REFERENCE CARD

**Beginning and Intermediate Algebra:
An Integrated Approach, Second Edition**
R. David Gustafson and Peter D. Frisk
ISBN 0-534-35943-4

 **Brooks/Cole Publishing Company**
ITP® An International Thomson Publishing Company
Visit Brooks/Cole on the Internet
http://www.brookscole.com

CHAPTER 1   REAL NUMBERS AND THEIR BASIC PROPERTIES

If n is a natural number, then $x^n = \overbrace{x \cdot x \cdot x \cdot \cdots \cdot x}^{n \text{ factors of } x}$

CHAPTER 4   POLYNOMIALS

If m and n are integers, then

$$x^m x^n = x^{m+n} \quad (x^m)^n = x^{m \cdot n} \quad (xy)^n = x^n y^n$$

$$\left(\frac{x}{y}\right)^n = \frac{x^n}{y^n} \ (y \neq 0) \quad \frac{x^m}{x^n} = x^{m-n} \ (x \neq 0)$$

$$x^0 = 1 \ (x \neq 0) \quad x^{-n} = \frac{1}{x^n} \ (x \neq 0)$$

CHAPTER 5   FACTORING POLYNOMIALS

$$x^2 - y^2 = (x + y)(x - y)$$
$$a^2 + 2ab + b^2 = (a + b)^2$$
$$a^2 - 2ab + b^2 = (a - b)^2$$

CHAPTER 7   MORE EQUATIONS, INEQUALITIES, AND FACTORING

If $k > 0$, then
$|x| = k$ is equivalent to $x = k$ or $x = -k$.
$|a| = |b|$ is equivalent to $a = b$ or $a = -b$.
$|x| < k$ is equivalent to $-k < x < k$.
$|x| > k$ is equivalent to $x < -k$ or $x > k$.

$$x^3 + y^3 = (x + y)(x^2 - xy + y^2)$$
$$x^3 - y^3 = (x - y)(x^2 + xy + y^2)$$

CHAPTER 8   WRITING EQUATIONS OF LINES; VARIATION

Midpoint formula: If $P(x_1, y_1)$ and $Q(x_2, y_2)$ are points on a line, the midpoint of segment PQ is

$$M\left(\frac{x_1 + x_2}{2}, \frac{y_1 + y_2}{2}\right)$$

Slope of a nonvertical line:

$$m = \frac{\Delta y}{\Delta x} = \frac{y_2 - y_1}{x_2 - x_1} \ (x_1 \neq x_2)$$

Equations of a line:

Point–slope form: $y - y_1 = m(x - x_1)$
Slope–intercept form: $y = mx + b$
General form: $Ax + By = C$
Horizontal line: $y = b$
Vertical line: $x = a$

CHAPTER 9   RATIONAL EXPONENTS AND RADICALS

The Pythagorean theorem: If a and b are the lengths of the legs of a right triangle and c is the length of the hypotenuse, then

$$a^2 + b^2 = c^2$$

The distance formula:

$$d(PQ) = \sqrt{(x_2 - x_1)^2 + (y_2 - y_1)^2}$$

Fractional exponents: If m and n are positive integers and $x > 0$, then

$$x^{m/n} = \sqrt[n]{x^m} = \left(\sqrt[n]{x}\right)^m \quad x^{-m/n} = \frac{1}{x^{m/n}} \qquad \frac{1}{x^{-m/n}} = x^{m/n}$$

Properties of radicals: If at least one of a or b is positive, then

$$\sqrt[n]{ab} = \sqrt[n]{a}\sqrt[n]{b} \quad \sqrt[n]{\frac{a}{b}} = \frac{\sqrt[n]{a}}{\sqrt[n]{b}} \ (b \neq 0)$$

CHAPTER 10   QUADRATIC FUNCTIONS, INEQUALITIES, AND ALGEBRA OF FUNCTIONS

The quadratic formula:

$$x = \frac{-b \pm \sqrt{b^2 - 4ac}}{2a} \ (a \neq 0)$$

Complex numbers: If a, b, c, and d are real numbers and $i^2 = -1$, then

$a + bi = c + di$ if and only if $a = c$ and $b = d$.

$(a + bi) + (c + di) = (a + c) + (b + d)i$

$(a + bi)(c + di) = (ac - bd) + (ad + bc)i$

$|a + bi| = \sqrt{a^2 + b^2}$

CHAPTER 11 EXPONENTIAL AND LOGARITHMIC FUNCTIONS

Properties of logarithms: If M, N, and b are positive numbers and $b \neq 1$, then

$\log_b 1 = 0 \qquad \log_b b = 1$

$\log_b b^x = x \qquad b^{\log_b x} = x$

$\log_b MN = \log_b M + \log_b N$

$\log_b \dfrac{M}{N} = \log_b M - \log_b N$

$\log_b M^p = p \log_b M$ \quad If $\log_b x = \log_b y$, then $x = y$.

Change-of-base formula: $\log_b y = \dfrac{\log_a y}{\log_a b}$

CHAPTER 12 MORE GRAPHING AND CONIC SECTIONS

Equations of a circle with radius r:

$(x - h)^2 + (y - k)^2 = r^2$ center at (h, k)

$x^2 + y^2 = r^2$ center at $(0, 0)$

Equations of a parabola: If $a > 0$, then

Parabola opening	Vertex at orgin	Vertex at (h, k)
Up	$y = ax^2$	$y = a(x - h)^2 + k$
Down	$y = -ax^2$	$y = -a(x - h)^2 + k$
Right	$x = ay^2$	$x = a(y - k)^2 + h$
Left	$x = -ay^2$	$x = -a(y - k)^2 + h$

Equations of an ellipse:

Center at $(0, 0)$ if $a > b > 0$

$$\dfrac{x^2}{a^2} + \dfrac{y^2}{b^2} = 1 \qquad \dfrac{x^2}{b^2} + \dfrac{y^2}{a^2} = 1$$

Center at (h, k)

$$\dfrac{(x - h)^2}{a^2} + \dfrac{(y - k)^2}{b^2} = 1 \quad (a > b > 0)$$

$$\dfrac{(x - h)^2}{b^2} + \dfrac{(y - k)^2}{a^2} = 1 \quad (a > b > 0)$$

Equations of a hyperbola:

Center at $(0, 0)$

$$\dfrac{x^2}{a^2} - \dfrac{y^2}{b^2} = 1 \qquad \dfrac{y^2}{a^2} - \dfrac{x^2}{b^2} = 1$$

Center at (h, k)

$$\dfrac{(x - h)^2}{a^2} - \dfrac{(y - k)^2}{b^2} = 1$$

$$\dfrac{(y - k)^2}{a^2} - \dfrac{(x - h)^2}{b^2} = 1$$

CHAPTER 14 MISCELLANEOUS TOPICS

The binomial theorem:

$$(a + b)^n = a^n + \dfrac{n!}{1!(n - 1)!} a^{n-1}b$$

$$+ \dfrac{n!}{2!(n - 2)!} a^{n-2}b^2 + \cdots + b^n$$

Arithmetic sequence:

$$S_n = \dfrac{n(a + l)}{2} \quad \text{(sum of the first } n \text{ terms)}$$

Geometric sequence:

$$S_n = \dfrac{a - ar^n}{1 - r} \quad (r \neq 1) \quad \text{(sum of the first } n \text{ terms)}$$

The sum of all the terms of the sequence is

$$S = \dfrac{a}{1 - r} \quad (|r| < 1)$$

Formulas for permutations and combinations:

$$P(n, r) = \dfrac{n!}{(n - r)!} \qquad P(n, n) = n!$$

$$P(n, 0) = 1 \qquad C(n, r) = \binom{n}{r} = \dfrac{n!}{r!(n - r)!}$$

$$C(n, n) = \binom{n}{n} = 1 \qquad C(n, 0) = \binom{n}{0} = 1$$

Turn the graphing calculator into a powerful tool for your success!

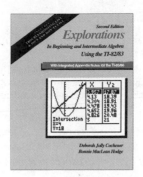

Explorations in Beginning and Intermediate Algebra Using the TI-82/83 with Integrated Appendix Notes for the TI-85/86, Second Edition

by Deborah J. Cochener and Bonnie M. Hodge,
both of Austin Peay State University

352 pages. Spiralbound. 8 1/2" x 11". ISBN: 0-534-36149-8.
©1999. Published by Brooks/Cole.

You can quickly learn to use the graphing calculator to develop problem-solving and critical-thinking skills that will improve your performance in beginning and intermediate algebra!

Designed to help you succeed in your algebra course, this unique and student-friendly workbook improves both your understanding *and* retention of algebra concepts—using the graphing calculator. By integrating technology into mathematics, the authors help you develop problem-solving and critical-thinking skills.

To guide you in your explorations, you'll find:

- hands-on applications with solutions
- correlation charts that relate course topics to the workbook units
- key charts (specific to the TI-82, TI-83, TI-85, and TI-86) that show which units introduce keys on the calculator
- a *Troubleshooting Section* to help you avoid common errors

Other primary features to help you succeed:

- *In Your Own Words*, a section at the end of each unit, gives you the opportunity to summarize main points from the unit and personalize the material for your own use
- Appendices that address the linking capabilities of the calculators and menu maps for each calculator that are designed to assist the you in locating a needed menu or function
- An optional unit that addresses the complex number operations available on the TI-83, TI-85,and TI-86
- *Extra for Experts* problems that extend the concepts within the unit
- *Accuracy Checks*—five-to-ten-question quizzes—that correspond to most of the units. The answers are posted at the Brooks/Cole web site.

Topics

This text contains 36 units divided into the following subsections:

Basic Calculator Operations
Graphically Solving Equations and Inequalities
Graphing and Applications of Equations in Two Variables
Stat Plots

Order your copy today!

To receive a sampler of *Explorations in Beginning and Intermediate Algebra Using the TI-82/83 with Integrated Appendix Notes for the TI-85/86, Second Edition*, simply mail in the order form on the other side of this page, call (800) 354-9706, or visit us on the Internet: http://www.brookscole.com

ORDER FORM

Call our toll-free number (800) 354-9706 to purchase, or use the form below and mail or fax to (831) 375-6414.

No risk. All of our books and software are backed by our 30-day, money-back guarantee. Major credit cards accepted. We accept purchase orders from your company or institution. The cost of shipping will be added to your bill. To order, simply fill out this coupon and return it to Brooks/Cole along with your check, money order, or credit card information.

_____ Yes! I would like to order *Explorations in Beginning and Intermediate Algebra Using the TI-82/83 with Integrated Appendix Notes for the TI-85/86, Second Edition*, by Deborah J. Cochener and Bonnie M. Hodge, ISBN: 0-534-36149-8 for $26.95. *Note: Pricing is subject to change. Please call (800) 354-9706 to receive current pricing information.*

Residents of AL, AZ, CA, CT, CO, FL, GA, IL, IN, KS, KY, LA, MA, MD, MI, MN, MO, NC, NJ, NY, OH, PA, RI, SC, TN, TX, UT, VA, WA, WI must add appropriate state sales tax.

Subtotal _____
Tax _____
Handling ___ $4.00 ___
TOTAL _____

Payment Options

_____ Payment enclosed (check or money order)

_____ Please charge the following credit card:

_____ VISA _____ MasterCard _____ American Express

Card #: _____ Expiration Date: _____

Contact Name: _____ Phone: _____

Signature: _____
(Note: Credit card billing and shipping address must be the same.)

All orders under 250 pounds will be shipped via UPS, unless the customer requests a specific carrier. Average shipping charges are estimated at 10% of the order. The cost of shipping will be added to your bill. Customer's requested shipper: _____

Prices are subject to change without notice. We will refund any pre-payments for unshipped, out-of-stock titles after 150 days and for not-yet-published titles after 180 days, unless an earlier date is requested in writing by you.

Please ship my order to: *(please print)*

Name _____

Street Address _____

City _____ State _____ Zip _____

Telephone (_____) _____ e-mail: _____

Mail to:

Brooks/Cole Publishing Company
Source Code 9BCMA
511 Forest Lodge Road
Pacific Grove, CA 93950-5098
Phone: (800) 354-9706 • Fax: (831) 375-6414

ITP **International Thomson Publishing**
Education Group

©1998 Brooks/Cole Publishing Company

SECOND EDITION

Beginning and Intermediate Algebra

An Integrated Approach

To
Craig, Jeremy,
Paula, Gary,
and Bob

Books in the Gustafson/Frisk Series

■ ■ ■ ■ ■ ■ ■ ■ ■ *SECOND EDITION*

Beginning and Intermediate Algebra ■ ■ ■ ■ ■ ■ ■ ■

An Integrated Approach

R. David Gustafson
Rock Valley College

Peter D. Frisk
Rock Valley College

Brooks/Cole Publishing Company

I**T**P® An International Thomson Publishing Company

Pacific Grove • Albany • Bonn • Boston • Cincinnati • Detroit • Johannesburg • London • Madrid • Melbourne
Mexico City • New York • Paris • San Francisco • Singapore • Tokyo • Toronto • Washington

A ROBERT W. PIRTLE BOOK

Publisher: *Robert W. Pirtle*
Marketing Team: *Jennifer Huber, Christine Davis, Debra Johnston*
Editorial Assistant: *Erin Wickersham*
Production Editor: *Ellen Brownstein*
Production Service: *Hoyt Publishing Services*
Manuscript Editor: *David Hoyt*
Permissions Editor: *Carline Haga*
Interior Design: *E. Kelly Shoemaker, Vernon Boes*

Interior Illustration: *Lori Heckelman*
Photo Editor: *Terry Powell*
Cover Design: *Roy Neuhaus*
Cover Photo: *John Paul Endress/The Stock Market*
Art Coordinator: *David Hoyt*
Typesetting: *The Clarinda Company*
Cover Printing: *Phoenix Color Corp.*
Printing and Binding: *World Color Book Services (Taunton)*

For more information, contact:

BROOKS/COLE PUBLISHING COMPANY
511 Forest Lodge Road
Pacific Grove, CA 93950
USA

International Thomson Publishing Europe
Berkshire House 168-173
High Holborn
London WC1V 7AA
England

Thomas Nelson Australia
102 Dodds Street
South Melbourne, 3205
Victoria, Australia

Nelson Canada
1120 Birchmount Road
Scarborough, Ontario
Canada M1K 5G4

International Thomson Editores
Seneca 53
Col. Polanco
11560 México, D. F., México

International Thomson Publishing GmbH
Königswinterer Strasse 418
53227 Bonn
Germany

International Thomson Publishing Asia
60 Albert Street #15-01
Albert Complex
Singapore 189969

International Thomson Publishing Japan
Hirakawacho Kyowa Building, 3F
2-2-1 Hirakawacho
Chiyoda-ku, Tokyo 102
Japan

Printed in the United States of America

10   9   8   7   6   5   4   3

Library of Congress Cataloging-in-Publication Data

Gustafson, R. David (Roy David), [date]
    Beginning and intermediate algebra: an integrated approach/
R. David Gustafson, Peter D. Frisk. — 2nd ed.
        p.   cm.
    Includes index.
    ISBN 0-534-35943-4 (hardcover: alk. paper)
    1. Algebra.   I. Frisk, Peter D., [date].   II. Title.
QA152.2.G86   1998
512.9—dc21   98-42081
                                                  CIP

Photo credits: p. 2, Philip Gould/Corbis; **p. 5,** The British Museum; **p. 91,** Robert Maass/Corbis; **p. 165,** Roger Ressmeyer/©Corbis; **p. 197,** Courtesy of IBM Corporation; **p. 258,** Leif Skoogfors/Corbis; **p. 336,** Richard T. Nowitz/Corbis; **p. 403,** Rob Rowan; Progressive Image/Corbis; **p. 475,** Tim Wright/Corbis; **p. 554,** Hulton/Duetsch Collection/Corbis; **p. 640,** Roger Ressmeyer/©Corbis; **p. 713,** Charles E. Rotkin/©Corbis; **p. 801,** Ed Young/Corbis; **p. 862,** Archaeological Consulting/ Gary Breschini & Trudy Haverstat; **p. 874,** David H. Wells/Corbis; **p. 929,** PhotoDisc; **p. 992,** Liba Taylor/Corbis.

To the Instructor

Beginning and Intermediate Algebra, second edition, combines the topics of beginning and intermediate algebra. This type of book has several advantages:

- By combining topics, much overlap and redundancy can be eliminated. The instructor will have time to teach for mastery of the material.
- For many students, the purchase of a single book will save money.
- A combination approach in one book will enable some colleges to cut back on the number of hours needed for mathematics remediation.

However, there are three concerns inherent in a combined approach:

- The first half of the book must include enough beginning algebra to ensure that students who complete the first half of the book and then transfer to another college will have the necessary prerequisites to enroll in an intermediate algebra course.
- The beginning algebra material should not get too difficult too fast.
- Intermediate algebra students beginning in the second half of the book must get some review of basic topics so that they can compete with students who are continuing from the first course.

Unlike many other texts, this book uses an *integrated approach*, which addresses each of the previous three concerns by

- including a full course in beginning algebra in the first six chapters,
- delaying the presentation of intermediate algebra topics until Chapter 7 or later, and
- providing a quick review of basic topics for those who begin in the second half of the book.

■ ORGANIZATION

The first six chapters of this second edition present all of the topics usually associated with a first course in algebra, except for a detailed discussion of manipulating radical expressions and the quadratic formula. These topics can be omitted because they will be carefully introduced and taught in any intermediate algebra course.

The first six chapters cover basic material at the beginning algebra level. Harder topics, such as absolute value inequalities and factoring the sum and difference of two cubes, are left until Chapter 7. Chapter 3 discusses graphs of linear equations and systems of two equations in two variables. Systems of three equations in three variables and the methods for solving them are left until Chapter 13.

Chapter 7 is the entry-level chapter for students enrolling in intermediate algebra. As such, it quickly reviews the topics taught in the first six chapters and extends those topics to the intermediate algebra level.

Chapters 8 through 14 are written at the intermediate algebra level and include a quick review of important topics as needed. For example, Chapter 8 begins with a review of the rectangular coordinate system and graphing linear equations, a topic first taught in Chapter 3. It then moves on to the topics of writing equations of lines, nonlinear functions, and variation. As another example, Chapter 12 begins with a review of solving simple systems of equations, which was first taught in Chapter 3. It then moves on to solving more difficult systems by matrices and determinants.

■ GOALS OF THE BOOK

In addition to using a truly integrated approach, our goal has been to write a book that

- is enjoyable to read.
- is easy to understand.
- is relevant.
- will develop the necessary skills for success in future academic courses or on the job.

Although the material has been extensively revised, this second edition retains the basic philosophy of the highly successful previous edition. The revisions include several improvements in line with the NCTM standards, the AMATYC Crossroads, and the current trends in mathematics reform. For example, more emphasis has been placed on graphing and problem solving.

■ GENERAL CHANGES IN THE SECOND EDITION

The overall effects of the changes made to the second edition are as follows:

- **To increase the emphasis on learning mathematics through graphing.** Although graphing calculators are used often, their use is not required. All of the topics are fully discussed in traditional ways. Of course, we recommend that instructors use the graphing calculator material.

Graphing calculator work ▶
appears throughout the book.

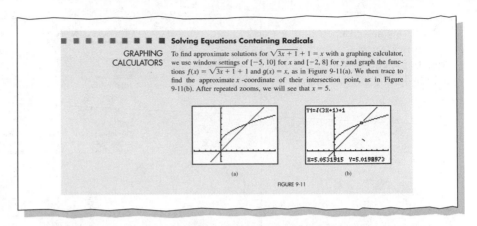

Solving Equations Containing Radicals

GRAPHING CALCULATORS To find approximate solutions for $\sqrt{3x + 1} + 1 = x$ with a graphing calculator, we use window settings of $[-5, 10]$ for x and $[-2, 8]$ for y and graph the functions $f(x) = \sqrt{3x + 1} + 1$ and $g(x) = x$, as in Figure 9-11(a). We then trace to find the approximate x-coordinate of their intersection point, as in Figure 9-11(b). After repeated zooms, we will see that $x = 5$.

FIGURE 9-11

- **To increase the emphasis on problem solving through realistic applications.** The variety of application problems has been increased significantly, and all application problems are labeled with special titles.

All applications have titles. ▶

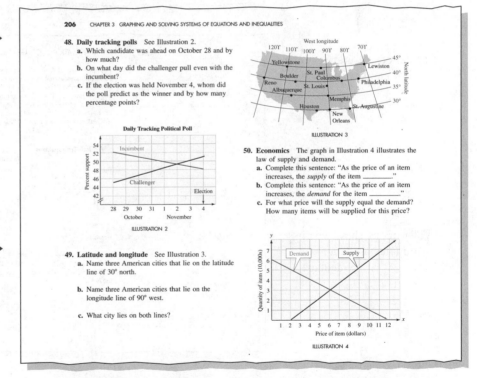

206   CHAPTER 3 GRAPHING AND SOLVING SYSTEMS OF EQUATIONS AND INEQUALITIES

48. Daily tracking polls See Illustration 2.
  a. Which candidate was ahead on October 28 and by how much?
  b. On what day did the challenger pull even with the incumbent?
  c. If the election was held November 4, whom did the poll predict as the winner and by how many percentage points?

ILLUSTRATION 3

Daily Tracking Political Poll

ILLUSTRATION 2

50. Economics The graph in Illustration 4 illustrates the law of supply and demand.
  a. Complete this sentence: "As the price of an item increases, the *supply* of the item _____."
  b. Complete this sentence: "As the price of an item increases, the *demand* for the item _____."
  c. For what price will the supply equal the demand? How many items will be supplied for this price?

Many new applications ▶
have been included.

49. Latitude and longitude See Illustration 3.
  a. Name three American cities that lie on the latitude line of 30° north.
  b. Name three American cities that lie on the longitude line of 90° west.
  c. What city lies on both lines?

ILLUSTRATION 4

- **To fine-tune the presentation of many topics** for better flow of ideas and for clarity.
- **To increase the visual interest by the use of color.** We include color not just as a design feature, but to highlight terms that instructors would point to in a classroom discussion.
- **To increase the emphasis on families of functions** and their graphs, including translations of their graphs.

Functions are grouped into ▶
families, such as squaring
functions, cubing functions,
and absolute value functions.

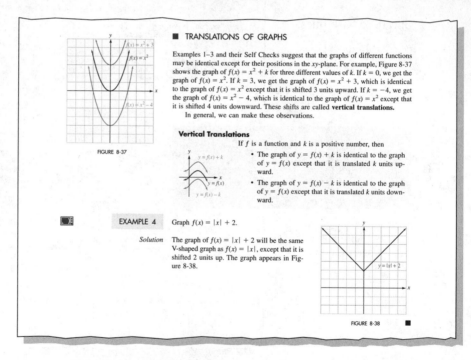

A third basic function is $f(x) = |x|$ (or $y = |x|$), often called the **absolute value function.**

EXAMPLE 3   Graph the function $f(x) = |x|$.

Solution   We substitute values for x in the equation and compute the corresponding values of $f(x)$. For example, if $x = -3$, we have

$$f(x) = |x|$$
$$f(-3) = |-3| \qquad \text{Substitute } -3 \text{ for } x.$$
$$= 3$$

The ordered pair $(-3, 3)$ satisfies the equation and will lie on the graph. We list this pair and others that satisfy the equation in the table shown in Figure 8-34. We plot the points and draw a V-shaped line through them to get the graph.

$$f(x) = |x|$$

x	y	$(x, f(x))$
-3	3	$(-3, 3)$
-2	2	$(-2, 2)$
-1	1	$(-1, 1)$
0	0	$(0, 0)$
1	1	$(1, 1)$
2	2	$(2, 2)$
3	3	$(3, 3)$

$f(x) = |x|$

FIGURE 8-34

Translations of graphs are ▶
thoroughly discussed.

FIGURE 8-37

■ TRANSLATIONS OF GRAPHS

Examples 1–3 and their Self Checks suggest that the graphs of different functions may be identical except for their positions in the xy-plane. For example, Figure 8-37 shows the graph of $f(x) = x^2 + k$ for three different values of k. If $k = 0$, we get the graph of $f(x) = x^2$. If $k = 3$, we get the graph of $f(x) = x^2 + 3$, which is identical to the graph of $f(x) = x^2$ except that it is shifted 3 units upward. If $k = -4$, we get the graph of $f(x) = x^2 - 4$, which is identical to the graph of $f(x) = x^2$ except that it is shifted 4 units downward. These shifts are called **vertical translations.**
    In general, we can make these observations.

Vertical Translations

If f is a function and k is a positive number, then

$y = f(x) + k$

$y = f(x)$

$y = f(x) - k$

• The graph of $y = f(x) + k$ is identical to the graph of $y = f(x)$ except that it is translated k units upward.
• The graph of $y = f(x) - k$ is identical to the graph of $y = f(x)$ except that it is translated k units downward.

Video icons show which ▶
examples are taught on
videotape.

EXAMPLE 4   Graph $f(x) = |x| + 2$.

Solution   The graph of $f(x) = |x| + 2$ will be the same V-shaped graph as $f(x) = |x|$, except that it is shifted 2 units up. The graph appears in Figure 8-38.

$y = |x| + 2$

FIGURE 8-38   ■

■ SPECIFIC CHANGES IN THE SECOND EDITION

To make the book more useful to students, we have

- **Added a Mathematics in the Workplace feature** at the start of each chapter, showing an application of mathematics to a specific career.
- **Added Getting Ready problems** to help students get ready for the section.

A Mathematics in the ▶
Workplace feature relates
mathematics to career
applications.

Each main section is broken ▶
into several subsections.

Getting Ready exercises ▶
prepare students for the
coming section.

■ ■ ■ ■ ■ ■ ■ ■ ■ **Economist**

MATHEMATICS IN
THE WORKPLACE

Economists study the way a society uses resources such as land, labor, raw materials, and machinery to provide goods and services. Some economists are theoreticians who use mathematical models to explain the causes of recession and inflation. Most economists, however, are concerned with practical applications of
economic policy in a particular area.

SAMPLE APPLICATION ■ An electronics firm manufactures tape recorders,
receiving $120 for each recorder it makes. If x represents the number of record-
ers produced, the income received is determined by the *revenue function*

$$R(x) = 120x$$

The manufacturer has fixed costs of $12,000 per month and variable costs of
$57.50 for each tape recorder manufactured. Thus, the *cost function* is

$$C(x) = variable\ costs + fixed\ costs$$
$$= 57.50x + 12,000$$

How many recorders must the company sell for revenue to equal cost?
(See Exercise 126 in Exercise 4.6.)

4.1 Natural-Number Exponents
■ EXPONENTS ■ POWERS OF EXPRESSIONS ■ THE PRODUCT RULE FOR EXPONENTS ■ THE POWER
RULES FOR EXPONENTS ■ THE QUOTIENT RULE FOR EXPONENTS

Getting Ready   *Evaluate each expression.*

1. 2^3 **2.** 3^2 **3.** $3(2)$ **4.** $2(3)$

5. $2^3 + 2^2$ **6.** $2^3 \cdot 2^2$ **7.** $3^3 - 3^2$ **8.** $\dfrac{3^3}{3^2}$

■ **EXPONENTS**

We have used natural-number exponents to indicate repeated multiplication. For example,

$$2^5 = 2 \cdot 2 \cdot 2 \cdot 2 \cdot 2 = 32 \qquad (-7)^3 = (-7)(-7)(-7) = -343$$
$$x^4 = x \cdot x \cdot x \cdot x \qquad -y^5 = -y \cdot y \cdot y \cdot y \cdot y$$

These examples suggest a definition for x^n, where n is a natural number.

258

- **Included Self Checks with most examples.**
- **Included Vocabulary and Concepts problems** in each exercise set. The
  Hints on Studying Algebra that appear in this preface recommend that
  students begin their study time with review, so review exercises are placed
  at the beginning of each exercise set. Most exercise sets follow this
  sequence:

  1. Review problems

  2. Vocabulary and Concepts problems

  3. Practice problems

  4. Application problems

  5. Writing problems

  6. Something to Think About problems

Most examples are followed by ▶
Self Checks.

Each exercise set is preceded ▶
by a set of oral exercises.

Review problems ▶

Vocabulary and Concepts ▶
problems

Practice problems ▶

Application problems ▶

Writing problems ▶

Something to Think About ▶
problems

Self Check   In Example 7, if a second inspector found 3 sweaters with faded colors in addition to the defectives found by inspector 1, what percent were defective?

Answer $7\frac{1}{2}\%$

Orals   *In Exercises 1–8, solve each equation.*

1. $3x = 3$ **2.** $5x = 5$ **3.** $-7x = 14$

4. $7.5x = 0$ **5.** $\frac{x}{5} = 2$ **6.** $\frac{x}{2} = -10$

7. $\frac{x}{-4} = 3$ **8.** $\frac{x}{8} = -3$

9. Change 30% to a decimal.   **10.** Change 0.08 to a percent.

EXERCISE 2.2

REVIEW   *Do the operations. Simplify the result when possible.*

1. $\frac{4}{5} + \frac{2}{3}$ **2.** $\frac{5}{6} \cdot \frac{12}{25}$ **3.** $\frac{5}{9} \div \frac{3}{5}$ **4.** $\frac{15}{7} - \frac{10}{3}$

5. $2 + 3 \cdot 4$ **6.** $3 \cdot 4^2$ **7.** $3 + 4^3(-5)$ **8.** $\frac{5(-4) - 3(-2)}{10 - (-4)}$

Find the area of each geometric figure.

9. A rectangle with dimensions of 3.5 feet by 7.2 feet.   **10.** A circle with a diameter of 12.45 inches. Give the result to the nearest hundredth.

VOCABULARY AND CONCEPTS   *Fill in each blank to make a true statement.*

11. If equal quantities are divided by the same nonzero quantity, the results are _____ quantities.

12. If $a = b$, then $\frac{a}{c} =$ ___, provided that $c \neq$ ___.

13. If $a = b$, then $ac =$ ___.

14. If _____ quantities are multiplied by the same nonzero quantity, the results will be equal quantities.

15. A percent is the numerator of a fraction whose denominator is _____.

16. Rate · _____ = percentage

PRACTICE   *In Exercises 17–52, use the division or multiplication property of equality to solve each equation.* **Check all solutions.**

17. $6x = 18$ **18.** $25x = 625$ **19.** $-4x = 36$ **20.** $-16y = 64$

APPLICATIONS   🔢   *In Exercises 69–78, solve each problem.*

69. Customer satisfaction   Two-thirds of a movie audience left the theater in disgust. If 78 angry patrons walked out, how many were there originally?

70. Stock split   After a three-for-two stock split, each shareholder will own 1.5 times as many shares as before. If 555 shares are owned after the split, how many were owned before?

71. Off-campus housing   Four-sevenths of the senior class is living in off-campus housing. If 868 students live off campus, how large is the senior class?

72. Union membership   The 2,484 union members represent 90% of a factory's workforce. How many are employed?

73. Shopper dissatisfaction   Refer to the survey results in Illustration 1. What percent of those surveyed were not pleased?

Shopper survey results	
First-time shoppers	1,731
Major purchase today	539
Shopped within previous month	1,823
Satisfied with service	4,140
Seniors	2,387
Total surveyed	9,200

ILLUSTRATION 1

WRITING

79. Explain how you would decide whether a number is a solution of an equation.

80. Distinguish between *percent* and *percentage*.

SOMETHING TO THINK ABOUT

81. The Ahmes Papyrus mentioned at the beginning of Chapter 1 contains this statement: *A circle nine units in diameter has the same area as a square eight units*

82. 🔢 Calculate the Egyptians' **percent of error:** What percent of the actual value of π is the difference between the values?

- **Changed the format of the chapter summaries.** Now the important concepts are listed in a left-hand column, with relevant review exercises beside them in the right-hand column.

The new format makes the ▶
Chapter Summary more useful
to students.

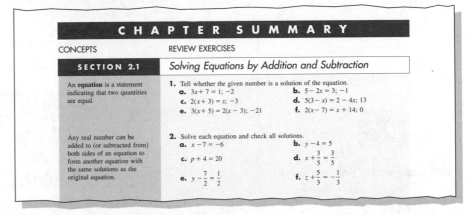

- **Used a consistent problem-solving approach** to real-life applications.
  1. Analyze the problem.
  2. Form an equation.
  3. Solve the equation.
  4. State the conclusion.
  5. Check the result.

Applications use a standard ▶
problem-solving approach.

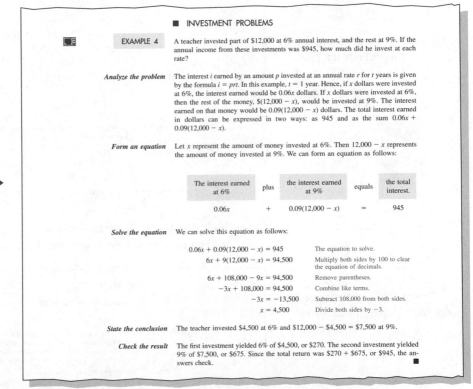

- 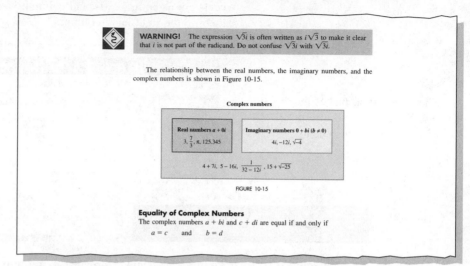 **Increased the number of warnings** about common errors.

Warnings are marked with a ▶
special icon.

Properties and definitions are ▶
boxed.

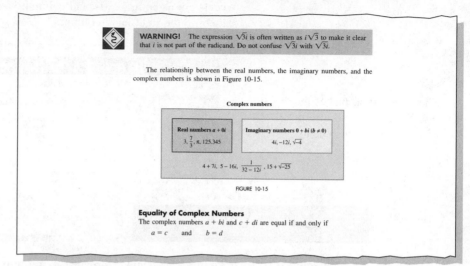

WARNING! The expression $\sqrt{3}i$ is often written as $i\sqrt{3}$ to make it clear that i is not part of the radicand. Do not confuse $\sqrt{3}i$ with $\sqrt{3i}$.

The relationship between the real numbers, the imaginary numbers, and the complex numbers is shown in Figure 10-15.

Complex numbers

Real numbers $a + 0i$	Imaginary numbers $0 + bi$ $(b \neq 0)$
$3, \frac{7}{3}, \pi, 125.345$	$4i, -12i, \sqrt{-4}$

$4 + 7i, \ 5 - 16i, \ \dfrac{1}{32 - 12i}, \ 15 + \sqrt{-25}$

FIGURE 10-15

Equality of Complex Numbers
The complex numbers $a + bi$ and $c + di$ are equal if and only if
$$a = c \quad \text{and} \quad b = d$$

- **Included more geometric content,** to integrate the subjects of algebra and geometry.

Geometric concepts appear ▶
throughout the text.

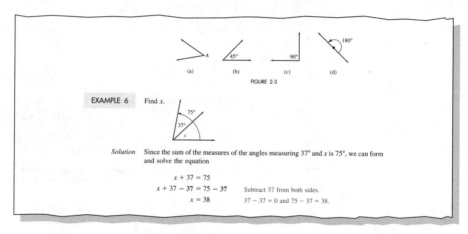

(a)   (b) 45°   (c) 90°   (d) 180°

FIGURE 2-3

EXAMPLE 6 Find x.

75°
37°
x

Solution Since the sum of the measures of the angles measuring 37° and x is 75°, we can form and solve the equation

$$x + 37 = 75$$
$$x + 37 - 37 = 75 - 37 \qquad \text{Subtract 37 from both sides.}$$
$$x = 38 \qquad \qquad 37 - 37 = 0 \text{ and } 75 - 37 = 38.$$

- **Included topics from statistics** as applications of algebra so as to provide background for students, who will encounter these ideas in the future.

Topics from statistics are ▶
discussed as applications of
algebra.

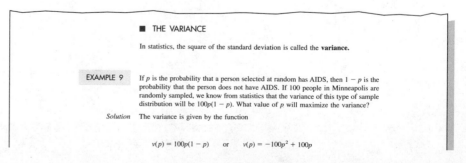

■ THE VARIANCE

In statistics, the square of the standard deviation is called the **variance.**

EXAMPLE 9 If p is the probability that a person selected at random has AIDS, then $1 - p$ is the probability that the person does not have AIDS. If 100 people in Minneapolis are randomly sampled, we know from statistics that the variance of this type of sample distribution will be $100p(1 - p)$. What value of p will maximize the variance?

Solution The variance is given by the function

$$v(p) = 100p(1 - p) \quad \text{or} \quad v(p) = -100p^2 + 100p$$

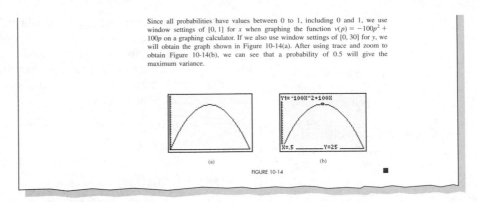

Since all probabilities have values between 0 to 1, including 0 and 1, we use window settings of $[0, 1]$ for x when graphing the function $v(p) = -100p^2 + 100p$ on a graphing calculator. If we also use window settings of $[0, 30]$ for y, we will obtain the graph shown in Figure 10-14(a). After using trace and zoom to obtain Figure 10-14(b), we can see that a probability of 0.5 will give the maximum variance.

(a)   (b)

FIGURE 10-14   ■

Specific changes made in the chapters are as follows.

Chapter 1   When number lines are introduced, both open and closed circles and parentheses and brackets are used to designate intervals on the number line. After students understand both notations, the book uses parentheses and brackets. The work on algebraic expressions has been expanded.

Chapter 2   The basic work in solving equations has been divided into two sections. Graphics like Figure 2-2 aid students in understanding the properties of equality. In applications involving markdown, markup, and geometry, simple equations are solved by the addition and subtraction properties of equality. In percent problems, equations are solved by the division and multiplication properties of equality. Applications have been eliminated from Section 2.4 so that students can concentrate on equation-solving techniques.

Interesting historical notes ▶
appear throughout the book.

■ ■ ■ ■ ■ ■ ■ ■ PERSPECTIVE

To find answers to such questions as How many? How far? How fast? and How heavy?, we often make use of mathematical statements called **equations.** The concept has a long history, and the techniques we will study in this chapter have been developed over many centuries.

The mathematical notation that we use today is the result of thousands of years of development. The ancient Egyptians used a word for variables, best translated as *heap.* Others used the word *res,* which is Latin for *thing.* In the fifteenth century, the letters *p:* and *m:* were used for *plus* and *minus.* What we would now write as $2x + 3 = 5$ might have been written by those early mathematicians as 2 *res p:* 3 *aequalis* 5.

We can think of the scale shown in Figure 2-2(a) as representing the equation $x + 4 = 9$. The weight on the left-hand side of the scale is $(x + 4)$ grams, and the weight on the right-hand side is 9 grams. Because these weights are equal, the scale is in balance. To find x, we need to isolate it by removing 4 grams from the left-hand side. To keep the scale in balance, we must also remove 4 grams from the right-hand side. In Figure 2-2(b), we can see that x grams will be balanced by 5 grams. We have found that the solution is 5.

Graphics help students ▶
understand concepts.

$x + 4$ grams   9 grams x grams   5 grams

(a)   (b)

FIGURE 2-2

Chapter 3 presents graphing linear equations and systems of equations and inequalities. A new section is devoted to the coordinate system, emphasizing reading information from graphs. Section 3.2 discusses graphing linear equations, and an application of graphing linear equations has been added to the section. There are many more applications that illustrate the use of intersecting graphs. Work with graphing calculators has been expanded.

A new section discusses ▶
reading information from
graphs.

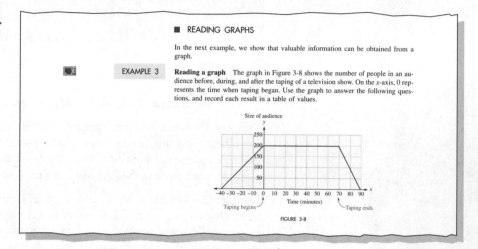

■ READING GRAPHS

In the next example, we show that valuable information can be obtained from a graph.

EXAMPLE 3   **Reading a graph**   The graph in Figure 3-8 shows the number of people in an audience before, during, and after the taping of a television show. On the x-axis, 0 represents the time when taping began. Use the graph to answer the following questions, and record each result in a table of values.

Size of audience

Taping begins   Time (minutes)   Taping ends

FIGURE 3-8

Chapter 4 covers exponents and polynomials. In Section 4.2, finding present value has been included as an application of negative exponents. In Section 4.4, functions are introduced in an informal way. Function notation is used instead of polynomial notation. Several simple polynomial functions are graphed. Again, graphing calculator material is included, as well as applications for adding, subtracting, multiplying, and dividing polynomials.

Polynomial functions are ▶
introduced in Chapter 4.

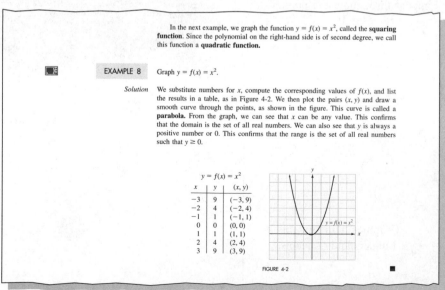

In the next example, we graph the function $y = f(x) = x^2$, called the **squaring function**. Since the polynomial on the right-hand side is of second degree, we call this function a **quadratic function.**

EXAMPLE 8   Graph $y = f(x) = x^2$.

Solution   We substitute numbers for x, compute the corresponding values of $f(x)$, and list the results in a table, as in Figure 4-2. We then plot the pairs (x, y) and draw a smooth curve through the points, as shown in the figure. This curve is called a **parabola**. From the graph, we can see that x can be any value. This confirms that the domain is the set of all real numbers. We can also see that y is always a positive number or 0. This confirms that the range is the set of all real numbers such that $y \geq 0$.

$y = f(x) = x^2$

x	y	(x, y)
-3	9	$(-3, 9)$
-2	4	$(-2, 4)$
-1	1	$(-1, 1)$
0	0	$(0, 0)$
1	1	$(1, 1)$
2	4	$(2, 4)$
3	9	$(3, 9)$

$y = f(x) = x^2$

FIGURE 4-2

Chapter 5 covers factoring. The method of factoring trinomials has been stream-lined, but this chapter remains a thorough treatment of traditional factoring techniques. However, factoring the sum and difference of two cubes is delayed until Chapter 7.

Chapter 6 presents proportion and rational expressions. The material on ratio has been extensively revised to show that ratios can be interpreted as unit costs and rates. Many applications involve comparison shopping. The material on proportions has also been extensively rewritten, including many more applications of proportions in everyday living. Similar triangles are thoroughly discussed. The skill work on manipulating rational expressions is placed toward the end of the chapter.

Similar triangles are fully ▶
discussed.

Author's notes (in color) explain ▶
the steps in the solution process.

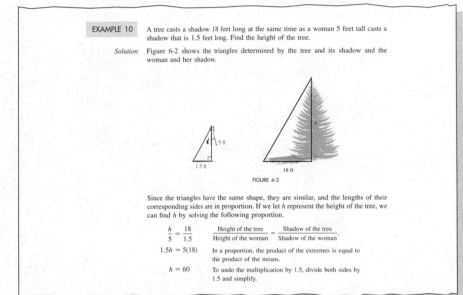

EXAMPLE 10   A tree casts a shadow 18 feet long at the same time as a woman 5 feet tall casts a shadow that is 1.5 feet long. Find the height of the tree.

Solution   Figure 6-2 shows the triangles determined by the tree and its shadow and the woman and her shadow.

FIGURE 6-2

Since the triangles have the same shape, they are similar, and the lengths of their corresponding sides are in proportion. If we let h represent the height of the tree, we can find h by solving the following proportion.

$$\frac{h}{5} = \frac{18}{1.5} \qquad \frac{\text{Height of the tree}}{\text{Height of the woman}} = \frac{\text{Shadow of the tree}}{\text{Shadow of the woman}}.$$

$$1.5h = 5(18) \qquad \text{In a proportion, the product of the extremes is equal to the product of the means.}$$

$$h = 60 \qquad \text{To undo the multiplication by 1.5, divide both sides by 1.5 and simplify.}$$

Chapter 7 is the entry-level chapter for students enrolling in intermediate algebra. As such, it quickly reviews the topics taught in the first six chapters and extends these topics to the intermediate algebra level.

Chapter 8 revisits the coordinate system and discusses slope, writing equations of lines, functions, and variation. The sections on slope and writing equations of lines have been reorganized for clarity. The work on functions reviews function notation and introduces the vertical line test. It also summarizes the basic types of functions, along with a discussion of their translations and reflections. Basic rational functions are discussed, and applications are given.

Chapter 9 now covers roots and radical expressions. Again, the practical material comes first, followed by the manipulation of radical expressions. Section 9.1 introduces square roots, cube roots, and nth roots, along with their corresponding functions. The Pythagorean theorem and the distance formula, along with many of their applications, are discussed in Section 9.2. Equations containing radicals, along with many applications, are covered in Section 9.3. The chapter concludes with a thorough treatment of rational exponents and manipulation of radical expressions.

Applications of the Pythagorean ▶
theorem are abundant.

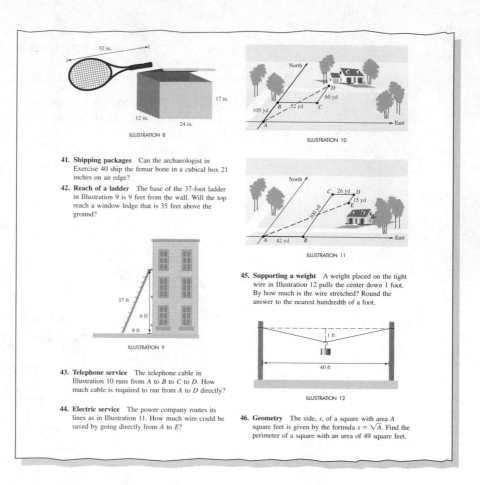

ILLUSTRATION 8

ILLUSTRATION 10

ILLUSTRATION 11

ILLUSTRATION 9

ILLUSTRATION 12

41. **Shipping packages**   Can the archaeologist in Exercise 40 ship the femur bone in a cubical box 21 inches on an edge?

42. **Reach of a ladder**   The base of the 37-foot ladder in Illustration 9 is 9 feet from the wall. Will the top reach a window ledge that is 35 feet above the ground?

43. **Telephone service**   The telephone cable in Illustration 10 runs from A to B to C to D. How much cable is required to run from A to D directly?

44. **Electric service**   The power company routes its lines as in Illustration 11. How much wire could be saved by going directly from A to E?

45. **Supporting a weight**   A weight placed on the tight wire in Illustration 12 pulls the center down 1 foot. By how much is the wire stretched? Round the answer to the nearest hundredth of a foot.

46. **Geometry**   The side, s, of a square with area A square feet is given by the formula $s = \sqrt{A}$. Find the perimeter of a square with an area of 49 square feet.

Chapter 10 presents quadratic equations and quadratic functions, complex numbers, and operations on functions. The chapter concludes with a discussion of the composition of functions and inverses of functions. The concepts of the domain and range of a function are thoroughly discussed.

Chapter 11 covers exponential and logarithmic functions. Special treatment is given to exponential expressions and logarithms with base e. This chapter is rich in applications.

Chapter 12 presents conic sections, piecewise-defined functions, and step functions. The work includes both conics centered at the origin and conics centered at (h, k). Completing the square is used to write equations of conics in standard form.

Chapter 13 covers solving systems of equations using matrices and determinants.

Chapter 14 includes a standard treatment of the binomial theorem, sequences, and permutations and combinations.

■ CALCULATORS

The use of calculators is assumed throughout the book. We believe that students should learn calculator skills in the mathematics classroom. They will then be prepared to use calculators in science and business classes and for nonacademic purposes. The directions within each exercise set indicate which exercises require calculators.

■ ANCILLARIES FOR THE INSTRUCTOR

Annotated Instructor's Edition
Free to professors when the text is adopted, the Instructor's Edition includes the complete text of the student edition, along with answers to all problems printed in blue next to the problems.

Complete Solutions Manual
This manual contains complete step-by-step solutions for all exercises in the text.

Test Manual
Includes printed test forms containing multiple-choice and free-response questions keyed to the text, as well as the answers for instructors.

Thomson World Class Learning^{TM} Testing Tools
This fully integrated suite of programs includes *World Class Test 1.0, World Class Test On-Line 1.0,* and *World Class Manager 1.0.* The program provides text-specific algorithmic testing options designed to offer instructors greater flexibility.

Thomson World Class Learning^{TM} Course
Using *World Class Course,* you can quickly and easily create and update a World Wide Web page specifically for your course or class—including what you plan to cover and when, assignments, grades, and even hot links to other resources on the Internet.

Text-Specific Video Tutorial Series
Free to schools when the text is adopted, this video series features worked-out examples from every section of the text, followed by supplementary examples that give students additional instruction and practice.

■ ANCILLARIES FOR THE STUDENT

Student Solutions Manual
The manual includes complete solutions to all odd-numbered exercises in the text.

Student Video
This *Greatest Hits* student video features the concepts and skills students traditionally have the most difficulty comprehending. The tape includes examples from each chapter.

Study Guide

Every copy of the text comes shrink-wrapped with a free, two-chapter sample of this text-specific Study Guide. Each chapter of the Study Guide contains chapter objectives, additional explanations of worked examples, exercises involving student participation, cautions, warnings, hints, and end-of-chapter tests.

Interactive Algebra 3.0

This extremely intuitive, text-specific tutorial provides explanations of concepts along with carefully graded, algorithmically generated examples and exercises. Hints are provided when students answer questions incorrectly. A management system provides a report of student progress upon completion of each unit. *Interactive Algebra* may be packaged with the text or sold as a stand-alone supplement.

To the Student

Congratulations. You now own a state-of-the-art textbook that has been written especially for you. We have tried to write a book that you can read and understand. The book includes carefully written narrative and an extensive number of worked examples with Self Checks.

To get the most out of this course, you must read and study the textbook properly. We recommend that you work the examples on paper first and then work the Self Checks. Only after you thoroughly understand the concepts taught in the examples should you attempt to work the exercises. Several ancillary materials may be helpful.

- A *Student Solutions Manual* contains the answers to the odd-numbered exercises.

- *Interactive Algebra 3.0* gives text-specific computer tutorials that cover every section of the text.

- A *Study Guide* contains chapter objectives, additional explanations of worked examples, exercises involving student participation, cautions, warnings, hints, and end-of-chapter tests.

- A complete set of videotapes is available, featuring worked-out examples from every section of the text, followed by supplementary examples.

Since the material presented in *Beginning and Intermediate Algebra*, Second Edition, will be of value to you in later years, we suggest that you keep this book. It will be a good source of reference and will keep at your fingertips the material that you have learned here.

We wish you well.

■ HINTS ON STUDYING ALGEBRA

The phrase "Practice makes perfect" is not quite true. It is *perfect* practice that makes perfect. For this reason, it is important that you learn how to study algebra to get the most out of this course.

Although we all learn differently, there are some hints on how to study algebra that most students find useful. Here are some things you should consider as you work on the material in this course.

Plan a strategy for success. To get where you want to be, you need a goal and a plan. Your goal should be to pass this course with a grade of A or B. To earn one of these grades, you must have a plan to achieve it. A good plan involves several points:

- Getting ready for class
- Attending class
- Doing homework
- Arranging for special help when you need it
- Having a strategy for taking tests

Getting ready for class. To get the most out of every class period, you will need to prepare for class. One of the best things you can do is to preview the material in the text that your instructor will be discussing. Perhaps you will not understand all of what you read, but you will understand it better when the instructor discusses the material in class.

Be sure to do your work every day. If you get behind and attend class without understanding previous material, you will be lost and will become frustrated and discouraged. Make a promise that you will always prepare for class, and then keep that promise.

Attending class. The classroom experience is your opportunity to learn from your instructor. Make the most of it by attending every class. Sit near the front of the room, where you can easily see and hear. It is easy to be distracted and lose interest if you sit in the back of the room. Remember that it is your responsibility to follow the discussion, even though that takes concentration and hard work.

Pay attention to your instructor, and jot down the important things that he or she says. However, do not spend so much time taking notes that you fail to concentrate on what your instructor is explaining. It is much better to listen and understand the big picture than just to copy solutions to problems.

Don't be afraid to ask questions when your instructor asks for them. If something is unclear to you, it is probably unclear to many other students as well. They will appreciate your willingness to ask. Besides, asking questions will make you an active participant in class. This will help you pay attention and keep you alert and involved.

Doing homework. It requires practice to excel at tennis, master a musical instrument, or learn a foreign language. In the same way, it requires practice to learn mathematics. Since practice in mathematics is the homework, homework is your opportunity to practice your skills and experiment with ideas.

It is very important for you to pick a definite time to study and do homework. Set a formal schedule and stick to it. Try to study in a place that is comfortable and quiet. If you can, do some homework shortly after class, or at least before you forget what was discussed in class. This quick follow-up will help you remember the skills and concepts your instructor taught that day.

Each formal study session should include three parts:

1. Begin every study session with a review period. Look over previous chapters and see if you can do a few problems from previous sections, chosen randomly. Keeping old skills alive will greatly reduce the time you will need to prepare for tests.

2. After reviewing, read the assigned material. Resist the temptation of diving into the exercises without reading and understanding the examples. Instead, work the examples and Self Checks with pencil and paper. Only after you completely understand the principles behind them should you try to work the exercises.

Once you begin to work the exercises, check your answers with those printed in the back of the book. If one of your answers differs from the printed answer, see if the two can be reconciled. Sometimes answers can have more than one form. If you decide that your answer is incorrect, compare your work to the example in the text that most closely resembles the exercise, and try to find your mistake. If you cannot find an error, consult the *Student Solutions Manual*. If nothing works, mark the problem and ask about it in your next class meeting.

3. After completing the written assignment, preview the next section. This preview will be helpful when you hear that material discussed during the next class period.

You probably know the general rule of thumb for college homework: two hours of practice for every hour in class. If mathematics is hard for you, plan on spending even more time on homework.

To make homework more enjoyable, study with one or more friends. The interaction will clarify ideas and help you remember them. If you must study alone, try talking to yourself. A good study technique is to explain the material to yourself out loud.

Arranging for special help. Take advantage of any special help that is available from your instructor. Often, the instructor can clear up difficulties in a very short time.

Find out whether your college has a free tutoring program. Peer tutors can often be of great help.

Taking tests. Students often get nervous before a test, because they are afraid that they will not do well. There are many different reasons for this fear, but the most common one is that students are not confident that they know the material.

To build confidence in your ability to work tests, rework many of the problems in the exercise sets, work the exercises in the Chapter Summaries, and take the Chapter Tests. Check all answers with those printed at the back of the text.

Then guess what the instructor will ask, build your own tests, and work them. Once you know your instructor, you will be surprised at how good you can get at picking test questions. With this preparation, you will have some idea of what will be on the test. You will have more confidence in your ability to do well. You should notice that you are far less nervous before tests, and this will also help your performance.

When you take a test, work slowly and deliberately. Scan the test and work the easy problems first. This will build confidence. Tackle the hardest problems last.

Acknowledgments

We are grateful to the following people, who reviewed the manuscript at various stages of its development. They all had valuable suggestions that have been incorporated into the text.

David Byrd
Enterprise State Junior College

Harold Farmer
Wallace Community College-Hanceville

Mark Foster
Santa Monica College

Dorothy K. Holtgrefe
Seminole Community College

Mike Judy
Fullerton College

Janet Mazzarella
Southwestern College

Daniel F. Mussa
Southern Illinois University

Joanne Peeples
El Paso Community College

Mary Ann Petruska
Pensacola Junior College

Janet Ritchie
SUNY-Old Westbury

Hattie White
St. Philip's College

George J. Witt
Glendale Community College

We are grateful to Diane Koenig and Robert Hessel, who read the entire manuscript and worked every problem. We also wish to thank Jerry Frang, Rob Clark, George Mader, Michael Welden, and Jennifer Dollar for their helpful comments and suggestions.

We are especially grateful to our editor, Robert Pirtle, for his encouragement and support. Finally, we express our thanks to Ellen Brownstein, who managed the production process; to David Hoyt, who skillfully edited the manuscript and guided the book to publication; to Lori Heckelman for her magnificent artwork; to Roy Neuhaus for his creative cover design; and to the Clarinda Company for outstanding composition.

R. David Gustafson
Peter D. Frisk

1 Real Numbers and Their Basic Properties

Computer Systems Analyst

Computer systems analysts help businesses and scientific research organizations develop computer systems to process and interpret data. Using techniques such as cost accounting, sampling, and mathematical model building, they analyze information and often present the results to management in the form of charts and diagrams.

SAMPLE APPLICATION ■ The process of sorting records into sequential order is a common task in electronic data processing. One sorting technique, called a *selection sort*, requires C comparisons to sort N records, where C and N are related by the formula

$$C = \frac{N^2 - N}{2}$$

How many comparisons are necessary to sort 10,000 records? (See Exercise 138 in Exercise 1.3.)

1.1 Real Numbers and Their Graphs

■ SETS OF NUMBERS ■ EQUALITY, INEQUALITY SYMBOLS, AND VARIABLES ■ THE NUMBER LINE
■ GRAPHING SUBSETS OF THE REAL NUMBERS ■ ABSOLUTE VALUE OF A NUMBER

Getting Ready

1. Give an example of a number that is used for counting.

2. Give an example of a number that is used when dividing a pizza.

3. Give an example of a number that is used for measuring very cold temperatures.

4. What other types of numbers can you think of?

■ SETS OF NUMBERS

A **set** is a collection of objects. For example, the set

{1, 2, 3, 4, 5}

contains the numbers 1, 2, 3, 4, and 5. The members, or **elements,** in a set are listed within braces {   }.

Two basic sets of numbers are the set of **natural numbers** (often called the **positive integers**) and the **whole numbers.**

> **Natural Numbers**
> The **natural numbers** (or the **positive integers**) are the numbers
> 1, 2, 3, 4, 5, 6, 7, 8, 9, 10, . . .

Whole Numbers

The **whole numbers** are the numbers

0, 1, 2, 3, 4, 5, 6, 7, 8, 9, 10, . . .

The three dots in the previous definitions, called an **ellipsis,** indicate that the lists of numbers continue on forever.

We can use whole numbers to describe many real-life situations. For example, some cars get 30 miles per gallon of gas, and some students might pay $1,750 in tuition.

Numbers that show a loss or a downward direction are called **negative integers,** denoted as -1, -2, -3, and so on. For example, a debt of $1,500 can be denoted as $-\$1,500$, and a temperature of $20°$ below zero can be denoted as $-20°$.

The set of negative integers and the set of whole numbers together form the set of **integers.**

Integers

The **integers** are the numbers

. . . , -5, -4, -3, -2, -1, 0, 1, 2, 3, 4, 5, . . .

Because the set of natural numbers and the set of whole numbers are included within the set of integers, we say that these sets are **subsets** of the set of integers.

Integers cannot describe every real-life situation. For example, a student might study $3\frac{1}{2}$ hours, or a television set might cost $217.37. To describe these situations, we need fractions, more formally called **rational numbers.**

Rational Numbers

A **rational number** is any number that can be written as a fraction with an integer in its numerator and a nonzero integer in its denominator.

Some examples of rational numbers are

$$\frac{3}{2}, \quad \frac{17}{12}, \quad -\frac{43}{8}, \quad 0.25, \quad \text{and} \quad -0.66666\ldots$$

The decimals 0.25 and $-0.66666\ldots$ are rational numbers, because 0.25 can be written as the fraction $\frac{1}{4}$, and $-0.66666\ldots$ can be written as the fraction $-\frac{2}{3}$.

Since every integer can be written as a fraction with a denominator of 1, every integer is also a rational number. Since every integer is a rational number, the set of integers is a subset of the rational numbers.

WARNING! Because division by 0 is undefined, expressions such as $\frac{6}{0}$ and $\frac{0}{0}$ do not represent any number.

Numbers such as $\sqrt{2}$ and π are not rational numbers, because they cannot be written as fractions with an integer numerator and a nonzero integer denominator.

Such numbers are called **irrational numbers.** We can find decimal approximations for irrational numbers by using a calculator. For example,

$$\sqrt{2} \approx 1.414213562\ldots \qquad \text{Press } \boxed{\sqrt{}}\,. \text{ Read} \approx \text{as "is approximately equal to."}$$
$$\pi \approx 3.141592654\ldots \qquad \text{Press } \boxed{\pi}\,.$$

If we combine the rational and the irrational numbers, we have the set of real numbers.

Real Numbers

A **real number** is any number that is either a rational number or an irrational number.

Figure 1-1 shows how the various sets of numbers are interrelated.

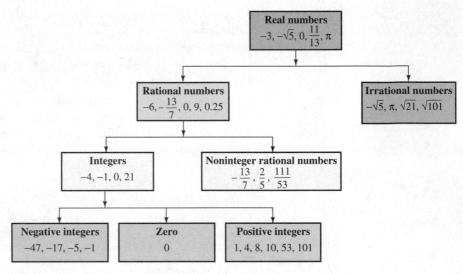

FIGURE 1-1

EXAMPLE 1   List the numbers in the set $\left\{-3, 0, \frac{1}{2}, 1.25, \sqrt{3}, 5\right\}$ that are   **a.** natural numbers, **b.** whole numbers,   **c.** negative integers,   **d.** rational numbers,   **e.** irrational numbers,   and   **f.** real numbers.

Solution   **a.** The only natural number is 5.

b. The whole numbers are 0 and 5.

c. The only negative integer is -3.

d. The rational numbers are $-3, 0, \frac{1}{2}, 1.25$, and 5. $\left(1.25 \text{ is rational, because } 1.25 \text{ can be written in the form } \frac{5}{4}.\right)$

e. The only irrational number is $\sqrt{3}$.

f. All of the numbers are real numbers.

Self Check : List the numbers in the set $\left\{-2, 0, 1.5, \sqrt{5}, 7\right\}$ that are **a.** positive integers and **b.** rational numbers.

Answers : **a.** 7, **b.** -2, 0, 1.5, 7

■ ■ ■ ■ ■ ■ ■ ■ ■ ■ PERSPECTIVE

Algebra is an extension of arithmetic. In algebra, the operations of addition, subtraction, multiplication, and division are performed on both numbers and letters, with the understanding that the letters represent numbers.

The origins of algebra are found in a papyrus written before 1600 B.C. by an Egyptian priest named Ahmes. This papyrus contains 84 algebra problems and their solutions. Because the Egyptians did not have a suitable system of notation, however, they were unable to develop algebra completely.

Further development of algebra occurred in the ninth century in the Middle East. In A.D. 830, an Arabian mathematician named al-Khowarazmi wrote a book called *Ihm al-jabr wa'l muqabalah*. This title was shortened to *al-Jabr*. We now know the subject as *algebra*. The French mathematician François Vieta (1540–1603) later simplified algebra by developing the symbolic notation that we use today.

The Ahmes Papyrus
(British Museum)

The natural numbers greater than 1 that can be divided evenly only by 1 and themselves are called **prime numbers.** The nonprime natural numbers greater than 1 are called **composite numbers.**

Prime numbers: {2, 3, 5, 7, 11, 13, 17, 19, 23, 29, . . .}
Composite numbers: {4, 6, 8, 9, 10, 12, 14, 15, 16, 18, 20, 21, 22, . . .}

Integers that can be divided evenly by 2 are called **even integers.** Integers that cannot be divided evenly by 2 are called **odd integers.**

Even integers: {. . . , $-10, -8, -6, -4, -2,$ 0, 2, 4, 6, 8, 10, . . .}
Odd integers: {. . . , $-9, -7, -5, -3, -1,$ 1, 3, 5, 7, 9, . . .}

EXAMPLE 2 : List the numbers in the set {$-3, -2,$ 0, 1, 2, 3, 4, 5, 9} that are   **a.** prime numbers, **b.** composite numbers,   **c.** even integers,   and   **d.** odd integers

Solution : **a.** The prime numbers are 2, 3, and 5.

b. The composite numbers and 4 and 9.

c. The even integers are -2, 0, 2, and 4.

d. The odd integers are -3, 1, 3, 5, and 9.   ■

List the numbers in the set $\{-5, 0, 1, 2, 4, 5\}$ that are   **a.** prime numbers and   **b.** even integers.

Answers   **a.** 2, 5,   **b.** 0, 2, 4

■ EQUALITY, INEQUALITY SYMBOLS, AND VARIABLES

To show that two expressions represent the same number, we use the **is equal to** sign ($=$). Since $4 + 5$ and 9 represent the same number, we can write

$4 + 5 = 9$ Read as "the sum of 4 and 5 is equal to 9."

Likewise, we can write

$5 - 3 = 2$ Read as "the difference between 5 and 3 equals 2," or "5 minus 3 equals 2."

$4 \cdot 5 = 20$ Read as "the product of 4 and 5 equals 20," or "4 times 5 equals 20."

and

$30 \div 6 = 5$ Read as "the quotient obtained when 30 is divided by 6 is 5," or "30 divided by 6 equals 5."

We can use **inequality symbols** to show that expressions are not equal.

Symbol	Read as
$\neq$	"is not equal to"
$<$	"is less than"
$>$	"is greater than"
$\leq$	"is less than or equal to"
$\geq$	"is greater than or equal to"

EXAMPLE 3   **Inequality symbols**

a. $6 \neq 9$ Read as "6 is not equal to 9."

b. $8 < 10$ Read as "8 is less than 10."

c. $12 > 1$ Read as "12 is greater than 1."

d. $5 \leq 5$ Read as "5 is less than or equal to 5." (Since $5 = 5$, this is a true statement.)

e. $9 \geq 7$ Read as "9 is greater than or equal to 7." (Since $9 > 7$, this is a true statement.)   ■

Self Check

Tell whether each statement is true or false: **a.** $12 \neq 12$, **b.** $7 \geq 7$, and **c.** $125 < 137$.

Answers

a. false, **b.** true, **c.** true

Inequality statements can be written so that the inequality symbol points in the opposite direction. For example,

$$5 < 7 \qquad \text{and} \qquad 7 > 5$$

both indicate that 5 is a smaller number than 7. Likewise,

$$12 \geq 3 \qquad \text{and} \qquad 3 \leq 12$$

both indicate that 12 is greater than or equal to 3.

In algebra, we use letters, called **variables,** to represent real numbers. For example,

- If x represents 4, then $x = 4$.
- If y represents any number greater than 3, then $y > 3$.
- If z represents any number less than or equal to -4, then $z \leq -4$.

 WARNING! In algebra, we usually do not use the times sign ($\times$) to indicate multiplication. It might be mistaken for the variable x.

■ THE NUMBER LINE

We can use the **number line** shown in Figure 1-2 to represent sets of numbers. The number line continues forever to the left and to the right. Numbers to the left of 0 are negative, and numbers to the right of 0 are positive.

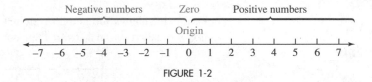

FIGURE 1-2

 WARNING! The number 0 is neither positive nor negative.

The number that corresponds to a point on the number line is called the **coordinate** of that point. For example, the coordinate of the **origin** is 0.

Many points on the number line do not have integer coordinates. For example, the point midway between 0 and 1 has the coordinate $\frac{1}{2}$, and the point midway between -3 and -2 has the coordinate $-\frac{5}{2}$ (see Figure 1-3).

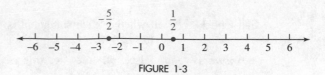

FIGURE 1-3

Numbers represented by points that lie on opposite sides of the origin and at equal distances from the origin are called **negatives** (or **opposites**) of each other. For example, 5 and −5 are negatives (or opposites). We need parentheses to express the opposite of a negative number. For example, −(−5) represents the opposite of −5, which we know to be 5. Thus,

$$-(-5) = 5$$

This suggests the following rule.

Double Negative Rule
If x represents a real number, then

$$-(-x) = x$$

If one point lies to the *right* of a second point on a number line, its coordinate is the *greater*. Since the point with coordinate 1 lies to the right of the point with co-ordinate −2 (see Figure 1.4(a)), it follows that $1 > -2$.

If one point lies to the *left* of another, its coordinate is the *smaller* (see Figure 1-4(b)). The point with coordinate −6 lies to the left of the point with coordinate −3, so it follows that $-6 < -3$.

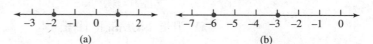

(a)   (b)

FIGURE 1-4

■ GRAPHING SUBSETS OF THE REAL NUMBERS

Figure 1-5 shows the graph of the natural numbers from 2 to 8. The points on the line are called the **graphs** of their corresponding coordinates.

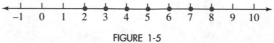

FIGURE 1-5

EXAMPLE 4   Graph the set of integers between −3 and 3.

Solution   The integers between −3 and 3 are −2, −1, 0, 1, and 2. The graph is shown in Fig-ure 1-6.

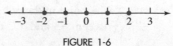

FIGURE 1-6

■

Self Check

Answer

Graph the set of integers between −4 and 0.

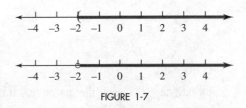

Graphs of many sets of real numbers are **intervals** on the number line. For example, two graphs of all real numbers x such that $x > -2$ are shown in Figure 1-7. The parenthesis or the open circle at −2 shows that this point is not included in the graph. The arrow pointing to the right shows that all numbers to the right of −2 are included.

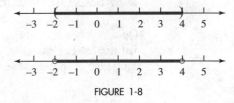

FIGURE 1-7

Figure 1-8 shows two graphs of the set of real numbers x between −2 and 4. This is the graph of all real numbers x such that $x > -2$ and $x < 4$. The parentheses or open circles at −2 and 4 show that these points are not included in the graph. However, all the numbers between −2 and 4 are included.

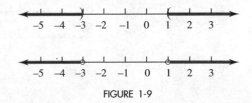

FIGURE 1-8

EXAMPLE 5

Graph all real numbers x such that $x < -3$ or $x > 1$.

Solution   The graph of all real numbers less than −3 includes all points on the number line that are to the left of −3. The graph of all real numbers greater than 1 includes all points that are to the right of 1. The two graphs are shown in Figure 1-9.

FIGURE 1-9   ■

Self Check

Answer

Graph all real numbers x such that $x < -1$ or $x > 0$. Use parentheses.

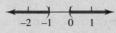

EXAMPLE 6   Graph the set of all real numbers from -5 to -1.

Solution   The set of all real numbers from -5 to -1 includes -5 and -1 and all the numbers in between. In the graphs shown in Figure 1-10, the brackets or the solid circles at -5 and -1 show that these points are included.

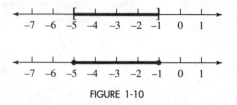

FIGURE 1-10   ■

Self Check   Graph the set of real numbers from -2 to 1. Use brackets.

Answer

■ ABSOLUTE VALUE OF A NUMBER

On a number line, the distance between a number x and 0 is called the **absolute value** of x. For example, the distance between 5 and 0 is 5 units (see Figure 1-11). Thus, the absolute value of 5 is 5:

$|5| = 5$ Read as "The absolute value of 5 is 5."

Since the distance between -6 and 0 is 6,

$|-6| = 6$ Read as "The absolute value of -6 is 6."

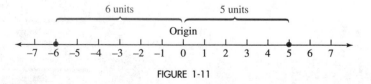

FIGURE 1-11

Because the absolute value of a real number represents that number's distance from 0 on the number line, the absolute value of every real number x is either positive or 0. In symbols, we say

$|x| \geq 0$ **for every real number x**

EXAMPLE 7 Evaluate **a.** $|6|$, **b.** $|-3|$, **c.** $|0|$, and **d.** $-|2+3|$.

Solution **a.** $|6| = 6$, because 6 is six units from 0.

b. $|-3| = 3$, because -3 is three units from 0.

c. $|0| = 0$, because 0 is zero units from 0.

d. $-|2+3| = -|5| = -5$ ∎

Self Check Evaluate **a.** $|8|$, **b.** $|-8|$, and **c.** $-|-8|$.
Answers **a.** 8, **b.** 8, **c.** -8

Orals *Describe each set of numbers in your own words.*

1. natural numbers 　　2. whole numbers
3. integers 　　4. rational numbers
5. real numbers 　　6. prime numbers
7. composite numbers 　　8. even integers
9. odd integers 　　10. irrational numbers

Find each value.

11. $-|15|$ 　　12. $|-25|$

EXERCISE 1.1

VOCABULARY AND CONCEPTS *Fill in each blank to make a true statement.*

1. A ___ is a collection of objects.

2. The numbers 1, 2, 3, 4, 5, . . . form the set of _____ numbers.

3. The set of _____ numbers is the set $\{0, 1, 2, 3, 4, 5, . . .\}$.

4. The set of _____ is the set $\{. . . , -3, -2, -1, 0, 1, 2, 3, . . .\}$.

5. Since every whole number is also an integer, the set of whole numbers is called a _____ of the set of integers.

6. $\sqrt{2}$ is an example of an _____ number.

7. If a natural number is greater than 1 and can be divided exactly only by 1 and itself, it is called a _____ number.

8. A composite number is a _____ number that is greater than 1 and is not _____.

9. The symbol $\neq$ means _____.

10. The symbol ___ means "is less than."

11. The symbol $\geq$ means _____.

12. The opposite of -7 is ___.

13. The figure

$$\xleftarrow{\quad|\quad|\quad|\quad|\quad|\quad|\quad|\quad}_{-3\;-2\;-1\;\;0\;\;1\;\;2\;\;3}\xrightarrow{}$$

is called a _____ line.

14. The distance between 8 and 0 on a number line is called the _____ of 8.

PRACTICE   *In Exercises 15–26, list the numbers in the set* $\left\{-3, -\frac{1}{2}, -1, 0, 1, 2, \frac{5}{3}, \sqrt{7}, 3.25, 6, 9\right\}$ *that are*

15. Natural numbers   **16.** Whole numbers   **17.** Positive integers   **18.** Negative integers

19. Integers   **20.** Rational numbers   **21.** Real numbers   **22.** Irrational numbers

23. Odd integers   **24.** Even integers   **25.** Composite numbers   **26.** Prime numbers

In Exercises 27–34, simplify each expression. Then classify the result as a natural number, an even integer, an odd integer, a prime number, a composite number, and/or a whole number.

27. $4 + 5$ **28.** $7 - 2$ **29.** $15 - 15$ **30.** $0 + 7$

31. $3 \cdot 8$ **32.** $8 \cdot 9$ **33.** $24 \div 8$ **34.** $3 \div 3$

In Exercises 35–48, place one of the symbols $=$, $<$, *and* $>$ *in each box to make a true statement.*

35. $5 \;\boxed{}\; 3 + 2$ **36.** $9 \;\boxed{}\; 7$ **37.** $25 \;\boxed{}\; 32$ **38.** $2 + 3 \;\boxed{}\; 17$

39. $5 + 7 \;\boxed{}\; 10$ **40.** $3 + 3 \;\boxed{}\; 9 - 3$ **41.** $3 + 9 \;\boxed{}\; 20 - 8$ **42.** $19 - 3 \;\boxed{}\; 8 + 6$

43. $4 \cdot 2 \;\boxed{}\; 2 \cdot 4$ **44.** $7 \cdot 9 \;\boxed{}\; 9 \cdot 6$ **45.** $8 \div 2 \;\boxed{}\; 4 + 2$ **46.** $0 \div 7 \;\boxed{}\; 1$

47. $3 + 2 + 5 \;\boxed{}\; 5 + 2 + 3$ **48.** $8 + 5 + 2 \;\boxed{}\; 5 + 2 + 8$

In Exercises 49–54, write each statement as a mathematical expression.

49. Seven is greater than three.   **50.** Five is less than thirty-two.

51. Eight is less than or equal to eight.   **52.** Twenty-five is not equal to twenty-three.

53. The result of adding three and four is equal to seven.   **54.** Thirty-seven is greater than or equal to the result of multiplying three and four.

In Exercises 55–66, rewrite each inequality statement as an equivalent inequality in which the inequality symbol points in the opposite direction.

55. $3 \leq 7$ **56.** $5 > 2$ **57.** $6 > 0$ **58.** $34 \leq 40$

59. $3 + 8 > 8$ **60.** $8 - 3 < 8$ **61.** $6 - 2 < 10 - 4$ **62.** $8 \cdot 2 \geq 8 \cdot 1$

63. $2 \cdot 3 < 3 \cdot 4$ **64.** $8 \div 2 \geq 9 \div 3$ **65.** $\dfrac{12}{4} < \dfrac{24}{6}$ **66.** $\dfrac{2}{3} \leq \dfrac{3}{4}$

In Exercises 67–74, graph each pair of numbers on a number line. In each pair, indicate which number is the greater and which number lies farther to the right.

67. 3, 6   **68.** 4, 7   **69.** 11, 6   **70.** 12, 10

71. 0, 2   **72.** 4, 10   **73.** 8, 0   **74.** 20, 30

In Exercises 75–86, graph each set of numbers on the number line.

75. The natural numbers between 2 and 8

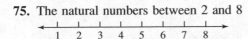

76. The prime numbers from 10 to 20

77. The even integers greater than 10 but less than 20

78. The even integers that are also prime numbers

79. The numbers that are whole numbers but not natural numbers

80. The prime numbers between 5 and 15

81. The natural numbers between 15 and 25 that are exactly divisible by 6

82. The odd integers between -5 and 5 that are exactly divisible by 3

83. The real numbers between 1 and 5

84. The real numbers greater than or equal to 8

85. The real numbers greater than or equal to 3 or less than or equal to -3

86. The real numbers greater than -2 and less than 3

In Exercises 87–94, find each absolute value.

87. $|36|$ **88.** $|-30|$ **89.** $|0|$ **90.** $|120|$

91. $|-230|$ **92.** $|18 - 12|$ **93.** $|12 - 20|$ **94.** $|100 - 100|$

WRITING

95. Explain why there is no greatest natural number.

96. Explain why 2 is the only even prime number.

97. Explain how to determine the absolute value of a number.

98. Explain why zero is an even integer.

SOMETHING TO THINK ABOUT   *Consider the following sets: the integers, natural numbers, even and odd integers, positive and negative numbers, prime and composite numbers, and rational numbers.*

99. Find a number that fits in as many of these categories as possible.

100. Find a number that fits in as few of these categories as possible.

1.2 Fractions

■ FRACTIONS ■ SIMPLIFYING FRACTIONS ■ MULTIPLYING FRACTIONS ■ DIVIDING FRACTIONS
■ ADDING FRACTIONS ■ SUBTRACTING FRACTIONS ■ MIXED NUMBERS ■ DECIMALS
■ ROUNDING DECIMALS ■ APPLICATIONS

Getting Ready

1. Add: 132
              45
              73

2. Subtract: 321
                   173

3. Multiply: 437
                   38

4. Divide: $37\overline{)3{,}885}$

■ FRACTIONS

In the **fractions**

$$\frac{1}{2}, \quad \frac{3}{5}, \quad \frac{2}{17}, \quad \text{and} \quad \frac{37}{7}$$

the number above the bar is called the **numerator,** and the number below the bar is called the **denominator.**

We often use fractions to indicate parts of a whole. In Figure 1-12(a), a rectangle has been divided into 5 equal parts, and 3 of the parts are shaded. The fraction $\frac{3}{5}$ indicates how much of the figure is shaded. In Figure 1-12(b), $\frac{5}{7}$ of the rectangle is shaded. In either example, the denominator of the fraction shows the total number of equal parts into which the whole is divided, and the numerator shows the number of these equal parts that are being considered.

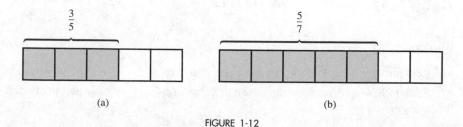

(a)   (b)

FIGURE 1-12

We can also use fractions to indicate division. For example, the fraction $\frac{8}{2}$ indicates that 8 is to be divided by 2:

$$\frac{8}{2} = 8 \div 2 = 4$$

WARNING! Note that $\frac{8}{2} = 4$, because $4 \cdot 2 = 8$, and that $\frac{0}{7} = 0$, because $0 \cdot 7 = 0$. However, $\frac{6}{0}$ is undefined, because no number multiplied by 0 gives 6. Since every number multiplied by 0 gives 0, $\frac{0}{0}$ is indeterminate. Remember that the denominator of a fraction cannot be 0.

■ SIMPLIFYING FRACTIONS

A fraction is in **lowest terms** when no integer other than 1 will divide both its numerator and its denominator exactly. The fraction $\frac{6}{11}$ is in lowest terms, because only 1 divides both 6 and 11 exactly. The fraction $\frac{6}{8}$ is not in lowest terms, because 2 divides both 6 and 8 exactly.

We can **simplify** a fraction that is not in lowest terms by dividing both its numerator and its denominator by the same number. For example, to simplify $\frac{6}{8}$, we divide both numerator and denominator by 2.

$$\frac{6}{8} = \frac{6 \div 2}{8 \div 2} = \frac{3}{4}$$

From Figure 1-13, we see that $\frac{6}{8}$ and $\frac{3}{4}$ are equal fractions, because each one represents the same part of the rectangle.

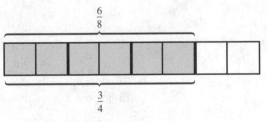

FIGURE 1-13

When a composite number has been written as the product of other natural numbers, we say that it has been **factored.** For example, 15 can be written as the product of 5 and 3.

$$15 = 5 \cdot 3$$

The numbers 5 and 3 are called **factors** of 15. When a composite number is written as the product of prime numbers, we say that it is written in **prime-factored form.**

EXAMPLE 1   Write 210 in prime-factored form.

Solution   We can write 210 as the product of 21 and 10 and proceed as follows:

$$210 = \mathbf{21} \cdot 10$$
$$210 = \mathbf{3 \cdot 7} \cdot \mathbf{2 \cdot 5} \qquad \text{Factor 21 as } 3 \cdot 7 \text{ and factor 10 as } 2 \cdot 5.$$

Since 210 is now written as the product of prime numbers, its prime-factored form is $210 = 2 \cdot 3 \cdot 5 \cdot 7$.   ∎

Self Check   Write 70 in prime-factored form.

Answer $2 \cdot 5 \cdot 7$

To simplify a fraction, we factor its numerator and its denominator and divide out all common factors that appear in both the numerator and denominator. For example,

$$\frac{6}{8} = \frac{3 \cdot 2}{4 \cdot 2} = \frac{3 \cdot \overset{1}{\cancel{2}}}{4 \cdot \underset{1}{\cancel{2}}} = \frac{3}{4} \qquad \text{and} \qquad \frac{15}{18} = \frac{5 \cdot 3}{6 \cdot 3} = \frac{5 \cdot \overset{1}{\cancel{3}}}{6 \cdot \underset{1}{\cancel{3}}} = \frac{5}{6}$$

WARNING!   Remember that a fraction is in lowest terms only when its numerator and denominator have no common factors.

EXAMPLE 2   Simplify each fraction, if possible.

a. To simplify $\frac{6}{30}$, we factor the numerator and denominator and divide out the common factor of 6.

$$\frac{6}{30} = \frac{6 \cdot 1}{6 \cdot 5} = \frac{\overset{1}{\cancel{6}} \cdot 1}{\underset{1}{\cancel{6}} \cdot 5} = \frac{1}{5}$$

b. To attempt to simplify $\frac{33}{40}$, we factor the numerator and denominator and hope to divide out any common factors.

$$\frac{33}{40} = \frac{3 \cdot 11}{2 \cdot 2 \cdot 2 \cdot 5}$$

Since the numerator and denominator have no common factors, $\frac{33}{40}$ is in lowest terms.   ∎

Self Check   Simplify $\frac{14}{35}$.

Answer $\frac{2}{5}$

The preceding examples illustrate the **fundamental property of fractions.**

The Fundamental Property of Fractions
If a, b, and x are real numbers, then

$$\frac{a \cdot x}{b \cdot x} = \frac{a}{b} \quad (b \neq 0 \text{ and } x \neq 0)$$

■ MULTIPLYING FRACTIONS

Multiplying Fractions
To multiply fractions, we multiply their numerators and multiply their denominators. In symbols, if a, b, c, and d are real numbers, then

$$\frac{a}{b} \cdot \frac{c}{d} = \frac{a \cdot c}{b \cdot d} \quad (b \neq 0 \text{ and } d \neq 0)$$

For example,

$$\frac{4}{7} \cdot \frac{2}{3} = \frac{4 \cdot 2}{7 \cdot 3} \quad \text{and} \quad \frac{4}{5} \cdot \frac{13}{9} = \frac{4 \cdot 13}{5 \cdot 9}$$

$$= \frac{8}{21} \qquad\qquad\qquad = \frac{52}{45}$$

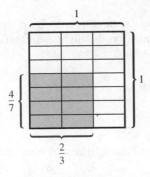

FIGURE 1-14

To justify the rule for multiplying fractions, we consider the square in Figure 1-14. Because the length of each side of the square is one unit and the area is the product of the lengths of two sides, the area is 1 square unit.

If this square is divided into 3 equal parts vertically and 7 equal parts horizontally, it is divided into 21 equal parts, and each represents $\frac{1}{21}$ of the total area. The area of the shaded rectangle in the square is $\frac{8}{21}$, because it contains 8 of the 21 parts. The width, w, of the shaded rectangle is $\frac{4}{7}$; its length, l, is $\frac{2}{3}$; and its area, A, is the product of l and w:

$$A = l \cdot w$$

$$\frac{8}{21} = \frac{2}{3} \cdot \frac{4}{7}$$

This suggests that we can find the product of

$$\frac{4}{7} \quad \text{and} \quad \frac{2}{3}$$

by multiplying their numerators and multiplying their denominators.

Fractions such as $\frac{8}{21}$ whose numerators are less than their denominators are called **proper fractions.** Fractions such as $\frac{52}{45}$ whose numerators are greater than their denominators are called **improper fractions.**

EXAMPLE 3   Do each multiplication.

a. $\dfrac{3}{7} \cdot \dfrac{13}{5} = \dfrac{3 \cdot 13}{7 \cdot 5}$ Multiply the numerators and multiply the denominators.

$= \dfrac{39}{35}$

b. $5 \cdot \dfrac{3}{15} = \dfrac{5}{1} \cdot \dfrac{3}{15}$ Write 5 as the improper fraction $\frac{5}{1}$.

$= \dfrac{5 \cdot 3}{1 \cdot 15}$ Multiply the numerators and multiply the denominators.

$= \dfrac{5 \cdot 3}{1 \cdot 5 \cdot 3}$ To attempt to simplify the fraction, factor the denominator.

$= \dfrac{\overset{1}{\cancel{5}} \cdot \overset{1}{\cancel{3}}}{1 \cdot \underset{1}{\cancel{5}} \cdot \underset{1}{\cancel{3}}}$ Divide out the common factors of 3 and 5.

$= 1$ $\frac{1 \cdot 1}{1 \cdot 1 \cdot 1} = 1.$ ■

Self Check   Multiply $\frac{5}{9} \cdot \frac{7}{10}$.
Answer
$\frac{7}{18}$

EXAMPLE 4   **European travel**   Out of 36 students in a history class, three-fourths have signed up for a trip to Europe. If there are 30 places available on the flight, will there be room for one more student?

Solution   We first find three-fourths of 36.

$\dfrac{3}{4} \cdot 36 = \dfrac{3}{4} \cdot \dfrac{36}{1}$ Write 36 as $\frac{36}{1}$.

$= \dfrac{3 \cdot 36}{4 \cdot 1}$ Multiply the numerators and multiply the denominators.

$= \dfrac{3 \cdot 4 \cdot 9}{4 \cdot 1}$ To simplify the fraction, factor the numerator.

$= \dfrac{3 \cdot \overset{1}{\cancel{4}} \cdot 9}{\underset{1}{\cancel{4}} \cdot 1}$ Divide out the common factor of 4.

$= \dfrac{27}{1}$

$= 27$

Twenty-seven students plan to go on the trip. Since there is room for 30 passengers, there is room for one more.   ■

Self Check   In Example 4, if five-sixths of the 36 students have signed up, will there be room for one more?

Answer   no

■ DIVIDING FRACTIONS

One number is called the **reciprocal** of another if their product is 1. For example, $\frac{3}{5}$ is the reciprocal of $\frac{5}{3}$, because

$$\frac{3}{5} \cdot \frac{5}{3} = \frac{15}{15} = 1$$

Dividing Fractions
To divide two fractions, we multiply the first fraction by the reciprocal of the second fraction. In symbols, if a, b, c, and d are real numbers, then

$$\frac{a}{b} \div \frac{c}{d} = \frac{a}{b} \cdot \frac{d}{c} = \frac{a \cdot d}{b \cdot c} \quad (b \neq 0, c \neq 0, \text{ and } d \neq 0)$$

EXAMPLE 5   Do each division.

a. $\dfrac{3}{5} \div \dfrac{6}{5} = \dfrac{3}{5} \cdot \dfrac{5}{6}$ Multiply $\frac{3}{5}$ by the reciprocal of $\frac{6}{5}$.

$= \dfrac{3 \cdot 5}{5 \cdot 6}$ Multiply the numerators and multiply the denominators.

$= \dfrac{3 \cdot 5}{5 \cdot 2 \cdot 3}$ Factor the denominator.

$= \dfrac{\overset{1}{\cancel{3}} \cdot \overset{1}{\cancel{5}}}{\underset{1}{\cancel{5}} \cdot 2 \cdot \underset{1}{\cancel{3}}}$ Divide out the common factors of 3 and 5.

$= \dfrac{1}{2}$

b. $\dfrac{15}{7} \div 10 = \dfrac{15}{7} \div \dfrac{10}{1}$ Write 10 as the improper fraction $\frac{10}{1}$.

$= \dfrac{15}{7} \cdot \dfrac{1}{10}$ Multiply $\frac{15}{7}$ by the reciprocal of $\frac{10}{1}$.

$= \dfrac{15 \cdot 1}{7 \cdot 10}$ Multiply the numerators and multiply the denominators.

$$= \frac{3 \cdot \overset{1}{\cancel{5}}}{7 \cdot 2 \cdot \underset{1}{\cancel{5}}}$$ Factor the numerator and the denominator, and divide out the common factor of 5.

$$= \frac{3}{14}$$ ∎

Self Check   Divide $\frac{13}{6} \div \frac{26}{8}$.

Answer $\frac{2}{3}$

■ ADDING FRACTIONS

Adding Fractions with the Same Denominator
To add fractions with the same denominator, we add the numerators and keep the common denominator. In symbols, if a, b, and d are real numbers, then

$$\frac{a}{d} + \frac{b}{d} = \frac{a + b}{d} \quad (d \neq 0)$$

For example,

$$\frac{3}{7} + \frac{2}{7} = \frac{3 + 2}{7}$$ Add the numerators and keep the common denominator.

$$= \frac{5}{7}$$

Figure 1-15 illustrates graphically why $\frac{3}{7} + \frac{2}{7} = \frac{5}{7}$.

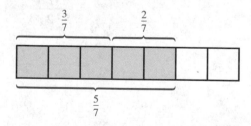

FIGURE 1-15

To add fractions with unlike denominators, we rewrite the fractions so that they have the same denominator. For example, we can multiply both the numerator and denominator of $\frac{1}{3}$ by 5 to obtain an equal fraction with a denominator of 15:

$$\frac{1}{3} = \frac{1 \cdot 5}{3 \cdot 5} = \frac{5}{15}$$

To rewrite $\frac{1}{5}$ as an equal fraction with a denominator of 15, we multiply the numerator and the denominator by 3:

$$\frac{1}{5} = \frac{1 \cdot 3}{5 \cdot 3} = \frac{3}{15}$$

Since 15 is the smallest number that can be used as a denominator for $\frac{1}{3}$ and $\frac{1}{5}$, it is called the **least** or **lowest common denominator** (the **LCD**).

To add the fractions $\frac{1}{3}$ and $\frac{1}{5}$, we rewrite each fraction as an equal fraction having a denominator of 15, and then we add the results:

$$\frac{1}{3} + \frac{1}{5} = \frac{1 \cdot 5}{3 \cdot 5} + \frac{1 \cdot 3}{5 \cdot 3}$$

$$= \frac{5}{15} + \frac{3}{15}$$

$$= \frac{5 + 3}{15}$$

$$= \frac{8}{15}$$

EXAMPLE 6   Add $\dfrac{3}{10} + \dfrac{5}{28}$.

Solution   To find the LCD, we find the prime factorization of each denominator and use each prime factor the greatest number of times it appears in either factorization:

$$\left.\begin{array}{l} 10 = 2 \cdot 5 \\ 28 = 2 \cdot 2 \cdot 7 \end{array}\right\} \text{LCD} = 2 \cdot 2 \cdot 5 \cdot 7 = 140$$

Since 140 is the smallest number that 10 and 28 divide exactly, we write both fractions as fractions with the LCD, 140.

$$\frac{3}{10} + \frac{5}{28} = \frac{3 \cdot 14}{10 \cdot 14} + \frac{5 \cdot 5}{28 \cdot 5} \qquad \text{Write each fraction as a fraction with a denominator of 140.}$$

$$= \frac{42}{140} + \frac{25}{140}$$

$$= \frac{42 + 25}{140} \qquad \text{Add the numerators and keep the denominator.}$$

$$= \frac{67}{140}$$

Since 67 is a prime number, it has no common factor with 140. Thus, $\frac{67}{140}$ is in lowest terms. ∎

Self Check   Add $\frac{3}{8} + \frac{5}{12}$.

Answer $\frac{19}{24}$

■ SUBTRACTING FRACTIONS

Subtracting Fractions with the Same Denominator
To subtract fractions with the same denominator, we subtract their numerators and keep their common denominator. In symbols, if a, b, and d are real numbers, then

$$\frac{a}{d} - \frac{b}{d} = \frac{a - b}{d} \quad (d \neq 0)$$

For example,

$$\frac{7}{9} - \frac{2}{9} = \frac{7 - 2}{9} = \frac{5}{9}$$

To subtract fractions with unlike denominators, we write them as equivalent fractions with a common denominator. For example, to subtract $\frac{2}{5}$ from $\frac{3}{4}$, we write $\frac{3}{4} - \frac{2}{5}$, find the LCD of 20, and proceed as follows:

$$\frac{3}{4} - \frac{2}{5} = \frac{3 \cdot 5}{4 \cdot 5} - \frac{2 \cdot 4}{5 \cdot 4}$$

$$= \frac{15}{20} - \frac{8}{20}$$

$$= \frac{15 - 8}{20}$$

$$= \frac{7}{20}$$

EXAMPLE 7 Subtract 5 from $\frac{23}{3}$.

Solution

$$\frac{23}{3} - 5 = \frac{23}{3} - \frac{5}{1}$$ Write 5 as the improper fraction $\frac{5}{1}$.

$$= \frac{23}{3} - \frac{5 \cdot 3}{1 \cdot 3}$$ Write $\frac{5}{1}$ as a fraction with a denominator of 3.

$$= \frac{23}{3} - \frac{15}{3}$$

$$= \frac{23 - 15}{3}$$ Subtract the numerators and keep the denominator.

$$= \frac{8}{3}$$

Self Check Subtract $\frac{5}{6} - \frac{3}{4}$.

Answer $\frac{1}{12}$

■ MIXED NUMBERS

The **mixed number** $3\frac{1}{2}$ represents the sum of 3 and $\frac{1}{2}$. We can write $3\frac{1}{2}$ as an improper fraction as follows:

$$3\frac{1}{2} = 3 + \frac{1}{2}$$

$$= \frac{6}{2} + \frac{1}{2} \qquad 3 = \frac{6}{2}.$$

$$= \frac{6 + 1}{2} \qquad \text{Add the numerators and keep the denominator.}$$

$$= \frac{7}{2}$$

To write the fraction $\frac{19}{5}$ as a mixed number, we divide 19 by 5 to get 3, with a remainder of 4.

$$\frac{19}{5} = 3 + \frac{4}{5} = 3\frac{4}{5}$$

EXAMPLE 8   Add $2\frac{1}{4} + 1\frac{1}{3}$.

Solution   We first change each mixed number to an improper fraction.

$$2\frac{1}{4} = 2 + \frac{1}{4} \qquad\qquad 1\frac{1}{3} = 1 + \frac{1}{3}$$

$$= \frac{8}{4} + \frac{1}{4} \qquad\qquad\quad = \frac{3}{3} + \frac{1}{3}$$

$$= \frac{9}{4} \qquad\qquad\qquad\quad = \frac{4}{3}$$

Then we add the fractions.

$$2\frac{1}{4} + 1\frac{1}{3} = \frac{9}{4} + \frac{4}{3}$$

$$= \frac{9 \cdot 3}{4 \cdot 3} + \frac{4 \cdot 4}{3 \cdot 4} \qquad \text{Change each fraction into a fraction with the LCD of 12.}$$

$$= \frac{27}{12} + \frac{16}{12}$$

$$= \frac{43}{12}$$

Finally, we change $\frac{43}{12}$ to a mixed number.

$$\frac{43}{12} = 3 + \frac{7}{12} = 3\frac{7}{12}$$

■

EXAMPLE 9

Fencing land The three sides of a triangular lot measure $33\frac{1}{4}$, $57\frac{3}{4}$, and $72\frac{1}{2}$ meters. How much fencing will be needed to enclose the area?

Solution We can find the sum of the lengths by adding the whole-number parts and the fractional parts of the dimensions separately:

$$33\frac{1}{4} + 57\frac{3}{4} + 72\frac{1}{2} = 33 + 57 + 72 + \frac{1}{4} + \frac{3}{4} + \frac{1}{2}$$

$$= 162 + \frac{1}{4} + \frac{3}{4} + \frac{2}{4} \qquad \text{Change } \tfrac{1}{2} \text{ to } \tfrac{2}{4} \text{ to obtain a common denominator.}$$

$$= 162 + \frac{6}{4} \qquad \text{Add the fractions by adding the numerators and keeping the common denominator.}$$

$$= 162 + \frac{3}{2} \qquad \tfrac{6}{4} = \tfrac{2\cdot3}{2\cdot2} = \tfrac{\cancel{2}\cdot3}{\cancel{2}\cdot2} = \tfrac{3}{2}.$$

$$= 162 + 1\frac{1}{2} \qquad \text{Change } \tfrac{3}{2} \text{ to a mixed number.}$$

$$= 163\frac{1}{2}$$

To enclose the area, $163\frac{1}{2}$ meters of fencing will be needed. ■

■ DECIMALS

Rational numbers can always be changed to decimal form. For example, to write $\frac{1}{4}$ and $\frac{5}{22}$ as decimals, we use long division:

```
    0.25              0.22727 . . .
 4)1.00           22)5.00000
    8                 4 4
   ――                 ――
   20                 60
   20                 44
   ――                ―――
                     160
                     154
                     ―――
                      60
                      44
                     ―――
                     160
```

The decimal 0.25 is called a **terminating decimal.** The decimal 0.2272727 . . . (often written as $0.2\overline{27}$) is called a **repeating decimal,** because it repeats the block of digits 27. Every rational number can be changed into either a **terminating** or a **repeating decimal.**

<table>
<tr><td colspan="2" align="center">*Terminating Decimals*</td><td colspan="2" align="center">*Repeating Decimals*</td></tr>
<tr><td>$\dfrac{1}{2} = 0.5$</td><td></td><td>$\dfrac{1}{3} = 0.33333$ . . .   or $0.\overline{3}$</td><td></td></tr>
<tr><td>$\dfrac{3}{4} = 0.75$</td><td></td><td>$\dfrac{1}{6} = 0.16666$ . . .   or $0.1\overline{6}$</td><td></td></tr>
<tr><td>$\dfrac{5}{8} = 0.625$</td><td></td><td>$\dfrac{5}{22} = 0.2272727$ . . .   or $0.2\overline{27}$</td><td></td></tr>
</table>

The decimal 0.5 has one **decimal place,** because it has one digit to the right of the decimal point. The decimal 0.75 has two decimal places, and 0.625 has three.

To *add* or *subtract* decimal fractions, we first align their decimal points and then add or subtract.

```
  25.568          25.568
 + 2.74          − 2.74
 -------         -------
  28.308          22.828
```

To do the previous operations with a scientific calculator, we would press these keys:

25.568 $\boxed{+}$ 2.74 $\boxed{=}$ and   25.568 $\boxed{-}$ 2.74 $\boxed{=}$

To *multiply* decimal fractions, we multiply the numbers and then place the decimal point so that the number of decimal places in the answer is equal to the sum of the decimal places in the factors.

```
     3.453        Here there are three decimal places.
  ×  9.25         Here there are two decimal places.
  --------
    17265
     6906
    31 077
  --------
  31.94025        The product has 3 + 2 = 5 decimal places.
```

To do this multiplication with a scientific calculator, we would press these keys:

3.453 $\boxed{\times}$ 9.25 $\boxed{=}$

To *divide* decimals, we move the decimal point in the divisor to the right to make the divisor a whole number. We then move the decimal point in the dividend the same number of places to the right.

$1.23\overline{)30.258}$ Move the decimal point in both the divisor and the dividend two places to the right.

We align the decimal point in the quotient with the repositioned decimal point in the dividend and then use long division.

$$
\begin{array}{r}
24.6 \\
123\overline{)3025.8} \\
\underline{246} \\
565 \\
\underline{492} \\
73\ 8 \\
\underline{73\ 8}
\end{array}
$$

To do the previous division with a scientific calculator, we would press these keys:

30.258 $\div$ 1.23 $=$

■ ROUNDING DECIMALS

When decimal fractions are long, we often **round** them to a specific number of decimal places. For example, the decimal fraction 25.36124 rounded to one place (or to the nearest tenth) is 25.4. Rounded to two places (or to the nearest one-hundredth), the decimal is 25.36. To round decimals, we use the following rules.

> **Rounding Decimals**
> 1. Determine to how many decimal places you wish to round.
> 2. Look at the first digit to the right of that decimal place.
> 3. If that digit is 4 or less, drop it and all digits that follow. If it is 5 or greater, add 1 to the digit in the position to which you wish to round, and drop all of the digits that follow.

■ APPLICATIONS

A **percent** is the numerator of a fraction with a denominator of 100. For example, $6\frac{1}{4}$ percent, written $6\frac{1}{4}\%$, is the fraction $\frac{6.25}{100}$, or the decimal 0.0625. In problems involving percent, the word *of* often indicates multiplication. For example, $6\frac{1}{4}\%$ of 8,500 is the product 0.0625(8,500).

EXAMPLE 10   **Auto loans**   Juan signs a one-year note to borrow $8,500 to buy a car. If the rate of interest is $6\frac{1}{4}\%$, how much interest will he pay?

Solution   For the privilege of using the bank's money for one year, Juan must pay $6\frac{1}{4}\%$ of $8,500. We calculate the interest, i, as follows:

$i = 6\frac{1}{4}\%$ of 8,500

$\quad = 0.0625 \cdot 8,500$ The word *of* means *times*.

$\quad = 531.25$

Juan will pay $531.25 interest.   ■

Self Check
Answer

In Example 10, how much interest will Juan pay if the rate is 9%?
$765

Orals *Simplify each fraction.*

1. $\dfrac{3}{6}$ **2.** $\dfrac{5}{10}$ **3.** $\dfrac{10}{20}$ **4.** $\dfrac{25}{75}$

Do each operation.

5. $\dfrac{5}{6} \cdot \dfrac{1}{2}$ **6.** $\dfrac{3}{4} \cdot \dfrac{3}{5}$ **7.** $\dfrac{2}{3} \div \dfrac{3}{2}$ **8.** $\dfrac{3}{5} \div \dfrac{5}{2}$

9. $\dfrac{4}{9} + \dfrac{7}{9}$ **10.** $\dfrac{6}{7} - \dfrac{3}{7}$ **11.** $\dfrac{2}{3} - \dfrac{1}{2}$ **12.** $\dfrac{3}{4} + \dfrac{1}{2}$

13. $2.5 + 0.36$ **14.** $3.45 - 2.21$

15. $0.2 \cdot 2.5$ **16.** $0.3 \cdot 13$

Round each decimal to two decimal places.

17. 3.244993   **18.** 3.24521

EXERCISE 1.2

REVIEW *Decide whether the following statements are true or false.*

1. 6 is an integer.   **2.** $\dfrac{1}{2}$ is a natural number.   **3.** 21 is a prime number.   **4.** No prime number is an even number.

5. $8 > -2$ **6.** $-3 < -2$ **7.** $9 \leq |-9|$ **8.** $|-11| \geq 10$

Place an appropriate symbol in each box to make the statement true.

9. $3 + 7 \,\boxed{}\, 10$ **10.** $\dfrac{3}{7} \,\boxed{}\, \dfrac{2}{7} = \dfrac{1}{7}$ **11.** $|-2| \,\boxed{}\, 2$ **12.** $4 + 8 \,\boxed{}\, 11$

VOCABULARY AND CONCEPTS *Fill in each blank to make a true statement.*

13. The number above the bar in a fraction is called the _____.

14. The number below the bar in a fraction is called the _____.

15. To _____ a fraction, we divide its numerator and denominator by the same number.

16. To write a number in prime-factored form, we write it as the product of _____ numbers.

17. If the numerator of a fraction is less than the denominator, the fraction is called a _____ fraction.

18. If the numerator of a fraction is greater than the denominator, the fraction is called an _____ fraction.

19. If the product of two numbers is ___, the numbers are called reciprocals.

20. $\dfrac{ax}{bx} = $ ___.

21. To multiply two fractions, _____ the numerators and multiply the denominators.

22. To divide two fractions, multiply the first fraction by the _____ of the second fraction.

23. To add fractions with a common denominator, add the _____ and keep the common _____.

24. To subtract fractions with a common denominator, _____ the numerators and keep the common _____.

25. $75\dfrac{2}{3}$ means 75 ___ $\dfrac{2}{3}$.

26. 0.75 is an example of a _____ decimal.

27. $5.3\overline{27}$ is an example of a _____ decimal.

28. A _____ is the numerator of a fraction whose denominator is 100.

PRACTICE   *In Exercises 29–36, write each fraction in lowest terms. If the fraction is already in lowest terms, so indicate.*

29. $\dfrac{6}{12}$

30. $\dfrac{3}{9}$

31. $\dfrac{15}{20}$

32. $\dfrac{22}{77}$

33. $\dfrac{24}{18}$

34. $\dfrac{35}{14}$

35. $\dfrac{72}{64}$

36. $\dfrac{26}{21}$

In Exercises 37–48, do each multiplication. Simplify each result when possible.

37. $\dfrac{1}{2} \cdot \dfrac{3}{5}$

38. $\dfrac{3}{4} \cdot \dfrac{5}{7}$

39. $\dfrac{4}{3} \cdot \dfrac{6}{5}$

40. $\dfrac{7}{8} \cdot \dfrac{6}{15}$

41. $\dfrac{5}{12} \cdot \dfrac{18}{5}$

42. $\dfrac{5}{4} \cdot \dfrac{12}{10}$

43. $\dfrac{17}{34} \cdot \dfrac{3}{6}$

44. $\dfrac{21}{14} \cdot \dfrac{3}{6}$

45. $12 \cdot \dfrac{5}{6}$

46. $9 \cdot \dfrac{7}{12}$

47. $\dfrac{10}{21} \cdot 14$

48. $\dfrac{5}{24} \cdot 16$

In Exercises 49–60, do each division. Simplify each result when possible.

49. $\dfrac{3}{5} \div \dfrac{2}{3}$

50. $\dfrac{4}{5} \div \dfrac{3}{7}$

51. $\dfrac{3}{4} \div \dfrac{6}{5}$

52. $\dfrac{3}{8} \div \dfrac{15}{28}$

53. $\dfrac{2}{13} \div \dfrac{8}{13}$

54. $\dfrac{4}{7} \div \dfrac{20}{21}$

55. $\dfrac{21}{35} \div \dfrac{3}{14}$

56. $\dfrac{23}{25} \div \dfrac{46}{5}$

57. $6 \div \dfrac{3}{14}$

58. $23 \div \dfrac{46}{5}$

59. $\dfrac{42}{30} \div 7$

60. $\dfrac{34}{8} \div 17$

In Exercises 61–84, do each addition or subtraction. Simplify each result when possible.

61. $\dfrac{3}{5} + \dfrac{3}{5}$

62. $\dfrac{4}{7} - \dfrac{2}{7}$

63. $\dfrac{4}{13} - \dfrac{3}{13}$

64. $\dfrac{2}{11} + \dfrac{9}{11}$

65. $\dfrac{1}{6} + \dfrac{1}{24}$

66. $\dfrac{17}{25} - \dfrac{2}{5}$

67. $\dfrac{3}{5} + \dfrac{2}{3}$

68. $\dfrac{4}{3} + \dfrac{7}{2}$

69. $\dfrac{9}{4} - \dfrac{5}{6}$

70. $\dfrac{2}{15} + \dfrac{7}{9}$

71. $\dfrac{7}{10} - \dfrac{1}{14}$

72. $\dfrac{7}{25} + \dfrac{3}{10}$

73. $3 - \dfrac{3}{4}$

74. $5 + \dfrac{21}{5}$

75. $\dfrac{17}{3} + 4$

76. $\dfrac{13}{9} - 1$

77. $4\dfrac{3}{5} + \dfrac{3}{5}$

78. $2\dfrac{1}{8} + \dfrac{3}{8}$

79. $3\dfrac{1}{3} - 1\dfrac{2}{3}$

80. $5\dfrac{1}{7} - 3\dfrac{2}{7}$

81. $3\dfrac{3}{4} - 2\dfrac{1}{2}$

82. $15\dfrac{5}{6} + 11\dfrac{5}{8}$

83. $8\dfrac{2}{9} - 7\dfrac{2}{3}$

84. $3\dfrac{4}{5} - 3\dfrac{1}{10}$

In Exercises 85–92, do each operation.

85. $23.45 + 135.2$

86. $345.213 - 27.35$

87. $67.235 - 22.45$

88. $12.17 + 3.457$

89. $3.4 \cdot 13.2$

90. $4.21 \cdot 2.73$

91. $0.23\overline{)1.0465}$

92. $4.7\overline{)10.857}$

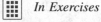 *In Exercises 93–100, use a calculator to do each operation. Round each answer to two decimal places.*

93. $323.24 + 27.2543$

94. $843.45213 - 712.765$

95. $55.77443 - 0.568245$

96. $0.62317 + 1.3316$

97. $25.25 \cdot 132.179$

98. $234.874 \cdot 242.46473$

99. $0.456\overline{)4.5694323}$

100. $43.225\overline{)32.465748}$

APPLICATIONS *Solve each problem.*

101. Buying fencing How many meters of fencing are needed to enclose the square field shown in Illustration 1?

$30\dfrac{2}{5}$ meters

ILLUSTRATION 1

102. Spring plowing A farmer has plowed $12\frac{1}{3}$ acres of a $43\frac{1}{2}$-acre field. How much more needs to be plowed?

103. Perimeter of a garden The four sides of a garden measure $7\frac{2}{3}$ feet, $15\frac{1}{4}$ feet, $19\frac{1}{2}$ feet, and $10\frac{3}{4}$ feet. Find the length of the fence needed to enclose the garden.

104. Making clothes A designer needs $3\frac{1}{4}$ yards of material for each dress he makes. How much material will he need to make 14 dresses?

105. Minority population 22% of the 11,431,000 citizens of Illinois are nonwhite. How many are nonwhite?

106. Quality control In the manufacture of active-matrix color LCD computer displays, many units must be rejected as defective. If 23% of a production run of 17,500 units is defective, how many units are acceptable?

107. Freeze-drying Almost all of the water must be removed when food is preserved by freeze-drying. Find the weight of the water removed from 750 pounds of a food that is 36% water.

108. Planning for growth This year, sales at Positronics Corporation totaled $18.7 million. If the projection of 12% annual growth is true, what will be next year's sales?

109. Speed skating In tryouts for the Olympics, a speed skater had times of 44.47, 43.24, 42.77, and 42.05 seconds. Find the average time. (*Hint:* Add the numbers and divide by 4.)

110. Cost of gasoline Otis drove his car 15,675.2 miles last year, averaging 25.5 miles per gallon of gasoline. If the average cost of gasoline was $1.27 per gallon, find the fuel cost to drive the car.

111. Paying taxes A woman earns $48,712.32 in taxable income. She must pay 15% tax on the first $23,000 and 28% on the rest. In addition, she must pay a Social Security tax of 15.4% on the total amount. How much tax will she need to pay?

112. Sealing asphalt A rectangular parking lot is 253.5 feet long and 178.5 feet wide. A 55-gallon drum of asphalt sealer covers 4,000 square feet and costs $97.50. Find the cost to seal the parking lot. (Sealer can only be purchased in full drums.)

113. Installing carpet What will it cost to carpet the area shown in Illustration 2 with carpet that costs $29.79 per square yard? (One square yard is 9 square feet.)

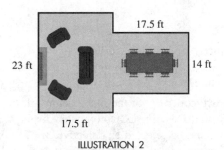

17.5 ft

23 ft   14 ft

17.5 ft

ILLUSTRATION 2

114. Inventory costs Each television a retailer buys costs $3.25 per day for warehouse storage. What does it cost to store 37 television sets for three weeks?

115. Manufacturing profits A manufacturer of computer memory boards has a profit of $37.50 on each standard-capacity memory board, and $57.35 on each high-capacity board. The sales department has orders for 2,530 standard boards and 1,670 high-capacity boards. Which order should production fill first, to receive the greater profit?

116. Dairy production A Holstein cow will produce 7,600 pounds of milk each year, with a $3\frac{1}{2}\%$ butterfat content. Each year, a Guernsey cow will produce about 6,500 pounds of milk that is 5% butterfat. Which cow produces more butterfat?

117. Feeding dairy cows Each year, a typical dairy cow will eat 12,000 pounds of food that is 57% silage. To feed 30 cows, how much silage will a farmer use in a year?

118. Comparing bids Two contractors bid on a home remodeling project. The first bids $9,350 for the entire job. The second contractor will work for $27.50 per hour, plus $4,500 for materials. He estimates that the job will take 150 hours. Which contractor has the lower bid?

119. Choosing a furnace A high-efficiency home heating system can be installed for $4,170, with an average monthly heating bill of $57.50. A regular furnace can be installed for $1,730, but monthly heating bills average $107.75. After three years, which system has cost more altogether?

120. Choosing a furnace Refer to Exercise 119. Decide which furnace system will have cost more after five years.

WRITING

121. Describe how you would find the common denominator of two fractions.

123. Explain how to convert a mixed number into an improper fraction.

122. Explain how to convert an improper fraction into a mixed number.

124. Explain how you would decide which of two decimal fractions is the larger.

SOMETHING TO THINK ABOUT

125. In what situations would it be better to leave an answer in the form of an improper fraction?

126. When would it be better to change an improper-fraction answer into a mixed number?

127. Can the product of two proper fractions be larger than either of the fractions?

128. How does the product of one proper and one improper fraction compare with the two factors?

1.3 Exponents and Order of Operations

■ EXPONENTS ■ ORDER OF OPERATIONS ■ GEOMETRY

Getting Ready *Do the operations.*

1. $2 \cdot 2$ **2.** $3 \cdot 3$ **3.** $3 \cdot 3 \cdot 3$ **4.** $2 \cdot 2 \cdot 2$

5. $\dfrac{1}{2} \cdot \dfrac{1}{2}$ **6.** $\dfrac{1}{3} \cdot \dfrac{1}{3} \cdot \dfrac{1}{3}$ **7.** $\dfrac{2}{5} \cdot \dfrac{2}{5} \cdot \dfrac{2}{5}$ **8.** $\dfrac{3}{10} \cdot \dfrac{3}{10} \cdot \dfrac{3}{10}$

■ EXPONENTS

To show how many times a number is to be used as a factor in a product, we use *exponents.* In the expression 2^3, 2 is called the **base** and 3 is called the **exponent.**

$$\text{Base} \longrightarrow 2^3 \longleftarrow \text{Exponent}$$

The exponent of 3 indicates that the base of 2 is to be used as a factor three times:

$$2^3 = \overbrace{2 \cdot 2 \cdot 2}^{3 \text{ factors of } 2} = 8$$

WARNING!   Note that $2^3 = 8$. This is not the same as $2 \cdot 3 = 6$.

In the expression x^5 (called an **exponential expression** or a **power of** x), 5 is the **exponent** and x is the **base.** The exponent of 5 indicates that a base of x is to be used as a factor five times.

$$x^5 = \overbrace{x \cdot x \cdot x \cdot x \cdot x}^{5 \text{ factors of } x}$$

In expressions such as x or y, the exponent is understood to be 1:

$$x = x^1 \quad \text{and} \quad y = y^1$$

In general, we have the following definition.

> **Natural-Number Exponents**
> If n is a natural number, then
> $$x^n = \overbrace{x \cdot x \cdot x \cdot \cdots \cdot x}^{n \text{ factors of } x}$$

EXAMPLE 1   Write each expression without using exponents.

   a. $4^2 = 4 \cdot 4 = 16$ Read 4^2 as "4 squared" or as "4 to the second power."

   b. $5^3 = 5 \cdot 5 \cdot 5 = 125$ Read 5^3 as "5 cubed" or as "5 to the third power."

   c. $6^4 = 6 \cdot 6 \cdot 6 \cdot 6 = 1{,}296$ Read 6^4 as "6 to the fourth power."

   d. $\left(\dfrac{2}{3}\right)^5 = \dfrac{2}{3} \cdot \dfrac{2}{3} \cdot \dfrac{2}{3} \cdot \dfrac{2}{3} \cdot \dfrac{2}{3} = \dfrac{32}{243}$ Read $\left(\dfrac{2}{3}\right)^5$ as "$\dfrac{2}{3}$ to the fifth power." ∎

Self Check   Evaluate   **a.** 7^2 and   **b.** $\left(\frac{3}{4}\right)^3$.
Answers   **a.** 49,   **b.** $\frac{27}{64}$

We can find powers using a calculator. For example, to find 2.35^4, we enter these numbers and press these keys:

2.35 y^x 4 $=$

The display will read 30.49800625 . Some calculators have a x^y key rather than a y^x key.

In the next example, the base of an exponential expression is a variable.

EXAMPLE 2   Write each expression without using exponents.

   a. $y^6 = y \cdot y \cdot y \cdot y \cdot y \cdot y$ Read y^6 as "y to the sixth power."

   b. $x^3 = x \cdot x \cdot x$ Read x^3 as "x cubed" or as "x to the third power."

   c. $z^2 = z \cdot z$ Read z^2 as "z squared" or as "z to the second power."

   d. $a^1 = a$ Read a^1 as "a to the first power." ∎

Self Check

Answers

Write each expression without using an exponent: **a.** a^3 and   **b.** b^4.
a. $a \cdot a \cdot a$,   **b.** $b \cdot b \cdot b \cdot b$

■ ORDER OF OPERATIONS

Suppose that you are asked to contact a friend if you see a Rolex watch for sale while traveling in Switzerland. After locating the watch, you send the following message to your friend.

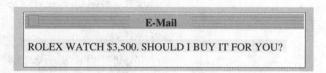

E-Mail

ROLEX WATCH $3,500. SHOULD I BUY IT FOR YOU?

The next day, you receive this response.

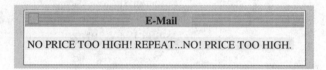

E-Mail

NO PRICE TOO HIGH! REPEAT...NO! PRICE TOO HIGH.

The first statement says to buy the watch at any price. The second says not to buy it, because it is too expensive. The placement of the exclamation point makes these statements read differently, resulting in different interpretations.

When reading a mathematical statement, the same kind of confusion is possible. To illustrate, we consider the expression $2 + 3 \cdot 4$, which contains the operations of addition and multiplication. We can calculate this expression in two different ways. We can do the addition first and then do the multiplication. Or we can do the multiplication first and then do the addition. However, we will get different results.

Method 1: Add First		*Method 2: Multiply First*	
$2 + 3 \cdot 4 = 5 \cdot 4$	Add 2 and 3.	$2 + 3 \cdot 4 = 2 + 12$	Multiply 3 and 4.
$= 20$	Multiply 5 and 4.	$= 14$	Add 2 and 12.

└──────────── Different results ────────────┘

To eliminate the possibility of getting different answers, we will agree to do multiplications before additions. The correct calculation of $2 + 3 \cdot 4$ is

$$2 + 3 \cdot 4 = 2 + 12$$
$$= 14$$

To indicate that additions should be done before multiplications, we must use **grouping symbols** such as parentheses ( ), brackets [ ], or braces { }. In the expression $(2 + 3)4$, the parentheses indicate that the addition is to be done first:

$$(\mathbf{2 + 3})4 = \mathbf{5} \cdot 4$$
$$= 20$$

To guarantee that calculations will have one correct result, we will always do calculations in the following order.

> ### Rules for the Order of Operations
> Use the following steps to do all calculations within each pair of grouping symbols, working from the innermost pair to the outermost pair.
> **1.** Find the values of any exponential expressions.
> **2.** Do all multiplications and divisions, working from left to right.
> **3.** Do all additions and subtractions, working from left to right.
> When all grouping symbols have been removed, repeat the rules above to finish the calculation.
>   In a fraction, simplify the numerator and the denominator separately. Then simplify the fraction, whenever possible.

> **WARNING!**   Note that $4(2)^3 \neq (4 \cdot 2)^3$:
> $$4(2)^3 = 4 \cdot 2 \cdot 2 \cdot 2 = 4(8) = 32 \quad \text{and} \quad (\mathbf{4 \cdot 2})^3 = \mathbf{8}^3 = 8 \cdot 8 \cdot 8 = 512$$
> Likewise, $4x^3 \neq (4x)^3$, because
> $$4x^3 = 4xxx \quad \text{and} \quad (4x)^3 = (4x)(4x)(4x) = 64x^3$$

EXAMPLE 3   Evaluate $5^3 + 2(8 - 3 \cdot 2)$.

Solution   We do the work within the parentheses first and then simplify.

$$5^3 + 2(8 - 3 \cdot 2) = 5^3 + 2(8 - 6)$$ Do the multiplication within the parentheses.

$$= 5^3 + 2(2)$$ Do the subtraction within the parentheses.

$$= 125 + 2(2)$$ Find the value of the exponential expression.

$$= 125 + 4$$ Do the multiplication.

$$= 129$$ Do the addition.   ∎

Self Check   Evaluate $5 + 4 \cdot 3^2$.

Answer   41

EXAMPLE 4   Evaluate $\dfrac{3(3 + 2) + 5}{17 - 3(4)}$.

Solution   We simplify the numerator and the denominator separately and then simplify the fraction.

$$\dfrac{3(3 + 2) + 5}{17 - 3(4)} = \dfrac{3(5) + 5}{17 - 3(4)} \qquad \text{Do the addition within the parentheses.}$$

$$= \dfrac{15 + 5}{17 - 12} \qquad \text{Do the multiplications.}$$

$$= \dfrac{20}{5} \qquad \text{Do the addition and the subtraction.}$$

$$= 4 \qquad \text{Do the division.} \qquad \blacksquare$$

Self Check   Evaluate $\dfrac{4 + 2(5 - 3)}{2 + 3(2)}$.

Answer   1

EXAMPLE 5   If $x = 3$ and $y = 4$, evaluate **a.** $3y + x^2$ and **b.** $3(y + x^2)$.

Solution   **a.** $3y + x^2 = 3(4) + 3^2$ Substitute 3 for x and 4 for y.

            $= 3(4) + 9$ Evaluate the exponential expression.

            $= 12 + 9$ Do the multiplication.

            $= 21$ Do the addition.

     b. $3(y + x^2) = 3(4 + 3^2)$ Substitute 3 for x and 4 for y.

                $= 3(4 + 9)$ Evaluate the exponential expression.

                $= 3(13)$ Do the addition in the parentheses.

                $= 39$ Do the multiplication. $\blacksquare$

Self Check   If $x = 2$ and $y = 3$, evaluate $2(x^2 + y^3)$.

Answer   62

EXAMPLE 6   If $x = 4$ and $y = 3$, evaluate $\dfrac{3x^2 - 2y}{2(x + y)}$.

Solution $\dfrac{3x^2 - 2y}{2(x + y)} = \dfrac{3(4^2) - 2(3)}{2(4 + 3)}$ Substitute 4 for x and 3 for y.

$= \dfrac{3(16) - 2(3)}{2(7)}$ Find the value of 4^2 in the numerator and do the addition in the denominator.

$= \dfrac{48 - 6}{14}$ Do the multiplications.

$= \dfrac{42}{14}$ Do the subtraction.

$= 3$ Do the division.   ■

Self Check   If $x = 2$ and $y = 5$, evaluate $\dfrac{x^2 + 6y}{2(x + y) + 3}$.

Answer   2

■ GEOMETRY

To find perimeters and areas of geometric figures, substituting numbers for variables is often required. The **perimeter** of a geometric figure is the distance around it, and the **area** of a figure is the amount of surface that it encloses. The perimeter of a circle is called its **circumference.**

EXAMPLE 7   **Circles** Find **a.** the circumference and **b.** the area of the circle shown in Figure 1-16.

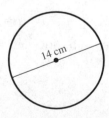

14 cm

FIGURE 1-16

Solution   **a.** The formula for the circumference of a circle is

$$C = \pi D$$

where C is the circumference, π is approximately $\frac{22}{7}$, and D is the diameter − the distance through the center of the circle. We can approximate the circumference by substituting $\frac{22}{7}$ for π and 14 for D in the formula and simplifying.

$$C = \pi D$$

$$C \approx \frac{22}{7} \cdot 14 \qquad \text{Read} \approx \text{as "is approximately equal to."}$$

$$C \approx \frac{22 \cdot \overset{2}{14}}{\underset{1}{7} \cdot 1} \qquad \text{Multiply the fractions and simplify.}$$

$$C \approx 44$$

The circumference is approximately 44 centimeters. To use a calculator, we would press these keys:

$$\boxed{\pi} \quad \boxed{\times} \quad \boxed{14} \quad \boxed{=}$$

The display will read 43.98229715. . . . The result is not 44, because a calculator uses a better approximation for π than $\frac{22}{7}$.

b. The formula for the area of a circle is

$$A = \pi r^2$$

where A is the area, $\pi \approx \frac{22}{7}$, and r is the **radius** of the circle. (The radius is one-half of the diameter.) We can approximate the area by substituting $\frac{22}{7}$ for π and 7 for r in the formula and simplifying.

$$A = \pi r^2$$

$$A \approx \frac{22}{7} \cdot 7^2$$

$$A \approx \frac{22}{7} \cdot \frac{49}{1} \qquad \text{Evaluate the exponential expression.}$$

$$A \approx \frac{22 \cdot \overset{7}{49}}{\underset{1}{7} \cdot 1} \qquad \text{Multiply the fractions and simplify.}$$

$$A \approx 154$$

The area is approximately 154 square centimeters. To use a calculator, we would press these keys:

$$\boxed{\pi} \quad \boxed{\times} \quad \boxed{7} \quad \boxed{x^2} \quad \boxed{=}$$

The display will read 153.93804.

Self Check   Find   **a.** the circumference   and   **b.** the area of a circle with a diameter of 28 meters. $\left(\text{Use } \frac{22}{7} \text{ as an estimate for } \pi.\right)$ Check your results with a calculator.

Answers   **a.** 88 m,   **b.** 616 m^2

Table 1-1 shows the formulas for the perimeter and area of several geometric figures.

TABLE 1-1

Figure	Name	Perimeter	Area
	Square	$P = 4s$	$A = s^2$
	Rectangle	$P = 2l + 2w$	$A = lw$
	Triangle	$P = a + b + c$	$A = \dfrac{1}{2}bh$
	Trapezoid	$P = a + b + c + d$	$A = \dfrac{1}{2}h(b + d)$
	Circle	$C = 2\pi r = \pi D$	$A = \pi r^2$

The **volume** of a three-dimensional geometric solid is the amount of space it encloses. Table 1-2 shows the formulas for the volume of several solids.

TABLE 1–2	Figure	Name	Volume
		Rectangular solid	$V = lwh$
		Cylinder	$V = Bh$, where B is the area of the base
		Pyramid	$V = \frac{1}{3}Bh$, where B is the area of the base
		Cone	$V = \frac{1}{3}Bh$, where B is the area of the base (If the base is a circle, then $B = \pi r^2$.)
		Sphere	$V = \frac{4}{3}\pi r^3$

EXAMPLE 8

Winter driving Find the number of cubic feet of road salt in the conical pile shown in Figure 1-17. Round the answer to two decimal places.

Solution We can find the area of the circular base by substituting $\frac{22}{7}$ for π and 14.30 for the radius.

$$A = \pi r^2$$

$$\approx \frac{22}{7}(14.3)^2$$

$$\approx 642.6828571 \qquad \text{Use a calculator.}$$

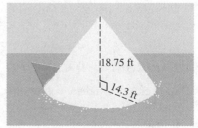

FIGURE 1-17

We then substitute 642.6828571 for B and 18.75 for h in the formula for the volume of a cone.

$$V = \frac{1}{3}Bh$$

$$\approx \frac{1}{3}(642.6828571)(18.75)$$

$$\approx 4{,}016.767857 \qquad \text{Use a calculator.}$$

To two decimal places, there are 4,016.77 cubic feet of salt in the pile. ∎

Self Check To the nearest hundredth, find the number of cubic feet of water that can be contained in a spherical tank that has a radius of 9 feet. $\left(\text{Use } \pi \approx \frac{22}{7}.\right)$

Answer 3,054.86 ft^3

Orals *Find the value of each expression.*

1. 2^5 **2.** 3^4 **3.** 4^3 **4.** 5^3

Simplify each expression.

5. $3(2)^3$ **6.** $(3 \cdot 2)^2$ **7.** $3 + 2 \cdot 4$
8. $10 - 3^2$ **9.** $4 + 2^2 \cdot 3$ **10.** $2 \cdot 3 + 2 \cdot 3^2$

EXERCISE 1.3

REVIEW

1. On the number line, graph the prime numbers between 10 and 20.

2. Write the inequality $7 \le 12$ as an inequality using the symbol $\ge$.

3. Classify the number 17 as a prime number or a composite number.

4. Evaluate $\dfrac{3}{5} - \dfrac{1}{2}$

VOCABULARY AND CONCEPTS *Fill in each blank to make a true statement.*

5. An _____ indicates how many times a base is to be used as a factor in a product.

6. In the expression x^5, x is called the _____ and 5 is called an _____.

7. In the expression $3 + 4 \cdot 5$, the _____ should be done first.

8. Parentheses, brackets, and braces are called _____ symbols.

Write the appropriate formula to find each quantity.

9. The perimeter of a square _____

10. The area of a square _____

11. The perimeter of a rectangle _____

12. The area of a rectangle _____

13. The perimeter of a triangle _____

14. The area of a triangle _____

15. The perimeter of a trapezoid _____

16. The area of a trapezoid _____

17. The circumference of a circle _____

18. The area of a circle _____

19. The volume of a rectangular solid _____

20. The volume of a cylinder _____

21. The volume of a pyramid _____

22. The volume of a cone _____

23. The volume of a sphere _____

24. In Exercises 20–22, B is the _____ of the base.

PRACTICE *In Exercises 25–30, find the value of each expression.*

25. 4^2

26. 5^2

27. 6^2

28. 7^3

29. $\left(\dfrac{1}{10}\right)^4$

30. $\left(\dfrac{1}{2}\right)^6$

In Exercises 31–34, use a calculator to find each power.

31. 7.9^3

32. 0.45^4

33. 25.3^2

34. 7.567^3

In Exercises 35–42, write each expression as the product of several factors.

35. x^2

36. y^3

37. $3z^4$

38. $5t^2$

39. $(5t)^2$

40. $(3z)^4$

41. $5(2x)^3$

42. $7(3t)^2$

In Exercises 43–50, find the value of each expression if $x = 3$ and $y = 2$.

43. $4x^2$

44. $4y^3$

45. $(5y)^3$

46. $(2y)^4$

47. $2x^y$

48. $3y^x$

49. $(3y)^x$

50. $(2x)^y$

In Exercises 51–78, simplify each expression by doing the operations.

51. $3 \cdot 5 - 4$

52. $4 \cdot 6 + 5$

53. $3(5 - 4)$

54. $4(6 + 5)$

55. $3 + 5^2$

56. $4^2 - 2^2$

57. $(3 + 5)^2$

58. $(5 - 2)^3$

59. $2 + 3 \cdot 5 - 4$

60. $12 + 2 \cdot 3 + 2$

61. $64 \div (3 + 1)$

62. $16 \div (5 + 3)$

63. $(7 + 9) \div (2 \cdot 4)$

64. $(7 + 9) \div 2 \cdot 4$

65. $(5 + 7) \div 3 \cdot 4$

66. $(5 + 7) \div (3 \cdot 4)$

67. $24 \div 4 \cdot 3 + 3$

68. $36 \div 9 \cdot 4 - 2$

69. $3^2 + 2(1 + 4) - 2$

70. $4 \cdot 3 + 2(5 - 2) - 2^3$

71. $5^2 - (7 - 3)^2$

72. $3^3 + (3 - 1)^3$

73. $(2 \cdot 3 - 4)^3$

74. $(3 \cdot 5 - 2 \cdot 6)^2$

75. $\dfrac{3}{5} \cdot \dfrac{10}{3} + \dfrac{1}{2} \cdot 12$

76. $\dfrac{15}{4}\left(1 + \dfrac{3}{5}\right)$

77. $\left[\dfrac{1}{3} - \left(\dfrac{1}{2}\right)^2\right]^2$

78. $\left[\left(\dfrac{2}{3}\right)^2 - \dfrac{1}{3}\right]^2$

In Exercises 79–86, use a calculator to simplify each fraction.

79. $\dfrac{(3+5)^2 + 2}{2(8-5)}$

80. $\dfrac{25 - (2 \cdot 3 - 1)}{2 \cdot 9 - 8}$

81. $\dfrac{(5-3)^2 + 2}{4^2 - (8+2)}$

82. $\dfrac{(4^2 - 2) + 7}{5(2+4) - 3^2}$

83. $\dfrac{2[4 + 2(3-1)]}{3[3(2 \cdot 3 - 4)]}$

84. $\dfrac{3[9 - 2(7-3)]}{(8-5)(9-7)}$

85. $\dfrac{3 \cdot 7 - 5(3 \cdot 4 - 11)}{4(3+2) - 3^2 + 5}$

86. $\dfrac{2 \cdot 5^2 - 2^2 + 3}{2(5-2)^2 - 11}$

In Exercises 87–110, evaluate each expression given that $x = 3$, $y = 2$, and $z = 4$.

87. $2x - y$

88. $2z + y$

89. $10 - 2x$

90. $15 - 3z$

91. $5z \div 2 + y$

92. $5x \div 3 + y$

93. $4x - 2z$

94. $5y - 3x$

95. $x + yz$

96. $3z + x - 2y$

97. $3(2x + y)$

98. $4(x + 3y)$

99. $(3 + x)y$

100. $(4 + z)y$

101. $(z + 1)(x + y)$

102. $3(z + 1) \div x$

103. $(x + y) \div (z + 1)$

104. $(2x + 2y) \div (3z - 2)$

105. $xyz + z^2 - 4x$

106. $zx + y^2 - 2z$

107. $3x^2 + 2y^2$

108. $3x^2 + (2y)^2$

109. $\dfrac{2x + y^2}{y + 2z}$

110. $\dfrac{2z^2 - y}{2x - y^2}$

In Exercises 111–114, insert parentheses in the expression $3 \cdot 8 + 5 \cdot 3$ to make its value equal to the given number.

111. 39

112. 117

113. 87

114. 69

In Exercises 115–118, find the perimeter of each figure.

115.

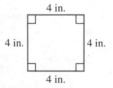
4 in.   4 in.   4 in.   4 in.

116.

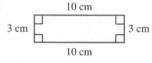
10 cm   3 cm   3 cm   10 cm

117.

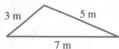
3 m   5 m   7 m

118.

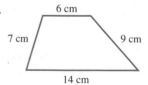
6 cm   7 cm   9 cm   14 cm

In Exercises 119–122, find the area of each figure.

119.

5 m   5m

120.

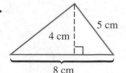
5 cm   4 cm   8 cm

121.

6 ft   10 ft

122.

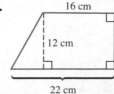
16 cm   12 cm   22 cm

In Exercises 123–124, find the circumference of each circle. Use $\pi \approx \frac{22}{7}$.

123.

14 m

124.

21 cm

In Exercises 125–126, find the area of each circle. Use $\pi \approx \frac{22}{7}$.

125.

42 ft

126.

7 m

In Exercises 127–132, find the volume of each solid. Use $\pi \approx \frac{22}{7}$.

127.

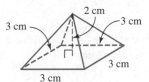

2 cm
3 cm
3 cm
3 cm
3 cm

128.

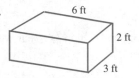

6 ft
2 ft
3 ft

129.

6 m

130.

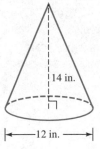

14 in.
12 in.

131.

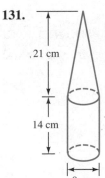

21 cm
14 cm
8 cm

132.

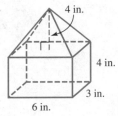

4 in.
4 in.
3 in.
6 in.

 In Exercises 133–138, use a calculator. For π, use the $\boxed{\pi}$ *key. Round to two decimal places.*

133. Volume of a tank   Find the number of cubic feet of water in a spherical tank with a radius of 21.35 feet.

134. Storing solvents   A hazardous solvent fills a rectangular tank with dimensions of 12 inches by 9.5 inches by 7.3 inches. For disposal, it must be transferred to a cylindrical canister 7.5 inches in diameter and 18 inches high. How much solvent will be left over?

135. Volume of a classroom   Thirty students are in a classroom with dimensions of 40 feet by 40 feet by 9 feet. How many cubic feet of air are there for each student?

136. Wallpapering One roll of wallpaper covers about 33 square feet. At $27.50 per roll, how much would it cost to paper two walls 8.5 feet high and 17.3 feet long? (*Hint:* Wallpaper can only be purchased in full rolls.)

137. Focal length The focal length f of a double-convex thin lens is given by the formula

$$f = \frac{rs}{(r + s)(n - 1)}$$

If $r = 8$, $s = 12$, and $n = 1.6$, find f.

138. Sorting records A selection sort requires C comparisons to sort N records, where C and N are related by the formula

$$C = \frac{N^2 - N}{2}$$

How many comparisons are needed to sort 10,000 records?

WRITING

139. Explain why the symbols $3x$ and x^3 have different meanings.

140. Students often say that x^n means "x multiplied by itself n times." Explain why this is not correct.

SOMETHING TO THINK ABOUT

141. If x were greater than 1, would raising x to higher and higher powers produce bigger numbers or smaller numbers?

142. What would happen in Exercise 141 if x were a positive number that was less than 1?

1.4 Adding and Subtracting Real Numbers

■ ADDING REAL NUMBERS WITH LIKE SIGNS ■ ADDING REAL NUMBERS WITH UNLIKE SIGNS
■ SUBTRACTING REAL NUMBERS ■ USING A CALCULATOR TO ADD AND SUBTRACT REAL NUMBERS

Getting Ready   *Do each operation.*

1. $14.32 + 3.2$ **2.** $5.54 - 2.6$ **3.** $4.2 - (3 - 0.8)$

4. $(5.42 - 4.22) - 0.2$ **5.** $(437 - 198) - 143$ **6.** $437 - (198 - 143)$

■ ADDING REAL NUMBERS WITH LIKE SIGNS

Since the positive direction on the number line is to the right, positive numbers can be represented by arrows pointing to the right. Negative numbers can be represented by arrows pointing to the left.

To add the integers $+2$ and $+3$, we can represent $+2$ with an arrow the length of 2, pointing to the right. We can then represent $+3$ with an arrow of length 3, also pointing to the right. To add the numbers, we place the arrows end to end, as in Figure 1-18. Since the endpoint of the second arrow is the point with coordinate $+5$, we have

$$(+2) + (+3) = +5$$

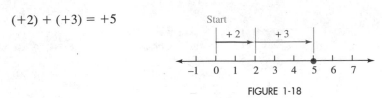

FIGURE 1-18

As a check, we can think of this problem in terms of money. If you had $2 and earned $3 more, you would have a total of $5.

The addition problem

$$(-2) + (-3)$$

can be represented by the arrows shown in Figure 1-19. Since the endpoint of the final arrow is the point with coordinate -5, we have

$$(-2) + (-3) = -5$$

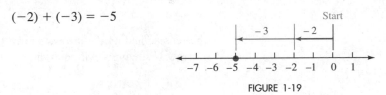

FIGURE 1-19

As a check, we can think of this problem in terms of money. If you lost $2 and then lost $3 more, you would have lost a total of $5.

Because two real numbers with the same sign can be represented by arrows pointing in the same direction, we have the following rule.

Adding Real Numbers with Like Signs

To find the sum of two real numbers with the same sign, add their absolute values and keep their common sign.

EXAMPLE 1   **Adding real numbers**

a. $(+4) + (+6) = +(4 + 6)$
$$= 10$$

b. $(-4) + (-6) = -(4 + 6)$
$$= -10$$

c. $+5 + (+10) = +(5 + 10)$
$$= 15$$

d. $-\dfrac{1}{2} + \left(-\dfrac{3}{2}\right) = -\left(\dfrac{1}{2} + \dfrac{3}{2}\right)$
$$= -\dfrac{4}{2}$$
$$= -2$$ ■

Add   **a.** $(+0.5) + (+1.2)$ and   **b.** $(-3.7) + (-2.3)$.
a. 1.7,   **b.** -6

WARNING! We do not need to write a + sign in front of a positive number:

$$+4 = 4 \quad \text{and} \quad +5 = 5$$

However, we must always remember to write a − sign in front of a negative number.

■ ADDING REAL NUMBERS WITH UNLIKE SIGNS

Real numbers with unlike signs can be represented by arrows on a number line that point in opposite directions. For example, the addition problem

$$(-6) + (+2)$$

can be represented by the arrows shown in Figure 1-20. Since the endpoint of the final arrow is the point with coordinate -4, we have

$$(-6) + (+2) = -4$$

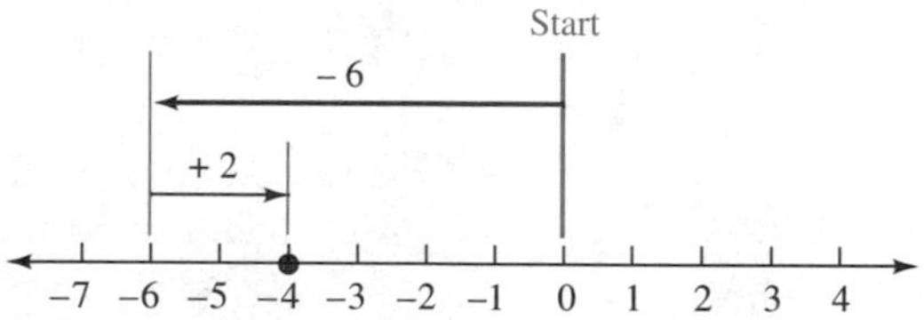

FIGURE 1-20

As a check, we can think of this problem in terms of money. If you lost $6 and then earned $2, you would still have a loss of $4.

The addition problem

$$(+7) + (-4)$$

can be represented by the arrows shown in Figure 1-21. Since the endpoint of the final arrow is the point with coordinate $+3$, we have

$$(+7) + (-4) = +3$$

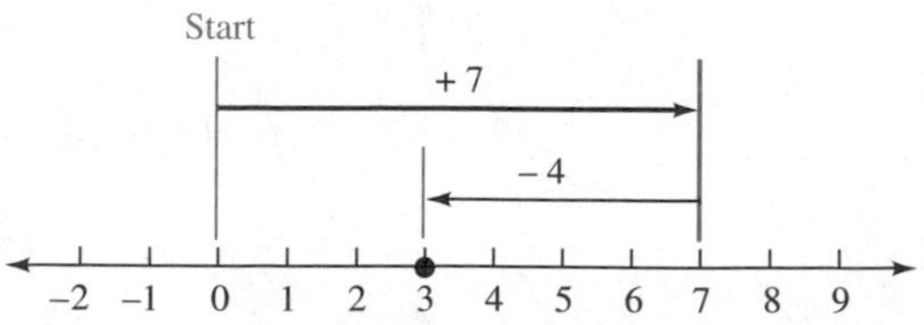

FIGURE 1-21

As a check, you can think of this problem in terms of money. If you had $7 and then lost $4, you would still have a gain of $3.

Because two real numbers with unlike signs can be represented by arrows pointing in opposite directions, we have the following rule.

Adding Real Numbers with Unlike Signs

To find the sum of two real numbers with unlike signs, subtract their absolute values (the smaller from the larger) and use the sign of the number with the greater absolute value.

EXAMPLE 2   **Adding real numbers**

a. $(+6) + (-5) = +(6 - 5)$
$$= 1$$

b. $(-2) + (+3) = +(3 - 2)$
$$= 1$$

c. $+6 + (-9) = -(9 - 6)$
$$= -3$$

d. $-\dfrac{2}{3} + \left(+\dfrac{1}{2}\right) = -\left(\dfrac{2}{3} - \dfrac{1}{2}\right)$
$$= -\left(\dfrac{4}{6} - \dfrac{3}{6}\right)$$
$$= -\dfrac{1}{6}$$ ∎

Self Check

Answers

Add   **a.** $(+3.5) + (-2.6)$ and   **b.** $(-7.2) + (+4.7)$.

a. 0.9,   **b.** -2.5

EXAMPLE 3   **Working with grouping symbols**

a. $[(+3) + (-7)] + (-4) = [-4] + (-4)$ Do the work within the brackets first.
$$= -8$$

b. $-3 + [(-2) + (-8)] = -3 + [-10]$ Do the work within the brackets first.
$$= -13$$ ∎

Self Check

Answer

Add $-2 + [(+5.2) + (-12.7)]$.

-9.5

EXAMPLE 4   If $x = -4$, $y = 5$, and $z = -13$, evaluate   **a.** $x + y$ and   **b.** $2y + z$.

Solution   We substitute -4 for x, 5 for y, and -13 for z. Then we simplify.

a. $x + y = (-4) + (5)$
$$= 1$$

b. $2y + z = 2 \cdot 5 + (-13)$
$$= 10 + (-13)$$
$$= -3$$ ∎

Self Check   If $x = 5$, $y = -3$, and $z = -4$, evaluate $2y + 3z + x$.
Answer -13

Sometimes numbers are added vertically, as shown in the next example.

EXAMPLE 5   **Adding numbers in a vertical format**

a.	$+5$	**b.**	$+5$	**c.**	-5	**d.**	-5
	$\underline{+2}$		$\underline{-2}$		$\underline{+2}$		$\underline{-2}$
	$+7$		$+3$		-3		-7	■

Self Check   Add   **a.** $+3.2$ and   **b.** -13.5
$\underline{-5.4}$ $\underline{-\ 4.3}$.

Answers   **a.** -2.2,   **b.** -17.8

Words and phrases such as *found, gain, credit, up, increase, forward, rises, in the future,* and *to the right* indicate a positive direction. Words and phrases such as *lost, loss, debit, down, backward, falls, in the past,* and *to the left* indicate a negative direction.

EXAMPLE 6   **Account balance**   The treasurer of a math club opens a checking account by depositing $350 in the bank. The bank debits the account $9 for check printing, and the treasurer writes a check for $22. Find the balance after these transactions.

Solution   The deposit can be represented by $+350$. The debit of $9 can be represented by -9, and the check written for $22 can be represented by -22. The balance in the account after these transactions is the sum of 350, -9, and -22.

$$350 + (-9) + (-22) = 341 + (-22) \qquad \text{Work from left to right.}$$
$$= 319$$

The balance is $319. ■

Self Check   Find the balance in Example 6 if another deposit of $17 is made.
Answer $336

■ SUBTRACTING REAL NUMBERS

In arithmetic, subtraction is a take-away process. For example,

$$7 - 4 = 3$$

can be thought of as taking 4 objects away from 7 objects, leaving 3 objects.
For algebra, a better approach treats the subtraction problem

$$7 - 4$$

as the equivalent addition problem:

$$7 + (-4)$$

In either case, the answer is 3.

$$7 - 4 = 3 \quad \text{and} \quad 7 + (-4) = 3$$

Thus, to subtract 4 from 7, we can add the negative (or opposite) of 4 to 7. In general, we have the following rule.

Subtracting Real Numbers

If a and b are two real numbers, then

$$a - b = a + (-b)$$

EXAMPLE 7 Evaluate **a.** $12 - 4$, **b.** $-13 - 5$, and **c.** $-14 - (-6)$.

Solution **a.** $12 - 4 = 12 + (-4)$ To subtract 4, add the opposite of 4.

$$= 8$$

b. $-13 - 5 = -13 + (-5)$ To subtract 5, add the opposite of 5.

$$= -18$$

c. $-14 - (-6) = -14 + [-(-6)]$ To subtract -6, add the opposite of -6.

$$= -14 + 6 \quad\quad\quad \text{The opposite of } -6 \text{ is } 6.$$

$$= -8$$

∎

Self Check Evaluate **a.** $-12.7 - 8.9$ and **b.** $15.7 - (-11.3)$.

Answers **a.** -21.6, **b.** 27

EXAMPLE 8 If $x = -5$ and $y = -3$, evaluate **a.** $\dfrac{y - x}{7 + x}$ and **b.** $\dfrac{6 + x}{y - x} - \dfrac{y - 4}{7 + x}$.

Solution We can substitute -5 for x and -3 for y into each expression and simplify.

a. $\dfrac{y - x}{7 + x} = \dfrac{-3 - (-5)}{7 + (-5)}$

$$= \dfrac{-3 + [-(-5)]}{2} \quad\quad\quad \text{To subtract } -5, \text{ add the opposite of } -5.$$

$$= \dfrac{-3 + 5}{2} \quad\quad\quad\quad\quad -(-5) = 5.$$

$$= \dfrac{2}{2}$$

$$= 1$$

b. $\dfrac{6 + x}{y - x} - \dfrac{y - 4}{7 + x} = \dfrac{6 + (-5)}{-3 - (-5)} - \dfrac{-3 - 4}{7 + (-5)}$

$$= \dfrac{1}{-3 + 5} - \dfrac{-3 + (-4)}{2} \qquad -(-5) = +5.$$

$$= \dfrac{1}{2} - \dfrac{-7}{2}$$

$$= \dfrac{1 - (-7)}{2}$$

$$= \dfrac{1 + [-(-7)]}{2} \qquad \text{To subtract } -7, \text{ add the opposite of } -7.$$

$$= \dfrac{1 + 7}{2} \qquad -(-7) = 7.$$

$$= \dfrac{8}{2}$$

$$= 4 \qquad\qquad\qquad\qquad\qquad\qquad \blacksquare$$

Self Check If $a = -3$ and $b = -5$, evaluate $\dfrac{7 - a}{b - a + 3}$.

Answer 10

To use a vertical format for subtracting real numbers, we add the opposite of the number that is to be subtracted by changing the sign of the lower number (called the **subtrahend**) and proceeding as in addition.

EXAMPLE 9 Do each subtraction by doing an equivalent addition.

a. The subtraction $-\;\dfrac{5}{\underline{-4}}$ becomes the addition $+\;\dfrac{5}{\underline{+4}}$
$$\qquad\qquad\qquad\qquad\qquad\qquad\qquad\qquad\qquad 9$$

b. The subtraction $-\;\dfrac{-8}{\underline{+3}}$ becomes the addition $+\;\dfrac{-8}{\underline{-3}}$
$$\qquad\qquad\qquad\qquad\qquad\qquad\qquad\qquad\quad -11 \qquad \blacksquare$$

Self Check Do the subtraction: $\dfrac{5.8}{\underline{-\;-4.6}}$.

Answer 10.4

EXAMPLE 10   Simplify   **a.** $3 - [4 + (-6)]$ and   **b.** $[-5 + (-3)] - [-2 - (+5)]$.

Solution   **a.** $3 - [4 + (-6)] = 3 - (-2)$ Do the addition within the brackets first.

$\qquad\qquad\qquad = 3 + [-(-2)]$ To subtract -2, add the opposite of -2.

$\qquad\qquad\qquad = 3 + 2$ $-(-2) = 2$.

$\qquad\qquad\qquad = 5$

b. $[-5 + (-3)] - [-2 - (+5)]$

$\qquad = [-5 + (-3)] - [-2 + (-5)]$ To subtract $+5$, add the opposite of 5.

$\qquad = -8 - (-7)$ Do the work within the brackets.

$\qquad = -8 + [-(-7)]$ To subtract -7, add the opposite of -7.

$\qquad = -8 + 7$ $-(-7) = 7$.

$\qquad = -1$ ■

Self Check   Simplify $[7.2 - (-3)] - [3.2 + (-1.7)]$.

Answer   8.7

EXAMPLE 11   **Temperature change**   At noon, the temperature was 7° above zero. At midnight, the temperature was 4° below zero. Find the difference between these two temperatures.

Solution   A temperature of 7° above zero can be represented as $+7$. A temperature of 4° below zero can be represented as -4. To find the difference between these temperatures, we can set up a subtraction problem and simplify.

$$7 - (-4) = 7 + [-(-4)] \qquad \text{To subtract } -4, \text{ add the opposite of } -4.$$

$$= 7 + 4 \qquad\qquad -(-4) = 4.$$

$$= 11$$

The difference between the temperatures is 11°. Figure 1-22 shows this difference.

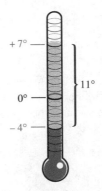

FIGURE 1-22   ■

Self Check   Find the difference between temperatures of 32° and −10°.

Answer   42°

■ USING A CALCULATOR TO ADD AND SUBTRACT REAL NUMBERS

A calculator can add positive and negative numbers.

- You do not have to do anything special to enter positive numbers. When you press 5, for example, a positive 5 is entered.

- To enter −5 into a scientific calculator, you must press the +/− key. This key is called the *plus–minus* or *change-of-sign* key.

To evaluate −345.678 + (−527.339), we enter these numbers and press these keys:

345.678 +/−   +   527.339 +/−   =

The display will read -873.017 .

Orals   *Find each value.*

1. $2 + 3$ **2.** $2 + (-5)$ **3.** $-4 + 7$

4. $-5 + (-6)$ **5.** $6 - 2$ **6.** $-8 - 4$

7. $-5 - (-7)$ **8.** $12 - (-4)$

9. $-5 + (3 - 4)$ **10.** $(-5 + 3) - 4$

EXERCISE 1.4

REVIEW   *If $x = 5$, $y = 7$, and $z = 2$, evaluate each expression.*

1. $x + 3(y - z)$ **2.** $(x + 3)(y - z)$ **3.** $x + 3y - z$ **4.** $(x + 3)y - z$

VOCABULARY AND CONCEPTS   *Fill in each blank to make a true statement.*

5. Positive and negative numbers can be represented by _____ on the number line.

6. To find the sum of two real numbers with like signs, ____ their absolute values and _____ their common sign.

7. To find the sum of two real numbers with unlike signs, _____ their absolute values and use the sign of the number with the _____ absolute value.

8. $a - b =$ _____

9. To subtract a number, we ____ its _____.

10. The subtraction

$$\begin{array}{r} 35 \\ -45 \\ \hline \end{array}$$

is equivalent to the subtraction 35 _____.

In Exercises 11–26, find each sum.

11. $4 + 8$ **12.** $(-4) + (-2)$ **13.** $(-3) + (-7)$ **14.** $(+4) + 11$

15. $6 + (-4)$ **16.** $5 + (-3)$ **17.** $9 + (-11)$ **18.** $10 + (-13)$

19. $(-0.4) + 0.9$ **20.** $(-1.2) + (-5.3)$ **21.** $\dfrac{1}{5} + \left(+\dfrac{1}{7}\right)$ **22.** $\dfrac{2}{3} + \left(-\dfrac{1}{4}\right)$

23. $\begin{array}{r} 5 \\ +\underline{-4} \end{array}$ **24.** $\begin{array}{r} -20 \\ +\underline{-17} \end{array}$ **25.** $\begin{array}{r} -1.3 \\ +\underline{3.5} \end{array}$ **26.** $\begin{array}{r} 1.3 \\ +\underline{-2.5} \end{array}$

In Exercises 27–38, evaluate each expression.

27. $5 + [4 + (-2)]$ **28.** $-6 + [(-3) + 8]$ **29.** $-2 + (-4 + 5)$ **30.** $5 + [-4 + (-6)]$

31. $[-4 + (-3)] + [2 + (-2)]$ **32.** $[3 + (-1)] + [-2 + (-3)]$

33. $-4 + (-3 + 2) + (-3)$ **34.** $5 + [2 + (-5)] + (-2)$

35. $-|8 + (-4)| + 7$ **36.** $\left| \dfrac{3}{5} + \left(-\dfrac{4}{5}\right) \right|$

37. $-5.2 + |-2.5 + (-4)|$ **38.** $6.8 + |8.6 + (-1.1)|$

In Exercises 39–52, let $x = 2$, $y = -3$, $z = -4$, and $u = 5$. Evaluate each expression.

39. $x + y$ **40.** $x + z$ **41.** $x + z + u$ **42.** $y + z + u$

43. $(x + u) + 3$ **44.** $(y + 5) + x$ **45.** $x + (-1 + z)$ **46.** $-7 + (z + x)$

47. $(x + z) + (u + z)$ **48.** $(z + u) + (x + y)$

49. $x + [5 + (y + u)]$ **50.** $y + \{[u + (z + (-6)]\} + y$

51. $|2x + y|$ **52.** $3|x + y + z|$

In Exercises 53–68, find each difference.

53. $8 - 4$ **54.** $-8 - 4$ **55.** $8 - (-4)$ **56.** $-9 - (-5)$

57. $0 - (-5)$ **58.** $0 - 75$ **59.** $\dfrac{5}{3} - \dfrac{7}{6}$ **60.** $-\dfrac{5}{9} - \dfrac{5}{3}$

61. $-3\dfrac{1}{2} - 5\dfrac{1}{4}$ **62.** $2\dfrac{1}{2} - \left(-3\dfrac{1}{2}\right)$ **63.** $-6.7 - (-2.5)$ **64.** $25.3 - 17.5$

65. $\begin{array}{r} 8 \\ \underline{-4} \end{array}$ **66.** $\begin{array}{r} 8 \\ -\underline{-3} \end{array}$ **67.** $\begin{array}{r} -10 \\ -\underline{3} \end{array}$ **68.** $\begin{array}{r} -13 \\ -\underline{5} \end{array}$

In Exercises 69–78, evaluate each quantity.

69. $+3 - [(-4) - 3]$ **70.** $-5 - [4 - (-2)]$ **71.** $(5 - 3) + (3 - 5)$ **72.** $(3 - 5) - [5 - (-3)]$

73. $5 - [4 + (-2) - 5]$ **74.** $3 - [-(-2) + 5]$

75. $\left(\dfrac{5}{2} - 3\right) - \left(\dfrac{3}{2} - 5\right)$ **76.** $\left(\dfrac{7}{3} - \dfrac{5}{6}\right) - \left[\dfrac{5}{6} - \left(-\dfrac{7}{3}\right)\right]$

77. $(5.2 - 2.5) - (5.25 - 5)$

78. $\left(3\frac{1}{2} - 2\frac{1}{2}\right) - \left[5\frac{1}{3} - \left(-5\frac{2}{3}\right)\right]$

In Exercises 79–86, let $x = -4$, $y = 5$, and $z = -6$. Evaluate each quantity.

79. $y - x$

80. $y - z$

81. $x - y - z$

82. $y + z - x$

83. $x - (y - z)$

84. $y + (z - x)$

85. $\dfrac{y - x}{3 - z}$

86. $\dfrac{y}{x - z} - \dfrac{x}{8 + z}$

In Exercises 87–90, let $a = 2$, $b = -3$, and $c = -4$. Evaluate each quantity.

87. $a + b - c$

88. $a - b + c$

89. $\dfrac{a + b}{b - c}$

90. $\dfrac{c - a}{-(a + b)}$

In Exercises 91–94, use a calculator to evaluate each quantity. Let $x = 2.34$, $y = 3.47$, and $z = 0.72$. Round the answers to one decimal place.

91. $x^3 - y + z^2$

92. $y - z^2 - x^2$

93. $x^2 - y^2 - z^2$

94. $z^3 - x^2 + y^3$

APPLICATIONS *Use signed numbers to solve each problem.*

95. College tuition A student owed $575 in tuition. If she earned a scholarship that would pay $400 of the bill, what did she still owe?

96. Dieting Scott weighed 212 pounds but lost 24 pounds during a diet. What does Scott weigh now?

97. Temperature The temperature rose 13 degrees in 1 hour and then dropped 4 degrees in the next hour. What signed number represents the net change in temperature?

98. Mountain climbing A team of mountaineers climbed 2,347 feet one day but then came down 597 feet to a good spot to make camp. What signed number represents their net change in altitude?

99. Temperature The temperature fell from zero to 14° below one night. By 5:00 P.M. the next day, the temperature had risen 10 degrees. What was the temperature at 5:00 P.M.?

100. History In 1897, Joseph Thompson discovered the electron. Fifty-four years later, the first fission reactor was built. Nineteen years before the reactor was erected, James Chadwick discovered the neutron. In what year was the neutron discovered?

101. History The Greek mathematician Euclid was alive in 300 B.C. The English mathematician Sir Isaac Newton was alive in A.D. 1700. How many years apart did they live?

102. Banking Abdul deposited $212 in a new checking account, wrote a check for $173, and deposited another $312. Find the balance in his account.

103. Military science An army retreated 2,300 meters. After regrouping, it moved forward 1,750 meters. The next day it gained another 1,875 meters. What was the army's net gain?

104. Football A football player gained and lost the following yardage on six consecutive plays: $+5$, $+7$, -5, $+1$, -2, and -6. How many yards were gained or lost?

105. Aviation A pilot flying at 32,000 feet is instructed to descend to 28,000 feet. How many feet must he descend?

106. Stock market Tuesday's high and low prices for Transitronics stock were $37\frac{1}{8}$ and $31\frac{5}{8}$. Find the range of prices for this stock.

107. Temperature Find the difference between a temperature of 32° above zero and a temperature of 27° above zero.

108. **Temperature**   Find the difference between a temperature of 3° below zero and a temperature of 21° below zero.

109. **Stock market**   At the opening bell on Monday, the Dow Jones Industrial Average was 9,153. At the close, the Dow was down 23 points, but news of a half-point drop in interest rates on Tuesday sent the market up 57 points. What was the Dow average after the market closed on Tuesday?

110. **Stock market**   On a Monday morning, the Dow Jones Industrial Average opened at 8,917. For the week, the Dow rose 29 points on Monday and 12 points on Wednesday. However, it fell 53 points on Tuesday and 27 points on both Thursday and Friday. Where did the Dow close on Friday?

111. **Stock splits**   A man owned 500 shares of Transitronics Corporation before the company declared a two-for-one stock split. After the split, he sold 300 shares. How many shares does the man now own?

112. **Small business**   Maria earned $2,532 in a part-time business. However, $633 of the earnings went for taxes. Find Maria's net earnings.

In Exercises 113–116, use a calculator.

113. **Balancing the books**   On January 1, Sally had $437.45 in the bank. During the month, she had deposits of $25.17, $37.93, and $45.26, and she had withdrawals of $17.13, $83.44, and $22.58. How much was in her account at the end of the month?

114. **Small business**   The owner of a small business has a gross income of $97,345.32. However, he paid $37,675.66 in expenses plus $7,537.45 in taxes, $3,723.41 in health care premiums, and $5,767.99 in pension payments. Find his profit.

115. **Closing a real estate transaction**   A woman sold her house for $115,000. Her fees at closing were $78 for preparing a deed, $446 for title work, $216 for revenue stamps, and a sales commission of $7,612.32. In addition, there was a deduction of $23,445.11 to pay off her old mortgage. As part of the deal, the buyer agreed to pay half of the title work. How much money did the woman receive after closing?

116. **Winning the lottery**   Mike won $500,000 in a state lottery. He will get $\frac{1}{20}$ of the sum each year for the next 20 years. After he receives his first installment, he plans to pay off a car loan of $7,645.12 and give his son $10,000 for college. By paying off the car loan, he will receive a rebate of 2% of the loan. If he must pay income tax of 28% on his first installment, how much will he have left to spend?

WRITING

117. Explain why the sum of two negative numbers is always negative, and the sum of two positive numbers is always positive.

118. Explain why the sum of a negative number and a positive number could be either negative or positive.

SOMETHING TO THINK ABOUT

119. Think of two numbers. First, add the absolute values of the two numbers, and write your answer. Second, add the two numbers, take the absolute value of that sum, and write that answer. Do the two answers agree? Can you find two numbers that produce different answers? When do you get answers that agree, and when don't you?

120. "Think of a very small number," requests the teacher. "One one-millionth," answers Charles. "Negative one million," responds Mia. Explain why either answer might be considered correct.

1.5 Multiplying and Dividing Real Numbers

■ MULTIPLYING REAL NUMBERS ■ DIVIDING REAL NUMBERS ■ USING A CALCULATOR TO MULTIPLY AND DIVIDE REAL NUMBERS

Getting Ready *Find each product or quotient.*

1. 8×7 **2.** 9×6 **3.** 8×9 **4.** 7×9

5. $\dfrac{81}{9}$ **6.** $\dfrac{48}{8}$ **7.** $\dfrac{64}{8}$ **8.** $\dfrac{56}{7}$

■ MULTIPLYING REAL NUMBERS

Because the times sign, $\times$, looks like the letter x, it is seldom used in algebra. Instead, a dot, parentheses, or no symbol at all is used to denote multiplication. Each of the following expressions indicates the **product** obtained when two real numbers x and y are multiplied.

$$x \cdot y \qquad (x)(y) \qquad x(y) \qquad (x)y \qquad xy$$

To develop rules for multiplying real numbers, we rely on the definition of multiplication. The expression $5 \cdot 4$ indicates that 4 is to be used as a term in a sum five times. That is,

$$5(4) = 4 + 4 + 4 + 4 + 4 = 20$$

Read 5(4) as "5 times 4."

Likewise, the expression $5(-4)$ indicates that -4 is to be used as a term in a sum five times. Thus,

$$5(-4) = (-4) + (-4) + (-4) + (-4) + (-4) = -20$$

Read $5(-4)$ as "5 times negative 4."

If multiplying by a positive number indicates repeated addition, it is reasonable that multiplication by a negative number indicates repeated subtraction. The expression $(-5)4$, for example, means that 4 is to be used as a term in a repeated subtraction five times. That is,

$$(-5)4 = -(4) - (4) - (4) - (4) - (4)$$
$$= (-4) + (-4) + (-4) + (-4) + (-4)$$
$$= -20$$

Likewise, the expression $(-5)(-4)$ indicates that -4 is to be used as a term in a repeated subtraction five times. Thus,

$$(-5)(-4) = -(-4) - (-4) - (-4) - (-4) - (-4)$$
$$= -(-4) + [-(-4)] + [-(-4)] + [-(-4)] + [-(-4)]$$
$$= 4 + 4 + 4 + 4 + 4$$
$$= 20$$

The expression $0(-2)$ indicates that -2 is to be used zero times as a term in a repeated addition. Thus,

$$0(-2) = 0$$

Finally, the expression $(-3)(1) = -3$ suggests that the product of any number and 1 is the number itself.

The previous results suggest the following rules.

Rules for Multiplying Signed Numbers

1. The product of two real numbers with like signs is the product of their absolute values.

2. The product of two real numbers with unlike signs is the negative of the product of their absolute values.

3. Any number multiplied by 0 is 0: $a \cdot 0 = 0 \cdot a = 0$.

4. Any number multiplied by 1 is that number itself: $a \cdot 1 = 1 \cdot a = a$.

EXAMPLE 1   Find each product: **a.** $4(-7)$,   **b.** $(-5)(-4)$,   **c.** $(-7)(6)$,   **d.** $8(6)$,
e. $(-3)(5)(-4)$,   and   **f.** $(-4)(-2)(-3)$.

Solution   **a.** $4(-7) = -(4 \cdot 7)$ **b.** $(-5)(-4) = +(5 \cdot 4)$
$= -28$ $= +20$

c. $(-7)(6) = -(7 \cdot 6)$ **d.** $8(6) = +(8 \cdot 6)$
$= -42$ $= +48$

e. $(-3)(5)(-4) = (-15)(-4)$ **f.** $(-4)(-2)(-3) = 8(-3)$
$= 60$ $= -24$ ■

Self Check   Find each product: **a.** $-7(5)$,   **b.** $-12(-7)$,   and   **c.** $-2(-4)(-9)$.
Answers   **a.** -35,   **b.** 84,   **c.** -72

EXAMPLE 2   If $x = -3$, $y = 2$, and $z = 4$, evaluate   **a.** $y + xz$ and   **b.** $x(y - z)$.

Solution   We substitute -3 for x, 2 for y, and 4 for z in each expression and simplify.

a. $y + xz = 2 + (-3)(4)$ **b.** $x(y - z) = -3[2 - 4]$
$= 2 + (-12)$ $= -3[2 + (-4)]$
$= -10$ $= -3(-2)$
$= 6$ ■

Self Check   If $x = -4$, $y = -3$, and $z = 5$, evaluate $x - yz$.
Answer   11

EXAMPLE 3   If $x = -2$ and $y = 3$, evaluate   **a.** $x^2 - y^2$ and   **b.** $-x^2$.

Solution   **a.** We substitute -2 for x and 3 for y and simplify.

$$x^2 - y^2 = (-2)^2 - 3^2$$
$$= 4 - 9 \qquad \text{Simplify the exponential expressions first.}$$
$$= -5 \qquad \text{Do the subtraction.}$$

b. We substitute -2 for x and simplify.

$$-x^2 = -(-2)^2$$
$$= -4 \qquad\qquad (-2)^2 = 4.$$

Self Check   If $a = -3.2$ and $b = -5$, evaluate $a^2 - 2b^3$.

Answer   260.24

EXAMPLE 4   Find each product: **a.** $\left(-\dfrac{2}{3}\right)\left(-\dfrac{6}{5}\right)$ and   **b.** $\left(\dfrac{3}{10}\right)\left(-\dfrac{5}{9}\right)$.

Solution   **a.** $\left(-\dfrac{2}{3}\right)\left(-\dfrac{6}{5}\right) = +\left(\dfrac{2}{3} \cdot \dfrac{6}{5}\right)$ **b.** $\left(\dfrac{3}{10}\right)\left(-\dfrac{5}{9}\right) = -\dfrac{3}{10} \cdot \dfrac{5}{9}$

$$= +\dfrac{2 \cdot 6}{3 \cdot 5} \qquad\qquad\qquad = -\dfrac{3 \cdot 5}{10 \cdot 9}$$

$$= +\dfrac{12}{15} \qquad\qquad\qquad = -\dfrac{15}{90}$$

$$= +\dfrac{4}{5} \qquad\qquad\qquad = -\dfrac{1}{6}$$

Self Check   Evaluate   **a.** $\frac{3}{5}\left(-\frac{10}{9}\right)$ and   **b.** $-\left(\frac{15}{8}\right)\left(-\frac{16}{5}\right)$.

Answers   **a.** $-\frac{2}{3}$,   **b.** 6

EXAMPLE 5   **Temperature change**   If the temperature is dropping $4°$ each hour, how much warmer was it 3 hours ago?

Solution   A temperature drop of $4°$ per hour can be represented by $-4°$ per hour. "Three hours ago" can be represented by -3. The temperature 3 hours ago is the product of -3 and -4.

$$(-3)(-4) = +12$$

The temperature was $12°$ warmer 3 hours ago.

Self Check   How much colder will it be after 5 hours?

Answer $20°$ colder

■ DIVIDING REAL NUMBERS

We know that 8 divided by 4 is 2 and 18 divided by 6 is 3.

$$\frac{8}{4} = 2, \text{ because } 2 \cdot 4 = 8 \qquad \frac{18}{6} = 3, \text{ because } 3 \cdot 6 = 18$$

These examples suggest that the following rule

$$\frac{a}{b} = c \quad \text{if and only if} \quad c \cdot b = a$$

is true for the division of any real number a by any nonzero real number b. For example,

$$\frac{+10}{+2} = +5, \text{ because } (+5)(+2) = +10$$

$$\frac{-10}{-2} = +5, \text{ because } (+5)(-2) = -10$$

$$\frac{+10}{-2} = -5, \text{ because } (-5)(-2) = +10$$

$$\frac{-10}{+2} = -5, \text{ because } (-5)(+2) = -10$$

These examples suggest the rules for dividing real numbers.

Rules for Dividing Signed Numbers

1. The quotient of two real numbers with like signs is the quotient of their absolute values.

2. The quotient of two real numbers with unlike signs is the negative of the quotient of their absolute values.

3. $\dfrac{a}{0}$ is undefined; $\dfrac{0}{0}$ is indeterminate.

4. If $a \neq 0$, then $\dfrac{0}{a} = 0$.

EXAMPLE 6   Find each quotient: **a.** $\dfrac{36}{18}$,   **b.** $\dfrac{-44}{11}$,   **c.** $\dfrac{27}{-9}$,   and   **d.** $\dfrac{-64}{-8}$.

Solution   **a.** $\dfrac{36}{18} = +\dfrac{36}{18} = 2$ The quotient of two numbers with like signs is the quotient of their absolute values.

b. $\dfrac{-44}{11} = -\dfrac{44}{11} = -4$ The quotient of two numbers with unlike signs is the negative of the quotient of their absolute values.

c. $\dfrac{27}{-9} = -\dfrac{27}{9} = -3$ The quotient of two numbers with unlike signs is the negative of the quotient of their absolute values.

d. $\dfrac{-64}{-8} = +\dfrac{64}{8} = 8$ The quotient of two numbers with like signs is the quotient of their absolute values.   ■

Self Check   Find each quotient: **a.** $\dfrac{-72.6}{12.1}$ and   **b.** $\dfrac{-24.51}{-4.3}$.

Answers   **a.** -6,   **b.** 5.7

EXAMPLE 7   If $x = -64$, $y = 16$, and $z = -4$, evaluate

a. $\dfrac{yz}{-x}$ and   **b.** $\dfrac{z^3 y}{x}$.

Solution   We substitute -64 for x, 16 for y, and -4 for z in each expression and simplify.

a. $\dfrac{yz}{-x} = \dfrac{16(-4)}{-(-64)}$ **b.** $\dfrac{z^3 y}{x} = \dfrac{(-4)^3(16)}{-64}$

$\qquad = \dfrac{-64}{+64}$ $\qquad = \dfrac{(-64)(16)}{(-64)}$

$\qquad = -1$ $\qquad = 16$ ■

Self Check   Evaluate $\dfrac{x + y}{-z^2}$, given the values in Example 7.

Answer   3

EXAMPLE 8   If $x = -50$, $y = 10$, and $z = -5$, evaluate

a. $\dfrac{xyz}{x - 5z}$ and   **b.** $\dfrac{3xy + 2yz}{2(x + y)}$.

Solution   We substitute -50 for x, 10 for y, and -5 for z in each expression and simplify.

a. $\dfrac{xyz}{x - 5z} = \dfrac{(-50)(10)(-5)}{-50 - 5(-5)}$ **b.** $\dfrac{3xy + 2yz}{2(x + y)} = \dfrac{3(-50)(10) + 2(10)(-5)}{2(-50 + 10)}$

$\qquad = \dfrac{(-500)(-5)}{-50 + 25}$ $\qquad = \dfrac{-150(10) + (20)(-5)}{2(-40)}$

$\qquad = \dfrac{2{,}500}{-25}$ $\qquad = \dfrac{-1{,}500 - 100}{-80}$

$\qquad = -100$ $\qquad = \dfrac{-1{,}600}{-80}$

$\qquad\qquad = 20$ ■

Self Check   Evaluate $\dfrac{2xy - 3z - 5}{3(y - z)}$, given the values in Example 8.

Answer -22

EXAMPLE 9   **Stock reports**   In its annual report, a corporation reports its performance on a per-share basis. When a company with 35 million shares outstanding loses $2.3 million, what will be the per-share loss?

Solution   A loss of $2.3 million can be represented by $-2,300,000$. Because there are 35 million shares, the per-share loss can be represented by the quotient $\frac{-2,300,000}{35,000,000}$.

$$\frac{-2,300,000}{35,000,000} \approx -0.065714285 \qquad \text{Use a calculator.}$$

The company lost about 6.6¢ per share.   ■

Self Check   If the company in Example 9 earns $1.5 million in the following year, find its per-share gain for that year.

Answer   about 4.3¢

Remember these facts about dividing real numbers.

Division

1. $\dfrac{a}{0}$ is undefined.

2. If $a \neq 0$, then $\dfrac{0}{a} = 0$.

3. $\dfrac{a}{1} = a$.

4. If $a \neq 0$, then $\dfrac{a}{a} = 1$.

■ USING A CALCULATOR TO MULTIPLY AND DIVIDE REAL NUMBERS

A calculator can be used to multiply and divide positive and negative numbers. To evaluate $(-345.678)(-527.339)$, we enter these numbers and press these keys:

345.678 +/−   ×   527.339 +/−   =

The display will read $\mathtt{182289.4908}$.
To evaluate $\frac{-345.678}{-527.339}$, we enter these numbers and press these keys:

345.678 +/−   ÷   527.339 +/−   =

The display will read $\mathtt{0.655513815}$.

Orals   *Find each product or quotient.*

1. $1(-3)$ **2.** $-2(-5)$ **3.** $-3(-6)$ **4.** $4(-6)$

5. $-2(3)(-4)$ **6.** $-2(-3)(-4)$ **7.** $\dfrac{-12}{6}$ **8.** $\dfrac{-10}{-5}$

9. $\dfrac{3(6)}{-2}$ **10.** $\dfrac{(-2)(-3)}{-6}$

EXERCISE 1.5

REVIEW

1. A concrete block weighs $37\frac{1}{2}$ pounds. How much will 30 of these blocks weigh?

2. If one brick weighs 1.3 pounds, how much will a skid of 500 bricks weigh?

3. If $x = 5$, $y = 8$, and $z = 3$, evaluate $x^3 - yz^2$.

4. Put $<$, $=$, or $>$ in the box to make a true statement:
$-2(-3 + 4)$ ▭ $-3[3 - (-4)]$

VOCABULARY AND CONCEPTS   *Fill in each blank to make a true statement.*

5. The product of two positive numbers is _____.

6. The product of a _____ number and a negative number is negative.

7. The product of two negative numbers is _____.

8. The quotient of a _____ number and a positive number is negative.

9. The quotient of two negative numbers is _____.

10. Any number multiplied by __ is 0.

11. $a \cdot 1 =$ __

12. The symbol $\dfrac{a}{0}$ is _____.

13. If $a \neq 0$, $\dfrac{0}{a} =$ __.

14. If $a \neq 0$, $\dfrac{a}{a} =$ __.

PRACTICE   *In Exercises 15–34, find each product.*

15. $(+6)(+8)$ **16.** $(-9)(-7)$ **17.** $(-8)(-7)$ **18.** $(9)(-6)$

19. $(+12)(-12)$ **20.** $(-9)(12)$ **21.** $\left(\dfrac{1}{2}\right)(-32)$ **22.** $\left(-\dfrac{3}{4}\right)(12)$

23. $\left(-\dfrac{3}{4}\right)\left(-\dfrac{8}{3}\right)$ **24.** $\left(-\dfrac{2}{5}\right)\left(\dfrac{15}{2}\right)$ **25.** $(-3)\left(-\dfrac{1}{3}\right)$ **26.** $(5)\left(-\dfrac{2}{5}\right)$

27. $(3)(-4)(-6)$ **28.** $(-1)(-3)(-6)$ **29.** $(-2)(3)(4)$ **30.** $(5)(0)(-3)$

31. $(2)(-5)(-6)(-7)$

32. $(-3)(-5)(-5)(-2)$

33. $(-2)(-2)(-2)(-3)(-4)$

34. $(-5)(4)(3)(-2)(-1)$

In Exercises 35–54, let $x = -1$, $y = 2$, and $z = -3$. Evaluate each expression.

35. y^2 **36.** x^2 **37.** $-z^2$ **38.** $-xz$

39. xy **40.** yz **41.** $y + xz$ **42.** $z - xy$

43. $(x + y)z$ **44.** $y(x - z)$ **45.** $(x - z)(x + z)$ **46.** $(y + z)(x - z)$

47. $xy + yz$ **48.** $zx - zy$ **49.** xyz **50.** $x^2 y$

51. $x^2(y - z)$ **52.** $y^2(x - z)$ **53.** $(-x)(-y) + z^2$ **54.** $(-x)(-z) - y^2$

In Exercises 55–66, simplify each expression.

55. $\dfrac{80}{-20}$ **56.** $\dfrac{-66}{33}$ **57.** $\dfrac{-110}{-55}$ **58.** $\dfrac{200}{40}$

59. $\dfrac{-160}{40}$ **60.** $\dfrac{-250}{-25}$ **61.** $\dfrac{320}{-16}$ **62.** $\dfrac{180}{-36}$

63. $\dfrac{8 - 12}{-2}$ **64.** $\dfrac{16 - 2}{2 - 9}$ **65.** $\dfrac{20 - 25}{7 - 12}$ **66.** $\dfrac{2(15)^2 - 2}{-2^3 + 1}$

In Exercises 67–74, evaluate each expression if $x = -2$, $y = 3$, $z = 4$, $t = 5$, and $w = -18$.

67. $\dfrac{yz}{x}$ **68.** $\dfrac{zt}{x}$ **69.** $\dfrac{tw}{y}$ **70.** $\dfrac{w}{xy}$

71. $\dfrac{z + w}{x}$ **72.** $\dfrac{xyz}{y - 1}$ **73.** $\dfrac{xtz}{y + 1}$ **74.** $\dfrac{x + y + z}{t}$

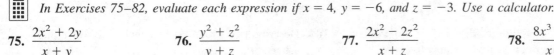 *In Exercises 75–82, evaluate each expression if $x = 4$, $y = -6$, and $z = -3$. Use a calculator.*

75. $\dfrac{2x^2 + 2y}{x + y}$ **76.** $\dfrac{y^2 + z^2}{y + z}$ **77.** $\dfrac{2x^2 - 2z^2}{x + z}$ **78.** $\dfrac{8x^3 - 8y^2}{x - z}$

79. $\dfrac{y^3 + 4z^3}{(x + y)^2}$ **80.** $\dfrac{x^2 - 2xz + z^2}{x - y + z}$ **81.** $\dfrac{xy^2 z + x^2 y}{2y - 2z}$ **82.** $\dfrac{(x^2 - 2y)z^2}{-xz}$

In Exercises 83–90, evaluate each expression if $x = \frac{1}{2}$, $y = -\frac{2}{3}$, and $z = -\frac{3}{4}$.

83. $x + y$ **84.** $y + z$ **85.** $x + y + z$ **86.** $y + x - z$

87. $(x + y)(x - y)$ **88.** $(x - z)(x + z)$ **89.** $(x + y + z)(xyz)$ **90.** $xyz(x - y - z)$

APPLICATIONS *Use signed numbers to solve each problem.*

91. Temperature change   If the temperature is increasing 2 degrees each hour for 3 hours, what product of signed numbers represents the temperature change?

92. Temperature change   If the temperature is decreasing 2 degrees each hour for 3 hours, what product of signed numbers represents the temperature change?

93. Gambling   In Las Vegas, Robert lost $30 per hour playing the slot machines for 15 hours. What product of signed numbers represents the change in his financial condition?

94. Draining a pool A pool is emptying at the rate of 12 gallons per minute. What product of signed numbers would represent how much more water was in the pool 2 hours ago?

95. Filling a pool Water from a pipe is filling a pool at the rate of 23 gallons per minute. What product of signed numbers represents the amount of water in the pool 2 hours ago?

96. Mowing lawns Rafael worked all day mowing lawns and was paid $8 per hour. If he had $94 at the end of an 8-hour day, how much did he have before he started working?

97. Temperature Suppose that the temperature is dropping at the rate of 3 degrees each hour. If the temperature has dropped 18 degrees, what signed number expresses how many hours the temperature has been falling?

98. Dieting A man lost 37.5 pounds. If he lost 2.5 pounds each week, how long has he been dieting?

Use a calculator and signed numbers to solve each problem.

99. Stock market Over a 7-day period, the Dow Jones Industrial Average had gains of 26, 35, and 17 points. In that period, there were also losses of 25, 31, 12, and 24 points. Find the average daily performance over the 7-day period.

100. Astronomy Light travels at the rate of 186,000 miles per second. How long will it take light to travel from the sun to Venus? (*Hint:* The distance from the sun to Venus is 67,000,000 miles.)

101. Saving for school A student has saved $15,000 to attend graduate school. If she estimates that her expenses will be $613.50 a month while in school, does she have enough to complete an 18-month master's degree program?

102. Earnings per share Over a five-year period, a corporation reported profits of $18 million, $21 million, and $33 million. It also reported losses of $5 million and $71 million. Find the average gain (or loss) each year.

WRITING

103. Explain how you would decide whether the product of several numbers is positive or negative.

104. Describe two situations in which negative numbers are useful.

SOMETHING TO THINK ABOUT

105. If the quotient of two numbers is undefined, what would their product be?

106. If the product of five numbers is negative, how many of the factors could be negative?

107. If x^5 is a negative number, can you decide whether x is negative too?

108. If x^6 is a positive number, can you decide whether x is positive too?

1.6 Algebraic Expressions

■ ALGEBRAIC EXPRESSIONS ■ EVALUATING ALGEBRAIC EXPRESSIONS ■ ALGEBRAIC TERMS

Getting Ready *Identify each of the following as a sum, difference, product, or quotient.*

1. $x + 3$　　　　**2.** $57x$

3. $\dfrac{x}{9}$

4. $19 - y$

5. $\dfrac{x - 7}{3}$

6. $x - \dfrac{7}{3}$

7. $5(x + 2)$

8. $5x + 10$

■ ALGEBRAIC EXPRESSIONS

Variables and numbers can be combined with the operations of arithmetic to produce **algebraic expressions.** For example, if x and y are variables, the algebraic expression $x + y$ represents the **sum** of x and y, and the algebraic expression $x - y$ represents their **difference.**

There are many other ways to express addition or subtraction with algebraic expressions, as shown in Tables 1-3 and 1-4.

The phrase	translates into the algebraic expression
the *sum* of *t* and 12	$t + 12$
5 *plus* *s*	$5 + s$
7 *added to* *a*	$a + 7$
10 *more than* *q*	$q + 10$
12 *greater than* *m*	$m + 12$
l increased by m	$l + m$
exceeds p by 50	$p + 50$

TABLE 1-3

The phrase	translates into the algebraic expression
the *difference* of 50 and *r*	$50 - r$
1,000 *minus* *q*	$1{,}000 - q$
15 *less than* *w*	$w - 15$
t decreased by q	$t - q$
12 *reduced by m*	$12 - m$
l subtracted from 250	$250 - l$
2,000 *less* *p*	$2{,}000 - p$

TABLE 1-4

EXAMPLE 1   Let x represent a certain number. Write an expression that represents   **a.** the number that is 5 more than x and   **b.** the number 12 decreased by x.

Solution   **a.** The number "5 more than x" is the number found by adding 5 to x. It is represented by $x + 5$.

b. The number "12 decreased by x" is the number found by subtracting x from 12. It is represented by $12 - x$.   ■

Self Check	Let y represent a certain number. Write an expression that represents y increased by 25.
Answer	$y + 25$

EXAMPLE 2   **Income taxes**   Bob worked x hours preparing his income tax return. He worked 3 hours less than that on his son's return. Write an expression that represents   **a.** the number of hours he spent preparing his son's return   and   **b.** the total number of hours he worked.

Solution   **a.** Because he worked x hours on his own return and 3 hours less on his son's return, he worked $(x - 3)$ hours on his son's return.

b. Because he worked x hours on his own return and $(x - 3)$ hours on his son's return, the total time he spent on taxes was $[x + (x - 3)]$ hours.   ■

Self Check	Javier deposited $\$d$ in a bank account. Later, he withdrew $\$500$. Write an expression that represents the difference of d and 500.
Answer	$d - 500$

There are several ways to indicate the **product** of two numbers with algebraic expressions, as shown in Table 1-5.

The phrase	translates into the algebraic expression
the *product* of 100 and a	$100a$
25 *times B*	$25B$
twice x	$2x$
$\dfrac{1}{2}$ *of z*	$\dfrac{1}{2}z$
12 *multiplied by m*	$12m$

TABLE 1-5

EXAMPLE 3   Let x represent a certain number. Denote a number that is   **a.** twice as large as x, **b.** 5 more than 3 times x,   and   **c.** 4 less than $\frac{1}{2}$ of x.

Solution   **a.** The number "twice as large as x" is found by multiplying x by 2. It is represented by $2x$.

b. The number "5 more than 3 times x" is found by adding 5 to the product of 3 and x. It is represented by $3x + 5$.

c. The number "4 less than $\frac{1}{2}$ of x" is found by subtracting 4 from the product of $\frac{1}{2}$ and x. It is represented by $\frac{1}{2}x - 4$. ∎

Self Check Find the product of 40 and t.

Answer $40t$

EXAMPLE 4

Stock valuation Jim owns x shares of Transitronic stock, valued at \$29 a share; y shares of Positone stock, valued at \$32 a share; and 300 shares of Baby Bell, valued at \$42 a share.

a. How many shares of stock does he own?

b. What is the value of his stock?

Solution **a.** Because there are x shares of Transitronic, y shares of Positone, and 300 shares of Baby Bell, his total number of shares is $x + y + 300$.

b. The value of x shares of Transitronic is \$$29x$, the value of y shares of Positone is \$$32y$, and the value of 300 shares of Baby Bell is \$$42(300)$. The total value of the stock is \$$(29x + 32y + 12{,}600)$. ∎

Self Check If water softener salt costs \$$p$ per bag, find the cost of 25 bags.

Answer $\$25p$

There are also several ways to indicate the **quotient** of two numbers with algebraic expressions, as shown in Table 1-6.

The phrase	translates into the algebraic expression
the *quotient* of 470 and A	$\dfrac{470}{A}$
B divided by C	$\dfrac{B}{C}$
the *ratio* of h to 5	$\dfrac{h}{5}$
x *split into* 5 equals parts	$\dfrac{x}{5}$

TABLE 1-6

EXAMPLE 5

Let x and y represent two numbers. Write an algebraic expression that represents the sum obtained when 3 times the first number is added to the quotient obtained when the second number is divided by 6.

Solution   Three times the first number x is denoted as $3x$. The quotient obtained when the second number y is divided by 6 is the fraction $\frac{y}{6}$. Their sum is expressed as $3x + \frac{y}{6}$. ∎

Self Check   If the cost c of a meal is split equally among 4 people, what is each person's share?

Answer $\frac{c}{4}$

EXAMPLE 6   **Cutting a rope**   A 5-foot section is cut from the end of a rope that is l feet long. If the remaining rope is divided into three equal pieces, find the length of each of the equal pieces.

Solution   After a 5-foot section is cut from one end of l feet of rope, the rope that remains is $(l - 5)$ feet long. When that remaining rope is cut into 3 equal pieces, each piece will be $\frac{l-5}{3}$ feet long. See Figure 1-23.

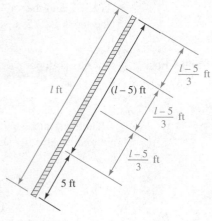

FIGURE 1-23 ∎

Self Check   If a 7-foot section is cut from a rope that is l feet long and the remaining rope is divided into two equal pieces, how long is each piece?

Answer $\frac{l-7}{2}$ ft

■ EVALUATING ALGEBRAIC EXPRESSIONS

Since variables represent numbers, algebraic expressions also represent numbers. We have seen that we can evaluate algebraic expressions when we know the values of the variables.

EXAMPLE 7   If $x = 8$ and $y = 10$, evaluate   **a.** $x + y$,   **b.** $y - x$,   **c.** $3xy$ and   **d.** $\dfrac{5x}{y - 5}$.

Solution   We substitute 8 for x and 10 for y in each expression and simplify.

a. $x + y = \mathbf{8} + \mathbf{10}$ **b.** $y - x = \mathbf{10} - \mathbf{8}$

$\qquad = 18$ $\qquad = 2$

c. $3xy = (3)(\mathbf{8})(\mathbf{10})$

$\qquad = (24)(10)$ Do the multiplications from left to right.

$\qquad = 240$

d. $\dfrac{5x}{y - 5} = \dfrac{5 \cdot \mathbf{8}}{\mathbf{10} - 5}$

$\qquad = \dfrac{40}{5}$ Simplify the numerator and the denominator separately.

$\qquad = 8$ Simplify the fraction.

WARNING!   After numbers are substituted for the variables in a product, it is often necessary to insert a dot or parentheses to show the multiplication. Otherwise (3)(8)(10), for example, might be mistaken for 3,810, and $5 \cdot 8$ might be mistaken for 58.

Self Check   If $a = -2$ and $b = 5$, evaluate $\dfrac{6b + 2}{a + 2b}$.

Answer   4

■ ALGEBRAIC TERMS

Numbers without variables, such as 7, 21, and 23, are called **constants.** Expressions such as 37, xyz, and $32t$, which are constants, variables, or products of constants and variables, are called **algebraic terms.**

- The expression $3x + 5y$ contains two terms. The first term is $3x$, and the second term is $5y$.

- The expression $xy + (-7)$ contains two terms. The first term is xy, and the second term is -7.

- The expression $3 + x + 2y$ contains three terms. The first term is 3, the second term is x, and the third term is $2y$.

Numbers and variables that are part of a product are called **factors.** For example,

- The product $7x$ has two factors, which are 7 and x.
- The product $-3xy$ has three factors, which are -3, x, and y.
- The product $\frac{1}{2}abc$ has four factors, which are $\frac{1}{2}$, a, b, and c.

The number factor of a product is called its **numerical coefficient.** The numerical coefficient (or just the *coefficient*) of $7x$ is 7. The coefficient of $-3xy$ is -3, and the coefficient of $\frac{1}{2}abc$ is $\frac{1}{2}$. The coefficient of terms such as x, ab, and rst is understood to be 1.

$$x = 1x, \qquad ab = 1ab, \qquad \text{and} \qquad rst = 1rst$$

EXAMPLE 8
a. The expression $5x + y$ has two terms. The numerical coefficient of its first term is 5. The numerical coefficient of its second term is 1.

b. The expression $-17wxyz$ has one term, which contains the five factors -17, w, x, y, and z. Its numerical coefficient is -17.

c. The expression 37 has one term, the constant 37. Its numerical coefficient is 37.

■

Self Check   How many terms does the expression $3x^2 - 2x + 7$ have? Find the sum of the coefficients.

Answers   3, 8

Orals   *If $x = -2$ and $y = 3$, find the value of each expression.*

1. $x + y$ 2. $7x$ 3. $7x + y$ 4. $7(x + y)$
5. $4x^2$ 6. $(4x)^2$ 7. $-3x^2$ 8. $(-3x)^2$

EXERCISE 1.6

REVIEW   *Evaluate each of the following.*

1. 14% of 3,800   2. $\frac{3}{5}$ of 4,765   3. $\dfrac{-4 + (7 - 9)}{(-9 - 7) + 4}$ 4. $\dfrac{5}{4}\left(1 - \dfrac{3}{5}\right)$

VOCABULARY AND CONCEPTS   *Fill in each blank to make a true statement.*

5. The answer to an addition problem is called a _____.

6. The answer to a _____ problem is called a difference.

7. The answer to a _____ problem is called a product.

8. The answer to a division problem is called a _____.

9. An _____ expression is a combination of variables, numbers, and the operation symbols for addition, subtraction, multiplication, or division.

10. To _____ an algebraic expression, we substitute values for the variables and simplify.

11. Letters that stand for numbers are called _____.

12. Terms that have no variables are called _____.

PRACTICE *In Exercises 13–30, let x, y, and z represent three real numbers. Write an algebraic expression to denote each quantity.*

13. The sum of x and y

14. The product of x and y

15. The product of x and twice y

16. The sum of twice x and twice y

17. The difference obtained when x is subtracted from y

18. The difference obtained when twice x is subtracted from y

19. The quotient obtained when y is divided by x

20. The quotient obtained when the sum of x and y is divided by z

21. The sum obtained when the quotient of x divided by y is added to z

22. y decreased by x

23. z less the product of x and y

24. z less than the product of x and y

25. The product of 3, x, and y

26. The quotient obtained when the product of 3 and z is divided by the product of 4 and x

27. The quotient obtained when the sum of x and y is divided by the sum of y and z

28. The quotient obtained when the product of x and y is divided by the sum of x and z

29. The sum of the product xy and the quotient obtained when y is divided by z

30. The number obtained when x decreased by 4 is divided by the product of 3 and y

In Exercises 31–42, write each algebraic expression as an English phrase.

31. $x + 3$

32. $y - 2$

33. $\dfrac{x}{y}$

34. xz

35. $2xy$

36. $\dfrac{x + y}{2}$

37. $\dfrac{5}{x + y}$

38. $\dfrac{3x}{y + z}$

39. $\dfrac{3 + x}{y}$

40. $3 + \dfrac{x}{y}$

41. $xy(x + y)$

42. $(x + y + z)(xyz)$

In Exercises 43–50, let x = 8, y = 4, and z = 2. Write each phrase as an algebraic expression, and evaluate it.

43. The sum of x and z

44. The product of x, y, and z

45. z less than y

46. The quotient obtained when y is divided by z

47. 3 less than the product of y and z

48. 7 less than the sum of x and y

49. The quotient obtained when the product of x and y is divided by z

50. The quotient obtained when 10 greater than x is divided by z

In Exercises 51–60, give the number of terms in each algebraic expression and also give the numerical coefficient of the first term.

51. $6d$

52. $-4c + 3d$

53. $-xy - 4t + 35$

54. xy

55. $3ab + bc - cd - ef$

56. $-2xyz + cde - 14$

57. $-4xyz + 7xy - z$

58. $5uvw - 4uv + 8uw$

59. $3x + 4y + 2z + 2$

60. $7abc - 9ab + 2bc + a - 1$

In Exercises 61–64, consider the algebraic expression $29xyz + 23xy + 19x$.

61. What are the factors of the third term?

62. What are the factors of the second term?

63. What are the factors of the first term?

64. What factor is common to all three terms?

In Exercises 65–68, consider the algebraic expression $3xyz + 5xy + 17xz$.

65. What are the factors of the first term?

66. What are the factors of the second term?

67. What are the factors of the third term?

68. What factor is common to all three terms?

In Exercises 69–72, consider the algebraic expression $5xy + yt + 8xyt$.

69. Find the numerical coefficients of each term.

70. What factor is common to all three terms?

71. What factors are common to the first and third terms?

72. What factors are common to the second and third terms?

In Exercises 73–76, consider the algebraic expression $3xy + y + 25xyz$.

73. Find the numerical coefficient of each term and find their product.

74. Find the numerical coefficient of each term and find their sum.

75. What factors are common to the first and third terms?

76. What factor is common to all three terms?

APPLICATIONS

77. Course load A man enrolls in college for c hours of credit, and his sister enrolls for 4 more hours than her brother. Write an expression that represents the number of hours the sister is taking.

78. Antique cars An antique Ford has 25,000 more miles on its odometer than a newer car. If the newer car has traveled m miles, find an expression that represents the mileage on the Ford.

79. T-bills Write an expression that represents the value of t T-bills, each worth \$9,987.

80. Real estate Write an expression that represents the value of a vacant lots if each lot is worth \$35,000.

81. Cutting rope A rope x feet long is cut into 5 equal pieces. Find an expression for the length of each piece.

82. Plumbing A plumber cuts a pipe that is 12 feet long into x equal pieces. Find an expression for the length of each piece.

83. Comparing assets A girl had d dollars, and her brother had $5 more than three times that amount. How much did the brother have?

84. Comparing investments Wendy has x shares of stock. Her sister has 2 fewer shares than twice Wendy's shares. How many shares does her sister have?

WRITING

85. Distinguish between the meanings of these two phrases: "3 less than x" and "3 is less than x."

87. What is the purpose of using variables? Why aren't ordinary numbers enough?

86. Distinguish between *factor* and *term*.

88. In words, xy is "the product of x and y." However, $\frac{x}{y}$ is "the quotient obtained when x is divided by y." Explain why the extra words are needed.

SOMETHING TO THINK ABOUT

89. If the value of x were doubled, what would happen to the value of $37x$?

90. If the values of both x and y were doubled, what would happen to the value of $5xy^2$?

1.7 Properties of Real Numbers

■ THE CLOSURE PROPERTIES ■ THE COMMUTATIVE PROPERTIES ■ THE ASSOCIATIVE PROPERTIES
■ THE DISTRIBUTIVE PROPERTY ■ THE IDENTITY ELEMENTS ■ INVERSES FOR ADDITION AND
MULTIPLICATION

Getting Ready *Do the operations.*

1. $3 + (5 + 9)$

2. $(3 + 5) + 9$

3. $23.7 + 14.9$

4. $14.9 + 23.7$

5. $7(5 + 3)$

6. $7 \cdot 5 + 7 \cdot 3$

7. $125.3 + (-125.3)$

8. $125.3\left(\dfrac{1}{125.3}\right)$

9. $777 + 0$

10. $777 \cdot 1$

■ THE CLOSURE PROPERTIES

The **closure properties** guarantee that the sum, difference, product, or quotient (except for division by zero) of any two real numbers is also a real number.

Closure Properties

If a and b are real numbers, then

$a + b$ is a real number $a - b$ is a real number

ab is a real number $\dfrac{a}{b}$ is a real number $(b \neq 0)$

EXAMPLE 1   Assume that $x = 8$ and $y = -4$. Find the real-number answer to show that **a.** $x + y$,   **b.** $x - y$,   **c.** xy,   and   **d.** $\frac{x}{y}$ all represent real numbers.

Solution   We substitute 8 for x and -4 for y in each expression and simplify.

a. $x + y = \mathbf{8} + (-\mathbf{4})$ **b.** $x - y = \mathbf{8} - (-\mathbf{4})$
$\qquad\quad = 4$ $\qquad\qquad\quad = 8 + 4$
$\qquad\qquad\qquad\qquad\qquad\qquad\quad = 12$

c. $xy = \mathbf{8}(-\mathbf{4})$ **d.** $\dfrac{x}{y} = \dfrac{\mathbf{8}}{-\mathbf{4}}$
$\qquad\quad = -32$ $\qquad\qquad\quad = -2$ ■

Self Check   Assume that $a = -6$ and $b = 3$. Find the real-number answer to show that **a.** $a - b$ and   **b.** $\frac{a}{b}$ are real numbers.

Answers   **a.** -9,   **b.** -2

■ **THE COMMUTATIVE PROPERTIES**

The **commutative properties** (from the word *commute,* which means to go back and forth) guarantee that addition or multiplication of two real numbers can be done in either order.

Commutative Properties

If a and b are real numbers, then

$a + b = b + a$ commutative property of addition

$ab = ba$ commutative property of multiplication

EXAMPLE 2   Assume that $x = -3$ and $y = 7$. Show that   **a.** $x + y = y + x$ and   **b.** $xy = yx$.

Solution   **a.** We can show that the sum $x + y$ is the same as the sum $y + x$ by substituting -3 for x and 7 for y in each expression and simplifying.

$$x + y = -\mathbf{3} + 7 = 4 \qquad \text{and} \qquad y + x = 7 + (-\mathbf{3}) = 4$$

b. We can show that the product xy is the same as the product yx by substituting -3 for x and 7 for y in each expression and simplifying.

$$xy = -\mathbf{3}(7) = -21 \qquad \text{and} \qquad yx = 7(-\mathbf{3}) = -21 \qquad ■$$

Self Check

Assume that $a = 6$ and $b = -5$. Show that **a.** $a + b = b + a$ and
b. $ab = ba$.

Answers

a. $a + b = 1$ and $b + a = 1$, **b.** $ab = -30$ and $ba = -30$

■ THE ASSOCIATIVE PROPERTIES

The **associative properties** guarantee that three real numbers can be regrouped in an addition or multiplication.

> **Associative Properties**
> If a, b, and c are real numbers, then
> $$(a + b) + c = a + (b + c) \quad \text{associative property of addition}$$
> $$(ab)c = a(bc) \quad \text{associative property of multiplication}$$

Because of the associative property of addition, we can group (or *associate*) the numbers in a sum in any way that we wish. For example,

$$(\mathbf{3 + 4}) + 5 = \mathbf{7} + 5 \quad \text{and} \quad 3 + (\mathbf{4 + 5}) = 3 + \mathbf{9}$$
$$= 12 \qquad\qquad\qquad\qquad = 12$$

The answer is 12 regardless of how we group the three numbers.

The associative property of multiplication permits us to group (or *associate*) the numbers in a product in any way that we wish. For example,

$$(\mathbf{3 \cdot 4}) \cdot 7 = \mathbf{12} \cdot 7 \quad \text{and} \quad 3 \cdot (\mathbf{4 \cdot 7}) = 3 \cdot \mathbf{28}$$
$$= 84 \qquad\qquad\qquad\qquad = 84$$

The answer is 84 regardless of how we group the three numbers.

■ THE DISTRIBUTIVE PROPERTY

The **distributive property** shows how to multiply the sum of two numbers by a third number. Because of this property, we can often add first and then multiply, or multiply first and then add.

For example, $2(3 + 7)$ can be calculated in two different ways. We can add and then multiply, or we can multiply each number within the parentheses by 2 and then add.

$$2(\mathbf{3 + 7}) = 2(\mathbf{10}) \quad \text{and} \quad 2(3 + 7) = \mathbf{2 \cdot 3 + 2 \cdot 7}$$
$$= 20 \qquad\qquad\qquad\qquad = 6 + 14$$
$$\qquad\qquad\qquad\qquad\qquad = 20$$

Either way, the result is 20.

In general, we have the following property.

Distributive Property

If a, b, and c are real numbers, then

$$a(b + c) = ab + ac$$

Because multiplication is commutative, the distributive property can also be written in the form

$$(b + c)a = ba + ca$$

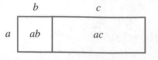

a | ab | ac

b c

FIGURE 1-24

We can interpret the distributive property geometrically. Since the area of the largest rectangle in Figure 1-24 is the product of its width a and its length $b + c$, its area is $a(b + c)$. The areas of the two smaller rectangles are ab and ac. Since the area of the largest rectangle is equal to the sum of the areas of the smaller rectangles, we have $a(b + c) = ab + ac$.

EXAMPLE 3   Evaluate each expression in two different ways:
a. $3(5 + 9)$ and   **b.** $-2(-7 + 3)$.

Solution   **a.** $3(\mathbf{5 + 9}) = 3(\mathbf{14})$ and $3(5 + 9) = \mathbf{3 \cdot 5 + 3 \cdot 9}$
$\qquad\qquad\qquad = 42$ $\qquad\qquad\qquad\qquad = 15 + 27$
$\qquad\qquad\qquad\qquad\qquad\qquad\qquad\qquad = 42$

b. $-2(\mathbf{-7 + 3}) = -2(\mathbf{-4})$ and $-2(-7 + 3) = \mathbf{-2}(-7) + (\mathbf{-2})3$
$\qquad\qquad\qquad\quad = 8$ $\qquad\qquad\qquad\qquad\qquad = 14 + (-6)$
$\qquad\qquad\qquad\qquad\qquad\qquad\qquad\qquad\qquad = 8$ ∎

Self Check   Evaluate $-5.2(2.7 + 3.5)$ in two different ways.
Answer -32.24

The distributive property can be extended to three or more terms. For example, if a, b, c, and d are real numbers, then

$$a(b + c + d) = ab + ac + ad$$

EXAMPLE 4   Write $3(x + y + 2)$ without using parentheses.

Solution $3(x + y + 2) = \mathbf{3}x + \mathbf{3}y + \mathbf{3} \cdot 2$ Distribute the multiplication by 3.
$\qquad\qquad\qquad\quad = 3x + 3y + 6$ ∎

Self Check   Write $-6.3(a + 2b + 3.7)$ without using parentheses.
Answer $-6.3a - 12.6b - 23.31$

■ THE IDENTITY ELEMENTS

The numbers 0 and 1 play special roles in arithmetic. The number 0 is the only number that can be added to another number (say, a) and give an answer of that same number a:

$$0 + a = a + 0 = a$$

The number 1 is the only number that can be multiplied by another number (say, a) and give an answer of that same number a:

$$1 \cdot a = a \cdot 1 = a$$

Because adding 0 to a number or multiplying a number by 1 leaves that number the same (identical), the numbers 0 and 1 are called **identity elements.**

> **Identity Elements**
>
> 0 is the **identity element for addition.**
>
> 1 is the **identity element for multiplication.**

■ INVERSES FOR ADDITION AND MULTIPLICATION

If the sum of two numbers is 0, the numbers are called *negatives,* or **additive inverses,** of each other. Since $3 + (-3) = 0$, the numbers 3 and -3 are negatives or additive inverses of each other. In general, because

$$a + (-a) = 0$$

the numbers represented by a and $-a$ are negatives or additive inverses of each other.

If the product of two numbers is 1, the numbers are called **reciprocals,** or **multiplicative inverses,** of each other. Since $7\left(\frac{1}{7}\right) = 1$, the numbers 7 and $\frac{1}{7}$ are reciprocals. Since $(-0.25)(-4) = 1$, the numbers -0.25 and -4 are reciprocals. In general, because

$$a\left(\frac{1}{a}\right) = 1 \qquad \text{provided } a \neq 0$$

the numbers represented by a and $\frac{1}{a}$ are reciprocals or multiplicative inverses of each other.

> **Additive and Multiplicative Inverses**
>
> Because $a + (-a) = 0$, the numbers a and $-a$ are called **negatives** or **additive inverses.**
>
> Because $a\left(\frac{1}{a}\right) = 1$ $(a \neq 0)$, the numbers a and $\frac{1}{a}$ are called **reciprocals** or **multiplicative inverses.**

EXAMPLE 5   The property in the right column justifies the statement in the left column.

$3 + 4$ is a real number	closure property of addition
$\dfrac{8}{3}$ is a real number	closure property of division
$3 + 4 = 4 + 3$	commutative property of addition
$-3 + (2 + 7) = (-3 + 2) + 7$	associative property of addition
$(5)(-4) = (-4)(5)$	commutative property of multiplication
$(ab)c = a(bc)$	associative property of multiplication
$3(a + 2) = 3a + 3 \cdot 2$	distributive property
$3 + 0 = 3$	additive identity property
$3(1) = 3$	multiplicative identity property
$2 + (-2) = 0$	additive inverse property
$\left(\dfrac{2}{3}\right)\left(\dfrac{3}{2}\right) = 1$	multiplicative inverse property

∎

Self Check   Which property justifies each statement?   **a.** $a + 7 = 7 + a$
b. $3(y + 2) = 3y + 3 \cdot 2$ **c.** $3 \cdot (2 \cdot p) = (3 \cdot 2) \cdot p$

Answers   **a.** commutative property of addition,   **b.** distributive property,   **c.** associative property of multiplication

The properties of the real numbers are summarized as follows.

Properties of Real Numbers

For all real numbers a, b, and c,

Closure properties	$a + b$ is a real number	$a \cdot b$ is a real number
	$a - b$ is a real number	$a \div b$ is a real number $(b \neq 0)$
	Addition	*Multiplication*
Commutative properties	$a + b = b + a$	$a \cdot b = b \cdot a$
Associative properties	$(a + b) + c = a + (b + c)$	$(ab)c = a(bc)$
Identity properties	$a + 0 = a$	$a \cdot 1 = a$
Inverse properties	$a + (-a) = 0$	$a \cdot \left(\dfrac{1}{a}\right) = 1$ $(a \neq 0)$
Distributive property	$a(b + c) = ab + ac$	

Orals   *Give an example of each property.*

1. The associative property of multiplication

2. The additive identity property

3. The distributive property

4. The inverse for multiplication

Provide an example to illustrate each statement.

5. Subtraction is not commutative.

6. Division is not associative.

EXERCISE 1.7

REVIEW

1. Write as a mathematical expression: The sum of x and the square of y is greater than or equal to z.

2. Write as an English phrase: $3(x + z)$.

In Exercises 3–4, fill each box with an appropriate symbol.

3. For any number x, $|x| \geq \boxed{}$.

4. $x - y = x + (\boxed{})$

In Exercises 5–6, fill in each blank to make a true statement.

5. The product of two negative numbers is a _____ number.

6. The sum of two negative numbers is a _____ number.

VOCABULARY AND CONCEPTS   *Fill in each blank to make a true statement.*

7. If a and b are real numbers, $a + b$ is a ____ number.

8. If a and b are real numbers, $\frac{a}{b}$ is a real number, provided that _____.

9. $a + b = b + \underline{}$

10. $a \cdot b = \underline{} \cdot a$

11. $(a + b) + c = a + \underline{}$

12. $(ab)c = \underline{} \cdot (bc)$

13. $a(b + c) = ab + \underline{}$

14. $0 + a = \underline{}$

15. $a \cdot 1 = \underline{}$

16. 0 is the _____ element for _____.

17. 1 is the identity _____ for _____.

18. If $a + (-a) = 0$, then a and $-a$ are called _____ inverses.

19. If $a\left(\dfrac{1}{a}\right) = 1$, then a and $\underline{}$ are called reciprocals.

20. $a(b + c + d) = ab + \underline{}$

PRACTICE   *In Exercises 21–28, assume that $x = 12$ and $y = -2$. Show that each expression represents a real number by finding the real-number answer.*

21. $x + y$

22. $y - x$

23. xy

24. $\dfrac{x}{y}$

25. x^2

26. y^2

27. $\dfrac{x}{y^2}$

28. $\dfrac{2x}{3y}$

In Exercises 29–34, assume that $x = 5$ and $y = 7$. Show that both given expressions have the same value.

29. $x + y;\ y + x$ **30.** $xy;\ yx$ **31.** $3x + 2y;\ 2y + 3x$ **32.** $3xy;\ 3yx$

33. $x(x + y);\ (x + y)x$ **34.** $xy + y^2;\ y^2 + xy$

In Exercises 35–40, assume that $x = 2$, $y = -3$, and $z = 1$. Show that the expressions have the same value.

35. $(x + y) + z;\ x + (y + z)$ **36.** $(xy)z;\ x(yz)$
37. $(xz)y;\ x(yz)$ **38.** $(x + y) + z;\ y + (x + z)$
39. $x^2(yz^2);\ (x^2y)z^2$ **40.** $x(y^2z^3);\ (xy^2)z^3$

In Exercises 41–52, use the distributive property to write each expression without parentheses. Simplify each result if possible.

41. $3(x + y)$ **42.** $4(a + b)$ **43.** $x(x + 3)$ **44.** $y(y + z)$
45. $-x(a + b)$ **46.** $a(x + y)$ **47.** $4(x^2 + x)$ **48.** $-2(a^2 + 3)$
49. $-5(t + 2)$ **50.** $2x(a - x)$ **51.** $-2a(x + a)$ **52.** $-p(p - q)$

In Exercises 53–64, give the additive and the multiplicative inverse of each number when possible.

53. 2 **54.** 3 **55.** $\dfrac{1}{3}$ **56.** $-\dfrac{1}{2}$

57. 0 **58.** -2 **59.** $-\dfrac{5}{2}$ **60.** 0.5

61. -0.2 **62.** 0.75 **63.** $\dfrac{4}{3}$ **64.** -1.25

In Exercises 65–76, state which property of real numbers justifies each statement.

65. $3 + x = x + 3$ **66.** $(3 + x) + y = 3 + (x + y)$
67. $xy = yx$ **68.** $(3)(2) = (2)(3)$
69. $-2(x + 3) = -2x + (-2)(3)$ **70.** $x(y + z) = (y + z)x$
71. $(x + y) + z = z + (x + y)$ **72.** $3(x + y) = 3x + 3y$

73. $5 \cdot 1 = 5$ **74.** $x + 0 = x$ **75.** $3 + (-3) = 0$ **76.** $9 \cdot \dfrac{1}{9} = 1$

In Exercises 77–86, use the given property to rewrite the expression in a different form.

77. $3(x + 2)$; distributive property **78.** $x + y$; commutative property of addition
79. y^2x; commutative property of multiplication **80.** $x + (y + z)$; associative property of addition

81. $(x + y)z$; commutative property of addition

82. $x(y + z)$; distributive property

83. $(xy)z$; associative property of multiplication

84. $1x$; multiplicative identity property

85. $0 + x$; additive identity property

86. $5 \cdot \dfrac{1}{5}$; multiplicative inverse property

WRITING

87. Explain why division is not commutative.

88. Describe two ways of calculating the value of $3(12 + 7)$.

SOMETHING TO THINK ABOUT

89. Suppose there were no other numbers than the odd integers.
- Would the closure property for addition still be true?
- Would the closure property for multiplication still be true?
- Would there still be an identity for addition?
- Would there still be an identity for multiplication?

90. Suppose there were no other numbers than the even integers. Answer the four parts of Exercise 89 again.

■ ■ ■ ■ ■ ■ ■ ■ ■ **PROJECTS**

PROJECT 1   The circumference of any circle (the distance around the circle) and the diameter of the circle (the distance across) are related. When you divide the circumference by the diameter, the quotient is always the same number, **pi**, denoted by the Greek letter π.

- Carefully measure the circumference of several circles—a quarter, a dinner plate, a bicycle tire—whatever you can find that is round. Then calculate approximations of π by dividing (with a calculator) each circle's circumference by its diameter.

- Press the $\boxed{\pi}$ button on the calculator to obtain a more accurate value of π. How close were your calculations?

PROJECT 2   **a.** The fraction $\frac{22}{7}$ is often used as an approximation of π. To how many decimal places is this approximation accurate?

b. Experiment with your calculator and try to do better. Find another fraction (with no more than three digits in either its numerator or its denominator) that is closer to π. Who in your class has done best?

C H A P T E R   S U M M A R Y

CONCEPTS

REVIEW EXERCISES

SECTION 1.1   *Real Numbers and Their Graphs*

Natural numbers:

1, 2, 3, 4, 5, . . .

Whole numbers:

0, 1, 2, 3, 4, 5, . . .

Integers:

. . . , −3, −2, −1, 0, 1, 2, 3, . . .

Rational numbers:

Fractions with integer numerators and nonzero integer denominators

Real numbers:

Rational numbers or irrational numbers

Prime numbers:

2, 3, 5, 7, 11, 13, 17, . . .

Composite numbers:

4, 6, 8, 9, 10, 12, 14, 15, . . .

Even integers:

. . . , −6, −4, −2, 0, 2, 4, 6, . . .

Odd integers:

. . . , −5, −3, −1, 1, 3, 5, . . .

1. Consider the set $\{0, 1, 2, 3, 4, 5\}$.

   a. Which numbers are natural numbers?

   b. Which numbers are prime numbers?

   c. Which numbers are odd natural numbers?

   d. Which numbers are composite numbers?

2. Consider the set $\left\{-6, -\frac{2}{3}, 0, \sqrt{2}, 2.6, \pi, 5\right\}$.

   a. Which numbers are integers?

   b. Which numbers are rational numbers?

   c. Which numbers are prime numbers?

   d. Which numbers are real numbers?

   e. Which numbers are even integers?

   f. Which numbers are odd integers?

   g. Which numbers are not rational?

3. Place one of the symbols $=$, $<$, or $>$ in each box to make a true statement.

   a. -5 ▢ $12 - 12$ **b.** $\dfrac{24}{6}$ ▢ 5

   c. $13 - 13$ ▢ $5 - \dfrac{25}{5}$ **d.** $\dfrac{21}{7}$ ▢ -33

Double negative rule:

$-(-x) = x$

4. Simplify each expression.

   a. $-(-8)$ **b.** $-(12 - 4)$

Sets of numbers can be graphed on the number line.

5. Draw a number line and graph each set of numbers.

   a. The composite numbers from 14 to 20

   b. The whole numbers between 19 and 25

c. The real numbers less than or equal to -3 or greater than 2

d. The real numbers greater than -4 and less than 3

The **absolute value** of x, denoted as $|x|$, is the distance between x and 0 on the number line.

$|x| \geq 0$

6. Find each absolute value.
  a. $|53 - 42|$ **b.** $|-31|$

SECTION 1.2   *Fractions*

To simplify a fraction, factor the numerator and the denominator. Then divide out all common factors.

7. Simplify each fraction.
  a. $\dfrac{45}{27}$ **b.** $\dfrac{121}{11}$

To multiply two fractions, multiply their numerators and multiply their denominators.

8. Do each operation and simplify the answer, if possible.
  a. $\dfrac{31}{15} \cdot \dfrac{10}{62}$ **b.** $\dfrac{25}{36} \cdot \dfrac{12}{15} \cdot \dfrac{3}{5}$

To divide two fractions, multiply the first by the reciprocal of the second.

  c. $\dfrac{18}{21} \div \dfrac{6}{7}$ **d.** $\dfrac{14}{24} \div \dfrac{7}{12} \div \dfrac{2}{5}$

To add (or subtract) two fractions with like denominators, add (or subtract) their numerators and keep their common denominator.

  e. $\dfrac{7}{12} + \dfrac{9}{12}$ **f.** $\dfrac{13}{24} - \dfrac{5}{24}$

To add (or subtract) two fractions with unlike denominators, rewrite the fractions with the same denominator, add (or subtract) their numerators, and use the common denominator.

  g. $\dfrac{1}{3} + \dfrac{1}{7}$ **h.** $\dfrac{5}{7} + \dfrac{4}{9}$

  i. $\dfrac{2}{3} - \dfrac{1}{7}$ **j.** $\dfrac{4}{5} - \dfrac{2}{3}$

Before working with mixed numbers, convert them to improper fractions.

  k. $3\dfrac{2}{3} + 5\dfrac{1}{4}$ **l.** $7\dfrac{5}{12} - 4\dfrac{1}{2}$

9. Do the operations.
  a. $32.71 + 15.9$ **b.** $27.92 - 14.93$
  c. $5.3 \cdot 3.5$ **d.** $21.83 \div 5.9$

10. Do each operation and round to two decimal places.
  a. $2.7(4.92 - 3.18)$ **b.** $\dfrac{3.3 + 2.5}{0.22}$

  c. $\dfrac{12.5}{14.7 - 11.2}$ **d.** $(3 - 0.7)(3.63 - 2)$

11. Average study time Four students recorded the time they spent working on a take-home exam: 5.2, 4.7, 9.5, and 8 hours. Find the average time spent. (*Hint:* Add the numbers and divide by 4.)

12. Absenteeism During the height of the flu season, 15% of the 380 university faculty members were sick. How many were ill?

13. Packaging Four steel bands surround the shipping crate in Illustration 1. Find the total length of strapping needed.

ILLUSTRATION 1

Exponents and Order of Operations

If n is a natural number, then

$$x^n = \overbrace{x \cdot x \cdot x \cdot x \cdot \cdots \cdot x}^{n \text{ factors of } x}$$

14. Find the value of each expression.
  a. 3^4 **b.** $\left(\dfrac{2}{3}\right)^2$

  c. $(0.5)^2$ **d.** $5^2 + 2^3$

15. Let $x = 2$ and $y = 3$ and evaluate each expression.
  a. y^4 **b.** x^y

16. **Petroleum storage** Find the volume of the cylindrical storage tank in Illustration 2. Round to one decimal place.

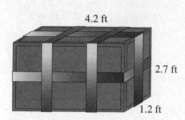

ILLUSTRATION 2

Order of operations

Within each pair of grouping symbols (working from the innermost pair to the outermost pair), do the following operations:

1. Evaluate all exponential expressions.
2. Do multiplications and divisions, working from left to right.
3. Do additions and subtractions, working from left to right.

When the grouping symbols are gone, repeat the above rules to finish the calculation.

In a fraction, simplify the numerator and denominator separately. Then simplify the fraction, if possible.

17. Simplify each expression.

a. $5 + 3^3$

b. $7 \cdot 2 - 7$

c. $4 + (8 \div 4)$

d. $(4 + 8) \div 4$

e. $5^3 - \dfrac{81}{3}$

f. $(5 - 2)^2 + 5^2 + 2^2$

g. $\dfrac{4 \cdot 3 + 3^4}{31}$

h. $\dfrac{4}{3} \cdot \dfrac{9}{2} + \dfrac{1}{2} \cdot 18$

18. Let $x = 6$ and $y = 8$ and evaluate each expression.

a. $y^2 - x$

b. $(y - x)^2$

c. $\dfrac{x + y}{x - 4}$

d. $\dfrac{xy - 12}{4 + y}$

19. Let $x = 2$ and $y = 3$ and evaluate each expression.

a. $x^2 + xy^2$

b. $\dfrac{x^2 + y}{x^3 - 1}$

SECTION 1.4

Adding and Subtracting Real Numbers

To find the sum of two real numbers with the same sign, add their absolute values and keep their common sign.

To add two real numbers with unlike signs, subtract their absolute values (the smaller from the larger) and use the sign of the number with the greater absolute value.

If x and y are two real numbers, then $x - y = x + (-y)$.

20. Evaluate each expression.

a. $(+7) + (+8)$

b. $(-25) + (-32)$

c. $(-2.7) + (-3.8)$

d. $\dfrac{1}{3} + \dfrac{1}{6}$

e. $(+12) + (-24)$

f. $(-44) + (+60)$

g. $3.7 + (-2.5)$

h. $-5.6 + (+2.06)$

i. $15 - (-4)$

j. $-12 - (-13)$

k. $[-5 + (-5)] - (-5)$

l. $1 - [5 - (-3)]$

m. $\dfrac{5}{6} - \left(-\dfrac{2}{3}\right)$

n. $\dfrac{2}{3} - \left(\dfrac{1}{3} - \dfrac{2}{3}\right)$

o. $\left|\dfrac{3}{7} - \left(-\dfrac{4}{7}\right)\right|$

p. $\dfrac{3}{7} - \left|-\dfrac{4}{7}\right|$

21. Let $x = 2$, $y = -3$, and $z = -1$ and evaluate each expression.

a. $y + z$

b. $x + y$

c. $x + (y + z)$

d. $x - y$

e. $x - (y - z)$

f. $(x - y) - z$

| **SECTION 1.5** | *Multiplying and Dividing Real Numbers* |

The product of two real numbers with like signs is the positive product of their absolute values.

The product of two real numbers with unlike signs is the negative of the product of their absolute values.

The quotient of two real numbers with like signs is the quotient of their absolute values.

The quotient of two real numbers with unlike signs is the negative of the quotient of their absolute values.

Division by zero is undefined.

22. Evaluate each expression.

a. $(+3)(+4)$

b. $(-5)(-12)$

c. $\left(-\dfrac{3}{14}\right)\left(-\dfrac{7}{6}\right)$

d. $(3.75)(0.37)$

e. $5(-7)$

f. $(-15)(7)$

g. $\left(-\dfrac{1}{2}\right)\left(\dfrac{4}{3}\right)$

h. $(-12.2)(3.7)$

i. $\dfrac{+25}{+5}$

j. $\dfrac{-14}{-2}$

k. $\dfrac{(-2)(-7)}{4}$

l. $\dfrac{-22.5}{-3.75}$

m. $\dfrac{-25}{5}$

n. $\dfrac{(-3)(-4)}{-6}$

o. $\left(\dfrac{-10}{2}\right)^2 - (-1)^3$

p. $\dfrac{[-3 + (-4)]^2}{10 + (-3)}$

q. $\left(\dfrac{-3 + (-3)}{3}\right)\left(\dfrac{-15}{5}\right)$

r. $\dfrac{-2 - (-8)}{5 + (-1)}$

23. Let $x = 2$, $y = -3$, and $z = -1$ and evaluate each expression.

a. xy

b. yz

c. $x(x + z)$

d. xyz

e. $y^2z + x$

f. $yz^3 + (xy)^2$

g. $\dfrac{xy}{z}$

h. $\dfrac{|xy|}{3z}$

| **SECTION 1.6** | *Algebraic Expressions* |

24. Let x, y, and z represent three real numbers. Write an algebraic expression that represents each quantity.

a. The product of x and z

b. The sum of x and twice y

c. Twice the sum of x and y

d. x decreased by the product of y and z

25. Write each algebraic expression as an English phrase.

  a. $3xy$

  b. $5 - yz$

  c. $yz - 5$

  d. $\dfrac{x + y + z}{2xyz}$

26. How many terms does the expression $3x + 4y + 9$ have?

27. What is the numerical coefficient of the term $7xy$?

28. What is the numerical coefficient of the term xy?

29. Find the sum of the numerical coefficients in $2x^3 + 4x^2 + 3x$.

| **SECTION 1.7** | *Properties of Real Numbers* |

The closure properties:

$x + y$ is a real number.

$x - y$ is a real number.

xy is a real number.

$\dfrac{x}{y}$ is a real number $(y \neq 0)$.

The commutative properties:

$x + y = y + x$

$xy = yx$

The associative properties:

$(x + y) + z = x + (y + z)$

$(xy)z = x(yz)$

The distributive property:

$x(y + z) = xy + xz$

The identity elements:

0 is the identity for addition.

1 is the identity for multiplication.

The additive and multiplicative inverse properties:

$x + (-x) = 0$

$x\left(\dfrac{1}{x}\right) = 1 \quad (x \neq 0)$

30. Tell which property of real numbers justifies each statement. Assume that all variables represent real numbers.

  a. $x + y$ is a real number

  b. $3 \cdot (4 \cdot 5) = (4 \cdot 5) \cdot 3$

  c. $3 + (4 + 5) = (3 + 4) + 5$

  d. $5(x + 2) = 5 \cdot x + 5 \cdot 2$

  e. $a + x = x + a$

  f. $3 \cdot (4 \cdot 5) = (3 \cdot 4) \cdot 5$

  g. $3 + (x + 1) = (x + 1) + 3$

  h. $x \cdot 1 = x$

  i. $17 + (-17) = 0$

  j. $x + 0 = x$

■ Chapter Test

1. List the prime numbers between 30 and 50.

2. What is the only even prime number?

3. Graph the composite numbers less than 10 on a number line.

4. Graph the real numbers from 5 to 15 on a number line.

5. Evaluate $-|23|$.

6. Evaluate $-|7| + |-7|$.

In Problems 7–10, place one of the symbols =, <, or > in each box to make a true statement.

7. $3(4 - 2)$ ▢ $-2(2 - 5)$

8. $1 + 4 \cdot 3$ ▢ $-2(-7)$

9. 25% of 136 ▢ $\dfrac{1}{2}$ of 66

10. -13.7 ▢ $-|-13.7|$

In Problems 11–16, simplify each expression.

11. $\dfrac{26}{40}$

12. $\dfrac{7}{8} \cdot \dfrac{24}{21}$

13. $\dfrac{18}{35} \div \dfrac{9}{14}$

14. $\dfrac{24}{16} + 3$

15. $\dfrac{17 - 5}{36} - \dfrac{2(13 - 5)}{12}$

16. $\dfrac{|-7 - (-6)|}{-7 - |-6|}$

17. Find 17% of 457 and round the answer to one decimal place.

18. Find the area of a rectangle 12.8 feet wide and 23.56 feet long. Round the answer to two decimal places.

19. Find the area of the figure in Illustration 1.

20. To the nearest cubic inch, find the volume of the solid in Illustration 2.

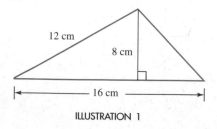

ILLUSTRATION 1

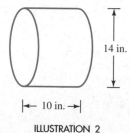

ILLUSTRATION 2

In Problems 21–26, let x = −2, y = 3, and z = 4. Evaluate each expression.

21. $xy + z$

22. $x(y + z)$

23. $\dfrac{z + 4y}{2x}$

24. $|x^y - z|$

25. $x^3 + y^2 + z$

26. $|x| - 3|y| - 4|z|$

27. Let x and y represent two real numbers. Write an algebraic expression to denote the quotient obtained when the product of the two numbers is divided by their sum.

28. Let x and y represent two real numbers. Write an algebraic expression to denote the difference obtained when the sum of x and y is subtracted from the product of 5 and y.

29. A man lives 12 miles from work and 7 miles from the grocery store. If he made x round trips to work and y round trips to the store, how many miles did he drive?

30. A baseball costs $\$a$ and a glove costs $\$b$. How much will it cost a community center to buy 12 baseballs and 8 gloves?

31. What is the numerical coefficient of the term $3xy^2$?

32. How many terms are in the expression $3x^2y + 5xy^2 + x + 7$?

33. What is the identity element for addition?

34. What is the multiplicative inverse of $\dfrac{1}{5}$?

In Problems 35–38, state which property of the real numbers justifies each statement.

35. $(xy)z = z(xy)$

36. $3(x + y) = 3x + 3y$

37. $2 + x = x + 2$

38. $7 \cdot \dfrac{1}{7} = 1$

2 Equations and Inequalities

MATHEMATICS IN THE WORKPLACE

Banker

Practically every bank has a group of officers who make decisions affecting bank operations: the president, who directs overall operations; one or more vice-presidents, who act as general managers or are in charge of bank departments, such as trust or credit; a comptroller, who is responsible for all bank property; and treasurers and other senior officers, who supervise sections within departments.

SAMPLE APPLICATION ■ A banker invested $24,000 for a client in two mutual funds, one earning 9% annual interest and the other earning 14%. After 1 year, the total interest earned was $3,135. How much was invested at each rate? (See Exercise 25 in Exercise 2.5.)

2.1   Solving Equations by Addition and Subtraction

■ EQUATIONS ■ SOLVING EQUATIONS ■ MARKDOWN AND MARKUP ■ GEOMETRY

Getting Ready   *Fill in each blank to make a true statement.*

1. $3 + \boxed{} = 0$ **2.** $(-7) + \boxed{} = 0$ **3.** $(-4) + \boxed{} = 0$

4. $7 - \boxed{} = 0$ **5.** $5 + 3(4) = \boxed{}$ **6.** $(-x) + \boxed{} = 0$

■ EQUATIONS

An **equation** is a statement indicating that two quantities are equal. Some examples of equations are

$$x + 5 = 21, \qquad 2x - 5 = 11, \qquad \text{and} \qquad 3x^2 - 4x + 5 = 0$$

The statement $3x + 2$ is not an equation, because it does not contain an $=$ sign.

In the equation $x + 5 = 21$, the expression $x + 5$ is called the **left-hand side,** and 21 is called the **right-hand side.** The letter x is called the **variable** (or the **unknown**).

An equation can be true or false. The equation $16 + 5 = 21$ is true, but the equation $10 + 5 = 21$ is false. The equation $2x - 5 = 11$ might be true or false, depending on the value of x. If $x = 8$, the equation is true, because when we substitute 8 for x, we get 11.

$$2(8) - 5 = 16 - 5$$
$$= 11$$

Any number that makes an equation true when substituted for its variable is said to *satisfy* the equation. All of the numbers that satisfy an equation are called its **solu-**

91

tions or **roots.** Since 8 is the only number that satisfies the equation $2x - 5 = 11$, it is the only solution.

EXAMPLE 1   Is 6 a solution of $3x - 5 = 2x$?

Solution   We substitute 6 for x and simplify.

$$3x - 5 = 2x$$
$$3 \cdot 6 - 5 \stackrel{?}{=} 2 \cdot 6 \qquad \text{Substitute 6 for } x.$$
$$18 - 5 \stackrel{?}{=} 12$$
$$13 = 12$$

Since $13 = 12$ is false, 6 is not a solution.   ∎

Self Check   Is 1 a solution of $2x + 3 = 5$?

Answer   yes

■ SOLVING EQUATIONS

To **solve an equation** means to find its solutions. To develop an understanding of how to solve equations, we refer to the scales shown in Figure 2-1. We can think of the scale shown in Figure 2-1(a) as representing the equation $x - 5 = 2$. The weight on the left-hand side of the scale is $(x - 5)$ grams, and the weight on the right-hand side is 2 grams. Because these weights are equal, the scale is in balance. To find x, we need to isolate it by adding 5 grams to the left-hand side of the scale. To keep the scale in balance, we must also add 5 grams to the right-hand side. After adding 5 grams to both sides of the scale, we can see from Figure 2-1(b) that x grams will

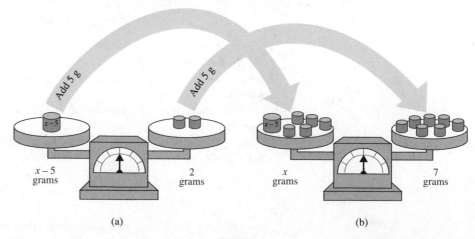

FIGURE 2-1

be balanced by 7 grams. We say that we have solved the equation and that the solution is 7.

The example suggests the following property of equality: *If the same quantity is added to equal quantities, the results will be equal quantities.* We can express this property in symbols.

Addition Property of Equality
Suppose that a, b, and c are real numbers. Then

If $a = b$, then $a + c = b + c$.

When we use this property, the resulting equation will have the same solutions as the original one. We say that the equations are *equivalent*.

Equivalent Equations
Two equations are **equivalent equations** when they have the same solutions.

In the previous example, we found that $x - 5 = 2$ is equivalent to $x = 7$. In the next example, we use the addition property of equality to solve the equation $x - 5 = 2$ algebraically.

EXAMPLE 2   Solve $x - 5 = 2$.

Solution   To isolate x on one side of the $=$ sign, we undo the subtraction of 5 by adding 5 to both sides of the equation.

$$x - 5 = 2$$
$$x - 5 + 5 = 2 + 5 \qquad \text{Add 5 to both sides of the equation.}$$
$$x = 7 \qquad \text{$-5 + 5 = 0$ and $2 + 5 = 7$.}$$

We check by substituting 7 for x in the original equation and simplifying.

$$x - 5 = 2$$
$$7 - 5 \overset{?}{=} 2 \qquad \text{Substitute 7 for x.}$$
$$2 = 2$$

Since $2 = 2$, the solution checks. ∎

Self Check   Solve $b - 21.8 = 13$.
Answer   34.8

■ ■ ■ ■ ■ ■ ■ ■ ■ ■ PERSPECTIVE

To find answers to such questions as How many? How far? How fast? and How heavy?, we often make use of mathematical statements called **equations.** The concept has a long history, and the techniques we will study in this chapter have been developed over many centuries.

The mathematical notation that we use today is the result of thousands of years of development. The ancient Egyptians used a word for variables, best translated as *heap.* Others used the word *res,* which is Latin for *thing.* In the fifteenth century, the letters *p:* and *m:* were used for *plus* and *minus.* What we would now write as $2x + 3 = 5$ might have been written by those early mathematicians as 2 *res p:* 3 *aequalis* 5.

We can think of the scale shown in Figure 2-2(a) as representing the equation $x + 4 = 9$. The weight on the left-hand side of the scale is $(x + 4)$ grams, and the weight on the right-hand side is 9 grams. Because these weights are equal, the scale is in balance. To find x, we need to isolate it by removing 4 grams from the left-hand side. To keep the scale in balance, we must also remove 4 grams from the right-hand side. In Figure 2-2(b), we can see that x grams will be balanced by 5 grams. We have found that the solution is 5.

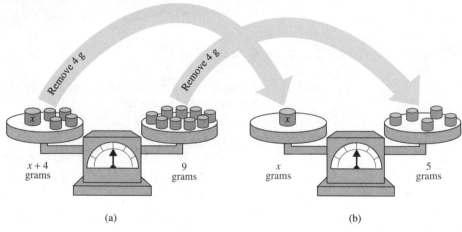

| $x + 4$ grams | 9 grams | x grams | 5 grams |

(a)   (b)

FIGURE 2-2

The previous example suggests the following property of equality: *If the same quantity is subtracted from equal quantities, the results will be equal quantities.* We can express this property in symbols.

Subtraction Property of Equality
Suppose that a, b, and c are real numbers. Then
  If $a = b$, then $a - c = b - c$.

When we use this property, the resulting equation will be equivalent to the original one.

In the next example, we use the subtraction property of equality to solve the equation $x + 4 = 9$ algebraically.

EXAMPLE 3   Solve $x + 4 = 9$.

Solution   To isolate x on one side of the $=$ sign, we undo the addition of 4 by subtracting 4 from both sides of the equation.

$$x + 4 = 9$$
$$x + 4 - 4 = 9 - 4 \qquad \text{Subtract 4 from both sides.}$$
$$x = 5 \qquad\qquad 4 - 4 = 0 \text{ and } 9 - 4 = 5.$$

We can check the solution by substituting 5 for x in the original equation and simplifying.

$$x + 4 = 9$$
$$5 + 4 \overset{?}{=} 9 \qquad\qquad \text{Substitute 5 for } x.$$
$$9 = 9$$

The solution checks.   ■

Self Check
Answer

Solve $a + 17.5 = 12.2$.

-5.3

■ MARKDOWN AND MARKUP

When the price of merchandise is reduced, the amount of the reduction is called the **markdown** or the **discount.** To find the sale price of an item, we subtract the markdown from the regular price.

Sale price	=	regular price	−	markdown

EXAMPLE 4   **Buying a sofa**   A sofa is on sale for $650. If it has been marked down $325, find its regular price.

Solution   We can let r represent the regular price and substitute 650 for the sale price and 325 for the markdown.

Sale price	=	regular price	−	markdown
650	=	r	−	325

We can use the addition property of equality to solve the equation.

$$650 = r - 325$$
$$650 + \mathbf{325} = r - 325 + \mathbf{325} \qquad \text{Add 325 to both sides.}$$
$$975 = r \qquad\qquad 650 + 325 = 975 \text{ and } -325 + 325 = 0.$$

The regular price is $975. ■

Self Check Find the regular price of the sofa in Example 4 if the discount were $275.

Answer $925

To make a profit, a merchant must sell an item for more than he or she paid for it. The retail price of the item is the sum of its wholesale cost and the **markup.**

| Retail price | = | wholesale cost | + | markup |

EXAMPLE 5 **Buying a car** A car with a sticker price of $17,500 has a markup of $3,500. Find the invoice price (the wholesale price) to the dealer.

Solution We can let w represent the wholesale price and substitute 17,500 for the retail price and 3,500 for the markup.

| Retail price | = | wholesale cost | + | markup |
| 17,500 | = | w | + | 3,500 |

We can use the subtraction property of equality to solve the equation.

$$17,500 = w + 3,500$$
$$17,500 - \mathbf{3,500} = w + 3,500 - \mathbf{3,500} \qquad \text{Subtract 3,500 from both sides.}$$
$$14,000 = w \qquad\qquad 17,500 - 3,500 = 14,000 \text{ and } 3,500 - 3,500 = 0.$$

The invoice price is $14,000. ■

Self Check Find the invoice price of the car in Example 5 if the markup is $6,700.

Answer $10,800

■ GEOMETRY

The geometric figure shown in Figure 2-3(a) is called an **angle.** Angles are measured in **degrees.** The angle shown in Figure 2-3(b) measures 45 degrees (denoted as 45°). If an angle measures 90°, as in Figure 2-3(c), it is called a **right angle.** If an angle measures 180°, it is called a **straight angle.**

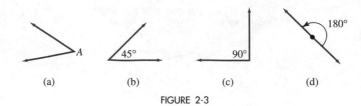

FIGURE 2-3

EXAMPLE 6   Find x.

Solution   Since the sum of the measures of the angles measuring 37° and x is 75°, we can form and solve the equation

$$x + 37 = 75$$
$$x + 37 - \mathbf{37} = 75 - \mathbf{37} \qquad \text{Subtract 37 from both sides.}$$
$$x = 38 \qquad\qquad 37 - 37 = 0 \text{ and } 75 - 37 = 38.$$

Thus, $x = 38°$.   ■

Self Check   Find x.

Answer   47°

EXAMPLE 7   Find x.

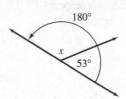

Solution   Since the sum of the measures of the angles measuring 53° and x is 180°, we can form and solve the equation

$$x + 53 = 180$$
$$x + 53 - \mathbf{53} = 180 - \mathbf{53} \qquad \text{Subtract 53 from both sides.}$$
$$x = 127 \qquad\qquad 53 - 53 = 0 \text{ and } 180 - 53 = 127.$$

Thus, $x = 127°$.   ■

Self Check   Find x.

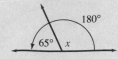

Answer   115°

If the sum of two angles is 90°, the angles are called **complementary.** If the sum of two angles is 180°, the angles are called **supplementary.**

EXAMPLE 8   Find   **a.** the complement of an angle measuring 30°   and   **b.** the supplement of an angle measuring 50°.

Solution   **a.** We can let x represent the complement of 30°. Since the sum of two complementary angles is 90°, we have

$$x + 30 = 90$$
$$x + 30 - \mathbf{30} = 90 - \mathbf{30} \qquad \text{Subtract 30 from both sides.}$$
$$x = 60 \qquad\qquad 30 - 30 = 0 \text{ and } 90 - 30 = 60.$$

The complement of a 30° angle is a 60° angle.

b. We can let x represent the supplement of 50°. Since the sum of two supplementary angles is 180°, we have

$$x + 50 = 180$$
$$x + 50 - \mathbf{50} = 180 - \mathbf{50} \qquad \text{Subtract 50 from both sides.}$$
$$x = 130 \qquad\qquad 50 - 50 = 0 \text{ and } 180 - 50 = 130.$$

The supplement of a 50° angle is a 130° angle.   ■

Self Check   Find   **a.** the supplement of 105°   and   **b.** the complement of 15°.
Answers   **a.** 75°,   **b.** 75°

Orals   *Solve each equation.*

1. $x - 9 = 11$ **2.** $x - 3 = 13$ **3.** $w + 5 = 7$

4. $x + 32 = 36$ **5.** $x - 2.5 = -2.5$ **6.** $x + 12.4 = 12.4$

7. $x + \dfrac{1}{5} = \dfrac{4}{5}$ **8.** $x - \dfrac{2}{7} = \dfrac{5}{7}$

9. Find the complement of a 10° angle.

10. Find the supplement of an 80° angle.

EXERCISE 2.1

REVIEW *Do the operations and classify the result as an integer, a prime number, or a composite number.*

1. $3[2 - (-3)]$

2. $(2 - 4)^4$

3. $\dfrac{2^3 - 14}{3^2 - 3}$

4. $\dfrac{3 + 5}{3} - \dfrac{5}{7 - 4}$

Tell which property of real numbers justifies each statement.

5. $3 + 31$ is a real number

6. $3(x + y) = 3x + 3y$

7. $a + (3 + b) = (3 + b) + a$

8. $a + (3 + b) = (a + 3) + b$

Evaluate each expression.

9. 4^3

10. $(-3)^4$

11. $-3(4^2 - 5^2)$

12. $-6^2 + 5 \cdot 4$

VOCABULARY AND CONCEPTS *Fill in each blank to make a true statement.*

13. An _____ is a statement that two quantities are equal.

14. A _____ of an equation is a number that satisfies the equation.

15. The answer to an equation is called a solution or a _____ of the equation.

16. A letter that represents a number is called a _____.

17. If two equations have the same solutions, they are called _____ equations.

18. To solve an equation, we isolate the _____ on one side of the equation.

19. The equation $3x - 2 = 7$ can be true or false, depending on the value of _.

20. If the same quantity is added to _____ quantities, the results will be equal quantities.

21. If the same quantity is subtracted from equal quantities, the results will be _____ quantities.

22. Sale price = _____ − markdown

23. Retail price = wholesale cost + _____

24. Another name for markdown is _____.

25. If the sum of two angles is 180°, the angles are called _____ angles.

26. If the sum of two angles is 90°, the angles are called _____ angles.

In Exercises 27–34, tell whether each statement is an equation.

27. $x = 2$

28. $y - 3$

29. $7x < 8$

30. $7 + x = 2$

31. $x + 7 = 0$

32. $3 - 3y > 2$

33. $1 + 1 = 3$

34. $5 = a + 2$

PRACTICE *In Exercises 35–46, tell whether the given number is a solution of the equation.*

35. $x + 2 = 3$; 1

36. $x - 2 = 4$; 6

37. $a - 7 = 0$; -7

38. $x + 4 = 4$; 0

39. $\dfrac{y}{7} = 4$; 28

40. $\dfrac{c}{-5} = -2$; -10

41. $\dfrac{x}{5} = x$; 0

42. $\dfrac{x}{7} = 7x$; 0

43. $3k + 5 = 5k - 1$; 3

44. $2s - 1 = s + 7$; 6

45. $\dfrac{5 + x}{10} - x = \dfrac{1}{2}$; 0

46. $\dfrac{x - 5}{6} = 12 - x$; 11

In Exercises 47–66, use the addition or the subtraction property of equality to solve each equation. **Check all solutions.**

47. $x + 7 = 13$

48. $y + 3 = 7$

49. $y - 7 = 12$

50. $c - 11 = 22$

51. $1 = y - 5$

52. $0 = r + 10$

53. $p - 404 = 115$

54. $41 = 45 + q$

55. $-37 + z = 37$

56. $-43 + a = -43$

57. $-57 = b - 29$

58. $-93 = 67 + y$

59. $\dfrac{4}{3} = -\dfrac{2}{3} + x$

60. $z + \dfrac{5}{7} = -\dfrac{2}{7}$

61. $d + \dfrac{2}{3} = \dfrac{3}{2}$

62. $s + \dfrac{2}{3} = \dfrac{1}{5}$

63. $-\dfrac{3}{5} = x - \dfrac{2}{5}$

64. $b + 7 = \dfrac{20}{3}$

65. $r - \dfrac{1}{5} = \dfrac{3}{10}$

66. $t + \dfrac{4}{7} = \dfrac{11}{14}$

APPLICATIONS   *Use an equation to solve each problem.*

67. Buying a boat   A boat is on sale for $7,995. Find its regular price if it has been marked down $1,350.

68. Buying a house   A house that was priced at $105,000 has been discounted $7,500. Find the new asking price.

69. Buying clothes   A sport jacket that sells for $175 has a markup of $85. Find the wholesale price.

70. Buying a vacuum cleaner   A vacuum that sells for $97 has a markup of $37. Find the wholesale price.

71. Banking   The amount A in an account is given by the formula

$$A = p + i$$

where p is the principal and i is the interest. How much interest has been earned if an original deposit (the principal) of $4,750 has grown to be $5,010?

72. Depreciation   The current value v of a car is given by the formula

$$v = c - d$$

where c is the original price and d is the depreciation. Find the original cost of a car that is worth $10,250 after depreciating $7,500.

73. Appreciation   The value v of a house is given by the formula

$$v = p + a$$

where p is the original purchase price and a is the appreciation. Find the original purchase price of a house that is worth $110,000 and has appreciated $57,000.

74. Taxes   The cost c of an item is given by the formula

$$c = p + t$$

where p is the price and t is the sales tax. Find the tax paid on an item that was priced at $37.10 and cost $39.32.

75. Buying carpet   The cost c of carpet is given by the formula

$$c = p + t$$

where p is the price and t is the cost of installation. How much did it cost to install $317 worth of carpet that cost $512?

76. Selling real estate   The money m the seller receives from selling a house is given by the formula

$$m = s - c$$

where s is the selling price and c is the agent's commission. Find the selling price of a house if the seller received $217,000 and the agent received $13,020.

77. Buying real estate   The cost of a condominium is $57,595 less than the cost of a house. If the house costs $202,744, find the cost of the condominium.

78. Buying paint   After reading the ad in Illustration 1, a decorator bought one gallon of primer, one gallon of paint, and a brush. The total cost was $30.44. Find the cost of the brush.

ILLUSTRATION 1

GEOMETRY   *In Exercises 79–84, find x.*

79.

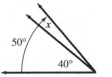

80.

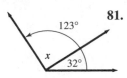

81.

82.

83.

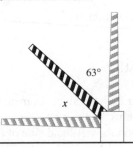

84.

85. Find the complement of 37°.

87. Find the supplement of the complement of 40°.

86. Find the supplement of 37°.

88. Find the complement of the supplement of 140°.

WRITING

89. Explain what it means for a number to satisfy an equation.

91. Explain what Figure 2-1 is trying to show.

90. How can you tell whether a number is the solution of an equation?

92. Explain what Figure 2-2 is trying to show.

SOMETHING TO THINK ABOUT

93. If two lines intersect as in Illustration 2, angles 1 and 2 and angles 3 and 4 are called **vertical angles.** Let the measure of angle 1 be various numbers and compute the values of the other three. What do you discover?

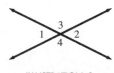

ILLUSTRATION 2

94. If two lines meet and form one right angle, the lines are said to be **perpendicular.** See Illustration 3. Find the measures of angles 1, 2, and 3. What do you discover?

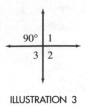

ILLUSTRATION 3

2.2 Solving Equations by Multiplication and Division

■ THE DIVISION PROPERTY OF EQUALITY ■ THE MULTIPLICATION PROPERTY OF EQUALITY ■ PERCENT
■ APPLICATIONS OF PERCENT

Getting Ready *Fill in each blank to make a true statement.*

1. $\dfrac{1}{3} \cdot 3 = $ ☐

2. $5 \cdot $ ☐ $= 1$

3. $\dfrac{-6}{-6} = $ ☐

4. $\dfrac{4(2)}{\text{☐}} = 2$

5. $5 \cdot \dfrac{4}{5} = $ ☐

6. $\dfrac{-5(3)}{-5} = $ ☐

7. $0.07 \cdot 900 = $ ☐

8. $0.09 \cdot 800 = $ ☐

■ THE DIVISION PROPERTY OF EQUALITY

We will now consider how to solve the equation $2x = 6$. Since $2x$ means $2 \cdot x$, the equation can be written as $2 \cdot x = 6$. We can think of the scale shown in Figure 2-4(a) as representing this equation. The weight on the left-hand side of the scale is $2 \cdot x$ grams, and the weight on the right-hand side is 6 grams. Because these weights

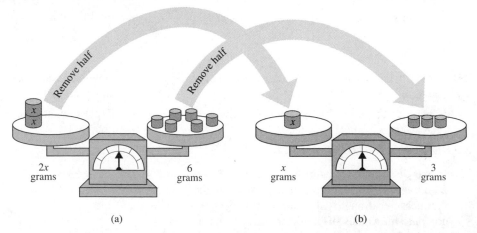

(a)   (b)

FIGURE 2-4

are equal, the scale is in balance. To find x, we remove half of the weight from each side. This is equivalent to dividing the weight on both sides by 2. When we do this, the scale will remain in balance. From the scale shown in Figure 2-4(b), we can see that x grams will be balanced by 3 grams. Thus, $x = 3$.

The previous example suggests the following property of equality: *If equal quantities are divided by the same nonzero quantity, the results will be equal quantities.* We can express this property in symbols.

Division Property of Equality
Suppose that a, b, and c are real numbers and that $c \neq 0$. Then

$$\text{If } a = b, \text{ then } \frac{a}{c} = \frac{b}{c}.$$

When we use the division property, the resulting equation will be equivalent to the original one.

To solve the equation $2x = 6$ algebraically, we proceed as in Example 1.

EXAMPLE 1   Solve $2x = 6$.

Solution   To isolate x on one side of the $=$ sign, we undo the multiplication by 2 by dividing both sides by 2.

$$2x = 6$$

$$\frac{2x}{2} = \frac{6}{2} \qquad \text{Divide both sides by 2.}$$

$$x = 3 \qquad \tfrac{2}{2} = 1 \text{ and } \tfrac{6}{2} = 3.$$

Verify that the solution is 3.   ■

Self Check   Solve $-5x = 15$.

Answer -3

■ THE MULTIPLICATION PROPERTY OF EQUALITY

We can think of the scale shown in Figure 2-5(a) as representing the equation $\frac{x}{3} = 12$. The weight on the left-hand side of the scale is $\frac{x}{3}$ grams, and the weight on the right-hand side is 12 grams. Because these weights are equal, the scale is in balance. To find x, we can triple (or multiply by 3) the weight on each side. When we do this, the scale will remain in balance. From the scale shown in Figure 2-5(b), we can see that x grams will be balanced by 36 grams. Thus, $x = 36$.

The previous example suggests the following property of equality: *If equal quantities are multiplied by the same nonzero quantity, the results will be equal quantities.* We can express this property in symbols.

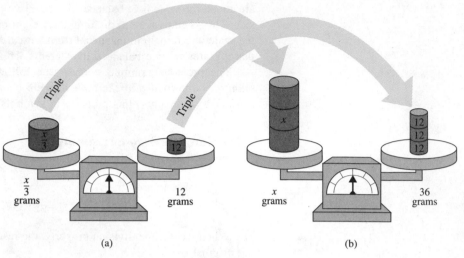

FIGURE 2-5

Multiplication Property of Equality

Suppose that a, b, and c are real numbers, and $c \neq 0$. Then

If $a = b$, then $ca = cb$.

When we use the multiplication property, the resulting equation will be equivalent to the original one.

To solve the equation $\frac{x}{3} = 12$ algebraically, we proceed as in Example 2.

EXAMPLE 2   Solve $\dfrac{x}{3} = 12$.

Solution   To find x, we undo the division by 3 by multiplying both sides of the equation by 3.

$$\frac{x}{3} = 12$$

$$\mathbf{3} \cdot \frac{x}{3} = \mathbf{3} \cdot 12 \qquad \text{Multiply both sides by 3.}$$

$$x = 36 \qquad 3 \cdot \tfrac{x}{3} = x \text{ and } 3 \cdot 12 = 36.$$

Verify that the solution checks.   ∎

Self Check   Solve $\frac{x}{5} = -7$.
Answer -35

■ PERCENT

A percent is the numerator of a fraction whose denominator is 100. For example, $6\frac{1}{4}$ percent (written as $6\frac{1}{4}\%$) is the fraction $\frac{6.25}{100}$, or the decimal 0.0625. In problems involving percent, the word *of* usually indicates multiplication. For example, $6\frac{1}{4}\%$ of 8,500 is the product of 0.0625 and 8,500.

$$6\frac{1}{4}\% \text{ of } 8{,}500 = 0.0625 \cdot 8{,}500$$
$$= 531.25$$

In the statement $6\frac{1}{4}\%$ of $8{,}500 = 531.25$, the percent $6\frac{1}{4}\%$ is called a **rate,** 8,500 is called the **base,** and their product, 531.25, is called a **percentage.** Every percent problem is based on the equation **rate · base = percentage.**

> **Percentage Formula**
> The product of a rate r and a base b is called a **percentage.** If p is the percentage, then
>
> $$rb = p$$

Percent problems involve questions such as

- What is 30% of 1,000?
- 45% of what number is 405?
- What percent of 400 is 60?

When we use equations, these problems are easy to solve.

EXAMPLE 3   What is 30% of 1,000?

Solution   In this problem, the rate r is 30%, and the base is 1,000.

Rate	·	base	=	percentage
30%	of	1,000	is	the percentage.

We can substitute these values into the percentage formula and solve for p.

$$rb = p$$
$$\mathbf{30\%} \cdot \mathbf{1{,}000} = p \qquad \text{Substitute 30\% for } r \text{ and 1,000 for } b.$$
$$0.30 \cdot 1{,}000 = p \qquad \text{Change 30\% to the decimal 0.30.}$$
$$300 = p \qquad \text{Multiply.}$$

Thus, 30% of 1,000 is 300.   ■

Self Check   Find 45% of 800.

Answer   360

EXAMPLE 4   45% of what number is 405?

Solution   In this problem, the rate r is 45%, and the percentage p is 405.

Rate	·	base	=	percentage
45%	of	what number	is	405?

We can substitute these values into the percentage formula and solve for b.

$$rb = p$$
$$45\% \cdot b = 405 \qquad \text{Substitute 45\% for } r \text{ and 405 for } p.$$
$$0.45 \cdot b = 405 \qquad \text{Change 45\% to a decimal.}$$
$$\frac{0.45b}{0.45} = \frac{405}{0.45} \qquad \text{To undo the multiplication by 0.45, divide both sides by 0.45.}$$
$$b = 900 \qquad \tfrac{0.45}{0.45} = 1 \text{ and } \tfrac{405}{0.45} = 900.$$

Thus, 45% of 900 is 405.   ■

Self Check   35% of what number is 306.25?
Answer   875

EXAMPLE 5   What percent of 400 is 60?

Solution   In this problem, the base b is 400, and the percentage p is 60.

Rate	·	base	=	percentage
What percent	of	400	is	60?

We can substitute these values in the percentage formula and solve for r.

$$rb = p$$
$$r \cdot 400 = 60 \qquad \text{Substitute 400 for } b \text{ and 60 for } p.$$
$$\frac{400r}{400} = \frac{60}{400} \qquad \text{To undo the multiplication by 400, divide both sides by 400.}$$
$$r = 0.15 \qquad \tfrac{400}{400} = 1 \text{ and } \tfrac{60}{400} = 0.15.$$
$$r = 15\% \qquad \text{To change the decimal into a percent, multiply by 100 and insert a \% sign.}$$

Thus, 15% of 400 is 60.   ■

Self Check

Answer

What percent of 600 is 150?

25%

■ APPLICATIONS OF PERCENT

EXAMPLE 6

Investing At a recent stockholders' meeting, 4.5 million shares of stock were voted in favor of a proposal for a mandatory retirement age for members of the board of directors. Since this represented 75% of the number of shares outstanding, the proposal passed. How many shares are outstanding?

Solution Let b represent the number of outstanding shares. Then 75% of b is 4.5 million. We can substitute 75% for r and 4.5 million for p in the formula for percentage and solve for b.

$$rb = p$$

$$75\% \cdot b = 4{,}500{,}000 \qquad \text{Substitute 75\% for } r \text{ and 4,500,000 for } p.$$

$$0.75b = 4{,}500{,}000 \qquad \text{Change 75\% to a decimal.}$$

$$\frac{0.75b}{0.75} = \frac{4{,}500{,}000}{0.75} \qquad \text{To undo the multiplication of 0.75, divide both sides by 0.75.}$$

$$b = 6{,}000{,}000 \qquad \frac{0.75}{0.75} = 1 \text{ and } \frac{4{,}500{,}000}{0.75} = 6{,}000{,}000.$$

There are 6 million shares outstanding.

Self Check

In Example 6, if 60% of the shares outstanding were voted in favor of the proposal, how many shares were voted in favor?

Answer 3.6 million

EXAMPLE 7

Quality control After examining 240 sweaters, a quality-control inspector found 5 with defective stitching, 8 with mismatched designs, and 2 with incorrect labels. What percent were defective?

Solution Let r represent the percent that are defective. Then the base b is 240, and the percentage p is the number of defective sweaters, which is $5 + 8 + 2 = 15$. We can find r by solving the equation

$$rb = p$$

$$r \cdot 240 = 15 \qquad \text{Substitute 240 for } b \text{ and 15 for } p.$$

$$\frac{240r}{240} = \frac{15}{240} \qquad \text{To undo the multiplication of 240, divide both sides by 240.}$$

$$r = 0.0625 \qquad \tfrac{240}{240} = 1 \text{ and } \tfrac{15}{240} = 0.0625.$$

$$r = 6.25\% \qquad \text{To change 0.0625 to a percent, multiply by 100 and add a \% sign.}$$

The defect rate is $6\tfrac{1}{4}\%$.

Self Check   In Example 7, if a second inspector found 3 sweaters with faded colors in addition to the defectives found by inspector 1, what percent were defective?

Answer $7\frac{1}{2}\%$

Orals   *In Exercises 1–8, solve each equation.*

1. $3x = 3$ **2.** $5x = 5$ **3.** $-7x = 14$

4. $7.5x = 0$ **5.** $\dfrac{x}{5} = 2$ **6.** $\dfrac{x}{2} = -10$

7. $\dfrac{x}{-4} = 3$ **8.** $\dfrac{x}{8} = -3$

9. Change 30% to a decimal.   **10.** Change 0.08 to a percent.

EXERCISE 2.2

REVIEW   *Do the operations. Simplify the result when possible.*

1. $\dfrac{4}{5} + \dfrac{2}{3}$ **2.** $\dfrac{5}{6} \cdot \dfrac{12}{25}$ **3.** $\dfrac{5}{9} \div \dfrac{3}{5}$ **4.** $\dfrac{15}{7} - \dfrac{10}{3}$

5. $2 + 3 \cdot 4$ **6.** $3 \cdot 4^2$ **7.** $3 + 4^3(-5)$ **8.** $\dfrac{5(-4) - 3(-2)}{10 - (-4)}$

Find the area of each geometric figure.

9. A rectangle with dimensions of 3.5 feet by 7.2 feet.   **10.** A circle with a diameter of 12.45 inches. Give the result to the nearest hundredth.

VOCABULARY AND CONCEPTS   *Fill in each blank to make a true statement.*

11. If equal quantities are divided by the same nonzero quantity, the results are _____ quantities.

12. If $a = b$, then $\dfrac{a}{c} = $ __, provided that $c \neq$ __.

13. If $a = b$, then $ac = $ __.

14. If _____ quantities are multiplied by the same nonzero quantity, the results will be equal quantities.

15. A percent is the numerator of a fraction whose denominator is ____.

16. Rate $\cdot$ _____ = percentage

PRACTICE   *In Exercises 17–52, use the division or multiplication property of equality to solve each equation.* **Check all solutions.**

17. $6x = 18$ **18.** $25x = 625$ **19.** $-4x = 36$ **20.** $-16y = 64$

21. $4t = 108$ **22.** $-66 = -6t$ **23.** $11x = -121$ **24.** $-9y = -9$

25. $\dfrac{x}{5} = 5$ **26.** $\dfrac{x}{15} = 3$ **27.** $\dfrac{x}{32} = -2$ **28.** $\dfrac{y}{16} = -5$

29. $\dfrac{b}{3} = 5$ **30.** $\dfrac{a}{5} = -3$ **31.** $-3 = \dfrac{s}{11}$ **32.** $\dfrac{s}{-12} = 4$

33. $-32z = 64$ **34.** $15 = \dfrac{r}{-5}$ **35.** $18z = -9$ **36.** $-12z = 3$

37. $\dfrac{z}{7} = 14$ **38.** $-19x = -57$ **39.** $\dfrac{w}{7} = \dfrac{5}{7}$ **40.** $-17z = -51$

41. $\dfrac{s}{-3} = -\dfrac{5}{6}$ **42.** $1,228 = \dfrac{x}{0.25}$ **43.** $0.25x = 1,228$ **44.** $-255y = 51$

45. $\dfrac{b}{3} = \dfrac{1}{3}$ **46.** $\dfrac{a}{13} = \dfrac{1}{26}$ **47.** $-0.2w = -17$ **48.** $1.5a = -14$

49. $\dfrac{u}{5} = -\dfrac{3}{10}$ **50.** $\dfrac{t}{-7} = \dfrac{1}{2}$ **51.** $\dfrac{p}{0.2} = 12$ **52.** $\dfrac{t}{0.3} = -36$

In Exercises 53–68, use the formula rb = p to find each value.

53. What number is 40% of 200?

54. What number is 35% of 520?

55. What number is 50% of 38?

56. What number is 25% of 300?

57. 15% of what number is 48?

58. 26% of what number is 78?

59. 133 is 35% of what number?

60. 13.3 is 3.5% of what number?

61. 28% of what number is 42?

62. 44% of what number is 143?

63. What percent of 357.5 is 71.5?

64. What percent of 254 is 13.208?

65. 0.32 is what percent of 4?

66. 3.6 is what percent of 28.8?

67. 34 is what percent of 17?

68. 39 is what percent of 13?

APPLICATIONS   *In Exercises 69–78, solve each problem.*

69. Customer satisfaction   Two-thirds of a movie audience left the theater in disgust. If 78 angry patrons walked out, how many were there originally?

70. Stock split   After a three-for-two stock split, each shareholder will own 1.5 times as many shares as before. If 555 shares are owned after the split, how many were owned before?

71. Off-campus housing   Four-sevenths of the senior class is living in off-campus housing. If 868 students live off campus, how large is the senior class?

72. Union membership   The 2,484 union members represent 90% of a factory's workforce. How many are employed?

73. Shopper dissatisfaction   Refer to the survey results in Illustration 1. What percent of those surveyed were not pleased?

Shopper survey results	
First-time shoppers	1,731
Major purchase today	539
Shopped within previous month	1,823
Satisfied with service	4,140
Seniors	2,387
Total surveyed	9,200

ILLUSTRATION 1

74. Charity overhead   Out of $237,000 donated to a certain charity, $5,925 was used to pay for fundraising expenses. What percent of donations was overhead?

75. Selling price of a microwave oven   The 5% sales tax on a microwave oven amounts to $13.50. What is the microwave's selling price?

76. Sales taxes   Sales tax on a $12 compact disc is $0.72. At what rate is sales tax computed?

77. Hospital occupancy   18% of hospital patients stay for less than one day. If 1,008 patients in January stayed for less than one day, what total number of patients did the hospital treat in January?

78. House prices   The average price of houses in one neighborhood decreased 8% since last year, a drop of $7,800. What was the average price of a house last year?

WRITING

79. Explain how you would decide whether a number is a solution of an equation.

80. Distinguish between *percent* and *percentage*.

SOMETHING TO THINK ABOUT

81. The Ahmes Papyrus mentioned at the beginning of Chapter 1 contains this statement: *A circle nine units in diameter has the same area as a square eight units on a side.* From this statement, determine the ancient Egyptians' approximation of π.

82. Calculate the Egyptians' **percent of error:** What percent of the actual value of π is the difference between the values?

2.3 Solving More Equations

■ SOLVING MORE COMPLICATED EQUATIONS   ■ MARKUP AND MARKDOWN

Getting Ready   *Do the operations.*

1. $7 + 3 \cdot 5$ **2.** $3(5 + 7)$ **3.** $\dfrac{3 + 7}{2}$ **4.** $3 + \dfrac{7}{2}$

5. $\dfrac{3(5 - 8)}{9}$ **6.** $3 \cdot \dfrac{5 - 8}{9}$ **7.** $\dfrac{3 \cdot 5 - 8}{9}$ **8.** $3 \cdot \dfrac{5}{9} - 8$

■ SOLVING MORE COMPLICATED EQUATIONS

We have solved equations by using the addition, subtraction, multiplication, and division properties of equality. To solve more complicated equations, we need to use several of these properties in succession.

EXAMPLE 1   Solve $-12x + 5 = 17$.

Solution   The left-hand side of the equation indicates that x is to be multiplied by -12 and then 5 is to be added to that product. To isolate x, we must undo these operations in the opposite order.

- To undo the addition of 5, we subtract 5 from both sides.
- To undo the multiplication by -12, we divide both sides by -12.

$$-12x + 5 = 17$$

$$-12x + 5 - \mathbf{5} = 17 - \mathbf{5} \qquad \text{To undo the addition of 5, subtract 5 from both sides.}$$

$$-12x = 12 \qquad 5 - 5 = 0 \text{ and } 17 - 5 = 12.$$

$$\frac{-12x}{\mathbf{-12}} = \frac{12}{\mathbf{-12}} \qquad \text{To undo the multiplication by } -12, \text{ divide both sides by } -12.$$

$$x = -1 \qquad \frac{-12}{-12} = 1 \text{ and } \frac{12}{-12} = -1.$$

Check: $-12x + 5 = 17$

$$-12(\mathbf{-1}) + 5 \stackrel{?}{=} 17 \qquad \text{Substitute } -1 \text{ for } x.$$

$$12 + 5 \stackrel{?}{=} 17 \qquad \text{Simplify.}$$

$$17 = 17$$

Because $17 = 17$, the solution checks.   ∎

Self Check   Solve $2x + 3 = 15$.

Answer   6

EXAMPLE 2   Solve $\dfrac{x}{3} - 7 = -3$.

Solution   The left-hand side of the equation indicates that x is to be divided by 3 and then 7 is to be subtracted from that quotient. To isolate x, we must undo these operations in the opposite order.

- To undo the subtraction of 7, we add 7 to both sides.
- To undo the division by 3, we multiply both sides by 3.

$$\frac{x}{3} - 7 = -3$$

$$\frac{x}{3} - 7 + \mathbf{7} = -3 + \mathbf{7} \qquad \text{To undo the subtraction of 7, add 7 to both sides.}$$

$$\frac{x}{3} = 4 \qquad -7 + 7 = 0 \text{ and } -3 + 7 = 4.$$

$$3 \cdot \frac{x}{3} = \mathbf{3} \cdot 4 \qquad \text{To undo the division by 3, multiply both sides by 3.}$$

$$x = 12 \qquad 3 \cdot \frac{1}{3} = 1 \text{ and } 3 \cdot 4 = 12.$$

$$\textit{Check: } \frac{x}{3} - 7 = -3$$

$$\frac{12}{3} - 7 \overset{?}{=} -3 \qquad \text{Substitute 12 for } x.$$

$$4 - 7 \overset{?}{=} -3 \qquad \text{Simplify.}$$

$$-3 = -3$$

Since $-3 = -3$, the solution checks. ∎

Self Check   Solve $\frac{x}{4} - 3 = 5$.

Answer   32

EXAMPLE 3   Solve $\dfrac{x-7}{3} = 9$.

Solution   The left-hand side of the equation indicates that 7 is to be subtracted from x and that the difference is to be divided by 3. To isolate x, we must undo these operations in the opposite order.

- To undo the division by 3, we multiply both sides by 3.
- To undo the subtraction of 7, we add 7 to both sides.

$$\frac{x-7}{3} = 9$$

$$\mathbf{3}\left(\frac{x-7}{3}\right) = \mathbf{3}(9) \qquad \text{To undo the division by 3, multiply both sides by 3.}$$

$$x - 7 = 27 \qquad 3 \cdot \tfrac{1}{3} = 1 \text{ and } 3(9) = 27.$$

$$x - 7 \mathbf{+ 7} = 27 \mathbf{+ 7} \qquad \text{To undo the subtraction of 7, add 7 to both sides.}$$

$$x = 34 \qquad -7 + 7 = 0 \text{ and } 27 + 7 = 34.$$

Verify that the solution checks. ∎

Self Check   Solve $\frac{a-3}{5} = -2$.

Answer -7

EXAMPLE 4	Solve $\dfrac{3x}{4} + 2 = -7$.

Solution   The left-hand side of the equation indicates that x is to be multiplied by 3, then $3x$ is to be divided by 4, and then 2 is to be added to that result. To isolate x, we must undo these operations in the opposite order.

- To undo the addition of 2, we subtract 2 from both sides.
- To undo the division by 4, we multiply both sides by 4.
- To undo the multiplication by 3, we divide both sides by 3.

$$\frac{3x}{4} + 2 = -7$$

$$\frac{3x}{4} + 2 - \mathbf{2} = -7 - \mathbf{2} \qquad \text{To undo the addition of 2, subtract 2 from both sides.}$$

$$\frac{3x}{4} = -9 \qquad 2 - 2 = 0 \text{ and } -7 - 2 = -9.$$

$$\mathbf{4}\left(\frac{3x}{4}\right) = \mathbf{4}(-9) \qquad \text{To undo the division by 4, multiply both sides by 4.}$$

$$3x = -36 \qquad 4 \cdot \tfrac{3}{4} = 3 \text{ and } 4(-9) = -36.$$

$$\frac{3x}{\mathbf{3}} = \frac{-36}{\mathbf{3}} \qquad \text{To undo the multiplication by 3, divide both sides by 3.}$$

$$x = -12 \qquad \tfrac{3}{3} = 1 \text{ and } \tfrac{-36}{3} = -12.$$

Verify that the solution checks.   ∎

Self Check	Solve $\frac{2x}{3} - 4 = 12$.
Answer	24

EXAMPLE 5	**Advertising** A store manager hires a student to distribute advertising circulars door to door. The student will be paid \$24 a day plus 12¢ for every ad distributed. How many circulars must she distribute to earn \$42 in one day?

Solution   We can let a represent the number of circulars that the student must distribute. Her earnings can be expressed in two ways: as \$24 plus the 12¢-apiece cost of distributing the circulars, and as \$42.

\$24	plus	a ads at \$0.12 each	is	\$42.	12¢ = \$0.12.
24	+	0.12a	=	42	

We can solve this equation as follows:

$$24 + 0.12a = 42$$

$$24 - \mathbf{24} + 0.12a = 42 - \mathbf{24}$$ To undo the addition of 24, subtract 24 from both sides.

$$0.12a = 18$$ $24 - 24 = 0$ and $42 - 24 = 18$.

$$\frac{0.12a}{\mathbf{0.12}} = \frac{18}{\mathbf{0.12}}$$ To undo the multiplication by 0.12, divide both sides by 0.12.

$$a = 150$$ $\frac{0.12}{0.12} = 1$ and $\frac{18}{0.12} = 150$.

The student must distribute 150 ads. Check the result. ∎

Self Check   How many circulars must the student in Example 5 deliver in one day to earn $48?

Answer   200

■ MARKUP AND MARKDOWN

We have seen that the retail price of an item is the sum of the cost and the markup.

$$\boxed{\text{Retail price}} = \boxed{\text{cost}} + \boxed{\text{markup}}$$

Often, the markup is expressed as a **percent of cost.**

$$\boxed{\text{Markup}} = \boxed{\text{percent of markup}} \cdot \boxed{\text{cost}}$$

Suppose a store manager buys toasters for $21 and sells them at a 17% markup. To find the retail price, the manager begins with his cost and adds 17% of that cost.

$$\boxed{\text{Retail price}} = \boxed{\text{cost}} + \boxed{\text{markup}}$$

$$= \boxed{\text{cost}} + \boxed{\text{percent of markup}} \cdot \boxed{\text{cost}}$$

$$= 21 + 0.17 \cdot 21$$

$$= 21 + 3.57$$

$$= 24.57$$

The retail price of a toaster is $24.57.

EXAMPLE 6   **Antique cars**   In 1956, a Chevrolet BelAir automobile sold for $4,000. Today, it is worth about $28,600. Find the **percent of increase.**

Solution   We let p represent the percent of increase, expressed as a decimal.

Current price	=	original price	+	p(original price)

$$28,600 \quad = \quad 4,000 \quad + \quad p(4,000)$$

$$28,600 - \mathbf{4,000} = 4,000 - \mathbf{4,000} + 4,000p$$ To undo the addition of 4,000, subtract 4,000 from both sides.

$$24,600 = 4,000p$$ $28,600 - 4,000 = 24,600$ and $4,000 - 4,000 = 0$.

$$\frac{24,600}{\mathbf{4,000}} = \frac{4,000p}{\mathbf{4,000}}$$ To undo the multiplication by 4,000, divide both sides by 4,000.

$$6.15 = p$$ Simplify.

To convert 6.15 to a percent, we multiply by 100 and insert a % sign. Since the percent of increase is 615%, the car has appreciated 615%.   ∎

Self Check   Find the percent of increase in Example 6 if the car sells for $30,000.

Answer   650%

We have seen that when the price of merchandise is reduced, the amount of reduction is the **markdown** (also called the **discount**).

Sale price	=	regular price	−	markdown

Usually, the markdown is expressed as a percent of the regular price.

Markdown	=	percent of markdown	·	regular price

Suppose that a television set that regularly sells for $570 has been marked down 25%. That means the customer will pay 25% less than the regular price. To find the sale price, we use the formula

Sale price	=	regular price	−	markdown		
	=	Regular price	−	percent of markdown	·	regular price
	=	$570	−	25%	of	$570

$$= \$570 - (0.25)(\$570) \qquad 25\% = 0.25.$$
$$= \$570 - \$142.50$$
$$= \$427.50$$

The television set is selling for $427.50.

EXAMPLE 7

Buying a camera   A camera that was originally priced at $452 is on sale for $384.20. Find the percent of markdown.

Solution   We let p represent the percent of discount, expressed as a decimal, and substitute $384.20 for the sale price and $452 for the regular price.

Sale price	=	Regular price	−	Percent of markdown	·	Regular price
384.20	=	452	−	p	·	452

$$384.20 - \mathbf{452} = 452 - \mathbf{452} - p(452)$$
To undo the addition of 452, subtract 452 from both sides.

$$-67.80 = -p(452)$$
$384.20 - 452 = -67.80$ and $452 - 452 = 0$.

$$\frac{-67.80}{\mathbf{-452}} = \frac{-p(452)}{\mathbf{-452}}$$
To undo the multiplication by -452, divide both sides by -452.

$$0.15 = p$$
$\frac{-67.80}{-452} = 0.15$ and $\frac{-452}{-452} = 1$.

The camera is on sale at a 15% markdown.   ∎

Self Check
Answer

If the camera in Example 7 is reduced another $23, find the percent of discount.
20%

WARNING!   When a price increases from $100 to $125, the percent of increase is 25%. When the price *decreases* from $125 to $100, the percent of decrease is 20%. These different results occur because the percent of increase is a percent of the original (smaller) price, $100. The percent of decrease is a percent of the original (larger) price, $125.

Orals   *What would you do first when solving each equation?*

1. $5x - 7 = -12$

2. $15 = \dfrac{x}{5} + 3$

3. $\dfrac{x}{7} - 3 = 0$

4. $\dfrac{x - 3}{7} = -7$

5. $5w - 5 = 5$ **6.** $5w + 5 = 5$

7. $\dfrac{x - 7}{3} = 5$ **8.** $\dfrac{3x - 5}{2} + 2 = 0$

Find the value of the variable in each equation.

9. $7z - 7 = 14$ **10.** $\dfrac{t - 1}{2} = 6$

EXERCISE 2.3

REVIEW *Refer to the formulas given in Section 1.3.*

1. Find the perimeter of a rectangle with sides measuring 8.5 cm and 16.5 cm.

2. Find the area of a rectangle with sides measuring 2.3 in. and 3.7 in.

3. Find the area of a trapezoid with a height of 8.5 in. and bases measuring 6.7 in. and 12.2 in.

4. Find the volume of a rectangular solid with dimensions of 8.2 cm by 7.6 cm by 10.2 cm.

VOCABULARY AND CONCEPTS *Fill in each blank to make a true statement.*

5. Retail price = _____ + markup

6. Markup = percent of markup · _____.

7. Markdown = _____ of markdown · regular price

8. Another word for markdown is _____.

PRACTICE *In Exercises 9–60, solve each equation.* ***Check all solutions.***

9. $5x - 1 = 4$ **10.** $5x + 3 = 8$ **11.** $6x + 2 = -4$ **12.** $4x - 4 = 4$

13. $3x - 8 = 1$ **14.** $7x - 19 = 2$ **15.** $11x + 17 = -5$ **16.** $13x - 29 = -3$

17. $43t + 72 = 158$ **18.** $96t + 23 = -265$ **19.** $-47 - 21s = 58$ **20.** $-151 + 13s = -229$

21. $2y - \dfrac{5}{3} = \dfrac{4}{3}$ **22.** $9y + \dfrac{1}{2} = \dfrac{3}{2}$ **23.** $-4y - 12 = -20$ **24.** $-8y + 64 = -32$

25. $\dfrac{x}{3} - 3 = -2$ **26.** $\dfrac{x}{7} + 3 = 5$ **27.** $\dfrac{z}{9} + 5 = -1$ **28.** $\dfrac{y}{5} - 3 = 3$

29. $\dfrac{b}{3} + 5 = 2$ **30.** $\dfrac{a}{5} - 3 = -4$ **31.** $\dfrac{s}{11} + 9 = 6$ **32.** $\dfrac{r}{12} + 2 = 4$

33. $\dfrac{k}{5} - \dfrac{1}{2} = \dfrac{3}{2}$ **34.** $\dfrac{y}{5} - \dfrac{8}{7} = -\dfrac{1}{7}$ **35.** $\dfrac{w}{16} + \dfrac{5}{4} = 1$ **36.** $\dfrac{m}{7} - \dfrac{1}{14} = \dfrac{1}{14}$

37. $\dfrac{b + 5}{3} = 11$ **38.** $\dfrac{2 + a}{13} = 3$ **39.** $\dfrac{r + 7}{3} = 4$ **40.** $\dfrac{t - 2}{7} = -3$

41. $\dfrac{u - 2}{5} = 1$ **42.** $\dfrac{v - 7}{3} = -1$ **43.** $\dfrac{x - 4}{4} = -3$ **44.** $\dfrac{3 + y}{5} = -3$

45. $\dfrac{3x}{2} - 6 = 9$ **46.** $\dfrac{5x}{7} + 3 = 8$ **47.** $\dfrac{3y}{2} + 5 = 11$ **48.** $\dfrac{5z}{3} + 3 = -2$

49. $\dfrac{3x - 12}{2} = 9$

50. $\dfrac{5x + 10}{7} = 0$

51. $\dfrac{5k - 8}{9} = 1$

52. $\dfrac{2x - 1}{3} = -5$

53. $\dfrac{3z + 2}{17} = 0$

54. $\dfrac{10t - 4}{2} = 1$

55. $\dfrac{17k - 28}{21} + \dfrac{4}{3} = 0$

56. $\dfrac{5a - 2}{3} = \dfrac{1}{6}$

57. $-\dfrac{x}{3} - \dfrac{1}{2} = -\dfrac{5}{2}$

58. $\dfrac{17 - 7a}{8} = 2$

59. $\dfrac{9 - 5w}{15} = \dfrac{2}{5}$

60. $\dfrac{3t - 5}{5} + \dfrac{1}{2} = -\dfrac{19}{2}$

APPLICATIONS

61. Integer problem Six less than 3 times a certain number is 9. Find the number.

62. Integer problem If a certain number is increased by 7 and that result is divided by 2, the number 5 is obtained. Find the original number.

63. Apartment rental A student moves into a bigger apartment that rents for $400 per month. That rent is $100 less than twice what she had been paying. Find her former rent.

64. Auto repair A mechanic charged $20 an hour to repair the water pump on a car, plus $95 for parts. If the total bill was $155, how many hours did the repair take?

65. Boarding dogs A sportsman boarded his dog at a kennel for $16 plus $12 a day. If the stay cost $100, how many days was the owner gone?

66. Water billing The city's water department charges $7 per month, plus 42¢ for every 100 gallons of water used. Last month, one homeowner used 1,900 gallons and received a bill for $17.98. Was the billing correct?

67. Telephone charges A call to Tucson from a pay phone in Chicago costs 85¢ for the first minute and 27¢ for each additional minute or portion of a minute. If a student has $8.50 in change, how long can she talk?

68. Monthly sales A clerk's sales in February were $2,000 less than three times her sales in January. If her February sales were $7,000, by what amount did her sales increase?

69. Ticket sales A music group charges $1,500 for each performance, plus 20% of the total ticket sales. After a concert, the group received $2,980. How much money did the ticket sales raise?

70. Getting an A To receive a grade of A, the average of four 100-point exams must be 90 or better. If a student received scores of 88, 83, and 92 on the first three exams, what score does he need on the fourth exam to earn an A?

71. Getting an A The grade in history class is based on the average of five 100-point exams. One student received scores of 85, 80, 95, and 78 on the first four exams. With an average of 90 needed, what chance does he have for an A?

72. Excess inventory From the portion of the ad shown in Illustration 1, determine the sale price of a shirt.

	Regularly	Sale
Sweaters	$45.95	$27.57
Shirts	$37.50	$

ILLUSTRATION 1

73. Clearance sales Sweaters already on sale for 20% off the regular price cost $36 when purchased with a promotional coupon that allows an additional 10% discount. Find the original price. (*Hint:* When you save 20%, you are paying 80%.)

74. Furniture sale   A $1,250 sofa is marked down to $900. Find the percent of markdown.

75. Value of coupons   The percent discount offered by the coupon in Illustration 2 depends on the amount purchased. Find the range of the percent discount.

76. Furniture pricing   A bedroom set selling for $1,900 cost $1,000 wholesale. Find the percent markup.

**Value coupon
Save $15**

on purchases of $100 to $250.

ILLUSTRATION 2

WRITING

77. In solving the equation $5x - 3 = 12$, explain why you would add 3 to both sides first, rather than dividing by 5 first.

78. To solve the equation $\frac{3x-4}{7} = 2$, what operations would you perform, and in what order?

SOMETHING TO THINK ABOUT

79. Suppose you must solve the following equation but you can't quite read one number. It reads

$$\frac{7x + \#}{22} = \frac{1}{2}$$

If the solution of the equation is 1, what is the equation?

80. A store manager first increases his prices by 30% and then advertises

SALE!! 30% savings!!

What is the real percent discount to customers?

2.4 Simplifying Expressions to Solve Equations

■ LIKE TERMS ■ COMBINING LIKE TERMS ■ SOLVING EQUATIONS ■ IDENTITIES AND IMPOSSIBLE EQUATIONS

Getting Ready   *Use the distributive property to remove parentheses.*

1. $(3 + 4)x$ **2.** $(7 + 2)x$

3. $(8 - 3)w$ **4.** $(10 - 4)y$

Simplify each expression by doing the operations within the parentheses.

5. $(3 + 4)x$ **6.** $(7 + 2)x$

7. $(8 - 3)w$ **8.** $(10 - 4)y$

■ LIKE TERMS

Recall that a *term* is either a number or the product of numbers and variables. Some examples of terms are $7x$, $-3xy$, y^2, and 8. The number part of each term is called its **numerical coefficient.**

- The numerical coefficient of $7x$ is 7.
- The numerical coefficient of $-3xy$ is -3.
- The numerical coefficient of y^2 is the understood factor of 1.
- The numerical coefficient of 8 is 8.

> ### Like Terms
> **Like terms,** or **similar terms,** are terms with exactly the same variables and exponents.

The terms $3x$ and $5x$ are **like terms,** as are $9x^2$ and $-3x^2$. The terms $4xy$ and $3x^2$ are **unlike terms,** because they have different variables. The terms $4x$ and $5x^2$ are unlike terms, because the variables have different exponents.

■ COMBINING LIKE TERMS

The distributive property can be used to combine terms of algebraic expressions that contain sums or differences of like terms. For example, the terms in $3x + 5x$ and $9xy^2 - 11xy^2$ can be combined as follows:

$$3x + 5x = (3 + 5)x \qquad\qquad 9xy^2 - 11xy^2 = (9 - 11)xy^2$$
$$= 8x \qquad\qquad\qquad\qquad\qquad = -2xy^2$$

These examples suggest the following rule.

> ### Combining Like Terms
> To combine like terms, add their numerical coefficients and keep the same variables and exponents.

> **WARNING!** If the terms of an expression are unlike terms, they cannot be combined. For example, since the terms in $9xy^2 - 11x^2y$ have variables with different exponents, they are unlike terms and cannot be combined.

EXAMPLE 1 Simplify $3(x + 2) + 2(x - 8)$.

Solution

$3(x + 2) + 2(x - 8)$

$= 3x + 3 \cdot 2 + 2x - 2 \cdot 8$ Use the distributive property to remove parentheses.

$= 3x + 6 + 2x - 16$ $3 \cdot 2 = 6$ and $2 \cdot 8 = 16$.

$= 3x + 2x + 6 - 16$ Use the commutative property of addition: $6 + 2x = 2x + 6$.

$= 5x - 10$ Combine like terms. ■

Self Check	Simplify $-5(a + 3) + 2(a - 5)$.
Answer	$-3a - 25$

EXAMPLE 2 Simplify $3(x - 3) - 5(x + 4)$.

Solution

$$3(x - 3) - 5(x + 4)$$

$= 3(x - 3) + (-5)(x + 4)$	$a - b = a + (-b)$.
$= 3x - 3 \cdot 3 + (-5)x + (-5)4$	Use the distributive property to remove parentheses.
$= 3x - 9 + (-5x) + (-20)$	$3 \cdot 3 = 9$ and $(-5)(4) = -20$.
$= -2x - 29$	Combine like terms. ■

Self Check	Simplify $-3(b - 2) - 4(b - 4)$.
Answer	$-7b + 22$

■ SOLVING EQUATIONS

To solve an equation, we must isolate the variable on one side. This is often a multistep process that may require combining like terms. As we solve equations, we will follow these steps.

Solving Equations

1. Clear the equation of fractions.
2. Use the distributive property to remove parentheses.
3. Combine like terms if necessary.
4. Undo the operations of addition and subtraction to get the variables on one side and the constants on the other.
5. Combine like terms and undo the operations of multiplication and division to isolate the variable.

EXAMPLE 3 Solve $3(x + 2) - 5x = 0$.

Solution

$3(x + 2) - 5x = 0$	
$3x + 3 \cdot 2 - 5x = 0$	Use the distributive property to remove parentheses.
$3x - 5x + 6 = 0$	Rearrange terms and simplify.
$-2x + 6 = 0$	Combine like terms.

$$-2x + 6 - 6 = 0 - 6 \qquad \text{Subtract 6 from both sides.}$$
$$-2x = -6 \qquad \text{Combine like terms.}$$
$$\frac{-2x}{-2} = \frac{-6}{-2} \qquad \text{Divide both sides by } -2.$$
$$x = 3 \qquad \text{Simplify.}$$

$$\text{Check: } 3(x + 2) - 5x = 0$$
$$3(3 + 2) - 5 \cdot 3 \stackrel{?}{=} 0 \qquad \text{Substitute 3 for } x.$$
$$3 \cdot 5 - 5 \cdot 3 \stackrel{?}{=} 0$$
$$15 - 15 \stackrel{?}{=} 0$$
$$0 = 0 \qquad\blacksquare$$

Self Check Solve $-2(y - 3) - 4y = 0$.

Answer 1

EXAMPLE 4 Solve $3(x - 5) = 4(x + 9)$.

Solution

$$3(x - 5) = 4(x + 9)$$
$$3x - 15 = 4x + 36 \qquad \text{Remove parentheses.}$$
$$3x - 15 - 3x = 4x + 36 - 3x \qquad \text{Subtract } 3x \text{ from both sides.}$$
$$-15 = x + 36 \qquad \text{Combine like terms.}$$
$$-15 - 36 = x + 36 - 36 \qquad \text{Subtract 36 from both sides.}$$
$$-51 = x \qquad \text{Combine like terms.}$$
$$x = -51$$

$$\text{Check: } 3(x - 5) = 4(x + 9)$$
$$3(-51 - 5) \stackrel{?}{=} 4(-51 + 9) \qquad \text{Substitute } -51 \text{ for } x.$$
$$3(-56) \stackrel{?}{=} 4(-42)$$
$$-168 = -168 \qquad\blacksquare$$

Self Check Solve $4(z + 3) = -3(z - 4)$.

Answer 0

EXAMPLE 5 Solve $\dfrac{3x + 11}{5} = x + 3$.

Solution We first multiply both sides by 5 to clear the equation of fractions. When we multiply the right-hand side by 5, we must multiply the *entire* right-hand side by 5.

$$\frac{3x + 11}{5} = x + 3$$

$$5\left(\frac{3x + 11}{5}\right) = 5(x + 3)$$ Multiply both sides by 5.

$$3x + 11 = 5x + 15$$ Remove parentheses.

$$3x + 11 - 11 = 5x + 15 - 11$$ Subtract 11 from both sides.

$$3x = 5x + 4$$ Combine like terms.

$$3x - 5x = 5x + 4 - 5x$$ Subtract 5x from both sides.

$$-2x = 4$$ Combine like terms.

$$\frac{-2x}{-2} = \frac{4}{-2}$$ Divide both sides by −2.

$$x = -2$$ Simplify.

Check: $\dfrac{3x + 11}{5} = x + 3$

$$\frac{3(-2) + 11}{5} \stackrel{?}{=} (-2) + 3$$ Substitute −2 for x.

$$\frac{-6 + 11}{5} \stackrel{?}{=} 1$$ Simplify.

$$\frac{5}{5} \stackrel{?}{=} 1$$

$$1 = 1$$ ■

Self Check   Solve $\frac{2x - 5}{4} = x - 2$.

Answer $\frac{3}{2}$

WARNING!   Remember that when you multiply one side of an equation by a nonzero number, you must multiply the other side of the equation by the same number.

EXAMPLE 6   Solve $0.2x + 0.4(50 - x) = 19$.

Solution   Since $0.2 = \frac{2}{10}$ and $0.4 = \frac{4}{10}$, this equation contains fractions. To clear the fractions, we multiply both sides by 10.

$$0.2x + 0.4(50 - x) = 19$$

$$10[0.2x + 0.4(50 - x)] = 10(19)$$ Multiply both sides by 10.

$$10[0.2x] + 10[0.4(50 - x)] = 10(19)$$ Use the distributive property on the left-hand side.

$$2x + 4(50 - x) = 190$$ Do the multiplications.

$$2x + 200 - 4x = 190$$ Remove parentheses.

$$-2x + 200 = 190 \qquad \text{Combine like terms.}$$
$$-2x = -10 \qquad \text{Subtract 200 from both sides.}$$
$$x = 5 \qquad \text{Divide both sides by } -2.$$

Verify that the solution checks. ∎

Self Check   Solve $0.3(20 - x) + 0.5x = 15$.

Answer   45

■ IDENTITIES AND IMPOSSIBLE EQUATIONS

An equation that is true for all values of its variable is called an **identity.** For example, the equation $x + x = 2x$ is an identity because it is true for all values of x.

Because no number can equal a number that is 1 larger than itself, the equation $x = x + 1$ is not true for any number x. Such equations are called **impossible equations** or **contradictions.**

The equations in Examples 3–6 are called **conditional equations.** For these equations, some values of x are solutions, but other values of x are not.

EXAMPLE 7   Solve $3(x + 8) + 5x = 2(12 + 4x)$.

Solution
$$3(x + 8) + 5x = 2(12 + 4x)$$
$$3x + 24 + 5x = 24 + 8x \qquad \text{Remove parentheses.}$$
$$8x + 24 = 24 + 8x \qquad \text{Combine like terms.}$$
$$8x + 24 - \mathbf{8x} = 24 + 8x - \mathbf{8x} \qquad \text{Subtract } 8x \text{ from both sides.}$$
$$24 = 24 \qquad \text{Combine like terms.}$$

Since the result $24 = 24$ is true for every number x, every number x is a solution of the original equation. This equation is an identity. ∎

Self Check   Solve $-2(x + 3) - 18x = 5(9 - 4x) - 51$.

Answer   all values of x

EXAMPLE 8   Solve $3(x + 7) - x = 2(x + 10)$.

Solution
$$3(x + 7) - x = 2(x + 10)$$
$$3x + 21 - x = 2x + 20 \qquad \text{Remove parentheses.}$$
$$2x + 21 = 2x + 20 \qquad \text{Combine like terms.}$$
$$2x + 21 - \mathbf{2x} = 2x + 20 - \mathbf{2x} \qquad \text{Subtract } 2x \text{ from both sides.}$$
$$21 = 20 \qquad \text{Combine like terms.}$$

Since the result $21 = 20$ is false, the original equation has no solution. It is an impossible equation. ∎

Self Check | Solve $5(x - 2) - 2x = 3(x + 7)$.
Answer | no values of x

Orals   *Simplify by combining like terms.*

1. $3x + 5x$ **2.** $-2y + 3y$ **3.** $3x + 2x - 5x$

4. $3y + 2y - 7y$ **5.** $3(x + 2) - 3x + 6$ **6.** $3(x + 2) + 3x - 6$

Solve each equation, when possible.

7. $5x = 4x + 3$ **8.** $2(x - 1) = 2(x + 1)$

9. $3x = 2(x + 1)$ **10.** $x + 2(x + 1) = 3$

EXERCISE 2.4

REVIEW   *Evaluate each expression when $x = -3$, $y = -5$, and $z = 0$.*

1. $x^2 z(y^3 - z)$ **2.** $z - y^3$ **3.** $\dfrac{x - y^2}{2y - 1 + x}$ **4.** $\dfrac{2y + 1}{x} - x$

Do the operations.

5. $\dfrac{6}{7} - \dfrac{5}{8}$ **6.** $\dfrac{6}{7} \cdot \dfrac{5}{8}$ **7.** $\dfrac{6}{7} \div \dfrac{5}{8}$ **8.** $\dfrac{6}{7} + \dfrac{5}{8}$

VOCABULARY AND CONCEPTS   *Fill in each blank to make a true statement.*

9. If terms have the same _____ with the same exponents, they are called _____ terms.

10. To combine like terms, _____ their numerical coefficients and _____ the same variables and exponents.

11. If an equation is true for all values of its variable, it is called an _____.

12. If an equation is true for some values of its variable, but not all, it is called a _____ equation.

PRACTICE   *In Exercises 13–34, simplify each expression, when possible.*

13. $3x + 17x$ **14.** $12y - 15y$ **15.** $8x^2 - 5x^2$ **16.** $17x^2 + 3x^2$

17. $9x + 3y$ **18.** $5x + 5y$ **19.** $3(x + 2) + 4x$ **20.** $9(y - 3) + 2y$

21. $5(z - 3) + 2z$ **22.** $4(y + 9) - 6y$

23. $12(x + 11) - 11$ **24.** $-3(3 + z) + 2z$

25. $8(y + 7) - 2(y - 3)$ **26.** $9(z + 2) + 5(3 - z)$

27. $2x + 4(y - x) + 3y$ **28.** $3y - 6(y + z) + y$

29. $(x + 2) - (x - y)$

30. $3z + 2(y - z) + y$

31. $2\left(4x + \dfrac{9}{2}\right) - 3\left(x + \dfrac{2}{3}\right)$

32. $7\left(3x - \dfrac{2}{7}\right) - 5\left(2x - \dfrac{3}{5}\right) + x$

33. $8x(x + 3) - 3x^2$

34. $2x + x(x + 3)$

In Exercises 35–72, solve each equation, when possible. **Check all solutions.**

35. $3x + 2 = 2x$

36. $5x + 7 = 4x$

37. $5x - 3 = 4x$

38. $4x + 3 = 5x$

39. $9y - 3 = 6y$

40. $8y + 4 = 4y$

41. $8y - 7 = y$

42. $9y - 8 = y$

43. $9 - 23w = 4w$

44. $y + 4 = -7y$

45. $22 - 3r = 8r$

46. $14 + 7s = s$

47. $3(a + 2) = 4a$

48. $4(a - 5) = 3a$

49. $5(b + 7) = 6b$

50. $8(b + 2) = 9b$

51. $2 + 3(x - 5) = 4(x - 1)$

52. $2 - (4x + 7) = 3 + 2(x + 2)$

53. $10x + 3(2 - x) = 5(x + 2) - 4$

54. $11x + 6(3 - x) = 3$

55. $3(a + 2) = 2(a - 7)$

56. $9(t - 1) = 6(t + 2) - t$

57. $9(x + 11) + 5(13 - x) = 0$

58. $3(x + 15) + 4(11 - x) = 0$

59. $\dfrac{3(t - 7)}{2} = t - 6$

60. $\dfrac{2(t + 9)}{3} = t - 8$

61. $\dfrac{5(2 - s)}{3} = s + 6$

62. $\dfrac{8(5 - s)}{5} = -2s$

63. $\dfrac{4(2x - 10)}{3} = 2(x - 4)$

64. $\dfrac{11(x - 12)}{2} = 9 - 2x$

65. $3.1(x - 2) = 1.3x + 2.8$

66. $0.6x - 0.8 = 0.8(2x - 1) - 0.7$

67. $2.7(y + 1) = 0.3(3y + 33)$

68. $1.5(5 - y) = 3y + 12$

69. $19.1x - 4(x + 0.3) = -46.5$

70. $18.6x + 7.2 = 1.5(48 - 2x)$

71. $14.3(x + 2) + 13.7(x - 3) = 15.5$

72. $1.25(x - 1) = 0.5(3x - 1) - 1$

In Exercises 73–84, solve each equation. If it is an identity or an impossible equation, so indicate.

73. $8x + 3(2 - x) = 5(x + 2) - 4$

74. $5(x + 2) = 5x - 2$

75. $2(s + 2) = 2(s + 1) + 3$

76. $21(b - 1) + 3 = 3(7b - 6)$

77. $\dfrac{2(t - 1)}{6} - 2 = \dfrac{t + 2}{6}$

78. $\dfrac{2(2r - 1)}{6} + 5 = \dfrac{3(r + 7)}{6}$

79. $2(3z + 4) = 2(3z - 2) + 13$

80. $x + 7 = \dfrac{2x + 6}{2} + 4$

81. $2(y - 3) - \dfrac{y}{2} = \dfrac{3}{2}(y - 4)$

82. $\dfrac{20 - a}{2} = \dfrac{3}{2}(a + 4)$

83. $\dfrac{3x + 14}{2} = x - 2 + \dfrac{x + 18}{2}$

84. $\dfrac{5(x + 3)}{3} - x = \dfrac{2(x + 8)}{3}$

WRITING

85. Explain why $3x^2y$ and $5x^2y$ are like terms.

86. Explain why $3x^2y$ and $3xy^2$ are unlike terms.

87. Discuss whether $7xxy^3$ and $5x^2yyy$ are like terms.

88. Discuss whether $\frac{3}{2}x$ and $\frac{3x}{2}$ are like terms.

SOMETHING TO THINK ABOUT

89. What number is equal to its own double?

90. What number is equal to one-half of itself?

2.5 Applications of Equations

■ PROBLEM SOLVING ■ NUMBER PROBLEMS ■ GEOMETRIC PROBLEMS ■ INVESTMENT PROBLEMS
■ MOTION PROBLEMS ■ LIQUID MIXTURE PROBLEMS ■ DRY MIXTURE PROBLEMS

Getting Ready

1. If one part of a pipe is x feet long and the other part is $(x + 2)$ feet long, find an expression that represents the length of the pipe.

2. If one part of a board is x feet long and the other part is three times as long, find an expression that represents the length of the board.

3. What is the formula for the perimeter of a rectangle?

4. Define a triangle.

5. Find 7% of $12,000.

6. At 55 miles per hour, how far would a car travel in 7 hours?

7. If 8 gallons of a mixture of water and alcohol is 70% alcohol, how many gallons of alcohol does the mixture contain?

8. At $7.50 per pound, how many pounds of chocolate would be worth $71.25?

■ PROBLEM SOLVING

The key to problem solving is to thoroughly understand the problem and then devise a plan to solve it. The following list of steps provides a strategy to follow.

Problem Solving

1. **Analyze the problem** by reading it several times to understand the given facts. What information is given? What are you asked to find? Often a sketch, chart, or diagram will help you visualize the facts of the problem.

2. **Form an equation** by picking a variable to represent the quantity to be found. Then express all other unknown quantities in the problem as expressions involving that variable. Finally, write an equation expressing a quantity in two different ways.

3. **Solve the equation.**

4. **State the conclusion.**

5. **Check the result.**

In this section, we will use this five-step strategy to solve many problems.

■ NUMBER PROBLEMS

EXAMPLE 1   A plumber wants to cut a 17-foot pipe into three parts. The longest part is to be 3 times as long as the shortest, and the middle-sized part is to be 2 feet longer than the shortest. How long should each part be?

Analyze the problem   The information is given in terms of the length of the shortest part. Therefore, we let a variable represent the length of the shortest part and express the other lengths in terms of that variable.

Form an equation   Let x represent the length of the shortest part. Then $3x$ represents the length of the longest part, and $x + 2$ represents the length of the middle-sized part. We sketch the pipe as shown in Figure 2-6.

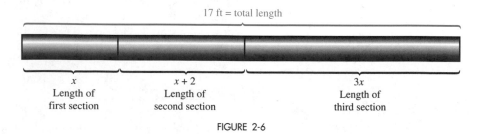

FIGURE 2-6

The sum of the lengths of these three parts equals the total length of the pipe.

The length of part 1	plus	the length of part 2	plus	the length of part 3	equals	the total length.
x	$+$	$x + 2$	$+$	$3x$	$=$	17

Solve the equation   We can solve this equation as follows:

$$x + x + 2 + 3x = 17 \qquad \text{The equation to solve.}$$
$$5x + 2 = 17 \qquad \text{Combine like terms.}$$
$$5x = 15 \qquad \text{Subtract 2 from both sides.}$$
$$x = 3 \qquad \text{Divide both sides by 5.}$$

State the conclusion   The shortest part is 3 feet long. Because the middle-sized part is 2 feet longer than the shortest, it is 5 feet long. Because the longest part is 3 times the shortest, it is 9 feet long.

Check the result   Because 3 feet, 5 feet, and 9 feet total 17 feet, the solution checks. ■

■ GEOMETRIC PROBLEMS

EXAMPLE 2   The length of a rectangle is 4 meters more than twice its width. If the perimeter of the rectangle is 26 meters, find its dimensions.

Analyze the problem   We can sketch the rectangle as in Figure 2-7. Since the formula for the perimeter of a rectangle is $P = 2l + 2w$, the perimeter of the rectangle in the figure is $2(4 + 2w) + 2w$. We are also told that the perimeter is 26.

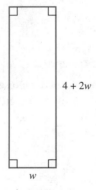

4 + 2w

w

FIGURE 2-7

Form an equation   Let w represent the width of the rectangle. Then $4 + 2w$ represents the length of the rectangle. We can form the equation

$2 \cdot$	the length	plus $2 \cdot$	the width	equals	the perimeter.
$2 \cdot$	$(4 + 2w)$	$+ \ 2 \cdot$	w	$=$	26

Solve the equation   We can solve this equation as follows:

$$2(4 + 2w) + 2w = 26 \qquad \text{The equation to solve.}$$
$$8 + 4w + 2w = 26 \qquad \text{Remove parentheses.}$$
$$6w + 8 = 26 \qquad \text{Combine like terms.}$$
$$6w = 18 \qquad \text{Subtract 8 from both sides.}$$
$$w = 3 \qquad \text{Divide both sides by 6.}$$

State the conclusion   The width of the rectangle is 3 meters, and the length, $4 + 2w$, is 10 meters.

Check the result   If a rectangle has a width of 3 meters and a length of 10 meters, then the length is 4 meters longer than twice the width ($4 + 2 \cdot 3 = 10$). The perimeter is $2 \cdot 10 + 2 \cdot 3 = 26$ meters. The solution checks.   ■

EXAMPLE 3   The vertex angle of an isosceles triangle is 56°. Find the measure of each base angle.

Analyze the problem   An **isosceles triangle** has two equal sides, which meet to form the **vertex angle.** See Figure 2-8. The angles opposite those sides, called **base angles,** are also equal. If we let x represent the measure of one base angle, then the measure of the other base angle is also x. In any triangle the sum of the three angles is 180°.

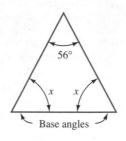

FIGURE 2-8

Form an equation   Let x represent the measure of one base angle. Then x also represents the measure of the other base angle. We can form the equation

One base angle	plus	the other base angle	plus	the vertex angle	equals	180°.
x	$+$	x	$+$	56	$=$	180

Solve the equation   We can solve this equation as follows:

$$x + x + 56 = 180 \qquad \text{The equation to solve.}$$
$$2x + 56 = 180 \qquad \text{Combine like terms.}$$
$$2x = 124 \qquad \text{Subtract 56 from both sides.}$$
$$x = 62 \qquad \text{Divide both sides by 2.}$$

State the conclusion   The measure of each base angle is 62°.

Check the result   The measure of each base angle is 62°, and the vertex angle measures 56°. These three angles total 180°. The solution checks.   ■

■ INVESTMENT PROBLEMS

EXAMPLE 4   A teacher invested part of $12,000 at 6% annual interest, and the rest at 9%. If the annual income from these investments was $945, how much did he invest at each rate?

Analyze the problem   The interest i earned by an amount p invested at an annual rate r for t years is given by the formula $i = prt$. In this example, $t = 1$ year. Hence, if x dollars were invested at 6%, the interest earned would be $0.06x$ dollars. If x dollars were invested at 6%, then the rest of the money, $\$(12{,}000 - x)$, would be invested at 9%. The interest earned on that money would be $0.09(12{,}000 - x)$ dollars. The total interest earned in dollars can be expressed in two ways: as 945 and as the sum $0.06x + 0.09(12{,}000 - x)$.

Form an equation   Let x represent the amount of money invested at 6%. Then $12{,}000 - x$ represents the amount of money invested at 9%. We can form an equation as follows:

The interest earned at 6%	plus	the interest earned at 9%	equals	the total interest.
$0.06x$	$+$	$0.09(12{,}000 - x)$	$=$	945

Solve the equation   We can solve this equation as follows:

$0.06x + 0.09(12{,}000 - x) = 945$	The equation to solve.
$6x + 9(12{,}000 - x) = 94{,}500$	Multiply both sides by 100 to clear the equation of decimals.
$6x + 108{,}000 - 9x = 94{,}500$	Remove parentheses.
$-3x + 108{,}000 = 94{,}500$	Combine like terms.
$-3x = -13{,}500$	Subtract 108,000 from both sides.
$x = 4{,}500$	Divide both sides by -3.

State the conclusion   The teacher invested $4,500 at 6% and $12{,}000 - \$4{,}500 = \$7{,}500$ at 9%.

Check the result   The first investment yielded 6% of $4,500, or $270. The second investment yielded 9% of $7,500, or $675. Since the total return was $270 + $675, or $945, the answers check.   ■

■ MOTION PROBLEMS

EXAMPLE 5

Chicago and Green Bay are about 200 miles apart. A car leaves Chicago traveling toward Green Bay at 55 mph at the same time as a truck leaves Green Bay bound for Chicago at 45 mph. How long will it take them to meet?

Analyze the problem

Motion problems are based on the formula $d = rt$, where d is the distance traveled, r is the rate, and t is the time. We can organize the information of this problem in chart form, as in Figure 2-9(a).

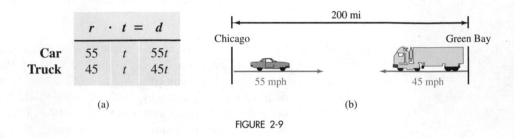

	r	$\cdot$	t	$=$	d
Car	55		t		$55t$
Truck	45		t		$45t$

(a)

(b)

FIGURE 2-9

We know that the two vehicles travel for the same amount of time—say, t hours. The faster car travels $55t$ miles, and the slower truck travels $45t$ miles. At the time they meet, the total distance can be expressed in two ways: as the sum $55t + 45t$ and as 200 miles.

Form an equation

Let t represent the time that each vehicle travels until they meet. Then $55t$ represents the distance traveled by the car, and $45t$ represents the distance traveled by the truck. After referring to Figure 2-9(b), we form the equation

The distance the car goes	plus	the distance the truck goes	equals	the total distance.
$55t$	$+$	$45t$	$=$	200

Solve the equation

We can solve the equation as follows:

$$55t + 45t = 200 \qquad \text{The equation to solve.}$$
$$100t = 200 \qquad \text{Combine like terms.}$$
$$t = 2 \qquad \text{Divide both sides by 100.}$$

State the conclusion

The vehicles meet after 2 hours.

Check the result

During those 2 hours, the car travels $55 \cdot 2 = 110$ miles, while the truck travels $45 \cdot 2 = 90$ miles. The total distance traveled is $110 + 90 = 200$ miles. Since this is the total distance between Chicago and Green Bay, the answer checks. ■

■ LIQUID MIXTURE PROBLEMS

EXAMPLE 6

A chemist has one solution that is 50% sulfuric acid and another that is 20% sulfuric acid. How much of each should she use to make 12 liters of a solution that is 30% acid?

Analyze the problem

The sulfuric acid present in the final mixture comes from the two solutions to be mixed. If x represents the number of liters of the 50% solution required for the mixture, then the rest of the mixture (($12 - x$) liters) must be the 20% solution. See Figure 2-10. Only 50% of the x liters, and only 20% of the ($12 - x$) liters, is pure sulfuric acid. The total of these amounts is also the amount of acid in the final mixture, which is 30% of 12 liters.

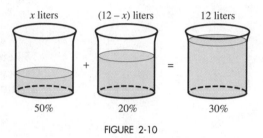

FIGURE 2-10

Form an equation

Let x represent the required number of liters of the 50% solution. Then $12 - x$ represents the required number of liters of the 20% solution. We can form the equation

The acid in the 50% solution	plus	the acid in the 20% solution	equals	the acid in the final mixture.
50% of x	+	20% of ($12 - x$)	=	30% of 12

Solve the equation

We can solve this equation as follows:

$$0.50x + 0.20(12 - x) = 0.30(12)$$ The equation to solve.

$$5x + 2(12 - x) = 3(12)$$ Multiply both sides by 10 to clear the equation of decimals.

$$5x + 24 - 2x = 36$$ Remove parentheses.

$$3x + 24 = 36$$ Combine like terms.

$$3x = 12$$ Subtract 24 from both sides.

$$x = 4$$ Divide both sides by 3.

State the conclusion

The chemist must mix 4 liters of the 50% solution and $12 - 4 = 8$ liters of the 20% solution.

Check the result.   ■

■ DRY MIXTURE PROBLEMS

EXAMPLE 7

Fancy cashews are not selling at $9 per pound, because they are too expensive. Filberts are selling at $6 per pound. How many pounds of filberts should be combined with 50 pounds of cashews to obtain a mixture that can be sold at $7 per pound?

Analyze the problem

Dry mixture problems are based on the formula $v = pn$, where v is the value of the mixture, p is the price per pound, and n is the number of pounds. Suppose x pounds of filberts are used in the mixture. At $6 per pound, they are worth $6x$. At $9 per pound, the 50 pounds of cashews are worth $9 \cdot 50$, or $450. The mixture will weigh $(50 + x)$ pounds, and at $7 per pound, it will be worth $7(50 + x)$. The value of the ingredients, $(6x + 450)$, is equal to the value of the mixture, $7(50 + x)$. See Figure 2-11.

	v	$= p$ $\cdot$	n
Filberts	$6x$	6	x
Cashews	$9(50)$	9	50
Mixture	$7(50 + x)$	7	$50 + x$

FIGURE 2-11

Form an equation

Let x represent the number of pounds of filberts in the mixture. We can form the equation

The value of the filberts	plus	the value of the cashews	equals	the value of the mixture.
$6x$	$+$	$9 \cdot 50$	$=$	$7(50 + x)$

Solve the equation

We can solve this equation as follows:

$6x + 9 \cdot 50 = 7(50 + x)$ The equation to solve.

$6x + 450 = 350 + 7x$ Remove parentheses and simplify.

$100 = x$ Subtract $6x$ and 350 from both sides.

State the conclusion

The storekeeper should use 100 pounds of filberts in the mixture.

Check the result

The value of 100 pounds of filberts at $6 per pound is $ 600
The value of 50 pounds of cashews at $9 per pound is $ 450
The value of the mixture is $1,050

The value of 150 pounds of mixture at $7 per pound is also $1,050.   ■

Orals

1. Express the value of 7 pounds of ground coffee worth $d per pound.
2. Express one year's interest on $18,000, invested at an annual rate r.

3. Express the length of a rectangle with area of A square feet and width 6 feet.

4. Express the length of a rectangle with perimeter of P feet and width of 9 feet.

EXERCISE 2.5

REVIEW   *Refer to the formulas in Section 1.3.*

1. Find the volume of a pyramid that has a height of 6 centimeters and a square base, 10 centimeters on each side.

2. Find the volume of a cone with a height of 6 centimeters and a circular base with radius 6 centimeters. Use $\pi \approx \frac{22}{7}$.

Simplify each expression.

3. $3(x + 2) + 4(x - 3)$

4. $4(x - 2) - 3(x + 1)$

5. $\frac{1}{2}(x + 1) - \frac{1}{2}(x + 4)$

6. $\frac{3}{2}\left(x + \frac{2}{3}\right) + \frac{1}{2}(x + 8)$

7. The amount A on deposit in a bank account bearing simple interest is given by the formula

$$A = P + Prt$$

Find A when $P = \$1,200$, $r = 0.08$, and $t = 3$.

8. The distance s that a certain object falls in t seconds is given by the formula

$$s = 350 - 16t^2 + vt$$

Determine s when $t = 4$ and $v = -3$.

VOCABULARY AND CONCEPTS   *Fill in each blank to make a true statement.*

9. The perimeter of a rectangle is given by the formula $P =$ _____.

10. An _____ triangle is a triangle with two sides of equal length.

11. The sides of equal length of an isosceles triangle meet to form the _____ angle.

12. The angles opposite the sides of equal length of an isosceles triangle are called _____ angles.

13. Motion problems are based on the formula _____.

14. The last step in the problem-solving process is to _____ the result.

APPLICATIONS

15. **Carpentry**   The 12-foot board in Illustration 1 has been cut into two parts, one twice as long as the other. How long is each part?

16. **Plumbing**   A 20-foot pipe has been cut into two parts, one 3 times as long as the other. How long is each part?

17. **Triangular bracing**   The outside perimeter of the triangular brace shown in Illustration 2 is 57 feet. If all three sides are of equal length, find the length of each side.

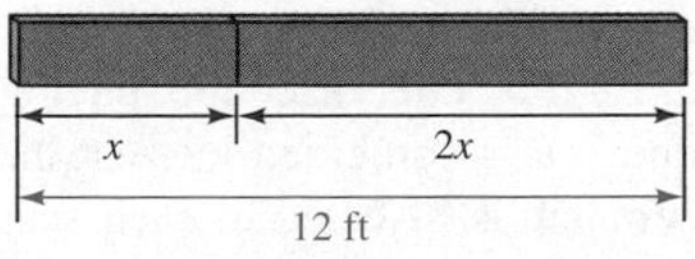

x $2x$

12 ft

ILLUSTRATION 1

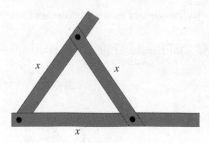

ILLUSTRATION 2

18. Circuit boards The perimeter of the circuit board in Illustration 3 is 90 centimeters. Find the dimensions of the board.

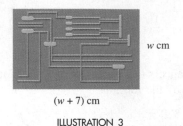

ILLUSTRATION 3

19. Swimming pool The width of a rectangular swimming pool is 11 meters less than the length, and the perimeter is 94 meters. Find its dimensions.

20. Wooden truss The truss in Illustration 4 is in the form of an isosceles triangle. Each of the two equal sides is 4 feet less than the third side. If the perimeter is 25 feet, find the length of each side.

ILLUSTRATION 4

21. Framing pictures The length of a rectangular picture is 5 inches greater than twice the width. If the perimeter is 112 inches, find the dimensions of the frame.

22. Guy wires The two guy wires in Illustration 5 form an isosceles triangle. One of the two equal angles of the triangle is four times the third angle (the vertex angle). Find the measure of the vertex angle.

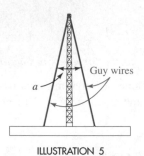

ILLUSTRATION 5

23. Equilateral triangles Find the measure of each angle of an equilateral triangle. (*Hint:* The three angles of an equilateral triangle are equal.)

24. Land areas The perimeter of a square piece of land is twice the perimeter of an equilateral (equal-sided) triangular lot. If one side of the square is 60 meters, find the length of a side of the triangle.

25. Investment problem A banker invested $24,000 for a client in two mutual funds, one earning 9% annual interest and the other earning 14%. After 1 year, the total interest earned was $3,135. How much was invested at each rate?

26. Investment problem A rollover IRA of $18,750 is invested in two mutual funds, one earning 12% interest and the other earning 10%. After 1 year, the combined interest income is $2,117. How much was invested at each rate?

27. Investment problem One investment pays 8%, and another pays 11%. If equal amounts are invested in each, the combined interest income for 1 year is $712.50. How much is invested at each rate?

28. Investment problem When equal amounts are invested in each of three accounts paying 5%, 6%, and 7%, one year's combined interest income is $882. How much is invested in each account?

29. Investment problem   A college professor wants to supplement her retirement income with investment interest. If she invests $15,000 at 6% annual interest, how much more would she have to invest at 7% to achieve a goal of $1,250 in supplemental income?

30. Investment problem   A teacher has a choice of two investment plans: an insured fund that has paid an average of 11% interest per year, or a riskier investment that has averaged a 13% return. If the same amount invested at the higher rate would generate an extra $150 per year, how much does the teacher have to invest?

31. Investment problem   A financial counselor recommends investing twice as much in CDs as in a bond fund. A client follows his advice and invests $21,000 in CDs paying 1% more interest than the fund. The CDs would generate $840 more interest than the fund. Find the two rates. (*Hint:* 1% = 0.01.)

32. Investment problem   The amount of annual interest earned by $8,000 invested at a certain rate is $200 less than $12,000 would earn at a 1% lower rate. At what rate is the $8,000 invested?

33. Travel time   Ashford and Bartlett are 315 miles apart. A car leaves Ashford bound for Bartlett at 50 mph. At the same time, another car leaves Bartlett and heads toward Ashford at 55 mph. In how many hours will the two cars meet?

34. Travel time   Granville and Preston are 535 miles apart. A car leaves Preston bound for Granville at 47 mph. At the same time, another car leaves Granville and heads toward Preston at 60 mph. How long will it take them to meet?

35. Travel time   Two cars leave Peoria at the same time, one heading east at 60 mph and the other west at 50 mph. (See Illustration 6.) How long will it take them to be 715 miles apart?

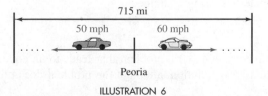

ILLUSTRATION 6

36. Boating   Two boats leave port at the same time, one heading north at 35 knots (nautical miles per hour), the other south at 47 knots. How long will it take them to be 738 nautical miles apart?

37. Travel time   Two cars start together and head east, one at 42 mph and the other at 53 mph. (See Illustration 7.) In how many hours will the cars be 82.5 miles apart?

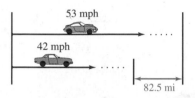

ILLUSTRATION 7

38. Speed of trains   Two trains are 330 miles apart, and their speeds differ by 20 mph. They travel toward each other and meet in 3 hours. Find the speed of each train.

39. Speed of an airplane   Two planes are 6,000 miles apart, and their speeds differ by 200 mph. They travel toward each other and meet in 5 hours. Find the speed of the slower plane.

40. Average speed   An automobile averaged 40 mph for part of a trip and 50 mph for the remainder. If the 5-hour trip covered 210 miles, for how long did the car average 40 mph?

41. Mixing fuels   How many gallons of fuel costing $1.15 per gallon must be mixed with 20 gallons of a fuel costing $.85 per gallon to obtain a mixture costing $1 per gallon? (See Illustration 8.)

42. Mixing paint   Paint costing $19 per gallon is to be mixed with 5 gallons of a $3-per-gallon thinner to make a paint that can be sold for $14 per gallon. How much paint will be produced?

43. Brine solution   How many gallons of a 3% salt solution must be mixed with 50 gallons of a 7% solution to obtain a 5% solution?

44. Making cottage cheese   To make low-fat cottage cheese, milk containing 4% butterfat is mixed with 10 gallons of milk containing 1% butterfat to obtain a mixture containing 2% butterfat. How many gallons of the richer milk must be used?

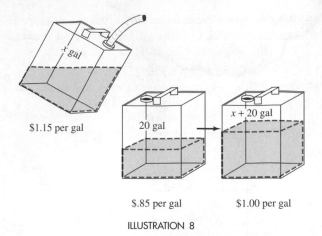

x gal

$1.15 per gal

20 gal

$x + 20$ gal

$.85 per gal $1.00 per gal

ILLUSTRATION 8

45. Antiseptic solutions   A nurse wishes to add water to 30 ounces of a 10% solution of benzalkonium chloride to dilute it to an 8% solution. How much water must she add?

46. Mixing photographic chemicals   A photographer wishes to mix 2 liters of a 5% acetic acid solution with a 10% solution to get a 7% solution. How many liters of 10% solution must be added?

47. Mixing candy   Lemon drops worth $1.90 per pound are to be mixed with jelly beans that cost $1.20 per pound to make 100 pounds of a mixture worth $1.48 per pound. How many pounds of each candy should be used?

48. Blending gourmet tea   One grade of tea, worth $3.20 per pound, is to be mixed with another grade worth $2 per pound to make 20 pounds that will sell for $2.72 per pound. How much of each grade of tea must be used?

49. Mixing nuts   A bag of peanuts is worth $.30 less than a bag of cashews. Equal amounts of peanuts and cashews are used to make 40 bags of a mixture that sells for $1.05 per bag. How much is a bag of cashews worth?

50. Mixing candy   Twenty pounds of lemon drops are to be mixed with cherry chews to make a mixture that will sell for $1.80 per pound. How much of the more expensive candy should be used? See Illustration 9.

	Price per pound
Peppermint patties	$1.35
Lemon drops	$1.70
Licorice lumps	$1.95
Cherry chews	$2.00

ILLUSTRATION 9

51. Coffee blends   A store sells regular coffee for $4 a pound and gourmet coffee for $7 a pound. To get rid of 40 pounds of the gourmet coffee, the shopkeeper plans to make a gourmet blend that he will put on sale for $5 a pound. How many pounds of regular coffee should be used?

52. Lawn seed blends   A garden store sells Kentucky bluegrass seed for $6 per pound and ryegrass seed for $3 per pound. How much rye must be mixed with 100 pounds of bluegrass to obtain a blend that will sell for $5 per pound?

WRITING

53. Describe the steps you would use to analyze and solve a problem.

55. Create a mixture problem of your own, and solve it.

54. Create a geometry problem that could be solved by solving the equation $2w + 2(w + 5) = 26$.

56. In mixture problems, explain why it is important to distinguish between the quantity and the value of the materials being combined.

SOMETHING TO THINK ABOUT

57. Is it possible for the equation of a problem to have a solution, but for the problem to have no solution? For example, is it possible to find two consecutive even integers whose sum is 16?

58. Invent a geometric problem that leads to an equation that has a solution, although the problem does not.

59. Consider the problem: How many gallons of a 10% and a 20% solution should be mixed to obtain a 30% solution? Without solving it, how do you know that the problem has no solution?

60. What happens if you try to solve Exercise 59?

2.6 Formulas

■ SOLVING FORMULAS

Getting Ready *Find the missing number.*

1. $\dfrac{3x}{\boxed{}} = x$

2. $\dfrac{-5y}{\boxed{}} = y$

3. $\dfrac{rx}{\boxed{}} = x$

4. $\dfrac{-ay}{\boxed{}} = y$

5. $\boxed{} \cdot \dfrac{x}{7} = x$

6. $\boxed{} \cdot \dfrac{y}{12} = y$

7. $\boxed{} \cdot \dfrac{x}{d} = x$

8. $\boxed{} \cdot \dfrac{y}{s} = y$

■ SOLVING FORMULAS

Equations with several variables are called **literal equations.** Often these equations are **formulas** such as $A = lw$, the formula for finding the area of a rectangle. Suppose that we wish to find the lengths of several rectangles whose areas and widths are known. It would be tedious to substitute values for A and w into the formula and then repeatedly solve the formula for l. It would be better to solve the formula $A = lw$ for l first and then substitute values for A and w and compute l directly.

To **solve an equation for a variable** means to isolate that variable on one side of the equation, with all other quantities on the opposite side. We can isolate the variable by using the usual equation-solving techniques.

EXAMPLE 1   Solve $A = lw$ for l.

Solution   To isolate l on the left-hand side, we undo the multiplication by w by dividing both sides of the equation by w.

$$A = lw$$

$$\frac{A}{w} = \frac{lw}{w} \qquad \text{To undo the multiplication by } w \text{, divide both sides by } w.$$

$$\frac{A}{w} = l \qquad \frac{w}{w} = 1.$$

$$l = \frac{A}{w}$$

■

Self Check

Solve $A = lw$ for w.

Answer

$w = \dfrac{A}{l}$

EXAMPLE 2

Recall that the formula $A = \frac{1}{2}bh$ gives the area of a triangle with base b and height h. Solve the formula for b.

Solution

$$A = \frac{1}{2}bh$$

$$2A = 2 \cdot \frac{1}{2}bh \qquad \text{To eliminate the fraction, multiply both sides by 2.}$$

$$2A = bh \qquad \qquad 2 \cdot \frac{1}{2} = 1.$$

$$\frac{2A}{h} = \frac{bh}{h} \qquad \text{To undo the multiplication by } h, \text{ divide both sides by } h.$$

$$\frac{2A}{h} = b \qquad \qquad \frac{h}{h} = 1.$$

$$b = \frac{2A}{h}$$

If the area A and the height h of a triangle are known, the base b is given by the formula $b = \frac{2A}{h}$. ∎

Self Check

Solve $A = \frac{1}{2}bh$ for h.

Answer

$h = \dfrac{2A}{b}$

EXAMPLE 3

The formula $C = \frac{5}{9}(F - 32)$ is used to convert Fahrenheit temperature readings into their Celsius equivalents. Solve the formula for F.

Solution

$$C = \frac{5}{9}(F - 32)$$

$$\frac{9}{5}C = \frac{9}{5} \cdot \frac{5}{9}(F - 32) \qquad \text{To eliminate } \frac{5}{9}, \text{ multiply both sides by } \frac{9}{5}.$$

$$\frac{9}{5}C = 1(F - 32) \qquad \frac{9}{5} \cdot \frac{5}{9} = \frac{9 \cdot 5}{5 \cdot 9} = 1.$$

$$\frac{9}{5}C = F - 32 \qquad \text{Remove parentheses.}$$

$$\frac{9}{5}C + 32 = F - 32 + 32 \qquad \text{To undo the subtraction of 32, add 32 to both sides.}$$

$$\frac{9}{5}C + 32 = F \qquad\qquad \text{Combine like terms.}$$

$$F = \frac{9}{5}C + 32$$

The formula $F = \frac{9}{5}C + 32$ is used to convert degrees Celsius to degrees Fahrenheit.

■

Self Check Solve $x = \frac{2}{3}(y + 5)$ for y.

Answer $y = \frac{3}{2}x - 5$

EXAMPLE 4 Recall that the area A of the trapezoid shown in Figure 2-12 is given by the formula

$$A = \frac{1}{2}h(B + b)$$

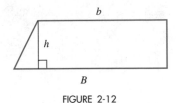

FIGURE 2-12

where B and b are its bases and h is its height. Solve the formula for b.

Solution **Method 1:** $A = \dfrac{1}{2}(B + b)h$

$$2A = 2 \cdot \frac{1}{2}(B + b)h \qquad \text{Multiply both sides by 2.}$$

$$2A = Bh + bh \qquad\qquad \text{Simplify and remove parentheses.}$$

$$2A - Bh = Bh + bh - Bh \qquad \text{Subtract } Bh \text{ from both sides.}$$

$$2A - Bh = bh \qquad\qquad \text{Combine like terms.}$$

$$\frac{2A - Bh}{h} = \frac{bh}{h} \qquad\qquad \text{Divide both sides by } h.$$

$$\frac{2A - Bh}{h} = b \qquad\qquad \tfrac{h}{h} = 1.$$

Method 2: $A = \dfrac{1}{2}(B + b)h$

$$2 \cdot A = 2 \cdot \frac{1}{2}(B + b)h \qquad \text{Multiply both sides by 2.}$$

$$2A = (B + b)h \qquad\qquad \text{Simplify.}$$

$$\frac{2A}{h} = \frac{(B + b)h}{h} \qquad\qquad \text{Divide both sides by } h.$$

$$\frac{2A}{h} = B + b \qquad\qquad \tfrac{h}{h} = 1.$$

$$\frac{2A}{h} - B = B + b - B \qquad \text{Subtract } B \text{ from both sides.}$$

$$\frac{2A}{h} - B = b \qquad \text{Combine like terms.}$$

Although they look different, the results of Methods 1 and 2 are equivalent. ■

Self Check   Solve $A = \frac{1}{2}h(B + b)$ for B.

Answer $B = \dfrac{2A - hb}{h}$ or $B = \dfrac{2A}{h} - b$

EXAMPLE 5   Solve the formula $P = 2l + 2w$ for l, and then find l when $P = 56$ and $w = 11$.

Solution   We first solve the formula $P = 2l + 2w$ for l.

$$P = 2l + 2w$$

$$P - 2w = 2l + 2w - 2w \qquad \text{Subtract } 2w \text{ from both sides.}$$

$$P - 2w = 2l \qquad \text{Combine like terms.}$$

$$\frac{P - 2w}{2} = \frac{2l}{2} \qquad \text{Divide both sides by 2.}$$

$$\frac{P - 2w}{2} = l \qquad \frac{2}{2} = 1.$$

$$l = \frac{P - 2w}{2}$$

We then substitute 56 for P and 11 for w and simplify.

$$l = \frac{P - 2w}{2}$$

$$l = \frac{56 - 2(11)}{2}$$

$$= \frac{56 - 22}{2}$$

$$= \frac{34}{2}$$

$$= 17$$

Thus, $l = 17$. ■

Self Check   Solve $P = 2l + 2w$ for w, and then find w when $P = 46$ and $l = 16$.

Answer $w = \frac{P - 2l}{2}$, 7

EXAMPLE 6   Recall that the volume V of the right-circular cone shown in Figure 2-13 is given by the formula

$$V = \frac{1}{3}Bh$$

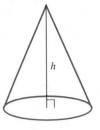

FIGURE 2-13

where B is the area of its circular base and h is its height. Solve the formula for h, and find the height of a right-circular cone with a volume of 64 cubic centimeters and a base area of 16 square centimeters.

Solution   We first solve the formula for h.

$$V = \frac{1}{3}Bh$$

$$3V = 3 \cdot \frac{1}{3}Bh \qquad \text{Multiply both sides by 3.}$$

$$3V = Bh \qquad 3 \cdot \frac{1}{3} = 1.$$

$$\frac{3V}{B} = \frac{Bh}{B} \qquad \text{Divide both sides by } B.$$

$$\frac{3V}{B} = h \qquad \frac{B}{B} = 1.$$

$$h = \frac{3V}{B}$$

We then substitute 64 for V and 16 for B and simplify.

$$h = \frac{3V}{B}$$

$$h = \frac{3(64)}{16}$$

$$= 3(4)$$

$$= 12$$

The height of the cone is 12 centimeters.   ∎

Self Check   Solve $V = \frac{1}{3}Bh$ for B, and find the area of the base when the volume is 42 cubic feet and the height is 6 feet.

Answer $B = \frac{3V}{h}$, 21 square feet

Orals   *Solve the equation* $ab + c - d = 0$,

1. for a **2.** for b

3. for c **4.** for d

Solve the equation $a + b = \dfrac{c}{d}$,

5. for a **6.** for b

7. for c **8.** for d

EXERCISE 2.6

REVIEW *Simplify each expression, if possible.*

1. $2x - 5y + 3x$

2. $2x^2y + 5x^2y^2$

3. $\dfrac{3}{5}(x + 5) - \dfrac{8}{5}(10 + x)$

4. $\dfrac{2}{11}(22x - y^2) + \dfrac{9}{11}y^2$

VOCABULARY AND CONCEPTS *Fill in each blank to make a true statement.*

5. Equations that contain several variables are called _____ equations.

6. The equation $A = lw$ is an example of a _____.

7. To solve a formula for a variable means to _____ the variable on one side of the formula.

8. To solve the formula $d = rt$ for t, divide both sides of the formula by __.

9. To solve $A = p + i$ for p, _____ i from both sides.

10. To solve $t = \dfrac{d}{r}$ for d, _____ both sides by r.

PRACTICE *In Exercises 11–34, solve each formula for the indicated variable.*

11. $E = IR$; for I

12. $i = prt$; for r

13. $V = lwh$; for w

14. $K = A + 32$; for A

15. $P = a + b + c$; for b

16. $P = 4s$; for s

17. $P = 2l + 2w$; for w

18. $d = rt$; for t

19. $A = P + Prt$; for t

20. $a = \dfrac{1}{2}(B + b)h$; for h

21. $C = 2\pi r$; for r

22. $I = \dfrac{E}{R}$; for R

23. $K = \dfrac{wv^2}{2g}$; for w

24. $V = \pi r^2 h$; for h

25. $P = I^2R$; for R

26. $V = \dfrac{1}{3}\pi r^2 h$; for h

27. $K = \dfrac{wv^2}{2g}$; for g

28. $P = \dfrac{RT}{mV}$; for V

29. $F = \dfrac{GMm}{d^2}$; for M

30. $C = 1 - \dfrac{A}{a}$; for A

31. $F = \dfrac{GMm}{d^2}$; for d^2

32. $y = mx + b$; for x

33. $G = 2(r - 1)b$; for r

34. $F = f(1 - M)$; for M

In Exercises 35–42, solve each formula for the indicated variable. Then substitute numbers to find the variable's value.

35. $d = rt$ Find t if $d = 135$ and $r = 45$.

36. $d = rt$ Find r if $d = 275$ and $t = 5$.

37. $i = prt$ Find t if $i = 12$, $p = 100$, and $r = 0.06$.

38. $i = prt$ Find r if $i = 120$, $p = 500$, and $t = 6$.

39. $P = a + b + c$ Find c if $P = 37$, $a = 15$, and $b = 19$.

40. $y = mx + b$ Find x if $y = 30$, $m = 3$, and $b = 0$.

41. $K = \dfrac{1}{2}h(a + b)$ Find h if $K = 48$, $a = 7$, and $b = 5$.

42. $\dfrac{x}{2} + y = z^2$ Find x if $y = 3$ and $z = 3$.

APPLICATIONS

43. Ohm's law   The formula $E = IR$, called **Ohm's law,** is used in electronics. Solve for I, and then calculate the current I if the voltage E is 48 volts and the resistance R is 12 ohms. Current has units of *amperes.*

44. Volume of a cone   The volume V of a cone is given by the formula $V = \frac{1}{3}\pi r^2 h$. Solve the formula for h, and then calculate the height h if V is 36π cubic inches and the radius r is 6 inches.

45. **Circumference of a circle**   The circumference C of a circle is given by $C = 2\pi r$, where r is the radius of the circle. Solve the formula for r, and then calculate the radius of a circle with a circumference of 14.32 feet. Round to the nearest hundredth of a foot.

46. **Growth of money**   At a simple interest rate r, an amount of money P grows to an amount A in t years according to the formula $A = P(1 + rt)$. Solve the formula for P. After $t = 3$ years, a girl has an amount $A = \$4{,}357$ on deposit. What amount P did she start with? Assume an interest rate of 6%.

47. **Power loss**   The power P lost when an electric current I passes through a resistance R is given by the formula $P = I^2 R$. Solve for R. If P is 2,700 watts and I is 14 amperes, calculate R to the nearest hundredth of an ohm.

48. **Geometry**   The perimeter P of a rectangle with length l and width w is given by the formula $P = 2l + 2w$. Solve this formula for w. If the perimeter of a certain rectangle is 58.37 meters and its length is 17.23 meters, find its width. Round to two decimal places.

49. Force of gravity   The masses of the two objects in Illustration 1 are m and M. The force of gravitation F between the masses is given by

$$F = \frac{GmM}{d^2}$$

where G is a constant and d is the distance between them. Solve for m.

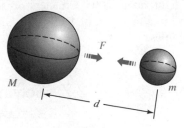

ILLUSTRATION 1

50. Thermodynamics   In thermodynamics, the Gibbs free-energy equation is given by

$$G = U - TS + pV$$

Solve this equation for the pressure, p.

51. Pulleys The approximate length L of a belt joining two pulleys of radii r and R feet with centers D feet apart is given by the formula

$$L = 2D + 3.25(r + R)$$

See Illustration 2. Solve the formula for D. If a 25-foot belt joins pulleys with radii of 1 foot and 3 feet, how far apart are the centers of the pulleys?

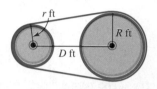

ILLUSTRATION 2

52. Geometry The measure a of an interior angle of a regular polygon with n sides is given by the formula $a = 180°\left(1 - \frac{2}{n}\right)$. See Illustration 3. Solve the formula for n. How many sides does a regular polygon have if an interior angle is 108°? (*Hint:* Distribute first.)

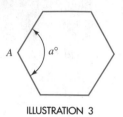

ILLUSTRATION 3

WRITING

53. The formula $P = 2l + 2w$ is also an equation, but an equation such as $2x + 3 = 5$ is not a formula. What equations do you think should be called formulas?

54. To solve the equation $s - A(s - 5) = r$ for the variable s, one student simply added $A(s - 5)$ to both sides to get $s = r + A(s - 5)$. Explain why this is not correct.

SOMETHING TO THINK ABOUT

55. The energy of an atomic bomb comes from the conversion of matter into energy, according to Einstein's formula $E = mc^2$. The constant c is the speed of light, about 300,000 meters per second. Find the energy in a mass, m, of 1 kilogram. Energy has units of **joules.**

56. When a car of mass m collides with a wall, the energy of the collision is given by the formula $E = \frac{1}{2}mv^2$. Compare the energy of two collisions: a car striking a wall at 30 mph, and at 60 mph.

2.7 Solving Inequalities

■ INEQUALITIES ■ COMPOUND INEQUALITIES ■ APPLICATIONS

Getting Ready *Graph each set on the number line.*

1. All real numbers greater than -1.

2. All real numbers less than or equal to 5.

3. All real numbers between -2 and 4.   **4.** All real numbers less than -2 or greater than or equal to 4.

■ INEQUALITIES

Recall the meaning of the following symbols.

Inequality Symbols
$<$ means   "is less than"
$>$ means   "is greater than"
$\leq$ means   "is less than or equal to"
$\geq$ means   "is greater than or equal to"

An **inequality** is a mathematical statement that indicates that two quantities are not necessarily equal. A **solution of an inequality** is any number that makes the inequality true. The number 2 is a solution of the inequality

$$x \leq 3$$

because $2 \leq 3$.

The inequality $x \leq 3$ has many more solutions, because any real number that is less than or equal to 3 will satisfy the inequality. We can use a graph on the number line to represent the solutions of the inequality $x \leq 3$. The colored arrow in Figure 2-14 indicates all those points with coordinates that satisfy the inequality $x \leq 3$.

The bracket at the point with coordinate 3 indicates that the number 3 is a solution of the inequality $x \leq 3$.

The graph of the inequality $x > 1$ appears in Figure 2-15. The colored arrow indicates all those points whose coordinates satisfy the inequality $x > 1$. The parenthesis at the point with coordinate 1 indicates that 1 is not a solution of the inequality $x > 1$.

To solve more complicated inequalities, we need to use the addition, subtraction, multiplication, and division properties of inequalities. When we use any of these properties, the resulting inequality will have the same solutions as the original one.

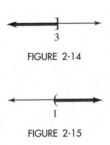

FIGURE 2-14

FIGURE 2-15

Addition Property of Inequality
If a, b, and c are real numbers, and
    If $a < b$, then $a + c < b + c$.
Similar statements can be made for the symbols $>$, $\leq$, and $\geq$.

The **addition property of inequality** can be stated this way: *If any quantity is added to both sides of an inequality, the resulting inequality has the same direction as the original inequality.*

> **Subtraction Property of Inequality**
> If a, b, and c are real numbers, and
>   If $a < b$, then $a - c < b - c$.
> Similar statements can be made for the symbols $>$, $\leq$, and $\geq$.

The **subtraction property of inequality** can be stated this way: *If any quantity is subtracted from both sides of an inequality, the resulting inequality has the same direction as the original inequality.*

The subtraction property of inequality is included in the addition property: To *subtract* a number a from both sides of an inequality, we could instead *add* the *negative* of a to both sides.

EXAMPLE 1   Solve $2x + 5 > x - 4$ and graph the solution on a number line.

Solution   To isolate the x on the left-hand side of the $>$ sign, we proceed as if we were solving equations.

$$2x + 5 > x - 4$$
$$2x + 5 - \mathbf{5} > x - 4 - \mathbf{5} \qquad \text{Subtract 5 from both sides.}$$
$$2x > x - 9 \qquad \text{Combine like terms.}$$
$$2x - \mathbf{x} > x - 9 - \mathbf{x} \qquad \text{Subtract } x \text{ from both sides.}$$
$$x > -9 \qquad \text{Combine like terms.}$$

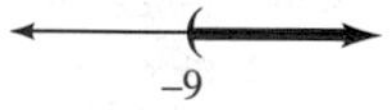

−9

FIGURE 2-16

The graph of the solution (see Figure 2-16) includes all points to the right of -9 but does not include -9 itself. For this reason, we use a parenthesis at -9.   ■

Self Check   Graph the solution of $3x - 2 < x + 4$.

Answer
3

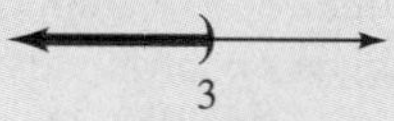

If both sides of the true inequality $2 < 5$ are multiplied by a *positive* number, such as 3, another true inequality results.

$$2 < 5$$
$$\mathbf{3} \cdot 2 < \mathbf{3} \cdot 5 \qquad \text{Multiply both sides by 3.}$$
$$6 < 15$$

The inequality $6 < 15$ is true. However, if both sides of $2 < 5$ are multiplied by a negative number, such as -3, the direction of the inequality symbol must be reversed to produce another true inequality.

$$2 < 5$$
$$\mathbf{-3} \cdot 2 > \mathbf{-3} \cdot 5 \qquad \text{Multiply both sides by the } \textit{negative} \text{ number } -3 \text{ and reverse the direction of the inequality.}$$
$$-6 > -15$$

The inequality $-6 > -15$ is true, because -6 lies to the right of -15 on the number line. These examples suggest the following properties.

Multiplication Property of Inequality

If a, b, and c are real numbers, and

If $a < b$ and $c > 0$, then $ac < bc$.

If $a < b$ and $c < 0$, then $ac > bc$.

Similar statements can be made for the symbols $>$, $\leq$, and $\geq$.

The multiplication property of inequality can be stated this way:

If unequal quantities are multiplied by the same positive quantity, the results will be unequal and in the same order.

If unequal quantities are multiplied by the same negative quantity, the results will be unequal but in the opposite order.

There is a similar property for division.

Division Property of Inequality

If a, b, and c are real numbers, and

If $a < b$ and $c > 0$, then $\dfrac{a}{c} < \dfrac{b}{c}$.

If $a < b$ and $c < 0$, then $\dfrac{a}{c} > \dfrac{b}{c}$.

Similar statements can be made for the symbols $>$, $\leq$, and $\geq$.

The division property of inequality can be stated this way:

If unequal quantities are divided by the same positive quantity, the results will be unequal and in the same order.

If unequal quantities are divided by the same negative quantity, the results will be unequal but in the opposite order.

To *divide* both sides of an inequality by a nonzero number c, we could instead *multiply* both sides by $\frac{1}{c}$.

WARNING!   If both sides of an inequality are multiplied by a *positive* number, the direction of the resulting inequality remains the same. However, if both sides of an inequality are multiplied by a *negative* number, the direction of the resulting inequality must be reversed.

EXAMPLE 2

Solve $3x + 7 \leq -5$ and graph the solution.

Solution

$$3x + 7 \leq -5$$
$$3x + 7 - 7 \leq -5 - 7 \qquad \text{Subtract 7 from both sides.}$$
$$3x \leq -12 \qquad \text{Combine like terms.}$$
$$\frac{3x}{3} \leq \frac{-12}{3} \qquad \text{Divide both sides by 3.}$$
$$x \leq -4$$

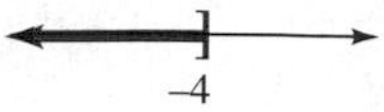

FIGURE 2-17

The solution consists of all real numbers that are less than or equal to -4. The bracket at -4 in the graph of Figure 2-17 indicates that -4 is one of the solutions. ∎

Self Check

Answer

Graph the solution of $2x - 5 \geq -3$.

EXAMPLE 3

Solve $5 - 3x \leq 14$ and graph the solution.

Solution

$$5 - 3x \leq 14$$
$$5 - 3x - 5 \leq 14 - 5 \qquad \text{Subtract 5 from both sides.}$$
$$-3x \leq 9 \qquad \text{Combine like terms.}$$
$$\frac{-3x}{-3} \geq \frac{9}{-3} \qquad \begin{array}{l}\text{Divide both sides by } -3 \text{ and reverse the direction of} \\ \text{the } \leq \text{ symbol.}\end{array}$$
$$x \geq -3$$

FIGURE 2-18

Since both sides of the inequality were divided by -3, the direction of the inequality was *reversed*. The graph of the solution appears in Figure 2-18. The bracket at -3 indicates that -3 is one of the solutions. ∎

Self Check

Answer

Graph the solution of $6 - 7x \geq -15$.

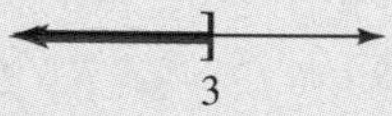

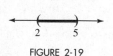

■ COMPOUND INEQUALITIES

Two inequalities can often be combined into a **double inequality** or **compound inequality** to indicate that numbers lie *between* two fixed values. For example, the inequality $2 < x < 5$ indicates that x is greater than 2 and that x is also less than 5. The solution of $2 < x < 5$ consists of all numbers that lie *between* 2 and 5. The graph of this set (called an **interval**) appears in Figure 2-19.

FIGURE 2-19

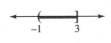

EXAMPLE 4   Solve $-4 < 2(x - 1) \leq 4$ and graph the solution.

Solution
$$-4 < 2(x - 1) \leq 4$$
$$-4 < 2x - 2 \leq 4 \qquad \text{Remove parentheses.}$$
$$-2 < 2x \leq 6 \qquad \text{Add 2 to all three parts.}$$
$$-1 < x \leq 3 \qquad \text{Divide all three parts by 2.}$$

FIGURE 2-20

The graph of the solution appears in Figure 2-20.   ■

Self Check   Graph the solution of $0 \leq 4(x + 5) < 26$.

Answer

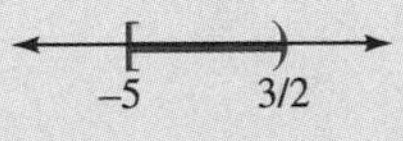

■ APPLICATIONS

EXAMPLE 5   A student has scores of 72%, 74%, and 78% on three mathematics examinations. What interval of scores does his last score need to fall in to earn a grade of B (80% or better)?

Solution   We can let x represent the score on the fourth (and last) exam. To find the average grade, we add the four scores and divide by 4. To earn a B, this average must be greater than or equal to 80%.

| The average of the four grades | $\geq$ | 80 |

$$\frac{72 + 74 + 78 + x}{4} \qquad \geq \qquad 80$$

We can solve this inequality for x.

$$\frac{224 + x}{4} \geq 80 \qquad 72 + 74 + 78 = 224.$$
$$224 + x \geq 320 \qquad \text{Multiply both sides by 4.}$$
$$x \geq 96 \qquad \text{Subtract 224 from both sides.}$$

FIGURE 2-21

A perfect score on the last exam is 100%. To earn a B, the student must score from 96% to 100%. This means that the student's score must be in the interval $96 \leq x \leq 100$. The graph of this interval appears in Figure 2-21.   ■

EXAMPLE 6   If the perimeter of an equilateral triangle is less than 15 feet, how long could each side be?

Solution   Recall that each side of an equilateral triangle is the same length and that the perimeter of a triangle is the sum of the lengths of its three sides. If we let x represent the length of one of the sides, then $x + x + x$ represents the perimeter. Since the perimeter is to be less than 15 feet, we have the following inequality:

$$x + x + x < 15$$
$$3x < 15 \qquad \text{Combine like terms.}$$
$$x < 5 \qquad \text{Divide both sides by 3.}$$

Each side of the triangle must be less than 5 feet long.   ∎

Orals   *Solve each inequality.*

1. $2x < 4$ **2.** $x + 5 \geq 6$

3. $-3x \leq -6$ **4.** $-x > 2$

5. $2x - 5 < 7$ **6.** $5 - 2x < 7$

EXERCISE 2.7

REVIEW   *Simplify each expression.*

1. $3x^2 - 2(y^2 - x^2)$

2. $5(xy + 2) - 3xy - 8$

3. $\dfrac{1}{3}(x + 6) - \dfrac{4}{3}(x - 9)$

4. $\dfrac{4}{5}x(y + 1) - \dfrac{9}{5}y(x - 1)$

VOCABULARY AND CONCEPTS   *Fill in each blank to make a true statement.*

5. The symbol $<$ means _____.

6. The symbol $>$ means _____.

7. The symbol ___ means "is greater than or equal to."

8. The symbol ___ means "is less than or equal to."

9. An _____ is a statement indicating that two quantities are not necessarily equal.

10. A _____ of an inequality is a number that makes the inequality true.

PRACTICE   *In Exercises 11–44, solve each inequality and graph the solution.*

11. $x + 2 > 5$ **12.** $x + 5 \geq 2$ **13.** $-x - 3 \leq 7$ **14.** $-x - 9 > 3$

15. $3 + x < 2$ **16.** $5 + x \geq 3$ **17.** $2x - 3 \leq 5$ **18.** $-3x - 5 < 4$

19. $-3x - 7 > -1$ **20.** $-5x + 7 \leq 12$ **21.** $-4x + 1 > 17$ **22.** $7x - 9 > 5$

23. $2x + 9 \le x + 8$ **24.** $3x + 7 \le 4x - 2$ **25.** $9x + 13 \ge 8x$ **26.** $7x - 16 < 6x$

27. $8x + 4 > 6x - 2$ **28.** $7x + 6 \ge 4x$ **29.** $5x + 7 < 2x + 1$ **30.** $7x + 2 > 4x - 1$

31. $7 - x \le 3x - 1$ **32.** $2 - 3x \ge 6 + x$

33. $9 - 2x > 24 - 7x$ **34.** $13 - 17x < 34 - 10x$

35. $3(x - 8) < 5x + 6$ **36.** $9(x - 11) > 13 + 7x$

37. $8(5 - x) \le 10(8 - x)$ **38.** $17(3 - x) \ge 3 - 13x$

39. $\dfrac{5}{2}(7x - 15) + x \ge \dfrac{13}{2}x - \dfrac{3}{2}$ **40.** $\dfrac{5}{3}(x + 1) \le -x + \dfrac{2}{3}$

41. $\dfrac{3x - 3}{2} < 2x + 2$ **42.** $\dfrac{x + 7}{3} \ge x - 3$

43. $\dfrac{2(x + 5)}{3} \le 3x - 6$ **44.** $\dfrac{3(x - 1)}{4} > x + 1$

In Exercises 45–62, solve each inequality and graph the solution.

45. $2 < x - 5 < 5$ **46.** $3 < x - 2 < 7$

47. $-5 < x + 4 \le 7$ **48.** $-9 \le x + 8 < 1$

49. $0 \le x + 10 \le 10$ **50.** $-8 < x - 8 < 8$

51. $4 < -2x < 10$ **52.** $-4 \le -4x < 12$

53. $-3 \le \dfrac{x}{2} \le 5$ **54.** $-12 \le \dfrac{x}{3} < 0$

55. $3 \le 2x - 1 < 5$ **56.** $4 < 3x - 5 \le 7$

57. $0 < 10 - 5x \le 15$ **58.** $1 \le -7x + 8 \le 15$

59. $-6 < 3(x + 2) < 9$ **60.** $-18 \le 9(x - 5) < 27$

61. $3 - x < 5 < 7 - x$

62. $x + 1 < 2x + 3 < x + 5$

APPLICATIONS *In Exercises 63–80, express each solution as an inequality.*

63. Calculating grades A student has test scores of 68%, 75%, and 79%. What must she score on the last exam to earn 80% or better?

64. Calculating grades A student has test scores of 70%, 74%, and 84%. What score does he need on the last exam to maintain 70% or better?

65. Fleet averages An automobile manufacturer produces three sedan models in equal quantities. One model has an economy rating of 17 miles per gallon, and the second model is rated for 19 mpg. If the manufacturer is required to have a fleet average of at least 21 mpg, what economy rating is required for the third model?

66. Avoiding a service charge When the average daily balance of a customer's checking account falls below $500 in any week, the bank assesses a $5 service charge. Bill's account balances for the week were as shown in Illustration 1.

Monday	$540.00
Tuesday	$435.50
Wednesday	$345.30
Thursday	$310.00

ILLUSTRATION 1

What must Friday's balance be to avoid the service charge?

67. Geometry The perimeter of an equilateral triangle is at most 57 feet. What could be the length of a side? (*Hint:* All three sides of an equilateral triangle are equal.)

68. Geometry The perimeter of a square is no less than 68 centimeters. How long can a side be?

69. Land elevations The land elevations in Nevada range from the 13,143-foot height of Boundary Peak to the Colorado River at 470 feet. To the nearest tenth, what is the range of these elevations in miles? (*Hint:* 1 mile is 5,280 feet.)

70. Doing homework A teacher requires that students do homework at least 2 hours a day. How many minutes should a student work each week?

71. Plane altitudes A pilot plans to fly at an altitude of between 17,500 and 21,700 feet. To the nearest tenth, what will be the range of altitudes in miles? (*Hint:* There are 5,280 feet in 1 mile.)

72. Getting exercise Doctors advise exercising at least 15 minutes but less than 30 minutes per day. Find the range of exercise time for one week.

73. Comparing temperatures To hold the temperature of a room between 19° and 22° Celsius, what Fahrenheit temperatures must be maintained? (*Hint:* Fahrenheit temperature (F) and Celsius temperature (C) are related by the formula $C = \frac{5}{9}(F - 32)$.)

74. Melting iron To melt iron, the temperature of a furnace must be at least $1{,}540°C$ but no more than $1{,}650°C$. What range of Fahrenheit temperatures must be maintained?

75. Phonograph records The radii of phonograph records must lie between 5.9 and 6.1 inches. What variation in circumference can occur? (*Hint:* The circumference of a circle is given by the formula $C = 2\pi r$, where r is the radius. Let $\pi = 3.14$.)

76. Pythons A large snake, the African Rock Python, can grow to a length of 25 feet. To the nearest hundredth, find the snake's range of lengths in meters. (*Hint:* There are about 3.281 feet in 1 meter.)

77. Comparing weights The normal weight of a 6 foot 2 inch man is between 150 and 190 pounds. To the nearest hundredth, what would such a person weigh in kilograms? (*Hint:* There are 2.2 pounds in 1 kilogram.)

78. Manufacturing   The time required to assemble a television set at the factory is 2 hours. A stereo receiver requires only 1 hour. The labor force at the factory can supply at least 640 and at most 810 hours of assembly time per week. When the factory is producing 3 times as many television sets as stereos, how many stereos could be manufactured in 1 week?

79. Geometry   A rectangle's length is 3 feet less than twice its width, and its perimeter is between 24 and 48 feet. What might be its width?

80. Geometry   A rectangle's width is 8 feet less than 3 times its length, and its perimeter is between 8 and 16 feet. What might be its length?

WRITING

81. Explain why multiplying both sides of an inequality by a negative constant reverses the direction of the inequality.

82. Explain the use of parentheses and brackets in the graphing of the solution of an inequality.

SOMETHING TO THINK ABOUT

83. To solve the inequality $1 < \frac{1}{x}$, one student multiplies both sides by x to get $x < 1$. Why is this not correct?

84. Find the solution of $1 < \frac{1}{x}$. (*Hint:* Will any negative values of x work?)

■ ■ ■ ■ ■ ■ ■ ■ ■ ■ **PROJECTS**

PROJECT 1   Build a scale similar to the one shown in Figure 2-1. Demonstrate to your class how you would use the scale to solve the following equations.

  a. $x - 4 = 6$ **b.** $x + 3 = 2$ **c.** $2x = 6$

  d. $\dfrac{x}{2} = 3$ **e.** $3x - 2 = 5$ **f.** $\dfrac{x}{3} + 1 = 2$

PROJECT 2   Magicians don't really saw people in half, or pull rabbits out of empty hats. Most magic tricks are just clever illusions, fooling the audience into seeing what isn't really there. The most successful magicians are very believable liars—and it takes a lot of practice to be good.

   Many magic tricks involve cutting ropes in various ways and then restoring them to their original lengths. For example, a magician holds up a long rope for all to see and then cuts it into three separate sections. He displays these to the audience; the sections are of three obviously different lengths, as in Illustration 1.

   The magician then folds the ropes, twists them, coils them around his fist and arm, and utters some magic words—all to distract and confuse the audience. When he holds up the three sections, as in Illustration 2, they are now the same length!

   The secret of the trick lies behind the magician's hand, hidden from the audience. What appears to be two equal lengths of rope in Illustration 3 is only one—the longest of the original three, folded in half. Those "two" sections are equal to the "third," which is just the middle-sized of the original three. What

(continued)

■ ■ ■ ■ ■ ■ ■ ■ ■ ■ PROJECTS (continued)

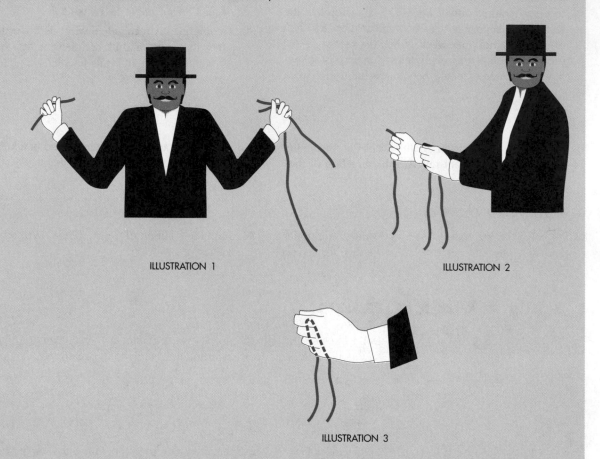

ILLUSTRATION 1

ILLUSTRATION 2

ILLUSTRATION 3

happened to the shortest of the original three? The magician disposed of it when the audience was distracted.

To prepare for this trick, the magician places two marks on an 8-foot rope, so that the two cuts can be made quickly and accurately. A third mark of a different color is the center of the largest section, the point where that rope is to be folded in half.

- If the shortest section is to be 1 foot long, where does the magician make the marks?

- Get some rope, cut an 8-foot piece, mark it as you have determined, and practice the trick. There are several ways to dispose of the shortest segment without being noticed. Try using a stretched rubber band to snap the rope up your sleeve, or fake a distracting sneeze while you slip it into your pocket. It is an easy trick to master, and an effective illusion.

C H A P T E R   S U M M A R Y

CONCEPTS	REVIEW EXERCISES

SECTION 2.1 — *Solving Equations by Addition and Subtraction*

An **equation** is a statement indicating that two quantities are equal.

1. Tell whether the given number is a solution of the equation.
   a. $3x + 7 = 1; -2$
   b. $5 - 2x = 3; -1$
   c. $2(x + 3) = x; -3$
   d. $5(3 - x) = 2 - 4x; 13$
   e. $3(x + 5) = 2(x - 3); -21$
   f. $2(x - 7) = x + 14; 0$

Any real number can be added to (or subtracted from) both sides of an equation to form another equation with the same solutions as the original equation.

2. Solve each equation and check all solutions.
   a. $x - 7 = -6$
   b. $y - 4 = 5$
   c. $p + 4 = 20$
   d. $x + \dfrac{3}{5} = \dfrac{3}{5}$
   e. $y - \dfrac{7}{2} = \dfrac{1}{2}$
   f. $z + \dfrac{5}{3} = -\dfrac{1}{3}$

Sale price
   = regular price − markdown

3. A necklace is on sale for $69.95. If it has been marked down $35.45, what is its regular price?

Retail price
   = wholesale cost + markup

4. A suit that has been marked up $115.25 sells for $212.95. Find its wholesale price.

If the sum of the measures of two angles is 90°, the angles are complementary.

5. Find the complement of an angle that measures 69°.

If the sum of the measures of two angles is 180°, the angles are supplementary.

6. Find the supplement of an angle that measures 69°.

SECTION 2.2 — *Solving Equations by Multiplication and Division*

Both sides of an equation can be multiplied (or divided) by any *nonzero* real number to form another equation with the same solutions as the original equation.

7. Solve each equation and check all solutions.
   a. $3x = 15$
   b. $8r = -16$
   c. $10z = 5$
   d. $14s = 21$
   e. $\dfrac{y}{3} = 6$
   f. $\dfrac{w}{7} = -5$
   g. $\dfrac{a}{-7} = \dfrac{1}{14}$
   h. $\dfrac{t}{12} = \dfrac{1}{2}$

Percentage = rate · base

8. Solve each problem.
  a. What number is 35% of 700?
  b. 72% of what number is 936?
  c. What percent of 2,300 is 851?
  d. 72 is what percent of 576?

9. Find the % average of a student with the following scores.

Test	Number of questions	Number correct
1	89	72
2	77	53
3	81	75

SECTION 2.3 *Solving More Equations*

10. Solve each equation and check all solutions.
  a. $5y + 6 = 21$ **b.** $5y - 9 = 1$
  c. $-12z + 4 = -8$ **d.** $17z + 3 = 20$
  e. $13 - 13t = 0$ **f.** $10 + 7t = -4$
  g. $23a - 43 = 3$ **h.** $84 - 21a = -63$
  i. $3x + 7 = 1$ **j.** $7 - 9x = 16$
  k. $\dfrac{b + 3}{4} = 2$ **l.** $\dfrac{b - 7}{2} = -2$
  m. $\dfrac{x - 8}{5} = 1$ **n.** $\dfrac{x + 10}{2} = -1$
  o. $\dfrac{2y - 2}{4} = 2$ **p.** $\dfrac{3y + 12}{11} = 3$
  q. $\dfrac{x}{2} + 7 = 11$ **r.** $\dfrac{r}{3} - 3 = 7$
  s. $\dfrac{a}{2} + \dfrac{9}{4} = 6$ **t.** $\dfrac{x}{8} - 2.3 = 3.2$

11. A compact disc player is on sale for $240, a 25% savings from the regular price. Find the regular price.

12. A $38 dictionary costs $40.47, with sales tax. Find the tax rate.

13. A Turkish rug was purchased for $560. If it is now worth $1,100, find the percent of increase.

14. A clock on sale for $215 was regularly priced at $465. Find the percent of discount.

| **SECTION 2.4** | *Simplifying Expressions to Solve Equations* |

Like terms can be combined by adding their numerical coefficients and using the same variables and exponents.

15. Simplify each expression, if possible.
- **a.** $5x + 9x$
- **b.** $7a + 12a$
- **c.** $18b - 13b$
- **d.** $21x - 23x$
- **e.** $5y - 7y$
- **f.** $19x - 19$
- **g.** $7(x + 2) + 2(x - 7)$
- **h.** $2(3 - x) + x - 6x$
- **i.** $y^2 + 3(y^2 - 2)$
- **j.** $2x^2 - 2(x^2 - 2)$

16. Solve each equation and check all solutions.
- **a.** $2x - 19 = 2 - x$
- **b.** $5b - 19 = 2b + 20$
- **c.** $3x + 20 = 5 - 2x$
- **d.** $0.9x + 10 = 0.7x + 1.8$
- **e.** $10(t - 3) = 3(t + 11)$
- **f.** $2(5x - 7) = 2(x - 35)$
- **g.** $\dfrac{3u - 6}{5} = 3$
- **h.** $\dfrac{5v - 35}{3} = -5$
- **i.** $\dfrac{7x - 28}{4} = -21$
- **j.** $\dfrac{27 + 9y}{5} = -27$

An **identity** is an equation that is true for all values of its variable.

An **impossible equation** or a **contradiction** is an equation that is true for no values of its variable.

17. Classify each equation as an identity or a contradiction.
- **a.** $2x - 5 = x - 5 + x$
- **b.** $-3(a + 1) - a = -4a + 3$
- **c.** $2(x - 1) + 4 = 4(1 + x) - (2x + 2)$

| **SECTION 2.5** | *Applications of Equations* |

Equations are useful in solving applied problems.

18. A carpenter wants to cut an 8-foot board into two pieces so that one piece is 7 feet shorter than twice the longer piece. Where should he make the cut?

19. If the length of the rectangular painting in Illustration 1 is 3 inches more than twice the width, how wide is the rectangle?

84 in.

ILLUSTRATION 1

20. A woman has $27,000. Part is invested for 1 year in a certificate of deposit paying 7% interest, and the remaining amount in a cash management fund paying 9%. The total interest on the two investments is $2,110. How much does she invest at each rate?

21. A bicycle path is 5 miles long. A man walks from one end at the rate of 3 mph. At the same time, a friend bicycles from the other end, traveling at 12 mph. In how many minutes will they meet?

22. A container is partly filled with 12 liters of whole milk containing 4% butterfat. How much 1% milk must be added to get a mixture that is 2% butterfat?

23. A store manager mixes candy worth 90¢ per pound with gumdrops worth $1.50 per pound to make 20 pounds of a mixture worth $1.20 per pound. How many pounds of each kind of candy must he use?

24. The electric company charges $17.50 per month, plus 18¢ for every kilowatt-hour of energy used. One resident's bill was $43.96. How many kilowatt-hours were used that month?

25. A contractor charges $35 for the installation of rain gutters, plus $1.50 per foot. If one installation cost $162.50, how many feet of gutter were required?

SECTION 2.6	*Formulas*

A literal equation, or formula, can often be solved for any of its variables.

26. Solve each equation for the indicated variable.

a. $E = IR$; for R **b.** $i = prt$; for t

c. $P = I^2R$; for R **d.** $d = rt$; for r

e. $V = lwh$; for h **f.** $y = mx + b$; for m

g. $V = \pi r^2 h$; for h **h.** $a = 2\pi rh$; for r

i. $F = \dfrac{GMm}{d^2}$; for G **j.** $P = \dfrac{RT}{mV}$; for m

SECTION 2.7	*Solving Inequalities*

Inequalities are solved by techniques similar to those used to solve equations, with this exception: *If both sides of an inequality are multiplied or divided by a negative number, the direction of the inequality must be reversed.*

The solution of an inequality can be graphed on the number line.

27. Graph the solution to each inequality.

a. $3x + 2 < 5$

b. $-5x - 8 > 7$

c. $5x - 3 \geq 2x + 9$

d. $7x + 1 \leq 8x - 5$

e. $5(3 - x) \leq 3(x - 3)$

f. $3(5 - x) \geq 2x$

g. $8 < x + 2 < 13$

h. $0 \leq 2 - 2x < 4$

■ Chapter Test

In Problems 1–4, state whether the given number is a solution of the equation.

1. $5x + 3 = -2; -1$

2. $3(x + 2) = 2x; -6$

3. $-3(2 - x) = 0; -2$

4. $3(x + 2) = 2x + 7; 1$

In Problems 5–14, solve each equation.

5. $x + 17 = -19$

6. $a - 15 = 32$

7. $12x = -144$

8. $\dfrac{x}{7} = -1$

9. $8x + 2 = -14$

10. $3 = 5 - 2x$

11. $\dfrac{2x - 5}{3} = 3$

12. $\dfrac{3x - 18}{2} = 6x$

13. $23 - 5(x + 10) = -12$

14. $\dfrac{7}{8}(x - 4) = 5x - \dfrac{7}{2}$

In Problems 15–18, simplify each expression.

15. $x + 5(x - 3)$

16. $3x - 5(2 - x)$

17. $-3x(x + 3) + 3x(x - 3)$

18. $-4x(2x - 5) - 7x(4x + 1)$

19. A car leaves Rockford at the rate of 65 mph bound for Madison. At the same time, a truck leaves Madison at the rate of 55 mph, bound for Rockford. If the cities are 72 miles apart, how long will it take for the car and the truck to meet?

20. How many liters of water must be added to 30 liters of a 10% brine solution to dilute it to an 8% solution?

In Problems 21–24, solve each equation for the variable indicated.

21. $d = rt$; for t

22. $P = 2l + 2w$; for l

23. $A = 2\pi rh$; for h

24. $A = P + Prt$; for r

In Problems 25–28, graph the solution of each inequality.

25. $8x - 20 \geq 4$

26. $x - 2(x + 7) > 14$

27. $-4 \leq 2(x + 1) < 10$

28. $-2 < 5(x - 1) \leq 10$

■ Cumulative Review Exercises

In Exercises 1–2, classify each number as an integer, a rational number, an irrational number, a real number, a positive number, or a negative number. Each number may be in several classifications.

1. $\dfrac{27}{9}$

2. -0.25

In Exercises 3–4, graph each set of numbers on the number line.

3. The natural numbers between 2 and 7

4. The real numbers between 2 and 7

In Exercises 5–8, simplify each expression.

5. $\dfrac{|-3| - |3|}{|-3 - 3|}$

6. $\dfrac{5}{7} \cdot \dfrac{14}{3}$

7. $2\dfrac{3}{5} + 5\dfrac{1}{2}$

8. $35.7 - 0.05$

In Exercises 9–12, let $x = -5$, $y = 3$, and $z = 0$. Evaluate each expression.

9. $(3x - 2y)z$

10. $\dfrac{x - 3y + |z|}{2 - x}$

11. $x^2 - y^2 + z^2$

12. $\dfrac{x}{y} + \dfrac{y + 2}{3 - z}$

13. What is $7\frac{1}{2}\%$ of 330?

14. 1,688 is 32% of what number?

In Exercises 15–16, consider the algebraic expression $3x^3 + 5x^2y + 37y$.

15. Find the coefficient of the second term.

16. List the factors of the third term.

In Exercises 17–20, simplify each expression.

17. $3x - 5x + 2y$

18. $3(x - 7) + 2(8 - x)$

19. $2x^2y^3 - xy(xy^2)$

20. $x^2(3 - y) + x(xy + x)$

In Exercises 21–24, solve each equation.

21. $3(x - 5) + 2 = 2x$

22. $\dfrac{x - 5}{3} - 5 = 7$

23. $\dfrac{2x - 1}{5} = \dfrac{1}{2}$

24. $2(a - 5) - (3a + 1) = 0$

25. Auto sales   An auto dealer's promotional ad appears in Illustration 1. One car is selling for $23,499. What was the dealer's invoice?

26. Furniture pricing   A sofa and a $300 chair are discounted 35%, and are priced at $780 for both. Find the original price of the sofa.

27. Cost of a car   The total cost of a new car, including an 8.5% sales tax, is $13,725.25. Find the cost before tax.

28. Manufacturing concrete   Concrete contains 3 times as much gravel as cement. How many pounds of cement are in 500 pounds of dry concrete mix?

29. Building construction   A 35-foot beam, 1 foot wide and 2 inches thick, is cut into three sections. One section is 14 feet long. Of the remaining two sections, one is twice as long as the other. Will the shortest section span an 8-foot-wide doorway?

700 cars to choose from!

Buy at

over dealer invoice!

ILLUSTRATION 1

30. Installing solar heating   One solar panel in Illustration 2 is 3.4 feet wider than the other. Find the width of each.

|← 18 ft →|

ILLUSTRATION 2

In Exercises 31–32, solve each formula for the variable indicated.

31. $A = \dfrac{1}{2}h(b + B)$; for h

32. $y = mx + b$; for x

In Exercises 33–36, evaluate each expression.

33. $4^2 - 5^2$

34. $(4 - 5)^2$

35. $5(4^3 - 2^3)$

36. $-2(5^4 - 7^3)$

In Exercises 37–38, graph the solutions of each inequality.

37. $8(4 + x) > 10(6 + x)$

38. $-9 < 3(x + 2) \le 3$

3 Graphing and Solving Systems of Equations and Inequalities

Electrical/Electronic Engineer

Electrical engineers design, develop, test, and supervise the manufacture of electronic equipment. Electrical engineers who work with electronic equipment are often called electronic engineers.

SAMPLE APPLICATION ■ In a radio, an inductor and a capacitor are used in a resonant circuit to select a wanted radio station at a frequency f and reject all others. The inductance L and the capacitance C determine the inductive reactance X_L and the capacitive reactance X_C of that circuit, where

$$X_L = 2\pi f L \qquad \text{and} \qquad X_C = \frac{1}{2\pi f C}$$

The radio station selected will be at the frequency f, where $X_L = X_C$. Write a formula for f^2 in terms of L and C.

(See Exercise 54 in Exercise 3.6.)

3.1 The Rectangular Coordinate System

■ THE RECTANGULAR COORDINATE SYSTEM ■ GRAPHING MATHEMATICAL RELATIONSHIPS
■ READING GRAPHS ■ STEP GRAPHS

Getting Ready *Graph each set of numbers on the number line.*

1. $-2, 1, 3$ **2.** All numbers greater than -2

3. All numbers less than or equal to 3   **4.** All numbers between -3 and 2

■ THE RECTANGULAR COORDINATE SYSTEM

When designing the Gateway Arch, shown in Figure 3-1(a), architects created a mathematical model of the arch called a **graph.** This graph, shown in Figure 3-1(b), is drawn on a grid called a **rectangular coordinate system.** This coordinate system is sometimes called a **Cartesian coordinate system** after the 17th-century French mathematician René Descartes.

165

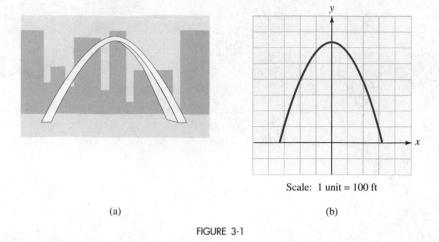

Scale: 1 unit = 100 ft

(a)   (b)

FIGURE 3-1

A rectangular coordinate system (see Figure 3-2) is formed by two perpendicular number lines.

- The horizontal number line is called the **x-axis.**
- The vertical number line is called the **y-axis.**

The positive direction on the x-axis is to the right, and the positive direction on the y-axis is upward. The scale on each axis should fit the data. For example, the axes of the graph of the arch shown in Figure 3-1(b) are scaled in units of 100 feet. If no scale is indicated on the axes, we assume that the axes are scaled in units of 1.

The point where the axes cross is called the **origin.** This is the 0 point on each axis. The two axes form a **coordinate plane** and divide it into four regions called **quadrants,** which are numbered as shown in Figure 3-2.

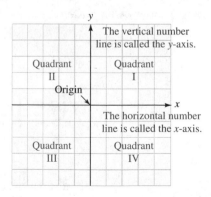

FIGURE 3-2

Each point in a coordinate plane can be identified by a pair of real numbers x and y, written as (x, y). The first number in the pair is the **x-coordinate,** and the second number is the **y-coordinate.** The numbers are called the **coordinates** of the point. Some examples of ordered pairs are $(3, -4)$, $\left(-1, -\frac{3}{2}\right)$, and $(0, 2.5)$.

$$(3, -4)$$

In an ordered pair, the   The y-coordinate
x-coordinate is listed first.   is listed second.

The process of locating a point in the coordinate plane is called **graphing** or **plotting** the point. In Figure 3-3(a), we show how to graph the point A with coordinates of $(3, -4)$. Since the x-coordinate is positive, we start at the origin and move 3 units to the right along the x-axis. Since the y-coordinate is negative, we then move down 4 units to locate point A. Point A is the **graph** of $(3, -4)$ and lies in quadrant IV.

To plot the point $B(-4, 3)$, we start at the origin, move 4 units to the left along the x-axis, and then move up 3 units to locate point B. Point B lies in quadrant II.

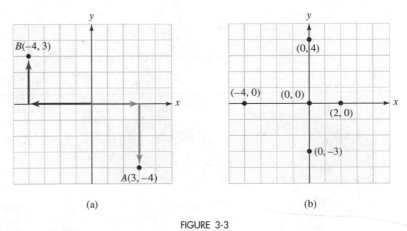

(a)   (b)

FIGURE 3-3

 WARNING!   Note that point A with coordinates of $(3, -4)$ is not the same as point B with coordinates $(-4, 3)$. Since the order of the coordinates of a point is important, we call the pairs **ordered pairs.**

In Figure 3-3(b), we see that the points $(-4, 0)$, $(0, 0)$, and $(2, 0)$ lie on the x-axis. In fact, all points with a y-coordinate of 0 will lie on the x-axis.

From Figure 3-3(b), we also see that the points $(0, -3)$, $(0, 0)$, and $(0, 4)$ lie on the y-axis. All points with an x-coordinate of 0 lie on the y-axis. From the figure, we can also see that the coordinates of the origin are $(0, 0)$.

EXAMPLE 1

Graphing points Plot the points   **a.** $A(-2, 3)$,   **b.** $B\left(-1, -\frac{3}{2}\right)$,   **c.** $C(0, 2.5)$, and   **d.** $D(4, 2)$.

Solution   **a.** To plot point A with coordinates $(-2, 3)$, we start at the origin, move 2 units to the *left* on the x-axis, and move 3 units *up*. Point A lies in quadrant II. (See Figure 3-4.)

b. To plot point B with coordinates of $\left(-1, -\frac{3}{2}\right)$, we start at the origin and move 1 unit to the *left* and $\frac{3}{2}$ $\left(\text{or } 1\frac{1}{2}\right)$ units *down*. Point B lies in quadrant III, as shown in Figure 3-4.

c. To graph point C with coordinates of $(0, 2.5)$, we start at the origin and move 0 units on the x-axis and 2.5 units *up*. Point C lies on the y-axis, as shown in Figure 3-4.

d. To graph point D with coordinates of $(4, 2)$, we start at the origin and move 4 units to the *right* and 2 units *up*. Point D lies in quadrant I, as shown in Figure 3-4.

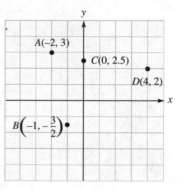

FIGURE 3-4

Self Check
Answer

Plot the points   **a.** $E(2, -2)$,   **b.** $F(-4, 0)$,   **c.** $G\left(1.5, \frac{5}{2}\right)$,   and   **d.** $H(0, 5)$.

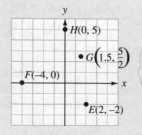

EXAMPLE 2   **Orbit of the earth**   The circle shown in Figure 3-5 is an approximate graph of the orbit of the earth. The graph is made up of infinitely many points, each with its own *x*- and *y*-coordinates. Use the graph to find the coordinates of the earth's position during the months of February, May, August, and December.

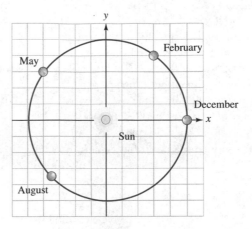

FIGURE 3-5

Solution   To find the coordinates of each position, we start at the origin and move left or right along the *x*-axis to find the *x*-coordinate and then up or down to find the *y*-coordinate. See Table 3-1.

Month	Position of earth on graph	Coordinates
February	3 units to the *right,* then 4 units *up*	(3, 4)
May	4 units to the *left,* then 3 units *up*	(−4, 3)
August	3.5 units to the *left,* then 3.5 units *down*	(−3.5, −3.5)
December	5 units *right,* then no units *up* or *down*	(5, 0)

TABLE 3-1

■ GRAPHING MATHEMATICAL RELATIONSHIPS

Every day, we deal with quantities that are related.

- The distance that we travel depends on how fast we are going.
- Our weight depends on how much we eat.
- The amount of water in a tub depends on how long the water has been running.

■ ■ ■ ■ ■ ■ ■ ■ ■ ■ P E R S P E C T I V E

As a child, René Descartes was frail and often sick. To improve his health, eight-year-old René was sent to a Jesuit school. The headmaster encouraged him to sleep in the morning as long as he wished. As a young man, Descartes spent several years as a soldier and world traveler, but his interests included mathematics and philosophy, as well as science, literature, writing, and taking it easy. The habit of sleeping late continued throughout his life. He claimed that his most productive thinking occurred when he was lying in bed. According to one story, Descartes first thought of analytic geometry as he watched a fly walking on his bedroom ceiling.

Descartes might have lived longer if he had stayed in bed. In 1649, Queen Christina of Sweden decided that she needed a tutor in philosophy, and she requested the services of Descartes. Tutoring would not have been difficult, except that the queen scheduled her lessons before dawn in her library with her windows open. The cold Stockholm mornings were too much for a man who was used to sleeping past noon. Within a few months, Descartes developed a fever and died, probably of pneumonia.

We can often use graphs to visualize relationships between two quantities. For example, suppose that we know the number of gallons of water that are in a tub at several time intervals after the water has been turned on. We can list that information in a **table of values** (see Figure 3-6).

Time (minutes)	Water in tub (gallons)	
0	0	→ (0, 0)
1	8	→ (1, 8)
3	24	→ (3, 24)
4	32	→ (4, 32)

At various times, the amount of water in the tub was measured and recorded in the table of values.

x-coordinate y-coordinate   The data in the table can be expressed as ordered pairs (x, y).

FIGURE 3-6

The information in the table can be used to construct a graph that shows the relationship between the amount of water in the tub and the time the water has been running. Since the amount of water in the tub depends on the time, we will associate *time* with the x-axis and the *amount of water* with the y-axis.

To construct the graph in Figure 3-7, we plot the four ordered pairs and draw a line through the resulting data points.

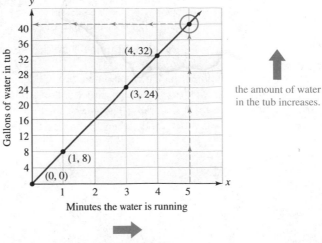

FIGURE 3-7

From the graph, we can see that the amount of water in the tub increases as the water is allowed to run. We can also use the graph to make observations about the amount of water in the tub at other times. For example, the dashed line on the graph shows that in 5 minutes, the tub will contain 40 gallons of water.

■ READING GRAPHS

In the next example, we show that valuable information can be obtained from a graph.

EXAMPLE 3   **Reading a graph**   The graph in Figure 3-8 shows the number of people in an audience before, during, and after the taping of a television show. On the x-axis, 0 represents the time when taping began. Use the graph to answer the following questions, and record each result in a table of values.

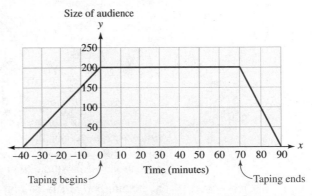

FIGURE 3-8

a. How many people were in the audience when taping began?

b. What was the size of the audience 10 minutes before taping began?

c. At what times were there exactly 100 people in the audience?

Solution   **a.** The time when taping began is represented by 0 on the x-axis. Since the point on the graph directly above 0 has a y-coordinate of 200, the point (0, 200) is on the graph. The y-coordinate of this point indicates that 200 people were in the audience when the taping began.

Time	Audience
0	200

b. Ten minutes before taping began is represented by -10 on the x-axis. Since the point on the graph directly above -10 has a y-coordinate of 150, the point $(-10, 150)$ is on the graph. The y-coordinate of this point indicates that 150 people were in the audience 10 minutes before the taping began.

Time	Audience
-10	150

c. We can draw a horizontal line passing through 100 on the y-axis. Since this line intersects the graph twice, there were two times when 100 people were in the audience. One time was 20 minutes before taping began, and the other was 80 minutes after taping began. So the points $(-20, 100)$ and $(80, 100)$ are on the graph. The y-coordinates of these points indicate that there were 100 people in the audience 20 minutes before and 80 minutes after taping began.

Time	Audience
-20	100
80	100

Self Check   Use the graph in Figure 3-8 to answer the following questions.   **a.** At what times were there exactly 50 people in the audience?   **b.** What was the size of the audience that watched the taping?   **c.** How long did it take for the audience to leave the studio after taping ended?

Answers   **a.** 30 min before and 85 min after taping began,   **b.** 200,   **c.** 20 min

■ STEP GRAPHS

The graph in Figure 3-9 shows the cost of renting a trailer for different periods of time. For example, the cost of renting the trailer for 4 days is $60, which is the y-coordinate of the point with coordinates of (4, 60). For renting the trailer for a period lasting over 4 and up to 5 days, the cost jumps to $70. Since the jumps in cost form steps in the graph, we call the graph a **step graph.**

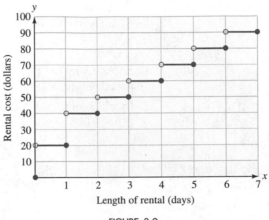

Length of rental (days)

FIGURE 3-9

EXAMPLE 4   Use the information in Figure 3-9 to answer the following questions. Write the results in a table of values.

a. Find the cost of renting the trailer for 2 days.

b. Find the cost of renting the trailer for $5\frac{1}{2}$ days.

c. How long can you rent the trailer if you have $50?

d. Is the rental cost per day the same?

Solution   **a.** We locate 2 days on the x-axis and move up to locate the point on the graph directly above the 2. Since the point has coordinates (2, 40), a two-day rental would cost $40. We enter this ordered pair in Table 3-2.

b. We locate $5\frac{1}{2}$ days on the x-axis and move straight up to locate the point on the graph with coordinates $\left(5\frac{1}{2}, 80\right)$, which indicates that a $5\frac{1}{2}$-day rental would cost $80. We enter this ordered pair in Table 3-2.

c. We draw a horizontal line through the point labeled 50 on the y-axis. Since this line intersects one step of the graph, we can look down to the x-axis to find the x-values that correspond to a y-value of 50. From the graph, we see that the trailer can be rented for more than 2 and up to 3 days for $50. We write (3, 50) in Table 3-2.

Length of rental (days)	Cost (dollars)
2	40
$5\frac{1}{2}$	80
3	50

TABLE 3-2

d. No, the cost per day is not the same. If we look at the y-coordinates, we see that for the first day, the rental fee is $20. For the second day, the cost jumps another $20. For the third day, and all subsequent days, the cost jumps only $10. ■

Orals **1.** Explain why the pair $(-2, 4)$ is called an ordered pair.

2. At what point do the coordinate axes intersect?

3. In which quadrant does the graph of $(3, -5)$ lie?

4. On which axis does the point $(0, 5)$ lie?

EXERCISE 3.1

REVIEW

1. Evaluate $-3 - 3(-5)$.

2. Evaluate $(-5)^2 + (-5)$.

3. What is the opposite of -8?

4. Simplify $|-1 - 9|$.

5. Solve $-4x + 7 = -21$.

6. Solve $P = 2l + 2w$ for w.

7. Evaluate $(x + 1)(x + y)^2$ for $x = -2$ and $y = -5$.

8. Simplify $-6(x - 3) - 2(1 - x)$.

VOCABULARY AND CONCEPTS Fill in each blank to make a true statement.

9. The pair of numbers $(-1, -5)$ is called an _____.

10. In the ordered pair $\left(-\frac{3}{2}, -5\right)$, -5 is called the ___ coordinate.

11. The point with coordinates $(0, 0)$ is the _____.

12. The x- and y-axes divide the coordinate plane into four regions called _____.

13. The point with coordinates $(4, 2)$ can be graphed on a _____ system.

14. The process of locating the position of a point on a coordinate plane is called _____ the point.

In Exercises 15–20, answer each question or fill in each blank to make a true statement.

15. Do $(3, 2)$ and $(2, 3)$ represent the same point?

16. In the ordered pair $(4, 5)$, is 4 associated with the horizontal or the vertical axis?

17. To plot the point with coordinates $(-5, 4.5)$, we start at the _____, move 5 units to the ____, and then move 4.5 units ____.

18. To plot the point with coordinates $\left(6, -\frac{3}{2}\right)$, we start at the _____, move 6 units to the _____, and then move $\frac{3}{2}$ units _____.

19. In which quadrant do points with a negative x-coordinate and a positive y-coordinate lie?

20. In which quadrant do points with a positive x-coordinate and a negative y-coordinate lie?

21. Use the graph to complete the table.

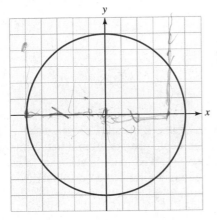

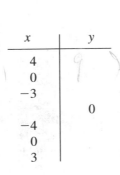

x	y
4	
0	
-3	
	0
-4	
0	
3	

22. Use the graph to complete the table.

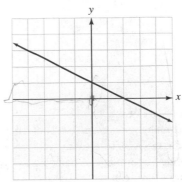

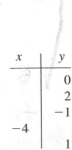

x	y
	0
	2
	-1
-4	
	1

The graph in Illustration 1 gives the heart rate of a woman before, during, and after an aerobic workout. In Exercises 23–30, use the graph to answer the following questions.

23. What information does the point $(-10, 60)$ give us?

24. After beginning the workout, how long did it take the woman to reach her training-zone heart rate?

25. What was her heart rate one-half hour after beginning the workout?

26. For how long did she work out at the training-zone level?

27. At what times was her heart rate 100 beats per minute?

28. How long was her cooldown period?

29. What was the difference in her heart rate before the workout and after the cooldown period?

30. What was her approximate heart rate 8 minutes after beginning?

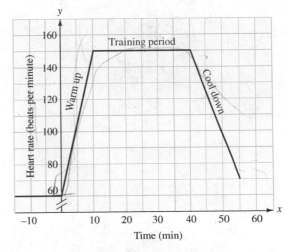

ILLUSTRATION 1

PRACTICE *In Exercises 31–32, graph each point on the coordinate grid.*

31. $A(-3, 4)$, $B(4, 3.5)$, $C\left(-2, -\frac{5}{2}\right)$, $D(0, -4)$, $E\left(\frac{3}{2}, 0\right)$, $F(3, -4)$

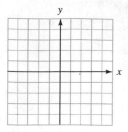

32. $G(4, 4)$, $H(0.5, -3)$, $I(-4, -4)$, $J(0, -1)$, $K(0, 0)$, $L(0, 3)$, $M(-2, 0)$

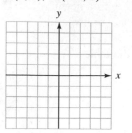

APPLICATIONS

33. Road maps Road maps usually have a coordinate system to help locate cities. Use the map in Illustration 2 to locate Carbondale, Champaign, Chicago, Peoria, Rockford, Springfield, and St. Louis. Express each answer in the form (number, letter).

34. Battleship In the game Battleship, players use coordinates to drop depth charges from a battleship to hit a hidden submarine. What coordinates should be used to make three hits on the exposed submarine shown in Illustration 3? Express each answer in the form (letter, number).

35. Water pressure The graphs in Illustration 4 show the paths of two streams of water from the same hose held at two different angles.

   a. At which angle does the stream of water shoot higher? How much higher?

   b. At which angle does the stream of water shoot out farther? How much farther?

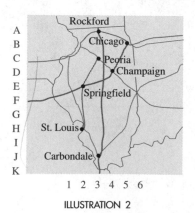

ILLUSTRATION 2

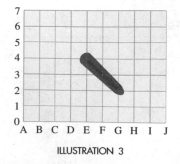

ILLUSTRATION 3

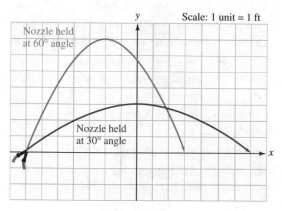

ILLUSTRATION 4

36. Golf swing   To correct her swing, a golfer was video-taped and then had her image displayed on a computer monitor so that it could be analyzed by a golf pro. See Illustration 5. Give the coordinates of the points that are highlighted on the arc of her swing.

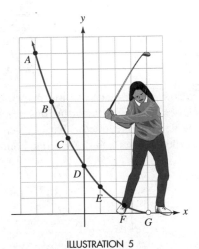

ILLUSTRATION 5

37. Video rental   The charges for renting a movie are shown in the graph in Illustration 6.

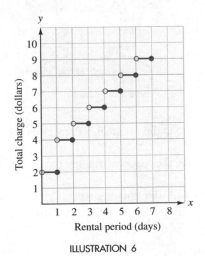

ILLUSTRATION 6

a. Find the charge for a 1-day rental.
b. Find the charge for a 2-day rental.
c. Find the charge if the tape is kept for 5 days.
d. Find the charge if the tape is kept for a week.

38. Postage rates   The graph shown in Illustration 7 gives the first-class postage rates for mailing parcels weighing up to 5 ounces.

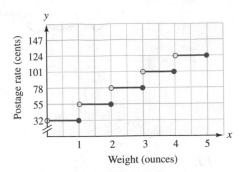

ILLUSTRATION 7

a. Find the cost of postage to mail each of the following letters first class: a 1-ounce letter, a 4-ounce letter, and a $2\frac{1}{2}$-ounce letter.
b. Find the difference in postage for a 3.75-ounce letter and a 4.75-ounce letter.
c. What is the heaviest letter than can be mailed first class for 55¢?

39. Gas mileage   The table in Illustration 8 gives the number of miles (y) that a truck can be driven on x gallons of gasoline. Plot the ordered pairs and draw a line connecting the points.

x	y
2	10
3	15
5	25

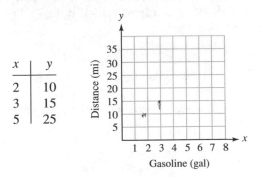

ILLUSTRATION 8

a. Estimate how far the truck can go on 7 gallons of gasoline.
b. How many gallons of gas are needed to travel a distance of 20 miles?
c. Estimate how far the truck can go on 6.5 gallons of gasoline.

40. Wages The table in Illustration 9 gives the amount y (in dollars) that a student can earn by working x hours. Plot the ordered pairs and draw a line connecting the points.

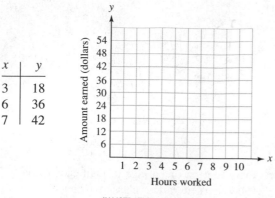

x	y
3	18
6	36
7	42

ILLUSTRATION 9

a. How much will the student earn in 5 hours?
b. How long would the student have to work to earn $12?
c. Estimate how much the student will earn in 3.5 hours.

41. Value of a car The table in Illustration 10 shows the value y (in thousands of dollars) of a car that is x years old. Plot the ordered pairs and draw a line connecting the points.
a. What does the point $(3, 7)$ on the graph tell you?

b. Estimate the value of the car when it is 7 years old.

c. After how many years will the car be worth $2,500?

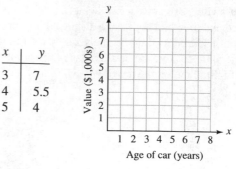

x	y
3	7
4	5.5
5	4

ILLUSTRATION 10

42. Depreciation As a piece of farm machinery gets older, it loses value. The table in Illustration 11 shows the value y of a tractor that is x years old. Plot the ordered pairs and draw a line connecting them.

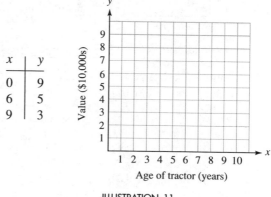

x	y
0	9
6	5
9	3

ILLUSTRATION 11

a. What does the point $(0, 9)$ on the graph tell you?

b. Estimate the value of the tractor in 3 years.

c. When will the tractor's value dip below $30,000?

WRITING

43. Explain why the point with coordinates $(-3, 3)$ is not the same as the point with coordinates $(3, -3)$.

45. Explain how to plot the point with coordinates $(-2, 5)$.

44. Explain what is meant when we say that the rectangular coordinate graph of the St. Louis Arch is made up of *infinitely many* points.

46. Explain why the coordinates of the origin are $(0, 0)$.

SOMETHING TO THINK ABOUT

47. Could you have a coordinate system where the coordinate axes were not perpendicular? How would it be different?

48. René Descartes is famous for saying, "I think. Therefore I am." What do you think he meant by that?

3.2 Graphing Linear Equations

■ EQUATIONS WITH TWO VARIABLES ■ CONSTRUCTING TABLES OF VALUES ■ GRAPHING EQUATIONS ■ THE INTERCEPT METHOD OF GRAPHING A LINE ■ GRAPHING HORIZONTAL AND VERTICAL LINES ■ AN APPLICATION OF LINEAR EQUATIONS

Getting Ready   *In Problems 1–4, let $y = 2x + 1$.*

1. Find y when $x = 0$.

2. Find y when $x = 2$.

3. Find y when $x = -2$.

4. Find y when $x = \dfrac{1}{2}$.

5. Find five pairs of numbers with a sum of 8.

6. Find five pairs of numbers with a difference of 5.

■ EQUATIONS WITH TWO VARIABLES

The equation $x + 2y = 5$ contains the two variables x and y. The solutions of such equations are ordered pairs of numbers. For example, the ordered pair $(1, 2)$ is a solution, because the equation is satisfied when $x = 1$ and $y = 2$.

$$x + 2y = 5$$
$$1 + 2(2) = 5 \qquad \text{Substitute 1 for } x \text{ and 2 for } y.$$
$$1 + 4 = 5$$
$$5 = 5$$

EXAMPLE 1   Is the pair $(-2, 4)$ a solution of $y = 3x + 9$?

Solution   We substitute -2 for x and 4 for y and see whether the resulting equation is true.

$$y = 3x + 9 \qquad \text{The original equation.}$$
$$4 \stackrel{?}{=} 3(-2) + 9 \qquad \text{Substitute } -2 \text{ for } x \text{ and 4 for } y.$$
$$4 \stackrel{?}{=} -6 + 9 \qquad \text{Do the multiplication: } 3(-2) = -6.$$
$$4 = 3 \qquad \text{Do the addition: } -6 + 9 = 3.$$

Since the equation $4 = 3$ is false, the pair $(-2, 4)$ is not a solution.   ■

Self Check   Is $(-1, -5)$ a solution of $y = 5x$?

Answer   yes

■ CONSTRUCTING TABLES OF VALUES

To find solutions of equations in x and y, we can pick numbers at random, substitute them for x, and find the corresponding values of y. For example, to find some ordered pairs that satisfy $y = 5 - x$, we can let $x = 1$ (called the **input value**), substitute 1 for x, and solve for y (called the **output value**).

$y = 5 - x$	The original equation.
$y = 5 - 1$	Substitute the input value of 1 for x.
$y = 4$	The output is 4.

$y = 5 - x$

x	y	(x, y)
1	4	$(1, 4)$

The ordered pair $(1, 4)$ is a solution. As we find solutions, we will list them in a **table of values** like the one shown at the left.

If $x = 2$, we have

$y = 5 - x$

x	y	(x, y)
1	4	$(1, 4)$
2	3	$(2, 3)$

$y = 5 - x$	The original equation.
$y = 5 - 2$	Substitute the input value of 2 for x.
$y = 3$	The output is 3.

A second solution is $(2, 3)$. We list it in the table of values at the left.

If $x = 5$, we have

$y = 5 - x$

x	y	(x, y)
1	4	$(1, 4)$
2	3	$(2, 3)$
5	0	$(5, 0)$

$y = 5 - x$	The original equation.
$y = 5 - 5$	Substitute the input value of 5 for x.
$y = 0$	The output is 0.

A third solution is $(5, 0)$. We list it in the table of values at the left.

If $x = -1$, we have

$y = 5 - x$

x	y	(x, y)
1	4	$(1, 4)$
2	3	$(2, 3)$
5	0	$(5, 0)$
-1	6	$(-1, 6)$

$y = 5 - x$	The original equation.
$y = 5 - (-1)$	Substitute the input value of -1 for x.
$y = 6$	The output is 6.

A fourth solution is $(-1, 6)$. We list it in the table of values at the left.

If $x = 6$, we have

$y = 5 - x$

x	y	(x, y)
1	4	$(1, 4)$
2	3	$(2, 3)$
5	0	$(5, 0)$
-1	6	$(-1, 6)$
6	-1	$(6, -1)$

$y = 5 - x$	The original equation.
$y = 5 - 6$	Substitute the input value 6 for x.
$y = -1$	The output is -1.

A fifth solution is $(6, -1)$. We list it in the table of values at the left.

Since we can choose any real number for x, and since any choice of x will give a corresponding value of y, it is apparent that the equation $y = 5 - x$ has *infinitely many solutions*.

■ GRAPHING EQUATIONS

To graph the equation $y = 5 - x$, we plot the ordered pairs listed in the table on a rectangular coordinate system, as in Figure 3-10. From the figure, we can see that the five points lie on a line.

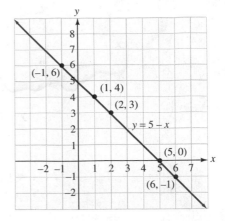

$$y = 5 - x$$

x	y	(x, y)
1	4	$(1, 4)$
2	3	$(2, 3)$
5	0	$(5, 0)$
-1	6	$(-1, 6)$
6	-1	$(6, -1)$

FIGURE 3-10

We draw a line through the points. The arrowheads on the line show that the graph continues forever in both directions. Since the graph of any solution of $y = 5 - x$ will lie on this line, the line is a picture of all of the solutions of the equation $y = 5 - x$. The line is said to be the **graph** of the equation.

Any equation, such as $y = 5 - x$, whose graph is a line is called a **linear equation in two variables.** Any point on the line has coordinates that satisfy the equation, and the graph of any pair (x, y) that satisfies the equation is a point on the line.

Since we will usually choose a number for x first and then find the corresponding value of y, the value of y depends on x. For this reason, we call y the **dependent variable** and x the **independent variable.** The value of the independent variable is the input value, and the value of the dependent variable is the output value.

Although only two points are needed to graph a linear equation, we often plot a third point as a check. If the three points do not lie on a line, at least one of them is in error.

Graphing Linear Equations

1. Find two pairs (x, y) that satisfy the equation by picking arbitrary input values for x and solving for the corresponding output values of y. A third point provides a check.

2. Plot each resulting pair (x, y) on a rectangular coordinate system. If they do not lie on a line, check your calculations.

3. Draw the line passing through the points.

EXAMPLE 2   Graph $y = 3x - 4$.

Solution   We find three ordered pairs that satisfy the equation.

If $x = 1$	**If $x = 2$**	**If $x = 3$**
$y = 3x - 4$	$y = 3x - 4$	$y = 3x - 4$
$y = 3(1) - 4$	$y = 3(2) - 4$	$y = 3(3) - 4$
$y = -1$	$y = 2$	$y = 5$

We enter the results in a table of values, plot the points, and draw a line through the points. The graph appears in Figure 3-11.

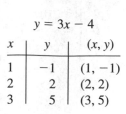

$$y = 3x - 4$$

x	y	(x, y)
1	-1	$(1, -1)$
2	2	$(2, 2)$
3	5	$(3, 5)$

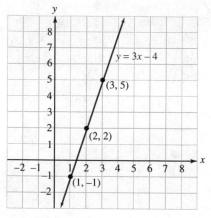

FIGURE 3-11

Self Check
Answer

Graph $y = 3x$.

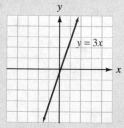

Note that the graph of $y = 3x$ is 4 units above the graph of $y = 3x - 4$.

EXAMPLE 3   Graph $y - 4 = \dfrac{1}{2}(x - 8)$.

Solution   We first solve for y and simplify.

$$y - 4 = \frac{1}{2}(x - 8)$$

$$y - 4 = \frac{1}{2}x - 4 \qquad \text{Use the distributive property to remove parentheses.}$$

$$y = \frac{1}{2}x \qquad \text{Add 4 to both sides.}$$

We now find three ordered pairs that satisfy the equation.

If $x = 0$	**If** $x = 2$	**If** $x = -4$
$y = \dfrac{1}{2}x$	$y = \dfrac{1}{2}x$	$y = \dfrac{1}{2}x$
$y = \dfrac{1}{2}(0)$	$y = \dfrac{1}{2}(2)$	$y = \dfrac{1}{2}(-4)$
$y = 0$	$y = 1$	$y = -2$

We enter the results in a table of values, plot the points, and draw a line through the points. The graph appears in Figure 3-12.

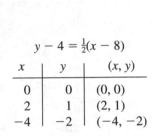

$$y - 4 = \tfrac{1}{2}(x - 8)$$

x	y	(x, y)
0	0	$(0, 0)$
2	1	$(2, 1)$
-4	-2	$(-4, -2)$

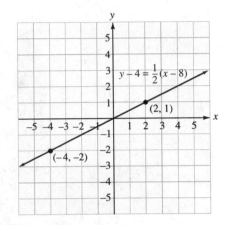

FIGURE 3-12

Self Check

Answer

Graph $y + 3 = \tfrac{1}{3}(x - 6)$.

■ THE INTERCEPT METHOD OF GRAPHING A LINE

The points where a line intersects the x- and y-axes are called **intercepts** of the line.

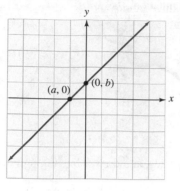

FIGURE 3-13

x- and y-Intercepts

The **x-intercept** of a line is a point $(a, 0)$ where the line intersects the x-axis. (See Figure 3-13.) To find a, substitute 0 for y in the equation of the line and solve for x.

A **y-intercept** of a line is a point $(0, b)$ where the line intersects the y-axis. To find b, substitute 0 for x in the equation of the line and solve for y.

Plotting the x- and y-intercepts and drawing a line through them is called the **intercept method of graphing a line.** This method is useful for graphing equations written in **general form.**

General Form of the Equation of a Line

If A, B, and C are real numbers and A and B are not both 0, then the equation

$$Ax + By = C$$

is called the **general form** of the equation of a line.

Whenever possible, we will write the general form $Ax + By = C$ so that A, B, and C are integers and $A \geq 0$.

EXAMPLE 4   Graph $3x + 2y = 6$.

Solution   To find the y-intercept, we let $x = 0$ and solve for y.

$$3x + 2y = 6$$
$$3(0) + 2y = 6 \qquad \text{Substitute 0 for } x.$$
$$2y = 6 \qquad \text{Simplify.}$$
$$y = 3 \qquad \text{Divide both sides by 2.}$$

The y-intercept is the pair $(0, 3)$. To find the x-intercept, we let $y = 0$ and solve for x.

$$3x + 2y = 6$$
$$3x + 2(0) = 6 \qquad \text{Substitute 0 for } y.$$
$$3x = 6 \qquad \text{Simplify.}$$
$$x = 2 \qquad \text{Divide both sides by 3.}$$

The x-intercept is the pair $(2, 0)$. As a check, we plot one more point. If $x = 4$, then

$$3x + 2y = 6$$
$$3(4) + 2y = 6 \qquad \text{Substitute 4 for } x.$$
$$12 + 2y = 6 \qquad \text{Simplify.}$$
$$2y = -6 \qquad \text{Add } -12 \text{ to both sides.}$$
$$y = -3 \qquad \text{Divide both sides by 2.}$$

The point $(4, -3)$ is on the graph. We plot these three points and join
line. The graph of $3x + 2y = 6$ is shown in Figure 3-14.

$3x + 2y = 6$

x	y	(x, y)
0	3	$(0, 3)$
2	0	$(2, 0)$
4	-3	$(4, -3)$

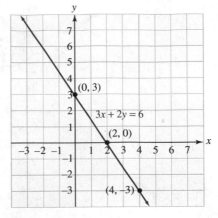

FIGURE 3-14

Self Check

Answer

Graph $4x + 3y = 6$.

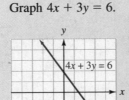

■ GRAPHING HORIZONTAL AND VERTICAL LINES

Equations such as $y = 3$ and $x = -2$ are linear equations, because they can be written in the general form $Ax + By = C$.

$y = 3$ is equivalent to $0x + 1y = 3$

$x = -2$ is equivalent to $1x + 0y = -2$

Next, we discuss how to graph these types of linear equations.

EXAMPLE 5   Graph   **a.** $y = 3$ and   **b.** $x = -2$.

Solution   **a.** We can write the equation $y = 3$ in general form as $0x + y = 3$. Since the coefficient of x is 0, the numbers chosen for x have no effect on y. The value of y is
always 3. For example, if we substitute -3 for x, we get

$$0x + y = 3$$
$$0(-3) + y = 3$$
$$0 + y = 3$$
$$y = 3$$

The table in Figure 3-15(a) gives several pairs that satisfy the equation $y = 3$. After plotting these pairs and joining them with a line, we see that the graph of $y = 3$ is a horizontal line that intersects the y-axis at 3. The y-intercept is (0, 3). There is no x-intercept.

b. We can write $x = -2$ in general form as $x + 0y = -2$. Since the coefficient of y is 0, the values of y have no effect on x. The number x is always -2. A table of values and the graph are shown in Figure 3-15(b). The graph of $x = -2$ is a vertical line that intersects the x-axis at -2. The x-intercept is $(-2, 0)$. There is no y-intercept.

$y = 3$

x	y	(x, y)
-3	3	$(-3, 3)$
0	3	$(0, 3)$
2	3	$(2, 3)$
4	3	$(4, 3)$

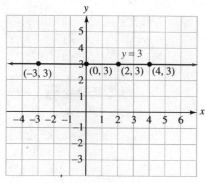

(a)

$x = -2$

x	y	(x, y)
-2	-2	$(-2, -2)$
-2	0	$(-2, 0)$
-2	2	$(-2, 2)$
-2	3	$(-2, 3)$

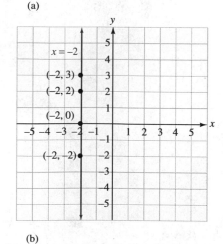

(b)

FIGURE 3-15

Self Check   Identify the graph of each equation as a horizontal or a vertical line: **a.** $x = 5$, **b.** $y = -3$,   and   **c.** $x = 0$.

Answers   **a.** vertical,   **b.** horizontal,   **c.** vertical

From the results of Example 5, we have the following facts.

Equations of Horizontal and Vertical Lines

The equation $y = b$ represents a horizontal line that intersects the y-axis at $(0, b)$. If $b = 0$, the line is the x-axis.

The equation $x = a$ represents a vertical line that intersects the x-axis at $(a, 0)$. If $a = 0$, the line is the y-axis.

■ AN APPLICATION OF LINEAR EQUATIONS

EXAMPLE 6

Birthday parties   A restaurant offers a party package that includes food, drinks, cake, and party favors for a cost of $25 plus $3 per child. Write a linear equation that will give the cost for a party of any size. Then graph the equation.

Solution   We can let c represent the cost of the party. Then the cost c will be the sum of the basic charge of $25 and the cost per child times the number of children attending. If the number of children attending is n, at $3 per child, the total cost for the children is $3n$.

The cost	is	the basic $25 charge	plus $3 times	the number of children.
c	$=$	25	$+\ 3\ \cdot$	n

For the equation $c = 25 + 3n$, the independent variable (input) is n, the number of children. The dependent variable (output) is c, the cost of the party. We will find three points on the graph of the equation by choosing n-values of 0, 5, and 10 and finding the corresponding c-values. The results are recorded in the table.

If $n = 0$	If $n = 5$	If $n = 10$
$c = 25 + 3(0)$	$c = 25 + 3(5)$	$c = 25 + 3(10)$
$c = 25$	$c = 25 + 15$	$c = 25 + 30$
	$c = 40$	$c = 55$

$T = 25 + 3n$

n	C
0	25
5	40
10	55

Next, we graph the points in Figure 3-16 and draw a line through them. We don't draw an arrowhead on the left, because it does not make sense to have a *negative* number of children attend a party. We can use the graph to determine the cost of a party of any size. For example, to find the cost of a party with 8 children, we locate 8 on the horizontal axis and then move up to find a point on the graph directly above the 8. Since the coordinates of that point are $(8, 49)$, the cost for 8 children would be $49.

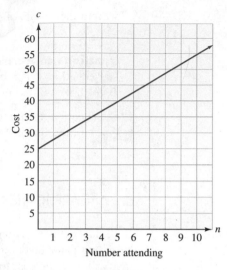

FIGURE 3-16

■

Making Tables and Graphs

GRAPHING CALCULATORS

TI-83 graphing calculator
(Courtesy of Texas Instruments)

So far, we have graphed equations by making tables of values and plotting points. This method is usually tedious and time-consuming. Fortunately, the task of making tables and graphing equations is much easier when we use a graphing calculator.

Several brands of calculators are available. Although we will use calculators to make tables and graph equations, we will not show complete keystrokes for any specific brand. For these details, please consult your owner's manual.

All graphing calculators have a **viewing window** that is used to display tables and graphs. We will first discuss how to make tables and then discuss how to draw graphs.

MAKING TABLES To construct a table of values for the equation $y = x^2$, simply press the Y = key, enter the expression x^2, and press the 2nd and TABLE keys to get a screen similar to Figure 3-17(a). You can use the up and down keys to scroll through the table to obtain a screen like Figure 3-17(b).

X	Y₁
0	0
1	1
2	4
3	9
4	16
5	25
6	36

X=0

X	Y₁
-5	25
-4	16
-3	9
-2	4
-1	1
0	0
1	1

X= -5

(a)   (b)

FIGURE 3-17

DRAWING GRAPHS   To see the proper picture of a graph, we must often set the minimum and maximum values for the x- and y-coordinates. The standard window settings of

$$\text{Xmin} = -10 \qquad \text{Xmax} = 10 \qquad \text{Ymin} = -10 \qquad \text{Ymax} = 10$$

indicate that -10 is the minimum x- and y-coordinate to be used in the graph, and that 10 is the maximum x- and y-coordinate to be used. We will usually express window values in interval notation. In this notation, the standard settings are

$$X = [-10, 10] \qquad Y = [-10, 10]$$

To graph the equation $2x - 3y = 14$ with a calculator, we must first solve the equation for y.

$$2x - 3y = 14$$
$$-3y = -2x + 14 \qquad \text{Subtract } 2x \text{ from both sides.}$$
$$y = \frac{2}{3}x - \frac{14}{3} \qquad \text{Divide both sides by } -3.$$

We now set the standard window values of $X = [-10, 10]$ and $Y = [-10, 10]$, press the $\boxed{Y=}$ key and enter the equation as $(2/3)x - 14/3$, and press $\boxed{\text{GRAPH}}$ to get the line shown in Figure 3-18.

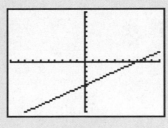

FIGURE 3-18

 WARNING!   To graph an equation with a graphing calculator, the equation must be solved for y.

USING THE TRACE AND ZOOM FEATURES   With the trace feature, we can find the coordinates of any point on a graph. For example, to find the x-intercept of the line shown in Figure 3-18, we press the $\boxed{\text{TRACE}}$ key and move the flashing cursor along the line with the cursor keys until we approach the x-intercept, as shown in Figure 3-19(a). The x- and y-coordinates of the flashing cursor appear at the bottom of the screen.

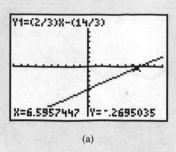

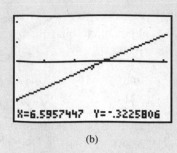

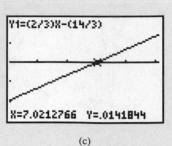

(a)   (b)   (c)

FIGURE 3-19

To get better results, we can press the ZOOM key to see a magnified picture of the line, as shown in Figure 3-19(b). We can trace again and move the cursor even closer to the *x*-intercept, as shown in Figure 3-19(c). Since the *y*-coordinate shown on the screen is close to 0, the *x*-coordinate shown on the screen is close to the *x*-value of the *x*-intercept. Repeated zooms will show that the *x*-intercept is $(7, 0)$.

Orals

1. How many points should be plotted to graph a line?

2. Define the intercepts of a line.

3. Find three pairs (x, y) that satisfy $x + y = 8$.

4. Find three pairs (x, y) that satisfy $x - y = 6$.

5. Which lines have no *y*-intercepts?

6. Which lines have no *x*-intercepts?

EXERCISE 3.2

REVIEW

1. Solve $\dfrac{x}{8} = -12$.

2. Combine like terms: $3t - 4T + 5T - 6t$.

3. Is $\dfrac{x + 5}{6}$ an expression or an equation?

4. Which formula is used to find the perimeter of a rectangle?

5. What number is 0.5% of 250?

6. Solve $-3x + 5 > 17$.

7. Find $-2.5 - (-2.6)$.

8. Evaluate $(-5)^3$.

VOCABULARY AND CONCEPTS   *Fill in each blank to make a true statement.*

9. The equation $y = x + 1$ is an equation in _____ variables.

10. An ordered pair is a _____ of an equation if the numbers in the ordered pair satisfy the equation.

11. In equations containing the variables x and y, x is called the _____ variable and y is called the _____ variable.

12. When constructing a _____ of values, the values of x are the _____ values and the values of y are the _____ values.

13. An equation whose graph is a line and whose variables are to the first power is called a _____ equation.

14. The equation $Ax + By = C$ is the _____ form of the equation of a line.

15. The _____ of a line is the point $(0, b)$, where the line intersects the y-axis.

16. The _____ of a line is the point $(a, 0)$, where the line intersects the x-axis.

PRACTICE *In Exercises 17–20, tell whether the ordered pair satisfies the equation.*

17. $x - 2y = -4$; $(4, 4)$

18. $y = 8x - 5$; $(4, 26)$

19. $y = \dfrac{2}{3}x + 5$; $(6, 12)$

20. $y = -\dfrac{1}{2}x - 2$; $(4, -4)$

$y = a + 4$

In Exercises 21–24, complete each table of values. Check your work with a graphing calculator.

21. $y = x - 3$

x	y
0	-3
1	-2
-2	-5

22. $y = x + 2$

x	$x + 2$
0	2
-1	
-2	
1	
3	

23. $y = -2x$

input	output
0	0
1	-2
3	-6
-1	2
-2	4

24. $y = \dfrac{x}{2}$

x	$\frac{x}{2}$
0	
1	
-2	
-4	

In Exercises 25–28, graph each equation. Check your work with a graphing calculator.

25. $y = 2x - 1$

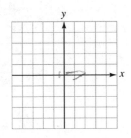

26. $y = 3x + 1$

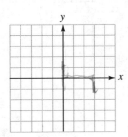

27. $y = \dfrac{x}{2} - 2$

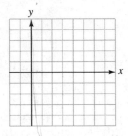

28. $y = \dfrac{x}{3} - 3$ $- 1$

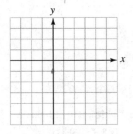

In Exercises 29–36, write each equation in general form, when necessary. Then graph it using the intercept method.

29. $x + y = 7$

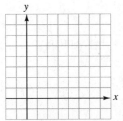

30. $x + y = -2$

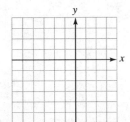

31. $x - y = 7$

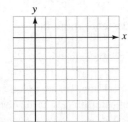

32. $x - y = -2$

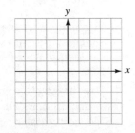

33. $y = -2x + 5$

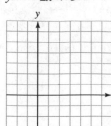

34. $y = -3x - 1$

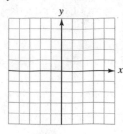

35. $2x + 3y = 12$

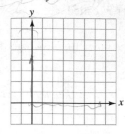

36. $3x - 2y = 6$

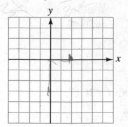

In Exercises 37–44, graph each equation.

37. $y = -5$

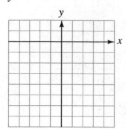

38. $x = 4$

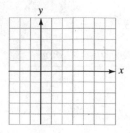

39. $x = 5$

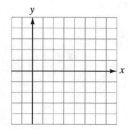

40. $y = 4$

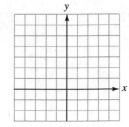

41. $y = 0$

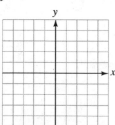

42. $x = 0$

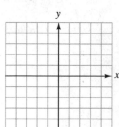

43. $2x = 5$

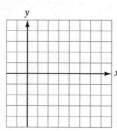

44. $3y = 7$

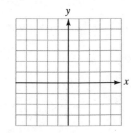

APPLICATIONS

45. Educational costs Each semester, a college charges a service fee of $50 plus $25 for each unit taken by a student.

   a. Write a linear equation that gives the total enrollment cost c for a student taking u units.

   b. Complete the table of values and graph the equation. See Illustration 1.

   c. What does the y-intercept of the line tell you?

   d. Use the graph to find the total cost for a student taking 18 units the first semester and 12 units the second semester.

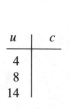

u	c
4	
8	
14	

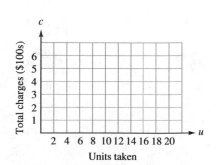

ILLUSTRATION 1

46. Group rates   To promote the sale of tickets for a cruise to Alaska, a travel agency reduces the regular ticket price of $3,000 by $5 for each individual traveling in the group.

  a. Write a linear equation that would find the ticket price T for the cruise if a group of p people travel together.

  b. Complete the table of values and then graph the equation. See Illustration 2.

  c. As the size of the group increases, what happens to the ticket price?

  d. Use the graph to determine the cost of an individual ticket if a group of 25 will be traveling together.

p	T
10	
30	
60	

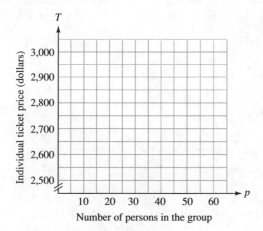

ILLUSTRATION 2

47. Physiology   Physiologists have found that a woman's height h in inches can be approximated using the linear equation $h = 3.9r + 28.9$, where r represents the length of her radius bone in inches. See Illustration 3.

  a. Complete the table of values (round to the nearest tenth), and then graph the equation.

  b. Complete this sentence: From the graph, we see that the longer the radius bone, the . . . .

  c. From the graph, estimate the height of a girl whose radius bone is 7.5 inches long.

r	H
7	
8.5	
9	

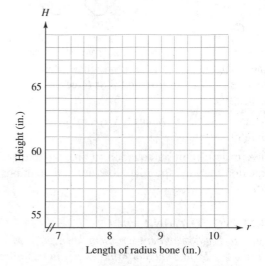

ILLUSTRATION 3

48. Research   A psychology major found that the time t in seconds that it took a white rat to complete a maze was related to the number of trials n the rat had been given by the equation $t = 25 - 0.25n$. See Illustration 4.

n	t
4	
12	
16	

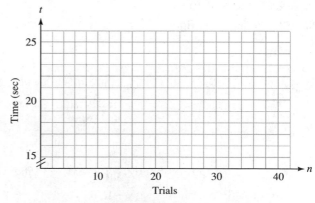

ILLUSTRATION 4

a. Complete the table of values and then graph the equation.

b. Complete this sentence: From the graph, we see that the more trials the rat had, the . . .

c. From the graph, estimate the time it will take the rat to complete the maze on its 32nd trial.

WRITING

49. From geometry, we know that two points determine a line. Explain why it is good practice when graphing linear equations to find and plot three points instead of just two.

50. Explain the process used to find the *x*- and *y*-intercepts of the graph of a line.

51. What is a table of values? Why is it often called a table of solutions?

52. When graphing an equation in two variables, how many solutions of the equation must be found?

53. Give examples of an equation in one variable and an equation in two variables. How do their solutions differ?

54. What does it mean when we say that an equation in two variables has infinitely many solutions?

SOMETHING TO THINK ABOUT   *If points P(a, b) and Q(c, d) are two points on a rectangular coordinate system and point M is midway between them, then point M is called the **midpoint** of the line segment joining P and Q. (See Illustration 5.) To find the coordinates of the midpoint M(x_M, y_M) of the segment PQ, we find the average of the x-coordinates and the average of the y-coordinates of P and Q.*

$$x_M = \frac{a + c}{2} \quad \text{and} \quad y_M = \frac{b + d}{2}$$

In Exercises 55–60, find the coordinates of the midpoint of the line segment with the given coordinates.

55. $P(5, 3)$ and $Q(7, 9)$

56. $P(5, 6)$ and $Q(7, 10)$

57. $P(2, -7)$ and $Q(-3, 12)$

58. $P(-8, 12)$ and $Q(3, -9)$

59. $A(4, 6)$ and $B(10, 6)$

60. $A(8, -6)$ and the origin

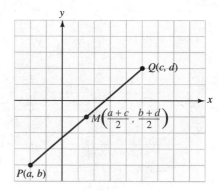

ILLUSTRATION 5

3.3 Solving Systems of Equations by Graphing

■ SYSTEMS OF EQUATIONS ■ THE GRAPHING METHOD ■ INCONSISTENT SYSTEMS ■ DEPENDENT EQUATIONS

Getting Ready   *If* $y = x^2 - 3$, *find y when x =*

1. 0   **2.** 1

3. −2   **4.** 3

■ SYSTEMS OF EQUATIONS

The lines graphed in Figure 3-20 approximate the per-person consumption of chicken and beef by Americans for the years 1990 to 1997. We can see that over this period, consumption of chicken increased, while that of beef decreased.

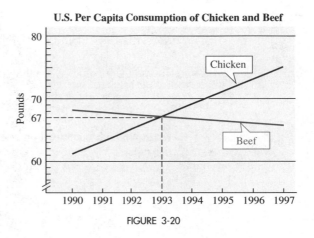

FIGURE 3-20

By graphing this *pair* of lines on the same coordinate system, it is apparent that Americans consumed equal amounts of chicken and beef in 1993—about 67 pounds each. In this section, we will work with pairs of linear equations whose graphs will be intersecting lines.

We have considered equations such as $x + y = 3$ that contain two variables. Because there are infinitely many pairs of numbers whose sum is 3, there are infinitely many pairs (x, y) that will satisfy this equation. Some of these pairs are listed in Table 3-3(a). Likewise, there are infinitely many pairs (x, y) that will satisfy the equation $3x - y = 1$. Some of these pairs are listed in Table 3-3(b).

$x + y = 3$			$3x - y = 1$	
x	y		x	y
0	3		0	−1
1	2		1	2
2	1		2	5
3	0		3	8
(a)			(b)	

TABLE 3-3

Although there are infinitely many pairs that satisfy each of these equations, only the pair (1, 2) satisfies both equations. The pair of equations

$$\begin{cases} x + y = 3 \\ 3x - y = 1 \end{cases}$$

is called a **system of equations.** Because the ordered pair (1, 2) satisfies both equations, it is called a **simultaneous solution** or just a **solution of the system of equations.** In this chapter, we will discuss three methods for finding the solution of a system of two equations, each with two variables. In this section, we consider the graphing method.

■ THE GRAPHING METHOD

To use the method of graphing to solve the system

$$\begin{cases} x + y = 3 \\ 3x - y = 1 \end{cases}$$

we graph both equations on one set of coordinate axes using the intercept method. See Figure 3-21.

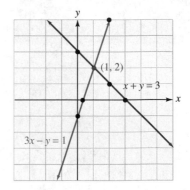

$x + y = 3$				$3x - y = 1$		
x	y	(x, y)		x	y	(x, y)
0	3	(0, 3)		0	-1	(0, -1)
3	0	(3, 0)		$\frac{1}{3}$	0	$(\frac{1}{3}, 0)$
2	1	(2, 1)		2	5	(2, 5)

FIGURE 3-21

Although there are infinitely many pairs (x, y) that satisfy $x + y = 3$ and infinitely many pairs (x, y) that satisfy $3x - y = 1$, only the coordinates of the point where their graphs intersect satisfy both equations. Thus, the solution of the system is $x = 1$ and $y = 2$, or just (1, 2).

To check the solution, we substitute 1 for x and 2 for y in each equation and verify that the pair (1, 2) satisfies each equation.

First equation   *Second equation*

$$x + y = 3 \qquad\qquad 3x - y = 1$$
$$1 + 2 \stackrel{?}{=} 3 \qquad\qquad 3(1) - 2 \stackrel{?}{=} 1$$
$$3 = 3 \qquad\qquad\quad 3 - 2 \stackrel{?}{=} 1$$
$$1 = 1$$

When the graphs of two equations in a system are different lines, the equations are called **independent equations.** When a system of equations has a solution, the system is called a **consistent system.**

To solve a system of equations in two variables by graphing, we follow these steps.

The Graphing Method

1. Carefully graph each equation.
2. When possible, find the coordinates of the point where the graphs intersect.
3. Check the solution in the equations of the original system.

■ ■ ■ ■ ■ ■ ■ ■ ■ ■ PERSPECTIVE

To schedule a company's workers, managers must consider several factors to match a worker's ability to the demands of various jobs and to match company resources to the requirements of the job. To design bridges or office buildings, engineers must analyze the effects of thousands of forces to ensure that structures won't collapse. A telephone switching network decides which of thousands of possible routes is the most efficient and then rings the correct telephone in seconds. Each of these tasks requires solving systems of equations—not just two equations in two variables, but hundreds of equations in hundreds of variables. These tasks are common in every business, industry, educational institution, and government in the world. All would be much more difficult without a computer.

One of the earliest computers in use was the Mark I, which resulted from a collaboration between IBM and a Harvard mathematician, Howard Aiken. The Mark I was started in 1939 and finished in 1944. It was 8 feet tall, 2 feet thick, and over 50 feet long. It contained over 750,000 parts and performed 3 calculations per second.

Ironically, Aiken could not envision the importance of his invention. He advised the National Bureau of Standards that there was no point in building a better computer, because "there will never be enough work for more than one or two of these machines."

Mark I Relay Computer (1944)
(Courtesy of IBM Corporation)

EXAMPLE 1   Use graphing to solve $\begin{cases} 2x + 3y = 2 \\ 3x = 2y + 16 \end{cases}$.

Solution   Using the intercept method, we graph both equations on one set of coordinate axes, as shown in Figure 3-22.

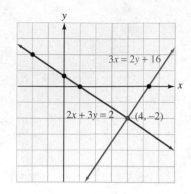

$2x + 3y = 2$		
x	y	(x, y)
0	$\frac{2}{3}$	$\left(0, \frac{2}{3}\right)$
1	0	$(1, 0)$
-2	2	$(-2, 2)$

$3x = 2y + 16$		
x	y	(x, y)
0	-8	$(0, -8)$
$\frac{16}{3}$	0	$\left(\frac{16}{3}, 0\right)$
4	-2	$(4, -2)$

FIGURE 3-22

Although there are infinitely many pairs (x, y) that satisfy $2x + 3y = 2$ and infinitely many pairs (x, y) that satisfy $3x = 2y + 16$, only the coordinates of the point where the graphs intersect satisfy both equations. The solution is $x = 4$ and $y = -2$, or just $(4, -2)$.

To check, we substitute 4 for x and -2 for y in each equation and verify that the pair $(4, -2)$ satisfies each equation.

$$2x + 3y = 2 \qquad\qquad 3x = 2y + 16$$
$$2(4) + 3(-2) \stackrel{?}{=} 2 \qquad\qquad 3(4) \stackrel{?}{=} 2(-2) + 16$$
$$8 - 6 \stackrel{?}{=} 2 \qquad\qquad 12 \stackrel{?}{=} -4 + 16$$
$$2 = 2 \qquad\qquad 12 = 12$$

The equations in this system are independent equations, and the system is a consistent system of equations. ∎

Self Check   Use graphing to solve $\begin{cases} 2x = y - 5 \\ x + y = -1 \end{cases}$.

Answer $(-2, 1)$

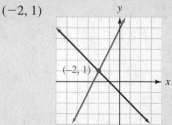

■ INCONSISTENT SYSTEMS

Sometimes a system of equations will have no solution. Such systems are called **inconsistent systems.**

EXAMPLE 2   Solve the system $\begin{cases} 2x + y = -6 \\ 4x + 2y = 8 \end{cases}$.

Solution   We graph both equations on one set of coordinate axes, as in Figure 3-23.

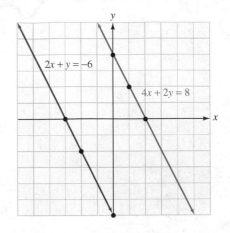

$2x + y = -6$

x	y	(x, y)
-3	0	$(-3, 0)$
0	-6	$(0, -6)$
-2	-2	$(-2, -2)$

$4x + 2y = 8$

x	y	(x, y)
2	0	$(2, 0)$
0	4	$(0, 4)$
1	2	$(1, 2)$

FIGURE 3-23

The lines in the figure are parallel. Because parallel lines do not intersect, the system has no solution, and the system is inconsistent. Since the graphs are different lines, the equations of the system are independent.   ■

Self Check   Solve $\begin{cases} 2y = 3x \\ 3x - 2y = 6 \end{cases}$.

Answer   Since the lines do not intersect, there is no solution.

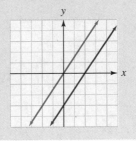

■ DEPENDENT EQUATIONS

Sometimes a system will have infinitely many solutions. In this case, we say that the equations of the system are **dependent equations.**

EXAMPLE 3   Solve the system $\begin{cases} y - 2x = 4 \\ 4x + 8 = 2y \end{cases}$.

Solution   We graph each equation on one set of axes, as in Figure 3-24.

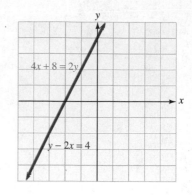

$y - 2x = 4$		
x	y	(x, y)
0	4	$(0, 4)$
-2	0	$(-2, 0)$
1	6	$(1, 6)$

$4x + 8 = 2y$		
x	y	(x, y)
0	4	$(0, 4)$
-2	0	$(-2, 0)$
-3	-2	$(-3, -2)$

FIGURE 3-24

The lines in the figure are the same line. Since the lines intersect at infinitely many points, there are infinitely many solutions. Any pair (x, y) that satisfies one of the equations satisfies the other also.

From the graph, we can see that some solutions are $(0, 4)$, $(1, 6)$, and $(-1, 2)$, since each of these points lies on the one line that is the graph of both equations. ∎

Self Check   Solve $\begin{cases} 6x - 2y = 4 \\ y + 2 = 3x \end{cases}$.

Answer   Since the graphs are the same line, there are infinitely many solutions.

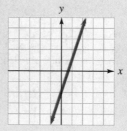

Table 3-4 summarizes the possibilities that can occur when two equations, each with two variables, are graphed.

EXAMPLE 4   Solve the system $\begin{cases} \frac{2}{3}x - \frac{1}{2}y = 1 \\ \frac{1}{10}x + \frac{1}{15}y = 1 \end{cases}$.

Solution   We can multiply both sides of the first equation by 6 to clear it of fractions.

$$\frac{2}{3}x - \frac{1}{2}y = 1$$

$$6\left(\frac{2}{3}x - \frac{1}{2}y\right) = 6(1)$$

1. $4x - 3y = 6$

Possible graph	If the	then
	lines are different and intersect,	the equations are independent and the system is consistent. One solution exists.
	lines are different and parallel,	the equations are independent and the system is inconsistent. No solutions exist.
	lines coincide (are the same line),	the equations are dependent and the system is consistent. Infinitely many solutions exist.

TABLE 3-4

We then multiply both sides of the second equation by 30 to clear it of fractions.

$$\frac{1}{10}x + \frac{1}{15}y = 1$$

$$30\left(\frac{1}{10}x + \frac{1}{15}y\right) = 30(1)$$

2. $\qquad 3x + 2y = 30$

Equations 1 and 2 form the following equivalent system of equations, which has the same solutions as the original system.

$$\begin{cases} 4x - 3y = 6 \\ 3x + 2y = 30 \end{cases}$$

We can graph each equation of the previous system (see Figure 3-25) and find that their point of intersection has coordinates of $(6, 6)$. The solution of the given system is $x = 6$ and $y = 6$, or just $(6, 6)$.

To verify that $(6, 6)$ satisfies each equation of the original system, we substitute 6 for x and 6 for y in each of the original equations and simplify.

$$\frac{2}{3}x - \frac{1}{2}y = 1 \qquad\qquad \frac{1}{10}x + \frac{1}{15}y = 1$$

$$\frac{2}{3}(6) - \frac{1}{2}(6) \stackrel{?}{=} 1 \qquad\qquad \frac{1}{10}(6) + \frac{1}{15}(6) \stackrel{?}{=} 1$$

$$4 - 3 \stackrel{?}{=} 1 \qquad\qquad \frac{3}{5} + \frac{2}{5} \stackrel{?}{=} 1$$

$$1 = 1 \qquad\qquad\qquad 1 = 1$$

The equations in this system are independent, and the system is consistent.

$4x - 3y = 6$

x	y	(x, y)
0	-2	$(0, -2)$
3	2	$(3, 2)$
6	6	$(6, 6)$

$3x + 2y = 30$

x	y	(x, y)
10	0	$(10, 0)$
8	3	$(8, 3)$
6	6	$(6, 6)$

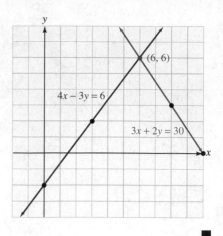

FIGURE 3-25

■

Self Check Solve $\begin{cases} -\frac{x}{2} = \frac{y}{4} \\ \frac{1}{4}x - \frac{3}{8}y = -2 \end{cases}$.

Answer $(-2, 4)$

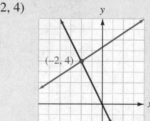

■ ■ ■ ■ ■ ■ ■ ■ ■ **Solving Systems of Equations**

GRAPHING
CALCULATORS

We can use a graphing calculator to solve the system $\begin{cases} 2x + y = 12 \\ 2x - y = -2 \end{cases}$. However, before we can enter the equations into the calculator, we must solve them for y.

$$2x + y = 12 \qquad\qquad 2x - y = -2$$
$$y = -2x + 12 \qquad\qquad -y = -2x - 2$$
$$\qquad\qquad\qquad\qquad\qquad y = 2x + 2$$

We can now enter the resulting equations into a calculator and graph them. If we use standard window settings of $x = [-10, 10]$ and $y = [-10, 10]$, their graphs will look like Figure 3-26(a). We can trace to see that the coordinates of the intersection point are approximately

$$x = 2.5531915 \qquad \text{and} \qquad y = 6.893617$$

See Figure 3-26(b). For better results, we can zoom in on the intersection point and trace again to find that

$$x = 2.5 \quad \text{and} \quad y = 7$$

See Figure 3-26(c). Check the solution.

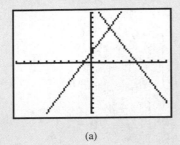
(a)

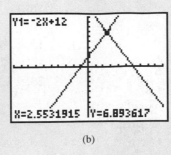
(b)

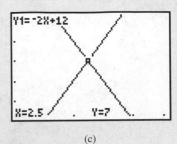
(c)

FIGURE 3-26

Orals *Tell whether the pair is a solution of the system.*

1. $(3, 2)$, $\begin{cases} x + y = 5 \\ x - y = 1 \end{cases}$

2. $(1, 2)$, $\begin{cases} x - y = -1 \\ x + y = 3 \end{cases}$

3. $(4, 1)$, $\begin{cases} x + y = 5 \\ x - y = 2 \end{cases}$

4. $(5, 2)$, $\begin{cases} x - y = 3 \\ x + y = 6 \end{cases}$

EXERCISE 3.3

REVIEW *Evaluate each expression. Assume that* $x = -3$.

1. $(-2)^4$

2. -2^4

3. $3x - x^2$

4. $\dfrac{-3 + 2x}{6x}$

VOCABULARY AND CONCEPTS *Fill in each blank to make a true statement.*

5. The pair of equations $\begin{cases} x - y = -1 \\ 2x - y = 1 \end{cases}$ is called a _____ of equations.

6. Because the ordered pair $(2, 3)$ satisfies both equations in Exercise 5, it is called a _____ of the system.

7. When the graphs of two equations in a system are different lines, the equations are called _____ equations.

8. When a system of equations has a solution, the system is called a _____.

9. Systems of equations that have no solution are called _____ systems.

10. When a system has infinitely many solutions, the equations of the system are said to be _____ equations.

In Exercises 11–22, tell whether the ordered pair is a solution of the given system.

11. $(1, 1),$ $\begin{cases} x + y = 2 \\ 2x - y = 1 \end{cases}$

12. $(1, 3),$ $\begin{cases} 2x + y = 5 \\ 3x - y = 0 \end{cases}$

13. $(3, -2),$ $\begin{cases} 2x + y = 4 \\ x + y = 1 \end{cases}$

14. $(-2, 4),$ $\begin{cases} 2x + 2y = 4 \\ x + 3y = 10 \end{cases}$

15. $(4, 5),$ $\begin{cases} 2x - 3y = -7 \\ 4x - 5y = 25 \end{cases}$

16. $(2, 3),$ $\begin{cases} 3x - 2y = 0 \\ 5x - 3y = -1 \end{cases}$

17. $(-2, -3),$ $\begin{cases} 4x + 5y = -23 \\ -3x + 2y = 0 \end{cases}$

18. $(-5, 1),$ $\begin{cases} -2x + 7y = 17 \\ 3x - 4y = -19 \end{cases}$

19. $\left(\dfrac{1}{2}, 3\right),$ $\begin{cases} 2x + y = 4 \\ 4x - 3y = 11 \end{cases}$

20. $\left(2, \dfrac{1}{3}\right),$ $\begin{cases} x - 3y = 1 \\ -2x + 6y = -6 \end{cases}$

21. $\left(-\dfrac{2}{5}, \dfrac{1}{4}\right),$ $\begin{cases} 5x - 4y = -6 \\ 8y = 10x + 12 \end{cases}$

22. $\left(-\dfrac{1}{3}, \dfrac{3}{4}\right),$ $\begin{cases} 3x + 4y = 2 \\ 12y = 3(2 - 3x) \end{cases}$

In Exercises 23–34, solve each system.

23. $\begin{cases} x + y = 2 \\ x - y = 0 \end{cases}$

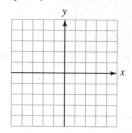

24. $\begin{cases} x + y = 4 \\ x - y = 0 \end{cases}$

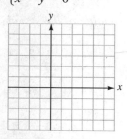

25. $\begin{cases} x + y = 2 \\ x - y = 4 \end{cases}$

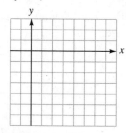

26. $\begin{cases} x + y = 1 \\ x - y = -5 \end{cases}$

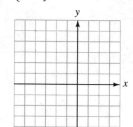

27. $\begin{cases} 3x + 2y = -8 \\ 2x - 3y = -1 \end{cases}$

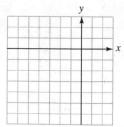

28. $\begin{cases} x + 4y = -2 \\ x + y = -5 \end{cases}$

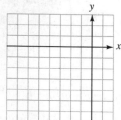

29. $\begin{cases} 4x - 2y = 8 \\ y = 2x - 4 \end{cases}$

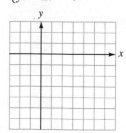

30. $\begin{cases} 3x - 6y = 18 \\ x = 2y + 3 \end{cases}$

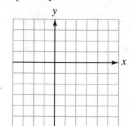

31. $\begin{cases} 2x - 3y = -18 \\ 3x + 2y = -1 \end{cases}$

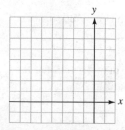

32. $\begin{cases} -x + 3y = -11 \\ 3x - y = 17 \end{cases}$

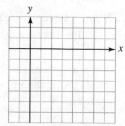

33. $\begin{cases} 4x = 3(4 - y) \\ 2y = 4(3 - x) \end{cases}$

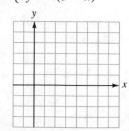

34. $\begin{cases} 2x = 3(2 - y) \\ 3y = 2(3 - x) \end{cases}$

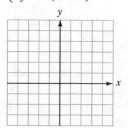

In Exercises 35–42, solve each system.

35. $\begin{cases} x + 2y = -4 \\ x - \dfrac{1}{2}y = 6 \end{cases}$

36. $\begin{cases} \dfrac{2}{3}x - y = -3 \\ 3x + y = 3 \end{cases}$

37. $\begin{cases} -\dfrac{3}{4}x + y = 3 \\ \dfrac{1}{4}x + y = -1 \end{cases}$

38. $\begin{cases} \dfrac{1}{3}x + y = 7 \\ \dfrac{2}{3}x - y = -4 \end{cases}$

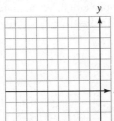 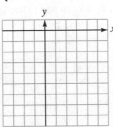 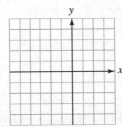 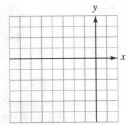

39. $\begin{cases} \dfrac{1}{2}x + \dfrac{1}{4}y = 0 \\ \dfrac{1}{4}x - \dfrac{3}{8}y = -2 \end{cases}$

40. $\begin{cases} \dfrac{1}{2}x + \dfrac{2}{3}y = -5 \\ \dfrac{3}{2}x - y = 3 \end{cases}$

41. $\begin{cases} \dfrac{1}{3}x - \dfrac{1}{2}y = \dfrac{1}{6} \\ \dfrac{2}{5}x + \dfrac{1}{2}y = \dfrac{13}{10} \end{cases}$

42. $\begin{cases} \dfrac{3}{4}x + \dfrac{2}{3}y = -\dfrac{19}{6} \\ y - x = -\dfrac{4x}{3} \end{cases}$

In Exercises 43–46, use a graphing calculator to solve each system, if possible. If answers are not exact, round to the nearest hundredth.

43. $\begin{cases} y = 4 - x \\ y = 2 + x \end{cases}$

44. $\begin{cases} y = x - 2 \\ y = x + 2 \end{cases}$

45. $\begin{cases} 3x - 6y = 4 \\ 2x + y = 1 \end{cases}$

46. $\begin{cases} 4x + 9y = 4 \\ 6x + 3y = -1 \end{cases}$

APPLICATIONS

47. Transplants See Illustration 1.
  a. What was the relationship between the number of donors and those awaiting a transplant in 1989?

  b. In what year was the number of donors and the number waiting for a transplant the same? Estimate the number.
  c. Explain the most recent trends.

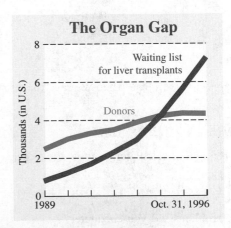

ILLUSTRATION 1

48. Daily tracking polls See Illustration 2.
   a. Which candidate was ahead on October 28 and by how much?
   b. On what day did the challenger pull even with the incumbent?
   c. If the election was held November 4, whom did the poll predict as the winner and by how many percentage points?

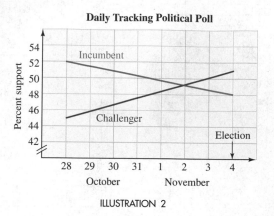

Daily Tracking Political Poll

ILLUSTRATION 2

ILLUSTRATION 3

49. Latitude and longitude See Illustration 3.
   a. Name three American cities that lie on the latitude line of 30° north.

   b. Name three American cities that lie on the longitude line of 90° west.

   c. What city lies on both lines?

50. Economics The graph in Illustration 4 illustrates the law of supply and demand.
   a. Complete this sentence: "As the price of an item increases, the *supply* of the item _____."
   b. Complete this sentence: "As the price of an item increases, the *demand* for the item _____."
   c. For what price will the supply equal the demand? How many items will be supplied for this price?

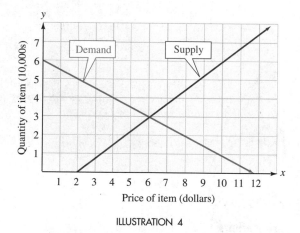

ILLUSTRATION 4

WRITING

51. Explain what we mean when we say "inconsistent system."

52. Explain what we mean when we say, "The equations of a system are dependent."

SOMETHING TO THINK ABOUT

53. Use a graphing calculator to solve the system

$$\begin{cases} 11x - 20y = 21 \\ -4x + 7y = 21 \end{cases}$$

What problems did you encounter?

54. Can the equations of an inconsistent system with two equations in two variables be dependent?

3.4 Solving Systems of Equations by Substitution

■ THE SUBSTITUTION METHOD ■ INCONSISTENT SYSTEMS ■ DEPENDENT EQUATIONS

Getting Ready   *Remove parentheses.*

1. $2(3x + 2)$ **2.** $5(-5 - 2x)$

Substitute $x - 2$ for y and remove parentheses.

3. $2y$ **4.** $3(y - 2)$

■ THE SUBSTITUTION METHOD

We now consider the **substitution method** for solving systems of equations.
    To solve the system

$$\begin{cases} y = 3x - 2 \\ 2x + y = 8 \end{cases}$$

by the substitution method, we note that $y = 3x - 2$. Because $y = 3x - 2$, we can
substitute $3x - 2$ for y in the equation $2x + y = 8$ to get

$$2x + y = 8$$
$$2x + (3x - 2) = 8$$

The resulting equation has only one variable and can be solved for x.

$$2x + (3x - 2) = 8$$
$$2x + 3x - 2 = 8 \qquad \text{Remove parentheses.}$$
$$5x - 2 = 8 \qquad \text{Combine like terms.}$$
$$5x = 10 \qquad \text{Add 2 to both sides.}$$
$$x = 2 \qquad \text{Divide both sides by 5.}$$

We can find y by substituting 2 for x in either equation of the given system. Because
$y = 3x - 2$ is already solved for y, it is easier to substitute in this equation.

$$y = 3x - 2$$
$$= 3(2) - 2$$
$$= 6 - 2$$
$$= 4$$

The solution of the given system is $x = 2$ and $y = 4$, or just $(2, 4)$.

Check: $\quad y = 3x - 2 \qquad\qquad\qquad 2x + y = 8$

$$4 \overset{?}{=} 3(2) - 2 \qquad\qquad 2(2) + 4 \overset{?}{=} 8$$
$$4 \overset{?}{=} 6 - 2 \qquad\qquad\quad 4 + 4 \overset{?}{=} 8$$
$$4 = 4 \qquad\qquad\qquad\qquad 8 = 8$$

Since the pair $x = 2$ and $y = 4$ is a solution, the lines represented by the equations of the given system intersect at the point $(2, 4)$. The equations of this system are independent, and the system is consistent.

To solve a system of equations in x and y by the substitution method, we follow these steps.

The Substitution Method

1. Solve one of the equations for x or y. (This step may not be necessary.)

2. Substitute the resulting expression for the variable obtained in Step 1 into the other equation, and solve that equation.

3. Find the value of the other variable by substituting the solution found in Step 2 into any equation containing both variables.

4. Check the solution in the equations of the original system.

EXAMPLE 1   Solve the system $\begin{cases} 2x + y = -5 \\ 3x + 5y = -4 \end{cases}$.

Solution   We solve one of the equations for one of its variables. Since the term y in the first equation has a coefficient of 1, we solve the first equation for y.

$$2x + y = -5$$
$$y = -5 - 2x \qquad \text{Subtract } 2x \text{ from both sides.}$$

We then substitute $-5 - 2x$ for y in the second equation and solve for x.

$$3x + 5y = -4$$
$$3x + 5(\mathbf{-5 - 2x}) = -4$$
$$3x - 25 - 10x = -4 \qquad \text{Remove parentheses.}$$
$$-7x - 25 = -4 \qquad \text{Combine like terms.}$$
$$-7x = 21 \qquad \text{Add 25 to both sides.}$$
$$x = -3 \qquad \text{Divide both sides by } -7.$$

We can find y by substituting -3 for x in the equation $y = -5 - 2x$.

$$y = -5 - 2x$$
$$= -5 - 2(\mathbf{-3})$$
$$= -5 + 6$$
$$= 1$$

The solution is $x = -3$ and $y = 1$, or just $(-3, 1)$.

Check:
$$2x + y = -5 \qquad\qquad\qquad 3x + 5y = -4$$
$$2(\mathbf{-3}) + 1 \overset{?}{=} -5 \qquad\qquad 3(\mathbf{-3}) + 5(\mathbf{1}) \overset{?}{=} -4$$
$$-6 + 1 \overset{?}{=} -5 \qquad\qquad\qquad -9 + 5 \overset{?}{=} -4$$
$$-5 = -5 \qquad\qquad\qquad\qquad -4 = -4$$

Self Check   Solve $\begin{cases} 2x - 3y = 13 \\ 3x + y = 3 \end{cases}$.

Answer $(2, -3)$

EXAMPLE 2   Solve the system $\begin{cases} 2x + 3y = 5 \\ 3x + 2y = 0 \end{cases}$.

Solution   We can solve the second equation for x:

$$3x + 2y = 0$$

$$3x = -2y \qquad \text{Subtract } 2y \text{ from both sides.}$$

$$x = \frac{-2y}{3} \qquad \text{Divide both sides by 3.}$$

We then substitute $\frac{-2y}{3}$ for x in the other equation and solve for y.

$$2x + 3y = 5$$

$$2\left(\frac{-2y}{3}\right) + 3y = 5$$

$$\frac{-4y}{3} + 3y = 5 \qquad \text{Remove parentheses.}$$

$$3\left(\frac{-4y}{3}\right) + 3(3y) = 3(5) \qquad \text{Multiply both sides by 3.}$$

$$-4y + 9y = 15 \qquad \text{Remove parentheses.}$$

$$5y = 15 \qquad \text{Combine like terms.}$$

$$y = 3 \qquad \text{Divide both sides by 5.}$$

We can find x by substituting 3 for y in the equation $x = \frac{-2y}{3}$.

$$x = \frac{-2y}{3}$$

$$= \frac{-2(3)}{3}$$

$$= -2$$

Check the solution $(-2, 3)$ in each equation of the system.

Self Check   Solve $\begin{cases} 3x - 2y = -19 \\ 2x + 5y = 0 \end{cases}$.

Answer $(-5, 2)$

EXAMPLE 3   Solve the system $\begin{cases} 3(x - y) = 5 \\ x + 3 = -\frac{5}{2}y \end{cases}$.

Solution   We begin by writing each equation in general form:

$$3(x - y) = 5 \qquad\qquad x + 3 = -\frac{5}{2}y$$

1. $3x - 3y = 5$ $\qquad\qquad$ $2x + 6 = -5y$ $\qquad$ Multiply both sides by 2.

2. $\qquad\qquad\qquad\qquad\qquad$ $2x + 5y = -6$ $\qquad$ Add $5y$ and subtract 6 from both sides.

To solve the system formed by Equations 1 and 2, we first solve Equation 1 for x.

1. $3x - 3y = 5$

$\qquad\qquad$ $3x = 5 + 3y$ $\qquad$ Add $3y$ to both sides.

3. $\qquad\qquad$ $x = \dfrac{5 + 3y}{3}$ $\qquad$ Divide both sides by 3.

We then substitute $\frac{5 + 3y}{3}$ for x in Equation 2 and proceed as follows:

2. $\qquad\qquad$ $2x + 5y = -6$

$$2\left(\frac{5 + 3y}{3}\right) + 5y = -6$$

$\qquad$ $2(5 + 3y) + 15y = -18$ $\qquad$ Multiply both sides by 3.

$\qquad\qquad$ $10 + 6y + 15y = -18$ $\qquad$ Remove parentheses.

$\qquad\qquad\qquad$ $10 + 21y = -18$ $\qquad$ Combine like terms.

$\qquad\qquad\qquad\qquad$ $21y = -28$ $\qquad$ Subtract 10 from both sides.

$\qquad\qquad\qquad\qquad$ $y = \dfrac{-28}{21}$ $\qquad$ Divide both sides by 21.

$\qquad\qquad\qquad\qquad$ $y = -\dfrac{4}{3}$ $\qquad$ Simplify $\frac{-28}{21}$.

To find x, we substitute $-\frac{4}{3}$ for y in Equation 3 and simplify.

$$x = \frac{5 + 3y}{3}$$

$$= \frac{5 + 3\left(-\frac{4}{3}\right)}{3}$$

$$= \frac{5 - 4}{3}$$

$$= \frac{1}{3}$$

Check the solution $\left(\frac{1}{3}, -\frac{4}{3}\right)$ in each equation. $\qquad\qquad$ ■

Self Check   Solve $\begin{cases} 2(x + y) = -5 \\ x + 2 = -\frac{3}{5}y \end{cases}$.

Answer $\left(-\frac{5}{4}, -\frac{5}{4}\right)$

■ INCONSISTENT SYSTEMS

EXAMPLE 4 Solve the system $\begin{cases} x = 4(3 - y) \\ 2x = 4(3 - 2y) \end{cases}$.

Solution Since $x = 4(3 - y)$, we can substitute $4(3 - y)$ for x in the second equation and solve for y.

$$2x = 4(3 - 2y)$$
$$2[\mathbf{4(3 - y)}] = 4(3 - 2y)$$
$$8(3 - y) = 4(3 - 2y) \qquad 2 \cdot 4 = 8.$$
$$24 - 8y = 12 - 8y \qquad \text{Remove parentheses.}$$
$$24 = 12 \qquad \text{Add } 8y \text{ to both sides.}$$

This impossible result indicates that the equations in this system are independent, but that the system is inconsistent. If each equation in this system were graphed, these graphs would be parallel lines. There are no solutions to this system. ■

Self Check Solve $\begin{cases} 0.1x - 0.4 = 0.1y \\ -2y = 2(2 - x) \end{cases}$.

Answer no solution

■ DEPENDENT EQUATIONS

EXAMPLE 5 Solve the system $\begin{cases} 3x = 4(6 - y) \\ 4y + 3x = 24 \end{cases}$.

Solution We can substitute $4(6 - y)$ for $3x$ in the second equation and proceed as follows:

$$4y + \mathbf{3x} = 24$$
$$4y + \mathbf{4(6 - y)} = 24$$
$$4y + 24 - 4y = 24 \qquad \text{Remove parentheses.}$$
$$24 = 24 \qquad \text{Combine like terms.}$$

Although $24 = 24$ is true, we did not find y. This result indicates that the equations of this system are dependent. If either equation were graphed, the same line would result.

Because any ordered pair that satisfies one equation satisfies the other also, the system has infinitely many solutions. To find some of them, we substitute 8, 0, and 4 for x in either equation and solve for y. The pairs $(8, 0)$, $(0, 6)$, and $(4, 3)$ are solutions. ■

Self Check Solve $\begin{cases} 3y = -3(x + 4) \\ 3x + 3y = -12 \end{cases}$.

Answer infinitely many solutions

Orals *Let $y = x + 1$. Find y after each quantity is substituted for x.*

1. $2z$

2. $z + 1$

3. $3t + 2$

4. $\dfrac{t}{3} + 3$

EXERCISE 3.4

REVIEW *Let $x = -2$ and $y = 3$ and evaluate each expression.*

1. $y^2 - x^2$

2. $-x^2 + y^3$

3. $\dfrac{3x - 2y}{2x + y}$

4. $-2x^2y^2$

5. $-x(3y - 4)$

6. $-2y(4x - y)$

VOCABULARY AND CONCEPTS *Fill in each blank to make a true statement.*

7. We say the equation $y = 2x + 4$ is solved for ___ or that y is expressed in _____ of x.

8. To _____ a solution of a system means to see whether the coordinates of the ordered pair satisfy both equations.

9. Consider $2(x - 6) = 2x - 12$. The distributive property was applied to _____ parentheses.

10. In mathematics, to _____ means to replace an expression with one that is equivalent to it.

11. A system with dependent equations has _____ solutions.

12. In the term y, the _____ is understood to be 1.

PRACTICE *In Exercises 13–54, use the substitution method to solve each system.*

13. $\begin{cases} y = 2x \\ x + y = 6 \end{cases}$

14. $\begin{cases} y = 3x \\ x + y = 4 \end{cases}$

15. $\begin{cases} y = 2x - 6 \\ 2x + y = 6 \end{cases}$

16. $\begin{cases} y = 2x - 9 \\ x + 3y = 8 \end{cases}$

17. $\begin{cases} y = 2x + 5 \\ x + 2y = -5 \end{cases}$

18. $\begin{cases} y = -2x \\ 3x + 2y = -1 \end{cases}$

19. $\begin{cases} 2a + 4b = -24 \\ a = 20 - 2b \end{cases}$

20. $\begin{cases} 3a + 6b = -15 \\ a = -2b - 5 \end{cases}$

21. $\begin{cases} 2a = 3b - 13 \\ b = 2a + 7 \end{cases}$

22. $\begin{cases} a = 3b - 1 \\ b = 2a + 2 \end{cases}$

23. $\begin{cases} r + 3s = 9 \\ 3r + 2s = 13 \end{cases}$

24. $\begin{cases} x - 2y = 2 \\ 2x + 3y = 11 \end{cases}$

25. $\begin{cases} 4x + 5y = 2 \\ 3x - y = 11 \end{cases}$

26. $\begin{cases} 5u + 3v = 5 \\ 4u - v = 4 \end{cases}$

27. $\begin{cases} 2x + y = 0 \\ 3x + 2y = 1 \end{cases}$

28. $\begin{cases} 3x - y = 7 \\ 2x + 3y = 1 \end{cases}$

29. $\begin{cases} 3x + 4y = -7 \\ 2y - x = -1 \end{cases}$

30. $\begin{cases} 4x + 5y = -2 \\ x + 2y = -2 \end{cases}$

31. $\begin{cases} 9x = 3y + 12 \\ 4 = 3x - y \end{cases}$

32. $\begin{cases} 8y = 15 - 4x \\ x + 2y = 4 \end{cases}$

33. $\begin{cases} 2x + 3y = 5 \\ 3x + 2y = 5 \end{cases}$

34. $\begin{cases} 3x - 2y = -1 \\ 2x + 3y = -5 \end{cases}$

35. $\begin{cases} 2x + 5y = -2 \\ 4x + 3y = 10 \end{cases}$

36. $\begin{cases} 3x + 4y = -6 \\ 2x - 3y = -4 \end{cases}$

37. $\begin{cases} 2x - 3y = -3 \\ 3x + 5y = -14 \end{cases}$

38. $\begin{cases} 4x - 5y = -12 \\ 5x - 2y = 2 \end{cases}$

39. $\begin{cases} 7x - 2y = -1 \\ -5x + 2y = -1 \end{cases}$

40. $\begin{cases} -8x + 3y = 22 \\ 4x + 3y = -2 \end{cases}$

41. $\begin{cases} 2a + 3b = 2 \\ 8a - 3b = 3 \end{cases}$

42. $\begin{cases} 3a - 2b = 0 \\ 9a + 4b = 5 \end{cases}$

43. $\begin{cases} y - x = 3x \\ 2(x + y) = 14 - y \end{cases}$

44. $\begin{cases} y + x = 2x + 2 \\ 2(3x - 2y) = 21 - y \end{cases}$

45. $\begin{cases} 3(x - 1) + 3 = 8 + 2y \\ 2(x + 1) = 4 + 3y \end{cases}$

46. $\begin{cases} 4(x - 2) = 19 - 5y \\ 3(x + 1) - 2y = 2y \end{cases}$

47. $\begin{cases} 6a = 5(3 + b + a) - a \\ 3(a - b) + 4b = 5(1 + b) \end{cases}$

48. $\begin{cases} 5(x + 1) + 7 = 7(y + 1) \\ 5(y + 1) = 6(1 + x) + 5 \end{cases}$

49. $\begin{cases} \dfrac{1}{2}x + \dfrac{1}{2}y = -1 \\ \dfrac{1}{3}x - \dfrac{1}{2}y = -4 \end{cases}$

50. $\begin{cases} \dfrac{2}{3}y + \dfrac{1}{5}z = 1 \\ \dfrac{1}{3}y - \dfrac{2}{5}z = 3 \end{cases}$

51. $\begin{cases} 5x = \dfrac{1}{2}y - 1 \\ \dfrac{1}{4}y = 10x - 1 \end{cases}$

52. $\begin{cases} \dfrac{2}{3}x = 1 - 2y \\ 2(5y - x) + 11 = 0 \end{cases}$

53. $\begin{cases} \dfrac{6x - 1}{3} - \dfrac{5}{3} = \dfrac{3y + 1}{2} \\ \dfrac{1 + 5y}{4} + \dfrac{x + 3}{4} = \dfrac{17}{2} \end{cases}$

54. $\begin{cases} \dfrac{5x - 2}{4} + \dfrac{1}{2} = \dfrac{3y + 2}{2} \\ \dfrac{7y + 3}{3} = \dfrac{x}{2} + \dfrac{7}{3} \end{cases}$

WRITING

55. Explain how to use substitution to solve a system of equations.

56. If the equations of a system are written in general form, why is it to your advantage to solve for a variable whose coefficient is 1?

SOMETHING TO THINK ABOUT

57. Could you use substitution to solve the system

$$\begin{cases} y = 2y + 4 \\ x = 3x - 5 \end{cases}$$

How would you solve it?

58. What are the advantages and disadvantages of
a. the graphing method?
b. the substitution method?

3.5 Solving Systems of Equations by Addition

■ THE ADDITION METHOD ■ INCONSISTENT SYSTEMS ■ DEPENDENT EQUATIONS

Getting Ready *Add the left-hand sides and the right-hand sides of the equations in each system.*

1. $\begin{cases} 2x + 3y = 4 \\ 3x - 3y = 6 \end{cases}$

2. $\begin{cases} 4x - 2y = 1 \\ -4x + 3y = 5 \end{cases}$

3. $\begin{cases} 6x - 5y = 23 \\ -4x + 5y = 10 \end{cases}$

4. $\begin{cases} -5x + 6y = 18 \\ 5x + 12y = 10 \end{cases}$

■ THE ADDITION METHOD

Another method used to solve systems of equations is the **addition method.** To solve the system

$$\begin{cases} x + y = 8 \\ x - y = -2 \end{cases}$$

by the addition method, we see that the coefficients of y are *opposites* and then add the left-hand sides and the right-hand sides of the equations to eliminate the variable y.

$\begin{array}{l} x + y = 8 \\ \underline{x - y = -2} \end{array}$ Equal quantities, $x - y$ and -2, are added to both sides of the equation $x + y = 8$. By the addition property of equality, the results will be equal.

Now, column by column, we add like terms.

Combine like terms.

$\begin{array}{l} x + y = 8 \\ \underline{x - y = -2} \\ 2x = 6 \end{array}$ ← Write each result here.

We can then solve the resulting equation for x.

$2x = 6$

$x = 3$ Divide both sides by 2.

To find y, we substitute 3 for x in either equation of the system and solve it for y.

$x + y = 8$ The first equation of the system.

$3 + y = 8$ Substitute 3 for x.

$ y = 5$ Subtract 3 from both sides.

We check the solution by verifying that the pair (3, 5) satisfies each equation of the original system.

To solve an equation in x and y by the addition method, we follow these steps.

The Addition Method

1. If necessary, write both equations in general form: $Ax + By = C$.
2. If necessary, multiply one or both of the equations by nonzero quantities to make the coefficients of x (or the coefficients of y) opposites.
3. Add the equations to eliminate the term involving x (or y).
4. Solve the equation resulting from Step 3.
5. Find the value of the other variable by substituting the solution found in Step 4 into any equation containing both variables.
6. Check the solution in the equations of the original system.

EXAMPLE 1   Solve the system $\begin{cases} 3y = 14 + x \\ x + 22 = 5y \end{cases}$.

Solution   We can write the equations in the form

$$\begin{cases} -x + 3y = 14 \\ x - 5y = -22 \end{cases}$$

When these equations are added, the terms involving x are eliminated. We solve the resulting equation for y.

$$\begin{array}{r} -x + 3y = 14 \\ \underline{x - 5y = -22} \\ -2y = -8 \\ y = 4 \end{array}$$ Divide both sides by -2.

To find x, we substitute 4 for y in either equation of the system. If we substitute 4 for y in the equation $-x + 3y = 14$, we have

$$\begin{array}{rl} -x + 3y = 14 & \\ -x + 3(\mathbf{4}) = 14 & \\ -x + 12 = 14 & \text{Simplify.} \\ -x = 2 & \text{Subtract 12 from both sides.} \\ x = -2 & \text{Divide both sides by } -1. \end{array}$$

Verify that $(-2, 4)$ satisfies each equation.   ∎

Self Check   Solve $\begin{cases} 3y = 7 - x \\ 2x - 3y = -22 \end{cases}$.

Answer $(-5, 4)$

Sometimes we need to multiply both sides of one equation in a system by a number to make the coefficients of one of the variables opposites.

EXAMPLE 2   Solve the system $\begin{cases} 3x + y = 7 \\ x + 2y = 4 \end{cases}$.

Solution   If we add the equations as they are, neither variable will be eliminated. We must write the equations so that the coefficients of one of the variables are opposites. To eliminate x, we can multiply both sides of the second equation by -3 to get

$$\begin{cases} 3x + y = 7 \\ -3(x + 2y) = -3(4) \end{cases} \longrightarrow \begin{cases} 3x + y = 7 \\ -3x - 6y = -12 \end{cases}$$

The coefficients of the terms $3x$ and $-3x$ are opposites. When the equations are added, x is eliminated.

$$\begin{array}{r} 3x + y = 7 \\ -3x - 6y = -12 \\ \hline -5y = -5 \\ y = 1 \end{array}$$ Divide both sides by -5.

To find x, we substitute 1 for y in the equation $3x + y = 7$.

$$\begin{aligned} 3x + y &= 7 \\ 3x + (1) &= 7 & &\text{Substitute 1 for } y. \\ 3x &= 6 & &\text{Subtract 1 from both sides.} \\ x &= 2 & &\text{Divide both sides by 3.} \end{aligned}$$

Check the solution $(2, 1)$ in the original system of equations.   ∎

Self Check   Solve $\begin{cases} 3x + 4y = 25 \\ 2x + y = 10 \end{cases}$.

Answer $(3, 4)$

In some instances, we must multiply both equations by nonzero quantities to make the coefficients of one of the variables opposites.

EXAMPLE 3   Solve the system $\begin{cases} 2a - 5b = 10 \\ 3a - 2b = -7 \end{cases}$.

Solution   The equations in the system must be written so that one of the variables will be eliminated when the equations are added.

To eliminate a, we can multiply the first equation by 3 and the second equation by -2 to get

$$\begin{cases} 3(2a - 5b) = 3(10) \\ -2(3a - 2b) = -2(-7) \end{cases} \longrightarrow \begin{cases} 6a - 15b = 30 \\ -6a + 4b = 14 \end{cases}$$

When these equations are added, the terms $6a$ and $-6a$ are eliminated.

$$
\begin{aligned}
6a - 15b &= 30 \\
\underline{-6a + 4b} &= \underline{14} \\
-11b &= 44 \\
b &= -4 \qquad \text{Divide both sides by } -11.
\end{aligned}
$$

To find a, we substitute -4 for b in the equation $2a - 5b = 10$.

$$
\begin{aligned}
2a - 5b &= 10 \\
2a - 5(-4) &= 10 \qquad \text{Substitute } -4 \text{ for } b. \\
2a + 20 &= 10 \qquad \text{Simplify.} \\
2a &= -10 \qquad \text{Subtract 20 from both sides.} \\
a &= -5 \qquad \text{Divide both sides by 2.}
\end{aligned}
$$

Check the solution $(-5, -4)$ in the original equations.   ■

Self Check   Solve $\begin{cases} 2a + 3b = 7 \\ 5a + 2b = 1 \end{cases}$.

Answer $(-1, 3)$

EXAMPLE 4   Solve $\begin{cases} \frac{5}{6}x + \frac{2}{3}y = \frac{7}{6} \\ \frac{10}{7}x - \frac{4}{9}y = \frac{17}{21} \end{cases}$.

Solution   To clear the equations of fractions, we multiply both sides of the first equation by 6 and both sides of the second equation by 63. This gives the system

1. $\begin{cases} 5x + 4y = 7 \\ 90x - 28y = 51 \end{cases}$
2.

We can solve for x by eliminating the terms involving y. To do so, we multiply Equation 1 by 7 and add the result to Equation 2.

$$
\begin{aligned}
35x + 28y &= 49 \\
\underline{90x - 28y} &= \underline{51} \\
125x &= 100 \\
x &= \frac{100}{125} \qquad \text{Divide both sides by 125.} \\
x &= \frac{4}{5} \qquad \text{Simplify.}
\end{aligned}
$$

To solve for y, we substitute $\frac{4}{5}$ for x in Equation 1 and simplify.

$$5x + 4y = 7$$

$$5\left(\frac{4}{5}\right) + 4y = 7$$

$$4 + 4y = 7 \qquad \text{Simplify.}$$

$$4y = 3 \qquad \text{Subtract 4 from both sides.}$$

$$y = \frac{3}{4} \qquad \text{Divide both sides by 4.}$$

Check the solution of $\left(\frac{4}{5}, \frac{3}{4}\right)$ in the original equations. ■

Self Check   Solve $\begin{cases} \frac{1}{3}x + \frac{1}{6}y = 1 \\ \frac{1}{2}x - \frac{1}{4}y = 0 \end{cases}$.

Answer $\left(\frac{3}{2}, 3\right)$

■ INCONSISTENT SYSTEMS

EXAMPLE 5   Solve $\begin{cases} x - \frac{2y}{3} = \frac{8}{3} \\ -\frac{3x}{2} + y = -6 \end{cases}$.

Solution   We can multiply both sides of the first equation by 3 and both sides of the second equation by 2 to clear the equations of fractions.

$$\begin{cases} 3\left(x - \dfrac{2y}{3}\right) = 3\left(\dfrac{8}{3}\right) \\ 2\left(-\dfrac{3x}{2} + y\right) = 2(-6) \end{cases} \longrightarrow \begin{cases} 3x - 2y = 8 \\ -3x + 2y = -12 \end{cases}$$

We can add the resulting equations to eliminate the term involving x.

$$\begin{array}{r} 3x - 2y = 8 \\ -3x + 2y = -12 \\ \hline 0 = -4 \end{array}$$

Here, the terms involving both x and y drop out, and a false result is obtained. This shows that the equations of the system are independent, but the system itself is inconsistent. This system has no solution. ■

Self Check   Solve $\begin{cases} x - \frac{y}{3} = \frac{10}{3} \\ 3x - y = \frac{5}{2} \end{cases}$.

Answer   no solution

■ DEPENDENT EQUATIONS

EXAMPLE 6

Solve $\begin{cases} \frac{2x-5y}{2} = \frac{19}{2} \\ -0.2x + 0.5y = -1.9 \end{cases}$.

Solution   We can multiply both sides of the first equation by 2 to clear it of fractions and both sides of the second equation by 10 to clear it of decimals.

$$\begin{cases} 2\left(\dfrac{2x-5y}{2}\right) = 2\left(\dfrac{19}{2}\right) \\ 10(-0.2x + 0.5y) = 10(-1.9) \end{cases} \longrightarrow \begin{cases} 2x - 5y = 19 \\ -2x + 5y = -19 \end{cases}$$

We add the resulting equations to get

$$\begin{array}{r} 2x - 5y = 19 \\ -2x + 5y = -19 \\ \hline 0 = 0 \end{array}$$

As in Example 5, both x and y drop out. However, this time a true result is obtained. This shows that the equations are dependent and the system has infinitely many solutions. Any ordered pair that satisfies one equation satisfies the other also. Some solutions are $(2, -3)$, $(12, 1)$, and $\left(0, -\frac{19}{5}\right)$.   ■

Self Check

Solve $\begin{cases} \frac{3x+y}{6} = \frac{1}{3} \\ -0.3x - 0.1y = -0.2 \end{cases}$.

Answer   infinitely many solutions

Orals   *Use addition to solve each system for x.*

1. $\begin{cases} x + y = 1 \\ x - y = 1 \end{cases}$ **2.** $\begin{cases} 2x + y = 4 \\ x - y = 2 \end{cases}$

Use addition to solve each system for y.

3. $\begin{cases} -x + y = 3 \\ x + y = 3 \end{cases}$ **4.** $\begin{cases} x + 2y = 4 \\ -x - y = 1 \end{cases}$

EXERCISE 3.5

REVIEW   *Solve each equation or inequality. For each inequality, give the answer in interval notation and graph the interval.*

1. $8(3x - 5) - 12 = 4(2x + 3)$

2. $5x - 13 = x - 1$

3. $x - 2 = \dfrac{x+2}{3}$

4. $\dfrac{3}{2}(y + 4) = \dfrac{20 - y}{2}$

5. $7x - 9 \le 5$

6. $-2x + 6 > 16$

VOCABULARY AND CONCEPTS *Fill in each blank to make a true statement.*

7. The numerical _____ of $-3x$ is -3.

8. The _____ of 4 is -4.

9. $Ax + By = C$ is the _____ form of the equation of a line.

10. When adding the equations

$$5x - 6y = 10$$
$$-3x + 6y = 24$$

the variable y will be _____.

11. To clear the equation $\frac{2}{3}x + 4y = -\frac{4}{5}$ of fractions, we must multiply both sides by ___.

12. To solve the system

$$\begin{cases} 3x + 12y = 4 \\ 6x - 4y = 8 \end{cases}$$

we would multiply the first equation by ___ and add to eliminate the x.

PRACTICE *In Exercises 13–24, use the addition method to solve each system.*

13. $\begin{cases} x + y = 5 \\ x - y = -3 \end{cases}$

14. $\begin{cases} x - y = 1 \\ x + y = 7 \end{cases}$

15. $\begin{cases} x - y = -5 \\ x + y = 1 \end{cases}$

16. $\begin{cases} x + y = 1 \\ x - y = 5 \end{cases}$

17. $\begin{cases} 2x + y = -1 \\ -2x + y = 3 \end{cases}$

18. $\begin{cases} 3x + y = -6 \\ x - y = -2 \end{cases}$

19. $\begin{cases} 2x - 3y = -11 \\ 3x + 3y = 21 \end{cases}$

20. $\begin{cases} 3x - 2y = 16 \\ -3x + 8y = -10 \end{cases}$

21. $\begin{cases} 2x + y = -2 \\ -2x - 3y = -6 \end{cases}$

22. $\begin{cases} 3x + 4y = 8 \\ 5x - 4y = 24 \end{cases}$

23. $\begin{cases} 4x + 3y = 24 \\ 4x - 3y = -24 \end{cases}$

24. $\begin{cases} 5x - 4y = 8 \\ -5x - 4y = 8 \end{cases}$

In Exercises 25–54, use the addition method to solve each system of equations. If the equations of a system are dependent or if a system is inconsistent, so indicate.

25. $\begin{cases} x + y = 5 \\ x + 2y = 8 \end{cases}$

26. $\begin{cases} x + 2y = 0 \\ x - y = -3 \end{cases}$

27. $\begin{cases} 2x + y = 4 \\ 2x + 3y = 0 \end{cases}$

28. $\begin{cases} 2x + 5y = -13 \\ 2x - 3y = -5 \end{cases}$

29. $\begin{cases} 3x + 29 = 5y \\ 4y - 34 = -3x \end{cases}$

30. $\begin{cases} 3x - 16 = 5y \\ 33 - 5y = 4x \end{cases}$

31. $\begin{cases} 2x = 3(y - 2) \\ 2(x + 4) = 3y \end{cases}$

32. $\begin{cases} 3(x - 2) = 4y \\ 2(2y + 3) = 3x \end{cases}$

33. $\begin{cases} -2(x + 1) = 3(y - 2) \\ 3(y + 2) = 6 - 2(x - 2) \end{cases}$

34. $\begin{cases} 5(x - 1) = 8 - 3(y + 2) \\ 4(x + 2) - 7 = 3(2 - y) \end{cases}$

35. $\begin{cases} 4(x + 1) = 17 - 3(y - 1) \\ 2(x + 2) + 3(y - 1) = 9 \end{cases}$

36. $\begin{cases} 3(x + 3) + 2(y - 4) = 5 \\ 3(x - 1) = -2(y + 2) \end{cases}$

37. $\begin{cases} 2x + y = 10 \\ x + 2y = 10 \end{cases}$

38. $\begin{cases} 3x + 2y = 0 \\ 2x - 3y = -13 \end{cases}$

39. $\begin{cases} 2x - y = 16 \\ 3x + 2y = 3 \end{cases}$

40. $\begin{cases} 3x + 4y = -17 \\ 4x - 3y = -6 \end{cases}$

41. $\begin{cases} 4x + 5y = -20 \\ 5x - 4y = -25 \end{cases}$

42. $\begin{cases} 3x - 5y = 4 \\ 7x + 3y = 68 \end{cases}$

43. $\begin{cases} 6x = -3y \\ 5y = 2x + 12 \end{cases}$

44. $\begin{cases} 3y = 4x \\ 5x = 4y - 2 \end{cases}$

45. $\begin{cases} 4(2x - y) = 18 \\ 3(x - 3) = 2y - 1 \end{cases}$

46. $\begin{cases} 2(2x + 3y) = 5 \\ 8x = 3(1 + 3y) \end{cases}$

47. $\begin{cases} \dfrac{3}{5}x + \dfrac{4}{5}y = 1 \\ -\dfrac{1}{4}x + \dfrac{3}{8}y = 1 \end{cases}$

48. $\begin{cases} \dfrac{1}{2}x - \dfrac{1}{4}y = 1 \\ \dfrac{1}{3}x + y = 3 \end{cases}$

49. $\begin{cases} \dfrac{3}{5}x + y = 1 \\ \dfrac{4}{5}x - y = -1 \end{cases}$

50. $\begin{cases} \dfrac{1}{2}x + \dfrac{4}{7}y = -1 \\ 5x - \dfrac{4}{5}y = -10 \end{cases}$

51. $\begin{cases} \dfrac{x}{2} - \dfrac{y}{3} = -2 \\ \dfrac{2x - 3}{2} + \dfrac{6y + 1}{3} = \dfrac{17}{6} \end{cases}$

52. $\begin{cases} \dfrac{x + 2}{4} + \dfrac{y - 1}{3} = \dfrac{1}{12} \\ \dfrac{x + 4}{5} - \dfrac{y - 2}{2} = \dfrac{5}{2} \end{cases}$

53. $\begin{cases} \dfrac{x - 3}{2} + \dfrac{y + 5}{3} = \dfrac{11}{6} \\ \dfrac{x + 3}{3} - \dfrac{5}{12} = \dfrac{y + 3}{4} \end{cases}$

54. $\begin{cases} \dfrac{x + 2}{3} = \dfrac{3 - y}{2} \\ \dfrac{x + 3}{2} = \dfrac{2 - y}{3} \end{cases}$

WRITING

55. Why is it usually to your advantage to write the equations of a system in general form before using the addition method to solve it?

56. How would you decide whether to use substitution or addition to solve a system of equations?

SOMETHING TO THINK ABOUT

57. If possible, find a solution to the system
$$\begin{cases} x + y = 5 \\ x - y = -3 \\ 2x - y = -2 \end{cases}$$

58. If possible, find a solution to the system
$$\begin{cases} x + y = 5 \\ x - y = -3 \\ x - 2y = 0 \end{cases}$$

3.6 Applications of Systems of Equations

■ SOLVING PROBLEMS WITH TWO VARIABLES

Getting Ready   *In Problems 1–4, let x and y represent two numbers. Use an algebraic expression to denote each phrase.*

1. The sum of x and y

2. The difference when y is subtracted from x

3. The product of x and y

4. The quotient x divided by y

5. Give the formula for the area of a rectangle.

6. Give the formula for the perimeter of a rectangle.

■ SOLVING PROBLEMS WITH TWO VARIABLES

We have previously set up equations involving one variable to solve problems. In this section, we consider ways to solve problems by using equations in two variables. The following steps are helpful when solving problems involving two unknown quantities.

Problem-Solving Strategy

1. Read the problem several times and *analyze* the facts. Occasionally, a sketch, chart, or diagram will help you visualize the facts of the problem.

2. Pick different variables to represent two unknown quantities. *Form two equations* involving each of the two variables. This will give a system of two equations in two variables.

3. *Solve the system* using the most convenient method: graphing, substitution, or addition.

4. *State the conclusion.*

5. *Check the solution* in the words of the problem.

EXAMPLE 1

Farming A farmer raises wheat and soybeans on 215 acres. If he wants to plant 31 more acres in wheat than in soybeans, how many acres of each should he plant?

Analyze the problem

The farmer plants two fields, one in wheat and one in soybeans. We know that the number of acres of wheat planted plus the number of acres of soybeans planted will equal a total of 215 acres.

Form two equations

If w represents the number of acres of wheat and s represents the number of acres of soybeans to be planted, we can form the two equations

The number of acres planted in wheat	+	the number of acres planted in soybeans	is	215 acres.
w	+	s	=	215

Since the farmer wants to plant 31 more acres in wheat than in soybeans, we have

The number of acres planted in wheat	−	the number of acres planted in soybeans	is	31 acres.
w	−	s	=	31

Solve the system   We can now solve the system

1. $\begin{cases} w + s = 215 \\ w - s = 31 \end{cases}$
2.

by the addition method.

$$
\begin{array}{r}
w + s = 215 \\
\underline{w - s = 31} \\
2w = 246 \\
w = 123 \qquad \text{Divide both sides by 2.}
\end{array}
$$

To find s, we substitute 123 for w in Equation 1.

$$
\begin{aligned}
w + s &= 215 \\
\mathbf{123} + s &= 215 \qquad \text{Substitute 123 for } w. \\
s &= 92 \qquad \text{Subtract 123 from both sides.}
\end{aligned}
$$

State the conclusion   The farmer should plant 123 acres of wheat and 92 acres of soybeans.

Check the result   The total acreage planted is $123 + 92$, or 215 acres. The area planted in wheat is 31 acres greater than that planted in soybeans, because $123 - 92 = 31$. The answers check. ■

EXAMPLE 2   **Lawn care**   An installer of underground irrigation systems wants to cut a 20-foot length of plastic tubing into two pieces. The longer piece is to be 2 feet longer than twice the shorter piece. Find the length of each piece.

Analyze the problem   Refer to Figure 3-27, which shows the pipe.

Form two equations   We can let s represent the length of the shorter piece and l represent the length of the longer piece. Then we can form the equations

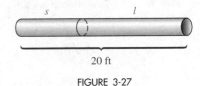

FIGURE 3-27

The length of the shorter piece	+	the length of the longer piece	is	20 feet.
s	+	l	=	20

Since the longer piece is 2 feet longer than twice the shorter piece, we have

The length of the longer piece	is	2	·	the length of the shorter piece	+	2 feet.
l	=	2	·	s	+	2

Solve the system   We can use the substitution method to solve the system

1. $\begin{cases} s + l = 20 \\ l = 2s + 2 \end{cases}$
2.

$$s + (2s + 2) = 20 \qquad \text{Substitute } 2s + 2 \text{ for } l \text{ in Equation 1.}$$
$$3s + 2 = 20 \qquad \text{Combine like terms.}$$
$$3s = 18 \qquad \text{Subtract 2 from both sides.}$$
$$s = 6 \qquad \text{Divide both sides by 3.}$$

State the conclusion   The shorter piece should be 6 feet long. To find the length of the longer piece, we substitute 6 for s in Equation 1 and solve for l.

$$s + l = 20$$
$$6 + l = 20 \qquad \text{Substitute 6 for } s.$$
$$l = 14 \qquad \text{Subtract 6 from both sides.}$$

The longer piece should be 14 feet long.

Check the result   The sum of 6 and 14 is 20. 14 is 2 more than twice 6. The answers check.   ∎

EXAMPLE 3

Gardening   Tom has 150 feet of fencing to enclose a rectangular garden. If the length is to be 5 feet less than 3 times the width, find the area of the garden.

Analyze the problem   To find the area of a rectangle, we need to know its length and width. See Figure 3-28.

Form two equations   We can let l represent the length of the garden and w represent the width. Since the perimeter of a rectangle is two lengths plus two widths, we can form the equations

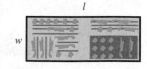

FIGURE 3-28

2	·	the length of the garden	+	2	·	the width of the garden	is	150 feet.
2	·	l	+	2	·	w	=	150

Since the length is 5 feet less than 3 times the width,

The length of the garden	is	3	·	the width of the garden	−	5 feet.
l	=	3	·	w	−	5

Solve the system   We can use the substitution method to solve this system.

1. $\begin{cases} 2l + 2w = 150 \\ l = 3w - 5 \end{cases}$
2.

$$2(3w - 5) + 2w = 150 \qquad \text{Substitute } 3w - 5 \text{ for } l \text{ in Equation 1.}$$
$$6w - 10 + 2w = 150 \qquad \text{Remove parentheses.}$$
$$8w - 10 = 150 \qquad \text{Combine like terms.}$$
$$8w = 160 \qquad \text{Add 10 to both sides.}$$
$$w = 20 \qquad \text{Divide both sides by 8.}$$

The width of the garden is 20 feet. To find the length, we substitute 20 for w in Equation 2 and simplify.

$$l = 3w - 5$$
$$= 3(\mathbf{20}) - 5 \qquad \text{Substitute 20 for } w.$$
$$= 60 - 5$$
$$= 55$$

Since the dimensions of the rectangle are 55 feet by 20 feet, and the area of a rectangle is given by the formula

$$A = l \cdot w \qquad \text{Area = length times width.}$$

we have

$$A = \mathbf{55} \cdot \mathbf{20}$$
$$= 1,100$$

State the conclusion   The garden covers an area of 1,100 square feet.

Check the result   Because the dimensions of the garden are 55 feet by 20 feet, the perimeter is

$$P = 2l + 2w$$
$$= 2(\mathbf{55}) + 2(\mathbf{20}) \qquad \text{Substitute for } l \text{ and } w.$$
$$= 110 + 40$$
$$= 150$$

It is also true that 55 feet is 5 feet less than 3 times 20 feet. The answers check.   ∎

EXAMPLE 4   **Manufacturing**   The setup cost of a machine that mills brass plates is $750. After setup, it costs $0.25 to mill each plate. Management is considering the purchase of a larger machine that can produce the same plate at a cost of $0.20 per plate. If the setup cost of the larger machine is $1,200, how many plates would the company have to produce to make the purchase worthwhile?

Analyze the problem   We begin by finding the number of plates (called the **break point**) that will cost equal amounts to produce on either machine.

Form two equations   We can let c represent the cost of milling p plates. If we call the machine currently being used machine 1, and the new one machine 2, we can form the two equations

The cost of making p plates on machine 1	is	the startup cost of machine 1	+	the cost per plate on machine 1	·	the number of plates p to be made.
c	$=$	750	+	0.25	·	p

The cost of making p plates on machine 2	is	the startup cost of machine 2	+	the cost per plate on machine 2	·	the number of plates p to be made.
c	$=$	1,200	+	0.20	·	p

Solve the system   Since the costs at the break point are equal, we can use the substitution method to solve the system

$$\begin{cases} c = \mathbf{750 + 0.25p} \\ c = 1{,}200 + 0.20p \end{cases}$$

$750 + 0.25p = 1{,}200 + 0.20p$	Substitute $750 + 0.25p$ for c in the second equation.
$0.25p = 450 + 0.20p$	Subtract 750 from both sides.
$0.05p = 450$	Subtract $0.20p$ from both sides.
$p = 9{,}000$	Divide both sides by 0.05.

State the conclusion   If 9,000 plates are milled, the cost will be the same on either machine. If more than 9,000 plates are milled, the cost will be cheaper on the newer machine, because it mills the plates less expensively than the smaller machine.

Check the solution   Figure 3-29 verifies that the break point is 9,000 plates. It also interprets the solution graphically. ∎

EXAMPLE 5   **Investing**   Terri and Juan earned $1,150 from a one-year investment of $15,000. If Terri invested some of the money at 8% interest and Juan invested the rest at 7%, how much did each invest?

Analyze the problem   We are told that Terri invested an unknown part of the $15,000 at 8% and Juan invested the rest at 7%. Together, these investments earned $1,150.

Form two equations   We can let x represent the amount invested by Terri and y represent the amount of money invested by Juan. Because the total investment is $15,000, we have

The amount invested by Terri	+	the amount invested by Juan	is	$15,000.
x	+	y	$=$	15,000

Current machine		New, larger machine	
$c = 750 + 0.25p$		$c = 1,200 + 0.20p$	
p	c	p	c
0	750	0	1,200
1,000	1,000	4,000	2,000
5,000	2,000	12,000	3,600

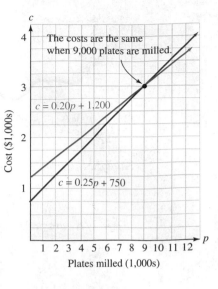

FIGURE 3-29

Since the income on x dollars invested at 8% is $0.08x$, the income on y dollars invested at 7% is $0.07y$, and the combined income is $1,150, we have

The income on the 8% investment	+	the income on the 7% investment	is	$1,150.
$0.08x$	+	$0.07y$	=	1,150

Thus, we have the system

1. $\begin{cases} x + y = 15,000 \\ 0.08x + 0.07y = 1,150 \end{cases}$
2.

Solve the system To solve the system, we use the addition method.

$$
\begin{array}{ll}
-8x - 8y = -120,000 & \text{Multiply both sides of Equation 1 by } -8. \\
\underline{8x + 7y = 115,000} & \text{Multiply both sides of Equation 2 by 100.} \\
-y = -5,000 & \text{Add the equations together.} \\
y = 5,000 & \text{Multiply both sides by } -1.
\end{array}
$$

To find x, we substitute 5,000 for y in Equation 1 and simplify.

$$
\begin{array}{ll}
x + y = 15,000 & \\
x + \mathbf{5,000} = 15,000 & \text{Substitute 5,000 for } y. \\
x = 10,000 & \text{Subtract 5,000 from both sides.}
\end{array}
$$

State the conclusion Terri invested $10,000, and Juan invested $5,000.

Check the result

$$\$10{,}000 + \$5{,}000 = \$15{,}000$$ The two investments total $15,000.

$$0.08(\$10{,}000) = \$800$$ Terri earned $800.

$$0.07(\$5{,}000) = \$350$$ Juan earned $350.

The combined interest is $800 + $350 = $1,150. The answers check.   ■

EXAMPLE 6   **Boating**   A boat traveled 30 kilometers downstream in 3 hours and made the return trip in 5 hours. Find the speed of the boat in still water.

Analyze the problem   Traveling downstream, the speed of the boat will be faster than it would be in still water. Traveling upstream, the speed of the boat will be less than it would be in still water.

Form two equations   We can let s represent the speed of the boat in still water and let c represent the speed of the current. Then the rate of speed of the boat while going downstream is $s + c$. The rate of the boat while going upstream is $s - c$. We can organize the information of the problem as in Figure 3-30.

	Distance	=	Rate	·	Time
Downstream	30		$s + c$		3
Upstream	30		$s - c$		5

FIGURE 3-30

Because $d = r \cdot t$, the information in the table gives two equations in two variables.

$$\begin{cases} 30 = 3(s + c) \\ 30 = 5(s - c) \end{cases}$$

After removing parentheses and rearranging terms, we have

1. $\begin{cases} 3s + 3c = 30 \\ 5s - 5c = 30 \end{cases}$
2.

Solve the system   To solve this system by addition, we multiply Equation 1 by 5, Equation 2 by 3, add the equations, and solve for s.

$$15s + 15c = 150$$
$$\underline{15s - 15c = 90}$$
$$30s = 240$$
$$s = 8 \qquad \text{Divide both sides by 30.}$$

State the conclusion   The speed of the boat in still water is 8 kilometers per hour.

Check the result   We leave the check to the reader.   ■

EXAMPLE 7   **Medical technology**   A laboratory technician has one batch of antiseptic that is 40% alcohol and a second batch that is 60% alcohol. She would like to make 8 liters of solution that is 55% alcohol. How many liters of each batch should she use?

Analyze the problem   Some 60%-alcohol solution must be added to some 40%-alcohol solution to make a 55%-alcohol solution.

Form two equations   We can let x represent the number of liters to be used from batch 1, let y represent the number of liters to be used from batch 2, and organize the information of the problem as in Figure 3-31.

	Fractional part that is alcohol	·	Number of liters of solution	=	Number of liters of alcohol
Batch 1	0.40		x		$0.40x$
Batch 2	0.60		y		$0.60y$
Mixture	0.55		8		$0.55(8)$

FIGURE 3-31

The information in Figure 3-31 provides two equations.

1. $x + y = 8$ The number of liters of batch 1 plus the number of liters of batch 2 equals the total number of liters in the mixture.

2. $0.40x + 0.60y = 0.55(8)$ The amount of alcohol in batch 1 plus the amount of alcohol in batch 2 equals the amount of alcohol in the mixture.

Solve the system   We can use addition to solve this system.

$$
\begin{array}{ll}
-40x - 40y = -320 & \text{Multiply both sides of Equation 1 by } -40. \\
\underline{40x + 60y = 440} & \text{Multiply both sides of Equation 2 by } 100. \\
20y = 120 & \\
y = 6 & \text{Divide both sides by 20.}
\end{array}
$$

To find x, we substitute 6 for y in Equation 1 and simplify:

$$
\begin{array}{ll}
x + y = 8 & \\
x + 6 = 8 & \text{Substitute 6 for } y. \\
x = 2 & \text{Subtract 6 from both sides.}
\end{array}
$$

State the conclusion   The technician should use 2 liters of the 40% solution and 6 liters of the 60% solution.

Check the result   The check is left to the reader.   ■

Orals   *If x and y are integers, express each quantity.*

1. Twice x **2.** One more than y

3. The sum of twice x and three times y

If a book costs $x and a calculator cost $y, find

4. The cost of 3 books and 2 calculators

5. The cost of 4 books and 5 calculators

EXERCISE 3.6

REVIEW   *In Exercises 1–4, graph each inequality.*

1. $x < 4$ **2.** $x \geq -3$ **3.** $-1 < x \leq 2$ **4.** $-2 \leq x \leq 0$

In Exercises 5–8, write each product using exponents.

5. $8 \cdot 8 \cdot 8 \cdot c$ **6.** $4(\pi)(r)(r)$ **7.** $a \cdot a \cdot b \cdot b$ **8.** $(-2)(-2)$

VOCABULARY AND CONCEPTS   *Fill in each blank to make a true statement.*

9. A _____ is a letter that stands for a number.

10. An _____ is a statement indicating that two quantities are equal.

11. $\begin{cases} a + b = 20 \\ a = 2b + 4 \end{cases}$ is a _____ of linear equations.

12. A _____ of a system of two linear equations satisfies both equations simultaneously.

PRACTICE   *In Exercises 13–16, use two equations in two variables to solve each problem.*

13. Integer problem   One integer is twice another, and their sum is 96. Find the integers.

14. Integer problem   The sum of two integers is 38, and their difference is 12. Find the integers.

15. Integer problem   Three times one integer plus another integer is 29. If the first integer plus twice the second is 18, find the integers.

16. Integer problem   Twice one integer plus another integer is 21. If the first integer plus 3 times the second is 33, find the integers.

APPLICATIONS   *In Exercises 17–54, use two equations in two variables to solve each problem.*

17. Raising livestock   A rancher raises five times as many cows as horses. If he has 168 animals, how many cows does he have?

18. Grass seed mixture   A landscaper used 100 pounds of grass seed containing twice as much bluegrass as rye. He added 15 more pounds of bluegrass to the mixture before seeding a lawn. How many pounds of bluegrass did he use?

19. Buying painting supplies Two partial receipts for paint supplies appear in Illustration 1. How much did each gallon of paint and each brush cost?

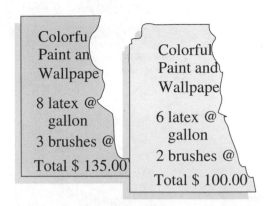

Colorfu
Paint an
Wallpape

8 latex @
gallon
3 brushes @
Total $ 135.00

Colorful
Paint and
Wallpape

6 latex @
gallon
2 brushes @
Total $ 100.00

ILLUSTRATION 1

20. Buying baseball equipment One catcher's mitt and ten outfielder's gloves cost $239.50. How much does each cost if one catcher's mitt and five outfielder's gloves cost $134.50?

21. Buying contact lens cleaner Two bottles of contact lens cleaner and three bottles of soaking solution cost $29.40, and three bottles of cleaner and two bottles of soaking solution cost $28.60. Find the cost of each.

22. Buying clothes Two pairs of shoes and four pairs of socks cost $109, and three pairs of shoes and five pairs of socks cost $160. Find the cost of a pair of socks.

23. Cutting pipe A plumber wants to cut the pipe shown in Illustration 2 into two pieces so that one piece is 5 feet longer than the other. How long should each piece be?

25 ft

ILLUSTRATION 2

24. Cutting lumber A carpenter wants to cut a 20-foot board into two pieces so that one piece is 4 times as long as the other. How long should each piece be?

25. Splitting the lottery Maria and Susan pool their resources to buy several lottery tickets. They win $250,000! They agree that Susan should get $50,000 more than Maria, because she gave most of the money. How much will Maria get?

26. Figuring inheritances In his will, a man left his older son $10,000 more than twice as much as he left his younger son. If the estate is worth $497,500, how much did the younger son get?

27. Television programming The producer of a 30-minute documentary about World War I divided it into two parts. Four times as much program time was devoted to the causes of the war as to the outcome. How long was each part of the documentary?

28. Government The salaries of the President and Vice President of the United States total $371,500 a year. If the President makes $28,500 more than the Vice President, find each of their salaries.

29. Causes of death In 1993, the number of Americans dying from cancer was six times the number that died from accidents. If the number of deaths from these two causes totaled 630,000, how many Americans died from each cause?

30. At the movies At an IMAX theater, the giant rectangular movie screen has a width 26 feet less than its length. If its perimeter is 332 feet, find the area of the screen.

31. Geometry The perimeter of the rectangle shown in Illustration 3 is 110 feet. Find its dimensions.

w

$l = w + 5$

ILLUSTRATION 3

32. Geometry A rectangle is 3 times as long as it is wide, and its perimeter is 80 centimeters. Find its dimensions.

33. Geometry The length of a rectangle is 2 feet more than twice its width. If its perimeter is 34 feet, find its area.

34. Geometry A 50-meter path surrounds the rectangular garden shown in Illustration 4. The width of the garden is two-thirds its length. Find its area.

ILLUSTRATION 4

35. Choosing a furnace A high-efficiency 90+ furnace costs $2,250 and costs an average of $412 per year to operate in Rockford, IL. An 80+ furnace costs only $1,715 but costs $466 per year to operate. Find the break point.

36. Making tires A company has two molds to form tires. One mold has a setup cost of $600 and the other a setup cost of $1,100. The cost to make each tire on the first machine is $15, and the cost per tire on the second machine is $13. Find the break point.

37. Choosing a furnace See Exercise 35. If you intended to live in a house for seven years, which furnace would you choose?

38. Making tires See Exercise 36. If you planned a production run of 500 tires, which mold would you use?

39. Investing money Bill invested some money at 5% annual interest, and Janette invested some at 7%. If their combined interest was $310 on a total investment of $5,000, how much did Bill invest?

40. Investing money Peter invested some money at 6% annual interest, and Martha invested some at 12%. If their combined investment was $6,000 and their combined interest was $540, how much money did Martha invest?

41. Buying tickets Students can buy tickets to a basketball game for $1. The admission for nonstudents is $2. If 350 tickets are sold and the total receipts are $450, how many student tickets are sold?

42. Buying tickets If receipts for the movie advertised in Illustration 5 were $720 for an audience of 190 people, how many senior citizens attended?

Admissions: $4
Seniors: $3
Showtimes: 7, 9, 11

ILLUSTRATION 5

43. Boating A boat can travel 24 miles downstream in 2 hours and can make the return trip in 3 hours. Find the speed of the boat in still water.

44. Aviation With the wind, a plane can fly 3,000 miles in 5 hours. Against the same wind, the trip takes 6 hours. Find the airspeed of the plane (the speed in still air).

45. Aviation An airplane can fly downwind a distance of 600 miles in 2 hours. However, the return trip against the same wind takes 3 hours. Find the speed of the wind.

46. Finding the speed of a current It takes a motorboat 4 hours to travel 56 miles down a river, and it takes 3 hours longer to make the return trip. Find the speed of the current.

47. Mixing chemicals A chemist has one solution that is 40% alcohol and another that is 55% alcohol. How much of each must she use to make 15 liters of a solution that is 50% alcohol?

48. Mixing pharmaceuticals A nurse has a solution that is 25% alcohol and another that is 50% alcohol. How much of each must he use to make 20 liters of a solution that is 40% alcohol?

49. Mixing nuts A merchant wants to mix the peanuts with the cashews shown in Illustration 6 to get 48 pounds of mixed nuts to sell at $4 per pound. How many pounds of each should the merchant use?

ILLUSTRATION 6

50. Mixing peanuts and candy A merchant wants to mix peanuts worth $3 per pound with jelly beans worth $1.50 per pound to make 30 pounds of a mixture worth $2.10 per pound. How many pounds of each should he use?

51. Selling radios An electronics store put two types of car radios on sale. One model sold for $87, and the other sold for $119. During the sale, the receipts for the 25 radios sold were $2,495. How many of the less expensive radios were sold?

52. Selling ice cream At a store, ice cream cones cost $.90 and sundaes cost $1.65. One day, the receipts for a total of 148 cones and sundaes were $180.45. How many cones were sold?

53. Investing money An investment of $950 at one rate of interest and $1,200 at a higher rate together generate an annual income of $205.50. If the investment rates differ by 1%, find the lower rate. (*Hint:* Treat 1% as .01.)

54. Selecting radio frequencies In a radio, an inductor and a capacitor are used in a resonant circuit to select a wanted radio station at a frequency f and reject all others. The inductance L and the capacitance C determine the inductive reactance X_L and the capacitive reactance X_C of that circuit, where

$$X_L = 2\pi fL \quad \text{and} \quad X_C = \frac{1}{2\pi fC}$$

The radio station selected will be at the frequency f where $X_L = X_C$. Write a formula for f^2 in terms of L and C.

WRITING

55. Which problem in the preceding set did you find the hardest? Why?

56. Which problem in the preceding set did you find the easiest? Why?

SOMETHING TO THINK ABOUT

57. How many nails will balance one nut in Illustration 7?

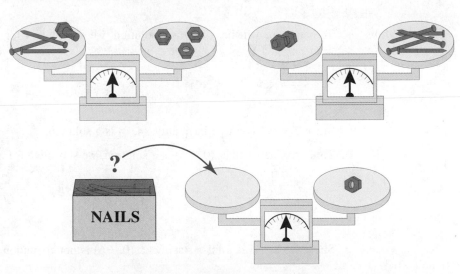

ILLUSTRATION 7

3.7   Systems of Linear Inequalities

■ SOLUTIONS OF LINEAR INEQUALITIES ■ GRAPHING LINEAR INEQUALITIES ■ AN APPLICATION OF
LINEAR INEQUALITIES ■ SOLVING SYSTEMS OF LINEAR INEQUALITIES ■ AN APPLICATION OF
SYSTEMS OF LINEAR INEQUALITIES

Getting Ready   *Graph $y = \frac{1}{3}x + 3$ and tell whether the given point lies on the line, above the line, or below the line.*

1. $(0, 0)$ **2.** $(0, 4)$ **3.** $(2, 2)$ **4.** $(6, 5)$

5. $(-3, 2)$ **6.** $(6, 8)$ **7.** $(-6, 0)$ **8.** $(-9, 5)$

■ SOLUTIONS OF LINEAR INEQUALITIES

A **linear inequality** in x and y is an inequality that can be written in one of the following forms:

$$Ax + By > C \qquad Ax + By < C \qquad Ax + By \geq C \qquad Ax + By \leq C$$

where A, B, and C are real numbers and A and B are not both 0. Some examples of linear inequalities are

$$2x - y > -3 \qquad y < 3 \qquad x + 4y \geq 6 \qquad x \leq -2$$

An ordered pair (x, y) is a solution of an inequality in x and y if a true statement results when the values of x and y are substituted into the inequality.

EXAMPLE 1   Determine whether each ordered pair is a solution of $y \geq x - 5$:   **a.** $(4, 2)$ and **b.** $(0, -6)$.

Solution   **a.** To determine whether $(4, 2)$ is a solution, we substitute 4 for x and 2 for y.

$$y \geq x - 5$$
$$2 \geq 4 - 5$$
$$2 \geq -1$$

Since $2 \geq -1$ is a true inequality, $(4, 2)$ is a solution.

b. To determine whether $(0, -6)$ is a solution, we substitute 0 for x and -6 for y.

$$y \geq x - 5$$
$$-6 \geq 0 - 5$$
$$-6 \geq -5$$

Since $-6 \geq -5$ is a false statement, $(0, -6)$ is not a solution.   ■

■ GRAPHING LINEAR INEQUALITIES

The graph of $y = x - 5$ is a line consisting of the points whose coordinates satisfy the equation. The graph of the inequality $y \geq x - 5$ is not a line but rather an area bounded by a line, called a **half-plane.** The half-plane consists of the points whose coordinates satisfy the inequality.

EXAMPLE 2   Graph the inequality $y \geq x - 5$.

Solution   Since $y \geq x - 5$ means that $y = x - 5$ or $y > x - 5$, we begin by graphing the equation $y = x - 5$. See Figure 3-32(a).

Because the graph of $y \geq x - 5$ also indicates that y can be greater than $x - 5$, the coordinates of points other than those shown in Figure 3-32(a) satisfy the inequality. For example, the coordinates of the origin satisfy the inequality. We can verify this by letting x and y be 0 in the given inequality:

$$y \geq x - 5$$
$$0 \geq 0 - 5 \qquad \text{Substitute 0 for } x \text{ and 0 for } y.$$
$$0 \geq -5$$

Because $0 \geq -5$ is true, the coordinates of the origin satisfy the original inequality. In fact, the coordinates of every point on the same side of the line as the origin satisfy the inequality. The graph of $y \geq x - 5$ is the half-plane that is shaded in Figure 3-32(b). Since the boundary line $y = x - 5$ is included, we draw it with a solid line.

$y = x - 5$

x	y	(x, y)
0	-5	$(0, -5)$
5	0	$(5, 0)$

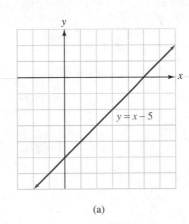

(a)

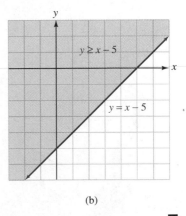

(b)

FIGURE 3-32

Self Check   Graph $y \geq -x - 2$.

Answer

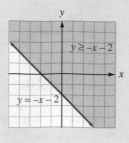

EXAMPLE 3   Graph $x + 2y < 6$.

Solution   We find the boundary by graphing the equation $x + 2y = 6$. Since the symbol $<$ does not include an $=$ sign, the points on the graph of $x + 2y = 6$ will not be a part of the graph. To show this, we draw the boundary line as a broken line. See Figure 3-33.

To determine which half-plane to shade, we substitute the coordinates of some point that lies on one side of the boundary line into $x + 2y < 6$. The origin is a convenient choice.

$$x + 2y < 6$$
$$\mathbf{0} + 2(\mathbf{0}) < 6 \qquad \text{Substitute 0 for } x \text{ and 0 for } y.$$
$$0 < 6$$

Since $0 < 6$ is true, we shade the side of the line that includes the origin. The graph is shown in Figure 3-33.

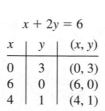

$x + 2y = 6$

x	y	(x, y)
0	3	(0, 3)
6	0	(6, 0)
4	1	(4, 1)

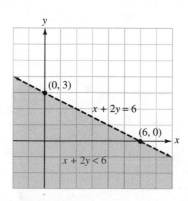

FIGURE 3-33

Self Check

Graph $2x - y < 4$.

Answer

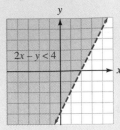

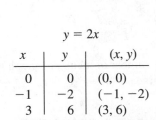

EXAMPLE 4

Graph $y > 2x$.

Solution

To find the boundary line, we graph the equation $y = 2x$. Since the symbol $>$ does not include an equal sign, the points on the graph of $y = 2x$ are not a part of the graph of $y > 2x$. To show this, we draw the boundary line as a broken line. See Figure 3-34(a).

To determine which half-plane to shade, we substitute the coordinates of some point that lies on one side of the boundary line into $y > 2x$. Point $T(2, 0)$, for example, is below the boundary line. See Figure 3-34(a). To see if point $T(2, 0)$ satisfies $y > 2x$, we substitute 2 for x and 0 for y in the inequality.

$$y > 2x$$
$$0 > 2(2) \qquad \text{Substitute 2 for } x \text{ and 0 for } y.$$
$$0 > 4$$

Since $0 > 4$ is false, the coordinates of point T do not satisfy the inequality, and point T is not on the side of the line we wish to shade. Instead, we shade the other side of the boundary line. The graph of the solution set of $y > 2x$ is shown in Figure 3-34(b).

$y = 2x$

x	y	(x, y)
0	0	$(0, 0)$
-1	-2	$(-1, -2)$
3	6	$(3, 6)$

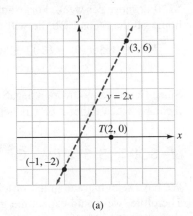

(a)

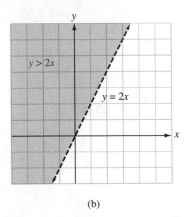

(b)

FIGURE 3-34

Self Check

Answer

Graph $y < 3x$.

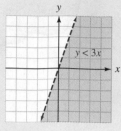

■ AN APPLICATION OF LINEAR INEQUALITIES

EXAMPLE 5

Earning money Carlos has two part-time jobs, one paying $5 per hour and the other paying $6 per hour. He must earn at least $120 per week to pay his expenses while attending college. Write an inequality that shows the various ways he can schedule his time to achieve his goal.

Solution If we let x represent the number of hours he works on the first job and y the number of hours he works on the second job, we have

The hourly rate on the first job	·	the hours worked on the first job	+	the hourly rate on the second job	·	the hours worked on the second job	is at least	$120.
$5	·	x	+	$6	·	y	$\geq$	$120

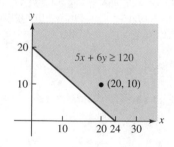

FIGURE 3-35

The graph of the inequality $5x + 6y \geq 120$ is shown in Figure 3-35. Any point in the shaded region indicates a possible way he can schedule his time and earn $120 or more per week. For example, if he works 20 hours on the first job and 10 hours on the second job, he will earn

$$\$5(20) + \$6(10) = \$100 + \$60$$
$$= \$160$$

Since Carlos cannot work a negative number of hours, the graph in the figure has no meaning when either x or y is negative. ■

■ SOLVING SYSTEMS OF LINEAR INEQUALITIES

We have seen that the graph of a linear inequality in two variables is a half-plane. Therefore, we would expect the graph of a system of two linear inequalities to be two overlapping half-planes. For example, to solve the system

$$\begin{cases} x + y \geq 1 \\ x - y \geq 1 \end{cases}$$

we graph each inequality and then superimpose the graphs on one set of coordinate axes.

The graph of $x + y \geq 1$ includes the graph of the equation $x + y = 1$ and all points above it. Because the boundary line is included, we draw it with a solid line. See Figure 3-36(a).

The graph of $x - y \geq 1$ includes the graph of the equation $x - y = 1$ and all points below it. Because the boundary line is included, we draw it with a solid line. See Figure 3-36(b).

$x + y = 1$

x	y	(x, y)
0	1	$(0, 1)$
1	0	$(1, 0)$
2	-1	$(2, -1)$

$x - y = 1$

x	y	(x, y)
0	-1	$(0, -1)$
1	0	$(1, 0)$
2	1	$(2, 1)$

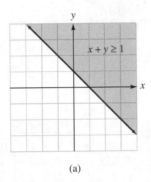

(a)

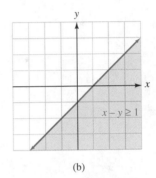

(b)

FIGURE 3-36

In Figure 3-37, we show the result when the graphs are superimposed on one coordinate system. The area that is shaded twice represents the set of solutions of the given system. Any point in the doubly shaded region has coordinates that satisfy both of the inequalities.

To see that this is true, we can pick a point, such as point A, that lies in the doubly shaded region and show that its coordinates satisfy both inequalities. Because point A has coordinates $(4, 1)$, we have

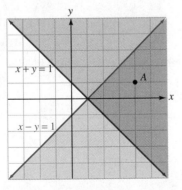

FIGURE 3-37

$$x + y \geq 1 \qquad \text{and} \qquad x - y \geq 1$$
$$4 + 1 \geq 1 \qquad\qquad 4 - 1 \geq 1$$
$$5 \geq 1 \qquad\qquad\quad 3 \geq 1$$

Since the coordinates of point A satisfy each equation, point A is a solution. If we pick a point that is not in the doubly shaded region, its coordinates will not satisfy both of the inequalities.

In general, to solve systems of linear inequalities, we will take the following steps.

Solving Systems of Inequalities

1. Graph each inequality in the system on the same coordinate axes.
2. Find the region where the graphs overlap.
3. Pick a test point from the region to verify the solution.

EXAMPLE 6   Graph the solution set of $\begin{cases} 2x + y < 4 \\ -2x + y > 2 \end{cases}$.

Solution   We graph each inequality on one set of coordinate axes, as in Figure 3-38.

- The graph of $2x + y < 4$ includes all points below the line $2x + y = 4$. Since the boundary is not included, we draw it as a broken line.
- The graph of $-2x + y > 2$ includes all points above the line $-2x + y = 2$. Since the boundary is not included, we draw it as a broken line.

The area that is shaded twice represents the set of solutions of the given system.

$2x + y = 4$		
x	y	(x, y)
0	4	$(0, 4)$
1	2	$(1, 2)$
2	0	$(2, 0)$

$-2x + y = 2$		
x	y	(x, y)
-1	0	$(-1, 0)$
0	2	$(0, 2)$
2	6	$(2, 6)$

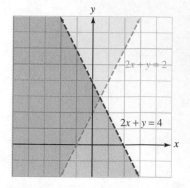

FIGURE 3-38

Pick a point in the doubly shaded region and show that it satisfies both inequalities. ∎

Self Check   Graph the solution of $\begin{cases} x + 3y < 6 \\ -x + 3y < 6 \end{cases}$.

Answer

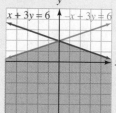

EXAMPLE 7   Graph the solution set of $\begin{cases} x \le 2 \\ y > 3 \end{cases}$.

Solution   We graph each inequality on one set of coordinate axes, as in Figure 3-39.

- The graph of $x \le 2$ includes all points on the line $x = 2$ and all points to the left of the line. Since the boundary line is included, we draw it as a solid line.
- The graph $y > 3$ includes all points above the line $y = 3$. Since the boundary is not included, we draw it as a broken line.

The area that is shaded twice represents the set of solutions of the given system.

$x = 2$

x	y	(x, y)
2	0	$(2, 0)$
2	2	$(2, 2)$
2	4	$(2, 4)$

$y = 3$

x	y	(x, y)
0	3	$(0, 3)$
1	3	$(1, 3)$
4	3	$(4, 3)$

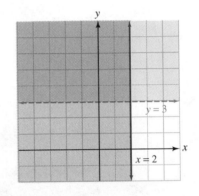

FIGURE 3-39

Pick a point in the doubly shaded region and show that this is true. ■

Self Check   Solve $\begin{cases} y \ge 1 \\ x > 2 \end{cases}$.

Answer

EXAMPLE 8   Graph the solution set of the system $\begin{cases} y < 3x - 1 \\ y \ge 3x + 1 \end{cases}$.

Solution   We graph each inequality, as in Figure 3-40.

- The graph of $y < 3x - 1$ includes all of the points below the broken line $y = 3x - 1$.

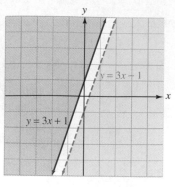

FIGURE 3-40

- The graph of $y \geq 3x + 1$ includes all of the points on and above the solid line $y = 3x + 1$.

Since the graphs of these inequalities do not intersect, there are no solutions. ■

Self Check   Solve $\begin{cases} y \geq -\frac{1}{2}x + 1 \\ y \leq -\frac{1}{2}x - 1 \end{cases}$.

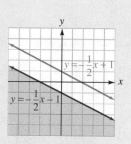

Answer   no solutions

■ AN APPLICATION OF SYSTEMS OF LINEAR INEQUALITIES

EXAMPLE 9   **Landscaping**   A homeowner budgets from $300 to $600 for trees and bushes to landscape his yard. After shopping around, he finds that good trees cost $150 and mature bushes cost $75. What combinations of trees and bushes can he afford to buy?

Analyze the problem   The homeowner wants to spend *at least* $300 but *not more than* $600 for trees and bushes.

Form two inequalities   We can let x represent the number of trees purchased and y the number of bushes purchased. We can then form the following system of inequalities.

The cost of a tree	·	the number of trees purchased	+	the cost of a bush	·	the number of bushes purchased	should be at least	$300.
$150	·	x	+	$75	·	y	≥	$300

The cost of a tree	·	the number of trees purchased	+	the cost of a bush	·	the number of bushes purchased	should not be more than	$600.
$150	·	x	+	$75	·	y	≤	$600

Solve the system   We graph the system

$$\begin{cases} 150x + 75y \geq 300 \\ 150x + 75y \leq 600 \end{cases}$$

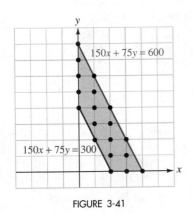

FIGURE 3-41

as in Figure 3-41. The coordinates of each point shown in the graph give a possible combination of the number of trees (x) and the number of bushes (y) that can be purchased. These possibilities are

$(0, 4)$, $(0, 5)$, $(0, 6)$, $(0, 7)$, $(0, 8)$

$(1, 2)$, $(1, 3)$, $(1, 4)$, $(1, 5)$, $(1, 6)$

$(2, 0)$, $(2, 1)$, $(2, 2)$, $(2, 3)$, $(2, 4)$

$(3, 0)$, $(3, 1)$, $(3, 2)$, $(4, 0)$

Only these points can be used, because the homeowner cannot buy part of a tree or part of a bush. ∎

Orals   *Tell whether the following coordinates satisfy $y > 3x + 2$.*

1. $(0, 0)$ **2.** $(5, 5)$ **3.** $(-2, 4)$ **4.** $(-3, -6)$

Tell whether the following coordinates satisfy the inequality $y \leq \frac{1}{2}x - 1$.

5. $(0, 0)$ **6.** $(2, 0)$ **7.** $(4, 3)$ **8.** $(-4, -3)$

EXERCISE 3.7

REVIEW

1. Solve $3x + 5 = 14$.

3. Solve $A = P + Prt$ for t.

2. Solve $2(x - 4) \leq -12$.

4. Does the graph of the line $y = -x$ pass through the origin?

Simplify each expression.

5. $2a + 5(a - 3)$

7. $4(b - a) + 3b + 2a$

6. $2t - 3(3 + t)$

8. $3p + 2(q - p) + q$

VOCABULARY AND CONCEPTS *Fill in each blank to make a true statement.*

9. $2x - y \leq 4$ is a linear _____ in x and y.

10. The symbol $\leq$ means _____ or _____.

11. In the accompanying graph, the line $2x - y = 4$ is the _____ of the graph $2x - y \leq 4$.

12. In the accompanying graph, the line $2x - y = 4$ divides the rectangular coordinate system into two _____.

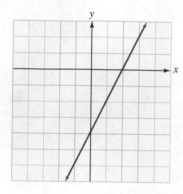

13. $\begin{cases} x + y > 2 \\ x + y < 4 \end{cases}$ This is a system of linear _____.

14. The _____ of a system of linear inequalities are all the ordered pairs that make all of the inequalities of the system true at the same time.

15. Any point in the _____ region of the graph of the solution of a system of two linear inequalities has coordinates that satisfy both of the inequalities of the system.

16. To graph a linear inequality such as $x + y > 2$, first graph the boundary. Then pick a test _____ to determine which half-plane to shade.

Answer each question.

17. Tell whether each ordered pair is a solution of $5x - 3y \geq 0$.
   a. $(1, 1)$ **b.** $(-2, -3)$
   c. $(0, 0)$ **d.** $\left(\dfrac{1}{5}, \dfrac{4}{3}\right)$

18. Tell whether each ordered pair is a solution of $x + 4y < -1$.
   a. $(3, 1)$ **b.** $(-2, 0)$
   c. $(0.5, 0.2)$ **d.** $\left(-2, \dfrac{1}{4}\right)$

19. Tell whether the graph of each linear inequality includes the boundary line.
   a. $y > -x$ **b.** $5x - 3y \leq -2$

20. If a false statement results when the coordinates of a test point are substituted into a linear inequality, which half-plane should be shaded to represent the solution of the inequality?

PRACTICE *In Exercises 21–28, complete the graph by shading the correct half-plane.*

21. $y \leq x + 2$

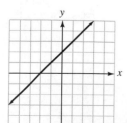

22. $y > x - 3$

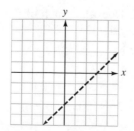

23. $y > 2x - 4$

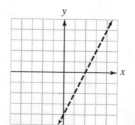

24. $y \leq -x + 1$

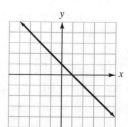

25. $x - 2y \le 4$

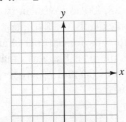

26. $3x + 2y \ge 12$

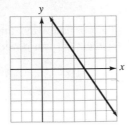

27. $y \le 4x$

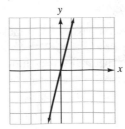

28. $y + 2x < 0$

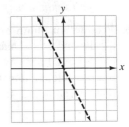

In Exercises 29–44, graph each inequality.

29. $y \ge 3 - x$

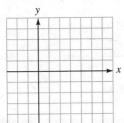

30. $y < 2 - x$

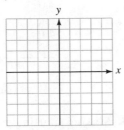

31. $y < 2 - 3x$

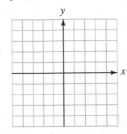

32. $y \ge 5 - 2x$

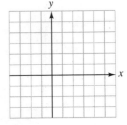

33. $y \ge 2x$

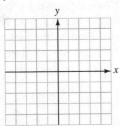

34. $y < 3x$

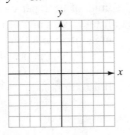

35. $2y - x < 8$

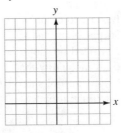

36. $y + 9x \ge 3$

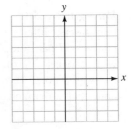

37. $3x - 4y > 12$

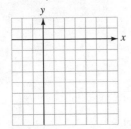

38. $4x + 3y \le 12$

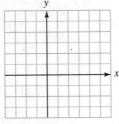

39. $5x + 4y \ge 20$

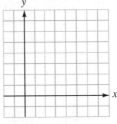

40. $7x - 2y < 21$

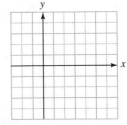

41. $x < 2$

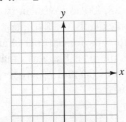

42. $y > -3$

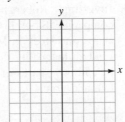

43. $y \le 1$

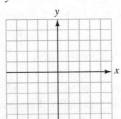

44. $x \ge -4$

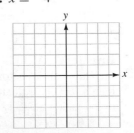

APPLICATIONS *In Exercises 45–50, graph each inequality for nonnegative values of x and y. Then give some ordered pairs that satisfy the inequality.*

45. Production planning It costs a bakery $3 to make a cake and $4 to make a pie. Production costs cannot exceed $120 per day. Find an inequality that shows the possible combinations of cakes (x) and pies (y) that can be made, and graph it in Illustration 1.

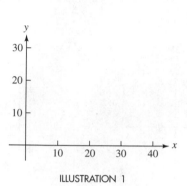

ILLUSTRATION 1

46. Hiring baby sitters Mary has a choice of two babysitters. Sitter 1 charges $6 per hour, and sitter 2 charges $7 per hour. Mary can afford no more than $42 per week for sitters. Find an inequality that shows the possible ways that she can hire sitter 1 (x) and sitter 2 (y), and graph it in Illustration 2.

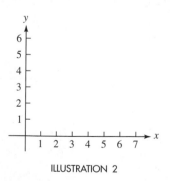

ILLUSTRATION 2

47. Inventory A clothing store advertises that it maintains an inventory of at least $4,400 worth of men's jackets. A leather jacket costs $100, and a nylon jacket costs $88. Find an inequality that shows the possible ways that leather jackets (x) and nylon jackets (y) can be stocked, and graph it in Illustration 3.

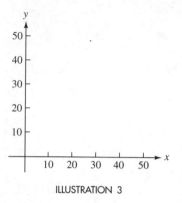

ILLUSTRATION 3

48. Making sporting goods To keep up with demand, a sporting goods manufacturer allocates at least 2,400 units of time per day to make baseballs and footballs. It takes 20 units of time to make a baseball and 30 units of time to make a football. Find an inequality that shows the possible ways to schedule the time to make baseballs (x) and footballs (y), and graph it in Illustration 4.

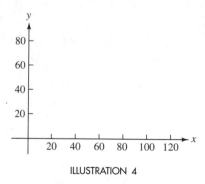

ILLUSTRATION 4

49. Investing Robert has up to $8,000 to invest in two companies. Stock in Robotronics sells for $40 per share, and stock in Macrocorp sells for $50 per share. Find an inequality that shows the possible ways that he can buy shares of Robotronics (x) and Macrocorp (y), and graph it in Illustration 5.

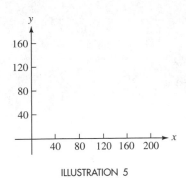

ILLUSTRATION 5

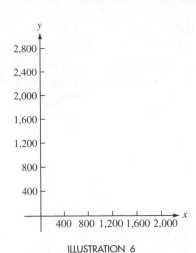

ILLUSTRATION 6

50. **Buying tickets**   Tickets to the Rockford Rox baseball games cost $6 for reserved seats and $4 for general admission. Nightly receipts must average at least $10,200 to meet expenses. Find an inequality that shows the possible ways that the Rox can sell reserved seats (x) and general admission tickets (y), and graph it in Illustration 6.

PRACTICE   *In Exercises 51–68, find the solution set of each system of inequalities, when possible.*

51. $\begin{cases} x + 2y \le 3 \\ 2x - y \ge 1 \end{cases}$

52. $\begin{cases} 2x + y \ge 3 \\ x - 2y \le -1 \end{cases}$

53. $\begin{cases} x + y < -1 \\ x - y > -1 \end{cases}$

54. $\begin{cases} x + y > 2 \\ x - y < -2 \end{cases}$

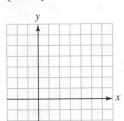
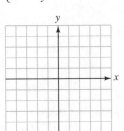
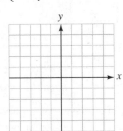
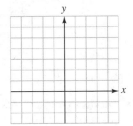

55. $\begin{cases} 2x - y < 4 \\ x + y \ge -1 \end{cases}$

56. $\begin{cases} x - y \ge 5 \\ x + 2y < -4 \end{cases}$

57. $\begin{cases} x > 2 \\ y \le 3 \end{cases}$

58. $\begin{cases} x \ge -1 \\ y > -2 \end{cases}$

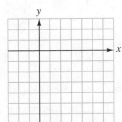
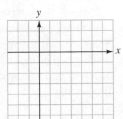
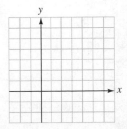
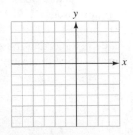

59. $\begin{cases} x + y < 1 \\ x + y > 3 \end{cases}$

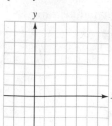

60. $\begin{cases} x \leq 0 \\ y < 0 \end{cases}$

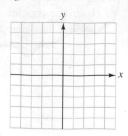

61. $\begin{cases} 3x + 4y > -7 \\ 2x - 3y \geq 1 \end{cases}$

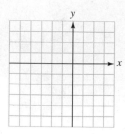

62. $\begin{cases} 3x + y \leq 1 \\ 4x - y > -8 \end{cases}$

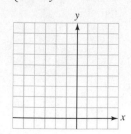

63. $\begin{cases} 2x - 4y > -6 \\ 3x + y \geq 5 \end{cases}$

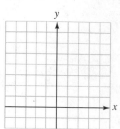

64. $\begin{cases} 2x - 3y < 0 \\ 2x + 3y \geq 12 \end{cases}$

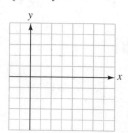

65. $\begin{cases} 3x - y \leq -4 \\ 3y > -2(x + 5) \end{cases}$

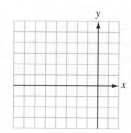

66. $\begin{cases} 3x + y < -2 \\ y > 3(1 - x) \end{cases}$

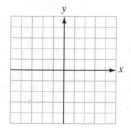

67. $\begin{cases} \dfrac{x}{2} + \dfrac{y}{3} \geq 2 \\ \dfrac{x}{2} - \dfrac{y}{2} < -1 \end{cases}$

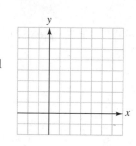

68. $\begin{cases} \dfrac{x}{3} - \dfrac{y}{2} < -3 \\ \dfrac{x}{3} + \dfrac{y}{2} > -1 \end{cases}$

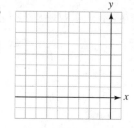

In Exercises 69–72, graph each system of inequalities and give two possible solutions to each problem.

69. Buying compact discs Melodic Music has compact discs on sale for either $10 or $15. A customer wants to spend at least $30 but no more than $60 on CDs. Find a system of inequalities whose graph will show the possible combinations of $10 CDs (x) and $15 CDs (y) that the customer can buy, and graph it in Illustration 7.

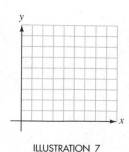

ILLUSTRATION 7

70. Buying boats Dry Boatworks wholesales aluminum boats for $800 and fiberglass boats for $600. Northland Marina wants to order at least $2,400 but no more than $4,800 worth of boats. Find a system of inequalities whose graph will show the possible combinations of aluminum boats (x) and fiberglass boats (y) that can be ordered, and graph it in Illustration 8.

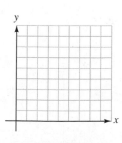

ILLUSTRATION 8

71. Buying furniture A distributor wholesales desk chairs for $150 and side chairs for $100. Best Furniture wants to order no more than $900 worth of chairs and wants to order more side chairs than desk chairs. Find a system of inequalities whose graph will show the possible combinations of desk chairs (x) and side chairs (y) that can be ordered, and graph it in Illustration 9.

72. Ordering furnace equipment J. Bolden Heating Company wants to order no more than $2,000 worth of electronic air cleaners and humidifiers from a wholesaler that charges $500 for aircleaners and $200 for humidifiers. Bolden wants more humidifiers than air cleaners. Find a system of inequalities whose graph will show the possible combinations of air cleaners (x) and humidifiers (y) that can be ordered, and graph it in Illustration 10.

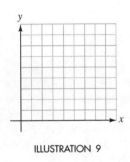

ILLUSTRATION 9

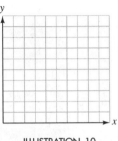

ILLUSTRATION 10

WRITING

73. Explain how to find the boundary for the graph of an inequality.

74. Explain how to decide which side of the boundary line to shade.

75. Explain how to use graphing to solve a system of inequalities.

76. Explain when a system of inequalities will have no solutions.

SOMETHING TO THINK ABOUT

77. What are some limitations of the graphing method for solving inequalities?

78. Graph $y = 3x + 1$, $y < 3x + 1$, and $y > 3x + 1$. What do you discover?

79. Can a system of inequalities have
   a. no solutions?
   b. exactly one solution?
   c. infinitely many solutions?

80. Find a system of two inequalities that has a solution of $(2, 0)$ but no solutions of the form (x, y) where $y < 0$.

■ ■ ■ ■ ■ ■ ■ ■ ■ **PROJECT**

The graphing method of solving a system of equations is not as accurate as algebraic methods, and some systems are more difficult than others to solve accurately. For example, the two lines in Illustration 1(a) could be drawn carelessly, and the point of intersection would not be far from the correct location. If the lines in Illustration 1(b) were drawn carelessly, the point of intersection could move substantially from its correct location.

■ ■ ■ ■ ■ ■ ■ ■ ■ ■ **PROJECT** *(continued)*

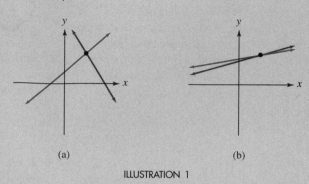

(a)   (b)

ILLUSTRATION 1

- Carefully solve each of these systems of equations graphically (by hand, not with a graphing calculator). Indicate your best estimate of the solution of each system.

$$\begin{cases} 2x - 4y = -7 \\ 4x + 2y = 11 \end{cases} \qquad \begin{cases} 5x - 4y = -1 \\ 12x - 10y = -3 \end{cases}$$

- Solve each system algebraically. How close were your graphical solutions to the actual solutions? Write a paragraph explaining any differences.

- Create a system of equations with the solutions $x = 3$, $y = 2$ for which an accurate solution could be obtained graphically.

- Create a system of equations with the solutions $x = 3$, $y = 2$ that is more difficult to solve accurately than the previous system, and write a paragraph explaining why.

CHAPTER SUMMARY

CONCEPTS

REVIEW EXERCISES

SECTION 3.1   *The Rectangular Coordinate System*

Any ordered pair of real numbers represents a point on the rectangular coordinate system.

1. Plot each point on the rectangular coordinate system in Illustration 1.
- **a.** $A(1, 3)$
- **b.** $B(1, -3)$
- **c.** $C(-3, 1)$
- **d.** $D(-3, -1)$
- **e.** $E(0, 5)$
- **f.** $F(-5, 0)$

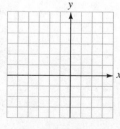

ILLUSTRATION 1

2. Find the coordinates of each point in Illustration 2.

   a. *A*   **b.** *B*
   c. *C*   **d.** *D*
   e. *E*   **f.** *F*
   g. *G*   **h.** *H*

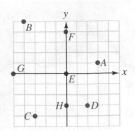

ILLUSTRATION 2

SECTION 3.2   *Graphing Linear Equations*

An ordered pair of real numbers is a **solution** if it satisfies the equation.

To graph a linear equation,
1. Find three pairs (x, y) that satisfy the equation.
2. Plot each pair on the rectangular coordinate system.
3. Draw a line passing through the three points.

3. Tell whether each pair satisfies the equation $3x - 4y = 12$.

   a. $(2, 1)$ **b.** $\left(3, -\dfrac{3}{4}\right)$

4. Graph each equation on a rectangular coordinate system.

   a. $y = x - 5$

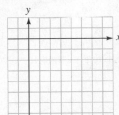

   b. $y = 2x + 1$

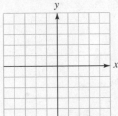

   c. $y = \dfrac{x}{2} + 2$

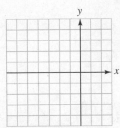

   d. $y = 3$

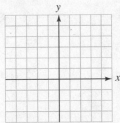

   e. $x + y = 4$

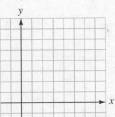

   f. $x - y = -3$

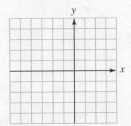

g. $3x + 5y = 15$

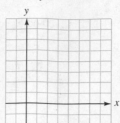

h. $7x - 4y = 28$

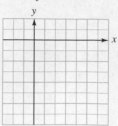

SECTION 3.3

Solving Systems of Equations by Graphing

To solve a system of equations graphically, carefully graph each equation of the system. If the lines intersect, the coordinates of the point of intersection give the solution of the system.

5. Tell whether the ordered pair is a solution of the system.

a. $(1, 5),\ \begin{cases} 3x - y = -2 \\ 2x + 3y = 17 \end{cases}$

b. $(-2, 4),\ \begin{cases} 5x + 3y = 2 \\ -3x + 2y = 16 \end{cases}$

c. $\left(14, \dfrac{1}{2}\right),\ \begin{cases} 2x + 4y = 30 \\ \dfrac{x}{4} - y = 3 \end{cases}$

d. $\left(\dfrac{7}{2}, -\dfrac{2}{3}\right),\ \begin{cases} 4x - 6y = 18 \\ \dfrac{x}{3} + \dfrac{y}{2} = \dfrac{5}{6} \end{cases}$

6. Use the graphing method to solve each system.

a. $\begin{cases} x + y = 7 \\ 2x - y = 5 \end{cases}$

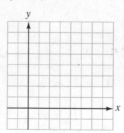

b. $\begin{cases} \dfrac{x}{3} + \dfrac{y}{5} = -1 \\ x - 3y = -3 \end{cases}$

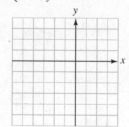

c. $\begin{cases} 3x + 6y = 6 \\ x + 2y = 2 \end{cases}$

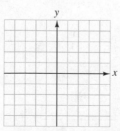

d. $\begin{cases} 6x + 3y = 12 \\ 2x + y = 2 \end{cases}$

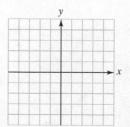

SECTION 3.4 — Solving Systems of Equations by Substitution

To solve a system of equations by substitution, solve one of the equations of the system for one of its variables, substitute the resulting expression into the other equation, and solve for the other variable.

7. Use the substitution method to solve each system.

a. $\begin{cases} x = 3y + 5 \\ 5x - 4y = 3 \end{cases}$

b. $\begin{cases} 3x - \dfrac{2y}{5} = 2(x - 2) \\ 2x - 3 = 3 - 2y \end{cases}$

c. $\begin{cases} 8x + 5y = 3 \\ 5x - 8y = 13 \end{cases}$

d. $\begin{cases} 6(x + 2) = y - 1 \\ 5(y - 1) = x + 2 \end{cases}$

SECTION 3.5 — Solving Systems of Equations by Addition

To solve a system of equations by addition, first multiply one or both of the equations by suitable constants, if necessary, to eliminate one of the variables when the equations are added. The equation that results can be solved for its single variable. Then substitute the value obtained back into one of the original equations and solve for the other variable.

8. Use the addition method to solve each system.

a. $\begin{cases} 2x + y = 1 \\ 5x - y = 20 \end{cases}$

b. $\begin{cases} x + 8y = 7 \\ x - 4y = 1 \end{cases}$

c. $\begin{cases} 5x + y = 2 \\ 3x + 2y = 11 \end{cases}$

d. $\begin{cases} x + y = 3 \\ 3x = 2 - y \end{cases}$

e. $\begin{cases} 11x + 3y = 27 \\ 8x + 4y = 36 \end{cases}$

f. $\begin{cases} 9x + 3y = 5 \\ 3x = 4 - y \end{cases}$

g. $\begin{cases} 9x + 3y = 5 \\ 3x + y = \dfrac{5}{3} \end{cases}$

h. $\begin{cases} \dfrac{x}{3} + \dfrac{y + 2}{2} = 1 \\ \dfrac{x + 8}{8} + \dfrac{y - 3}{3} = 0 \end{cases}$

SECTION 3.6 — Applications of Systems of Equations

Systems of equations are useful in solving many different types of problems.

9. **Integer problem** One number is 5 times another, and their sum is 18. Find the numbers.

10. **Geometry** The length of a rectangle is 3 times its width, and its perimeter is 24 feet. Find its dimensions.

11. **Buying grapefruit** A grapefruit costs 15 cents more than an orange. Together, they cost 85 cents. Find the cost of a grapefruit.

12. **Utility bills** A man's electric bill for January was $23 less than his gas bill. The two utilities cost him a total of $109. Find the amount of his gas bill.

13. **Buying groceries** Two gallons of milk and 3 dozen eggs cost $6.80. Three gallons of milk and 2 dozen eggs cost $7.35. How much does each gallon of milk cost?

14. **Investing money**   Carlos invested part of $3,000 in a 10% certificate of deposit account and the rest in a 6% passbook account. If the total annual interest from both accounts is $270, how much did he invest at 6%?

SECTION 3.7   *Systems of Linear Inequalities*

To graph a system of inequalities, first graph the individual inequalities of the system. The final solution, if one exists, is that region where all the individual graphs intersect.

15. Graph each inequality.

    a. $y \geq x + 2$

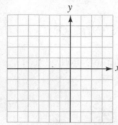

    b. $x < 3$

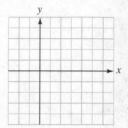

16. Solve each system of inequalities.

    a. $\begin{cases} 5x + 3y < 15 \\ 3x - y > 3 \end{cases}$

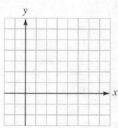

    b. $\begin{cases} 5x - 3y \geq 5 \\ 3x + 2y \geq 3 \end{cases}$

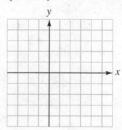

    c. $\begin{cases} x \geq 3y \\ y < 3x \end{cases}$

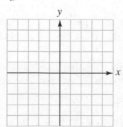

    d. $\begin{cases} x > 0 \\ x \leq 3 \end{cases}$

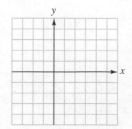

■ Chapter Test

In Problems 1–4, graph each equation.

1. $y = \dfrac{x}{2} + 1$

2. $2(x + 1) - y = 4$

3. $x = 1$

4. $2y = 8$

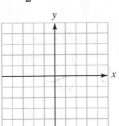   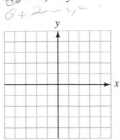   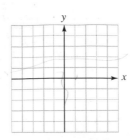   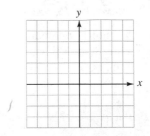

In Problems 5–6, tell whether the given ordered pair is a solution of the given system.

5. $(2, -3)$, $\begin{cases} 3x - 2y = 12 \\ 2x + 3y = -5 \end{cases}$

6. $(-2, -1)$, $\begin{cases} 4x + y = -9 \\ 2x - 3y = -7 \end{cases}$

In Problems 7–8, solve each system by graphing.

7. $\begin{cases} 3x + y = 7 \\ x - 2y = 0 \end{cases}$

8. $\begin{cases} x + \dfrac{y}{2} = 1 \\ y = 1 - 3x \end{cases}$

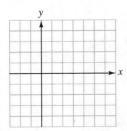   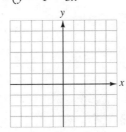

In Problems 9–10, solve each system by substitution.

9. $\begin{cases} y = x - 1 \\ x + y = -7 \end{cases}$

10. $\begin{cases} \dfrac{x}{6} + \dfrac{y}{10} = 3 \\ \dfrac{5x}{16} - \dfrac{3y}{16} = \dfrac{15}{8} \end{cases}$

In Problems 11–12, solve each system by addition.

11. $\begin{cases} 3x - y = 2 \\ 2x + y = 8 \end{cases}$

12. $\begin{cases} 4x + 3 = -3y \\ \dfrac{-x}{7} + \dfrac{4y}{21} = 1 \end{cases}$

In Problems 13–14, classify each system as consistent or inconsistent.

13. $\begin{cases} 2x + 3(y - 2) = 0 \\ -3y = 2(x - 4) \end{cases}$

14. $\begin{cases} \dfrac{x}{3} + y - 4 = 0 \\ -3y = x - 12 \end{cases}$

In Problems 15–16, use a system of equations in two variables to solve each problem.

15. The sum of two numbers is -18. One number is 2 greater than 3 times the other. Find the product of the numbers.

16. A woman invested some money at 8% and some at 9%. The interest on the combined investment of $10,000 was $840. How much was invested at 9%?

In Problems 17–18, solve each system of inequalities by graphing.

17. $\begin{cases} x + y < 3 \\ x - y < 1 \end{cases}$

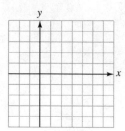

18. $\begin{cases} 2x + 3y \le 6 \\ x \ge 2 \end{cases}$

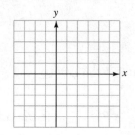

4 Polynomials

MATHEMATICS IN THE WORKPLACE

Economist

Economists study the way a society uses resources such as land, labor, raw materials, and machinery to provide goods and services. Some economists are theoreticians who use mathematical models to explain the causes of recession and inflation. Most economists, however, are concerned with practical applications of economic policy in a particular area.

SAMPLE APPLICATION ■ An electronics firm manufactures tape recorders, receiving $120 for each recorder it makes. If x represents the number of recorders produced, the income received is determined by the *revenue function*

$$R(x) = 120x$$

The manufacturer has fixed costs of $12,000 per month and variable costs of $57.50 for each tape recorder manufactured. Thus, the *cost function* is

$$C(x) = \textit{variable costs} + \textit{fixed costs}$$
$$= 57.50x + 12,000$$

How many recorders must the company sell for revenue to equal cost? (See Exercise 126 in Exercise 4.6.)

4.1 Natural-Number Exponents

■ EXPONENTS ■ POWERS OF EXPRESSIONS ■ THE PRODUCT RULE FOR EXPONENTS ■ THE POWER RULES FOR EXPONENTS ■ THE QUOTIENT RULE FOR EXPONENTS

Getting Ready *Evaluate each expression.*

1. 2^3 **2.** 3^2 **3.** $3(2)$ **4.** $2(3)$

5. $2^3 + 2^2$ **6.** $2^3 \cdot 2^2$ **7.** $3^3 - 3^2$ **8.** $\dfrac{3^3}{3^2}$

■ EXPONENTS

We have used natural-number exponents to indicate repeated multiplication. For example,

$$2^5 = 2 \cdot 2 \cdot 2 \cdot 2 \cdot 2 = 32 \qquad\qquad (-7)^3 = (-7)(-7)(-7) = -343$$
$$x^4 = x \cdot x \cdot x \cdot x \qquad\qquad\qquad -y^5 = -y \cdot y \cdot y \cdot y \cdot y$$

These examples suggest a definition for x^n, where n is a natural number.

Natural-Number Exponents

If n is a natural number, then

$$\overbrace{x^n = x \cdot x \cdot x \cdot \cdots \cdot x}^{n \text{ factors of } x}$$

In the exponential expression x^n, x is called the **base** and n is called the **exponent**. The entire expression is called a **power of x.**

$$\text{Base} \longrightarrow x^n \longleftarrow \text{Exponent}$$

■ POWERS OF EXPRESSIONS

If an exponent is a natural number, it tells how many times its base is to be used as a factor. An exponent of 1 indicates that its base is to be used one time as a factor, an exponent of 2 indicates that its base is to be used two times as a factor, and so on.

$$3^1 = 3 \qquad (-y)^1 = -y \qquad (-4z)^2 = (-4z)(-4z) \qquad \text{and} \qquad (t^2)^3 = t^2 \cdot t^2 \cdot t^2$$

EXAMPLE 1   Show that **a.** -2^4 and **b.** $(-2)^4$ have different values.

Solution   We find each power and show that the results are different.

$$-2^4 = -(2^4) \qquad\qquad\qquad (-2)^4 = (-2)(-2)(-2)(-2)$$
$$= -(2 \cdot 2 \cdot 2 \cdot 2) \qquad\qquad\qquad = 16$$
$$= -16$$

Since $-16 \neq 16$, if follows that $-2^4 \neq (-2)^4$.   ■

Self Check   Show that **a.** $(-4)^3$ and **b.** -4^3 have the same value.

Answers   **a.** -64, **b.** -64

EXAMPLE 2   Write each expression without using exponents: **a.** r^3, **b.** $(-2s)^4$, and **c.** $\left(\frac{1}{3}ab\right)^5$.

Solution   **a.** $r^3 = r \cdot r \cdot r$

b. $(-2s)^4 = (-2s)(-2s)(-2s)(-2s)$

c. $\left(\frac{1}{3}ab\right)^5 = \left(\frac{1}{3}ab\right)\left(\frac{1}{3}ab\right)\left(\frac{1}{3}ab\right)\left(\frac{1}{3}ab\right)\left(\frac{1}{3}ab\right)$ ■

Self Check   Write each expression without using exponents: **a.** x^4 and **b.** $\left(-\frac{1}{2}xy\right)^3$.

Answers   **a.** $x \cdot x \cdot x \cdot x$, **b.** $\left(-\frac{1}{2}\right)\left(-\frac{1}{2}\right)\left(-\frac{1}{2}\right)x \cdot x \cdot x \cdot y \cdot y \cdot y$

■ THE PRODUCT RULE FOR EXPONENTS

To develop a rule for multiplying exponential expressions with the same base, we consider the product $x^2 \cdot x^3$. Since the expression x^2 means that x is to be used as a factor two times and the expression x^3 means that x is to be used as a factor three times, we have

$$x^2x^3 = \overbrace{x \cdot x}^{2 \text{ factors of } x} \cdot \overbrace{x \cdot x \cdot x}^{3 \text{ factors of } x}$$

$$= \overbrace{x \cdot x \cdot x \cdot x \cdot x}^{5 \text{ factors of } x}$$

$$= x^5$$

In general,

$$x^m \cdot x^n = \overbrace{x \cdot x \cdot x \cdot \cdots \cdot x}^{m \text{ factors of } x} \cdot \overbrace{x \cdot x \cdot x \cdot x \cdot \cdots \cdot x}^{n \text{ factors of } x}$$

$$= \overbrace{x \cdot x \cdot x \cdot x \cdot x \cdot x \cdot \cdots \cdot x \cdot x \cdot x}^{m + n \text{ factors of } x}$$

$$= x^{m+n}$$

This discussion suggests the following rule: *To multiply two exponential expressions with the same base, keep the base and add the exponents.*

> **Product Rule for Exponents**
> If m and n are natural numbers, then
> $$x^m x^n = x^{m+n}$$

EXAMPLE 3   Simplify each expression.

a. $x^3x^4 = x^{3+4}$ Keep the base and add the exponents.

     $= x^7$ $3 + 4 = 7$.

b. $y^2y^4y = (y^2y^4)y$ Use the associative property to group y^2 and y^4 together.

     $= (y^{2+4})y$ Keep the base and add the exponents.

     $= y^6y$ $2 + 4 = 6$.

     $= y^{6+1}$ Keep the base and add the exponents; $y = y^1$.

     $= y^7$ $6 + 1 = 7$.   ■

Self Check   Simplify each expression: **a.** zz^3 and   **b.** $x^2x^3x^6$.

Answers   **a.** z^4,   **b.** x^{11}

EXAMPLE 4   Simplify $(2y^3)(3y^2)$.

Solution $(2y^3)(3y^2) = 2(3)y^3y^2$ Use the commutative and associative properties to group the numbers together and the variables together.

$\qquad\qquad = 6y^{3+2}$ Multiply the coefficients. Keep the base and add the exponents.

$\qquad\qquad = 6y^5$ $3 + 2 = 5$. ∎

Self Check   Simplify $(4x)(-3x^2)$.
Answer $-12x^3$

WARNING!   The product rule for exponents applies only to exponential expressions with the same base. An expression such as x^2y^3 cannot be simplified, because x^2 and y^3 have different bases.

■ THE POWER RULES FOR EXPONENTS

To find another rule of exponents, we consider the expression $(x^3)^4$, which can be written as $x^3 \cdot x^3 \cdot x^3 \cdot x^3$. Because each of the four factors of x^3 contains three factors of x, there are $4 \cdot 3$ (or 12) factors of x. Thus, the product can be written as x^{12}.

$$(x^3)^4 = x^3 \cdot x^3 \cdot x^3 \cdot x^3$$

$$= \overbrace{\underbrace{x \cdot x \cdot x}_{x^3} \cdot \underbrace{x \cdot x \cdot x}_{x^3} \cdot \underbrace{x \cdot x \cdot x}_{x^3} \cdot \underbrace{x \cdot x \cdot x}_{x^3}}^{12 \text{ factors of } x}$$

$$= x^{12}$$

In general,

$$(x^m)^n = \overbrace{x^m \cdot x^m \cdot x^m \cdot \cdots \cdot x^m}^{n \text{ factors of } x^m}$$

$$= \overbrace{x \cdot x \cdot x \cdot x \cdot x \cdot x \cdot x \cdot \cdots \cdot x}^{m \cdot n \text{ factors of } x}$$

$$= x^{m \cdot n}$$

This discussion suggests the following rule: *To raise an exponential expression to a power, keep the base and multiply the exponents.*

First Power Rule for Exponents
If m and n are natural numbers, then

$$(x^m)^n = x^{m \cdot n}$$

EXAMPLE 5 Write each expression using one exponent.

a. $(2^3)^7 = 2^{3 \cdot 7}$ Keep the base and multiply the exponents.

$\quad\quad = 2^{21}$ $3 \cdot 7 = 21.$

b. $(z^7)^7 = z^{7 \cdot 7}$ Keep the base and multiply the exponents.

$\quad\quad = z^{49}$ $7 \cdot 7 = 49.$ ■

Self Check Write each expression using one exponent: **a.** $(y^5)^2$ and **b.** $(u^x)^y$.

Answers   **a.** y^{10}, **b.** u^{xy}

In the next example, the product and power rules of exponents are both used.

EXAMPLE 6 Write each expression using one exponent.

a. $(x^2 x^5)^2 = (x^7)^2$ **b.** $(y^6 y^2)^3 = (y^8)^3$

$\quad\quad\quad = x^{14}$ $\quad\quad\quad = y^{24}$

c. $(z^2)^4 (z^3)^3 = z^8 z^9$ **d.** $(x^3)^2 (x^5 x^2)^3 = x^6 (x^7)^3$

$\quad\quad\quad = z^{17}$ $\quad\quad\quad\quad = x^6 x^{21}$

$\quad\quad\quad\quad\quad\quad\quad\quad\quad\quad\quad\quad = x^{27}$ ■

Self Check Write each expression using one exponent: **a.** $(a^4 a^3)^3$ and **b.** $(a^3)^3 (a^4)^2$.

Answers   **a.** a^{21}, **b.** a^{17}

To find two more rules for exponents, we consider the expressions $(2x)^3$ and $\left(\frac{2}{x}\right)^3$.

$$(2x)^3 = (2x)(2x)(2x) \qquad\qquad \left(\frac{2}{x}\right)^3 = \left(\frac{2}{x}\right)\left(\frac{2}{x}\right)\left(\frac{2}{x}\right) \quad (x \neq 0)$$

$$= (2 \cdot 2 \cdot 2)(x \cdot x \cdot x) \qquad\qquad = \frac{2 \cdot 2 \cdot 2}{x \cdot x \cdot x}$$

$$= 2^3 x^3 \qquad\qquad\qquad\qquad = \frac{2^3}{x^3}$$

$$= 8x^3 \qquad\qquad\qquad\qquad = \frac{8}{x^3}$$

These examples suggest the following rules: *To raise a product to a power, we raise each factor of the product to that power,* and *to raise a fraction to a power, we raise both the numerator and denominator to that power.*

More Power Rules for Exponents

If n is a natural number, then

$$(xy)^n = x^n y^n \qquad \text{and if } y \neq 0, \text{ then} \qquad \left(\frac{x}{y}\right)^n = \frac{x^n}{y^n}$$

EXAMPLE 7 Write each expression without using parentheses. Assume there are no divisions by zero.

a. $(ab)^4 = a^4 b^4$

b. $(3c)^3 = 3^3 c^3$
$$= 27c^3$$

c. $(x^2 y^3)^5 = (x^2)^5 (y^3)^5$
$$= x^{10} y^{15}$$

d. $(-2x^3 y)^2 = (-2)^2 (x^3)^2 y^2$
$$= 4x^6 y^2$$

e. $\left(\dfrac{4}{k}\right)^3 = \dfrac{4^3}{k^3}$
$$= \dfrac{64}{k^3}$$

f. $\left(\dfrac{3x^2}{2y^3}\right)^5 = \dfrac{3^5 (x^2)^5}{2^5 (y^3)^5}$
$$= \dfrac{243x^{10}}{32y^{15}}$$ ∎

Self Check Write each expression without using parentheses: **a.** $(3x^2 y)^2$ and
b. $\left(\dfrac{2x^3}{3y^2}\right)^4$.

Answers **a.** $9x^4 y^2$, **b.** $\dfrac{16x^{12}}{81y^8}$

■ THE QUOTIENT RULE FOR EXPONENTS

To find a rule for dividing exponential expressions, we consider the fraction $\frac{4^5}{4^2}$, where the exponent in the numerator is greater than the exponent in the denominator. We can simplify the fraction as follows:

$$\frac{4^5}{4^2} = \frac{4 \cdot 4 \cdot 4 \cdot 4 \cdot 4}{4 \cdot 4}$$

$$= \frac{\overset{1}{\cancel{4}} \cdot \overset{1}{\cancel{4}} \cdot 4 \cdot 4 \cdot 4}{\underset{1}{\cancel{4}} \cdot \underset{1}{\cancel{4}}}$$

$$= 4^3$$

The result of 4^3 has a base of 4 and an exponent of $5 - 2$ (or 3). This suggests that *to divide exponential expressions with the same base, we keep the base and subtract the exponents.*

Quotient Rule for Exponents
If m and n are natural numbers, $m > n$, and $x \neq 0$, then
$$\frac{x^m}{x^n} = x^{m-n}$$

b,d

EXAMPLE 8

Simplify each expression. Assume that there are no divisions by 0.

a. $\dfrac{x^4}{x^3} = x^{4-3}$

$= x^1$

$= x$

b. $\dfrac{8y^2y^6}{4y^3} = \dfrac{8y^8}{4y^3}$

$= \dfrac{8}{4}y^{8-3}$

$= 2y^5$

c. $\dfrac{a^3a^5a^7}{a^4a} = \dfrac{a^{15}}{a^5}$

$= a^{15-5}$

$= a^{10}$

d. $\dfrac{(a^3b^4)^2}{ab^5} = \dfrac{a^6b^8}{ab^5}$

$= a^{6-1}b^{8-5}$

$= a^5b^3$

∎

Self Check

Simplify a. $\dfrac{a^5}{a^3}$, b. $\dfrac{6b^2b^3}{2b^4}$, and c. $\dfrac{(x^2y^3)^2}{x^3y^4}$.

Answers

a. a^2, b. $3b$, c. xy^2

We summarize the rules for positive exponents as follows.

Properties of Exponents

If n is a natural number, then

$$n \text{ factors of } x$$
$$x^n = x \cdot x \cdot x \cdot \cdots \cdot x$$

If m and n are natural numbers and there are no divisions by 0, then

$$x^m x^n = x^{m+n} \qquad (x^m)^n = x^{m \cdot n} \qquad (xy)^n = x^n y^n \qquad \left(\dfrac{x}{y}\right)^n = \dfrac{x^n}{y^n}$$

$$\dfrac{x^m}{x^n} = x^{m-n} \quad \text{provided } m > n.$$

Orals

Find the base and the exponent in each expression.

1. x^3 **2.** 3^x **3.** ab^c **4.** $(ab)^c$

Evaluate each expression.

5. 6^2 **6.** $(-6)^2$ **7.** $2^3 + 1^3$ **8.** $(2 + 1)^3$

EXERCISE 4.1

REVIEW

1. Graph the real numbers -3, 0, 2, and $-\frac{3}{2}$ on a number line.

$$\xleftarrow{\hspace{1em}} \underset{-4\;\;-3\;\;-2\;\;-1\;\;\;0\;\;\;1\;\;\;2\;\;\;3}{\rule{0pt}{1em}} \xrightarrow{\hspace{1em}}$$

2. Graph the interval $(-2, 3]$ on a number line.

$$\xleftarrow{\hspace{1em}} \underset{-3\;\;-2\;\;-1\;\;\;0\;\;\;1\;\;\;2\;\;\;3}{\rule{0pt}{1em}} \xrightarrow{\hspace{1em}}$$

Write each algebraic expression as an English phrase.

3. $3(x + y)$

4. $3x + y$

Write each English phrase as an algebraic expression.

5. Three greater than the absolute value of twice x

6. The sum of the numbers y and z decreased by the sum of their squares

VOCABULARY AND CONCEPTS   *Fill in each blank to make a true statement.*

7. The base of the exponential expression $(-5)^3$ is ____. The exponent is ___.

8. The base of the exponential expression -5^3 is ___. The exponent is ___.

9. $(3x)^4$ means _____.

10. Write $(-3y)(-3y)(-3y)$ as a power. _____

11. $y^5 =$ _____

12. $x^m x^n =$ _____

13. $(xy)^n =$ _____

14. $\left(\dfrac{a}{b}\right)^n =$ ____

15. $(a^b)^c =$ ____

16. $\dfrac{x^m}{x^n} =$ _____

17. The area of the square shown in Illustration 1 is $s \cdot s$. Why do you think the symbol s^2 is called "s squared"?

18. The volume of the cube shown in Illustration 2 is $s \cdot s \cdot s$. Why do you think the symbol s^3 is called "s cubed"?

s

s

ILLUSTRATION 1

s

s

s

ILLUSTRATION 2

In Exercises 19–30, identify the base and the exponent in each expression.

19. 4^3

20. $(-5)^2$

21. x^5

22. y^8

23. $(2y)^3$

24. $(-3x)^2$

25. $-x^4$

26. $(-x)^4$

27. x

28. $(xy)^3$

29. $2x^3$

30. $-3y^6$

In Exercises 31–38, write each expression without using exponents.

31. 5^3

32. -4^5

33. x^7

34. $3x^3$

35. $-4x^5$

36. $(-2y)^4$

37. $(3t)^5$

38. a^3b^2

In Exercises 39–46, write each expression using exponents.

39. $2 \cdot 2 \cdot 2$

40. $5 \cdot 5$

41. $x \cdot x \cdot x \cdot x$

42. $y \cdot y \cdot y \cdot y \cdot y$

43. $(2x)(2x)(2x)$

44. $(-4y)(-4y)$

45. $-4t \cdot t \cdot t \cdot t$

46. $5 \cdot u \cdot u$

PRACTICE *In Exercises 47–54, evaluate each expression.*

47. 5^4

48. $(-3)^3$

49. $2^2 + 3^2$

50. $2^3 - 2^2$

51. $5^4 - 4^3$

52. $2(4^3 + 3^2)$

53. $-5(3^4 + 4^3)$

54. $-5^2(4^3 - 2^6)$

In Exercises 55–70, write each expression as an expression involving only one exponent.

55. x^4x^3

56. y^5y^2

57. x^5x^5

58. yy^3

59. tt^2

60. w^3w^5

61. $a^3a^4a^5$

62. $b^2b^3b^5$

63. $y^3(y^2y^4)$

64. $(y^4y)y^6$

65. $4x^2(3x^5)$

66. $-2y(y^3)$

67. $(-y^2)(4y^3)$

68. $(-4x^3)(-5x)$

69. $6x^3(-x^2)(-x^4)$

70. $-2x(-x^2)(-3x)$

In Exercises 71–86, write each expression as an expression involving only one exponent.

71. $(3^2)^4$

72. $(4^3)^3$

73. $(y^5)^3$

74. $(b^3)^6$

75. $(a^3)^7$

76. $(b^2)^3$

77. $(x^2x^3)^5$

78. $(y^3y^4)^4$

79. $(3zz^2z^3)^5$

80. $(4t^3t^6t^2)^2$

81. $(x^5)^2(x^7)^3$

82. $(y^3y)^2(y^2)^2$

83. $(r^3r^2)^4(r^3r^5)^2$

84. $(s^2)^3(s^3)^2(s^4)^4$

85. $(s^3)^3(s^2)^2(s^5)^4$

86. $(yy^3)^3(y^2y^3)^4(y^3y^3)^2$

In Exercises 87–102, write each expression without using parentheses.

87. $(xy)^3$

88. $(uv^2)^4$

89. $(r^3s^2)^2$

90. $(a^3b^2)^3$

91. $(4ab^2)^2$

92. $(3x^2y)^3$

93. $(-2r^2s^3t)^3$

94. $(-3x^2y^4z)^2$

95. $\left(\dfrac{a}{b}\right)^3$

96. $\left(\dfrac{r^2}{s}\right)^4$

97. $\left(\dfrac{x^2}{y^3}\right)^5$

98. $\left(\dfrac{u^4}{v^2}\right)^6$

99. $\left(\dfrac{-2a}{b}\right)^5$

100. $\left(\dfrac{2t}{3}\right)^4$

101. $\left(\dfrac{b^2}{3a}\right)^3$

102. $\left(\dfrac{a^3b}{c^4}\right)^5$

In Exercises 103–118, simplify each expression.

103. $\dfrac{x^5}{x^3}$

104. $\dfrac{a^6}{a^3}$

105. $\dfrac{y^3y^4}{yy^2}$

106. $\dfrac{b^4b^5}{b^2b^3}$

107. $\dfrac{12a^2a^3a^4}{4(a^4)^2}$

108. $\dfrac{16(aa^2)^3}{2a^2a^3}$

109. $\dfrac{(ab^2)^3}{(ab)^2}$

110. $\dfrac{(m^3n^4)^3}{(mn^2)^3}$

111. $\dfrac{20(r^4s^3)^4}{6(rs^3)^3}$

112. $\dfrac{15(x^2y^5)^5}{21(x^3y)^2}$

113. $\dfrac{17(x^4y^3)^8}{34(x^5y^2)^4}$

114. $\dfrac{35(r^3s^2)^2}{49r^2s^4}$

115. $\left(\dfrac{y^3y}{2yy^2}\right)^3$

116. $\left(\dfrac{3t^3t^4t^5}{4t^2t^6}\right)^3$

117. $\left(\dfrac{-2r^3r^3}{3r^4r}\right)^3$

118. $\left(\dfrac{-6y^4y^5}{5y^3y^5}\right)^2$

APPLICATIONS

119. Bouncing ball When a certain ball is dropped, it always rebounds to one-half of its previous height. If the ball is dropped from a height of 32 feet, explain why the expression $32\left(\frac{1}{2}\right)^4$ represents the height of the ball on the fourth bounce. Find the height of the fourth bounce.

120. Having babies The probability that a couple will have n baby boys in a row is given by the formula $\left(\frac{1}{2}\right)^n$. Find the probability that a couple will have four baby boys in a row.

121. Investing If an investment of $1,000 doubles every seven years, find the value of the investment after 28 years.

122. Investing Guess the answer to the following problem. Then use a calculator to find the correct answer. Were you close?
If the value of 1¢ is to double every day, what will the penny be worth after 31 days?

WRITING

123. Describe how you would multiply two exponential expressions with like bases.

124. Describe how you would divide two exponential expressions with like bases.

SOMETHING TO THINK ABOUT

125. Is the operation of raising to a power commutative? That is, is $a^b = b^a$? Explain.

126. Is the operation of raising to a power associative? That is, is $(a^b)^c = a^{(b^c)}$? Explain.

4.2 Zero and Negative-Integer Exponents

■ ZERO EXPONENTS ■ NEGATIVE-INTEGER EXPONENTS ■ EXPONENTS WITH VARIABLES
■ FINDING PRESENT VALUE

Getting Ready   *Simplify by dividing out common factors.*

1. $\dfrac{3\cdot3\cdot3}{3\cdot3\cdot3\cdot3}$

2. $\dfrac{2yy}{2yyy}$

3. $\dfrac{3xx}{3xx}$

4. $\dfrac{xxy}{xxxyy}$

■ ZERO EXPONENTS

When we discussed the quotient rule for exponents in the previous section, the exponent in the numerator was always greater than the exponent in the denominator. We now consider what happens when the exponents are equal.

If we apply the quotient rule to the fraction $\frac{5^3}{5^3}$, where the exponents in the numerator and denominator are equal, we obtain 5^0. However, because any nonzero number divided by itself equals 1, we also obtain 1.

$$\frac{5^3}{5^3} = 5^{3-3} = 5^0 \qquad\qquad \frac{5^3}{5^3} = \frac{\overset{1}{\cancel{5}} \cdot \overset{1}{\cancel{5}} \cdot \overset{1}{\cancel{5}}}{\underset{1}{\cancel{5}} \cdot \underset{1}{\cancel{5}} \cdot \underset{1}{\cancel{5}}} = 1$$

These are equal.

For this reason, we will define 5^0 to be equal to 1. In general, the following is true.

> **Zero Exponents**
> If x is any nonzero real number, then
> $$x^0 = 1$$
> Since $x \neq 0$, 0^0 is undefined.

EXAMPLE 1   Write each expression without using exponents.

a. $\left(\dfrac{1}{13}\right)^0 = 1$ $\qquad\qquad$ **b.** $\dfrac{x^5}{x^5} = x^{5-5}$ $\quad (x \neq 0)$

$$= x^0$$
$$= 1$$

c. $3x^0 = 3(1)$ $\qquad\qquad$ **d.** $(3x)^0 = 1$
$$= 3$$

e. $\dfrac{6^n}{6^n} = 6^{n-n}$ $\qquad\qquad$ **f.** $\dfrac{y^m}{y^m} = y^{m-m}$ $\quad (y \neq 0)$

$$= 6^0 \qquad\qquad\qquad\qquad = y^0$$
$$= 1 \qquad\qquad\qquad\qquad = 1$$

Parts **c** and **d** point out that $3x^0 \neq (3x)^0$. ∎

Self Check   Write each expression without using exponents: **a.** $(-0.115)^0$,
b. $\dfrac{4^2}{4^2}$,   and   **c.** $\dfrac{x^m}{x^m}$, $(x \neq 0)$.

Answers   **a.** 1,   **b.** 1,   **c.** 1

■ NEGATIVE-INTEGER EXPONENTS

If we apply the quotient rule to $\frac{6^2}{6^5}$, where the exponent in the numerator is less than the exponent in the denominator, we obtain 6^{-3}. However, by dividing out two factors of 6, we also obtain $\frac{1}{6^3}$.

$$\frac{6^2}{6^5} = 6^{2-5} = 6^{-3}$$

$$\frac{6^2}{6^5} = \frac{\overset{1}{\cancel{6}} \cdot \overset{1}{\cancel{6}}}{\cancel{6} \cdot \cancel{6} \cdot 6 \cdot 6 \cdot 6} = \frac{1}{6^3}$$

These are equal.

For these reasons, we define 6^{-3} to be equal to $\frac{1}{6^3}$. In general, the following is true.

Negative Exponents
If x is any nonzero number and n is a natural number, then

$$x^{-n} = \frac{1}{x^n}$$

EXAMPLE 2 b,c,e   Express each quantity without using negative exponents or parentheses. Assume that no denominators are zero.

a. $3^{-5} = \dfrac{1}{3^5}$

$= \dfrac{1}{243}$

b. $x^{-4} = \dfrac{1}{x^4}$

c. $(2x)^{-2} = \dfrac{1}{(2x)^2}$

$= \dfrac{1}{4x^2}$

d. $2x^{-2} = 2\left(\dfrac{1}{x^2}\right)$

$= \dfrac{2}{x^2}$

e. $(-3a)^{-4} = \dfrac{1}{(-3a)^4}$

$= \dfrac{1}{81a^4}$

f. $(x^3x^2)^{-3} = (x^5)^{-3}$

$= \dfrac{1}{(x^5)^3}$

$= \dfrac{1}{x^{15}}$ ■

Self Check Write each expression without using negative exponents or parentheses:
a. a^{-5}, **b.** $(3y)^{-3}$, and **c.** $(a^4a^3)^{-2}$.

Answers **a.** $\dfrac{1}{a^5}$, **b.** $\dfrac{1}{27y^3}$, **c.** $\dfrac{1}{a^{14}}$

Because of the definitions of negative and zero exponents, the product, power, and quotient rules are true for all integer exponents.

Properties of Exponents

If m and n are integers and there are no divisions by 0, then

$$x^m x^n = x^{m+n} \qquad (x^m)^n = x^{m \cdot n} \qquad (xy)^n = x^n y^n \qquad \left(\frac{x}{y}\right)^n = \frac{x^n}{y^n}$$

$$x^0 = 1 \quad (x \neq 0) \qquad x^{-n} = \frac{1}{x^n} \qquad \frac{x^m}{x^n} = x^{m-n}$$

c,e

EXAMPLE 3   Simplify and write the result without using negative exponents. Assume that no denominators are zero.

a. $(x^{-3})^2 = x^{-6}$

$\qquad\qquad = \dfrac{1}{x^6}$

b. $\dfrac{x^3}{x^7} = x^{3-7}$

$\qquad = x^{-4}$

$\qquad = \dfrac{1}{x^4}$

c. $\dfrac{y^{-4}y^{-3}}{y^{-20}} = \dfrac{y^{-7}}{y^{-20}}$

$\qquad\qquad = y^{-7-(-20)}$

$\qquad\qquad = y^{-7+20}$

$\qquad\qquad = y^{13}$

d. $\dfrac{12a^3 b^4}{4a^5 b^2} = 3a^{3-5}b^{4-2}$

$\qquad\qquad = 3a^{-2}b^2$

$\qquad\qquad = \dfrac{3b^2}{a^2}$

e. $\left(-\dfrac{x^3 y^2}{xy^{-3}}\right)^{-2} = (-x^{3-1}y^{2-(-3)})^{-2}$

$\qquad\qquad\qquad = (-x^2 y^5)^{-2}$

$\qquad\qquad\qquad = \dfrac{1}{(-x^2 y^5)^2}$

$\qquad\qquad\qquad = \dfrac{1}{x^4 y^{10}}$

■

Self Check   Simplify and write the result without using negative exponents:

a. $(x^4)^{-3}$, **b.** $\dfrac{a^4}{a^8}$, **c.** $\dfrac{a^{-4}a^{-5}}{a^{-3}}$, and **d.** $\dfrac{20x^5 y^3}{5x^3 y^6}$.

Answers   **a.** $\dfrac{1}{x^{12}}$, **b.** $\dfrac{1}{a^4}$, **c.** $\dfrac{1}{a^6}$, **d.** $\dfrac{4x^2}{y^3}$

■ EXPONENTS WITH VARIABLES

These properties of exponents are also true when the exponents are algebraic expressions.

EXAMPLE 4   Simplify each expression.

a. $x^{2m}x^{3m} = x^{2m+3m}$
$= x^{5m}$

b. $\dfrac{y^{2m}}{y^{4m}} = y^{2m-4m}$ $(y \neq 0)$
$= y^{-2m}$
$= \dfrac{1}{y^{2m}}$

c. $a^{2m-1}a^{2m} = a^{2m-1+2m}$
$= a^{4m-1}$

d. $(b^{m+1})^{2m} = b^{(m+1)2m}$
$= b^{2m^2+2m}$ ∎

Self Check   Simplify each expression: **a.** $z^{3n}z^{2n}$,   **b.** $\dfrac{z^{3n}}{z^{5n}}$,   and   **c.** $(x^{m+2})^{3m}$.

Answers   **a.** z^{5n},   **b.** $\dfrac{1}{z^{2n}}$,   **c.** x^{3m^2+6m}

■ **FINDING PRESENT VALUE**

■ ■ ■ ■ ■ ■ ■ ■ ■   **Finding Present Value**

CALCULATORS   To find out how much money P must be invested at an annual rate i (expressed as a decimal) to have A in n years, we use the formula $P = A(1 + i)^{-n}$. To find out how much we must invest at 6% to have $50,000 in 10 years, we substitute 50,000 for A, 0.06 (6%) for i, and 10 for n to get

$$P = A(1 + i)^{-n}$$
$$P = 50{,}000(1 + 0.06)^{-10}$$

To evaluate P with a scientific calculator, we enter these numbers and press these keys:

Keystrokes

(1 + .06) y^x 10   +/−   ×   50000   **27919.73885**

We must invest $27,919.74 to have $50,000 in 10 years.

Orals   *Simplify each quantity.*

1. 2^{-1}

2. 2^{-2}

3. $\left(\dfrac{1}{2}\right)^{-1}$

4. $\left(\dfrac{7}{9}\right)^{0}$

5. $x^{-1}x^{2}$

6. $y^{-2}y^{-5}$

7. $\dfrac{x^{5}x^{2}}{x^{7}}$

8. $\left(\dfrac{x}{y}\right)^{-1}$

EXERCISE 4.2

REVIEW

1. If $a = -2$ and $b = 3$, evaluate $\dfrac{3a^2 + 4b + 8}{a + 2b^2}$.

2. Evaluate $|-3 + 5 \cdot 2|$.

Solve each equation.

3. $5\left(x - \dfrac{1}{2}\right) = \dfrac{7}{2}$

4. $\dfrac{5(2 - x)}{6} = \dfrac{x + 6}{2}$

5. Solve $P = L + \dfrac{s}{f}i$ for s.

6. Solve $P = L + \dfrac{s}{f}i$ for i.

VOCABULARY AND CONCEPTS   *Fill in each blank to make a true statement.*

7. If x is any nonzero real number, then $x^0 = \underline{\quad}$.

8. If x is any nonzero real number, then $x^{-n} = \underline{\quad}$.

9. Since $\dfrac{6^4}{6^4} = 6^{4-4} = 6^0$ and $\dfrac{6^4}{6^4} = 1$, we define 6^0 to be $\underline{\quad}$.

10. Since $\dfrac{8^3}{8^5} = 8^{3-5} = 8^{-2}$ and $\dfrac{8^3}{8^5} = \dfrac{8 \cdot 8 \cdot 8}{8 \cdot 8 \cdot 8 \cdot 8 \cdot 8} = \dfrac{1}{8^2}$, we define 8^{-2} to be $\underline{\quad}$.

PRACTICE   *In Exercises 11–74, simplify each expression. Write each answer without using parentheses or negative exponents.*

11. $2^5 \cdot 2^{-2}$

12. $10^2 \cdot 10^{-4} \cdot 10^5$

13. $4^{-3} \cdot 4^{-2} \cdot 4^5$

14. $3^{-4} \cdot 3^5 \cdot 3^{-3}$

15. $\dfrac{3^5 \cdot 3^{-2}}{3^3}$

16. $\dfrac{6^2 \cdot 6^{-3}}{6^{-2}}$

17. $\dfrac{2^5 \cdot 2^7}{2^6 \cdot 2^{-3}}$

18. $\dfrac{5^{-2} \cdot 5^{-4}}{5^{-6}}$

19. $2x^0$

20. $(2x)^0$

21. $(-x)^0$

22. $-x^0$

23. $\left(\dfrac{a^2b^3}{ab^4}\right)^0$

24. $\dfrac{2}{3}\left(\dfrac{xyz}{x^2y}\right)^0$

25. $\dfrac{x^0 - 5x^0}{2x^0}$

26. $\dfrac{4a^0 + 2a^0}{3a^0}$

27. x^{-2}

28. y^{-3}

29. b^{-5}

30. c^{-4}

31. $(2y)^{-4}$

32. $(-3x)^{-1}$

33. $(ab^2)^{-3}$

34. $(m^2n^3)^{-2}$

35. $\dfrac{y^4}{y^5}$

36. $\dfrac{t^7}{t^{10}}$

37. $\dfrac{(r^2)^3}{(r^3)^4}$

38. $\dfrac{(b^3)^4}{(b^5)^4}$

39. $\dfrac{y^4y^3}{y^4y^{-2}}$

40. $\dfrac{x^{12}x^{-7}}{x^3x^4}$

41. $\dfrac{a^4a^{-2}}{a^2a^0}$

42. $\dfrac{b^0b^3}{b^{-3}b^4}$

43. $(ab^2)^{-2}$

44. $(c^2d^3)^{-2}$

45. $(x^2y)^{-3}$

46. $(-xy^2)^{-4}$

47. $(x^{-4}x^3)^3$

48. $(y^{-2}y)^3$

49. $(y^3y^{-2})^{-2}$

50. $(x^{-3}x^{-2})^2$

51. $(a^{-2}b^{-3})^{-4}$

52. $(y^{-3}z^5)^{-6}$

53. $(-2x^3y^{-2})^{-5}$

54. $(-3u^{-2}v^3)^{-3}$

55. $\left(\dfrac{a^3}{a^{-4}}\right)^2$

56. $\left(\dfrac{a^4}{a^{-3}}\right)^3$

57. $\left(\dfrac{b^5}{b^{-2}}\right)^{-2}$

58. $\left(\dfrac{b^{-2}}{b^3}\right)^{-3}$

59. $\left(\dfrac{4x^2}{3x^{-5}}\right)^4$

60. $\left(\dfrac{-3r^4r^{-3}}{r^{-3}r^7}\right)^3$

61. $\left(\dfrac{12y^3z^{-2}}{3y^{-4}z^3}\right)^2$

62. $\left(\dfrac{6xy^3}{3x^{-1}y}\right)^3$

63. $\left(\dfrac{2x^3y^{-2}}{4xy^2}\right)^7$

64. $\left(\dfrac{9u^2v^3}{18u^{-3}v}\right)^4$

65. $\left(\dfrac{14u^{-2}v^3}{21u^{-3}v}\right)^4$

66. $\left(\dfrac{-27u^{-5}v^{-3}w}{18u^3v^{-2}}\right)^4$

67. $\left(\dfrac{6a^2b^3}{2ab^2}\right)^{-2}$

68. $\left(\dfrac{15r^2s^{-2}t}{3r^{-3}s^3}\right)^{-3}$

69. $\left(\dfrac{18a^2b^3c^{-4}}{3a^{-1}b^2c}\right)^{-3}$

70. $\left(\dfrac{21x^{-2}y^2z^{-2}}{7x^3y^{-1}}\right)^{-2}$

71. $\dfrac{(2x^{-2}y)^{-3}}{(4x^2y^{-1})^3}$

72. $\dfrac{(ab^{-2}c)^2}{(a^{-2}b)^{-3}}$

73. $\dfrac{(17x^5y^{-5}z)^{-3}}{(17x^{-5}y^3z^2)^{-4}}$

74. $\dfrac{16(x^{-2}yz)^{-2}}{(2x^{-3}z^0)^4}$

In Exercises 75–90, write each expression with a single exponent.

75. $x^{2m}x^m$

76. $y^{3m}y^{2m}$

77. $u^{2m}v^{3n}u^{3m}v^{-3n}$

78. $r^{2m}s^{-3}r^{3m}s^3$

79. $y^{3m+2}y^{-m}$

80. $x^{m+1}x^m$

81. $\dfrac{y^{3m}}{y^{2m}}$

82. $\dfrac{z^{4m}}{z^{2m}}$

83. $\dfrac{x^{3n}}{x^{6n}}$

84. $\dfrac{x^m}{x^{5m}}$

85. $(x^{m+1})^2$

86. $(y^2)^{m+1}$

87. $(x^{3-2n})^{-4}$

88. $(y^{1-n})^{-3}$

89. $(y^{2-n})^{-4}$

90. $(x^{3-4n})^{-2}$

APPLICATIONS

91. **Present value** How much money must be invested at 7% to have $100,000 in 40 years?

92. **Present value** How much money must be invested at 8% to have $100,000 in 40 years?

93. **Present value** How much money must be invested at 9% to have $100,000 in 40 years?

94. **Biology** During bacterial reproduction, the time required for a population to double is called the **generation time.** If b bacteria are introduced into a medium, then after the generation time has elapsed, there will be $2b$ bacteria. After n generations, there will be $b \cdot 2^n$ bacteria. Give the meaning of this expression when $n = 0$.

WRITING

95. Tell how you would help a friend understand that 2^{-3} is not equal to -8.

96. Describe how you would verify on a calculator that

$$2^{-3} = \frac{1}{2^3}$$

SOMETHING TO THINK ABOUT

97. If a positive number x is raised to a negative power, is the result greater than, equal to, or less than x? Explore the possibilities.

98. We know that $x^{-n} = \dfrac{1}{x^n}$. Is it also true that $x^n = \dfrac{1}{x^{-n}}$? Explain.

4.3 Scientific Notation

■ SCIENTIFIC NOTATION ■ WRITING NUMBERS IN SCIENTIFIC NOTATION ■ CHANGING FROM SCIENTIFIC NOTATION TO STANDARD NOTATION ■ USING SCIENTIFIC NOTATION TO SIMPLIFY COMPUTATIONS

Getting Ready   *Evaluate each expression.*

1. 10^2 **2.** 10^3 **3.** 10^1 **4.** 10^{-2}

5. $5(10^2)$ **6.** $8(10^3)$ **7.** $3(10^1)$ **8.** $7(10^{-2})$

■ SCIENTIFIC NOTATION

Scientists often deal with extremely large and extremely small numbers. For example,

- The distance from the earth to the sun is approximately 150,000,000 kilometers.

- Ultraviolet light emitted from a mercury arc has a wavelength of approximately 0.000025 centimeter.

The large number of zeros in these numbers makes them difficult to read and hard to remember. In this section, we will discuss a notation that will make these numbers easier to work with.

Scientific notation provides a compact way of writing large and small numbers.

> **Scientific Notation**
> A number is written in **scientific notation** if it is written as the product of a number between 1 (including 1) and 10 and an integer power of 10.

Each of the following numbers is written in scientific notation.

$$3.67 \times 10^6 \qquad 2.24 \times 10^{-4} \qquad \text{and} \qquad 9.875 \times 10^{22}$$

Every number that is written in scientific notation has the following form:

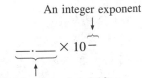

An integer exponent

$$\underbrace{\underline{}.\underline{}}_{\text{A decimal between 1 and 10}} \times 10^{-}$$

■ WRITING NUMBERS IN SCIENTIFIC NOTATION

EXAMPLE 1   Change 150,000,000 to scientific notation.

Solution   We note that 1.5 lies between 1 and 10. To obtain 150,000,000, the decimal point in 1.5 must be moved eight places to the right. Because multiplying a number by 10 moves the decimal point one place to the right, we can accomplish this by multiplying 1.5 by 10 eight times.

$$1.5\underbrace{0\,0\,0\,0\,0\,0\,0}_{\text{8 places to the right}}$$

150,000,000 written in scientific notation is 1.5×10^8.   ■

Self Check   Change 93,000,000 to scientific notation.
Answer 9.3×10^7

EXAMPLE 2   Change 0.000025 to scientific notation.

Solution   We note that 2.5 is between 1 and 10. To obtain 0.000025, the decimal point in 2.5 must be moved five places to the left. We can accomplish this by dividing 2.5 by 10^5, which is equivalent to multiplying 2.5 by $\frac{1}{10^5}$ (or by 10^{-5}).

$$\underbrace{0\,0\,0\,0}_{\text{5 places to the left}}2.5$$

0.000025 written in scientific notation is 2.5×10^{-5}.   ■

Self Check	Write 0.00125 in scientific notation.
Answer	1.25×10^{-3}

EXAMPLE 3   Write   **a.** 235,000   and   **b.** 0.00000235   in scientific notation.

Solution   **a.** $235,000 = 2.35 \times 10^5$, because $2.35 \times 10^5 = 235,000$ and 2.35 is between 1 and 10.

b. $0.00000235 = 2.35 \times 10^{-6}$, because $2.35 \times 10^{-6} = 0.00000235$ and 2.35 is between 1 and 10. ∎

Self Check	Write **a.** 17,500 and **b.** 0.657 in scientific notation.
Answers	**a.** 1.75×10^4, **b.** 6.57×10^{-1}

■ ■ ■ ■ ■ ■ ■ ■ ■ ■ P E R S P E C T I V E

The Metric System

A common metric unit of length is the kilometer, which is 1,000 meters. Because 1,000 is 10^3, we can write 1 km = 10^3 m. Similarly, 1 centimeter is one-hundredth of a meter: 1 cm = 10^{-2} m. In the metric system, prefixes such as *kilo* and *centi* refer to powers of 10. Other prefixes are used in the metric system, as shown in the table.

Prefix	Symbol	Meaning	
peta	P	10^{15} =	1,000,000,000,000,000.
tera	T	10^{12} =	1,000,000,000,000.
giga	G	10^{9} =	1,000,000,000.
mega	M	10^{6} =	1,000,000.
kilo	k	10^{3} =	1,000.
deci	d	10^{-1} =	0.1
centi	c	10^{-2} =	0.01
milli	m	10^{-3} =	0.001
micro	μ	10^{-6} =	0.000 001
nano	n	10^{-9} =	0.000 000 001
pico	p	10^{-12} =	0.000 000 000 001
femto	f	10^{-15} =	0.000 000 000 000 001
atto	a	10^{-18} =	0.000 000 000 000 000 001

To appreciate the magnitudes involved, consider these facts: Light, which travels 186,000 miles every second, will travel about one foot in one nanosecond. The distance to the nearest star is 43 petameters, and the diameter of an atom is about 10 nanometers. To measure some quantities, however, even these units are inadequate. The sun, for example, radiates 5×10^{26} watts. That's a lot of light bulbs!

EXAMPLE 4   Write 432.0×10^5 in scientific notation.

Solution   The number 432.0×10^5 is not written in scientific notation, because 432.0 is not a number between 1 and 10. To write the number in scientific notation, we proceed as follows:

$$432.0 \times 10^5 = 4.32 \times 10^2 \times 10^5 \qquad \text{Write 432.0 in scientific notation.}$$
$$= 4.32 \times 10^7 \qquad\qquad 10^2 \times 10^5 = 10^7. \qquad\blacksquare$$

Self Check   Write 85×10^{-3} in scientific notation.
Answer 8.5×10^{-2}

■ CHANGING FROM SCIENTIFIC NOTATION TO STANDARD NOTATION

We can change a number written in scientific notation to **standard notation.** For example, to write 9.3×10^7 in standard notation, we multiply 9.3 by 10^7.

$$9.3 \times 10^7 = 9.3 \times 10,000,000$$
$$= 93,000,000$$

EXAMPLE 5   Write   **a.** 3.4×10^5 and   **b.** 2.1×10^{-4} in standard notation.

Solution   **a.** $3.4 \times 10^5 = 3.4 \times 100,000$ **b.** $2.1 \times 10^{-4} = 2.1 \times \dfrac{1}{10^4}$
$$= 340,000$$

$$= 2.1 \times \frac{1}{10,000}$$

$$= 0.00021 \qquad\blacksquare$$

Self Check   Write   **a.** 4.76×10^5 and   **b.** 9.8×10^{-3} in standard notation.
Answers   **a.** 476,000,   **b.** 0.0098

Each of the following numbers is written in both scientific and standard notation. In each case, the exponent gives the number of places that the decimal point moves, and the sign of the exponent indicates the direction that it moves.

$5.32 \times 10^5 = 5\,3\,2\,0\,0\,0.$ 5 places to the right.

$2.37 \times 10^6 = 2\,3\,7\,0\,0\,0\,0.$ 6 places to the right.

$8.95 \times 10^{-4} = 0.0\,0\,0\,8\,9\,5$ 4 places to the left.

$8.375 \times 10^{-3} = 0.0\,0\,8\,3\,7\,5$ 3 places to the left.

$9.77 \times 10^0 = 9.77$ No movement of the decimal point.

■ USING SCIENTIFIC NOTATION TO SIMPLIFY COMPUTATIONS

Another advantage of scientific notation becomes apparent when we simplify fractions such as

$$\frac{(0.0032)(25{,}000)}{0.00040}$$

that contain very large or very small numbers. Although we can simplify this fraction by using arithmetic, scientific notation provides an easier way. First, we write each number in scientific notation; then we do the arithmetic on the numbers and the exponential expressions separately. Finally, we write the result in standard form, if desired.

$$\frac{(0.0032)(25{,}000)}{0.00040} = \frac{(3.2 \times 10^{-3})(2.5 \times 10^4)}{4.0 \times 10^{-4}}$$

$$= \frac{(3.2)(2.5)}{4.0} \times \frac{10^{-3}10^4}{10^{-4}}$$

$$= \frac{8.0}{4.0} \times 10^{-3+4-(-4)}$$

$$= 2.0 \times 10^5$$

$$= 200{,}000$$

■ ■ ■ ■ ■ ■ ■ ■ ■ ■   **Finding Powers of Decimals**

CALCULATORS   To find the value of $(453.46)^5$, we can use a scientific calculator and enter these numbers and press these keys:

Keystrokes

453.46 y^x 5 =   `1.917321395 13`

So we have $(453.46)^5 = 1.917321395 \times 10^{13}$. Since this number is too large to show on the calculator display, the calculator gives the result in scientific notation.

As the previous example shows, scientific calculators can do operations using scientific notation. Consult your owner's manual to see how to enter numbers written in scientific notation into the calculator.

EXAMPLE 6   **Speed of light**   In a vacuum, light travels 1 meter in approximately 0.000000003 second. How long does it take for light to travel 500 kilometers?

Solution   Because 1 kilometer = 1,000 meters, the length of time for light to travel 500 kilometers (500 · 1,000 meters) is given by

$$(0.000000003)(500)(1,000) = (3 \times 10^{-9})(5 \times 10^2)(1 \times 10^3)$$
$$= 3(5) \times 10^{-9+2+3}$$
$$= \mathbf{15 \times 10^{-4}}$$
$$= \mathbf{1.5 \times 10^1} \times 10^{-4}$$
$$= 1.5 \times 10^{-3}$$
$$= 0.0015$$

Light travels 500 kilometers in approximately 0.0015 second.   ■

Orals   *Tell which number of each pair is the larger.*

1. 37.2 or 3.72×10^2 **2.** 37.2 or 3.72×10^{-1}

3. 3.72×10^3 or 4.72×10^3 **4.** 3.72×10^3 or 4.72×10^2

5. 3.72×10^{-1} or 4.72×10^{-2} **6.** 3.72×10^{-3} or 2.72×10^{-2}

EXERCISE 4.3

REVIEW

1. If $y = -1$, find the value of $-5y^{55}$.

2. Evaluate $\dfrac{3a^2 - 2b}{2a + 2b}$ if $a = 4$ and $b = 3$.

Tell which property of real numbers justifies each statement.

3. $5 + z = z + 5$

4. $7(u + 3) = 7u + 7 \cdot 3$

Solve each equation.

5. $3(x - 4) - 6 = 0$

6. $8(3x - 5) - 4(2x + 3) = 12$

VOCABULARY AND CONCEPTS   *Fill in each blank to make a true statement.*

7. A number is written in _____ when it is written as the product of a number between 1 (including 1) and 10 and an integer power of 10.

8. The number 125,000 is written in _____ notation.

PRACTICE   *In Exercises 9–20, write each number in scientific notation.*

9. 23,000   **10.** 4,750   **11.** 1,700,000   **12.** 290,000

13. 0.062   **14.** 0.00073   **15.** 0.0000051   **16.** 0.04

17. 42.5×10^2 **18.** 0.3×10^3 **19.** 0.25×10^{-2} **20.** 25.2×10^{-3}

In Exercises 21–32, write each number in standard notation.

21. 2.3×10^2 **22.** 3.75×10^4 **23.** 8.12×10^5 **24.** 1.2×10^3

25. 1.15×10^{-3} **26.** 4.9×10^{-2} **27.** 9.76×10^{-4} **28.** 7.63×10^{-5}

29. 25×10^6 **30.** 0.07×10^3 **31.** 0.51×10^{-3} **32.** 617×10^{-2}

In Exercises 33–38, use scientific notation to simplify each expression. Give all answers in standard notation.

33. $(3.4 \times 10^2)(2.1 \times 10^3)$ **34.** $(4.1 \times 10^{-3})(3.4 \times 10^4)$

35. $\dfrac{9.3 \times 10^2}{3.1 \times 10^{-2}}$ **36.** $\dfrac{7.2 \times 10^6}{1.2 \times 10^8}$

37. $\dfrac{96{,}000}{(12{,}000)(0.00004)}$ **38.** $\dfrac{(0.48)(14{,}400{,}000)}{96{,}000{,}000}$

APPLICATIONS

39. Distance to Alpha Centauri The distance from the earth to the nearest star outside our solar system is approximately 25,700,000,000,000 miles. Write this number in scientific notation.

40. Speed of sound The speed of sound in air is 33,100 centimeters per second. Write this number in scientific notation.

41. Distance to Mars The distance from Mars to the sun is approximately 1.14×10^8 miles. Write this number in standard notation.

42. Distance to Venus The distance from Venus to the sun is approximately 6.7×10^7 miles. Write this number in standard notation.

43. Length of one meter One meter is approximately 0.00622 mile. Use scientific notation to express this number.

44. Angstrom One angstrom is 1×10^{-7} millimeter. Write this number in standard notation.

45. Distance between Mercury and the sun The distance from Mercury to the sun is approximately 3.6×10^7 miles. Use scientific notation to express this distance in feet. (*Hint:* 5,280 feet = 1 mile.)

46. Mass of a proton The mass of one proton is approximately 1.7×10^{-24} gram. Use scientific notation to express the mass of 1 million protons.

47. Speed of sound The speed of sound in air is approximately 3.3×10^4 centimeters per second. Use scientific notation to express this speed in kilometers per second. (*Hint:* 100 centimeters = 1 meter and 1,000 meters = 1 kilometer.)

48. Light year One light year is approximately 5.87×10^{12} miles. Use scientific notation to express this distance in feet. (*Hint:* 5,280 feet = 1 mile.)

WRITING

49. In what situations would scientific notation be more convenient than standard notation?

50. To multiply a number by a power of 10, we move the decimal point. Which way, and how far? Explain.

SOMETHING TO THINK ABOUT

51. Two positive numbers are written in scientific notation. How could you decide which is larger, without converting either to standard notation?

52. The product $1 \cdot 2 \cdot 3 \cdot 4 \cdot 5$, or 120, is called **5 factorial,** written 5!. Similarly, the number $6! = 6 \cdot 5 \cdot 4 \cdot 3 \cdot 2 \cdot 1 = 620$. Factorials get large very quickly. Calculate 30!, and write the number in standard notation. (*Hint:* Experiment with the $x!$ key on a calculator.) How large a factorial can you compute with a calculator?

4.4 Polynomials

■ POLYNOMIALS ■ MONOMIALS, BINOMIALS, AND TRINOMIALS ■ DEGREE OF A POLYNOMIAL
■ EVALUATING POLYNOMIALS ■ FUNCTIONS ■ FUNCTION NOTATION ■ GRAPHING POLYNOMIAL
FUNCTIONS

Getting Ready *Write each expression using exponents.*

1. $2xxyyy$

2. $3xyyy$

3. $2xx + 3yy$

4. $xxx + yyy$

5. $(3xxy)(2xyy)$

6. $(5xyzzz)(xyz)$

7. $3(5xy)\left(\dfrac{1}{3}xy\right)$

8. $(xy)(xz)(yz)(xyz)$

■ POLYNOMIALS

Recall that expressions such as

$$3x \qquad 4y^2 \qquad -8x^2y^3 \qquad \text{and} \qquad 25$$

with constant and/or variable factors are called **algebraic terms.** The numerical coefficients of the first three of these terms are 3, 4, and -8, respectively. Because $25 = 25x^0$, 25 is considered to be the numerical coefficient of the term 25.

> **Polynomials**
> A **polynomial** is an algebraic expression that is the sum of one or more terms containing whole-number exponents on the variables.

Here are some examples of polynomials:

$$8xy^2t \qquad 3x + 2 \qquad 4y^2 - 2y + 3 \qquad \text{and} \qquad 3a - 4b - 4c + 8d$$

 WARNING! The expression $2x^3 - 3y^{-2}$ is not a polynomial, because the second term contains a negative exponent on a variable base.

EXAMPLE 1

Tell whether each expression is a polynomial.

a. $x^2 + 2x + 1$ Yes.

b. $3x^{-1} - 2x - 3$ No. The first term has a negative exponent on a variable base.

c. $\dfrac{1}{2}x^3 - 2.3x + 5$ Yes.   ∎

Self Check

Tell whether each expression is a polynomial: **a.** $3x^{-4} + 2x^2 - 3$ and **b.** $7.5x^3 - 4x^2 - 3x$

Answers   **a.** no,   **b.** yes

■ MONOMIALS, BINOMIALS, AND TRINOMIALS

A polynomial with one term is called a **monomial.** A polynomial with two terms is called a **binomial.** A polynomial with three terms is called a **trinomial.** Here are some examples.

Monomials	Binomials	Trinomials
$5x^2y$	$3u^3 - 4u^2$	$-5t^2 + 4t + 3$
$-6x$	$18a^2b + 4ab$	$27x^3 - 6x - 2$
29	$-29z^{17} - 1$	$-32r^6 + 7y^3 - z$

EXAMPLE 2

Classify each polynomial as a monomial, a binomial, or a trinomial.

a. $5x^4 + 3x$ Since the polynomial has two terms, it is a binomial.

b. $7x^4 - 5x^3 - 2$ Since the polynomial has three terms, it is a trinomial.

c. $-5x^2y^3$ Since the polynomial has one term, it is a monomial.   ∎

Self Check

Classify each polynomial as a monomial, a binomial, or a trinomial: **a.** $5x$, **b.** $-5x^2 + 2x - 5$,   and   **c.** $16x^2 - 9y^2$.

Answers   **a.** monomial,   **b.** trinomial,   **c.** binomial

■ DEGREE OF A POLYNOMIAL

The monomial $7x^6$ is called a **monomial of sixth degree** or a **monomial of degree 6,** because the variable x occurs as a factor six times. The monomial $3x^3y^4$ is a monomial of seventh degree, because the variables x and y occur as factors a total of seven times. Other examples are

$-2x^3$ is a monomial of degree 3.

$47x^2y^3$ is a monomial of degree 5.

$18x^4y^2z^8$ is a monomial of degree 14.

8 is a monomial of degree 0, because $8 = 8x^0$.

These examples illustrate the following definition.

Degree of a Monomial

If a is a nonzero constant, the **degree of the monomial** ax^n is n.

The **degree of a monomial with several variables** is the sum of the exponents on those variables.

WARNING! Note that the degree of ax^n is not defined when $a = 0$. Since $ax^n = 0$ when $a = 0$, the constant 0 has no defined degree.

Because each term of a polynomial is a monomial, we define the degree of a polynomial by considering the degree of each of its terms.

Degree of a Polynomial

The **degree of a polynomial** is the same as the degree of its term with largest degree.

For example,

- $x^2 + 2x$ is a binomial of degree 2, because the degree of its first term is 2 and the degree of its other term is less than 2.
- $3x^3y^2 + 4x^4y^4 - 3x^3$ is a trinomial of degree 8, because the degree of its second term is 8 and the degree of each of its other terms is less than 8.
- $25x^4y^3z^7 - 15xy^8z^{10} - 32x^8y^8z^3 + 4$ is a polynomial of degree 19, because its second and third terms are of degree 19. Its other terms have degrees less than 19.

EXAMPLE 3   Find the degree of each polynomial.

a. $-4x^3 - 5x^2 + 3x$ 3

b. $5x^4y^2 + 7xy^2 - 16x^3y^5$ 8

c. $-17a^2b^3c^4 + 12a^3b^4c$ 9

■ EVALUATING POLYNOMIALS

When a number is substituted for the variable in a polynomial, the polynomial takes
on a numerical value. Finding that value is called **evaluating the polynomial.**

EXAMPLE 4   Evaluate the polynomial $3x^2 + 2$ when   **a.** $x = 0$,   **b.** $x = 2$,   **c.** $x = -3$,   and
d. $x = -\frac{1}{5}$.

Solution   **a.** $3x^2 + 2 = 3(0)^2 + 2$ **b.** $3x^2 + 2 = 3(2)^2 + 2$
$$= 3(0) + 2$$ $$= 3(4) + 2$$
$$= 0 + 2$$ $$= 12 + 2$$
$$= 2$$ $$= 14$$

c. $3x^2 + 2 = 3(-3)^2 + 2$ **d.** $3x^2 + 2 = 3\left(-\frac{1}{5}\right)^2 + 2$
$$= 3(9) + 2$$ $$= 3\left(\frac{1}{25}\right) + 2$$
$$= 27 + 2$$
$$= 29$$ $$= \frac{3}{25} + \frac{50}{25}$$
$$= \frac{53}{25}$$

When we evaluate a polynomial function for several values of its variable, we
often write the results in a table.

EXAMPLE 5   Evaluate the polynomial $x^3 + 1$ when   **a.** $x = -2$,   **b.** $x = -1$,   **c.** $x = 0$,
d. $x = 1$,   and   **e.** $x = 2$. Write the results in a table.

Solution

x	$x^3 + 1$	
a. -2	-7	$x^3 + 1 = (-2)^3 + 1 = -7.$
b. -1	0	$x^3 + 1 = (-1)^3 + 1 = 0.$
c. 0	1	$x^3 + 1 = (0)^3 + 1 = 1.$
d. 1	2	$x^3 + 1 = (1)^3 + 1 = 2.$
e. 2	9	$x^3 + 1 = (2)^3 + 1 = 9.$

Self Check

Consider the polynomial $-x^3 + 1$. Complete the following table.

x	$-x^3 + 1$
-2	
-1	
0	
1	
2	

Answers   9, 2, 1, 0, -7

■ FUNCTIONS

The results of Examples 4 and 5 illustrate that for every input value x that we substitute into a polynomial with the variable x, there is exactly one output value. Whenever we consider a polynomial equation such as $y = 3x^2 + 2$, where each input value x determines a single output value y, we say that y is a *function* of x.

Function
Any equation in x and y where each value of x (the input) determines one value of y (the output) is called a **function.** In this case, we say that y is a function of x.

The set of all input values x is called the **domain** of the function, and the set of all output values y is called the **range.**

Since each output value y depends on some input value of x, we call y the **dependent variable** and x the **independent variable.** Here are some equations that define y to be a function of x.

1. $y = 2x - 3$ Note that each input value of x determines a single output value of y. For example, if $x = 4$, then $y = 5$. Since any real number can be substituted for x, the domain is the set of real numbers. We will soon show that the range is also the set of real numbers.

2. $y = x^2$ Note that each input value of x determines a single output value of y. For example, if $x = 3$, then $y = 9$. Since any real number can be substituted for x, the domain is the set of real numbers. Since the square of any real number is positive or 0, the range is the set of all numbers y such that $y \geq 0$.

3. $y = x^3$ Note that each input value of x determines a single output value of y. For example, if $x = -2$, then $y = -8$. Since any number can be substituted for x, the domain is the set of real numbers. We will soon show that the range is also the set of real numbers.

■ FUNCTION NOTATION

There is a special notation for functions that uses the symbol $f(x)$, read as "f of x."

> **Function Notation**
> The notation $y = f(x)$ denotes that the variable y is a function of x.

 WARNING!   The notation $f(x)$ does not mean "f times x."

The notation $y = f(x)$ provides a way to denote the values of y in a function that correspond to individual values of x. For example, if $y = f(x)$, the value of y that is determined by $x = 3$ is denoted as $f(3)$. Similarly, $f(-1)$ represents the value of y that corresponds to $x = -1$.

EXAMPLE 6   Let $y = f(x) = 2x - 3$ and find   **a.** $f(3)$,   **b.** $f(-1)$,   **c.** $f(0)$,   and   **d.** $f(0.2)$.

Solution   **a.** We replace x with 3.   **b.** We replace x with -1.

$$f(x) = 2x - 3$$
$$f(3) = 2(3) - 3$$
$$= 6 - 3$$
$$= 3$$

$$f(x) = 2x - 3$$
$$f(-1) = 2(-1) - 3$$
$$= -2 - 3$$
$$= -5$$

c. We replace x with 0.   **d.** We replace x with 0.2.

$$f(x) = 2x - 3$$
$$f(0) = 2(0) - 3$$
$$= 0 - 3$$
$$= -3$$

$$f(x) = 2x - 3$$
$$f(0.2) = 2(0.2) - 3$$
$$= 0.4 - 3$$
$$= -2.6$$ ■

Self Check   Using the function of Example 6, find   **a.** $f(-2)$ and   **b.** $f\left(\frac{3}{2}\right)$.
Answers   **a.** -7,   **b.** 0

■ ■ ■ ■ ■ ■ ■ ■ ■ **Height of a Rocket**

CALCULATORS   The height h (in feet) of a toy rocket launched straight up into the air with an initial velocity of 64 feet per second is given by the polynomial function

$$h = f(t) = -16t^2 + 64t$$

In this case, the height h is the dependent variable, and the time t is the independent variable. To find the height of the rocket 3.5 seconds after launch, we substitute 3.5 for t and evaluate h.

$$h = -16t^2 + 64t$$
$$h = -16(3.5)^2 + 64(3.5)$$

To evaluate h using a scientific calculator, we enter these numbers and press these keys:

Keystrokes

$$16 \quad +/- \quad \times \quad 3.5 \quad x^2 \quad + \quad (\quad 64 \quad \times \quad 3.5 \quad) \quad =$$

The display will read $28.$
After 3.5 seconds, the rocket will be 28 feet above the ground.

■ GRAPHING POLYNOMIAL FUNCTIONS

Since the right-hand sides of the functions $y = f(x) = 2x - 3$, $y = f(x) = x^2$, and $y = f(x) = x^3$ are polynomials, they are called **polynomial functions.** We can graph these functions as we graphed equations in Section 3.2. We make a table of values, plot points, and draw the line or curve that passes through those points.

   In the next example, we graph the function $y = f(x) = 2x - 3$. Since its graph is a line, we call this function a **linear function.**

EXAMPLE 7   Graph $y = f(x) = 2x - 3$.

Solution   We substitute numbers for x, compute the corresponding values of $f(x)$, and list the results in a table, as in Figure 4-1. We then plot the pairs (x, y) and draw a line

$$y = f(x) = 2x - 3$$

x	y	(x, y)
-3	-9	$(-3, -9)$
-2	-7	$(-2, -7)$
-1	-5	$(-1, -5)$
0	-3	$(0, -3)$
1	-1	$(1, -1)$
2	1	$(2, 1)$
3	3	$(3, 3)$

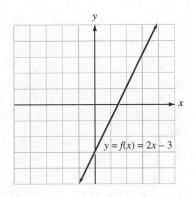

FIGURE 4-1

through the points, as shown in the figure. From the graph, we can see that x can be any value. This confirms that the domain is the set of all real numbers. We can also see that y can be any value. This confirms that the range is also the set of all real numbers.   ∎

Self Check   Graph $y = f(x) = \frac{1}{2}x + 3$ and tell whether it is a linear function.

Answer   It is a linear function.

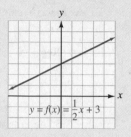

In the next example, we graph the function $y = f(x) = x^2$, called the **squaring function**. Since the polynomial on the right-hand side is of second degree, we call this function a **quadratic function.**

EXAMPLE 8   Graph $y = f(x) = x^2$.

Solution   We substitute numbers for x, compute the corresponding values of $f(x)$, and list the results in a table, as in Figure 4-2. We then plot the pairs (x, y) and draw a smooth curve through the points, as shown in the figure. This curve is called a **parabola.** From the graph, we can see that x can be any value. This confirms that the domain is the set of all real numbers. We can also see that y is always a positive number or 0. This confirms that the range is the set of all real numbers such that $y \geq 0$.

$$y = f(x) = x^2$$

x	y	(x, y)
-3	9	$(-3, 9)$
-2	4	$(-2, 4)$
-1	1	$(-1, 1)$
0	0	$(0, 0)$
1	1	$(1, 1)$
2	4	$(2, 4)$
3	9	$(3, 9)$

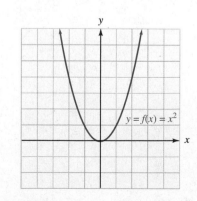

FIGURE 4-2   ∎

Self Check   Graph $y = f(x) = x^2 - 3$ and compare the graph to the graph of $y = f(x) = x^2$ shown in Figure 4-2.

Answer   The graph has the same shape but is 3 units lower.

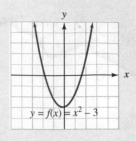

In the next example, we graph the function $y = f(x) = x^3$, called the **cubing function**.

EXAMPLE 9   Graph $y = f(x) = x^3$.

Solution   We substitute numbers for x, compute the corresponding values of $f(x)$, and list the results in a table, as in Figure 4-3. We then plot the pairs (x, y) and draw a smooth curve through the points, as shown in the figure.

$$y = f(x) = x^3$$

x	y	(x, y)
-2	-8	$(-2, -8)$
-1	-1	$(-1, -1)$
0	0	$(0, 0)$
1	1	$(1, 1)$
2	8	$(2, 8)$

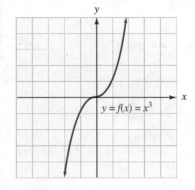

FIGURE 4-3   ■

Self Check   Graph $y = f(x) = x^3 + 3$ and compare the graph to the graph of $y = f(x) = x^3$ shown in Figure 4-3.

Answer   The graph has the same shape but is 3 units higher.

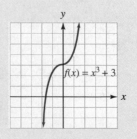

Graphing Polynomial Functions

GRAPHING CALCULATORS

It is possible to use a graphing calculator to generate tables and graphs for polynomial functions. For example, Figure 4-4 shows calculator tables and graphs of $y = f(x) = 2x - 3$, $y = f(x) = x^2$, and $y = f(x) = x^3$.

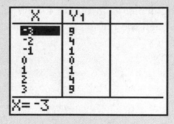 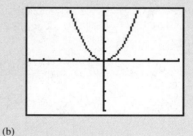

(a)

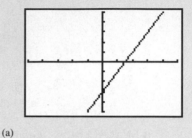

(b)

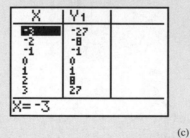 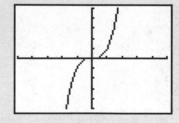

(c)

FIGURE 4-4

EXAMPLE 10

Graph $y = f(x) = x^2 - 2x$.

Solution

We substitute numbers for x, compute the corresponding values of $f(x)$, and list the results in a table, as in Figure 4-5. We then plot the pairs (x, y) and draw a smooth curve through the points, as shown in the figure.

$$y = f(x) = x^2 - 2x$$

x	y	(x, y)
-2	8	$(-2, 8)$
-1	3	$(-1, 3)$
0	0	$(0, 0)$
1	-1	$(1, -1)$
2	0	$(2, 0)$
3	3	$(3, 3)$
4	8	$(4, 8)$

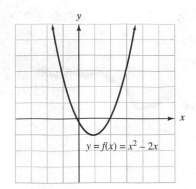

FIGURE 4-5

Self Check
Answer

Use a graphing calculator to graph $y = f(x) = x^2 - 2x$.

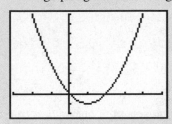

Orals *Give an example of a polynomial that is . . .*

1. a binomial

2. a monomial

3. a trinomial

4. not a monomial, a binomial, or a trinomial

5. of degree 3

6. of degree 1

7. of degree 0

8. has no defined degree

EXERCISE 4.4

REVIEW *Solve each equation.*

1. $5(u - 5) + 9 = 2(u + 4)$

2. $8(3a - 5) - 12 = 4(2a + 3)$

Solve each inequality and graph the solution set.

3. $-4(3y + 2) \leq 28$

4. $-5 < 3t + 4 \leq 13$

Write each expression without using parentheses or negative exponents.

5. $(x^2 x^4)^3$

6. $(a^2)^3 (a^3)^2$

7. $\left(\dfrac{y^2 y^5}{y^4} \right)^3$

8. $\left(\dfrac{2t^3}{t} \right)^{-4}$

VOCABULARY AND CONCEPTS   *Fill in each blank to make a true statement.*

9. An expression with a constant and/or a variable is called an _____ term.

10. The numerical coefficient of the term $-25x^2y^3$ is _____.

11. A _____ is an algebraic expression that is the sum of one or more terms containing whole-number exponents.

12. A _____ is a polynomial with two terms.

13. A _____ is a polynomial with three terms.

14. A _____ is a polynomial with one term.

15. If $a \neq 0$, the _____ of ax^n is n.

16. The degree of a monomial with several variables is the _____ of the exponents on those variables.

17. Any equation in x and y where each input value x determines exactly one output value y is called a _____.

18. $f(x)$ is read as _____.

19. In a function, the set of all input values is called the _____.

20. In a function, the set of all output values is called the _____.

Tell whether each expression is a polynomial.

21. $x^3 - 5x^2 - 2$

22. $x^{-4} - 5x$

23. $\frac{1}{2}x^3 + 3$

24. $x^3 - 1$

In Exercises 25–36, classify each polynomial as a monomial, a binomial, a trinomial, or none of these.

25. $3x + 7$

26. $3y - 5$

27. $3y^2 + 4y + 3$

28. $3xy$

29. $3z^2$

30. $3x^4 - 2x^3 + 3x - 1$

31. $5t - 32$

32. $9x^2y^3z^4$

33. $s^2 - 23s + 31$

34. $12x^3 - 12x^2 + 36x - 3$

35. $3x^5 - 2x^4 - 3x^3 + 17$

36. x^3

In Exercises 37–48, give the degree of each polynomial.

37. $3x^4$

38. $3x^5 - 4x^2$

39. $-2x^2 + 3x^3$

40. $-5x^5 + 3x^2 - 3x$

41. $3x^2y^3 + 5x^3y^5$

42. $-2x^2y^3 + 4x^3y^2z$

43. $-5r^2s^2t - 3r^3st^2 + 3$

44. $4r^2s^3t^3 - 5r^2s^8$

45. $x^{12} + 3x^2y^3z^4$

46. 17^2x

47. 38

48. -25

PRACTICE   *In Exercises 49–52, evaluate $5x - 3$ for each value.*

49. $x = 2$

50. $x = 0$

51. $x = -1$

52. $x = -2$

In Exercises 53–56, evaluate $-x^2 - 4$ for each value.

53. $x = 0$

54. $x = 1$

55. $x = -1$

56. $x = -2$

In Exercises 57–60, evaluate $x^2 - 2x + 3$ for each value.

57. $x = 0$

58. $x = 3$

59. $x = -2$

60. $x = -1$

In Exercises 61–64, complete each table.

61.

x	$x^2 - 3$
-2	
-1	
0	
1	
2	

62.

x	$-x^2 + 3$
-2	
-1	
0	
1	
2	

63.

x	$x^3 + 2$
-2	
-1	
0	
1	
2	

64.

x	$-x^3 + 2$
-2	
-1	
0	
1	
2	

In Exercises 65–68, graph each polynomial function. Check your work with a graphing calculator.

65. $f(x) = x^2 - 1$

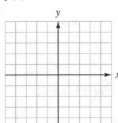

66. $f(x) = x^2 + 2$

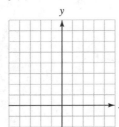

67. $f(x) = x^3 + 2$

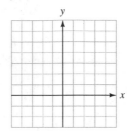

68. $f(x) = x^3 - 2$

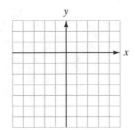

APPLICATIONS *Use a calculator to help solve each problem.*

69. Height of a rocket See the Calculators section on page 289. Find the height of the rocket 2 seconds after launch.

70. Height of a rocket Again referring to page 289, make a table of values to find the rocket's height at various times. For what values of t will the height of the rocket be 0?

71. Stopping distance The number of feet that a car travels before stopping depends on the driver's reaction time and the braking distance. (See Illustration 1.) For one driver, the stopping distance d is given by the function $d = f(v) = 0.04v^2 + 0.9v$, where v is the velocity of the car. Find the stopping distance when the driver is traveling at 30 mph.

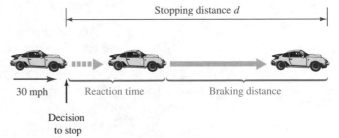

ILLUSTRATION 1

72. Stopping distance Find the stopping distance of the car discussed in Exercise 71 when the driver is going 70 mph.

WRITING

73. Describe how to determine the degree of a polynomial.

74. Describe how to classify a polynomial as a monomial, a binomial, a trinomial, or none of these.

SOMETHING TO THINK ABOUT

75. Find a polynomial whose value will be 1 if you substitute $\frac{3}{2}$ for x.

76. Graph the function $y = f(x) = -x^2$. What do you discover?

4.5 Adding and Subtracting Polynomials

■ ADDING MONOMIALS ■ SUBTRACTING MONOMIALS ■ ADDING POLYNOMIALS ■ SUBTRACTING POLYNOMIALS ■ ADDING AND SUBTRACTING MULTIPLES OF POLYNOMIALS ■ AN APPLICATION OF ADDING POLYNOMIALS

Getting Ready   *Combine like terms and simplify, if possible.*

1. $3x + 2x$ **2.** $5y - 3y$ **3.** $19x + 6x$ **4.** $8z - 3z$

5. $9r + 3r$ **6.** $4r - 3s$ **7.** $7r - 7r$ **8.** $17r - 17r^2$

■ ADDING MONOMIALS

Recall that like terms have the same variables with the same exponents. For example,

$3xyz^2$ and $-2xyz^2$ are like terms.

$\dfrac{1}{2}ab^2c$ and $\dfrac{1}{3}a^2bd^2$ are unlike terms.

Also recall that to combine like terms, we add (or subtract) their coefficients and keep the same variables with the same exponents. For example,

$$2y + 5y = (2 + 5)y \qquad \text{and} \qquad -3x^2 + 7x^2 = (-3 + 7)x^2$$
$$= 7y \qquad\qquad\qquad\qquad\qquad = 4x^2$$

Likewise,

$$4x^3y^2 + 9x^3y^2 = 13x^3y^2 \qquad \text{and} \qquad 4r^2s^3t^4 + 7r^2s^3t^4 = 11r^2s^3t^4$$

These examples suggest that to add like monomials, we simply combine like terms.

⬛ a,c   **EXAMPLE 1**

a. $5xy^3 + 7xy^3 = 12xy^3$

b. $-7x^2y^2 + 6x^2y^2 + 3x^2y^2 = -x^2y^2 + 3x^2y^2$
$$= 2x^2y^2$$

c. $(2x^2)^2 + 81x^4 = 4x^4 + 81x^4 \qquad (2x^2)^2 = (2x^2)(2x^2) = 4x^4.$
$$= 85x^4$$

■

Self Check   Do the following additions: **a.** $6a^3b^2 + 5a^3b^2$,   **b.** $-2pq^2 + 5pq^2 + 8pq^2$, and   **c.** $27x^6 + (2x^2)^3$.

Answers   **a.** $11a^3b^2$,   **b.** $11pq^2$,   **c.** $35x^6$

■ SUBTRACTING MONOMIALS

To subtract one monomial from another, we add the negative of the monomial that is to be subtracted. In symbols, $x - y = x + (-y)$.

EXAMPLE 2

a. $8x^2 - 3x^2 = 8x^2 + (-3x^2)$
$$= 5x^2$$

b. $6x^3y^2 - 9x^3y^2 = 6x^3y^2 + (-9x^3y^2)$
$$= -3x^3y^2$$

c. $-3r^2st^3 - 5r^2st^3 = -3r^2st^3 + (-5r^2st^3)$
$$= -8r^2st^3$$ ■

Self Check

Answers

Find each difference: **a.** $12m^3 - 7m^3$ and **b.** $-4p^3q^2 - 8p^3q^2$.
a. $5m^3$, **b.** $-12p^3q^2$

■ ADDING POLYNOMIALS

Because of the distributive property, we can remove parentheses enclosing several terms when the sign preceding the parentheses is a + sign. We simply drop the parentheses.

$$+(3x^2 + 3x - 2) = +1(3x^2 + 3x - 2)$$
$$= 1(3x^2) + 1(3x) + 1(-2)$$
$$= 3x^2 + 3x + (-2)$$
$$= 3x^2 + 3x - 2$$

We can add polynomials by removing parentheses, if necessary, and then combining any like terms that are contained within the polynomials.

EXAMPLE 3   Add $(3x^2 - 3x + 2) + (2x^2 + 7x - 4)$.

Solution

$$(3x^2 - 3x + 2) + (2x^2 + 7x - 4)$$
$$= 3x^2 - 3x + 2 + 2x^2 + 7x - 4$$
$$= 3x^2 + 2x^2 - 3x + 7x + 2 + (-4)$$
$$= 5x^2 + 4x - 2$$ ■

Self Check

Answer

Add $(2a^2 - a + 4) + (5a^2 + 6a - 5)$.
$7a^2 + 5a - 1$

Problems such as Example 3 are often written with like terms aligned vertically. We can then add column by column.

$$3x^2 - 3x + 2$$
$$\underline{2x^2 + 7x - 4}$$
$$5x^2 + 4x - 2$$

EXAMPLE 4   Add: $4x^2y + 8x^2y^2 - 3x^2y^3$
$$\underline{3x^2y - 8x^2y^2 + 8x^2y^3}$$
$$7x^2y \qquad\quad + 5x^2y^3$$ ■

Self Check   Add: $4pq^2 + 6pq^3 - 7pq^4$
$$\underline{2pq^2 - 8pq^3 + 9pq^4}$$

Answer $6pq^2 - 2pq^3 + 2pq^4$

■ SUBTRACTING POLYNOMIALS

Because of the distributive property, we can remove parentheses enclosing several terms when the sign preceding the parentheses is a − sign. We simply drop the minus sign and the parentheses, and *change the sign of every term within the parentheses.*

$$-(3x^2 + 3x - 2) = -1(3x^2 + 3x - 2)$$
$$= -1(3x^2) + (-1)(3x) + (-1)(-2)$$
$$= -3x^2 + (-3x) + 2$$
$$= -3x^2 - 3x + 2$$

This suggests that the way to subtract polynomials is to remove parentheses and combine like terms.

EXAMPLE 5   **a.** $(3x - 4) - (5x + 7) = 3x - 4 - 5x - 7$
$$= -2x - 11$$

b. $(3x^2 - 4x - 6) - (2x^2 - 6x + 12) = 3x^2 - 4x - 6 - 2x^2 + 6x - 12$
$$= x^2 + 2x - 18$$

c. $(-4rt^3 + 2r^2t^2) - (-3rt^3 + 2r^2t^2) = -4rt^3 + 2r^2t^2 + 3rt^3 - 2r^2t^2$
$$= -rt^3$$ ■

Self Check   Find the difference: $(-2a^2b + 5ab^2) - (-5a^2b - 7ab^2)$.

Answer $3a^2b + 12ab^2$

To subtract polynomials in vertical form, we add the negative of the **subtrahend** (the bottom polynomial) to the **minuend** (the top polynomial).

EXAMPLE 6   Subtract $3x^2y - 2xy^2$ from $2x^2y + 4xy^2$.

Solution   We write the subtraction in vertical form, change the signs of the terms of the subtrahend, and add:

$$\begin{array}{r} 2x^2y + 4xy^2 \\ -\ 3x^2y - 2xy^2 \end{array} \longrightarrow \begin{array}{r} 2x^2y + 4xy^2 \\ +\ -3x^2y + 2xy^2 \\ \hline -\ x^2y + 6xy^2 \end{array}$$

In horizontal form, the solution is

$$2x^2y + 4xy^2 - (3x^2y - 2xy^2) = 2x^2y + 4xy^2 - 3x^2y + 2xy^2$$
$$= -x^2y + 6xy^2$$

Self Check   Find the difference: $\begin{array}{r} 5p^2q - 6pq + 7q \\ -\ 2p^2q + 2pq - 8q \end{array}$

Answer $3p^2q - 8pq + 15q$

EXAMPLE 7   Subtract $6xy^2 + 4x^2y^2 - x^3y^2$ from $-2xy^2 - 3x^3y^2$.

Solution

$$\begin{array}{r} -2xy^2 \qquad -3x^3y^2 \\ -\ 6xy^2 + 4x^2y^2 - x^3y^2 \end{array} \longrightarrow \begin{array}{r} -2xy^2 \qquad -3x^3y^2 \\ +\ -6xy^2 - 4x^2y^2 + x^3y^2 \\ \hline -8xy^2 - 4x^2y^2 - 2x^3y^2 \end{array}$$

In horizontal form, the solution is

$$-2xy^2 - 3x^3y^2 - (6xy^2 + 4x^2y^2 - x^3y^2)$$
$$= -2xy^2 - 3x^3y^2 - 6xy^2 - 4x^2y^2 + x^3y^2$$
$$= -8xy^2 - 4x^2y^2 - 2x^3y^2$$

Self Check   Subtract $-2pq^2 - 2p^2q^2 + 3p^3q^2$ from $5pq^2 + 3p^2q^2 - p^3q^2$.
Answer $7pq^2 + 5p^2q^2 - 4p^3q^2$

■ **ADDING AND SUBTRACTING MULTIPLES OF POLYNOMIALS**

Because of the distributive property, we can remove parentheses enclosing several terms when a monomial precedes the parentheses. We simply multiply every term within the parentheses by that monomial. For example, to add $3(2x + 5)$ and $2(4x - 3)$, we proceed as follows:

$$3(2x + 5) + 2(4x - 3) = 6x + 15 + 8x - 6$$
$$= 6x + 8x + 15 - 6 \qquad \text{$15 + 8x = 8x + 15$.}$$
$$= 14x + 9 \qquad \text{Combine like terms.}$$

EXAMPLE 8   **a.** $3(x^2 + 4x) + 2(x^2 - 4) = 3x^2 + 12x + 2x^2 - 8$
$$= 5x^2 + 12x - 8$$

b. $8(y^2 - 2y + 3) - 4(2y^2 + y - 3) = 8y^2 - 16y + 24 - 8y^2 - 4y + 12$
$$= -20y + 36$$

c. $-4x(xy^2 - xy + 3) - x(xy^2 - 2) + 3(x^2y^2 + 2x^2y)$
$$= -4x^2y^2 + 4x^2y - 12x - x^2y^2 + 2x + 3x^2y^2 + 6x^2y$$
$$= -2x^2y^2 + 10x^2y - 10x \qquad ■$$

Self Check   Remove parentheses and simplify: **a.** $2(a^3 - 3a) + 5(a^3 + 2a)$ and
b. $5x(xy + 2x) - x^2(y - 3)$.

Answers   **a.** $7a^3 + 4a$,   **b.** $4x^2y + 13x^2$

■ AN APPLICATION OF ADDING POLYNOMIALS

EXAMPLE 9   **Property values**   A house purchased for \$95,000 is expected to appreciate according to the formula $y = 2{,}500x + 95{,}000$, where y is the value of the house after x years. A second house purchased for \$125,000 is expected to appreciate according to the formula $y = 4{,}500x + 125{,}000$. Find one formula that will give the value of both properties after x years.

Solution   The value of the first house after x years is given by the polynomial $2{,}500x + 95{,}000$. The value of the second house after x years is given by the polynomial $4{,}500x + 125{,}000$. The value of both houses will be the sum of these two polynomials.

$$2{,}500x + 95{,}000 + 4{,}500x + 125{,}000 = 7{,}000x + 220{,}000$$

The total value of the properties is given by the formula $7{,}000x + 220{,}000$.   ■

Orals   *Simplify.*

1. $x^3 + 3x^3$
2. $3xy + xy$
3. $(x + 3y) - (x + y)$
4. $5(1 - x) + 3(x - 1)$
5. $(2x - y^2) - (2x + y^2)$
6. $5(x^2 + y) + (x^2 - y)$
7. $3x^2 + 2y + x^2 - y$
8. $2x^2y + y - (2x^2y - y)$

EXERCISE 4.5

REVIEW *Let $a = 3$, $b = -2$, $c = -1$, and $d = 2$. Evaluate each expression.*

1. $ab + cd$

2. $ad + bc$

3. $a(b + c)$

4. $d(b + a)$

5. Solve the inequality $-4(2x - 9) \geq 12$ and graph the solution set.

6. The **kinetic energy** of a moving object is given by the formula

$$K = \frac{mv^2}{2}$$

Solve the formula for m.

VOCABULARY AND CONCEPTS *Fill in each blank to make a true statement.*

7. A _____ is a polynomial with one term.

8. If two polynomials are subtracted in vertical form, the bottom polynomial is called the _____, and the top polynomial is called the _____.

9. To add like monomials, add the numerical _____ and keep the _____.

10. $a - b = a +$ _____

11. To add two polynomials, combine any _____ contained in the polynomials.

12. To subtract polynomials, remove parentheses and combine _____.

In Exercises 13–24, tell whether the terms are like or unlike terms. If they are like terms, add them.

13. $3y$, $4y$

14. $3x^2$, $5x^2$

15. $3x$, $3y$

16. $3x^2$, $6x$

17. $3x^3$, $4x^3$, $6x^3$

18. $-2y^4$, $-6y^4$, $10y^4$

19. $-5x^3y^2$, $13x^3y^2$

20. 23, $12x$

21. $-23t^6$, $32t^6$, $56t^6$

22. $32x^5y^3$, $-21x^5y^3$, $-11x^5y^3$

23. $-x^2y$, xy, $3xy^2$

24. $4x^3y^2z$, $-6x^3y^2z$, $2x^3y^2z$

PRACTICE *In Exercises 25–42, simplify each expression if possible.*

25. $4y + 5y$

26. $-2x + 3x$

27. $-8t^2 - 4t^2$

28. $15x^2 + 10x^2$

29. $32u^3 - 16u^3$

30. $25xy^2 - 7xy^2$

31. $18x^5y^2 - 11x^5y^2$

32. $17x^6y - 22x^6y$

33. $3rst + 4rst + 7rst$

34. $-2ab + 7ab - 3ab$

35. $-4a^2bc + 5a^2bc - 7a^2bc$

36. $(xy)^2 + 4x^2y^2 - 2x^2y^2$

37. $(3x)^2 - 4x^2 + 10x^2$

38. $(2x)^4 - (3x^2)^2$

39. $5x^2y^2 + 2(xy)^2 - (3x^2)y^2$

40. $-3x^3y^6 + 2(xy^2)^3 - (3x)^3y^6$

41. $(-3x^2y)^4 + (4x^4y^2)^2 - 2x^8y^4$

42. $5x^5y^{10} - (2xy^2)^5 + (3x)^5y^{10}$

In Exercises 43–74, do the operations and simplify.

43. $(3x + 7) + (4x - 3)$

44. $(2y - 3) + (4y + 7)$

45. $(4a + 3) - (2a - 4)$

46. $(5b - 7) - (3b + 5)$

47. $(2x + 3y) + (5x - 10y)$

48. $(5x - 8y) - (2x + 5y)$

49. $(-8x - 3y) - (11x + y)$

50. $(-4a + b) + (5a - b)$

51. $(3x^2 - 3x - 2) + (3x^2 + 4x - 3)$

52. $(3a^2 - 2a + 4) - (a^2 - 3a + 7)$

53. $(2b^2 + 3b - 5) - (2b^2 - 4b - 9)$

54. $(4c^2 + 3c - 2) + (3c^2 + 4c + 2)$

55. $(2x^2 - 3x + 1) - (4x^2 - 3x + 2) + (2x^2 + 3x + 2)$

56. $(-3z^2 - 4z + 7) + (2z^2 + 2z - 1) - (2z^2 - 3z + 7)$

57. $2(x + 3) + 3(x + 3)$

58. $5(x + y) + 7(x + y)$

59. $-8(x - y) + 11(x - y)$

60. $-4(a - b) - 5(a - b)$

61. $2(x^2 - 5x - 4) - 3(x^2 - 5x - 4) + 6(x^2 - 5x - 4)$

62. $7(x^2 + 3x + 1) + 9(x^2 + 3x + 1) - 5(x^2 + 3x + 1)$

63. Add: $\quad 3x^2 + 4x + 5$
$$\underline{2x^2 - 3x + 6}$$

64. Add: $\quad 2x^3 + 2x^2 - 3x + 5$
$$\underline{3x^3 - 4x^2 - x - 7}$$

65. Add: $\quad 2x^3 - 3x^2 + 4x - 7$
$$\underline{-9x^3 - 4x^2 - 5x + 6}$$

66. Add: $\quad -3x^3 + 4x^2 - 4x + 9$
$$\underline{2x^3 + 9x - 3}$$

67. Add: $\quad -3x^2y + 4xy + 25y^2$
$$\underline{5x^2y - 3xy - 12y^2}$$

68. Add: $\quad -6x^3z - 4x^2z^2 + 7z^3$
$$\underline{-7x^3z + 9x^2z^2 - 21z^3}$$

69. Subtract: $\quad 3x^2 + 4x - 5$
$$\underline{-2x^2 - 2x + 3}$$

70. Subtract: $\quad 3y^2 - 4y + 7$
$$\underline{6y^2 - 6y - 13}$$

71. Subtract: $\quad 4x^3 + 4x^2 - 3x + 10$
$$\underline{5x^3 - 2x^2 - 4x - 4}$$

72. Subtract: $\quad 3x^3 + 4x^2 + 7x + 12$
$$\underline{-4x^3 + 6x^2 + 9x - 3}$$

73. Subtract: $\quad -2x^2y^2 - 4xy + 12y^2$
$$\underline{10x^2y^2 + 9xy - 24y^2}$$

74. Subtract: $\quad 25x^3 - 45x^2z + 31xz^2$
$$\underline{12x^3 + 27x^2z - 17xz^2}$$

75. Find the sum when $x^2 + x - 3$ is added to the sum of $2x^2 - 3x + 4$ and $3x^2 - 2$.

76. Find the sum when $3y^2 - 5y + 7$ is added to the sum of $-3y^2 - 7y + 4$ and $5y^2 + 5y - 7$.

77. Find the difference when $t^3 - 2t^2 + 2$ is subtracted from the sum of $3t^3 + t^2$ and $-t^3 + 6t - 3$.

78. Find the difference when $-3z^3 - 4z + 7$ is subtracted from the sum of $2z^2 + 3z - 7$ and $-4z^3 - 2z - 3$.

79. Find the sum when $3x^2 + 4x - 7$ is added to the sum of $-2x^2 - 7x + 1$ and $-4x^2 + 8x - 1$.

80. Find the difference when $32x^2 - 17x + 45$ is subtracted from the sum of $23x^2 - 12x - 7$ and $-11x^2 + 12x + 7$.

In Exercises 81–90, simplify each expression.

81. $2(x + 3) + 4(x - 2)$

82. $3(y - 4) - 5(y + 3)$

83. $-2(x^2 + 7x - 1) - 3(x^2 - 2x + 7)$

84. $-5(y^2 - 2y - 6) + 6(2y^2 + 2y - 5)$

85. $2(2y^2 - 2y + 2) - 4(3y^2 - 4y - 1) + 4y(y^2 - y - 1)$

86. $-4(z^2 - 5z) - 5(4z^2 - 1) + 6(2z - 3)$

87. $2a(ab^2 - b) - 3b(a + 2ab) + b(b - a + a^2b)$

88. $3y(xy + y) - 2y^2(x - 4 + y) + 2(y^3 + y^2)$

89. $-4xy^2(x + y + z) - 2x(xy^2 - 4y^2z) - 2y(8xy^2 - 1)$

90. $-3uv(u - v^2 + w) + 4w(uv + w) - 3w(w + uv)$

APPLICATIONS *In Exercises 91–96, consider the following information: If a house was purchased for $105,000 and is expected to appreciate $900 per year, its value y after x years is given by the formula* $y = 900x + 105,000$.

91. Value of a house Find the expected value of the house in 10 years.

92. Value of a house A second house was purchased for $120,000 and was expected to appreciate $1,000 per year. Find a polynomial equation that will give the value y of the house in x years.

93. Value of a house Find the value of the house discussed in Exercise 92 after 12 years.

94. Value of a house Find one polynomial equation that will give the combined value y of both houses after x years.

95. Value of two houses Find the value of the two houses after 20 years by
   a. substituting 20 into the polynomial equations $y = 900x + 105,000$ and $y = 1,000x + 120,000$ and adding the results.
   b. substituting into the result of Exercise 94.

96. Value of two houses Find the value of the two houses after 25 years by
   a. substituting 25 into the polynomial equations $y = 900x + 105,000$ and $y = 1,000x + 120,000$ and adding the results.
   b. substituting into the result of Exercise 94.

In Exercises 97–100, consider the following information: A business bought two computers, one for $6,600 and the other for $9,200. The first computer is expected to depreciate $1,100 per year and the second $1,700 per year.

97. Value of a computer Write a polynomial equation that will give the value of the first computer after x years.

98. Value of a computer Write a polynomial equation that will give the value of the second computer after x years.

99. Value of two computers Find one polynomial equation that will give the value of both computers after x years.

100. Value of two computers In two ways, find the value of the computers after 3 years.

WRITING

101. How do you recognize like terms?

102. How do you add like terms?

SOMETHING TO THINK ABOUT *In Exercises 103–104, let $P(x) = 3x - 5$. Find each value.*

103. $P(x + h) + P(x)$

104. $P(x + h) - P(x)$

105. If $P(x) = x^{23} + 5x^2 + 73$ and $Q(x) = x^{23} + 4x^2 + 73$, find $P(7) - Q(7)$.

106. If two numbers written in scientific notation have the same power of 10, they can be added as similar terms:

$$2 \times 10^3 + 3 \times 10^3 = 5 \times 10^3$$

Without converting to standard form, how could you add

$$2 \times 10^3 + 3 \times 10^4$$

4.6 Multiplying Polynomials

■ MULTIPLYING MONOMIALS ■ MULTIPLYING A POLYNOMIAL BY A MONOMIAL ■ MULTIPLYING A BINOMIAL BY A BINOMIAL ■ THE FOIL METHOD ■ MULTIPLYING A POLYNOMIAL BY A BINOMIAL ■ MULTIPLYING BINOMIALS TO SOLVE EQUATIONS ■ AN APPLICATION OF MULTIPLYING POLYNOMIALS

Getting Ready *Simplify.*

1. $(2x)(3)$ **2.** $(3xxx)(x)$ **3.** $5x^2 \cdot x$ **4.** $8x^2x^3$

Use the distributive property to remove parentheses.

5. $3(x + 5)$ **6.** $x(x + 5)$ **7.** $4(y - 3)$ **8.** $2y(y - 3)$

■ MULTIPLYING MONOMIALS

We have previously multiplied monomials by other monomials. For example, to multiply $4x^2$ by $-2x^3$, we use the commutative and associative properties of multiplication to group the numerical factors together and the variable factors together. Then we multiply the numerical factors and multiply the variable factors.

$$4x^2(-2x^3) = 4(-2)x^2x^3$$
$$= -8x^5$$

This example suggests the following rule.

Multiplying Monomials
To multiply two monomials, multiply the numerical factors and then multiply the variable factors.

a,c

EXAMPLE 1   Multiply **a.** $3x^5(2x^5)$, **b.** $-2a^2b^3(5ab^2)$, and **c.** $-4y^5z^2(2y^3z^3)(3yz)$.

Solution   **a.** $3x^5(2x^5) = 3(2)x^5x^5$
$\qquad\qquad = 6x^{10}$

b. $-2a^2b^3(5ab^2) = -2(5)a^2ab^3b^2$
$\qquad\qquad\qquad = -10a^3b^5$

c. $-4y^5z^2(2y^3z^3)(3yz) = -4(2)(3)y^5y^3yz^2z^3z$
$\qquad\qquad\qquad\qquad = -24y^9z^6$ ■

Self Check   Multiply **a.** $(5a^2b^3)(6a^3b^4)$ and **b.** $(-15p^3q^2)(5p^3q^2)$.
Answers   **a.** $30a^5b^7$, **b.** $-75p^6q^4$

■ MULTIPLYING A POLYNOMIAL BY A MONOMIAL

To find the product of a monomial and a polynomial with more than one term, we use the distributive property. To multiply $2x + 4$ by $5x$, for example, we proceed as follows:

$$5x(2x + 4) = 5x \cdot 2x + 5x \cdot 4 \qquad\qquad \text{Use the distributive property.}$$

$$= 10x^2 + 20x \qquad\qquad \text{Multiply the monomials: } 5x \cdot 2x = 10x^2 \text{ and } 5x \cdot 4 = 20x.$$

This example suggests the following rule.

Multiplying Polynomials by Monomials
To multiply a polynomial with more than one term by a monomial, use the distributive property to remove parentheses and simplify.

EXAMPLE 2   Multiply **a.** $3a^2(3a^2 - 5a)$ and **b.** $-2xz^2(2x - 3z + 2z^2)$.

Solution   **a.** $3a^2(3a^2 - 5a) = 3a^2 \cdot 3a^2 - 3a^2 \cdot 5a$ Use the distributive property.

$$= 9a^4 - 15a^3 \qquad\qquad \text{Multiply: } 3a^2 \cdot 3a^2 = 9a^4 \text{ and } 3a^2 \cdot 5a = 15a^3.$$

b. $-2xz^2(2x - 3z + 2z^2)$

$$= -2xz^2 \cdot 2x - (-2xz^2) \cdot 3z + (-2xz^2) \cdot 2z^2$$

Use the distributive property.

$$= -4x^2z^2 - (-6xz^3) + (-4xz^4)$$

Multiply:
$-2xz^2 \cdot 2x = -4x^2z^2$,
$-2xz^2 \cdot 3z = -6xz^3$,
and $-2xz^2 \cdot 2z^2 = -4xz^4$.

$$= -4x^2z^2 + 6xz^3 - 4xz^4$$

■

Self Check

Answers

Multiply   **a.** $2p^3(3p^2 - 5p)$ and   **b.** $-5a^2b(3a + 2b - 4ab)$.

a. $6p^5 - 10p^4$,   **b.** $-15a^3b - 10a^2b^2 + 20a^3b^2$

■ MULTIPLYING A BINOMIAL BY A BINOMIAL

To multiply two binomials, we must use the distributive property more than once. For example, to multiply $2a - 4$ by $3a + 5$, we proceed as follows.

$$(2a - 4)(3a + 5) = (2a - 4) \cdot 3a + (2a - 4) \cdot 5$$

Use the distributive property.

$$= 3a(2a - 4) + 5(2a - 4)$$

Use the commutative property of multiplication.

$$= 3a \cdot 2a - 3a \cdot 4 + 5 \cdot 2a - 5 \cdot 4$$

Use the distributive property.

$$= 6a^2 - 12a + 10a - 20$$

Do the multiplications.

$$= 6a^2 - 2a - 20$$

Combine like terms.

This example suggests the following rule.

Multiplying Two Binomials

To multiply two binomials, multiply each term of one binomial by each term of the other binomial and combine like terms.

■ THE FOIL METHOD

We can use a shortcut method, called the **FOIL** method, to multiply binomials. FOIL is an acronym for **F**irst terms, **O**uter terms, **I**nner terms, and **L**ast terms. To use the FOIL method to multiply $2a - 4$ by $3a + 5$, we

1. multiply the **F**irst terms $2a$ and $3a$ to obtain $6a^2$,

2. multiply the **O**uter terms $2a$ and 5 to obtain $10a$,

3. multiply the **Inner** terms -4 and $3a$ to obtain $-12a$, and

4. multiply the **Last** terms -4 and 5 to obtain -20.

Then we simplify the resulting polynomial, if possible.

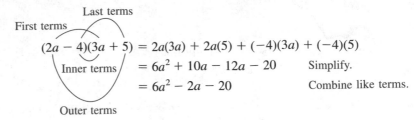

$$
\begin{aligned}
(2a - 4)(3a + 5) &= 2a(3a) + 2a(5) + (-4)(3a) + (-4)(5) \\
&= 6a^2 + 10a - 12a - 20 \qquad \text{Simplify.} \\
&= 6a^2 - 2a - 20 \qquad \text{Combine like terms.}
\end{aligned}
$$

EXAMPLE 3   Find each product.

a.
$$
\begin{aligned}
(3x + 4)(2x - 3) &= 3x(2x) + 3x(-3) + 4(2x) + 4(-3) \\
&= 6x^2 - 9x + 8x - 12 \\
&= 6x^2 - x - 12
\end{aligned}
$$

b.
$$
\begin{aligned}
(2y - 7)(5y - 4) &= 2y(5y) + 2y(-4) + (-7)(5y) + (-7)(-4) \\
&= 10y^2 - 8y - 35y + 28 \\
&= 10y^2 - 43y + 28
\end{aligned}
$$

c.
$$
\begin{aligned}
(2r - 3s)(2r + t) &= 2r(2r) + 2r(t) - 3s(2r) - 3s(t) \\
&= 4r^2 + 2rt - 6rs - 3st
\end{aligned}
$$

∎

Self Check
Answers

Find each product: **a.** $(2a - 1)(3a + 2)$ and   **b.** $(5y - 2z)(2y + 3z)$.
a. $6a^2 + a - 2$,   **b.** $10y^2 + 11yz - 6z^2$

EXAMPLE 4   Simplify each expression.

a. $3(2x - 3)(x + 1)$

$$
\begin{aligned}
&= 3(2x^2 + 2x - 3x - 3) && \text{Use FOIL to multiply the binomials.} \\
&= 3(2x^2 - x - 3) && \text{Combine like terms.} \\
&= 6x^2 - 3x - 9 && \text{Use the distributive property to remove parentheses.}
\end{aligned}
$$

b. $(x + 1)(x - 2) - 3x(x + 3)$

$$= x^2 - 2x + x - 2 - 3x^2 - 9x$$
$$= -2x^2 - 10x - 2 \qquad \text{Combine like terms.} \quad \blacksquare$$

Self Check Simplify $(x + 3)(2x - 1) + 2x(x - 1)$.

Answer $4x^2 + 3x - 3$

The products discussed in Example 5 are called **special products**.

EXAMPLE 5 Find each product.

a. $(x + y)^2 = (x + y)(x + y)$

$$= x^2 + xy + xy + y^2$$
$$= x^2 + 2xy + y^2$$

The square of the sum of two quantities has three terms: the square of the first quantity, plus twice the product of the quantities, plus the square of the second quantity.

b. $(x - y)^2 = (x - y)(x - y)$

$$= x^2 - xy - xy + y^2$$
$$= x^2 - 2xy + y^2$$

The square of the difference of two quantities has three terms: the square of the first quantity, minus twice the product of the quantities, plus the square of the second quantity.

c. $(x + y)(x - y) = x^2 - xy + xy - y^2$
$$= x^2 - y^2$$

The product of a sum and a difference of two quantities is a binomial. It is the product of the first quantities minus the product of the second quantities. Binomials that have the same terms, but different signs, are often called **conjugate binomials.** $\blacksquare$

Self Check Find each product: **a.** $(p + 2)^2$, **b.** $(p - 2)^2$, and **c.** $(p + 2q)(p - 2q)$.

Answers **a.** $p^2 + 4p + 4$, **b.** $p^2 - 4p + 4$, **c.** $p^2 - 4q^2$

Because the products discussed in Example 5 occur so often, it is wise to learn their forms.

Special Products

$$(x + y)^2 = x^2 + 2xy + y^2$$
$$(x - y)^2 = x^2 - 2xy + y^2$$
$$(x + y)(x - y) = x^2 - y^2$$

WARNING! Note that $(x + y)^2 \neq x^2 + y^2$ and $(x - y)^2 \neq x^2 - y^2$.

■ MULTIPLYING A POLYNOMIAL BY A BINOMIAL

We must use the distributive property more than once to multiply a polynomial by a binomial. For example, to multiply $3x^2 + 3x - 5$ by $2x + 3$, we proceed as follows:

$$
\begin{aligned}
\mathbf{(2x + 3)}(3x^2 + 3x - 5) &= \mathbf{(2x + 3)}3x^2 + \mathbf{(2x + 3)}3x - \mathbf{(2x + 3)}5 \\
&= 3x^2(2x + 3) + 3x(2x + 3) - 5(2x + 3) \\
&= 6x^3 + 9x^2 + 6x^2 + 9x - 10x - 15 \\
&= 6x^3 + 15x^2 - x - 15
\end{aligned}
$$

This example suggests the following rule.

Multiplying Polynomials

To multiply one polynomial by another, multiply each term of one polynomial by each term of the other polynomial and combine like terms.

It is often convenient to organize the work vertically.

EXAMPLE 6

a. Multiply:

$$
\begin{array}{r}
3a^2 - 4a + 7 \\
2a + 5 \\
\hline
\end{array}
$$

$2a(3a^2 - 4a + 7) \longrightarrow \quad 6a^3 - 8a^2 + 14a$

$5(3a^2 - 4a + 7) \longrightarrow \quad \underline{+ 15a^2 - 20a + 35}$

$\qquad\qquad\qquad\quad 6a^3 + 7a^2 - 6a + 35$

b. Multiply:

$$
\begin{array}{r}
3y^2 - 5y + 4 \\
- 4y^2 - 3 \\
\hline
\end{array}
$$

$-4y^2(3y^2 - 5y + 4) \longrightarrow -12y^4 + 20y^3 - 16y^2$

$-3(3y^2 - 5y + 4) \longrightarrow \quad \underline{\qquad - 9y^2 + 15y - 12}$

$\qquad\qquad\qquad\quad -12y^4 + 20y^3 - 25y^2 + 15y - 12$

■

Self Check   Multiply   **a.** $(3x + 2)(2x^2 - 4x + 5)$ and   **b.** $(-2x^2 + 3)(2x^2 - 4x - 1)$.

Answers   **a.** $6x^3 - 8x^2 + 7x + 10$,   **b.** $-4x^4 + 8x^3 + 8x^2 - 12x - 3$

■ MULTIPLYING BINOMIALS TO SOLVE EQUATIONS

To solve an equation such as $(x + 2)(x + 3) = x(x + 7)$, we can first use the FOIL method to remove the parentheses on the left-hand side, use the distributive property to remove parentheses on the right-hand side, and proceed as follows:

$$(x + 2)(x + 3) = x(x + 7)$$
$$x^2 + 3x + 2x + 6 = x^2 + 7x$$

$3x + 2x + 6 = 7x$ Subtract x^2 from both sides.

$5x + 6 = 7x$ Combine like terms.

$6 = 2x$ Subtract $5x$ from both sides.

$3 = x$ Divide both sides by 2.

Check: $(x + 2)(x + 3) = x(x + 7)$

$(3 + 2)(3 + 3) \overset{?}{=} 3(3 + 7)$ Replace x with 3.

$5(6) \overset{?}{=} 3(10)$ Do the additions within parentheses.

$30 = 30$

EXAMPLE 7   Solve $(x + 5)(x + 4) = (x + 9)(x + 10)$.

Solution   We use the FOIL method to remove parentheses on both sides of the equation. Then we proceed as follows:

$$(x + 5)(x + 4) = (x + 9)(x + 10)$$
$$x^2 + 4x + 5x + 20 = x^2 + 10x + 9x + 90$$

$9x + 20 = 19x + 90$ Subtract x^2 from both sides and combine like terms.

$20 = 10x + 90$ Subtract $9x$ from both sides.

$-70 = 10x$ Subtract 90 from both sides.

$-7 = x$ Divide both sides by 10.

Check: $(x + 5)(x + 4) = (x + 9)(x + 10)$

$(-7 + 5)(-7 + 4) \overset{?}{=} (-7 + 9)(-7 + 10)$ Replace x with -7.

$(-2)(-3) \overset{?}{=} (2)(3)$ Do the additions within parentheses.

$6 = 6$ ■

■ AN APPLICATION OF MULTIPLYING POLYNOMIALS

EXAMPLE 8

Dimensions of a painting   A square painting is surrounded by a border 2 inches wide. If the area of the border is 96 square inches, find the dimensions of the painting.

Analyze the problem

Refer to Figure 4-6, which shows a square painting surrounded by a border 2 inches wide. We know that the area of this border is 96 square inches, and we are to find the dimensions of the painting.

FIGURE 4-6

Form an equation

Let x represent the length of each side of the square painting. The outer rectangle is also a square, and its dimensions are $(x + 4)$ by $(x + 4)$ inches. Since the area of a square is the product of its length and width, the area of the larger square is $(x + 4)(x + 4)$, and the area of the painting is $x \cdot x$. If we subtract the area of the painting from the area of the larger square, the difference is 96 (the area of the border).

The area of the large square	minus	the area of the square painting	=	the area of the border.
$(x + 4)(x + 4)$	$-$	$x \cdot x$	$=$	96

Solve the equation

$(x + 4)(x + 4) - x^2 = 96$ $x \cdot x = x^2.$

$x^2 + 8x + 16 - x^2 = 96$ $(x + 4)(x + 4) = x^2 + 8x + 16.$

$8x + 16 = 96$ Combine like terms.

$8x = 80$ Subtract 16 from both sides.

$x = 10$ Divide both sides by 8.

State the conclusion

The dimensions of the painting are 10 inches by 10 inches.

Check the result   Check the result.

■

Orals   *Find each product.*

1. $2x^2(3x - 1)$ 2. $5y(2y^2 - 3)$ 3. $7xy(x + y)$ 4. $-2y(2x - 3y)$

5. $(x + 3)(x + 2)$ 6. $(x - 3)(x + 2)$
7. $(2x + 3)(x + 2)$ 8. $(3x - 1)(3x + 1)$
9. $(x + 3)^2$ 10. $(x - 5)^2$

EXERCISE 4.6

REVIEW   *In Exercises 1–4, tell which property of real numbers justifies each statement.*

1. $3(x + 5) = 3x + 3 \cdot 5$

2. $(x + 3) + y = x + (3 + y)$

3. $3(ab) = (ab)3$

4. $a + 0 = a$

5. Solve $\dfrac{5}{3}(5y + 6) - 10 = 0$.

6. Solve $F = \dfrac{GMm}{d^2}$ for m.

VOCABULARY AND CONCEPTS   *Fill in each blank to make a true statement.*

7. A polynomial with one term is called a _____.

8. A _____ is a polynomial with two terms.

9. A polynomial with three terms is called a _____.

10. In the acronym FOIL, F stands for _____, O stands for _____, I stands for _____, and L stands for _____.

In Exercises 11–14, consider the product $(2x + 5)(3x - 4)$.

11. The product of the first terms is ____ .

12. The product of the outer terms is ____ .

13. The product of the inner terms is ____ .

14. The product of the last terms is ____ .

PRACTICE   *In Exercises 15–26, find each product.*

15. $(3x^2)(4x^3)$

16. $(-2a^3)(3a^2)$

17. $(3b^2)(-2b)(4b^3)$

18. $(3y)(2y^2)(-y^4)$

19. $(2x^2y^3)(3x^3y^2)$

20. $(-x^3y^6z)(x^2y^2z^7)$

21. $(x^2y^5)(x^2z^5)(-3y^2z^3)$

22. $(-r^4st^2)(2r^2st)(rst)$

23. $(x^2y^3)^5$

24. $(a^3b^2c)^4$

25. $(a^3b^2c)(abc^3)^2$

26. $(xyz^3)(xy^2z^2)^3$

In Exercises 27–44, find each product.

27. $3(x + 4)$

28. $-3(a - 2)$

29. $-4(t + 7)$

30. $6(s^2 - 3)$

31. $3x(x - 2)$

32. $4y(y + 5)$

33. $-2x^2(3x^2 - x)$

34. $4b^3(2b^2 - 2b)$

35. $3xy(x + y)$

36. $-4x^2(3x^2 - x)$

37. $2x^2(3x^2 + 4x - 7)$

38. $3y^3(2y^2 - 7y - 8)$

39. $\frac{1}{4}x^2(8x^5 - 4)$

40. $\frac{4}{3}a^2b(6a - 5b)$

41. $-\frac{2}{3}r^2t^2(9r - 3t)$

42. $-\frac{4}{5}p^2q(10p + 15q)$

43. $(3xy)(-2x^2y^3)(x + y)$

44. $(-2a^2b)(-3a^3b^2)(3a - 2b)$

In Exercises 45–62, use the FOIL method to find each product.

45. $(a + 4)(a + 5)$

46. $(y - 3)(y + 5)$

47. $(3x - 2)(x + 4)$

48. $(t + 4)(2t - 3)$

49. $(2a + 4)(3a - 5)$

50. $(2b - 1)(3b + 4)$

51. $(3x - 5)(2x + 1)$

52. $(2y - 5)(3y + 7)$

53. $(x + 3)(2x - 3)$

54. $(2x + 3)(2x - 5)$

55. $(2s + 3t)(3s - t)$

56. $(3a - 2b)(4a + b)$

57. $(x + y)(x + z)$

58. $(a - b)(x + y)$

59. $(u + v)(u + 2t)$

60. $(x - 5y)(a + 2y)$

61. $(-2r - 3s)(2r + 7s)$

62. $(-4a + 3)(-2a - 3)$

In Exercises 63–70, find each product.

63. $4x + 3$
$\underline{\quad x + 2}$

64. $5r + 6$
$\underline{\quad 2r - 1}$

65. $4x - 2y$
$\underline{3x + 5y}$

66. $5r + 6s$
$\underline{2r - \ \ s}$

67. $x^2 + x + 1$
$\underline{\qquad x - 1}$

68. $4x^2 - 2x + 1$
$\underline{\qquad 2x + 1}$

69. $(2x + 1)(x^2 + 3x - 1)$

70. $(3x - 2)(2x^2 - x + 2)$

In Exercises 71–88, find each special product.

71. $(x + 4)(x + 4)$

72. $(a + 3)(a + 3)$

73. $(t - 3)(t - 3)$

74. $(z - 5)(z - 5)$

75. $(r + 4)(r - 4)$

76. $(b + 2)(b - 2)$

77. $(x + 5)^2$

78. $(y - 6)^2$

79. $(2s + 1)(2s + 1)$

80. $(3t - 2)(3t - 2)$

81. $(4x + 5)(4x - 5)$

82. $(5z + 1)(5z - 1)$

83. $(x - 2y)^2$

84. $(3a + 2b)^2$

85. $(2a - 3b)^2$

86. $(2x + 5y)^2$

87. $(4x + 5y)(4x - 5y)$

88. $(6p + 5q)(6p - 5q)$

In Exercises 89–98, find each product.

89. $2(x - 4)(x + 1)$

90. $-3(2x + 3y)(3x - 4y)$

91. $3a(a + b)(a - b)$

92. $-2r(r + s)(r + s)$

93. $(4t + 3)(t^2 + 2t + 3)$

94. $(3x + y)(2x^2 - 3xy + y^2)$

95. $(-3x + y)(x^2 - 8xy + 16y^2)$

96. $(3x - y)(x^2 + 3xy - y^2)$

97. $(x - 2y)(x^2 + 2xy + 4y^2)$

98. $(2m + n)(4m^2 - 2mn + n^2)$

In Exercises 99–108, simplify each expression.

99. $2t(t + 2) + 3t(t - 5)$

100. $3y(y + 2) + (y + 1)(y - 1)$

101. $3xy(x + y) - 2x(xy - x)$

102. $(a + b)(a - b) - (a + b)(a + b)$

103. $(x + y)(x - y) + x(x + y)$

104. $(2x - 1)(2x + 1) + x(2x + 1)$

105. $(x + 2)^2 - (x - 2)^2$

106. $(x - 3)^2 - (x + 3)^2$

107. $(2s - 3)(s + 2) + (3s + 1)(s - 3)$

108. $(3x + 4)(2x - 2) - (2x + 1)(x + 3)$

In Exercises 109–118, solve each equation.

109. $(s - 4)(s + 1) = s^2 + 5$

110. $(y - 5)(y - 2) = y^2 - 4$

111. $z(z + 2) = (z + 4)(z - 4)$

112. $(z + 3)(z - 3) = z(z - 3)$

113. $(x + 4)(x - 4) = (x - 2)(x + 6)$

114. $(y - 1)(y + 6) = (y - 3)(y - 2) + 8$

115. $(a - 3)^2 = (a + 3)^2$

116. $(b + 2)^2 = (b - 1)^2$

117. $4 + (2y - 3)^2 = (2y - 1)(2y + 3)$

118. $7s^2 + (s - 3)(2s + 1) = (3s - 1)^2$

APPLICATIONS

119. Millstones The radius of one millstone in Illustration 1 is 3 meters greater than the radius of the other, and their areas differ by 15π square meters. Find the radius of the larger millstone.

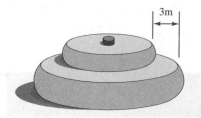

3m

ILLUSTRATION 1

120. Bookbinding Two square sheets of cardboard used for making book covers differ in area by 44 square inches. An edge of the larger square is 2 inches greater than an edge of the smaller square. Find the length of an edge of the smaller square.

121. Baseball In major league baseball, the distance between bases is 30 feet greater than it is in softball. The bases in major league baseball mark the corners of a square that has an area 4,500 square feet greater than for softball. Find the distance between the bases in baseball.

122. Pulley design The radius of one pulley in Illustration 2 is 1 inch greater than the radius of the second pulley, and their areas differ by 4π square inches. Find the radius of the smaller pulley.

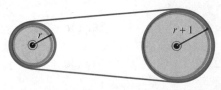

r $r + 1$

ILLUSTRATION 2

WRITING

123. Describe the steps involved in finding the product of a binomial and its conjugate.

124. Writing the expression $(x + y)^2$ as $x^2 + y^2$ illustrates a common error. Explain.

SOMETHING TO THINK ABOUT

125. The area of the square in Illustration 3 is the total of the areas of the four smaller regions. The picture illustrates the product $(x + y)^2$. Explain.

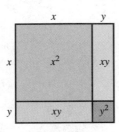

ILLUSTRATION 3

126. Selling tape recorders   An electronics firm manufactures tape recorders, receiving $120 for each recorder it makes. If x represents the number of recorders produced, the income received is determined by the *revenue function* $R(x) = 120x$. The manufacturer has fixed costs of $12,000 per month and variable costs of $57.50 for each recorder manufactured. Thus, the *cost function* is $C(x) = 57.50x + 12,000$. How many recorders must the company sell for revenue to equal cost?

4.7   Dividing Polynomials by Monomials

■ DIVIDING A MONOMIAL BY A MONOMIAL ■ DIVIDING A POLYNOMIAL BY A MONOMIAL ■ AN APPLICATION OF DIVIDING A POLYNOMIAL BY A MONOMIAL

Getting Ready   *Simplify each fraction.*

1. $\dfrac{4x^2y^3}{2xy}$ **2.** $\dfrac{9xyz}{9xz}$ **3.** $\dfrac{15x^2y}{10x}$ **4.** $\dfrac{6x^2y}{6xy^2}$

5. $\dfrac{(2x^2)(5y^2)}{10xy}$ **6.** $\dfrac{(5x^3y)(6xy^3)}{10x^4y^4}$

■ DIVIDING A MONOMIAL BY A MONOMIAL

We have seen that dividing by a number is equivalent to multiplying by its recipro-cal. For example, dividing the number 8 by 2 gives the same answer as multiplying 8 by $\frac{1}{2}$.

$$\frac{8}{2} = 4 \qquad \text{and} \qquad \frac{1}{2} \cdot 8 = 4$$

In general, the following is true.

Division

$$\frac{a}{b} = \frac{1}{b} \cdot a \quad (b \neq 0)$$

Recall that to simplify a fraction, we write both its numerator and denominator as the product of several factors and then divide out all common factors. For example,

$$\frac{4}{6} = \frac{2 \cdot 2}{2 \cdot 3}$$
Factor: $4 = 2 \cdot 2$ and $6 = 2 \cdot 3$.

$$\frac{20}{25} = \frac{4 \cdot 5}{5 \cdot 5}$$
Factor: $20 = 4 \cdot 5$ and $25 = 5 \cdot 5$.

$$= \frac{\overset{1}{\cancel{2}} \cdot 2}{\underset{1}{\cancel{2}} \cdot 3}$$
Divide out the common factor of 2.

$$= \frac{4 \cdot \overset{1}{\cancel{5}}}{\underset{1}{\cancel{5}} \cdot 5}$$
Divide out the common factor of 5.

$$= \frac{2}{3}$$
$\frac{2}{2} = 1$.

$$= \frac{4}{5}$$
$\frac{5}{5} = 1$.

We can use the same method to simplify algebraic fractions that contain variables.

$$\frac{3p^2 q}{6pq^3} = \frac{3 \cdot p \cdot p \cdot q}{2 \cdot 3 \cdot p \cdot q \cdot q \cdot q}$$
Factor: $p^2 = p \cdot p$, $6 = 2 \cdot 3$, and $q^3 = q \cdot q \cdot q$.

$$= \frac{\overset{1}{\cancel{3}} \cdot \overset{1}{\cancel{p}} \cdot p \cdot \overset{1}{\cancel{q}}}{2 \cdot \underset{1}{\cancel{3}} \cdot \underset{1}{\cancel{p}} \cdot \underset{1}{\cancel{q}} \cdot q \cdot q}$$
Divide out the common factors of 3, p, and q.

$$= \frac{p}{2q^2}$$
$\frac{3}{3} = 1$, $\frac{p}{p} = 1$, and $\frac{q}{q} = 1$.

To divide monomials, we can either use the previous method used for simplifying arithmetic fractions or use the rules of exponents.

EXAMPLE 1 Simplify **a.** $\dfrac{x^2 y}{xy^2}$ and **b.** $\dfrac{-8a^3 b^2}{4ab^3}$.

Solution

Using Fractions

a. $\dfrac{x^2 y}{xy^2} = \dfrac{x \cdot x \cdot y}{x \cdot y \cdot y}$

$$= \frac{\overset{1}{\cancel{x}} \cdot x \cdot \overset{1}{\cancel{y}}}{\underset{1}{\cancel{x}} \cdot y \cdot \underset{1}{\cancel{y}}}$$

$$= \frac{x}{y}$$

Using the Rules of Exponents

$$\frac{x^2 y}{xy^2} = x^{2-1} y^{1-2}$$

$$= x^1 y^{-1}$$

$$= \frac{x}{y}$$

	Using Fractions	*Using the Rules of Exponents*

b. $\dfrac{-8a^3b^2}{4ab^3} = \dfrac{-2\cdot 4\cdot a\cdot a\cdot a\cdot b\cdot b}{4\cdot a\cdot b\cdot b\cdot b}$ $\dfrac{-8a^3b^2}{4ab^3} = \dfrac{(-1)2^3a^3b^2}{2^2ab^3}$

$$= \dfrac{-2\cdot \overset{1}{\cancel{4}}\cdot \overset{1}{\cancel{a}}\cdot a\cdot a\cdot \overset{1}{\cancel{b}}\cdot \overset{1}{\cancel{b}}}{\underset{1}{\cancel{4}}\cdot \underset{1}{\cancel{a}}\cdot \underset{1}{\cancel{b}}\cdot \underset{1}{\cancel{b}}\cdot b}$$ $= (-1)2^{3-2}a^{3-1}b^{2-3}$

$$\qquad\qquad\qquad\qquad = (-1)2^1a^2b^{-1}$$

$$= \dfrac{-2a^2}{b}\qquad\qquad\qquad = \dfrac{-2a^2}{b}\qquad\blacksquare$$

Self Check Simplify $\dfrac{-5p^2q^3}{10pq^4}$.

Answer $\dfrac{-p}{2q}$

■ DIVIDING A POLYNOMIAL BY A MONOMIAL

To divide a polynomial with more than one term by a monomial, we write the division as a product, use the distributive property to remove parentheses, and simplify each resulting fraction.

EXAMPLE 2 Simplify $\dfrac{9x + 6y}{3xy}$.

Solution
$$\dfrac{9x + 6y}{3xy} = \dfrac{1}{3xy}(9x + 6y)$$

$$= \dfrac{9x}{3xy} + \dfrac{6y}{3xy} \qquad\text{Remove parentheses.}$$

$$= \dfrac{3}{y} + \dfrac{2}{x} \qquad\text{Simplify each fraction.}\qquad\blacksquare$$

Self Check Simplify $\dfrac{4a - 8b}{4ab}$.

Answer $\dfrac{1}{b} - \dfrac{2}{a}$

EXAMPLE 3 Simplify $\dfrac{6x^2y^2 + 4x^2y - 2xy}{2xy}$.

Solution
$$\dfrac{6x^2y^2 + 4x^2y - 2xy}{2xy}$$

$$= \dfrac{1}{2xy}(6x^2y^2 + 4x^2y - 2xy)$$

$$= \frac{6x^2y^2}{2xy} + \frac{4x^2y}{2xy} - \frac{2xy}{2xy} \qquad \text{Remove parentheses.}$$

$$= 3xy + 2x - 1 \qquad \text{Simplify each fraction.} \quad \blacksquare$$

Self Check Simplify $\dfrac{9a^2b - 6ab^2 + 3ab}{3ab}$.

Answer $3a - 2b + 1$

EXAMPLE 4   Simplify $\dfrac{12a^3b^2 - 4a^2b + a}{6a^2b^2}$.

Solution $\dfrac{12a^3b^2 - 4a^2b + a}{6a^2b^2}$

$$= \frac{1}{6a^2b^2}(12a^3b^2 - 4a^2b + a)$$

$$= \frac{12a^3b^2}{6a^2b^2} - \frac{4a^2b}{6a^2b^2} + \frac{a}{6a^2b^2} \qquad \text{Remove parentheses.}$$

$$= 2a - \frac{2}{3b} + \frac{1}{6ab^2} \qquad \text{Simplify each fraction.} \quad \blacksquare$$

Self Check   Simplify $\dfrac{14p^3q + pq^2 - p}{7p^2q}$.

Answer $2p + \frac{q}{7p} - \frac{1}{7pq}$

EXAMPLE 5   Simplify $\dfrac{(x - y)^2 - (x + y)^2}{xy}$.

Solution $\dfrac{(x - y)^2 - (x + y)^2}{xy}$

$$= \frac{x^2 - 2xy + y^2 - (x^2 + 2xy + y^2)}{xy} \qquad \begin{array}{l}\text{Multiply the binomials in the}\\ \text{numerator.}\end{array}$$

$$= \frac{x^2 - 2xy + y^2 - x^2 - 2xy - y^2}{xy} \qquad \text{Remove parentheses.}$$

$$= \frac{-4xy}{xy} \qquad \text{Combine like terms.}$$

$$= -4 \qquad \text{Divide out } xy. \quad \blacksquare$$

Self Check   Simplify $\dfrac{(x + y)^2 - (x - y)^2}{xy}$.

Answer   4

■ AN APPLICATION OF DIVIDING A POLYNOMIAL BY A MONOMIAL

The area of the trapezoidal drainage ditch shown in Figure 4-7 is given by the formula $A = \frac{1}{2}h(B + b)$, where B and b are its bases and h is its height. To solve the formula for b, we proceed as follows.

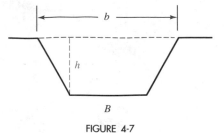

FIGURE 4-7

$$A = \frac{1}{2}h(B + b)$$

$$2A = 2 \cdot \frac{1}{2}h(B + b) \qquad \text{Multiply both sides by 2.}$$

$$2A = h(B + b) \qquad \text{Simplify: } 2 \cdot \frac{1}{2} = \frac{2}{2} = 1.$$

$$2A = hB + hb \qquad \text{Use the distributive property to remove parentheses.}$$

$$2A - hB = hB - hB + hb \qquad \text{Subtract } hB \text{ from both sides.}$$

$$2A - hB = hb \qquad \text{Combine like terms: } hB - hB = 0.$$

$$\frac{2A - hB}{h} = \frac{hb}{h} \qquad \text{Divide both sides by } h.$$

$$\frac{2A - hB}{h} = b \qquad \frac{hb}{h} = b.$$

EXAMPLE 6 Another student worked the previous problem in a different way and got a result of $b = \frac{2A}{h} - B$. Is this result correct?

Solution To show that this result is correct, we must show that $\frac{2A - hB}{h} = \frac{2A}{h} - B$. We can do this by dividing $2A - hB$ by h.

$$\frac{2A - hB}{h} = \frac{1}{h}(2A - hB)$$

$$= \frac{2A}{h} - \frac{hB}{h} \qquad \text{Use the distributive property to remove parentheses.}$$

$$= \frac{2A}{h} - B \qquad \text{Simplify } \frac{hB}{h} = B.$$

The results are the same. ■

Self Check In Example 6, suppose another student got $2A - B$. Is this result correct?

Answer no

Orals *Simplify each fraction.*

1. $\dfrac{4x^3y}{2xy}$ **2.** $\dfrac{6x^3y^2}{3x^3y}$ **3.** $\dfrac{35ab^2c^3}{7abc}$ **4.** $\dfrac{-14p^2q^5}{7pq^4}$

5. $\dfrac{(x+y)+(x-y)}{2x}$ **6.** $\dfrac{(2x^2-z)+(x^2+z)}{x}$

EXERCISE 4.7

REVIEW *In Exercises 1–4, identify each polynomial as a monomial, a binomial, a trinomial, or none of these.*

1. $5a^2b + 2ab^2$ **2.** $-3x^3y$

3. $-2x^3 + 3x^2 - 4x + 12$ **4.** $17t^2 - 15t + 27$

5. What is the degree of the trinomial $3x^2 - 2x + 4$? **6.** What is the numerical coefficient of the second term of the trinomial $-7t^2 - 5t + 17$?

VOCABULARY AND CONCEPTS *Fill in each blank to make a true statement.*

7. A _____ is an algebraic expression in which the exponents on the variables are whole numbers. **8.** A _____ is a polynomial with one algebraic term.

9. A binomial is a polynomial with ___ terms. **10.** A trinomial is a polynomial with ___ terms.

11. $\dfrac{1}{b} \cdot a =$ ___ **12.** $\dfrac{15x - 6y}{6xy} =$ ___ $\cdot (15x - 6y)$

PRACTICE *In Exercises 13–24, simplify each fraction.*

13. $\dfrac{5}{15}$ **14.** $\dfrac{64}{128}$ **15.** $\dfrac{-125}{75}$ **16.** $\dfrac{-98}{21}$

17. $\dfrac{120}{160}$ **18.** $\dfrac{70}{420}$ **19.** $\dfrac{-3,612}{-3,612}$ **20.** $\dfrac{-288}{-112}$

21. $\dfrac{-90}{360}$ **22.** $\dfrac{8,423}{-8,423}$ **23.** $\dfrac{5,880}{2,660}$ **24.** $\dfrac{-762}{366}$

In Exercises 25–52, do each division by simplifying each fraction. Write all answers without using negative or zero exponents.

25. $\dfrac{xy}{yz}$ **26.** $\dfrac{a^2b}{ab^2}$ **27.** $\dfrac{r^3s^2}{rs^3}$ **28.** $\dfrac{y^4z^3}{y^2z^2}$

29. $\dfrac{8x^3y^2}{4xy^3}$ **30.** $\dfrac{-3y^3z}{6yz^2}$ **31.** $\dfrac{12u^5v}{-4u^2v^3}$ **32.** $\dfrac{16rst^2}{-8rst^3}$

33. $\dfrac{-16r^3y^2}{-4r^2y^4}$ **34.** $\dfrac{35xyz^2}{-7x^2yz}$ **35.** $\dfrac{-65rs^2t}{15r^2s^3t}$ **36.** $\dfrac{112u^3z^6}{-42u^3z^6}$

37. $\dfrac{x^2x^3}{xy^6}$

38. $\dfrac{(xy)^2}{x^2y^3}$

39. $\dfrac{(a^3b^4)^3}{ab^4}$

40. $\dfrac{(a^2b^3)^3}{a^6b^6}$

41. $\dfrac{15(r^2s^3)^2}{-5(rs^5)^3}$

42. $\dfrac{-5(a^2b)^3}{10(ab^2)^3}$

43. $\dfrac{-32(x^3y)^3}{128(x^2y^2)^3}$

44. $\dfrac{68(a^6b^7)^2}{-96(abc^2)^3}$

45. $\dfrac{(5a^2b)^3}{(2a^2b^2)^3}$

46. $\dfrac{-(4x^3y^3)^2}{(x^2y^4)^8}$

47. $\dfrac{-(3x^3y^4)^3}{-(9x^4y^5)^2}$

48. $\dfrac{(2r^3s^2t)^2}{-(4r^2s^2t^2)^2}$

49. $\dfrac{(a^2a^3)^4}{(a^4)^3}$

50. $\dfrac{(b^3b^4)^5}{(bb^2)^2}$

51. $\dfrac{(z^3z^{-4})^3}{(z^{-3})^2}$

52. $\dfrac{(t^{-3}t^5)}{(t^2)^{-3}}$

In Exercises 53–66, do each division.

53. $\dfrac{6x + 9y}{3xy}$

54. $\dfrac{8x + 12y}{4xy}$

55. $\dfrac{5x - 10y}{25xy}$

56. $\dfrac{2x - 32}{16x}$

57. $\dfrac{3x^2 + 6y^3}{3x^2y^2}$

58. $\dfrac{4a^2 - 9b^2}{12ab}$

59. $\dfrac{15a^3b^2 - 10a^2b^3}{5a^2b^2}$

60. $\dfrac{9a^4b^3 - 16a^3b^4}{12a^2b}$

61. $\dfrac{4x - 2y + 8z}{4xy}$

62. $\dfrac{5a^2 + 10b^2 - 15ab}{5ab}$

63. $\dfrac{12x^3y^2 - 8x^2y - 4x}{4xy}$

64. $\dfrac{12a^2b^2 - 8a^2b - 4ab}{4ab}$

65. $\dfrac{-25x^2y + 30xy^2 - 5xy}{-5xy}$

66. $\dfrac{-30a^2b^2 - 15a^2b - 10ab^2}{-10ab}$

In Exercises 67–76, simplify each numerator and do the division.

67. $\dfrac{5x(4x - 2y)}{2y}$

68. $\dfrac{9y^2(x^2 - 3xy)}{3x^2}$

69. $\dfrac{(-2x)^3 + (3x^2)^2}{6x^2}$

70. $\dfrac{(-3x^2y)^3 + (3xy^2)^3}{27x^3y^4}$

71. $\dfrac{4x^2y^2 - 2(x^2y^2 + xy)}{2xy}$

72. $\dfrac{-5a^3b - 5a(ab^2 - a^2b)}{10a^2b^2}$

73. $\dfrac{(3x - y)(2x - 3y)}{6xy}$

74. $\dfrac{(2m - n)(3m - 2n)}{-3m^2n^2}$

75. $\dfrac{(a + b)^2 - (a - b)^2}{2ab}$

76. $\dfrac{(x - y)^2 + (x + y)^2}{2x^2y^2}$

APPLICATIONS

77. Reconciling formulas   Are the formulas

$$l = \dfrac{P - 2w}{2} \quad \text{and} \quad l = \dfrac{P}{2} - w$$

the same?

78. Reconciling formulas   Are the formulas

$$r = \dfrac{G + 2b}{2b} \quad \text{and} \quad r = \dfrac{G}{2b} + b$$

the same?

79. Phone bills   On a phone bill, the following formulas are given to compute the average cost per minute of x minutes of phone usage. Are they equivalent?

$$C = \frac{0.15x + 12}{x} \quad \text{and} \quad C = 0.15 + \frac{12}{x}$$

80. Electric bills   On an electric bill, the following formulas are given to compute the average cost of x kwh of electricity. Are they equivalent?

$$C = \frac{0.08x + 5}{x} \quad \text{and} \quad C = 0.08x + \frac{5}{x}$$

WRITING

81. Describe how you would simplify the fraction

$$\frac{4x^2y + 8xy^2}{4xy}$$

82. A fellow student attempts to simplify the fraction $\frac{3x + 5}{x + 5}$ by dividing out the $x + 5$:

$$\frac{3x + 5}{x + 5} = \frac{3x\!\!\!\!\diagup + 5}{x\!\!\!\!\diagup + 5} = 3$$

What would you say to him?

SOMETHING TO THINK ABOUT

83. If $x = 501$, evaluate $\dfrac{x^{500} - x^{499}}{x^{499}}$.

84. An exercise reads as follows:

$$\text{Simplify } \frac{3x^3y + 6xy^2}{3xy^3}$$

It contains a misprint: one mistyped letter or digit. The correct answer is $\frac{x^2}{y} + 2$. Fix the exercise.

4.8   Dividing Polynomials by Polynomials

■ DIVIDING POLYNOMIALS BY POLYNOMIALS ■ WRITING POWERS IN DESCENDING ORDER ■ THE CASE OF THE MISSING TERMS

Getting Ready   *Divide.*

1. $12)\overline{156}$ **2.** $17)\overline{357}$ **3.** $13)\overline{247}$ **4.** $19)\overline{247}$

■ DIVIDING POLYNOMIALS BY POLYNOMIALS

To divide one polynomial by another, we use a method similar to long division in arithmetic. We will illustrate the method with several examples.

EXAMPLE 1   Divide $x^2 + 5x + 6$ by $x + 2$.

Solution   Here the divisor is $x + 2$, and the dividend is $x^2 + 5x + 6$.

Step 1:

$$\begin{array}{r} x \phantom{{}+5x+6} \\ x + 2 \overline{)x^2 + 5x + 6} \end{array}$$

How many times does x divide x^2? $x^2/x = x$. Place the x above the division symbol.

Step 2:
$$\begin{array}{r} x \\ x + 2\overline{)x^2 + 5x + 6} \\ x^2 + 2x \end{array}$$

Multiply each term in the divisor by x. Place the product under $x^2 + 5x$ and draw a line.

Step 3:
$$\begin{array}{r} x \\ x + 2\overline{)x^2 + 5x + 6} \\ x^2 + 2x \\ 3x + 6 \end{array}$$

Subtract $x^2 + 2x$ from $x^2 + 5x$ by adding the negative of $x^2 + 2x$ to $x^2 + 5x$.

Bring down the 6.

Step 4:
$$\begin{array}{r} x + 3 \\ x + 2\overline{)x^2 + 5x + 6} \\ x^2 + 2x \\ 3x + 6 \end{array}$$

How many times does x divide $3x$? $3x/x = +3$. Place the $+3$ above the division symbol.

Step 5:
$$\begin{array}{r} x + 3 \\ x + 2\overline{)x^2 + 5x + 6} \\ x^2 + 2x \\ 3x + 6 \\ 3x + 6 \end{array}$$

Multiply each term in the divisor by 3. Place the product under the $3x + 6$ and draw a line.

Step 6:
$$\begin{array}{r} x + 3 \\ x + 2\overline{)x^2 + 5x + 6} \\ x^2 + 2x \\ 3x + 6 \\ 3x + 6 \\ \hline 0 \end{array}$$

Subtract $3x + 6$ from $3x + 6$ by adding the negative of $3x + 6$.

The quotient is $x + 3$, and the remainder is 0.

Step 7: Check the work by verifying that $x + 2$ times $x + 3$ is $x^2 + 5x + 6$.

$$(x + 2)(x + 3) = x^2 + 3x + 2x + 6$$
$$= x^2 + 5x + 6$$

The answer checks.   ∎

Self Check
Answer

Divide $x^2 + 7x + 12$ by $x + 3$.

$x + 4$

EXAMPLE 2   Divide $\dfrac{6x^2 - 7x - 2}{2x - 1}$.

Solution   Here the divisor is $2x - 1$, and the dividend is $6x^2 - 7x - 2$.

Step 1:
$$\begin{array}{r} 3x \\ 2x - 1\overline{)6x^2 - 7x - 2} \end{array}$$

How many times does $2x$ divide $6x^2$? $6x^2/2x = 3x$. Place the $3x$ above the division symbol.

Step 2:

$$\begin{array}{r} 3x \\ 2x - 1 \overline{)6x^2 - 7x - 2} \\ 6x^2 - 3x \end{array}$$

Multiply each term in the divisor by $3x$. Place the product under $6x^2 - 7x$ and draw a line.

Step 3:

$$\begin{array}{r} 3x \\ 2x - 1 \overline{)6x^2 - 7x - 2} \\ 6x^2 - 3x \\ \hline - 4x - 2 \end{array}$$

Subtract $6x^2 - 3x$ from $6x^2 - 7x$ by adding the negative of $6x^2 - 3x$ to $6x^2 - 7x$.

Bring down the -2.

Step 4:

$$\begin{array}{r} 3x - 2 \\ 2x - 1 \overline{)6x^2 - 7x - 2} \\ 6x^2 - 3x \\ \hline - 4x - 2 \end{array}$$

How many times does $2x$ divide $-4x$? $-4x/2x = -2$. Place the -2 above the division symbol.

Step 5:

$$\begin{array}{r} 3x - 2 \\ 2x - 1 \overline{)6x^2 - 7x - 2} \\ 6x^2 - 3x \\ \hline - 4x - 2 \\ - 4x + 2 \end{array}$$

Multiply each term in the divisor by -2. Place the product under the $-4x - 2$ and draw a line.

Step 6:

$$\begin{array}{r} 3x - 2 \\ 2x - 1 \overline{)6x^2 - 7x - 2} \\ 6x^2 - 3x \\ \hline - 4x - 2 \\ - 4x + 2 \\ \hline - 4 \end{array}$$

Subtract $-4x + 2$ from $-4x - 2$ by adding the negative of $-4x + 2$.

Here the quotient is $3x - 2$, and the remainder is -4. It is common to write the answer in quotient $+ \frac{\text{remainder}}{\text{divisor}}$ form:

$$3x - 2 + \frac{-4}{2x - 1}$$

where the fraction $\dfrac{-4}{2x - 1}$ is formed by dividing the remainder by the divisor.

Step 7: To check the answer, we multiply $3x - 2 + \frac{-4}{2x-1}$ by $2x - 1$. The product should be the dividend.

$$
\begin{aligned}
(2x - 1)\left(3x - 2 + \frac{-4}{2x - 1}\right) &= (2x - 1)(3x - 2) + (2x - 1)\left(\frac{-4}{2x - 1}\right) \\
&= (2x - 1)(3x - 2) - 4 \\
&= 6x^2 - 4x - 3x + 2 - 4 \\
&= 6x^2 - 7x - 2
\end{aligned}
$$

Because the result is the dividend, the answer checks.   ∎

Self Check Divide $\dfrac{8x^2 + 6x - 3}{2x + 3}$.

Answer $4x - 3 + \dfrac{6}{2x+3}$

■ WRITING POWERS IN DESCENDING ORDER

The division method works best when exponents of the terms in the divisor and the dividend are written in descending order. This means that the term involving the highest power of x appears first, the term involving the second-highest power of x appears second, and so on. For example, the terms in

$$3x^3 + 2x^2 - 7x + 5$$

have their exponents written in descending order.

If the powers in the dividend or divisor are not in descending order, we can use the commutative property of addition to write them that way.

EXAMPLE 3 Divide $4x^2 + 2x^3 + 12 - 2x$ by $x + 3$.

Solution We write the dividend so that the exponents are in descending order and divide.

$$
\begin{array}{r}
2x^2 - 2x\ + 4 \\
x + 3\overline{)2x^3 + 4x^2 - 2x + 12} \\
\underline{2x^3 + 6x^2} \\
-2x^2 - 2x \\
\underline{-2x^2 - 6x} \\
+4x + 12 \\
\underline{+4x + 12}
\end{array}
$$

Check: $(x + 3)(2x^2 - 2x + 4) = 2x^3 - 2x^2 + 4x + 6x^2 - 6x + 12$
$$= 2x^3 + 4x^2 - 2x + 12 \qquad ■$$

Self Check Divide $x^2 - 10x + 6x^3 + 4$ by $2x - 1$.

Answer $3x^2 + 2x - 4$

■ THE CASE OF THE MISSING TERMS

When we write the terms of a dividend in descending powers of x, we may notice that some powers of x are missing. For example, in the dividend of

$$x + 1\overline{)3x^4 - 7x^2 - 3x + 15}$$

the term involving x^3 is missing. When this happens, we should either write the term with a coefficient of 0 or leave a blank space for it. In this case, we would write the dividend as

$$3x^4 + 0x^3 - 7x^2 - 3x + 15 \quad \text{or} \quad 3x^4 \qquad - 7x^2 - 3x + 15$$

EXAMPLE 4   Divide $\dfrac{x^2 - 4}{x + 2}$.

Solution   Since $x^2 - 4$ does not have a term involving x, we must either include the term $0x$ or leave a space for it.

$$
\begin{array}{r}
x - 2 \\
x + 2 \overline{)\, x^2 + 0x - 4} \\
\underline{x^2 + 2x} \\
-2x - 4 \\
\underline{-2x - 4}
\end{array}
$$

Check: $(x + 2)(x - 2) = x^2 - 2x + 2x - 4$
$$= x^2 - 4$$ ∎

Self Check   Divide $\dfrac{x^2 - 9}{x - 3}$.

Answer $x + 3$

EXAMPLE 5   Divide $x^3 + y^3$ by $x + y$.

Solution   We write $x^3 + y^3$ leaving spaces for the missing terms and proceed as follows.

$$
\begin{array}{r}
x^2 - xy + y^2 \\
x + y \overline{)\, x^3 + y^3} \\
\underline{x^3 + x^2y} \\
-x^2y \\
\underline{-x^2y - xy^2} \\
+ xy^2 + y^3 \\
\underline{xy^2 + y^3}
\end{array}
$$

Check: $(x + y)(x^2 - xy + y^2) = x^3 - x^2y + xy^2 + x^2y - xy^2 + y^3$
$$= x^3 + y^3$$ ∎

Self Check   Divide $x^3 - y^3$ by $x - y$.
Answer $x^2 + xy + y^2$

Orals   *Divide, and give the answer in* quotient $+ \frac{\text{remainder}}{\text{divisor}}$ *form.*

1. $x \overline{)2x + 3}$ **2.** $x \overline{)3x - 5}$ **3.** $x + 1 \overline{)2x + 3}$

4. $x + 1 \overline{)3x + 5}$ **5.** $x + 1 \overline{)x^2 + x}$ **6.** $x + 2 \overline{)x^2 + 2x}$

EXERCISE 4.8

REVIEW

1. List the composite numbers between 20 and 30.

2. Graph the set of prime numbers between 10 and 20 on a number line.

Let $a = -2$ and $b = 3$. Evaluate each expression.

3. $|a - b|$ **4.** $|a + b|$ **5.** $-|a^2 - b^2|$ **6.** $a - |-b|$

Simplify each expression.

7. $3(2x^2 - 4x + 5) + 2(x^2 + 3x - 7)$

8. $-2(y^3 + 2y^2 - y) - 3(3y^3 + y)$

VOCABULARY AND CONCEPTS   *Fill in each blank to make a true statement.*

9. In the division $x + 1 \overline{)x^2 + 2x + 1}$, $x + 1$ is called the _____, and $x^2 + 2x + 1$ is called the _____.

10. The answer to a division problem is called the _____.

11. If a division does not come out even, the leftover part is called a _____.

12. The exponents in $2x^4 + 3x^3 + 4x^2 - 7x - 2$ are said to be written in _____ order.

Write each polynomial with the powers in descending order.

13. $4x^3 + 7x - 2x^2 + 6$

14. $5x^2 + 7x^3 - 3x - 9$

15. $9x + 2x^2 - x^3 + 6x^4$

16. $7x^5 + x^3 - x^2 + 2x^4$

Identify the missing terms in each polynomial.

17. $5x^4 + 2x^2 - 1$

18. $-3x^5 - 2x^3 + 4x - 6$

PRACTICE   *In Exercises 19–24, do each division.*

19. Divide $x^2 + 4x + 4$ by $x + 2$.

20. Divide $x^2 - 5x + 6$ by $x - 2$.

21. Divide $y^2 + 13y + 12$ by $y + 1$.

22. Divide $z^2 - 7z + 12$ by $z - 3$.

23. Divide $a^2 + 2ab + b^2$ by $a + b$.

24. Divide $a^2 - 2ab + b^2$ by $a - b$.

In Exercises 25–30, do each division.

25. $\dfrac{6a^2 + 5a - 6}{2a + 3}$

26. $\dfrac{8a^2 + 2a - 3}{2a - 1}$

27. $\dfrac{3b^2 + 11b + 6}{3b + 2}$

28. $\dfrac{3b^2 - 5b + 2}{3b - 2}$

29. $\dfrac{2x^2 - 7xy + 3y^2}{2x - y}$

30. $\dfrac{3x^2 + 5xy - 2y^2}{x + 2y}$

In Exercises 31–42, write the powers of x in descending order and do each division.

31. $5x + 3 \overline{)11x + 10x^2 + 3}$

32. $2x - 7 \overline{)-x - 21 + 2x^2}$

33. $4 + 2x \overline{)-10x - 28 + 2x^2}$

34. $1 + 3x \overline{)9x^2 + 1 + 6x}$

35. $2x - y \overline{)xy - 2y^2 + 6x^2}$

36. $2y + x \overline{)3xy + 2x^2 - 2y^2}$

37. $x + 3y \overline{)2x^2 - 3y^2 + 5xy}$

38. $2x - 3y \overline{)2x^2 - 3y^2 - xy}$

39. $3x - 2y \overline{)-10y^2 + 13xy + 3x^2}$

40. $2x + 3y \overline{)-12y^2 + 10x^2 + 7xy}$

41. $4x + y \overline{)-19xy + 4x^2 - 5y^2}$

42. $x - 4y \overline{)5x^2 - 4y^2 - 19xy}$

In Exercises 43–48, do each division.

43. $2x + 3 \overline{)2x^3 + 7x^2 + 4x - 3}$

44. $2x - 1 \overline{)2x^3 - 3x^2 + 5x - 2}$

45. $3x + 2 \overline{)6x^3 + 10x^2 + 7x + 2}$

46. $4x + 3 \overline{)4x^3 - 5x^2 - 2x + 3}$

47. $2x + y \overline{)2x^3 + 3x^2y + 3xy^2 + y^3}$

48. $3x - 2y \overline{)6x^3 - x^2y + 4xy^2 - 4y^3}$

In Exercises 49–58, do each division. If there is a remainder, leave the answer in quotient $+ \frac{\text{remainder}}{\text{divisor}}$ *form.*

49. $\dfrac{2x^2 + 5x + 2}{2x + 3}$

50. $\dfrac{3x^2 - 8x + 3}{3x - 2}$

51. $\dfrac{4x^2 + 6x - 1}{2x + 1}$

52. $\dfrac{6x^2 - 11x + 2}{3x - 1}$

53. $\dfrac{x^3 + 3x^2 + 3x + 1}{x + 1}$

54. $\dfrac{x^3 + 6x^2 + 12x + 8}{x + 2}$

55. $\dfrac{2x^3 + 7x^2 + 4x + 3}{2x + 3}$

56. $\dfrac{6x^3 + x^2 + 2x + 1}{3x - 1}$

57. $\dfrac{2x^3 + 4x^2 - 2x + 3}{x - 2}$

58. $\dfrac{3y^3 - 4y^2 + 2y + 3}{y + 3}$

In Exercises 59–68, do each division.

59. $\dfrac{x^2 - 1}{x - 1}$

60. $\dfrac{x^2 - 9}{x + 3}$

61. $\dfrac{4x^2 - 9}{2x + 3}$

62. $\dfrac{25x^2 - 16}{5x - 4}$

63. $\dfrac{x^3 + 1}{x + 1}$

64. $\dfrac{x^3 - 8}{x - 2}$

65. $\dfrac{a^3 + a}{a + 3}$

66. $\dfrac{y^3 - 50}{y - 5}$

67. $3x - 4 \overline{)15x^3 - 23x^2 + 16x}$

68. $2y + 3 \overline{)21y^2 + 6y^3 - 20}$

WRITING

69. Distinguish among *dividend, divisor, quotient,* and *remainder.*

70. How would you check the results of a division?

SOMETHING TO THINK ABOUT

71. What's wrong here?

$$
\begin{array}{r}
x + 1 \\
x - 2\overline{)x^2 + 3x - 2} \\
\underline{x^2 - 2x} \\
x - 2 \\
\underline{x - 2} \\
0
\end{array}
$$

72. What's wrong here?

$$
\begin{array}{r}
3x \\
x + 2\overline{)3x^2 + 10x + 7} \\
\underline{3x^2 + 9x} \\
x + 7
\end{array}
$$

The quotient is $3x$ and the remainder is $x + 7$.

■ ■ ■ ■ ■ ■ ■ ■ ■ ■ **PROJECT**

There is a pattern in the behavior of polynomials. To discover it, consider the polynomial $2x^2 - 3x - 5$. First, evaluate the polynomial at $x = 1$ and $x = 3$. Then divide the polynomial by $x - 1$ and again by $x - 3$.

1. What do you notice about the remainders of these divisions?

2. Try others. For example, evaluate the polynomial at $x = 2$ and then divide by $x - 2$.

3. Can you make the pattern hold when you evaluate the polynomial at $x = -2$?

4. Does the pattern hold for other polynomials? Try some polynomials of your own, experiment, and report your conclusions.

C H A P T E R   S U M M A R Y

CONCEPTS

REVIEW EXERCISES

SECTION 4.1   *Natural-Number Exponents*

If n is a natural number, then

$$
x^n = \overbrace{x \cdot x \cdot x \cdot \cdots \cdot x}^{n \text{ factors of } x}
$$

1. Write each expression without using exponents.
  a. $(-3x)^4$
  b. $\left(\dfrac{1}{2}pq\right)^3$

2. Evaluate each expression.
  a. 5^3 **b.** 3^5
  c. $(-8)^2$ **d.** -8^2
  e. $3^2 + 2^2$ **f.** $(3 + 2)^2$

If m and n are integers, then

$$x^m x^n = x^{m+n}$$

$$(x^m)^n = x^{m \cdot n}$$

$$(xy)^n = x^n y^n$$

$$\left(\frac{x}{y}\right)^n = \frac{x^n}{y^n} \quad (y \neq 0)$$

$$\frac{x^m}{x^n} = x^{m-n} \quad (x \neq 0)$$

3. Do the operations and simplify.

a. $x^3 x^2$ **b.** $x^2 x^7$

c. $(y^7)^3$ **d.** $(x^{21})^2$

e. $(ab)^3$ **f.** $(3x)^4$

g. $b^3 b^4 b^5$ **h.** $-z^2(z^3 y^2)$

i. $(16s)^2 s$ **j.** $-3y(y^5)$

k. $(x^2 x^3)^3$ **l.** $(2x^2 y)^2$

m. $\dfrac{x^7}{x^3}$ **n.** $\left(\dfrac{x^2 y}{xy^2}\right)^2$

o. $\dfrac{8(y^2 x)^2}{4(yx^2)^2}$ **p.** $\dfrac{(5y^2 z^3)^3}{25(yz)^5}$

SECTION 4.2   *Zero and Negative-Integer Exponents*

$$x^0 = 1 \quad (x \neq 0)$$

$$x^{-n} = \frac{1}{x^n} \quad (x \neq 0)$$

4. Write each expression without using negative exponents or parentheses.

a. x^0 **b.** $(3x^2 y^2)^0$

c. $(3x^0)^2$ **d.** $(3x^2 y^0)^2$

e. x^{-3} **f.** $x^{-2} x^3$

g. $y^4 y^{-3}$ **h.** $\dfrac{x^3}{x^{-7}}$

i. $(x^{-3} x^4)^{-2}$ **j.** $(a^{-2} b)^{-3}$

k. $\left(\dfrac{x^2}{x}\right)^{-5}$ **l.** $\left(\dfrac{15z^4}{5z^3}\right)^{-2}$

SECTION 4.3   *Scientific Notation*

A number is written in scientific notation if it is written as the product of a number between 1 (including 1) and 10 and an integer power of 10.

5. Write each number in scientific notation.

a. 728   **b.** 9,370

c. 0.0136   **d.** 0.00942

e. 7.73   **f.** 753×10^3

g. 0.018×10^{-2} **h.** 600×10^2

6. Write each number in standard notation.

a. 7.26×10^5 **b.** 3.91×10^{-4}

c. 2.68×10^0 **d.** 5.76×10^1

e. 739×10^{-2} **f.** 0.437×10^{-3}

g. $\dfrac{(0.00012)(0.00004)}{0.00000016}$ **h.** $\dfrac{(4,800)(20,000)}{600,000}$

| SECTION 4.4 | *Polynomials* |

7. Find the degree of each polynomial and classify it as a monomial, a binomial, or a trinomial.

    a. $13x^7$ **b.** $5^3x + x^2$

    c. $-3x^5 + x - 1$ **d.** $9xy + 21x^3y^2$

When a number is substituted for the variable in a polynomial, the polynomial takes on a numerical value.

8. Evaluate $3x + 2$ for each value of x.

    a. $x = 3$ **b.** $x = 0$

    c. $x = -2$ **d.** $x = \dfrac{2}{3}$

9. Evaluate $5x^4 - x$ for each value of x.

    a. $x = 3$ **b.** $x = 0$

    c. $x = -2$ **d.** $x = (-0.3)$

Any equation in x and y where each value of x determines a single value of y is a **function.** We say that y is a function of x.

10. If $y = f(x) = x^2 - 4$, find each value.

    a. $f(0)$ **b.** $f(5)$

    c. $f(-2)$ **d.** $f\left(\dfrac{1}{2}\right)$

11. Graph each polynomial function.

    a. $y = f(x) = x^2 - 5$ **b.** $y = f(x) = x^3 - 2$

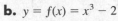

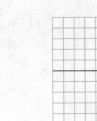

| SECTION 4.5 | *Adding and Subtracting Polynomials* |

When adding or subtracting polynomials, combine like terms by adding or subtracting the numerical coefficients and using the same variables and the same exponents.

12. Simplify each expression.

    a. $3x + 5x - x$ **b.** $3x + 2y$

    c. $(xy)^2 + 3x^2y^2$ **d.** $-2x^2yz + 3yx^2z$

    e. $(3x^2 + 2x) + (5x^2 - 8x)$

    f. $(7a^2 + 2a - 5) - (3a^2 - 2a + 1)$

    g. $3(9x^2 + 3x + 7) - 2(11x^2 - 5x + 9)$

    h. $4(4x^3 + 2x^2 - 3x - 8) - 5(2x^3 - 3x + 8)$

| SECTION 4.6 | *Multiplying Polynomials* |

To multiply two monomials, first multiply the numerical factors and then multiply the variable factors using the properties of exponents.

13. Find each product.
   a. $(2x^2y^3)(5xy^2)$ **b.** $(xyz^3)(x^3z)^2$

To multiply a polynomial with more than one term by a monomial, multiply each term of the polynomial by the monomial and simplify.

14. Find each product.
   a. $5(x + 3)$ **b.** $3(2x + 4)$
   c. $x^2(3x^2 - 5)$ **d.** $2y^2(y^2 + 5y)$
   e. $-x^2y(y^2 - xy)$
   f. $-3xy(xy - x)$

To multiply two binomials, use the **FOIL method.**

15. Find each product.
   a. $(x + 3)(x + 2)$
   b. $(2x + 1)(x - 1)$
   c. $(3a - 3)(2a + 2)$
   d. $6(a - 1)(a + 1)$
   e. $(a - b)(2a + b)$
   f. $(3x - y)(2x + y)$

Special products:
$$(x + y)^2 = x^2 + 2xy + y^2$$
$$(x - y)^2 = x^2 - 2xy + y^2$$
$$(x + y)(x - y) = x^2 - y^2$$

16. Find each product.
   a. $(x + 3)(x + 3)$ **b.** $(x + 5)(x - 5)$
   c. $(y - 2)(y + 2)$ **d.** $(x + 4)^2$
   e. $(x - 3)^2$ **f.** $(y - 1)^2$
   g. $(2y + 1)^2$ **h.** $(y^2 + 1)(y^2 - 1)$

To multiply one polynomial by another, multiply each term of one polynomial by each term of the other polynomial, and simplify.

17. Find each product.
   a. $(3x + 1)(x^2 + 2x + 1)$
   b. $(2a - 3)(4a^2 + 6a + 9)$

18. Solve each equation.
   a. $x^2 + 3 = x(x + 3)$
   b. $x^2 + x = (x + 1)(x + 2)$
   c. $(x + 2)(x - 5) = (x - 4)(x - 1)$
   d. $(x - 1)(x - 2) = (x - 3)(x + 1)$
   e. $x^2 + x(x + 2) = x(2x + 1) + 1$
   f. $(x + 5)(3x + 1) = x^2 + (2x - 1)(x - 5)$

SECTION 4.7	*Dividing Polynomials by Monomials*

To divide a polynomial by a monomial, write the division as a product, use the distributive property to remove parentheses, and simplify each resulting fraction.

19. Do each division.

a. $\dfrac{3x + 6y}{2xy}$

b. $\dfrac{14xy - 21x}{7xy}$

c. $\dfrac{15a^2bc + 20ab^2c - 25abc^2}{-5abc}$

d. $\dfrac{(x + y)^2 + (x - y)^2}{-2xy}$

SECTION 4.8	*Dividing Polynomials by Polynomials*

Use long division to divide one polynomial by another.

20. Do each division.

a. $x + 2 \overline{)x^2 + 3x + 5}$

b. $x - 1 \overline{)x^2 - 6x + 5}$

c. $x + 3 \overline{)2x^2 + 7x + 3}$

d. $3x - 1 \overline{)3x^2 + 14x - 2}$

e. $2x - 1 \overline{)6x^3 + x^2 + 1}$

f. $3x + 1 \overline{)-13x - 4 + 9x^3}$

■ Chapter Test

1. Use exponents to rewrite $2xxxyyyy$.

2. Evaluate $3^2 + 5^3$.

In Problems 3–6, write each expression as an expression containing only one exponent.

3. $y^2(yy^3)$

4. $(-3b^2)(2b^3)(-b^2)$

5. $(2x^3)^5(x^2)^3$

6. $(2rr^2r^3)^3$

In Problems 7–10, simplify each expression. Write answers without using parentheses or negative exponents.

7. $3x^0$

8. $2y^{-5}y^2$

9. $\dfrac{y^2}{yy^{-2}}$

10. $\left(\dfrac{a^2b^{-1}}{4a^3b^{-2}}\right)^{-3}$

11. Write 28,000 in scientific notation.

12. Write 0.0025 in scientific notation.

13. Write 7.4×10^3 in standard notation.

14. Write 9.3×10^{-5} in standard notation.

15. Classify $3x^2 + 2$ as a monomial, a binomial, or a trinomial.

16. Find the degree of the polynomial $3x^2y^3z^4 + 2x^3y^2z - 5x^2y^3z^5$

17. Evaluate $x^2 + x - 2$ when $x = -2$.

18. Graph the polynomial function $y = f(x) = x^2 + 2$.

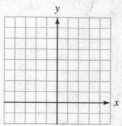

19. Simplify $-6(x - y) + 2(x + y) - 3(x + 2y)$

20. Simplify $-2(x^2 + 3x - 1) - 3(x^2 - x + 2) + 5(x^2 + 2)$

21. Add $\quad 3x^3 + 4x^2 - x - 7$
$\qquad\quad \underline{2x^3 - 2x^2 + 3x + 2}$

22. Subtract $\quad 2x^2 - 7x + 3$
$\qquad\qquad\quad \underline{3x^2 - 2x - 1}$

In Problems 23–26, find each product.

23. $(-2x^3)(2x^2y)$

24. $3y^2(y^2 - 2y + 3)$

25. $(2x - 5)(3x + 4)$

26. $(2x - 3)(x^2 - 2x + 4)$

27. Solve the equation $(a + 2)^2 = (a - 3)^2$.

28. Simplify $\dfrac{8x^2y^3z^4}{16x^3y^2z^4}$.

29. Simplify $\dfrac{6a^2 - 12b^2}{24ab}$.

30. Divide $2x + 3\overline{)2x^2 - x - 6}$.

■ Cumulative Review Exercises

In Exercises 1–4, evaluate each expression. Assume that $x = 2$ and $y = -5$.

1. $5 + 3 \cdot 2$

2. $3 \cdot 5^2 - 4$

3. $\dfrac{3x - y}{xy}$

4. $\dfrac{x^2 - y^2}{x + y}$

In Exercises 5–8, solve each equation.

5. $\dfrac{4}{5}x + 6 = 18$

6. $x - 2 = \dfrac{x + 2}{3}$

7. $2(5x + 2) = 3(3x - 2)$

8. $4(y + 1) = -2(4 - y)$

In Exercises 9–12, graph the solution of each inequality.

9. $5x - 3 > 7$

10. $7x - 9 < 5$

11. $-2 < -x + 3 < 5$

12. $0 \leq \dfrac{4 - x}{3} \leq 2$

In Exercises 13–14, solve each formula for the indicated variable.

13. $A = p + prt$, for r

14. $A = \dfrac{1}{2} bh$, for h

In Exercises 15–16, graph each equation.

15. $3x - 4y = 12$

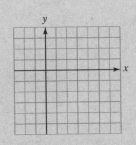

16. $y - 2 = \dfrac{1}{2}(x - 4)$

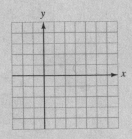

In Exercises 17–18, solve each system by graphing.

17. $\begin{cases} x - y = 4 \\ 2x + y = 5 \end{cases}$

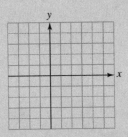

18. $\begin{cases} 3x + 2y \geq 6 \\ x + 3y \leq 6 \end{cases}$

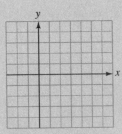

In Exercises 19–20, solve each system of equations by an algebraic method.

19. $\begin{cases} x + y = 1 \\ x - y = 7 \end{cases}$

20. $\begin{cases} 4x + 9y = 8 \\ 2x - 6y = -3 \end{cases}$

In Exercises 21–24, write each expression as an expression using only one exponent.

21. $(y^3 y^5)y^6$

22. $\dfrac{x^3 y^4}{x^2 y^3}$

23. $\dfrac{a^4 b^{-3}}{a^{-3} b^3}$

24. $\left(\dfrac{-x^{-2} y^3}{x^{-3} y^2} \right)^2$

In Exercises 25–28, do each operation.

25. $(3x^2 + 2x - 7) - (2x^2 - 2x + 7)$

26. $(3x - 7)(2x + 8)$

27. $(x - 2)(x^2 + 2x + 4)$

28. $x - 3 \overline{)2x^2 - 5x - 3}$

29. Astronomy   The **parsec,** a unit of distance used in astronomy, is 3×10^{16} meters. The distance to Betelgeuse, a star in the constellation Orion, is 1.6×10^2 parsecs. Use scientific notation to express this distance in meters.

30. Surface area   The total surface area A of a box with dimensions l, w, and d (see Illustration 1) is given by the formula

$$A = 2lw + 2wd + 2ld$$

If $A = 202$ square inches, $l = 9$ inches, and $w = 5$ inches, find d.

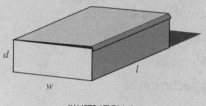

ILLUSTRATION 1

31. Concentric circles   The area of the ring between the two concentric circles of radius r and R (see Illustration 2) is given by the formula

$$A = \pi(R + r)(R - r)$$

If $r = 3$ inches and $R = 17$ inches, find A to the nearest tenth.

ILLUSTRATION 2

32. Employee discounts   Employees at an appliance store can purchase merchandise at 25% less than the regular price. An employee buys a color TV set for $414.72, including 8% sales tax. Find the regular price of the TV.

5 Factoring Polynomials

Computer Programmer

Computers process vast quantities of information rapidly and accurately when they are given programs to follow. Computer programmers write those programs, which logically list the steps the machine must follow to organize data, solve a problem, or do other tasks. Applications programmers are usually oriented toward business, engineering, or science. System programmers maintain the software that controls a computer system.

SAMPLE APPLICATION ■ Computers take more time to do multiplications than additions. To make a program run as quickly as possible, a computer programmer wants to write the polynomial $3x^4 + 2x^3 + 5x^2 + 7x + 1$ in a form that requires fewer multiplications. Write the polynomial so that it contains only four multiplications.
(See Exercise 82 in Exercise 5.2.)

5.1 Factoring Out the Greatest Common Factor

■ FACTORING NATURAL NUMBERS ■ FACTORING MONOMIALS ■ FACTORING OUT A COMMON MONOMIAL ■ FACTORING OUT A NEGATIVE FACTOR ■ QUADRATIC EQUATIONS

Getting Ready   *Simplify each expression by removing parentheses.*

1. $5(x + 3)$ **2.** $7(y - 8)$ **3.** $x(3x - 2)$ **4.** $y(5y + 9)$

5. $a(b + 9)$ **6.** $x(3 + x + y)$ **7.** $xy(x - 4)$ **8.** $xy^2(2x - 5y)$

■ FACTORING NATURAL NUMBERS

In this chapter, we shall reverse the operation of multiplication and show how to find the factors of a known product. The process of finding the individual factors of a product is called **factoring.**

Because 4 divides 12 exactly, 4 is called a **factor** of 12. The numbers 1, 2, 3, 4, 6, and 12 are the natural-number factors of 12, because each one divides 12 exactly. Recall that a natural number greater than 1 whose only factors are 1 and the number itself is called a **prime number.** For example, 19 is a prime number, because

1. 19 is a natural number greater than 1, and

2. The only two natural number factors of 19 are 1 and 19.

The prime numbers less than 50 are

2, 3, 5, 7, 11, 13, 17, 19, 23, 29, 31, 37, 41, 43, and 47

A natural number is said to be in **prime-factored form** if it is written as the product of factors that are prime numbers.

To find the prime-factored form of a natural number, we can use a **factoring tree.** For example, to find the prime-factored form of 60, we proceed as follows:

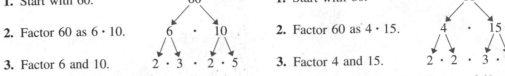

Solution 1	*Solution 2*
1. Start with 60.	**1.** Start with 60.
2. Factor 60 as $6 \cdot 10$.	**2.** Factor 60 as $4 \cdot 15$.
3. Factor 6 and 10.	**3.** Factor 4 and 15.

We stop when only prime numbers appear. In either case, the prime factors of 60 are $2 \cdot 2 \cdot 3 \cdot 5$. Thus, the prime-factored form of 60 is $2^2 \cdot 3 \cdot 5$. This illustrates the **fundamental theorem of arithmetic,** which states that there is only one prime factorization for any natural number greater than 1.

The right-hand sides of the equations

$$42 = 2 \cdot 3 \cdot 7$$
$$60 = 2^2 \cdot 3 \cdot 5$$
$$90 = 2 \cdot 3^2 \cdot 5$$

show the prime-factored forms (or **prime factorizations**) of 42, 60, and 90. The largest natural number that divides each of these numbers is called their **greatest common factor (GCF).** The GCF of 42, 60, and 90 is 6, because 6 is the largest natural number that divides each of these numbers:

$$\frac{42}{6} = 7 \qquad \frac{60}{6} = 10 \qquad \text{and} \qquad \frac{90}{6} = 15$$

■ FACTORING MONOMIALS

Algebraic monomials can also have a greatest common factor. The right-hand sides of the equations

$$6a^2b^3 = 2 \cdot 3 \cdot a \cdot a \cdot b \cdot b \cdot b$$
$$4a^3b^2 = 2 \cdot 2 \cdot a \cdot a \cdot a \cdot b \cdot b$$
$$18a^2b = 2 \cdot 3 \cdot 3 \cdot a \cdot a \cdot b$$

show the prime factorizations of $6a^2b^3$, $4a^3b^2$, and $18a^2b$. Since all three of these monomials have one factor of 2, two factors of a, and one factor of b, the GCF is

$$2 \cdot a \cdot a \cdot b \qquad \text{or} \qquad 2a^2b$$

To find the GCF of several monomials, we follow these steps.

Strategy for Finding the Greatest Common Factor (GCF)

1. Find the prime factorization of each monomial.

2. List each common factor the least number of times it appears in any one monomial.

3. Find the product of the factors found in the list to obtain the GCF.

EXAMPLE 1   Find the GCF of $10x^3y^2$, $60x^2y$, and $30xy^2$.

Solution   **1.** Find the prime factorization of each monomial.

$$10x^3y^2 = \mathbf{2 \cdot 5 \cdot x \cdot} x \cdot x \cdot \mathbf{y} \cdot y$$
$$60x^2y = \mathbf{2} \cdot 2 \cdot 3 \cdot \mathbf{5 \cdot x \cdot} x \cdot \mathbf{y}$$
$$30xy^2 = \mathbf{2} \cdot 3 \cdot \mathbf{5 \cdot x \cdot y \cdot} y$$

2. List each common factor the least number of times it appears in any one monomial: 2, 5, x, and y.

3. Find the product of the factors in the list:

$$2 \cdot 5 \cdot x \cdot y = 10xy$$ ∎

Self Check   Find the GCF of $20a^2b^3$, $12ab^4$, and $8a^3b^2$.
Answer $4ab^2$

■ ■ ■ ■ ■ ■ ■ ■ ■ ■   PERSPECTIVE

Much of the mathematics that we have inherited from earlier times is the result of teamwork. In a battle early in the 12th century, control of the Spanish city of Toledo was taken from the Mohammedans, who had ruled there for four centuries. Libraries in this great city contained many books written in Arabic, full of knowledge that was unknown in Europe.

The Archbishop of Toledo wanted to share this knowledge with the rest of the world. He knew that these books should be translated into Latin, the universal language of scholarship. But what European scholar could read Arabic? The citizens of Toledo knew both Arabic and Spanish, and most scholars of Europe could read Spanish.

Teamwork saved the day. A citizen of Toledo read the Arabic text aloud, in Spanish. The scholars listened to the Spanish version and wrote it down in Latin. One of these scholars was an Englishman, Robert of Chester. It was he who translated al-Khowarazmi's book, *Ihm al-jabr wa'l muqabalah*, the beginning of the subject we now know as algebra.

■ FACTORING OUT A COMMON MONOMIAL

Recall that the distributive property provides a way to multiply a polynomial by a monomial. For example,

$$3x^2(2x - 3y) = 3x^2 \cdot 2x - 3x^2 \cdot 3y$$
$$= 6x^3 - 9x^2y$$

To reverse this process and factor the product $6x^3 - 9x^2y$, we can find the GCF of each monomial (which is $3x^2$) and then use the distributive property in reverse.

$$6x^3 - 9x^2y = 3x^2 \cdot 2x - 3x^2 \cdot 3y$$
$$= 3x^2(2x - 3y)$$

This process is called **factoring out the greatest common factor.**

EXAMPLE 2   Factor $12y^2 + 20y$.

Solution   To find the GCF, we find the prime factorization of $12y^2$ and $20y$.

$$\left.\begin{array}{l} 12y^2 = \mathbf{2 \cdot 2} \cdot 3 \cdot \mathbf{y} \cdot y \\ 20y = \mathbf{2 \cdot 2} \cdot 5 \cdot \mathbf{y} \end{array}\right\} \text{GCF} = 4y$$

We can use the distributive property to factor out the GCF of $4y$.

$$\begin{aligned} 12y^2 + 20y &= \mathbf{4y} \cdot 3y + \mathbf{4y} \cdot 5 \\ &= \mathbf{4y}(3y + 5) \end{aligned}$$

Check by verifying that $4y(3y + 5) = 12y^2 + 20y$.   ■

Self Check   Factor $15x^3 - 20x^2$.

Answer $5x^2(3x - 4)$

EXAMPLE 3   Factor $35a^3b^2 - 14a^2b^3$.

Solution   To find the GCF, we find the prime factorization of $35a^3b^2$ and $-14a^2b^3$.

$$\left.\begin{array}{l} 35a^3b^2 = 5 \cdot \mathbf{7} \cdot \mathbf{a \cdot a} \cdot a \cdot \mathbf{b} \cdot b \\ -14a^2b^3 = -2 \cdot \mathbf{7} \cdot \mathbf{a \cdot a} \cdot \mathbf{b} \cdot b \cdot b \end{array}\right\} \text{GCF} = 7a^2b^2$$

We factor out the GCF of $7a^2b^2$.

$$\begin{aligned} 35a^3b^2 - 14a^2b^3 &= \mathbf{7a^2b^2} \cdot 5a - \mathbf{7a^2b^2} \cdot 2b \\ &= \mathbf{7a^2b^2}(5a - 2b) \end{aligned}$$

Check by verifying that $7a^2b^2(5a - 2b) = 35a^3b^2 - 14a^2b^3$.   ■

Self Check   Factor $40x^2y^3 + 15x^3y^2$.

Answer $5x^2y^2(8y + 3x)$

EXAMPLE 4   Factor $a^2b^2 - ab$.

Solution   We factor out the GCF, which is ab.

$$\begin{aligned} a^2b^2 - ab &= \mathbf{ab} \cdot ab - \mathbf{ab} \cdot 1 \\ &= \mathbf{ab}(ab - 1) \end{aligned}$$

WARNING!   The last term of $a^2b^2 - ab$ has an implied coefficient of 1. When ab is factored out, we must write the coefficient of 1.

We check by verifying that $ab(ab - 1) = a^2b^2 - ab$.   ■

Self Check
Answer

Factor $x^3y^5 + x^2y^3$.
$x^2y^3(xy^2 + 1)$

EXAMPLE 5   Factor $12x^3y^2z + 6x^2yz - 3xz$.

Solution   We factor out the GCF, which is $3xz$.

$$12x^3y^2z + 6x^2yz - 3xz = \mathbf{3xz} \cdot 4x^2y^2 + \mathbf{3xz} \cdot 2xy - \mathbf{3xz} \cdot 1$$
$$= \mathbf{3xz}(4x^2y^2 + 2xy - 1)$$

Check by verifying that

$$3xz(4x^2y^2 + 2xy - 1) = 12x^3y^2z + 6x^2yz - 3xz \qquad \blacksquare$$

Self Check
Answer

Factor $6ab^2c - 12a^2bc + 3ab$.
$3ab(2bc - 4ac + 1)$

■ FACTORING OUT A NEGATIVE FACTOR

It is often useful to factor out the negative of a monomial.

EXAMPLE 6   Factor -1 out of $-a^3 + 2a^2 - 4$.

Solution

$$-a^3 + 2a^2 - 4$$
$$= (\mathbf{-1})a^3 + (\mathbf{-1})(-2a^2) + (\mathbf{-1})4 \qquad (-1)(-2a^2) = +2a^2.$$
$$= \mathbf{-1}(a^3 - 2a^2 + 4) \qquad \text{Factor out } -1.$$
$$= -(a^3 - 2a^2 + 4) \qquad \begin{array}{l}\text{The coefficient of 1 need not be}\\\text{written.}\end{array}$$

Check by verifying that

$$-(a^3 - 2a^2 + 4) = -a^3 + 2a^2 - 4 \qquad \blacksquare$$

Self Check
Answer

Factor -1 out of $-b^4 - 3b^2 + 2$.
$-(b^4 + 3b^2 - 2)$

EXAMPLE 7   Factor out the negative of the GCF: $-18a^2b + 6ab^2 - 12a^2b^2$.

Solution   The GCF is $6ab$. To factor out its negative, we factor out $-6ab$.

$$-18a^2b + 6ab^2 - 12a^2b^2 = (\mathbf{-6ab})3a - (\mathbf{-6ab})b + (\mathbf{-6ab})2ab$$
$$= \mathbf{-6ab}(3a - b + 2ab)$$

Check by verifying that

$$-6ab(3a - b + 2ab) = -18a^2b + 6ab^2 - 12a^2b^2 \qquad \blacksquare$$

Self Check

Answer

Factor out the negative of the GCF: $-25xy^2 - 15x^2y + 30x^2y^2$.

$-5xy(5y + 3x - 6xy)$

■ QUADRATIC EQUATIONS

Equations such as $9x - 6 = 0$ that involve first-degree polynomials are called **linear equations.** Equations such as $9x^2 - 6x = 0$ that involve second-degree polynomials are called **quadratic equations.**

Quadratic Equations

A **quadratic equation** is an equation of the form

$$ax^2 + bx + c = 0$$

where a, b, and c are real numbers, and $a \neq 0$.

The techniques that we have used to solve linear equations cannot be used to solve quadratic equations. For example, these techniques cannot be used to isolate x on one side of the equation $9x^2 - 6x = 0$. However, we can often solve quadratic equations by factoring and using the following property of real numbers.

Zero-Factor Property of Real Numbers

Suppose a and b represent two real numbers. Then

If $ab = 0$, then $a = 0$ or $b = 0$.

We already know that if either of two numbers is 0, their product is 0. The zero-factor property says that if the product of two numbers is 0, then at least one of them must be 0.

For example, the equation $(x - 4)(x + 5) = 0$ indicates that a product is equal to 0. By the zero-factor property, one of the factors must be 0:

$$x - 4 = 0 \qquad \text{or} \qquad x + 5 = 0$$

We can solve each of these linear equations to get

$$x = 4 \qquad \text{or} \qquad x = -5$$

The equation $(x - 4)(x + 5) = 0$ has two solutions: 4 and -5.

EXAMPLE 8

Solve $9x^2 - 6x = 0$.

Solution

We begin by factoring the left-hand side of the equation.

$$9x^2 - 6x = 0$$
$$3x(3x - 2) = 0$$

By the zero-factor theorem, we have

$$3x = 0 \qquad \text{or} \qquad 3x - 2 = 0$$

We can solve each of these equations to get

$$x = 0 \qquad \text{or} \qquad x = \frac{2}{3}$$

Check: To check, we substitute these results for x in the original equation, and simplify.

For $x = 0$	**For $x = \frac{2}{3}$**
$9x^2 - 6x = 0$	$9x^2 - 6x = 0$
$9(0)^2 - 6(0) \stackrel{?}{=} 0$	$9\left(\dfrac{2}{3}\right)^2 - 6\left(\dfrac{2}{3}\right) \stackrel{?}{=} 0$
$0 - 0 \stackrel{?}{=} 0$	$9\left(\dfrac{4}{9}\right) - 6\left(\dfrac{2}{3}\right) \stackrel{?}{=} 0$
$0 = 0$	$4 - 4 \stackrel{?}{=} 0$
	$0 = 0$

Both solutions check.

Self Check

Solve $5y^2 + 10y = 0$.

Answer 0, −2

Orals *Find the prime factorization of each number.*

1. 36   **2.** 27   **3.** 81   **4.** 45

Find the greatest common factor:

5. 3, 6, and 9   **6.** $3a^2b$, $6ab$, and $9ab^2$

Factor out the greatest common factor:

7. $15xy + 10$ **8.** $15xy + 10xy^2$

EXERCISE 5.1

REVIEW *Solve each equation and check all solutions.*

1. $3x - 2(x + 1) = 5$

2. $5(y - 1) + 1 = y$

3. $\dfrac{2x - 7}{5} = 3$

4. $2x - \dfrac{x}{2} = 5x$

VOCABULARY AND CONCEPTS   *Fill in each blank to make a true statement.*

5. A natural number greater than 1 whose only factors are 1 and itself is called a _____ number.

6. If a natural number is written as the product of prime numbers, it is written in _____ form.

7. The GCF of several natural numbers is the _____ number that divides each of the numbers.

8. An equation of the form $ax^2 + bx + c = 0$, where $a \neq 0$, is called a _____ equation.

9. If $ab = 0$, then $a = $ __ or $b = $ __.

10. A quadratic equation contains a _____-degree polynomial.

PRACTICE   *In Exercises 11–22, find the prime factorization of each number.*

11. 12

12. 24

13. 15

14. 20

15. 40

16. 62

17. 98

18. 112

19. 225

20. 144

21. 288

22. 968

In Exercises 23–28, complete each factorization.

23. $4a + 12 = (a + 3)$

24. $3t - 27 = 3\left(t - \right)$

25. $r^4 + r^2 = r^2\left(+ 1\right)$

26. $a^3 - a^2 = (a - 1)$

27. $4y^2 + 8y - 2xy = 2y\left(2y + - \right)$

28. $3x^2 - 6xy + 9xy^2 = \left(- 2y + 3y^2\right)$

In Exercises 29–56, factor out the greatest common factor.

29. $3x + 6$

30. $2y - 10$

31. $xy - xz$

32. $uv + ut$

33. $t^3 + 2t^2$

34. $b^3 - 3b^2$

35. $r^4 - r^2$

36. $a^3 + a^2$

37. $a^3b^3z^3 - a^2b^3z^2$

38. $r^3s^6t^9 + r^2s^2t^2$

39. $24x^2y^3z^4 + 8xy^2z^3$

40. $3x^2y^3 - 9x^4y^3z$

41. $12uvw^3 - 18uv^2w^2$

42. $14xyz - 16x^2y^2z$

43. $3x + 3y - 6z$

44. $2x - 4y + 8z$

45. $ab + ac - ad$

46. $rs - rt + ru$

47. $4y^2 + 8y - 2xy$

48. $3x^2 - 6xy + 9xy^2$

49. $12r^2 - 3rs + 9r^2s^2$

50. $6a^2 - 12a^3b + 36ab$

51. $abx - ab^2x + abx^2$

52. $a^2b^2x^2 + a^3b^2x^2 - a^3b^3x^3$

53. $4x^2y^2z^2 - 6xy^2z^2 + 12xyz^2$

54. $32xyz + 48x^2yz + 36xy^2z$

55. $70a^3b^2c^2 + 49a^2b^3c^3 - 21a^2b^2c^2$

56. $8a^2b^2 - 24ab^2c + 9b^2c^2$

In Exercises 57–68, factor out -1 from each polynomial.

57. $-a - b$

58. $-x - 2y$

59. $-2x + 5y$

60. $-3x + 8z$

61. $-2a + 3b$

62. $-2x + 5y$

63. $-3m - 4n + 1$

64. $-3r + 2s - 3$

65. $-3xy + 2z + 5w$

66. $-4ab + 3c - 5d$

67. $-3ab - 5ac + 9bc$

68. $-6yz + 12xz - 5xy$

In Exercises 69–78, factor out the greatest common factor, including −1.

69. $-3x^2y - 6xy^2$

70. $-4a^2b^2 + 6ab^2$

71. $-4a^2b^3 + 12a^3b^2$

72. $-25x^4y^3z^2 + 30x^2y^3z^4$

73. $-4a^2b^2c^2 + 14a^2b^2c - 10ab^2c^2$

74. $-10x^4y^3z^2 + 8x^3y^2z - 20x^2y$

75. $-14a^6b^6 + 49a^2b^3 - 21ab$

76. $-35r^9s^9t^9 + 25r^6s^6t^6 + 75r^3s^3t^3$

77. $-5a^2b^3c + 15a^3b^4c^2 - 25a^4b^3c$

78. $-7x^5y^4z^3 + 49x^5y^5z^4 - 21x^6y^4z^3$

In Exercises 79–86, solve each equation.

79. $(x - 2)(x + 3) = 0$

80. $(x - 3)(x - 2) = 0$

81. $(x - 4)(x + 1) = 0$

82. $(x + 5)(x + 2) = 0$

83. $(2x - 5)(3x + 6) = 0$

84. $(3x - 4)(x + 1) = 0$

85. $(x - 1)(x + 2)(x - 3) = 0$

86. $(x + 2)(x + 3)(x - 4) = 0$

In Exercises 87–98, solve each equation.

87. $x^2 - 3x = 0$

88. $x^2 + 5x = 0$

89. $2x^2 - 5x = 0$

90. $5x^2 + 7x = 0$

91. $x^2 - 7x = 0$

92. $x^2 - 8x = 0$

93. $3x^2 + 8x = 0$

94. $5x^2 - x = 0$

95. $8x^2 - 16x = 0$

96. $15x^2 - 20x = 0$

97. $10x^2 + 2x = 0$

98. $5x^2 + x = 0$

WRITING

99. When we add $5x$ and $7x$, we combine like terms: $5x + 7x = 12x$. Explain how this is related to factoring out a common factor.

100. One student summarized the zero-factor property of real numbers by saying, "Anything times zero is zero." This answer is true, but it does not describe the zero-factor property. Explain.

SOMETHING TO THINK ABOUT

101. Think of two positive integers. Divide their product by their greatest common factor. Why do you think the result is called the **lowest common multiple** of the two integers? (*Hint:* The **multiples** of an integer such as 5 are 5, 10, 15, 20, 25, 30, and so on.)

102. Two integers are **relatively prime** if their greatest common factor is 1. For example, 6 and 25 are relatively prime, but 6 and 15 are not. If the greatest common factor of three integers is 1, must any two of them be relatively prime? Explain.

5.2 Factoring by Grouping

■ FACTORING OUT A POLYNOMIAL ■ FACTORING BY GROUPING

Getting Ready   *Remove parentheses and simplify.*

1. $3(x + y) + a(x + y)$ **2.** $x(y + 1) + 5(y + 1)$

3. $5(x + 1) - y(x + 1)$ **4.** $x(x + 2) - y(x + 2)$

5. $(3x - y)x + (3x - y)y$ **6.** $5(y - 7) - y(y - 7)$

■ FACTORING OUT A POLYNOMIAL

If the GCF of several terms is a polynomial, we can factor out the common polynomial factor. For example, since $a + b$ is a common factor of $(a + b)x$ and $(a + b)y$, we can factor out the $a + b$.

$$(a + b)x + (a + b)y = (a + b)(x + y)$$

We can check by verifying that $(a + b)(x + y) = (a + b)x + (a + b)y$.

EXAMPLE 1   Factor $a + 3$ out of $(a + 3) + (a + 3)^2$.

Solution   Recall that $a + 3$ is equal to $(a + 3)1$ and that $(a + 3)^2$ is equal to $(a + 3)(a + 3)$. We can factor out $a + 3$ and simplify.

$$(a + 3) + (a + 3)^2 = (a + 3)1 + (a + 3)(a + 3)$$
$$= (a + 3)[1 + (a + 3)]$$
$$= (a + 3)(a + 4) \qquad ■$$

Self Check   Factor out $y + 2$: $(y + 2)^2 - 3(y + 2)$.
Answer $(y + 2)(y - 1)$

EXAMPLE 2   Factor $6a^2b^2(x + 2y) - 9ab(x + 2y)$.

Solution   The GCF of $6a^2b^2$ and $9ab$ is $3ab$. We can factor out this GCF as well as $(x + 2y)$.

$$6a^2b^2(x + 2y) - 9ab(x + 2y)$$
$$= 3ab \cdot 2ab(x + 2y) - 3ab \cdot 3(x + 2y)$$
$$= 3ab(x + 2y)(2ab - 3) \qquad \text{Factor out } 3ab(x + 2y). \qquad ■$$

Self Check

Factor $4p^3q^2(2a + b) + 8p^2q^3(2a + b)$.

Answer

$4p^2q^2(2a + b)(p + 2q)$

■ FACTORING BY GROUPING

Suppose we wish to factor

$$ax + ay + cx + cy$$

Although no factor is common to all four terms, there is a common factor of a in $ax + ay$ and a common factor of c in $cx + cy$. We can factor out the a and the c to obtain

$$ax + ay + cx + cy = a(x + y) + c(x + y)$$
$$= (x + y)(a + c) \qquad \text{Factor out } x + y.$$

We can check the result by multiplication.

$$(x + y)(a + c) = ax + cx + ay + cy$$
$$= ax + ay + cx + cy$$

Thus, $ax + ay + cx + cy$ factors as $(x + y)(a + c)$. This type of factoring is called **factoring by grouping.**

EXAMPLE 3

Factor $2c + 2d - cd - d^2$.

Solution

$$2c + 2d - cd - d^2 = 2(c + d) - d(c + d) \qquad \begin{array}{l}\text{Factor out 2 from } 2c + 2d \text{ and} \\ -d \text{ from } -cd - d^2.\end{array}$$

$$= (c + d)(2 - d) \qquad \text{Factor out } c + d.$$

$$\text{Check: } (c + d)(2 - d) = 2c - cd + 2d - d^2$$
$$= 2c + 2d - cd - d^2 \qquad \blacksquare$$

Self Check

Factor $3a + 3b - ac - bc$.

Answer

$(a + b)(3 - c)$

EXAMPLE 4

Factor $x^2y - ax - xy + a$.

Solution

$$x^2y - ax - xy + a = x(xy - a) - 1(xy - a) \qquad \begin{array}{l}\text{Factor out } x \text{ from } x^2y - ax \\ \text{and } -1 \text{ from } -xy + a.\end{array}$$

$$= (xy - a)(x - 1) \qquad \text{Factor out } xy - a.$$

Check by multiplication. $\qquad \blacksquare$

Self Check

Factor $pq^2 + tq + 2pq + 2t$.

Answer

$(pq + t)(q + 2)$

> **WARNING!** When factoring expressions such as those in the previous two examples, don't think that $2(c + d) - d(c + d)$ or $x(xy - a) - 1(xy - a)$ are in factored form. To be in factored form, the final result must be a product.

Factoring by grouping often works on polynomials with more than four terms.

EXAMPLE 5 Factor $6am - 6bm + 6cm + 5an - 5bn + 5cn$.

Solution Factor $6m$ from the first three terms and $5n$ from the last three terms to obtain

$$6am - 6bm + 6cm + 5an - 5bn + 5cn = 6m(a - b + c) + 5n(a - b + c)$$

Then factor out the common factor of $(a - b + c)$.

$$6am - 6bm + 6cm + 5an - 5bn + 5cn = (a - b + c)(6m + 5n)$$

Check by multiplication. ∎

Self Check Factor $2ap + 2aq - 2at - bp - bq + bt$.
Answer $(p + q - t)(2a - b)$

 b

EXAMPLE 6 Factor **a.** $a(c - d) + b(d - c)$ and **b.** $ac + bd - ad - bc$.

Solution **a.**

$$
\begin{aligned}
a(c - d) + b(d - c) &= a(c - d) - b(-d + c) && \text{Factor } -1 \text{ from } d - c. \\
&= a(c - d) - b(c - d) && -d + c = c - d. \\
&= (c - d)(a - b) && \text{Factor out } (c - d).
\end{aligned}
$$

b. In this example, we cannot factor anything from the first two terms or the last two terms. However, if we rearrange the terms, the factoring is routine:

$$
\begin{aligned}
ac + bd - ad - bc &= ac - ad + bd - bc && bd - ad = -ad + bd. \\
&= a(c - d) + b(d - c) && \text{Factor } a \text{ from } ac - ad \text{ and} \\
& && b \text{ from } bd - bc. \\
&= (c - d)(a - b) && \text{See part a.}
\end{aligned}
$$
∎

Self Check Factor $ax - by - ay + bx$.
Answer $(a + b)(x - y)$

Orals *Find the common factor of the given terms.*

1. $a(x + 3)$ and $3(x + 3)$
2. $5(a - 1)$ and $xy(a - 1)$
3. $b(x - 2)$ and $(x - 2)^2$
4. $(y + 5)$ and $(y + 5)^2$
5. $a(x - 7)$, $9(x - 7)$, and $x(x - 7)$
6. $5(2y + 9)$, $y(2y + 9)$, and $y^2(2y + 9)$

EXERCISE 5.2

REVIEW *Simplify each expression and write all results without using negative exponents.*

1. $u^3u^2u^4$

2. $\dfrac{y^6}{y^8}$

3. $\dfrac{a^3b^4}{a^2b^5}$

4. $(3x^5)^0$

VOCABULARY AND CONCEPTS *Fill in each blank to make a true statement.*

5. The GCF of $x(a + b) - y(a + b)$ is _____.

6. Check the results of a factoring problem by _____.

In Exercises 7–10, complete each factorization.

7. $a(x + y) + b(x + y) = (x + y)$_____

8. $p(m - n) - q(m - n) =$ _____$(p - q)$

9. $(r - s)p - (r - s)q = (r - s)$_____

10. $ax + bx + ap + bp = x$_____ $+ p$_____

$=$ _____$(x + p)$

PRACTICE *In Exercises 11–30, factor each expression.*

11. $(x + y)2 + (x + y)b$

12. $(a - b)c + (a - b)d$

13. $3(x + y) - a(x + y)$

14. $x(y + 1) - 5(y + 1)$

15. $3(r - 2s) - x(r - 2s)$

16. $x(a + 2b) + y(a + 2b)$

17. $(x - 3)^2 + (x - 3)$

18. $(3t + 5)^2 - (3t + 5)$

19. $2x(a^2 + b) + 2y(a^2 + b)$

20. $3x(c - 3d) + 6y(c - 3d)$

21. $3x^2(r + 3s) - 6y^2(r + 3s)$

22. $9a^2b^2(3x - 2y) - 6ab(3x - 2y)$

23. $3x(a + b + c) - 2y(a + b + c)$

24. $2m(a - 2b + 3c) - 21xy(a - 2b + 3c)$

25. $14x^2y(r + 2s - t) - 21xy(r + 2s - t)$

26. $15xy^3(2x - y + 3z) + 25xy^2(2x - y + 3z)$

27. $(x + 3)(x + 1) - y(x + 1)$ •

28. $x(x^2 + 2) - y(x^2 + 2)$

29. $(3x - y)(x^2 - 2) + (x^2 - 2)$

30. $(x - 5y)(a + 2) - (x - 5y)$

In Exercises 31–50, factor each expression.

31. $2x + 2y + ax + ay$

32. $bx + bz + 5x + 5z$

33. $7r + 7s - kr - ks$

34. $9p - 9q + mp - mq$

35. $xr + xs + yr + ys$

36. $pm - pn + qm - qn$

37. $2ax + 2bx + 3a + 3b$

38. $3xy + 3xz - 5y - 5z$

39. $2ab + 2ac + 3b + 3c$

40. $3ac + a + 3bc + b$

41. $2x^2 + 2xy - 3x - 3y$

42. $3ab + 9a - 2b - 6$

43. $3tv - 9tw + uv - 3uw$

44. $ce - 2cf + 3de - 6df$

45. $9mp + 3mq - 3np - nq$

46. $ax + bx - a - b$

47. $mp - np - m + n$

48. $6x^2u - 3x^2v + 2yu - yv$

49. $x(a - b) + y(b - a)$

50. $p(m - n) - q(n - m)$

In Exercises 51–58, factor each expression. Factor out all common factors first, if they exist.

51. $ax^3 + bx^3 + 2ax^2y + 2bx^2y$

52. $x^3y^2 - 2x^2y^2 + 3xy^2 - 6y^2$

53. $4a^2b + 12a^2 - 8ab - 24a$

54. $-4abc - 4ac^2 + 2bc + 2c^2$

55. $x^3 + 2x^2 + x + 2$

56. $y^3 - 3y^2 - 5y + 15$

57. $x^3y - x^2y - xy^2 + y^2$

58. $2x^3z - 4x^2z + 32xz - 64z$

In Exercises 59–66, factor each expression completely.

59. $x^2 + xy + x + 2x + 2y + 2$

60. $ax + ay + az + bx + by + bz$

61. $am + bm + cm - an - bn - cn$

62. $x^2 + xz - x - xy - yz + y$

63. $ad - bd - cd + 3a - 3b - 3c$

64. $ab + ac - ad - b - c + d$

65. $ax^2 - ay + bx^2 - by + cx^2 - cy$

66. $a^2x - bx - a^2y + by + a^2z - bz$

In Exercises 67–78, factor each expression completely. You may have to rearrange some terms first.

67. $2r - bs - 2s + br$

68. $5x + ry + rx + 5y$

69. $ax + by + bx + ay$

70. $mr + ns + ms + nr$

71. $ac + bd - ad - bc$

72. $sx - ry + rx - sy$

73. $ar^2 - brs + ars - br^2$

74. $a^2bc + a^2c + abc + ac$

75. $ba + 3 + a + 3b$

76. $xy + 7 + y + 7x$

77. $pr + qs - ps - qr$

78. $ac - bd - ad + bc$

WRITING

79. Explain why $a - b$ and $b - a$ are negatives of each other.

80. Explain how you would factor $x(a - b) + y(b - a)$.

SOMETHING TO THINK ABOUT

81. Factor $ax + ay + bx + by$ by grouping the first two terms and the last two terms. Then rearrange the terms as $ax + bx + ay + by$, and factor again by grouping the first two and the last two. Do the results agree?

82. The polynomial $3x^4 + 2x^3 + 5x^2 + 7x + 1$ can be written as

$3 \cdot x \cdot x \cdot x \cdot x + 2 \cdot x \cdot x \cdot x + 5 \cdot x \cdot x + 7 \cdot x + 1$

to illustrate that it involves 10 multiplications and 4 additions. Since computers take more time to do multiplications than additions, computer programmers write polynomials in a way that requires the fewest multiplications possible. Use factoring by grouping to write the previous polynomial so that it contains only four multiplications.

5.3 Factoring the Difference of Two Squares

■ FACTORING THE DIFFERENCE OF TWO SQUARES ■ MULTISTEP FACTORING ■ SOLVING EQUATIONS

Getting Ready *Multiply the binomials.*

1. $(a + b)(a - b)$ **2.** $(2r + s)(2r - s)$

3. $(3x + 2y)(3x - 2y)$ **4.** $(4x^2 + 3)(4x^2 - 3)$

■ FACTORING THE DIFFERENCE OF TWO SQUARES

Whenever we multiply a binomial of the form $x + y$ by a binomial of the form $x - y$, we obtain a binomial of the form $x^2 - y^2$.

$$(x + y)(x - y) = x^2 - xy + xy - y^2$$
$$= x^2 - y^2$$

The binomial $x^2 - y^2$ is called the **difference of two squares,** because x^2 is the square of x and y^2 is the square of y. The difference of the squares of two quantities always factors into the sum of those two quantities multiplied by the difference of those two quantities.

> **Factoring the Difference of Two Squares**
> $$x^2 - y^2 = (x + y)(x - y)$$

If we think of the difference of two squares as the square of a **First** quantity minus the square of a **Last** quantity, we have the formula

$$F^2 - L^2 = (F + L)(F - L)$$

and we say, *To factor the square of a First quantity minus the square of a Last quantity, we multiply the First plus the Last by the First minus the Last.*

To factor $x^2 - 9$, we note that it can be written in the form $x^2 - 3^2$ and use the formula for factoring the difference of two squares:

$$\mathbf{F^2 - L^2} = \mathbf{(F + L)(F - L)}$$
$$x^2 - 3^2 = (x + 3)(x - 3)$$

We can check by verifying that $(x + 3)(x - 3) = x^2 - 9$.

To factor the difference of two squares, it is helpful to know the integers that are perfect squares. The number 400, for example, is a perfect square, because $20^2 = 400$. The perfect integer squares less than 400 are

1, 4, 9, 16, 25, 36, 49, 64, 81, 100, 121, 144, 169, 196, 225, 256, 289, 324, 361

Expressions containing variables such as x^4y^2 are also perfect squares, because they can be written as the square of a quantity:

$$x^4y^2 = (x^2y)^2$$

EXAMPLE 1 Factor $25x^2 - 49$.

Solution We can write $25x^2 - 49$ in the form $(5x)^2 - 7^2$ and use the formula for factoring the difference of two squares:

$$\mathbf{F}^2 \ - \mathbf{L}^2 = (\mathbf{F} + \mathbf{L})(\mathbf{F} - \mathbf{L})$$
$$\downarrow \qquad \downarrow \qquad \downarrow \quad \downarrow \quad \downarrow \quad \downarrow$$
$$(\mathbf{5x})^2 - 7^2 = (\mathbf{5x} + 7)(\mathbf{5x} - 7) \qquad \text{Substitute } 5x \text{ for F and 7 for L.}$$

We can check by multiplying $5x + 7$ and $5x - 7$.

$$(5x + 7)(5x - 7) = 25x^2 - 35x + 35x - 49$$
$$= 25x^2 - 49 \qquad \blacksquare$$

Self Check Factor $16a^2 - 81$.
Answer $(4a + 9)(4a - 9)$

EXAMPLE 2 Factor $4y^4 - 25z^2$.

Solution We can write $4y^4 - 25z^2$ in the form $(2y^2)^2 - (5z)^2$ and use the formula for factoring the difference of two squares:

$$\mathbf{F}^2 \quad - \ \mathbf{L}^2 \ = (\ \mathbf{F} \ + \ \mathbf{L})(\ \mathbf{F} \ - \ \mathbf{L})$$
$$\downarrow \qquad \downarrow \qquad \downarrow \quad \downarrow \quad \downarrow \quad \downarrow$$
$$(\mathbf{2y^2})^2 - (5z)^2 = (\mathbf{2y^2} + 5z)(\mathbf{2y^2} - 5z)$$

Check by multiplication. $\qquad \blacksquare$

Self Check Factor $9m^2 - 64n^4$.
Answer $(3m + 8n^2)(3m - 8n^2)$

■ MULTISTEP FACTORING

We can often factor out a greatest common factor before factoring the difference of two squares. To factor $8x^2 - 32$, for example, we factor out the GCF of 8 and then factor the resulting difference of two squares.

$$8x^2 - 32 = 8(x^2 - 4) \qquad \text{Factor out 8.}$$
$$= 8(x^2 - 2^2) \qquad \text{Write 4 as } 2^2.$$
$$= 8(x + 2)(x - 2) \qquad \text{Factor the difference of two squares.}$$

We can check by multiplication:

$$8(x + 2)(x - 2) = 8(x^2 - 4)$$
$$= 8x^2 - 32$$

EXAMPLE 3   Factor $2a^2x^3y - 8b^2xy$.

Solution   We factor out the GCF of $2xy$ and then factor the resulting difference of two squares.

$$2a^2x^3y - 8b^2xy$$
$$= \mathbf{2xy} \cdot a^2x^2 - \mathbf{2xy} \cdot 4b^2 \qquad \text{The GCF is } 2xy.$$
$$= \mathbf{2xy}(a^2x^2 - 4b^2) \qquad \text{Factor out } 2xy.$$
$$= 2xy[(ax)^2 - (2b)^2] \qquad \text{Write } a^2x^2 \text{ as } (ax)^2 \text{ and } 4b^2 \text{ as } (2b)^2.$$
$$= 2xy(ax + 2b)(ax - 2b) \qquad \text{Factor the difference of two squares.}$$

We check by multiplication.   ∎

Self Check   Factor $2p^2q^2s - 18r^2s$.

Answer $2s(pq + 3r)(pq - 3r)$

Sometimes we must factor a difference of two squares more than once to factor a polynomial. For example, the binomial $625a^4 - 81b^4$ can be written in the form $(25a^2)^2 - (9b^2)^2$, which factors as

$$625a^4 - 81b^4 = (25a^2)^2 - (9b^2)^2$$
$$= (25a^2 + 9b^2)(\mathbf{25a^2 - 9b^2})$$

Since the factor $25a^2 - 9b^2$ can be written in the form $(5a)^2 - (3b)^2$, it is the difference of two squares and can be factored as $(5a + 3b)(5a - 3b)$. Thus,

$$625a^4 - 81b^4 = (25a^2 + 9b^2)(\mathbf{5a + 3b})(\mathbf{5a - 3b})$$

WARNING! The binomial $25a^2 + 9b^2$ is the **sum of two squares,** because it can be written in the form $(5a)^2 + (3b)^2$. If we are limited to integer coefficients, binomials that are the sum of two squares cannot be factored.

Polynomials that do not factor over the integers are called **prime polynomials.**

EXAMPLE 4   Factor $2x^4y - 32y$.

Solution
$$2x^4y - 32y = \mathbf{2y} \cdot x^4 - \mathbf{2y} \cdot 16$$
$$= \mathbf{2y}(x^4 - 16) \qquad \text{Factor out the GCF of } 2y.$$
$$= 2y(x^2 + 4)(\mathbf{x^2 - 4}) \qquad \text{Factor } x^4 - 16.$$
$$= 2y(x^2 + 4)(\mathbf{x + 2})(\mathbf{x - 2}) \qquad \text{Factor } x^2 - 4. \text{ Note that } x^2 + 4 \text{ does not factor.}$$ ∎

Self Check

Answer

Factor $48a^5 - 3ab^4$.

$3a(4a^2 + b^2)(2a + b)(2a - b)$

Example 5 requires the techniques of factoring out a common factor, factoring by grouping, and factoring the difference of two squares.

EXAMPLE 5   Factor $2x^3 - 8x + 2yx^2 - 8y$.

Solution

$$2x^3 - 8x + 2yx^2 - 8y = 2(x^3 - 4x + yx^2 - 4y)$$ Factor out 2.

$$= 2[x(x^2 - 4) + y(x^2 - 4)]$$ Factor out x from $x^3 - 4x$ and y from $yx^2 - 4y$.

$$= 2[(x^2 - 4)(x + y)]$$ Factor out $x^2 - 4$.

$$= 2(x + 2)(x - 2)(x + y)$$ Factor $x^2 - 4$.

Check by multiplication.   ∎

Self Check

Answer

Factor $3a^3 - 12a + 3a^2b - 12b$.

$3(a + 2)(a - 2)(a + b)$

WARNING!   To *factor* an expression means to factor the expression *completely.*

■ SOLVING EQUATIONS

We can use factoring the difference of two squares to solve many quadratic equations.

EXAMPLE 6   Solve $4x^2 = 36$.

Solution   Before we can use the zero-factor theorem, we must subtract 36 from both sides to make the right-hand side 0.

$$4x^2 = 36$$

$$4x^2 - 36 = 0$$ Subtract 36 from both sides.

$$x^2 - 9 = 0$$ Divide both sides by 4.

$$(x + 3)(x - 3) = 0$$ Factor $x^2 - 9$.

$$x + 3 = 0 \quad \text{or} \quad x - 3 = 0$$ Set each factor equal to 0.

$$x = -3 \qquad\qquad x = 3$$ Solve each linear equation.

Check each solution.

For x = −3	*For x = 3*
$4x^2 = 36$	$4x^2 = 36$
$4(-3)^2 \stackrel{?}{=} 36$	$4(3)^2 \stackrel{?}{=} 36$
$4(9) \stackrel{?}{=} 36$	$4(9) \stackrel{?}{=} 36$
$36 = 36$	$36 = 36$

Both solutions check.   ■

Self Check   Solve $9p^2 = 64$.

Answer $\frac{8}{3}, -\frac{8}{3}$

Orals   *Factor each binomial.*

1. $x^2 - 9$ **2.** $y^2 - 36$

3. $z^2 - 4$ **4.** $p^2 - q^2$

5. $25 - t^2$ **6.** $36 - r^2$

7. $100 - y^2$ **8.** $100 - y^4$

EXERCISE 5.3

REVIEW

1. In the study of the flow of fluids, Bernoulli's law is given by the equation

$$\frac{p}{w} + \frac{v^2}{2g} + h = k$$

Solve the equation for p.

2. Solve Bernoulli's law for h. (See Exercise 1.)

$$h = k - \frac{p}{w} - \frac{v^2}{2g}$$

VOCABULARY AND CONCEPTS   *Fill in each blank to make a true statement.*

3. A binomial of the form $a^2 - b^2$ is called the

_____.

4. A binomial of the form $a^2 + b^2$ is called the

_____.

5. $p^2 - q^2 = (p + q)$_____

6. The _____ of two squares cannot be factored by using only integer coefficients.

Complete each factorization.

7. $x^2 - 9 = (x + 3)$_____

8. $p^2 - q^2 =$ _____$(p - q)$

9. $4m^2 - 9n^2 = (2m + 3n)$_____

10. $16p^2 - 25q^2 =$ _____$(4p - 5q)$

PRACTICE *In Exercises 11–30, factor each expression, if possible.*

11. $x^2 - 16$

12. $x^2 - 25$

13. $y^2 - 49$

14. $y^2 - 81$

15. $4y^2 - 49$

16. $9z^2 - 4$

17. $9x^2 - y^2$

18. $4x^2 - z^2$

19. $25t^2 - 36u^2$

20. $49u^2 - 64v^2$

21. $16a^2 - 25b^2$

22. $36a^2 - 121b^2$

23. $a^2 + b^2$

24. $121a^2 - 144b^2$

25. $a^4 - 4b^2$

26. $9y^2 + 16z^2$

27. $49y^2 - 225z^4$

28. $25x^2 + 36y^2$

29. $196x^4 - 169y^2$

30. $144a^4 + 169b^4$

In Exercises 31–46, factor each expression.

31. $8x^2 - 32y^2$

32. $2a^2 - 200b^2$

33. $2a^2 - 8y^2$

34. $32x^2 - 8y^2$

35. $3r^2 - 12s^2$

36. $45u^2 - 20v^2$

37. $x^3 - xy^2$

38. $a^2b - b^3$

39. $4a^2x - 9b^2x$

40. $4b^2y - 16c^2y$

41. $3m^3 - 3mn^2$

42. $2p^2q - 2q^3$

43. $4x^4 - x^2y^2$

44. $9xy^2 - 4xy^4$

45. $2a^3b - 242ab^3$

46. $50c^4d^2 - 8c^2d^4$

In Exercises 47–58, factor each expression.

47. $x^4 - 81$

48. $y^4 - 625$

49. $a^4 - 16$

50. $b^4 - 256$

51. $a^4 - b^4$

52. $m^4 - 16n^4$

53. $81r^4 - 256s^4$

54. $x^8 - y^4$

55. $a^4 - b^8$

56. $16y^8 - 81z^4$

57. $x^8 - y^8$

58. $x^8y^8 - 1$

In Exercises 59–78, factor each expression.

59. $2x^4 - 2y^4$

60. $a^5 - ab^4$

61. $a^4b - b^5$

62. $m^5 - 16mn^4$

63. $48m^4n - 243n^5$

64. $2x^4y - 512y^5$

65. $3a^5y + 6ay^5$

66. $2p^{10}q - 32p^2q^5$

67. $3a^{10} - 3a^2b^4$

68. $2x^9y + 2xy^9$

69. $2x^8y^2 - 32y^6$

70. $3a^8 - 243a^4b^8$

71. $a^6b^2 - a^2b^6c^4$

72. $a^2b^3c^4 - a^2b^3d^4$

73. $a^2b^7 - 625a^2b^3$

74. $16x^3y^4z - 81x^3y^4z^5$

75. $243r^5s - 48rs^5$

76. $1,024m^5n - 324mn^5$

77. $16(x - y)^2 - 9$

78. $9(x + 1)^2 - y^2$

In Exercises 79–88, factor each expression.

79. $a^3 - 9a + 3a^2 - 27$

80. $b^3 - 25b - 2b^2 + 50$

81. $y^3 - 16y - 3y^2 + 48$

82. $a^3 - 49a + 2a^2 - 98$

83. $3x^3 - 12x + 3x^2 - 12$

84. $2x^3 - 18x - 6x^2 + 54$

85. $3m^3 - 3mn^2 + 3am^2 - 3an^2$

86. $ax^3 - axy^2 - bx^3 + bxy^2$

87. $2m^3n^2 - 32mn^2 + 8m^2 - 128$

88. $2x^3y + 4x^2y - 98xy - 196y$

In Exercises 89–100, solve each equation.

89. $x^2 - 25 = 0$

90. $x^2 - 36 = 0$

91. $y^2 - 49 = 0$

92. $z^2 - 121 = 0$

93. $4x^2 - 1 = 0$

94. $9y^2 - 1 = 0$

95. $9y^2 - 4 = 0$

96. $16z^2 - 25 = 0$

97. $x^2 = 49$

98. $z^2 = 25$

99. $4x^2 = 81$

100. $9y^2 = 64$

WRITING

101. Explain how to factor the difference of two squares.

102. Explain why $x^4 - y^4$ is not completely factored as $(x^2 + y^2)(x^2 - y^2)$.

SOMETHING TO THINK ABOUT

103. It is easy to multiply 399 by 401 without a calculator: The product is $400^2 - 1$, or 159,999. Explain.

104. Use the method in the previous exercise to find $498 \cdot 502$ without a calculator.

5.4 Factoring Trinomials with Lead Coefficients of 1

■ FACTORING TRINOMIALS OF THE FORM $x^2 + bx + c$ ■ FACTORING OUT -1 ■ PRIME TRINOMIALS
■ MULTISTEP FACTORING ■ FACTORING PERFECT-SQUARE TRINOMIALS ■ SOLVING EQUATIONS

Getting Ready   *Multiply the binomials.*

1. $(x + 6)(x + 6)$

2. $(y - 7)(y - 7)$

3. $(a - 3)(a - 3)$

4. $(x + 4)(x + 5)$ **5.** $(r - 2)(r - 5)$ **6.** $(m + 3)(m - 7)$

7. $(a - 3b)(a + 4b)$ **8.** $(u - 3v)(u - 5v)$ **9.** $(x + 4y)(x - 6y)$

■ FACTORING TRINOMIALS OF THE FORM $x^2 + bx + c$

The product of two binomials is often a trinomial. For example,

$$(x + 3)(x + 3) = x^2 + 6x + 9 \quad \text{and} \quad (x - 4y)(x - 4y) = x^2 - 8xy + 16y^2$$

For this reason, we should not be surprised that many trinomials factor into the product of two binomials. To develop a method for factoring trinomials, we multiply $(x + a)$ and $(x + b)$.

$$\begin{aligned}
(x + a)(x + b) &= x^2 + bx + ax + ab && \text{Use the FOIL method.} \\
&= x^2 + ax + bx + ab && \text{Write } bx + ax \text{ as } ax + bx. \\
&= x^2 + (a + b)x + ab && \text{Factor } x \text{ out of } ax + bx.
\end{aligned}$$

From the result, we can see that

- the coefficient of the middle term is the sum of a and b, and
- the last term is the product of a and b

We can use these facts to factor trinomials with lead coefficients of 1.

EXAMPLE 1   Factor $x^2 + 5x + 6$.

Solution   To factor this trinomial, we will write it as the product of two binomials. Since the first term of the trinomial is x^2, the first term of each binomial factor must be x. To fill in the following blanks, we must find two integers whose product is $+6$ and whose sum is $+5$.

$$x^2 + 5x + 6 = (x \qquad)(x \qquad)$$

The positive factorizations of 6 and the sums of the factors are shown in the following table.

Product of the factors	Sum of the factors
$1(6) = 6$	$1 + 6 = 7$
$2(3) = 6$	$2 + 3 = 5$

The last row contains the integers $+2$ and $+3$, whose product is $+6$ and whose sum is $+5$. So we can fill in the blanks with $+2$ and $+3$.

$$x^2 + 5x + 6 = (x + 2)(x + 3)$$

358 CHAPTER 5 FACTORING POLYNOMIALS

To check the result, we verify that $(x + 2)$ times $(x + 3)$ is $x^2 + 5x + 6$.

$$(x + 2)(x + 3) = x^2 + 3x + 2x + 2 \cdot 3$$
$$= x^2 + 5x + 6$$

Self Check Factor $y^2 + 5y + 4$.

Answer $(y + 1)(y + 4)$

In Example 1, the factors can be written in either order. An equivalent factorization is $x^2 + 5x + 6 = (x + 3)(x + 2)$.

EXAMPLE 2 Factor $y^2 - 7y + 12$.

Solution Since the first term of the trinomial is y^2, the first term of each binomial factor must be y. To fill in the following blanks, we must find two integers whose product is $+12$ and whose sum is -7.

$$y^2 - 7y + 12 = (y \quad\quad)(y \quad\quad)$$

The two-integer factorizations of 12 and the sums of the factors are shown in the following table.

Product of the factors	Sum of the factors
$1(12) = 12$	$1 + 12 = 13$
$2(6) = 12$	$2 + 6 = 8$
$3(4) = 12$	$3 + 4 = 7$
$-1(-12) = 12$	$-1 + (-12) = -13$
$-2(-6) = 12$	$-2 + (-6) = -8$
$-3(-4) = 12$	$-3 + (-4) = -7$

The last row contains the integers -3 and -4, whose product is $+12$ and whose sum is -7. So we can fill in the blanks with -3 and -4.

$$y^2 - 7y + 12 = (y - 3)(y - 4)$$

To check the result, we verify that $(y - 3)$ times $(y - 4)$ is $y^2 - 7y + 12$.

$$(y - 4)(y - 3) = y^2 - 3y - 4y + 12$$
$$= y^2 - 7y + 12$$

Self Check Factor $p^2 - 5p + 6$.

Answer $(p - 3)(p - 2)$

EXAMPLE 3   Factor $a^2 + 2a - 15$.

Solution   Since the first term is a^2, the first term of each binomial factor must be a. To fill in the following blanks, we must find two integers whose product is -15 and whose sum is $+2$.

$$a^2 + 2a - 15 = (a \qquad)(a \qquad)$$

The possible factorizations of -15 and the sums of the factors are shown in the following table.

Product of the factors	Sum of the factors
$1(-15) = -15$	$1 + (-15) = -14$
$3(-5) = -15$	$3 + (-5) = -2$
$5(-3) = -15$	$5 + (-3) = 2$
$15(-1) = -15$	$15 + (-1) = 14$

The third row contains the integers $+5$ and -3, whose product is -15 and whose sum is $+2$. So we can fill in the blanks with $+5$ and -3.

$$a^2 + 2a - 15 = (a + 5)(a - 3)$$

We can check by multiplying $a + 5$ and $a - 3$.

$$(a + 5)(a - 3) = a^2 - 3a + 5a - 15$$
$$= a^2 + 2a - 15 \qquad ■$$

Self Check   Factor $p^2 + 3p - 18$.
Answer $(p + 6)(p - 3)$

EXAMPLE 4   Factor $z^2 - 4z - 21$.

Solution   Since the first term is z^2, the first term of each binomial factor must be z. To fill in the following blanks, we must find two integers whose product is -21 and whose sum is -4.

$$z^2 - 4y - 21 = (z \qquad)(z \qquad)$$

The factorizations of -21 and the sums of the factors are shown in the following table.

Product of the factors	Sum of the factors
$1(-21) = -21$	$1 + (-21) = -20$
$3(-7) = -21$	$3 + (-7) = -4$
$7(-3) = -21$	$7 + (-3) = 4$
$21(-1) = -21$	$21 + (-1) = 20$

The second row contains the integers $+3$ and -7, whose product is -21 and whose sum is -4. So we can fill in the blanks with $+3$ and -7.

$$z^2 - 4z - 21 = (z + 3)(z - 7)$$

To check, we multiply $z + 3$ and $z - 7$.

$$(z + 3)(z - 7) = z^2 - 7z + 3z - 21$$
$$= z^2 - 4z - 21 \qquad \blacksquare$$

Self Check   Factor $q^2 - 2q - 24$.

Answer $(q + 4)(q - 6)$

The next example has two variables.

EXAMPLE 5   Factor $x^2 + xy - 6y^2$.

Solution   Since the first term is x^2, the first term of each binomial factor must be x. Since the last term is $-6y^2$, the second term of each binomial factor has a factor of y. To fill in the following blanks, we must find coefficients whose product is -6 that will give a middle term of xy.

$$x^2 + xy - 6y^2 = (x \quad\quad y\,)(x \quad\quad y\,)$$

The possible factorizations of -6 and the sums of the factors are shown in the following table.

Product of the factors	Sum of the factors
$1(-6) = -6$	$1 + (-6) = -5$
$2(-3) = -6$	$2 + (-3) = -1$
$3(-2) = -6$	$3 + (-2) = 1$
$6(-1) = -6$	$6 + (-1) = 5$

The third row contains the integers 3 and -2. These are the only integers whose product is -6 and will give the correct middle term of xy. So we can fill in the blanks with 3 and -2.

$$x^2 + xy - 6y^2 = (x + 3y)(x - 2y)$$

We can check by multiplying $x + 3y$ and $x - 2y$.

$$(x + 3y)(x - 2y) = x^2 - 2xy + 3xy - 6y^2$$
$$= x^2 + xy - 6y^2 \qquad \blacksquare$$

Self Check   Factor $a^2 + ab - 12b^2$.

Answer $(a - 3b)(a + 4b)$

■ FACTORING OUT −1

When the coefficient of the first term is −1, we begin by factoring out −1.

EXAMPLE 6   Factor $-x^2 + 2x + 15$.

Solution   We factor out −1 and then factor the trinomial.

$$-x^2 + 2x + 15 = -(x^2 - 2x - 15) \qquad \text{Factor out } -1.$$
$$= -(x - 5)(x + 3) \qquad \text{Factor } x^2 - 2x - 15.$$

We check by multiplying −1, $x - 5$, and $x + 3$.

$$-(x - 5)(x + 3) = -(x^2 + 3x - 5x - 15)$$
$$= -(x^2 - 2x - 15)$$
$$= -x^2 + 2x + 15 \qquad ■$$

Self Check   Factor $-x^2 + 11x - 18$.
Answer $-(x - 9)(x - 2)$

■ PRIME TRINOMIALS

If a trinomial cannot be factored using only integers, it is called a **prime polynomial.**

EXAMPLE 7   Factor $x^2 + 2x + 3$, if possible.

Solution   To factor the trinomial, we must find two integers whose product is +3 and whose sum is +2. The possible factorizations of 3 and the sums of the factors are shown in the following table.

Product of the factors	Sum of the factors
$1(3) = 3$	$1 + 3 = 4$
$-1(-3) = 3$	$-1 + (-3) = -4$

Since two integers whose product is +3 and whose sum is +2 do not exist, $x^2 + 2x + 3$ cannot be factored. It is a prime trinomial.   ■

Self Check   Factor $x^2 - 4x + 6$, if possible.
Answer   It is prime.

■ MULTISTEP FACTORING

The following examples require more than one step.

EXAMPLE 8

Factor $-3ax^2 + 9a - 6ax$.

Solution   We write the trinomial in descending powers of x and factor out the common factor of $-3a$.

$$-3ax^2 + 9a - 6ax = -3ax^2 - 6ax + 9a$$
$$= -3a(x^2 + 2x - 3)$$

Finally, we factor the trinomial $x^2 + 2x - 3$.

$$-3ax^2 + 9a - 6ax = -3a(x + 3)(x - 1)$$

We can check by multiplying.

$$-3a(x + 3)(x - 1) = -3a(x^2 + 2x - 3)$$
$$= -3ax^2 - 6ax + 9a$$
$$= -3ax^2 + 9a - 6ax$$ ■

Self Check   Factor $-2pq^2 + 6p - 4pq$.
Answer $-2p(q + 3)(q - 1)$

EXAMPLE 9

Factor $m^2 - 2mn + n^2 - 64a^2$.

Solution   We group the first three terms together and factor the resulting trinomial.

$$m^2 - 2mn + n^2 - 64a^2 = (m - n)(m - n) - 64a^2$$
$$= (m - n)^2 - (8a)^2$$

Then we factor the resulting difference of two squares:

$$m^2 - 2mn + n^2 - 64a^2 = (m - n)^2 - (8a)^2$$
$$= (m - n + 8a)(m - n - 8a)$$ ■

Self Check   Factor $p^2 + 4pq + 4q^2 - 25y^2$.
Answer $(p + 2q + 5y)(p + 2q - 5y)$

■ FACTORING PERFECT-SQUARE TRINOMIALS

We have discussed the following special product formulas used to square binomials.

Special Product Formulas

$$(x + y)^2 = x^2 + 2xy + y^2$$
$$(x - y)^2 = x^2 - 2xy + y^2$$

These formulas can be used in reverse order to factor perfect-square trinomials.

1. $x^2 + 2xy + y^2 = (x + y)^2$
2. $x^2 - 2xy + y^2 = (x - y)^2$

In words, Formula 1 states that *if a trinomial is the square of one quantity, plus twice the product of two quantities, plus the square of the second quantity, it factors into the square of the sum of the quantities.*

Formula 2 states that *if a trinomial is the square of one quantity, minus twice the product of two quantities, plus the square of the second quantity, it factors into the square of the difference of the quantities.*

The trinomials on the left-hand sides of the previous equations are called **perfect-square trinomials,** because they are the results of squaring a binomial. Although we can factor perfect-square trinomials by using the techniques discussed earlier in this section, we can usually factor them by inspecting their terms. For example, $x^2 + 8x + 16$ is a perfect-square trinomial, because

- The first term x^2 is the square of x.
- The last term 16 is the square of 4.
- The middle term $8x$ is twice the product of x and 4.

Thus,

$$x^2 + 8x + 16 = x^2 + 2(x)(4) + 4^2$$
$$= (x + 4)^2$$

EXAMPLE 10   Factor $x^2 - 10x + 25$.

Solution $x^2 - 10x + 25$ is a perfect-square trinomial, because

- The first term x^2 is the square of x.
- The last term 25 is the square of 5.
- The middle term $-10x$ is the negative of twice the product of x and 5.

Thus,

$$x^2 - 10x + 25 = x^2 - 2(x)(5) + 5^2$$
$$= (x - 5)^2$$ ■

Self Check   Factor $x^2 + 10x + 25$.
Answer $(x + 5)^2$

■ SOLVING EQUATIONS

We can use the factoring of trinomials and the zero-factor property to solve many equations.

EXAMPLE 11   Solve $x^3 - 2x^2 - 63x = 0$.

Solution

$$x^3 - 2x^2 - 63x = 0$$

$$x(x^2 - 2x - 63) = 0 \qquad \text{Factor out } x.$$

$$x(x + 7)(x - 9) = 0 \qquad \text{Factor the trinomial.}$$

$$x = 0 \quad \text{or} \quad x + 7 = 0 \quad \text{or} \quad x - 9 = 0 \qquad \text{Set each factor equal to 0.}$$

$$x = -7 \qquad\qquad x = 9 \qquad \text{Solve each linear equation.}$$

The solutions are 0, -7, and 9. Check each one.   ■

Self Check   Solve $x^3 - x^2 - 2x = 0$.

Answers $0, -1, 2$

Orals   *Finish each factoring problem.*

1. $x^2 + 5x + 4 = (x + 1)(x + \quad)$
2. $x^2 - 5x + 6 = (x \quad 2)(x \quad 3)$
3. $x^2 + x - 6 = (x \quad 2)(x + \quad)$
4. $x^2 - x - 6 = (x \quad 3)(x + \quad)$
5. $x^2 + 5x - 6 = (x + \quad)(x - \quad)$
6. $x^2 - 7x + 6 = (x - \quad)(x - \quad)$

EXERCISE 5.4

REVIEW   *Graph the solution of each inequality on a number line.*

1. $x - 3 > 5$
2. $x + 4 \le 3$
3. $-3x - 5 \ge 4$
4. $2x - 3 < 7$

5. $\dfrac{3(x - 1)}{4} < 12$
6. $\dfrac{-2(x + 3)}{3} \ge 9$
7. $-2 < x \le 4$
8. $-5 \le x + 1 < 5$

VOCABULARY AND CONCEPTS   *Complete each formula.*

9. $x^2 + 2xy + y^2 = $ _____
10. $x^2 - 2xy + y^2 = $ _____

Complete each factorization.

11. $y^2 + 6y + 8 = (y + \quad)(y + \quad)$
12. $z^2 - 3z - 10 = (z + \quad)(z - \quad)$
13. $x^2 - xy - 2y^2 = (x + \quad)(x - \quad)$
14. $a^2 + ab - 6b^2 = (a + \quad)(a - \quad)$

In Exercises 15–42, factor each trinomial, if possible. Use the FOIL method to check each result.

15. $x^2 + 3x + 2$

16. $y^2 + 4y + 3$

17. $z^2 + 12z + 11$

18. $x^2 + 7x + 10$

19. $a^2 - 4a - 5$

20. $b^2 + 6b - 7$

21. $t^2 - 9t + 14$

22. $c^2 - 9c + 8$

23. $u^2 + 10u + 15$

24. $v^2 + 9v + 15$

25. $y^2 - y - 30$

26. $x^2 - 3x - 40$

27. $a^2 + 6a - 16$

28. $x^2 + 5x - 24$

29. $t^2 - 5t - 50$

30. $a^2 - 10a - 39$

31. $r^2 - 9r - 12$

32. $s^2 + 11s - 26$

33. $y^2 + 2yz + z^2$

34. $r^2 - 2rs + 4s^2$

35. $x^2 + 4xy + 4y^2$

36. $a^2 + 10ab + 9b^2$

37. $m^2 + 3mn - 10n^2$

38. $m^2 - mn - 12n^2$

39. $a^2 - 4ab - 12b^2$

40. $p^2 + pq - 6q^2$

41. $u^2 + 2uv - 15v^2$

42. $m^2 + 3mn - 10n^2$

In Exercises 43–54, factor each trinomial. Factor out -1 first.

43. $-x^2 - 7x - 10$

44. $-x^2 + 9x - 20$

45. $-y^2 - 2y + 15$

46. $-y^2 - 3y + 18$

47. $-t^2 - 15t + 34$

48. $-t^2 - t + 30$

49. $-r^2 + 14r - 40$

50. $-r^2 + 14r - 45$

51. $-a^2 - 4ab - 3b^2$

52. $-a^2 - 6ab - 5b^2$

53. $-x^2 + 6xy + 7y^2$

54. $-x^2 - 10xy + 11y^2$

In Exercises 55–66, write each trinomial in descending powers of one variable, and then factor.

55. $4 - 5x + x^2$

56. $y^2 + 5 + 6y$

57. $10y + 9 + y^2$

58. $x^2 - 13 - 12x$

59. $c^2 - 5 + 4c$

60. $b^2 - 6 - 5b$

61. $-r^2 + 2s^2 + rs$

62. $u^2 - 3v^2 + 2uv$

63. $4rx + r^2 + 3x^2$

64. $-a^2 + 5b^2 + 4ab$

65. $-3ab + a^2 + 2b^2$

66. $-13yz + y^2 - 14z^2$

In Exercises 67–78, completely factor each trinomial. Factor out any common monomials first (including -1, if necessary).

67. $2x^2 + 10x + 12$

68. $3y^2 - 21y + 18$

69. $3y^3 + 6y^2 + 3y$

70. $4x^4 + 16x^3 + 16x^2$

71. $-5a^2 + 25a - 30$

72. $-2b^2 + 20b - 18$

73. $3z^2 - 15tz + 12t^2$

74. $5m^2 + 45mn - 50n^2$

75. $12xy + 4x^2y - 72y$

76. $48xy + 6xy^2 + 96x$

77. $-4x^2y - 4x^3 + 24xy^2$

78. $3x^2y^3 + 3x^3y^2 - 6xy^4$

In Exercises 79–86, completely factor each expression.

79. $ax^2 + 4ax + 4a + bx + 2b$

80. $mx^2 + mx - 6m + nx - 2n$

81. $a^2 + 8a + 15 + ab + 5b$

82. $x^2 + 2xy + y^2 + 2x + 2y$

83. $a^2 + 2ab + b^2 - 4$

84. $a^2 + 6a + 9 - b^2$

85. $b^2 - y^2 - 4y - 4$

86. $c^2 - a^2 + 8a - 16$

In Exercises 87–98, factor each perfect square trinomial.

87. $x^2 + 6x + 9$

88. $x^2 + 10x + 25$

89. $y^2 - 8y + 16$

90. $z^2 - 2z + 1$

91. $t^2 + 20t + 100$

92. $r^2 + 24r + 144$

93. $u^2 - 18u + 81$

94. $v^2 - 14v + 49$

95. $x^2 + 4xy + 4y^2$

96. $a^2 + 6ab + 9b^2$

97. $r^2 - 10rs + 25s^2$

98. $m^2 - 12mn + 36n^2$

In Exercises 99–116, solve each equation.

99. $x^2 - 13x + 12 = 0$

100. $x^2 + 7x + 6 = 0$

101. $x^2 - 2x - 15 = 0$

102. $x^2 - x - 20 = 0$

103. $-4x - 21 + x^2 = 0$

104. $2x + x^2 - 15 = 0$

105. $x^2 + 8 - 9x = 0$

106. $45 + x^2 - 14x = 0$

107. $a^2 + 8a = -15$

108. $a^2 - a = 56$

109. $2y - 8 = -y^2$

110. $-3y + 18 = y^2$

111. $x^3 + 3x^2 + 2x = 0$

112. $x^3 - 7x^2 + 10x = 0$

113. $x^3 - 27x - 6x^2 = 0$

114. $x^3 - 22x - 9x^2 = 0$

115. $(x - 1)(x^2 + 5x + 6) = 0$

116. $(x - 2)(x^2 - 8x + 7) = 0$

WRITING

117. Explain how you would write a trinomial in descending order.

118. Explain how to use the FOIL method to check the factoring of a trinomial.

SOMETHING TO THINK ABOUT

119. Two students factor $2x^2 + 20x + 42$ and get two different answers: $(2x + 6)(x + 7)$, and $(x + 3)(2x + 14)$. Do both answers check? Why don't they agree? Is either completely correct?

120. Find the error:

$x = y$	
$x^2 = xy$	Multiply both sides by x.
$x^2 - y^2 = xy - y^2$	Subtract y^2 from both sides.
$(x + y)(x - y) = y(x - y)$	Factor.
$x + y = y$	Divide both sides by $(x - y)$.
$y + y = y$	Substitute y for its equal, x.
$2y = y$	Combine like terms.
$2 = 1$	Divide both sides by y.

5.5 Factoring General Trinomials

■ FACTORING TRINOMIALS OF THE FORM $ax^2 + bx + c$ ■ FACTORING PERFECT-SQUARE TRINOMIALS ■ SOLVING EQUATIONS

Getting Ready   *Multiply and combine like terms.*

1. $(2x + 1)(3x + 2)$ **2.** $(3y - 2)(2y - 5)$ **3.** $(4t - 3)(2t + 3)$

4. $(2r + 5)(2r - 3)$ **5.** $(2m - 3)(3m - 2)$ **6.** $(4a + 3)(4a + 1)$

■ FACTORING TRINOMIALS OF THE FORM $ax^2 + bx + c$

We must consider more combinations of factors when we factor trinomials with lead coefficients other than 1.

EXAMPLE 1   Factor $2x^2 + 5x + 3$.

Solution   Since the first term is $2x^2$, the first terms of the binomial factors must be $2x$ and x. To fill in the following blanks, we must find two factors of $+3$ that will give a middle term of $+5x$.

$$(2x \qquad)(x \qquad)$$

Since the sign of each term of the trinomial is $+$, we need to consider only positive factors of the last term (3). Since the positive factors of 3 are 1 and 3, there are two possible factorizations.

$$(2x + 1)(x + 3) \qquad \text{or} \qquad (2x + 3)(x + 1)$$

The first possibility is incorrect, because it gives a middle term of $7x$. The second possibility is correct, because it gives a middle term of $5x$. Thus,

$$2x^2 + 5x + 3 = (2x + 3)(x + 1)$$

Check by multiplication.   ■

Self Check   Factor $3x^2 + 7x + 2$.
Answer $(3x + 1)(x + 2)$

EXAMPLE 2   Factor $6x^2 - 17x + 5$.

Solution   Since the first term is $6x^2$, the first terms of the binomial factors must be $6x$ and x or $3x$ and $2x$. To fill in the following blanks, we must find two factors of $+5$ that will give a middle term of $-17x$.

$$(6x \qquad)(x \qquad) \qquad \text{or} \qquad (3x \qquad)(2x \qquad)$$

Since the sign of the third term is $+$ and the sign of the middle term is $-$, we need to consider only negative factors of the last term (5). Since the negative factors of 5 are -1 and -5, there are four possible factorizations.

$$(6x - 1)(x - 5) \qquad (6x - 5)(x - 1)$$
$$(3x - 1)(2x - 5) \qquad (3x - 5)(2x - 1)$$

Only the possibility printed in color gives the correct middle term of $-17x$. Thus,

$$6x^2 - 17x + 5 = (3x - 1)(2x - 5)$$

Check by multiplication. ∎

Self Check

Answer

Factor $6x^2 - 7x + 2$.

$(3x - 2)(2x - 1)$

EXAMPLE 3 Factor $3y^2 - 4y - 4$.

Solution Since the first term is $3y^2$, the first terms of the binomial factors must be $3y$ and y. To fill in the following blanks, we must find two factors of -4 that will give a middle term of $-4y$.

$$(3y \quad\quad)(y \quad\quad)$$

Since the sign of the third term is $-$, the signs inside the binomial factors will be different. Because the factors of the last term (4) are 1, 2, and 4, there are six possibilities to consider.

$$(3y + 1)(y - 4) \qquad (3y + 4)(y - 1)$$
$$(3y - 1)(y + 4) \qquad (3y - 4)(y + 1)$$
$$(3y - 2)(y + 2) \qquad (3y + 2)(y - 2)$$

Only the possibility printed in color gives the correct middle term of $-4y$. Thus,

$$3y^2 - 4y - 4 = (3y + 2)(y - 2)$$

Check by multiplication. ∎

Self Check

Answer

Factor $5a^2 - 7a - 6$.

$(5a + 3)(a - 2)$

EXAMPLE 4 Factor $6b^2 + 7b - 20$.

Solution Since the first term is $6b^2$, the first terms of the binomial factors must be $6b$ and b or $3b$ and $2b$. To fill in the following blanks, we must find two factors of -20 that will give a middle term of $+7b$.

$$(6b \quad\quad)(b \quad\quad) \qquad \text{or} \qquad (3b \quad\quad)(2b \quad\quad)$$

Since the sign of the third term is $-$, the signs inside the binomial factors will be different. Because the factors of the last term (20) are 1, 2, 5, 10, and 20, there are many possible combinations for the last terms. We must try to find one that will give a last term of -20 and a sum of the products of the outer terms and inner terms of $+7b$.

If we pick factors of $6b$ and b for the first terms and -5 and $+4$ for the last terms, we have

$$(6b - 5)(b + 4)$$

$$\begin{array}{r} -5b \\ \underline{24b} \\ 19b \end{array}$$

which gives a wrong middle term of $19b$.

If we pick factors of $3b$ and $2b$ for the first terms and -4 and $+5$ for the last terms, we have

$$(3b - 4)(2b + 5)$$

$$\begin{array}{r} -8b \\ \underline{15b} \\ 7b \end{array}$$

which gives the correct middle term of $+7b$ and the correct last term of -20. Thus,

$$6b^2 + 7b - 20 = (3b - 4)(2b + 5)$$

Check by multiplication. ∎

Self Check

Factor $4x^2 + 4x - 3$.

Answer

$(2x + 3)(2x - 1)$

The next example has two variables.

EXAMPLE 5

Factor $2x^2 + 7xy + 6y^2$.

Solution

Since the first term is $2x^2$, the first terms of the binomial factors must be $2x$ and x. To fill in the following blanks, we must find two factors of $6y^2$ that will give a middle term of $+7xy$.

$$(2x \quad\quad)(x \quad\quad)$$

Since the sign of each term is $+$, the signs inside the binomial factors will be $+$. The possible factors of the last term ($6y^2$) are y, $2y$, $3y$, and $6y$. We must try to find one that will give a last term of $+6y^2$ and a sum of the products of the outer terms and inner terms of $+7xy$.

If we pick factors of $6y$ and y, we have

$$(2x + y)(x + 6y)$$

$$\begin{array}{c} xy \\ \underline{12xy} \\ 13xy \end{array}$$

which gives a wrong middle term of $13xy$.

If we pick factors of $3y$ and $2y$, we have

$$(2x + 3y)(x + 2y)$$

$$\begin{array}{c} 3xy \\ \underline{4xy} \\ 7xy \end{array}$$

which gives a correct middle term of $7xy$. Thus,

$$2x^2 + 7xy + 6y^2 = (2x + 3y)(x + 2y)$$

Check by multiplication. ∎

Self Check

Answer

Factor $4x^2 + 8xy + 3y^2$.

$(2x + 3y)(2x + y)$

Because some guesswork is often necessary, it is difficult to give specific rules for factoring trinomials. However, the following hints are often helpful.

Factoring General Trinomials

1. Write the trinomial in descending powers of one variable.

2. Factor out any GCF (including -1 if that is necessary to make the coefficient of the first term positive).

3. If the sign of the third term is $+$, the signs between the terms of the binomial factors are the same as the sign of the middle term. If the sign of the third term is $-$, the signs between the terms of the binomial factors are opposite.

4. Try combinations of first terms and last terms until you find one that works, or until you exhaust all the possibilities. If no combination works, the trinomial is prime.

5. Check the factorization by multiplication.

EXAMPLE 6 Factor $2x^2y - 8x^3 + 3xy^2$.

Solution *Step 1:* Write the trinomial in descending powers of x.

$$-8x^3 + 2x^2y + 3xy^2$$

Step 2: Factor out the negative of the GCF, which is $-x$.

$$-8x^3 + 2x^2y + 3xy^2 = -x(8x^2 - 2xy - 3y^2)$$

Step 3: Because the sign of the third term of the trinomial factor is $-$, the signs within its binomial factors will be opposites.

Step 4: Find the binomial factors of the trinomial.

$$-8x^3 + 2x^2y + 3xy^2 = -x(\mathbf{8x^2 - 2xy - 3y^2})$$
$$= -x(\mathbf{2x + y})(\mathbf{4x - 3y})$$

Step 5: Check by multiplication.

$$-x(2x + y)(4x - 3y) = -x(8x^2 - 6xy + 4xy - 3y^2)$$
$$= -x(8x^2 - 2xy - 3y^2)$$
$$= -8x^3 + 2x^2y + 3xy^2$$
$$= 2x^2y - 8x^3 + 3xy^2$$

■

Self Check Factor $12y - 2y^3 - 2y^2$.
Answer $-2y(y + 3)(y - 2)$

■ FACTORING PERFECT-SQUARE TRINOMIALS

As before, we can factor perfect-square trinomials by inspection.

EXAMPLE 7 Factor $4x^2 - 20x + 25$.

Solution $4x^2 - 20x + 25$ is a perfect-square trinomial, because

- The first term $4x^2$ is the square of $2x$: $(2x)^2 = 4x^2$.
- The last term 25 is the square of 5: $5^2 = 25$.
- The middle term $-20x$ is the negative of twice the product of $2x$ and 5.

Thus,

$$4x^2 - 20x + 25 = (2x)^2 - 2(2x)(5) + 5^2$$
$$= (2x - 5)^2$$ ■

Self Check Factor $9x^2 - 12x + 4$.

Answer $(3x - 2)^2$

The next example combines the techniques of factoring by grouping, factoring a perfect-square trinomial, and factoring the difference of two squares.

EXAMPLE 8 Factor $4x^2 - 4xy + y^2 - 9$.

Solution

$$4x^2 - 4xy + y^2 - 9$$

$= (4x^2 - 4xy + y^2) - 9$	Group the first three terms.
$= (2x - y)^2 - 9$	Factor the perfect-square trinomial.
$= [(2x - y) + 3][(2x - y) - 3]$	Factor the difference of two squares.
$= (2x - y + 3)(2x - y - 3)$	Remove parentheses.

Check by multiplication. ■

Self Check Factor $x^2 + 4x + 4 - y^2$.

Answer $(x + 2 + y)(x + 2 - y)$

EXAMPLE 9 Factor $9 - 4x^2 - 4xy - y^2$.

Solution

$9 - 4x^2 - 4xy - y^2 = 9 - (4x^2 + 4xy + y^2)$	Factor -1 from the trinomial.
$= 9 - (2x + y)(2x + y)$	Factor the perfect-square trinomial.
$= 9 - (2x + y)^2$	$(2x + y)(2x + y) = (2x + y)^2$.
$= [3 + (2x + y)][(3 - (2x + y)]$	Factor the difference of two squares.
$= (3 + 2x + y)(3 - 2x - y)$	Remove parentheses.

Check by multiplication. ■

Self Check Factor $16 - a^2 - 2a - 1$.

Answer $(a + 5)(3 - a)$

■ SOLVING EQUATIONS

EXAMPLE 10   Solve $2x^2 + 3x = 2$.

Solution   We write the equation in the form $ax^2 + bx + c = 0$ and solve for x.

$$2x^2 + 3x = 2$$
$$2x^2 + 3x - 2 = 0 \qquad \text{Add } -2 \text{ to both sides.}$$
$$(2x - 1)(x + 2) = 0 \qquad \text{Factor } 2x^2 + 3x - 2.$$
$$2x - 1 = 0 \quad \text{or} \quad x + 2 = 0 \qquad \text{Set each factor equal to 0.}$$
$$2x = 1 \qquad\qquad x = -2 \qquad \text{Solve each linear equation.}$$
$$x = \frac{1}{2}$$

Check each solution.   ■

Self Check   Solve $3x^2 - 5x - 2 = 0$.
Answers $2, -\frac{1}{3}$

EXAMPLE 11   Solve $6x^3 + 12x = 17x^2$.

Solution
$$6x^3 + 12x = 17x^2$$
$$6x^3 - 17x^2 + 12x = 0 \qquad \text{Subtract } 17x^2 \text{ from both sides.}$$
$$x(6x^2 - 17x + 12) = 0 \qquad \text{Factor out } x.$$
$$x(2x - 3)(3x - 4) = 0 \qquad \text{Factor } 6x^2 - 17x + 12.$$
$$x = 0 \quad \text{or} \quad 2x - 3 = 0 \quad \text{or} \quad 3x - 4 = 0 \qquad \text{Set each factor equal to 0.}$$
$$x = 0 \qquad\qquad 2x = 3 \qquad\qquad 3x = 4 \qquad \text{Solve the linear equations.}$$
$$x = \frac{3}{2} \qquad\qquad x = \frac{4}{3}$$

Check each solution.   ■

Self Check   Solve $6x^3 + 7x^2 = 5x$.
Answers $0, \frac{1}{2}, -\frac{5}{3}$

Orals   *Finish factoring each problem.*

1. $2x^2 + 5x + 3 = (x +)(x + 1)$ **2.** $6x^2 + 5x + 1 = (x + 1)(3x + 1)$

3. $6x^2 + 5x - 1 = (x 1)(6x 1)$ **4.** $6x^2 + x - 1 = (2x 1)(3x 1)$

5. $4x^2 + 4x - 3 = (2x +)(2x -)$ **6.** $4x^2 - x - 3 = (4x +)(x -)$

EXERCISE 5.5

REVIEW

1. The nth term l of an arithmetic sequence is

$$l = f + (n - 1)d$$

where f is the first term and d is the common difference. Remove the parentheses and solve the equation for n.

2. The sum S of n consecutive terms of an arithmetic sequence is

$$S = \frac{n}{2}(f + l)$$

where f is the first term and l is the nth term. Solve for f.

VOCABULARY AND CONCEPTS   Fill in each blank to make a true statement.

3. To factor a general trinomial, first write the trinomial in _____ powers of one variable.

4. If the sign of the first and third terms of a trinomial are $+$, the signs within the binomial factors are _____ as the sign of the middle term.

5. If the sign of the first term of a trinomial is $+$ and the sign of the third term is $-$, the signs within the binomial factors are _____.

6. Always check factorizations by _____.

Complete each factorization.

7. $6x^2 + x - 2 = (3x + \quad)(2x - \quad)$

8. $15x^2 - 7x - 4 = (5x - \quad)(3x + \quad)$

9. $12x^2 - 7xy + y^2 = (3x - \quad)(4x - \quad)$

10. $6x^2 + 5xy - 6y^2 = (2x + \quad)(3x - \quad)$

PRACTICE   In Exercises 11–34, factor each trinomial.

11. $2x^2 - 3x + 1$

12. $2y^2 - 7y + 3$

13. $3a^2 + 13a + 4$

14. $2b^2 + 7b + 6$

15. $4z^2 + 13z + 3$

16. $4t^2 - 4t + 1$

17. $6y^2 + 7y + 2$

18. $4x^2 + 8x + 3$

19. $6x^2 - 7x + 2$

20. $4z^2 - 9z + 2$

21. $3a^2 - 4a - 4$

22. $8u^2 - 2u - 15$

23. $2x^2 - 3x - 2$

24. $12y^2 - y - 1$

25. $2m^2 + 5m - 12$

26. $10u^2 - 13u - 3$

27. $10y^2 - 3y - 1$

28. $6m^2 + 19m + 3$

29. $12y^2 - 5y - 2$

30. $10x^2 + 21x - 10$

31. $5t^2 + 13t + 6$

32. $16y^2 + 10y + 1$

33. $16m^2 - 14m + 3$

34. $16x^2 + 16x + 3$

In Exercises 35–46, factor each trinomial.

35. $3x^2 - 4xy + y^2$

36. $2x^2 + 3xy + y^2$

37. $2u^2 + uv - 3v^2$

38. $2u^2 + 3uv - 2v^2$

39. $4a^2 - 4ab + b^2$

40. $2b^2 - 5bc + 2c^2$

41. $6r^2 + rs - 2s^2$

42. $3m^2 + 5mn + 2n^2$

43. $4x^2 + 8xy + 3y^2$

44. $4b^2 + 15bc - 4c^2$

45. $4a^2 - 15ab + 9b^2$

46. $12x^2 + 5xy - 3y^2$

In Exercises 47–62, write the terms of each trinomial in descending powers of one variable. Then factor the trinomial, if possible.

47. $-13x + 3x^2 - 10$

48. $-14 + 3a^2 - a$

49. $15 + 8a^2 - 26a$

50. $16 - 40a + 25a^2$

51. $12y^2 + 12 - 25y$

52. $12t^2 - 1 - 4t$

53. $3x^2 + 6 + x$

54. $25 + 2u^2 + 3u$

55. $2a^2 + 3b^2 + 5ab$

56. $11uv + 3u^2 + 6v^2$

57. $pq + 6p^2 - q^2$

58. $-11mn + 12m^2 + 2n^2$

59. $b^2 + 4a^2 + 16ab$

60. $3b^2 + 3a^2 - ab$

61. $12x^2 + 10y^2 - 23xy$

62. $5ab + 25a^2 - 2b^2$

In Exercises 63–78, factor each polynomial.

63. $4x^2 + 10x - 6$

64. $9x^2 + 21x - 18$

65. $y^3 + 13y^2 + 12y$

66. $2xy^2 + 8xy - 24x$

67. $6x^3 - 15x^2 - 9x$

68. $9y^3 + 3y^2 - 6y$

69. $30r^5 + 63r^4 - 30r^3$

70. $6s^5 - 26s^4 - 20s^3$

71. $4a^2 - 4ab - 8b^2$

72. $6x^2 + 3xy - 18y^2$

73. $8x^2 - 12xy - 8y^2$

74. $24a^2 + 14ab + 2b^2$

75. $-16m^3n - 20m^2n^2 - 6mn^3$

76. $-84x^4 - 100x^3y - 24x^2y^2$

77. $-28u^3v^3 + 26u^2v^4 - 6uv^5$

78. $-16x^4y^3 + 30x^3y^4 + 4x^2y^5$

In Exercises 79–84, factor each perfect-square trinomial.

79. $4x^2 + 12x + 9$

80. $4x^2 - 4x + 1$

81. $9x^2 + 12x + 4$

82. $4x^2 - 20x + 25$

83. $16x^2 - 8xy + y^2$

84. $25x^2 + 20xy + 4y^2$

In Exercises 85–90, factor each polynomial.

85. $4x^2 + 4xy + y^2 - 16$

86. $9x^2 - 6x + 1 - d^2$

87. $9 - a^2 - 4ab - 4b^2$

88. $25 - 9a^2 + 6ac - c^2$

89. $4x^2 + 4xy + y^2 - a^2 - 2ab - b^2$

90. $a^2 - 2ab + b^2 - x^2 + 2x - 1$

In Exercises 91–106, solve each equation.

91. $2x^2 - 5x + 2 = 0$

92. $2x^2 + x - 3 = 0$

93. $5x^2 - 6x + 1 = 0$

94. $6x^2 - 5x + 1 = 0$

95. $3x^2 - 8x = 3$

96. $2x^2 - 11x = 21$

97. $15x^2 - 2 = 7x$

98. $8x^2 + 10x = 3$

99. $x(6x + 5) = 6$ **100.** $x(2x - 3) = 14$ **101.** $(x + 1)(8x + 1) = 18x$ **102.** $4x(3x + 2) = x + 12$

103. $2x(3x^2 + 10x) = -6x$ **104.** $2x^3 = 2x(x + 2)$ **105.** $x^3 + 7x^2 = x^2 - 9x$ **106.** $x^2(x + 10) = 2x(x - 8)$

WRITING

107. Describe an organized approach to finding all of the possibilities when you attempt to factor $12x^2 - 4x + 9$.

108. Explain how to determine whether a trinomial is prime.

SOMETHING TO THINK ABOUT

109. For what values of b will the trinomial $6x^2 + bx + 6$ be factorable?

110. Create a quadratic equation with the two solutions $x = 3$ and $x = \frac{3}{2}$.

5.6   Summary of Factoring Techniques

■ IDENTIFYING FACTORING TYPES

Getting Ready   *Factor each polynomial.*

1. $3ax^2 + 3a^2x$

2. $x^2 - 9y^2$

3. $x^2 - 3x - 10$

4. $6x^2 - 13x + 6$

5. $6x^2 - 14x + 4$

6. $ax^2 + bx^2 - ay^2 - by^2$

■ IDENTIFYING FACTORING TYPES

In this brief section, we will discuss ways to approach a randomly chosen factoring problem. For example, suppose we wish to factor the trinomial

$$x^4y + 7x^3y - 18x^2y$$

We begin by attempting to identify the problem type. The first type we look for is **factoring out a common factor.** Because the trinomial has a common factor of x^2y, we factor it out:

$$x^4y + 7x^3y - 18x^2y = x^2y(x^2 + 7x - 18)$$

We can factor the remaining trinomial $x^2 + 7x - 18$ as $(x + 9)(x - 2)$. Thus,

$$x^4y + 7x^3y - 18x^2y = x^2y(x^2 + 7x - 18)$$
$$= x^2y(x + 9)(x - 2)$$

To identify the type of factoring problem, we follow these steps.

> **Factoring a Polynomial**
> 1. Factor out all common factors.
> 2. If an expression has two terms, check to see if the problem type is the **difference of two squares:** $x^2 - y^2 = (x + y)(x - y)$.
> 3. If an expression has three terms, check to see if it is a **perfect trinomial square:**
> $$x^2 + 2xy + y^2 = (x + y)(x + y)$$
> $$x^2 - 2xy + y^2 = (x - y)(x - y)$$
>   If the trinomial is not a trinomial square, attempt to factor the trinomial as a **general trinomial.**
> 4. If an expression has four or more terms, try to factor the expression by **grouping.**
> 5. Continue factoring until each individual factor is prime.
> 6. Check the results by multiplying.

EXAMPLE 1   Factor $x^5 y^2 - xy^6$.

Solution   We begin by factoring out the common factor of xy^2:

$$x^5 y^2 - xy^6 = xy^2(x^4 - y^4)$$

Since the expression $x^4 - y^4$ has two terms, we check to see if it is the difference of two squares, which it is. As the difference of two squares, it factors as $(x^2 + y^2)(x^2 - y^2)$.

$$x^5 y^2 - xy^6 = xy^2(x^4 - y^4)$$
$$= xy^2(x^2 + y^2)(x^2 - y^2)$$

The binomial $x^2 + y^2$ is the sum of two squares and cannot be factored. However, $x^2 - y^2$ is the difference of two squares and factors as $(x + y)(x - y)$.

$$x^5 y^2 - xy^6 = xy^2(x^4 - y^4)$$
$$= xy^2(x^2 + y^2)(x^2 - y^2)$$
$$= xy^2(x^2 + y^2)(x + y)(x - y)$$

Since each individual factor is prime, the given expression is in completely factored form. ∎

Self Check   Factor $-a^5 b + ab^5$.

Answer $-ab(a^2 + b^2)(a + b)(a - b)$

EXAMPLE 2 Factor $x^6 - x^4y^2 - 2x^3y^3 + 2xy^5$.

Solution We begin by factoring out the common factor of x.

$$x^6 - x^4y^2 - 2x^3y^3 + 2xy^5 = x(x^5 - x^3y^2 - 2x^2y^3 + 2y^5)$$

Since $x^5 - x^3y^2 - 2x^2y^3 + 2y^5$ has four terms, we try factoring it by grouping:

$$\begin{aligned} x^6 - x^4y^2 - 2x^3y^3 + 2xy^5 &= x(x^5 - x^3y^2 - 2x^2y^3 + 2y^5) \\ &= x[x^3(x^2 - y^2) - 2y^3(x^2 - y^2)] \\ &= x(x^2 - y^2)(x^3 - 2y^3) \qquad \text{Factor out } x^2 - y^2. \end{aligned}$$

Finally, we factor the difference of two squares:

$$x^6 - x^4y^2 - 2x^3y^3 + 2xy^5 = x(x + y)(x - y)(x^3 - 2y^3)$$

Since each factor is prime, the given expression is in prime-factored form. ∎

Self Check Factor $2a^5 - 4a^2b^3 - 8a^3 + 16b^3$.
Answer $2(a + 2)(a - 2)(a^3 - 2b^3)$

Orals *Indicate which factoring technique you would use first, if any.*

1. $2x^2 - 4x$ **2.** $16 - 25y^2$ **3.** $ax + ay - x - y$

4. $x^2 + 4$ **5.** $8x^2 - 50$ **6.** $25r^2 - s^4$

E X E R C I S E 5.6

REVIEW *Solve each equation, if possible.*

1. $2(t - 5) + t = 3(2 - t)$ **2.** $5 + 3(2x - 1) = 2(4 + 3x) - 24$
3. $5x^2 - 35x = 0$ **4.** $6x^2 - x = 35$

VOCABULARY AND CONCEPTS *Fill in each blank to make a true statement.*

5. The first step in any factoring problem is to factor out all common _____, if possible.
6. If a polynomial has two terms, check to see if it is the _____.
7. If a polynomial has three terms, try to factor it as the product of two _____.
8. If a polynomial has four or more terms, try factoring by _____.

PRACTICE *In Exercises 9–58, factor each expression.*

9. $6x + 3$ **10.** $x^2 - 9$ **11.** $x^2 - 6x - 7$ **12.** $a^2 - b^2$

13. $6t^2 + 7t - 3$ **14.** $3rs^2 - 6r^2st$ **15.** $4x^2 - 25$ **16.** $ac + ad + bc + bd$

17. $t^2 - 2t + 1$ **18.** $6p^2 - 3p - 2$ **19.** $3a^2 - 12$ **20.** $2x^2 - 32$

21. $x^2y^2 - 2x^2 - y^2 + 2$
23. $70p^4q^3 - 35p^4q^2 + 49p^5q^2$

22. $a^2c + a^2d^2 + bc + bd^2$
24. $a^2 + 2ab + b^2 - x^2 - 2xy - y^2$

25. $2ab^2 + 8ab - 24a$ **26.** $t^4 - 16$ **27.** $-8p^3q^7 - 4p^2q^3$ **28.** $8m^2n^3 - 24mn^4$

29. $4a^2 - 4ab + b^2 - 9$ **30.** $3rs + 6r^2 - 18s^2$ **31.** $x^2 + 7x + 1$ **32.** $3a^2 + 27b^2$

33. $-2x^4 + 32x^2$ **34.** $16 - 40z + 25z^2$ **35.** $14t^3 - 40t^2 + 6t^4$ **36.** $6x^2 + 7x - 20$

37. $a^2(x - a) - b^2(x - a)$ **38.** $5x^3y^3z^4 + 25x^2y^3z^2 - 35x^3y^2z^5$

39. $2x^4 - 32$
41. $x^3 + 2x^2 - 9x - 18$
43. $-16x^4y^2z + 24x^5y^3z^4 - 15x^2y^3z^7$

40. $2c^2 - 5cd - 3d^2$
42. $8a^2x^3y - 2b^2xy$
44. $2ac + 4ad + bc + 2bd$

45. $81p^4 - 16q^4$
47. $4x^2 + 9y^2$
49. $x^4 - 4x^2 - x^2y^2 + 4y^2$
51. $10r^2 - 13r - 4$
53. $21t^3 - 10t^2 + t$
55. $x^2 - y^2 - 4y - 4$
57. $2a^2c - 2b^2c + 4a^2d - 4b^2d$

46. $6x^2 - x - 16$
48. $30a^4 + 5a^3 - 200a^2$
50. $6a^3 + 35a^2 - 6a$
52. $4x^2 + 4x + 1 - y^2$
54. $16x^2 - 40x^3 + 25x^4$
56. $a^2 - 4b^2 - 4b - 1$
58. $3a^2x^2 + 6a^2x + 3a^2 - 6b^2x^2 - 12b^2x - 6b^2$

WRITING

59. Explain how to identify the type of factoring required to factor a polynomial.

60. Which factoring technique do you find most difficult? Why?

SOMETHING TO THINK ABOUT

61. A test for factorability A trinomial of the form $ax^2 + bx + c$, with integer coefficients and $a \neq 0$, will factor into two binomials with integer coefficients if the value of

$$b^2 - 4ac$$

is a perfect square. If $b^2 - 4ac = 0$, the factors will be the same. Calculate $b^2 - 4ac$ and tell whether each trinomial will factor.
     a. $3p^2 - 4p - 4$ **b.** $4t^2 - 3t - 5$

62. Because it is the difference of two squares, $x^2 - q^2$ always factors. Does the test for factorability predict this?

5.7 Problem Solving

■ INTEGER PROBLEMS ■ BALLISTICS ■ GEOMETRIC PROBLEMS

Getting Ready

1. One side of a square is s inches long. Find an expression that represents its area.

2. The length of a rectangle is 4 centimeters more than twice the width. If w represents the width, find an expression that represents the length.

3. If x represents the smaller of two consecutive integers, find an expression that represents their product.

4. The length of a rectangle is 3 inches greater than the width. If w represents the width of the rectangle, find an expression that represents the area.

■ **INTEGER PROBLEMS**

EXAMPLE 1   One negative integer is 5 less than another, and their product is 84. Find the integers.

Analyze the problem   Let x represent the larger number. Then $x - 5$ represents the smaller number. We know that the product of the negative integers is 84.

Form an equation   Since their product is 84, we can form the equation $x(x - 5) = 84$.

Solve the equation   To solve the equation, we proceed as follows.

$$
\begin{aligned}
x(x - 5) &= 84 & &\text{Remove parentheses.}\\
x^2 - 5x &= 84 & &\text{Subtract 84 from both sides.}\\
x^2 - 5x - 84 &= 0 & &\text{Factor.}\\
(x - 12)(x + 7) &= 0\\
x - 12 = 0 \quad &\text{or} \quad x + 7 = 0 & &\text{Set each factor equal to 0.}\\
x = 12 \qquad & \qquad x = -7 & &\text{Solve each linear equation.}
\end{aligned}
$$

State the conclusion   Since we need two negative numbers, we discard the result $x = 12$. The two negative integers are

$$x = -7 \qquad \text{and} \qquad x - 5 = -7 - 5$$
$$= -12$$

Check the result   The number -12 is five less than -7, and $(-12)(-7) = 84$.   ■

■ BALLISTICS

EXAMPLE 2   If an object is thrown straight up into the air with an initial velocity of 112 feet per second, its height after t seconds is given by the formula

$$h = 112t - 16t^2$$

where h represents the height of the object in feet. After this object has been thrown, in how many seconds will it hit the ground?

Analyze the problem   Before the object is thrown, its height above the ground is 0. When it is thrown, it will go up and then come down. When it hits the ground, its height will again be 0.

Form an equation   Thus, we set h equal to 0 in the formula $h = 112t - 16t^2$ to form the equation $0 = 112t - 16t^2$.

$$\boldsymbol{h} = 112t - 16t^2$$
$$\boldsymbol{0} = 112t - 16t^2$$

Solve the equation   We then solve the equation as follows.

$$0 = 112t - 16t^2$$
$$0 = 16t(7 - t) \qquad \text{Factor out } 16t.$$
$$16t = 0 \quad \text{or} \quad 7 - t = 0 \qquad \text{Set each factor equal to 0.}$$
$$t = 0 \qquad\qquad t = 7 \qquad \text{Solve each linear equation.}$$

State the conclusion   When $t = 0$, the object's height above the ground is 0 feet, because it has just been released. When $t = 7$, the height is again 0 feet. The object has hit the ground. The solution is 7 seconds.

Check the result.   ■

■ GEOMETRIC PROBLEMS

Recall that the area of a rectangle is given by the formula

$$A = lw$$

where A represents the area, l the length, and w the width of the rectangle. The perimeter of a rectangle is given by the formula

$$P = 2l + 2w$$

where P represents the perimeter of the rectangle, l the length, and w the width of the rectangle.

EXAMPLE 3   Assume that the rectangle in Figure 5-1 has an area of 52 square centimeters and that its length is 1 centimeter more than 3 times its width. Find the perimeter of the rectangle.

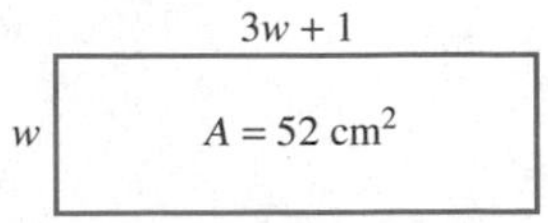

FIGURE 5-1

Analyze the problem   Let w represent the width of the rectangle. Then $3w + 1$ represents its length. Its area is 52 square centimeters. We can use this fact to find the values of its width and length. Then we can find the perimeter.

Form and solve an equation   Because the area is 52 square centimeters, we substitute 52 for A and $3w + 1$ for l in the formula $A = lw$ and solve for w.

$$A = lw$$
$$52 = (3w + 1)w$$
$$52 = 3w^2 + w \qquad \text{Remove parentheses.}$$
$$0 = 3w^2 + w - 52 \qquad \text{Subtract 52 from both sides.}$$
$$0 = (3w + 13)(w - 4) \qquad \text{Factor.}$$
$$3w + 13 = 0 \quad \text{or} \quad w - 4 = 0 \qquad \text{Set each factor equal to 0.}$$
$$3w = -13 \qquad\qquad w = 4 \qquad \text{Solve each linear equation.}$$
$$w = -\frac{13}{3}$$

Because the width of a rectangle cannot be negative, we discard the result $w = -\frac{13}{3}$. Thus, the width of the rectangle is 4, and the length is given by

$$3w + 1 = 3(4) + 1$$
$$= 12 + 1$$
$$= 13$$

The dimensions of the rectangle are 4 centimeters by 13 centimeters. We find the perimeter by substituting 13 for l and 4 for w in the formula for the perimeter.

$$P = 2l + 2w$$
$$= 2(13) + 2(4)$$
$$= 26 + 8$$
$$= 34$$

State the conclusion   The perimeter of the rectangle is 34 centimeters.

Check the result   A rectangle with dimensions of 13 centimeters by 4 centimeters does have an area of 52 square centimeters, and the length is 1 centimeter more than 3 times the width. A rectangle with these dimensions has a perimeter of 34 centimeters. ■

EXAMPLE 4   The triangle in Figure 5-2 has an area of 10 square centimeters and a height that is 3 centimeters less than twice the length of its base. Find the length of the base and the height of the triangle.

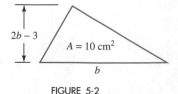

FIGURE 5-2

Analyze the problem Let b represent the length of the base of the triangle. Then $2b - 3$ represents the height. Because the area is 10 square centimeters, we can substitute 10 for A and $2b - 3$ for h in the formula $A = \frac{1}{2}bh$ and solve for b.

Form and solve an equation

$$A = \frac{1}{2}bh$$

$$10 = \frac{1}{2}b(2b - 3)$$

$20 = b(2b - 3)$	Multiply both sides by 2.
$20 = 2b^2 - 3b$	Remove parentheses.
$0 = 2b^2 - 3b - 20$	Subtract 20 from both sides.
$0 = (2b + 5)(b - 4)$	Factor.
$2b + 5 = 0$ or $\quad b - 4 = 0$	Set both factors equal to 0.
$2b = -5 \qquad\qquad b = 4$	Solve each linear equation.
$b = -\dfrac{5}{2}$	

State the conclusion Because a triangle cannot have a negative number for the length of its base, we discard the result $b = -\frac{5}{2}$. The length of the base of the triangle is 4 centimeters. Its height is $2(4) - 3$, or 5 centimeters.

Check the result If the base of the triangle has a length of 4 centimeters and the height of the triangle is 5 centimeters, its height is 3 centimeters less than twice the length of its base. Its area is 10 centimeters.

$$A = \frac{1}{2}bh$$

$$= \frac{1}{2}(4)(5)$$

$$= 2(5)$$

$$= 10 \qquad\qquad\blacksquare$$

Orals *Give the formula for . . .*

1. The area of a rectangle
2. The area of a triangle
3. The area of a square
4. The area of a rectangular solid

5. The perimeter of a rectangle
6. The perimeter of a square

EXERCISE 5.7

REVIEW   *Solve each equation.*

1. $-2(5z + 2) = 3(2 - 3z)$

2. $3(2a - 1) - 9 = 2a$

3. A rectangle is 3 times as long as it is wide, and its perimeter is 120 centimeters. Find its area.

4. A woman invested $15,000, part at 7% annual interest and part at 8% annual interest. If she receives $1,100 interest per year, how much did she invest at 7%?

VOCABULARY AND CONCEPTS   *Fill in each blank to make a true statement.*

5. The first step in the problem-solving process is to _____ the problem.

6. The last step in the problem-solving process is to _____.

PRACTICE   *Solve each problem.*

7. Integer problem   One positive integer is 2 more than another. Their product is 35. Find the integers.

8. Integer problem   One positive integer is 5 less than 4 times another. Their product is 21. Find the integers.

9. Integer problem   If 4 is added to the square of a composite integer, the result is 5 less than 10 times that integer. Find the integer.

10. Integer problem   If 3 times the square of a certain natural number is added to the number itself, the result is 14. Find the number.

In Exercises 11–14, an object has been thrown straight up into the air. The formula $h = vt - 16t^2$ gives the height h of the object above the ground after t seconds when it is thrown upward with an initial velocity v.

11. Time of flight   After how many seconds will an object hit the ground if it was thrown with a velocity of 144 feet per second?

12. Time of flight   After how many seconds will an object hit the ground if it was thrown with a velocity of 160 feet per second?

13. Ballistics   If a cannonball is fired with an upward velocity of 220 feet per second, at what times will it be at a height of 600 feet?

14. Ballistics   A cannonball's initial upward velocity is 128 feet per second. At what times will it be 192 feet above the ground?

APPLICATIONS   *Solve each problem.*

15. Exhibition diving   At a resort, tourists watch swimmers dive from a cliff to the water 64 feet below. A diver's height h above the water t seconds after diving is given by $h = -16t^2 + 64$. How long does a dive last?

16. Forensic medicine   The kinetic energy E of a moving object is given by $E = \frac{1}{2}mv^2$, where m is the mass of the object (in kilograms) and v is the object's velocity (in meters per second). Kinetic energy is measured in Joules. By the damage done to a victim, a police pathologist determines that the energy of a 3-kilogram mass at impact was 54 Joules. Find the velocity at impact.

17. Insulation The area of the rectangular slab of foam insulation in Illustration 1 is 36 square meters. Find the dimensions of the slab.

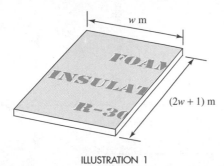

ILLUSTRATION 1

18. Shipping pallets The length of a rectangular shipping pallet is 2 feet less than 3 times its width. Its area is 21 square feet. Find the dimensions of the pallet.

19. Carpentry A room containing 143 square feet is 2 feet longer than it is wide. How long a crown molding is needed to trim the perimeter of the ceiling?

20. Designing a tent The length of the base of the triangular sheet of canvas above the door of the tent in Illustration 2 is 2 feet more than twice its height. The area is 30 square feet. Find the height and the length of the base of the triangle.

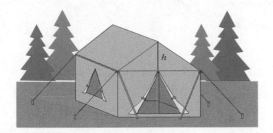

ILLUSTRATION 2

21. Dimensions of a triangle The height of a triangle is 2 inches less than 5 times the length of its base. The area is 36 square inches. Find the length of the base and the height of the triangle.

22. Area of a triangle The base of a triangle is numerically 3 less than its area, and the height is numerically 6 less than its area. Find the area of the triangle.

23. Area of a triangle The length of the base and the height of a triangle are numerically equal. Their sum is 6 less than the number of units in the area of the triangle. Find the area of the triangle.

24. Dimensions of a parallelogram The formula for the area of a parallelogram is $A = bh$. The area of the parallelogram in Illustration 3 is 200 square centimeters. If its base is twice its height, how long is the base?

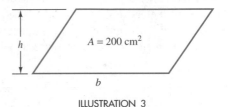

ILLUSTRATION 3

25. Swimming pool border The owners of the rectangular swimming pool in Illustration 4 want to surround the pool with a crushed-stone border of uniform width. They have enough stone to cover 74 square meters. How wide should they make the border? (*Hint:* The area of the larger rectangle minus the area of the smaller is the area of the border.)

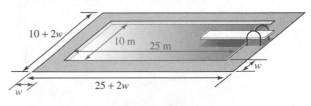

ILLUSTRATION 4

26. House construction The formula for the area of a trapezoid is $A = \frac{h(B + b)}{2}$. The area of the trapezoidal truss in Illustration 5 is 24 square meters. Find the height of the trapezoid if one base is 8 meters and the other base is the same as the height.

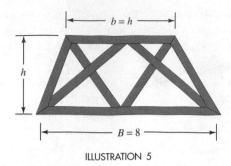

ILLUSTRATION 5

27. Volume of a solid The volume of a rectangular solid is given by the formula $V = lwh$, where l is the length, w is the width, and h is the height. The volume of the rectangular solid in Illustration 6 is 210 cubic centimeters. Find the width of the rectangular solid if its length is 10 centimeters and its height is 1 centimeter longer than twice its width.

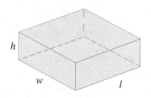

ILLUSTRATION 6

28. Volume of a pyramid The volume of a pyramid is given by the formula $V = \frac{Bh}{3}$, where B is the area of its base and h is its height. The volume of the pyramid in

Illustration 7 is 192 cubic centimeters. Find the dimensions of its rectangular base if one edge of the base is 2 centimeters longer than the other, and the height of the pyramid is 12 centimeters.

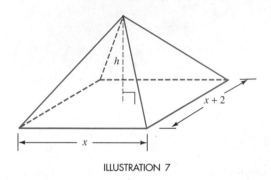

ILLUSTRATION 7

29. Volume of a pyramid The volume of a pyramid is 84 cubic centimeters. Its height is 9 centimeters, and one side of its rectangular base is 3 centimeters shorter than the other. Find the dimensions of its base. (See Exercise 28.)

30. Volume of a solid The volume of a rectangular solid is 72 cubic centimeters. Its height is 4 centimeters, and its width is 3 centimeters shorter than its length. Find the sum of its length and width. (See Exercise 27.)

WRITING

31. Explain the steps you would use to set up and solve an application problem.

32. Explain how you should check the solution to an application problem.

SOMETHING TO THINK ABOUT

33. Here is an easy-sounding problem:
The length of a rectangle is 2 feet greater than the width, and the area is 18 square feet. Find the width of the rectangle.
Set up the equation. Can you solve it? Why not?

34. Does the equation in Exercise 33 have a solution, even if you can't find it? If it does, find an estimate of the solution.

■ ■ ■ ■ ■ ■ ■ ■ ■ PROJECT

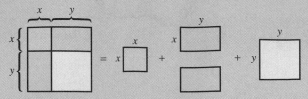

FIGURE 5-3

Because the length of each side of the largest square in Figure 5-3 is $x + y$, its area is $(x + y)^2$. This area is also the sum of four smaller areas, which illustrates the factorization

$$x^2 + 2xy + y^2 = (x + y)^2$$

What factorization is illustrated by each of the following figures?

1.
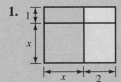

3.
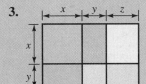

2.
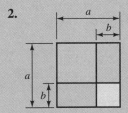

4.
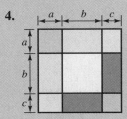

5. Factor the expression

$$a^2 + ac + 2a + ab + bc + 2b$$

and draw a figure that illustrates the factorization.

6. Verify the factorization

$$x^3 + 3x^2y + 3xy^2 + y^3 = (x + y)^3$$

Hint: Expand the right-hand side: $(x + y)^3 = (x + y)(x + y)(x + y)$

Then draw a figure that illustrates the factorization.

CHAPTER SUMMARY

CONCEPTS

REVIEW EXERCISES

SECTION 5.1

Factoring Out the Greatest Common Factor

A natural number is in **prime-factored form** if it is written as the product of prime-number factors.

The **greatest common factor (GCF)** of several monomials is found by taking each common prime factor and variable factor the fewest number of times it appears in any one monomial.

1. Find the prime factorization of each number.

   a. 35   **b.** 45

   c. 96   **d.** 102

   e. 87   **f.** 99

   g. 2,050   **h.** 4,096

2. Factor each expression completely.

   a. $3x + 9y$ **b.** $5ax^2 + 15a$

   c. $7x^2 + 14x$ **d.** $3x^2 - 3x$

   e. $2x^3 + 4x^2 - 8x$

   f. $ax + ay - az$

   g. $ax + ay - a$

   h. $x^2yz + xy^2z$

Zero-factor property:
If a and b represent two real numbers and if $ab = 0$, then $a = 0$ or $b = 0$.

3. Solve each equation.

   a. $x^2 + 2x = 0$ **b.** $2x^2 - 6x = 0$

SECTION 5.2

Factoring by Grouping

If a polynomial has four or more terms, consider factoring it by grouping.

4. Factor each polynomial.

   a. $(x + y)a + (x + y)b$

   b. $(x + y)^2 + (x + y)$

   c. $2x^2(x + 2) + 6x(x + 2)$

   d. $3x(y + z) - 9x(y + z)^2$

   e. $3p + 9q + ap + 3aq$

   f. $ar - 2as + 7r - 14s$

   g. $x^2 + ax + bx + ab$

   h. $xy + 2x - 2y - 4$

   i. $xa + yb + ya + xb$

SECTION 5.3	*Factoring the Difference of Two Squares*

To factor the difference of two squares, use the pattern

$$x^2 - y^2 = (x + y)(x - y)$$

5. Factor each expression.
   a. $x^2 - 9$
   b. $x^2y^2 - 16$
   c. $(x + 2)^2 - y^2$
   d. $z^2 - (x + y)^2$
   e. $6x^2y - 24y^3$
   f. $(x + y)^2 - z^2$

6. Solve each equation.
   a. $x^2 - 9 = 0$
   b. $x^2 - 25 = 0$

SECTIONS 5.4–5.5	*Factoring Trinomials*

Factor trinomials by trying these steps:

1. Write the trinomial with the exponents of one variable in descending order.

2. Factor out any greatest common factor (including -1 if that is necessary to make the coefficient of the first term positive).

3. If the sign of the third term of the trinomial is plus ($+$), the signs between the terms of each binomial factor are the same as the sign of the trinomial's second term. If the sign of the third term is minus ($-$), the signs between the terms of the binomials are opposite.

4. Mentally try various combinations of first terms and last terms until you find the one that works or you exhaust all the possibilities. In that case, the trinomial is prime.

5. Check by multiplication.

7. Factor each polynomial.
   a. $x^2 + 10x + 21$
   b. $x^2 + 4x - 21$
   c. $x^2 + 2x - 24$
   d. $x^2 - 4x - 12$

8. Factor each polynomial.
   a. $2x^2 - 5x - 3$
   b. $3x^2 - 14x - 5$
   c. $6x^2 + 7x - 3$
   d. $6x^2 + 3x - 3$
   e. $6x^3 + 17x^2 - 3x$
   f. $4x^3 - 5x^2 - 6x$

9. Solve each equation.
   a. $a^2 - 7a + 12 = 0$
   b. $x^2 - 2x - 15 = 0$
   c. $2x - x^2 + 24 = 0$
   d. $16 + x^2 - 10x = 0$
   e. $2x^2 - 5x - 3 = 0$
   f. $2x^2 + x - 3 = 0$
   g. $4x^2 = 1$
   h. $9x^2 = 4$
   i. $x^3 - 7x^2 + 12x = 0$
   j. $x^3 + 5x^2 + 6x = 0$
   k. $2x^3 + 5x^2 = 3x$
   l. $3x^3 - 2x = x^2$

Summary of Factoring Techniques

Steps for factoring polynomials:

1. Factor out all common factors.

2. If an expression has two terms, check to see if the problem type is the **difference of two squares:**

$$a^2 - b^2 = (a + b)(a - b)$$

3. If an expression has three terms, check to see if the problem type is a **perfect trinomial square:**

$$a^2 + 2ab + b^2 = (a + b)(a + b)$$
$$a^2 - 2ab + b^2 = (a - b)(a - b)$$

If the trinomial is not a trinomial square, attempt to factor the trinomial as a **general trinomial.**

4. If an expression has four or more terms, try to factor it by **grouping.**

10. Factor each polynomial.

   a. $3x^2y - xy^2 - 6xy + 2y^2$

   b. $5x^2 + 10x - 15xy - 30y$

   c. $2a^2x + 2abx + a^3 + a^2b$

   d. $x^2 + 2ax + a^2 - y^2$

   e. $ax^2 + 4ax + 3a - bx - b$

Problem Solving

11. Number problem   The sum of two numbers is 12, and their product is 35. Find the numbers.

12. Number problem   If 3 times the square of a positive number is added to 5 times the number, the result is 2. Find the number.

13. Dimensions of a rectangle   A rectangle is 2 feet longer than it is wide, and its area is 48 square feet. Find its dimensions.

14. Gardening   A rectangular flower bed is 3 feet longer than twice its width, and its area is 27 square feet. Find its dimensions.

15. Geometry   A rectangle is 3 feet longer than it is wide. Its area is numerically equal to its perimeter. Find its dimensions.

■ Chapter Test

1. Find the prime factorization of 196.

2. Find the prime factorization of 111.

In Problems 3–4, factor out the greatest common factor.

3. $60ab^2c^3 + 30a^3b^2c - 25a$

4. $3x^2(a + b) - 6xy(a + b)$

In Problems 5–18, factor each expression.

5. $ax + ay + bx + by$

6. $x^2 - 25$

7. $3a^2 - 27b^2$

8. $16x^4 - 81y^4$

9. $x^2 + 4x + 3$

10. $x^2 - 9x - 22$

11. $x^2 + 10xy + 9y^2$

12. $6x^2 - 30xy + 24y^2$

13. $3x^2 + 13x + 4$

14. $2a^2 + 5a - 12$

15. $2x^2 + 3xy - 2y^2$

16. $12 - 25x + 12x^2$

17. $12a^2 + 6ab - 36b^2$

18. $x^4 - 16y^4$

In Problems 19–26, solve each equation.

19. $x^2 + 3x = 0$

20. $2x^2 + 5x + 3 = 0$

21. $9y^2 - 81 = 0$

22. $-3(y - 6) + 2 = y^2 + 2$

23. $10x^2 - 13x = 9$

24. $10x^2 - x = 9$

25. $10x^2 + 43x = 9$

26. $10x^2 - 89x = 9$

27. Cannon fire A cannonball is fired straight up into the air with a velocity of 192 feet per second. In how many seconds will it hit the ground? (Its height above the ground is given by the formula $h = vt - 16t^2$, where v is the velocity and t is the time in seconds.)

28. Base of a triangle The base of a triangle with an area of 40 square meters is 2 meters longer than it is high. Find the base of the triangle.

6 Proportion and Rational Expressions

Mechanical Engineer

MATHEMATICS IN
THE WORKPLACE

Mechanical engineers design and develop power-producing machines such as internal combustion engines, steam and gas turbines, and jet rocket engines, as well as power-using machines such as refrigeration and air-conditioning equipment, machine tools, printing presses, and steel rolling mills. Many mechanical engineers do research, test, and design work. Others work in maintenance, technical sales, and production operations. Many teach in colleges and universities or work as consultants.

SAMPLE APPLICATION ■ The stiffness of the shaft shown in Illustration 1 is given by the formula

$$k = \cfrac{1}{\cfrac{1}{k_1} + \cfrac{1}{k_2}}$$

Section 1   Section 2

ILLUSTRATION 1

where k_1 and k_2 are the individual stiffnesses of each section. If the stiffness, k_2, of Section 2 is 4,200,000 in. lb/rad and design specifications require that the overall stiffness, k, of the entire shaft be 1,900,000 in. lb/rad, what must the stiffness of Section 1 be?
(See Exercise 60 in Exercise 6.6.)

6.1 Ratios

■ RATIOS ■ UNIT COSTS ■ RATES

Getting Ready   *Simplify each fraction.*

1. $\dfrac{2}{4}$ **2.** $\dfrac{8}{12}$ **3.** $-\dfrac{20}{25}$ **4.** $\dfrac{-45}{81}$

■ RATIOS

Ratios appear often in real-life situations. For example,

- To prepare fuel for a Lawnboy lawnmower, gasoline must be mixed with oil in the ratio of 50 to 1.

- To make 14-karat jewelry, gold is mixed with other metals in the ratio of 14 to 10.

- In the stock market, winning stocks might outnumber losing stocks in the ratio of 7 to 4.

- At Rock Valley College, the ratio of students to faculty is 16 to 1.

Ratios give us a way to compare numerical quantities.

Ratios

A **ratio** is the comparison of two numbers by their indicated quotient. In symbols,

If a and b are two numbers, the ratio of a to b is $\dfrac{a}{b}$.

WARNING!   The denominator b cannot be 0 in the fraction $\frac{a}{b}$, but b can be 0 in the ratio $\frac{a}{b}$. For example, the ratio of women to men on a women's softball team could be 25 to 0. However, these applications are rare.

Some examples of ratios are

$$\frac{7}{9}, \quad \frac{21}{27}, \quad \text{and} \quad \frac{2{,}290}{1{,}317}$$

- The fraction $\frac{7}{9}$ can be read as "the ratio of 7 to 9."
- The fraction $\frac{21}{27}$ can be read as "the ratio of 21 to 27."
- The ratio $\frac{2{,}290}{1{,}317}$ can be read as "the ratio of 2,290 to 1,317."

Because $\frac{7}{9}$ and $\frac{21}{27}$ represent equal numbers, they are **equal ratios.**

EXAMPLE 1   Express each phrase as a ratio in lowest terms: **a.** the ratio of 15 to 12   and   **b.** the ratio of 0.3 to 1.2.

Solution   **a.** The ratio of 15 to 12 can be written as the fraction $\frac{15}{12}$. After simplifying, the ratio is $\frac{5}{4}$.

b. The ratio of 0.3 to 1.2 can be written as the fraction $\frac{0.3}{1.2}$. We can simplify this fraction as follows:

$$\frac{0.3}{1.2} = \frac{0.3 \cdot \mathbf{10}}{1.2 \cdot \mathbf{10}}$$
To clear the decimal, multiply both numerator and denominator by 10.

$$= \frac{3}{12}$$
Multiply: $0.3 \cdot 10 = 3$ and $1.2 \cdot 10 = 12$.

$$= \frac{1}{4}$$
Simplify the fraction: $\dfrac{3}{12} = \dfrac{\overset{1}{\cancel{3}} \cdot 1}{\underset{1}{\cancel{3}} \cdot 4} = \dfrac{1}{4}$. ∎

Self Check   Express each ratio in lowest terms: **a.** the ratio of 8 to 12   and   **b.** the ratio of 3.2 to 16.

Answers   **a.** $\frac{2}{3}$,   **b.** $\frac{1}{5}$

EXAMPLE 2

Express each phrase as a ratio in lowest terms: **a.** the ratio of 3 meters to 8 meters and **b.** the ratio of 4 ounces to 1 pound.

Solution **a.** The ratio of 3 meters to 8 meters can be written as the fraction $\frac{3 \text{ meters}}{8 \text{ meters}}$, or just $\frac{3}{8}$.

b. When possible, we should express ratios in the same units. Since there are 16 ounces in 1 pound, the proper ratio is $\frac{4 \text{ ounces}}{16 \text{ ounces}}$, which simplifies to $\frac{1}{4}$. ∎

Self Check

Express each ratio in lowest terms: **a.** the ratio of 8 ounces to 2 pounds and **b.** the ratio of 1 foot to 2 yards. (*Hint:* 3 feet = 1 yard.)

Answers **a.** $\frac{1}{4}$, **b.** $\frac{1}{6}$

EXAMPLE 3

At a college, there are 2,772 students and 154 faculty members. Write a fraction in simplified form that expresses the ratio of students per faculty member.

Solution The ratio of students to faculty is 2,772 to 154. We can write this ratio as the fraction $\frac{2,772}{154}$ and simplify it.

$$\frac{2,772}{154} = \frac{18 \cdot \overset{1}{\cancel{154}}}{1 \cdot \underset{1}{\cancel{154}}}$$

$$= \frac{18}{1} \qquad \tfrac{154}{154} = 1.$$

The ratio of students to faculty is 18 to 1. ∎

Self Check

In a college graduating class, 224 students out of 632 went on to graduate school. Write a fraction in simplified form that expresses the ratio of the number of students going on to the number in the graduating class.

Answer $\frac{28}{79}$

■ UNIT COSTS

The *unit cost* of an item is the ratio of its cost to its quantity. For example, the unit cost (the cost per pound) of 5 pounds of coffee priced at $20.75 is given by the ratio

$$\frac{\$20.75}{5 \text{ pounds}} = \frac{\$2,075}{500 \text{ pounds}} \qquad \text{To eliminate the decimal, multiply numerator and denominator by 100.}$$

$$= \$4.15 \text{ per pound} \qquad \$2,070 \div 500 = \$4.15.$$

The unit cost is $4.15 per pound.

EXAMPLE 4   Olives come packaged in a 12-ounce jar, which sells for $3.09, or in a 6-ounce jar, which sells for $1.53. Which is the better buy?

Solution   To find the better buy, we must find each unit cost. The unit cost of the 12-ounce jar is

$$\frac{\$3.09}{12 \text{ ounces}} = \frac{309\text{¢}}{12 \text{ ounces}} \qquad \text{Change } \$3.09 \text{ to } 309 \text{ cents.}$$

$$= 25.75\text{¢ per ounce}$$

The unit cost of the 6-ounce jar is

$$\frac{\$1.53}{6 \text{ ounces}} = \frac{153\text{¢}}{6 \text{ ounces}} \qquad \text{Change } \$1.53 \text{ to } 153 \text{ cents.}$$

$$= 25.5\text{¢ per ounce}$$

Since the unit cost is less when olives are packaged in 6-ounce jars, that is the better buy.   ∎

Self Check   A fast-food restaurant sells a 12-ounce soft drink for 79¢ and a 16-ounce soft drink for 99¢. Which is the better buy?

Answer   the 16-oz drink

■ RATES

When ratios are used to compare quantities with different units, they are called *rates*. For example, if we drive 413 miles in 7 hours, the average rate of speed is the ratio of the miles driven to the length of time of the trip.

$$\text{Average rate of speed} = \frac{413 \text{ miles}}{7 \text{ hours}} = \frac{59 \text{ miles}}{1 \text{ hour}} \qquad \frac{413}{7} = \frac{\cancel{7} \cdot 59}{\cancel{7} \cdot 1} = \frac{59}{1}.$$

The ratio $\frac{59 \text{ miles}}{1 \text{ hour}}$ can be expressed in any of the following forms:

$$59 \frac{\text{miles}}{\text{hour}}, \qquad 59 \text{ miles per hour}, \qquad 59 \text{ miles/hour}, \qquad \text{or} \qquad 59 \text{ mph}$$

EXAMPLE 5   Find the hourly rate of pay for a student who earns $370 for working 40 hours.

Solution   We can write the rate of pay as the ratio

$$\text{Rate of pay} = \frac{\$370}{40 \text{ hours}}$$

and simplify by dividing 370 by 40.

$$\text{Rate of pay} = 9.25 \frac{\text{dollars}}{\text{hour}}$$

The rate is $9.25 per hour.   ∎

Self Check

Joan earns $316 per 40-hour week managing a dress shop. Set up a ratio and find her hourly rate of pay.

Answer

$7.90 per hour

EXAMPLE 6

One household used 813.75 kilowatt hours of electricity during a 31-day period. Find the rate of energy consumption in kilowatt hours per day.

Solution

We can write the rate of energy consumption as the ratio

$$\text{Rate of energy consumption} = \frac{813.75 \text{ kilowatt hours}}{31 \text{ days}}$$

and simplify by dividing 813.75 by 31.

$$\text{Rate of energy consumption} = 26.25 \frac{\text{kilowatt hours}}{\text{day}}$$

The rate of consumption is 26.25 kilowatt hours per day. ■

Self Check

To heat a house for 30 days, a furnace burned 72 therms of natural gas. Find the rate of gas consumption in therms per day.

Answer

2.4 therms per day

EXAMPLE 7

A textbook costs $49.22, including sales tax. If the tax was $3.22, find the sales tax rate.

Solution

Since the tax was $3.22, the cost of the book alone was

$49.22 − $3.22 = $46.00

We can write the sales tax rate as the ratio

$$\text{Sales tax rate} = \frac{\text{amount of sales tax}}{\text{cost of the book, without tax}}$$

$$= \frac{\$3.22}{\$46}$$

and simplify by dividing 3.22 by 46.

Sales tax rate = 0.07

The tax rate is 0.07, or 7%. ■

Self Check

A sport coat costs $160.50, including sales tax. If the cost of the coat without tax is $150, find the sales tax rate.

Answer

7%

■ ■ ■ ■ ■ ■ ■ ■ ■   **Computing Gas Mileage**

CALCULATORS   A man drove a total of 775 miles. Along the way, he stopped for gas three times, pumping 10.5, 11.3, and 8.75 gallons of gas. He started with the tank half-full and ended with the tank half-full. To find how many miles he got per gallon, we need to divide the total distance by the total number of gallons of gas consumed.

$$\frac{775}{10.5 + 11.3 + 8.75}$$ ← Total distance
                                        ← Total number of gallons consumed

We can make this calculation by pressing these keys on a scientific calculator.

Keystrokes

775 ÷ (10.5 + 11.3 + 8.75) =

The display will read 25.36824877. To the nearest one-hundredth, he got 25.37 mpg.

Orals   *Express as a ratio in lowest terms.*

1. 5 to 7   **2.** 50 to 1   **3.** 3 to 9   **4.** 7 to 10

EXERCISE 6.1

REVIEW   *Solve each equation.*

1. $2x + 4 = 38$ **2.** $\dfrac{x}{2} - 4 = 38$ **3.** $3(x + 2) = 24$ **4.** $\dfrac{x - 6}{3} = 20$

Factor each expression.

5. $2x + 6$ **6.** $x^2 - 49$ **7.** $2x^2 - x - 6$ **8.** $x^3 + 27$

VOCABULARY AND CONCEPTS   *Fill in each blank to make a true statement.*

9. A ratio is a _____ of two numbers by their indicated _____.

10. The _____ of an item is the ratio of its cost to its quantity.

11. The ratios $\frac{2}{3}$ and $\frac{4}{6}$ are _____ ratios.

12. The ratio $\frac{500 \text{ miles}}{15 \text{ hours}}$ is called a _____.

13. Give three examples of ratios that you have encountered this past week.

14. Suppose that a basketball player made 8 free throws out of 12 tries. The ratio of $\frac{8}{12}$ can be simplified as $\frac{2}{3}$. Interpret this result.

PRACTICE   *In Exercises 15–30, express each phrase as a ratio in lowest terms.*

15. 5 to 7   **16.** 3 to 5   **17.** 17 to 34   **18.** 19 to 38

19. 22 to 33   **20.** 14 to 21   **21.** 7 to 24.5   **22.** 0.65 to 0.15

23. 4 ounces to 12 ounces

24. 3 inches to 15 inches

25. 12 minutes to 1 hour

26. 8 ounces to 1 pound

27. 3 days to 1 week

28. 4 inches to 2 yards

29. 18 months to 2 years

30. 8 feet to 4 yards

In Exercises 31–34, refer to the monthly family budget shown in Illustration 1. Give each ratio in lowest terms.

31. Find the total amount of the budget.

32. Find the ratio of the amount budgeted for rent to the total budget.

33. Find the ratio of the amount budgeted for entertainment to the total budget.

34. Find the ratio of the amount budgeted for phone to the amount budgeted for entertainment.

Item	Amount
Rent	$750
Food	$652
Gas and electric	$188
Phone	$125
Entertainment	$110

ILLUSTRATION 1

In Exercises 35–38, refer to the tax deductions listed in Illustration 2. Give each ratio in lowest terms.

35. Find the total amount of deductions.

36. Find the ratio of real estate tax deductions to the total deductions.

37. Find the ratio of the contributions to the total deductions.

38. Find the ratio of the mortgage interest deduction to the union dues deduction.

Item	Amount
Medical	$ 995
Real estate tax	$1,245
Contributions	$1,680
Mortgage interest	$4,580
Union dues	$ 225

ILLUSTRATION 2

APPLICATIONS *In Exercises 39–56, find each ratio and express it in lowest terms. You may use a calculator when it is helpful.*

39. Faculty-to-student ratio At a college, there are 125 faculty members and 2,000 students. Find the faculty-to-student ratio.

40. Ratio of men to women In a state senate, there are 94 men and 24 women. Find the ratio of men to women.

41. Unit cost of gasoline A driver pumped 17 gallons of gasoline into his tank at a cost of $21.59. Write a ratio of dollars to gallons, and give the unit cost of gasoline.

42. Unit cost of grass seed A 50-pound bag of grass seed costs $222.50. Write a ratio of dollars to pounds, and give the unit cost of grass seed.

43. Unit cost of cranberry juice A 12-ounce can of cranberry juice sells for 84¢. Give the unit cost in cents per ounce.

44. Unit cost of beans A 24-ounce package of green beans sells for $1.29. Give the unit cost in cents per ounce.

45. Comparative shopping A 6-ounce can of orange juice sells for 89¢, and an 8-ounce can sells for $1.19. Which is the better buy?

46. Comparing speeds A car travels 345 miles in 6 hours, and a truck travels 376 miles in 6.2 hours. Which vehicle travels faster?

47. Comparing reading speeds One seventh-grader read a 54-page book in 40 minutes, and another read an 80-page book in 62 minutes. If the books were equally difficult, which student read faster?

48. Comparative shopping A 30-pound bag of fertilizer costs $12.25, and an 80-pound bag costs $30.25. Which is the better buy?

49. Emptying a tank An 11,880-gallon tank can be emptied in 27 minutes. Write a ratio of gallons to minutes, and give the rate of flow in gallons per minute.

50. Rate of pay Ricardo worked for 27 hours to help insulate a hockey arena. For his work, he received $337.50. Write a ratio of dollars to hours, and find his hourly rate of pay.

51. Sales tax A sweater cost $36.75 after sales tax had been added. Find the tax rate as a percent if the sweater retailed for $35.

52. Real estate taxes The real estate taxes on a summer home assessed at $75,000 were $1,500. Find the tax rate as a percent.

53. Rate of speed A car travels 325 miles in 5 hours. Find its rate of speed in miles per hour.

54. Rate of speed An airplane travels from Chicago to San Francisco, a distance of 1,883 miles, in 3.5 hours. Find the average rate of speed of the plane.

55. Comparing gas mileage One car went 1,235 miles on 51.3 gallons of gasoline, and another went 1,456 miles on 55.78 gallons. Which car had the better mpg rating?

56. Comparing electric rates In one community, a bill for 575 kilowatt hours (kwh) of electricity was $38.81. In a second community, a bill for 831 kwh was $58.10. In which community is electricity cheaper?

WRITING

57. Some people think that the word *ratio* comes from the words *rational number.* Explain why this may be true.

58. In the fraction $\frac{a}{b}$, b cannot be 0. Explain why. In the ratio $\frac{a}{b}$, b can be 0. Explain why.

SOMETHING TO THINK ABOUT

59. Which ratio is the larger? How can you tell?

$$\frac{17}{19} \quad \text{or} \quad \frac{19}{21}$$

60. Which ratio is the smaller? How can you tell?

$$-\frac{13}{29} \quad \text{or} \quad -\frac{17}{31}$$

6.2 Proportions and Similar Triangles

■ PROPORTIONS ■ MEANS AND EXTREMES OF A PROPORTION ■ SOLVING PROPORTIONS
■ PROBLEM SOLVING ■ SIMILAR TRIANGLES

Getting Ready *Solve each equation.*

1. $\dfrac{5}{2} = \dfrac{x}{4}$ **2.** $\dfrac{7}{9} = \dfrac{y}{3}$ **3.** $\dfrac{y}{10} = \dfrac{2}{7}$ **4.** $\dfrac{1}{x} = \dfrac{8}{40}$

5. $\dfrac{w}{14} = \dfrac{7}{21}$ **6.** $\dfrac{c}{12} = \dfrac{5}{12}$ **7.** $\dfrac{3}{q} = \dfrac{1}{7}$ **8.** $\dfrac{16}{3} = \dfrac{8}{z}$

■ PROPORTIONS

Consider Table 6-1, in which we are given the costs of various numbers of gallons of gasoline.

Number of gallons	Cost
2	$ 2.72
5	$ 6.80
8	$10.88
12	$16.32
20	$27.20

TABLE 6-1

If we find the ratios of the costs to the numbers of gallons purchased, we will see that they are equal. In this example, each ratio represents the cost of 1 gallon of gasoline, which is $1.36 per gallon.

$$\frac{\$2.72}{2} = \$1.36, \qquad \frac{\$6.80}{5} = \$1.36, \qquad \frac{\$10.88}{8} = \$1.36, \qquad \frac{\$16.32}{12} = \$1.36,$$

and $\qquad \dfrac{\$27.20}{20} = \$1.36$

When two ratios such as $\frac{\$2.72}{2}$ and $\frac{\$6.80}{5}$ are equal, they form a *proportion*. In this section, we will discuss proportions and use them to solve problems.

> **Proportions**
> A **proportion** is a statement that two ratios are equal.

Some examples of proportions are

$$\frac{1}{2} = \frac{3}{6}, \qquad \frac{7}{3} = \frac{21}{9}, \qquad \frac{8x}{1} = \frac{40x}{5}, \qquad \text{and} \qquad \frac{a}{b} = \frac{c}{d}$$

- The proportion $\dfrac{1}{2} = \dfrac{3}{6}$ can be read as "1 is to 2 as 3 is to 6."

- The proportion $\dfrac{7}{3} = \dfrac{21}{9}$ can be read as "7 is to 3 as 21 is to 9."

- The proportion $\dfrac{8x}{1} = \dfrac{40x}{5}$ can be read as "8x is to 1 as 40x is to 5."

- The proportion $\dfrac{a}{b} = \dfrac{c}{d}$ can be read as "a is to b as c is to d."

The terms of the proportion $\frac{a}{b} = \frac{c}{d}$ are numbered as follows:

$$\text{First term} \longrightarrow \quad \frac{a}{b} = \frac{c}{d} \quad \longleftarrow \text{Third term}$$
$$\text{Second term} \longrightarrow \qquad\qquad \longleftarrow \text{Fourth term}$$

■ MEANS AND EXTREMES OF A PROPORTION

In the proportion $\frac{1}{2} = \frac{3}{6}$, the numbers 1 and 6 are called the **extremes,** and the numbers 2 and 3 are called the **means.**

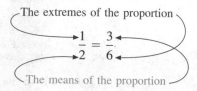

The extremes of the proportion

$$\frac{1}{2} = \frac{3}{6}$$

The means of the proportion

In this proportion, the product of the extremes is equal to the product of the means.

$$1 \cdot 6 = 6 \quad \text{and} \quad 2 \cdot 3 = 6$$

This illustrates a fundamental property of proportions.

> **Fundamental Property of Proportions**
> In any proportion, the product of the extremes is equal to the product of the means.

In the proportion $\frac{a}{b} = \frac{c}{d}$, a and d are the extremes, and b and c are the means. We can show that the product of the extremes (ad) is equal to the product of the means (bc) by multiplying both sides of the proportion by bd and observing that $ad = bc$.

$$\frac{a}{b} = \frac{c}{d}$$

$$\frac{bd}{1} \cdot \frac{a}{b} = \frac{bd}{1} \cdot \frac{c}{d} \qquad \text{To eliminate the fractions, multiply both sides by } \frac{bd}{1}.$$

$$\frac{abd}{b} = \frac{bcd}{d} \qquad \text{Multiply the numerators and multiply the denominators.}$$

$$ad = bc \qquad \text{Divide out the common factors: } \frac{b}{b} = 1 \text{ and } \frac{d}{d} = 1.$$

Since $ad = bc$, the product of the extremes equals the product of the means.

To determine whether an equation is a proportion, we can check to see whether the product of the extremes is equal to the product of the means.

EXAMPLE 1   Determine whether each equation is a proportion: **a.** $\dfrac{3}{7} = \dfrac{9}{21}$ and

b. $\dfrac{8}{3} = \dfrac{13}{5}$.

Solution   In each case, we check to see whether the product of the extremes is equal to the product of the means.

a. The product of the extremes is $3 \cdot 21 = 63$. The product of the means is $7 \cdot 9 = 63$. Since the products are equal, the equation is a proportion: $\frac{3}{7} = \frac{9}{21}$.

b. The product of the extremes is $8 \cdot 5 = 40$. The product of the means is $3 \cdot 13 = 39$. Since the products are not equal, the equation is not a proportion: $\frac{8}{3} \neq \frac{13}{5}$.   ∎

Self Check   Determine whether the equation is a proportion: $\frac{6}{13} = \frac{24}{53}$.

Answer   no

When two pairs of numbers such as 2, 3 and 8, 12 form a proportion, we say that they are **proportional.** To show that 2, 3, 8, and 12 are proportional, we check to see whether the equation

$$\frac{2}{3} = \frac{8}{12}$$

is a proportion. To do so, we find the product of the extremes and the product of the means:

$$2 \cdot 12 = 24 \qquad\qquad 3 \cdot 8 = 24$$

Since the products are equal, the equation is a proportion, and the numbers are proportional.

EXAMPLE 2   Determine whether 3, 7, 36, and 91 are proportional.

Solution   We check to see whether $\frac{3}{7} = \frac{36}{91}$ is a proportion by finding two products:

$$3 \cdot 91 = 273 \qquad \text{The product of the extremes.}$$
$$7 \cdot 36 = 252 \qquad \text{The product of the means.}$$

Since the products are not equal, the numbers are not proportional.   ∎

Self Check   Determine whether 6, 11, 54, and 99 are proportional.

Answer   yes

■ SOLVING PROPORTIONS

Suppose that we know three terms in the proportion

$$\frac{x}{5} = \frac{24}{20}$$

To find the unknown term, we multiply the extremes and multiply the means, set them equal, and solve for x:

$$\frac{x}{5} = \frac{24}{20}$$

$20x = 5 \cdot 24$ In a proportion, the product of the extremes is equal to the product of the means.

$20x = 120$ Multiply: $5 \cdot 24 = 120$.

$$\frac{20x}{20} = \frac{120}{20}$$ To undo the multiplication by 20, divide both sides by 20.

$x = 6$ Simplify: $\frac{20}{20} = 1$ and $\frac{120}{20} = 6$.

The first term is 6.

EXAMPLE 3   Solve $\dfrac{12}{18} = \dfrac{3}{x}$.

Solution

$$\frac{12}{18} = \frac{3}{x}$$

$12 \cdot x = 18 \cdot 3$ In a proportion, the product of the extremes equals the product of the means.

$12x = 54$ Multiply: $18 \cdot 3 = 54$.

$$\frac{12x}{12} = \frac{54}{12}$$ To undo the multiplication by 12, divide both sides by 12.

$x = \dfrac{9}{2}$ Simplify: $\frac{12}{12} = 1$ and $\frac{54}{12} = \frac{9}{2}$.

Thus, $x = \dfrac{9}{2}$.

■

Self Check   Solve $\frac{15}{x} = \frac{25}{40}$.

Answer   24

EXAMPLE 4   Find the third term of the proportion $\dfrac{3.5}{7.2} = \dfrac{x}{15.84}$.

Solution

$$\frac{3.5}{7.2} = \frac{x}{15.84}$$

$$3.5(15.84) = 7.2x \qquad \text{In a proportion, the product of the extremes equals the product of the means.}$$

$$55.44 = 7.2x \qquad \text{Multiply: } 3.5 \cdot 15.84 = 55.44.$$

$$\frac{55.44}{7.2} = \frac{7.2x}{7.2} \qquad \text{To undo the multiplication by 7.2, divide both sides by 7.2.}$$

$$7.7 = x \qquad \text{Simplify: } \frac{55.44}{7.2} = 7.7 \text{ and } \frac{7.2}{7.2} = 1.$$

The third term is 7.7. ■

Self Check   Find the second term of the proportion $\dfrac{6.7}{x} = \dfrac{33.5}{38}$.

Answer   7.6

■ ■ ■ ■ ■ ■ ■ ■ ■   **Solving Equations with a Calculator**

CALCULATORS   To solve the equation in Example 4 with a calculator, we can proceed as follows.

$$\frac{3.5}{7.2} = \frac{x}{15.84}$$

$$\frac{3.5(15.84)}{7.2} = x \qquad \text{Multiply both sides by 15.84.}$$

We can find x by pressing these keys on a scientific calculator.

Keystrokes

    3.5 × 15.84 ÷ 7.2 =

The display will read 7.7 . Thus, $x = 7.7$.

EXAMPLE 5   Solve $\dfrac{2x + 1}{4} = \dfrac{10}{8}$.

Solution

$$\frac{2x + 1}{4} = \frac{10}{8}$$

$8(2x + 1) = 40$ — In a proportion, the product of the extremes equals the product of the means.

$16x + 8 = 40$ — Use the distributive property to remove parentheses.

$16x + 8 - \mathbf{8} = 40 - \mathbf{8}$ — To undo the addition of 8, subtract 8 from both sides.

$16x = 32$ — Simplify: $8 - 8 = 0$ and $40 - 8 = 32$.

$$\frac{16x}{\mathbf{16}} = \frac{32}{\mathbf{16}}$$ — To undo the multiplication by 16, divide both sides by 16.

$x = 2$ — Simplify: $\frac{16}{16} = 1$ and $\frac{32}{16} = 2$.

Thus, $x = 2$. ■

Self Check Solve $\dfrac{3x - 1}{2} = \dfrac{12.5}{5}$.

Answer 2

■ PROBLEM SOLVING

We can use proportions to solve problems.

EXAMPLE 6 If 6 apples cost $1.38, how much will 16 apples cost?

Solution Let c represent the cost of 16 apples. The ratios of the numbers of apples to their costs are equal.

6 apples is to $1.38 as 16 apples is to $$c$.

6 apples → $\dfrac{6}{1.38}$ = $\dfrac{16}{c}$ ← 16 apples
Cost of 6 apples →   ← Cost of 16 apples

$6 \cdot c = 1.38(16)$ — In a proportion, the product of the extremes is equal to the product of the means.

$6c = 22.08$ — Do the multiplication: $1.38 \cdot 16 = 22.08$.

$$\frac{6c}{6} = \frac{22.08}{6}$$ — To undo the multiplication by 6, divide both sides by 6.

$c = 3.68$ — Simplify: $\frac{6}{6} = 1$ and $\frac{22.08}{6} = 3.68$.

Sixteen apples will cost $3.68. ■

Self Check If 9 tickets to a concert cost $112.50, how much will 15 tickets cost?

Answer $187.50

EXAMPLE 7   A solution contains 2 quarts of antifreeze and 5 quarts of water. How many quarts of antifreeze must be mixed with 18 quarts of water to have the same concentration?

Solution   Let q represent the number of quarts of antifreeze to be mixed with the water. The ratios of the quarts of antifreeze to the quarts of water are equal.

2 quarts antifreeze is to 5 quarts water as q quarts antifreeze is to 18 quarts water.

2 quarts antifreeze → $\dfrac{2}{5} = \dfrac{q}{18}$ ← q quarts of antifreeze
5 quarts water →   ← 18 quarts of water

$2 \cdot 18 = 5q$ In a proportion, the product of the extremes is equal to the product of the means.

$36 = 5q$ Do the multiplication: $2 \cdot 18 = 36$.

$\dfrac{36}{5} = \dfrac{5q}{5}$ To undo the multiplication by 5, divide both sides by 5.

$\dfrac{36}{5} = q$ Simplify: $\frac{5}{5} = 1$.

The mixture should contain $\frac{36}{5}$ or 7.2 quarts of antifreeze. ∎

Self Check   A solution should contain 2 ounces of alcohol for every 7 ounces of water. How much alcohol should be added to 20 ounces of water to get the proper concentration?

Answer $\frac{40}{7}$ oz

EXAMPLE 8   A recipe for rhubarb cake calls for $1\frac{1}{4}$ cups of sugar for every $2\frac{1}{2}$ cups of flour. How many cups of flour are needed if the baker intends to use 3 cups of sugar?

Solution   Let f represent the number of cups of flour to be mixed with the sugar. The ratios of the cups of sugar to the cups of flour are equal.

$1\frac{1}{4}$ cups sugar is to $2\frac{1}{2}$ cups flour as 3 cups sugar is to f cups flour.

$1\frac{1}{4}$ cups sugar → $\dfrac{1\frac{1}{4}}{2\frac{1}{2}} = \dfrac{3}{f}$ ← 3 cups sugar
$2\frac{1}{2}$ cups flour →   ← f cups flour

$\dfrac{1.25}{2.5} = \dfrac{3}{f}$ Change the fractions to decimals.

$1.25f = 2.5 \cdot 3$ In a proportion, the product of the extremes is equal to the product of the means.

$1.25f = 7.5$ Do the multiplication: $2.5 \cdot 3 = 7.5$.

$\dfrac{1.25f}{1.25} = \dfrac{7.5}{1.25}$ To undo the multiplication by 1.25, divide both sides by 1.25.

$f = 6$ Divide: $\frac{1.25}{1.25} = 1$ and $\frac{7.5}{1.25} = 6$.

The baker should use 6 cups of flour. ∎

EXAMPLE 9   In a manufacturing process, 15 parts out of 90 were found to be defective. How many defective parts will be expected in a run of 120 parts?

Solution   Let d represent the expected number of defective parts. In each run, the ratio of the defective parts to the total number of parts should be the same.

15 defective parts is to 90 as d defective parts is to 120.

15 defective parts → $\dfrac{15}{90}$ = $\dfrac{d}{120}$ ← d defective parts

90 parts → ⎯⎯ 120 ← 120 parts

$$15 \cdot 120 = 90d \qquad \text{In a proportion, the product of the extremes is equal to the product of the means.}$$

$$1{,}800 = 90d \qquad \text{Do the multiplication: } 15 \cdot 120 = 1{,}800.$$

$$\frac{1{,}800}{90} = \frac{90d}{90} \qquad \text{To undo the multiplication by 90, divide both sides by 90.}$$

$$20 = d \qquad \text{Divide: } \frac{1{,}800}{90} = 20 \text{ and } \frac{90}{90} = 1.$$

The expected number of defective parts is 20.   ∎

■ SIMILAR TRIANGLES

If two angles of one triangle have the same measure as two angles of a second triangle, the triangles will have the same shape. Triangles with the same shape are called **similar triangles.** In Figure 6-1, $\triangle ABC \sim \triangle DEF$ (read the symbol $\sim$ as "is similar to.")

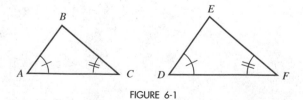

FIGURE 6-1

Property of Similar Triangles
If two triangles are similar, all pairs of corresponding sides are in proportion.

In the similar triangles shown in Figure 6-1, the following proportions are true.

$$\frac{AB}{DE} = \frac{BC}{EF}, \qquad \frac{BC}{EF} = \frac{CA}{FD}, \qquad \text{and} \qquad \frac{CA}{FD} = \frac{AB}{DE}$$

EXAMPLE 10   A tree casts a shadow 18 feet long at the same time as a woman 5 feet tall casts a shadow that is 1.5 feet long. Find the height of the tree.

Solution   Figure 6-2 shows the triangles determined by the tree and its shadow and the woman and her shadow.

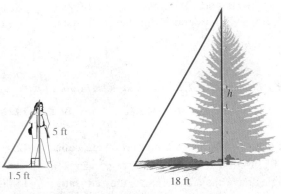

5 ft

1.5 ft

18 ft

FIGURE 6-2

Since the triangles have the same shape, they are similar, and the lengths of their corresponding sides are in proportion. If we let h represent the height of the tree, we can find h by solving the following proportion.

$$\frac{h}{5} = \frac{18}{1.5} \qquad \frac{\text{Height of the tree}}{\text{Height of the woman}} = \frac{\text{Shadow of the tree}}{\text{Shadow of the woman}}.$$

$$1.5h = 5(18) \qquad \text{In a proportion, the product of the extremes is equal to the product of the means.}$$

$$h = 60 \qquad \text{To undo the multiplication by 1.5, divide both sides by 1.5 and simplify.}$$

The tree is 60 feet tall.   ∎

Self Check   Find the height of the tree in Example 10 if the woman is 5 feet 6 inches tall and her shadow is still 1.5 feet long.

Answer   66 ft

Orals   *Indicate which are proportions.*

1. $\dfrac{3}{5} = \dfrac{6}{10}$ **2.** $\dfrac{1}{2} = \dfrac{1}{3}$ **3.** $\dfrac{1}{2} + \dfrac{2}{4}$ **4.** $\dfrac{1}{x} = \dfrac{2}{2x}$

EXERCISE 6.2

REVIEW

1. Change $\dfrac{9}{10}$ to a percent.

2. Change $\dfrac{7}{8}$ to a percent.

3. Change $33\frac{1}{3}\%$ to a fraction.

4. Change 75% to a fraction.

5. Find 30% of 1,600.

6. Find $\dfrac{1}{2}\%$ of 520.

7. **Shopping** If Maria bought a dress for 25% off the original price of $98, how much did the dress cost?

8. **Shopping** Bill purchased a shirt on sale for $17.50. Find the original cost of the shirt if it was marked down 30%.

VOCABULARY AND CONCEPTS   *Fill in each blank to make a true statement.*

9. A _____ is a statement that two _____ are equal.

10. The first and fourth terms of a proportion are called the _____ of the proportion.

11. The second and third terms of a proportion are called the _____ of the proportion.

12. When two pairs of numbers form a proportion, we say that the numbers are _____.

13. If two triangles have the same _____, they are said to be similar.

14. If two triangles are similar, the lengths of their corresponding sides are in _____.

15. The equation $\dfrac{a}{b} = \dfrac{c}{d}$ is a proportion if the product ____ is equal to the product ____.

16. If $3 \cdot 10 = 17 \cdot x$, then _____ is a proportion. (Note that answers may differ.)

17. Read $\triangle ABC$ as _____ ABC.

18. The symbol $\sim$ is read as _____.

PRACTICE   *In Exercises 19–26, tell whether each statement is a proportion.*

19. $\dfrac{9}{7} = \dfrac{81}{70}$

20. $\dfrac{5}{2} = \dfrac{20}{8}$

21. $\dfrac{-7}{3} = \dfrac{14}{-6}$

22. $\dfrac{13}{-19} = \dfrac{-65}{95}$

23. $\dfrac{9}{19} = \dfrac{38}{80}$

24. $\dfrac{40}{29} = \dfrac{29}{22}$

25. $\dfrac{10.4}{3.6} = \dfrac{41.6}{14.4}$

26. $\dfrac{13.23}{3.45} = \dfrac{39.96}{11.35}$

In Exercises 27–42, solve for the variable in each proportion.

27. $\dfrac{2}{3} = \dfrac{x}{6}$

28. $\dfrac{3}{6} = \dfrac{x}{8}$

29. $\dfrac{5}{10} = \dfrac{3}{c}$

30. $\dfrac{7}{14} = \dfrac{2}{b}$

31. $\dfrac{-6}{x} = \dfrac{8}{4}$

32. $\dfrac{4}{x} = \dfrac{2}{8}$

33. $\dfrac{x}{3} = \dfrac{9}{3}$

34. $\dfrac{x}{2} = \dfrac{-18}{6}$

35. $\dfrac{x+1}{5} = \dfrac{3}{15}$

36. $\dfrac{x-1}{7} = \dfrac{2}{21}$

37. $\dfrac{x+3}{12} = \dfrac{-7}{6}$

38. $\dfrac{x+7}{-4} = \dfrac{3}{12}$

39. $\dfrac{4-x}{13} = \dfrac{11}{26}$

40. $\dfrac{5-x}{17} = \dfrac{13}{34}$

41. $\dfrac{2x+1}{18} = \dfrac{14}{3}$

42. $\dfrac{2x-1}{18} = \dfrac{9}{54}$

APPLICATIONS   ⊞   *In Exercises 43–62, set up and solve a proportion. Use a calculator when it is helpful.*

43. Grocery shopping   If 3 pints of yogurt cost $1, how much will 51 pints cost?

44. Shopping for clothes   If shirts are on sale at two for $25, how much will 5 shirts cost?

45. Gardening   Garden seed is on sale at 3 packets for 50¢. How much will 39 packets cost?

46. Cooking   A recipe for spaghetti sauce requires four 16-ounce bottles of catsup to make two gallons of sauce. How many bottles of catsup are needed to make 10 gallons of sauce?

47. Mixing perfume   A perfume is to be mixed in the ratio of 3 drops of pure essence to 7 drops of alcohol. How many drops of pure essence should be mixed with 56 drops of alcohol?

48. Making cologne   A cologne can be made by mixing 2 drops of pure essence with 5 drops of distilled water. How many drops of water should be used with 15 drops of pure essence?

49. Making cookies   A recipe for chocolate chip cookies calls for $1\frac{1}{4}$ cups of flour and 1 cup of sugar. The recipe will make $3\frac{1}{2}$ dozen cookies. How many cups of flour will be needed to make 12 dozen cookies?

50. Making brownies   A recipe for brownies calls for 4 eggs and $1\frac{1}{2}$ cups of flour. If the recipe make 15 brownies, how many cups of flour will be needed to make 130 brownies?

51. Quality control   In a manufacturing process, 95% of the parts made are to be within specifications. How many defective parts would be expected in a run of 940 pieces?

52. Quality control   Out of a sample of 500 men's shirts, 17 were rejected because of crooked collars. How many crooked collars would you expect to find in a run of 15,000 shirts?

53. Gas consumption   If a car can travel 42 miles on 1 gallon of gas, how much gas will it need to travel 315 miles?

54. Gas consumption   If a truck gets 12 miles per gallon of gas, how far can it go on 17 gallons?

55. Computing paychecks   Bill earns $412 for a 40-hour week. If he missed 10 hours of work last week, how much did he get paid?

56. Model railroading   An HO-scale model railroad engine is 9 inches long. If HO scale is 87 feet to 1 foot, how long is a real engine?

57. Model railroading   An N-scale model railroad caboose is 3.5 inches long. If N scale is 169 feet to 1 foot, how long is a real caboose?

58. Model houses   A model house is built to a scale of 1 inch to 8 inches. If a model house is 36 inches wide, how wide is the real house?

59. Staffing   A school board determined that there should be 3 teachers for every 50 students. How many teachers are needed for an enrollment of 2,700 students?

60. Drafting   In a scale drawing, a 280-foot antenna tower is drawn 7 inches high. The building next to it is drawn 2 inches high. How tall is the actual building?

61. Mixing fuel   The instructions on a can of oil intended to be added to lawnmower gasoline read:

Recommended	Gasoline	Oil
50 to 1	6 gal	16 oz

Are these instructions correct? (*Hint:* There are 128 ounces in 1 gallon.)

62. Mixing fuel   See Exercise 61. How much oil should be mixed with 28 gallons of gas?

In Exercises 63–70, use similar triangles to solve each problem.

63. Height of a tree   A tree casts a shadow of 26 feet at the same time as a 6-foot man casts a shadow of 4 feet. (See Illustration 1.) Find the height of the tree.

64. Height of a flagpole   A man places a mirror on the ground and sees the reflection of the top of a flagpole, as in Illustration 2. The two triangles in the illustration are similar. Find the height *h* of the flagpole.

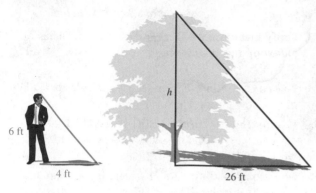

ILLUSTRATION 2

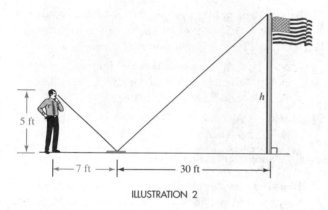

ILLUSTRATION 2

65. Width of a river Use the dimensions in Illustration 3 to find w, the width of the river. The two triangles in the illustration are similar.

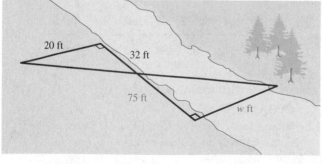

ILLUSTRATION 3

66. Flight path An airplane ascends 100 feet as it flies a horizontal distance of 1,000 feet. How much altitude will it gain as it flies a horizontal distance of 1 mile? See Illustration 4. (*Hint:* 5,280 feet = 1 mile.)

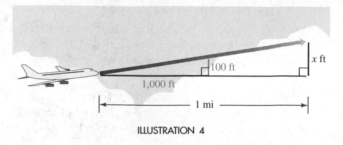

ILLUSTRATION 4

67. Flight path An airplane descends 1,350 feet as it flies a horizontal distance of 1 mile. How much altitude is lost as it flies a horizontal distance of 5 miles?

68. Ski runs A ski course falls 100 feet in every 300 feet of horizontal run. If the total horizontal run is $\frac{1}{2}$ mile, find the height of the hill.

69. Mountain travel A road ascends 750 feet in every 2,500 feet of travel. By how much will the road rise in a trip of 10 miles?

70. Photo enlargements The 3-by-5 photo in Illustration 5 is to be blown up to the larger size. Find x.

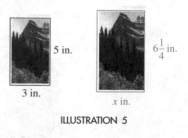

ILLUSTRATION 5

WRITING

71. Explain the difference between a ratio and a proportion.

72. Explain how to tell whether the equation $\frac{3.2}{3.7} = \frac{5.44}{6.29}$ is a proportion.

SOMETHING TO THINK ABOUT

73. Verify that $\frac{3}{5} = \frac{12}{20} = \frac{3+12}{5+20}$. Is the following rule always true?

$$\frac{a}{b} = \frac{c}{d} = \frac{a+c}{b+d}$$

74. Verify that since $\frac{3}{5} = \frac{9}{15}$, then $\frac{3+5}{5} = \frac{9+15}{15}$. Is the following rule always true?

$$\text{If } \frac{a}{b} = \frac{c}{d}, \text{ then } \frac{a+b}{b} = \frac{c+d}{d}.$$

6.3 Simplifying Fractions

■ SIMPLIFYING FRACTIONS   ■ DIVISION BY 1   ■ DIVIDING POLYNOMIALS THAT ARE NEGATIVES

Getting Ready   *Simplify.*

1. $\dfrac{12}{16}$ **2.** $\dfrac{16}{8}$ **3.** $\dfrac{25}{55}$ **4.** $\dfrac{36}{72}$

■ SIMPLIFYING FRACTIONS

Ratios such as $\frac{1}{2}$ and $\frac{3}{4}$ that are the quotient of two integers are *rational numbers.* Expressions such as

$$\frac{x}{x+2} \quad \text{and} \quad \frac{5a^2 + b^2}{3a - b}$$

where the numerators and denominators are polynomials, are called **rational expressions.**

We have seen that a fraction can be simplified by dividing out common factors shared by its numerator and denominator. For example,

$$\frac{18}{30} = \frac{3 \cdot 6}{5 \cdot 6} = \frac{3 \cdot \overset{1}{\cancel{6}}}{5 \cdot \underset{1}{\cancel{6}}} = \frac{3}{5} \quad \text{and} \quad -\frac{6}{15} = -\frac{3 \cdot 2}{3 \cdot 5} = -\frac{\overset{1}{\cancel{3}} \cdot 2}{\underset{1}{\cancel{3}} \cdot 5} = -\frac{2}{5}$$

To simplify the fraction $\frac{ac}{bc}$, we can divide out the common factor of c to obtain

$$\frac{ac}{bc} = \frac{a\overset{1}{\cancel{c}}}{b\underset{1}{\cancel{c}}} = \frac{a}{b}$$

This fact establishes the fundamental property of fractions.

Fundamental Property of Fractions

If a is a real number and b and c are nonzero real numbers, then

$$\frac{ac}{bc} = \frac{a}{b}$$

The fundamental property of fractions implies that factors common to both the numerator and denominator of a fraction can be divided out. When all common factors have been divided out, we say that the fraction has been **expressed in lowest terms.** To **simplify a fraction** means to write it in lowest terms.

EXAMPLE 1   Simplify $\dfrac{21x^2y}{14xy^2}$.

Solution   To simplify a fraction means to write it in lowest terms.

$$\frac{21x^2y}{14xy^2} = \frac{3 \cdot 7 \cdot x \cdot x \cdot y}{2 \cdot 7 \cdot x \cdot y \cdot y} \qquad \text{Factor the numerator and denominator.}$$

$$= \frac{3 \cdot \overset{1}{\cancel{7}} \cdot \overset{1}{\cancel{x}} \cdot x \cdot \overset{1}{\cancel{y}}}{2 \cdot \underset{1}{\cancel{7}} \cdot \underset{1}{\cancel{x}} \cdot y \cdot \underset{1}{\cancel{y}}} \qquad \text{Divide out the common factors of 7, } x\text{, and } y.$$

$$= \frac{3x}{2y}$$

This fraction can also be simplified by using the rules of exponents:

$$\frac{21x^2y}{14xy^2} = \frac{3 \cdot 7}{2 \cdot 7}x^{2-1}y^{1-2} \qquad \frac{x^2}{x} = x^{2-1}; \frac{y}{y^2} = y^{1-2}.$$

$$= \frac{3}{2}xy^{-1} \qquad\qquad 2 - 1 = 1; 1 - 2 = -1.$$

$$= \frac{3}{2} \cdot \frac{x}{y} \qquad\qquad y^{-1} = \frac{1}{y}.$$

$$= \frac{3x}{2y} \qquad\qquad \text{Multiply.}$$

∎

Self Check   Simplify $\dfrac{32a^3b^2}{24ab^4}$.

Answer $\dfrac{4a^2}{3b^2}$

■ ■ ■ ■ ■ ■ ■ ■ ■ PERSPECTIVE

The fraction $\frac{8}{4}$ is equal to 2, because $4 \cdot 2 = 8$. The expression $\frac{8}{0}$ is undefined, because there is no number x for which $0 \cdot x = 8$. The expression $\frac{0}{0}$ presents a different problem, however, because $\frac{0}{0}$ seems to equal any number. For example, $\frac{0}{0} = 17$, because $0 \cdot 17 = 0$. Similarly, $\frac{0}{0} = \pi$, because $0 \cdot \pi = 0$. Since "no answer" and "any answer" are both unacceptable, division by 0 is not allowed.

Although $\frac{0}{0}$ represents many numbers, there is often one best answer. In the 17th century, mathema-ticians such as Sir Isaac Newton (1642–1727) and Gottfried Wilhelm von Leibniz (1646–1716) began to look more closely at expressions related to the fraction $\frac{0}{0}$. They discovered that under certain conditions, there was one best answer. Expressions related to $\frac{0}{0}$ are called **indeterminate forms.** One of these expressions, called a **derivative,** is the foundation of **calculus,** an important area of mathematics discovered independently by both Newton and Leibniz.

EXAMPLE 2 Write $\dfrac{x^2 + 3x}{3x + 9}$ in lowest terms.

Solution

$$\frac{x^2 + 3x}{3x + 9} = \frac{x(x + 3)}{3(x + 3)}$$

Factor the numerator and the denominator.

$$= \frac{x\cancel{(x + 3)}}{3\cancel{(x + 3)}}$$

Divide out the common factor of $x + 3$.

$$= \frac{x}{3}$$

■

Self Check Simplify $\dfrac{x^2 - 5x}{5x - 25}$.

Answer $\frac{x}{5}$

■ DIVISION BY 1

Any number divided by the number 1 remains unchanged. For example,

$$\frac{37}{1} = 37, \qquad \frac{5x}{1} = 5x, \qquad \text{and} \qquad \frac{3x + y}{1} = 3x + y$$

In general, for any real number a, the following is true.

Division by 1

$$\frac{a}{1} = a$$

EXAMPLE 3   Simplify $\dfrac{x^3 + x^2}{x + 1}$.

Solution $\dfrac{x^3 + x^2}{x + 1} = \dfrac{x^2(x + 1)}{x + 1}$ Factor the numerator.

$= \dfrac{x^2\cancel{(x + 1)}^{\,1}}{\cancel{x + 1}_{\,1}}$ Divide out the common factor of $x + 1$.

$= \dfrac{x^2}{1}$

$= x^2$ Denominators of 1 need not be written.   ∎

Self Check   Simplify $\dfrac{x^2 - x}{x - 1}$.

Answer x

■ DIVIDING POLYNOMIALS THAT ARE NEGATIVES

If the terms of two polynomials are the same, except for sign, the polynomials are called **negatives** of each other. For example,

$x - y$ and $-x + y$ are negatives,

$2a - 1$ and $-2a + 1$ are negatives, and

$3x^2 - 2x + 5$ and $-3x^2 + 2x - 5$ are negatives.

Example 4 shows why the quotient of two binomials that are negatives is always -1.

EXAMPLE 4   Simplify   **a.** $\dfrac{x - y}{y - x}$ and   **b.** $\dfrac{2a - 1}{1 - 2a}$.

Solution   We can rearrange terms in each numerator, factor out -1, and proceed as follows:

a. $\dfrac{x - y}{y - x} = \dfrac{-y + x}{y - x}$ **b.** $\dfrac{2a - 1}{1 - 2a} = \dfrac{-1 + 2a}{1 - 2a}$

$= \dfrac{-(y - x)}{y - x}$ $= \dfrac{-(1 - 2a)}{1 - 2a}$

$= \dfrac{-\cancel{(y - x)}^{\,1}}{\cancel{y - x}_{\,1}}$ $= \dfrac{-\cancel{(1 - 2a)}^{\,1}}{\cancel{1 - 2a}_{\,1}}$

$= -1$ $= -1$ ∎

Self Check   Simplify $\dfrac{3p - 2q}{2q - 3p}$.

Answer -1

In general, we have this important result.

Division of Negatives

The quotient of any nonzero expression and its negative is -1.

EXAMPLE 5   Simplify $\dfrac{x^2 + 13x + 12}{x^2 - 144}$.

Solution

$$\frac{x^2 + 13x + 12}{x^2 - 144} = \frac{(x + 1)(x + 12)}{(x + 12)(x - 12)}$$ Factor the numerator and denominator.

$$= \frac{(x + 1)\overset{1}{\cancel{(x + 12)}}}{\underset{1}{\cancel{(x + 12)}}(x - 12)}$$ Divide out the common factor of $x + 12$.

$$= \frac{x + 1}{x - 12}$$ ∎

Self Check   Simplify $\dfrac{x^2 - 9}{x^3 - 3x^2}$.

Answer $\dfrac{x + 3}{x^2}$

WARNING!   Remember that only *factors* that are common to the *entire numerator* and the *entire denominator* can be divided out. *Terms* that are common to both the numerator and denominator *cannot* be divided out. For example, consider the correct simplification

$$\frac{5 + 8}{5} = \frac{13}{5}$$

It would be incorrect to divide out the common *term* of 5 in the above simplification. Doing so gives an incorrect answer.

$$\frac{5 + 8}{5} \neq \frac{\overset{1}{\cancel{5}} + 8}{\underset{1}{\cancel{5}}} = \frac{1 + 8}{1} = 9$$

EXAMPLE 6   Write $\dfrac{5(x + 3) - 5}{7(x + 3) - 7}$ in lowest terms.

Solution   We cannot divide out $x + 3$, because it is not a factor of the entire numerator, nor is it a factor of the entire denominator. Instead, we simplify the numerator and denominator, factor them, and then divide out any common factors.

$$\frac{5(x + 3) - 5}{7(x + 3) - 7} = \frac{5x + 15 - 5}{7x + 21 - 7}$$ Remove parentheses.

$$= \frac{5x + 10}{7x + 14}$$ Combine like terms.

$$= \frac{5(x + 2)}{7(x + 2)}$$ Factor the numerator and denominator.

$$= \frac{5\cancel{(x + 2)}}{7\cancel{(x + 2)}}$$ Divide out the common factor of $x + 2$.

$$= \frac{5}{7}$$ ∎

Self Check   Simplify $\dfrac{4(x - 2) + 4}{3(x - 2) + 3}$.

Answer $\frac{4}{3}$

EXAMPLE 7   Simplify $\dfrac{x(x + 3) - 3(x - 1)}{x^2 + 3}$.

Solution $$\frac{x(x + 3) - 3(x - 1)}{x^2 + 3} = \frac{x^2 + 3x - 3x + 3}{x^2 + 3}$$ Remove parentheses in the numerator.

$$= \frac{x^2 + 3}{x^2 + 3}$$ Combine like terms in the numerator.

$$= \frac{\cancel{x^2 + 3}}{\cancel{x^2 + 3}}$$ Divide out the common factor of $x^2 + 3$.

$$= 1$$ ∎

Self Check   Simplify $\dfrac{a(a + 2) - 2(a - 1)}{a^2 + 2}$.

Answer   1

Sometimes a fraction does not simplify. Such a fraction is already in lowest terms. For example, to attempt to simplify

$$\frac{x^2 + x - 2}{x^2 + x}$$

we factor the numerator and denominator.

$$\frac{x^2 + x - 2}{x^2 + x} = \frac{(x + 2)(x - 1)}{x(x + 1)}$$

Because there are no factors common to the numerator and denominator, this fraction is already in lowest terms.

Orals *Simplify each fraction.*

1. $\dfrac{14}{21}$ 2. $\dfrac{34}{17}$ 3. $\dfrac{xyz}{wxy}$ 4. $\dfrac{8x^2}{4x}$

5. $\dfrac{6x^2y}{6xy^2}$ 6. $\dfrac{x^2y^3}{x^2y^4}$ 7. $\dfrac{x + y}{y + x}$ 8. $\dfrac{x - y}{y - x}$

EXERCISE 6.3

REVIEW

1. State the associative property of addition.

2. State the distributive property.

3. What is the additive identity?

4. What is the multiplicative identity?

5. Find the additive inverse of $-\dfrac{5}{3}$.

6. Find the multiplicative inverse of $-\dfrac{5}{3}$.

VOCABULARY AND CONCEPTS   *Fill in each blank to make a true statement.*

7. In a fraction, the part above the fraction bar is called the _____.

8. In a fraction, the part below the fraction bar is called the _____.

9. The denominator of a fraction cannot be __.

10. A fraction that has polynomials in its numerator and denominator is called a _____ expression.

11. $x - 2$ and $2 - x$ are called _____ of each other.

12. To *simplify* a fraction means to write it in _____ terms.

13. The fundamental property of fractions states that

$$\frac{ac}{bc} = \underline{\quad}.$$

14. Any number x divided by 1 is __.

15. To simplify a fraction, we _____ the numerator and denominator and divide out _____ factors.

16. A fraction cannot be simplified when it is written in _____.

PRACTICE   *In Exercises 17–84, write each fraction in lowest terms. If a fraction is already in lowest terms, so indicate. Assume that no denominators are 0.*

17. $\dfrac{8}{10}$

18. $\dfrac{16}{28}$

19. $\dfrac{28}{35}$

20. $\dfrac{14}{20}$

21. $\dfrac{8}{52}$

22. $\dfrac{15}{21}$

23. $\dfrac{10}{45}$

24. $\dfrac{21}{35}$

25. $\dfrac{-18}{54}$

26. $\dfrac{16}{40}$

27. $\dfrac{4x}{2}$

28. $\dfrac{2x}{4}$

29. $\dfrac{-6x}{18}$

30. $\dfrac{-25y}{5}$

31. $\dfrac{45}{9a}$

32. $\dfrac{48}{16y}$

33. $\dfrac{7+3}{5z}$

34. $\dfrac{(3-18)k}{25}$

35. $\dfrac{(3+4)a}{24-3}$

36. $\dfrac{x+x}{2}$

37. $\dfrac{2x}{3x}$

38. $\dfrac{5y}{7y}$

39. $\dfrac{6x^2}{4x^2}$

40. $\dfrac{9xy}{6xy}$

41. $\dfrac{2x^2}{3y}$

42. $\dfrac{5y^2}{2y^2}$

43. $\dfrac{15x^2y}{5xy^2}$

44. $\dfrac{12xz}{4xz^2}$

45. $\dfrac{28x}{32y}$

46. $\dfrac{14xz^2}{7x^2z^2}$

47. $\dfrac{x+3}{3(x+3)}$

48. $\dfrac{2(x+7)}{x+7}$

49. $\dfrac{5x+35}{x+7}$

50. $\dfrac{x-9}{3x-27}$

51. $\dfrac{x^2+3x}{2x+6}$

52. $\dfrac{xz-2x}{yz-2y}$

53. $\dfrac{15x-3x^2}{25y-5xy}$

54. $\dfrac{3y+xy}{3x+xy}$

55. $\dfrac{6a-6b+6c}{9a-9b+9c}$

56. $\dfrac{3a-3b-6}{2a-2b-4}$

57. $\dfrac{x-7}{7-x}$

58. $\dfrac{d-c}{c-d}$

59. $\dfrac{6x-3y}{3y-6x}$

60. $\dfrac{3c-4d}{4c-3d}$

61. $\dfrac{a+b-c}{c-a-b}$

62. $\dfrac{x-y-z}{z+y-x}$

63. $\dfrac{x^2+3x+2}{x^2+x-2}$

64. $\dfrac{x^2+x-6}{x^2-x-2}$

65. $\dfrac{x^2-8x+15}{x^2-x-6}$

66. $\dfrac{x^2-6x-7}{x^2+8x+7}$

67. $\dfrac{2x^2-8x}{x^2-6x+8}$

68. $\dfrac{3y^2-15y}{y^2-3y-10}$

69. $\dfrac{xy+2x^2}{2xy+y^2}$

70. $\dfrac{3x+3y}{x^2+xy}$

71. $\dfrac{x^2+3x+2}{x^3+x^2}$

72. $\dfrac{6x^2-13x+6}{3x^2+x-2}$

73. $\dfrac{x^2-8x+16}{x^2-16}$

74. $\dfrac{3x+15}{x^2-25}$

75. $\dfrac{2x^2-8}{x^2-3x+2}$

76. $\dfrac{3x^2-27}{x^2+3x-18}$

77. $\dfrac{x^2-2x-15}{x^2+2x-15}$

78. $\dfrac{x^2+4x-77}{x^2-4x-21}$

79. $\dfrac{x^2-3(2x-3)}{9-x^2}$

80. $\dfrac{x(x-8)+16}{16-x^2}$

81. $\dfrac{4(x+3)+4}{3(x+2)+6}$

82. $\dfrac{4+2(x-5)}{3x-5(x-2)}$

83. $\dfrac{x^2 - 9}{(2x + 3) - (x + 6)}$

84. $\dfrac{x^2 + 5x + 4}{2(x + 3) - (x + 2)}$

WRITING

85. Explain why $\dfrac{x - 7}{7 - x} = -1$.

86. Explain why $\dfrac{x + 7}{7 + x} = 1$.

SOMETHING TO THINK ABOUT

87. Exercise 79,

$$\dfrac{x^2 - 3(2x - 3)}{9 - x^2}$$

has two possible answers: $\dfrac{3 - x}{3 + x}$ and $-\dfrac{x - 3}{x + 3}$.

Why is either answer correct?

88. Find two different-looking but correct answers for the following problem.

Simplify $\dfrac{y^2 + 5(2y + 5)}{25 - y^2}$.

6.4 Multiplying and Dividing Fractions

■ MULTIPLYING FRACTIONS ■ MULTIPLYING A FRACTION BY A POLYNOMIAL ■ DIVIDING FRACTIONS ■ DIVIDING A FRACTION BY A POLYNOMIAL ■ COMBINED OPERATIONS

Getting Ready *Multiply the fractions and simplify.*

1. $\dfrac{3}{7} \cdot \dfrac{14}{9}$ **2.** $\dfrac{21}{15} \cdot \dfrac{10}{3}$ **3.** $\dfrac{19}{38} \cdot 6$ **4.** $42 \cdot \dfrac{3}{21}$

5. $\dfrac{4}{9} \cdot \dfrac{45}{8}$ **6.** $\dfrac{11}{7} \cdot \dfrac{14}{22}$ **7.** $\dfrac{75}{12} \cdot \dfrac{6}{50}$ **8.** $\dfrac{13}{5} \cdot \dfrac{20}{26}$

■ MULTIPLYING FRACTIONS

Recall that to multiply fractions, we multiply their numerators and multiply their denominators. For example, to find the product of $\frac{4}{7}$ and $\frac{3}{5}$, we proceed as follows.

$$\dfrac{4}{7} \cdot \dfrac{3}{5} = \dfrac{4 \cdot 3}{7 \cdot 5} \qquad \text{Multiply the numerators and multiply the denominators.}$$

$$= \dfrac{12}{35} \qquad 4 \cdot 3 = 12 \text{ and } 7 \cdot 5 = 35.$$

In general, the following is true.

Rule for Multiplying Fractions

If a, b, c, and d are real numbers and $b \neq 0$ and $d \neq 0$, then

$$\frac{a}{b} \cdot \frac{c}{d} = \frac{ac}{bd}$$

EXAMPLE 1 Multiply **a.** $\dfrac{1}{3} \cdot \dfrac{2}{5}$, **b.** $\dfrac{7}{9} \cdot \dfrac{-5}{3x}$, **c.** $\dfrac{x^2}{2} \cdot \dfrac{3}{y^2}$, and **d.** $\dfrac{t+1}{t} \cdot \dfrac{t-1}{t-2}$.

Solution **a.** $\dfrac{1}{3} \cdot \dfrac{2}{5} = \dfrac{1 \cdot 2}{3 \cdot 5}$ **b.** $\dfrac{7}{9} \cdot \dfrac{-5}{3x} = \dfrac{7(-5)}{9 \cdot 3x}$

$\qquad\qquad = \dfrac{2}{15}$ $\qquad\qquad\qquad\qquad = \dfrac{-35}{27x}$

c. $\dfrac{x^2}{2} \cdot \dfrac{3}{y^2} = \dfrac{x^2 \cdot 3}{2 \cdot y^2}$ **d.** $\dfrac{t+1}{t} \cdot \dfrac{t-1}{t-2} = \dfrac{(t+1)(t-1)}{t(t-2)}$

$\qquad\qquad = \dfrac{3x^2}{2y^2}$ ∎

Self Check Multiply $\dfrac{3x}{4} \cdot \dfrac{p-3}{y}$.

Answer $\dfrac{3x(p-3)}{4y}$

EXAMPLE 2 Multiply $\dfrac{35x^2y}{7y^2z} \cdot \dfrac{z}{5xy}$.

Solution $\dfrac{35x^2y}{7y^2z} \cdot \dfrac{z}{5xy} = \dfrac{35x^2y \cdot z}{7y^2z \cdot 5xy}$ Multiply the numerators and multiply the denominators.

$\qquad = \dfrac{5 \cdot 7 \cdot x \cdot x \cdot y \cdot z}{7 \cdot y \cdot y \cdot z \cdot 5 \cdot x \cdot y}$ Factor.

$\qquad = \dfrac{\overset{1}{\cancel{5}} \cdot \overset{1}{\cancel{7}} \cdot \overset{1}{\cancel{x}} \cdot x \cdot \overset{1}{\cancel{y}} \cdot \overset{1}{\cancel{z}}}{\underset{1}{\cancel{7}} \cdot \underset{1}{\cancel{y}} \cdot y \cdot \underset{1}{\cancel{z}} \cdot \underset{1}{\cancel{5}} \cdot \underset{1}{\cancel{x}} \cdot y}$ Divide out common factors.

$\qquad = \dfrac{x}{y^2}$ ∎

Self Check Multiply $\dfrac{a^2b^2}{2a} \cdot \dfrac{9a^3}{3b^3}$.

Answer $\dfrac{3a^4}{2b}$

EXAMPLE 3

Multiply $\dfrac{x^2 - x}{2x + 4} \cdot \dfrac{x + 2}{x}$.

Solution

$$\dfrac{x^2 - x}{2x + 4} \cdot \dfrac{x + 2}{x} = \dfrac{(x^2 - x)(x + 2)}{(2x + 4)(x)}$$

Multiply the numerators and multiply the denominators.

$$= \dfrac{x(x - 1)(x + 2)}{2(x + 2)x}$$

Factor.

$$= \dfrac{\overset{1}{\cancel{x}}(x - 1)\overset{1}{\cancel{(x + 2)}}}{2\underset{1}{\cancel{(x + 2)}}\underset{1}{\cancel{x}}}$$

Divide out common factors.

$$= \dfrac{x - 1}{2}$$

∎

Self Check

Multiply $\dfrac{x^2 + x}{3x + 6} \cdot \dfrac{x + 2}{x + 1}$.

Answer $\frac{x}{3}$

EXAMPLE 4

Multiply $\dfrac{x^2 - 3x}{x^2 - x - 6}$ and $\dfrac{x^2 + x - 2}{x^2 - x}$.

Solution

$$\dfrac{x^2 - 3x}{x^2 - x - 6} \cdot \dfrac{x^2 + x - 2}{x^2 - x}$$

$$= \dfrac{(x^2 - 3x)(x^2 + x - 2)}{(x^2 - x - 6)(x^2 - x)}$$

Multiply the numerators and multiply the denominators.

$$= \dfrac{x(x - 3)(x + 2)(x - 1)}{(x + 2)(x - 3)x(x - 1)}$$

Factor.

$$= \dfrac{\overset{1}{\cancel{x}}\overset{1}{\cancel{(x - 3)}}\overset{1}{\cancel{(x + 2)}}\overset{1}{\cancel{(x - 1)}}}{\underset{1}{\cancel{(x + 2)}}\underset{1}{\cancel{(x - 3)}}\underset{1}{\cancel{x}}\underset{1}{\cancel{(x - 1)}}}$$

Divide out common factors.

$$= 1$$

∎

Self Check

Multiply $\dfrac{a^2 + a}{a^2 - 4} \cdot \dfrac{a^2 - a - 2}{a^2 + 2a + 1}$.

Answer $\frac{a}{a + 2}$

■ MULTIPLYING A FRACTION BY A POLYNOMIAL

Since any number divided by 1 remains unchanged, we can write any polynomial as a fraction by inserting a denominator of 1.

EXAMPLE 5   Multiply $\dfrac{x^2 + x}{x^2 + 8x + 7} \cdot x + 7.$

Solution $\dfrac{x^2 + x}{x^2 + 8x + 7} \cdot (x + 7) = \dfrac{x^2 + x}{x^2 + 8x + 7} \cdot \dfrac{x + 7}{1}$ Write $x + 7$ as a fraction with a denominator of 1.

$= \dfrac{x(x + 1)(x + 7)}{(x + 1)(x + 7)1}$ Multiply the fractions and factor where possible.

$= \dfrac{x\cancel{(x + 1)}\cancel{(x + 7)}}{1\cancel{(x + 1)}\cancel{(x + 7)}}$ Divide out all common factors.

$= x$ ∎

Self Check   Multiply $a - 7 \cdot \dfrac{a^2 - a}{a^2 - 8a + 7}.$

Answer a

■ DIVIDING FRACTIONS

Recall that division by a nonzero number is equivalent to multiplying by the reciprocal of that number. Thus, to divide two fractions, we can invert the **divisor** (the fraction following the ÷ sign) and multiply. For example, to divide $\frac{4}{7}$ by $\frac{3}{5}$, we proceed as follows:

$\dfrac{4}{7} \div \dfrac{3}{5} = \dfrac{4}{7} \cdot \dfrac{5}{3}$ Invert $\frac{3}{5}$ and change the division to a multiplication.

$= \dfrac{20}{21}$ Multiply the numerators and multiply the denominators.

In general, the following is true.

Division of Fractions
If a is a real number and b, c, and d are nonzero real numbers, then

$$\dfrac{a}{b} \div \dfrac{c}{d} = \dfrac{a}{b} \cdot \dfrac{d}{c} = \dfrac{ad}{bc}$$

EXAMPLE 6   Do the divisions: **a.** $\dfrac{7}{13} \div \dfrac{21}{26}$ and   **b.** $\dfrac{-9x}{35y} \div \dfrac{15x^2}{14}.$

Solution **a.** $\dfrac{7}{13} \div \dfrac{21}{26} = \dfrac{7}{13} \cdot \dfrac{26}{21}$ Invert the divisor and multiply.

$$= \dfrac{7 \cdot 2 \cdot 13}{13 \cdot 3 \cdot 7}$$ Multiply the fractions and factor where possible.

$$= \dfrac{\overset{1}{\cancel{7}} \cdot 2 \cdot 13}{13 \cdot 3 \cdot \underset{1}{\cancel{7}}}$$ Divide out common factors.

$$= \dfrac{2}{3}$$

b. $\dfrac{-9x}{35y} \div \dfrac{15x^2}{14} = \dfrac{-9x}{35y} \cdot \dfrac{14}{15x^2}$ Invert the divisor and multiply.

$$= \dfrac{-3 \cdot 3 \cdot x \cdot 2 \cdot 7}{5 \cdot 7 \cdot y \cdot 3 \cdot 5 \cdot x \cdot x}$$ Multiply the fractions and factor where possible.

$$= \dfrac{-3 \cdot \overset{1}{\cancel{3}} \cdot \overset{1}{\cancel{x}} \cdot 2 \cdot \overset{1}{\cancel{7}}}{5 \cdot \underset{1}{\cancel{7}} \cdot y \cdot \underset{1}{\cancel{3}} \cdot 5 \cdot \underset{1}{\cancel{x}} \cdot x}$$ Divide out common factors.

$$= -\dfrac{6}{25xy}$$ Multiply the remaining factors. ∎

Self Check   Divide $\dfrac{-8a}{3b} \div \dfrac{16a^2}{9b^2}$.

Answer $-\frac{3b}{2a}$

EXAMPLE 7   Divide $\dfrac{x^2 + x}{3x - 15} \div \dfrac{x^2 + 2x + 1}{6x - 30}$.

Solution $\dfrac{x^2 + x}{3x - 15} \div \dfrac{x^2 + 2x + 1}{6x - 30}$

$$= \dfrac{x^2 + x}{3x - 15} \cdot \dfrac{6x - 30}{x^2 + 2x + 1}$$ Invert the divisor and multiply.

$$= \dfrac{x(x + 1) \cdot 2 \cdot 3(x - 5)}{3(x - 5)(x + 1)(x + 1)}$$ Multiply the fractions and factor.

$$= \dfrac{x\cancel{(x + 1)} \cdot 2 \cdot \cancel{3}\cancel{(x - 5)}}{\cancel{3}\cancel{(x - 5)}\cancel{(x + 1)}(x + 1)}$$ Divide out all common factors.

$$= \dfrac{2x}{x + 1}$$ ∎

Self Check   Divide $\dfrac{a^2 - 1}{a^2 + 4a + 3} \div \dfrac{a - 1}{a^2 + 2a - 3}$.

Answer $a - 1$

■ DIVIDING A FRACTION BY A POLYNOMIAL

To divide a fraction by a polynomial, we write the polynomial as a fraction by inserting a denominator of 1 and then divide the fractions.

EXAMPLE 8   Divide $\dfrac{2x^2 - 3x - 2}{2x + 1} \div (4 - x^2)$.

Solution $\dfrac{2x^2 - 3x - 2}{2x + 1} \div (4 - x^2)$

$= \dfrac{2x^2 - 3x - 2}{2x + 1} \div \dfrac{4 - x^2}{1}$ Write $4 - x^2$ as a fraction with a denominator of 1.

$= \dfrac{2x^2 - 3x - 2}{2x + 1} \cdot \dfrac{1}{4 - x^2}$ Invert the divisor and multiply.

$= \dfrac{(2x + 1)(x - 2) \cdot 1}{(2x + 1)(2 + x)(2 - x)}$ Multiply the fractions and factor where possible.

$= \dfrac{\overset{1}{\cancel{(2x + 1)}}\overset{-1}{\cancel{(x - 2)}} \cdot 1}{\underset{1}{\cancel{(2x + 1)}}(2 + x)\underset{1}{\cancel{(2 - x)}}}$ Divide out common factors: $\frac{x - 2}{2 - x} = -1$.

$= \dfrac{-1}{2 + x}$

$= -\dfrac{1}{2 + x}$ ■

Self Check   Divide $(b - a) \div \dfrac{a^2 - b^2}{a^2 + ab}$.

Answer $-a$

■ COMBINED OPERATIONS

Unless parentheses indicate otherwise, we do multiplications and divisions in order from left to right.

EXAMPLE 9   Simplify $\dfrac{x^2 - x - 6}{x - 2} \div \dfrac{x^2 - 4x}{x^2 - x - 2} \cdot \dfrac{x - 4}{x^2 + x}$.

Solution   Since there are no parentheses to indicate otherwise, we do the division first.

$$\dfrac{x^2 - x - 6}{x - 2} \div \dfrac{x^2 - 4x}{x^2 - x - 2} \cdot \dfrac{x - 4}{x^2 + x}$$

$$= \dfrac{x^2 - x - 6}{x - 2} \cdot \dfrac{x^2 - x - 2}{x^2 - 4x} \cdot \dfrac{x - 4}{x^2 + x} \qquad \text{Invert the divisor and multiply.}$$

$$= \dfrac{(x + 2)(x - 3)(x + 1)(x - 2)(x - 4)}{(x - 2)x(x - 4)x(x + 1)} \qquad \begin{array}{l}\text{Multiply the fractions and}\\\text{factor.}\end{array}$$

$$= \dfrac{(x + 2)(x - 3)\overset{1}{\cancel{(x + 1)}}\overset{1}{\cancel{(x - 2)}}\overset{1}{\cancel{(x - 4)}}}{\underset{1}{\cancel{(x - 2)}}x\underset{1}{\cancel{(x - 4)}}x\underset{1}{\cancel{(x + 1)}}} \qquad \text{Divide out all common factors.}$$

$$= \dfrac{(x + 2)(x - 3)}{x^2} \qquad \blacksquare$$

Self Check   Simplify $\dfrac{a^2 + ab}{ab - b^2} \cdot \dfrac{a^2 - b^2}{a^2 + ab} \div \dfrac{a + b}{b}$.

Answer   1

EXAMPLE 10   Simplify $\dfrac{x^2 + 6x + 9}{x^2 - 2x}\left(\dfrac{x^2 - 4}{x^2 + 3x} \div \dfrac{x + 2}{x} \right)$.

Solution   We do the division within the parentheses first.

$$\dfrac{x^2 + 6x + 9}{x^2 - 2x}\left(\dfrac{x^2 - 4}{x^2 + 3x} \div \dfrac{x + 2}{x} \right)$$

$$= \dfrac{x^2 + 6x + 9}{x^2 - 2x}\left(\dfrac{x^2 - 4}{x^2 + 3x} \cdot \dfrac{x}{x + 2} \right) \qquad \text{Invert the divisor and multiply.}$$

$$= \dfrac{(x + 3)(x + 3)(x + 2)(x - 2)x}{x(x - 2)x(x + 3)(x + 2)} \qquad \begin{array}{l}\text{Multiply the fractions and factor}\\\text{where possible.}\end{array}$$

$$= \dfrac{\overset{1}{\cancel{(x + 3)}}(x + 3)\overset{1}{\cancel{(x + 2)}}\overset{1}{\cancel{(x - 2)}}\overset{1}{\cancel{x}}}{\underset{1}{\cancel{x}}\underset{1}{\cancel{(x - 2)}}x\underset{1}{\cancel{(x + 3)}}\underset{1}{\cancel{(x + 2)}}} \qquad \text{Divide out all common factors.}$$

$$= \dfrac{x + 3}{x} \qquad \blacksquare$$

Self Check   Simplify $\dfrac{x^2 - 2x}{x^2 + 6x + 9} \div \left(\dfrac{x^2 - 4}{x^2 + 3x} \cdot \dfrac{x}{x + 2}\right)$.

Answer $\frac{x}{x+3}$

Orals   *Do the operations and simplify.*

1. $\dfrac{x}{2} \cdot \dfrac{3}{x}$

2. $\dfrac{x+1}{5} \cdot \dfrac{7}{x+1}$

3. $\dfrac{5}{x+7} \cdot (x+7)$

4. $\dfrac{3}{7} \div \dfrac{3}{7}$

5. $\dfrac{3}{4} \div 3$

6. $(x+1) \div \dfrac{x+1}{x}$

EXERCISE 6.4

REVIEW   *Simplify each expression. Write all answers without using negative exponents.*

1. $2x^3 y^2 (-3x^2 y^4 z)$

2. $\dfrac{8x^4 y^5}{-2x^3 y^2}$

3. $(3y)^{-4}$

4. $(a^{-2} a)^{-3}$

5. $\dfrac{x^{3m}}{x^{4m}}$

6. $(3x^2 y^3)^0$

Do the operations and simplify.

7. $-4(y^3 - 4y^2 + 3y - 2) + 6(-2y^2 + 4) - 4(-2y^3 - y)$

8. $y - 5 \overline{)5y^3 - 3y^2 + 4y - 1}$

VOCABULARY AND CONCEPTS   *Fill in each blank to make a true statement.*

9. In a fraction, the part above the fraction bar is called the _____.

10. In a fraction, the part below the fraction bar is called the _____.

11. To multiply fractions, we multiply their _____ and multiply their _____.

12. $\dfrac{a}{b} \cdot \dfrac{c}{d} = \underline{\quad}$

13. To write a polynomial in fractional form, we insert a denominator of __.

14. $\dfrac{a}{b} \div \dfrac{c}{d} = \dfrac{a}{b} \cdot \underline{\quad}$

15. To divide two fractions, invert the _____ and _____.

16. Unless parentheses indicate otherwise, do multiplications and divisions in order from ____ to _____.

PRACTICE   *In Exercises 17–62, do the multiplications. Simplify answers if possible.*

17. $\dfrac{5}{7} \cdot \dfrac{9}{13}$

18. $\dfrac{2}{7} \cdot \dfrac{5}{11}$

19. $\dfrac{25}{35} \cdot \dfrac{-21}{55}$

20. $-\dfrac{27}{24} \cdot \left(-\dfrac{56}{35}\right)$

21. $\dfrac{2}{3} \cdot \dfrac{15}{2} \cdot \dfrac{1}{7}$

22. $\dfrac{2}{5} \cdot \dfrac{10}{9} \cdot \dfrac{3}{2}$

23. $\dfrac{3x}{y} \cdot \dfrac{y}{2}$

24. $\dfrac{2y}{z} \cdot \dfrac{z}{3}$

25. $\dfrac{5y}{7} \cdot \dfrac{7x}{5z}$

26. $\dfrac{4x}{3y} \cdot \dfrac{3y}{7x}$

27. $\dfrac{7z}{9z} \cdot \dfrac{4z}{2z}$

28. $\dfrac{8z}{2x} \cdot \dfrac{16x}{3x}$

29. $\dfrac{2x^2y}{3xy} \cdot \dfrac{3xy^2}{2}$

30. $\dfrac{2x^2z}{z} \cdot \dfrac{5x}{z}$

31. $\dfrac{8x^2y^2}{4x^2} \cdot \dfrac{2xy}{2y}$

32. $\dfrac{9x^2y}{3x} \cdot \dfrac{3xy}{3y}$

33. $\dfrac{-2xy}{x^2} \cdot \dfrac{3xy}{2}$

34. $\dfrac{-3x}{x^2} \cdot \dfrac{2xz}{3}$

35. $\dfrac{ab^2}{a^2b} \cdot \dfrac{b^2c^2}{abc} \cdot \dfrac{abc^2}{a^3c^2}$

36. $\dfrac{x^3y}{z} \cdot \dfrac{xz^3}{x^2y^2} \cdot \dfrac{yz}{xyz}$

37. $\dfrac{10r^2st^3}{6rs^2} \cdot \dfrac{3r^3t}{2rst} \cdot \dfrac{2s^3t^4}{5s^2t^3}$

38. $\dfrac{3a^3b}{25cd^3} \cdot \dfrac{-5cd^2}{6ab} \cdot \dfrac{10abc^2}{2bc^2d}$

39. $\dfrac{z+7}{7} \cdot \dfrac{z+2}{z}$

40. $\dfrac{a-3}{a} \cdot \dfrac{a+3}{5}$

41. $\dfrac{x-2}{2} \cdot \dfrac{2x}{x-2}$

42. $\dfrac{y+3}{y} \cdot \dfrac{3y}{y+3}$

43. $\dfrac{x+5}{5} \cdot \dfrac{x}{x+5}$

44. $\dfrac{y-9}{y+9} \cdot \dfrac{y}{9}$

45. $\dfrac{(x+1)^2}{x+1} \cdot \dfrac{x+2}{x+1}$

46. $\dfrac{(y-3)^2}{y-3} \cdot \dfrac{y-3}{y-3}$

47. $\dfrac{2x+6}{x+3} \cdot \dfrac{3}{4x}$

48. $\dfrac{3y-9}{y-3} \cdot \dfrac{y}{3y^2}$

49. $\dfrac{x^2-x}{x} \cdot \dfrac{3x-6}{3x-3}$

50. $\dfrac{5z-10}{z+2} \cdot \dfrac{3}{3z-6}$

51. $\dfrac{7y-14}{y-2} \cdot \dfrac{x^2}{7x}$

52. $\dfrac{y^2+3y}{9} \cdot \dfrac{3x}{y+3}$

53. $\dfrac{x^2+x-6}{5x} \cdot \dfrac{5x-10}{x+3}$

54. $\dfrac{z^2+4z-5}{5z-5} \cdot \dfrac{5z}{z+5}$

55. $\dfrac{m^2-2m-3}{2m+4} \cdot \dfrac{m^2-4}{m^2+3m+2}$

56. $\dfrac{p^2-p-6}{3p-9} \cdot \dfrac{p^2-9}{p^2+6p+9}$

57. $\dfrac{x^2+7xy+12y^2}{x^2+2xy-8y^2} \cdot \dfrac{x^2-xy-2y^2}{x^2+4xy+3y^2}$

58. $\dfrac{m^2+9mn+20n^2}{m^2-25n^2} \cdot \dfrac{m^2-9mn+20n^2}{m^2-16n^2}$

59. $\dfrac{abc^2}{a+1} \cdot \dfrac{c}{a^2b^2} \cdot \dfrac{a^2+a}{ac}$

60. $\dfrac{x^3yz^2}{4x+8} \cdot \dfrac{x^2-4}{2x^2y^2z^2} \cdot \dfrac{8yz}{x-2}$

61. $\dfrac{3x^2+5x+2}{x^2-9} \cdot \dfrac{x-3}{x^2-4} \cdot \dfrac{x^2+5x+6}{6x+4}$

62. $\dfrac{x^2-25}{3x+6} \cdot \dfrac{x^2+x-2}{2x+10} \cdot \dfrac{6x}{3x^2-18x+15}$

In Exercises 63–92, do each division. Simplify answers when possible.

63. $\dfrac{1}{3} \div \dfrac{1}{2}$

64. $\dfrac{3}{4} \div \dfrac{1}{3}$

65. $\dfrac{21}{14} \div \dfrac{5}{2}$

66. $\dfrac{14}{3} \div \dfrac{10}{3}$

67. $\dfrac{2}{y} \div \dfrac{4}{3}$

68. $\dfrac{3}{a} \div \dfrac{a}{9}$

69. $\dfrac{3x}{2} \div \dfrac{x}{2}$

70. $\dfrac{y}{6} \div \dfrac{2}{3y}$

71. $\dfrac{3x}{y} \div \dfrac{2x}{4}$

72. $\dfrac{3y}{8} \div \dfrac{2y}{4y}$

73. $\dfrac{4x}{3x} \div \dfrac{2y}{9y}$

74. $\dfrac{14}{7y} \div \dfrac{10}{5z}$

75. $\dfrac{x^2}{3} \div \dfrac{2x}{4}$

76. $\dfrac{z^2}{z} \div \dfrac{z}{3z}$

77. $\dfrac{x^2y}{3xy} \div \dfrac{xy^2}{6y}$

78. $\dfrac{2xz}{z} \div \dfrac{4x^2}{z^2}$

79. $\dfrac{x+2}{3x} \div \dfrac{x+2}{2}$

80. $\dfrac{z-3}{3z} \div \dfrac{z+3}{z}$

81. $\dfrac{(z-2)^2}{3z^2} \div \dfrac{z-2}{6z}$

82. $\dfrac{(x+7)^2}{x+7} \div \dfrac{(x-3)^2}{x+7}$

83. $\dfrac{(z-7)^2}{z+2} \div \dfrac{z(z-7)}{5z^2}$

84. $\dfrac{y(y+2)}{y^2(y-3)} \div \dfrac{y^2(y+2)}{(y-3)^2}$

85. $\dfrac{x^2-4}{3x+6} \div \dfrac{x-2}{x+2}$

86. $\dfrac{x^2-9}{5x+15} \div \dfrac{x-3}{x+3}$

87. $\dfrac{x^2-1}{3x-3} \div \dfrac{x+1}{3}$

88. $\dfrac{x^2-16}{x-4} \div \dfrac{3x+12}{x}$

89. $\dfrac{5x^2+13x-6}{x+3} \div \dfrac{5x^2-17x+6}{x-2}$

90. $\dfrac{x^2-x-6}{2x^2+9x+10} \div \dfrac{x^2-25}{2x^2+15x+25}$

91. $\dfrac{2x^2+8x-42}{x-3} \div \dfrac{2x^2+14x}{x^2+5x}$

92. $\dfrac{x^2-2x-35}{3x^2+27x} \div \dfrac{x^2+7x+10}{6x^2+12x}$

In Exercises 93–106, do the operations.

93. $\dfrac{x}{3} \cdot \dfrac{9}{4} \div \dfrac{x^2}{6}$

94. $\dfrac{y^2}{2} \div \dfrac{4}{y} \cdot \dfrac{y^2}{8}$

95. $\dfrac{x^2}{18} \div \dfrac{x^3}{6} \div \dfrac{12}{x^2}$

96. $\dfrac{y^3}{3y} \cdot \dfrac{3y^2}{4} \div \dfrac{15}{20}$

97. $\dfrac{x^2-1}{x^2-9} \cdot \dfrac{x+3}{x+2} \div \dfrac{5}{x+2}$

98. $\dfrac{2}{3x-3} \div \dfrac{2x+2}{x-1} \cdot \dfrac{5}{x+1}$

99. $\dfrac{x^2-4}{2x+6} \div \dfrac{x+2}{4} \cdot \dfrac{x+3}{x-2}$

100. $\dfrac{x^2-5x}{x+1} \cdot \dfrac{x+1}{x^2+3x} \div \dfrac{x-5}{x-3}$

101. $\dfrac{x-x^2}{x^2-4}\left(\dfrac{2x+4}{x+2} \div \dfrac{5}{x+2}\right)$

102. $\dfrac{2}{3x-3} \div \left(\dfrac{2x+2}{x-1} \cdot \dfrac{5}{x+1}\right)$

103. $\dfrac{y^2}{x+1} \cdot \dfrac{x^2+2x+1}{x^2-1} \div \dfrac{3y}{xy-y}$

104. $\dfrac{x^2-y^2}{x^4-x^3} \div \dfrac{x-y}{x^2} \div \dfrac{x^2+2xy+y^2}{x+y}$

105. $\dfrac{x^2+x-6}{x^2-4} \cdot \dfrac{x^2+2x}{x-2} \div \dfrac{x^2+3x}{x+2}$

106. $\dfrac{x^2-x-6}{x^2+6x-7} \cdot \dfrac{x^2+x-2}{x^2+2x} \div \dfrac{x^2+7x}{x^2-3x}$

WRITING

107. Explain how to multiply two fractions and how to simplify the result.

108. Explain why any mathematical expression can be written as a fraction.

109. To divide fractions, you must first know how to multiply fractions. Explain.

110. Explain how to do the division $\dfrac{a}{b} \div \dfrac{c}{d} \div \dfrac{e}{f}$.

SOMETHING TO THINK ABOUT

111. Let x equal a number of your choosing. Without simplifying first, use a calculator to evaluate

$$\frac{x^2 + x - 6}{x^2 + 3x} \cdot \frac{x^2}{x - 2}$$

Try again, with a different value of x. If you were to simplify the expression, what do you think you would get?

112. Simplify the expression in Exercise 111 to determine whether your guess was correct.

6.5 Adding and Subtracting Fractions

■ ADDING FRACTIONS WITH LIKE DENOMINATORS ■ SUBTRACTING FRACTIONS WITH LIKE DENOMINATORS ■ COMBINED OPERATIONS ■ THE LCD ■ ADDING FRACTIONS WITH UNLIKE DENOMINATORS ■ SUBTRACTING FRACTIONS WITH UNLIKE DENOMINATORS ■ COMBINED OPERATIONS

Getting Ready *Add the fractions and simplify.*

1. $\dfrac{1}{5} + \dfrac{3}{5}$ **2.** $\dfrac{3}{7} + \dfrac{4}{7}$ **3.** $\dfrac{3}{8} + \dfrac{4}{8}$ **4.** $\dfrac{18}{19} + \dfrac{20}{19}$

Subtract the fractions and simplify.

5. $\dfrac{5}{9} - \dfrac{4}{9}$ **6.** $\dfrac{7}{12} - \dfrac{1}{12}$ **7.** $\dfrac{7}{13} - \dfrac{9}{13}$ **8.** $\dfrac{20}{10} - \dfrac{7}{10}$

■ ADDING FRACTIONS WITH LIKE DENOMINATORS

To add fractions with a common denominator, we add their numerators and keep the common denominator. For example,

$$\frac{2x}{7} + \frac{3x}{7} = \frac{2x + 3x}{7} \qquad \text{Add the numerators and keep the common denominator.}$$

$$= \frac{5x}{7} \qquad 2x + 3x = 5x.$$

In general, we have the following result.

Adding Fractions with Like Denominators

If a, b, and d represent real numbers, then

$$\frac{a}{d} + \frac{b}{d} = \frac{a+b}{d} \quad (d \neq 0)$$

EXAMPLE 1

Do each addition.

a. $\dfrac{xy}{8z} + \dfrac{3xy}{8z} = \dfrac{xy + 3xy}{8z}$ Add the numerators and keep the common denominator.

$= \dfrac{4xy}{8z}$ Combine like terms.

$= \dfrac{xy}{2z}$ $\frac{4xy}{8z} = \frac{4xy}{4 \cdot 2z} = \frac{xy}{2z}$, because $\frac{4}{4} = 1$.

b. $\dfrac{3x+y}{5x} + \dfrac{x+y}{5x} = \dfrac{3x + y + x + y}{5x}$ Add the numerators and keep the common denominator.

$= \dfrac{4x + 2y}{5x}$ Combine like terms.

∎

Self Check   Add **a.** $\frac{x}{7} + \frac{y}{7}$ and **b.** $\frac{3x}{7y} + \frac{4x}{7y}$.

Answers   **a.** $\frac{x+y}{7}$, **b.** $\frac{x}{y}$

EXAMPLE 2   Add $\dfrac{3x + 21}{5x + 10} + \dfrac{8x + 1}{5x + 10}$.

Solution   Because the fractions have the same denominator, we add their numerators and keep the common denominator.

$$\frac{3x + 21}{5x + 10} + \frac{8x + 1}{5x + 10} = \frac{3x + 21 + 8x + 1}{5x + 10}$$ Add the fractions.

$$= \frac{11x + 22}{5x + 10}$$ Combine like terms.

$$= \frac{11\overset{1}{\cancel{(x+2)}}}{5\underset{1}{\cancel{(x+2)}}}$$ Factor and divide out the common factor of $x + 2$.

$$= \frac{11}{5}$$

∎

Self Check   Add $\frac{x+4}{6x-12} + \frac{x-8}{6x-12}$.

Answer $\frac{1}{3}$

■ SUBTRACTING FRACTIONS WITH LIKE DENOMINATORS

To subtract fractions with a common denominator, we subtract their numerators and keep the common denominator.

> **Subtracting Fractions with Like Denominators**
> If a, b, and d represent real numbers, then
> $$\frac{a}{d} - \frac{b}{d} = \frac{a - b}{d} \quad (d \neq 0)$$

b

EXAMPLE 3   Subtract   **a.** $\dfrac{5x}{3} - \dfrac{2x}{3}$ and   **b.** $\dfrac{5x + 1}{x - 3} - \dfrac{4x - 2}{x - 3}$.

Solution   In each part, the fractions have the same denominator. To subtract them, we subtract their numerators and keep the common denominator.

a. $\dfrac{5x}{3} - \dfrac{2x}{3} = \dfrac{5x - 2x}{3}$ Subtract the fractions.

$\phantom{\dfrac{5x}{3} - \dfrac{2x}{3}} = \dfrac{3x}{3}$ Combine like terms.

$\phantom{\dfrac{5x}{3} - \dfrac{2x}{3}} = \dfrac{x}{1}$ $\frac{3}{3} = 1$.

$\phantom{\dfrac{5x}{3} - \dfrac{2x}{3}} = x$ Denominators of 1 need not be written.

b. $\dfrac{5x + 1}{x - 3} - \dfrac{4x - 2}{x - 3} = \dfrac{(5x + 1) - (4x - 2)}{x - 3}$ Subtract the fractions.

$\phantom{\dfrac{5x + 1}{x - 3} - \dfrac{4x - 2}{x - 3}} = \dfrac{5x + 1 - 4x + 2}{x - 3}$ Remove parentheses.

$\phantom{\dfrac{5x + 1}{x - 3} - \dfrac{4x - 2}{x - 3}} = \dfrac{x + 3}{x - 3}$ Combine like terms.   ■

Self Check   Subtract $\dfrac{2y + 1}{y + 5} - \dfrac{y - 4}{y + 5}$.

Answer   1

■ COMBINED OPERATIONS

To add and/or subtract three or more fractions, we follow the rules for order of operations.

EXAMPLE 4   Simplify $\dfrac{3x + 1}{x - 7} - \dfrac{5x + 2}{x - 7} + \dfrac{2x + 1}{x - 7}$.

Solution   This example involves both addition and subtraction of fractions. Unless parentheses indicate otherwise, we do additions and subtractions from left to right.

$$\frac{3x+1}{x-7} - \frac{5x+2}{x-7} + \frac{2x+1}{x-7}$$

$$= \frac{(3x+1)-(5x+2)+(2x+1)}{x-7}$$ Combine the numerators and keep the common denominator.

$$= \frac{3x+1-5x-2+2x+1}{x-7}$$ Remove parentheses.

$$= \frac{0}{x-7}$$ Combine like terms.

$$= 0$$ Simplify.   ∎

Self Check   Simplify $\frac{2a-3}{a-5} + \frac{3a+2}{a-5} - \frac{24}{a-5}$.

Answer   5

Example 4 illustrates that if the numerator of a fraction is 0, its value is 0.

■ THE LCD

Since the denominators of the fractions in the addition $\frac{4}{7} + \frac{3}{5}$ are different, we cannot add the fractions in their present form.

four-sevenths   +   three-fifths
└── Different denominators ──┘

To add these fractions, we need to find a common denominator. The smallest common denominator (called the **least** or **lowest common denominator**) is the easiest one to work with.

> **Least Common Denominator**
> The **least common denominator (LCD)** for a set of fractions is the smallest number that each denominator will divide exactly.

In the addition $\frac{4}{7} + \frac{3}{5}$, the denominators are 7 and 5. The smallest number that 7 and 5 will divide evenly is 35. This is the LCD. We now build each fraction into a fraction with a denominator of 35.

$$\frac{4}{7} + \frac{3}{5} = \frac{4 \cdot 5}{7 \cdot 5} + \frac{3 \cdot 7}{5 \cdot 7}$$ Multiply numerator and denominator of $\frac{4}{7}$ by 5, and multiply numerator and denominator of $\frac{3}{5}$ by 7.

$$= \frac{20}{35} + \frac{21}{35}$$ Do the multiplications.

Now that the fractions have a common denominator, we can add them.

$$\frac{20}{35} + \frac{21}{35} = \frac{20 + 21}{35} = \frac{41}{35}$$

EXAMPLE 5   Change   **a.** $\dfrac{1}{2y}$,   **b.** $\dfrac{3y}{5}$,   and   **c.** $\dfrac{7x}{10y}$ into fractions with a common denominator of $30y$.

Solution   To build each fraction, we multiply the numerator and denominator by what it takes to make the denominator $30y$.

    a. $\dfrac{1}{2y} = \dfrac{1 \cdot \mathbf{15}}{2y \cdot \mathbf{15}} = \dfrac{15}{30y}$

    b. $\dfrac{3y}{5} = \dfrac{3y \cdot \mathbf{6y}}{5 \cdot \mathbf{6y}} = \dfrac{18y^2}{30y}$

    c. $\dfrac{7x}{10y} = \dfrac{7x \cdot \mathbf{3}}{10y \cdot \mathbf{3}} = \dfrac{21x}{30y}$ ∎

Self Check   Change $\frac{5a}{6b}$ into a fraction with a denominator of $30ab$.

Answer $\dfrac{25a^2}{30ab}$

There is a process that we can use to find the least common denominator of several fractions.

Finding the Least Common Denominator (LCD)

1. List the different denominators that appear in the fractions.

2. Completely factor each denominator.

3. Form a product using each different factor obtained in Step 2. Use each different factor the *greatest* number of times it appears in any one factorization. The product formed by multiplying these factors is the LCD.

EXAMPLE 6   Find the LCD of $\dfrac{5a}{24b}$, $\dfrac{11a}{18b}$, and $\dfrac{35a}{36b}$.

Solution   We list and factor each denominator into the product of prime numbers.

$$24b = 2 \cdot 2 \cdot 2 \cdot 3 \cdot b = 2^3 \cdot 3 \cdot b$$
$$18b = 2 \cdot 3 \cdot 3 \cdot b = 2 \cdot 3^2 \cdot b$$
$$36b = 2 \cdot 2 \cdot 3 \cdot 3 \cdot b = 2^2 \cdot 3^2 \cdot b$$

We then form a product with factors of 2, 3, and b. To find the LCD, we use each of these factors the greatest number of times it appears in any one factorization. We use 2 three times, because it appears three times as a factor of 24. We use 3 twice, because it occurs twice as a factor of 18 and 36. We use b once.

$$\begin{aligned} \text{LCD} &= 2 \cdot 2 \cdot 2 \cdot 3 \cdot 3 \cdot b \\ &= 8 \cdot 9 \cdot b \\ &= 72b \end{aligned}$$
∎

Self Check   Find the LCD of $\frac{3y}{28z}$ and $\frac{5x}{21z}$.

Answer $84z$

■ ADDING FRACTIONS WITH UNLIKE DENOMINATORS

The following list of steps summarizes how to add fractions that have unlike denominators.

> **Adding Fractions with Unlike Denominators**
> To add fractions with different denominators,
> **1.** Find the LCD.
> **2.** Write each fraction as a fraction with a denominator that is the LCD.
> **3.** Add the resulting fractions and simplify the result, if possible.

To add $\frac{4x}{7}$ and $\frac{3x}{5}$, we first find the LCD, which is 35. We then build the fractions so that each one has a denominator of 35. Finally, we add the resulting fractions.

$$\begin{aligned} \frac{4x}{7} + \frac{3x}{5} &= \frac{4x \cdot 5}{7 \cdot 5} + \frac{3x \cdot 7}{5 \cdot 7} \qquad && \text{Multiply numerator and denominator of } \tfrac{4x}{7} \text{ by 5} \\ && &\text{and numerator and denominator of } \tfrac{3x}{5} \text{ by 7.} \\ &= \frac{20x}{35} + \frac{21x}{35} && \text{Do the multiplications.} \\ &= \frac{41x}{35} && \text{Add the numerators and keep the common} \\ &&& \text{denominator.} \end{aligned}$$

    EXAMPLE 7   Add $\dfrac{5a}{24b}, \dfrac{11a}{18b},$ and $\dfrac{35a}{36b}$.

Solution   In Example 6, we saw that the LCD of these fractions is $2 \cdot 2 \cdot 2 \cdot 3 \cdot 3 \cdot b = 72b$. To add the fractions, we first factor each denominator:

$$\frac{5a}{24b} + \frac{11a}{18b} + \frac{35a}{36b} = \frac{5a}{2 \cdot 2 \cdot 2 \cdot 3 \cdot b} + \frac{11a}{2 \cdot 3 \cdot 3 \cdot b} + \frac{35a}{2 \cdot 2 \cdot 3 \cdot 3 \cdot b}$$

In each resulting fraction, we multiply the numerator and the denominator by whatever it takes to build the denominator to the lowest common denominator of $2 \cdot 2 \cdot 2 \cdot 3 \cdot 3 \cdot b$.

$$= \frac{5a \cdot 3}{2 \cdot 2 \cdot 2 \cdot 3 \cdot b \cdot 3} + \frac{11a \cdot 2 \cdot 2}{2 \cdot 3 \cdot 3 \cdot b \cdot 2 \cdot 2} + \frac{35a \cdot 2}{2 \cdot 2 \cdot 3 \cdot 3 \cdot b \cdot 2}$$

$$= \frac{15a + 44a + 70a}{72b} \qquad \text{Do the multiplications and add the fractions.}$$

$$= \frac{129a}{72b} \qquad \text{Simplify.} \qquad \blacksquare$$

Self Check   Add $\frac{3y}{28z} + \frac{5x}{21z}$.

Answer $\frac{9y + 20x}{84z}$

EXAMPLE 8   Add $\dfrac{5y}{14x} + \dfrac{2y}{21x}$.

Solution   We first find the LCD.

$$\left. \begin{array}{l} 14x = 2 \cdot 7 \cdot x \\ 21x = 3 \cdot 7 \cdot x \end{array} \right\} \; \text{LCD} = 2 \cdot 3 \cdot 7 \cdot x = 42x$$

We then build the fractions so that each one has a denominator of $42x$.

$$\frac{5y}{14x} + \frac{2y}{21x} = \frac{5y \cdot 3}{14x \cdot 3} + \frac{2y \cdot 2}{21x \cdot 2} \qquad \begin{array}{l} \text{Multiply the numerator and denominator} \\ \text{of } \frac{5y}{14x} \text{ by 3 and those of } \frac{2y}{21x} \text{ by 2.} \end{array}$$

$$= \frac{15y}{42x} + \frac{4y}{42x} \qquad \text{Do the multiplications.}$$

$$= \frac{19y}{42x} \qquad \text{Add the fractions.} \qquad \blacksquare$$

Self Check   Add $\frac{3y}{4x} + \frac{2y}{3x}$.

Answer $\frac{17y}{12x}$

EXAMPLE 9   Add $\dfrac{1}{x} + \dfrac{x}{y}$.

Solution   By inspection, the LCD is xy.

$$\dfrac{1}{x} + \dfrac{x}{y} = \dfrac{1(y)}{x(y)} + \dfrac{(x)x}{(x)y}$$ Build the fractions to get the common denominator of xy.

$$= \dfrac{y}{xy} + \dfrac{x^2}{xy}$$ Do the multiplications.

$$= \dfrac{y + x^2}{xy}$$ Add the fractions.   ∎

Self Check   Add $\dfrac{a}{b} + \dfrac{3}{a}$.

Answer $\dfrac{a^2 + 3b}{ab}$

■ SUBTRACTING FRACTIONS WITH UNLIKE DENOMINATORS

To subtract fractions with unlike denominators, we first change them into fractions with the same denominator.

EXAMPLE 10   Subtract $\dfrac{x}{x + 1} - \dfrac{3}{x}$.

Solution   By inspection, the least common denominator is $(x + 1)x$.

$$\dfrac{x}{x + 1} - \dfrac{3}{x} = \dfrac{x(x)}{(x + 1)x} - \dfrac{3(x + 1)}{x(x + 1)}$$ Build the fractions to get the common denominator.

$$= \dfrac{x(x) - 3(x + 1)}{(x + 1)x}$$ Subtract the numerators and keep the common denominator.

$$= \dfrac{x^2 - 3x - 3}{(x + 1)x}$$ Do the multiplications in the numerator.   ∎

Self Check   Subtract $\dfrac{a}{a - 1} - \dfrac{5}{b}$.

Answer $\dfrac{ab - 5a + 5}{(a - 1)b}$

EXAMPLE 11   Subtract $\dfrac{a}{a - 1} - \dfrac{2}{a^2 - 1}$.

Solution   We factor $a^2 - 1$ and discover that the LCD is $(a + 1)(a - 1)$.

$$\frac{a}{a - 1} - \frac{2}{a^2 - 1}$$

$$= \frac{a(a + 1)}{(a - 1)(a + 1)} - \frac{2}{(a + 1)(a - 1)}$$ Build the first fraction and factor the denominator of the second fraction.

$$= \frac{a(a + 1) - 2}{(a - 1)(a + 1)}$$ Subtract the numerators and keep the common denominator.

$$= \frac{a^2 + a - 2}{(a - 1)(a + 1)}$$ Remove parentheses.

$$= \frac{(a + 2)\overset{1}{\cancel{(a - 1)}}}{\underset{1}{\cancel{(a - 1)}}(a + 1)}$$ Factor.

$$= \frac{a + 2}{a + 1}$$ Divide out the common factor of $a - 1$.   ∎

Self Check   Subtract $\dfrac{b}{b + 1} - \dfrac{3}{b^2 - 1}$.

Answer $\dfrac{b^2 - b - 3}{(b + 1)(b - 1)}$

EXAMPLE 12   Subtract $\dfrac{3}{x - y} - \dfrac{x}{y - x}$.

Solution   We note that the second denominator is the negative of the first. So we can multiply the numerator and denominator of the second fraction by -1 to get

$$\frac{3}{x - y} - \frac{x}{y - x} = \frac{3}{x - y} - \frac{-1x}{-1(y - x)}$$ Multiply numerator and denominator by -1.

$$= \frac{3}{x - y} - \frac{-x}{-y + x}$$ Remove parentheses.

$$= \frac{3}{x - y} - \frac{-x}{x - y}$$ $-y + x = x - y$.

$$= \frac{3 - (-x)}{x - y}$$ Subtract the numerators and keep the common denominator.

$$= \frac{3 + x}{x - y}$$ $-(-x) = x$.   ∎

Self Check   Subtract $\frac{5}{a - b} - \frac{2}{b - a}$.

Answer $\frac{7}{a - b}$

■ COMBINED OPERATIONS

To add and/or subtract three or more fractions, we follow the rules for order of operations.

EXAMPLE 13 Do the operations: $\dfrac{3}{x^2y} + \dfrac{2}{xy} - \dfrac{1}{xy^2}$.

Solution Find the least common denominator.

$$\left.\begin{array}{l} x^2y = x \cdot x \cdot y \\ xy = x \cdot y \\ xy^2 = x \cdot y \cdot y \end{array}\right\} \quad \text{Factor each denominator.}$$

In any one of these denominators, the factor x occurs at most twice, and the factor y occurs at most twice. Thus,

$$\begin{aligned} \text{LCD} &= x \cdot x \cdot y \cdot y \\ &= x^2y^2 \end{aligned}$$

We build each fraction into a fraction with a denominator of x^2y^2.

$$\dfrac{3}{x^2y} + \dfrac{2}{xy} - \dfrac{1}{xy^2}$$

$$= \dfrac{3 \cdot y}{x \cdot x \cdot y \cdot y} + \dfrac{2 \cdot x \cdot y}{x \cdot y \cdot x \cdot y} - \dfrac{1 \cdot x}{x \cdot y \cdot y \cdot x} \qquad \begin{array}{l}\text{Factor each denominator}\\\text{and build each fraction.}\end{array}$$

$$= \dfrac{3y + 2xy - x}{x^2y^2} \qquad \begin{array}{l}\text{Do the multiplications and}\\\text{combine the numerators.} ■\end{array}$$

Self Check Combine $\dfrac{5}{ab^2} - \dfrac{b}{a} + \dfrac{a}{b}$.

Answer $\dfrac{5 - b^3 + a^2b}{ab^2}$

EXAMPLE 14 Do the operations: $\dfrac{3}{x^2 - y^2} + \dfrac{2}{x - y} - \dfrac{1}{x + y}$.

Solution Find the least common denominator.

$$\left.\begin{array}{l} x^2 - y^2 = (x - y)(x + y) \\ x - y = x - y \\ x + y = x + y \end{array}\right\} \quad \text{Factor each denominator, where possible.}$$

Since the least common denominator is $(x - y)(x + y)$, we build each fraction into a new fraction with that common denominator.

$$\frac{3}{x^2 - y^2} + \frac{2}{x - y} - \frac{1}{x + y}$$

$$= \frac{3}{(x - y)(x + y)} + \frac{2}{x - y} - \frac{1}{x + y} \qquad \text{Factor.}$$

$$= \frac{3}{(x - y)(x + y)} + \frac{2(x + y)}{(x - y)(x + y)} - \frac{1(x - y)}{(x + y)(x - y)} \qquad \begin{array}{l}\text{Build each} \\ \text{fraction to get} \\ \text{a common} \\ \text{denominator.}\end{array}$$

$$= \frac{3 + 2(x + y) - 1(x - y)}{(x - y)(x + y)} \qquad \begin{array}{l}\text{Combine the} \\ \text{numerators} \\ \text{and keep the} \\ \text{common} \\ \text{denominator.}\end{array}$$

$$= \frac{3 + 2x + 2y - x + y}{(x - y)(x + y)} \qquad \begin{array}{l}\text{Remove} \\ \text{parentheses.}\end{array}$$

$$= \frac{3 + x + 3y}{(x - y)(x + y)} \qquad \begin{array}{l}\text{Combine like} \\ \text{terms.} \quad \blacksquare\end{array}$$

Self Check Combine $\dfrac{5}{a^2 - b^2} - \dfrac{3}{a + b} + \dfrac{4}{a - b}$.

Answer $\frac{a + 7b + 5}{(a + b)(a - b)}$

Orals *Indicate whether the fractions are equal.*

1. $\dfrac{1}{2}, \dfrac{6}{12}$ 2. $\dfrac{3}{8}, \dfrac{15}{40}$ 3. $\dfrac{7}{9}, \dfrac{14}{27}$ 4. $\dfrac{5}{10}, \dfrac{15}{30}$

5. $\dfrac{x}{3}, \dfrac{3x}{9}$ 6. $\dfrac{5}{3}, \dfrac{5x}{3y}$ 7. $\dfrac{5}{3}, \dfrac{5x}{3x}$ 8. $\dfrac{5y}{10}, \dfrac{y}{2}$

EXERCISE 6.5

REVIEW *Write each number in prime-factored form.*

1. 49   2. 64   3. 136   4. 242

5. 102   6. 315   7. 144   8. 145

VOCABULARY AND CONCEPTS   *Fill in each blank to make a true statement.*

9. The _____ for a set of fractions is the smallest number that each denominator divides exactly.

10. When we multiply the numerator and denominator of a fraction by some number to get a common denominator, we say that we are _____ the fraction.

11. To add two fractions with like denominators, we add their _____ and keep the _____.

12. To subtract two fractions with _____ denominators, we need to find a common denominator.

PRACTICE   *In Exercises 13–24, do each addition. Simplify answers, if possible.*

13. $\dfrac{1}{3} + \dfrac{1}{3}$

14. $\dfrac{3}{4} + \dfrac{3}{4}$

15. $\dfrac{2}{9} + \dfrac{1}{9}$

16. $\dfrac{5}{7} + \dfrac{9}{7}$

17. $\dfrac{2x}{y} + \dfrac{2x}{y}$

18. $\dfrac{4y}{3x} + \dfrac{2y}{3x}$

19. $\dfrac{4}{7y} + \dfrac{10}{7y}$

20. $\dfrac{x^2}{4y} + \dfrac{x^2}{4y}$

21. $\dfrac{y+2}{5z} + \dfrac{y+4}{5z}$

22. $\dfrac{x+3}{x^2} + \dfrac{x+5}{x^2}$

23. $\dfrac{3x-5}{x-2} + \dfrac{6x-13}{x-2}$

24. $\dfrac{8x-7}{x+3} + \dfrac{2x+37}{x+3}$

In Exercises 25–36, do each subtraction. Simplify answers, if possible.

25. $\dfrac{5}{7} - \dfrac{4}{7}$

26. $\dfrac{5}{9} - \dfrac{3}{9}$

27. $\dfrac{35}{72} - \dfrac{44}{72}$

28. $\dfrac{35}{99} - \dfrac{13}{99}$

29. $\dfrac{2x}{y} - \dfrac{x}{y}$

30. $\dfrac{7y}{5} - \dfrac{4y}{5}$

31. $\dfrac{9y}{3x} - \dfrac{6y}{3x}$

32. $\dfrac{5r^2}{2r} - \dfrac{r^2}{2r}$

33. $\dfrac{6x-5}{3xy} - \dfrac{3x-5}{3xy}$

34. $\dfrac{7x+7}{5y} - \dfrac{2x+7}{5y}$

35. $\dfrac{3y-2}{y+3} - \dfrac{2y-5}{y+3}$

36. $\dfrac{5x+8}{x+5} - \dfrac{3x-2}{x+5}$

In Exercises 37–44, do the operations. Simplify answers, if possible.

37. $\dfrac{13x}{15} + \dfrac{12x}{15} - \dfrac{5x}{15}$

38. $\dfrac{13y}{32} + \dfrac{13y}{32} - \dfrac{10y}{32}$

39. $\dfrac{x}{3y} + \dfrac{2x}{3y} - \dfrac{x}{3y}$

40. $\dfrac{5y}{8x} + \dfrac{4y}{8x} - \dfrac{y}{8x}$

41. $\dfrac{3x}{y+2} - \dfrac{3y}{y+2} + \dfrac{x+y}{y+2}$

42. $\dfrac{3y}{x-5} + \dfrac{x}{x-5} - \dfrac{y-x}{x-5}$

43. $\dfrac{x+1}{x-2} - \dfrac{2(x-3)}{x-2} + \dfrac{3(x+1)}{x-2}$

44. $\dfrac{3xy}{x-y} - \dfrac{x(3y-x)}{x-y} - \dfrac{x(x-y)}{x-y}$

In Exercises 45–56, build each fraction into an equivalent fraction with the indicated denominator.

45. $\dfrac{25}{4}$; 20

46. $\dfrac{5}{y}$; xy

47. $\dfrac{8}{x}$; x^2y

48. $\dfrac{7}{y}$; xy^2

49. $\dfrac{3x}{x+1}$; $(x+1)^2$

50. $\dfrac{5y}{y-2}$; $(y-2)^2$

51. $\dfrac{2y}{x}$; x^2+x

52. $\dfrac{3x}{y}$; y^2-y

53. $\dfrac{z}{z-1}$; $z^2 - 1$ **54.** $\dfrac{y}{y+2}$; $y^2 - 4$ **55.** $\dfrac{2}{x+1}$; $x^2 + 3x + 2$ **56.** $\dfrac{3}{x-1}$; $x^2 + x - 2$

In Exercises 57–66, several denominators are given. Find the LCD.

57. $2x$, $6x$

58. $3y$, $9y$

59. $3x$, $6y$, $9xy$

60. $2x^2$, $6y$, $3xy$

61. $x^2 - 1$, $x + 1$

62. $y^2 - 9$, $y - 3$

63. $x^2 + 6x$, $x + 6$, x

64. $xy^2 - xy$, xy, $y - 1$

65. $x^2 - 4x - 5$, $x^2 - 25$

66. $x^2 - x - 6$, $x^2 - 9$

In Exercises 67–96, do the operations. Simplify answers, if possible.

67. $\dfrac{1}{2} + \dfrac{2}{3}$

68. $\dfrac{2}{3} - \dfrac{5}{6}$

69. $\dfrac{2y}{9} + \dfrac{y}{3}$

70. $\dfrac{8a}{15} - \dfrac{5a}{12}$

71. $\dfrac{21x}{14} - \dfrac{5x}{21}$

72. $\dfrac{7y}{6} + \dfrac{10y}{9}$

73. $\dfrac{4x}{3} + \dfrac{2x}{y}$

74. $\dfrac{2y}{5x} - \dfrac{y}{2}$

75. $\dfrac{2}{x} - 3x$

76. $14 + \dfrac{10}{y^2}$

77. $\dfrac{y+2}{5y} + \dfrac{y+4}{15y}$

78. $\dfrac{x+3}{x^2} + \dfrac{x+5}{2x}$

79. $\dfrac{x+5}{xy} - \dfrac{x-1}{x^2y}$

80. $\dfrac{y-7}{y^2} - \dfrac{y+7}{2y}$

81. $\dfrac{x}{x+1} + \dfrac{x-1}{x}$

82. $\dfrac{3x}{xy} + \dfrac{x+1}{y-1}$

83. $\dfrac{x-1}{x} + \dfrac{y+1}{y}$

84. $\dfrac{a+2}{b} + \dfrac{b-2}{a}$

85. $\dfrac{x}{x-2} + \dfrac{4+2x}{x^2-4}$

86. $\dfrac{y}{y+3} - \dfrac{2y-6}{y^2-9}$

87. $\dfrac{x+1}{x-1} + \dfrac{x-1}{x+1}$

88. $\dfrac{2x}{x+2} + \dfrac{x+1}{x-3}$

89. $\dfrac{2x+2}{x-2} - \dfrac{2x}{2-x}$

90. $\dfrac{y+3}{y-1} - \dfrac{y+4}{1-y}$

91. $\dfrac{2x}{x^2-3x+2} + \dfrac{2x}{x-1} - \dfrac{x}{x-2}$

92. $\dfrac{4a}{a-2} - \dfrac{3a}{a-3} + \dfrac{4a}{a^2-5a+6}$

93. $\dfrac{2x}{x-1} + \dfrac{3x}{x+1} - \dfrac{x+3}{x^2-1}$

94. $\dfrac{a}{a-1} - \dfrac{2}{a+2} + \dfrac{3(a-2)}{a^2+a-2}$

95. $\dfrac{x+1}{2x+4} - \dfrac{x^2}{2x^2-8}$

96. $\dfrac{x+1}{x+2} - \dfrac{x^2+1}{x^2-x-6}$

WRITING

97. Explain how to add fractions with the same denominator.

98. Explain how to subtract fractions with the same denominator.

99. Explain how to find a lowest common denominator.

100. Explain how to add two fractions with different denominators.

SOMETHING TO THINK ABOUT

101. Find the mistake:

$$\frac{2x+3}{x+5} - \frac{x+2}{x+5} = \frac{2x+3-x+2}{x+5}$$

$$= \frac{x+5}{x+5}$$

$$= 1$$

102. Find the mistake:

$$\frac{5x-4}{y} + \frac{x}{y} = \frac{5x-4+x}{y+y}$$

$$= \frac{6x-4}{2y}$$

$$= \frac{3x-2}{y}$$

In Exercises 103–104, show that each formula is true.

103. $\dfrac{a}{b} + \dfrac{c}{d} = \dfrac{ad+bc}{bd}$

104. $\dfrac{a}{b} - \dfrac{c}{d} = \dfrac{ad-bc}{bd}$

6.6 Complex Fractions

■ SIMPLIFYING COMPLEX FRACTIONS ■ SIMPLIFYING FRACTIONS WITH TERMS CONTAINING NEGATIVE EXPONENTS

Getting Ready   *Use the distributive property to remove parentheses, and simplify.*

1. $3\left(1 + \dfrac{1}{3}\right)$ **2.** $10\left(\dfrac{1}{5} - 2\right)$ **3.** $4\left(\dfrac{3}{2} + \dfrac{1}{4}\right)$ **4.** $14\left(\dfrac{3}{7} - 1\right)$

5. $x\left(\dfrac{3}{x} + 3\right)$ **6.** $y\left(\dfrac{2}{y} - 1\right)$ **7.** $4x\left(3 - \dfrac{1}{2x}\right)$ **8.** $6xy\left(\dfrac{1}{2x} + \dfrac{1}{3y}\right)$

■ SIMPLIFYING COMPLEX FRACTIONS

Fractions such as

$$\frac{\frac{1}{3}}{4}, \qquad \frac{\frac{5}{3}}{\frac{2}{9}}, \qquad \frac{x + \frac{1}{2}}{3 - x}, \qquad \text{and} \qquad \frac{\frac{x+1}{2}}{x + \frac{1}{x}}$$

that contain fractions in their numerators or denominators are called **complex fractions.** Complex fractions can often be simplified. For example, we can simplify the complex fraction

$$\frac{\frac{5x}{3}}{\frac{2y}{9}}$$

by doing the division:

$$\frac{\dfrac{5x}{3}}{\dfrac{2y}{9}} = \frac{5x}{3} \div \frac{2y}{9} = \frac{5x}{3} \cdot \frac{9}{2y} = \frac{5x \cdot 3 \cdot \overset{1}{\cancel{3}}}{\cancel{3} \cdot 2y} = \frac{15x}{2y}$$

There are two methods that we can use to simplify complex fractions.

Simplifying Complex Fractions
Method 1
Write the numerator and the denominator of the complex fraction as single fractions. Then divide the fractions and simplify.

Method 2
Multiply the numerator and denominator of the complex fraction by the LCD of the fractions in its numerator and denominator. Then simplify the results, if possible.

To simplify the complex fraction $\dfrac{\dfrac{3x}{5} + 1}{2 - \dfrac{x}{5}}$ by using Method 1, we proceed as follows:

$$\frac{\dfrac{3x}{5} + 1}{2 - \dfrac{x}{5}} = \frac{\dfrac{3x}{5} + \dfrac{5}{5}}{\dfrac{10}{5} - \dfrac{x}{5}} \qquad \text{Change 1 to } \tfrac{5}{5} \text{ and 2 to } \tfrac{10}{5}.$$

$$= \frac{\dfrac{3x + 5}{5}}{\dfrac{10 - x}{5}} \qquad \text{Add the fractions in the numerator and subtract the fractions in the denominator.}$$

$$= \frac{3x + 5}{5} \div \frac{10 - x}{5} \qquad \text{Write the complex fraction as an equivalent division problem.}$$

$$= \frac{3x + 5}{5} \cdot \frac{5}{10 - x} \qquad \text{Invert the divisor and multiply.}$$

$$= \frac{(3x + 5)5}{5(10 - x)} \qquad \text{Multiply the fractions.}$$

$$= \frac{3x + 5}{10 - x} \qquad \text{Divide out the common factor of 5; } \tfrac{5}{5} = 1.$$

To use Method 2, we proceed as follows:

$$\frac{\dfrac{3x}{5}+1}{2-\dfrac{x}{5}} = \frac{5\left(\dfrac{3x}{5}+1\right)}{5\left(2-\dfrac{x}{5}\right)}$$ Multiply both numerator and denominator by 5, the LCD of $\frac{3x}{5}$ and $\frac{x}{5}$.

$$= \frac{5\cdot\dfrac{3x}{5}+5\cdot 1}{5\cdot 2 - 5\cdot \dfrac{x}{5}}$$ Remove parentheses.

$$= \frac{3x+5}{10-x}$$ Do the multiplications.

In this example, Method 2 is easier than Method 1. Any complex fraction can be simplified by using either method. With practice, you will be able to see which method is best to use in any given situation.

EXAMPLE 1   Simplify $\dfrac{\dfrac{x}{3}}{\dfrac{y}{3}}$.

Solution

Method 1

$$\frac{\dfrac{x}{3}}{\dfrac{y}{3}} = \frac{x}{3}\div\frac{y}{3}$$

$$= \frac{x}{3}\cdot\frac{3}{y}$$

$$= \frac{3x}{3y}$$

$$= \frac{x}{y}$$

Method 2

$$\frac{\dfrac{x}{3}}{\dfrac{y}{3}} = \frac{3\left(\dfrac{x}{3}\right)}{3\left(\dfrac{y}{3}\right)}$$

$$= \frac{\dfrac{x}{1}}{\dfrac{y}{1}}$$

$$= \frac{x}{y}$$

■

Self Check   Simplify $\dfrac{\dfrac{a}{4}}{\dfrac{5}{b}}$.

Answer $\frac{ab}{20}$

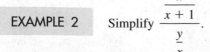

EXAMPLE 2   Simplify $\dfrac{\dfrac{x}{x+1}}{\dfrac{y}{x}}$.

Solution

Method 1

$$\frac{\dfrac{x}{x+1}}{\dfrac{y}{x}} = \frac{x}{x+1} \div \frac{y}{x}$$

$$= \frac{x}{x+1} \cdot \frac{x}{y}$$

$$= \frac{x^2}{y(x+1)}$$

Method 2

$$\frac{\dfrac{x}{x+1}}{\dfrac{y}{x}} = \frac{x(x+1)\left(\dfrac{x}{x+1}\right)}{x(x+1)\left(\dfrac{y}{x}\right)}$$

$$= \frac{\dfrac{x^2}{1}}{\dfrac{y(x+1)}{1}}$$

$$= \frac{x^2}{y(x+1)}$$ ∎

Self Check   Simplify $\dfrac{\dfrac{x}{y}}{\dfrac{x}{y+1}}$.

Answer $\dfrac{y+1}{y}$

EXAMPLE 3   Simplify $\dfrac{1+\dfrac{1}{x}}{1-\dfrac{1}{x}}$.

Solution

Method 1

$$\frac{1+\dfrac{1}{x}}{1-\dfrac{1}{x}} = \frac{\dfrac{x}{x}+\dfrac{1}{x}}{\dfrac{x}{x}-\dfrac{1}{x}}$$

$$= \frac{\dfrac{x+1}{x}}{\dfrac{x-1}{x}}$$

$$= \frac{x+1}{x} \div \frac{x-1}{x}$$

$$= \frac{x+1}{x} \cdot \frac{x}{x-1}$$

$$= \frac{(x+1)x}{x(x-1)}$$

$$= \frac{x+1}{x-1}$$

Method 2

$$\frac{1+\dfrac{1}{x}}{1-\dfrac{1}{x}} = \frac{x\left(1+\dfrac{1}{x}\right)}{x\left(1-\dfrac{1}{x}\right)}$$

$$= \frac{x+1}{x-1}$$ ∎

Self Check Simplify $\dfrac{\frac{1}{x} + 1}{\frac{1}{x} - 1}$.

Answer $\dfrac{1+x}{1-x}$

EXAMPLE 4 Simplify $\dfrac{1}{1 + \dfrac{1}{x+1}}$.

Solution Use Method 2.

$$\dfrac{1}{1 + \dfrac{1}{x+1}} = \dfrac{(x+1) \cdot 1}{(x+1)\left(1 + \dfrac{1}{x+1}\right)}$$

Multiply numerator and denominator by $x+1$.

$$= \dfrac{x+1}{(x+1)1 + 1}$$

Simplify.

$$= \dfrac{x+1}{x+2}$$

Simplify.

■

Self Check Simplify $\dfrac{2}{\dfrac{1}{x+2} - 2}$.

Answer $\dfrac{2(x+2)}{-2x - 3}$

■ SIMPLIFYING FRACTIONS WITH TERMS CONTAINING NEGATIVE EXPONENTS

Many fractions with terms containing negative exponents are complex fractions in disguise.

EXAMPLE 5 Simplify $\dfrac{x^{-1} + y^{-2}}{x^{-2} - y^{-1}}$.

Solution Write the fraction as a complex fraction and simplify:

$$\dfrac{x^{-1} + y^{-2}}{x^{-2} - y^{-1}} = \dfrac{\dfrac{1}{x} + \dfrac{1}{y^2}}{\dfrac{1}{x^2} - \dfrac{1}{y}}$$

$$= \frac{x^2y^2\left(\dfrac{1}{x} + \dfrac{1}{y^2}\right)}{x^2y^2\left(\dfrac{1}{x^2} - \dfrac{1}{y}\right)}$$ Multiply numerator and denominator by x^2y^2.

$$= \frac{xy^2 + x^2}{y^2 - x^2y}$$ Remove parentheses.

$$= \frac{x(y^2 + x)}{y(y - x^2)}$$ Attempt to simplify the fraction by factoring the numerator and denominator.

The result cannot be simplified. ∎

Self Check Simplify $\dfrac{x^{-2} - y^{-1}}{x^{-1} + y^{-2}}$.

Answer $\dfrac{y(y - x^2)}{x(y^2 + x)}$

Orals *Simplify each complex fraction.*

1. $\dfrac{\dfrac{2}{3}}{\dfrac{1}{2}}$ 2. $\dfrac{\dfrac{2}{1}}{\dfrac{1}{2}}$ 3. $\dfrac{\dfrac{1}{2}}{2}$ 4. $\dfrac{1 + \dfrac{1}{2}}{\dfrac{1}{2}}$

EXERCISE 6.6

REVIEW *Write each expression as an expression involving only one exponent.*

1. $t^3t^4t^2$ 2. $(a^0a^2)^3$ 3. $-2r(r^3)^2$ 4. $(s^3)^2(s^4)^0$

Write each expression without using parentheses or negative exponents.

5. $\left(\dfrac{3r}{4r^3}\right)^4$ 6. $\left(\dfrac{12y^{-3}}{3y^2}\right)^{-2}$ 7. $\left(\dfrac{6r^{-2}}{2r^3}\right)^{-2}$ 8. $\left(\dfrac{4x^3}{5x^{-3}}\right)^{-2}$

VOCABULARY AND CONCEPTS *Fill in each blank to make a true statement.*

9. If a fraction has a fraction in its numerator or denominator, it is called a _____.

10. The denominator of the complex fraction
$$\dfrac{\dfrac{3}{x} + \dfrac{x}{y}}{\dfrac{1}{x} + 2}$$
is _____.

11. In Method 1, we write the numerator and denominator of a complex fraction as _____ fractions and then _____.

12. In Method 2, we multiply the numerator and denominator of the complex fraction by the _____ of the fractions in its numerator and denominator.

PRACTICE *In Exercises 13–46, simplify each complex fraction.*

13. $\dfrac{\dfrac{2}{3}}{\dfrac{3}{4}}$

14. $\dfrac{\dfrac{3}{5}}{\dfrac{2}{7}}$

15. $\dfrac{\dfrac{4}{5}}{\dfrac{32}{15}}$

16. $\dfrac{\dfrac{7}{8}}{\dfrac{49}{4}}$

17. $\dfrac{\dfrac{2}{3}+1}{\dfrac{1}{3}+1}$

18. $\dfrac{\dfrac{3}{5}-2}{\dfrac{2}{5}-2}$

19. $\dfrac{\dfrac{1}{2}+\dfrac{3}{4}}{\dfrac{3}{2}+\dfrac{1}{4}}$

20. $\dfrac{\dfrac{2}{3}-\dfrac{5}{2}}{\dfrac{2}{3}-\dfrac{3}{2}}$

21. $\dfrac{\dfrac{x}{y}}{\dfrac{1}{x}}$

22. $\dfrac{\dfrac{y}{x}}{\dfrac{x}{xy}}$

23. $\dfrac{\dfrac{5t^2}{9x^2}}{\dfrac{3t}{x^2t}}$

24. $\dfrac{\dfrac{5w^2}{4tz}}{\dfrac{15wt}{z^2}}$

25. $\dfrac{\dfrac{1}{x}-3}{\dfrac{5}{x}+2}$

26. $\dfrac{\dfrac{1}{y}+3}{\dfrac{3}{y}-2}$

27. $\dfrac{\dfrac{2}{x}+2}{\dfrac{4}{x}+2}$

28. $\dfrac{\dfrac{3}{x}-3}{\dfrac{9}{x}-3}$

29. $\dfrac{\dfrac{3y}{x}-y}{y-\dfrac{y}{x}}$

30. $\dfrac{\dfrac{y}{x}+3y}{y+\dfrac{2y}{x}}$

31. $\dfrac{\dfrac{1}{x+1}}{1+\dfrac{1}{x+1}}$

32. $\dfrac{\dfrac{1}{x-1}}{1-\dfrac{1}{x-1}}$

33. $\dfrac{\dfrac{x}{x+2}}{\dfrac{x}{x+2}+x}$

34. $\dfrac{\dfrac{2}{x-2}}{\dfrac{2}{x-2}-1}$

35. $\dfrac{1}{\dfrac{1}{x}+\dfrac{1}{y}}$

36. $\dfrac{1}{\dfrac{b}{a}-\dfrac{a}{b}}$

37. $\dfrac{\dfrac{2}{x}}{\dfrac{2}{y}-\dfrac{4}{x}}$

38. $\dfrac{\dfrac{2y}{3}}{\dfrac{2y}{3}-\dfrac{8}{y}}$

39. $\dfrac{3+\dfrac{3}{x-1}}{3-\dfrac{3}{x}}$

40. $\dfrac{2-\dfrac{2}{x+1}}{2+\dfrac{2}{x}}$

41. $\dfrac{\dfrac{3}{x}+\dfrac{4}{x+1}}{\dfrac{2}{x+1}-\dfrac{3}{x}}$

42. $\dfrac{\dfrac{5}{y-3}-\dfrac{2}{y}}{\dfrac{1}{y}+\dfrac{2}{y-3}}$

43. $\dfrac{\dfrac{2}{x}-\dfrac{3}{x+1}}{\dfrac{2}{x+1}-\dfrac{3}{x}}$

44. $\dfrac{\dfrac{5}{y}+\dfrac{4}{y+1}}{\dfrac{4}{y}-\dfrac{5}{y+1}}$

45. $\dfrac{\dfrac{1}{y^2+y}-\dfrac{1}{xy+x}}{\dfrac{1}{xy+x}-\dfrac{1}{y^2+y}}$

46. $\dfrac{\dfrac{2}{b^2-1}-\dfrac{3}{ab-a}}{\dfrac{3}{ab-a}-\dfrac{2}{b^2-1}}$

In Exercises 47–56, simplify each fraction.

47. $\dfrac{x^{-2}}{y^{-1}}$

48. $\dfrac{a^{-4}}{b^{-2}}$

49. $\dfrac{1 + x^{-1}}{x^{-1} - 1}$

50. $\dfrac{y^{-2} + 1}{y^{-2} - 1}$

51. $\dfrac{a^{-2} + a}{a + 1}$

52. $\dfrac{t - t^{-2}}{1 - t^{-1}}$

53. $\dfrac{2x^{-1} + 4x^{-2}}{2x^{-2} + x^{-1}}$

54. $\dfrac{x^{-2} - 3x^{-3}}{3x^{-2} - 9x^{-3}}$

55. $\dfrac{1 - 25y^{-2}}{1 + 10y^{-1} + 25y^{-2}}$

56. $\dfrac{1 - 9x^{-2}}{1 - 6x^{-1} + 9x^{-2}}$

WRITING

57. Explain how to use Method 1 to simplify

$$\dfrac{1 + \dfrac{1}{x}}{3 - \dfrac{1}{x}}$$

58. Explain how to use Method 2 to simplify the expression in Exercise 57.

SOMETHING TO THINK ABOUT

59. Simplify these four complex fractions:

$$\dfrac{1}{1 + 1}, \quad \dfrac{1}{1 + \dfrac{1}{2}}, \quad \dfrac{1}{1 + \dfrac{1}{1 + \dfrac{1}{2}}}, \quad \text{and} \quad \dfrac{1}{1 + \dfrac{1}{1 + \dfrac{1}{1 + \dfrac{1}{2}}}}$$

60. Engineering The stiffness of the shaft shown in Illustration 1 is given by the formula

$$k = \dfrac{1}{\dfrac{1}{k_1} + \dfrac{1}{k_2}}$$

ILLUSTRATION 1

where k_1 and k_2 are the individual stiffnesses of each section. If the stiffness k_2 of Section 2 is 4,200,000 in. lb/rad, and the design specifications require that the stiffness k of the entire shaft be 1,900,000 in. lb/rad, what must the stiffness k_1 of Section 1 be?

6.7 Solving Equations That Contain Fractions

■ SOLVING EQUATIONS THAT CONTAIN FRACTIONS ■ EXTRANEOUS SOLUTIONS ■ FORMULAS

Getting Ready   *Simplify.*

1. $3\left(x + \dfrac{1}{3}\right)$

2. $8\left(x - \dfrac{1}{8}\right)$

3. $x\left(\dfrac{3}{x} + 2\right)$

4. $3y\left(\dfrac{1}{3} - \dfrac{2}{y}\right)$

5. $6x\left(\dfrac{5}{2x} + \dfrac{2}{3x}\right)$ **6.** $9x\left(\dfrac{7}{9} + \dfrac{2}{3x}\right)$

7. $(y - 1)\left(\dfrac{1}{y - 1} + 1\right)$ **8.** $(x + 2)\left(3 - \dfrac{1}{x + 2}\right)$

■ SOLVING EQUATIONS THAT CONTAIN FRACTIONS

To solve equations containing fractions, it is usually best to eliminate those fractions. To do so, we multiply both sides of the equation by the LCD of the fractions that appear in the equation. For example, to solve $\frac{x}{3} + 1 = \frac{x}{6}$, we multiply both sides of the equation by 6:

$$\frac{x}{3} + 1 = \frac{x}{6}$$

$$6\left(\frac{x}{3} + 1\right) = 6\left(\frac{x}{6}\right)$$

We then use the distributive property to remove parentheses, simplify, and solve the resulting equation for x.

$$6 \cdot \frac{x}{3} + 6 \cdot 1 = 6 \cdot \frac{x}{6}$$

$$2x + 6 = x$$

$$x + 6 = 0 \qquad \text{Subtract } x \text{ from both sides.}$$

$$x = -6 \qquad \text{Subtract 6 from both sides.}$$

Check: $\dfrac{x}{3} + 1 = \dfrac{x}{6}$

$$\frac{-6}{3} + 1 \overset{?}{=} \frac{-6}{6} \qquad \text{Substitute } -6 \text{ for } x.$$

$$-2 + 1 \overset{?}{=} -1 \qquad \text{Simplify.}$$

$$-1 = -1$$

EXAMPLE 1   Solve $\dfrac{4}{x} + 1 = \dfrac{6}{x}$.

Solution   To clear the equation of fractions, we multiply both sides by the LCD of $\frac{4}{x}$ and $\frac{6}{x}$, which is x.

$$\frac{4}{x} + 1 = \frac{6}{x}$$

$$x\left(\frac{4}{x} + 1\right) = x\left(\frac{6}{x}\right) \qquad \text{Multiply both sides by } x.$$

$$x \cdot \frac{4}{x} + x \cdot 1 = x \cdot \frac{6}{x} \qquad \text{Remove parentheses.}$$

$$4 + x = 6 \qquad \text{Simplify.}$$
$$x = 2 \qquad \text{Subtract 4 from both sides.}$$

Check: $\dfrac{4}{x} + 1 = \dfrac{6}{x}$

$$\dfrac{4}{2} + 1 \overset{?}{=} \dfrac{6}{2} \qquad \text{Substitute 2 for } x.$$

$$2 + 1 \overset{?}{=} 3 \qquad \text{Simplify.}$$

$$3 = 3 \qquad\qquad\qquad \blacksquare$$

Self Check

Solve $\dfrac{6}{x} - 1 = \dfrac{3}{x}$.

Answer   3

■ EXTRANEOUS SOLUTIONS

If we multiply both sides of an equation by an expression that involves a variable, as we did in Example 1, we *must* check the apparent solutions. The next example shows why.

EXAMPLE 2   Solve $\dfrac{x + 3}{x - 1} = \dfrac{4}{x - 1}$.

Solution   To clear the equation of fractions, we multiply both sides by $x - 1$, the LCD of the fractions contained in the equation.

$$\frac{x + 3}{x - 1} = \frac{4}{x - 1}$$

$$(x - 1)\frac{x + 3}{x - 1} = (x - 1)\frac{4}{x - 1} \qquad \text{Multiply both sides by } x - 1.$$

$$x + 3 = 4 \qquad\qquad \text{Simplify.}$$

$$x = 1 \qquad\qquad \text{Subtract 3 from both sides.}$$

Because both sides were multiplied by an expression containing a variable, we must check the apparent solution.

$$\frac{x + 3}{x - 1} = \frac{4}{x - 1}$$

$$\frac{1 + 3}{1 - 1} \overset{?}{=} \frac{4}{1 - 1} \qquad \text{Substitute 1 for } x.$$

$$\frac{4}{0} \overset{\times}{=} \frac{4}{0} \qquad \text{Division by 0 is undefined.}$$

Since zeros appear in the denominators, the fractions are undefined. Thus, 1 is a false solution, and the equation has no solutions. Such false solutions are often called **extraneous solutions.**   ■

Self Check

Solve $\frac{x+5}{x-2} = \frac{7}{x-2}$.

Answer

2 is extraneous.

EXAMPLE 3

Solve $\dfrac{3x+1}{x+1} - 2 = \dfrac{3(x-3)}{x+1}$.

Solution

To clear the equation of fractions, we multiply both sides by $x + 1$, the LCD of the fractions contained in the equation.

$$\frac{3x+1}{x+1} - 2 = \frac{3(x-3)}{x+1}$$

$$(x+1)\left[\frac{3x+1}{x+1} - 2\right] = (x+1)\left[\frac{3(x-3)}{x+1}\right]$$

$$3x + 1 - 2(x+1) = 3(x-3)$$
Use the distributive property to remove parentheses.

$$3x + 1 - 2x - 2 = 3x - 9$$
Remove parentheses.

$$x - 1 = 3x - 9$$
Combine like terms.

$$-2x = -8$$
On both sides, subtract $3x$ and add 1.

$$x = 4$$
Divide both sides by -2.

Check: $\dfrac{3x+1}{x+1} - 2 = \dfrac{3(x-3)}{x+1}$

$$\frac{3(4)+1}{4+1} - 2 \stackrel{?}{=} \frac{3(4-3)}{4+1}$$
Substitute 4 for x.

$$\frac{13}{5} - \frac{10}{5} \stackrel{?}{=} \frac{3(1)}{5}$$

$$\frac{3}{5} = \frac{3}{5}$$

∎

Self Check

Solve $\frac{12}{x+1} - 5 = \frac{2}{x+1}$.

Answer

1

Many times, we will have to factor a denominator to find the LCD.

EXAMPLE 4

Solve $\dfrac{x+2}{x+3} + \dfrac{1}{x^2+2x-3} = 1$.

Solution

To find the LCD, we must factor the second denominator.

$$\frac{x+2}{x+3} + \frac{1}{x^2+2x-3} = 1$$

$$\frac{x+2}{x+3} + \frac{1}{(x+3)(x-1)} = 1$$
Factor $x^2 + 2x - 3$.

To clear the equation of fractions, we multiply both sides by $(x + 3)(x - 1)$, the LCD of the fractions contained in the equation.

$$(x + 3)(x - 1)\left[\frac{x + 2}{x + 3} + \frac{1}{(x + 3)(x - 1)}\right] = (x + 3)(x - 1)1 \qquad \text{Multiply both sides by } (x + 3)(x - 1).$$

$$(x + 3)(x - 1)\frac{x + 2}{x + 3} + (x + 3)(x - 1)\frac{1}{(x + 3)(x - 1)} = (x + 3)(x - 1)1 \qquad \text{Remove brackets.}$$

$$(x - 1)(x + 2) + 1 = (x + 3)(x - 1) \qquad \text{Simplify.}$$

$$x^2 + x - 2 + 1 = x^2 + 2x - 3 \qquad \text{Remove parentheses.}$$

$$x - 2 + 1 = 2x - 3 \qquad \text{Subtract } x^2 \text{ from both sides.}$$

$$x - 1 = 2x - 3 \qquad \text{Combine like terms.}$$

$$-x - 1 = -3 \qquad \text{Subtract } 2x \text{ from both sides.}$$

$$-x = -2 \qquad \text{Add 1 to both sides.}$$

$$x = 2 \qquad \text{Divide both sides by } -1.$$

Verify that 2 is a solution of the given equation. ■

EXAMPLE 5 Solve $\dfrac{4}{5} + y = \dfrac{4y - 50}{5y - 25}$.

Solution

$$\frac{4}{5} + y = \frac{4y - 50}{5y - 25}$$

$$\frac{4}{5} + y = \frac{4y - 50}{5(y - 5)} \qquad \text{Factor } 5y - 25.$$

$$5(y - 5)\left[\frac{4}{5} + y\right] = 5(y - 5)\left[\frac{4y - 50}{5(y - 5)}\right] \qquad \text{Multiply both sides by } 5(y - 5).$$

$$4(y - 5) + 5y(y - 5) = 4y - 50 \qquad \text{Remove brackets.}$$

$$4y - 20 + 5y^2 - 25y = 4y - 50 \qquad \text{Remove parentheses.}$$

$$5y^2 - 25y - 20 = -50 \qquad \text{Subtract } 4y \text{ from both sides and rearrange terms.}$$

$$5y^2 - 25y + 30 = 0 \qquad \text{Add 50 to both sides.}$$

$$y^2 - 5y + 6 = 0 \qquad \text{Divide both sides by 5.}$$

$$(y - 3)(y - 2) = 0 \qquad \text{Factor } y^2 - 5y + 6.$$

$$y - 3 = 0 \quad \text{or} \quad y - 2 = 0 \qquad \text{Set each factor equal to 0.}$$

$$y = 3 \qquad \qquad y = 2$$

Verify that 3 and 2 both satisfy the original equation. ■

Self Check Solve $\frac{x-6}{3x-9} - \frac{1}{3} = \frac{x}{2}$.

Answer 1, 2

■ FORMULAS

Many formulas are equations that contain fractions.

EXAMPLE 6 The formula $\frac{1}{r} = \frac{1}{r_1} + \frac{1}{r_2}$ is used in electronics to calculate parallel resistances. Solve the formula for r.

Solution Clear the equation of fractions by multiplying both sides by the LCD, which is rr_1r_2.

$$\frac{1}{r} = \frac{1}{r_1} + \frac{1}{r_2}$$

$$rr_1r_2\left(\frac{1}{r}\right) = rr_1r_2\left(\frac{1}{r_1} + \frac{1}{r_2}\right) \qquad \text{Multiply both sides by } rr_1r_2.$$

$$\frac{rr_1r_2}{r} = \frac{rr_1r_2}{r_1} + \frac{rr_1r_2}{r_2} \qquad \text{Remove parentheses.}$$

$$r_1r_2 = rr_2 + rr_1 \qquad \text{Simplify.}$$

$$r_1r_2 = r(r_2 + r_1) \qquad \text{Factor out an } r.$$

$$\frac{r_1r_2}{r_2 + r_1} = r \qquad \text{Divide both sides by } r_2 + r_1.$$

or

$$r = \frac{r_1r_2}{r_2 + r_1}$$ ■

Self Check Solve the formula in Example 6 for r_1.

Answer $r_1 = \dfrac{rr_2}{r_2 - r}$

Orals *Indicate your first step in solving each equation.*

1. $\dfrac{x-3}{5} = \dfrac{x}{2}$ **2.** $\dfrac{1}{x-1} = \dfrac{8}{x}$

3. $\dfrac{y}{9} + 5 = \dfrac{y+1}{3}$ **4.** $\dfrac{5x-8}{3} + 3x = \dfrac{x}{5}$

EXERCISE 6.7

REVIEW *Factor each expression.*

1. $x^2 + 4x$ **2.** $x^2 - 16y^2$ **3.** $2x^2 + x - 3$

4. $6a^2 - 5a - 6$ **5.** $x^4 - 16$ **6.** $4x^2 + 10x - 6$

VOCABULARY AND CONCEPTS *Fill in each blank to make a true statement.*

7. False solutions that result from multiplying both sides of an equation by a variable are called _____ solutions.

8. If the product of two numbers is 1, the numbers are called _____.

9. To clear an equation of fractions, we multiply both sides by the _____ of the fractions in the equation.

10. If you multiply both sides of an equation by an expression that involves a variable, you must _____ the solution.

11. To clear the equation $\frac{1}{x} + \frac{2}{y} = 5$ of fractions, we multiply both sides by ___.

12. To clear the equation $\frac{x}{x-2} - \frac{x}{x-1} = 5$ of fractions, we multiply both sides by _____.

PRACTICE *In Exercises 13–70, solve each equation and check the solution. If an equation has no solution, so indicate.*

13. $\dfrac{x}{2} + 4 = \dfrac{3x}{2}$

14. $\dfrac{y}{3} + 6 = \dfrac{4y}{3}$

15. $\dfrac{2y}{5} - 8 = \dfrac{4y}{5}$

16. $\dfrac{3x}{4} - 6 = \dfrac{x}{4}$

17. $\dfrac{x}{3} + 1 = \dfrac{x}{2}$

18. $\dfrac{x}{2} - 3 = \dfrac{x}{5}$

19. $\dfrac{x}{5} - \dfrac{x}{3} = -8$

20. $\dfrac{2}{3} + \dfrac{x}{4} = 7$

21. $\dfrac{3a}{2} + \dfrac{a}{3} = -22$

22. $\dfrac{x}{2} + x = \dfrac{9}{2}$

23. $\dfrac{x-3}{3} + 2x = -1$

24. $\dfrac{x+2}{2} - 3x = x + 8$

25. $\dfrac{z-3}{2} = z + 2$

26. $\dfrac{b+2}{3} = b - 2$

27. $\dfrac{5(x+1)}{8} = x + 1$

28. $\dfrac{3(x-1)}{2} + 2 = x$

29. $\dfrac{c-4}{4} = \dfrac{c+4}{8}$

30. $\dfrac{t+3}{2} = \dfrac{t-3}{3}$

31. $\dfrac{x+1}{3} + \dfrac{x-1}{5} = \dfrac{2}{15}$

32. $\dfrac{y-5}{7} + \dfrac{y-7}{5} = \dfrac{-2}{5}$

33. $\dfrac{3x-1}{6} - \dfrac{x+3}{2} = \dfrac{3x+4}{3}$

34. $\dfrac{2x+3}{3} + \dfrac{3x-4}{6} = \dfrac{x-2}{2}$

35. $\dfrac{3}{x} + 2 = 3$

36. $\dfrac{2}{x} + 9 = 11$

37. $\dfrac{5}{a} - \dfrac{4}{a} = 8 + \dfrac{1}{a}$

38. $\dfrac{11}{b} + \dfrac{13}{b} = 12$

39. $\dfrac{2}{y+1} + 5 = \dfrac{12}{y+1}$

40. $\dfrac{1}{t-3} = \dfrac{-2}{t-3} + 1$

41. $\dfrac{1}{x-1} + \dfrac{3}{x-1} = 1$

42. $\dfrac{3}{p+6} - 2 = \dfrac{7}{p+6}$

43. $\dfrac{a^2}{a+2} - \dfrac{4}{a+2} = a$

44. $\dfrac{z^2}{z+1} + 2 = \dfrac{1}{z+1}$

45. $\dfrac{x}{x-5} - \dfrac{5}{x-5} = 3$

46. $\dfrac{3}{y-2} + 1 = \dfrac{3}{y-2}$

47. $\dfrac{3r}{2} - \dfrac{3}{r} = \dfrac{3r}{2} + 3$

48. $\dfrac{2p}{3} - \dfrac{1}{p} = \dfrac{2p-1}{3}$

49. $\dfrac{1}{3} + \dfrac{2}{x-3} = 1$

50. $\dfrac{3}{5} + \dfrac{7}{x+2} = 2$

51. $\dfrac{u}{u-1} + \dfrac{1}{u} = \dfrac{u^2+1}{u^2-u}$

52. $\dfrac{v}{v+2} + \dfrac{1}{v-1} = 1$

53. $\dfrac{3}{x-2} + \dfrac{1}{x} = \dfrac{2(3x+2)}{x^2 - 2x}$

54. $\dfrac{5}{x} + \dfrac{3}{x+2} = \dfrac{-6}{x(x+2)}$

55. $\dfrac{7}{q^2 - q - 2} + \dfrac{1}{q+1} = \dfrac{3}{q-2}$

56. $\dfrac{-5}{s^2 + s - 2} + \dfrac{3}{s+2} = \dfrac{1}{s-1}$

57. $\dfrac{3y}{3y-6} + \dfrac{8}{y^2 - 4} = \dfrac{2y}{2y+4}$

58. $\dfrac{x-3}{4x-4} + \dfrac{1}{9} = \dfrac{x-5}{6x-6}$

59. $y + \dfrac{2}{3} = \dfrac{2y-12}{3y-9}$

60. $y + \dfrac{3}{4} = \dfrac{3y-50}{4y-24}$

61. $\dfrac{5}{4y+12} - \dfrac{3}{4} = \dfrac{5}{4y+12} - \dfrac{y}{4}$

62. $\dfrac{3}{5x-20} + \dfrac{4}{5} = \dfrac{3}{5x-20} - \dfrac{x}{5}$

63. $\dfrac{x}{x-1} - \dfrac{12}{x^2 - x} = \dfrac{-1}{x-1}$

64. $1 - \dfrac{3}{b} = \dfrac{-8b}{b^2 + 3b}$

65. $\dfrac{z-4}{z-3} = \dfrac{z+2}{z+1}$

66. $\dfrac{a+2}{a+8} = \dfrac{a-3}{a-2}$

67. $\dfrac{n}{n^2 - 9} + \dfrac{n+8}{n+3} = \dfrac{n-8}{n-3}$

68. $\dfrac{x-3}{x-2} - \dfrac{1}{x} = \dfrac{x-3}{x}$

69. $\dfrac{b+2}{b+3} + 1 = \dfrac{-7}{b-5}$

70. $\dfrac{x-4}{x-3} + \dfrac{x-2}{x-3} = x - 3$

71. Solve the formula $\dfrac{1}{a} + \dfrac{1}{b} = 1$ for a.

72. Solve the formula $\dfrac{1}{a} - \dfrac{1}{b} = 1$ for b.

73. Optics   The local length f of a lens is given by the formula

$$\dfrac{1}{f} = \dfrac{1}{d_1} + \dfrac{1}{d_2}$$

where d_1 is the distance from the object to the lens and d_2 is the distance from the lens to the image.

Solve the formula for f.

74. Solve the formula in Exercise 73 for d_1.

WRITING

75. Explain how you would decide what to do first when you solve an equation that involves fractions.

76. Explain why it is important to check your solutions to an equation that contains fractions with variables in the denominator.

SOMETHING TO THINK ABOUT

77. What number is equal to its own reciprocal?

78. Solve $x^{-2} + x^{-1} = 0$.

6.8 Applications of Equations That Contain Fractions

■ PROBLEM SOLVING

Getting Ready

1. If it takes 5 hours to fill a pool, what part could be filled in 1 hour?

2. $x is invested at 5% annual interest. Write an expression for the interest earned in one year.

3. Write an expression for the amount of an investment that earns $y interest in one year at 5%.

4. Express how long it takes to travel y miles at 52 mph.

■ PROBLEM SOLVING

EXAMPLE 1

Number problem   If the same number is added to both the numerator and denominator of the fraction $\frac{3}{5}$, the result is $\frac{4}{5}$. Find the number.

Analyze the problem   We are asked to find a number. If we add it to both the numerator and denominator of a fraction, we will get $\frac{4}{5}$.

Form an equation   Let n represent the unknown number and add n to both the numerator and denominator of $\frac{3}{5}$. Then set the result equal to $\frac{4}{5}$ to get the equation

$$\frac{3 + n}{5 + n} = \frac{4}{5}$$

Solve the equation   To solve the equation, we proceed as follows:

$$\frac{3 + n}{5 + n} = \frac{4}{5}$$

$$5(5 + n)\frac{3 + n}{5 + n} = 5(5 + n)\frac{4}{5} \qquad \text{Multiply both sides by } 5(5 + n).$$

$$5(3 + n) = (5 + n)4 \qquad \text{Simplify.}$$

$$15 + 5n = 20 + 4n \qquad \text{Use the distributive property to remove parentheses.}$$

$$5n = 5 + 4n \qquad \text{Subtract 15 from both sides.}$$

$$n = 5 \qquad \text{Subtract } 4n \text{ from both sides.}$$

State the conclusion   The number is 5.

Check the result   Add 5 to both the numerator and denominator of $\frac{3}{5}$ and get

$$\frac{3 + 5}{5 + 5} = \frac{8}{10} = \frac{4}{5}$$

The result checks.

■

EXAMPLE 2

Draining an oil tank   An inlet pipe can fill an oil tank in 7 days, and a second inlet pipe can fill the same tank in 9 days. If both pipes are used, how long will it take to fill the tank?

Analyze the problem   The key is to note what each pipe can do in 1 day. If you add what the first pipe can do in 1 day to what the second pipe can do in 1 day, the sum is what they can do together in 1 day. Since the first pipe can fill the tank in 7 days, it can do $\frac{1}{7}$ of the job in 1 day. Since the second pipe can fill the tank in 9 days, it can do $\frac{1}{9}$ of the job in 1 day. If it takes x days for both pipes to fill the tank, together they can do $\frac{1}{x}$ of the job in 1 day.

Form an equation   Let x represent the number of days it will take to fill the tank if both inlet pipes are used. Then form the equation

What the first inlet pipe can do in 1 day	$+$	What the second inlet pipe can do in 1 day	$=$	what they can do together in 1 day.
$\dfrac{1}{7}$	$+$	$\dfrac{1}{9}$	$=$	$\dfrac{1}{x}$

Solve the equation   To solve the equation, we proceed as follows:

$$\frac{1}{7} + \frac{1}{9} = \frac{1}{x}$$

$$63x\left(\frac{1}{7} + \frac{1}{9}\right) = 63x\left(\frac{1}{x}\right) \qquad \text{Multiply both sides by } 63x.$$

$$9x + 7x = 63 \qquad \text{Use the distributive property to remove parentheses and simplify.}$$

$$16x = 63 \qquad \text{Combine like terms.}$$

$$x = \frac{63}{16} \qquad \text{Divide both sides by 16.}$$

State the conclusion   It will take $\frac{63}{16}$ or $3\frac{15}{16}$ days for both inlet pipes to fill the tank.

Check the result   In $\frac{63}{16}$ days, the first pipe fills $\frac{1}{7}\left(\frac{63}{16}\right)$ of the tank, and the second pipe fills $\frac{1}{9}\left(\frac{63}{16}\right)$ of the tank. The sum of these efforts, $\frac{9}{16} + \frac{7}{16}$, is equal to one full tank.   ∎

EXAMPLE 3

Track and field   A coach can run 10 miles in the same amount of time that his best student athlete can run 12 miles. If the student can run 1 mph faster than the coach, how fast can the student run?

Analyze the problem   This is a uniform motion problem, which is based on the formula $d = rt$, where d is the distance traveled, r is the rate, and t is the time. If we solve this formula for t, we obtain

$$t = \frac{d}{r}$$

If the coach runs 10 miles at some unknown rate of r mph, it will take $\frac{10}{r}$ hours. If the student runs 12 miles at some unknown rate of $(r + 1)$ mph, it will take $\frac{12}{r+1}$ hours. We can organize the information of the problem as in Figure 6-3.

	d	$=$ r	$\cdot$ t
Student	12	$r + 1$	$\dfrac{12}{r+1}$
Coach	10	r	$\dfrac{10}{r}$

FIGURE 6-3

Because the times are given to be equal, we know that $\dfrac{12}{r+1} = \dfrac{10}{r}$.

Form an equation   Let r be the rate that the coach can run. Then $r + 1$ is the rate that the student can run. We can form the equation

The time it takes the student to run 12 miles	$=$	the time it takes the coach to run 10 miles.
$\dfrac{12}{r+1}$	$=$	$\dfrac{10}{r}$

Solve the equation   We can solve the equation as follows:

$$\frac{12}{r+1} = \frac{10}{r}$$

$$r(r+1)\frac{12}{r+1} = r(r+1)\frac{10}{r} \qquad \text{Multiply both sides by } r(r+1).$$

$$12r = 10(r+1) \qquad \text{Simplify.}$$

$$12r = 10r + 10 \qquad \text{Use the distributive property to remove parentheses.}$$

$$2r = 10 \qquad \text{Subtract } 10r \text{ from both sides.}$$

$$r = 5 \qquad \text{Divide both sides by 2.}$$

State the conclusion   The coach can run 5 mph. The student, running 1 mph faster, can run 6 mph.

Check the results   Verify that this result checks.   ■

EXAMPLE 4   **Comparing investments**   At one bank, a sum of money invested for one year will earn $96 interest. If invested in bonds, that same money would earn $108, because the interest rate paid by the bonds is 1% greater than that paid by the bank. Find the bank's rate of interest.

Analyze the problem   This interest problem is based on the formula $i = pr$, where i is the interest, p is the principal (the amount invested), and r is the annual rate of interest. If we solve this formula for p, we obtain

$$p = \frac{i}{r}$$

If we let r represent the bank's rate of interest, then $r + .01$ represents the rate paid by the bonds. If an investment at a bank earns $96 interest at some unknown rate r, the principal invested is $\frac{96}{r}$. If an investment in bonds earns $108 interest at some unknown rate $(r + .01)$, the principal invested is $\frac{108}{r + .01}$. We can organize the information of the problem as in Figure 6-4.

	Interest	= Principal	·	Rate
Bank	96	$\dfrac{96}{r}$		r
Bonds	108	$\dfrac{108}{r + .01}$		$r + .01$

FIGURE 6-4

Form an equation   Because the same principal would be invested in either account, we can set up the following equation:

$$\frac{96}{r} = \frac{108}{r + .01}$$

Solve the equation   We can solve the equation as follows:

$$\frac{96}{r} = \frac{108}{r + .01}$$

$$r(r + .01) \cdot \frac{96}{r} = \frac{108}{r + .01} \cdot r(r + .01) \qquad \text{Multiply both sides by } r(r + .01).$$

$$96(r + .01) = 108r$$

$$96r + .96 = 108r \qquad \text{Remove parentheses.}$$

$$.96 = 12r \qquad \text{Subtract } 96r \text{ from both sides.}$$

$$.08 = r \qquad \text{Divide both sides by 12.}$$

State the conclusion   The bank's interest rate is .08, or 8%. The bonds pay 9% interest, a rate 1% greater than that paid by the bank.

Check the results   Verify that these rates check.   ■

Orals   **1.** What is the formula that relates the principal p that is invested, the earned interest i, and the rate r for 1 year?

2. What is the formula that relates the distance d traveled at a speed r, for a time t?

3. What is the formula that relates the cost C of purchasing q items that cost \$$d$ each?

EXERCISE 6.8

REVIEW *Solve each equation.*

1. $x^2 - 5x - 6 = 0$

2. $x^2 - 25 = 0$

3. $(t + 2)(t^2 + 7t + 12) = 0$ **4.** $2(y - 4) = -y^2$

5. $y^3 - y^2 = 0$

6. $5a^3 - 125a = 0$

7. $(x^2 - 1)(x^2 - 4) = 0$ **8.** $6t^3 + 35t^2 = 6t$

VOCABULARY AND CONCEPTS

9. List the five steps used in problem solving.

10. Write 6% as a decimal.

PRACTICE

11. Number problem If the denominator of $\frac{3}{4}$ is increased by a number and the numerator is doubled, the result is 1. Find the number.

12. Number problem If a number is added to the numerator of $\frac{7}{8}$ and the same number is subtracted from the denominator, the result is 2. Find the number.

13. Number problem If a number is added to the numerator of $\frac{3}{4}$ and twice as much is added to the denominator, the result is $\frac{4}{7}$. Find the number.

14. Number problem If a number is added to the numerator of $\frac{5}{7}$ and twice as much is subtracted from the denominator, the result is 8. Find the number.

15. Number problem The sum of a number and its reciprocal is $\frac{13}{6}$. Find the numbers.

16. Number problem The sum of the reciprocals of two consecutive even integers is $\frac{7}{24}$. Find the integers.

17. Filling a pool An inlet pipe can fill an empty swimming pool in 5 hours, and another inlet pipe can fill the pool in 4 hours. How long will it take both pipes to fill the pool?

18. Filling a pool One inlet pipe can fill an empty pool in 4 hours, and a drain can empty the pool in 8 hours. How long will it take the pipe to fill the pool if the drain is left open?

19. Roofing a house A homeowner estimates that it will take 7 days to roof his house. A professional roofer estimates that he could roof the house in 4 days. How long will it take if the homeowner helps the roofer?

20. Sewage treatment A sludge pool is filled by two inlet pipes. One pipe can fill the pool in 15 days and the other pipe can fill it in 21 days. However, if no sewage is added, waste removal will empty the pool in 36 days. How long will it take the two inlet pipes to fill an empty pool?

21. Touring A tourist can bicycle 28 miles in the same time as he can walk 8 miles. If he can ride 10 mph faster than he can walk, how much time should he allow to walk a 30-mile trail? See Illustration 1. (*Hint:* How fast can he walk?)

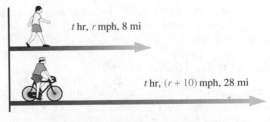

t hr, r mph, 8 mi

t hr, $(r + 10)$ mph, 28 mi

ILLUSTRATION 1

22. Comparing travel A plane can fly 300 miles in the same time as it takes a car to go 120 miles. If the car travels 90 mph slower than the plane, find the speed of the plane.

23. Boating A boat that can travel 18 mph in still water can travel 22 miles downstream in the same amount of time that it can travel 14 miles upstream. Find the speed of the current in the river. (See Illustration 2.)

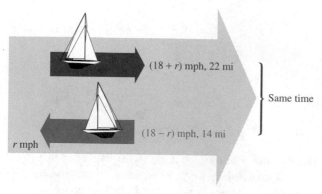

$(18 + r)$ mph, 22 mi

Same time

$(18 - r)$ mph, 14 mi

r mph

ILLUSTRATION 2

24. Wind speed A plane can fly 300 miles downwind in the same amount of time as it can travel 210 miles upwind. Find the velocity of the wind if the plane can fly 255 mph in still air.

25. Comparing investments Two certificates of deposit pay interest at rates that differ by 1%. Money invested for one year in the first CD earns $175 interest. The same principal invested in the other CD earns $200. Find the two rates of interest.

26. Comparing interest rates Two bond funds pay interest at rates that differ by 2%. Money invested for one year in the first fund earns $315 interest. The same amount invested in the other fund earns $385. Find the lower rate of interest.

27. Sharing costs Some office workers bought a $35 gift for their boss. If there had been two more employees to contribute, everyone's cost would have been $2 less. How many workers contributed to the gift?

28. Sales A dealer bought some radios for a total of $1,200. She gave away 6 radios as gifts, sold each of the rest for $10 more than she paid for each radio, and broke even. How many radios did she buy?

29. Sales A bookstore can purchase several calculators for a total cost of $120. If each calculator cost $1 less, the bookstore could purchase 10 additional calculators at the same total cost. How many calculators can be purchased at the regular price?

30. Furnace repair A repairman purchased several furnace-blower motors for a total cost of $210. If his cost per motor had been $5 less, he could have purchased 1 additional motor. How many motors did he buy at the regular rate?

31. River tours A river boat tour begins by going 60 miles upstream against a 5 mph current. There, the boat turns around and returns with the current. What still-water speed should the captain use to complete the tour in 5 hours?

32. Travel time A company president flew 680 miles in a corporate jet but returned in a smaller plane that could fly only half as fast. If the total travel time was 6 hours, find the speeds of the planes.

WRITING

33. The key to solving shared work problems is to ask, "How much of the job could be done in 1 unit of time?" Explain.

34. It is difficult to check the solution of a shared work problem. Explain how you could decide if the answer is at least reasonable.

SOMETHING TO THINK ABOUT

35. Create a problem, involving either investment income or shared work, that can be solved by an equation that contains fractions.

36. Solve the problem you created in Exercise 35.

■ ■ ■ ■ ■ ■ ■ ■ ■ PROJECTS

1. If the sides of two similar triangles are in the ratio of 1 to 1, the triangles are said to be **congruent.** Congruent triangles have the same shape and the same size (area).

   a. Draw several triangles with sides of length 1, 1.5, and 2 inches. Are the triangles all congruent? What general rule could you make?

   b. Draw several triangles with the dimensions shown in Illustration 1. Are the triangles all congruent? What general rule could you make?

   c. Draw several triangles with the dimensions shown in Illustration 2. Are the triangles all congruent? What general rule could you make?

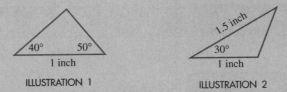

ILLUSTRATION 1   ILLUSTRATION 2

   d. If three angles of one triangle have the same measure as three angles of a second triangle, are the triangles congruent? Explain your answer.

2. Our solar system consists of nine planets and their moons, and some assorted asteroids, comets, and other debris, all orbiting the sun. If the sizes of the planets and their distances from the sun were reduced proportionally so that the sun was the size of an orange, earth would be a grain of sand, and the farthest planet, Pluto, would be half a mile away.

    The diameters of the planets and their distances from the sun are given in Table 6-2.

	Diameter (km)	Distance from sun (AU)*
Sun	1.5×10^6	0
Mercury	4.9×10^3	0.39
Venus	1.2×10^4	0.72
Earth	1.3×10^4	1.0
Mars	6.8×10^3	1.5
Jupiter	1.4×10^5	5.2
Saturn	1.2×10^5	9.5
Uranus	5.1×10^5	19
Neptune	4.9×10^5	30
Pluto	2.3×10^3	39

*One AU (astronomical unit) is the distance from the earth to the sun, about 93 million miles.

TABLE 6-2

a. Use the information in Table 6-2 to draw a scale diagram of the relative *positions* of the sun and the planets. You will need a large sheet of paper, or perhaps the classroom chalkboard. From your diagram, which planets do you think are called the *inner planets,* and which are the *outer planets?*

b. Draw a scale diagram that shows the relative *sizes* of the sun and planets.

c. What difficulty would you have in drawing a scale diagram that shows both relative sizes and distances? Could you draw a scale diagram if you disregarded the enormous size of the sun? Write your observations in a brief paragraph.

C H A P T E R   S U M M A R Y

CONCEPTS

REVIEW EXERCISES

SECTION 6.1 · *Ratios*

A **ratio** is the comparison of two numbers by their indicated quotient.

1. Write each ratio as a fraction in lowest terms.
  a. 3 to 6   **b.** $12x$ to $15x$
  c. 2 feet to 1 yard   **d.** 5 pints to 3 quarts

The **unit cost** of an item is the ratio of its cost to its quantity.

2. If three pounds of coffee cost $8.79, find the unit cost (the cost per pound).

Rates are ratios that are used to compare quantities with different units.

3. If a factory used 2,275 kwh of electricity in February, what was the rate of energy consumption in kwh per week?

SECTION 6.2 · *Proportions and Similar Triangles*

A **proportion** is a statement that two ratios are equal.

4. Determine whether the following equations are proportions.
  a. $\dfrac{4}{7} = \dfrac{20}{34}$ **b.** $\dfrac{5}{7} = \dfrac{30}{42}$

In any proportion, the product of the extremes is equal to the product of the means.

5. Solve each proportion.
  a. $\dfrac{3}{x} = \dfrac{6}{9}$ **b.** $\dfrac{x}{3} = \dfrac{x}{5}$
  c. $\dfrac{x-2}{5} = \dfrac{x}{7}$ **d.** $\dfrac{4x-1}{18} = \dfrac{x}{6}$

The measures of corresponding sides of similar triangles are in proportion.

6. A telephone pole casts a shadow 12 feet long at the same time that a man 6 feet tall casts a shadow of 3.6 feet. How tall is the pole?

SECTION 6.3 — Simplifying Fractions

If b and c are not 0, then

$$\frac{a}{b} = \frac{a \cdot c}{b \cdot c}$$

$$\frac{a}{1} = a$$

$$\frac{a}{0} \text{ is undefined.}$$

7. Write each fraction in lowest terms. If a fraction is already in lowest terms, so indicate.

a. $\dfrac{10}{25}$ **b.** $-\dfrac{12}{18}$

c. $-\dfrac{51}{153}$ **d.** $\dfrac{105}{45}$

e. $\dfrac{3x^2}{6x^3}$ **f.** $\dfrac{5xy^2}{2x^2y^2}$

g. $\dfrac{x^2}{x^2 + x}$ **h.** $\dfrac{x + 2}{x^2 + 2x}$

i. $\dfrac{6xy}{3xy}$ **j.** $\dfrac{8x^2y}{2x(4xy)}$

k. $\dfrac{3p - 2}{2 - 3p}$ **l.** $\dfrac{x^2 - x - 56}{x^2 - 5x - 24}$

m. $\dfrac{2x^2 - 16x}{2x^2 - 18x + 16}$ **n.** $\dfrac{x^2 + x - 2}{x^2 - x - 2}$

SECTION 6.4 — Multiplying and Dividing Fractions

$$\frac{a}{b} \cdot \frac{c}{d} = \frac{a \cdot c}{b \cdot d} \quad (b, d \neq 0)$$

8. Do each multiplication and simplify.

a. $\dfrac{3xy}{2x} \cdot \dfrac{4x}{2y^2}$ **b.** $\dfrac{3x}{x^2 - x} \cdot \dfrac{2x - 2}{x^2}$

c. $\dfrac{x^2 + 3x + 2}{x^2 + 2x} \cdot \dfrac{x}{x + 1}$ **d.** $\dfrac{x^2 + x}{3x - 15} \cdot \dfrac{6x - 30}{x^2 + 2x + 1}$

$$\frac{a}{b} \div \frac{c}{d} = \frac{a}{b} \cdot \frac{d}{c} \quad (b, c, d \neq 0)$$

9. Do each division and simplify.

a. $\dfrac{3x^2}{5x^2y} \div \dfrac{6x}{15xy^2}$ **b.** $\dfrac{x^2 + 5x}{x^2 + 4x - 5} \div \dfrac{x^2}{x - 1}$

c. $\dfrac{x^2 - x - 6}{2x - 1} \div \dfrac{x^2 - 2x - 3}{2x^2 + x - 1}$

d. $\dfrac{x^2 - 3x}{x^2 - x - 6} \div \dfrac{x^2 - x}{x^2 + x - 2}$

e. $\dfrac{x^2 + 4x + 4}{x^2 + x - 6} \left(\dfrac{x - 2}{x - 1} \div \dfrac{x + 2}{x^2 + 2x - 3} \right)$

| SECTION 6.5 | *Adding and Subtracting Fractions* |

$\dfrac{a}{d} + \dfrac{b}{d} = \dfrac{a+b}{d} \quad (d \neq 0)$

10. Do each operation. Simplify all answers.

a. $\dfrac{x}{x+y} + \dfrac{y}{x+y}$ **b.** $\dfrac{3x}{x-7} - \dfrac{x-2}{x-7}$

$\dfrac{a}{d} - \dfrac{b}{d} = \dfrac{a-b}{d} \quad (d \neq 0)$

c. $\dfrac{x}{x-1} + \dfrac{1}{x}$ **d.** $\dfrac{1}{7} - \dfrac{1}{x}$

To add or subtract fractions with unlike denominators, first find the LCD of the fractions. Then express each fraction in equivalent form with a common denominator. Finally, add or subtract the fractions.

e. $\dfrac{3}{x+1} - \dfrac{2}{x}$ **f.** $\dfrac{x+2}{2x} - \dfrac{2-x}{x^2}$

g. $\dfrac{x}{x+2} + \dfrac{3}{x} - \dfrac{4}{x^2+2x}$

h. $\dfrac{2}{x-1} - \dfrac{3}{x+1} + \dfrac{x-5}{x^2-1}$

| SECTION 6.6 | *Complex Fractions* |

To simplify a complex fraction, use either of these methods:

11. Simplify each complex fraction.

1. Write the numerator and denominator of the complex fraction as single fractions, do the division of the fractions, and simplify.

a. $\dfrac{\frac{3}{2}}{\frac{2}{3}}$ **b.** $\dfrac{\frac{3}{2}+1}{\frac{2}{3}+1}$

c. $\dfrac{\frac{1}{x}+1}{\frac{1}{x}-1}$ **d.** $\dfrac{1+\frac{3}{x}}{2-\frac{1}{x^2}}$

2. Multiply both the numerator and the denominator of the complex fraction by the LCD of the fractions that appear in the numerator and the denominator; then simplify.

e. $\dfrac{\frac{2}{x-1}+\frac{x-1}{x+1}}{\frac{1}{x^2-1}}$ **f.** $\dfrac{\frac{a}{b}+c}{\frac{b}{a}+c}$

| **SECTION 6.7** | *Solving Equations That Contain Fractions* |

To solve an equation that contains fractions, change it to another equation without fractions. Do so by multiplying both sides by the LCD of the fractions. Check all solutions.

12. Solve each equation and check all answers.

a. $\dfrac{3}{x} = \dfrac{2}{x-1}$

b. $\dfrac{5}{x+4} = \dfrac{3}{x+2}$

c. $\dfrac{2}{3x} + \dfrac{1}{x} = \dfrac{5}{9}$

d. $\dfrac{2x}{x+4} = \dfrac{3}{x-1}$

e. $\dfrac{2}{x-1} + \dfrac{3}{x+4} = \dfrac{-5}{x^2+3x-4}$

f. $\dfrac{4}{x+2} - \dfrac{3}{x+3} = \dfrac{6}{x^2+5x+6}$

13. Solve for r_1: $\dfrac{1}{r} = \dfrac{1}{r_1} + \dfrac{1}{r_2}$.

| **SECTION 6.8** | *Applications of Equations That Contain Fractions* |

14. The efficiency E of a Carnot engine is given by the formula

$$E = 1 - \dfrac{T_2}{T_1}$$

Solve the formula for T_1.

15. Radioactive tracers are used for diagnostic work in nuclear medicine. The **effective half-life** H of a radioactive material in a biological organism is given by the formula

$$H = \dfrac{RB}{R+B}$$

where R is the radioactive half-life and B is the biological half-life of the tracer. Solve the formula for R.

16. Pumping a basement   If one pump can empty a flooded basement in 18 hours and a second pump can empty the basement in 20 hours, how long will it take to empty the basement when both pumps are used?

17. Painting houses   If a homeowner can paint a house in 14 days and a professional painter can paint it in 10 days, how long will it take if they work together?

18. Exercise   A jogger can bicycle 30 miles in the same time as he can jog 10 miles. If he can ride 10 miles per hour faster than he can jog, how fast can he jog?

19. Wind speed   A plane can fly 400 miles downwind in the same amount of time as it can travel 320 miles upwind. If the plane can fly at 360 miles per hour in still air, find the velocity of the wind.

■ Chapter Test

1. Express as a ratio in lowest terms: 6 feet to 3 yards.

2. Is the equation $\dfrac{3xy}{5xy} = \dfrac{3xt}{5xt}$ a proportion?

3. Solve the proportion for y: $\dfrac{y}{y-1} = \dfrac{y-2}{y}$.

4. A tree casts a shadow that is 30 feet long when a 6-foot-tall man casts a shadow that is 4 feet long. How tall is the tree?

5. Simplify $\dfrac{48x^2y}{54xy^2}$.

6. Simplify $\dfrac{2x^2 - x - 3}{4x^2 - 9}$.

7. Simplify $\dfrac{3(x+2) - 3}{2x - 4 - (x - 5)}$.

8. Multiply and simplify $\dfrac{12x^2y}{15xyz} \cdot \dfrac{25y^2z}{16xt}$.

9. Multiply and simplify $\dfrac{x^2 + 3x + 2}{3x + 9} \cdot \dfrac{x + 3}{x^2 - 4}$.

10. Divide and simplify $\dfrac{8x^2y}{25xt} \div \dfrac{16x^2y^3}{30xyt^3}$.

11. Divide and simplify $\dfrac{x^2 - x}{3x^2 + 6x} \div \dfrac{3x - 3}{3x^3 + 6x^2}$.

12. Simplify $\dfrac{x^2 + xy}{x - y} \cdot \dfrac{x^2 - y^2}{x^2 - 2x} \div \dfrac{x^2 + 2xy + y^2}{x^2 - 4}$.

13. Add $\dfrac{5x - 4}{x - 1} + \dfrac{5x + 3}{x - 1}$.

14. Subtract $\dfrac{3y + 7}{2y + 3} - \dfrac{3(y - 2)}{2y + 3}$.

15. Add $\dfrac{x + 1}{x} + \dfrac{x - 1}{x + 1}$.

16. Subtract $\dfrac{5x}{x - 2} - 3$.

17. Simplify $\dfrac{\dfrac{8x^2}{xy^3}}{\dfrac{4y^3}{x^2y^3}}$.

18. Simplify $\dfrac{1 + \dfrac{y}{x}}{\dfrac{y}{x} - 1}$.

19. Solve for x: $\dfrac{x}{10} - \dfrac{1}{2} = \dfrac{x}{5}$.

20. Solve for x: $3x - \dfrac{2(x+3)}{3} = 16 - \dfrac{x+2}{2}$.

21. Solve for x: $\dfrac{7}{x+4} - \dfrac{1}{2} = \dfrac{3}{x+4}$.

22. Solve for B: $H = \dfrac{RB}{R+B}$.

23. Cleaning highways One highway worker could pick up all the trash on a strip of highway in 7 hours, and his helper could pick up the trash in 9 hours. How long will it take them if they work together?

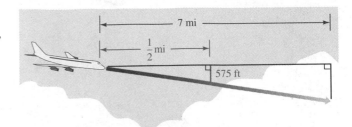

24. Boating A boat can motor 28 miles downstream in the same amount of time as it can motor 18 miles upstream. Find the speed of the current if the boat can motor at 23 mph in still water.

25. Flight path A plane drops 575 feet as it flies a horizontal distance of $\frac{1}{2}$ mile. How much altitude will it lose as it flies a horizontal distance of 7 miles? (See Illustration 1.)

ILLUSTRATION 1

■ Cumulative Review Exercises

In Exercises 1–4, simplify each expression.

1. $x^2 x^5$

2. $(x^2)^5$

3. $\dfrac{x^5}{x^2}$

4. $(3x^5)^0$

In Exercises 5–8, simplify each expression.

5. $(3x^2 - 2x) + (6x^3 - 3x^2 - 1)$

6. $(4x^3 - 2x) - (2x^3 - 2x^2 - 3x + 1)$

7. $3(5x^2 - 4x + 3) + 2(-x^2 + 2x - 4)$

8. $4(3x^2 - 4x - 1) - 2(-2x^2 + 4x - 3)$

In Exercises 9–12, do each multiplication.

9. $(3x^3 y^2)(-4x^2 y^3)$

10. $-5x^2(7x^3 - 2x^2 - 2)$

11. $(3x + 1)(2x + 4)$

12. $(5x - 4y)(3x + 2y)$

In Exercises 13–14, do each division.

13. $x + 3\overline{)x^2 + 7x + 12}$

14. $2x - 3\overline{)2x^3 - x^2 - x - 3}$

In Exercises 15–24, factor each expression.

15. $3x^2 y - 6xy^2$

16. $3(a + b) + x(a + b)$

17. $2a + 2b + ab + b^2$

18. $25p^4 - 16q^2$

19. $x^2 - 11x - 12$

20. $x^2 - xy - 6y^2$

21. $6a^2 - 7a - 20$

22. $8m^2 - 10mn - 3n^2$

23. $x^4 - 81$

24. $x^2 + 2x + 1 - y^2$

In Exercises 25–30, solve each equation.

25. $\dfrac{4}{5}x + 6 = 18$ **26.** $5 - \dfrac{x+2}{3} = 7 - x$ **27.** $6x^2 - x - 2 = 0$ **28.** $5x^2 = 10x$

29. $x^2 + 3x + 2 = 0$

30. $2y^2 + 5y - 12 = 0$

In Exercises 31–34, solve each inequality and graph the solution set.

31. $5x - 3 > 7$

32. $7x - 9 < 5$

33. $-2 < -x + 3 < 5$

34. $0 \le \dfrac{4-x}{3} \le 2$

In Exercises 35–36, graph each equation.

35. Graph the equation $4x - 3y = 12$.

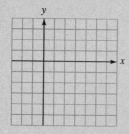

36. Graph the equation $3x + 4y = 4y + 12$.

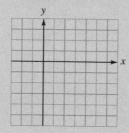

In Exercises 37–38, solve each system of equations.

37. $\begin{cases} x + y = 1 \\ x - y = 7 \end{cases}$

38. $\begin{cases} 4x + 9y = 8 \\ 2x - 6y = -3 \end{cases}$

In Exercises 39–40, solve each system by graphing.

39. $\begin{cases} x - y = 4 \\ 2x + y = 5 \end{cases}$

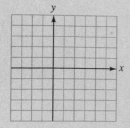

40. $\begin{cases} 3x + 2y \ge 6 \\ x + 3y \le 6 \end{cases}$

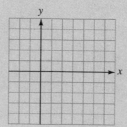

In Exercises 41–44, $y = f(x) = 2x^2 - 3$. Find each value.

41. $f(0)$ **42.** $f(3)$ **43.** $f(-2)$ **44.** $f(2x)$

In Exercises 45–46, simplify each fraction.

45. $\dfrac{x^2 + 2x + 1}{x^2 - 1}$ **46.** $\dfrac{x^2 + 2x - 15}{x^2 + 3x - 10}$

In Exercises 47–52, do the operation(s) and simplify when possible.

47. $\dfrac{x^2 + x - 6}{5x - 5} \cdot \dfrac{5x - 10}{x + 3}$ **48.** $\dfrac{p^2 - p - 6}{3p - 9} \div \dfrac{p^2 + 6p + 9}{p^2 - 9}$

49. $\dfrac{3x}{x + 2} + \dfrac{5x}{x + 2} - \dfrac{7x - 2}{x + 2}$ **50.** $\dfrac{x - 1}{x + 1} + \dfrac{x + 1}{x - 1}$

51. $\dfrac{a + 1}{2a + 4} - \dfrac{a^2}{2a^2 - 8}$ **52.** $\dfrac{\dfrac{1}{x} + \dfrac{1}{y}}{\dfrac{1}{x} - \dfrac{1}{y}}$

7 More Equations, Inequalities, and Factoring

MATHEMATICS IN THE WORKPLACE

Pilots are highly trained professionals who fly airplanes and helicopters to carry out a variety of tasks. Most nonmilitary pilots transport passengers and cargo, but many are needed for crop dusting, spreading seed, testing aircraft, firefighting, police work, traffic reporting, and rescue and evacuation. Of course, many pilots are employed by the military.

SAMPLE APPLICATION ■ The air passing over the curved top of the wing shown in Illustration 1 moves faster than the air beneath, because it has a greater distance to travel. This faster-moving air exerts less pressure on the wing than the air beneath, causing *lift*. Two factors that determine lift are the *velocity* of the plane and the plane's *angle of attack*.

If the velocity of a 2,050-pound plane is 130 feet per second, find the proper angle of attack.
(See Project 2.)

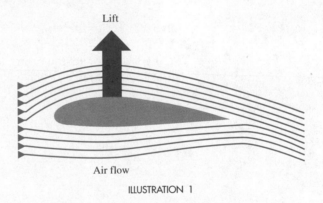

Lift

Air flow

ILLUSTRATION 1

In this chapter, we will review many of the ideas covered in the first six chapters and then extend them to the intermediate algebra level. If you have trouble with any topic, review the sections of the text that first discussed that topic.

7.1   Review of Equations and Inequalities

■ SOLVING EQUATIONS ■ IDENTITIES AND IMPOSSIBLE EQUATIONS ■ SOLVING FORMULAS
■ PROBLEM SOLVING ■ INEQUALITIES ■ COMPOUND INEQUALITIES

Getting Ready   *Find the value of x that will make each statement true.*

1. $x + 3 = 5$ **2.** $x - 5 = 3$ **3.** $\dfrac{3x}{5} = 6$ **4.** $2x + 3 = x - 4$

475

■ SOLVING EQUATIONS

Recall that an *equation* is a statement indicating that two mathematical expressions are equal. The set of numbers that satisfy an equation is called its *solution set,* and the elements in the solution set are called *solutions* or *roots* of the equation. Finding the solution set of an equation is called *solving the equation.*

To solve an equation, we will use the following two properties of equality to replace the equation with simpler equivalent equations that have the same solution set. We continue this process until we have isolated the variable on one side of the = sign.

1. If any quantity is added to (or subtracted from) both sides of an equation, a new equation is formed that is equivalent to the original equation.

2. If both sides of an equation are multiplied (or divided) by the same nonzero constant, a new equation is formed that is equivalent to the original equation.

EXAMPLE 1   Solve $3(2x - 1) = 2x + 9$.

Solution   We use the distributive property to remove parentheses and then isolate x on the left-hand side of the equation.

$$3(2x - 1) = 2x + 9$$

$$6x - 3 = 2x + 9 \qquad \text{To remove parentheses, use the distributive property.}$$

$$6x - 3 + 3 = 2x + 9 + 3 \qquad \text{To undo the subtraction by 3, add 3 to both sides.}$$

$$6x = 2x + 12 \qquad \text{Combine like terms.}$$

$$6x - 2x = 2x + 12 - 2x \qquad \text{To eliminate } 2x \text{ from the right-hand side, subtract } 2x \text{ from both sides.}$$

$$4x = 12 \qquad \text{Combine like terms.}$$

$$x = 3 \qquad \text{To undo the multiplication by 4, divide both sides by 4.}$$

Check: We substitute 3 for x in the original equation to see whether it satisfies the equation.

$$3(2x - 1) = 2x + 9$$

$$3(2 \cdot 3 - 1) \stackrel{?}{=} 2 \cdot 3 + 9$$

$$3(5) \stackrel{?}{=} 6 + 9 \qquad \text{On the left-hand side, do the work in parentheses first.}$$

$$15 = 15$$

Since 3 satisfies the original equation, it is a solution. The solution set is $\{3\}$. ■

Self Check   Solve $2(3x - 2) = 3x - 13$.

Answer -3

To solve more complicated linear equations, we will follow these steps.

Solving Equations

1. If an equation contains fractions, multiply both sides of the equation by their least common denominator (LCD) to eliminate the denominators.

2. Use the distributive property to remove all grouping symbols and combine like terms.

3. Use the addition and subtraction properties to get all variables on one side of the equation and all numbers on the other side. Combine like terms, if necessary.

4. Use the multiplication and division properties to make the coefficient of the variable equal to 1.

5. Check the result by replacing the variable with the possible solution and verifying that the number satisfies the equation.

EXAMPLE 2   Solve $\dfrac{5}{3}(x - 3) = \dfrac{3}{2}(x - 2) + 2$.

Solution   *Step 1:* Since 6 is the smallest number that can be divided by both 2 and 3, we multiply both sides of the equation by 6 to eliminate the fractions:

$$\frac{5}{3}(x - 3) = \frac{3}{2}(x - 2) + 2$$

$$6\left[\frac{5}{3}(x - 3)\right] = 6\left[\frac{3}{2}(x - 2) + 2\right]$$

$$10(x - 3) = 9(x - 2) + 12 \qquad 6 \cdot \tfrac{5}{3} = 10, \quad 6 \cdot \tfrac{3}{2} = 9, \quad 6 \cdot 2 = 12.$$

Step 2: We use the distributive property to remove parentheses and then combine like terms.

$$10x - 30 = 9x - 18 + 12$$
$$10x - 30 = 9x - 6$$

Step 3: We use the addition and subtraction properties by adding 30 to both sides and subtracting $9x$ from both sides.

$$10x - 30 - 9x + 30 = 9x - 6 - 9x + 30$$
$$x = 24 \qquad\qquad \text{Combine like terms.}$$

Since the coefficient of x in the above equation is 1, Step 4 is unnecessary.

Step 5: We check by substituting 24 for x in the original equation and simplifying:

$$\frac{5}{3}(x - 3) = \frac{3}{2}(x - 2) + 2$$

$$\frac{5}{3}(\mathbf{24} - 3) \overset{?}{=} \frac{3}{2}(\mathbf{24} - 2) + 2$$

$$\frac{5}{3}(21) \overset{?}{=} \frac{3}{2}(22) + 2$$

$$5(7) \overset{?}{=} 33 + 2$$

$$35 = 35$$

Since 24 satisfies the equation, it is a solution. The solution set is {24}. ■

Self Check Solve $\frac{2}{3}(x - 2) = \frac{5}{2}(x - 1) + 3$.

Answer -1

■ IDENTITIES AND IMPOSSIBLE EQUATIONS

The equations discussed so far have been **conditional equations.** For these equations, some numbers x are solutions and others are not. An **identity** is an equation that is satisfied by every number x for which both sides of the equation are defined.

EXAMPLE 3 Solve $2(x - 1) + 4 = 4(1 + x) - (2x + 2)$.

Solution

$2(x - 1) + 4 = 4(1 + x) - (2x + 2)$	
$2x - 2 + 4 = 4 + 4x - 2x - 2$	Use the distributive property to remove parentheses.
$2x + 2 = 2x + 2$	Combine like terms.
$2 = 2$	Subtract $2x$ from both sides.

Since $2 = 2$, the equation is true for every number x. Since every number x satisfies this equation, it is an identity. ■

Self Check Solve $3(x + 1) - (20 + x) = 5(x - 1) - 3(x + 4)$.

Answer an identity

An **impossible equation** or a **contradiction** is an equation that has no solution.

EXAMPLE 4 Solve $\dfrac{x - 1}{3} + 4x = \dfrac{3}{2} + \dfrac{13x - 2}{3}$.

Solution

$$\frac{x-1}{3} + 4x = \frac{3}{2} + \frac{13x - 2}{3}$$

$$6\left(\frac{x-1}{3} + 4x\right) = 6\left(\frac{3}{2} + \frac{13x-2}{3}\right) \qquad \text{To eliminate the fractions, multiply both sides by 6.}$$

$$2(x-1) + 6(4x) = 9 + 2(13x - 2) \qquad \text{Use the distributive property to remove parentheses.}$$

$$2x - 2 + 24x = 9 + 26x - 4 \qquad \text{Remove parentheses.}$$

$$26x - 2 = 26x + 5 \qquad \text{Combine like terms.}$$

$$-2 = 5 \qquad \text{Subtract } 26x \text{ from both sides.}$$

Since $-2 = 5$ is false, no number x satisfies the equation. The solution set is $\varnothing$, called the **empty set.** ∎

Self Check   Solve $\frac{x-2}{3} - 3 = \frac{1}{5} + \frac{x+1}{3}$.

Answer   no solution

■ SOLVING FORMULAS

To solve a formula for a variable means to isolate that variable on one side of the = sign and place all other quantities on the other side.

EXAMPLE 5   **Wages and commissions**   A sales clerk earns $200 per week plus a 5% commission on the value of the merchandise she sells. What dollar volume must she sell each week to earn $250, $300, and $350 in three successive weeks?

Solution   The weekly earnings e are computed using the formula

1. $e = 200 + 0.05v$

where v represents the value of the merchandise sold. To find v for the three values of e, we first solve Equation 1 for v.

$$e = 200 + 0.05v$$

$$e - 200 = 0.05v \qquad \text{Subtract 200 from both sides.}$$

$$\frac{e - 200}{0.05} = v \qquad \text{Divide both sides by 0.05.}$$

We can now substitute $250, $300, and $350 for e and compute v.

$$v = \frac{e - 200}{0.05} \qquad\qquad v = \frac{e - 200}{0.05} \qquad\qquad v = \frac{e - 200}{0.05}$$

$$v = \frac{250 - 200}{0.05} \qquad\qquad v = \frac{300 - 200}{0.05} \qquad\qquad v = \frac{350 - 200}{0.05}$$

$$v = 1{,}000 \qquad\qquad\qquad v = 2{,}000 \qquad\qquad\qquad v = 3{,}000$$

She must sell $1,000 worth of merchandise the first week, $2,000 worth in the second week, and $3,000 worth in the third week. ■

■ PROBLEM SOLVING

EXAMPLE 6   **Building a dog run** A man has 28 meters of fencing to make a rectangular dog run. If he wants the dog run to be 6 meters longer than it is wide, find its dimensions.

Analyze the problem   The perimeter P of a rectangle is the distance around it. If w is chosen to represent the width of the dog run, then $w + 6$ represents its length. (See Figure 7-1.) The perimeter can be expressed either as $2w + 2(w + 6)$ or as 28.

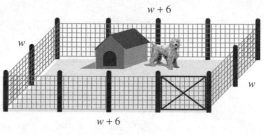

FIGURE 7-1

Form an equation   We let w represent the width of the dog run. Then $w + 6$ represents its length.

Two widths	+	two lengths	=	the perimeter.
$2 \cdot w$	+	$2 \cdot (w + 6)$	=	28

Solve the equation   We can solve this equation as follows:

$$2w + 2(w + 6) = 28$$
$$2w + 2w + 12 = 28 \qquad \text{Use the distributive property to remove parentheses.}$$
$$4w + 12 = 28 \qquad \text{Combine like terms.}$$
$$4w = 16 \qquad \text{Subtract 12 from both sides.}$$
$$w = 4 \qquad \text{Divide both sides by 4.}$$
$$w + 6 = 10$$

State the conclusion   The dimensions of the dog run are 4 meters by 10 meters.

Check the result   If a dog run has a width of 4 meters and a length of 10 meters, its length is 6 meters longer than its width, and the perimeter is $2(4) + 2(10) = 28$. ■

For more examples of using equations to solve problems, review Chapter 2, Section 2.5.

■ INEQUALITIES

Recall that *inequalities* are statements indicating that quantities are unequal.

- $a < b$ means "a is less than b."
- $a > b$ means "a is greater than b."
- $a \leq b$ means "a is less than or equal to b."
- $a \geq b$ means "a is greater than or equal to b."

In Chapter 1, we saw that many inequalities can be graphed as regions on the number line, called **intervals.** For example, the graph of the inequality $-4 < x < 2$ is shown in Figure 7-2(a). Since neither endpoint is included, we say that the graph is an **open interval.** In **interval notation,** this interval is denoted as $(-4, 2)$, where the parentheses indicate that the endpoints are not included.

The graph of the inequality $-2 \leq x \leq 5$ is shown in Figure 7-2(b). Since both endpoints are included, we say that the graph is a **closed interval.** This interval is denoted as $[-2, 5]$, where the brackets indicate that the endpoints are included.

Since one endpoint is included and one is not in the interval shown in Figure 7-2(c), we call the interval a **half-open interval.** This interval is denoted as $[-10, 10)$. Since the interval shown in Figure 7-2(d) extends forever in one direction, it is called an **unbounded interval.** This interval is denoted as $[-6, \infty)$, where the symbol ∞ is read as "infinity."

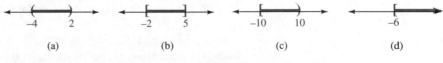

FIGURE 7-2

If a and b are real numbers, Table 7-1 shows the different types of intervals that can occur.

Kind of Interval	Inequality	Graph	Interval
Open interval	$a < x < b$		(a, b)
Half-open interval	$a \leq x < b$		$[a, b)$
	$a < x \leq b$		$(a, b]$
Closed interval	$a \leq x \leq b$		$[a, b]$
Unbounded interval	$x > a$		(a, ∞)
	$x \geq a$		$[a, \infty)$
	$x < a$		$(-\infty, a)$
	$x \leq a$		$(-\infty, a]$
	$-\infty < x < \infty$		$(-\infty, \infty)$

TABLE 7-1

Inequalities that are true for all numbers x, such as $x + 1 > x$, are called **absolute inequalities.** Inequalities that are true for some numbers x, but not all numbers x, such as $3x + 2 < 8$, are called **conditional inequalities.**

If a and b are two real numbers, then $a < b$, $a = b$, or $a > b$. This property, called the **trichotomy property,** indicates that one and only one of three statements is true about any two real numbers. Either

- the first number is less than the second,
- the first number is equal to the second, or
- the first number is greater than the second.

If a, b, and c are real numbers with $a < b$ and $b < c$, then $a < c$. This property, called the **transitive property,** indicates that if we have three numbers and

- the first number is less than the second and
- the second number is less than the third, then
- the first number is less than the third.

To solve an inequality, we use the following properties of inequalities.

Properties of Inequalities

1. Any real number can be added to (or subtracted from) both sides of an inequality to produce another inequality with the same direction.
2. If both sides of an inequality are multiplied (or divided) by a positive number, another inequality results with the same direction as the original inequality.
3. If both sides of an inequality are multiplied (or divided) by a negative number, another inequality results, but with the opposite direction from the original inequality.

Property 1 indicates that any number can be added to both sides of a true inequality to get another true inequality with the same direction. For example, if 4 is added to both sides of the inequality $3 < 12$, we get

$$3 + 4 < 12 + 4$$
$$7 < 16$$

and the $<$ symbol remains an $<$ symbol. Adding 4 to both sides does not change the direction (sometimes called the **order**) of the inequality.

Subtracting 4 from both sides of $3 < 12$ does not change the direction of the inequality either.

$$3 - 4 < 12 - 4$$
$$-1 < 8$$

Property 2 indicates that both sides of a true inequality can be multiplied by any positive number to get another true inequality with the same direction. For example, if both sides of the true inequality $-4 < 6$ are multiplied by 2, we get

$$2(-4) < 2(6)$$
$$-8 < 12$$

and the $<$ symbol remains an $<$ symbol. Multiplying both sides by 2 does not change the direction of the inequality.

Dividing both sides by 2 does not change the direction of the inequality either.

$$\frac{-4}{2} < \frac{6}{2}$$
$$-2 < 3$$

Property 3 indicates that if both sides of a true inequality are multiplied by any negative number, another true inequality results, but with the opposite direction. For example, if both sides of the true inequality $-4 < 6$ are multiplied by -2, we get

$$-4 < 6$$
$$-2(-4) > -2(6)$$
$$8 > -12$$

and the $<$ symbol becomes an $>$ symbol. Multiplying both sides by -2 reverses the direction of the inequality.

Dividing both sides by -2 also reverses the direction of the inequality.

$$-4 < 6$$

$$\frac{-4}{-2} > \frac{6}{-2}$$

$$2 > -3$$

WARNING! We must remember to reverse the inequality symbol every time we multiply or divide both sides by a negative number.

A **linear inequality** is any inequality that can be expressed in the form

$$ax + c < 0 \qquad ax + c > 0 \qquad ax + c \leq 0 \qquad \text{or} \qquad ax + c \geq 0 \quad (a \neq 0)$$

We can solve linear inequalities by using the same steps that we use for solving linear equations, with one exception. If we multiply or divide both sides by a *negative* number, we must reverse the direction of the inequality.

EXAMPLE 7   Solve   **a.** $3(2x - 9) < 9$ and   **b.** $-4(3x + 2) \leq 16$.

Solution   **a.** We solve the inequality as if it were an equation:

$$3(2x - 9) < 9$$

$6x - 27 < 9$	Use the distributive property to remove parentheses.
$6x < 36$	Add 27 to both sides.
$x < 6$	Divide both sides by 6.

The solution set is the interval $(-\infty, 6)$. The graph of the solution set is shown in Figure 7-3(a).

b. We solve the inequality as if it were an equation:

$$-4(3x + 2) \leq 16$$

$-12x - 8 \leq 16$	Use the distributive property to remove parentheses.
$-12x \leq 24$	Add 8 to both sides.
$x \geq -2$	Divide both sides by -12 and reverse the $\leq$ symbol.

The solution set is the interval $[-2, \infty)$. The graph of the solution set is shown in Figure 7-3(b).

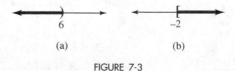

(a)   (b)

FIGURE 7-3

Self Check

Solve $-3(2x + 1) > 9$.

Answer

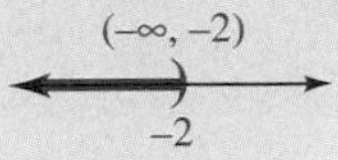

$(-\infty, -2)$

-2

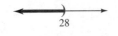

EXAMPLE 8

Solve $\dfrac{2}{3}(x + 2) > \dfrac{4}{5}(x - 3)$.

Solution

$$\dfrac{2}{3}(x + 2) > \dfrac{4}{5}(x - 3)$$

$$\mathbf{15} \cdot \dfrac{2}{3}(x + 2) > \mathbf{15} \cdot \dfrac{4}{5}(x - 3)$$ To eliminate the fractions, multiply both sides by 15.

$$10(x + 2) > 12(x - 3)$$ $15 \cdot \frac{2}{3} = 10$ and $15 \cdot \frac{4}{5} = 12$.

$$10x + 20 > 12x - 36$$ Use the distributive property to remove parentheses.

$$-2x + 20 > -36$$ Add $-12x$ to both sides.

$$-2x > -56$$ Subtract 20 from both sides.

$$x < 28$$ Divide both sides by -2 and reverse the $>$ symbol.

28

FIGURE 7-4

The solution set is the interval $(-\infty, 28)$, whose graph is shown in Figure 7-4. ■

Self Check

Solve $\frac{1}{2}(x - 1) \leq \frac{2}{3}(x + 1)$.

Answer

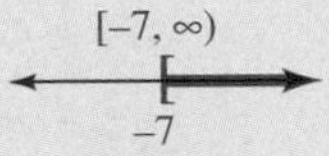

$[-7, \infty)$

-7

■ COMPOUND INEQUALITIES

To say that x is between -3 and 8, we write a double inequality:

$$-3 < x < 8$$ Read as "-3 is less than x and x is less than 8."

This double inequality contains two different linear inequalities:

$$-3 < x \qquad \text{and} \qquad x < 8$$

These two inequalities mean that $x > -3$ and $x < 8$. The word *and* indicates that these two inequalities are true at the same time.

Double Inequalities
The double inequality $c < x < d$ is equivalent to $c < x$ and $x < d$.

WARNING!   The inequality $c < x < d$ cannot be expressed as

$$c < x \quad \text{or} \quad x < d$$

EXAMPLE 9   Solve $-3 \leq 2x + 5 < 7$.

Solution   This inequality means that $2x + 5$ is between -3 and 7. We can solve it by isolating x between the inequality symbols:

$$-3 \leq 2x + 5 < 7$$
$$-8 \leq 2x < 2 \qquad \text{Subtract 5 from all three parts.}$$
$$-4 \leq x < 1 \qquad \text{Divide all three parts by 2.}$$

FIGURE 7-5

The solution set is the interval $[-4, 1)$. Its graph is shown in Figure 7-5.   ■

Self Check   Solve $-3 < 2x - 5 \leq 9$.

Answer

$(1, 7]$

EXAMPLE 10   Solve $x + 3 < 2x - 1 < 4x - 3$.

Solution   Since it is impossible to isolate x between the inequality symbols, we solve each of its linear inequalities separately.

$$x + 3 < 2x - 1 \qquad \text{and} \qquad 2x - 1 < 4x - 3$$
$$4 < x \qquad\qquad\qquad 2 < 2x$$
$$\qquad\qquad\qquad 1 < x$$

FIGURE 7-6

Only those numbers x where $x > 4$ and $x > 1$ are in the solution set. Since all numbers greater than 4 are also greater than 1, the solutions are the numbers x where $x > 4$. The solution set is the interval $(4, \infty)$. The graph is shown in Figure 7-6.   ■

Self Check   Solve $x - 5 \leq 3x - 1 \leq 5x + 5$.

Answer

$[-2, \infty)$

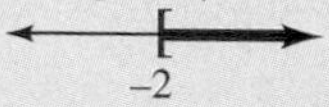

EXAMPLE 11   Solve the compound inequality $x \leq -3$ or $x \geq 8$.

Solution   The graph of the compound inequality $x \leq -3$ or $x \geq 8$ is the union of two intervals:

$$(-\infty, -3] \cup [8, \infty)$$

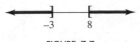

FIGURE 7-7

Its graph is shown in Figure 7-7.

The word *or* in the statement $x \le -3$ or $x \ge 8$ indicates that only one of the inequalities needs to be true to make the statement true. ∎

Self Check

Solve $x < -2$ or $x > 4$.

Answer

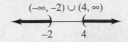

WARNING! In the statement $x \le -3$ or $x \ge 8$, it is incorrect to string the equalities together as $8 \le x \le -3$, because that would imply that $8 \le -3$, which is false.

Orals *Solve each equation or inequality.*

1. $2x + 4 = 6$

2. $3x - 4 = 8$

3. $2x < 4$

4. $3x + 1 \ge 10$

5. $-3x > 12$

6. $-\dfrac{x}{2} \le 4$

EXERCISE 7.1

REVIEW *Simplify each expression.*

1. $\left(\dfrac{t^3 t^5 t^{-6}}{t^2 t^{-4}} \right)^{-3}$

2. $\left(\dfrac{a^{-2} b^3 a^5 b^{-2}}{a^6 b^{-5}} \right)^{-4}$

3. A man invested $1,200 in baking equipment to make pies. Each pie requires $3.40 in ingredients. If the man can sell all the pies he can make for $5.95 each, how many pies will he have to make to earn a profit?

4. A woman invested $15,000, part at 7% annual interest and the rest at 8%. If she earned $2,200 in income over a two-year period, how much did she invest at 7%?

VOCABULARY AND CONCEPTS *Fill in each blank to make a true statement.*

5. An _____ is a statement indicating that two mathematical expressions are equal.

6. If any quantity is _____ to both sides of an equation, a new equation is formed that is equivalent to the original equation.

7. If both sides of an equation are _____ (or _____) by the same nonzero number, a new equation is formed that is equivalent to the original equation.

8. An _____ is an equation that is true for all values of its variable.

9. An _____ equation is an equation that is true for no values of its variable.

10. The symbol $<$ is read as "_____."

11. The symbol $\geq$ is read as "_____ or equal to."

12. An open interval has no _____.

13. A _____ interval has one endpoint.

14. If $a < b$ and $b < c$, then _____.

15. If both sides of an inequality are multiplied by a _____ number, a new inequality is formed that has the same direction as the first.

16. If both sides of an inequality are multiplied by a _____ number, a new inequality is formed that has the opposite direction from the first.

PRACTICE   *In Exercises 17–36, solve each equation.*

17. $2x + 1 = 13$

18. $2x - 4 = 16$

19. $3(x + 1) = 15$

20. $-2(x + 5) = 30$

21. $2r - 5 = 1 - r$

22. $5s - 13 = s - 1$

23. $3(2y - 4) - 6 = 3y$

24. $2x + (2x - 3) = 5$

25. $5(5 - a) = 37 - 2a$

26. $4a + 17 = 7(a + 2)$

27. $4(y + 1) = -2(4 - y)$

28. $5(r + 4) = -2(r - 3)$

29. $2(a - 5) - (3a + 1) = 0$

30. $8(3a - 5) - 4(2a + 3) = 12$

31. $\dfrac{x}{2} - \dfrac{x}{3} = 4$

32. $\dfrac{x}{2} + \dfrac{x}{3} = 10$

33. $\dfrac{x}{6} + 1 = \dfrac{x}{3}$

34. $\dfrac{3}{2}(y + 4) = \dfrac{20 - y}{2}$

35. $\dfrac{a + 1}{3} + \dfrac{a - 1}{5} = \dfrac{2}{15}$

36. $\dfrac{2z + 3}{3} + \dfrac{3z - 4}{6} = \dfrac{z - 2}{2}$

In Exercises 37–44, solve each equation. If the equation is an identity or an impossible equation, so indicate.

37. $4(2 - 3t) + 6t = -6t + 8$

38. $2x - 6 = -2x + 4(x - 2)$

39. $\dfrac{a + 1}{4} + \dfrac{2a - 3}{4} = \dfrac{a}{2} - 2$

40. $\dfrac{y - 8}{5} + 2 = \dfrac{2}{5} - \dfrac{y}{3}$

41. $3(x - 4) + 6 = -2(x + 4) + 5x$

42. $2(x - 3) = \dfrac{3}{2}(x - 4) + \dfrac{x}{2}$

43. $y(y + 2) = (y + 1)^2 - 1$

44. $x(x - 3) = (x - 1)^2 - (5 + x)$

In Exercises 45–54, solve each formula for the indicated variable.

45. $V = \dfrac{1}{3}Bh$ for B

46. $A = \dfrac{1}{2}bh$ for b

47. $p = 2l + 2w$ for w

48. $p = 2l + 2w$ for l

49. $z = \dfrac{x - \mu}{\sigma}$ for x

50. $z = \dfrac{x - \mu}{\sigma}$ for μ

51. $y = mx + b$ for x

52. $y = mx + b$ for m

53. $P = L + \dfrac{s}{f}i$ for s

54. $P = L + \dfrac{s}{f}i$ for f

In Exercises 55–80, solve each inequality. Give the result in interval notation and graph the solution set.

55. $5x - 3 > 7$

56. $7x - 9 < 5$

57. $-3x - 1 \leq 5$

58. $-2x + 6 \geq 16$

59. $8 - 9y \geq -y$

60. $4 - 3x \leq x$

61. $-3(a + 2) > 2(a + 1)$

62. $-4(y - 1) < y + 8$

63. $\dfrac{1}{2}y + 2 \geq \dfrac{1}{3}y - 4$

64. $\dfrac{1}{4}x - \dfrac{1}{3} \leq x + 2$

65. $-2 < -b + 3 < 5$

66. $4 < -t - 2 < 9$

67. $15 > 2x - 7 > 9$

68. $25 > 3x - 2 > 7$

69. $-6 < -3(x - 4) \leq 24$

70. $-4 \leq -2(x + 8) < 8$

71. $0 \geq \dfrac{1}{2}x - 4 > 6$

72. $-6 \leq \dfrac{1}{3}a + 1 < 0$

73. $0 \leq \dfrac{4 - x}{3} \leq 2$

74. $-2 \leq \dfrac{5 - 3x}{2} \leq 2$

75. $3x + 2 < 8$ or $2x - 3 > 11$

76. $3x + 4 < -2$ or $3x + 4 > 10$

77. $-4(x + 2) \geq 12$ or $3x + 8 < 11$

78. $5(x - 2) \geq 0$ and $-3x < 9$

79. $x < -3$ and $x > 3$

80. $x < 3$ or $x > -3$

APPLICATIONS

81. Cutting a board   The carpenter in Illustration 1 saws a board into two pieces. He wants one piece to be 1 foot longer than twice the length of the shorter piece. Find the length of each piece.

82. Cutting a beam   A 30-foot steel beam is to be cut into two pieces. The longer piece is to be 2 feet more than 3 times as long as the shorter piece. Find the length of each piece.

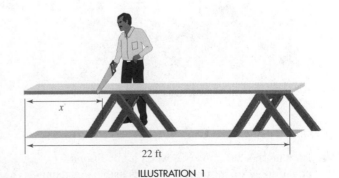

ILLUSTRATION 1

83. Finding dimensions   The rectangular garden shown in Illustration 2 is twice as long as it is wide. Find its dimensions.

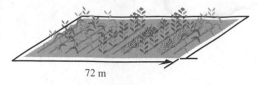

72 m

ILLUSTRATION 2

84. Fencing a pasture   A farmer has 624 feet of fencing to enclose the pasture shown in Illustration 3. Because a river runs along one side, fencing will be needed on only three sides. Find the dimensions of the pasture if its length is parallel to the river and is double its width.

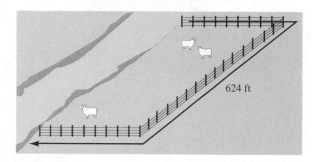

624 ft

ILLUSTRATION 3

85. Fencing a pen   A man has 150 feet of fencing to build the pen shown in Illustration 4. If one end is a square, find the outside dimensions of the entire pen.

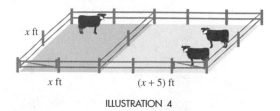

x ft

x ft $(x+5)$ ft

ILLUSTRATION 4

86. Enclosing a swimming pool   A woman wants to enclose the swimming pool shown in Illustration 5 and have a walkway of uniform width all the way around. How wide will the walkway be if the woman uses 180 feet of fencing?

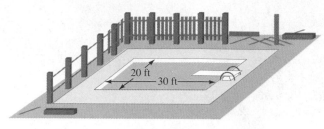

20 ft   30 ft

ILLUSTRATION 5

87. Finding profit   The wholesale cost of a radio is $27. A store owner knows that for the radio to sell, it must be priced under $42. If p is the profit, express the possible profit as an inequality.

88. Investing money   If a woman invests $10,000 at 8% annual interest, how much more must she invest at 9% so that her annual income will exceed $1,250?

89. Buying compact discs   A student can afford to spend up to $330 on a stereo system and some compact discs. If the stereo costs $175 and the discs are $8.50 each, find the greatest number of discs the student can buy.

90. Grade averages   A student has scores of 70, 77, and 85 on three exams. What score is needed on a fourth exam to make the student's average 80 or better?

WRITING

91. Explain the difference between a conditional equation, an identity, and an impossible equation.

92. The techniques for solving linear equations and linear inequalities are similar, yet different. Explain.

SOMETHING TO THINK ABOUT

93. Find the error:

$$4(x + 3) = 16$$
$$4x + 3 = 16$$
$$4x = 13$$
$$x = \frac{13}{4}$$

94. Which of these relations is transitive?
a. $=$ **b.** $\leq$ **c.** $\not=$ **d.** $\neq$

7.2 Equations Containing Absolute Values

■ ABSOLUTE VALUE ■ ABSOLUTE VALUE EQUATIONS ■ EQUATIONS WITH TWO ABSOLUTE VALUES

Getting Ready   *Simplify each expression.*

1. $-(-6)$ **2.** $-(-5)$ **3.** $-(x - 2)$ **4.** $-(2 - \pi)$

■ ABSOLUTE VALUE

We begin this section by reviewing the definition of the absolute value of x.

Absolute Value
If $x \geq 0$, then $|x| = x$.

If $x < 0$, then $|x| = -x$.

This definition provides a way for associating a nonnegative real number with any real number.

- If $x \geq 0$, then x is its own absolute value.
- If $x < 0$, then $-x$ (which is positive) is the absolute value.

Either way, $|x|$ is positive or 0:

$|x| \geq 0$ **for all real numbers** x

EXAMPLE 1   Find   **a.** $|9|$,   **b.** $|-5|$,   **c.** $|0|$,   and   **d.** $|2 - \pi|$.

Solution   **a.** Since $9 \geq 0$, 9 is its own absolute value: $|9| = 9$.

b. Since $-5 < 0$, the negative of -5 is the absolute value:

$$|-5| = -(-5) = 5$$

c. Since $0 \geq 0$, 0 is its own absolute value: $|0| = 0$.

d. Since $\pi \approx 3.14$, it follows that $2 - \pi < 0$. Thus,

$$|2 - \pi| = -(2 - \pi) = \pi - 2$$ ■

Self Check

Find **a.** $|-12|$ and **b.** $|4 - \pi|$.

Answers

a. 12,   **b.** $4 - \pi$

 WARNING! The placement of a $-$ sign in an expression containing an absolute value symbol is important. For example, $|-19| = 19$, but $-|19| = -19$.

EXAMPLE 2

Find **a.** $-|-10|$,   **b.** $-|13|$,   and   **c.** $-(-|-3|)$.

Solution

a. $-|-10| = -(10) = -10$

b. $-|13| = -13$

c. $-(-|-3|) = -(-3) = 3$ ■

Self Check

Find $-|-15|$.

Answer

-15

■ ABSOLUTE VALUE EQUATIONS

In the equation $|x| = 5$, x can be either 5 or -5, because

$$|5| = 5 \qquad \text{and} \qquad |-5| = 5$$

Thus, if $|x| = 5$, then $x = 5$ or $x = -5$. In general, the following is true.

Absolute Value Equations
If $k > 0$, then

$$|x| = k \qquad \text{is equivalent to} \qquad x = k \text{ or } x = -k$$

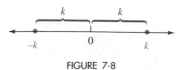

FIGURE 7-8

The absolute value of x can be interpreted as the distance on the number line from a point to the origin. The solutions of $|x| = k$ are represented by the two points that lie exactly k units from the origin. (See Figure 7-8.)

The equation $|x - 3| = 7$ indicates that a point on the number line with a co-ordinate of $x - 3$ is 7 units from the origin. Thus, $x - 3$ can be either 7 or -7.

$$x - 3 = 7 \qquad \text{or} \qquad x - 3 = -7$$
$$x = 10 \qquad\qquad x = -4$$

The solutions of the equation $|x - 3| = 7$ are 10 and -4. See Figure 7-9. If either of these numbers is substituted for x in the equation $|x - 3| = 7$, the equation is satisfied:

$$|x - 3| = 7 \qquad\qquad |x - 3| = 7$$
$$|\mathbf{10} - 3| \overset{?}{=} 7 \qquad\qquad |\mathbf{-4} - 3| \overset{?}{=} 7$$
$$|7| \overset{?}{=} 7 \qquad\qquad |-7| \overset{?}{=} 7$$
$$7 = 7 \qquad\qquad 7 = 7$$

FIGURE 7-9

EXAMPLE 3   Solve $|3x - 2| = 5$.

Solution   We can write $|3x - 2| = 5$ as

$$3x - 2 = 5 \quad \text{or} \quad 3x - 2 = -5$$

and solve each equation for x:

$$3x - 2 = 5 \qquad \text{or} \qquad 3x - 2 = -5$$
$$3x = 7 \qquad\qquad 3x = -3$$
$$x = \frac{7}{3} \qquad\qquad x = -1$$

Verify that both solutions check.   ■

Self Check   Solve $|2x + 3| = 5$.

Answer $1, -4$

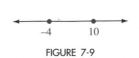

EXAMPLE 4   Solve $\left| \dfrac{2}{3}x + 3 \right| + 4 = 10$.

Solution   We first isolate the absolute value on the left-hand side.

$$\left| \frac{2}{3}x + 3 \right| + 4 = 10$$

1. $\left| \dfrac{2}{3}x + 3 \right| = 6$ Subtract 4 from both sides.

We can now write Equation 1 as

$$\frac{2}{3}x + 3 = 6 \qquad \text{or} \qquad \frac{2}{3}x + 3 = -6$$

and solve each equation for x:

$$\frac{2}{3}x + 3 = 6 \qquad \text{or} \qquad \frac{2}{3}x + 3 = -6$$

$$\frac{2}{3}x = 3 \qquad\qquad\qquad \frac{2}{3}x = -9$$

$$2x = 9 \qquad\qquad\qquad 2x = -27$$

$$x = \frac{9}{2} \qquad\qquad\qquad x = \frac{-27}{2}$$

Verify that both solutions check. ∎

Self Check Solve $\left|\frac{3}{2}x - 3\right| + 1 = 7$.

Answer 6, −2

EXAMPLE 5 Solve $\left| 7x + \frac{1}{2} \right| = -4$, if possible.

Solution Since the absolute value of a number cannot be negative, no value of x can make $\left| 7x + \frac{1}{2} \right| = -4$. Since this equation has no solutions, its solution set is $\varnothing$. ∎

Self Check Solve $-|3x + 2| = 4$, if possible.

Answer no solution

EXAMPLE 6 Solve $\left| \frac{1}{2}x - 5 \right| - 4 = -4$.

Solution We first isolate the absolute value on the left-hand side.

$$\left| \frac{1}{2}x - 5 \right| - 4 = -4$$

$$\left| \frac{1}{2}x - 5 \right| = 0 \qquad \text{Add 4 to both sides.}$$

Since 0 is the only number whose absolute value is 0, the binomial $\frac{1}{2}x - 5$ must be 0, and we have

$$\frac{1}{2}x - 5 = 0$$

$$\frac{1}{2}x = 5 \qquad \text{Add 5 to both sides.}$$

$$x = 10 \qquad \text{Multiply both sides by 2.}$$

Verify that 10 satisfies the original equation. ∎

Self Check	Solve $\left|\frac{2}{3}x - 4\right| + 2 = 2$.
Answer	6

■ EQUATIONS WITH TWO ABSOLUTE VALUES

The equation $|a| = |b|$ is true when $a = b$ or when $a = -b$. For example,

$$|3| = |3| \qquad \text{or} \qquad |3| = |-3|$$
$$3 = 3 \qquad\qquad\qquad 3 = 3$$

Thus, we have the following result.

Equations with Two Absolute Values
If a and b represent algebraic expressions, the equation $|a| = |b|$ is equivalent to the pair of equations

$$a = b \qquad \text{or} \qquad a = -b$$

EXAMPLE 7   Solve $|5x + 3| = |3x + 25|$.

Solution   This equation is true when $5x + 3 = 3x + 25$, or when $5x + 3 = -(3x + 25)$. We solve each equation for x.

$$
\begin{array}{rcl}
5x + 3 = 3x + 25 & \text{or} & 5x + 3 = -(3x + 25) \\
2x = 22 & & 5x + 3 = -3x - 25 \\
x = 11 & & 8x = -28 \\
& & x = -\dfrac{28}{8} \\
& & x = -\dfrac{7}{2}
\end{array}
$$

Verify that both solutions check.   ■

Self Check	Solve $|4x - 3| = |2x + 5|$.
Answer	$4, -\frac{1}{3}$

Orals   *Find each absolute value.*

1. $|-5|$ 2. $-|5|$ 3. $-|-6|$ 4. $-|4|$

Solve each equation.

5. $|x| = 8$ 6. $|x| = -5$
7. $|x - 5| = 0$ 8. $|x + 1| = 1$

EXERCISE 7.2

REVIEW   *Solve each equation.*

1. $3(2a - 1) = 2a$

2. $\dfrac{t}{6} - \dfrac{t}{3} = -1$

3. $\dfrac{5x}{2} - 1 = \dfrac{x}{3} + 12$

4. $4b - \dfrac{b + 9}{2} = \dfrac{b + 2}{5} - \dfrac{8}{5}$

VOCABULARY AND CONCEPTS   *Fill in each blank to make a true statement.*

5. If $x \geq 0$, then $|x| = $ __.

6. If $x < 0$, then $|x| = $ ____.

7. $|x| \geq$ __ for all real numbers x

8. If $k > 0$, then $|x| = k$ is equivalent to _____.

9. If $|a| = |b|$, then $a = b$ or _____.

10. If $k > 0$, the equation $|x| = k$ has ____ solutions.

PRACTICE   *In Exercises 11–22, find the value of each expression.*

11. $|8|$

12. $|-18|$

13. $|-12|$

14. $|15|$

15. $-|2|$

16. $-|-20|$

17. $-|-30|$

18. $-|25|$

19. $-(-|50|)$

20. $-(-|-20|)$

21. $|\pi - 4|$

22. $|2\pi - 4|$

In Exercises 23–34, select the smaller of the two numbers.

23. $|2|, |5|$

24. $|-6|, |2|$

25. $|5|, |-8|$

26. $|6|, |3|$

27. $|-2|, |10|$

28. $|-6|, -|6|$

29. $|-3|, -|-4|$

30. $|-3|, |-2|$

31. $-|-5|, -|-7|$

32. $-|-8|, -|20|$

33. $-x, |x + 1|$ $(x > 0)$

34. $y, |y - 1|$ $(y > 0)$

In Exercises 35–68, solve each equation, if possible.

35. $|x| = 8$

36. $|x| = 9$

37. $|x - 3| = 6$

38. $|x + 4| = 8$

39. $|2x - 3| = 5$

40. $|4x - 4| = 20$

41. $|3x + 2| = 16$

42. $|5x - 3| = 22$

43. $\left| \dfrac{7}{2}x + 3 \right| = -5$

44. $|2x + 10| = 0$

45. $\left| \dfrac{x}{2} - 1 \right| = 3$

46. $\left| \dfrac{4x - 64}{4} \right| = 32$

47. $|3 - 4x| = 5$

48. $|8 - 5x| = 18$

49. $|3x + 24| = 0$

50. $|x - 21| = -8$

51. $\left| \dfrac{3x + 48}{3} \right| = 12$

52. $\left| \dfrac{x}{2} + 2 \right| = 4$

53. $|x + 3| + 7 = 10$

54. $|2 - x| + 3 = 5$

55. $\left| \dfrac{3}{5}x - 4 \right| - 2 = -2$

56. $\left| \dfrac{3}{4}x + 2 \right| + 4 = 4$

57. $|2x + 1| = |3x + 3|$

58. $|5x - 7| = |4x + 1|$

59. $|3x - 1| = |x + 5|$

60. $|3x + 1| = |x - 5|$

61. $|2 - x| = |3x + 2|$

62. $|4x + 3| = |9 - 2x|$

63. $\left| \dfrac{x}{2} + 2 \right| = \left| \dfrac{x}{2} - 2 \right|$

64. $|7x + 12| = |x - 6|$

65. $\left| x + \dfrac{1}{3} \right| = |x - 3|$

66. $\left| x - \dfrac{1}{4} \right| = |x + 4|$

67. $|3x + 7| = -|8x - 2|$

68. $-|17x + 13| = |3x - 14|$

WRITING

69. Explain how to find the absolute value of a given number.

70. Explain why the equation $|x| + 5 = 0$ has no solution.

SOMETHING TO THINK ABOUT

71. For what values of k does $|x| + k = 0$ have exactly two solutions?

72. For what value of k does $|x| + k = 0$ have exactly one solution?

73. Construct several examples to show that $|a \cdot b| = |a| \cdot |b|$.

74. Construct several examples to show that $\left| \dfrac{a}{b} \right| = \dfrac{|a|}{|b|}$.

75. Construct several examples to show that $|a + b| \neq |a| + |b|$.

76. Construct several examples to show that $|a - b| \neq |a| - |b|$.

7.3 Inequalities Containing Absolute Values

■ INEQUALITIES OF THE FORM $|x| < k$ ■ INEQUALITIES OF THE FORM $|x| > k$

Getting Ready   *Solve each inequality.*

1. $2x + 3 > 5$ **2.** $-3x - 1 < 5$ **3.** $2x - 5 \leq 9$

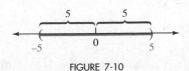

FIGURE 7-10

■ INEQUALITIES OF THE FORM $|x| < k$

The inequality $|x| < 5$ indicates that a point with coordinate x is less than 5 units from the origin. (See Figure 7-10.)

Thus, x is between -5 and 5, and

$$|x| < 5 \qquad \text{is equivalent to} \qquad -5 < x < 5$$

The solution to the inequality $|x| < k$ ($k > 0$) includes the coordinates of the points on the number line that are less than k units from the origin. (See Figure 7-11.)

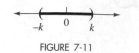

FIGURE 7-11

We have the following facts.

$$|x| < k \qquad \text{is equivalent to} \qquad -k < x < k \quad (k > 0)$$
$$|x| \leq k \qquad \text{is equivalent to} \qquad -k \leq x \leq k \quad (k \geq 0)$$

EXAMPLE 1   Solve $|2x - 3| < 9$.

Solution   We write the inequality as the double inequality

$$-9 < 2x - 3 < 9$$

and solve for x:

$$-9 < 2x - 3 < 9$$
$$-6 < 2x < 12 \qquad \text{Add 3 to all three parts.}$$
$$-3 < x < 6 \qquad \text{Divide all parts by 2.}$$

Any number between -3 and 6, not including either -3 or 6, is in the solution set. This is the interval $(-3, 6)$. The graph is shown in Figure 7-12.   ■

FIGURE 7-12

Self Check   Solve $|3x + 1| < 5$.

Answer

$(-2, 4/3)$

EXAMPLE 2   Solve $|3x + 2| \leq 5$.

Solution   We write the expression as the double inequality

$$-5 \leq 3x + 2 \leq 5$$

and solve for x:

$$-5 \leq 3x + 2 \leq 5$$
$$-7 \leq 3x \leq 3 \qquad \text{Subtract 2 from all three parts.}$$
$$-\frac{7}{3} \leq x \leq 1 \qquad \text{Divide all three parts by 3.}$$

FIGURE 7-13

The solution set is the interval $\left[-\frac{7}{3}, 1\right]$, whose graph is shown in Figure 7-13.   ■

Self Check
Answer

Solve $|2x - 3| \leq 5$.

[-1, 4]

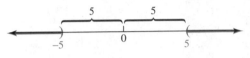

■ INEQUALITIES OF THE FORM $|x| > k$

The inequality $|x| > 5$ can be interpreted to mean that a point with coordinate x is more than 5 units from the origin. (See Figure 7-14.)

FIGURE 7-14

Thus, $x < -5$ or $x > 5$.

In general, the inequality $|x| > k$ $(k > 0)$ can be interpreted to mean that a point with coordinate x is more than k units from the origin. (See Figure 7-15.) Thus,

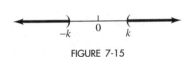

FIGURE 7-15

$$|x| > k \quad \text{is equivalent to} \quad x < -k \text{ or } x > k$$

The *or* indicates an either/or situation. It is necessary for x to satisfy only one of the two conditions to be in the solution set.

If k is a nonnegative constant, then

$$|x| > k \quad \text{is equivalent to} \quad x < -k \text{ or } x > k$$
$$|x| \geq k \quad \text{is equivalent to} \quad x \leq -k \text{ or } x \geq k$$

EXAMPLE 3

Solve $|5x - 10| > 20$.

Solution

We write the inequality as two separate inequalities

$$5x - 10 < -20 \quad \text{or} \quad 5x - 10 > 20$$

and solve each one for x:

$$
\begin{array}{llll}
5x - 10 < -20 & \text{or} & 5x - 10 > 20 & \\
5x < -10 & & 5x > 30 & \text{Add 10 to both sides.} \\
x < -2 & & x > 6 & \text{Divide both sides by 5.}
\end{array}
$$

Thus, x is either less than -2 or greater than 6:

$$x < -2 \text{ or } x > 6$$

FIGURE 7-16

This is the interval $(-\infty, -2) \cup (6, \infty)$. The graph appears in Figure 7-16. ■

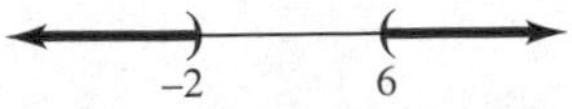

Self Check   Solve $|3x - 2| > 4$.
Answer $(-\infty, -2/3) \cup (2, \infty)$

EXAMPLE 4   Solve $\left| \dfrac{3 - x}{5} \right| \geq 6$.

Solution   We write the inequality as two separate inequalities

$$\frac{3 - x}{5} \leq -6 \qquad \text{or} \qquad \frac{3 - x}{5} \geq 6$$

and solve each one for x:

$$\frac{3 - x}{5} \leq -6 \qquad \text{or} \qquad \frac{3 - x}{5} \geq 6$$

$3 - x \leq -30$	$3 - x \geq 30$	Multiply both sides by 5.
$-x \leq -33$	$-x \geq 27$	Subtract 3 from both sides.
$x \geq 33$	$x \leq -27$	Divide both sides by -1 and reverse the direction of the inequality symbol.

The solution set is the interval $(-\infty, -27] \cup [33, \infty)$, whose graph appears in Figure 7-17. ■

FIGURE 7-17

Self Check   Solve $\left| \frac{4 - x}{3} \right| \geq 2$.
Answer $(-\infty, -2] \cup [10, \infty)$

EXAMPLE 5   Solve $\left| \dfrac{2}{3}x - 2 \right| - 3 > 6$.

Solution   We begin by adding 3 to both sides to isolate the absolute value on the left-hand side. We then proceed as follows:

$$\left| \frac{2}{3}x - 2 \right| - 3 > 6$$

$$\left| \frac{2}{3}x - 2 \right| > 9 \qquad\qquad \text{Add 3 to both sides.}$$

$$\frac{2}{3}x - 2 < -9 \qquad \text{or} \qquad \frac{2}{3}x - 2 > 9$$

$$\frac{2}{3}x < -7 \qquad\qquad \frac{2}{3}x > 11 \qquad\qquad \text{Add 2 to both sides.}$$

$$2x < -21 \qquad\qquad 2x > 33 \qquad\qquad \text{Multiply both sides by 3.}$$

$$x < -\frac{21}{2} \qquad\qquad x > \frac{33}{2} \qquad\qquad \text{Divide both sides by 2.}$$

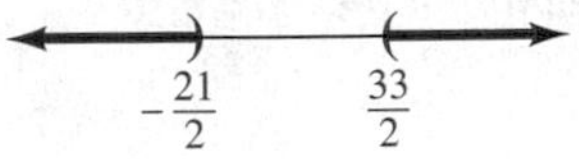

FIGURE 7-18

The solution set is $\left(-\infty, -\frac{21}{2}\right) \cup \left(\frac{33}{2}, \infty\right)$, whose graph appears in Figure 7-18. ■

Self Check   Solve $\left|\frac{3}{2}x + 1\right| - 2 > 1$.

Answer $(-\infty, -8/3) \cup (4/3, \infty)$

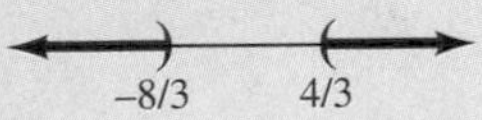

$-8/3 \qquad 4/3$

EXAMPLE 6   Solve $|3x - 5| \geq -2$.

Solution   Since the absolute value of any number is non-negative, and since any nonnegative number is larger than -2, the inequality is true for all x. The solution set is the interval $(-\infty, \infty)$, whose graph appears in Figure 7-19.

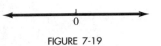

0

FIGURE 7-19

■

Self Check   Solve $|2x + 3| > -5$.

Answer $(-\infty, \infty)$

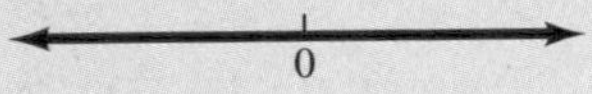

0

Orals   *Solve each inequality.*

1. $|x| < 8$
2. $|x| > 8$
3. $|x| \geq 4$
4. $|x| \leq 7$
5. $|x + 1| < 2$
6. $|x + 1| > 2$

EXERCISE 7.3

REVIEW   *Solve each formula for the given variable.*

1. $A = p + prt$ for t
2. $A = p + prt$ for r
3. $P = 2w + 2l$ for l
4. $V = \frac{1}{3}Bh$ for B

VOCABULARY AND CONCEPTS   *Fill in each blank to make a true statement.*

5. If $k > 0$, then $|x| < k$ is equivalent to _____.

6. If $k > 0$, then _____ is equivalent to $-k \leq x \leq k$.

7. If k is a nonnegative constant, then $|x| > k$ is equivalent to _____.

8. If k is a nonnegative constant, then _____ is equivalent to $x \leq -k$ or $x \geq k$.

PRACTICE   *In Exercises 9–52, solve each inequality. Write the solution set in interval notation and graph it.*

9. $|2x| < 8$

10. $|3x| < 27$

11. $|x + 9| \leq 12$

12. $|x - 8| \leq 12$

13. $|3x + 2| \leq -3$

14. $|3x - 2| \leq 10$

15. $|4x - 1| \leq 7$

16. $|5x - 12| < -5$

17. $|3 - 2x| < 7$

18. $|4 - 3x| \leq 13$

19. $|5x| > 5$

20. $|7x| > 7$

21. $|x - 12| > 24$

22. $|x + 5| \geq 7$

23. $|3x + 2| > 14$

24. $|2x - 5| \geq 25$

25. $|4x + 3| > -5$

26. $|4x + 3| > 0$

27. $|2 - 3x| \geq 8$

28. $|-1 - 2x| > 5$

29. $-|2x - 3| < -7$

30. $-|3x + 1| < -8$

31. $|8x - 3| > 0$

32. $|7x + 2| > -8$

33. $\left| \dfrac{x - 2}{3} \right| \leq 4$

34. $\left| \dfrac{x - 2}{3} \right| > 4$

35. $|3x + 1| + 2 < 6$

36. $|3x - 2| + 2 \geq 0$

37. $3|2x + 5| \geq 9$

38. $-2|3x - 4| < 16$

39. $|5x - 1| + 4 \le 0$

40. $-|5x - 1| + 2 < 0$

41. $\left| \dfrac{1}{3}x + 7 \right| + 5 > 6$

42. $\left| \dfrac{1}{2}x - 3 \right| - 4 < 2$

43. $\left| \dfrac{1}{5}x - 5 \right| + 4 > 4$

44. $\left| \dfrac{1}{6}x + 6 \right| + 2 < 2$

45. $\left| \dfrac{3}{5}x + \dfrac{7}{3} \right| < 2$

46. $\left| \dfrac{7}{3}x - \dfrac{3}{5} \right| \ge 1$

47. $\left| 3\left(\dfrac{x + 4}{4} \right) \right| > 0$

48. $3\left| \dfrac{1}{3}(x - 2) \right| + 2 \le 3$

49. $\left| \dfrac{1}{7}x + 1 \right| \le 0$

50. $|2x + 1| + 2 \le 2$

51. $\left| \dfrac{x - 5}{10} \right| \le 0$

52. $\left| \dfrac{3}{5}x - 2 \right| + 3 \le 3$

WRITING

53. Explain how parentheses and brackets are used when graphing inequalities.

54. If $k > 0$, explain the differences between the solution sets of $|x| < k$ and $|x| > k$.

SOMETHING TO THINK ABOUT

55. Under what conditions is $|x| + |y| > |x + y|$?

56. Under what conditions is $|x| + |y| = |x + y|$?

7.4 Review of Factoring

■ FACTORING OUT THE GREATEST COMMON FACTOR ■ FACTORING BY GROUPING ■ FORMULAS
■ FACTORING THE DIFFERENCE OF TWO SQUARES ■ FACTORING TRINOMIALS ■ TEST FOR
FACTORABILITY ■ USING GROUPING TO FACTOR TRINOMIALS

Getting Ready   *Do each multiplication.*

1. $3x^2y(2x - y)$

2. $(x + 2)(x - 2)$

3. $(x + 2)(x - 3)$

4. $(2x + 3)(3x - 1)$

In this section, we will review the basic types of factoring discussed in Chapter 5.

■ FACTORING OUT THE GREATEST COMMON FACTOR

EXAMPLE 1   Factor $3xy^2z^3 + 6xz^2 - 9xyz^4$.

Solution   We begin by factoring each monomial:

$$3xy^2z^3 = \mathbf{3} \cdot \mathbf{x} \cdot y \cdot y \cdot \mathbf{z} \cdot \mathbf{z} \cdot z$$
$$6xz^2 = \mathbf{3} \cdot 2 \cdot \mathbf{x} \cdot \mathbf{z} \cdot \mathbf{z}$$
$$-9xyz^4 = -\mathbf{3} \cdot 3 \cdot \mathbf{x} \cdot y \cdot \mathbf{z} \cdot \mathbf{z} \cdot z \cdot z$$

Since each term has one factor of 3, one factor of x, and two factors of z, and there are no other common factors, $3xz^2$ is the greatest common factor of the three terms. We can use the distributive property to factor out $3xz^2$.

$$3xy^2z^3 + 6xz^2 - 9xyz^4 = \mathbf{3xz^2} \cdot y^2z + \mathbf{3xz^2} \cdot 2 - \mathbf{3xz^2} \cdot 3yz^2$$
$$= \mathbf{3xz^2}(y^2z + 2 - 3yz^2)$$ ■

Self Check   Factor $4ab^3 - 6a^2b^2$.
Answer $2ab^2(2b - 3a)$

EXAMPLE 2   Factor the negative of the greatest common factor from $-6u^2v^3 + 8u^3v^2$.

Solution   Because the greatest common factor of the two terms is $2u^2v^2$, the negative of the greatest common factor is $-2u^2v^2$. To factor out $-2u^2v^2$, we proceed as follows:

$$-6u^2v^3 + 8u^3v^2 = -2u^2v^2 \cdot 3v + 2u^2v^2 \cdot 4u$$
$$= -\mathbf{2u^2v^2} \cdot 3v - (-\mathbf{2u^2v^2})4u$$
$$= -\mathbf{2u^2v^2}(3v - 4u)$$ ■

Self Check   Factor the negative of the greatest common factor from $-3p^3q + 6p^2q^2$.
Answer $-3p^2q(p - 2q)$

We can also factor out a common factor with a variable exponent.

EXAMPLE 3   Factor x^{2n} from $x^{4n} + x^{3n} + x^{2n}$.

Solution   We can write the trinomial in the form

$$\mathbf{x^{2n}} \cdot x^{2n} + \mathbf{x^{2n}} \cdot x^n + \mathbf{x^{2n}} \cdot 1$$

and factor out x^{2n}.

$$x^{4n} + x^{3n} + x^{2n} = \mathbf{x^{2n}} \cdot x^{2n} + \mathbf{x^{2n}} \cdot x^n + \mathbf{x^{2n}} \cdot 1$$
$$= \mathbf{x^{2n}}(x^{2n} + x^n + 1)$$ ■

Self Check

Factor $2a^n$ from $6a^{2n} - 4a^{n+1}$.

Answer

$2a^n(3a^n - 2a)$

■ FACTORING BY GROUPING

Suppose we wish to factor

$$ac + ad + bc + bd$$

Although there is no factor common to all four terms, there is a factor of a in the first two terms and a factor of b in the last two terms. We can factor out these common factors to get

$$ac + ad + bc + bd = a(c + d) + b(c + d)$$

We can now factor out the common factor of $c + d$ on the right-hand side:

$$ac + ad + bc + bd = (c + d)(a + b)$$

The grouping in this type of problem is not always unique. For example, if we write the expression $ac + ad + bc + bd$ in the form

$$ac + bc + ad + bd$$

and factor c from the first two terms and d from the last two terms, we obtain

$$ac + bc + ad + bd = c(a + b) + d(a + b)$$
$$= (a + b)(c + d)$$

EXAMPLE 4

Factor $3ax^2 + 3bx^2 + a + 5bx + 5ax + b$.

Solution

Although there is no factor common to all six terms, $3x^2$ can be factored out of the first two terms, and $5x$ can be factored out of the fourth and fifth terms to get

$$3ax^2 + 3bx^2 + a + 5bx + 5ax + b = 3x^2(a + b) + a + 5x(b + a) + b$$

This result can be written in the form

$$3ax^2 + 3bx^2 + a + 5bx + 5ax + b = 3x^2(a + b) + 5x(a + b) + (a + b)$$

Since $a + b$ is common to all three terms, it can be factored out to get

$$3ax^2 + 3bx^2 + a + 5bx + 5ax + b = (a + b)(3x^2 + 5x + 1)$$ ■

Self Check

Factor $2mp - np + 2mq - nq$.

Answer

$(p + q)(2m - n)$

■ FORMULAS

Factoring is often required to solve a formula for one of its variables.

EXAMPLE 5   The formula $r_1 r_2 = rr_2 + rr_1$ is used in electronics to relate the combined resistance r of two resistors wired in parallel. The variable r_1 represents the resistance of the first resistor, and the variable r_2 represents the resistance of the second resistor. Solve the formula for r_2.

Solution   To isolate r_2 on one side of the equation, we get all terms involving r_2 on the left-hand side and all terms not involving r_2 on the right-hand side. We then proceed as follows:

$$r_1 r_2 = rr_2 + rr_1$$
$$r_1 r_2 - rr_2 = rr_1 \qquad \text{Subtract } rr_2 \text{ from both sides.}$$
$$r_2(r_1 - r) = rr_1 \qquad \text{Factor out } r_2 \text{ on the left-hand side.}$$
$$r_2 = \frac{rr_1}{r_1 - r} \qquad \text{Divide both sides by } r_1 - r. \qquad ■$$

Self Check   Solve $f_1 f_2 = ff_1 + ff_2$ for f_1.

Answer $f_1 = \dfrac{ff_2}{f_2 - f}$

■ FACTORING THE DIFFERENCE OF TWO SQUARES

Recall the formula for factoring the difference of two squares.

> **Factoring the Difference of Two Squares**
> $$x^2 - y^2 = (x + y)(x - y)$$

If we think of the difference of two squares as the square of a **First** quantity minus the square of a **Last** quantity, we have the formula

$$F^2 - L^2 = (F + L)(F - L)$$

In words, we say

> *To factor the square of a **First** quantity minus the square of a **Last** quantity, we multiply the **First** plus the **Last** by the **First** minus the **Last**.*

To factor $49x^2 - 16$, for example, we write $49x^2 - 16$ in the form $(7x)^2 - (4)^2$ and use the formula for factoring the difference of two squares:

$$49x^2 - 16 = (7x)^2 - (4)^2$$
$$= (7x + 4)(7x - 4)$$

We can verify this result by multiplying $7x + 4$ and $7x - 4$.

$$
\begin{array}{c}
\overset{\text{F}\qquad\quad\text{L}}{(7x + 4)(7x - 4)} = 49x^2 - 28x + 28x - 16 \\
\qquad\qquad\qquad\quad = 49x^2 - 16
\end{array}
$$

We note that if $49x^2 - 16$ is divided by $7x + 4$, the quotient is $7x - 4$, and that if $49x^2 - 16$ is divided by $7x - 4$, the quotient is $7x + 4$.

> **WARNING!** Expressions such as $(7x)^2 + (4)^2$ that represent the sum of two squares cannot be factored in the real number system. The binomial $49x^2 + 16$ is a prime binomial.

EXAMPLE 6   Factor $(x + y)^4 - z^4$.

Solution   This expression is the difference of two squares that can be factored:

$$
\begin{aligned}
(x + y)^4 - z^4 &= [(x + y)^2]^2 - (z^2)^2 \\
&= [(x + y)^2 + z^2][(x + y)^2 - z^2]
\end{aligned}
$$

The factor $(x + y)^2 + z^2$ is the sum of two squares and is prime. However, the factor $(x + y)^2 - z^2$ is the difference of two squares and can be factored as $(x + y + z)(x + y - z)$. Thus,

$$
\begin{aligned}
(x + y)^4 - z^4 &= [(x + y)^2 + z^2][(x + y)^2 - z^2] \\
&= [(x + y)^2 + z^2](x + y + z)(x + y - z) \qquad \blacksquare
\end{aligned}
$$

Self Check   Factor $a^4 - (b + c)^4$.
Answer $[a^2 + (b + c)^2](a + b + c)(a - b - c)$

When possible, we will always factor out a common factor before factoring the difference of two squares. The factoring process is easier when all common factors are factored out first.

EXAMPLE 7   Factor $2x^4y - 32y$.

Solution

$$
\begin{array}{ll}
2x^4y - 32y = 2y(x^4 - 16) & \text{Factor out } 2y. \\
\qquad\quad = 2y(x^2 + 4)(x^2 - 4) & \text{Factor } x^4 - 16. \\
\qquad\quad = 2y(x^2 + 4)(x + 2)(x - 2) & \text{Factor } x^2 - 4. \qquad \blacksquare
\end{array}
$$

Self Check   Factor $3ap^4 - 243a$.
Answer $3a(p^2 + 9)(p + 3)(p - 3)$

■ FACTORING TRINOMIALS

Recall that to factor trinomials with lead coefficients of 1, we follow these steps.

> **Factoring Trinomials**
> 1. Write the trinomial in descending powers of one variable.
> 2. List the factorizations of the third term of the trinomial.
> 3. Pick the factorization where the sum of the factors is the coefficient of the middle term.

EXAMPLE 8 Factor $x^2 - 6x + 8$.

Solution Since this trinomial is already written in descending powers of x, we can move to Step 2 and list the possible factorizations of the third term, which is 8.

The one to choose
↓

$$8(1) \qquad 4(2) \qquad -8(-1) \qquad \mathbf{-4(-2)}$$

In this trinomial, the coefficient of the middle term is -6. The only factorization where the sum of the factors is -6 is $-4(-2)$. Thus,

$$x^2 - 6x + 8 = (x - 4)(x - 2)$$

The factorization of $x^2 - 6x + 8$ is $(x - 4)(x - 2)$. Because of the commutative property of multiplication, the order of the factors is not important. We can verify this result by multiplication:

$$(x - 4)(x - 2) = x^2 - 2x - 4x + 8$$
$$= x^2 - 6x + 8 \qquad ■$$

Self Check Factor $x^2 + 5x + 6$.
Answer $(x + 3)(x + 2)$

EXAMPLE 9 Factor $30x - 4xy - 2xy^2$.

Solution We begin by writing the trinomial in descending powers of y:

$$30x - 4xy - 2xy^2 = -2xy^2 - 4xy + 30x$$

Each term in this trinomial has a common monomial factor of $-2x$, which can be factored out.

$$30x - 4xy - 2xy^2 = -2x(y^2 + 2y - 15)$$

To factor $y^2 + 2y - 15$, we list the factors of -15 and find the pair whose sum is 2.

The one to choose
↓

$$15(-1) \qquad \mathbf{5(-3)} \qquad 1(-15) \qquad 3(-5)$$

The only factorization where the sum of the factors is 2 (the coefficient of the middle term of $y^2 + 2y - 15$) is $5(-3)$. Thus,

$$30x - 4xy - 2xy^2 = -2x(y^2 + 2y - 15)$$
$$= -2x(y + 5)(y - 3)$$

Verify this result by multiplication. ∎

Self Check

Factor $16a - 2ap^2 - 4ap$.

Answer

$-2a(p + 4)(p - 2)$

WARNING! In Example 9, be sure to include all factors in the final answer. It is a common error to forget to write the $-2x$.

There are more combinations of factors to consider when factoring trinomials with lead coefficients other than 1. To factor $5x^2 + 7x + 2$, for example, we must find two binomials of the form $ax + b$ and $cx + d$ such that

$$5x^2 + 7x + 2 = (ax + b)(cx + d)$$

Since the first term of the trinomial $5x^2 + 7x + 2$ is $5x^2$, the first terms of the binomial factors must be $5x$ and x.

$$5x^2 + 7x + 2 = \overbrace{(5x + b)(x}^{5x^2} + d)$$

Since the product of the last terms must be 2, and the sum of the products of the outer and inner terms must be $7x$, we must find two numbers whose product is 2 that will give a middle term of $7x$.

$$5x^2 + 7x + 2 = \underbrace{(5x + b)(x}_{\mathrm{O} + \mathrm{I} = 7x} + \overbrace{d)}^{2}$$

Because both $2(1)$ and $(-2)(-1)$ give a product of 2, there are four possible combinations to consider:

$$(5x + 2)(x + 1) \qquad (5x - 2)(x - 1)$$
$$(5x + 1)(x + 2) \qquad (5x - 1)(x - 2)$$

Of these possibilities, only the first one gives the proper middle term of $7x$.

1. $5x^2 + 7x + 2 = (5x + 2)(x + 1)$

We can verify this result by multiplication:

$$(5x + 2)(x + 1) = 5x^2 + 5x + 2x + 2$$
$$= 5x^2 + 7x + 2$$

■ TEST FOR FACTORABILITY

If a trinomial has the form $ax^2 + bx + c$, with integer coefficients and $a \neq 0$, we can test to see whether it is factorable. If the value of $b^2 - 4ac$ is a perfect square, the trinomial can be factored using only integers. If the value is not a perfect square, the trinomial is prime and cannot be factored using only integers.

For example, $5x^2 + 7x + 2$ is a trinomial in the form $ax^2 + bx + c$ with

$$a = 5, \qquad b = 7, \qquad \text{and} \qquad c = 2$$

For this trinomial, the value of $b^2 - 4ac$ is

$$b^2 - 4ac = 7^2 - 4(5)(2)$$
$$= 49 - 40$$
$$= 9$$

Since 9 is a perfect square, the trinomial is factorable. Its factorization is shown in Equation 1 on page 509.

> **Test for Factorability**
> A trinomial of the form $ax^2 + bx + c$, with integer coefficients and $a \neq 0$, will factor into two binomials with integer coefficients if the value of
> $$b^2 - 4ac$$
> is a perfect square. If $b^2 - 4ac = 0$, the factors will be the same.
> If $b^2 - 4ac$ is not a perfect square, the trinomial is prime.

EXAMPLE 10   Factor $3p^2 - 4p - 4$.

Solution   In this trinomial, $a = 3$, $b = -4$, and $c = -4$. To see whether it factors, we evaluate $b^2 - 4ac$.

$$b^2 - 4ac = (-4)^2 - 4(3)(-4)$$
$$= 16 + 48$$
$$= 64$$

Since 64 is a perfect square, the trinomial is factorable.

To factor the trinomial, we note that the first terms of the binomial factors must be $3p$ and p to give the first term of $3p^2$.

$$3p^2 - 4p - 4 = (3p + ?)(p + ?)$$

The product of the last terms must be -4, and the sum of the products of the outer terms and the inner terms must be $-4p$.

$$3p^2 - 4p - 4 = (3p + \text{?})(p + \text{?})$$

$$\text{O} + \text{I} = -4p$$

Because $1(-4)$, $-1(4)$, and $-2(2)$ all give a product of -4, there are six possible combinations to consider:-

$(3p + 1)(p - 4)$ $(3p - 4)(p + 1)$

$(3p - 1)(p + 4)$ $(3p + 4)(p - 1)$

$(3p - 2)(p + 2)$ $\mathbf{(3p + 2)(p - 2)}$

Of these possibilities, only the last gives the required middle term of $-4p$. Thus,

$$3p^2 - 4p - 4 = (3p + 2)(p - 2)$$

Verify this result by multiplying.　　　　　　　　　　　　　　　■

Self Check

Answer

Factor $2m^2 - 3m - 9$, if possible.

$(2m + 3)(m - 3)$

Recall the following hints for factoring trinomials.

Factoring a General Trinomial

1. Write the trinomial in descending powers of one variable.

2. Test the trinomial for factorability.

3. Factor out any greatest common factor (including -1 if that is necessary to make the coefficient of the first term positive).

4. When the sign of the first term of a trinomial is $+$ and the sign of the third term is $+$, the signs between the terms of each binomial factor are the same as the sign of the middle term of the trinomial.

    When the sign of the first term is $+$ and the sign of the third term is $-$, the signs between the terms of the binomial are opposite.

5. Try various combinations of first terms and last terms until you find one that works.

6. Check the factorization by multiplication.

EXAMPLE 11　　Factor $24y + 10xy - 6x^2y$.

Solution　　We write the trinomial in descending powers of x and factor out the common factor of $-2y$:

$$24y + 10xy - 6x^2y = -6x^2y + 10xy + 24y$$
$$= -2y(3x^2 - 5x - 12)$$

In the trinomial $3x^2 - 5x - 12$, $a = 3$, $b = -5$, and $c = -12$. Thus,

$$b^2 - 4ac = (-5)^2 - 4(3)(-12)$$
$$= 25 + 144$$
$$= 169$$

Since 169 is a perfect square, the trinomial will factor.

Because the sign of the third term of $3x^2 - 5x - 12$ is $-$, the signs between the binomial factors will be opposite. Because the first term is $3x^2$, the first terms of the binomial factors must be $3x$ and x.

$$24y + 10xy - 6x^2y = -2y(3x \qquad)(x \qquad)$$

$3x^2$

The product of the last terms must be -12, and the sum of the outer terms and the inner terms must be $-5x$.

$$24y + 10xy - 6x^2y = -2y(3x \quad ?)(x \quad ?)$$

-12

$$O + I = -5x$$

Because $1(-12)$, $2(-6)$, $3(-4)$, $12(-1)$, $6(-2)$, and $4(-3)$ all give a product of -12, there are 12 possible combinations to consider.

$(3x + 1)(x - 12)$	$(3x - 12)(x + 1)$
$(3x + 2)(x - 6)$	$(3x - 6)(x + 2)$
$(3x + 3)(x - 4)$	$(3x - 4)(x + 3)$
$(3x + 12)(x - 1)$	$(3x - 1)(x + 12)$
$(3x + 6)(x - 2)$	$(3x - 2)(x + 6)$
The one to choose → $(3x + 4)(x - 3)$	$(3x - 3)(x + 4)$

The six combinations marked in blue cannot work, because one of the factors has a common factor. This implies that $3x^2 - 5x - 12$ would have a common factor, which it doesn't.

After mentally trying the remaining combinations, we find that only $(3x + 4)(x - 3)$ gives the proper middle term of $-5x$.

$$24y + 10xy - 6x^2y = -2y(3x^2 - 5x - 12)$$
$$= -2y(3x + 4)(x - 3)$$

Verify this result by multiplication.   ∎

Self Check

Factor $18a - 6ap^2 + 3ap$.

Answer

$-3a(2p + 3)(p - 2)$

EXAMPLE 12   Factor $x^{2n} + x^n - 2$.

Solution   Since the first term is x^{2n}, the first terms of the binomial factors must be x^n and x^n.

$$x^{2n} + x^n - 2 = (x^n \qquad)(x^n \qquad)$$

Since the third term is -2, the last terms of the binomial factors must have **opposite** signs, have a product of -2, and lead to a middle term of x^n. The only combination that works is

$$x^{2n} + x^n - 2 = (x^n + 2)(x^n - 1)$$

Verify this result by multiplication.   ■

Self Check   Factor $a^{2n} + 2a^n - 3$.
Answer $(a^n + 3)(a^n - 1)$

EXAMPLE 13   Factor $x^2 + 6x + 9 - z^2$.

Solution   We group the first three terms together and factor the trinomial to get

$$x^2 + 6x + 9 - z^2 = (x + 3)(x + 3) - z^2$$
$$= (x + 3)^2 - z^2$$

We can now factor the difference of two squares to get

$$x^2 + 6x + 9 - z^2 = (x + 3 + z)(x + 3 - z)$$ ■

Self Check   Factor $y^2 + 4y + 4 - t^2$.
Answer $(y + 2 + t)(y + 2 - t)$

■ **USING GROUPING TO FACTOR TRINOMIALS**

The method of factoring by grouping can be used to help factor trinomials of the form $ax^2 + bx + c$. For example, to factor the trinomial $6x^2 + 7x - 3$, we proceed as follows:

1. First determine the product ac: $6(-3) = -18$. This number is called the **key number.**

2. Find two factors of the key number -18 whose sum is $b = 7$:

$$9(-2) = -18 \qquad \text{and} \qquad 9 + (-2) = 7$$

3. Use the factors 9 and -2 as coefficients of terms to be placed between $6x^2$ and -3:

$$6x^2 + 7x - 3 = 6x^2 + 9x - 2x - 3$$

4. Factor by grouping:

$$6x^2 + 9x - 2x - 3 = 3x(2x + 3) - (2x + 3)$$
$$= (2x + 3)(3x - 1) \qquad \text{Factor out } 2x + 3.$$

We can verify this factorization by multiplication.

Orals   *Factor each expression, if possible.*

1. $2x^2 + 4x$ 2. $3xy^2 - 6x^2y$

3. $x^2 - 1$ 4. $a^4 - 16$

5. $x^2 + 5x - 6$ 6. $2x^2 - x - 1$

EXERCISE 7.4

REVIEW   *Do each multiplication.*

1. $(x + 1)(x^2 - x + 1)$ 2. $(m + 3)(m^2 - 3m + 9)$
3. $(r - 2)(r^2 + 2r + 4)$ 4. $(a - 5)(a^2 + 5a + 25)$

Solve each equation.

5. $\dfrac{2}{3}(5t - 3) = 38$ 6. $2q^2 - 9 = q(q + 3) + q^2$

VOCABULARY AND CONCEPTS   *Fill in each blank to make a true statement.*

7. Factoring out a common monomial is based on the distributive property, which is $a(b + c) = $ _____.

8. Factoring the difference of two squares is based on the formula $F^2 - L^2 = $ _____.

9. A trinomial of the form $ax^2 + bx + c$, with integer coefficients and $a \neq 0$, will factor if the value of $b^2 - 4ac$ is a _____ square.

10. If $b^2 - 4ac$ in Exercise 9 is not a perfect square, the trinomial is _____.

PRACTICE   *In Exercises 11–24, factor each expression.*

11. $2x + 8$ 12. $3y - 9$ 13. $2x^2 - 6x$ 14. $3y^3 + 3y^2$
15. $15x^2y - 10x^2y^2$ 16. $63x^3y^2 + 81x^2y^4$
17. $13ab^2c^3 - 26a^3b^2c$ 18. $4x^2yz^2 + 4xy^2z^2$
19. $27z^3 + 12z^2 + 3z$ 20. $25t^6 - 10t^3 + 5t^2$
21. $24s^3 - 12s^2t + 6st^2$ 22. $18y^2z^2 + 12y^2z^3 - 24y^4z^3$
23. $45x^{10}y^3 - 63x^7y^7 + 81x^{10}y^{10}$ 24. $48u^6v^6 - 16u^4v^4 - 3u^6v^3$

In Exercises 25–30, factor out the negative of the greatest common factor.

25. $-3a - 6$

26. $-6b + 12$

27. $-6x^2 - 3xy$

28. $-15y^3 + 25y^2$

29. $-63u^3v^6z^9 + 28u^2v^7z^2 - 21u^3v^3z^4$

30. $-56x^4y^3z^2 - 72x^3y^4z^5 + 80xy^2z^3$

In Exercises 31–34, factor out the designated common factor.

31. x^2 from $x^{n+2} + x^{n+3}$

32. y^3 from $y^{n+3} + y^{n+5}$

33. y^n from $2y^{n+2} - 3y^{n+3}$

34. x^n from $4x^{n+3} - 5x^{n+5}$

In Exercises 35–40, factor by grouping.

35. $ax + bx + ay + by$

36. $ar - br + as - bs$

37. $x^2 + yx + 2x + 2y$

38. $2c + 2d - cd - d^2$

39. $3c - cd + 3d - c^2$

40. $x^2 + 4y - xy - 4x$

In Exercises 41–44, solve for the indicated variable.

41. $r_1r_2 = rr_2 + rr_1$ for r_1

42. $r_1r_2 = rr_2 + rr_1$ for r

43. $S(1 - r) = a - lr$ for r

44. $Sn = (n - 2)180°$ for n

In Exercises 45–62, factor each expression. Factor out any common factors first.

45. $x^2 - 4$

46. $y^2 - 9$

47. $9y^2 - 64$

48. $16x^4 - 81y^2$

49. $81a^4 - 49b^2$

50. $64r^6 - 121s^2$

51. $(x + y)^2 - z^2$

52. $a^2 - (b - c)^2$

53. $x^4 - y^4$

54. $16a^4 - 81b^4$

55. $2x^2 - 288$

56. $8x^2 - 72$

57. $2x^3 - 32x$

58. $3x^3 - 243x$

59. $r^2s^2t^2 - t^2x^4y^2$

60. $16a^4b^3c^4 - 64a^2bc^6$

61. $x^{2m} - y^{4n}$

62. $a^{4m} - b^{8n}$

In Exercises 63–66, factor each expression by grouping.

63. $a^2 - b^2 + a + b$

64. $x^2 - y^2 - x - y$

65. $2x + y + 4x^2 - y^2$

66. $m - 2n + m^2 - 4n^2$

In Exercises 67–72, test each trinomial for factorability and factor it, if possible.

67. $x^2 + 5x + 6$

68. $y^2 + 7y + 6$

69. $x^2 - 7x + 10$

70. $c^2 - 7c + 12$

71. $a^2 + 5a - 52$

72. $b^2 + 9b - 38$

In Exercises 73–80, factor each trinomial. If the coefficient of the first term is negative, begin by factoring out −1.

73. $3x^2 + 12x - 63$

74. $2y^2 + 4y - 48$

75. $a^2b^2 - 13ab + 22b^2$

76. $a^2b^2x^2 - 18a^2b^2x + 81a^2b^2$

77. $-a^2 + 4a + 32$

78. $-x^2 - 2x + 15$

79. $-3x^2 + 15x - 18$

80. $-2y^2 - 16y + 40$

In Exercises 81–100, factor each trinomial. Factor out all common monomials first (including −1 if the first term is negative). If a trinomial is prime, so indicate.

81. $6y^2 + 7y + 2$

82. $6x^2 - 11x + 3$

83. $8a^2 + 6a - 9$

84. $15b^2 + 4b - 4$

85. $5x^2 + 4x + 1$

86. $6z^2 + 17z + 12$

87. $8x^2 - 10x + 3$

88. $4a^2 + 20a + 3$

89. $a^2 - 3ab - 4b^2$

90. $b^2 + 2bc - 80c^2$

91. $2y^2 + yt - 6t^2$

92. $3x^2 - 10xy - 8y^2$

93. $-3a^2 + ab + 2b^2$

94. $-2x^2 + 3xy + 5y^2$

95. $3x^3 - 10x^2 + 3x$

96. $6y^3 + 7y^2 + 2y$

97. $-4x^3 - 9x + 12x^2$

98. $6x^2 + 4x + 9x^3$

99. $8x^2z + 6xyz + 9y^2z$

100. $x^3 - 60xy^2 + 7x^2y$

In Exercises 101–106, factor each trinomial.

101. $x^4 + 8x^2 + 15$

102. $x^4 + 11x^2 + 24$

103. $y^4 - 13y^2 + 30$

104. $y^4 - 13y^2 + 42$

105. $a^4 - 13a^2 + 36$

106. $b^4 - 17b^2 + 16$

In Exercises 107–114, factor each expression. Assume that n is a natural number.

107. $x^{2n} + 2x^n + 1$

108. $x^{4n} - 2x^{2n} + 1$

109. $2a^{6n} - 3a^{3n} - 2$

110. $b^{2n} - b^n - 6$

111. $x^{4n} + 2x^{2n}y^{2n} + y^{4n}$

112. $y^{6n} + 2y^{3n}z + z^2$

113. $6x^{2n} + 7x^n - 3$

114. $12y^{4n} + 10y^{2n} + 2$

In Exercises 115–120, factor each expression.

115. $x^2 + 4x + 4 - y^2$

116. $x^2 - 6x + 9 - 4y^2$

117. $x^2 + 2x + 1 - 9z^2$

118. $x^2 + 10x + 25 - 16z^2$

119. $c^2 - 4a^2 + 4ab - b^2$

120. $4c^2 - a^2 - 6ab - 9b^2$

In Exercises 121–128, use factoring by grouping to help factor each trinomial.

121. $a^2 - 17a + 16$

122. $b^2 - 4b - 21$

123. $2u^2 + 5u + 3$

124. $6y^2 + 5y - 6$

125. $20r^2 - 7rs - 6s^2$

126. $6s^2 + st - 12t^2$

127. $20u^2 + 19uv + 3v^2$

128. $12m^2 + mn - 6n^2$

WRITING

129. Explain how you would factor -1 from a trinomial.

130. Explain how you would test the polynomial $ax^2 + bx + c$ for factorability.

SOMETHING TO THINK ABOUT

131. Because it is the difference of two squares, $x^2 - q^2$ always factors. Does the test for factorability predict this?

132. The polynomial $ax^2 + ax + a$ factors: a is a common factor. Does the test for factorability predict this? Is there something wrong with the test? Explain.

133. Find the error.

$$x = y$$
$$x^2 = xy$$
$$x^2 - y^2 = xy - y^2$$
$$(x + y)(x - y) = y(x - y)$$
$$\frac{(x + y)(x - y)}{x - y} = \frac{y(x - y)}{x - y}$$
$$x + y = y$$
$$y + y = y$$
$$2y = y$$
$$\frac{2y}{y} = \frac{y}{y}$$
$$2 = 1$$

134. Factor $x^{32} - y^{32}$.

7.5 Factoring the Sum and Difference of Two Cubes

■ FACTORING THE SUM OF TWO CUBES ■ FACTORING THE DIFFERENCE OF TWO CUBES
■ MULTISTEP FACTORING

Getting Ready   *Find each product.*

1. $(x - 3)(x^2 + 3x + 9)$ **2.** $(x + 2)(x^2 - 2x + 4)$
3. $(y + 4)(y^2 - 4y + 16)$ **4.** $(r - 5)(r^2 + 5r + 25)$
5. $(a - b)(a^2 + ab + b^2)$ **6.** $(a + b)(a^2 - ab + b^2)$

Recall that the difference of the squares of two quantities factors into the product of two binomials. One binomial is the sum of the quantities, and the other is the difference of the quantities.

$$x^2 - y^2 = (x + y)(x - y) \qquad \text{or} \qquad F^2 - L^2 = (F + L)(F - L)$$

There are similar formulas for factoring the sum of two cubes and the difference of two cubes.

■ FACTORING THE SUM OF TWO CUBES

To find the formula for factoring the sum of two cubes, we need to find the following product:

$$(x + y)(x^2 - xy + y^2) = (x + y)x^2 - (x + y)xy + (x + y)y^2$$

Use the distributive property.

$$= x^3 + x^2y - x^2y - xy^2 + xy^2 + y^3$$
$$= x^3 + y^3$$

This result justifies the formula for factoring the **sum of two cubes.**

> **Factoring the Sum of Two Cubes**
>
> $$x^3 + y^3 = (x + y)(x^2 - xy + y^2)$$

If we think of the sum of two cubes as the cube of a **F**irst quantity plus the cube of a **L**ast quantity, we have the formula

$$F^3 + L^3 = (F + L)(F^2 - FL + L^2)$$

In words, we say, *To factor the cube of a First quantity plus the cube of a Last quantity, we multiply the First plus the Last by*

- *the First squared*
- *minus the First times the Last*
- *plus the Last squared.*

To factor the sum of two cubes, it is helpful to know the cubes of the numbers from 1 to 10:

1, 8, 27, 64, 125, 216, 343, 512, 729, 1,000

Expressions containing variables such as x^6y^3 are also perfect cubes, because they can be written as the cube of a quantity:

$$x^6y^3 = (x^2y)^3$$

EXAMPLE 1   Factor $x^3 + 8$.

Solution   The binomial $x^3 + 8$ is the sum of two cubes, because

$$x^3 + 8 = x^3 + 2^3$$

Thus, $x^3 + 8$ factors as $x + 2$ times the trinomial $x^2 - 2x + 2^2$.

$$F^3 + L^3 = (F + L)(F^2 - F\,L + L^2)$$
$$\downarrow \quad \downarrow \qquad \downarrow \quad \downarrow\downarrow \qquad \downarrow\downarrow \qquad \downarrow$$
$$x^3 + 2^3 = (x + 2)(x^2 - x \cdot 2 + 2^2)$$
$$= (x + 2)(x^2 - 2x + 4)$$

Check by multiplication.

$$(x + 2)(x^2 - 2x + 4) = (x + 2)x^2 - (x + 2)2x + (x + 2)4$$
$$= x^3 + 2x^2 - 2x^2 - 4x + 4x + 8$$
$$= x^3 + 8$$ ∎

Self Check Factor $p^3 + 64$.

Answer $(p + 4)(p^2 - 4p + 16)$

EXAMPLE 2 Factor $8b^3 + 27c^3$.

Solution The binomial $8b^3 + 27c^3$ is the sum of two cubes, because

$$8b^3 + 27c^3 = (2b)^3 + (3c)^3$$

Thus, $8b^3 + 27c^3$ factors as $2b + 3c$ times the trinomial $(2b)^2 - (2b)(3c) + (3c)^2$.

$$\mathbf{F}^3 \ + \ \mathbf{L}^3 \ = (\mathbf{F} + \mathbf{L})\,(\mathbf{F}^2 \ - \ \mathbf{F}\ \mathbf{L} \ + \ \mathbf{L}^2)$$
$$\downarrow \qquad \downarrow \qquad \downarrow \quad \downarrow \quad \downarrow \qquad \downarrow \quad \downarrow \qquad \downarrow$$
$$(2b)^3 + (3c)^3 = (2b + 3c)[(2b)^2 - (2b)(3c) + (3c)^2]$$
$$= (2b + 3c)(4b^2 - 6bc + 9c^2)$$

Check by multiplication.

$$(2b + 3c)(4b^2 - 6bc + 9c^2)$$
$$= (2b + 3c)4b^2 - (2b + 3c)6bc + (2b + 3c)9c^2$$
$$= 8b^3 + 12b^2c - 12b^2c - 18bc^2 + 18bc^2 + 27c^3$$
$$= 8b^3 + 27c^3$$ ∎

Self Check Factor $1{,}000p^3 + q^3$.

Answer $(10p + q)(100p^2 - 10pq + q^2)$

■ **FACTORING THE DIFFERENCE OF TWO CUBES**

To find the formula for factoring the difference of two cubes, we need to find the following product:

$$(x - y)(x^2 + xy + y^2) = (x - y)x^2 + (x - y)xy + (x - y)y^2 \qquad \text{Use the distributive property.}$$

$$= x^3 - x^2y + x^2y - xy^2 + xy^2 - y^3$$
$$= x^3 - y^3$$

This result justifies the formula for factoring the **difference of two cubes.**

Factoring the Difference of Two Cubes

$$x^3 - y^3 = (x - y)(x^2 + xy + y^2)$$

If we think of the difference of two cubes as the cube of a **First** quantity minus the cube of a **Last** quantity, we have the formula

$$F^3 - L^3 = (F - L)(F^2 + FL + L^2)$$

In words, we say, *To factor the cube of a First quantity minus the cube of a Last quantity, we multiply the First minus the Last by*

- *the First squared*
- *plus the First times the Last*
- *plus the Last squared.*

EXAMPLE 3   Factor $a^3 - 64b^3$.

Solution   The binomial $a^3 - 64b^3$ is the difference of two cubes.

$$a^3 - 64b^3 = a^3 - (4b)^3$$

Thus, its factors are the difference $a - 4b$ and the trinomial $a^2 + a(4b) + (4b)^2$.

$$\mathbf{F}^3 - \mathbf{L}^3 = (\mathbf{F} - \mathbf{L})(\mathbf{F}^2 + \mathbf{F}\ \mathbf{L} + \mathbf{L}^2)$$

$$a^3 - (4b)^3 = (a - 4b)[a^2 + a(4b) + (4b)^2]$$
$$= (a - 4b)(a^2 + 4ab + 16b^2)$$

Check by multiplication.

$$(a - 4b)(a^2 + 4ab + 16b^2)$$
$$= (a - 4b)a^2 + (a - 4b)4ab + (a - 4b)16b^2$$
$$= a^3 - 4a^2b + 4a^2b - 16ab^2 + 16ab^2 - 64b^3$$
$$= a^3 - 64b^3$$

■

Self Check

Answer

Factor $27p^3 - 8$.

$(3p - 2)(9p^2 + 6p + 4)$

■ MULTISTEP FACTORING

Sometimes we must factor out a greatest common factor before factoring a sum or difference of two cubes.

EXAMPLE 4   Factor $-2t^5 + 128t^2$.

Solution $\qquad$ $\begin{aligned} -2t^5 + 128t^2 &= -2t^2(t^3 - 64) & & \text{Factor out } -2t^2. \\ &= -2t^2(t - 4)(t^2 + 4t + 16) & & \text{Factor } t^3 - 64. \end{aligned}$

Verify this factorization by multiplication.   ■

Self Check $\qquad$ Factor $-3p^4 + 81p$.

Answer $\qquad$ $-3p(p - 3)(p^2 + 3p + 9)$

EXAMPLE 5   Factor $x^6 - 64$.

Solution $\qquad$ The binomial $x^6 - 64$ is both the difference of two squares and the difference of two cubes. Since it is easier to factor the difference of two squares first, the expression factors into the product of a sum and a difference.

$$\begin{aligned} x^6 - 64 &= (x^3)^2 - 8^2 \\ &= (x^3 + 8)(x^3 - 8) \end{aligned}$$

Because $x^3 + 8$ is the sum of two cubes and $x^3 - 8$ is the difference of two cubes, each of these binomials can be factored.

$$\begin{aligned} x^6 - 64 &= (\mathbf{x^3 + 8})(\mathbf{x^3 - 8}) \\ &= (\mathbf{x + 2})(\mathbf{x^2 - 2x + 4})(x - 2)(x^2 + 2x + 4) \end{aligned}$$

Verify this factorization by multiplication.   ■

Self Check $\qquad$ Factor $a^6 - 1$.

Answer $\qquad$ $(a + 1)(a^2 - a + 1)(a - 1)(a^2 + a + 1)$

Orals $\qquad$ *Factor each sum or difference of two cubes.*

1. $x^3 - y^3$ 2. $x^3 + y^3$

3. $a^3 + 8$ 4. $b^3 - 27$

5. $1 + 8x^3$ 6. $8 - r^3$

7. $x^3y^3 + 1$ 8. $125 - 8t^3$

EXERCISE 7.5

REVIEW

1. The length of one Fermi is 1×10^{-13} centimeter, approximately the radius of a proton. Express this number in standard notation.

2. In the 14th century, the Black Plague killed about 25,000,000 people, which was 25% of the population of Europe. Find the population at that time, expressed in scientific notation.

VOCABULARY AND CONCEPTS *Complete each formula.*

3. $x^3 + y^3 = (x + y)$_____

4. $x^3 - y^3 = (x - y)$_____

PRACTICE *In Exercises 5–24, factor each expression.*

5. $y^3 + 1$ **6.** $x^3 - 8$ **7.** $a^3 - 27$ **8.** $b^3 + 125$

9. $8 + x^3$ **10.** $27 - y^3$ **11.** $s^3 - t^3$ **12.** $8u^3 + w^3$

13. $27x^3 + y^3$ **14.** $x^3 - 27y^3$ **15.** $a^3 + 8b^3$ **16.** $27a^3 - b^3$

17. $64x^3 - 27$ **18.** $27x^3 + 125$

19. $27x^3 - 125y^3$ **20.** $64x^3 + 27y^3$

21. $a^6 - b^3$ **22.** $a^3 + b^6$

23. $x^9 + y^6$ **24.** $x^3 - y^9$

In Exercises 25–40, factor each expression. Factor out any greatest common factors first.

25. $2x^3 + 54$ **26.** $2x^3 - 2$ **27.** $-x^3 + 216$ **28.** $-x^3 - 125$

29. $64m^3x - 8n^3x$ **30.** $16r^4 + 128rs^3$

31. $x^4y + 216xy^4$ **32.** $16a^5 - 54a^2b^3$

33. $81r^4s^2 - 24rs^5$ **34.** $4m^5n + 500m^2n^4$

35. $125a^6b^2 + 64a^3b^5$ **36.** $216a^4b^4 - 1,000ab^7$

37. $y^7z - yz^4$ **38.** $x^{10}y^2 - xy^5$

39. $2mp^4 + 16mpq^3$ **40.** $24m^5n - 3m^2n^4$

In Exercises 41–44, factor each expression completely. Factor a difference of two squares first.

41. $x^6 - 1$ **42.** $x^6 - y^6$

43. $x^{12} - y^6$ **44.** $a^{12} - 64$

In Exercises 45–52, factor each expression completely.

45. $3(x^3 + y^3) - z(x^3 + y^3)$ **46.** $x(8a^3 - b^3) + 4(8a^3 - b^3)$

47. $(m^3 + 8n^3) + (m^3x + 8n^3x)$ **48.** $(a^3x + b^3x) - (a^3y + b^3y)$

49. $(a^4 + 27a) - (a^3b + 27b)$ **50.** $(x^4 + xy^3) - (x^3y + y^4)$

51. $y^3(y^2 - 1) - 27(y^2 - 1)$ **52.** $z^3(y^2 - 4) + 8(y^2 - 4)$

WRITING

53. Explain how to factor $a^3 + b^3$.

54. Explain the difference between $x^3 - y^3$ and $(x - y)^3$.

SOMETHING TO THINK ABOUT

55. Use a calculator to verify that

$$a^3 - b^3 = (a - b)(a^2 + ab + b^2)$$

when $a = 11$ and $b = 7$.

56. What difficulty do you encounter when you solve $x^3 - 8 = 0$ by factoring?

7.6 Review of Rational Expressions

■ SIMPLIFYING RATIONAL EXPRESSIONS ■ MULTIPLYING AND DIVIDING RATIONAL EXPRESSIONS
■ ADDING AND SUBTRACTING RATIONAL EXPRESSIONS ■ COMPLEX FRACTIONS

Getting Ready   *Do each operation.*

1. $\dfrac{2}{3} \cdot \dfrac{5}{2}$ **2.** $\dfrac{2}{3} \div \dfrac{5}{2}$ **3.** $\dfrac{2}{3} + \dfrac{5}{2}$ **4.** $\dfrac{2}{3} - \dfrac{5}{2}$

■ SIMPLIFYING RATIONAL EXPRESSIONS

Recall that rational expressions are algebraic fractions with polynomial numerators and polynomial denominators. To manipulate rational expressions, we use the same rules as we use to simplify, multiply, divide, add, and subtract arithmetic fractions.

EXAMPLE 1   Simplify $\dfrac{-8y^3z^5}{6y^4z^3}$.

Solution   We factor the numerator and denominator and divide out all common factors:

$$\frac{-8y^3z^5}{6y^4z^3} = \frac{-2 \cdot 4 \cdot y \cdot y \cdot y \cdot z \cdot z \cdot z \cdot z \cdot z}{2 \cdot 3 \cdot y \cdot y \cdot y \cdot y \cdot z \cdot z \cdot z}$$

$$= \frac{-\overset{1}{\cancel{2}} \cdot 4 \cdot \overset{1}{\cancel{y}} \cdot \overset{1}{\cancel{y}} \cdot \overset{1}{\cancel{y}} \cdot \overset{1}{\cancel{z}} \cdot \overset{1}{\cancel{z}} \cdot \overset{1}{\cancel{z}} \cdot z \cdot z}{\underset{1}{\cancel{2}} \cdot 3 \cdot \underset{1}{\cancel{y}} \cdot \underset{1}{\cancel{y}} \cdot \underset{1}{\cancel{y}} \cdot y \cdot \underset{1}{\cancel{z}} \cdot \underset{1}{\cancel{z}} \cdot \underset{1}{\cancel{z}}}$$

$$= -\frac{4z^2}{3y}$$

■

Self Check   Simplify $\dfrac{10k}{25k^2}$.

Answer $\dfrac{2}{5k}$

The fractions in Example 1 and the Self Check can be simplified by using the rules of exponents:

$$\frac{10k}{25k^2} = \frac{5 \cdot 2}{5 \cdot 5} k^{1-2} \qquad\qquad \frac{-8y^3z^5}{6y^4z^3} = \frac{-2 \cdot 4}{2 \cdot 3} y^{3-4}z^{5-3}$$

$$= \frac{2}{5} \cdot k^{-1} \qquad\qquad\qquad = \frac{-4}{3} \cdot y^{-1}z^2$$

$$= \frac{2}{5k} \qquad\qquad\qquad\qquad = -\frac{4}{3} \cdot \frac{1}{y} \cdot \frac{z^2}{1}$$

$$\qquad\qquad\qquad\qquad\qquad\qquad = -\frac{4z^2}{3y}$$

EXAMPLE 2   Simplify $\dfrac{2x^2 + 11x + 12}{3x^2 + 11x - 4}$.

Solution   We factor the numerator and denominator and divide out all common factors:

$$\frac{2x^2 + 11x + 12}{3x^2 + 11x - 4} = \frac{(2x + 3)\overset{1}{\cancel{(x + 4)}}}{(3x - 1)\underset{1}{\cancel{(x + 4)}}}$$

$$= \frac{2x + 3}{3x - 1} \qquad\qquad \tfrac{x+4}{x+4} = 1.$$

WARNING!   Do not divide out the x's in the fraction $\frac{2x+3}{3x-1}$. The x in the numerator is a factor of the first term only. It is not a factor of the entire numerator. Likewise, the x in the denominator is not a factor of the entire denominator.

Self Check   Simplify $\dfrac{2x^2 + 5x + 2}{3x^2 + 5x - 2}$.

Answer $\dfrac{2x+1}{3x-1}$

EXAMPLE 3   Simplify $\dfrac{3x^2 - 10xy - 8y^2}{4y^2 - xy}$.

Solution   We factor the numerator and denominator and proceed as follows:

$$\frac{3x^2 - 10xy - 8y^2}{4y^2 - xy} = \frac{(3x + 2y)\overset{-1}{\cancel{(x - 4y)}}}{y\underset{1}{\cancel{(4y - x)}}}$$

Because $x - 4y$ and $4y - x$ are negatives, their quotient is -1.

$$= \frac{-(3x + 2y)}{y}$$

$$= \frac{-3x - 2y}{y}$$ ∎

Self Check   Simplify $\dfrac{-2a^2 - ab + 3b^2}{a^2 - ab}$.

Answer $\dfrac{-2a - 3b}{a}$

■ MULTIPLYING AND DIVIDING RATIONAL EXPRESSIONS

To multiply two fractions, we multiply the numerators and multiply the denominators.

EXAMPLE 4   Multiply $\dfrac{x^2 - 6x + 9}{x} \cdot \dfrac{x^2}{x - 3}$.

Solution   We multiply the numerators and multiply the denominators. Then we simplify the resulting fraction.

$$\frac{x^2 - 6x + 9}{x} \cdot \frac{x^2}{x - 3} = \frac{(x^2 - 6x + 9)(x^2)}{x(x - 3)}$$

Multiply the numerators and multiply the denominators.

$$= \frac{(x - 3)(x - 3)xx}{x(x - 3)}$$

Factor the numerator and the denominator.

$$= \frac{\overset{1}{\cancel{(x - 3)}}(x - 3)\cancel{x}x}{\underset{1}{\cancel{x}\underset{1}{\cancel{(x - 3)}}}}$$

Divide out common factors.

$$= x(x - 3)$$ ∎

Self Check   Multiply $\dfrac{a^2 - 2a + 1}{a} \cdot \dfrac{a^3}{a - 1}$.

Answer $a^2(a - 1)$

EXAMPLE 5 Multiply $\dfrac{6x^2 + 5x - 4}{2x^2 + 5x + 3} \cdot \dfrac{8x^2 + 6x - 9}{12x^2 + 7x - 12}$.

Solution We multiply the fractions, factor each polynomial, and simplify.

$$\dfrac{6x^2 + 5x - 4}{2x^2 + 5x + 3} \cdot \dfrac{8x^2 + 6x - 9}{12x^2 + 7x - 12}$$

$$= \dfrac{(6x^2 + 5x - 4)(8x^2 + 6x - 9)}{(2x^2 + 5x + 3)(12x^2 + 7x - 12)}$$ Multiply the numerators and multiply the denominators.

$$= \dfrac{(3x + 4)(2x - 1)(4x - 3)(2x + 3)}{(2x + 3)(x + 1)(3x + 4)(4x - 3)}$$ Factor the polynomials.

$$= \dfrac{(3x + 4)(2x - 1)(4x - 3)(2x + 3)}{(2x + 3)(x + 1)(3x + 4)(4x - 3)}$$ Divide out the common factors.

$$= \dfrac{2x - 1}{x + 1}$$ ∎

Self Check Multiply $\dfrac{2x^2 + 5x + 3}{3x^2 + 5x + 2} \cdot \dfrac{2x^2 - 5x + 3}{4x^2 - 9}$.

Answer $\dfrac{x - 1}{3x + 2}$

In Examples 4 and 5, we would obtain the same answers if we factored first and divided out the common factors before we multiplied.

To divide two fractions, we invert the divisor and multiply.

EXAMPLE 6 Divide $\dfrac{x^3 + 8}{x + 1} \div \dfrac{x^2 - 2x + 4}{2x^2 - 2}$.

Solution Using the rule for division of fractions, we invert the divisor and multiply.

$$\dfrac{x^3 + 8}{x + 1} \div \dfrac{x^2 - 2x + 4}{2x^2 - 2}$$

$$= \dfrac{x^3 + 8}{x + 1} \cdot \dfrac{2x^2 - 2}{x^2 - 2x + 4}$$

$$= \dfrac{(x^3 + 8)(2x^2 - 2)}{(x + 1)(x^2 - 2x + 4)}$$

$$= \dfrac{(x + 2)(x^2 - 2x + 4)2(x + 1)(x - 1)}{(x + 1)(x^2 - 2x + 4)}$$ $2x^2 - 2 = 2(x^2 - 1)$
$= 2(x + 1)(x - 1)$.

$$= 2(x + 2)(x - 1)$$ ∎

Self Check   Divide $\dfrac{x^3 + 27}{x^2 - 4} \div \dfrac{x^2 - 3x + 9}{x + 2}$.

Answer $\dfrac{x + 3}{x - 2}$

EXAMPLE 7   Simplify $\dfrac{x^2 + 2x - 3}{6x^2 + 5x + 1} \div \dfrac{2x^2 - 2}{2x^2 - 5x - 3} \cdot \dfrac{6x^2 + 4x - 2}{x^2 - 2x - 3}$.

Solution   We change the division to a multiplication. Since multiplications and divisions are done from left to right, only the middle fraction should be inverted. Finally, we multiply the fractions, factor each polynomial, and divide out the common factors.

$$\frac{x^2 + 2x - 3}{6x^2 + 5x + 1} \div \frac{2x^2 - 2}{2x^2 - 5x - 3} \cdot \frac{6x^2 + 4x - 2}{x^2 - 2x - 3}$$

$$= \frac{x^2 + 2x - 3}{6x^2 + 5x + 1} \cdot \frac{2x^2 - 5x - 3}{2x^2 - 2} \cdot \frac{6x^2 + 4x - 2}{x^2 - 2x - 3}$$

$$= \frac{(x^2 + 2x - 3)(2x^2 - 5x - 3)(6x^2 + 4x - 2)}{(6x^2 + 5x + 1)(2x^2 - 2)(x^2 - 2x - 3)}$$

$$= \frac{\overset{1}{(x + 3)}\overset{1}{(x - 1)}\overset{1}{(2x + 1)}\overset{1}{(x - 3)}2\overset{1}{(3x - 1)(x + 1)}}{(3x + 1)\underset{1}{(2x + 1)}2\underset{1}{(x + 1)}\underset{1}{(x - 1)}\underset{1}{(x - 3)}\underset{1}{(x + 1)}}$$

$$= \frac{(x + 3)(3x - 1)}{(3x + 1)(x + 1)}$$ ■

■ ADDING AND SUBTRACTING RATIONAL EXPRESSIONS

To add or subtract fractions with like denominators, we add or subtract the numerators and keep the same denominator. Whenever possible, we should simplify the result.

EXAMPLE 8   Simplify $\dfrac{4x}{x + 2} + \dfrac{7x}{x + 2}$.

Solution $\dfrac{4x}{x + 2} + \dfrac{7x}{x + 2} = \dfrac{4x + 7x}{x + 2}$

$$= \frac{11x}{x + 2}$$ ■

Self Check Simplify $\frac{4a}{a+3} + \frac{2a}{a+3}$.

Answer $\frac{6a}{a+3}$

To add or subtract fractions with unlike denominators, we must convert them to fractions with the same denominator.

EXAMPLE 9 Simplify $\dfrac{4x}{x+2} - \dfrac{7x}{x-2}$.

Solution $\dfrac{4x}{x+2} - \dfrac{7x}{x-2} = \dfrac{4x(x-2)}{(x+2)(x-2)} - \dfrac{(x+2)7x}{(x+2)(x-2)}$ $\frac{x-2}{x-2} = 1, \frac{x+2}{x+2} = 1.$

$= \dfrac{(4x^2 - 8x) - (7x^2 + 14x)}{(x+2)(x-2)}$ Subtract the numerators and keep the common denominator.

$= \dfrac{4x^2 - 8x - 7x^2 - 14x}{(x+2)(x-2)}$ To remove parentheses, use the distributive property.

$= \dfrac{-3x^2 - 22x}{(x+2)(x-2)}$ Combine like terms.

 WARNING! The $-$ sign between the fractions in Step 1 applies to both terms of $7x^2 + 14x$.

■

Self Check Simplify $\frac{3a}{a+3} - \frac{2a}{a-3}$.

Answer $\dfrac{a^2 - 15a}{(a+3)(a-3)}$

EXAMPLE 10 Add $\dfrac{x}{x^2 - 2x + 1} + \dfrac{3}{x^2 - 1}$.

Solution We factor each denominator and find the LCD:

$$x^2 - 2x + 1 = (x-1)(x-1) = (x-1)^2$$
$$x^2 - 1 = (x+1)(x-1)$$

The LCD is $(x-1)^2(x+1)$.

We now write each fraction with its denominator in factored form and convert the fractions to fractions with an LCD of $(x-1)^2(x+1)$. Finally, we add the fractions.

$$\frac{x}{x^2 - 2x + 1} + \frac{3}{x^2 - 1} = \frac{x}{(x - 1)(x - 1)} + \frac{3}{(x + 1)(x - 1)}$$

$$= \frac{x(x + 1)}{(x - 1)(x - 1)(x + 1)} + \frac{3(x - 1)}{(x + 1)(x - 1)(x - 1)}$$

$$= \frac{x^2 + x + 3x - 3}{(x - 1)(x - 1)(x + 1)}$$

$$= \frac{x^2 + 4x - 3}{(x - 1)^2(x + 1)} \qquad \text{This result does not simplify.} \qquad ■$$

Self Check   Add $\dfrac{3}{a^2 + a} + \dfrac{2}{a^2 - 1}$.

Answer $\dfrac{5a - 3}{a(a + 1)(a - 1)}$

■ COMPLEX FRACTIONS

Recall that a *complex fraction* is a fraction with a fraction in its numerator or its denominator or both. Examples of complex fractions are

$$\frac{\dfrac{3}{5}}{\dfrac{6}{7}}, \qquad \frac{\dfrac{x + 2}{3}}{x - 4}, \qquad \text{and} \qquad \frac{\dfrac{3x^2 - 2}{2x}}{3x - \dfrac{2}{y}}$$

EXAMPLE 11   Simplify $\dfrac{\dfrac{3a}{b}}{\dfrac{6ac}{b^2}}$.

Solution   *Method 1:* We write the complex fraction as a division and proceed as follows:

$$\frac{\dfrac{3a}{b}}{\dfrac{6ac}{b^2}} = \frac{3a}{b} \div \frac{6ac}{b^2}$$

$$= \frac{3a}{b} \cdot \frac{b^2}{6ac} \qquad \text{Invert the divisor and multiply.}$$

$$= \frac{b}{2c} \qquad \text{Multiply the fractions and simplify.}$$

Method 2: We multiply the numerator and denominator by b^2, the LCD of $\frac{3a}{b}$ and $\frac{6ac}{b^2}$, and simplify:

$$\frac{\dfrac{3a}{b}}{\dfrac{6ac}{b^2}} = \frac{\dfrac{3a}{b} \cdot b^2}{\dfrac{6ac}{b^2} \cdot b^2} \qquad \frac{b^2}{b^2} = 1.$$

$$= \frac{\dfrac{3ab^2}{b}}{\dfrac{6ab^2c}{b^2}}$$

$$= \frac{3ab}{6ac} \qquad \text{Simplify the fractions in the numerator and denominator.}$$

$$= \frac{b}{2c} \qquad \text{Divide out the common factor of } 3a. \qquad \blacksquare$$

Self Check Simplify $\dfrac{\dfrac{2x}{y^2}}{\dfrac{6xz}{y}}$.

Answer $\dfrac{1}{3yz}$

EXAMPLE 12 Simplify $\dfrac{\dfrac{1}{x} + \dfrac{1}{y}}{\dfrac{1}{x} - \dfrac{1}{y}}$.

Solution *Method 1:* We add the fractions in the numerator and in the denominator and proceed as follows:

$$\frac{\dfrac{1}{x} + \dfrac{1}{y}}{\dfrac{1}{x} - \dfrac{1}{y}} = \frac{\dfrac{1y}{xy} + \dfrac{x1}{xy}}{\dfrac{1y}{xy} - \dfrac{x1}{xy}}$$

$$= \frac{\dfrac{y + x}{xy}}{\dfrac{y - x}{xy}}$$

$$= \frac{y + x}{xy} \div \frac{y - x}{xy}$$

$$= \frac{y + x}{xy} \cdot \frac{xy}{y - x}$$

$$= \frac{y + x}{y - x} \qquad \text{Multiply and then divide out the factor of } xy.$$

Method 2: We multiply the numerator and denominator by xy (the LCD of the fractions appearing in the complex fraction) and simplify.

$$\frac{\dfrac{1}{x} + \dfrac{1}{y}}{\dfrac{1}{x} - \dfrac{1}{y}} = \frac{xy\left(\dfrac{1}{x} + \dfrac{1}{y}\right)}{xy\left(\dfrac{1}{x} - \dfrac{1}{y}\right)} \qquad \frac{xy}{xy} = 1.$$

$$= \frac{\dfrac{xy}{x} + \dfrac{xy}{y}}{\dfrac{xy}{x} - \dfrac{xy}{y}}$$

$$= \frac{y + x}{y - x} \qquad \text{Simplify the fractions.} \qquad \blacksquare$$

Self Check Simplify $\dfrac{\dfrac{1}{x} - \dfrac{1}{y}}{\dfrac{1}{x} + \dfrac{1}{y}}$.

Answer $\frac{y - x}{y + x}$

EXAMPLE 13 Simplify $\dfrac{x^{-1} + y^{-1}}{x^{-2} - y^{-2}}$.

Solution *Method 1:* We proceed as follows:

$$\frac{x^{-1} + y^{-1}}{x^{-2} - y^{-2}} = \frac{\dfrac{1}{x} + \dfrac{1}{y}}{\dfrac{1}{x^2} - \dfrac{1}{y^2}} \qquad \text{Write the fraction without using negative exponents.}$$

$$= \frac{\dfrac{y}{xy} + \dfrac{x}{xy}}{\dfrac{y^2}{x^2 y^2} - \dfrac{x^2}{x^2 y^2}} \qquad \text{Get a common denominator in the numerator and denominator.}$$

$$= \frac{\dfrac{y + x}{xy}}{\dfrac{y^2 - x^2}{x^2 y^2}} \qquad \text{Add the fractions in the numerator and denominator.}$$

$$= \frac{y + x}{xy} \div \frac{y^2 - x^2}{x^2 y^2} \qquad \text{Write the fraction as a division.}$$

$$= \frac{y + x}{xy} \cdot \frac{xxyy}{(y - x)(y + x)} \qquad \text{Invert and multiply.}$$

$$= \frac{(y + x)xxyy}{xy(y - x)(y + x)} \qquad \begin{array}{l} \text{Multiply the numerators and the} \\ \text{denominators.} \end{array}$$

$$= \frac{xy}{y - x} \qquad \begin{array}{l} \text{Divide out the common factors of } x, \\ y, \text{ and } y + x \text{ in the numerator and} \\ \text{denominator.} \end{array}$$

Method 2: We multiply both numerator and denominator by $x^2 y^2$, the LCD of the fractions in the problem, and proceed as follows:

$$\frac{x^{-1} + y^{-1}}{x^{-2} - y^{-2}} = \frac{\dfrac{1}{x} + \dfrac{1}{y}}{\dfrac{1}{x^2} - \dfrac{1}{y^2}} \qquad \begin{array}{l} \text{Write the fraction without negative} \\ \text{exponents.} \end{array}$$

$$= \frac{x^2 y^2 \left(\dfrac{1}{x} + \dfrac{1}{y} \right)}{x^2 y^2 \left(\dfrac{1}{x^2} - \dfrac{1}{y^2} \right)} \qquad \dfrac{x^2 y^2}{x^2 y^2} = 1.$$

$$= \frac{xy^2 + x^2 y}{y^2 - x^2} \qquad \begin{array}{l} \text{Use the distributive property to} \\ \text{remove parentheses, and simplify.} \end{array}$$

$$= \frac{xy(y + x)}{(y + x)(y - x)} \qquad \text{Factor the numerator and denominator.}$$

$$= \frac{xy}{y - x} \qquad \text{Divide out } y + x. \qquad \blacksquare$$

Self Check   Simplify $\dfrac{x^{-1} - y^{-1}}{x^{-2}}$.

Answer $\dfrac{xy - x^2}{y}$

WARNING! $x^{-1} + y^{-1}$ means $\frac{1}{x} + \frac{1}{y}$, and $(x + y)^{-1}$ means $\frac{1}{x + y}$. Thus,

$$x^{-1} + y^{-1} \neq \frac{1}{x + y} \qquad \text{and} \qquad (x + y)^{-1} \neq x^{-1} + y^{-1}$$

■ ■ ■ ■ ■ ■ ■ ■ ■ PERSPECTIVE

Each of the complex fractions in the list

$$1 + \frac{1}{2},\ 1 + \cfrac{1}{1+\frac{1}{2}},\ 1 + \cfrac{1}{1+\cfrac{1}{1+\frac{1}{2}}},\ 1 + \cfrac{1}{1+\cfrac{1}{1+\cfrac{1}{1+\frac{1}{2}}}},\ \ldots$$

can be simplified by using the value of the expression preceding it. For example, to simplify the second expression in the list, replace $1 + \frac{1}{2}$ with $\frac{3}{2}$.

$$1 + \cfrac{1}{1+\frac{1}{2}} = 1 + \cfrac{1}{\frac{3}{2}} = 1 + \frac{2}{3} = \frac{5}{3}$$

To simplify the third expression, replace

$1 + \cfrac{1}{1+\frac{1}{2}}$ with $\frac{5}{3}$:

$$1 + \cfrac{1}{1+\cfrac{1}{1+\frac{1}{2}}} = 1 + \cfrac{1}{\frac{5}{3}} = 1 + \frac{3}{5} = \frac{8}{5}$$

Can you show that the expressions in the list simplify to the fractions $\frac{3}{2}, \frac{5}{3}, \frac{8}{5}, \frac{13}{8}, \frac{21}{13}, \frac{34}{21} \ldots$?

Do you see a pattern, and can you predict the next fraction?

Use a calculator to write each of these fractions as a decimal. The values produced get closer and closer to the irrational number $1.61803398875\ldots$, which is known as the **golden ratio.** This number often appears in the architecture of the ancient Greeks and Egyptians. The width of the stairs in front of the Greek Parthenon (Illustration 1), divided by the building's height, is the golden ratio. The height of the triangular face of the Great Pyramid of Cheops (Illustration 2), divided by the pyramid's width, is also the golden ratio.

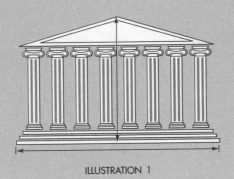

ILLUSTRATION 1

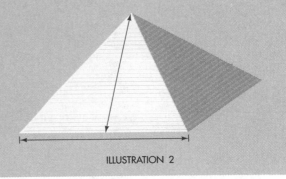

ILLUSTRATION 2

EXAMPLE 14   Simplify $\cfrac{\dfrac{2x}{1-\frac{1}{x}} + 3}{3 - \dfrac{2}{x}}$.

Solution   We begin by multiplying the numerator and denominator of the fraction

$$\frac{2x}{1 - \dfrac{1}{x}}$$

by x. This will eliminate the complex fraction in the numerator of the given fraction.

$$\frac{\dfrac{2x}{1 - \dfrac{1}{x}} + 3}{3 - \dfrac{2}{x}} = \frac{\dfrac{x2x}{x\left(1 - \dfrac{1}{x}\right)} + 3}{3 - \dfrac{2}{x}} \qquad \frac{x}{x} = 1.$$

$$= \frac{\dfrac{2x^2}{x - 1} + 3}{3 - \dfrac{2}{x}}$$

We then multiply the numerator and denominator of the previous fraction by $x(x - 1)$, the LCD of $\frac{2x^2}{x-1}$, 3, and $\frac{2}{x}$, and simplify:

$$\frac{\dfrac{2x}{1 - \dfrac{1}{x}} + 3}{3 - \dfrac{2}{x}} = \frac{x(x - 1)\left(\dfrac{2x^2}{x - 1} + 3\right)}{x(x - 1)\left(3 - \dfrac{2}{x}\right)} \qquad \frac{x(x-1)}{x(x-1)} = 1.$$

$$= \frac{2x^3 + 3x(x - 1)}{3x(x - 1) - 2(x - 1)}$$

$$= \frac{2x^3 + 3x^2 - 3x}{3x^2 - 5x + 2}$$

This result does not simplify.   ■

Self Check   Simplify $\dfrac{\dfrac{3}{1 - \dfrac{2}{x}} + 1}{2 - \dfrac{1}{x}}$.

Answer $\dfrac{4x^2 - 2x}{2x^2 - 5x + 2}$

Orals   *Simplify each fraction.*

1. $\dfrac{4}{6}$

2. $\dfrac{10}{15}$

3. $-\dfrac{25}{30}$

4. $-\dfrac{22}{55}$

5. $\dfrac{x^2}{xy}$

6. $\dfrac{2x-4}{x-2}$

7. $\dfrac{x-2}{2-x}$

8. $\dfrac{x^2-1}{x+1}$

EXERCISE 7.6

REVIEW   *Graph each interval.*

1. $(-\infty, -4) \cup [5, \infty)$

2. $(4, 8]$

Solve each formula for the indicated letter.

3. $P = 2l + 2w$; for w

4. $S = \dfrac{a - lr}{1 - r}$; for a

Solve each equation.

5. $a^4 - 13a^2 + 36 = 0$

6. $|2x - 1| = 9$

VOCABULARY AND CONCEPTS   *Fill in each blank to make a true statement.*

7. $\dfrac{ax}{bx} = \underline{\quad}$ $(b, x \neq 0)$

8. $\dfrac{a}{b} \cdot \dfrac{c}{d} = \underline{\quad}$ $(b, d \neq 0)$

9. $\dfrac{a}{b} \div \dfrac{c}{d} = \underline{\quad}$ $(b, c, d \neq 0)$

10. $\dfrac{a}{b} + \dfrac{c}{b} = \underline{\quad}$ $(b \neq 0)$

PRACTICE   *In Exercises 11–34, simplify each fraction.*

11. $\dfrac{12x^3}{3x}$

12. $-\dfrac{15a^2}{25a^3}$

13. $\dfrac{-24x^3y^4}{18x^4y^3}$

14. $\dfrac{15a^5b^4}{21b^3c^2}$

15. $\dfrac{9y^2(y-z)}{21y(y-z)^2}$

16. $\dfrac{-3ab^2(a-b)}{9ab(b-a)}$

17. $\dfrac{(a-b)(b-c)(c-d)}{(c-d)(b-c)(a-b)}$

18. $\dfrac{(p+q)(p-r)(r+s)}{(r-p)(r+s)(p+q)}$

19. $\dfrac{x+y}{x^2-y^2}$

20. $\dfrac{x-y}{x^2-y^2}$

21. $\dfrac{12-3x^2}{x^2-x-2}$

22. $\dfrac{x^2+2x-15}{x^2-25}$

23. $\dfrac{x^3+8}{x^2-2x+4}$

24. $\dfrac{x^2+3x+9}{x^3-27}$

25. $\dfrac{x^2+2x+1}{x^2+4x+3}$

26. $\dfrac{6x^2+x-2}{8x^2+2x-3}$

27. $\dfrac{3m-6n}{3n-6m}$

28. $\dfrac{ax+by+ay+bx}{a^2-b^2}$

29. $\dfrac{4x^2+24x+32}{16x^2+8x-48}$

30. $\dfrac{a^2-4}{a^3-8}$

31. $\dfrac{3x^2 - 3y^2}{x^2 + 2y + 2x + yx}$

32. $\dfrac{x^2 + 2xy}{x + 2y + x^2 - 4y^2}$

33. $\dfrac{x - y}{x^3 - y^3 - x + y}$

34. $\dfrac{2x^2 + 2x - 12}{x^3 + 3x^2 - 4x - 12}$

In Exercises 35–104, do the operations and simplify.

35. $\dfrac{x^2y^2}{cd} \cdot \dfrac{c^{-2}d^2}{x}$

36. $\dfrac{a^{-2}b^2}{x^{-1}y} \cdot \dfrac{a^4b^4}{x^2y^3}$

37. $\dfrac{-x^2y^{-2}}{x^{-1}y^{-3}} \div \dfrac{x^{-3}y^2}{x^4y^{-1}}$

38. $\dfrac{(a^3)^2}{b^{-1}} \div \dfrac{(a^3)^{-2}}{b^{-1}}$

39. $\dfrac{x^2 + 2x + 1}{x} \cdot \dfrac{x^2 - x}{x^2 - 1}$

40. $\dfrac{a + 6}{a^2 - 16} \cdot \dfrac{3a - 12}{3a + 18}$

41. $\dfrac{2x^2 - x - 3}{x^2 - 1} \cdot \dfrac{x^2 + x - 2}{2x^2 + x - 6}$

42. $\dfrac{9x^2 + 3x - 20}{3x^2 - 7x + 4} \cdot \dfrac{3x^2 - 5x + 2}{9x^2 + 18x + 5}$

43. $\dfrac{x^2 - 16}{x^2 - 25} \div \dfrac{x + 4}{x - 5}$

44. $\dfrac{a^2 - 9}{a^2 - 49} \div \dfrac{a + 3}{a + 7}$

45. $\dfrac{a^2 + 2a - 35}{12x} \div \dfrac{ax - 3x}{a^2 + 4a - 21}$

46. $\dfrac{x^2 - 4}{2b - bx} \div \dfrac{x^2 + 4x + 4}{2b + bx}$

47. $\dfrac{3t^2 - t - 2}{6t^2 - 5t - 6} \cdot \dfrac{4t^2 - 9}{2t^2 + 5t + 3}$

48. $\dfrac{2p^2 - 5p - 3}{p^2 - 9} \cdot \dfrac{2p^2 + 5p - 3}{2p^2 + 5p + 2}$

49. $\dfrac{3n^2 + 5n - 2}{12n^2 - 13n + 3} \div \dfrac{n^2 + 3n + 2}{4n^2 + 5n - 6}$

50. $\dfrac{8y^2 - 14y - 15}{6y^2 - 11y - 10} \div \dfrac{4y^2 - 9y - 9}{3y^2 - 7y - 6}$

51. $(2x^2 - 15x + 25) \div \dfrac{2x^2 - 3x - 5}{x + 1}$

52. $(x^2 - 6x + 9) \div \dfrac{x^2 - 9}{x + 3}$

53. $\dfrac{x^3 + y^3}{x^3 - y^3} \div \dfrac{x^2 - xy + y^2}{x^2 + xy + y^2}$

54. $\dfrac{x^2 - 6x + 9}{4 - x^2} \div \dfrac{x^2 - 9}{x^2 - 8x + 12}$

55. $\dfrac{m^2 - n^2}{2x^2 + 3x - 2} \cdot \dfrac{2x^2 + 5x - 3}{n^2 - m^2}$

56. $\dfrac{x^2 - y^2}{2x^2 + 2xy + x + y} \cdot \dfrac{2x^2 - 5x - 3}{yx - 3y - x^2 + 3x}$

57. $\dfrac{ax + ay + bx + by}{x^3 - 27} \cdot \dfrac{x^2 + 3x + 9}{xc + xd + yc + yd}$

58. $\dfrac{x^2 + 3x + xy + 3y}{x^2 - 9} \cdot \dfrac{x - 3}{x + 3}$

59. $\dfrac{x^2 - x - 6}{x^2 - 4} \cdot \dfrac{x^2 - x - 2}{9 - x^2}$

60. $\dfrac{2x^2 - 7x - 4}{20 - x - x^2} \div \dfrac{2x^2 - 9x - 5}{x^2 - 25}$

61. $\dfrac{2x^2 + 3xy + y^2}{y^2 - x^2} \div \dfrac{6x^2 + 5xy + y^2}{2x^2 - xy - y^2}$

62. $\dfrac{p^3 - q^3}{q^2 - p^2} \cdot \dfrac{q^2 + pq}{p^3 + p^2q + pq^2}$

63. $\dfrac{3x^2y^2}{6x^3y} \cdot \dfrac{-4x^7y^{-2}}{18x^{-2}y} \div \dfrac{36x}{18y^{-2}}$

64. $\dfrac{9ab^3}{7xy} \cdot \dfrac{14xy^2}{27z^3} \div \dfrac{18a^2b^2x}{3z^2}$

65. $(4x + 12) \cdot \dfrac{x^2}{2x - 6} \div \dfrac{2}{x - 3}$

66. $(4x^2 - 9) \div \dfrac{2x^2 + 5x + 3}{x + 2} \div (2x - 3)$

67. $\dfrac{2x^2 - 2x - 4}{x^2 + 2x - 8} \cdot \dfrac{3x^2 + 15x}{x + 1} \div \dfrac{4x^2 - 100}{x^2 - x - 20}$

68. $\dfrac{6a^2 - 7a - 3}{a^2 - 1} \div \dfrac{4a^2 - 12a + 9}{a^2 - 1} \cdot \dfrac{2a^2 - a - 3}{3a^2 - 2a - 1}$

69. $\dfrac{2x^2 + 5x - 3}{x^2 + 2x - 3} \div \left(\dfrac{x^2 + 2x - 35}{x^2 - 6x + 5} \div \dfrac{x^2 - 9x + 14}{2x^2 - 5x + 2} \right)$

70. $\dfrac{x^2 - 4}{x^2 - x - 6} \div \left(\dfrac{x^2 - x - 2}{x^2 - 8x + 15} \cdot \dfrac{x^2 - 3x - 10}{x^2 + 3x + 2} \right)$

71. $\dfrac{x^2 - x - 12}{x^2 + x - 2} \div \dfrac{x^2 - 6x + 8}{x^2 - 3x - 10} \cdot \dfrac{x^2 - 3x + 2}{x^2 - 2x - 15}$

72. $\dfrac{4x^2 - 10x + 6}{x^4 - 3x^3} \div \dfrac{2x - 3}{2x^3} \cdot \dfrac{x - 3}{2x - 2}$

73. $\dfrac{3}{a + b} - \dfrac{a}{a + b}$

74. $\dfrac{x}{x + 4} + \dfrac{5}{x + 4}$

75. $\dfrac{3x}{2x + 2} + \dfrac{x + 4}{2x + 2}$

76. $\dfrac{4y}{y - 4} - \dfrac{16}{y - 4}$

77. $\dfrac{5x}{x + 1} + \dfrac{3}{x + 1} - \dfrac{2x}{x + 1}$

78. $\dfrac{4}{a + 4} - \dfrac{2a}{a + 4} + \dfrac{3a}{a + 4}$

79. $\dfrac{3(x^2 + x)}{x^2 - 5x + 6} + \dfrac{-3(x^2 - x)}{x^2 - 5x + 6}$

80. $\dfrac{2x + 4}{x^2 + 13x + 12} - \dfrac{x + 3}{x^2 + 13x + 12}$

81. $\dfrac{a}{2} + \dfrac{2a}{5}$

82. $\dfrac{b}{6} + \dfrac{3a}{4}$

83. $\dfrac{3}{4x} + \dfrac{2}{3x}$

84. $\dfrac{2}{5a} + \dfrac{3}{2b}$

85. $\dfrac{a + b}{3} + \dfrac{a - b}{7}$

86. $\dfrac{x - y}{2} + \dfrac{x + y}{3}$

87. $\dfrac{3}{x + 2} + \dfrac{5}{x - 4}$

88. $\dfrac{2}{a + 4} - \dfrac{6}{a + 3}$

89. $\dfrac{x + 2}{x + 5} - \dfrac{x - 3}{x + 7}$

90. $\dfrac{7}{x + 3} + \dfrac{4x}{x + 6}$

91. $x + \dfrac{1}{x}$

92. $2 - \dfrac{1}{x + 1}$

93. $\dfrac{x + 8}{x - 3} - \dfrac{x - 14}{3 - x}$

94. $\dfrac{3 - x}{2 - x} + \dfrac{x - 1}{x - 2}$

95. $\dfrac{x}{x^2 + 5x + 6} + \dfrac{x}{x^2 - 4}$

96. $\dfrac{x}{3x^2 - 2x - 1} + \dfrac{4}{3x^2 + 10x + 3}$

97. $\dfrac{8}{x^2 - 9} + \dfrac{2}{x - 3} - \dfrac{6}{x}$

98. $\dfrac{x}{x^2 - 4} - \dfrac{x}{x + 2} + \dfrac{2}{x}$

99. $1 + x - \dfrac{x}{x - 5}$

100. $2 - x + \dfrac{3}{x - 9}$

101. $\dfrac{3}{x + 1} - \dfrac{2}{x - 1} + \dfrac{x + 3}{x^2 - 1}$

102. $\dfrac{2}{x - 2} + \dfrac{3}{x + 2} - \dfrac{x - 1}{x^2 - 4}$

103. $\dfrac{x - 2}{x^2 - 3x} + \dfrac{2x - 1}{x^2 + 3x} - \dfrac{2}{x^2 - 9}$

104. $\dfrac{2}{x - 1} - \dfrac{2x}{x^2 - 1} - \dfrac{x}{x^2 + 2x + 1}$

In Exercises 105–128, simplify each complex fraction.

105. $\dfrac{\dfrac{4x}{y}}{\dfrac{6xz}{y^2}}$

106. $\dfrac{\dfrac{5t^4}{9x}}{\dfrac{2t}{18x}}$

107. $\dfrac{\dfrac{x-y}{xy}}{\dfrac{y-x}{x}}$

108. $\dfrac{\dfrac{x^2+5x+6}{3xy}}{\dfrac{x^2-9}{6xy}}$

109. $\dfrac{\dfrac{1}{a}+\dfrac{1}{b}}{\dfrac{1}{a}}$

110. $\dfrac{\dfrac{1}{b}}{\dfrac{1}{a}-\dfrac{1}{b}}$

111. $\dfrac{\dfrac{y}{x}-\dfrac{x}{y}}{\dfrac{1}{x}+\dfrac{1}{y}}$

112. $\dfrac{\dfrac{y}{x}-\dfrac{x}{y}}{\dfrac{1}{y}-\dfrac{1}{x}}$

113. $\dfrac{\dfrac{1}{a}-\dfrac{1}{b}}{\dfrac{a}{b}-\dfrac{b}{a}}$

114. $\dfrac{\dfrac{1}{a}+\dfrac{1}{b}}{\dfrac{a}{b}-\dfrac{b}{a}}$

115. $\dfrac{1+\dfrac{6}{x}+\dfrac{8}{x^2}}{1+\dfrac{1}{x}-\dfrac{12}{x^2}}$

116. $\dfrac{1-x-\dfrac{2}{x}}{\dfrac{6}{x^2}+\dfrac{1}{x}-1}$

117. $\dfrac{\dfrac{1}{a+1}+1}{\dfrac{3}{a-1}+1}$

118. $\dfrac{2+\dfrac{3}{x+1}}{\dfrac{1}{x}+x+x^2}$

119. $\dfrac{x^{-1}+y^{-1}}{x^{-1}-y^{-1}}$

120. $\dfrac{(x+y)^{-1}}{x^{-1}+y^{-1}}$

121. $\dfrac{x+y}{x^{-1}+y^{-1}}$

122. $\dfrac{x-y}{x^{-1}-y^{-1}}$

123. $\dfrac{x-y^{-2}}{y-x^{-2}}$

124. $\dfrac{x^{-2}-y^{-2}}{x^{-1}-y^{-1}}$

125. $\dfrac{1+\dfrac{a}{b}}{1-\dfrac{a}{1-\dfrac{a}{b}}}$

126. $\dfrac{1+\dfrac{2}{1+\dfrac{a}{b}}}{1-\dfrac{a}{b}}$

127. $a+\dfrac{a}{1+\dfrac{a}{a+1}}$

128. $b+\dfrac{b}{1-\dfrac{b+1}{b}}$

WRITING

129. Explain how to simplify a rational expression.

130. Explain how to multiply two rational expressions.

131. Explain how to divide two rational expressions.

132. Explain how to add two rational expressions.

SOMETHING TO THINK ABOUT

133. A student compared his answer, $\frac{a-3b}{2b-a}$, with the answer, $\frac{3b-a}{a-2b}$, in the back of the text. Is the student's work correct?

134. Another student shows this work:

$$\frac{3x^2 + 6}{3y} = \frac{\cancel{3}x^2 + \overset{2}{\cancel{6}}}{\cancel{3}y} = \frac{x^2 + 2}{y}$$

Is the student's work correct?

135. In which parts can you divide out the 4's?

a. $\dfrac{4x}{4y}$

b. $\dfrac{4x}{x + 4}$

c. $\dfrac{4 + x}{4 + y}$

d. $\dfrac{4x}{4 + 4y}$

136. In which parts can you divide out the 3's?

a. $\dfrac{3x + 3y}{3z}$

b. $\dfrac{3(x + y)}{3x + y}$

c. $\dfrac{x + 3}{3y}$

d. $\dfrac{3x + 3y}{3a - 3b}$

7.7 Synthetic Division

■ SYNTHETIC DIVISION ■ THE REMAINDER THEOREM ■ THE FACTOR THEOREM

Getting Ready   *Do each division and find P(2).*

1. $x - 2 \overline{)x^2 - x - 1}$

2. $x - 2 \overline{)x^2 + x + 3}$

■ SYNTHETIC DIVISION

There is a shortcut method, called **synthetic division,** that we can use to divide a polynomial by a binomial of the form $x - r$. To see how it works, we consider the division of $4x^3 - 5x^2 - 11x + 20$ by $x - 2$.

$$
\begin{array}{r}
4x^2 + 3x \ - \ 5 \\
x - 2 \overline{)4x^3 - 5x^2 - 11x + 20} \\
\underline{4x^3 - 8x^2} \\
3x^2 - 11x \\
\underline{3x^2 - \ 6x} \\
- \ 5x + 20 \\
\underline{- \ 5x + 10} \\
10 \quad \text{(remainder)}
\end{array}
$$

$$
\begin{array}{r}
4 \quad 3 - 5 \\
1 - 2 \overline{)4 - 5 - 11 \quad 20} \\
\underline{4 - 8} \\
3 - 11 \\
\underline{3 - \ 6} \\
- \ 5 \quad 20 \\
\underline{- \ 5 \quad 10} \\
10 \quad \text{(remainder)}
\end{array}
$$

On the left is the long division, and on the right is the same division with the variables and their exponents removed. The various powers of x can be remembered without actually writing them, because the exponents of the terms in the divisor, dividend, and quotient were written in descending order.

We can further shorten the version on the right. The numbers printed in color need not be written, because they are duplicates of the numbers above them. Thus, we can write the division in the following form:

$$
\begin{array}{r}
4 \quad 3 \quad -5 \\
1-2\overline{)4 \quad -5 \quad -11 \quad 20} \\
\underline{-8} \\
3 \\
\underline{-6} \\
-5 \\
\underline{10} \\
10
\end{array}
$$

We can shorten the process further by compressing the work vertically and eliminating the 1 (the coefficient of x in the divisor):

$$
\begin{array}{r}
4 \quad 3 \quad -5 \\
-2\overline{)4 \quad -5 \quad -11 \quad 20} \\
\underline{-8 \quad -6 \quad 10} \\
3 \quad -5 \quad 10
\end{array}
$$

If we write the 4 in the quotient on the bottom line, the bottom line gives the coefficients of the quotient and the remainder. If we eliminate the top line, the division appears as follows:

$$
\begin{array}{r}
-2 \;|\; 4 \quad -5 \quad -11 \quad 20 \\
\underline{-8 \quad -6 \quad 10} \\
4 \quad 3 \quad -5 \quad 10
\end{array}
$$

The bottom line was obtained by subtracting the middle line from the top line. If we replace the -2 in the divisor by $+2$, the division process will reverse the signs of every entry in the middle line, and then the bottom line can be obtained by addition. This gives the final form of the synthetic division.

$$
\begin{array}{r}
+2 \;|\; 4 \quad -5 \quad -11 \quad 20 \\
\underline{8 \quad 6 \quad -10} \\
4 \quad 3 \quad -5 \quad 10
\end{array}
$$

The coefficients of the dividend.

The coefficients of the quotient and the remainder.

Thus,

$$
\frac{4x^3 - 5x^2 - 11x + 20}{x - 2} = 4x^2 + 3x - 5 + \frac{10}{x - 2}
$$

EXAMPLE 1   Divide $6x^2 + 5x - 2$ by $x - 5$.

Solution   We write the coefficients in the dividend and the 5 in the divisor in the following form:

$$
\begin{array}{r}
5 \;|\; 6 \quad 5 \quad -2 \\
\hline
\end{array}
$$

Then we follow these steps:

$$5\,|\quad 6\quad 5\quad -2 \qquad \text{Begin by bringing down the 6.}$$

$$\underline{}$$
$$6$$

$$5\,|\quad 6\quad 5\quad -2 \qquad \text{Multiply 5 by 6 to get 30.}$$
$$\underline{\ \ 30}$$
$$6$$

$$5\,|\quad 6\quad 5\quad -2 \qquad \text{Add 5 and 30 to get 35.}$$
$$\underline{\ \ 30}$$
$$6\quad 35$$

$$5\,|\quad 6\quad 5\quad -2 \qquad \text{Multiply 35 by 5 to get 175.}$$
$$\underline{\ \ 30\quad 175}$$
$$6\quad 35$$

$$5\,|\quad 6\quad 5\quad -2 \qquad \text{Add } -2 \text{ and 175 to get 173.}$$
$$\underline{\ \ 30\quad 175}$$
$$6\quad 35\quad 173$$

The numbers 6 and 35 represent the quotient $6x + 35$, and 173 is the remainder. Thus,

$$\frac{6x^2 + 5x - 2}{x - 5} = 6x + 35 + \frac{173}{x - 5}$$

■

EXAMPLE 2 Divide $5x^3 + x^2 - 3$ by $x - 2$.

Solution We begin by writing

$$2\,|\quad 5\quad 1\quad \mathbf{0}\quad -3 \qquad \text{Write 0 for the coefficient of } x, \text{ the missing term.}$$

and complete the division as follows:

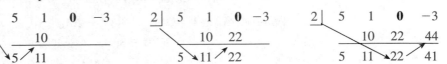

Thus,

$$\frac{5x^3 + x^2 - 3}{x - 2} = 5x^2 + 11x + 22 + \frac{41}{x - 2}$$

■

EXAMPLE 3 Divide $5x^2 + 6x^3 + 2 - 4x$ by $x + 2$.

Solution First, we write the dividend with the exponents in descending order.

$$6x^3 + 5x^2 - 4x + 2$$

Then we write the divisor in $x - r$ form: $x - (-2)$. Using synthetic division, we begin by writing

$$-2\,\underline{|\;\;6\quad 5\quad -4\quad 2}$$

and complete the division.

$$
\begin{array}{r|rrrr}
-2 & 6 & 5 & -4 & 2 \\
   &   & -12 & 14 & -20 \\
\hline
   & 6 & -7 & 10 & -18
\end{array}
$$

Thus,

$$\frac{5x^2 + 6x^3 + 2 - 4x}{x + 2} = 6x^2 - 7x + 10 + \frac{-18}{x + 2}$$

■

Self Check   Divide $2x - 4x^2 + 3x^3 - 3$ by $x - 1$.

Answer $3x^2 - x + 1 + \frac{-2}{x-1}$

■ THE REMAINDER THEOREM

Synthetic division is important in mathematics because of the **remainder theorem.**

> **Remainder Theorem**
>
> If a polynomial $P(x)$ is divided by $x - r$, the remainder is $P(r)$.

We illustrate the remainder theorem in the next example.

EXAMPLE 4   Let $P(x) = 2x^3 - 3x^2 - 2x + 1$. Find   **a.** $P(3)$ and   **b.** the remainder when $P(x)$ is divided by $x - 3$.

Solution   **a.** $P(3) = 2(3)^3 - 3(3)^2 - 2(3) + 1$ Substitute 3 for x.

$= 2(27) - 3(9) - 6 + 1$

$= 54 - 27 - 6 + 1$

$= \mathbf{22}$

b. We use synthetic division to find the remainder when $P(x) = 2x^3 - 3x^2 - 2x + 1$ is divided by $x - 3$.

$$
\begin{array}{r|rrrr}
3 & 2 & -3 & -2 & 1 \\
  &   & 6 & 9 & 21 \\
\hline
  & 2 & 3 & 7 & 22
\end{array}
$$

The remainder is 22.

The results of parts **a** and **b** show that when $P(x)$ is divided by $x - 3$, the remainder is $P(3)$. ∎

It is often easier to find $P(r)$ by using synthetic division than by substituting r for x in $P(x)$. This is especially true if r is a decimal.

■ THE FACTOR THEOREM

Recall that if two quantities are multiplied, each is called a **factor** of the product. Thus, $x - 2$ is one factor of $6x - 12$, because $6(x - 2) = 6x - 12$. A theorem, called the **factor theorem,** tells us how to find one factor of a polynomial if the remainder of a certain division is 0.

> **Factor Theorem**
> If $P(x)$ is a polynomial in x, then
> $$P(r) = 0 \quad \text{if and only if} \quad x - r \text{ is a factor of } P(x)$$

If $P(x)$ is a polynomial in x and if $P(r) = 0$, r is a **zero** of the polynomial.

EXAMPLE 5 Let $P(x) = 3x^3 - 5x^2 + 3x - 10$. Show that **a.** $P(2) = 0$ and **b.** $x - 2$ is a factor of $P(x)$.

Solution **a.** We can use the remainder theorem to evaluate $P(2)$ by dividing $P(x) = 3x^3 - 5x^2 + 3x - 10$ by $x - 2$.

$$
\begin{array}{r|rrrr}
2 & 3 & -5 & 3 & -10 \\
  &   & 6 & 2 & 10 \\
\hline
  & 3 & 1 & 5 & 0 \\
\end{array}
$$

The remainder in this division is 0. By the remainder theorem, the remainder is $P(2)$. Thus, $P(2) = 0$, and 2 is a zero of the polynomial.

b. Because the remainder is 0, the numbers 3, 1, and 5 in the synthetic division in part **a** represent the quotient $3x^2 + x + 5$. Thus,

$$\underbrace{(x - 2)}_{\text{Divisor} \cdot} \cdot \underbrace{(3x^2 + x + 5)}_{\text{quotient}} + \underbrace{0}_{+ \text{ remainder}} = \underbrace{3x^3 - 5x^2 + 3x - 10}_{\text{the dividend, } P(x)}$$

or

$$(x - 2)(3x^2 + x + 5) = 3x^3 - 5x^2 + 3x - 10$$

Thus, $x - 2$ is a factor of $P(x)$. ∎

The result in Example 5 is true, because the remainder, $P(2)$, is 0. If the remainder had not been 0, then $x - 2$ would not have been a factor of $P(x)$.

■ ■ ■ ■ ■ ■ ■ ■ ■   **Approximating Zeros of Polynomials**

GRAPHING
CALCULATORS

We can use a graphing calculator to approximate the real zeros of a polynomial function $f(x)$. For example, to find the real zeros of $f(x) = 2x^3 - 6x^2 + 7x - 21$, we graph the function as in Figure 7-20.

It is clear from the figure that the function f has a zero at $x = 3$.

$$f(3) = 2(3)^3 - 6(3)^2 + 7(3) - 21 \qquad \text{Substitute 3 for } x.$$
$$= 2(27) - 6(9) + 21 - 21$$
$$= 0$$

From the factor theorem, we know that $x - 3$ is a factor of the polynomial. To find the other factor, we can synthetically divide by 3.

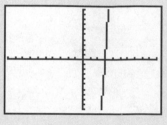

FIGURE 7-20

$$
\begin{array}{r|rrrr}
3 & 2 & -6 & 7 & -21 \\
  &   & 6 & 0 & 21 \\
\hline
  & 2 & 0 & 7 & 0
\end{array}
$$

Thus, $f(x) = (x - 3)(2x^2 + 7)$. Since $2x^2 + 7$ cannot be factored over the real numbers, we can conclude that 3 is the only real zero of the polynomial function.

Orals   *Find the remainder in each division.*

1. $(x^2 + 2x + 1) \div (x - 2)$ **2.** $(x^2 - 4) \div (x + 1)$

Tell whether $x - 2$ is a factor of each polynomial.

3. $x^3 - 2x^2 + x - 2$ **4.** $x^3 + 4x^2 - 1$

EXERCISE 7.7

REVIEW   *Let $f(x) = 3x^2 + 2x - 1$ and find each value.*

1. $f(1)$ **2.** $f(-2)$ **3.** $f(2a)$ **4.** $f(-t)$

Remove parentheses and simplify.

5. $2(x^2 + 4x - 1) + 3(2x^2 - 2x + 2)$
6. $-2(3y^3 - 2y + 7) - 3(y^2 + 2y - 4) + 4(y^3 + 2y - 1)$

VOCABULARY AND CONCEPTS   *Fill in each blank to make a true statement.*

7. If a polynomial $P(x)$ is divided by $x - r$, the remainder is ____.

8. If $P(x)$ is a polynomial in x, then $P(r) = 0$ if and only if ____ is a factor of $P(x)$.

PRACTICE *In Exercises 9–22, use synthetic division to do each division.*

9. $(x^2 + x - 2) \div (x - 1)$ **10.** $(x^2 + x - 6) \div (x - 2)$

11. $(x^2 - 7x + 12) \div (x - 4)$ **12.** $(x^2 - 6x + 5) \div (x - 5)$

13. $(x^2 + 8 + 6x) \div (x + 4)$ **14.** $(x^2 - 15 - 2x) \div (x + 3)$

15. $(x^2 - 5x + 14) \div (x + 2)$ **16.** $(x^2 + 13x + 42) \div (x + 6)$

17. $(3x^3 - 10x^2 + 5x - 6) \div (x - 3)$ **18.** $(2x^3 - 9x^2 + 10x - 3) \div (x - 3)$

19. $(2x^3 - 5x - 6) \div (x - 2)$ **20.** $(4x^3 + 5x^2 - 1) \div (x + 2)$

21. $(5x^2 + 6x^3 + 4) \div (x + 1)$ **22.** $(4 - 3x^2 + x) \div (x - 4)$

In Exercises 23–28, use a calculator and synthetic division to do each division.

23. $(7.2x^2 - 2.1x + 0.5) \div (x - 0.2)$ **24.** $(8.1x^2 + 3.2x - 5.7) \div (x - 0.4)$

25. $(2.7x^2 + x - 5.2) \div (x + 1.7)$ **26.** $(1.3x^2 - 0.5x - 2.3) \div (x + 2.5)$

27. $(9x^3 - 25) \div (x + 57)$ **28.** $(0.5x^3 + x) \div (x - 2.3)$

In Exercises 29–36, let $P(x) = 2x^3 - 4x^2 + 2x - 1$. Evaluate $P(x)$ by substituting the given value of x into the polynomial and simplifying. Then evaluate the polynomial by using the remainder theorem and synthetic division.

29. $P(1)$ **30.** $P(2)$ **31.** $P(-2)$ **32.** $P(-1)$

33. $P(3)$ **34.** $P(-4)$ **35.** $P(0)$ **36.** $P(4)$

In Exercises 37–44, let $Q(x) = x^4 - 3x^3 + 2x^2 + x - 3$. Evaluate $Q(x)$ by substituting the given value of x into the polynomial and simplifying. Then evaluate the polynomial by using the remainder theorem and synthetic division.

37. $Q(-1)$ **38.** $Q(1)$ **39.** $Q(2)$ **40.** $Q(-2)$

41. $Q(3)$ **42.** $Q(0)$ **43.** $Q(-3)$ **44.** $Q(-4)$

In Exercises 45–52, use the remainder theorem and synthetic division to find $P(r)$.

45. $P(x) = x^3 - 4x^2 + x - 2;\ r = 2$ **46.** $P(x) = x^3 - 3x^2 + x + 1;\ r = 1$

47. $P(x) = 2x^3 + x + 2;\ r = 3$ **48.** $P(x) = x^3 + x^2 + 1;\ r = -2$

49. $P(x) = x^4 - 2x^3 + x^2 - 3x + 2;\ r = -2$ **50.** $P(x) = x^5 + 3x^4 - x^2 + 1;\ r = -1$

51. $P(x) = 3x^5 + 1;\ r = -\dfrac{1}{2}$ **52.** $P(x) = 5x^7 - 7x^4 + x^2 + 1;\ r = 2$

In Exercises 53–56, use the factor theorem and tell whether the first expression is a factor of $P(x)$.

53. $x - 3;\ P(x) = x^3 - 3x^2 + 5x - 15$ **54.** $x + 1;\ P(x) = x^3 + 2x^2 - 2x - 3$ (*Hint:* Write $x + 1$ as $x - (-1)$.)

55. $x + 2;\ P(x) = 3x^2 - 7x + 4$ (*Hint:* Write $x + 2$ as $x - (-2)$.)   **56.** $x;\ P(x) = 7x^3 - 5x^2 - 8x$ (*Hint:* $x = x - 0$.)

In Exercises 57–58, use a calculator to work each problem.

57. Find 2^6 by using synthetic division to evaluate the polynomial $P(x) = x^6$ at $x = 2$. Then check the answer by evaluating 2^6 with a calculator.

58. Find $(-3)^5$ by using synthetic division to evaluate the polynomial $P(x) = x^5$ at $x = -3$. Then check the answer by evaluating $(-3)^5$ with a calculator.

WRITING

59. If you are given $P(x)$, explain how to use synthetic division to calculate $P(a)$.

60. Explain the factor theorem.

SOMETHING TO THINK ABOUT *Suppose that* $P(x) = x^{100} - x^{99} + x^{98} - x^{97} + \cdots + x^2 - x + 1.$

61. Find the remainder when $P(x)$ is divided by $x - 1$.

62. Find the remainder when $P(x)$ is divided by $x + 1$.

■ ■ ■ ■ ■ ■ ■ ■ ■ ■ **PROJECTS**

PROJECT 1

The expression $1 + x + x^2 + x^3$ is a polynomial of degree 3. The polynomial $1 + x + x^2 + x^3 + x^4$ has the same pattern, but one more term. Its degree is 4. As the pattern continues and more terms are added, the degree of the polynomial increases. If there were no end to the number of terms, the "polynomial" would have infinitely many terms, and no defined degree:

$$1 + x + x^2 + x^3 + x^4 + x^5 + x^6 + \cdots$$

Such "unending polynomials," called **power series,** are studied in calculus. However, this particular series is the result of a division of polynomials:

- Consider the division $\frac{1}{1-x}$. Find the quotient by filling in more steps of this long division:

Step 1

$$
\begin{array}{r}
1 \\
1 - x {\overline{\smash{\big)}\,1 + 0x + 0x^2 +}} \\
\underline{1 - x} \\
x
\end{array}
$$

Step 2

$$
\begin{array}{r}
1 + x \\
1 - x {\overline{\smash{\big)}\,1 + 0x + 0x^2 +}} \\
\underline{1 - x} \\
x + 0x^2 \\
\underline{x - x^2} \\
x^2
\end{array}
$$

To determine how the fraction $\frac{1}{1-x}$ and the series $1 + x + x^2 + x^3 + x^4 + x^5 + x^6 + \cdots$ could be equal, try this experiment.

- Let $x = \frac{1}{2}$ and evaluate $\frac{1}{1-x}$.
- Again, let $x = \frac{1}{2}$ and evaluate the series. Because you cannot add infinitely many numbers, just add the first 3, or 4, or 5 terms and see if you find a pattern. Use a calculator to complete this chart:

Polynomial	Value at $x = \frac{1}{2}$
$1 + x + x^2$	
$1 + x + x^2 + x^3$	
$1 + x + x^2 + x^3 + x^4$	
$1 + x + x^2 + x^3 + x^4 + x^5$	
$1 + x + x^2 + x^3 + x^4 + x^5 + x^6$	

What number do the values in the second column seem to be approaching? That number is called the *sum* of the series.

- Explain why the nonterminating decimal 1.1111111. . . represents the infinite series

$$1 + \left(\frac{1}{10}\right) + \left(\frac{1}{10}\right)^2 + \left(\frac{1}{10}\right)^3 + \left(\frac{1}{10}\right)^4 + \left(\frac{1}{10}\right)^5 + \left(\frac{1}{10}\right)^6 + \cdots$$

- Using the fraction $\frac{1}{1-x}$, explain why $1.11111\ldots = \frac{10}{9}$.
- Verify that $\frac{10}{9} = 1.11111\ldots$ by dividing 10 by 9.

PROJECT 2 We began this chapter by discussing *lift* provided by the wing of an airplane. We learned that two factors that determine lift are controlled by the pilot. One is the speed (or velocity) of the plane, and the other is the *angle of attack,* which is the angle between the direction the plane is aimed and the direction it is actually moving, as shown in Illustration 1.

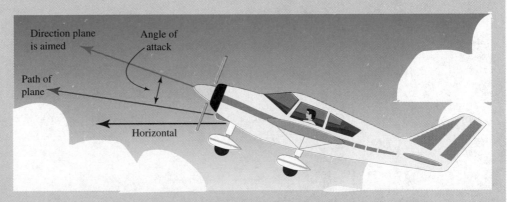

ILLUSTRATION 1

For one particular plane weighing 2,050 pounds, the lift, velocity, and angle of attack are related by the equation

$$L = (0.017a + 0.023)V^2$$

where L is the lift in pounds, a is the angle of attack in degrees, and V is the velocity in feet per second. To support the plane, the lift must equal the plane's weight.

a. Find the correct angle of attack when the velocity of the plane is 88.64 mph. (*Hint:* You must change the velocity to units of feet per second.)

b. As the angle of attack approaches 17°, the plane begins to stall. With more cargo on the return trip, the same plane weighs 2,325 pounds. If the pilot allows the velocity to drop to 80 feet per second (about 55 mph), will the plane stall?

C H A P T E R   S U M M A R Y

CONCEPTS

REVIEW EXERCISES

SECTION 7.1

Review of Equations and Inequalities

If $a = b$, then
$$a + c = b + c$$
$$a - c = b - c$$
$$ac = bc$$
$$\frac{a}{c} = \frac{b}{c} \quad (c \neq 0)$$

1. Solve and check each equation.
  a. $4(y - 1) = 28$
  b. $3(x + 7) = 42$
  c. $13(x - 9) - 2 = 7x - 5$
  d. $\dfrac{8(x - 5)}{3} = 2(x - 4)$
  e. $2x + 4 = 2(x + 3) - 2$
  f. $(3x - 2) - x = 2(x - 4)$

2. Solve for the indicated variable.
  a. $V = \dfrac{1}{3}\pi r^2 h$ for h

  b. $V = \dfrac{1}{6}ab(x + y)$ for x

3. Carpentry A carpenter wants to cut a 20-foot rafter so that one piece is 3 times as long as the other. Where should he cut the board?

4. Geometry A rectangle is 4 meters longer than it is wide. If the perimeter of the rectangle is 28 meters, find its area.

If $a < b$, then
$$a + c < b + c$$
$$a - c < b - c$$
$$ac < bc \quad (c > 0)$$
$$\frac{a}{c} < \frac{b}{c} \quad (c > 0)$$
$$ac > bc \quad (c < 0)$$
$$\frac{a}{c} > \frac{b}{c} \quad (c < 0)$$

$a < x < b$ means
$a < x$ and $x < b$

5. Solve each inequality. Give each solution set in interval notation and graph it.
  a. $\dfrac{1}{3}y - 2 \geq \dfrac{1}{2}y + 2$

  b. $\dfrac{7}{4}(x + 3) < \dfrac{3}{8}(x - 3)$

  c. $3 < 3x + 4 < 10$

  d. $4x > 3x + 2 > x - 3$

SECTION 7.2

Equations Containing Absolute Values

If $x \geq 0$, then $|x| = x$.
If $x < 0$, then $|x| = -x$.
$|x| = k$ means $x = k$ or
  $x = -k$

$|a| = |b|$ means $a = b$ or
  $a = -b$

6. Solve and check each equation.
  a. $|3x + 1| = 10$
  b. $\left|\dfrac{3}{2}x - 4\right| = 9$
  c. $|3x + 2| = |2x - 3|$
  d. $|5x - 4| = |4x - 5|$

| **SECTION 7.3** | *Inequalities Containing Absolute Values* |

If $k > 0$, then
   $|x| < k$ means $-k < x < k$
   $|x| \leq k$ means $-k \leq x \leq k$

If k is a nonnegative constant, then
   $|x| > k$ means $x < -k$ or
     $x > k$
   $|x| \geq k$ means $x \leq -k$ or
     $x \geq k$

7. Solve each inequality. Give each solution in interval notation and graph it.
  a. $|2x + 7| < 3$

  b. $|3x - 8| \geq 4$

  c. $\left| \dfrac{3}{2}x - 14 \right| \geq 0$

  d. $\left| \dfrac{2}{3}x + 14 \right| < 0$

| **SECTION 7.4** | *Review of Factoring* |

$x^2 - y^2 = (x + y)(x - y)$

8. Factor each polynomial.
  a. $4x + 8$
                              b. $5x^2y^3 - 10xy^2$
  c. $-8x^2y^3z^4 - 12x^4y^3z^2$
  d. $12a^6b^4c^2 + 15a^2b^4c^6$
  e. $xy + 2y + 4x + 8$
  f. $ac + bc + 3a + 3b$

9. Factor x^n from $x^{2n} + x^n$.

10. Factor y^{2n} from $y^{2n} - y^{4n}$.

11. Factor each polynomial.
  a. $x^4 + 4y + 4x^2 + x^2y$
  b. $a^5 + b^2c + a^2c + a^3b^2$
  c. $z^2 - 16$
  d. $y^2 - 121$
  e. $2x^4 - 98$
  f. $3x^6 - 300x^2$
  g. $y^2 + 21y + 20$
  h. $z^2 - 11z + 30$
  i. $-x^2 - 3x + 28$
  j. $-y^2 + 5y + 24$
  k. $y^3 + y^2 - 2y$
  l. $2a^4 + 4a^3 - 6a^2$
  m. $15x^2 - 57xy - 12y^2$
  n. $30x^2 + 65xy + 10y^2$
  o. $x^2 + 4x + 4 - 4p^4$
  p. $y^2 + 3y + 2 + 2x + xy$

SECTION 7.5	Factoring the Sum and Difference of Two Cubes

$x^3 + y^3 = (x + y)(x^2 - xy + y^2)$
$x^3 - y^3 = (x - y)(x^2 + xy + y^2)$

12. Factor each polynomial.

   a. $x^3 + 343$

   b. $a^3 - 125$

   c. $8y^3 - 512$

   d. $4x^3y + 108yz^3$

SECTION 7.6	Review of Rational Expressions

Use the same rules to manipulate rational expressions as you would use to manipulate arithmetic fractions.

13. Simplify each fraction.

   a. $\dfrac{248x^2y}{576xy^2}$ **b.** $\dfrac{x^2 - 49}{x^2 + 14x + 49}$

14. Do the operations and simplify.

   a. $\dfrac{x^2 + 4x + 4}{x^2 - x - 6} \cdot \dfrac{x^2 - 9}{x^2 + 5x + 6}$

   b. $\dfrac{x^3 - 64}{x^2 + 4x + 16} \div \dfrac{x^2 - 16}{x + 4}$

   c. $\dfrac{5y}{x - y} - \dfrac{3}{x - y}$

   d. $\dfrac{3x - 1}{x^2 + 2} + \dfrac{3(x - 2)}{x^2 + 2}$

   e. $\dfrac{3}{x + 2} + \dfrac{2}{x + 3}$

   f. $\dfrac{4x}{x - 4} - \dfrac{3}{x + 3}$

   g. $\dfrac{x^2 + 3x + 2}{x^2 - x - 6} \cdot \dfrac{3x^2 - 3x}{x^2 - 3x - 4} \div \dfrac{x^2 + 3x + 2}{x^2 - 2x - 8}$

   h. $\dfrac{x^2 - x - 6}{x^2 - 3x - 10} \div \dfrac{x^2 - x}{x^2 - 5x} \cdot \dfrac{x^2 - 4x + 3}{x^2 - 6x + 9}$

   i. $\dfrac{2x}{x + 1} + \dfrac{3x}{x + 2} + \dfrac{4x}{x^2 + 3x + 2}$

   j. $\dfrac{5x}{x - 3} + \dfrac{5}{x^2 - 5x + 6} + \dfrac{x + 3}{x - 2}$

   k. $\dfrac{3(x + 2)}{x^2 - 1} - \dfrac{2}{x + 1} + \dfrac{4(x + 3)}{x^2 - 2x + 1}$

   l. $\dfrac{x}{x^2 + 4x + 4} + \dfrac{2x}{x^2 - 4} - \dfrac{x^2 - 4}{x - 2}$

15. Simplify each complex fraction.

a. $\dfrac{\dfrac{3}{x} - \dfrac{2}{y}}{xy}$

b. $\dfrac{\dfrac{1}{x} + \dfrac{2}{y}}{\dfrac{2}{x} - \dfrac{1}{y}}$

c. $\dfrac{2x + 3 + \dfrac{1}{x}}{x + 2 + \dfrac{1}{x}}$

d. $\dfrac{x^{-1} - y^{-1}}{x^{-1} + y^{-1}}$

SECTION 7.7 — *Synthetic division*

The remainder theorem:
If a polynomial $P(x)$ is divided by $x - r$, the remainder is $P(r)$.

16. Use synthetic division to find the remainder in each division.

a. $x - 2\overline{)3x^3 + 2x^2 - 7x + 2}$

b. $x + 2\overline{)2x^3 - 4x^2 - 14x + 3}$

The factor theorem:
If $P(x)$ is a polynomial in x, then $P(r) = 0$ if and only if $x - r$ is a factor of $P(x)$.

17. Use the factor theorem to decide whether the first expression is a factor of $P(x)$.

a. $x - 5$; $P(x) = x^3 - 3x^2 - 8x - 10$

b. $x + 5$; $P(x) = x^3 + 4x^2 - 5x + 5$ (*Hint:* Write $x + 5$ as $x - (-5)$.)

■ Chapter Test

In Problems 1–2, solve each equation.

1. $9(x + 4) + 4 = 4(x - 5)$

2. $\dfrac{y - 1}{5} + 2 = \dfrac{2y - 3}{3}$

3. Solve $P = L + \dfrac{s}{f}i$ for i.

4. Solve $n = \dfrac{360}{180 - a}$ for a.

5. A 20-foot pipe is to be cut into three pieces. One piece is to be twice as long as another, and the third piece is to be six times as long as the shortest. Find the length of the longest piece.

6. A rectangle with a perimeter of 26 centimeters is 5 centimeters longer than it is wide. Find its area.

In Problems 7–12, solve each equation or inequality.

7. $-2(2x + 3) \geq 14$

8. $-2 < \dfrac{x - 4}{3} < 4$

9. $|2x + 3| = 11$

10. $|3x + 4| = |x + 12|$

11. $|x + 3| \leq 4$

12. $|2x - 4| > 22$

In Problems 13–14, factor each polynomial.

13. $3xy^2 + 6x^2y$

14. $12a^3b^2c - 3a^2b^2c^2 + 6abc^3$

In Problems 15–26, factor each polynomial.

15. $ax - xy + ay - y^2$

16. $ax + ay + bx + by - cx - cy$

17. $x^2 - 49$

18. $2x^2 - 32$

19. $4y^4 - 64$

20. $b^3 + 125$

21. $b^3 - 27$

22. $3u^3 - 24$

23. $x^2 + 8x + 15$

24. $6b^2 + b - 2$

25. $6u^2 + 9u - 6$

26. $x^2 + 6x + 9 - y^2$

In Problems 27–28, simplify each fraction.

27. $\dfrac{-12x^2y^3z^2}{18x^3y^4z^2}$

28. $\dfrac{2x^2 + 7x + 3}{4x + 12}$

In Problems 29–32, do the operations and simplify, if necessary. Write all answers without negative exponents.

29. $\dfrac{x^2y^{-2}}{x^3z^2} \cdot \dfrac{x^2z^4}{y^2z}$

30. $\dfrac{u^2 + 5u + 6}{u^2 - 4} \cdot \dfrac{u^2 - 5u + 6}{u^2 - 9}$

31. $\dfrac{x^3 + y^3}{4} \div \dfrac{x^2 - xy + y^2}{2x + 2y}$

32. $\dfrac{x + 2}{x + 1} - \dfrac{x + 1}{x + 2}$

In Problems 33–34, simplify each complex fraction.

33. $\dfrac{\dfrac{2u^2w^3}{v^2}}{\dfrac{4uw^4}{uv}}$

34. $\dfrac{\dfrac{x}{y} + \dfrac{1}{2}}{\dfrac{x}{2} - \dfrac{1}{y}}$

35. Find the remainder: $\dfrac{x^3 - 4x^2 + 5x + 3}{x + 1}$.

36. Use synthetic division to find the remainder when $4x^3 + 3x^2 + 2x - 1$ is divided by $x - 2$.

$$\underline{2\rfloor} \quad 4 \quad 3 \quad 2 \quad -1$$

8 Writing Equations of Lines; Variation

MATHEMATICS IN THE WORKPLACE

Stable Owner

Stable owners raise and train horses for show and for racing. Usually, the stable provides horses that people can use during riding lessons. The lessons may be in the English or western styles of riding. Many stables also board horses for private individuals.

SAMPLE APPLICATION ■ A woman wishes to purchase a horse for $5,000. If she plans to keep the horse at a stable that charges $350 per month for board, what will be the average cost per month if she keeps the horse for ten years? (See Exercise 74 in Exercise 8.5.)

We will begin this chapter by reviewing how to graph a known linear equation. We will then consider the reverse problem of writing the equation of a line with a known graph. In Sections 8.4 and 8.5, we will continue our discussion of functions, one of the most important topics in mathematics, and then conclude the chapter by discussing variation.

8.1 A Review of the Rectangular Coordinate System

■ THE COORDINATE SYSTEM ■ GRAPHING EQUATIONS ■ GRAPHING HORIZONTAL AND VERTICAL LINES ■ THE MIDPOINT FORMULA

Getting Ready   *In the equation $2x + y = 5$, find y when x has the following values.*

1. $x = 2$ **2.** $x = -2$ **3.** $x = 0$ **4.** $x = \dfrac{3}{2}$

■ THE COORDINATE SYSTEM

René Descartes (1596–1650) is credited with the idea of associating ordered pairs of real numbers with points in the geometric plane. His idea is based on two perpendicular number lines, one horizontal and one vertical, that divide the plan into four quadrants, numbered as in Figure 8-1.

The horizontal number line is the **x-axis,** and the vertical number line is the **y-axis.** The point where the axes intersect, called the **origin,** is the 0 point on each number line.

The positive direction on the *x*-axis is to the right, the positive direction on the *y*-axis is upward, and the unit distance on each axis is the same. This **xy-plane** is called a **rectangular coordinate system** or a **Cartesian coordinate system.**

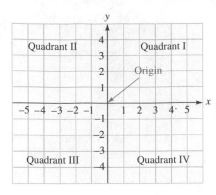

FIGURE 8-1

To plot the point associated with the pair of real numbers (2, 3), we start at the origin and count 2 units to the right and then 3 units up, as in Figure 8-2. The point P, which lies in the first quadrant, is the **graph** of the pair (2, 3). The pair (2, 3) gives the **coordinates** of point P.

To plot point Q with coordinates $(-4, 6)$, we start at the origin and count 4 units to the left and then 6 units up. Point Q lines in the second quadrant. Point R with coordinates $(6, -4)$ lies in the fourth quadrant.

> **WARNING!** The pairs $(-4, 6)$ and $(6, -4)$ represent different points. One is in the second quadrant, and one is in the fourth quadrant. Since order is important when graphing pairs of real numbers, such pairs are called **ordered pairs.**

In the ordered pair (a, b), a is called the **x-coordinate,** and b is called the **y-coordinate.**

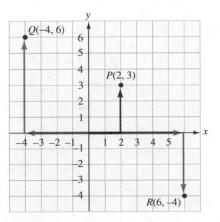

FIGURE 8-2

■ GRAPHING EQUATIONS

The **graph of an equation** in the variables x and y is the set of all points on a rectangular coordinate system with coordinates (x, y) that satisfy the equation.

EXAMPLE 1   Graph the equation $3x + 2y = 12$.

Solution   We pick values for either x or y, substitute them in the equation, and solve for the other variable. For example, if $x = 2$, then

$$3x + 2y = 12$$
$$3(2) + 2y = 12 \qquad \text{Substitute 2 for } x.$$
$$6 + 2y = 12 \qquad \text{Simplify.}$$
$$2y = 6 \qquad \text{Subtract 6 from both sides.}$$
$$y = 3 \qquad \text{Divide both sides by 2.}$$

One ordered pair that satisfies the equation is $(2, 3)$. If $y = 6$, we have

$$3x + 2y = 12$$
$$3x + 2(6) = 12 \qquad \text{Substitute 6 for } y.$$
$$3x + 12 = 12 \qquad \text{Simplify.}$$
$$3x = 0 \qquad \text{Subtract 12 from both sides.}$$
$$x = 0 \qquad \text{Divide both sides by 3.}$$

A second ordered pair that satisfies the equation is $(0, 6)$.

The pairs $(2, 3)$ and $(0, 6)$ and others that satisfy the equation are shown in the table in Figure 8-3. We plot each of these pairs on a rectangular coordinate system and join them to get the line shown in the figure. This line is the graph of the equation.

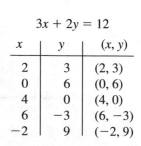

$$3x + 2y = 12$$

x	y	(x, y)
2	3	$(2, 3)$
0	6	$(0, 6)$
4	0	$(4, 0)$
6	-3	$(6, -3)$
-2	9	$(-2, 9)$

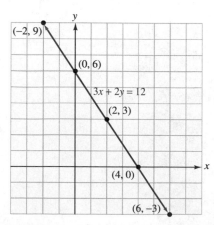

FIGURE 8-3

Self Check
Answer

Graph $2x + 3y = 6$.

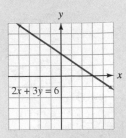

Intercepts of a Line

The **y-intercept** of a line is the point $(0, b)$ where the line intersects the y-axis. To find b, substitute 0 for x in the equation of the line and solve for y.

The **x-intercept** of a line is the point $(a, 0)$ where the line intersects the x-axis. To find a, substitute 0 for y in the equation of a line and solve for x.

In Example 1, the y-intercept of the line is the point with coordinates of $(0, 6)$, and the the x-intercept is the point with coordinates of $(4, 0)$.

EXAMPLE 2 Use the x- and y-intercepts to graph $2x + 5y = 10$.

Solution To find the y-intercept, we substitute 0 for x and solve for y:

$$2x + 5y = 10$$
$$2(0) + 5y = 10 \qquad \text{Substitute 0 for } x.$$
$$5y = 10 \qquad \text{Simplify.}$$
$$y = 2 \qquad \text{Divide both sides by 5.}$$

The y-intercept is the point $(0, 2)$.

To find the x-intercept, we substitute 0 for y and solve for x:

$$2x + 5y = 10$$
$$2x + 5(0) = 10 \qquad \text{Substitute 0 for } y.$$
$$2x = 10 \qquad \text{Simplify.}$$
$$x = 5 \qquad \text{Divide both sides by 2.}$$

The x-intercept is the point $(5, 0)$.

Although two points are sufficient to draw the line, it is a good idea to find and plot a third point as a check. To find the coordinates of a third point, we can substitute any convenient number (such as -5) for x and solve for y:

$$2x + 5y = 10$$
$$2(-5) + 5y = 10 \qquad \text{Substitute } -5 \text{ for } x.$$
$$-10 + 5y = 10 \qquad \text{Simplify.}$$
$$5y = 20 \qquad \text{Add 10 to both sides.}$$
$$y = 4 \qquad \text{Divide both sides by 5.}$$

The line will also pass through the point $(-5, 4)$. The graph is shown in Figure 8-4.

$$2x + 5y = 10$$

x	y	(x, y)
0	2	$(0, 2)$
5	0	$(5, 0)$
-5	4	$(-5, 4)$

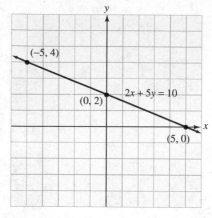

FIGURE 8-4

Self Check
Answer

Graph $5x - 2y = 10$.

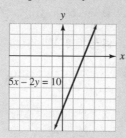

EXAMPLE 3

Graph $y = 3x + 4$.

Solution   We find the y- and x-intercepts: If $x = 0$, then

$$y = 3x + 4$$
$$y = 3(0) + 4 \qquad \text{Substitute 0 for } x.$$
$$y = 4 \qquad\qquad \text{Simplify.}$$

The y-intercept is the point $(0, 4)$.
  If $y = 0$, then

$$y = 3x + 4$$
$$0 = 3x + 4 \qquad \text{Substitute 0 for } y.$$
$$-4 = 3x \qquad\quad \text{Subtract 4 from both sides.}$$
$$-\frac{4}{3} = x \qquad\quad \text{Divide both sides by 3.}$$

The x-intercept is the point $\left(-\frac{4}{3}, 0\right)$.

To find the coordinates of a third point, we can substitute 1 for x and solve for y:

$$y = 3x + 4$$
$$y = 3(\mathbf{1}) + 4 \qquad \text{Substitute 1 for } x.$$
$$y = 7 \qquad\qquad \text{Simplify.}$$

The point $(1, 7)$ lies on the graph, as shown in Figure 8-5.

$y = 3x + 4$

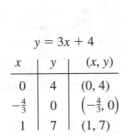

x	y	(x, y)
0	4	$(0, 4)$
$-\frac{4}{3}$	0	$\left(-\frac{4}{3}, 0\right)$
1	7	$(1, 7)$

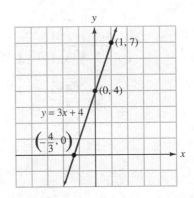

FIGURE 8-5

Self Check
Answer

Graph $y = -2x + 3$.

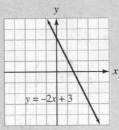

■ GRAPHING HORIZONTAL AND VERTICAL LINES

EXAMPLE 4   Graph   **a.** $y = 3$ and   **b.** $x = -2$.

Solution   **a.** Since the equation $y = 3$ does not contain x, the numbers chosen for x have no effect on y. The value of y is always 3.

After plotting the pairs (x, y) shown in Figure 8-6 and joining them with a straight line, we see that the graph is a horizontal line, parallel to the x-axis, with a y-intercept of $(0, 3)$. The line has no x-intercept.

b. Since the equation $x = -2$ does not contain y, the value of y can be any number. The value of x is always -2.

After plotting the pairs (x, y) shown in Figure 8-6 and joining them with a straight line, we see that the graph is a vertical line, parallel to the y-axis, with an x-intercept of $(-2, 0)$. The line has no y-intercept.

$y = 3$

x	y	(x, y)
-3	3	$(-3, 3)$
0	3	$(0, 3)$
2	3	$(2, 3)$
4	3	$(4, 3)$

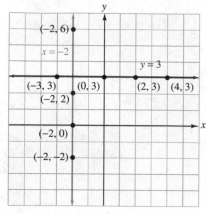

$x = -2$

x	y	(x, y)
-2	-2	$(-2, -2)$
-2	0	$(-2, 0)$
-2	2	$(-2, 2)$
-2	6	$(-2, 6)$

FIGURE 8-6

Self Check

Answer

Graph $x = 4$ and $y = -2$.

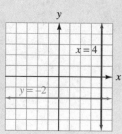

The results of Example 4 suggest the following facts.

Equations of Horizontal and Vertical Lines

If a and b are real numbers, then

The graph of $y = b$ is a horizontal line with y-intercept at $(0, b)$. If $b = 0$, the line is the x-axis.

The graph of $x = a$ is a vertical line with x-intercept at $(a, 0)$. If $a = 0$, the line is the y-axis.

■ THE MIDPOINT FORMULA

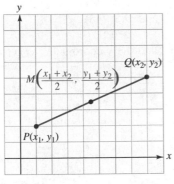

FIGURE 8-7

If point M in Figure 8-7 lies midway between points $P(x_1, y_1)$ and $Q(x_2, y_2)$, point M is called the **midpoint** of segment PQ. To find the coordinates of M, we average the x-coordinates and average the y-coordinates of P and Q.

The Midpoint Formula
The midpoint of the line segment $P(x_1, y_1)$ and $Q(x_2, y_2)$ is the point M with coordinates of

$$\left(\frac{x_1 + x_2}{2}, \frac{y_1 + y_2}{2} \right)$$

■ ■ ■ ■ ■ ■ ■ ■ ■ ■   P E R S P E C T I V E

In an xy-coordinate system, graphs of equations containing the two variables x and y are lines or curves. Other equations have more than two variables, and graphing them often requires some ingenuity and perhaps the aid of a computer. Graphs of equations with the three variables x, y, and z are viewed in a three-dimensional coordinate system with three axes. The coordinates of points in a three-dimensional coordinate system are ordered triples, (x, y, z).

For example, the points $P(2, 3, 4)$ and $Q(-1, 2, 3)$ are plotted in Illustration 1.

Graphs of equations in three variables are not lines or curves, but flat planes or curved surfaces. Only the simplest of these equations can be conveniently graphed by hand; a computer provides the best images of others. The graph in Illustration 2 is called a *paraboloid*. Illustration 3 models a portion of the vibrating surface of a drum head.

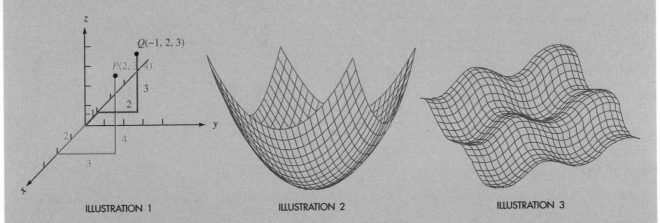

ILLUSTRATION 1   ILLUSTRATION 2   ILLUSTRATION 3

Computer programs for producing three-dimensional graphs are readily available. Some of the more powerful are Maple, Mathematica, Math-Cad, and Derive. Perhaps your school has such a program available for student use. With a brief introduction on the program's use, you can easily create several interesting graphs.

EXAMPLE 5   Find the midpoint of the segment joining $P(-2, 3)$ and $Q(3, -5)$.

Solution   To find the midpoint, we average the x-coordinates and the y-coordinates to get

$$\frac{x_1 + x_2}{2} = \frac{-2 + 3}{2} \quad \text{and} \quad \frac{y_1 + y_2}{2} = \frac{3 + (-5)}{2}$$

$$= \frac{1}{2} \qquad\qquad = -1$$

The midpoint of segment PQ is the point $M\left(\frac{1}{2}, -1\right)$.   ■

Self Check   Find the midpoint of $P(5, -3)$ and $Q(-2, 5)$.

Answer $\left(\frac{3}{2}, 1\right)$

Orals   *Find the x- and y-intercepts of each line.*

1. $x + y = 3$ **2.** $3x + y = 6$

3. $x + 4y = 8$ **4.** $3x - 4y = 12$

Find the midpoint of a line segment with endpoints at

5. $(2, 4), (6, 8)$ **6.** $(-4, 6), (4, -8)$

EXERCISE 8.1

REVIEW   *Graph each interval on the number line.*

1. $(-\infty, -2) \cup [2, \infty)$ **2.** $(-2, 4]$

Factor each expression.

3. $x^2 - x$ **4.** $x^2 - 1$

5. $x^3 - 1$ **6.** $x^4 - 1$

VOCABULARY AND CONCEPTS   *Fill in each blank to make a true statement.*

7. The point where the x- and y-axes intersect is called the _____.

8. The x-coordinate of a point is the first number in an ordered _____.

9. The _____ of a point is the second number in an ordered pair.

10. The y-intercept of a line is the point where the line intersects the _____.

11. The x-intercept of a line is the point where the line intersects the _____.

12. The graph of any equation of the form $x = a$, where a is a constant, is a _____ line.

13. The graph of any equation of the form $y = b$, where b is a constant, is a _____ line.

14. The midpoint of a segment with endpoints at $P(a, b)$ and $Q(c, d)$ has coordinates of _____.

In Exercises 15–22, plot each point on the rectangular coordinate system shown in Illustration 1.

15. $A(4, 3)$

16. $B(-2, 1)$

17. $C(3, -2)$

18. $D(-2, -3)$

19. $E(0, 5)$

20. $F(-4, 0)$

21. $G(2, 0)$

22. $H(0, 3)$

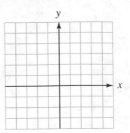

ILLUSTRATION 1

In Exercises 23–30, give the coordinates of each point shown in Illustration 2.

23. A

24. B

25. C

26. D

27. E

28. F

29. G

30. H

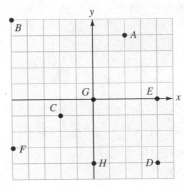

ILLUSTRATION 2

In Exercises 31–46, graph each equation.

31. $x + y = 4$

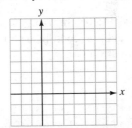

32. $x - y = 2$

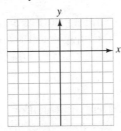

33. $2x - y = 3$

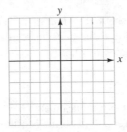

34. $x + 2y = 5$

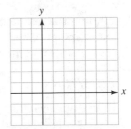

35. $3x + 4y = 12$

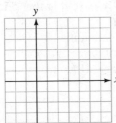

36. $4x - 3y = 12$

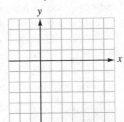

37. $y = -3x + 2$

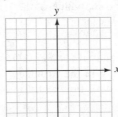

38. $y = 2x - 3$

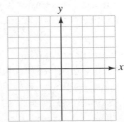

39. $3y = 6x - 9$

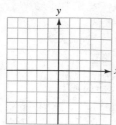

40. $2x = 4y - 10$

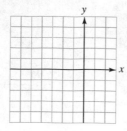

41. $3x + 4y - 8 = 0$

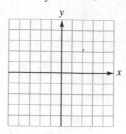

42. $-2y - 3x + 9 = 0$

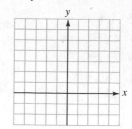

43. $x = 3$

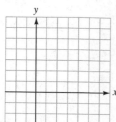

44. $y = -4$

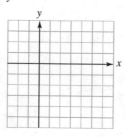

45. $-3y + 2 = 5$

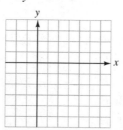

46. $-2x + 3 = 11$

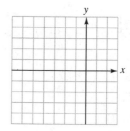

In Exercises 47–60, find the midpoint of segment PQ.

47. $P(0, 0)$, $Q(6, 8)$

48. $P(10, 12)$, $Q(0, 0)$

49. $P(6, 8)$, $Q(12, 16)$

50. $P(10, 4)$, $Q(2, -2)$

51. $P(2, 4)$, $Q(5, 8)$

52. $P(5, 9)$, $Q(8, 13)$

53. $P(-2, -8)$, $Q(3, 4)$

54. $P(-5, -2)$, $Q(7, 3)$

55. $Q(-3, 5)$, $P(-5, -5)$

56. $Q(2, -3)$, $P(4, -8)$

57. $Q(a, b)$, $P(4a, 3b)$

58. $Q(a + b, b)$, $P(-b, -a)$

59. $P(a - b, b)$, $Q(a + b, 3b)$

60. $P(3a, a + b)$, $Q(a + 2b, a - b)$

61. Finding the endpoint of a segment If $M(-2, 3)$ is the midpoint of segment PQ and the coordinates of P are $(-8, 5)$, find the coordinates of Q.

62. Finding the endpoint of a segment If $M(6, -5)$ is the midpoint of segment PQ and the coordinates of Q are $(-5, -8)$, find the coordinates of P.

APPLICATIONS

63. House appreciation A house purchased for $125,000 is expected to appreciate according to the formula $y = 7,500x + 125,000$, where y is the value of the house after x years. Find the value of the house 5 years later and 10 years later.

64. Car depreciation A car purchased for $17,000 is expected to depreciate according to the formula $y = -1,360x + 17,000$. When will the car be worthless?

65. Demand equation The number of television sets that consumers buy depends on price. The higher the price, the fewer people will buy. The equation that relates price to the number of TVs sold at that price is called a **demand equation.**

For a 13-inch TV, this equation is $p = -\frac{1}{10}q + 170$, where p is the price and q is the number of TVs sold at that price. How many TVs will be sold at a price of $150?

66. Supply equation The number of TVs that manufacturers produce depends on price. The higher the price, the more TVs manufacturers will produce. The equation that relates price to the number of TVs produced at that price is called a **supply equation.**

For a 13-inch TV, the supply equation is $p = \frac{1}{10}q + 130$, where p is the price and q is the number of TVs produced for sale at that price. How many TVs will be produced if the price is $150?

67. Meshing gears The rotational speed V of a large gear (with N teeth) is related to the speed v of the smaller gear (with n teeth) by the equation $V = \frac{nv}{N}$. If the larger gear in Illustration 3 is making 60 revolutions per minute, how fast is the smaller gear spinning?

ILLUSTRATION 3

68. Crime prevention The number n of incidents of family violence requiring police response appears to be related to d, the money spent on crisis intervention, by the equation

$$n = 430 - 0.005d$$

What expenditure would reduce the number of incidents to 350?

WRITING

69. Explain how to graph a line using the intercept method.

70. Explain how to determine in which quadrant the point $P(a, b)$ lies.

SOMETHING TO THINK ABOUT

71. If the line $y = ax + b$ passes only through quadrants I and II, what can be known about a and b?

72. What are the coordinates of the three points that divide the segment joining $P(a, b)$ and $Q(c, d)$ into four equal parts?

8.2 Slope of a Nonvertical Line

■ SLOPE OF A LINE ■ INTERPRETATION OF SLOPE ■ HORIZONTAL AND VERTICAL LINES ■ SLOPES OF PARALLEL LINES ■ SLOPES OF PERPENDICULAR LINES

Getting Ready *Evaluate each expression.*

1. $\dfrac{6 - 3}{8 - 5}$ **2.** $\dfrac{10 - 4}{2 - 8}$ **3.** $\dfrac{25 - 12}{9 - (-5)}$ **4.** $\dfrac{-9 - (-6)}{-4 - 10}$

■ SLOPE OF A LINE

A service offered by a computer online company costs $2 per month plus $3 for each hour of connect time. The table shown in Figure 8-8(a) gives the cost y for certain numbers of hours x of connect time. If we construct a graph from this data, we get the line shown in Figure 8-8(b).

Hours of connect time						
x	0	1	2	3	4	5
y	2	5	8	11	14	17
Cost						

(a)

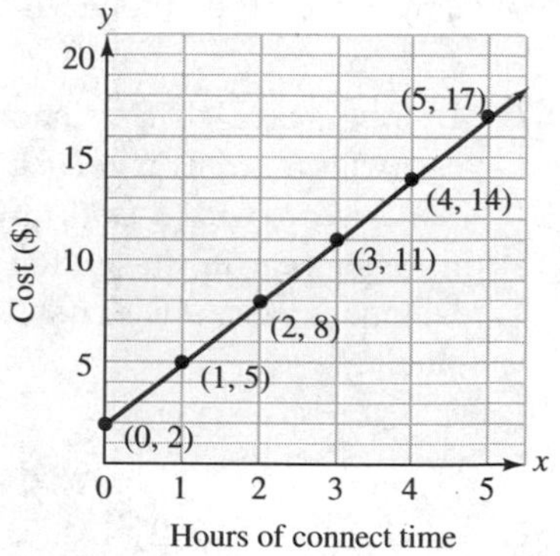

(b)

FIGURE 8-8

From the graph, we can see that if x changes from 0 to 1, y changes from 2 to 5. As x changes from 1 to 2, y changes from 5 to 8, and so on. The ratio of the change in y divided by the change in x is the constant 3.

$$\frac{\text{Change in } y}{\text{Change in } x} = \frac{5-2}{1-0} = \frac{8-5}{2-1} = \frac{11-8}{3-2} = \frac{14-11}{4-3} = \frac{17-14}{5-4} = \frac{3}{1} = 3$$

The ratio of the change in y divided by the change in x between any two points on any line is always a constant. This constant rate of change is called the **slope** of the line.

Slope of a Nonvertical Line

The **slope of the nonvertical line** passing through points $P(x_1, y_1)$ and $Q(x_2, y_2)$ is

$$m = \frac{\text{change in } y}{\text{change in } x} = \frac{y_2 - y_1}{x_2 - x_1} \quad (x_2 \neq x_1)$$

EXAMPLE 1   Find the slope of the line shown in Figure 8-9.

Solution   We can let $P(x_1, y_1) = P(-2, 4)$ and $Q(x_2, y_2) = Q(3, -4)$. Then

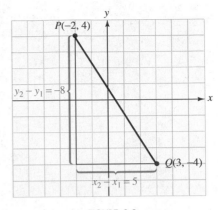

$$m = \frac{\text{change in } y}{\text{change in } x}$$

$$= \frac{y_2 - y_1}{x_2 - x_1}$$

$$= \frac{-4 - 4}{3 - (-2)}$$

Substitute -4 for y_2, 4 for y_1, 3 for x_2, and -2 for x_1.

$$= \frac{-8}{5}$$

$$= -\frac{8}{5}$$

FIGURE 8-9

The slope of the line is $-\frac{8}{5}$. We would obtain the same result if we let $P(x_1, y_1) = P(3, -4)$ and $Q(x_2, y_2) = Q(-2, 4)$. ∎

Self Check   Find the slope of the line joining the points $(-3, 6)$ and $(4, -8)$.

Answer -2

WARNING!   When calculating slope, always subtract the y values and the x values in the same order.

$$m = \frac{y_2 - y_1}{x_2 - x_1} \qquad \text{or} \qquad m = \frac{y_1 - y_2}{x_1 - x_2}$$

However,

$$m \neq \frac{y_2 - y_1}{x_1 - x_2} \qquad \text{and} \qquad m \neq \frac{y_1 - y_2}{x_2 - x_1}$$

The change in y (often denoted as Δy) is the **rise** of the line between points P and Q. The change in x (often denoted as Δx) is the **run.** Using this terminology, we can define slope to be the ratio of the rise to the run:

$$m = \frac{\Delta y}{\Delta x} = \frac{\text{rise}}{\text{run}} \qquad (\Delta x \neq 0)$$

EXAMPLE 2 Find the slope of the line determined by $3x - 4y = 12$.

Solution We first find the coordinates of two points on the line.

- If $x = 0$, then $y = -3$. The point $(0, -3)$ is on the line.
- If $y = 0$, then $x = 4$. The point $(4, 0)$ is on the line.

We then refer to Figure 8-10 and find the slope of the line between $P(0, -3)$ and $Q(4, 0)$ by substituting 0 for y_2, -3 for y_1, 4 for x_2, and 0 for x_1 in the formula for slope.

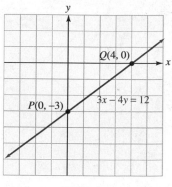

$$m = \frac{\text{change in } y}{\text{change in } x}$$

$$= \frac{y_2 - y_1}{x_2 - x_1}$$

$$= \frac{0 - (-3)}{4 - 0}$$

$$= \frac{3}{4}$$

FIGURE 8-10

The slope of the line is $\frac{3}{4}$.

Self Check Find the slope of the line determined by $2x + 5y = 12$.

Answer $-\frac{2}{5}$

■ INTERPRETATION OF SLOPE

Many applied problems involve equations of lines and their slopes.

EXAMPLE 3 **Cost of carpet** A store sells a high-quality carpet for $25 per square yard, plus a $20 delivery charge. The total cost c of n square yards is given by the following formula.

$c =$	cost per square yard	$\cdot$	the number of square yards	$+$	the delivery charge.
$c =$	25	$\cdot$	n	$+$	20

Graph this equation and interpret the slope of the line.

Solution   We can graph the equation on a coordinate system with a vertical c-axis and a horizontal n-axis. Figure 8-11 shows a table of ordered pairs and the graph.

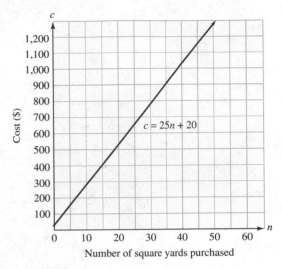

$$c = 25n + 20$$

x	y	(x, y)
10	270	(10, 270)
20	520	(20, 520)
30	770	(30, 770)
40	1,020	(40, 1,020)
50	1,270	(50, 1,270)

FIGURE 8-11

If we pick the points $(30, 770)$ and $(50, 1,270)$ to find the slope, we have

$$m = \frac{\Delta c}{\Delta n}$$

$$= \frac{c_2 - c_1}{n_2 - n_1}$$

$$= \frac{1{,}270 - 770}{50 - 30} \qquad \text{Substitute 1,270 for } c_2, \text{ 770 for } c_1, \text{ 50 for } n_2, \text{ and 30 for } n_1.$$

$$= \frac{500}{20}$$

$$= 25$$

The slope of 25 is the cost of the carpet in dollars per square yard. ∎

Self Check   Interpret the y-intercept of the graph in Figure 8-11.

Answer   The y-coordinate of the y-intercept is the delivery charge.

EXAMPLE 4   **Rate of descent**   It takes a skier 25 minutes to complete the course shown in Figure 8-12. Find his average rate of descent in feet per minute.

Solution   To find the average rate of descent, we must find the ratio of the change in altitude to the change in time. To find this ratio, we calculate the slope of the line passing through the points $(0, 12{,}000)$ and $(25, 8{,}500)$.

$$\text{Average rate of descent} = \frac{12{,}000 - 8{,}500}{0 - 25}$$

$$= \frac{3{,}500}{-25}$$

$$= -140$$

The average rate of descent is -140 ft/min.

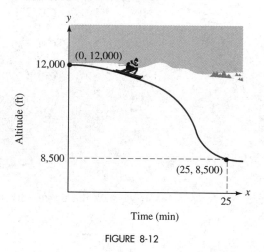

FIGURE 8-12

■ HORIZONTAL AND VERTICAL LINES

If $P(x_1, y_1)$ and $Q(x_2, y_2)$ are points on the horizontal line shown in Figure 8-13(a), then $y_1 = y_2$, and the numerator of the fraction

$$\frac{y_2 - y_1}{x_2 - x_1} \qquad \text{On a horizontal line, } x_2 \neq x_1.$$

is 0. Thus, the value of the fraction is 0, and the slope of the horizontal line is 0.

If $P(x_1, y_1)$ and $Q(x_2, y_2)$ are two points on the vertical line shown in Figure 8-13(b), then $x_1 = x_2$, and the denominator of the fraction

$$\frac{y_2 - y_1}{x_2 - x_1} \qquad \text{On a vertical line, } y_2 \neq y_1.$$

is 0. Since the denominator of a fraction cannot be 0, a vertical line has no defined slope.

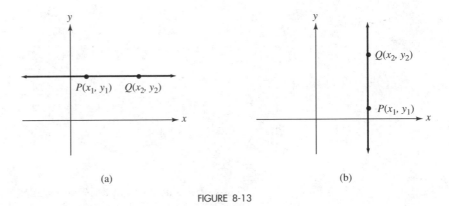

(a)

(b)

FIGURE 8-13

Slopes of Horizontal and Vertical Lines

All horizontal lines (lines with equations of the form $y = b$) have a slope of 0.

All vertical lines (lines with equations of the form $x = a$) have no defined slope.

If a line rises as we follow it from left to right, as in Figure 8-14(a), its slope is positive. If a line drops as we follow it from left to right, as in Figure 8-14(b), its slope is negative. If a line is horizontal, as in Figure 8-14(c), its slope is 0. If a line is vertical, as in Figure 8-14(d), it has no defined slope.

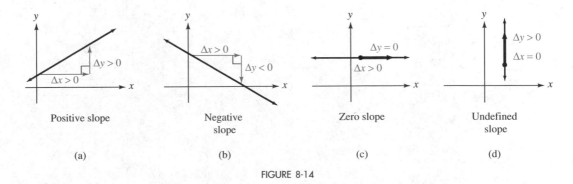

FIGURE 8-14

■ SLOPES OF PARALLEL LINES

To see a relationship between parallel lines and their slopes, we refer to the parallel lines l_1 and l_2 shown in Figure 8-15, with slopes of m_1 and m_2, respectively. Because right triangles ABC and DEF are similar, it follows that

$$m_1 = \frac{\Delta y \text{ of } l_1}{\Delta x \text{ of } l_1}$$

$$= \frac{\Delta y \text{ of } l_2}{\Delta x \text{ of } l_2}$$

$$= m_2$$

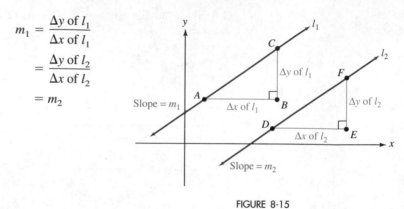

FIGURE 8-15

Thus, if two nonvertical lines are parallel, they have the same slope. It is also true that when two lines have the same slope, they are parallel.

> **Slopes of Parallel Lines**
> Nonvertical parallel lines have the same slope, and lines having the same slope are parallel.
>
> Since vertical lines are parallel, lines with no defined slope are parallel.

EXAMPLE 5   The lines in Figure 8-16 are parallel. Find y.

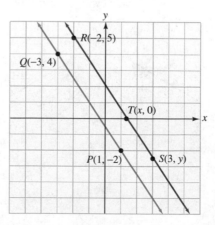

FIGURE 8-16

Solution   Since the lines are parallel, they have equal slopes. To find y, we find the slope of each line, set them equal, and solve the resulting equation.

$$\underbrace{\frac{-2-4}{1-(-3)}}_{\textit{Slope of PQ}} = \underbrace{\frac{y-5}{3-(-2)}}_{\textit{Slope of RS}}$$

$$\frac{-6}{4} = \frac{y-5}{5}$$

$-30 = 4(y-5)$ Multiply both sides by 20.

$-30 = 4y - 20$ Use the distributive property.

$-10 = 4y$ Add 20 to both sides.

$-\dfrac{5}{2} = y$ Divide both sides by 4 and simplify.

Thus, $y = -\frac{5}{2}$. ∎

Self Check

In Figure 8-16, find x.

Answer $\frac{4}{3}$

■ SLOPES OF PERPENDICULAR LINES

Two real numbers a and b are called **negative reciprocals** if $ab = -1$. For example,

$$-\frac{4}{3} \quad \text{and} \quad \frac{3}{4}$$

are negative reciprocals, because $-\frac{4}{3}\left(\frac{3}{4}\right) = -1$.

The following theorem relates perpendicular lines and their slopes.

Slopes of Perpendicular Lines

If two nonvertical lines are perpendicular, their slopes are negative reciprocals.

If the slopes of two lines are negative reciprocals, the lines are perpendicular.

Because a horizontal line is perpendicular to a vertical line, a line with a slope of 0 is perpendicular to a line with no defined slope.

EXAMPLE 6   Are the lines shown in Figure 8-17 perpendicular?

Solution   We find the slopes of the lines and see whether they are negative reciprocals.

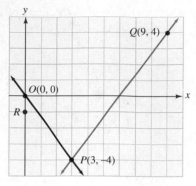

FIGURE 8-17

$$\text{Slope of } OP = \frac{\Delta y}{\Delta x} \quad \text{and} \quad \text{Slope of } PQ = \frac{\Delta y}{\Delta x}$$

$$= \frac{y_2 - y_1}{x_2 - x_1} \qquad\qquad = \frac{y_2 - y_1}{x_2 - x_1}$$

$$= \frac{-4 - 0}{3 - 0} \qquad\qquad = \frac{4 - (-4)}{9 - 3}$$

$$= -\frac{4}{3} \qquad\qquad = \frac{8}{6}$$

$$\qquad\qquad\qquad = \frac{4}{3}$$

Since their slopes are not negative reciprocals, the lines are not perpendicular. ∎

Self Check
Answer

In Figure 8-17, is PR perpendicular to PQ?

no

Orals   *Find the slope of the line passing through*

1. $(0, 0)$, $(1, 3)$ 2. $(0, 0)$, $(3, 6)$
3. Are lines with slopes of -2 and $\frac{8}{-4}$ parallel?
4. Find the negative reciprocal of -0.2.
5. Are lines with slopes of -2 and $\frac{1}{2}$ perpendicular?

EXERCISE 8.2

REVIEW   *Simplify each expression. Write all answers without negative exponents.*

1. $(x^3 y^2)^3$

2. $\left(\dfrac{x^5}{x^3}\right)^3$

3. $(x^{-3} y^2)^{-4}$

4. $\left(\dfrac{x^{-6}}{y^3}\right)^{-4}$

5. $\left(\dfrac{3x^2 y^3}{8}\right)^0$

6. $\left(\dfrac{x^3 x^{-7} y^{-6}}{x^4 y^{-3} y^{-2}}\right)^{-2}$

VOCABULARY AND CONCEPTS   *Fill in each blank to make a true statement.*

7. Slope is defined as the change in ___ divided by the change in ___.

8. A slope is a rate of _____.

9. The formula to compute slope is $m =$ _____.

10. The change in y (denoted as Δy) is the _____ of the line between points P and Q.

11. The change in x (denoted as Δx) is the _____ of the line between points P and Q.

12. The slope of a _____ line is 0.

13. The slope of a _____ line is undefined.

14. If a line rises as x increases, its slope is _____.

15. _____ lines have the same slope.

16. The slopes of _____ lines are negative _____.

PRACTICE *In Exercises 17–28, find the slope of the line that passes through the given points, if possible.*

17. $(0, 0)$, $(3, 9)$
18. $(9, 6)$, $(0, 0)$
19. $(-1, 8)$, $(6, 1)$
20. $(-5, -8)$, $(3, 8)$

21. $(3, -1)$, $(-6, 2)$
22. $(0, -8)$, $(-5, 0)$
23. $(7, 5)$, $(-9, 5)$
24. $(2, -8)$, $(3, -8)$

25. $(-7, -5)$, $(-7, -2)$
26. $(3, -5)$, $(3, 14)$

27. (a, b), (b, a)
28. (a, b), $(-b, -a)$

In Exercises 29–36, find the slope of the line determined by each equation.

29. $3x + 2y = 12$
30. $2x - y = 6$
31. $3x = 4y - 2$
32. $x = y$

33. $y = \dfrac{x - 4}{2}$
34. $x = \dfrac{3 - y}{4}$
35. $4y = 3(y + 2)$
36. $x + y = \dfrac{2 - 3y}{3}$

In Exercises 37–42, tell whether the slope of the line in each graph is positive, negative, 0, or undefined.

37.

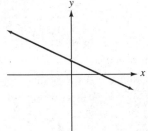

38.

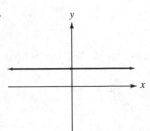

39.

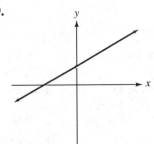

40.

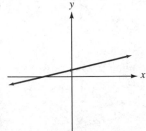

41.

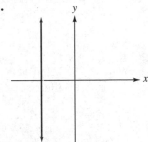

42.

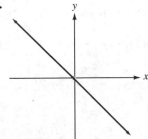

In Exercises 43–48, tell whether the lines with the given slopes are parallel, perpendicular, or neither.

43. $m_1 = 3$, $m_2 = -\dfrac{1}{3}$

44. $m_1 = \dfrac{1}{4}$, $m_2 = 4$

45. $m_1 = 4$, $m_2 = 0.25$

46. $m_1 = -5$, $m_2 = \dfrac{1}{-0.2}$

47. $m_1 = \dfrac{a}{b}$, $m_2 = \left(\dfrac{b}{a}\right)^{-1}$

48. $m_1 = \dfrac{c}{d}$, $m_2 = \dfrac{d}{c}$

In Exercises 49–54, tell whether the line PQ is parallel or perpendicular (or neither) to a line with a slope of -2.

49. $P(3, 4)$, $Q(4, 2)$

50. $P(6, 4)$, $Q(8, 5)$

51. $P(-2, 1)$, $Q(6, 5)$

52. $P(3, 4)$, $Q(-3, -5)$

53. $P(5, 4)$, $Q(6, 6)$

54. $P(-2, 3)$, $Q(4, -9)$

In Exercises 55–60, find the slopes of lines PQ and PR and tell whether the points P, Q, and R lie on the same line. (Hint: Two lines with the same slope and a point in common must be the same line.)

55. $P(-2, 4)$, $Q(4, 8)$, $R(8, 12)$

56. $P(6, 10)$, $Q(0, 6)$, $R(3, 8)$

57. $P(-4, 10)$, $Q(-6, 0)$, $R(-1, 5)$

58. $P(-10, -13)$, $Q(-8, -10)$, $R(-12, -16)$

59. $P(-2, 4)$, $Q(0, 8)$, $R(2, 12)$

60. $P(8, -4)$, $Q(0, -12)$, $R(8, -20)$

61. Find the equation of the *x*-axis and its slope.

62. Find the equation of the *y*-axis and its slope, if any.

In Exercises 63–68, work each geometry problem.

63. Show that points with coordinates of $(-3, 4)$, $(4, 1)$, and $(-1, -1)$ are the vertices of a right triangle.

64. Show that a triangle with vertices at $(0, 0)$, $(12, 0)$, and $(13, 12)$ is not a right triangle.

65. A square has vertices at points $(a, 0)$, $(0, a)$, $(-a, 0)$, and $(0, -a)$, where $a \neq 0$. Show that its adjacent sides are perpendicular.

66. If a and b are not both 0, show that the points $(2b, a)$, (b, b), and $(a, 0)$ are the vertices of a right triangle.

67. Show that the points $(0, 0)$, $(0, a)$, (b, c), and $(b, a + c)$ are the vertices of a parallelogram. (*Hint:* Opposite sides of a parallelogram are parallel.)

68. If $b \neq 0$, show that the points $(0, 0)$, $(0, b)$, $(8, b + 2)$, and $(12, 3)$ are the vertices of a trapezoid. (*Hint:* A **trapezoid** is a four-sided figure with exactly two sides parallel.)

APPLICATIONS

69. Grade of a road   Find the slope of the road shown in Illustration 1. (*Hint:* 1 mi = 5,280 ft.)

70. Pitch of a roof   Find the pitch of the roof shown in Illustration 2.

32 ft

1 mi

ILLUSTRATION 1

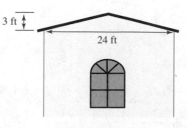

3 ft

24 ft

ILLUSTRATION 2

71. Slope of a ladder   A ladder reaches 18 feet up the side of a building with its base 5 feet from the building. Find the slope of the ladder.

72. Rate of growth   When a college started an aviation program, the administration agreed to predict enrollments using a straight-line method. If the enrollment during the first year was 8, and the enrollment during the fifth year was 20, find the rate of growth per year (the slope of the line). (See Illustration 3.)

73. Rate of growth   A small business predicts sales according to a straight-line method. If sales were $85,000 in the first year and $125,000 in the third year, find the rate of growth in sales per year (the slope of the line).

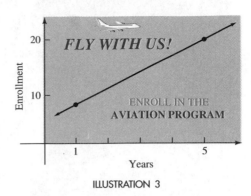

ILLUSTRATION 3

74. Rate of decrease   The price of computer technology has been dropping steadily for the past ten years. If a desktop PC cost $5,700 ten years ago, and the same computing power cost $1,499 two years ago, find the rate of decrease per year. (Assume a straight-line model.)

WRITING

75. Explain why a vertical line has no defined slope.

76. Explain how to determine from their slopes whether two lines are parallel, perpendicular, or neither.

SOMETHING TO THINK ABOUT

77. Find the slope of the line $Ax + By = C$. Follow the procedure of Example 2.

79. The points $(3, a)$, $(5, 7)$, and $(7, 10)$ lie on a line. Find a.

78. Follow Example 2 to find the slope of the line $y = mx + b$.

80. The line passing through points $A(1, 3)$ and $B(-2, 7)$ is perpendicular to the line passing through points $C(4, b)$ and $D(8, -1)$. Find b.

8.3   Writing Equations of Lines

■ POINT–SLOPE FORM OF THE EQUATION OF A LINE ■ SLOPE–INTERCEPT FORM OF THE EQUATION OF A LINE ■ USING SLOPE AS AN AID IN GRAPHING ■ GENERAL FORM OF THE EQUATION OF A LINE ■ STRAIGHT-LINE DEPRECIATION ■ CURVE FITTING

Getting Ready   *Solve each equation.*

1. $3 = \dfrac{x - 2}{4}$

2. $-2 = 3(x + 1)$

3. Solve $y - 2 = 3(x - 2)$ for y.

4. Solve $Ax + By + 3 = 0$ for x.

■ POINT–SLOPE FORM OF THE EQUATION OF A LINE

Suppose that line l shown in Figure 8-18 has a slope of m and passes through the point $P(x_1, y_1)$. If $Q(x, y)$ is a second point on line l, we have

$$m = \frac{y - y_1}{x - x_1}$$

or if we multiply both sides by $x - x_1$, we have

1. $y - y_1 = m(x - x_1)$

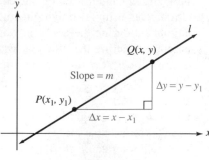

FIGURE 8-18

Because Equation 1 displays the coordinates of the point (x_1, y_1) on the line and the slope m of the line, it is called the **point–slope form** of the equation of a line.

> **Point–Slope Form**
> The equation of the line passing through $P(x_1, y_1)$ and with slope m is
> $$y - y_1 = m(x - x_1)$$

EXAMPLE 1   Write the equation of the line with a slope of $-\frac{2}{3}$ and passing through $P(-4, 5)$.

Solution   We substitute $-\frac{2}{3}$ for m, -4 for x_1, and 5 for y_1 into the point–slope form and simplify.

$$y - \mathbf{y_1} = m(x - x_1)$$

$$y - \mathbf{5} = -\frac{2}{3}[x - (-4)] \qquad \text{Substitute } -\tfrac{2}{3} \text{ for } m, -4 \text{ for } x_1, \text{ and 5 for } y_1.$$

$$y - 5 = -\frac{2}{3}(x + 4) \qquad -(-4) = 4.$$

$$y - 5 = -\frac{2}{3}x - \frac{8}{3} \qquad \text{Use the distributive property to remove parentheses.}$$

$$y = -\frac{2}{3}x + \frac{7}{3} \qquad \text{Add 5 to both sides and simplify.}$$

The equation of the line is $y = -\dfrac{2}{3}x + \dfrac{7}{3}$. ■

Self Check   Write the equation of the line with slope of $\frac{5}{4}$ and passing through $Q(2, 5)$.

Answer $y = \frac{5}{4}x + \frac{5}{2}$

EXAMPLE 2   Write the equation of the line passing through $P(-5, 4)$ and $Q(8, -6)$.

Solution   First we find the slope of the line.

$$m = \frac{y_2 - y_1}{x_2 - x_1}$$

$$= \frac{-6 - 4}{8 - (-5)} \qquad \text{Substitute } -6 \text{ for } y_2, 4 \text{ for } y_1, 8 \text{ for } x_2, \text{ and } -5 \text{ for } x_1.$$

$$= -\frac{10}{13}$$

Because the line passes through both P and Q, we can choose either point and substitute its coordinates into the point–slope form. If we choose $P(-5, 4)$, we substitute -5 for x_1, 4 for y_1, and $-\frac{10}{13}$ for m and proceed as follows.

$$y - y_1 = m(x - x_1)$$

$$y - 4 = -\frac{10}{13}[x - (-5)] \qquad \text{Substitute } -\tfrac{10}{13} \text{ for } m, -5 \text{ for } x_1, \text{ and } 4 \text{ for } y_1.$$

$$y - 4 = -\frac{10}{13}(x + 5) \qquad -(-5) = 5.$$

$$y - 4 = -\frac{10}{13}x - \frac{50}{13} \qquad \text{Remove parentheses.}$$

$$y = -\frac{10}{13}x + \frac{2}{13} \qquad \text{Add 4 to both sides and simplify.}$$

The equation of the line is $y = -\dfrac{10}{13}x + \dfrac{2}{13}$.   ∎

Self Check   Write the equation of the line passing through $R(-2, 5)$ and $S(4, -3)$.

Answer $y = -\frac{4}{3}x + \frac{7}{3}$

■ SLOPE–INTERCEPT FORM OF THE EQUATION OF A LINE

Since the y-intercept of the line shown in Figure 8-19 is the point $(0, b)$, we can write the equation of the line by substituting 0 for x_1 and b for y_1 in the point–slope form and simplifying.

$$y - y_1 = m(x - x_1)$$

$$y - b = m(x - 0)$$

$$y - b = mx$$

2. $$y = mx + b$$

Because Equation 2 displays the slope m and the y-coordinate b of the y-intercept, it is called the **slope–intercept form** of the equation of a line.

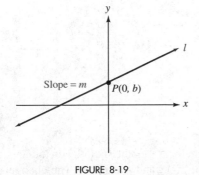

FIGURE 8-19

> **Slope-Intercept Form**
> The equation of the line with slope m and y-intercept $(0, b)$ is
>
> $$y = mx + b$$

EXAMPLE 3

Use the slope–intercept form to write the equation of the line with slope 4 that passes through the point $P(5, 9)$.

Solution

Since we are given that $m = 4$ and that the ordered pair $(5, 9)$ satisfies the equation, we can substitute 5 for x, 9 for y, and 4 for m in the equation $y = mx + b$ and solve for b.

$$y = mx + b$$
$$9 = 4(5) + b \qquad \text{Substitute 9 for } y, \text{ 4 for } m, \text{ and 5 for } x.$$
$$9 = 20 + b \qquad \text{Simplify.}$$
$$-11 = b \qquad \text{Subtract 20 from both sides.}$$

Because $m = 4$ and $b = -11$, the equation is $y = 4x - 11$.   ∎

Self Check

Write the equation of the line with slope -2 that passes through the point $Q(-2, 8)$.

Answer $y = -2x + 4$

■ USING SLOPE AS AN AID IN GRAPHING

It is easy to graph a linear equation when it is written in slope–intercept form. For example, to graph $y = \frac{4}{3}x - 2$, we note that $b = -2$ and that the y-intercept is $(0, b) = (0, -2)$. (See Figure 8-20.)

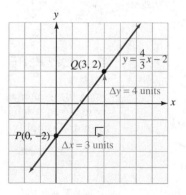

FIGURE 8-20

Because the slope of the line is $\frac{\Delta y}{\Delta x} = \frac{4}{3}$, we can locate another point Q on the line by starting at point P and counting 3 units to the right and 4 units up. The change in x from point P to point Q is $\Delta x = 3$, and the corresponding change in y is $\Delta y = 4$. The line joining points P and Q is the graph of the equation.

EXAMPLE 4   Find the slope and the y-intercept of the line with the equation $2(x - 3) = -3(y + 5)$. Then graph the line.

Solution   We write the equation in the form $y = mx + b$ to find the slope m and the y-intercept $(0, b)$.

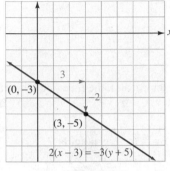

FIGURE 8-21

$$2(x - 3) = -3(y + 5)$$

$$2x - 6 = -3y - 15 \qquad \text{Use the distributive property to remove parentheses.}$$

$$2x + 3y - 6 = -15 \qquad \text{Add } 3y \text{ to both sides.}$$

$$3y - 6 = -2x - 15 \qquad \text{Subtract } 2x \text{ from both sides.}$$

$$3y = -2x - 9 \qquad \text{Add 6 to both sides.}$$

$$y = -\frac{2}{3}x - 3 \qquad \text{Divide both sides by 3.}$$

The slope is $-\frac{2}{3}$, and the y-intercept is $(0, -3)$. To draw the graph, we plot the y-intercept $(0, -3)$ and then locate a second point on the line by moving 3 units to the right and 2 units down. We draw a line through the two points to obtain the graph shown in Figure 8-21.   ■

Self Check   Find the slope and the y-intercept of the line with the equation $2(y - 1) = 3x + 2$, and graph the line.

Answer $m = \frac{3}{2}$, $(0, 2)$

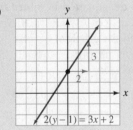

EXAMPLE 5   Show that the lines represented by $4x + 8y = 10$ and $2x = 12 - 4y$ are parallel.

Solution   We solve each equation for y to see that the lines are distinct and that their slopes are equal.

$$4x + 8y = 10 \qquad\qquad 2x = 12 - 4y$$

$$8y = -4x + 10 \qquad\quad 4y = -2x + 12$$

$$y = -\frac{1}{2}x + \frac{5}{4} \qquad\quad y = -\frac{1}{2}x + 3$$

Since the values of b in these equations are different, the lines are distinct. Since the slope of each line is $-\frac{1}{2}$, they are parallel.   ■

Self Check

Are lines represented by $3x - 2y = 4$ and $6x = 4(y + 1)$ parallel?

Answer   yes

EXAMPLE 6   Show that the lines represented by $4x + 8y = 10$ and $4x - 2y = 21$ are perpendicular.

Solution   We solve each equation for y to see that the slopes of their straight-line graphs are negative reciprocals.

$$4x + 8y = 10 \qquad\qquad 4x - 2y = 21$$
$$8y = -4x + 10 \qquad\qquad -2y = -4x + 21$$
$$y = -\frac{1}{2}x + \frac{5}{4} \qquad\qquad y = 2x - \frac{21}{2}$$

Since the slopes are $-\frac{1}{2}$ and 2 (which are negative reciprocals), the lines are perpendicular. ∎

Self Check

Are lines represented by $3x + 2y = 6$ and $2x - 3y = 6$ perpendicular?

Answer   yes

EXAMPLE 7   Write the equation of the line passing through $P(-2, 5)$ and parallel to the line $y = 8x - 3$.

Solution   Since the equation is solved for y, the slope of the line given by $y = 8x - 3$ is the coefficient of x, which is 8. Since the desired equation is to have a graph that is parallel to the graph of $y = 8x - 3$, its slope must also be 8.

We substitute -2 for x_1, 5 for y_1, and 8 for m in the point–slope form and simplify.

$$y - y_1 = m(x - x_1)$$
$$y - 5 = 8[x - (-2)] \qquad \text{Substitute 5 for } y_1, \text{ 8 for } m, \text{ and } -2 \text{ for } x_1.$$
$$y - 5 = 8(x + 2) \qquad -(-2) = 2.$$
$$y - 5 = 8x + 16 \qquad \text{Use the distributive property to remove parentheses.}$$
$$y = 8x + 21 \qquad \text{Add 5 to both sides.}$$

The equation is $y = 8x + 21$. ∎

Self Check

Write the equation of the line that is parallel to the line $y = 8x - 3$ and passes through the origin.

Answer $y = 8x$

EXAMPLE 8   Write the equation of the line passing through $P(-2, 5)$ and perpendicular to the line $y = 8x - 3$.

Solution   The slope of the given line is 8. Thus, the slope of the desired line must be $-\frac{1}{8}$, which is the negative reciprocal of 8.

We substitute -2 for x_1, 5 for y_1, and $-\frac{1}{8}$ for m into the point–slope form and simplify:

$$y - y_1 = m(x - x_1)$$

$$y - 5 = -\frac{1}{8}[x - (-2)] \qquad \text{Substitute 5 for } y_1, -\frac{1}{8} \text{ for } m, \text{ and } -2 \text{ for } x_1.$$

$$y - 5 = -\frac{1}{8}(x + 2) \qquad -(-2) = 2.$$

$$y - 5 = -\frac{1}{8}x - \frac{1}{4} \qquad \text{Remove parentheses.}$$

$$y = -\frac{1}{8}x - \frac{1}{4} + 5 \qquad \text{Add 5 to both sides.}$$

$$y = -\frac{1}{8}x + \frac{19}{4} \qquad \text{Combine terms: } -\frac{1}{4} + \frac{20}{4} = \frac{19}{4}.$$

The equation is $y = -\dfrac{1}{8}x + \dfrac{19}{4}$. ■

Self Check   Write the equation of the line that is perpendicular to the line $y = 8x - 3$ and passes through $Q(2, 4)$.

Answer $y = -\frac{1}{8}x + \frac{17}{4}$

■ GENERAL FORM OF THE EQUATION OF A LINE

Any linear equation that is written in the form $Ax + By = C$, where A, B, and C are constants, is said to be written in **general form.**

WARNING!   When writing equations in general form, we usually clear the equation of fractions and make A positive. We will also make A, B, and C as small as possible. For example, the equation $6x + 12y = 24$ can be changed to $x + 2y = 4$ by dividing both sides by 6.

Finding the Slope and y-Intercept from the General Form

If A, B, and C are real numbers and $B \ne 0$, the graph of the equation

$$Ax + By = C$$

is a nonvertical line with slope of $-\dfrac{A}{B}$ and a y-intercept of $\left(0, \dfrac{C}{B}\right)$.

You will be asked to justify the previous results in the exercises. You will also be asked to show that if $B = 0$, the equation $Ax + By = C$ represents a vertical line with x-intercept of $\left(\frac{C}{A}, 0\right)$.

EXAMPLE 9   Show that the lines represented by $4x + 3y = 7$ and $3x - 4y = 12$ are perpendicular.

Solution   To show that the lines are perpendicular, we will show that their slopes are negative reciprocals. The first equation, $4x + 3y = 7$, is written in general form, with $A = 4$, $B = 3$, and $C = 7$. By the previous result, the slope of the line is

$$m_1 = -\frac{A}{B} = -\frac{4}{3}$$

The second equation, $3x - 4y = 12$, is also written in general form, with $A = 3$, $B = -4$, and $C = 12$. The slope of this line is

$$m_2 = -\frac{A}{B} = -\frac{3}{-4} = \frac{3}{4}$$

Since the slopes are negative reciprocals, the lines are perpendicular.   ∎

Self Check   Are the lines $4x + 3y = 7$ and $y = -\frac{4}{3}x + 2$ parallel?

Answer   yes

We summarize the various forms for the equation of a line in Table 8-1.

General form of a linear equation	$Ax + By = C$ A and B cannot both be 0.
Slope–intercept form of a linear equation	$y = mx + b$ The slope is m, and the y-intercept is $(0, b)$.
Point–slope form of a linear equation	$y - y_1 = m(x - x_1)$ The slope is m, and the line passes through (x_1, y_1).
A **horizontal line**	$y = b$ The slope is 0, and the y-intercept is $(0, b)$.
A **vertical line**	$x = a$ There is no defined slope, and the x-intercept is $(a, 0)$.

TABLE 8-1

■ STRAIGHT-LINE DEPRECIATION

For tax purposes, many businesses use *straight-line depreciation* to find the declining value of aging equipment.

EXAMPLE 10   **Value of a lathe**   The owner of a machine shop buys a lathe for $1,970 and expects it to last for ten years. It can then be sold as scrap for an estimated *salvage value* of $270. If y represents the value of the lathe after x years of use, and y and x are related by the equation of a line,

a. Find the equation of the line.

b. Find the value of the lathe after $2\frac{1}{2}$ years.

c. Find the economic meaning of the y-intercept of the line.

d. Find the economic meaning of the slope of the line.

Solution   **a.** To find the equation of the line, we find its slope and use point–slope form to find its equation.

When the lathe is new, its age x is 0, and its value y is $1,970. When the lathe is 10 years old, $x = 10$ and its value is $y = $270. Since the line passes through the points $(0, 1,970)$ and $(10, 270)$, as shown in Figure 8-22, the slope of the line is

$$m = \frac{y_2 - y_1}{x_2 - x_1}$$

$$= \frac{270 - 1,970}{10 - 0}$$

$$= \frac{-1,700}{10}$$

$$= -170$$

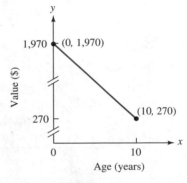

FIGURE 8-22

To find the equation of the line, we substitute -170 for m, 0 for x_1, and 1,970 for y_1 into the point–slope form and simplify.

$$y - y_1 = m(x - x_1)$$
$$y - 1{,}970 = -170(x - 0)$$
3. $$y = -170x + 1{,}970$$

The current value y of the lathe is related to its age x by the equation $y = -170x + 1{,}970$.

b. To find the value of the lathe after $2\frac{1}{2}$ years, we substitute 2.5 for x in Equation 3 and solve for y.

$$y = -170x + 1{,}970$$
$$= -170(2.5) + 1{,}970$$
$$= -425 + 1{,}970$$
$$= 1{,}545$$

In $2\frac{1}{2}$ years, the lathe will be worth $1,545.

c. The y-intercept of the graph is $(0, b)$, where b is the value of y when $x = 0$.

$$y = -170x + 1{,}970$$
$$y = -170(\mathbf{0}) + 1{,}970$$
$$y = 1{,}970$$

Thus, b is the value of a 0-year-old lathe, which is the lathe's original cost, $1,970.

d. Each year, the value of the lathe decreases by $170, because the slope of the line is -170. The slope of the depreciation line is the **annual depreciation rate.**

■

■ CURVE FITTING

In statistics, the process of using one variable to predict another is called **regression.** For example, if we know a man's height, we can make a good prediction about his weight, because taller men usually weigh more than shorter men.

Figure 8-23 shows the results of sampling ten men at random and finding their heights and weights. The graph of the ordered pairs (h, w) is called a **scattergram.**

Man	Height (h) in inches	Weight (w) in pounds
1	66	140
2	68	150
3	68	165
4	70	180
5	70	165
6	71	175
7	72	200
8	74	190
9	75	210
10	75	215

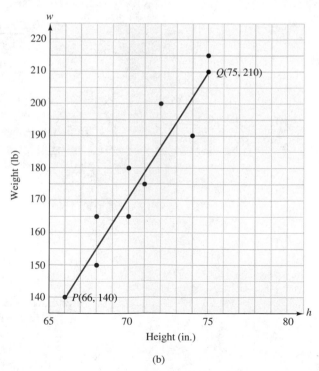

(a)

(b)

FIGURE 8-23

To write a **prediction equation** (sometimes called a **regression equation**), we must find the equation of the line that comes closer to all of the points in the scattergram than any other possible line. There are exact methods to find this equation, but we can only approximate it here.

To write an approximation of the regression equation, we place a straightedge on the scattergram shown in Figure 8-23 and draw the line joining two points that seems to best fit all of the points. In the figure, line PQ is drawn, where point P has coordinates of (66, 140) and point Q has coordinates of (75, 210).

Our approximation of the regression equation will be the equation of the line passing through points P and Q. To find the equation of this line, we first find its slope.

$$m = \frac{y_2 - y_1}{x_2 - x_1}$$
$$= \frac{210 - 140}{75 - 66}$$
$$= \frac{70}{9}$$

We can then use point–slope form to find the equation of the line.

$$y - y_1 = m(x - x_1)$$

$$y - 140 = \frac{70}{9}(x - 66) \qquad \text{Choose (66, 140) for } (x_1, y_1).$$

$$y = \frac{70}{9}x - \frac{4{,}620}{9} + 140 \qquad \begin{array}{l}\text{Remove parentheses and add 140 to both} \\ \text{sides.}\end{array}$$

4. $$y = \frac{70}{9}x - \frac{1{,}120}{3} \qquad -\frac{4{,}620}{9} + 140 = \frac{-1{,}120}{3}.$$

Our approximation of the regression equation is $y = \frac{70}{9}x - \frac{1{,}120}{3}$.

To predict the weight of a man who is 73 inches tall, for example, we substitute 73 for x in Equation 4 and simplify.

$$y = \frac{70}{9}x - \frac{1{,}120}{3}$$

$$y = \frac{70}{9}(73) - \frac{1{,}120}{3}$$

$$y \approx 194.4$$

We would predict that a 73-inch-tall man chosen at random will weigh about 194 pounds.

Orals   *Write the point–slope form of the equation of a line with* $m = 2$, *passing through the given point.*

1. (2, 3)   **2.** (−3, 8)

Write the equation of a line with $m = -3$ *and y-intercept of*

3. (0, 5)   **4.** (0, −7)

Tell whether the lines are parallel, perpendicular, or neither.

5. $y = 3x - 4$, $y = 3x + 5$ **6.** $y = -3x + 7$, $x = 3y - 1$

EXERCISE 8.3

REVIEW EXERCISES   *Solve each equation.*

1. $3(x + 2) + x = 5x$

2. $12b + 6(3 - b) = b + 3$

3. $\dfrac{5(2 - x)}{3} - 1 = x + 5$

4. $\dfrac{r - 1}{3} = \dfrac{r + 2}{6} + 2$

5. Mixing alloys   In 60 ounces of alloy for watch cases, there are 20 ounces of gold. How much copper must be added to the alloy so that a watch case weighing 4 ounces, made from the new alloy, will contain exactly 1 ounce of gold?

6. Mixing coffee   To make a mixture of 80 pounds of coffee worth $272, a grocer mixes coffee worth $3.25 a pound with coffee worth $3.85 a pound. How many pounds of the cheaper coffee should the grocer use?

VOCABULARY AND CONCEPTS   *Fill in each blank to make a true statement.*

7. The point–slope form of the equation of a line is _____.

8. The slope–intercept form of the equation of a line is _____.

9. The general form of the equation of a line is _____.

10. Two lines are parallel when they have the _____ slope.

11. Two lines are _____ when their slopes are negative reciprocals.

12. The process that recognizes that equipment loses value with age is called _____.

PRACTICE   *In Exercises 13–16, use point–slope form to write the equation of the line with the given properties. Write each equation in general form.*

13. $m = 5$, passing through $P(0, 7)$

14. $m = -8$, passing through $P(0, -2)$

15. $m = -3$, passing through $P(2, 0)$

16. $m = 4$, passing through $P(-5, 0)$

In Exercises 17–18, use point–slope form to write the equation of each line. Write the equation in general form.

17.

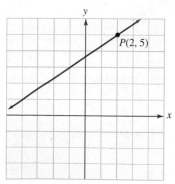

$P(2, 5)$

18.

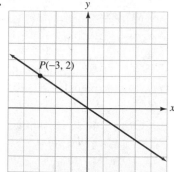

$P(-3, 2)$

In Exercises 19–22, use point–slope form to write the equation of the line passing through the two given points. Write each equation in slope–intercept form.

19. $P(0, 0)$, $Q(4, 4)$

20. $P(-5, -5)$, $Q(0, 0)$

21. $P(3, 4)$, $Q(0, -3)$

22. $P(4, 0)$, $Q(6, -8)$

In Exercises 23–24, use point–slope form to write the equation of each line. Write each answer in slope–intercept form.

23.

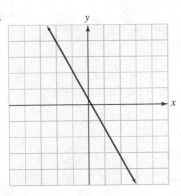

24.

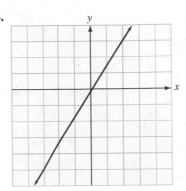

In Exercises 25–32, use the slope–intercept form to write the equation of the line with the given properties. Write each equation in slope–intercept form.

25. $m = 3$, $b = 17$

26. $m = -2$, $b = 11$

27. $m = -7$, passing through $P(7, 5)$

28. $m = 3$, passing through $P(-2, -5)$

29. $m = 0$, passing through $P(2, -4)$

30. $m = -7$, passing through the origin

31. Passing through $P(6, 8)$ and $Q(2, 10)$

32. Passing through $P(-4, 5)$ and $Q(2, -6)$

In Exercises 33–38, write each equation in slope–intercept form to find the slope and the y-intercept. Then use the slope and y-intercept to draw the line.

33. $y + 1 = x$

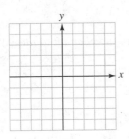

34. $x + y = 2$

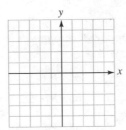

35. $x = \dfrac{3}{2}y - 3$

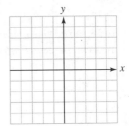

36. $x = -\dfrac{4}{5}y + 2$

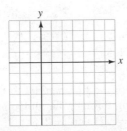

37. $3(y - 4) = -2(x - 3)$

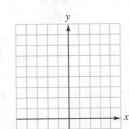

38. $-4(2x + 3) = 3(3y + 8)$

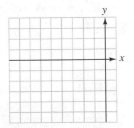

In Exercises 39–44, find the slope and the y-intercept of the line determined by the given equation.

39. $3x - 2y = 8$

40. $-2x + 4y = 12$

41. $-2(x + 3y) = 5$

42. $5(2x - 3y) = 4$

43. $x = \dfrac{2y - 4}{7}$

44. $3x + 4 = -\dfrac{2(y - 3)}{5}$

In Exercises 45–56, tell whether the graphs of each pair of equations are parallel, perpendicular, or neither.

45. $y = 3x + 4$, $y = 3x - 7$

46. $y = 4x - 13$, $y = \dfrac{1}{4}x + 13$

47. $x + y = 2$, $y = x + 5$

48. $x = y + 2$, $y = x + 3$

49. $y = 3x + 7$, $2y = 6x - 9$

50. $2x + 3y = 9$, $3x - 2y = 5$

51. $x = 3y + 4$, $y = -3x + 7$

52. $3x + 6y = 1$, $y = \dfrac{1}{2}x$

53. $y = 3$, $x = 4$

54. $y = -3$, $y = -7$

55. $x = \dfrac{y - 2}{3}$, $3(y - 3) + x = 0$

56. $2y = 8$, $3(2 + x) = 2(x + 2)$

In Exercises 57–62, write the equation of the line that passes through the given point and is parallel to the given line. Write the answer in slope–intercept form.

57. $P(0, 0)$, $y = 4x - 7$

58. $P(0, 0)$, $x = -3y - 12$

59. $P(2, 5)$, $4x - y = 7$

60. $P(-6, 3)$, $y + 3x = -12$

61. $P(4, -2)$, $x = \dfrac{5}{4}y - 2$

62. $P(1, -5)$, $x = -\dfrac{3}{4}y + 5$

In Exercises 63–68, write the equation of the line that passes through the given point and is perpendicular to the given line. Write the answer in slope–intercept form.

63. $P(0, 0)$, $y = 4x - 7$

64. $P(0, 0)$, $x = -3y - 12$

65. $P(2, 5)$, $4x - y = 7$

66. $P(-6, 3)$, $y + 3x = -12$

67. $P(4, -2)$, $x = \dfrac{5}{4}y - 2$

68. $P(1, -5)$, $x = -\dfrac{3}{4}y + 5$

In Exercises 69–72, use the method of Example 9 to find whether the graphs determined by each pair of equations are parallel, perpendicular, or neither.

69. $4x + 5y = 20$, $5x - 4y = 20$

70. $9x - 12y = 17$, $3x - 4y = 17$

71. $2x + 3y = 12$, $6x + 9y = 32$

72. $5x + 6y = 30$, $6x + 5y = 24$

73. Find the equation of the line perpendicular to the line $y = 3$ and passing through the midpoint of the segment joining $(2, 4)$ and $(-6, 10)$.

74. Find the equation of the line parallel to the line $y = -8$ and passing through the midpoint of the segment joining $(-4, 2)$ and $(-2, 8)$.

75. Find the equation of the line parallel to the line $x = 3$ and passing through the midpoint of the segment joining $(2, -4)$ and $(8, 12)$.

76. Find the equation of the line perpendicular to the line $x = 3$ and passing through the midpoint of the segment joining $(-2, 2)$ and $(4, -8)$.

77. Solve $Ax + By = C$ for y and thereby show that the slope of its graph is $-\frac{A}{B}$ and its y-intercept is $\left(0, \frac{C}{B}\right)$.

78. Show that the x-intercept of the graph of $Ax + By = C$ is $\left(\frac{C}{A}, 0\right)$.

APPLICATIONS *Assume straight-line depreciation or straight-line appreciation.*

79. Finding a depreciation equation A truck was purchased for $19,984. Its salvage value at the end of 8 years is expected to be $1,600. Find the depreciation equation.

80. Finding a depreciation equation A business purchased the computer shown in Illustration 1. It will be depreciated over a 5-year period, when it will probably be worth $200. Find the depreciation equation.

$2,350

ILLUSTRATION 1

81. Finding an appreciation equation A famous oil painting was purchased for $250,000 and is expected to double in value in 5 years. Find the appreciation equation.

82. Finding an appreciation equation A house purchased for $142,000 is expected to double in value in 8 years. Find its appreciation equation.

83. Finding a depreciation equation Find the depreciation equation for the TV in the want ad in Illustration 2.

> *For Sale*: 3-year-old 54-inch TV, $1,750 new. Asking $800. Call 875-5555. Ask for Mike.

ILLUSTRATION 2

84. Depreciating a lawn mower A lawn mower cost $450 when new and is expected to last 10 years. What will it be worth in $6\frac{1}{2}$ years?

85. Salvage value A copy machine that cost $1,750 when new will be depreciated at the rate of $180 per year. If the useful life of the copier is 7 years, find its salvage value.

86. Annual rate of depreciation A machine that cost $47,600 when new will have a salvage value of $500 after its useful life of 15 years. Find its annual rate of depreciation.

87. Real estate A vacation home is expected to appreciate about $4,000 per year. If the home will be worth $122,000 in 2 years, what will it be worth in 10 years?

88. Car repair A garage charges a fixed amount, plus an hourly rate, to service a car. Use the information in Illustration 3 to find the hourly rate.

A1 Car Repair	
Typical Charges	
2 hours	$143
5 hours	$320

ILLUSTRATION 3

89. Printer charges A printer charges a fixed setup cost, plus $15 for every 100 copies. If 300 copies cost $75, how much will 1,000 copies cost?

90. Predicting burglaries A police department knows that city growth and the number of burglaries are related by a linear equation. City records show that 575 burglaries were reported in a year when the local population was 77,000, and 675 were reported in a year when the population was 87,000. How many burglaries can be expected when the population reaches 110,000?

WRITING

91. Explain how to find the equation of a line passing through two given points.

92. In straight-line depreciation, explain why the slope of the line is called the *rate of depreciation*.

SOMETHING TO THINK ABOUT

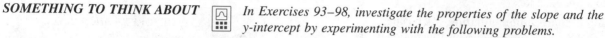 *In Exercises 93–98, investigate the properties of the slope and the y-intercept by experimenting with the following problems.*

93. Graph $y = mx + 2$ for several positive values of m. What do you notice?

94. Graph $y = mx + 2$ for several negative values of m. What do you notice?

95. Graph $y = 2x + b$ for several increasing positive values of b. What do you notice?

96. Graph $y = 2x + b$ for several decreasing negative values of b. What do you notice?

97. How will the graph of $y = \frac{1}{2}x + 5$ compare to the graph of $y = \frac{1}{2}x - 5$?

98. How will the graph of $y = \frac{1}{2}x - 5$ compare to the graph of $y = \frac{1}{2}x$?

99. If the graph of $y = ax + b$ passes through quadrants I, II, and IV, what can be known about the constants a and b?

100. The graph of $Ax + By = C$ passes only through the quadrants I and IV. What is known about the constants A, B, and C?

8.4 A Review of Functions

■ FUNCTIONS ■ FUNCTION NOTATION ■ FINDING DOMAINS AND RANGES OF FUNCTIONS
■ THE VERTICAL LINE TEST ■ LINEAR FUNCTIONS

Getting Ready   *If $y = \frac{3}{2}x - 2$, find the value of y for each value of x.*

1. $x = 2$ **2.** $x = 6$ **3.** $x = -12$ **4.** $x = -\dfrac{1}{2}$

■ FUNCTIONS

In Chapter 3, we saw that if x and y are real numbers, an equation in x and y determines a correspondence between the values of x and y. To see how, we consider the equation $y = \frac{1}{2}x + 3$. To find the value of y (called an **output value**) that corresponds to $x = 4$ (called an **input value**), we substitute 4 for x and simplify.

$$y = \frac{1}{2}x + 3$$

$$y = \frac{1}{2}(4) + 3 \qquad \text{Substitute the input value of 4 for } x.$$

$$= 2 + 3$$

$$= 5$$

The ordered pair $(4, 5)$ satisfies the equation and shows that a y-value of 5 corresponds to an x-value of 4. This ordered pair and others that satisfy the equation appear in the table shown in Figure 8-24. The graph of the equation also appears in the figure.

To see how the table determines the correspondence, we simply find an input in the x-column and then read across to find the corresponding output in the y-column. For example, if we select 2 as an input value, we get 4 as an output value. Thus, a y value of 4 corresponds to an x value of 2.

To see how the graph determines the correspondence, we draw a vertical and a horizontal line through any point (say, point P) on the graph, as shown in Figure 8-24. Because these lines intersect the x-axis at 4 and the y-axis at 5, the point $P(4, 5)$ associates 5 on the y-axis with 4 on the x-axis. This shows that a y value of 5 corresponds to an x value of 4.

$y = \frac{1}{2}x + 3$

x	y	
-2	2	A y value of 2 corresponds to an x value of -2.
0	3	A y value of 3 corresponds to an x value of 0.
2	4	A y value of 4 corresponds to an x value of 2.
4	5	A y value of 5 corresponds to an x value of 4.
6	6	A y value of 6 corresponds to an x value of 6.

Inputs   Outputs

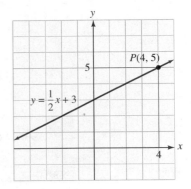

FIGURE 8-24

In this example, the set of all inputs x is the set of real numbers. The set of all outputs y is also the set of real numbers.

Recall that when a correspondence is set up by an equation, a table, or a graph, in which only one y value corresponds to each x value, we call the correspondence a **function.** The set of input values x is called the **domain** of the function, and the set of output values y is called the **range.** Since the value of y usually depends on the number x, we call y the **dependent variable** and x the **independent variable.**

Functions

A **function** is a correspondence between a set of input values x (called the **domain**) and a set of output values y (called the **range**), where exactly one y value in the range corresponds to each number x in the domain.

EXAMPLE 1   Does $y = 2x - 3$ define y to be a function of x? If so, find its domain and range and illustrate the function with a table and graph.

Solution   For a function to exist, every number x must determine one value of y. To find y in the equation $y = 2x - 3$, we multiply x by 2 and then subtract 3. Since this arithmetic gives one result, each choice of x determines one value of y. Thus, the equation does define y to be a function of x.

Since the input x can be any real number, the domain of the function is the set of real numbers, denoted by the interval $(-\infty, \infty)$. Since the output y can be any real number, the range is also the set of real numbers, denoted as $(-\infty, \infty)$.

A table of values and the graph appear in Figure 8-25.

$$y = 2x - 3$$

x	y
-4	-11
-2	-7
0	-3
2	1
4	5
6	9

The inputs can be any real number.   The outputs can be any real number.

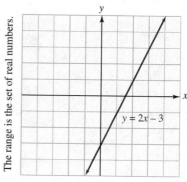

The range is the set of real numbers.

The domain is the set of real numbers.

FIGURE 8-25

Self Check   Does $y = -2x + 3$ define y to be a function of x?

Answer   yes

EXAMPLE 2   Does $y^2 = x$ define y to be a function of x?

Solution   For a function to exist, each number x must determine one value of y. If we let $x = 16$, for example, y could be either 4 or -4, because $4^2 = 16$ and $(-4)^2 = 16$. Since more than one value of y is determined when $x = 16$, the equation does not represent a function.

Self Check   Does $|y| = x$ define y to be a function of x?

Answer   no

■ FUNCTION NOTATION

In Chapter 4, we introduced the following special notation, which is used to denote functions.

Function Notation

The notation $y = f(x)$ denotes that the variable y is a function of x.

The notation $y = f(x)$ is read as "y equals f of x." Note that y and $f(x)$ are two notations for the same quantity. Thus, the equations $y = 4x + 3$ and $f(x) = 4x + 3$ are equivalent.

WARNING!   The notation $f(x)$ does not means "f times x."

The notation $y = f(x)$ provides a way of denoting the value of y (the dependent variable) that corresponds to some number x (the independent variable). For example, if $y = f(x)$, the value of y that is determined by $x = 3$ is denoted by $f(3)$.

EXAMPLE 3   Let $f(x) = 4x + 3$. Find   **a.** $f(3)$,   **b.** $f(-1)$,   **c.** $f(0)$,   and   **d.** $f(r)$.

Solution   **a.** We replace x with 3:   **b.** We replace x with -1:

$$f(x) = 4x + 3$$
$$f(3) = 4(3) + 3$$
$$= 12 + 3$$
$$= 15$$

$$f(x) = 4x + 3$$
$$f(-1) = 4(-1) + 3$$
$$= -4 + 3$$
$$= -1$$

c. We replace x with 0:   **d.** We replace x with r:

$$f(x) = 4x + 3$$
$$f(0) = 4(0) + 3$$
$$= 3$$

$$f(x) = 4x + 3$$
$$f(r) = 4r + 3$$

Self Check   If $f(x) = -2x - 1$, find   **a.** $f(2)$ and   **b.** $f(-3)$.
Answer   **a.** -5,   **b.** 5

To see why function notation is helpful, consider the following sentences:

1. In the equation $y = 4x + 3$, find the value of y when x is 3.
2. In the equation $f(x) = 4x + 3$, find $f(3)$.

Statement 2, which uses $f(x)$ notation, is much more concise.

We can think of a function as a machine that takes some input x and turns it into some output $f(x)$, as shown in Figure 8-26(a). The machine shown in Figure 8-26(b) turns the input number 2 into the output value -3 and turns the input number 6 into the output value -11. The set of numbers that we can put into the machine is the domain of the function, and the set of numbers that comes out is the range.

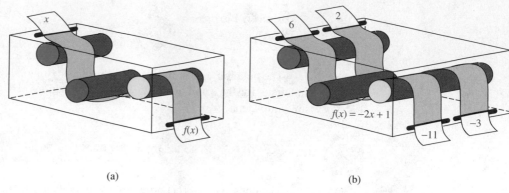

(a)　　　　　　　　　　　(b)

FIGURE 8-26

The letter f used in the notation $y = f(x)$ represents the word *function*. However, other letters can be used to represent functions. The notations $y = g(x)$ and $y = h(x)$ also denote functions involving the independent variable x.

In Example 4, the equation $y = g(x) = x^2 - 2x$ determines a function, because every possible value of x gives a single value of $g(x)$.

 a,d

EXAMPLE 4　　Let $g(x) = x^2 - 2x$. Find　**a.** $g\left(\frac{2}{5}\right)$,　**b.** $g(s)$,　**c.** $g(s^2)$,　and　**d.** $g(-t)$.

Solution　**a.** We replace x with $\frac{2}{5}$:

$$g(x) = x^2 - 2x$$

$$g\left(\frac{2}{5}\right) = \left(\frac{2}{5}\right)^2 - 2\left(\frac{2}{5}\right)$$

$$= \frac{4}{25} - \frac{4}{5}$$

$$= -\frac{16}{25}$$

b. We replace x with s:

$$g(x) = x^2 - 2x$$

$$g(s) = s^2 - 2s$$

c. We replace x with s^2:

$$g(x) = x^2 - 2x$$

$$g(s^2) = (s^2)^2 - 2s^2$$

$$= s^4 - 2s^2$$

d. We replace x with $-t$:

$$g(x) = x^2 - 2x$$

$$g(-t) = (-t)^2 - 2(-t)$$

$$= t^2 + 2t$$ ■

Self Check　Let $h(x) = -x^2 + 3$. Find　**a.** $h(2)$　and　**b.** $h(-a)$.
Answers　**a.** -1,　**b.** $-a^2 + 3$

EXAMPLE 5　　Let $f(x) = 4x - 1$. Find　**a.** $f(3) + f(2)$　and　**b.** $f(a) - f(b)$.

Solution　**a.** We find $f(3)$ and $f(2)$ separately.

$$f(x) = 4x - 1 \qquad f(x) = 4x - 1$$
$$f(3) = 4(3) - 1 \qquad f(2) = 4(2) - 1$$
$$= 12 - 1 \qquad\qquad = 8 - 1$$
$$= 11 \qquad\qquad\quad = 7$$

We then add the results to obtain $f(3) + f(2) = 11 + 7 = 18$.

b. We find $f(a)$ and $f(b)$ separately.

$$f(x) = 4x - 1 \qquad f(x) = 4x - 1$$
$$f(a) = 4a - 1 \qquad f(b) = 4b - 1$$

We then subtract the results to obtain

$$f(a) - f(b) = (4a - 1) - (4b - 1)$$
$$= 4a - 1 - 4b + 1$$
$$= 4a - 4b$$

Self Check   Let $g(x) = -2x + 3$. Find   **a.** $g(-2) + g(3)$ and   **b.** $g(\frac{1}{2}) - g(2)$.

Answers   **a.** 4,   **b.** 3

■ FINDING DOMAINS AND RANGES OF FUNCTIONS

EXAMPLE 6   Find the domains and ranges of the functions defined by   **a.** the ordered pairs $(-2, 4), (0, 6), (2, 8)$ and   **b.** the equation $y = \frac{1}{x-2}$.

Solution   **a.** The ordered pairs can be placed in a table to show a correspondence between x and y where a single value of y corresponds to each x.

x	y	
-2	4	4 corresponds to -2.
0	6	6 corresponds to 0.
2	8	8 corresponds to 2.

The domain is the set of numbers x: $\{-2, 0, 2\}$. The range is the set of values y: $\{4, 6, 8\}$.

b. The number 2 cannot be substituted for x, because that would make the denominator equal to zero. Since any real number except 2 can be substituted for x in the equation $y = \frac{1}{x-2}$, the domain is the set of all real numbers but 2. This is the interval $(-\infty, 2) \cup (2, \infty)$.

Since a fraction with a numerator of 1 cannot be 0, the range is the set of all real numbers but 0. This is the interval $(-\infty, 0) \cup (0, \infty)$.

Self Check   Find the domains and ranges of the functions defined by   **a.** the ordered pairs $(-3, 5), (-2, 7)$, and $(1, 11)$ and   **b.** the equation $y = \frac{2}{x+3}$.

Answers   **a.** $\{-3, -2, 1\}$, $\{5, 7, 11\}$,   **b.** $(-\infty, -3) \cup (-3, \infty)$, $(-\infty, 0) \cup (0, \infty)$

The *graph of a function* is the graph of the ordered pairs $(x, f(x))$ that define the function. For the graph in Figure 8-27, the domain is shown on the x-axis, and the range is shown on the y-axis. For any x in the domain, there corresponds one value $y = f(x)$ in the range.

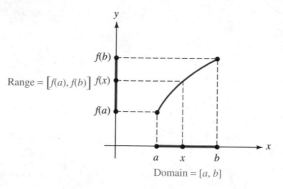

FIGURE 8-27

EXAMPLE 7   Find the domain and range of the function defined by $y = -2x + 1$.

Solution   We graph the equation as in Figure 8-28. Since every real number x on the x-axis determines a corresponding value of y, the domain is the interval $(-\infty, \infty)$ shown on the x-axis. Since the values of y can be any real number, the range is the interval $(-\infty, \infty)$ shown on the y-axis.

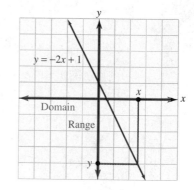

FIGURE 8-28   ∎

∎ THE VERTICAL LINE TEST

The **vertical line test** can be used to determine whether the graph of an equation represents a function. If any vertical line intersects a graph more than once, the graph cannot represent a function, because to one number x there would correspond more than one value of y.

The graph in Figure 8-29(a) represents a function, because every vertical line that intersects the graph does so exactly once. The graph in Figure 8-29(b) does not represent a function, because some vertical lines intersect the graph more than once.

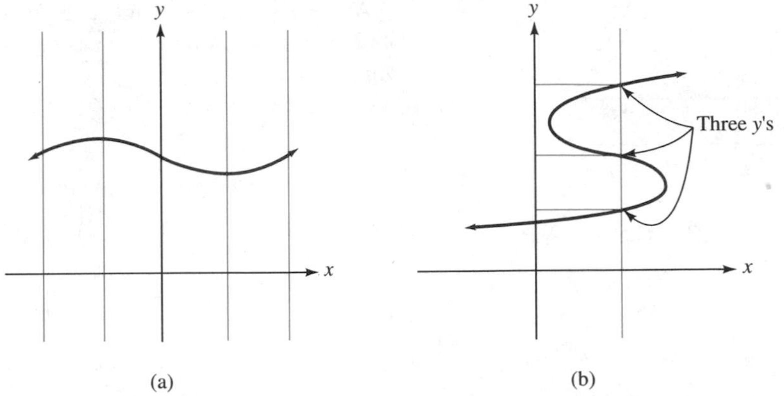

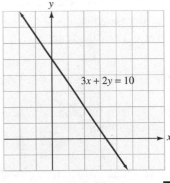

(a)   (b)

FIGURE 8-29

■ LINEAR FUNCTIONS

In Section 3.2, we graphed equations whose graphs were lines. These equations define basic functions, called **linear functions.**

Linear Functions
A **linear function** is a function defined by an equation that can be written in the form

$$f(x) = mx + b \quad \text{or} \quad y = mx + b$$

where m is the slope of the line graph and $(0, b)$ is the y-intercept.

EXAMPLE 8

Solve the equation $3x + 2y = 10$ for y to show that it defines a linear function. Then graph it to find its domain and range.

Solution

We solve the equation for y as follows:

$$3x + 2y = 10$$

$$2y = -3x + 10 \qquad \text{Subtract } 3x \text{ from both sides.}$$

$$y = -\frac{3}{2}x + 5 \qquad \text{Divide both sides by 2.}$$

Because the given equation is written in the form $y = mx + b$, it defines a linear function. The slope of its line graph is $-\frac{3}{2}$, and the y-intercept is $(0, 5)$. The graph appears in Figure 8-30. From the graph, we can see that both the domain and the range are the interval $(-\infty, \infty)$.

$3x + 2y = 10$

FIGURE 8-30

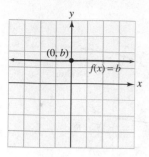

Constant function
Domain: $(-\infty, \infty)$
Range: $\{b\}$

FIGURE 8-31

A special case of a linear function is the **constant function,** defined by the equation $f(x) = b$, where b is a constant. Its graph, domain, and range are shown in Figure 8-31.

Orals   *Tell whether each equation or inequality determines y to be a function of x.*

1. $y = 2x + 1$ **2.** $y \geq 2x$ **3.** $y^2 = x$

If $f(x) = 2x + 1$, find

4. $f(0)$ **5.** $f(1)$ **6.** $f(-2)$

EXERCISE 8.4

REVIEW   *Solve each equation.*

1. $\dfrac{y + 2}{2} = 4(y + 2)$

2. $\dfrac{3z - 1}{6} - \dfrac{3z + 4}{3} = \dfrac{z + 3}{2}$

3. $\dfrac{2a}{3} + \dfrac{1}{2} = \dfrac{6a - 1}{6}$

4. $\dfrac{2x + 3}{5} - \dfrac{3x - 1}{3} = \dfrac{x - 1}{15}$

VOCABULARY AND CONCEPTS   *In Exercises 5–12, consider the function $y = f(x) = 5x - 4$. Fill in each blank to make a true statement.*

5. Any substitution for x is called an _____ value.

6. The value of __ is called the output value.

7. The independent variable is __.

8. The dependent variable is __.

9. A _____ is a correspondence between a set of input values and a set of output values, where each _____ value determines one _____ value.

10. In a function, the set of all inputs is called the _____ of the function.

11. In a function, the set of all output values is called the _____ of the function.

12. The notation $f(3)$ is the value of __ when $x = 3$.

13. The denominator of a fraction can never be __.

14. If a vertical line intersects a graph more than once, the graph _____ represent a function.

15. A linear function is any function that can be written in the form $y =$ _____.

16. In the function $f(x) = mx + b$, m is the _____ of its graph, and b is the y-coordinate of the _____.

PRACTICE   *In Exercises 17–24, tell whether the equation determines y to be a function of x.*

17. $y = 2x + 3$ **18.** $y = -1$ **19.** $y = 2x^2$ **20.** $y^2 = x + 1$

21. $y = 3 + 7x^2$ **22.** $y^2 = 3 - 2x$ **23.** $x = |y|$ **24.** $y = |x|$

In Exercises 25–32, find f(3) and f(−1).

25. $f(x) = 3x$

26. $f(x) = -4x$

27. $f(x) = 2x - 3$

28. $f(x) = 3x - 5$

29. $f(x) = 7 + 5x$

30. $f(x) = 3 + 3x$

31. $f(x) = 9 - 2x$

32. $f(x) = 12 + 3x$

In Exercises 33–40, find f(2) and f(3).

33. $f(x) = x^2$

34. $f(x) = x^2 - 2$

35. $f(x) = x^3 - 1$

36. $f(x) = x^3$

37. $f(x) = (x + 1)^2$

38. $f(x) = (x - 3)^2$

39. $f(x) = 2x^2 - x$

40. $f(x) = 5x^2 + 2x$

In Exercises 41–48, find f(2) and f(−2).

41. $f(x) = |x| + 2$

42. $f(x) = |x| - 5$

43. $f(x) = x^2 - 2$

44. $f(x) = x^2 + 3$

45. $f(x) = \dfrac{1}{x + 3}$

46. $f(x) = \dfrac{3}{x - 4}$

47. $f(x) = \dfrac{x}{x - 3}$

48. $f(x) = \dfrac{x}{x^2 + 2}$

In Exercises 49–52, find g(w) and g(w + 1).

49. $g(x) = 2x$

50. $g(x) = -3x$

51. $g(x) = 3x - 5$

52. $g(x) = 2x - 7$

In Exercises 53–60, f(x) = 2x + 1. Find each value.

53. $f(3) + f(2)$

54. $f(1) - f(-1)$

55. $f(b) - f(a)$

56. $f(b) + f(a)$

57. $f(b) - 1$

58. $f(b) - f(1)$

59. $f(0) + f\left(-\tfrac{1}{2}\right)$

60. $f(a) + f(2a)$

In Exercises 61–64, find the domain and range of each function.

61. $\{(-2, 3), (4, 5), (6, 7)\}$

62. $\{(0, 2), (1, 2), (3, 4)\}$

63. $f(x) = \dfrac{1}{x - 4}$

64. $f(x) = \dfrac{5}{x + 1}$

In Exercises 65–68, each graph represents a correspondence between x and y. Tell whether the correspondence is a function. If it is, give its domain and range.

65.

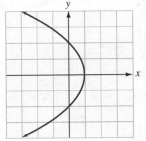

66.

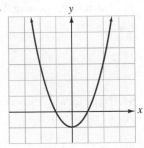

67.

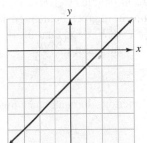

68.

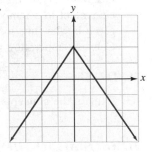

In Exercises 69–72, draw the graph of each linear function. Give the domain and range.

69. $f(x) = 2x - 1$

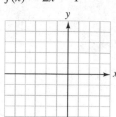

70. $f(x) = -x + 2$

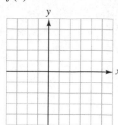

71. $2x - 3y = 6$

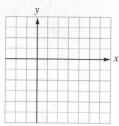

72. $3x + 2y = -6$

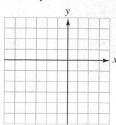

In Exercises 73–76, tell whether each equation defines a linear function.

73. $y = 3x^2 + 2$

74. $y = \dfrac{x - 3}{2}$

75. $x = 3y - 4$

76. $x = \dfrac{8}{y}$

APPLICATIONS

77. Ballistics A bullet shot straight upward is s feet above the ground after t seconds, where $s = f(t) = -16t^2 + 256t$. Find the height of the bullet 3 seconds after it is shot.

78. Artillery fire A mortar shell is s feet above the ground after t seconds, where $s = f(t) = -16t^2 + 512t + 64$. Find the height of the shell 20 seconds after it is fired.

79. Conversion from degrees Celsius to degrees Fahrenheit The temperature in degrees Fahrenheit that is equivalent to a temperature in degrees Celsius is given by the function $F(C) = \frac{9}{5}C + 32$. Find the Fahrenheit temperature that is equivalent to 25° C.

80. Conversion from degrees Fahrenheit to degrees Celsius The temperature in degrees Celsius that is equivalent to a temperature in degrees Fahrenheit is given by the function $C(F) = \frac{5}{9}F - \frac{160}{9}$. Find the Celsius temperature that is equivalent to 14°F.

81. Selling tape recorders An electronics firm manufactures tape recorders, receiving $120 for each recorder it makes. If x represents the number of recorders produced, the income received is determined by the *revenue function* $R(x) = 120x$. The manufacturer has fixed costs of $12,000 per month and variable costs of $57.50 for each recorder manufactured. Thus, the *cost function* is $C(x) = 57.50x + 12,000$. How many recorders must the company sell for revenue to equal cost?

82. Selling tires A tire company manufactures premium tires, receiving $130 for each tire it makes. If the manufacturer has fixed costs of $15,512.50 per month and variable costs of $93.50 for each tire manufactured, how many tires must the company sell for revenue to equal cost? (*Hint:* See Exercise 81.)

WRITING

83. Explain the concepts of function, domain, and range.

84. Explain why the constant function is a special case of a linear function.

SOMETHING TO THINK ABOUT Let $f(x) = 2x + 1$ and $g(x) = x^2$. Assume that $f(x) \neq 0$ and $g(x) \neq 0$.

85. Is $f(x) + g(x)$ equal to $g(x) + f(x)$?

86. Is $f(x) - g(x)$ equal to $g(x) - f(x)$?

8.5 Graphs of Nonlinear Functions

■ GRAPHS OF NONLINEAR FUNCTIONS ■ TRANSLATIONS OF GRAPHS ■ REFLECTIONS OF GRAPHS
■ RATIONAL FUNCTIONS ■ FINDING THE DOMAIN AND RANGE OF A RATIONAL FUNCTION

Getting Ready   *Give the slope and the y-intercept of each linear function.*

1. $f(x) = 2x - 3$ **2.** $f(x) = -3x + 4$

Find the value of f(x) when x = 2 and x = −1.

3. $f(x) = 5x - 4$ **4.** $f(x) = \dfrac{1}{2}x + 3$

■ GRAPHS OF NONLINEAR FUNCTIONS

If f is a function whose domain and range are sets of real numbers, its graph is the set of all points $(x, f(x))$ in the xy-plane. In other words, the graph of f is the graph of the equation $y = f(x)$. In this section, we will draw the graphs of many basic functions whose graphs are not straight lines.

The first basic function is $f(x) = x^2$ (or $y = x^2$), often called the **squaring function.**

EXAMPLE 1   Graph the function $f(x) = x^2$.

Solution   We substitute values for x in the equation and compute the corresponding values of $f(x)$. For example, if $x = -3$, we have

$$f(x) = x^2$$
$$f(-3) = (-3)^2 \qquad \text{Substitute } -3 \text{ for } x.$$
$$= 9$$

The ordered pair $(-3, 9)$ satisfies the equation and will lie on the graph. We list this pair and others that satisfy the equation in the table shown in Figure 8-32. We plot the points and draw a smooth curve through them to get the graph, called a **parabola.**

$$f(x) = x^2$$

x	y	$(x, f(x))$
-3	9	$(-3, 9)$
-2	4	$(-2, 4)$
-1	1	$(-1, 1)$
0	0	$(0, 0)$
1	1	$(1, 1)$
2	4	$(2, 4)$
3	9	$(3, 9)$

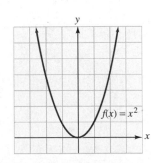

FIGURE 8-32

From the graph, we see that x can be any real number. This indicates that the domain of the squaring function is the set of real numbers, which is the interval $(-\infty, \infty)$. We can also see that y is always positive or zero. This indicates that the range is the set of nonnegative real numbers, which is the interval $[0, \infty)$.   ■

Self Check

Answer

Graph $f(x) = x^2 - 2$ and compare the graph to the graph of $f(x) = x^2$.

The graph has the same shape but is 2 units lower.

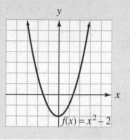

A second basic function is $f(x) = x^3$ (or $y = x^3$), often called the **cubing function.**

EXAMPLE 2

Graph the function $f(x) = x^3$.

Solution

We substitute values for x in the equation and compute the corresponding values of $f(x)$. For example, if $x = -2$, we have

$$f(x) = x^3$$
$$f(-2) = (-2)^3 \qquad \text{Substitute } -2 \text{ for } x.$$
$$= -8$$

The ordered pair $(-2, -8)$ satisfies the equation and will lie on the graph. We list this pair and others that satisfy the equation in the table shown in Figure 8-33. We plot the points and draw a smooth curve through them to get the graph.

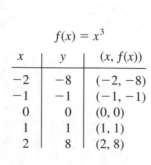

$f(x) = x^3$

x	y	$(x, f(x))$
-2	-8	$(-2, -8)$
-1	-1	$(-1, -1)$
0	0	$(0, 0)$
1	1	$(1, 1)$
2	8	$(2, 8)$

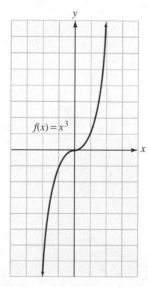

FIGURE 8-33

From the graph, we can see that x can be any real number. This indicates that the domain of the cubing function is the set of real numbers, which is the interval $(-\infty, \infty)$. We can also see that y can be any real number. This indicates that the range is the set of real numbers, which is the interval $(-\infty, \infty)$. ∎

Self Check   Graph $f(x) = x^3 + 1$ and compare the graph to the graph of $f(x) = x^3$.

Answer   The graph has the same shape but is 1 unit higher.

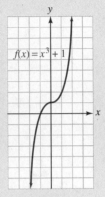

A third basic function is $f(x) = |x|$ (or $y = |x|$), often called the **absolute value function.**

EXAMPLE 3   Graph the function $f(x) = |x|$.

Solution   We substitute values for x in the equation and compute the corresponding values of $f(x)$. For example, if $x = -3$, we have

$$f(x) = |x|$$
$$f(-3) = |-3| \qquad \text{Substitute } -3 \text{ for } x.$$
$$= 3$$

The ordered pair $(-3, 3)$ satisfies the equation and will lie on the graph. We list this pair and others that satisfy the equation in the table shown in Figure 8-34. We plot the points and draw a V-shaped line through them to get the graph.

$$f(x) = |x|$$

x	y	$(x, f(x))$
-3	3	$(-3, 3)$
-2	2	$(-2, 2)$
-1	1	$(-1, 1)$
0	0	$(0, 0)$
1	1	$(1, 1)$
2	2	$(2, 2)$
3	3	$(3, 3)$

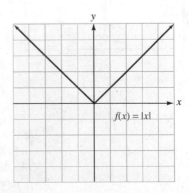

FIGURE 8-34

From the graph, we see that x can be any real number. This indicates that the domain of the absolute value function is the set of real numbers, which is the interval $(-\infty, \infty)$. We can also see that y is always positive or zero. This indicates that the range is the set of nonnegative real numbers, which is the interval $[0, \infty)$.   ■

Self Check   Graph $f(x) = |x - 2|$ and compare the graph to the graph of $f(x) = |x|$.

Answer   The graph has the same shape but is 2 units to the right.

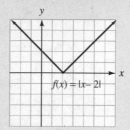

$f(x) = |x - 2|$

■ ■ ■ ■ ■ ■ ■ ■ ■   **Graphing Functions**

**GRAPHING
CALCULATORS**
We can graph nonlinear functions with a graphing calculator. For example, to graph $f(x) = x^2$ in a standard window of $[-10, 10]$ for x and $[-10, 10]$ for y, we enter the function by typing x ˆ 2 and press the GRAPH key. We will obtain the graph shown in Figure 8-35(a).

To graph $f(x) = x^3$, we enter the function by typing x ˆ 3 and press the GRAPH key to obtain the graph in Figure 8-35(b). To graph $f(x) = |x|$, we enter the function by pressing the ABS key (or selecting "abs" from a menu), typing x, and pressing the GRAPH key to obtain the graph in Figure 8-35(c).

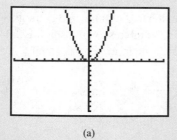

(a)

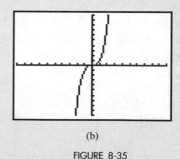

(b)

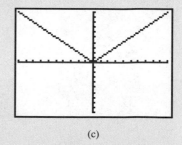

(c)

FIGURE 8-35

When using a graphing calculator, we must be sure that the viewing window does not show a misleading graph. For example, if we graph $f(x) = |x|$ in the window $[0, 10]$ for x and $[0, 10]$ for y, we will obtain a misleading graph that looks like a line. (See Figure 8-36.) This is not true. The proper graph is the V-shaped graph shown in Figure 8-35(c).

One of the challenges of using graphing calculators is finding an appropriate viewing window.

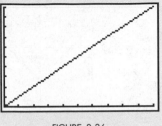

FIGURE 8-36

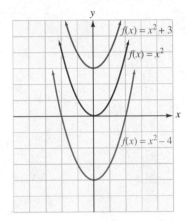

FIGURE 8-37

■ TRANSLATIONS OF GRAPHS

Examples 1–3 and their Self Checks suggest that the graphs of different functions may be identical except for their positions in the xy-plane. For example, Figure 8-37 shows the graph of $f(x) = x^2 + k$ for three different values of k. If $k = 0$, we get the graph of $f(x) = x^2$. If $k = 3$, we get the graph of $f(x) = x^2 + 3$, which is identical to the graph of $f(x) = x^2$ except that it is shifted 3 units upward. If $k = -4$, we get the graph of $f(x) = x^2 - 4$, which is identical to the graph of $f(x) = x^2$ except that it is shifted 4 units downward. These shifts are called **vertical translations.**

In general, we can make these observations.

Vertical Translations

If f is a function and k is a positive number, then

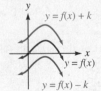

- The graph of $y = f(x) + k$ is identical to the graph of $y = f(x)$ except that it is translated k units upward.

- The graph of $y = f(x) - k$ is identical to the graph of $y = f(x)$ except that it is translated k units downward.

EXAMPLE 4

Graph $f(x) = |x| + 2$.

Solution   The graph of $f(x) = |x| + 2$ will be the same V-shaped graph as $f(x) = |x|$, except that it is shifted 2 units up. The graph appears in Figure 8-38.

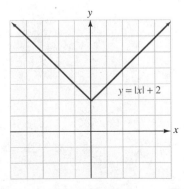

FIGURE 8-38

Self Check

Answer.

Graph $f(x) = |x| - 3$.

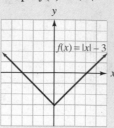

Figure 8-39 shows the graph of $f(x) = (x + h)^2$ for three different values of h. If $h = 0$, we get the graph of $f(x) = x^2$. The graph of $f(x) = (x - 3)^2$ is identical to the graph of $f(x) = x^2$, except that it is shifted 3 units to the right. The graph of $f(x) = (x + 2)^2$ is identical to the graph of $f(x) = x^2$, except that it is shifted 2 units to the left. These shifts are called **horizontal translations.**

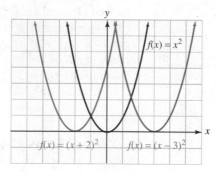

FIGURE 8-39

In general, we can make these observations.

Horizontal Translations

If f is a function and k is a positive number, then

- The graph of $y = f(x - k)$ is identical to the graph of $y = f(x)$ except that it is translated k units to the right.

- The graph of $y = f(x + k)$ is identical to the graph of $y = f(x)$ except that it is translated k units to the left.

EXAMPLE 5

Graph $f(x) = (x - 2)^2$.

Solution

The graph of $f(x) = (x - 2)^2$ will be the same shape as the graph of $f(x) = x^2$ except that it is shifted 2 units to the right. The graph appears in Figure 8-40.

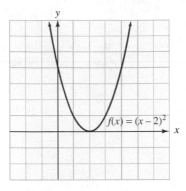

FIGURE 8-40

Self Check

Answer

Graph $f(x) = (x + 3)^3$.

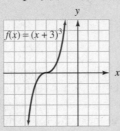

EXAMPLE 6   Graph $f(x) = (x - 3)^2 + 2$.

Solution   We can graph this equation by translating the graph of $f(x) = x^2$ to the right 3 units and then up 2 units, as shown in Figure 8-41.

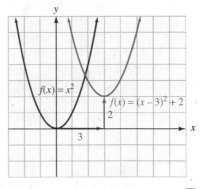

FIGURE 8-41

Self Check

Answer

Graph $f(x) = |x + 2| - 3$.

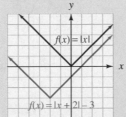

■ REFLECTIONS OF GRAPHS

We now consider the graph of $y = f(x) = -|x|$. To graph this function, we can make a table of values, plot each point, and draw the graph, as in Figure 8-42.

$$y = f(x) = -|x|$$

x	y	(x, y)
-3	-3	$(-3, -3)$
-2	-2	$(-2, -2)$
-1	-1	$(-1, -1)$
0	0	$(0, 0)$
1	-1	$(1, -1)$
2	-2	$(2, -2)$
3	-3	$(3, -3)$

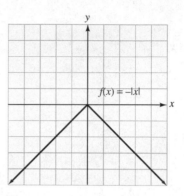

FIGURE 8-42

As we can see from the graph, its shape is the same as the graph of $y = f(x) = |x|$, except that it has been flipped upside down. We say that the graph of $y = f(x) = |x|$ has been *reflected* about the x-axis.

In general, we can make the following statement.

Reflections about the x-Axis

The graph of $y = -f(x)$ is identical to the graph of $y = f(x)$ except that it is reflected about the x-axis.

EXAMPLE 7   Graph the absolute value function $y = f(x) = -|x - 1| + 3$.

Solution   We graph this function by translating the graph of $y = f(x) = -|x|$ to the right 1 unit and up 3 units, as shown in Figure 8-43.

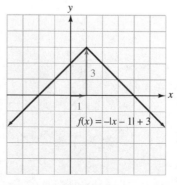

FIGURE 8-43

Self Check
Answer

Graph $y = f(x) = -|x + 2| - 3$.

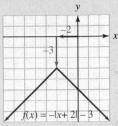

■ RATIONAL FUNCTIONS

Rational expressions often define functions. For example, if the cost of subscribing to an on-line information network is $6 per month plus $1.50 per hour of access time, the average (mean) hourly cost of the service is the total monthly cost, divided by the number of hours of access time:

$$\bar{c} = \frac{C}{n} = \frac{1.50n + 6}{n}$$ $\bar{c}$ is the mean hourly cost, C is the total monthly cost, and n is the number of hours the service is used.

The function

1. $\bar{c} = f(n) = \dfrac{1.50n + 6}{n}$ $(n > 0)$

gives the mean hourly cost of using the information network for n hours per month.

Figure 8-44 shows the graph of the rational function $\bar{c} = f(n) = \frac{1.50n + 6}{n}$ $(n > 0)$. Since $n > 0$, the domain of this function is the interval $(0, \infty)$.

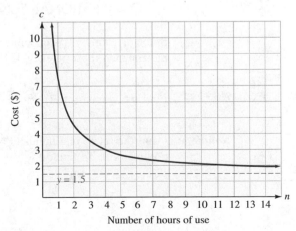

FIGURE 8-44

From the graph, we can see that the mean hourly cost decreases as the number of hours of access time increases. Since the cost of each extra hour of access time is $1.50, the mean hourly cost can approach $1.50 but never drop below it. Thus, the graph of the function approaches the line $y = 1.5$ as n increases without bound. When a graph approaches a line as the dependent variable gets large, we call the line an **asymptote**. The line $y = 1.5$ is a **horizontal asymptote** of the graph.

As n gets smaller and approaches 0, the graph approaches the y-axis but never touches it. The y-axis is a **vertical asymptote** of the graph.

EXAMPLE 8   Find the mean hourly cost when the network described above is used for **a.** 3 hours and **b.** 70.4 hours.

Solution   **a.** To find the mean hourly cost for 3 hours of access time, we substitute 3 for n in Equation 1 and simplify:

$$\bar{c} = f(3) = \frac{1.50(3) + 6}{3} = 3.5$$

The mean hourly cost for 3 hours of access time is $3.50.

b. To find the mean hourly cost for 70.4 hours of access time, we substitute 70.4 for n in Equation 1 and simplify:

$$\bar{c} = f(70.4) = \frac{1.50(70.4) + 6}{70.4} = 1.585227273$$

The mean hourly cost for 70.4 hours of access time is approximately $1.59.   ∎

Self Check   In Example 8, find the mean hourly cost when the network is used for 5 hours.
Answer $2.70

■ FINDING THE DOMAIN AND RANGE OF A RATIONAL FUNCTION

Since division by 0 is undefined, any values that make the denominator 0 in a rational function must be excluded from the domain of the function.

EXAMPLE 9   Find the domain of $f(x) = \dfrac{3x + 2}{x^2 + x - 6}$.

Solution   From the set of real numbers, we must exclude any values of x that make the denominator 0. To find these values, we set $x^2 + x - 6$ equal to 0 and solve for x.

$$x^2 + x - 6 = 0$$
$$(x + 3)(x - 2) = 0 \qquad \text{Factor.}$$
$$x + 3 = 0 \quad \text{or} \quad x - 2 = 0 \qquad \text{Set each factor equal to 0.}$$
$$x = -3 \qquad\qquad x = 2 \qquad \text{Solve each linear equation.}$$

Thus, the domain of the function is the set of all real numbers except -3 and 2. In interval notation, the domain is $(-\infty, -3) \cup (-3, 2) \cup (2, \infty)$. ∎

Self Check   Find the domain of $f(x) = \dfrac{x^2 + 1}{x - 2}$.

Answer $(-\infty, 2) \cup (2, \infty)$

■ ■ ■ ■ ■ ■ ■ ■ ■ ■   **Finding the Domain and Range of a Function**

GRAPHING   We can find the domain and range of the function in Example 9 by looking at its
CALCULATORS   graph. If we use window settings of $[-10, 10]$ for x and $[-10, 10]$ for y and graph the function

$$f(x) = \frac{3x + 2}{x^2 + x - 6}$$

we will obtain the graph in Figure 8-45(a).

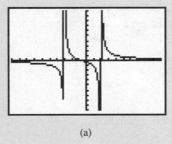   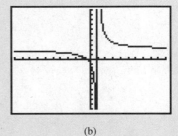

(a)   (b)

FIGURE 8-45

From the figure, we can see that

- As x approaches -3 from the left, the values of y decrease, and the graph approaches the vertical line $x = -3$.

- As x approaches -3 from the right, the values of y increase, and the graph approaches the vertical line $x = -3$.

From the figure, we can also see that

- As x approaches 2 from the left, the values of y decrease, and the graph approaches the vertical line $x = 2$.

- As x approaches 2 from the right, the values of y increase, and the graph approaches the vertical line $x = 2$.

The lines $x = -3$ and $x = 2$ are vertical asymptotes. Although the vertical lines in the graph appear to be the graphs of $x = -3$ and $x = 2$, they are not. Graphing calculators draw graphs by connecting dots whose x-coordinates are close together. Often, when two such points straddle a vertical asymptote and their y-coordinates are far apart, the calculator draws a line between them anyway, producing what appears to be a vertical asymptote. If you set your calculator to dot mode instead of connected mode, the vertical lines will not appear.

From Figure 8-45(a), we can also see that

- As x increases to the right of 2, the values of y decrease and approach the value $y = 0$.
- As x decreases to the left of -3, the values of y increase and approach the value $y = 0$.

The line $y = 0$ (the x-axis) is a horizontal asymptote. Graphing calculators do not draw lines that appear to be horizontal asymptotes.

From the graph, we can see that all real numbers x, except -3 and 2, give a value of y. This confirms that the domain of the function is $(-\infty, -3) \cup (-3, 2) \cup (2, \infty)$. We can also see that y can be any value. Thus, the range is $(-\infty, \infty)$.

To find the domain and range of the function $f(x) = \frac{2x+1}{x-1}$, we use a calculator to draw the graph shown in Figure 8-45(b). From this graph, we can see that the line $x = 1$ is a vertical asymptote and that the line $y = 2$ is a horizontal asymptote. Since x can be any real number except 1, the domain is the interval $(-\infty, 1) \cup (1, \infty)$. Since y can be any value except 2, the range is $(-\infty, 2) \cup (2, \infty)$.

Orals   1. Describe a parabola.

2. Describe the graph of $f(x) = |x| + 3$.

3. Describe the graph of $f(x) = x^3 - 4$.

4. Tell why the choice of a viewing window is important when using a graphing calculator.

EXERCISE 8.5

REVIEW

1. List the prime numbers between 40 and 50.

2. State the associative property of addition.

3. State the commutative property of multiplication.

4. What is the additive identity element?

5. What is the multiplicative identity element?

6. Find the multiplicative inverse of $\frac{5}{3}$.

VOCABULARY AND CONCEPTS   *Fill in each blank to make a true statement.*

7. The function $f(x) = x^2$ is called the _____ function.

8. The function $f(x) = x^3$ is called the _____ function.

9. The function $f(x) = |x|$ is called the _____ function.

10. Shifting the graph of an equation up or down is called a _____ translation.

11. Shifting the graph of an equation to the left or to the right is called a _____ translation.

12. The graph of $f(x) = x^2 + 5$ is the same as the graph of $f(x) = x^2$ except that it is shifted ___ units ___.

13. The graph of $f(x) = x^3 - 2$ is the same as the graph of $f(x) = x^3$ except that it is shifted ___ units _____.

14. The graph of $f(x) = (x - 5)^3$ is the same as the graph of $f(x) = x^3$ except that it is shifted ___ units _____.

15. The graph of $f(x) = (x + 4)^3$ is the same as the graph of $f(x) = x^3$ except that it is shifted ___ units _____.

16. To solve an equation with a graphing calculator, graph _____ sides of the equation and find the _____ of the point where the graphs intersect.

17. If a fraction is the quotient of two polynomials, it is called a _____ expression.

18. If a graph approaches a line but never touches it, the line is called an _____.

PRACTICE *In Exercises 19–26, graph each function by plotting points. Check your work with a graphing calculator.*

19. $f(x) = x^2 - 3$

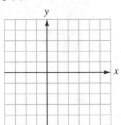

20. $f(x) = x^2 + 2$

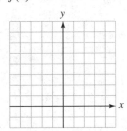

21. $f(x) = (x - 1)^3$

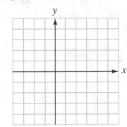

22. $f(x) = (x + 1)^3$

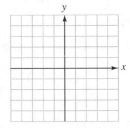

23. $f(x) = |x| - 2$

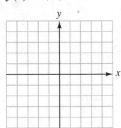

24. $f(x) = |x| + 1$

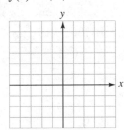

25. $f(x) = |x - 1|$

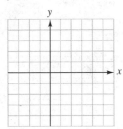

26. $f(x) = |x + 2|$

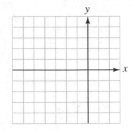

In Exercises 27–34, use a graphing calculator to graph each function, using values of $[-4, 4]$ for x and $[-4, 4]$ for y. The graph is not what it appears to be. Pick a better viewing window and find the true graph.

27. $f(x) = x^2 + 8$

28. $f(x) = x^3 - 8$

29. $f(x) = |x + 5|$

30. $f(x) = |x - 5|$

31. $f(x) = (x - 6)^2$

32. $f(x) = (x + 9)^2$

33. $f(x) = x^3 + 8$

34. $f(x) = x^3 - 12$

In Exercises 35–42, draw each graph using a translation of the graph of $f(x) = x^2$, $f(x) = x^3$, or $f(x) = |x|$.

35. $f(x) = x^2 - 5$

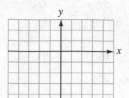

36. $f(x) = x^3 + 4$

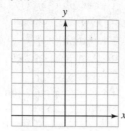

37. $f(x) = (x - 1)^3$

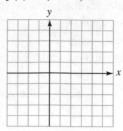

38. $f(x) = (x + 4)^2$

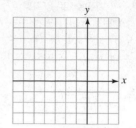

39. $f(x) = |x - 2| - 1$

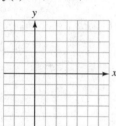

40. $f(x) = (x + 2)^2 - 1$

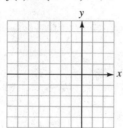

41. $f(x) = (x + 1)^3 - 2$

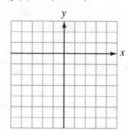

42. $f(x) = |x + 4| + 3$

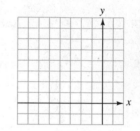

In Exercises 43–46, graph each function.

43. $f(x) = -x^2$

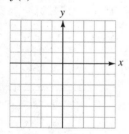

44. $f(x) = -x^3 + 2$

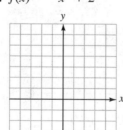

45. $f(x) = -(x - 2)^2 - 3$

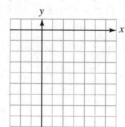

46. $f(x) = -|x - 2| + 3$

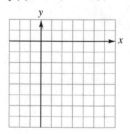

In Exercises 47–50, the time t it takes to travel 600 miles is a function of the mean rate of speed r: $t = f(r) = \frac{600}{r}$. Find t for each value of r.

47. 30 mph

48. 40 mph

49. 50 mph

50. 60 mph

In Exercises 51–54, suppose the cost (in dollars) of removing p% of the pollution in a river is given by the function $c = f(p) = \frac{50,000p}{100 - p}$ $(0 \le p < 100)$. Find the cost of removing each percent of pollution.

51. 10%

52. 30%

53. 50%

54. 80%

In Exercises 55–58, use a graphing calculator to graph each rational function. From the graph, determine its domain and range.

55. $f(x) = \dfrac{x}{x-2}$

56. $f(x) = \dfrac{x+2}{x}$

57. $f(x) = \dfrac{x+1}{x^2-4}$

58. $f(x) = \dfrac{x-2}{x^2-3x-4}$

APPLICATIONS   In Exercises 59–64, a service club wants to publish a directory of its members. Some investigation shows that the cost of typesetting and photography will be $700, and the cost of printing each directory will be $1.25.

59. Find a function that gives the total cost c of printing x directories.

60. Find a function that gives the mean cost per directory $\bar{c}$ of printing x directories.

61. Find the total cost of printing 500 directories.

62. Find the mean cost per directory if 500 directories are printed.

63. Find the mean cost per directory if 1,000 directories are printed.

64. Find the mean cost per directory if 2,000 directories are printed.

In Exercises 65–70, an electric company charges $7.50 per month plus 9¢ for each kilowatt hour (kwh) of electricity used.

65. Find a function that gives the total cost c of n kwh of electricity.

66. Find a function that gives the mean cost per kwh $\bar{c}$ when using n kwh.

67. Find the total cost for using 775 kwh.

68. Find the mean cost per kwh when 775 kwh are used.

69. Find the mean cost per kwh when 1,000 kwh are used.

70. Find the mean cost per kwh when 1,200 kwh are used.

In Exercises 71–74, assume that a person buys a horse for $5,000 and plans to pay $350 per month to board the horse.

71. Find a function that will give the total cost of owning the horse for x months.

72. Find a function that will give the mean cost per month $\bar{c}$ after owning the horse for x months.

73. Find the total cost of owning the horse for 10 years.

74. Find the mean cost per month if the horse is owned for 10 years.

WRITING

75. Explain how to graph an equation by plotting points.

76. Explain how the graphs of $y = (x-4)^2 - 3$ and $y = x^2$ are related.

8.6 Variation

■ DIRECT VARIATION ■ INVERSE VARIATION ■ JOINT VARIATION ■ COMBINED VARIATION

Getting Ready   *Solve each equation.*

1. $\dfrac{x}{2} = \dfrac{3}{4}$ **2.** $\dfrac{5}{7} = \dfrac{x}{2}$

3. $8 = 2k$ **4.** $12 = \dfrac{k}{3}$

Recall that the quotient of two numbers is often called a **ratio.** For example, the fraction $\frac{2}{3}$ can be read as "the ratio of 2 to 3." An equation indicating that two ratios are equal is called a **proportion.** Two examples of proportions are

$$\frac{1}{4} = \frac{2}{8} \qquad \text{and} \qquad \frac{4}{7} = \frac{12}{21}$$

In the proportion $\frac{a}{b} = \frac{c}{d}$, the terms a and d are called the **extremes** of the proportion, and the terms b and c are called the **means.**

To develop a fundamental property of proportions, we suppose that

$$\frac{a}{b} = \frac{c}{d}$$

is a proportion and multiply both sides by bd to obtain

$$bd\left(\frac{a}{b}\right) = bd\left(\frac{c}{d}\right)$$

$$\frac{bda}{b} = \frac{bdc}{d}$$

$$ad = bc$$

Thus, if $\frac{a}{b} = \frac{c}{d}$, then $ad = bc$. In a proportion, *the product of the extremes equals the product of the means.*

EXAMPLE 1   Solve the proportion $\dfrac{x+1}{x} = \dfrac{x}{x+2}$ for x.

Solution

$$\frac{x+1}{x} = \frac{x}{x+2}$$

$$(x+1)(x+2) = x \cdot x \qquad \text{The product of the extremes equals the product of the means.}$$

$$x^2 + 3x + 2 = x^2$$

$$3x + 2 = 0 \qquad \text{Subtract } x^2 \text{ from both sides.}$$

$$x = -\frac{2}{3} \qquad \text{Subtract 2 from both sides and divide by 3.}$$

Self Check Solve $\dfrac{x-2}{x} = \dfrac{x}{x-3}$.

Answer $\dfrac{6}{5}$

■ DIRECT VARIATION

To introduce direct variation, we consider the formula

$$C = \pi D$$

for the circumference of a circle, where C is the circumference, D is the diameter, and $\pi \approx 3.14159$. If we double the diameter of a circle, we determine another circle with a larger circumference C_1 such that

$$C_1 = \pi(2D) = 2\pi D = 2C$$

Thus, doubling the diameter results in doubling the circumference. Likewise, if we triple the diameter, we triple the circumference.

In this formula, we say that the variables C and D **vary directly,** or that they are **directly proportional.** This is because as one variable gets larger, so does the other, in a predictable way. In this example, the constant π is called the **constant of variation** or the **constant of proportionality.**

> **Direct Variation**
> The words "y varies directly with x" or "y is directly proportional to x" mean that $y = kx$ for some nonzero constant k. The constant k is called the **constant of variation** or the **constant of proportionality.**

Since the formula for direct variation ($y = kx$) defines a linear function, its graph is always a line with a y-intercept at the origin. The graph of $y = kx$ appears in Figure 8-46 for three positive values of k.

One example of direct variation is Hooke's law from physics. Hooke's law states that the distance a spring will stretch varies directly with the force that is applied to it.

If d represents a distance and f represents a force, Hooke's law is expressed mathematically as

$$d = kf$$

where k is the constant of variation. If the spring stretches 10 inches when a weight of 6 pounds is attached, k can be found as follows:

$d = kf$

$\mathbf{10} = k(\mathbf{6})$ Substitute 10 for d and 6 for f.

$\dfrac{5}{3} = k$ Divide both sides by 6 and simplify.

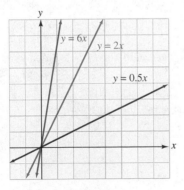

FIGURE 8-46

To find the force required to stretch the spring a distance of 35 inches, we can solve the equation $d = kf$ for f, with $d = 35$ and $k = \frac{5}{3}$.

$$d = kf$$

$$35 = \frac{5}{3}f \qquad \text{Substitute 35 for } d \text{ and } \frac{5}{3} \text{ for } k.$$

$$105 = 5f \qquad \text{Multiply both sides by 3.}$$

$$21 = f \qquad \text{Divide both sides by 5.}$$

The force required to stretch the spring a distance of 35 inches is 21 pounds.

EXAMPLE 2

Direct variation   The distance traveled in a given time is directly proportional to the speed. If a car travels 70 miles at 30 mph, how far will it travel in the same time at 45 mph?

Solution   The words *distance is directly proportional to speed* can be expressed by the equation

1. $d = ks$

where d is distance, k is the constant of variation, and s is the speed. To find k, we substitute 70 for d and 30 for s, and solve for k.

$$d = ks$$

$$70 = k(30)$$

$$k = \frac{7}{3}$$

To find the distance traveled at 45 mph, we substitute $\frac{7}{3}$ for k and 45 for s in Equation 1 and simplify.

$$d = ks$$

$$d = \frac{7}{3}(45)$$

$$= 105$$

In the time it took to go 70 miles at 30 mph, the car could travel 105 miles at 45 mph.   ■

Self Check   In Example 2, how far will the car travel in the same time at 60 mph?

Answer   140 mi

■ INVERSE VARIATION

In the formula $w = \frac{12}{l}$, w gets smaller as l gets larger, and w gets larger as l gets smaller. Since these variables vary in opposite directions in a predictable way, we say that the variables **vary inversely,** or that they are **inversely proportional.** The constant 12 is the constant of variation.

Inverse Variation
The words "y varies inversely with x" or "y is inversely proportional to x" mean that $y = \frac{k}{x}$ for some nonzero constant k. The constant k is called the **constant of variation.**

The formula for inverse variation $\left(y = \frac{k}{x} \right)$ defines a rational function. The graph of $y = \frac{k}{x}$ appears in Figure 8-47 for three positive values of k.

Because of gravity, an object in space is attracted to the earth. The force of this attraction varies inversely with the square of the distance from the object to the center of the earth.

If f represents the force and d represents the distance, this information can be expressed by the equation

$$f = \frac{k}{d^2}$$

If we know that an object 4,000 miles from the center of the earth is attracted to the earth with a force of 90 pounds, we can find k.

$$f = \frac{k}{d^2}$$

$$90 = \frac{k}{4{,}000^2} \qquad \text{Substitute 90 for } f \text{ and 4,000 for } d.$$

$$k = 90(4{,}000^2)$$

$$= 1.44 \times 10^9$$

To find the force of attraction when the object is 5,000 miles from the center of the earth, we proceed as follows:

$$f = \frac{k}{d^2}$$

$$f = \frac{1.44 \times 10^9}{5{,}000^2} \qquad \text{Substitute } 1.44 \times 10^9 \text{ for } k \text{ and 5,000 for } d.$$

$$= 57.6$$

The object will be attracted to the earth with a force of 57.6 pounds when it is 5,000 miles from the earth's center.

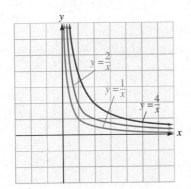

FIGURE 8-47

EXAMPLE 3   **Light intensity**   The intensity I of light received from a light source varies inversely with the square of the distance from the light source. If the intensity of a light source 4 feet from an object is 8 candelas, find the intensity at a distance of 2 feet.

Solution The words *intensity varies inversely with the square of the distance d* can be expressed by the equation

$$I = \frac{k}{d^2}$$

To find k, we substitute 8 for I and 4 for d and solve for k.

$$I = \frac{k}{d^2}$$

$$8 = \frac{k}{4^2}$$

$$128 = k$$

To find the intensity when the object is 2 feet from the light source, we substitute 2 for d and 128 for k and simplify.

$$I = \frac{k}{d^2}$$

$$I = \frac{128}{2^2}$$

$$= 32$$

The intensity at 2 feet is 32 candelas. ■

Self Check

Answer

In Example 3, find the intensity at a distance of 8 feet.

2 candelas

■ JOINT VARIATION

There are times when one variable varies with the product of several variables. For example, the area of a triangle varies directly with the product of its base and height:

$$A = \frac{1}{2}bh$$

Such variation is called **joint variation.**

Joint Variation

If one variable varies directly with the product of two or more variables, the relationship is called **joint variation.** If y varies jointly with x and z, then $y = kxz$. The nonzero constant k is called the **constant of variation.**

EXAMPLE 4   The volume V of a cone varies jointly with its height h and the area of its base B. If $V = 6$ cm^3 when $h = 3$ cm and $B = 6$ cm^2, find V when $h = 2$ cm and $B = 8$ cm^2.

Solution   The words V *varies jointly with h and B* can be expressed by the equation

$$V = khB$$ The relationship can also be read as "V is directly proportional to the product of h and B."

We can find k by substituting 6 for V, 3 for h, and 6 for B.

$$V = khB$$
$$6 = k(3)(6)$$
$$6 = k(18)$$
$$\frac{1}{3} = k$$ Divide both sides by 18; $\frac{6}{18} = \frac{1}{3}$.

To find V when $h = 2$ and $B = 8$, we substitute these values into the formula $V = \frac{1}{3}hB$.

$$V = \frac{1}{3}hB$$
$$V = \left(\frac{1}{3}\right)(2)(8)$$
$$= \frac{16}{3}$$

When $h = 2$ and $B = 8$, the volume is $5\frac{1}{3}$ cm^3.   ■

■ COMBINED VARIATION

Many applied problems involve a combination of direct and inverse variation. Such variation is called **combined variation.**

EXAMPLE 5   **Building highways**   The time it takes to build a highway varies directly with the length of the road, but inversely with the number of workers. If it takes 100 workers 4 weeks to build 2 miles of highway, how long will it take 80 workers to build 10 miles of highway?

Solution   We can let t represent the time in weeks, l represent the length in miles, and w represent the number of workers. The relationship between these variables can be expressed by the equation

$$t = \frac{kl}{w}$$

We substitute 4 for t, 100 for w, and 2 for l to find k:

$$4 = \frac{k(2)}{100}$$

$$400 = 2k \qquad \text{Multiply both sides by 100.}$$

$$200 = k \qquad \text{Divide both sides by 2.}$$

We now substitute 80 for w, 10 for l, and 200 for k in the equation $t = \frac{kl}{w}$ and simplify:

$$t = \frac{kl}{w}$$

$$t = \frac{200(10)}{80}$$

$$= 25$$

It will take 25 weeks for 80 workers to build 10 miles of highway. ■

Self Check Answer

In Example 5, how long will it take 60 workers to build 6 miles of highway?

20 weeks

Orals *Solve each proportion.*

1. $\dfrac{x}{2} = \dfrac{3}{6}$ **2.** $\dfrac{3}{x} = \dfrac{4}{12}$ **3.** $\dfrac{5}{7} = \dfrac{2}{x}$

Express each sentence with a formula.

4. a varies directly with b.   **5.** a varies inversely with b.

6. a varies jointly with b and c.   **7.** a varies directly with b and inversely with c.

EXERCISE 8.6

REVIEW *Simplify each expression.*

1. $(x^2x^3)^2$ **2.** $\left(\dfrac{a^3a^5}{a^{-2}}\right)^3$ **3.** $\dfrac{b^0 - 2b^0}{b^0}$ **4.** $\left(\dfrac{2r^{-2}r^{-3}}{4r^{-5}}\right)^{-3}$

5. Write 35,000 in scientific notation.

6. Write 0.00035 in scientific notation.

7. Write 2.5×10^{-3} in standard notation.

8. Write 2.5×10^4 in standard notation.

VOCABULARY AND CONCEPTS *Fill in each blank to make a true statement.*

9. An equation stating that two ratios are equal is called a _____.

10. In a proportion, the product of the _____ is equal to the product of the _____.

11. The equation $y = kx$ indicates _____ variation.

12. The equation $y = \frac{k}{x}$ indicates _____ variation.

13. Inverse variation is represented by a _____ function.

14. Direct variation is represented by a _____ function whose graph passes through the origin.

15. The equation $y = kxz$ indicates _____ variation.

16. The equation $y = \frac{kx}{z}$ indicates _____ variation.

Tell whether the graph represents direct variation, inverse variation, or neither.

17.

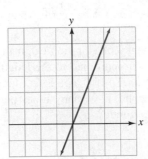

18.

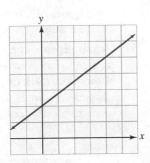

19.

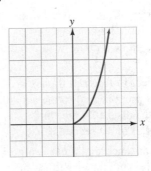

20.

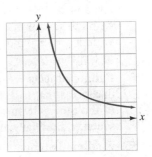

PRACTICE *In Exercises 21–36, solve each proportion for the variable, if possible.*

21. $\dfrac{x}{5} = \dfrac{15}{25}$

22. $\dfrac{4}{y} = \dfrac{6}{27}$

23. $\dfrac{r-2}{3} = \dfrac{r}{5}$

24. $\dfrac{x+1}{x-1} = \dfrac{6}{4}$

25. $\dfrac{3}{n} = \dfrac{2}{n+1}$

26. $\dfrac{4}{x+3} = \dfrac{3}{5}$

27. $\dfrac{5}{5z+3} = \dfrac{2z}{2z^2+6}$

28. $\dfrac{9t+6}{t(t+3)} = \dfrac{7}{t+3}$

29. $\dfrac{2}{c} = \dfrac{c-3}{2}$

30. $\dfrac{y}{4} = \dfrac{4}{y}$

31. $\dfrac{2}{3x} = \dfrac{6x}{36}$

32. $\dfrac{2}{x+6} = \dfrac{-2x}{5}$

33. $\dfrac{2(x+3)}{3} = \dfrac{4(x-4)}{5}$

34. $\dfrac{x+4}{5} = \dfrac{3(x-2)}{3}$

35. $\dfrac{1}{x+3} = \dfrac{-2x}{x+5}$

36. $\dfrac{x-1}{x+1} = \dfrac{2}{3x}$

In Exercises 37–44, express each sentence as a formula.

37. A varies directly with the square of p.

38. z varies inversely with the cube of t.

39. v varies inversely with the cube of r.

40. r varies directly with the square of s.

41. B varies jointly with m and n.

42. C varies jointly with x, y, and z.

43. P varies directly with the square of a, and inversely with the cube of j.

44. M varies inversely with the cube of n, and jointly with x and the square of z.

In Exercises 45–52, express each formula in words. In each formula, k is the constant of variation.

45. $L = kmn$

46. $P = \dfrac{km}{n}$

47. $E = kab^2$

48. $U = krs^2t$

49. $X = \dfrac{kx^2}{y^2}$ **50.** $Z = \dfrac{kw}{xy}$ **51.** $R = \dfrac{kL}{d^2}$ **52.** $e = \dfrac{kPL}{A}$

APPLICATIONS

53. Area of a circle   The area of a circle varies directly with the square of its radius. The constant of variation is π. Find the area of a circle with a radius of 6 inches.

54. Falling objects   An object in free fall travels a distance s that is directly proportional to the square of the time t. If an object falls 1,024 feet in 8 seconds, how far will it fall in 10 seconds?

55. Finding distance   The distance that a car can go is directly proportional to the number of gallons of gasoline it consumes. If a car can go 288 miles on 12 gallons of gasoline, how far can it go on a full tank of 18 gallons?

56. Farming   A farmer's harvest in bushels varies directly with the number of acres planted. If 8 acres can produce 144 bushels, how many acres are required to produce 1,152 bushels?

57. Farming   The length of time that a given number of bushels of corn will last when feeding cattle varies inversely with the number of animals. If x bushels will feed 25 cows for 10 days, how long will the feed last for 10 cows?

58. Geometry   For a fixed area, the length of a rectangle is inversely proportional to its width. A rectangle has a width of 18 feet and a length of 12 feet. If the length is increased to 16 feet, find the width.

59. Gas pressure   Under constant temperature, the volume occupied by a gas is inversely proportional to the pressure applied. If the gas occupies a volume of 20 cubic inches under a pressure of 6 pounds per square inch, find the volume when the gas is subjected to a pressure of 10 pounds per square inch.

60. Value of a car   The value of a car usually varies inversely with its age. If a car is worth $7,000 when

it is 3 years old, how much will it be worth when it is 7 years old?

61. Organ pipes   The frequency of vibration of air in an organ pipe is inversely proportional to the length of the pipe. (See Illustration 1.) If a pipe 2 feet long vibrates 256 times per second, how many times per second will a 6-foot pipe vibrate?

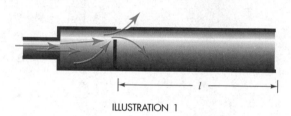

ILLUSTRATION 1

62. Geometry   The area of a rectangle varies jointly with its length and width. If both the length and the width are tripled, by what factor is the area multiplied?

63. Geometry   The volume of a rectangular solid varies jointly with its length, width, and height. If the length is doubled, the width is tripled, and the height is doubled, by what factor is the volume multiplied?

64. Costs of a trucking company   The costs incurred by a trucking company vary jointly with the number of trucks in service and the number of hours they are used. When 4 trucks are used for 6 hours each, the costs are $1,800. Find the costs of using 10 trucks, each for 12 hours.

65. Storing oil   The number of gallons of oil that can be stored in a cylindrical tank varies jointly with the height of the tank and the square of the radius of its base. The constant of proportionality is 23.5. Find the number of gallons that can be stored in the cylindrical tank in Illustration 2.

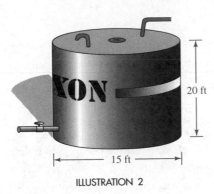

ILLUSTRATION 2

66. Finding the constant of variation A quantity l varies jointly with x and y and inversely with z. If the value of l is 30 when $x = 15$, $y = 5$, and $z = 10$, find k.

67. Electronics The voltage (in volts) measured across a resistor is directly proportional to the current (in amperes) flowing through the resistor. The constant of variation is the **resistance** (in ohms). If 6 volts is measured across a resistor carrying a current of 2 amperes, find the resistance.

68. Electronics The power (in watts) lost in a resistor (in the form of heat) is directly proportional to the square of the current (in amperes) passing through it. The constant of proportionality is the resistance (in ohms). What power is lost in a 5-ohm resistor carrying a 3-ampere current?

69. Building construction The deflection of a beam is inversely proportional to its width and the cube of its depth. If the deflection of a 4-inch-by-4-inch beam is 1.1 inches, find the deflection of a 2-inch-by-8-inch beam positioned as in Illustration 3.

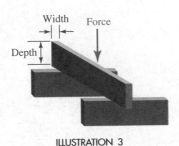

ILLUSTRATION 3

70. Building construction Find the deflection of the beam in Exercise 69 when the beam is positioned as in Illustration 4.

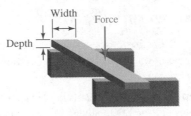

ILLUSTRATION 4

71. Gas pressure The pressure of a certain amount of gas is directly proportional to the temperature (measured in degrees Kelvin) and inversely proportional to the volume. A sample of gas at a pressure of 1 atmosphere occupies a volume of 1 cubic meter at a temperature of 273 Kelvin. When heated, the gas expands to twice its volume, but the pressure remains constant. To what temperature is it heated?

72. Tension A stone, twirled at the end of a string, is kept in its circular path by the tension of the string. The tension T is directly proportional to the square of the speed s and inversely proportional to the radius r of the circle. In Illustration 5, the tension is 32 pounds when the speed is 8 ft/sec and the radius is 6 feet. Find the tension when the speed is 4 ft/sec and the radius is 3 feet.

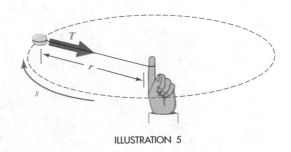

ILLUSTRATION 5

WRITING

73. Explain the terms *means* and *extremes*.

75. Explain the term *joint variation*.

74. Distinguish between a *ratio* and a *proportion*.

76. Explain why the equation $\frac{y}{x} = k$ indicates that y varies directly with x.

SOMETHING TO THINK ABOUT

77. As temperature increases on the Fahrenheit scale, it also increases on the Celsius scale. Is this direct variation? Explain.

79. Is a proportion useful for solving this problem?
   A water bill for 1,000 gallons was $15, and a bill for 2,000 gallons was $25. Find the bill for 3,000 gallons.
   Explain.

78. As the cost of a purchase (less than $5) increases, the amount of change received from a five-dollar bill decreases. Is this inverse variation? Explain.

80. How would you solve the problem in Exercise 79?

■ ■ ■ ■ ■ ■ ■ ■ ■   **PROJECTS**

PROJECT 1   The graph of a line is determined by two pieces of information. If we know the line's slope and its y-intercept, we would use the *slope–intercept* form to find the equation of the line. If we know the slope of the line and the coordinates of some point on that line, we would use the *point–slope* form. We have studied several standard forms of the equation of a line. Here is one more standard form that is useful when we know a line's x- and y-intercepts.

> **The Intercept Form of the Equation of a Line**
> The equation of a line with x-intercept $(a, 0)$ and y-intercept $(0, b)$ is
> $$\frac{x}{a} + \frac{y}{b} = 1$$

- Derive the intercept form of the equation of a line. (*Hint:* You know two points on the line.)
- Find the x- and y-intercepts of the line $\frac{x}{5} + \frac{y}{9} = 1$.
- Find the equation of the line with x-intercept $(3, 0)$ and y-intercept $(0, 7)$.
- Find the x- and y-intercepts of the line $4x + 5y = 20$ by writing the equation in intercept form.
- Graph the line $\frac{x}{k} + \frac{y}{k} = 1$ for five different values of k (your choice). What do these lines have in common?
- Graph the line $\frac{x}{3} + \frac{y}{k} = 1$ for five different values of k. What do these lines have in common?
- Can the equation of *every* line be written in intercept form? Discuss which lines can and which ones can't.

PROJECT 2   You are representing your branch of the large Buy-from-Us Corporation at the company's regional meeting, and you are looking forward to presenting your revenue and cost reports to the other branch representatives. But now disaster strikes! The graphs you had planned to present, containing cost and revenue information for this year and last year, are unlabeled! You cannot immediately recognize which graphs represent costs, which represent revenues, and which represent which year. Without these graphs, your presentation will not be effective.

   The only other information you have with you is in the notes you made for your talk. From these you are able to glean the following financial data about your branch.

1. All cost and revenue figures on the graphs are rounded to the nearest $50,000.

2. Costs for the fourth quarter of last year were $400,000.

3. Revenue was not above $400,000 for any quarter last year.

4. Last year, your branch lost money during the first quarter.

5. This year, your branch made money during three of the four quarters.

6. Profit during the second quarter of this year was $150,000.

And, of course, you know that profit = revenue − cost.

   With this information, you must match each of the graphs (Illustrations 1–4) with one of the following titles:

   Costs, This Year   Costs, Last Year

   Revenues, This Year   Revenues, Last Year

You should be sure to have sound reasons for your choices—reasons ensuring that no other arrangement of the titles will fit the data. The *last* thing you want to do is present incorrect information to the company bigwigs!

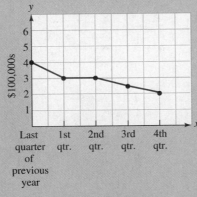

ILLUSTRATION 1

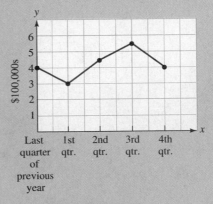

ILLUSTRATION 2

(continued)

■ ■ ■ ■ ■ ■ ■ ■ ■ **PROJECTS** *(continued)*

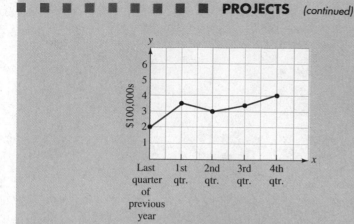

ILLUSTRATION 3

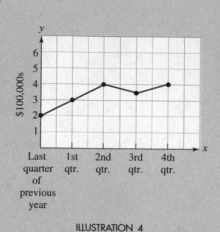

ILLUSTRATION 4

C H A P T E R   S U M M A R Y

CONCEPTS

REVIEW EXERCISES

SECTION 8.1

A Review of the Rectangular Coordinate System

The y-intercept of a line is the point $(0, b)$ where the line intersects the y-axis.

The x-intercept of a line is the point $(a, 0)$ where the line intersects the x-axis.

1. Graph each equation.

a. $x + y = 4$

b. $2x - y = 8$

c. $y = 3x + 4$

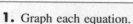

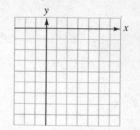

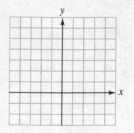

Graph of a vertical line:
  $x = a$
  x-intercept at $(a, 0)$

Graph of a horizontal line:
  $y = b$
  y-intercept at $(0, b)$

d. $x = 4 - 2y$

e. $y = 4$

f. $x = -2$

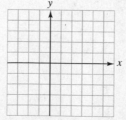

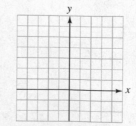

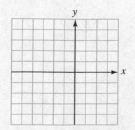

g. $2(x + 3) = x + 2$

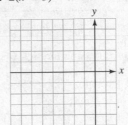

h. $3y = 2(y - 1)$

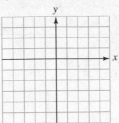

Midpoint formula:
If $P(x_1, y_1)$ and $Q(x_2, y_2)$ are two points on a line, the midpoint of PQ is

$$M\left(\frac{x_1 + x_2}{2}, \frac{y_1 + y_2}{2}\right)$$

2. Find the midpoint of the line segment joining $P(-3, 5)$ and $Q(6, 11)$.

SECTION 8.2

Slope of a Nonvertical Line

Slope of a nonvertical line:
If $x_2 \neq x_1$,

$$m = \frac{\Delta y}{\Delta x} = \frac{y_2 - y_1}{x_2 - x_1}$$

Horizontal lines have a slope of 0. Vertical lines have no defined slope.

3. Find the slope of the line passing through points P and Q, if possible.
   a. $P(2, 5)$ and $Q(5, 8)$
   b. $P(-3, -2)$ and $Q(6, 12)$
   c. $P(-3, 4)$ and $Q(-5, -6)$
   d. $P(5, -4)$ and $Q(-6, -9)$
   e. $P(-2, 4)$ and $Q(8, 4)$
   f. $P(-5, -4)$ and $Q(-5, 8)$

4. Find the slope of the graph of each equation, if one exists.
   a. $2x - 3y = 18$
   b. $2x + y = 8$
   c. $-2(x - 3) = 10$
   d. $3y + 1 = 7$

Parallel lines have the same slope. The slopes of two nonvertical perpendicular lines are negative reciprocals.

5. Tell whether the lines with the given slopes are parallel, perpendicular, or neither.
   a. $m_1 = 4$, $m_2 = -\dfrac{1}{4}$

   b. $m_1 = 0.5$, $m_2 = \dfrac{1}{2}$

   c. $m_1 = 0.5$, $m_2 = -\dfrac{1}{2}$

   d. $m_1 = 5$, $m_2 = -0.2$

6. If the sales of a new business were \$65,000 in its first year and \$130,000 in its fourth year, find the rate of growth in sales per year.

| **SECTION 8.3** | *Writing Equations of Lines* |

Equations of a line:

Point–slope form:

$$y - y_1 = m(x - x_1)$$

Slope–intercept form:

$$y = mx + b$$

General form:

$$Ax + By = C$$

7. Write the equation of the line with the given properties. Write the equation in general form.
   a. Slope of 3; passing through $P(-8, 5)$
   b. Passing through $(-2, 4)$ and $(6, -9)$
   c. Passing through $(-3, -5)$; parallel to the graph of $3x - 2y = 7$
   d. Passing through $(-3, -5)$; perpendicular to the graph of $3x - 2y = 7$

8. Use the slope of a line to help graph the function $y = \dfrac{2}{3}x + 1$.

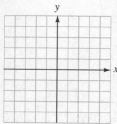

9. Are the lines represented by $2x + 3y = 8$ and $3x - 2y = 10$ parallel or perpendicular?

10. A business purchased a copy machine for \$8,700 and will depreciate it on a straight-line basis over the next 5 years. At the end of its useful life, it will be sold as scrap for \$100. Find its depreciation equation.

| **SECTION 8.4** | *A Review of Functions* |

A **function** is a correspondence between a set of input values x and a set of output values y, where exactly one value of y in the range corresponds to each number x in the domain.

$f(k)$ represents the value of $f(x)$ when $x = k$.

11. Tell whether each equation determines y to be a function of x.
   a. $y = 6x - 4$ **b.** $y = 4 - x$
   c. $y^2 = x$ **d.** $|y| = x^2$

12. Assume that $f(x) = 3x + 2$ and $g(x) = x^2 - 4$ and find each value.
   a. $f(-3)$ **b.** $g(8)$
   c. $g(-2)$ **d.** $f(5)$

The **domain** of a function is the set of input values. The **range** is the set of output values.

13. Find the domain and range of each function.
  a. $f(x) = 4x - 1$
  b. $f(x) = 3x - 10$
  c. $f(x) = x^2 + 1$
  d. $f(x) = \dfrac{4}{2 - x}$
  e. $f(x) = \dfrac{7}{x - 3}$
  f. $y = 7$

The **vertical line test** can be used to determine whether a graph represents a function.

14. Use the vertical line test to determine whether each graph represents a function.
  a.

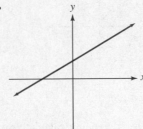

  b.

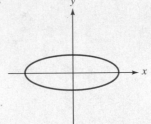

  c.

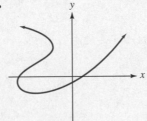

  d.

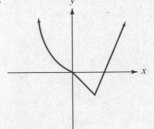

SECTION 8.5   *Graphs of Nonlinear Functions*

Graphs of nonlinear equations are not lines.

15. Graph each function.
  a. $f(x) = x^2 - 3$

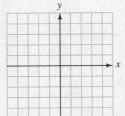

  b. $f(x) = |x| - 4$

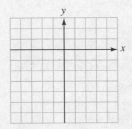

c. $f(x) = (x - 2)^3$

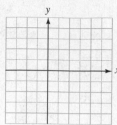

d. $f(x) = (x + 4)^2 - 3$

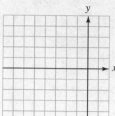

e. $f(x) = -x^3 - 2$

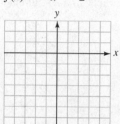

f. $f(x) = -|x - 1| + 2$

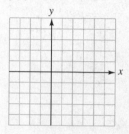

16. Use a graphing calculator to graph each function. Compare the results in Problem 15.

a. $f(x) = x^2 - 3$

b. $f(x) = |x| - 4$

c. $f(x) = (x - 2)^3$

d. $f(x) = (x + 4)^2 - 3$

e. $f(x) = -x^3 - 2$

f. $f(x) = -|x - 1| + 2$

17. Use a graphing calculator to graph each rational function and find its domain and range.

a. $f(x) = \dfrac{2}{x - 2}$

b. $f(x) = \dfrac{x}{x + 3}$

SECTION 8.6   *Variation*

In any proportion, the product of the extremes is equal to the product of the means.

18. Solve each proportion.

a. $\dfrac{x + 1}{8} = \dfrac{4x - 2}{23}$

b. $\dfrac{1}{x + 6} = \dfrac{x + 10}{12}$

Direct variation:
$$y = kx$$

Inverse variation:
$$y = \frac{k}{x}$$

Joint variation:
$$y = kxz$$

19. Assume that y varies directly with x. If $x = 12$ when $y = 2$, find the value of y when $x = 12$.

20. Assume that y varies inversely with x. If $x = 24$ when $y = 3$, find the value of y when $x = 12$.

21. Assume that y varies jointly with x and z. Find the constant of variation if $x = 24$ when $y = 3$ and $z = 4$.

22. Assume that y varies directly with t and inversely with x. Find the constant of variation if $x = 2$ when $t = 8$ and $y = 64$.

■ Chapter Test

1. Graph the equation $2x - 5y = 10$.

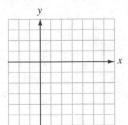

2. Find the midpoint of the line segment shown in Illustration 1.

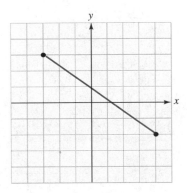

ILLUSTRATION 1

3. Find the x- and y-intercepts of the graph of $y = \frac{x-3}{5}$.

4. Is the graph of $x - 7 = 0$ a horizontal or a vertical line?

In Problems 5–8, find the slope of each line, if possible.

5. The line through $P(-2, 4)$ and $Q(6, 8)$

6. The graph of $2x - 3y = 8$

7. The graph of $x = 12$.

8. The graph of $y = 12$.

9. Write the equation of the line with slope of $\frac{2}{3}$ that passes through $P(4, -5)$. Give the answer in slope–intercept form.

10. Write the equation of the line that passes through $P(-2, 6)$ and $Q(-4, -10)$. Give the answer in general form.

11. Find the slope and the y-intercept of the graph of $-2(x - 3) = 3(2y + 5)$.

12. Determine whether the graphs of $4x - y = 12$ and $y = \frac{1}{4}x + 3$ are parallel, perpendicular, or neither.

13. Determine whether the graphs of $y = -\frac{2}{3}x + 4$ and $2y = 3x - 3$ are parallel, perpendicular, or neither.

14. Write the equation of the line that passes through the origin and is parallel to the graph of $y = \frac{3}{2}x - 7$.

15. Write the equation of the line that passes through $P(-3, 6)$ and is perpendicular to the graph of $y = -\frac{2}{3}x - 7$.

16. Does $|y| = x$ define y to be a function of x?

17. Find the domain and range of the function $f(x) = |x|$.

18. Find the domain and range of the function $f(x) = x^3$.

In Problems 19–22, $f(x) = 3x + 1$ and $g(x) = x^2 - 2$. Find each value.

19. $f(3)$

20. $g(0)$

21. $f(a)$

22. $g(-x)$

In Problems 23–24, tell whether each graph represents a function.

23.

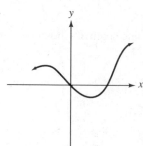

24.

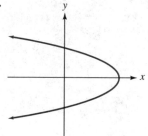

25. Graph $f(x) = x^2 - 1$.

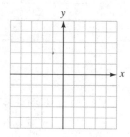

26. Graph $f(x) = -|x + 2|$.

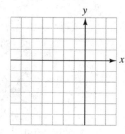

27. Solve the proportion $\dfrac{3}{x - 2} = \dfrac{x + 3}{2x}$.

28. Assume that y varies directly with x. If $x = 30$ when $y = 4$, find y when $x = 9$.

29. Assume that V varies inversely with t. If $V = 55$ when $t = 20$, find t when $V = 75$.

30. Does the graph define a function?

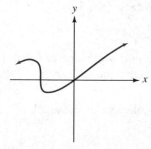

■ Cumulative Review Exercises

In Exercises 1–10, tell which numbers in the set $\left\{-2, 0, 1, 2, \frac{13}{12}, 6, 7, \sqrt{5}, \pi\right\}$ are in each category.

1. Natural numbers

2. Whole numbers

3. Rational numbers

4. Irrational numbers

5. Negative numbers

6. Real numbers

7. Prime numbers

8. Composite numbers

9. Even numbers

10. Odd numbers

In Exercises 11–12, graph each interval on the number line.

11. $-2 < x \le 5$

12. $[-5, 0) \cup [3, 6]$

In Exercises 13–14, simplify each expression.

13. $-|5| + |-3|$

14. $\dfrac{|-5| + |-3|}{-|4|}$

In Exercises 15–18, do the operations.

15. $2 + 4 \cdot 5$

16. $\dfrac{8 - 4}{2 - 4}$

17. $20 \div (-10 \div 2)$

18. $\dfrac{6 + 3(6 + 4)}{2(3 - 9)}$

In Exercises 19–20, $x = 2$ and $y = -3$. Evaluate each expression.

19. $-x - 2y$

20. $\dfrac{x^2 - y^2}{2x + y}$

In Exercises 21–24, tell which property of real numbers justifies each statement.

21. $(a + b) + c = a + (b + c)$

22. $3(x + y) = 3x + 3y$

23. $(a + b) + c = c + (a + b)$

24. $(ab)c = a(bc)$

In Exercises 25–28, simplify each expression. Assume that all variables are positive numbers and write all answers without negative exponents.

25. $(x^2 y^3)^4$

26. $\dfrac{c^4 c^8}{(c^5)^2}$

27. $\left(-\dfrac{a^3 b^{-2}}{ab}\right)^{-1}$

28. $\left(\dfrac{-3a^3 b^{-2}}{6a^{-2} b^3}\right)^0$

29. Change 0.00000497 to scientific notation.

30. Change 9.32×10^8 to standard notation.

In Exercises 31–34, solve each equation.

31. $2x - 5 = 11$

32. $\dfrac{2x - 6}{3} = x + 7$

33. $4(y - 3) + 4 = -3(y + 5)$

34. $2x - \dfrac{3(x - 2)}{2} = 7 - \dfrac{x - 3}{3}$

In Exercises 35–36, solve each formula for the indicated variable.

35. $S = \dfrac{n(a + l)}{2}$ for a

36. $A = \dfrac{1}{2}h(b_1 + b_2)$ for h

37. The sum of three consecutive even integers is 90. Find the integers.

38. A rectangle is three times as long as it is wide. If its perimeter is 112 centimeters, find its dimensions.

39. Tell whether the graph of $2x - 3y = 6$ defines a function.

40. Find the slope of a line passing through $P(-2, 5)$ and $Q(8, -9)$.

41. Write the equation of the line passing through $P(-2, 5)$ and $Q(8, -9)$.

42. Write the equation of the line passing through $P(-2, 3)$ and parallel to the graph of $3x + y = 8$.

In Exercises 43–46, $f(x) = 3x^2 + 2$ and $g(x) = 2x - 1$. Evaluate each expression.

43. $f(-1)$

44. $g(0)$

45. $g(t)$

46. $f(-r)$

In Exercises 47–48, graph each equation and tell whether it is a function. If it is a function, give the domain and range.

47. $y = -x^2 + 1$

48. $y = \left| \dfrac{1}{2}x - 3 \right|$

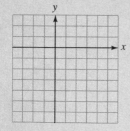

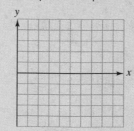

9 Rational Exponents and Radicals

MATHEMATICS IN
THE WORKPLACE

Photographer

Photographers use cameras and film to portray people, objects, places, and events. Some specialize in scientific, medical, or engineering photography and provide illustrations and documentation for publications and research reports. Others specialize in portrait, fashion, or industrial photography and provide pictures for catalogs and other publications. Photojournalists capture newsworthy events, people, and places, and their work is seen in newspapers and magazines, as well as on television.

SAMPLE APPLICATION ■ Many cameras have an adjustable lens opening called an *aperture*. The aperture controls the amount of light passing through the lens. Various lenses—wide-angle, close-up, and telephoto—are distinguished by their *focal length*. The *f-number* of a lens is its focal length divided by the diameter of its circular aperture:

$$f\text{-number} = \frac{f}{d} \qquad f \text{ is the focal length, and } d \text{ is the diameter of the aperture.}$$

A lens with a focal length of 12 centimeters and an aperture with a diameter of 6 centimeters has an *f*-number of $\frac{12}{6}$ and is called an *f*/2 lens. If the area of the aperture is reduced to admit half as much light, the *f*-number of the lens will change. Find the new *f*-number.
See Section 9.6, Example 5.

9.1 Radical Expressions

■ SQUARE ROOTS ■ SQUARE ROOTS OF EXPRESSIONS WITH VARIABLES ■ THE SQUARE ROOT FUNCTION ■ CUBE ROOTS ■ nTH ROOTS ■ STANDARD DEVIATION

Getting Ready   *Find each power.*

1. 0^2 **2.** 4^2 **3.** $(-4)^2$ **4.** -4^2

5. $\left(\dfrac{2}{5}\right)^3$ **6.** $\left(-\dfrac{3}{4}\right)^4$ **7.** $(7xy)^2$ **8.** $(7xy)^3$

In this chapter, we will reverse the squaring process and learn how to find **square roots** of numbers. We will also discuss how to find other roots of numbers.

■ SQUARE ROOTS

When solving problems, we must often find what number must be squared to obtain a second number a. If such a number can be found, it is called a **square root of a**. For example,

- 0 is a square root of 0, because $0^2 = 0$.
- 4 is a square root of 16, because $4^2 = 16$.

- -4 is a square root of 16, because $(-4)^2 = 16$.
- $7xy$ is a square root of $49x^2y^2$, because $(7xy)^2 = 49x^2y^2$.
- $-7xy$ is a square root of $49x^2y^2$, because $(-7xy)^2 = 49x^2y^2$.

The preceding examples illustrate the following definition.

Square Root of a
The number b is a **square root of a** if $b^2 = a$.

All positive numbers have two real number square roots: one that is positive and one that is negative.

EXAMPLE 1 Find the two square roots of 121.

Solution The two square roots of 121 are 11 and -11, because

$$11^2 = 121 \quad \text{and} \quad (-11)^2 = 121 \qquad \blacksquare$$

Self Check Find the square roots of 144.
Answers $12, -12$

In the following definition, the symbol $\sqrt{}$ is called a **radical sign,** and the number x within the radical sign is called the **radicand.**

Principal Square Root
If $x > 0$, the **principal square root of x** is the positive square root of x, denoted as $\sqrt{x}$.
The principal square root of 0 is 0: $\sqrt{0} = 0$.

By definition, the principal square root of a positive number is always positive. Although 5 and -5 are both square roots of 25, only 5 is the principal square root. The radical $\sqrt{25}$ represents 5. The radical $-\sqrt{25}$ represents -5.

EXAMPLE 2 Simplify each radical.

a. $\sqrt{1} = 1$ **b.** $\sqrt{81} = 9$
c. $-\sqrt{81} = -9$ **d.** $-\sqrt{225} = -15$
e. $\sqrt{\dfrac{1}{4}} = \dfrac{1}{2}$ **f.** $-\sqrt{\dfrac{16}{121}} = -\dfrac{4}{11}$
g. $\sqrt{0.04} = 0.2$ **h.** $-\sqrt{0.0009} = -0.03$ $\qquad \blacksquare$

Numbers such as 1, 4, 9, 16, 49, and 1,600 are called **integer squares,** because each one is the square of an integer. The square root of every integer square is a rational number.

$$\sqrt{1} = 1, \qquad \sqrt{4} = 2, \qquad \sqrt{9} = 3, \qquad \sqrt{16} = 4, \qquad \sqrt{49} = 7, \qquad \sqrt{1,600} = 40$$

The square roots of many positive integers are not rational numbers. For example, $\sqrt{11}$ is an **irrational number.** To find an approximate value of $\sqrt{11}$, we enter 11 into a scientific calculator and press the $\boxed{\sqrt{}}$ key.

$$\sqrt{11} \approx 3.31662479 \qquad \text{Read} \approx \text{as "is approximately equal to."}$$

Square roots of negative numbers are not real numbers. For example, $\sqrt{-9}$ is not a real number, because no real number squared equals -9. Square roots of negative numbers come from a set called the **imaginary numbers,** which we will discuss in the next chapter.

■ **SQUARE ROOTS OF EXPRESSIONS WITH VARIABLES**

If $x \neq 0$, the positive number x^2 has x and $-x$ for its two square roots. To denote the positive square root of $\sqrt{x^2}$, we must know whether x is positive or negative.

If $x > 0$, we can write

$$\sqrt{x^2} = x \qquad \sqrt{x^2} \text{ represents the positive square root of } x^2, \text{ which is } x.$$

If x is negative, then $-x > 0$, and we can write

$$\sqrt{x^2} = -x \qquad \sqrt{x^2} \text{ represents the positive square root of } x^2, \text{ which is } -x.$$

If we don't know whether x is positive or negative, we must use absolute value symbols to guarantee that $\sqrt{x^2}$ is positive.

Definition of $\sqrt{x^2}$
If x can be any real number, then
$$\sqrt{x^2} = |x|$$

EXAMPLE 3 Simplify each expression.

If x can be any real number, we have

a. $\sqrt{16x^2} = \sqrt{(4x)^2}$ Write $16x^2$ as $(4x)^2$.

$\qquad\quad = |4x|$ Because $(|4x|)^2 = 16x^2$. Since x could be negative, absolute value symbols are needed.

$\qquad\quad = 4|x|$ Since 4 is a positive constant in the product $4x$, we can write it outside the absolute value symbols.

b. $\sqrt{x^2 + 2x + 1}$

$\qquad = \sqrt{(x + 1)^2}$ Factor $x^2 + 2x + 1$.

$\qquad = |x + 1|$ Because $(x + 1)^2 = x^2 + 2x + 1$. Since $x + 1$ can be negative (for example, when $x = -5$), absolute value symbols are needed.

c. $\sqrt{x^4} = x^2$ Because $(x^2)^2 = x^4$. Since $x^2 \geq 0$, no absolute value symbols are needed. ∎

Self Check Simplify **a.** $\sqrt{25a^2}$ and **b.** $\sqrt{16a^4}$.
Answers **a.** $5|a|$, **b.** $4a^2$

■ THE SQUARE ROOT FUNCTION

Since there is one principal square root for every nonnegative real number x, the equation $y = f(x) = \sqrt{x}$ determines a function, called the **square root function.**

EXAMPLE 4 Graph $y = f(x) = \sqrt{x}$ and find its domain and range.

Solution We can make a table of values and plot points to get the graph shown in Figure 9-1(a), or we can use a graphing calculator with window settings of $[-1, 9]$ for x and $[-2, 5]$ for y to get the graph shown in Figure 9-1(b). Since the equation defines a function, its graph passes the vertical line test.

$y = f(x) = \sqrt{x}$

x	$f(x)$	$(x, f(x))$
0	0	(0, 0)
1	1	(1, 1)
4	2	(4, 2)
9	3	(9, 3)

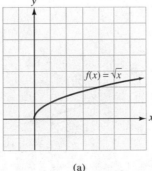

(a)

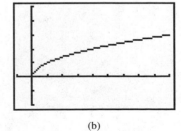

(b)

FIGURE 9-1

From either graph, we can see that the domain and the range are the set of non-negative real numbers, which is the interval $[0, \infty)$.   ■

Self Check   Graph $y = f(x) = \sqrt{x} + 2$ and compare the graph to the graph of $y = f(x) = \sqrt{x}$.

Answer   It is 2 units higher.

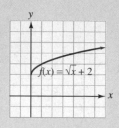

The graphs of many functions are translations or reflections of the square root function. For example, if $k > 0$

- The graph of $f(x) = \sqrt{x} + k$ is the graph of $f(x) = \sqrt{x}$ translated k units up.
- The graph of $f(x) = \sqrt{x} - k$ is the graph of $f(x) = \sqrt{x}$ translated k units down.
- The graph of $f(x) = \sqrt{x + k}$ is the graph of $f(x) = \sqrt{x}$ translated k units to the left.
- The graph of $f(x) = \sqrt{x - k}$ is the graph of $f(x) = \sqrt{x}$ translated k units to the right.
- The graph of $f(x) = -\sqrt{x}$ is the graph of $f(x) = \sqrt{x}$ reflected about the x-axis.

EXAMPLE 5   Graph $y = f(x) = -\sqrt{x + 4} - 2$ and find its domain and range.

Solution   This graph will be the reflection of $f(x) = \sqrt{x}$ about the x-axis, translated 4 units to the left and 2 units down. See Figure 9-2(a). We can confirm this graph by using a graphing calculator with window settings of $[-5, 6]$ for x and $[-6, 2]$ for y to get the graph shown in Figure 9-2(b).

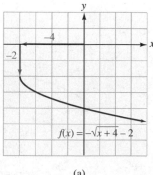

(a)   (b)

FIGURE 9-2

From either graph, we can see that the domain is the interval $[-4, \infty)$ and that the range is the interval $(-\infty, -2]$. ∎

Self Check

Graph $f(x) = \sqrt{x-2} - 4$.

Answer

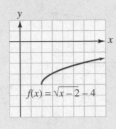

EXAMPLE 6

Period of a pendulum The **period of a pendulum** is the time required for the pendulum to swing back and forth to complete one cycle. (See Figure 9-3.) The period t (in seconds) is a function of the pendulum's length l, which is defined by the formula

$$t = f(l) = 2\pi \sqrt{\frac{l}{32}}$$

Find the period of a pendulum that is 5 feet long.

Solution   We substitute 5 for l in the formula and simplify.

$$t = 2\pi \sqrt{\frac{l}{32}}$$

$$t = 2\pi \sqrt{\frac{5}{32}}$$

$$\approx 2.483647066$$

To the nearest tenth, the period is 2.5 seconds. ∎

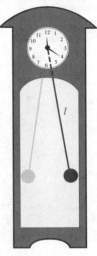

FIGURE 9-3

Self Check

To the nearest hundredth, find the period of a pendulum that is 3 feet long.

Answer   1.92 sec

■ ■ ■ ■ ■ ■ ■ ■ ■ ■

GRAPHING CALCULATORS

To solve Example 6 with a graphing calculator with window settings of $[-2, 10]$ for x and $[-2, 10]$ for y, we graph the function $f(x) = 2\pi \sqrt{\frac{x}{32}}$, as in Figure 9-4(a). We then trace and move the cursor toward an x value of 5 until we see the coordinates shown in Figure 9-4(b). The period is given by the y value shown in the screen. By zooming in, we can get better results.

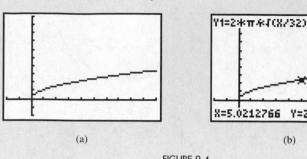

(a)   (b)

FIGURE 9-4

■ CUBE ROOTS

The **cube root of x** is any number whose cube is x. For example,

4 is a cube root of 64, because $4^3 = 64$.

$3x^2y$ is a cube root of $27x^6y^3$, because $(3x^2y)^3 = 27x^6y^3$.

$-2y$ is a cube root of $-8y^3$, because $(-2y)^3 = -8y^3$.

> **Cube Roots**
> The **cube root of x** is denoted as $\sqrt[3]{x}$ and is defined by
> $$\sqrt[3]{x} = y \quad \text{if} \quad y^3 = x$$

We note that 64 has two real-number square roots, 8 and -8. However, 64 has only one real-number cube root, 4, because 4 is the only real number whose cube is 64. Since every real number has exactly one real cube root, it is unnecessary to use absolute value symbols when simplifying cube roots.

> **Definition of $\sqrt[3]{x^3}$**
> If x is any real number, then
> $$\sqrt[3]{x^3} = x$$

EXAMPLE 7   Simplify each radical.

a. $\sqrt[3]{125} = 5$ Because $5^3 = 5 \cdot 5 \cdot 5 = 125$.

b. $\sqrt[3]{\dfrac{1}{8}} = \dfrac{1}{2}$ Because $\left(\dfrac{1}{2}\right)^3 = \dfrac{1}{2} \cdot \dfrac{1}{2} \cdot \dfrac{1}{2} = \dfrac{1}{8}$.

c. $\sqrt[3]{-27x^3} = -3x$ Because $(-3x)^3 = (-3x)(-3x)(-3x) = -27x^3$.

d. $\sqrt[3]{-\dfrac{8a^3}{27b^3}} = -\dfrac{2a}{3b}$ Because $\left(-\dfrac{2a}{3b}\right)^3 = \left(-\dfrac{2a}{3b}\right)\left(-\dfrac{2a}{3b}\right)\left(-\dfrac{2a}{3b}\right) = -\dfrac{8a^3}{27b^3}$.

e. $\sqrt[3]{0.216x^3y^6} = 0.6xy^2$ Because $(0.6xy^2)^3 = (0.6xy^2)(0.6xy^2)(0.6xy^2) = 0.216x^3y^6$.   ■

b,d

The equation $y = f(x) = \sqrt[3]{x}$ defines a **cube root function.** From the graph shown in Figure 9-5(a), we can see that the domain and range of the function $f(x) = \sqrt[3]{x}$ are the set of real numbers. Note that the graph of $f(x) = \sqrt[3]{x}$ passes the vertical line test. Figures 9-5(b) and 9-5(c) show several translations of the cube root function.

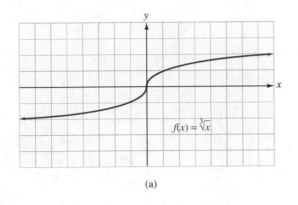

(a)

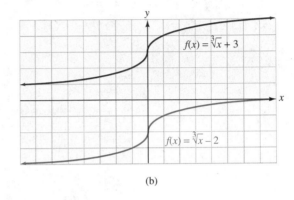

(b)

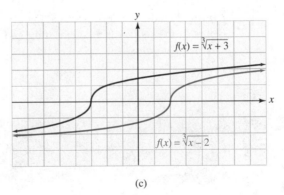

(c)

FIGURE 9-5

■ nTH ROOTS

Just as there are square roots and cube roots, there are fourth roots, fifth roots, sixth roots, and so on.

When n is an odd natural number, the radical $\sqrt[n]{x}$ $(n > 1)$ represents an **odd root.** Since every real number has just one real nth root when n is odd, we don't need to worry about absolute value symbols when finding odd roots. For example,

$$\sqrt[5]{243} = \sqrt[5]{3^5} = 3 \quad \text{because} \quad 3^5 = 243$$

$$\sqrt[7]{-128x^7} = \sqrt[7]{(-2x)^7} = -2x \quad \text{because} \quad (-2x)^7 = -128x^7$$

When n is an even natural number, the radical $\sqrt[n]{x}$ $(n > 1)$ represents an **even root.** In this case, there will be one positive and one negative real nth root. For example, the two real sixth roots of 729 are 3 and -3, because $3^6 = 729$ and $(-3)^6 = 729$. When finding even roots, we often use absolute value symbols to guarantee that the principal nth root is positive.

$$\sqrt[4]{(-3)^4} = |-3| = 3$$

We could also simplify this as follows: $\sqrt[4]{(-3)^4} = \sqrt[4]{81} = 3$.

$$\sqrt[6]{729x^6} = \sqrt[6]{(3x)^6} = |3x| = 3|x|$$

The absolute value symbols guarantee that the sixth root is positive.

In general, we have the following rules.

Rules for $\sqrt[n]{x^n}$

If x is a real number and $n > 1$, then

If n is an odd natural number, then $\sqrt[n]{x^n} = x$.

If n is an even natural number, then $\sqrt[n]{x^n} = |x|$.

In the radical $\sqrt[n]{x}$, n is called the **index** (or **order**) of the radical. When the index is 2, the radical is a square root, and we usually do not write the index.

$$\sqrt[2]{x} = \sqrt{x}$$

WARNING! When n is even $(n > 1)$ and $x < 0$, $\sqrt[n]{x}$ is not a real number. For example, $\sqrt[4]{-81}$ is not a real number, because no real number raised to the 4th power is -81.

EXAMPLE 8   Simplify each radical.

a. $\sqrt[4]{625} = 5$, because $5^4 = 625$

Read $\sqrt[4]{625}$ as "the fourth root of 625."

b. $\sqrt[5]{-32} = -2$, because $(-2)^5 = -32$

Read $\sqrt[5]{-32}$ as "the fifth root of -32."

c. $\sqrt[6]{\dfrac{1}{64}} = \dfrac{1}{2}$, because $\left(\dfrac{1}{2}\right)^6 = \dfrac{1}{64}$

Read $\sqrt[6]{\frac{1}{64}}$ as "the sixth root of $\frac{1}{64}$."

d. $\sqrt[7]{10^7} = 10$, because $10^7 = 10^7$

Read $\sqrt[7]{10^7}$ as "the seventh root of 10^7."

b,d

EXAMPLE 9 Simplify each radical. Assume that x can be any real number.

Solution **a.** $\sqrt[5]{x^5} = x$ Since n is odd, absolute value symbols aren't needed.

b. $\sqrt[4]{16x^4} = |2x| = 2|x|$ Since n is even and x can be negative, absolute value symbols are needed to guarantee that the result is positive.

c. $\sqrt[6]{(x + 4)^6} = |x + 4|$ Absolute value symbols are needed to guarantee that the result is positive.

d. $\sqrt[3]{(x + 1)^3} = x + 1$ Since n is odd, absolute value symbols aren't needed.

e. $\sqrt{(x^2 + 4x + 4)^2} = \sqrt{[(x + 2)^2]^2}$ Factor $x^2 + 4x + 4$.
$$= \sqrt{(x + 2)^4}$$
$$= (x + 2)^2$$ Since $(x + 2)^2$ is always positive, absolute value symbols aren't needed. ■

We summarize the definitions concerning $\sqrt[n]{x}$ as follows.

Summary of the Definitions of $\sqrt[n]{x}$

If n is a natural number greater than 1 and x is a real number, then

If $x > 0$, then $\sqrt[n]{x}$ is the positive number such that $\left(\sqrt[n]{x}\right)^n = x$.

If $x = 0$, then $\sqrt[n]{x} = 0$.

If $x < 0$ $\begin{cases} \text{and } n \text{ is odd, then } \sqrt[n]{x} \text{ is the real number such that } \left(\sqrt[n]{x}\right)^n = x. \\ \text{and } n \text{ is even, then } \sqrt[n]{x} \text{ is not a real number.} \end{cases}$

■ STANDARD DEVIATION

In statistics, the *standard deviation* is used to tell which of a set of distributions is the most variable. To see how to compute the standard deviation of a distribution, we consider the distribution 4, 5, 5, 8, 13 and construct the following table.

Original terms	Mean of the distribution	Differences (original term minus mean)	Squares of the differences from the mean
4	7	−3	9
5	7	−2	4
5	7	−2	4
8	7	1	1
13	7	6	36

The **standard deviation** of the distribution is the positive square root of the mean of the numbers shown in column 4 of the table.

$$\text{Standard deviation} = \sqrt{\frac{\text{sum of the squares of the differences from the mean}}{\text{number of differences}}}$$

$$= \sqrt{\frac{9 + 4 + 4 + 1 + 36}{5}}$$

$$= \sqrt{\frac{54}{5}}$$

$$\approx 3.286335345$$

To the nearest hundredth, the standard deviation of the given distribution is 3.29. The symbol for standard deviation is σ, the lowercase Greek letter *sigma*.

EXAMPLE 10   Which of the following distributions has the most variability: **a.** 3, 5, 7, 8, 12   or **b.** 1, 4, 6, 11?

Solution   We compute the standard deviation of each distribution.

a.

Original terms	Mean of the distribution	Differences (original term minus mean)	Squares of the differences from the mean
3	7	−4	16
5	7	−2	4
7	7	0	0
8	7	1	1
12	7	5	25

$$\sigma = \sqrt{\frac{16 + 4 + 0 + 1 + 25}{5}} = \sqrt{\frac{46}{5}} \approx 3.03$$

b.

Original terms	Mean of the distribution	Differences (original term minus mean)	Squares of the differences from the mean
1	5.5	−4.5	20.25
4	5.5	−1.5	2.25
6	5.5	0.5	0.25
11	5.5	5.5	30.25

$$\sigma = \sqrt{\frac{20.25 + 2.25 + 0.25 + 30.25}{4}} = \sqrt{\frac{53}{4}} \approx 3.64$$

Since the standard deviation for the second distribution is greater than the standard deviation for the first distribution, the second distribution has the greater variability. ∎

Orals *Simplify each radical, if possible.*

1. $\sqrt{9}$ **2.** $-\sqrt{16}$ **3.** $\sqrt[3]{-8}$ **4.** $\sqrt[5]{32}$

5. $\sqrt{64x^2}$ **6.** $\sqrt[3]{-27x^3}$

7. $\sqrt{-3}$ **8.** $\sqrt[4]{(x+1)^8}$

EXERCISE 9.1

REVIEW *Simplify each fraction.*

1. $\dfrac{x^2 + 7x + 12}{x^2 - 16}$

2. $\dfrac{a^3 - b^3}{b^2 - a^2}$

Do the operations.

3. $\dfrac{x^2 - x - 6}{x^2 - 2x - 3} \cdot \dfrac{x^2 - 1}{x^2 + x - 2}$

4. $\dfrac{x^2 - 3x - 4}{x^2 - 5x + 6} \div \dfrac{x^2 - 2x - 3}{x^2 - x - 2}$

5. $\dfrac{3}{m+1} + \dfrac{3m}{m-1}$

6. $\dfrac{2x+3}{3x-1} - \dfrac{x-4}{2x+1}$

VOCABULARY AND CONCEPTS *Fill in each blank to make a true statement.*

7. $5x^2$ is the square root of $25x^4$, because _____ $= 25x^4$.

8. b is a square root of a if _____.

9. The principal square root of x $(x > 0)$ is the _____ square root of x.

10. $\sqrt{x^2} =$ ____

11. The graph of $f(x) = \sqrt{x} + 3$ is the graph of $f(x) = \sqrt{x}$ translated ___ units ____.

12. The graph of $f(x) = \sqrt{x + 5}$ is the graph of $f(x) = \sqrt{x}$ translated ___ units to the ____.

13. $\sqrt[3]{x} = y$ if _____.

14. $\sqrt[3]{x^3} =$ ___

15. When n is an odd number, $\sqrt[n]{x}$ represents an _____ root.

16. When n is an _____ number, $\sqrt[n]{x}$ represents an even root.

17. $\sqrt{0} =$ ___

18. The _____ deviation of a set of numbers is the positive square root of the squares of the differences of the numbers and their mean.

In Exercises 19–22, identify the radicand in each expression.

19. $\sqrt{3x^2}$

20. $5\sqrt{x}$

21. $ab^2\sqrt{a^2 + b^3}$

22. $\dfrac{1}{2}x\sqrt{\dfrac{x}{y}}$

PRACTICE   *In Exercises 23–38, find each square root, if possible.*

23. $\sqrt{121}$

24. $\sqrt{144}$

25. $-\sqrt{64}$

26. $-\sqrt{1}$

27. $\sqrt{\dfrac{1}{9}}$

28. $-\sqrt{\dfrac{4}{25}}$

29. $-\sqrt{\dfrac{25}{49}}$

30. $\sqrt{\dfrac{49}{81}}$

31. $\sqrt{-25}$

32. $\sqrt{0.25}$

33. $\sqrt{0.16}$

34. $\sqrt{-49}$

35. $\sqrt{(-4)^2}$

36. $\sqrt{(-9)^2}$

37. $\sqrt{-36}$

38. $-\sqrt{-4}$

⊞ *In Exercises 39–42, use a calculator to find each square root. Give the answer to four decimal places.*

39. $\sqrt{12}$

40. $\sqrt{340}$

41. $\sqrt{679.25}$

42. $\sqrt{0.0063}$

In Exercises 43–50, find each square root. Assume that all variables are unrestricted, and use absolute value symbols when necessary.

43. $\sqrt{4x^2}$

44. $\sqrt{16y^4}$

45. $\sqrt{(t + 5)^2}$

46. $\sqrt{(a + 6)^2}$

47. $\sqrt{(-5b)^2}$

48. $\sqrt{(-8c)^2}$

49. $\sqrt{a^2 + 6a + 9}$

50. $\sqrt{x^2 + 10x + 25}$

In Exercises 51–54, find each value given that $f(x) = \sqrt{x - 4}$.

51. $f(4)$

52. $f(8)$

53. $f(20)$

54. $f(29)$

⊞ *In Exercises 55–58, find each value given that $f(x) = \sqrt{x^2 + 1}$. Give each answer to four decimal places.*

55. $f(4)$

56. $f(6)$

57. $f(2.35)$

58. $f(21.57)$

In Exercises 59–62, graph each function and find its domain and range.

59. $f(x) = \sqrt{x} + 4$

60. $f(x) = -\sqrt{x} - 2$

61. $f(x) = -\sqrt{x} - 3$

62. $f(x) = \sqrt[3]{x} - 1$

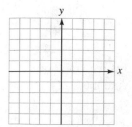
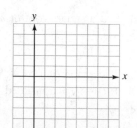
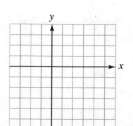
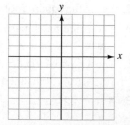

In Exercises 63–78, simplify each cube root.

63. $\sqrt[3]{1}$

64. $\sqrt[3]{-8}$

65. $\sqrt[3]{-125}$

66. $\sqrt[3]{512}$

67. $\sqrt[3]{-\dfrac{8}{27}}$

68. $\sqrt[3]{\dfrac{125}{216}}$

69. $\sqrt[3]{0.064}$

70. $\sqrt[3]{0.001}$

71. $\sqrt[3]{8a^3}$

72. $\sqrt[3]{-27x^6}$

73. $\sqrt[3]{-1,000p^3q^3}$

74. $\sqrt[3]{343a^6b^3}$

75. $\sqrt[3]{-\dfrac{1}{8}m^6n^3}$

76. $\sqrt[3]{\dfrac{27}{1,000}a^6b^6}$

77. $\sqrt[3]{0.008z^9}$

78. $\sqrt[3]{0.064s^9t^6}$

In Exercises 79–90, simplify each radical, if possible.

79. $\sqrt[4]{81}$

80. $\sqrt[6]{64}$

81. $-\sqrt[5]{243}$

82. $-\sqrt[4]{625}$

83. $\sqrt[5]{-32}$

84. $\sqrt[6]{729}$

85. $\sqrt[4]{\dfrac{16}{625}}$

86. $\sqrt[5]{-\dfrac{243}{32}}$

87. $-\sqrt[5]{-\dfrac{1}{32}}$

88. $\sqrt[6]{-729}$

89. $\sqrt[4]{-256}$

90. $-\sqrt[4]{\dfrac{81}{256}}$

In Exercises 91–102, simplify each radical. Assume that all variables are unrestricted, and use absolute value symbols where necessary.

91. $\sqrt[4]{16x^4}$

92. $\sqrt[5]{32a^5}$

93. $\sqrt[3]{8a^3}$

94. $\sqrt[6]{64x^6}$

95. $\sqrt[4]{\dfrac{1}{16}x^4}$

96. $\sqrt[4]{\dfrac{1}{81}x^8}$

97. $\sqrt[4]{x^{12}}$

98. $\sqrt[8]{x^{24}}$

99. $\sqrt[5]{-x^5}$

100. $\sqrt[3]{-x^6}$

101. $\sqrt[3]{-27a^6}$

102. $\sqrt[5]{-32x^5}$

In Exercises 103–106, simplify each radical. Assume that all variables are unrestricted, and use absolute value symbols when necessary.

103. $\sqrt[25]{(x+2)^{25}}$

104. $\sqrt[44]{(x+4)^{44}}$

105. $\sqrt[8]{0.00000001x^{16}y^8}$

106. $\sqrt[5]{0.00032x^{10}y^5}$

107. Find the standard deviation of the following distribution to the nearest hundredth: 2, 5, 5, 6, 7.

108. Find the standard deviation of the following distribution to the nearest hundredth: 3, 6, 7, 9, 11, 12.

109. Statistics In statistics, the formula

$$s_{\bar{x}} = \dfrac{s}{\sqrt{N}}$$

gives an estimate of the standard error of the mean. Find $s_{\bar{x}}$ to four decimal places when $s = 65$ and $N = 30$.

110. Statistics In statistics, the formula

$$\sigma_{\bar{x}} = \dfrac{\sigma}{\sqrt{N}}$$

gives the standard deviation of means of samples of size N. Find $\sigma_{\bar{x}}$ to four decimal places when $\sigma = 12.7$ and $N = 32$.

APPLICATIONS ⊞ *Use a calculator to solve each problem.*

111. Radius of a circle The radius r of a circle is given by the formula $r = \sqrt{\frac{A}{\pi}}$, where A is its area. Find the radius of a circle whose area is 9π square units.

112. Diagonal of a baseball diamond The diagonal d of a square is given by the formula $d = \sqrt{2s^2}$, where s is the length of each side. Find the diagonal of the baseball diamond shown in Illustration 1.

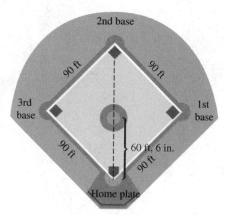

2nd base

90 ft 90 ft

3rd base

1st base

90 ft

60 ft, 6 in.

90 ft

Home plate

ILLUSTRATION 1

113. Falling objects The time t (in seconds) that it will take for an object to fall a distance of s feet is given by the formula $t = \sqrt{s}/4$. If a stone is dropped down a 256-foot well, how long will it take it to hit bottom?

114. Law enforcement Police sometimes use the formula $s = k\sqrt{l}$ to estimate the speed s (in mph) of a car involved in an accident. In this formula, l is the length of the skid in feet, and k is a constant depending on the condition of the pavement. For wet pavement, $k \approx 3.24$. How fast was a car going if its skid was 400 feet on wet pavement?

115. Electronics When the resistance in a circuit is 18 ohms, the current I (measured in amperes) and the power P (measured in watts) are related by the formula $I = \sqrt{P/18}$. Find the current used by an electrical appliance that is rated at 980 watts.

116. Medicine The approximate pulse rate p (in beats per minute) of an adult who is t inches tall is given by the formula

$$p = \frac{590}{\sqrt{t}}$$

Find the approximate pulse rate of an adult who is 71 inches tall.

WRITING

117. If x is any real number, then $\sqrt{x^2} = x$ is not correct. Explain

118. If x is any real number, then $\sqrt[3]{x^3} = |x|$ is not correct. Explain.

SOMETHING TO THINK ABOUT

119. Is $\sqrt{x^2 - 4x + 4} = x - 2$? What are the exceptions?

120. When is $\sqrt{x^2} \neq x$?

9.2 Applications of Radicals

■ THE PYTHAGOREAN THEOREM ■ THE DISTANCE FORMULA

Getting Ready *Evaluate each expression.*

1. $3^2 + 4^2$

2. $5^2 + 12^2$

3. $(5 - 2)^2 + (2 + 1)^2$

4. $(111 - 21)^2 + (60 - 4)^2$

■ THE PYTHAGOREAN THEOREM

If we know the lengths of two legs of a right triangle, we can always find the length of the **hypotenuse** (the side opposite the 90° angle) by using the **Pythagorean theorem.**

Pythagorean Theorem
If a and b are the lengths of two legs of a right triangle and c is the length of the hypotenuse, then
$$a^2 + b^2 = c^2$$

In words, the Pythagorean theorem says,

In any right triangle, the square of the hypotenuse is equal to the sum of the squares of the two legs.

Suppose the right triangle shown in Figure 9-6 has legs of length 3 and 4 units. To find the length of the hypotenuse, we use the Pythagorean theorem.

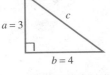

$a = 3$

c

$b = 4$

FIGURE 9-6

$$a^2 + b^2 = c^2$$
$$3^2 + 4^2 = c^2$$
$$9 + 16 = c^2$$
$$25 = c^2$$
$$\sqrt{25} = \sqrt{c^2} \qquad \text{Take the positive square root of both sides.}$$
$$5 = c$$

The length of the hypotenuse is 5 units.

EXAMPLE 1

Fighting fires   To fight a forest fire, the forestry department plans to clear a rectangular fire break around the fire, as shown in Figure 9-7. Crews are equipped with mobile communications with a 3,000-yard range. Can crews at points A and B remain in radio contact?

Solution   Points A, B, and C form a right triangle. To find the distance c from point A to point B, we can use the Pythagorean theorem, substituting 2,400 for a and 1,000 for b and solving for c.

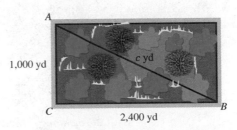

A

1,000 yd

c yd

C

B

2,400 yd

FIGURE 9-7

$$a^2 + b^2 = c^2$$
$$2,400^2 + 1,000^2 = c^2$$
$$5,760,000 + 1,000,000 = c^2$$
$$6,760,000 = c^2$$
$$\sqrt{6,760,000} = \sqrt{c^2} \qquad \text{Take the positive square root of both sides.}$$
$$2,600 = c \qquad \text{Use a calculator to find the square root.}$$

The two crews are 2,600 yards apart. Because this distance is less than the range of the radios, they can communicate. ■

Self Check   Can the crews communicate if $b = 1,500$ yards?
Answer   yes

■ ■ ■ ■ ■ ■ ■ ■ ■ ■ PERSPECTIVE

Pythagoras was a teacher. Although it was unusual for schools at that time, his classes were coeducational. According to some legends, Pythagoras married one of his students. He and his followers formed a secret society with two rules: membership was for life, and members could not reveal the secrets they knew.

Much of their teaching was good mathematics, but some ideas were strange. To them, numbers were sacred. Because beans were used as counters to represent numbers, Pythagoreans refused to eat beans. They also believed that the *only* numbers were the whole numbers. To them, fractions were not numbers; $\frac{2}{3}$ was just a way of comparing the whole numbers 2 and 3. They believed that whole numbers were the building blocks of the universe, just as atoms are to us. The basic Pythagorean doctrine was, "All things are number," and they meant *whole* number.

The Pythagorean theorem was an important discovery of the Pythagorean school, yet it caused some division in the ranks. The right triangle in Illustration 1 has two legs of length 1. By the Pythagorean theo-

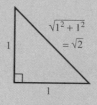

ILLUSTRATION 1

rem, the length of the hypotenuse is $\sqrt{2}$. One of their own group, Hippasus of Metapontum, discovered that $\sqrt{2}$ is an irrational number: There are *no* whole numbers a and b that make the fraction $\frac{a}{b}$ exactly equal to $\sqrt{2}$. This discovery was not appreciated by the other Pythagoreans. How could everything in the universe be described with whole numbers, when the side of this simple triangle couldn't? The Pythagoreans had a choice. Either revise and expand their beliefs, or cling to the old. According to legend, the group was at sea at the time of the discovery. Rather than upset the system, they threw Hippasus overboard.

■ THE DISTANCE FORMULA

With the *distance formula,* we can find the distance between any two points that are graphed on a rectangular coordinate system.

To find the distance d between points $P(x_1, y_1)$ and $Q(x_2, y_2)$ shown in Figure 9-8, we construct the right triangle PRQ. The distance between P and R is $|x_2 - x_1|$, and the distance between R and Q is $|y_2 - y_1|$. We apply the Pythagorean theorem to the right triangle PRQ to get

$$[d(PQ)]^2 = |x_2 - x_1|^2 + |y_2 - y_1|^2 \qquad \text{Read } d(PQ) \text{ as "the distance between } P \text{ and } Q\text{."}$$

$$= (x_2 - x_1)^2 + (y_2 - y_1)^2 \qquad \text{Because } |x_2 - x_1|^2 = (x_2 - x_1)^2 \text{ and } |y_2 - y_1|^2 = (y_2 - y_1)^2.$$

or

1. $d(PQ) = \sqrt{(x_2 - x_1)^2 + (y_2 - y_1)^2}$

Equation 1 is called the **distance formula.**

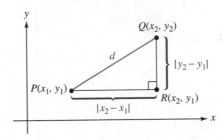

FIGURE 9-8

Distance Formula
The distance between two points $P(x_1, y_1)$ and $Q(x_2, y_2)$ is given by the formula
$$d(PQ) = \sqrt{(x_2 - x_1)^2 + (y_2 - y_1)^2}$$

EXAMPLE 2 Find the distance between points $P(-2, 3)$ and $Q(4, -5)$.

Solution To find the distance, we can use the distance formula by substituting 4 for x_2, -2 for x_1, -5 for y_2, and 3 for y_1.

$$d(PQ) = \sqrt{(x_2 - x_1)^2 + (y_2 - y_1)^2}$$
$$= \sqrt{[4 - (-2)]^2 + (-5 - 3)^2}$$
$$= \sqrt{(4 + 2)^2 + (-5 - 3)^2}$$
$$= \sqrt{6^2 + (-8)^2}$$
$$= \sqrt{36 + 64}$$
$$= \sqrt{100}$$
$$= 10$$

The distance between P and Q is 10 units.

| Self Check | Find the distance between $P(-2, -2)$ and $Q(3, 10)$. |
| Answer | 13 |

EXAMPLE 3

Building a freeway   In a city, streets run north and south, and avenues run east and west. Streets and avenues are 850 feet apart. The city plans to construct a straight freeway from the intersection of 25th Street and 8th Avenue to the intersection of 115th Street and 64th Avenue. How long will the freeway be?

Solution   We can represent the roads of the city by the coordinate system in Figure 9-9, where the units on each axis represent 850 feet. We represent the end of the freeway at 25th Street and 8th Avenue by the point $(x_1, y_1) = (25, 8)$. The other end is $(x_2, y_2) = (115, 64)$.

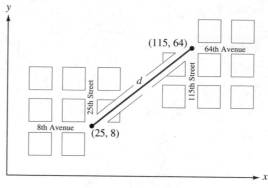

FIGURE 9-9

We can use the distance formula to find the length of the freeway.

$$d = \sqrt{(x_2 - x_1)^2 + (y_2 - y_1)^2}$$
$$d = \sqrt{(115 - 25)^2 + (64 - 8)^2}$$
$$= \sqrt{90^2 + 56^2}$$
$$= \sqrt{8{,}100 + 3{,}136}$$
$$= \sqrt{11{,}236}$$
$$= 106 \qquad \text{Use a calculator to find the square root.}$$

Because each unit represents 850 feet, the length of the freeway is $106(850) = 90{,}100$ feet. Since 5,280 feet = 1 mile, we can divide 90,100 by 5,280 to convert 90,100 feet to 17.064394 miles. Thus, the freeway will be about 17 miles long.   ■

EXAMPLE 4

Bowling   The velocity, v, of an object after it has fallen d feet is given by the equation $v^2 = 64d$. An inexperienced bowler lofts the ball 4 feet. With what velocity does it strike the alley?

Solution   We find the velocity by substituting 4 for d in the equation $v^2 = 64d$ and solving for v.

$$v^2 = 64d$$
$$v^2 = 64(4)$$
$$v^2 = 256$$
$$v = \sqrt{256}$$ Take the square root of both sides. Only the positive square root is meaningful.
$$= 16$$

The ball strikes the alley with a velocity of 16 feet per second.   ■

Orals   *Evaluate each expression.*

1. $\sqrt{25}$ **2.** $\sqrt{100}$ **3.** $\sqrt{169}$

4. $\sqrt{3^2 + 4^2}$ **5.** $\sqrt{8^2 + 6^2}$ **6.** $\sqrt{5^2 + 12^2}$

7. $\sqrt{5^2 - 3^2}$ **8.** $\sqrt{5^2 - 4^2}$ **9.** $\sqrt{169 - 12^2}$

EXERCISE 9.2

REVIEW   *Find each product.*

1. $(4x + 2)(3x - 5)$

2. $(3y - 5)(2y + 3)$

3. $(5t + 4s)(3t - 2s)$

4. $(4r - 3)(2r^2 + 3r - 4)$

VOCABULARY AND CONCEPTS   *Fill in each blank to make a true statement.*

5. In a right triangle, the side opposite the 90° angle is called the _____.

6. In a right triangle, the two shorter sides are called ____.

7. If a and b are the lengths of two legs of a right triangle and c is the length of the hypotenuse, then _____.

8. In any right triangle, the square of the hypotenuse is equal to the ____ of the squares of the two ____.

9. With the _____ formula, we can find the distance between two points on a rectangular coordinate system.

10. $d(PQ) = $ _____

PRACTICE   *In Exercises 11–14, the lengths of two legs of the right triangle ABC shown in Illustration 1 are given. Find the length of the missing side.*

11. $a = 6$ ft and $b = 8$ ft

12. $a = 10$ cm and $c = 26$ cm

13. $b = 18$ m and $c = 82$ m

14. $a = 14$ in. and $c = 50$ in.

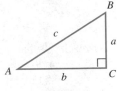

ILLUSTRATION 1

APPLICATIONS   *In Exercises 15–16, give each answer to the nearest tenth.*

15. Geometry   Find the length of the diagonal of one of the faces of the cube shown in Illustration 2.

16. Geometry   Find the length of the diagonal of the cube shown in Illustration 2.

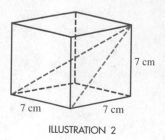

7 cm

7 cm   7 cm

ILLUSTRATION 2

In Exercises 17–26, find the distance between P and Q. If an answer is not exact, give the answer to the nearest tenth.

17. $Q(0, 0)$, $P(3, -4)$

18. $Q(0, 0)$, $P(-6, 8)$

19. $P(2, 4)$, $Q(5, 8)$

20. $P(5, 9)$, $Q(8, 13)$

21. $P(-2, -8)$, $Q(3, 4)$

22. $P(-5, -2)$, $Q(7, 3)$

23. $P(6, 8)$, $Q(12, 16)$

24. $P(10, 4)$, $Q(2, -2)$

25. $Q(-3, 5)$, $P(-5, -5)$

26. $Q(2, -3)$, $P(4, -8)$

27. Geometry   Show that a triangle with vertices at $(-2, 4)$, $(2, 8)$, and $(6, 4)$ is isosceles.

28. Geometry   Show that a triangle with vertices at $(-2, 13)$, $(-8, 9)$, and $(-2, 5)$ is isosceles.

29. Finding the equation of a line   Every point on the line *CD* in Illustration 3 is equidistant from points *A* and *B*. Use the distance formula to find the equation of line *CD*.

31. Geometry   Find the coordinates of the two points on the *x*-axis that are $\sqrt{5}$ units from the point $(5, 1)$.

32. Geometry   The square in Illustration 4 has an area of 18 square units, and its diagonals lie on the *x*- and *y*-axes. Find the coordinates of each corner of the square.

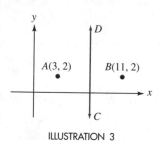

ILLUSTRATION 3

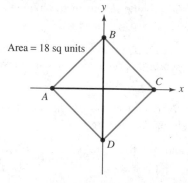

ILLUSTRATION 4

30. Geometry   Show that a triangle with vertices at $(2, 3)$, $(-3, 4)$, and $(1, -2)$ is a right triangle. (*Hint:* If the Pythagorean relation holds, the triangle is a right triangle.)

APPLICATIONS

33. Sailing Refer to the sailboat in Illustration 5. How long must a rope be to fasten the top of the mast to the bow?

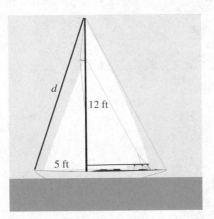

ILLUSTRATION 5

34. Carpentry The gable end of the roof shown in Illustration 6 is divided in half by a vertical brace. Find the distance from eaves to peak.

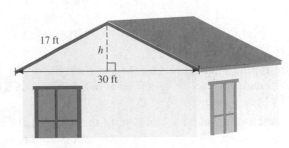

ILLUSTRATION 6

 In Exercises 35–38, use a calculator. The baseball diamond shown in Illustration 7 is a square, 90 feet on a side.

35. Baseball How far must a catcher throw the ball to throw out a runner stealing second base?

36. Baseball In baseball, the pitcher's mound is 60 feet, 6 inches from home plate. How far from the mound is second base?

37. Baseball If the third baseman fields a ground ball 10 feet directly behind third base, how far must he throw the ball to throw a runner out at first base?

38. Baseball The shortstop fields a grounder at a point one-third of the way from second base to third base. How far will he have to throw the ball to make an out at first base?

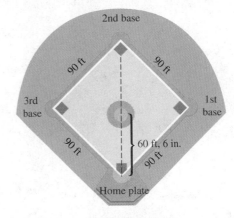

ILLUSTRATION 7

 In Exercises 39–48, use a calculator.

39. Packing a tennis racket The diagonal d of a rectangular box with dimensions $a \times b \times c$ is given by

$$d = \sqrt{a^2 + b^2 + c^2}$$

See Illustration 8. Will the racket fit in the shipping carton?

40. Shipping packages A delivery service won't accept a package for shipping if any dimension exceeds 21 inches. An archaeologist wants to ship a 36-inch femur bone. Will it fit in a 3-inch-tall box that has a 21-inch-square base?

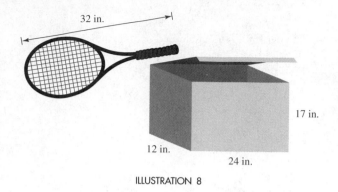

ILLUSTRATION 8

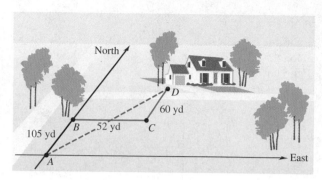

ILLUSTRATION 10

41. Shipping packages Can the archaeologist in Exercise 40 ship the femur bone in a cubical box 21 inches on an edge?

42. Reach of a ladder The base of the 37-foot ladder in Illustration 9 is 9 feet from the wall. Will the top reach a window ledge that is 35 feet above the ground?

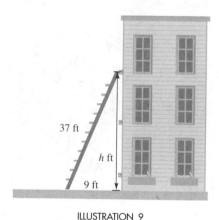

ILLUSTRATION 9

43. Telephone service The telephone cable in Illustration 10 runs from A to B to C to D. How much cable is required to run from A to D directly?

44. Electric service The power company routes its lines as in Illustration 11. How much wire could be saved by going directly from A to E?

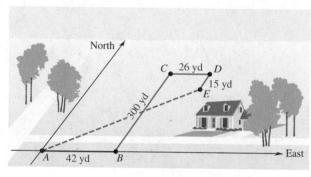

ILLUSTRATION 11

45. Supporting a weight A weight placed on the tight wire in Illustration 12 pulls the center down 1 foot. By how much is the wire stretched? Round the answer to the nearest hundredth of a foot.

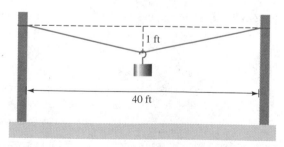

ILLUSTRATION 12

46. Geometry The side, s, of a square with area A square feet is given by the formula $s = \sqrt{A}$. Find the perimeter of a square with an area of 49 square feet.

47. Volume of a cube   The total surface area, A, of a cube is related to its volume, V, by the formula $A = 6\sqrt[3]{V^2}$. Find the volume of a cube with a surface area of 24 square centimeters.

48. Area of many cubes   A grain of table salt is a cube with a volume of approximately 6×10^{-6} cubic inches, and there are about 1.5 million grains of salt in one cup. Find the total surface area of the salt in one cup. (See Exercise 47.)

WRITING

49. State the Pythagorean theorem.

50. Explain the distance formula.

SOMETHING TO THINK ABOUT

51. The formula

$$I = \frac{703w}{h^2}$$

(where w is weight in pounds and h is height in inches) can be used to estimate body mass index, I. The scale shown in Illustration 13 can be used to judge a person's risk of heart attack. A girl weighing 104 pounds has a body mass index of 25. How tall is she?

20–26	normal
27–29	higher risk
30 and above	very high risk

ILLUSTRATION 13

52. What is the risk of a heart attack for a man who is 6 feet tall and weighs 220 pounds?

9.3 Radical Equations

■ EQUATIONS CONTAINING ONE RADICAL ■ EQUATIONS CONTAINING TWO RADICALS
■ EQUATIONS CONTAINING THREE RADICALS

Getting Ready   *Find each power.*

1. $\left(\sqrt{a}\right)^2$ **2.** $\left(\sqrt{5x}\right)^2$ **3.** $\left(\sqrt{x+4}\right)^2$ **4.** $\left(\sqrt[4]{y-3}\right)^4$

As we will see, the solutions of many problems involve equations containing radicals. To solve these equations, we will use the **power rule.**

> **The Power Rule**
> If x, y, and n are real numbers and $x = y$, then
> $$x^n = y^n$$

If we raise both sides of an equation to the same power, the resulting equation might not be equivalent to the original equation. For example, if we square both sides of the equation

1. $x = 3$ With a solution set of {3}.

we obtain the equation

2. $x^2 = 9$ With a solution set of $\{3, -3\}$.

Equations 1 and 2 are not equivalent, because they have different solution sets, and the solution -3 of Equation 2 does not satisfy Equation 1. Since raising both sides of an equation to the same power can produce an equation with roots that don't satisfy the original equation, we must always check each suspected solution in the original equation.

■ EQUATIONS CONTAINING ONE RADICAL

EXAMPLE 1   Solve $\sqrt{x + 3} = 4$.

Solution   To eliminate the radical, we apply the power rule by squaring both sides of the equation, and proceed as follows:

$$\sqrt{x + 3} = 4$$
$$\left(\sqrt{x + 3}\right)^2 = (4)^2 \qquad \text{Square both sides.}$$
$$x + 3 = 16$$
$$x = 13 \qquad \text{Subtract 3 from both sides.}$$

We must check the apparent solution of 13 to see whether it satisfies the original equation.

Check: $\sqrt{x + 3} = 4$
$$\sqrt{13 + 3} \overset{?}{=} 4 \qquad \text{Substitute 13 for } x.$$
$$\sqrt{16} \overset{?}{=} 4$$
$$4 = 4$$

Since 13 satisfies the original equation, it is a solution.   ■

Self Check   Solve $\sqrt{a - 2} = 3$.

Answer   11

To solve an equation with radicals, we follow these steps.

Solving an Equation Containing Radicals

1. Isolate one radical expression on one side of the equation.

2. Raise both sides of the equation to the power that is the same as the index of the radical.

3. Solve the resulting equation. If it still contains a radical, go back to Step 1.

4. Check the possible solutions to eliminate the ones that do not satisfy the original equation.

EXAMPLE 2   **Height of a bridge**   The distance d (in feet) that an object will fall in t seconds is given by the formula

$$t = \sqrt{\frac{d}{16}}$$

To find the height of a bridge, a man drops a stone into the water (see Figure 9-10). If it takes the stone 3 seconds to hit the water, how far above the river is the bridge?

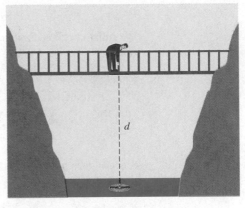

FIGURE 9-10

Solution   We substitute 3 for t in the formula and solve for d.

$$t = \sqrt{\frac{d}{16}}$$

$$\textbf{3} = \sqrt{\frac{d}{16}}$$

$$9 = \frac{d}{16} \qquad \text{Square both sides.}$$

$$144 = d \qquad \text{Multiply both sides by 16.}$$

The bridge is 144 feet above the river.   ∎

Self Check   In Example 2, how high is the bridge if it takes 4 seconds for the stone to hit the water?

Answer   256 ft

EXAMPLE 3   Solve $\sqrt{3x + 1} + 1 = x$.

Solution   We first subtract 1 from both sides to isolate the radical. Then, to eliminate the radical, we square both sides of the equation and proceed as follows:

$$\sqrt{3x + 1} + 1 = x$$

$$\sqrt{3x + 1} = x - 1 \qquad \text{Subtract 1 from both sides.}$$

$$\left(\sqrt{3x + 1}\right)^2 = (x - 1)^2 \qquad \text{Square both sides to eliminate the square root.}$$

$$3x + 1 = x^2 - 2x + 1 \qquad (x-1)^2 \neq x^2 - 1. \text{ Instead, } (x-1)^2 = (x-1)(x-1) = x^2 - x - x + 1 = x^2 - 2x + 1.$$

$$0 = x^2 - 5x \qquad \text{Subtract } 3x \text{ and 1 from both sides.}$$

$$0 = x(x - 5) \qquad \text{Factor } x^2 - 5x.$$

$$x = 0 \quad \text{or} \quad x - 5 = 0 \qquad \text{Set each factor equal to 0.}$$

$$x = 0 \qquad\qquad x = 5$$

We must check each apparent solution to see whether it satisfies the original equation.

Check:
$$\sqrt{3x + 1} + 1 = x \qquad\qquad \sqrt{3x + 1} + 1 = x$$
$$\sqrt{3(0) + 1} + 1 \stackrel{?}{=} 0 \qquad\qquad \sqrt{3(5) + 1} + 1 \stackrel{?}{=} 5$$
$$\sqrt{1} + 1 \stackrel{?}{=} 0 \qquad\qquad \sqrt{16} + 1 \stackrel{?}{=} 5$$
$$2 \neq 0 \qquad\qquad\qquad 5 = 5$$

Since 0 does not check, it must be discarded. The only solution of the original equation is 5. ■

Self Check Solve $\sqrt{4x + 1} + 1 = x$.

Answer 6; 0 is extraneous

■ ■ ■ ■ ■ ■ ■ ■ ■ **Solving Equations Containing Radicals**

GRAPHING CALCULATORS To find approximate solutions for $\sqrt{3x + 1} + 1 = x$ with a graphing calculator, we use window settings of $[-5, 10]$ for x and $[-2, 8]$ for y and graph the functions $f(x) = \sqrt{3x + 1} + 1$ and $g(x) = x$, as in Figure 9-11(a). We then trace to find the approximate x-coordinate of their intersection point, as in Figure 9-11(b). After repeated zooms, we will see that $x = 5$.

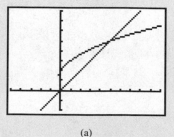

(a)

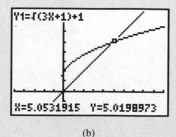

(b)

FIGURE 9-11

EXAMPLE 4 Solve $\sqrt[3]{x^3 + 7} = x + 1$.

Solution To eliminate the radical, we cube both sides of the equation and proceed as follows:

$$\sqrt[3]{x^3 + 7} = x + 1$$
$$\left(\sqrt[3]{x^3 + 7}\right)^3 = (x + 1)^3 \qquad \text{Cube both sides to eliminate the cube root.}$$
$$x^3 + 7 = x^3 + 3x^2 + 3x + 1$$
$$0 = 3x^2 + 3x - 6 \qquad \text{Subtract } x^3 \text{ and 7 from both sides.}$$
$$0 = x^2 + x - 2 \qquad \text{Divide both sides by 3.}$$
$$0 = (x + 2)(x - 1)$$
$$x + 2 = 0 \quad \text{or} \quad x - 1 = 0$$
$$x = -2 \qquad\qquad x = 1$$

We check each apparent solution to see whether it satisfies the original equation.

Check:
$$\sqrt[3]{x^3 + 7} = x + 1 \qquad\qquad \sqrt[3]{x^3 + 7} = x + 1$$
$$\sqrt[3]{(-2)^3 + 7} \stackrel{?}{=} -2 + 1 \qquad \sqrt[3]{1 + 7} \stackrel{?}{=} 1 + 1$$
$$\sqrt[3]{-8 + 7} \stackrel{?}{=} -1 \qquad\qquad \sqrt[3]{8} \stackrel{?}{=} 2$$
$$\sqrt[3]{-1} \stackrel{?}{=} -1 \qquad\qquad\qquad 2 = 2$$
$$-1 = -1$$

Both solutions satisfy the original equation. ∎

Self Check
Answer

Solve $\sqrt[3]{x^3 + 8} = x + 2$.

$0, -2$

■ EQUATIONS CONTAINING TWO RADICALS

When more than one radical appears in an equation, it is often necessary to apply the power rule more than once.

EXAMPLE 5 Solve $\sqrt{x} + \sqrt{x + 2} = 2$.

Solution To remove the radicals, we must square both sides of the equation. This is easier to do if one radical is on each side of the equation. So we subtract $\sqrt{x}$ from both sides to isolate one radical on one side of the equation.

$$\sqrt{x} + \sqrt{x + 2} = 2$$
$$\sqrt{x + 2} = 2 - \sqrt{x} \qquad \text{Subtract } \sqrt{x} \text{ from both sides.}$$
$$\left(\sqrt{x + 2}\right)^2 = \left(2 - \sqrt{x}\right)^2 \qquad \text{Square both sides to eliminate the square root.}$$

$$x + 2 = 4 - 4\sqrt{x} + x \qquad \left(2 - \sqrt{x}\right)\left(2 - \sqrt{x}\right) =$$
$$4 - 2\sqrt{x} - 2\sqrt{x} + x = 4 - 4\sqrt{x} + x$$

$$2 = 4 - 4\sqrt{x} \qquad \text{Subtract } x \text{ from both sides.}$$

$$-2 = -4\sqrt{x} \qquad \text{Subtract 4 from both sides.}$$

$$\frac{1}{2} = \sqrt{x} \qquad \text{Divide both sides by } -4.$$

$$\frac{1}{4} = x \qquad \text{Square both sides.}$$

Check: $\sqrt{x} + \sqrt{x + 2} = 2$

$$\sqrt{\frac{1}{4}} + \sqrt{\frac{1}{4} + 2} \overset{?}{=} 2$$

$$\frac{1}{2} + \sqrt{\frac{9}{4}} \overset{?}{=} 2$$

$$\frac{1}{2} + \frac{3}{2} \overset{?}{=} 2$$

$$2 = 2$$

The solution checks.   ■

Self Check   Solve $\sqrt{a} + \sqrt{a + 3} = 3$.

Answer   1

■ ■ ■ ■ ■ ■ ■ ■ ■ ■   **Solving Equations Containing Radicals**

GRAPHING CALCULATORS   To find approximate solutions for $\sqrt{x} + \sqrt{x + 2} = 5$ with a graphing calculator, we use window settings of $[-2, 10]$ for x and $[-2, 8]$ for y and graph the functions $f(x) = \sqrt{x} + \sqrt{x + 2}$ and $g(x) = 5$, as in Figure 9-12(a). We then trace to find an approximation of the x-coordinate of their intersection point, as in Figure 9-12(b). From the figure, we can see that $x \approx 5.15$. We can zoom to get better results.

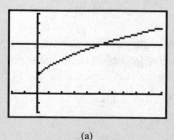
(a)

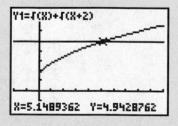
(b)

FIGURE 9-12

■ EQUATIONS CONTAINING THREE RADICALS

EXAMPLE 6   Solve $\sqrt{x+2} + \sqrt{2x} = \sqrt{18-x}$.

Solution   In this case, it is impossible to isolate one radical on each side of the equation, so we begin by squaring both sides. Then we proceed as follows.

$$\sqrt{x+2} + \sqrt{2x} = \sqrt{18-x}$$

$$\left(\sqrt{x+2} + \sqrt{2x}\right)^2 = \left(\sqrt{18-x}\right)^2 \qquad \text{Square both sides to eliminate one square root.}$$

$$x + 2 + 2\sqrt{x+2}\sqrt{2x} + 2x = 18 - x$$

$$2\sqrt{x+2}\sqrt{2x} = 16 - 4x \qquad \text{Subtract } 3x \text{ and } 2 \text{ from both sides.}$$

$$\sqrt{x+2}\sqrt{2x} = 8 - 2x \qquad \text{Divide both sides by 2.}$$

$$\left(\sqrt{x+2}\sqrt{2x}\right)^2 = (8-2x)^2 \qquad \text{Square both sides to eliminate the other square roots.}$$

$$(x+2)2x = 64 - 32x + 4x^2$$

$$2x^2 + 4x = 64 - 32x + 4x^2$$

$$0 = 2x^2 - 36x + 64 \qquad \text{Write the equation in quadratic form.}$$

$$0 = x^2 - 18x + 32 \qquad \text{Divide both sides by 2.}$$

$$0 = (x - 16)(x - 2) \qquad \text{Factor the trinomial.}$$

$$x - 16 = 0 \quad \text{or} \quad x - 2 = 0 \qquad \text{Set each factor equal to 0.}$$

$$x = 16 \qquad\qquad x = 2$$

Verify that 2 satisfies the equation, but 16 does not. Thus, the only solution is 2.   ■

Self Check   Solve $\sqrt{3x+4} + \sqrt{x+9} = \sqrt{x+25}$.
Answer   0

Orals   *Solve each equation.*

1. $\sqrt{x+2} = 3$ **2.** $\sqrt{x-2} = 1$
3. $\sqrt[3]{x+1} = 1$ **4.** $\sqrt[3]{x-1} = 2$
5. $\sqrt[4]{x-1} = 2$ **6.** $\sqrt[5]{x+1} = 2$

EXERCISE 9.3

REVIEW   *If $f(x) = 3x^2 - 4x + 2$, find each quantity.*

1. $f(0)$ **2.** $f(-3)$ **3.** $f(2)$ **4.** $f\left(\dfrac{1}{2}\right)$

VOCABULARY AND CONCEPTS   *Fill in each blank to make a true statement.*

5. If x, y, and n are real numbers and $x = y$, then
_____ .

6. When solving equations containing radicals, try to
_____ one radical expression on one side of the
equation.

7. To solve the equation $\sqrt{x + 4} = 5$, we first _____
both sides.

8. To solve the equation $\sqrt[3]{x + 4} = 2$, we first _____
both sides.

9. Squaring both sides of an equation can introduce
_____ solutions.

10. Always remember to _____ the solutions of an
equation containing radicals to eliminate any
_____ solutions.

PRACTICE   *In Exercises 11–64, solve each equation. Write all solutions and cross out those that are extraneous.*

11. $\sqrt{5x - 6} = 2$

12. $\sqrt{7x - 10} = 12$

13. $\sqrt{6x + 1} + 2 = 7$

14. $\sqrt{6x + 13} - 2 = 5$

15. $2\sqrt{4x + 1} = \sqrt{x + 4}$

16. $\sqrt{3(x + 4)} = \sqrt{5x - 12}$

17. $\sqrt[3]{7n - 1} = 3$

18. $\sqrt[3]{12m + 4} = 4$

19. $\sqrt[4]{10p + 1} = \sqrt[4]{11p - 7}$

20. $\sqrt[4]{10y + 2} = 2\sqrt[4]{2}$

21. $x = \dfrac{\sqrt{12x - 5}}{2}$

22. $x = \dfrac{\sqrt{16x - 12}}{2}$

23. $\sqrt{x + 2} = \sqrt{4 - x}$

24. $\sqrt{6 - x} = \sqrt{2x + 3}$

25. $2\sqrt{x} = \sqrt{5x - 16}$

26. $3\sqrt{x} = \sqrt{3x + 12}$

27. $r - 9 = \sqrt{2r - 3}$

28. $-s - 3 = 2\sqrt{5 - s}$

29. $\sqrt{-5x + 24} = 6 - x$

30. $\sqrt{-x + 2} = x - 2$

31. $\sqrt{y + 2} = 4 - y$

32. $\sqrt{22y + 86} = y + 9$

33. $\sqrt{x}\sqrt{x + 16} = 15$

34. $\sqrt{x}\sqrt{x + 6} = 4$

35. $\sqrt[3]{x^3 - 7} = x - 1$

36. $\sqrt[3]{x^3 + 56} - 2 = x$

37. $\sqrt[4]{x^4 + 4x^2 - 4} = -x$

38. $\sqrt[4]{8x - 8} + 2 = 0$

39. $\sqrt[4]{12t + 4} + 2 = 0$

40. $u = \sqrt[4]{u^4 - 6u^2 + 24}$

41. $\sqrt{2y + 1} = 1 - 2\sqrt{y}$

42. $\sqrt{u + 3} = \sqrt{u - 3}$

43. $\sqrt{y + 7} + 3 = \sqrt{y + 4}$

44. $1 + \sqrt{z} = \sqrt{z + 3}$

45. $\sqrt{v} + \sqrt{3} = \sqrt{v + 3}$

46. $\sqrt{x} + 2 = \sqrt{x + 4}$

47. $2 + \sqrt{u} = \sqrt{2u + 7}$

48. $5r + 4 = \sqrt{5r + 20} + 4r$

49. $\sqrt{6t + 1} - 3\sqrt{t} = -1$

50. $\sqrt{4s + 1} - \sqrt{6s} = -1$

51. $\sqrt{2x + 5} + \sqrt{x + 2} = 5$

52. $\sqrt{2x + 5} + \sqrt{2x + 1} + 4 = 0$

53. $\sqrt{z - 1} + \sqrt{z + 2} = 3$

54. $\sqrt{16v + 1} + \sqrt{8v + 1} = 12$

55. $\sqrt{x - 5} - \sqrt{x + 3} = 4$

56. $\sqrt{x + 8} - \sqrt{x - 4} = -2$

57. $\sqrt{x + 1} + \sqrt{3x} = \sqrt{5x + 1}$

58. $\sqrt{3x} - \sqrt{x + 1} = \sqrt{x - 2}$

59. $\sqrt{\sqrt{a} + \sqrt{a + 8}} = 2$

60. $\sqrt{\sqrt{2y} - \sqrt{y - 1}} = 1$

61. $\dfrac{6}{\sqrt{x + 5}} = \sqrt{x}$

62. $\dfrac{\sqrt{2x}}{\sqrt{x + 2}} = \sqrt{x - 1}$

63. $\sqrt{x + 2} + \sqrt{2x - 3} = \sqrt{11 - x}$

64. $\sqrt{8 - x} - \sqrt{3x - 8} = \sqrt{x - 4}$

APPLICATIONS

65. Highway design   A curve banked at 8° will accommodate traffic traveling s mph if the radius of the curve is r feet, according to the formula $s = 1.45 \sqrt{r}$. If engineers expect 65-mph traffic, what radius should they specify? (See Illustration 1.)

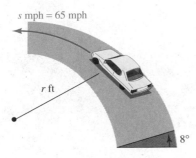

s mph = 65 mph

r ft

8°

ILLUSTRATION 1

66. Horizon distance   The higher a lookout tower is built, the farther an observer can see. (See Illustration 2.) That distance d (called the *horizon distance,* measured in miles) is related to the height h of the observer (measured in feet) by the formula $d = 1.4 \sqrt{h}$. How tall must a lookout tower be to see the edge of the forest, 25 miles away?

67. Generating power   The power generated by a windmill is related to the velocity of the wind by the formula

$$v = \sqrt[3]{\frac{P}{0.02}}$$

where P is the power (in watts) and v is the velocity of the wind (in mph). Find the speed of the wind when the windmill is generating 500 watts of power.

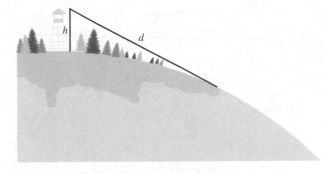

ILLUSTRATION 2

68. Carpentry   During construction, carpenters often brace walls as shown in Illustration 3, where the length of the brace is given by the formula

$$l = \sqrt{f^2 + h^2}$$

If a carpenter nails a 10-ft brace to the wall 6 feet above the floor, how far from the base of the wall should he nail the brace to the floor?

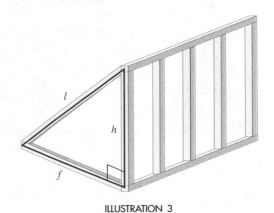

ILLUSTRATION 3

In Exercises 69–70, use a graphing calculator.

69. Marketing   The number of wrenches that will be produced at a given price can be predicted by the formula $s = \sqrt{5x}$, where s is the supply (in thousands) and x is the price (in dollars). If the demand, d, for wrenches can be predicted by the formula $d = \sqrt{100 - 3x^2}$, find the equilibrium price.

70. Marketing   The number of footballs that will be produced at a given price can be predicted by the formula $s = \sqrt{23x}$, where s is the supply (in thousands) and x is the price (in dollars). If the demand, d, for footballs can be predicted by the formula $d = \sqrt{312 - 2x^2}$, find the equilibrium price.

WRITING

71. If both sides of an equation are raised to the same power, the resulting equation might not be equivalent to the original equation. Explain.

72. Explain why you must check each apparent solution of a radical equation.

SOMETHING TO THINK ABOUT

73. Solve $\sqrt[3]{2x} = \sqrt{x}$. (*Hint:* Square and then cube both sides.)

74. Solve $\sqrt[4]{x} = \sqrt{\dfrac{x}{4}}$.

9.4 Rational Exponents

■ RATIONAL EXPONENTS ■ EXPONENTIAL EXPRESSIONS WITH VARIABLES IN THEIR BASES
■ FRACTIONAL EXPONENTS WITH NUMERATORS OTHER THAN 1 ■ NEGATIVE FRACTIONAL
EXPONENTS ■ SIMPLIFYING RADICAL EXPRESSIONS

Getting Ready   *Simplify each expression.*

1. $x^3 x^4$ **2.** $(a^3)^4$ **3.** $\dfrac{a^8}{a^4}$ **4.** a^0

5. x^{-4} **6.** $(ab^2)^3$ **7.** $\left(\dfrac{b^2}{c^3}\right)^3$ **8.** $(a^2 a^3)^2$

■ RATIONAL EXPONENTS

We have seen that positive integer exponents indicate the number of times that a base is to be used as a factor in a product. For example, x^5 means that x is to be used as a factor five times.

$$x^5 = \overbrace{x \cdot x \cdot x \cdot x \cdot x}^{5 \text{ factors of } x}$$

Furthermore, we recall the following properties of exponents.

Rules of Exponents
If there are no divisions by 0, then for all integers m and n,

1. $x^m x^n = x^{m+n}$ **2.** $(x^m)^n = x^{mn}$ **3.** $(xy)^n = x^n y^n$ **4.** $\left(\dfrac{x}{y}\right)^n = \dfrac{x^n}{y^n}$

5. $x^0 = 1 \quad (x \neq 0)$ **6.** $x^{-n} = \dfrac{1}{x^n}$ **7.** $\dfrac{x^m}{x^n} = x^{m-n}$ **8.** $\left(\dfrac{x}{y}\right)^{-n} = \left(\dfrac{y}{x}\right)^n$

It is possible to raise many bases to fractional powers. Since we want fractional exponents to obey the same rules as integer exponents, the square of $10^{1/2}$ must be 10, because

$$(10^{1/2})^2 = 10^{(1/2)2} \qquad \text{Keep the base and multiply the exponents.}$$
$$= 10^1 \qquad \qquad \tfrac{1}{2} \cdot 2 = 1.$$
$$= 10 \qquad \qquad 10^1 = 10.$$

However, we have seen that

$$\left(\sqrt{10}\right)^2 = 10$$

Since $(10^{1/2})^2$ and $\left(\sqrt{10}\right)^2$ both equal 10, we define $10^{1/2}$ to be $\sqrt{10}$. Likewise, we define

$$10^{1/3} \text{ to be } \sqrt[3]{10} \qquad \text{and} \qquad 10^{1/4} \text{ to be } \sqrt[4]{10}$$

Rational Exponents
If n ($n > 1$) is a natural number and $\sqrt[n]{x}$ is a real number, then
$$x^{1/n} = \sqrt[n]{x}$$

EXAMPLE 1   Simplify each expression.

a. $9^{1/2} = \sqrt{9} = 3$

b. $-\left(\dfrac{16}{9}\right)^{1/2} = -\sqrt{\dfrac{16}{9}} = -\dfrac{4}{3}$

c. $(-64)^{1/3} = \sqrt[3]{-64} = -4$

d. $16^{1/4} = \sqrt[4]{16} = 2$

e. $\left(\dfrac{1}{32}\right)^{1/5} = \sqrt[5]{\dfrac{1}{32}} = \dfrac{1}{2}$

f. $0^{1/8} = \sqrt[8]{0} = 0$

g. $-(32x^5)^{1/5} = -\sqrt[5]{32x^5} = -2x$

h. $(xyz)^{1/4} = \sqrt[4]{xyz}$ ∎

Self Check   Assume that $x > 0$. Simplify   **a.** $16^{1/2}$,   **b.** $\left(\tfrac{27}{8}\right)^{1/3}$,   and   **c.** $-(16x^4)^{1/4}$.

Answers   **a.** 4,   **b.** $\tfrac{3}{2}$,   **c.** $-2x$

EXAMPLE 2   Write each radical as an expression with a fractional exponent: **a.** $\sqrt[4]{5xyz}$ and
b. $\sqrt[5]{\dfrac{xy^2}{15}}$.

Solution   **a.** $\sqrt[4]{5xyz} = (5xyz)^{1/4}$ **b.** $\sqrt[5]{\dfrac{xy^2}{15}} = \left(\dfrac{xy^2}{15}\right)^{1/5}$ ∎

Self Check   Write the radical with a fractional exponent: $\sqrt[6]{4ab}$.

Answer $(4ab)^{1/6}$

■ EXPONENTIAL EXPRESSIONS WITH VARIABLES IN THEIR BASES

As with radicals, when n is odd in the expression $x^{1/n}$ $(n > 1)$, there is exactly one real nth root, and we don't have to worry about absolute value symbols.

When n is even, there are two nth roots. Since we want the expression $x^{1/n}$ to represent the positive nth root, we must often use absolute value symbols to guarantee that the simplified result is positive. Thus, if n is even,

$$(x^n)^{1/n} = |x|$$

When n is even and x is negative, the expression $x^{1/n}$ is not a real number.

c,e

EXAMPLE 3   Assume that all variables can be any real number, and simplify each expression.

a. $(-27x^3)^{1/3} = -3x$ Because $(-3x)^3 = -27x^3$. Since n is odd, no absolute value symbols are needed.

b. $(49x^2)^{1/2} = |7x|$ Because $(|7x|)^2 = 49x^2$. Since $7x$ can be
$\phantom{(49x^2)^{1/2}} = 7|x|$ negative, absolute value symbols are needed.

c. $(256a^8)^{1/8} = 2|a|$ Because $(2|a|)^8 = 256a^8$. Since a can be any real number, $2a$ can be negative. Thus, absolute value symbols are needed.

d. $[(y + 1)^2]^{1/2} = |y + 1|$ Because $|y + 1|^2 = (y + 1)^2$. Since y can be any real number, $y + 1$ can be negative, and the absolute value symbols are needed.

e. $(25b^4)^{1/2} = 5b^2$ Because $(5b^2)^2 = 25b^4$. Since $b^2 \geq 0$, no absolute value symbols are needed.

f. $(-256x^4)^{1/4}$ is not a real number.   Because no real number raised to the 4th power is $-256x^4$. ■

Self Check   Simplify each expression: **a.** $(625a^4)^{1/4}$ and   **b.** $(b^4)^{1/2}$.
Answer   **a.** $5|a|$,   **b.** b^2

We summarize the cases as follows.

Summary of the Definitions of $x^{1/n}$

If n is a natural number greater than 1 and x is a real number, then

If $x > 0$, then $x^{1/n}$ is the positive number such that $(x^{1/n})^n = x$.

If $x = 0$, then $x^{1/n} = 0$.

If $x < 0$ $\begin{cases} \text{and } n \text{ is odd, then } x^{1/n} \text{ is the real number such that } (x^{1/n})^n = x. \\ \text{and } n \text{ is even, then } x^{1/n} \text{ is not a real number.} \end{cases}$

■ FRACTIONAL EXPONENTS WITH NUMERATORS OTHER THAN 1

We can extend the definition of $x^{1/n}$ to include fractional exponents with numerators other than 1. For example, since $4^{3/2}$ can be written as $(4^{1/2})^3$, we have

$$4^{3/2} = (4^{1/2})^3 = (\sqrt{4})^3 = 2^3 = 8$$

Thus, we can simplify $4^{3/2}$ by cubing the square root of 4. We can also simplify $4^{3/2}$ by taking the square root of 4 cubed.

$$4^{3/2} = (4^3)^{1/2} = 64^{1/2} = \sqrt{64} = 8$$

In general, we have the following rule.

Changing from Rational Exponents to Radicals
If m and n are positive integers, $x \geq 0$, and $\frac{m}{n}$ is in simplified form, then
$$x^{m/n} = \left(\sqrt[n]{x}\right)^m = \sqrt[n]{x^m}$$

Because of the previous definition, we can interpret $x^{m/n}$ in two ways:

1. $x^{m/n}$ means the mth power of the nth root of x.
2. $x^{m/n}$ means the nth root of the mth power of x.

EXAMPLE 4   Simplify each expression.

a. $27^{2/3} = \left(\sqrt[3]{27}\right)^2$ or $27^{2/3} = \sqrt[3]{27^2}$
$\phantom{27^{2/3}} = 3^2$ $ = \sqrt[3]{729}$
$\phantom{27^{2/3}} = 9$ $ = 9$

b. $\left(\dfrac{1}{16}\right)^{3/4} = \left(\sqrt[4]{\dfrac{1}{16}}\right)^3$ or $\left(\dfrac{1}{16}\right)^{3/4} = \sqrt[4]{\left(\dfrac{1}{16}\right)^3}$
$ = \left(\dfrac{1}{2}\right)^3$ $ = \sqrt[4]{\dfrac{1}{4,096}}$
$ = \dfrac{1}{8}$ $ = \dfrac{1}{8}$

c. $(-8x^3)^{4/3} = \left(\sqrt[3]{-8x^3}\right)^4$ or $(-8x^3)^{4/3} = \sqrt[3]{(-8x^3)^4}$
$ = (-2x)^4$ $ = \sqrt[3]{4,096x^{12}}$
$ = 16x^4$ $ = 16x^4$ ■

Self Check   Simplify   **a.** $16^{3/2}$ and   **b.** $(-27x^6)^{2/3}$.
Answers   **a.** 64,   **b.** $9x^4$

To avoid large numbers, it is usually better to find the root of the base first, as shown in Example 4.

■ NEGATIVE FRACTIONAL EXPONENTS

To be consistent with the definition of negative integer exponents, we define $x^{-m/n}$ as follows.

> **Definition of $x^{-m/n}$**
>
> If m and n are positive integers, $\frac{m}{n}$ is in simplified form, and $x^{1/n}$ is a real number, then
>
> $$x^{-m/n} = \frac{1}{x^{m/n}} \qquad \text{and} \qquad \frac{1}{x^{-m/n}} = x^{m/n} \quad (x \neq 0)$$

 a

EXAMPLE 5   Write each expression without using negative exponents, if possible.

a. $64^{-1/2} = \dfrac{1}{64^{1/2}}$

$= \dfrac{1}{8}$

b. $16^{-3/2} = \dfrac{1}{16^{3/2}}$

$= \dfrac{1}{(16^{1/2})^3}$

$= \dfrac{1}{64} \qquad \begin{array}{l} (16^{1/2})^3 = \\ 4^3 = 64. \end{array}$

c. $(-32x^5)^{-2/5} = \dfrac{1}{(-32x^5)^{2/5}} \quad (x \neq 0)$

$= \dfrac{1}{[(-32x^5)^{1/5}]^2}$

$= \dfrac{1}{(-2x)^2}$

$= \dfrac{1}{4x^2}$

d. $(-16)^{-3/4}$ is not a real number, because $(-16)^{1/4}$ is not a real number.

■

Self Check   Write without using negative exponents: **a.** $25^{-3/2}$ and   **b.** $(-27a^3)^{-2/3}$.

Answers   **a.** $\dfrac{1}{125}$,   **b.** $\dfrac{1}{9a^2}$

> **WARNING!**   By definition, 0^0 is undefined. A base of 0 raised to a negative power is also undefined, because 0^{-2} would equal $\frac{1}{0^2}$, which is undefined since we cannot divide by 0.

We can use the laws of exponents to simplify many expressions with fractional exponents. If all variables represent positive numbers, no absolute value symbols are necessary.

EXAMPLE 6   Assume that all variables represent positive numbers. Write all answers without using negative exponents.

a. $5^{2/7}5^{3/7} = 5^{2/7+3/7}$ Use the rule $x^m x^n = x^{m+n}$.

       $= 5^{5/7}$ Add: $\frac{2}{7} + \frac{3}{7} = \frac{5}{7}$.

b. $(5^{2/7})^3 = 5^{(2/7)(3)}$ Use the rule $(x^m)^n = x^{mn}$.

       $= 5^{6/7}$ Multiply: $\frac{2}{7}(3) = \frac{6}{7}$.

c. $(a^{2/3}b^{1/2})^6 = (a^{2/3})^6(b^{1/2})^6$ Use the rule $(xy)^n = x^n y^n$.

       $= a^{12/3}b^{6/2}$ Use the rule $(x^m)^n = x^{mn}$ twice.

       $= a^4 b^3$ Simplify the exponent.

d. $\dfrac{a^{8/3}a^{1/3}}{a^2} = a^{8/3+1/3-2}$ Use the rules $x^m x^n = x^{m+n}$ and $\frac{x^m}{x^n} = x^{m-n}$.

       $= a^{8/3+1/3-6/3}$ $2 = \frac{6}{3}$.

       $= a^{3/3}$ $\frac{8}{3} + \frac{1}{3} - \frac{6}{3} = \frac{3}{3}$.

       $= a$ $\frac{3}{3} = 1$.   ∎

Self Check   Simplify   **a.** $(x^{1/3}y^{3/2})^6$ and   **b.** $\dfrac{x^{5/3}x^{2/3}}{x^{1/3}}$.

Answers   **a.** $x^2 y^9$,   **b.** x^2

EXAMPLE 7   Assume that all variables represent positive numbers, and do the operations. Write all answers without using negative exponents.

a. $a^{4/5}(a^{1/5} + a^{3/5}) = a^{4/5}a^{1/5} + a^{4/5}a^{3/5}$ Use the distributive property.

       $= a^{4/5+1/5} + a^{4/5+3/5}$ Use the rule $x^m x^n = x^{m+n}$.

       $= a^{5/5} + a^{7/5}$ Simplify the exponents.

       $= a + a^{7/5}$

 WARNING!   Note that $a + a^{7/5} \neq a^{1+7/5}$. The expression $a + a^{7/5}$ cannot be simplified, because a and $a^{7/5}$ are not like terms.

b. $x^{1/2}(x^{-1/2} + x^{1/2}) = x^{1/2}x^{-1/2} + x^{1/2}x^{1/2}$ Use the distributive property.

       $= x^{1/2-1/2} + x^{1/2+1/2}$ Use the rule $x^m x^n = x^{m+n}$.

       $= x^0 + x^1$ Simplify.

       $= 1 + x$ $x^0 = 1$.

c. $(x^{2/3} + 1)(x^{2/3} - 1) = x^{4/3} - x^{2/3} + x^{2/3} - 1$ Use the FOIL method.

$$= x^{4/3} - 1$$

d. $(x^{1/2} + y^{1/2})^2 = (x^{1/2} + y^{1/2})(x^{1/2} + y^{1/2})$

$$= x + 2x^{1/2}y^{1/2} + y$$ Use the FOIL method.   ∎

■ SIMPLIFYING RADICAL EXPRESSIONS

We can simplify many radical expressions by using the following steps.

> ### Using Fractional Exponents to Simplify Radicals
> 1. Change the radical expression into an exponential expression with rational exponents.
> 2. Simplify the rational exponents.
> 3. Change the exponential expression back into a radical.

EXAMPLE 8   Simplify **a.** $\sqrt[4]{3^2}$, **b.** $\sqrt[8]{x^6}$, and **c.** $\sqrt[9]{27x^6y^3}$.

Solution   **a.** $\sqrt[4]{3^2} = (3^2)^{1/4}$ Change the radical to an exponential expression.

$\qquad\quad = 3^{2/4}$ Use the rule $(x^m)^n = x^{mn}$.

$\qquad\quad = 3^{1/2}$ $\frac{2}{4} = \frac{1}{2}$.

$\qquad\quad = \sqrt{3}$ Change back to radical notation.

b. $\sqrt[8]{x^6} = (x^6)^{1/8}$ Change the radical to an exponential expression.

$\qquad\quad = x^{6/8}$ Use the rule $(x^m)^n = x^{mn}$.

$\qquad\quad = x^{3/4}$ $\frac{6}{8} = \frac{3}{4}$.

$\qquad\quad = (x^3)^{1/4}$ $\frac{3}{4} = 3\left(\frac{1}{4}\right)$.

$\qquad\quad = \sqrt[4]{x^3}$ Change back to radical notation.

c. $\sqrt[9]{27x^6y^3} = (3^3x^6y^3)^{1/9}$ Write 27 as 3^3 and change the radical to an exponential expression.

$\qquad\quad = 3^{3/9}x^{6/9}y^{3/9}$ Raise each factor to the $\frac{1}{9}$ power by multiplying the fractional exponents.

$\qquad\quad = 3^{1/3}x^{2/3}y^{1/3}$ Simplify each fractional exponent.

$\qquad\quad = (3x^2y)^{1/3}$ Use the rule $(xy)^n = x^n y^n$.

$\qquad\quad = \sqrt[3]{3x^2y}$ Change back to radical notation.   ∎

Self Check   Simplify **a.** $\sqrt[6]{3^3}$ and **b.** $\sqrt[4]{64x^2y^2}$.

Answers   **a.** $\sqrt{3}$, **b.** $\sqrt{8xy}$

Orals *Simplify each expression.*

1. $4^{1/2}$ **2.** $9^{1/2}$ **3.** $27^{1/3}$ **4.** $1^{1/4}$

5. $4^{3/2}$ **6.** $8^{2/3}$ **7.** $\left(\dfrac{1}{4}\right)^{1/2}$ **8.** $\left(\dfrac{1}{4}\right)^{-1/2}$

9. $(8x^3)^{1/3}$ **10.** $(16x^8)^{1/4}$

EXERCISE 9.4

REVIEW *Solve each inequality.*

1. $5x - 4 < 11$ **2.** $2(3t - 5) \geq 8$ **3.** $\dfrac{4}{5}(r - 3) > \dfrac{2}{3}(r + 2)$ **4.** $-4 < 2x - 4 \leq 8$

5. How much water must be added to 5 pints of a 20% alcohol solution to dilute it to a 15% solution?

6. A grocer bought some boxes of apples for $70. However, 4 boxes were spoiled. The grocer sold the remaining boxes at a profit of $2 each. How many boxes did the grocer sell if she managed to break even?

VOCABULARY AND CONCEPTS *Fill in each blank to make a true statement.*

7. $a^4 =$ _____ **8.** $a^m a^n =$ _____

9. $(a^m)^n =$ ____ **10.** $(ab)^n =$ _____

11. $\left(\dfrac{a}{b}\right)^n =$ ___ **12.** $a^0 =$ __, provided $a \neq$ __

13. $a^{-n} =$ ___, provided $a \neq$ __ **14.** $\dfrac{a^m}{a^n} =$ _____

15. $\left(\dfrac{a}{b}\right)^{-n} =$ ____ **16.** $x^{1/n} =$ ____

17. $(x^n)^{1/n} =$ ____, provided n is even **18.** $x^{m/n} = \sqrt[n]{x^m} =$ _____

PRACTICE *In Exercises 19–26, change each expression into radical notation.*

19. $7^{1/3}$ **20.** $26^{1/2}$ **21.** $(3x)^{1/4}$ **22.** $(4ab)^{1/6}$

23. $\left(\dfrac{1}{2}x^3 y\right)^{1/4}$ **24.** $\left(\dfrac{3}{4}a^2 b^2\right)^{1/5}$ **25.** $(x^2 + y^2)^{1/2}$ **26.** $(x^3 + y^3)^{1/3}$

In Exercises 27–34, change each radical to an exponential expression.

27. $\sqrt{11}$ **28.** $\sqrt[3]{12}$ **29.** $\sqrt[4]{3a}$ **30.** $3\sqrt[5]{a}$

31. $\sqrt[6]{\dfrac{1}{7}abc}$ **32.** $\sqrt[7]{\dfrac{3}{8}p^2 q}$ **33.** $\sqrt[3]{a^2 - b^2}$ **34.** $\sqrt{x^2 + y^2}$

In Exercises 35–54, simplify each expression, if possible.

35. $4^{1/2}$ **36.** $25^{1/2}$ **37.** $8^{1/3}$ **38.** $125^{1/3}$

39. $16^{1/4}$ **40.** $625^{1/4}$ **41.** $32^{1/5}$ **42.** $0^{1/5}$

43. $\left(\dfrac{1}{4}\right)^{1/2}$ **44.** $\left(\dfrac{1}{16}\right)^{1/2}$ **45.** $\left(\dfrac{1}{8}\right)^{1/3}$ **46.** $\left(\dfrac{1}{16}\right)^{1/4}$

47. $-16^{1/4}$ **48.** $-125^{1/3}$ **49.** $(-27)^{1/3}$ **50.** $(-125)^{1/3}$

51. $(-64)^{1/2}$ **52.** $(-243)^{1/5}$ **53.** $0^{1/3}$ **54.** $(-216)^{1/2}$

In Exercises 55–62, simplify each expression, if possible. Assume that all variables are unrestricted, and use absolute value symbols when necessary.

55. $(25y^2)^{1/2}$ **56.** $(-27x^3)^{1/3}$ **57.** $(16x^4)^{1/4}$ **58.** $(-16x^4)^{1/2}$

59. $(243x^5)^{1/5}$ **60.** $[(x+1)^4]^{1/4}$ **61.** $(-64x^8)^{1/4}$ **62.** $[(x+5)^3]^{1/3}$

In Exercises 63–74, simplify each expression.

63. $36^{3/2}$ **64.** $27^{2/3}$ **65.** $81^{3/4}$ **66.** $100^{3/2}$

67. $144^{3/2}$ **68.** $1{,}000^{2/3}$ **69.** $\left(\dfrac{1}{8}\right)^{2/3}$ **70.** $\left(\dfrac{4}{9}\right)^{3/2}$

71. $(25x^4)^{3/2}$ **72.** $(27a^3b^3)^{2/3}$ **73.** $\left(\dfrac{8x^3}{27}\right)^{2/3}$ **74.** $\left(\dfrac{27}{64y^6}\right)^{2/3}$

In Exercises 75–90, write each expression without using negative exponents. Assume that all variables represent positive numbers.

75. $4^{-1/2}$ **76.** $8^{-1/3}$ **77.** $(4)^{-3/2}$ **78.** $25^{-5/2}$

79. $(16x^2)^{-3/2}$ **80.** $(81c^4)^{-3/2}$ **81.** $(-27y^3)^{-2/3}$ **82.** $(-8z^9)^{-2/3}$

83. $(-32p^5)^{-2/5}$ **84.** $(16q^6)^{-5/2}$ **85.** $\left(\dfrac{1}{4}\right)^{-3/2}$ **86.** $\left(\dfrac{4}{25}\right)^{-3/2}$

87. $\left(\dfrac{27}{8}\right)^{-4/3}$ **88.** $\left(\dfrac{25}{49}\right)^{-3/2}$ **89.** $\left(-\dfrac{8x^3}{27}\right)^{-1/3}$ **90.** $\left(\dfrac{16}{81y^4}\right)^{-3/4}$

In Exercises 91–102, do the operations. Write answers without negative exponents. Assume that all variables represent positive numbers.

91. $5^{4/9}5^{4/9}$ **92.** $4^{2/5}4^{2/5}$ **93.** $(4^{1/5})^3$ **94.** $(3^{1/3})^5$

95. $\dfrac{9^{4/5}}{9^{3/5}}$ **96.** $\dfrac{7^{2/3}}{7^{1/2}}$ **97.** $\dfrac{7^{1/2}}{7^0}$ **98.** $5^{1/3}5^{-5/3}$

99. $6^{-2/3}6^{-4/3}$ **100.** $\dfrac{3^{4/3}3^{1/3}}{3^{2/3}}$ **101.** $\dfrac{2^{5/6}2^{1/3}}{2^{1/2}}$ **102.** $\dfrac{5^{1/3}5^{1/2}}{5^{1/3}}$

In Exercises 103–114, do the operations. Assume that all variables are positive, and write all answers without using negative exponents.

103. $a^{2/3}a^{1/3}$

104. $b^{3/5}b^{1/5}$

105. $(a^{2/3})^{1/3}$

106. $(t^{4/5})^{10}$

107. $(a^{1/2}b^{1/3})^{3/2}$

108. $(a^{3/5}b^{3/2})^{2/3}$

109. $(mn^{-2/3})^{-3/5}$

110. $(r^{-2}s^3)^{1/3}$

111. $\dfrac{(4x^3y)^{1/2}}{(9xy)^{1/2}}$

112. $\dfrac{(27x^3y)^{1/3}}{(8xy^2)^{2/3}}$

113. $(27x^{-3})^{-1/3}$

114. $(16a^{-2})^{-1/2}$

In Exercises 115–126, do the multiplications. Assume that all variables represent positive numbers, and write all answers without using negative exponents.

115. $y^{1/3}(y^{2/3} + y^{5/3})$

116. $y^{2/5}(y^{-2/5} + y^{3/5})$

117. $x^{3/5}(x^{7/5} - x^{2/5} + 1)$

118. $x^{4/3}(x^{2/3} + 3x^{5/3} - 4)$

119. $(x^{1/2} + 2)(x^{1/2} - 2)$

120. $(x^{1/2} + y^{1/2})(x^{1/2} - y^{1/2})$

121. $(x^{2/3} - x)(x^{2/3} + x)$

122. $(x^{1/3} + x^2)(x^{1/3} - x^2)$

123. $(x^{2/3} + y^{2/3})^2$

124. $(a^{1/2} - b^{2/3})^2$

125. $(a^{3/2} - b^{3/2})^2$

126. $(x^{-1/2} - x^{1/2})^2$

In Exercises 127–130, use rational exponents to simplify each radical. Assume that all variables represent positive numbers.

127. $\sqrt[6]{p^3}$

128. $\sqrt[8]{q^2}$

129. $\sqrt[4]{25b^2}$

130. $\sqrt[9]{-8x^6}$

WRITING

131. Explain how you would decide whether $a^{1/n}$ is a real number.

132. The expression $(a^{1/2} + b^{1/2})^2$ is not equal to $a + b$. Explain.

SOMETHING TO THINK ABOUT

133. The fraction $\frac{2}{4}$ is equal to $\frac{1}{2}$. Is $16^{2/4}$ equal to $16^{1/2}$? Explain.

134. How would you evaluate an expression with a mixed-number exponent? For example, what is $8^{1\frac{1}{3}}$? What is $25^{2\frac{1}{2}}$? Discuss.

9.5 Simplifying and Combining Radical Expressions

■ PROPERTIES OF RADICALS ■ SIMPLIFYING RADICAL EXPRESSIONS ■ ADDING AND SUBTRACTING RADICAL EXPRESSIONS ■ SOME SPECIAL TRIANGLES

Getting Ready *Simplify each radical. Assume that all variables represent positive numbers.*

1. $\sqrt{225}$

2. $\sqrt{576}$

3. $\sqrt[3]{125}$

4. $\sqrt[3]{343}$

5. $\sqrt{16x^4}$

6. $\sqrt{\dfrac{64}{121}x^6}$

7. $\sqrt[3]{27a^3b^9}$

8. $\sqrt[3]{-8a^{12}}$

■ PROPERTIES OF RADICALS

Many properties of exponents have counterparts in radical notation. For example, because $a^{1/n}b^{1/n} = (ab)^{1/n}$, we have

1. $\sqrt[n]{a}\sqrt[n]{b} = \sqrt[n]{ab}$

For example,

$$\sqrt{5}\sqrt{5} = \sqrt{5 \cdot 5} = \sqrt{5^2} = 5$$
$$\sqrt[3]{7x}\sqrt[3]{49x^2} = \sqrt[3]{7x \cdot 7^2x^2} = \sqrt[3]{7^3 \cdot x^3} = 7x$$
$$\sqrt[4]{2x^3}\sqrt[4]{8x} = \sqrt[4]{2x^3 \cdot 2^3x} = \sqrt[4]{2^4 \cdot x^4} = 2|x|$$

If we rewrite Equation 1, we have the following rule.

> **Multiplication Property of Radicals**
> If $\sqrt[n]{a}$ and $\sqrt[n]{b}$ are real numbers, then
> $$\sqrt[n]{ab} = \sqrt[n]{a}\sqrt[n]{b}$$

As long as all radicals represent real numbers, *the nth root of the product of two numbers is equal to the product of their nth roots.*

WARNING! The multiplication property of radicals applies to the *n*th root of the product of two numbers. There is no such property for sums or differences. For example,

$$\sqrt{9 + 4} \neq \sqrt{9} + \sqrt{4} \qquad \sqrt{9 - 4} \neq \sqrt{9} - \sqrt{4}$$
$$\sqrt{13} \neq 3 + 2 \qquad\qquad \sqrt{5} \neq 3 - 2$$
$$\sqrt{13} \neq 5 \qquad\qquad\qquad \sqrt{5} \neq 1$$

Thus, $\sqrt{a + b} \neq \sqrt{a} + \sqrt{b}$ and $\sqrt{a - b} \neq \sqrt{a} - \sqrt{b}$.

A second property of radicals involves quotients. Because

$$\frac{a^{1/n}}{b^{1/n}} = \left(\frac{a}{b}\right)^{1/n}$$

it follows that

2. $\dfrac{\sqrt[n]{a}}{\sqrt[n]{b}} = \sqrt[n]{\dfrac{a}{b}}$ $(b \neq 0)$

For example,

$$\frac{\sqrt{8x^3}}{\sqrt{2x}} = \sqrt{\frac{8x^3}{2x}} = \sqrt{4x^2} = 2x \quad (x > 0)$$
$$\frac{\sqrt[3]{54x^5}}{\sqrt[3]{2x^2}} = \sqrt[3]{\frac{54x^5}{2x^2}} = \sqrt[3]{27x^3} = 3x$$

If we rewrite Equation 2, we have the following rule.

> **Quotient Property of Radicals**
> If $\sqrt[n]{a}$ and $\sqrt[n]{b}$ are real numbers, then
> $$\sqrt[n]{\frac{a}{b}} = \frac{\sqrt[n]{a}}{\sqrt[n]{b}} \quad (b \neq 0)$$

As long as all radicals represent real numbers, *the nth root of the quotient of two numbers is equal to the quotient of their nth roots.*

■ SIMPLIFYING RADICAL EXPRESSIONS

A radical expression is said to be in simplest form when each of the following statements is true.

> **Simplified Form of a Radical Expression**
> A radical expression is in simplest form when
> 1. No radicals appear in the denominator of a fraction.
> 2. The radicand contains no fractions or negative numbers.
> 3. Each prime and variable factor in the radicand appears to a power that is less than the index of the radical.

EXAMPLE 1   Simplify   **a.** $\sqrt{12}$,   **b.** $\sqrt{98}$,   and   **c.** $\sqrt[3]{54}$.

Solution   **a.** Recall that numbers that are squares of integers, such as 1, 4, 9, 16, 25, and 36, are *perfect squares*. To simplify $\sqrt{12}$, we first factor 12 so that one factor is the largest perfect square that divides 12. Since 4 is the largest perfect square factor of 12, we write 12 as $4 \cdot 3$, use the multiplication property of radicals, and simplify.

$$\begin{aligned} \sqrt{12} &= \sqrt{4 \cdot 3} && \text{Write 12 as } 4 \cdot 3. \\ &= \sqrt{4}\sqrt{3} && \sqrt{4 \cdot 3} = \sqrt{4}\sqrt{3}. \\ &= 2\sqrt{3} && \sqrt{4} = 2. \end{aligned}$$

b. The largest perfect square factor of 98 is 49. Thus,

$$\begin{aligned} \sqrt{98} &= \sqrt{49 \cdot 2} && \text{Write 98 as } 49 \cdot 2. \\ &= \sqrt{49}\sqrt{2} && \sqrt{49 \cdot 2} = \sqrt{49}\sqrt{2}. \\ &= 7\sqrt{2} && \sqrt{49} = 7. \end{aligned}$$

c. Numbers that are cubes of integers, such as 1, 8, 27, 64, 125, and 216, are called *perfect cubes*. Since the largest perfect cube factor of 54 is 27, we have

$$\begin{aligned} \sqrt[3]{54} &= \sqrt[3]{27 \cdot 2} && \text{Write 54 as } 27 \cdot 2. \\ &= \sqrt[3]{27}\sqrt[3]{2} && \sqrt[3]{27 \cdot 2} = \sqrt[3]{27}\sqrt[3]{2}. \\ &= 3\sqrt[3]{2} && \sqrt[3]{27} = 3. \end{aligned}$$

■

EXAMPLE 2 Simplify **a.** $\sqrt{\dfrac{15}{49x^2}}$ $(x > 0)$ and **b.** $\sqrt[3]{\dfrac{10x^2}{27y^6}}$ $(y \neq 0)$.

Solution **a.** We can write the square root of the quotient as the quotient of the square roots and simplify the denominator. Since $x > 0$, we have

$$\sqrt{\frac{15}{49x^2}} = \frac{\sqrt{15}}{\sqrt{49x^2}}$$

$$= \frac{\sqrt{15}}{7x}$$

b. We can write the cube root of the quotient as the quotient of two cube roots. Since $y \neq 0$, we have

$$\sqrt[3]{\frac{10x^2}{27y^6}} = \frac{\sqrt[3]{10x^2}}{\sqrt[3]{27y^6}}$$

$$= \frac{\sqrt[3]{10x^2}}{3y^2}$$

EXAMPLE 3 Simplify each expression. Assume that all variables represent positive numbers.

a. $\sqrt{128a^5}$, **b.** $\sqrt[3]{24x^5}$, **c.** $\dfrac{\sqrt{45xy^2}}{\sqrt{5x}}$, and **d.** $\dfrac{\sqrt[3]{-432x^5}}{\sqrt[3]{8x}}$.

Solution **a.** We write $128a^5$ as $64a^4 \cdot 2a$ and use the multiplication property of radicals.

$$\sqrt{128a^5} = \sqrt{64a^4 \cdot 2a} \qquad \text{$64a^4$ is the largest perfect square that divides $128a^5$.}$$

$$= \sqrt{64a^4}\sqrt{2a} \qquad \text{Use the multiplication property of radicals.}$$

$$= 8a^2\sqrt{2a} \qquad \sqrt{64a^4} = 8a^2.$$

b. We write $24x^5$ as $8x^3 \cdot 3x^2$ and use the multiplication property of radicals.

$$\sqrt[3]{24x^5} = \sqrt[3]{8x^3 \cdot 3x^2} \qquad \text{$8x^3$ is the largest perfect cube that divides $24x^5$.}$$

$$= \sqrt[3]{8x^3}\sqrt[3]{3x^2} \qquad \text{Use the multiplication property of radicals.}$$

$$= 2x\sqrt[3]{3x^2} \qquad \sqrt[3]{8x^3} = 2x.$$

c. We can write the quotient of the square roots as the square root of a quotient.

$$\frac{\sqrt{45xy^2}}{\sqrt{5x}} = \sqrt{\frac{45xy^2}{5x}}$$ Use the quotient property of radicals.

$$= \sqrt{9y^2}$$ Simplify the fraction.

$$= 3y$$

d. We can write the quotient of the cube roots as the cube root of a quotient.

$$\frac{\sqrt[3]{-432x^5}}{\sqrt[3]{8x}} = \sqrt[3]{\frac{-432x^5}{8x}}$$ Use the quotient property of radicals.

$$= \sqrt[3]{-54x^4}$$ Simplify the fraction.

$$= \sqrt[3]{-27x^3 \cdot 2x}$$ $-27x^3$ is the largest perfect cube that divides $-54x^4$.

$$= \sqrt[3]{-27x^3}\sqrt[3]{2x}$$ Use the multiplication property of radicals.

$$= -3x\sqrt[3]{2x}$$ ■

Self Check   Simplify **a.** $\sqrt{98b^3}$, **b.** $\sqrt[3]{54y^5}$, and **c.** $\dfrac{\sqrt{50ab^2}}{\sqrt{2a}}$. Assume that all variables represent positive numbers.

Answers   **a.** $7b\sqrt{2b}$, **b.** $3y\sqrt[3]{2y^2}$, **c.** $5b$

To simplify more complicated radicals, we can use the prime factorization of the radicand to find its perfect square factors. For example, to simplify $\sqrt{3,168x^5y^7}$, we first find the prime factorization of $3,168x^5y^7$.

$$3,168x^5y^7 = 2^5 \cdot 3^2 \cdot 11 \cdot x^5 \cdot y^7$$

Then we have

$$\sqrt{3,168x^5y^7} = \sqrt{2^4 \cdot 3^2 \cdot x^4 \cdot y^6 \cdot 2 \cdot 11 \cdot x \cdot y}$$

$$= \sqrt{2^4 \cdot 3^2 \cdot x^4 \cdot y^6}\sqrt{2 \cdot 11 \cdot x \cdot y}$$ Write each perfect square under the left radical and each nonperfect square under the right radical.

$$= 2^2 \cdot 3x^2y^3\sqrt{22xy}$$

$$= 12x^2y^3\sqrt{22xy}$$

■ ADDING AND SUBTRACTING RADICAL EXPRESSIONS

Radical expressions with the same index and the same radicand are called **like** or **similar radicals.** For example, $3\sqrt{2}$ and $2\sqrt{2}$ are like radicals. However,

$3\sqrt{5}$ and $4\sqrt{2}$ are not like radicals, because the radicands are different.

$3\sqrt{5}$ and $2\sqrt[3]{5}$ are not like radicals, because the indexes are different.

We can often combine like terms. For example, to simplify the expression $3\sqrt{2} + 2\sqrt{2}$, we use the distributive property to factor out $\sqrt{2}$ and simplify.

$$3\sqrt{2} + 2\sqrt{2} = (3 + 2)\sqrt{2}$$
$$= 5\sqrt{2}$$

Radicals with the same index but different radicands can often be written as like radicals. For example, to simplify the expression $\sqrt{27} - \sqrt{12}$, we simplify both radicals and then combine the like radicals.

$$\sqrt{27} - \sqrt{12} = \sqrt{9 \cdot 3} - \sqrt{4 \cdot 3}$$
$$= \sqrt{9}\sqrt{3} - \sqrt{4}\sqrt{3} \qquad \sqrt{ab} = \sqrt{a}\sqrt{b}$$
$$= 3\sqrt{3} - 2\sqrt{3} \qquad \sqrt{9} = 3 \text{ and } \sqrt{4} = 2.$$
$$= (3 - 2)\sqrt{3} \qquad \text{Factor out } \sqrt{3}.$$
$$= \sqrt{3}$$

As the previous examples suggest, we can use the following rule to add or subtract radicals.

Adding and Subtracting Radicals

To add or subtract radicals, simplify each radical and combine all like radicals. To combine like radicals, add the coefficients and keep the common radical.

EXAMPLE 4   Simplify $2\sqrt{12} - 3\sqrt{48} + 3\sqrt{3}$.

Solution   We simplify each radical separately and combine like radicals.

$$2\sqrt{12} - 3\sqrt{48} + 3\sqrt{3} = 2\sqrt{4 \cdot 3} - 3\sqrt{16 \cdot 3} + 3\sqrt{3}$$
$$= 2\sqrt{4}\sqrt{3} - 3\sqrt{16}\sqrt{3} + 3\sqrt{3}$$
$$= 2(2)\sqrt{3} - 3(4)\sqrt{3} + 3\sqrt{3}$$
$$= 4\sqrt{3} - 12\sqrt{3} + 3\sqrt{3}$$
$$= (4 - 12 + 3)\sqrt{3}$$
$$= -5\sqrt{3}$$

Self Check   Simplify $3\sqrt{75} - 2\sqrt{12} + 2\sqrt{48}$.
Answer $19\sqrt{3}$

EXAMPLE 5   Simplify $\sqrt[3]{16} - \sqrt[3]{54} + \sqrt[3]{24}$.

Solution   We simplify each radical separately and combine like radicals:

$$\sqrt[3]{16} - \sqrt[3]{54} + \sqrt[3]{24} = \sqrt[3]{8 \cdot 2} - \sqrt[3]{27 \cdot 2} + \sqrt[3]{8 \cdot 3}$$
$$= \sqrt[3]{8}\sqrt[3]{2} - \sqrt[3]{27}\sqrt[3]{2} + \sqrt[3]{8}\sqrt[3]{3}$$
$$= 2\sqrt[3]{2} - 3\sqrt[3]{2} + 2\sqrt[3]{3}$$
$$= -\sqrt[3]{2} + 2\sqrt[3]{3}$$

WARNING! We cannot combine $-\sqrt[3]{2}$ and $2\sqrt[3]{3}$, because the radicals have different radicands.

Self Check   Simplify $\sqrt[3]{24} - \sqrt[3]{16} + \sqrt[3]{54}$.

Answer $2\sqrt[3]{3} + \sqrt[3]{2}$

EXAMPLE 6   Simplify $\sqrt[3]{16x^4} + \sqrt[3]{54x^4} - \sqrt[3]{-128x^4}$ $(x > 0)$.

Solution   We simplify each radical separately, factor out $\sqrt[3]{2x}$, and simplify.

$$\sqrt[3]{16x^4} + \sqrt[3]{54x^4} - \sqrt[3]{-128x^4}$$
$$= \sqrt[3]{8x^3 \cdot 2x} + \sqrt[3]{27x^3 \cdot 2x} - \sqrt[3]{-64x^3 \cdot 2x}$$
$$= \sqrt[3]{8x^3}\sqrt[3]{2x} + \sqrt[3]{27x^3}\sqrt[3]{2x} - \sqrt[3]{-64x^3}\sqrt[3]{2x}$$
$$= 2x\sqrt[3]{2x} + 3x\sqrt[3]{2x} + 4x\sqrt[3]{2x}$$
$$= (2x + 3x + 4x)\sqrt[3]{2x}$$
$$= 9x\sqrt[3]{2x}$$

Self Check   Simplify $\sqrt{32x^3} + \sqrt{50x^3} - \sqrt{18x^3}$ $(x > 0)$.

Answer $6x\sqrt{2x}$

■ SOME SPECIAL TRIANGLES

An **isosceles right triangle** is a right triangle with two legs of equal length. If we know the length of one leg of an isosceles right triangle, we can use the Pythagorean theorem to find the length of the hypotenuse. Since the triangle shown in Figure 9-13 is a right triangle, we have

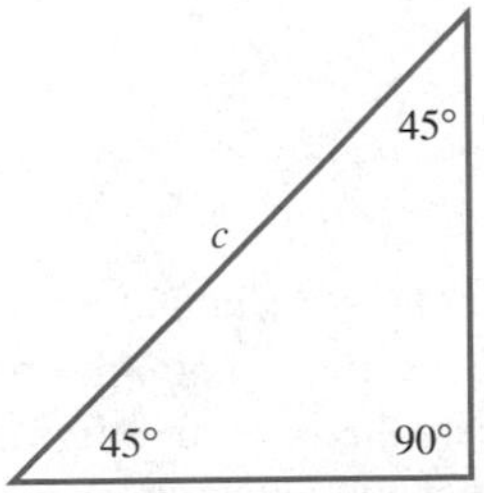

$$c^2 = a^2 + a^2 \qquad \text{Use the Pythagorean theorem.}$$
$$c^2 = 2a^2 \qquad \text{Combine like terms.}$$
$$\sqrt{c^2} = \sqrt{2a^2} \qquad \text{Take the positive square root of both sides.}$$
$$c = a\sqrt{2} \qquad \sqrt{2a^2} = \sqrt{2}\sqrt{a^2} = \sqrt{2}a = a\sqrt{2}. \text{ No absolute value symbols are needed, because } a \text{ and } c \text{ are positive.}$$

FIGURE 9-13

Thus, *in an isosceles right triangle, the length of the hypotenuse is the length of one leg times* $\sqrt{2}$.

EXAMPLE 7   If one leg of the isosceles right triangle shown in Figure 9-13 is 10 feet long, find the length of the hypotenuse.

Solution   Since the length of the hypotenuse is the length of a leg times $\sqrt{2}$, we have

$$c = 10\sqrt{2}$$

The length of the hypotenuse is $10\sqrt{2}$ feet. To two decimal places, the length is 14.14 feet.   ■

Self Check   Find the length of the hypotenuse of an isosceles right triangle if one leg is 12 meters long.

Answer $12\sqrt{2}$ m

If the length of the hypotenuse of an isosceles right triangle is known, we can use the Pythagorean theorem to find the length of each leg.

EXAMPLE 8   Find the length of each leg of the isosceles right triangle shown in Figure 9-14.

Solution   We use the Pythagorean theorem.

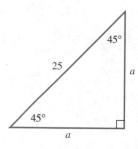

FIGURE 9-14

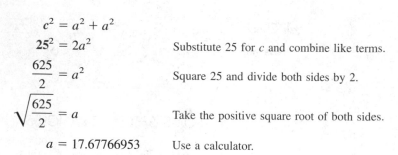

$$c^2 = a^2 + a^2$$

$$25^2 = 2a^2 \qquad \text{Substitute 25 for } c \text{ and combine like terms.}$$

$$\frac{625}{2} = a^2 \qquad \text{Square 25 and divide both sides by 2.}$$

$$\sqrt{\frac{625}{2}} = a \qquad \text{Take the positive square root of both sides.}$$

$$a = 17.67766953 \qquad \text{Use a calculator.}$$

To two decimal places, the length is 17.68 units.   ■

From geometry, we know that an **equilateral triangle** is a triangle with three sides of equal length and three 60° angles. If an **altitude** is drawn upon the base of an equilateral triangle, as shown in Figure 9-15, it bisects the base and divides the triangle into two 30°–60°–90° triangles. We can see that the shortest leg of each 30°–60°–90° triangle is a units long. Thus,

The shorter leg of a 30°–60°–90° right triangle is half as long as its hypotenuse.

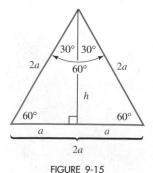

FIGURE 9-15

We can find the length of the altitude, h, by using the Pythagorean theorem.

$$a^2 + h^2 = (2a)^2$$
$$a^2 + h^2 = 4a^2 \qquad (2a)^2 = (2a)(2a) = 4a^2.$$
$$h^2 = 3a^2 \qquad \text{Subtract } a^2 \text{ from both sides.}$$
$$h = \sqrt{3a^2} \qquad \text{Take the positive square root of both sides.}$$
$$h = a\sqrt{3} \qquad \sqrt{3a^2} = \sqrt{3}\sqrt{a^2} = a\sqrt{3}. \text{ No absolute value symbols are needed, because } a \text{ is positive.}$$

Thus,

The length of the longer leg is the length of the shorter leg times $\sqrt{3}$.

EXAMPLE 9   Find the length of the hypotenuse and the longer leg of the right triangle shown in Figure 9-16.

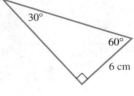

FIGURE 9-16

Solution   Since the shorter leg of a 30°–60°–90° right triangle is half as long as its hypotenuse, the hypotenuse is 12 centimeters long.
   Since the length of the longer leg is the length of the shorter leg times $\sqrt{3}$, the longer leg is $6\sqrt{3}$ (about 10.39) centimeters long.   ∎

Self Check   Find the length of the hypotenuse and the longer leg of a 30°–60°–90° right triangle if the shorter leg is 8 centimeters long.

Answers   16 cm, $8\sqrt{3}$ cm

EXAMPLE 10   Find the length of each leg of the triangle shown in Figure 9-17.

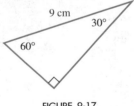

FIGURE 9-17

Solution   Since the shorter leg of a 30°–60°–90° right triangle is half as long as its hypotenuse, the shorter leg is $\frac{9}{2}$ centimeters long.
   Since the length of the longer leg is the length of the shorter leg times $\sqrt{3}$, the longer leg is $\frac{9}{2}\sqrt{3}$ (or about 7.79) centimeters long.   ∎

Orals   *Simplify.*

1. $\sqrt{7}\sqrt{7}$ **2.** $\sqrt[3]{4^2}\sqrt[3]{4}$ **3.** $\dfrac{\sqrt[3]{54}}{\sqrt[3]{2}}$

Simplify each expression. Assume that $b \neq 0$.

4. $\sqrt{18}$ **5.** $\sqrt[3]{16}$ **6.** $\sqrt[3]{\dfrac{3x^2}{64b^6}}$

Combine like terms.

7. $3\sqrt{3} + 4\sqrt{3}$

8. $5\sqrt{7} - 2\sqrt{7}$

9. $2\sqrt[3]{9} + 3\sqrt[3]{9}$

10. $10\sqrt[5]{4} - 2\sqrt[5]{4}$

EXERCISE 9.5

REVIEW *Do each operation.*

1. $3x^2y^3(-5x^3y^{-4})$

2. $-2a^2b^{-2}(4a^{-2}b^4 - 2a^2b + 3a^3b^2)$

3. $(3t + 2)^2$

4. $(5r - 3s)(5r + 2s)$

5. $2p - 5\overline{)6p^2 - 7p - 25}$

6. $3m + n\overline{)6m^3 - m^2n + 2mn^2 + n^3}$

VOCABULARY AND CONCEPTS *Fill in each blank to make a true statement.*

7. $\sqrt[n]{ab} = $ _____

8. $\sqrt[n]{\dfrac{a}{b}} = $ ___

PRACTICE *In Exercises 9–24, simplify each expression. Assume that all variables represent positive numbers.*

9. $\sqrt{6}\sqrt{6}$

10. $\sqrt{11}\sqrt{11}$

11. $\sqrt{t}\sqrt{t}$

12. $-\sqrt{z}\sqrt{z}$

13. $\sqrt[3]{5x^2}\sqrt[3]{25x}$

14. $\sqrt[4]{25a}\sqrt[4]{25a^3}$

15. $\dfrac{\sqrt{500}}{\sqrt{5}}$

16. $\dfrac{\sqrt{128}}{\sqrt{2}}$

17. $\dfrac{\sqrt{98x^3}}{\sqrt{2x}}$

18. $\dfrac{\sqrt{75y^5}}{\sqrt{3y}}$

19. $\dfrac{\sqrt{180ab^4}}{\sqrt{5ab^2}}$

20. $\dfrac{\sqrt{112ab^3}}{\sqrt{7ab}}$

21. $\dfrac{\sqrt[3]{48}}{\sqrt[3]{6}}$

22. $\dfrac{\sqrt[3]{64}}{\sqrt[3]{8}}$

23. $\dfrac{\sqrt[3]{189a^4}}{\sqrt[3]{7a}}$

24. $\dfrac{\sqrt[3]{243x^7}}{\sqrt[3]{9x}}$

In Exercises 25–44, simplify each radical.

25. $\sqrt{20}$

26. $\sqrt{8}$

27. $-\sqrt{200}$

28. $-\sqrt{250}$

29. $\sqrt[3]{80}$

30. $\sqrt[3]{270}$

31. $\sqrt[3]{-81}$

32. $\sqrt[3]{-72}$

33. $\sqrt[4]{32}$

34. $\sqrt[4]{48}$

35. $\sqrt[5]{96}$

36. $\sqrt[7]{256}$

37. $\sqrt{\dfrac{7}{9}}$

38. $\sqrt{\dfrac{3}{4}}$

39. $\sqrt[3]{\dfrac{7}{64}}$

40. $\sqrt[3]{\dfrac{4}{125}}$

41. $\sqrt[4]{\dfrac{3}{10,000}}$

42. $\sqrt[5]{\dfrac{4}{243}}$

43. $\sqrt[5]{\dfrac{3}{32}}$

44. $\sqrt[6]{\dfrac{5}{64}}$

In Exercises 45–64, simplify each radical. Assume that all variables represent positive numbers.

45. $\sqrt{50x^2}$

46. $\sqrt{75a^2}$

47. $\sqrt{32b}$

48. $\sqrt{80c}$

49. $-\sqrt{112a^3}$

50. $\sqrt{147a^5}$

51. $\sqrt{175a^2b^3}$

52. $\sqrt{128a^3b^5}$

53. $-\sqrt{300xy}$

54. $\sqrt{200x^2y}$

55. $\sqrt[3]{-54x^6}$

56. $-\sqrt[3]{-81a^3}$

57. $\sqrt[3]{16x^{12}y^3}$

58. $\sqrt[3]{40a^3b^6}$

59. $\sqrt[4]{32x^{12}y^4}$

60. $\sqrt[5]{64x^{10}y^5}$

61. $\sqrt{\dfrac{z^2}{16x^2}}$

62. $\sqrt{\dfrac{b^4}{64a^8}}$

63. $\sqrt[4]{\dfrac{5x}{16z^4}}$

64. $\sqrt[3]{\dfrac{11a^2}{125b^6}}$

In Exercises 65–104, simplify and combine like radicals. All variables represent positive numbers.

65. $4\sqrt{2x} + 6\sqrt{2x}$

66. $6\sqrt[3]{5y} + 3\sqrt[3]{5y}$

67. $8\sqrt[5]{7a^2} - 7\sqrt[5]{7a^2}$

68. $10\sqrt[6]{12xyz} - \sqrt[6]{12xyz}$

69. $\sqrt{3} + \sqrt{27}$ $4\sqrt{3}$

70. $\sqrt{8} + \sqrt{32}$

71. $\sqrt{2} - \sqrt{8}$

72. $\sqrt{20} - \sqrt{125}$

73. $\sqrt{98} - \sqrt{50}$

74. $\sqrt{72} - \sqrt{200}$

75. $3\sqrt{24} + \sqrt{54}$

76. $\sqrt{18} + 2\sqrt{50}$

77. $\sqrt[3]{24} + \sqrt[3]{3}$

78. $\sqrt[3]{16} + \sqrt[3]{128}$

79. $\sqrt[3]{32} - \sqrt[3]{108}$

80. $\sqrt[3]{80} - \sqrt[3]{10{,}000}$

81. $2\sqrt[3]{125} - 5\sqrt[3]{64}$

82. $3\sqrt[3]{27} + 12\sqrt[3]{216}$

83. $14\sqrt[4]{32} - 15\sqrt[4]{162}$

84. $23\sqrt[4]{768} + \sqrt[4]{48}$

85. $3\sqrt[4]{512} + 2\sqrt[4]{32}$

86. $4\sqrt[4]{243} - \sqrt[4]{48}$

87. $\sqrt{98} - \sqrt{50} - \sqrt{72}$

88. $\sqrt{20} + \sqrt{125} - \sqrt{80}$

89. $\sqrt{18} + \sqrt{300} - \sqrt{243}$

90. $\sqrt{80} - \sqrt{128} + \sqrt{288}$

91. $2\sqrt[3]{16} - \sqrt[3]{54} - 3\sqrt[3]{128}$

92. $\sqrt[4]{48} - \sqrt[4]{243} - \sqrt[4]{768}$

93. $\sqrt{25y^2z} - \sqrt{16y^2z}$

94. $\sqrt{25yz^2} + \sqrt{9yz^2}$

95. $\sqrt{36xy^2} + \sqrt{49xy^2}$

96. $3\sqrt{2x} - \sqrt{8x}$

97. $2\sqrt[3]{64a} + 2\sqrt[3]{8a}$

98. $3\sqrt[4]{x^4y} - 2\sqrt[4]{x^4y}$

99. $\sqrt{y^5} - \sqrt{9y^5} - \sqrt{25y^5}$

100. $\sqrt{8y^7} + \sqrt{32y^7} - \sqrt{2y^7}$

101. $\sqrt[5]{x^6y^2} + \sqrt[5]{32x^6y^2} + \sqrt[5]{x^6y^2}$

102. $\sqrt[3]{xy^4} + \sqrt[3]{8xy^4} - \sqrt[3]{27xy^4}$

103. $\sqrt{x^2 + 2x + 1} + \sqrt{x^2 + 2x + 1}$

104. $\sqrt{4x^2 + 12x + 9} + \sqrt{9x^2 + 6x + 1}$

In Exercises 105–112, find the missing lengths in each triangle. Give each answer to two decimal places.

105.

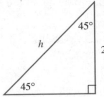

106.

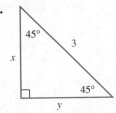

107.

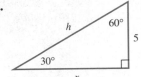

108.

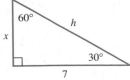

109.
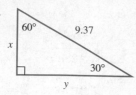

110.
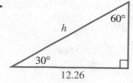

111.
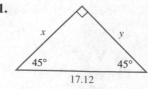

112.
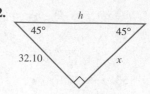

WRITING

113. Explain how to recognize like radicals.

114. Explain how to combine like radicals.

SOMETHING TO THINK ABOUT

115. Can you find any numbers a and b such that $\sqrt{a + b} = \sqrt{a} + \sqrt{b}$?

116. Find the sum: $\sqrt{3} + \sqrt{3^2} + \sqrt{3^3} + \sqrt{3^4} + \sqrt{3^5}$.

9.6 Multiplying and Dividing Radical Expressions

■ MULTIPLYING A MONOMIAL BY A MONOMIAL ■ MULTIPLYING A POLYNOMIAL BY A MONOMIAL
■ MULTIPLYING A POLYNOMIAL BY A POLYNOMIAL ■ PROBLEM SOLVING ■ RATIONALIZING
DENOMINATORS ■ RATIONALIZING NUMERATORS

Getting Ready *Do each operation and simplify, if possible.*

1. $a^3 a^4$

2. $\dfrac{b^5}{b^2}$

3. $a(a - 2)$

4. $3b^2(2b + 3)$

5. $(a + 2)(a - 5)$

6. $(2a + 3b)(2a - 3b)$

■ MULTIPLYING A MONOMIAL BY A MONOMIAL

Radical expressions with the same index can be multiplied and divided.

EXAMPLE 1   Multiply $3\sqrt{6}$ by $2\sqrt{3}$.

Solution   We use the commutative and associative properties of multiplication to multiply the coefficients and the radicals separately. Then we simplify any radicals in the product, if possible.

$$3\sqrt{6} \cdot 2\sqrt{3} = 3(2)\sqrt{6}\sqrt{3} \qquad \text{Multiply the coefficients and multiply the radicals.}$$
$$= 6\sqrt{18} \qquad 3(2) = 6 \text{ and } \sqrt{6}\sqrt{3} = \sqrt{18}.$$
$$= 6\sqrt{9}\sqrt{2} \qquad \sqrt{18} = \sqrt{9 \cdot 2} = \sqrt{9}\sqrt{2}.$$
$$= 6(3)\sqrt{2} \qquad \sqrt{9} = 3.$$
$$= 18\sqrt{2}$$

■

Self Check

Answer

Multiply $-2\sqrt{7}$ by $5\sqrt{2}$.

$-10\sqrt{14}$

■ MULTIPLYING A POLYNOMIAL BY A MONOMIAL

To multiply a polynomial by a monomial, we use the distributive property to remove parentheses and then simplify each resulting term, if possible.

EXAMPLE 2   Multiply $3\sqrt{3}\left(4\sqrt{8} - 5\sqrt{10}\right)$.

Solution

$$3\sqrt{3}\left(4\sqrt{8} - 5\sqrt{10}\right)$$
$$= 3\sqrt{3} \cdot 4\sqrt{8} - 3\sqrt{3} \cdot 5\sqrt{10} \qquad \text{Use the distributive property.}$$
$$= 12\sqrt{24} - 15\sqrt{30} \qquad \text{Multiply the coefficients and multiply the radicals.}$$
$$= 12\sqrt{4}\sqrt{6} - 15\sqrt{30}$$
$$= 12(2)\sqrt{6} - 15\sqrt{30}$$
$$= 24\sqrt{6} - 15\sqrt{30}$$

Self Check

Answer

Multiply $4\sqrt{2}\left(3\sqrt{5} - 2\sqrt{8}\right)$.

$12\sqrt{10} - 32$

■ MULTIPLYING A POLYNOMIAL BY A POLYNOMIAL

To multiply a binomial by a binomial, we use the FOIL method.

EXAMPLE 3   Multiply $\left(\sqrt{7} + \sqrt{2}\right)\left(\sqrt{7} - 3\sqrt{2}\right)$.

Solution

$$\left(\sqrt{7} + \sqrt{2}\right)\left(\sqrt{7} - 3\sqrt{2}\right)$$

$$= \left(\sqrt{7}\right)^2 - 3\sqrt{7}\sqrt{2} + \sqrt{2}\sqrt{7} - 3\sqrt{2}\sqrt{2}$$
$$= 7 - 3\sqrt{14} + \sqrt{14} - 3(2)$$
$$= 7 - 2\sqrt{14} - 6$$
$$= 1 - 2\sqrt{14}$$

Self Check

Answer

Multiply $\left(\sqrt{5} + 2\sqrt{3}\right)\left(\sqrt{5} - \sqrt{3}\right)$.

$-1 + \sqrt{15}$

Technically, the expression $\sqrt{3x} - \sqrt{5}$ is not a polynomial, because the variable does not have a whole-number exponent $\left(\sqrt{3x} = 3^{1/2}x^{1/2}\right)$. However, we will multiply such expressions as if they were polynomials.

EXAMPLE 4   Multiply $\left(\sqrt{3x} - \sqrt{5}\right)\left(\sqrt{2x} + \sqrt{10}\right)$.

Solution
$$\left(\sqrt{3x} - \sqrt{5}\right)\left(\sqrt{2x} + \sqrt{10}\right)$$

$$= \sqrt{3x}\sqrt{2x} + \sqrt{3x}\sqrt{10} - \sqrt{5}\sqrt{2x} - \sqrt{5}\sqrt{10}$$
$$= \sqrt{6x^2} + \sqrt{30x} - \sqrt{10x} - \sqrt{50}$$
$$= \sqrt{6}\sqrt{x^2} + \sqrt{30x} - \sqrt{10x} - \sqrt{25}\sqrt{2}$$
$$= \sqrt{6}x + \sqrt{30x} - \sqrt{10x} - 5\sqrt{2}$$
■

Self Check   Multiply $\left(\sqrt{x} + 1\right)\left(\sqrt{x} - 3\right)$.
Answer $x - 2\sqrt{x} - 3$

WARNING!   It is important to draw radical signs carefully so that they completely cover the radicand, but no more than the radicand. To avoid confusion, we often write an expression such as $\sqrt{6}x$ in the form $x\sqrt{6}$.

■ **PROBLEM SOLVING**

EXAMPLE 5   **Photography**   Many camera lenses (see Figure 9-18) have an adjustable opening called the *aperture,* which controls the amount of light passing through the lens. The *f-number* of a lens is its *focal length* divided by the diameter of its circular aperture:

$$f\text{-number} = \frac{f}{d} \qquad \begin{array}{l} f \text{ is the focal length,} \\ \text{and } d \text{ is the diameter of} \\ \text{the aperture.} \end{array}$$

FIGURE 9-18

A lens with a focal length of 12 centimeters and an aperture with a diameter of 6 centimeters has an *f*-number of $\frac{12}{6}$ and is an *f*/2 lens. If the area of the aperture is reduced to admit half as much light, the *f*-number of the lens will change. Find the new *f*-number.

Solution   We first find the area of the aperture when its diameter is 6 centimeters.

$$A = \pi r^2 \qquad \text{The formula for the area of a circle.}$$
$$A = \pi(3)^2 \qquad \text{Since a radius is half the diameter, substitute 3 for } r.$$
$$A = 9\pi$$

When the size of the aperture is reduced to admit half as much light, the area of the aperture will be $\frac{9\pi}{2}$ square centimeters. To find the diameter of a circle with this area, we proceed as follows:

$$A = \pi r^2 \qquad \text{The formula for the area of a circle.}$$

$$\frac{9\pi}{2} = \pi\left(\frac{d}{2}\right)^2 \qquad \text{Substitute } \frac{9\pi}{2} \text{ for } A \text{ and } \frac{d}{2} \text{ for } r.$$

$$\frac{9\pi}{2} = \frac{\pi d^2}{4} \qquad \left(\frac{d}{2}\right)^2 = \frac{d^2}{4}.$$

$$18 = d^2 \qquad \text{Multiply both sides by 4, and divide both sides by } \pi.$$

$$d = 3\sqrt{2} \qquad \sqrt{18} = \sqrt{9}\sqrt{2} = 3\sqrt{2}.$$

Since the focal length of the lens is still 12 centimeters and the diameter is now $3\sqrt{2}$ centimeters, the new f-number of the lens is

$$f\text{-number} = \frac{f}{d} = \frac{12}{3\sqrt{2}} \qquad \text{Substitute 12 for } f \text{ and } 3\sqrt{2} \text{ for } d.$$

$$\approx 2.828427125 \qquad \text{Use a calculator.}$$

The lens is now an $f/2.8$ lens. ■

■ RATIONALIZING DENOMINATORS

To divide radical expressions, we **rationalize the denominator** of a fraction to replace the denominator with a rational number. For example, to divide $\sqrt{70}$ by $\sqrt{3}$, we write the division as the fraction

$$\frac{\sqrt{70}}{\sqrt{3}}$$

To eliminate the radical in the denominator, we multiply the numerator and the denominator by a number that will give a perfect square under the radical in the denominator. Because $3 \cdot 3 = 9$ and 9 is a perfect square, $\sqrt{3}$ is such a number.

$$\frac{\sqrt{70}}{\sqrt{3}} = \frac{\sqrt{70} \cdot \sqrt{3}}{\sqrt{3} \cdot \sqrt{3}} \qquad \text{Multiply numerator and denominator by } \sqrt{3}.$$

$$= \frac{\sqrt{210}}{3} \qquad \text{Multiply the radicals.}$$

Since there is no radical in the denominator and $\sqrt{210}$ cannot be simplified, the expression $\sqrt{210}/3$ is in simplest form, and the division is complete.

EXAMPLE 6   Rationalize the denominator: **a.** $\sqrt{\dfrac{20}{7}}$ and   **b.** $\dfrac{4}{\sqrt[3]{2}}$.

Solution   **a.** We first write the square root of the quotient as the quotient of two square roots:

$$\sqrt{\frac{20}{7}} = \frac{\sqrt{20}}{\sqrt{7}}$$

We then proceed as follows:

$$\frac{\sqrt{20}}{\sqrt{7}} = \frac{\sqrt{20} \cdot \sqrt{7}}{\sqrt{7} \cdot \sqrt{7}} \qquad \text{Multiply numerator and denominator by } \sqrt{7}.$$

$$= \frac{\sqrt{140}}{7} \qquad \text{Multiply the radicals.}$$

$$= \frac{2\sqrt{35}}{7} \qquad \text{Simplify } \sqrt{140}: \sqrt{140} = \sqrt{4 \cdot 35} = \sqrt{4}\sqrt{35} = 2\sqrt{35}.$$

b. Since the denominator is a cube root, we multiply the numerator and the denominator by a number that will give a perfect cube under the radical sign. Since $2 \cdot 4 = 8$ is a perfect cube, $\sqrt[3]{4}$ is such a number.

$$\frac{4}{\sqrt[3]{2}} = \frac{4 \cdot \sqrt[3]{4}}{\sqrt[3]{2} \cdot \sqrt[3]{4}} \qquad \text{Multiply numerator and denominator by } \sqrt[3]{4}.$$

$$= \frac{4\sqrt[3]{4}}{\sqrt[3]{8}} \qquad \text{Multiply the radicals in the denominator.}$$

$$= \frac{4\sqrt[3]{4}}{2} \qquad \sqrt[3]{8} = 2.$$

$$= 2\sqrt[3]{4} \qquad \text{Simplify.} \qquad\qquad ■$$

Self Check   Rationalize the denominator: $\dfrac{5}{\sqrt[4]{3}}$.

Answer $\dfrac{5\sqrt[4]{27}}{3}$

EXAMPLE 7   Rationalize the denominator of $\dfrac{\sqrt[3]{5}}{\sqrt[3]{18}}$.

Solution   We multiply the numerator and the denominator by a number that will result in a perfect cube under the radical sign in the denominator.

Since 216 is the smallest perfect cube that is divisible by 18 ($216 \div 18 = 12$), multiplying the numerator and the denominator by $\sqrt[3]{12}$ will give the smallest possible perfect cube under the radical in the denominator.

$$\frac{\sqrt[3]{5}}{\sqrt[3]{18}} = \frac{\sqrt[3]{5} \cdot \sqrt[3]{12}}{\sqrt[3]{18} \cdot \sqrt[3]{12}} \qquad \text{Multiply numerator and denominator by } \sqrt[3]{12}.$$

$$= \frac{\sqrt[3]{60}}{\sqrt[3]{216}} \qquad \text{Multiply the radicals.}$$

$$= \frac{\sqrt[3]{60}}{6} \qquad \sqrt[3]{216} = 6.$$

Self Check Rationalize the denominator: $\dfrac{\sqrt{5}}{\sqrt{18}}$.

Answer $\dfrac{\sqrt{10}}{6}$

EXAMPLE 8 Rationalize the denominator of $\dfrac{\sqrt{5xy^2}}{\sqrt{xy^3}}$ (x and y are positive numbers).

Solution

Method 1	*Method 2*
$\dfrac{\sqrt{5xy^2}}{\sqrt{xy^3}} = \sqrt{\dfrac{5xy^2}{xy^3}}$	$\dfrac{\sqrt{5xy^2}}{\sqrt{xy^3}} = \sqrt{\dfrac{5xy^2}{xy^3}}$
$= \sqrt{\dfrac{5}{y}}$	$= \sqrt{\dfrac{5}{y}}$
$= \dfrac{\sqrt{5}}{\sqrt{y}}$	$= \sqrt{\dfrac{5 \cdot y}{y \cdot y}}$
$= \dfrac{\sqrt{5}\sqrt{y}}{\sqrt{y}\sqrt{y}}$	$= \dfrac{\sqrt{5y}}{\sqrt{y^2}}$
$= \dfrac{\sqrt{5y}}{y}$	$= \dfrac{\sqrt{5y}}{y}$

Self Check Rationalize the denominator: $\dfrac{\sqrt{4ab^3}}{\sqrt{2a^2b^2}}$.

Answer $\dfrac{\sqrt{2ab}}{a}$

To rationalize the denominator of a fraction with square roots in a binomial denominator, we multiply its numerator and denominator by the *conjugate* of its denominator. Conjugate binomials are binomials with the same terms but with opposite signs between their terms.

> **Conjugate Binomials**
> The **conjugate** of the binomial $a + b$ is $a - b$, and the conjugate of $a - b$ is $a + b$.

EXAMPLE 9   Rationalize the denominator of $\dfrac{1}{\sqrt{2} + 1}$.

Solution   We multiply the numerator and denominator of the fraction by $\sqrt{2} - 1$, which is the conjugate of the denominator.

$$\frac{1}{\sqrt{2} + 1} = \frac{1(\sqrt{2} - 1)}{(\sqrt{2} + 1)(\sqrt{2} - 1)}$$

$$= \frac{\sqrt{2} - 1}{(\sqrt{2})^2 - 1} \qquad (\sqrt{2} + 1)(\sqrt{2} - 1) = (\sqrt{2})^2 - 1.$$

$$= \frac{\sqrt{2} - 1}{2 - 1} \qquad (\sqrt{2})^2 = 2.$$

$$= \sqrt{2} - 1 \qquad \frac{\sqrt{2} - 1}{2 - 1} = \frac{\sqrt{2} - 1}{1} = \sqrt{2} - 1. \qquad ■$$

Self Check   Rationalize the denominator of $\dfrac{2}{\sqrt{3} + 1}$.

Answer $\sqrt{3} - 1$

EXAMPLE 10   Rationalize the denominator of $\dfrac{\sqrt{x} + \sqrt{2}}{\sqrt{x} - \sqrt{2}}$ $(x > 0)$.

Solution   We multiply the numerator and denominator by $\sqrt{x} + \sqrt{2}$, which is the conjugate of $\sqrt{x} - \sqrt{2}$, and simplify.

$$\frac{\sqrt{x} + \sqrt{2}}{\sqrt{x} - \sqrt{2}} = \frac{(\sqrt{x} + \sqrt{2})(\sqrt{x} + \sqrt{2})}{(\sqrt{x} - \sqrt{2})(\sqrt{x} + \sqrt{2})}$$

$$= \frac{x + \sqrt{2x} + \sqrt{2x} + 2}{x - 2} \qquad \text{Use the FOIL method.}$$

$$= \frac{x + 2\sqrt{2x} + 2}{x - 2} \qquad ■$$

Self Check   Rationalize the denominator of $\dfrac{\sqrt{x} - \sqrt{2}}{\sqrt{x} + \sqrt{2}}$.

Answer $\dfrac{x - 2\sqrt{2x} + 2}{x - 2}$

■ RATIONALIZING NUMERATORS

In calculus, we sometimes have to rationalize a numerator by multiplying the numerator and denominator of the fraction by the conjugate of the numerator.

EXAMPLE 11   Rationalize the numerator of $\dfrac{\sqrt{x} - 3}{\sqrt{x}}$ $(x > 0)$.

Solution   We multiply the numerator and denominator by $\sqrt{x} + 3$, which is the conjugate of the numerator.

$$\frac{\sqrt{x} - 3}{\sqrt{x}} = \frac{(\sqrt{x} - 3)(\sqrt{x} + 3)}{\sqrt{x}(\sqrt{x} + 3)}$$

$$= \frac{x + 3\sqrt{x} - 3\sqrt{x} - 9}{x + 3\sqrt{x}}$$

$$= \frac{x - 9}{x + 3\sqrt{x}}$$

The final expression is not in simplified form. However, this nonsimplified form is sometimes desirable in calculus.   ■

Self Check   Rationalize the numerator of $\dfrac{\sqrt{x} + 3}{\sqrt{x}}$.

Answer $\dfrac{x - 9}{x - 3\sqrt{x}}$

Orals   *Multiply and simplify.*

1. $\sqrt{3}\sqrt{3}$ 2. $\sqrt[3]{2}\sqrt[3]{2}\sqrt[3]{2}$ 3. $\sqrt{3}\sqrt{9}$
4. $\sqrt{a^3 b}\sqrt{ab}$ 5. $3\sqrt{2}(\sqrt{2} + 1)$ 6. $(\sqrt{2} + 1)(\sqrt{2} - 1)$

7. $\dfrac{1}{\sqrt{2}}$ 8. $\dfrac{1}{\sqrt{3} - 1}$

EXERCISE 9.6

REVIEW   *Solve each equation.*

1. $\dfrac{2}{3 - a} = 1$

2. $5(s - 4) = -5(s - 4)$

3. $\dfrac{8}{b - 2} + \dfrac{3}{2 - b} = -\dfrac{1}{b}$

4. $\dfrac{2}{x - 2} + \dfrac{1}{x + 1} = \dfrac{1}{(x + 1)(x - 2)}$

VOCABULARY AND CONCEPTS *Fill in each blank to make a true statement.*

5. To multiply $2\sqrt{7}$ by $3\sqrt{5}$, we multiply ___ by 3 and then multiply ____ by ____.

6. To multiply $2\sqrt{5}\left(3\sqrt{8} + \sqrt{3}\right)$, we use the _____ property to remove parentheses and simplify each resulting term.

7. To multiply $\left(\sqrt{3} + \sqrt{2}\right)\left(\sqrt{3} - 2\sqrt{2}\right)$, we can use the _____ method.

8. The conjugate of $\sqrt{x} + 1$ is _____.

9. To rationalize the denominator of $\dfrac{1}{\sqrt{3} - 1}$, multiply both the numerator and denominator by the _____ of the denominator.

10. To rationalize the numerator of $\dfrac{\sqrt{5} + 2}{\sqrt{5} - 2}$, multiply both the numerator and denominator by _____.

PRACTICE *In Exercises 11–34, do each multiplication and simplify, if possible. All variables represent positive numbers.*

11. $\sqrt{2}\sqrt{8}$

12. $\sqrt{3}\sqrt{27}$

13. $\sqrt{5}\sqrt{10}$

14. $\sqrt{7}\sqrt{35}$

15. $2\sqrt{3}\sqrt{6}$

16. $3\sqrt{11}\sqrt{33}$

17. $\sqrt[3]{5}\sqrt[3]{25}$

18. $\sqrt[3]{7}\sqrt[3]{49}$

19. $\left(3\sqrt[3]{9}\right)\left(2\sqrt[3]{3}\right)$

20. $\left(2\sqrt[3]{16}\right)\left(-\sqrt[3]{4}\right)$

21. $\sqrt[3]{2}\sqrt[3]{12}$

22. $\sqrt[3]{3}\sqrt[3]{18}$

23. $\sqrt{ab^3}\sqrt{ab}$

24. $\sqrt{8x}\sqrt{2x^3y}$

25. $\sqrt{5ab}\sqrt{5a}$

26. $\sqrt{15rs^2}\sqrt{10r}$

27. $\sqrt[3]{5r^2s}\sqrt[3]{2r}$

28. $\sqrt[3]{3xy^2}\sqrt[3]{9x^3}$

29. $\sqrt[3]{a^5b}\sqrt[3]{16ab^5}$

30. $\sqrt[3]{3x^4y}\sqrt[3]{18x}$

31. $\sqrt{x(x + 3)}\sqrt{x^3(x + 3)}$

32. $\sqrt{y^2(x + y)}\sqrt{(x + y)^3}$

33. $\sqrt[3]{6x^2(y + z)^2}\sqrt[3]{18x(y + z)}$

34. $\sqrt[3]{9x^2y(z + 1)^2}\sqrt[3]{6xy^2(z + 1)}$

In Exercises 35–54, do each multiplication and simplify. All variables represent positive numbers.

35. $3\sqrt{5}\left(4 - \sqrt{5}\right)$

36. $2\sqrt{7}\left(3\sqrt{7} - 1\right)$

37. $3\sqrt{2}\left(4\sqrt{3} + 2\sqrt{7}\right)$

38. $-\sqrt{3}\left(\sqrt{7} - \sqrt{5}\right)$

39. $-2\sqrt{5x}\left(4\sqrt{2x} - 3\sqrt{3}\right)$

40. $3\sqrt{7t}\left(2\sqrt{7t} + 3\sqrt{3t^2}\right)$

41. $\left(\sqrt{2} + 1\right)\left(\sqrt{2} - 3\right)$

42. $\left(2\sqrt{3} + 1\right)\left(\sqrt{3} - 1\right)$

43. $\left(4\sqrt{x} + 3\right)\left(2\sqrt{x} - 5\right)$

44. $\left(7\sqrt{y} + 2\right)\left(3\sqrt{y} - 5\right)$

45. $\left(\sqrt{5z} + \sqrt{3}\right)\left(\sqrt{5z} + \sqrt{3}\right)$

46. $\left(\sqrt{3p} - \sqrt{2}\right)\left(\sqrt{3p} + \sqrt{2}\right)$

47. $\left(\sqrt{3x} - \sqrt{2y}\right)\left(\sqrt{3x} + \sqrt{2y}\right)$

48. $\left(\sqrt{3m} + \sqrt{2n}\right)\left(\sqrt{3m} + \sqrt{2n}\right)$

49. $\left(2\sqrt{3a} - \sqrt{b}\right)\left(\sqrt{3a} + 3\sqrt{b}\right)$

50. $\left(5\sqrt{p} - \sqrt{3q}\right)\left(\sqrt{p} + 2\sqrt{3q}\right)$

51. $\left(3\sqrt{2r} - 2\right)^2$

52. $\left(2\sqrt{3t} + 5\right)^2$

53. $-2\left(\sqrt{3x} + \sqrt{3}\right)^2$

54. $3\left(\sqrt{5x} - \sqrt{3}\right)^2$

In Exercises 55–78, rationalize each denominator. All variables represent positive numbers.

55. $\sqrt{\dfrac{1}{7}}$

56. $\sqrt{\dfrac{5}{3}}$

57. $\sqrt{\dfrac{2}{3}}$

58. $\sqrt{\dfrac{3}{2}}$

59. $\dfrac{\sqrt{5}}{\sqrt{8}}$

60. $\dfrac{\sqrt{3}}{\sqrt{50}}$

61. $\dfrac{\sqrt{8}}{\sqrt{2}}$

62. $\dfrac{\sqrt{27}}{\sqrt{3}}$

63. $\dfrac{1}{\sqrt[3]{2}}$ **64.** $\dfrac{2}{\sqrt[3]{6}}$ **65.** $\dfrac{3}{\sqrt[3]{9}}$ **66.** $\dfrac{2}{\sqrt[3]{a}}$

67. $\dfrac{\sqrt[3]{2}}{\sqrt[3]{9}}$ **68.** $\dfrac{\sqrt[3]{9}}{\sqrt[3]{54}}$ **69.** $\dfrac{\sqrt{8x^2y}}{\sqrt{xy}}$ **70.** $\dfrac{\sqrt{9xy}}{\sqrt{3x^2y}}$

71. $\dfrac{\sqrt{10xy^2}}{\sqrt{2xy^3}}$ **72.** $\dfrac{\sqrt{5ab^2c}}{\sqrt{10abc}}$ **73.** $\dfrac{\sqrt[3]{4a^2}}{\sqrt[3]{2ab}}$ **74.** $\dfrac{\sqrt[3]{9x}}{\sqrt[3]{3xy}}$

75. $\dfrac{1}{\sqrt[4]{4}}$ **76.** $\dfrac{1}{\sqrt[5]{2}}$ **77.** $\dfrac{1}{\sqrt[5]{16}}$ **78.** $\dfrac{4}{\sqrt[4]{32}}$

In Exercises 79–94, do each division by rationalizing the denominator and simplifying. All variables represent positive numbers.

79. $\dfrac{1}{\sqrt{2}-1}$ **80.** $\dfrac{3}{\sqrt{3}-1}$ **81.** $\dfrac{\sqrt{2}}{\sqrt{5}+3}$ **82.** $\dfrac{\sqrt{3}}{\sqrt{3}-2}$

83. $\dfrac{\sqrt{3}+1}{\sqrt{3}-1}$ **84.** $\dfrac{\sqrt{2}-1}{\sqrt{2}+1}$ **85.** $\dfrac{\sqrt{7}-\sqrt{2}}{\sqrt{2}+\sqrt{7}}$ **86.** $\dfrac{\sqrt{3}+\sqrt{2}}{\sqrt{3}-\sqrt{2}}$

87. $\dfrac{2}{\sqrt{x}+1}$ **88.** $\dfrac{3}{\sqrt{x}-2}$ **89.** $\dfrac{x}{\sqrt{x}-4}$ **90.** $\dfrac{2x}{\sqrt{x}+1}$

91. $\dfrac{2z-1}{\sqrt{2z}-1}$ **92.** $\dfrac{3t-1}{\sqrt{3t}+1}$

93. $\dfrac{\sqrt{x}-\sqrt{y}}{\sqrt{x}+\sqrt{y}}$ **94.** $\dfrac{\sqrt{x}+\sqrt{y}}{\sqrt{x}-\sqrt{y}}$

In Exercises 95–100, rationalize each numerator. All variables represent positive numbers.

95. $\dfrac{\sqrt{3}+1}{2}$ **96.** $\dfrac{\sqrt{5}-1}{2}$ **97.** $\dfrac{\sqrt{x}+3}{x}$ **98.** $\dfrac{2+\sqrt{x}}{5x}$

99. $\dfrac{\sqrt{x}+\sqrt{y}}{\sqrt{x}}$ **100.** $\dfrac{\sqrt{x}-\sqrt{y}}{\sqrt{x}+\sqrt{y}}$

APPLICATIONS

101. Photography In Example 5, we saw that a lens with a focal length of 12 centimeters and an aperture $3\sqrt{2}$ centimeters in diameter is an *f*/2.8 lens. Find the *f*-number if the area of the aperture is again cut in half.

102. Photography In Exercise 101, we saw that a lens with a focal length of 12 centimeters and an aperture 3 centimeters in diameter is an *f*/4 lens. Find the *f*-number if the area of the aperture is again cut in half.

WRITING

103. Explain how to simplify a fraction with the monomial denominator $\sqrt[3]{3}$.

104. Explain how to simplify a fraction with the monomial denominator $\sqrt[3]{9}$.

SOMETHING TO THINK ABOUT *Use the comcepts in the following example to simplify each expression. Assume that all variables represent positive numbers.*

$$\sqrt{x^3}\sqrt[3]{x^2} = x^{3/2} \cdot x^{2/3} = x^{9/6+4/6} = x^{13/6} = x^{12/6} \cdot x^{1/6} = x^2\sqrt[6]{x}$$

105. $\sqrt{x^3}\sqrt[3]{x}$

106. $\sqrt{y}\sqrt[4]{y}$

107. $\sqrt[5]{p^3}\sqrt{p}$

108. $\sqrt[3]{a}\sqrt[4]{a^3}$

109. $\dfrac{\sqrt{b^3}}{\sqrt[3]{b^2}}$

110. $\dfrac{\sqrt[3]{x^2y^2}}{\sqrt[5]{x^3y}}$

■ ■ ■ ■ ■ ■ ■ ■ ■   **PROJECTS**

PROJECT 1   Tom and Brian arrange to have a bicycle race. Each leaves his own house at the same time and rides to the other's house, whereupon the winner of the race calls his own house and leaves a message for the loser. A map of the race is shown in Illustration 1. Brian stays on the highway, averaging 21 mph. Tom knows that he and Brian are evenly matched when biking on the highway, so he cuts across country for the first part of his trip, averaging 15 mph. When Tom reaches the highway at point A, he turns right and follows the highway, averaging 21 mph.

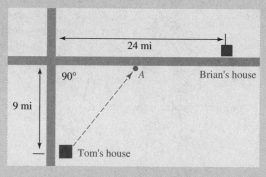

ILLUSTRATION 1

Tom and Brian never meet during the race, and amazingly, the race is a tie. Each of them calls the other at exactly the same moment!

a. How long (to the nearest second) did it take each person to complete the race?

b. How far from the intersection of the two highways is point A? (*Hint:* Set the travel times for Brian and Tom equal to each other. You may find two answers, but only one of them matches all of the information.)

(continued)

■ ■ ■ ■ ■ ■ ■ ■ ■ ■   **PROJECTS**   *(continued)*

c. Show that if Tom had started straight across country for Brian's house (in order to minimize the distance he had to travel), he would have lost the race. By how much time (to the nearest second) would he have lost? Then show that if Tom had biked across country to a point 9 miles from the intersection of the two highways, he would have won the race. By how much time (to the nearest second) would he have won?

PROJECT 2   You have been hired by the accounts receivable department of Widget Industries. Your boss, I. Tally, the head of accounting, informs you that those who purchase from Widget Industries are billed according to a pricing structure given by the following function:

$$\text{Price per widget in an order of } x \text{ widgets} = p(x) = p_0 - \frac{p_0}{2}\left(1 - \sqrt{1 - \frac{x^2}{k^2}}\right)$$

where x is the number of widgets ordered, p_0 is the standard price of a widget, and k is the maximum number of widgets that can be ordered at one time.

The formula was developed because it allows for changes in the standard price and the maximum order size without requiring that the entire pricing scheme be redone. Note that the total cost of an order of x widgets is $x \cdot p(x)$.

Tally is quite a taskmaster. On your first day, he assigns the following problems to you.

a. The pricing function is designed to benefit those who order the largest number of widgets. To see this, calculate the price (as a percentage of p_0) that a customer would pay for each widget if the customer ordered 100, 500, 800, or 1,000 widgets. Assume that $k = 1,000$. For what value of x is the cost per widget a minimum, and what is that minimum? Looking at the graph of $p(x)$ (using, say, 8 as p_0) on a graphing calculator will help you see how buying more widgets means that each widget costs less.

b. Tally tells you that the company lost money in the past because someone sent around a memo with a simplified, but incorrect, pricing function. Find the error(s) in the simplification given below.

$$p(x) = p_0 - \frac{p_0}{2}\left(1 - \sqrt{1 - \frac{x^2}{k^2}}\right)$$

$$= p_0 - \frac{p_0}{2}\left[1 - \left(1 - \frac{x}{k}\right)\right]$$

$$= p_0 - \frac{p_0}{2}\left(\frac{x}{k}\right)$$

$$= p_0\left(1 - \frac{x}{2k}\right)$$

Show that $x = 0$ and 1,000 are the *only* values of x for which the two formulas agree. (*Hint:* Set the two formulas equal to each other and solve for x.)

Now calculate the amount of money lost by using the incorrect formula, rather than the correct formula, on orders of 100, 500, 800, and 1,000 widgets. For these calculations, assume that $k = 1,000$ and $p_0 = \$5.60$.

c. Now is your chance to show Tally a thing or two. The pricing function *can* be simplified. Demonstrate to Tally that his pricing function is equivalent to

$$p(x) = \frac{p_0}{2}\left(1 + \frac{\sqrt{k^2 - x^2}}{k}\right)$$

Use this function to help a customer who wants to know how many widgets she must order to receive a 30% discount off the standard price p_0. Again, assume that $k = 1,000$ when doing this calculation.

CHAPTER SUMMARY

CONCEPTS

REVIEW EXERCISES

SECTION 9.1

Radical Expressions

If n is an even natural number,

$$\sqrt[n]{a^n} = |a|$$

If n is an odd natural number,

$$\sqrt[n]{a^n} = a$$

If n is a natural number greater than 1 and x is a real number, then

If $x > 0$, then $\sqrt[n]{x}$ is the positive number such that $\left(\sqrt[n]{x}\right)^n = x$.

If $x = 0$, then $\sqrt[n]{x} = 0$.

If $x < 0$, and n is odd, $\sqrt[n]{x}$ is the real number such that $\left(\sqrt[n]{x}\right)^n = x$.

If $x < 0$, and n is even, $\sqrt[n]{x}$ is not a real number.

1. Simplify each radical. Assume that x can be any number.

a. $\sqrt{49}$ **b.** $-\sqrt{121}$

c. $-\sqrt{36}$ **d.** $\sqrt{225}$

e. $\sqrt[3]{-27}$ **f.** $-\sqrt[3]{216}$

g. $\sqrt[4]{625}$ **h.** $\sqrt[5]{-32}$

i. $\sqrt{25x^2}$ **j.** $\sqrt{x^2 + 4x + 4}$

k. $\sqrt[3]{27a^6b^3}$ **l.** $\sqrt[4]{256x^8y^4}$

2. Graph each function.

a. $y = f(x) = \sqrt{x} + 2$

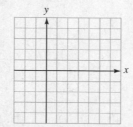

b. $y = f(x) = -\sqrt{x} - 1$

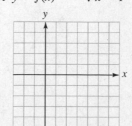

c. $y = f(x) = -\sqrt{x} + 2$

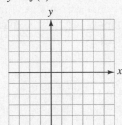

d. $y = f(x) = -\sqrt[3]{x} + 3$

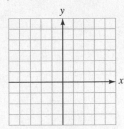

3. Consider the distribution 4, 8, 12, 16, 20.

    a. Find the mean of the distribution.

    b. Find the standard deviation.

| **SECTION 9.2** | *Applications of Radicals* |

The Pythagorean theorem:
If *a* and *b* are the lengths of the legs of a right triangle and *c* is the length of the hypotenuse, then
$a^2 + b^2 = c^2$.

In Exercises 4–5, the horizon distance d (measured in miles) is related to the height h (measured in feet) of the observer by the formula $d = 1.4\sqrt{h}$.

4. View from a submarine   A submarine's periscope extends 4.7 feet above the surface. How far away is the horizon?

5. View from a submarine   How far out of the water must a submarine periscope extend to provide a 4-mile horizon?

6. Sailing   A technique called *tacking* allows a sailboat to make progress into the wind. A sailboat follows the course in Illustration 1. Find *d*, the distance the boat advances into the wind.

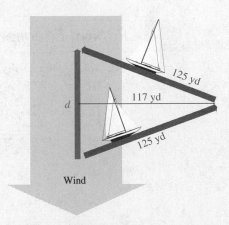

ILLUSTRATION 1

7. Communications   Some campers 3,900 yards from a highway are chatting with truckers on a citizen's band radio with an 8,900-yard range. Over what length of highway can these conversations take place? (See Illustration 2.)

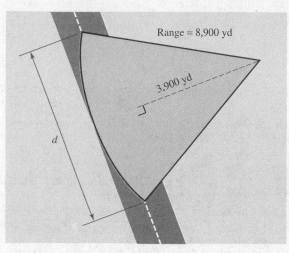

Range = 8,900 yd

3,900 yd

d

ILLUSTRATION 2

The distance formula:

$d(PQ) =$
$\sqrt{(x_2 - x_1)^2 + (y_2 - y_1)^2}$

8. Find the distance between points $P(0, 0)$ and $Q(5, -12)$.

9. Find the distance between points $P(-4, 6)$ and $Q(-2, 8)$. Give the result to the nearest hundredth.

SECTION 9.3   *Radical Equations*

The power rule:

If $x = y$, then $x^n = y^n$.

Raising both sides of an equation to the same power can lead to extraneous solutions. Be sure to check all suspected solutions.

10. Solve each equation.

a. $\sqrt{y + 3} = \sqrt{2y - 19}$

b. $u = \sqrt{25u - 144}$

c. $r = \sqrt{12r - 27}$

d. $\sqrt{z + 1} + \sqrt{z} = 2$

e. $\sqrt{2x + 5} - \sqrt{2x} = 1$

f. $\sqrt[3]{x^3 + 8} = x + 2$

SECTION 9.4 — *Rational Exponents*

If n ($n > 1$) is a natural number and $\sqrt[n]{x}$ is a real number, then $x^{1/n} = \sqrt[n]{x}$.

If n is even, $(x^n)^{1/n} = |x|$.

If n is a natural number greater than 1 and x is a real number, then

If $x > 0$, then $x^{1/n}$ is the positive number such that $(x^{1/n})^n = x$.

If $x = 0$, then $x^{1/n} = 0$.

If $x < 0$, and n is odd, then $x^{1/n}$ is the real number such that $(x^{1/n})^n = x$.

If $x < 0$ and n is even, then $x^{1/n}$ is not a real number.

If m and n are positive integers and $x > 0$,

$$x^{m/n} = \sqrt[n]{x^m} = \left(\sqrt[n]{x}\right)^m$$

$$x^{-m/n} = \frac{1}{x^{m/n}}$$

$$\frac{1}{x^{-m/n}} = x^{m/n} \quad (x \neq 0)$$

11. Simplify each expression, if possible. Assume that all variables represent positive numbers.

  a. $25^{1/2}$ **b.** $-36^{1/2}$

  c. $9^{3/2}$ **d.** $16^{3/2}$

  e. $(-8)^{1/3}$ **f.** $-8^{2/3}$

  g. $8^{-2/3}$ **h.** $8^{-1/3}$

  i. $-49^{5/2}$ **j.** $\dfrac{1}{25^{5/2}}$

  k. $\left(\dfrac{1}{4}\right)^{-3/2}$ **l.** $\left(\dfrac{4}{9}\right)^{-3/2}$

  m. $(27x^3 y)^{1/3}$ **n.** $(81x^4 y^2)^{1/4}$

  o. $(25x^3 y^4)^{3/2}$ **p.** $(8u^2 v^3)^{-2/3}$

12. Do the multiplications. Assume that all variables represent positive numbers and write all answers without negative exponents.

  a. $5^{1/4} 5^{1/2}$ **b.** $a^{3/7} a^{2/7}$

  c. $u^{1/2}(u^{1/2} - u^{-1/2})$ **d.** $v^{2/3}(v^{1/3} + v^{4/3})$

  e. $(x^{1/2} + y^{1/2})^2$

  f. $(a^{2/3} + b^{2/3})(a^{2/3} - b^{2/3})$

13. Simplify each expression. Assume that all variables are positive.

  a. $\sqrt[6]{5^2}$ **b.** $\sqrt[8]{x^4}$

  c. $\sqrt[9]{27a^3 b^6}$ **d.** $\sqrt[4]{25a^2 b^2}$

SECTION 9.5 — *Simplifying and Combining Radical Expressions*

Properties of radicals:

$$\sqrt[n]{ab} = \sqrt[n]{a}\sqrt[n]{b}$$

$$\sqrt[n]{\frac{a}{b}} = \frac{\sqrt[n]{a}}{\sqrt[n]{b}} \quad (b \neq 0)$$

14. Simplify each expression.

  a. $\sqrt{240}$ **b.** $\sqrt[3]{54}$

  c. $\sqrt[4]{32}$ **d.** $\sqrt[5]{96}$

  e. $\sqrt{8x^3}$ **f.** $\sqrt{18x^4 y^3}$

  g. $\sqrt[3]{16x^5 y^4}$ **h.** $\sqrt[3]{54x^7 y^3}$

  i. $\dfrac{\sqrt{32x^3}}{\sqrt{2x}}$ **j.** $\dfrac{\sqrt[3]{16x^5}}{\sqrt[3]{2x^2}}$

  k. $\sqrt[3]{\dfrac{2a^2 b}{27x^3}}$ **l.** $\sqrt{\dfrac{17xy}{64a^4}}$

Like radicals can be combined by addition and subtraction:
$$3\sqrt{2} + 5\sqrt{2} = 8\sqrt{2}$$

Radicals that are not similar can often be simplified to radicals that are similar and then combined:
$$\sqrt{2} + \sqrt{8} = \sqrt{2} + \sqrt{4}\sqrt{2}$$
$$= \sqrt{2} + 2\sqrt{2}$$
$$= 3\sqrt{2}$$

15. Simplify and combine like radicals. Assume that all variables represent positive numbers.

  a. $\sqrt{2} + \sqrt{8}$ **b.** $\sqrt{20} - \sqrt{5}$

  c. $2\sqrt[3]{3} - \sqrt[3]{24}$ **d.** $\sqrt[4]{32} + 2\sqrt[4]{162}$

  e. $2x\sqrt{8} + 2\sqrt{200x^2} + \sqrt{50x^2}$

  f. $3\sqrt{27a^3} - 2a\sqrt{3a} + 5\sqrt{75a^3}$

  g. $\sqrt[3]{54} - 3\sqrt[3]{16} + 4\sqrt[3]{128}$

  h. $2\sqrt[4]{32x^5} + 4\sqrt[4]{162x^5} - 5x\sqrt[4]{512x}$

In an isosceles right triangle, the length of the hypotenuse is the length of one leg times $\sqrt{2}$.

16. Find the length of the hypotenuse of an isosceles right triangle whose legs measure 7 meters.

The shorter leg of a 30°–60°–90° triangle is half as long as the hypotenuse. The longer leg is the length of the shorter leg times $\sqrt{3}$.

17. The hypotenuse of a 30°–60°–90° triangle measures $12\sqrt{3}$ centimeters. Find the length of each leg.

18. Find x to two decimal places.

  a.

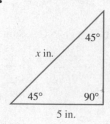

5 in.

  b.

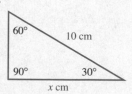

SECTION 9.6   *Multiplying and Dividing Radical Expressions*

If two radicals have the same index, they can be multiplied:
$$\sqrt{3x}\sqrt{6x} = \sqrt{18x^2} \ (x > 0)$$
$$= 3x\sqrt{2}$$

19. Simplify each expression. Assume that all variables represent positive numbers.

  a. $(2\sqrt{5})(3\sqrt{2})$ **b.** $2\sqrt{6}\sqrt{216}$

  c. $\sqrt{9x}\sqrt{x}$ **d.** $\sqrt[3]{3}\sqrt[3]{9}$

  e. $-\sqrt[3]{2x^2}\sqrt[3]{4x}$

  f. $-\sqrt[4]{256x^5y^{11}}\sqrt[4]{625x^9y^3}$

  g. $\sqrt{2}(\sqrt{8} - 3)$

  h. $\sqrt{2}(\sqrt{2} + 3)$

i. $\sqrt{5}(\sqrt{2} - 1)$

j. $\sqrt{3}(\sqrt{3} + \sqrt{2})$

k. $(\sqrt{2} + 1)(\sqrt{2} - 1)$

l. $(\sqrt{3} + \sqrt{2})(\sqrt{3} + \sqrt{2})$

m. $(\sqrt{x} + \sqrt{y})(\sqrt{x} - \sqrt{y})$

n. $(2\sqrt{u} + 3)(3\sqrt{u} - 4)$

To rationalize the binomial denominator of a fraction, multiply the numerator and the denominator by the conjugate of the binomial in the denominator.

20. Rationalize each denominator.

a. $\dfrac{1}{\sqrt{3}}$

b. $\dfrac{\sqrt{3}}{\sqrt{5}}$

c. $\dfrac{x}{\sqrt{xy}}$

d. $\dfrac{\sqrt[3]{uv}}{\sqrt[3]{u^5 v^7}}$

e. $\dfrac{2}{\sqrt{2} - 1}$

f. $\dfrac{\sqrt{2}}{\sqrt{3} - 1}$

g. $\dfrac{2x - 32}{\sqrt{x} + 4}$

h. $\dfrac{\sqrt{a} + 1}{\sqrt{a} - 1}$

21. Rationalize each numerator.

a. $\dfrac{\sqrt{3}}{5}$

b. $\dfrac{\sqrt[3]{9}}{3}$

c. $\dfrac{3 - \sqrt{x}}{2}$

d. $\dfrac{\sqrt{a} - \sqrt{b}}{\sqrt{a}}$

■ Chapter Test

In Problems 1–4, find each root.

1. $\sqrt{49}$

2. $\sqrt[3]{64}$

3. $\sqrt{4x^2}$

4. $\sqrt[3]{8x^3}$

In Problems 5–6, graph each function and find its domain and range.

5. $f(x) = \sqrt{x - 2}$

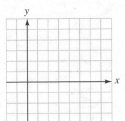

6. $f(x) = \sqrt[3]{x} + 3$

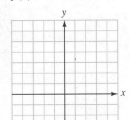

In Problems 7–8, consider the distribution 7, 8, 12, 13.

7. Find the mean of the distribution.

8. Find the standard deviation.

⊞   *In Problems 9–10, use a calculator.*

9. Shipping crates   The diagonal brace on the shipping crate in Illustration 1 is 53 inches. Find the height, h, of the crate.

10. Pendulum   The 2-meter pendulum in Illustration 2 rises 0.1 meter at the extremes of its swing. Find the width, w, of the swing.

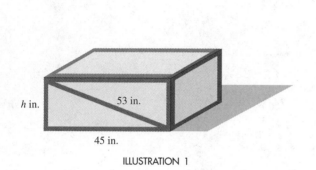

ILLUSTRATION 1

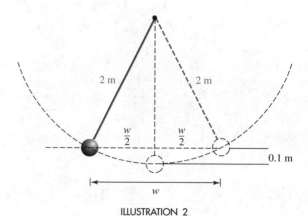

ILLUSTRATION 2

In Problems 11–12, find the distance between points P and Q.

11. $P(6, 8)$ and $Q(0, 0)$

12. $P(-2, 5)$ and $Q(22, 12)$

In Problems 13–14, solve and check each equation.

13. $\sqrt[3]{6n + 4} - 4 = 0$

14. $1 - \sqrt{u} = \sqrt{u - 3}$

In Problems 15–20, simplify each expression. Assume that all variables represent positive numbers, and write answers without using negative exponents.

15. $16^{1/4}$

16. $27^{2/3}$

17. $36^{-3/2}$

18. $\left(-\dfrac{8}{27}\right)^{-2/3}$

19. $\dfrac{2^{5/3}2^{1/6}}{2^{1/2}}$

20. $\dfrac{(8x^3y)^{1/2}(8xy^5)^{1/2}}{(x^3y^6)^{1/3}}$

In Problems 21–24, simplify each expression. Assume that all variables represent positive numbers.

21. $\sqrt{48}$

22. $\sqrt{250x^3y^5}$

23. $\dfrac{\sqrt[3]{24x^{15}y^4}}{\sqrt[3]{y}}$

24. $\sqrt{\dfrac{3a^5}{48a^7}}$

In Problems 25–28, simplify each expression. Assume that the variables are unrestricted.

25. $\sqrt{12x^2}$ **26.** $\sqrt{8x^6}$ **27.** $\sqrt[3]{81x^3}$ **28.** $\sqrt{18x^4y^9}$

In Problems 29–32, simplify and combine like radicals. Assume that all variables represent positive numbers.

29. $\sqrt{12} - \sqrt{27}$ **30.** $2\sqrt[3]{40} - \sqrt[3]{5,000} + 4\sqrt[3]{625}$

31. $2\sqrt{48y^5} - 3y\sqrt{12y^3}$ **32.** $\sqrt[4]{768z^5} + z\sqrt[4]{48z}$

In Problems 33–34, find x to two decimal places.

33.

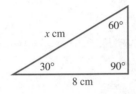

34.

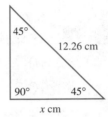

In Problems 35–36, do each operation and simplify, if possible. All variables represent positive numbers.

35. $-2\sqrt{xy}\left(3\sqrt{x} + \sqrt{xy^3}\right)$ **36.** $\left(3\sqrt{2} + \sqrt{3}\right)\left(2\sqrt{2} - 3\sqrt{3}\right)$

In Problems 37–40, rationalize each denominator.

37. $\dfrac{1}{\sqrt{5}}$ **38.** $\dfrac{6}{\sqrt[3]{9}}$ **39.** $\dfrac{-4\sqrt{2}}{\sqrt{5} + 3}$ **40.** $\dfrac{3t - 1}{\sqrt{3t} - 1}$

In Problems 41–42, rationalize each numerator.

41. $\dfrac{\sqrt{3}}{\sqrt{7}}$ **42.** $\dfrac{\sqrt{a} + \sqrt{b}}{\sqrt{a} - \sqrt{b}}$

10

Quadratic Functions, Inequalities, and Algebra of Functions

Chemist

Chemists search for knowledge about substances and put it to practical use. Most chemists work in research and development. In basic research, chemists investigate the properties, composition, and structure of matter and the laws that govern the combination of elements and the reactions of substances. Their research has resulted in the development of a tremendous variety of synthetic materials, of ingredients that have improved other substances, and of processes that help save energy and reduce pollution. In applied research and development, they create new products or improve existing ones.

SAMPLE APPLICATION ■ A weak acid (0.1 M concentration) breaks down into free cations (the hydrogen ion, H^+) and anions (A^-). When this acid dissociates, the following equilibrium equation is established.

1. $$\frac{[H^+][A^-]}{[HA]} = 4 \times 10^{-4}$$

where $[H^+]$, the hydrogen ion concentration, is equal to $[A^-]$, the anion concentration. $[HA]$ is the concentration of the undissociated acid itself. Find $[H^+]$ at equilibrium.
(See Exercise 85 in Exercise 10.1.)

We have discussed how to solve linear equations and certain quadratic equations in which the quadratic expression is factorable. In this chapter, we will discuss more general methods for solving quadratic equations, and we will consider their graphs.

10.1 Completing the Square and the Quadratic Formula

■ SOLVING QUADRATIC EQUATIONS BY FACTORING ■ THE SQUARE ROOT PROPERTY
■ COMPLETING THE SQUARE ■ SOLVING EQUATIONS BY COMPLETING THE SQUARE ■ THE
QUADRATIC FORMULA ■ PROBLEM SOLVING

Getting Ready   *Factor each expression.*

1. $6x^2 + x - 2$ **2.** $4x^2 - 4x - 3$

Evaluate $\sqrt{b^2 - 4ac}$ for the following values.

3. $a = 6, b = 1, c = -2$ **4.** $a = 4, b = -4, c = -3$

■ SOLVING QUADRATIC EQUATIONS BY FACTORING

A *quadratic equation* is an equation of the form $ax^2 + bx + c = 0$ $(a \neq 0)$, where a, b, and c are real numbers. We have discussed how to solve quadratic equations by using factoring. For example, to solve $6x^2 - 7x - 3 = 0$, we proceed as follows:

$$6x^2 - 7x - 3 = 0$$
$$(2x - 3)(3x + 1) = 0 \qquad \text{Factor.}$$
$$2x - 3 = 0 \quad \text{or} \quad 3x + 1 = 0 \qquad \text{Set each factor equal to 0.}$$
$$x = \frac{3}{2} \qquad\qquad x = -\frac{1}{3} \qquad \text{Solve each linear equation.}$$

However, many quadratic expressions do not factor easily. For example, it would be difficult to solve $2x^2 + 4x + 1 = 0$ by factoring, because $2x^2 + 4x + 1$ cannot be factored by using only integers.

■ THE SQUARE ROOT PROPERTY

To develop general methods for solving quadratic equations, we first consider the equation $x^2 = c$. If $c > 0$, we can find the real solutions of $x^2 = c$ as follows:

$$x^2 = c$$
$$x^2 - c = 0 \qquad\qquad \text{Subtract } c \text{ from both sides.}$$
$$x^2 - \left(\sqrt{c}\right)^2 = 0 \qquad\qquad c = \left(\sqrt{c}\right)^2.$$
$$\left(x + \sqrt{c}\right)\left(x - \sqrt{c}\right) = 0 \qquad \text{Factor the difference of two squares.}$$
$$x + \sqrt{c} = 0 \quad \text{or} \quad x - \sqrt{c} = 0 \qquad \text{Set each factor equal to 0.}$$
$$x = -\sqrt{c} \qquad\qquad x = \sqrt{c} \qquad \text{Solve each linear equation.}$$

The two solutions of $x^2 = c$ are $x = \sqrt{c}$ and $x = -\sqrt{c}$.

> **Square Root Property**
> If $c > 0$, the equation $x^2 = c$ has two real solutions:
> $$x = \sqrt{c} \qquad \text{and} \qquad x = -\sqrt{c}$$

EXAMPLE 1   Solve $x^2 - 12 = 0$.

Solution   We can write the equation as $x^2 = 12$ and use the square root property.

$$x^2 - 12 = 0$$
$$x^2 = 12 \qquad\qquad \text{Add 12 to both sides.}$$
$$x = \sqrt{12} \quad \text{or} \quad x = -\sqrt{12} \qquad \text{Use the square root property.}$$
$$x = 2\sqrt{3} \qquad\qquad x = -2\sqrt{3} \qquad \sqrt{12} = \sqrt{4}\sqrt{3} = 2\sqrt{3}.$$

Verify that each solution satisfies the equation.   ■

Self Check

Solve $x^2 - 18 = 0$.

Answer

$3\sqrt{2}, -3\sqrt{2}$

EXAMPLE 2   Solve $(x - 3)^2 = 16$.

Solution

$$(x - 3)^2 = 16$$

$x - 3 = \sqrt{16}$ or $x - 3 = -\sqrt{16}$ Use the square root property.

$x - 3 = 4$ $\qquad\qquad$ $x - 3 = -4$ $\sqrt{16} = 4$.

$\qquad x = 3 + 4$ $\qquad\qquad$ $x = 3 - 4$ Add 3 to both sides.

$\qquad x = 7$ $\qquad\qquad\quad$ $x = -1$ Simplify.

Verify that each solution satisfies the equation.   ■

Self Check

Solve $(x + 2)^2 = 9$.

Answer

$1, -5$

■ COMPLETING THE SQUARE

All quadratic equations can be solved by **completing the square.** This method is based on the special products

$$x^2 + 2ax + a^2 = (x + a)^2 \qquad \text{and} \qquad x^2 - 2ax + a^2 = (x - a)^2$$

The trinomials $x^2 + 2ax + a^2$ and $x^2 - 2ax + a^2$ are perfect square trinomials, because each one factors as the square of a binomial. In each case, the coefficient of the first term is 1, and if we take one-half of the coefficient of x in the middle term and square it, we obtain the third term.

$$\left[\frac{1}{2}(2a)\right]^2 = a^2$$

$$\left[\frac{1}{2}(-2a)\right]^2 = (-a)^2 = a^2$$

EXAMPLE 3   Add a number to make each binomial a perfect square trinomial: **a.** $x^2 + 10x$, **b.** $x^2 - 6x$, and **c.** $x^2 - 11x$.

Solution   **a.** To make $x^2 + 10x$ a perfect square trinomial, we find one-half of 10 to get 5, square 5 to get 25, and add 25 to $x^2 + 10x$.

$$x^2 + 10x + \left[\frac{1}{2}(10)\right]^2 = x^2 + 10x + (5)^2$$

$$= x^2 + 10x + 25 \qquad \text{Note that } x^2 + 10x + 25 =$$
$$(x + 5)^2.$$

b. To make $x^2 - 6x$ a perfect square trinomial, we find one-half of -6 to get -3, square -3 to get 9, and add 9 to $x^2 - 6x$.

$$x^2 - 6x + \left[\frac{1}{2}(-6)\right]^2 = x^2 - 6x + (-3)^2$$

$$= x^2 - 6x + 9 \qquad \text{Note that } x^2 - 6x + 9 = (x - 3)^2.$$

c. To make $x^2 - 11x$ a perfect square trinomial, we find one-half of -11 to get $-\frac{11}{2}$, square $-\frac{11}{2}$ to get $\frac{121}{4}$, and add $\frac{121}{4}$ to $x^2 - 11x$.

$$x^2 - 11x + \left[\frac{1}{2}(-11)\right]^2 = x^2 - 11x + \left(-\frac{11}{2}\right)^2$$

$$= x^2 - 11x + \frac{121}{4} \qquad \text{Note that } x^2 - 11x + \frac{121}{4} = \left(x - \frac{11}{2}\right)^2. \qquad ■$$

Self Check   Change $a^2 - 5a$ into a perfect trinomial square.

Answer $a^2 - 5a + \frac{25}{4}$

■ SOLVING EQUATIONS BY COMPLETING THE SQUARE

To solve an equation of the form $ax^2 + bx + c = 0$ $(a \ne 0)$ by completing the square, we use the following steps.

Completing the Square

1. Make sure that the coefficient of x^2 is 1. If it is not, make it 1 by dividing both sides of the equation by the coefficient of x^2.

2. If necessary, add a number to both sides of the equation to place the constant term on the right-hand side of the equal sign.

3. Complete the square:

   a. Find one-half of the coefficient of x and square it.

   b. Add the square to both sides of the equation.

4. Factor the trinomial square and combine like terms.

5. Solve the resulting equation by using the square root property.

EXAMPLE 4   Use completing the square to solve $x^2 + 8x + 7 = 0$.

Solution   *Step 1:* In this example, the coefficient of x^2 is already 1.

Step 2: We add -7 to both sides to place the constant on the right-hand side of the equal sign:

$$x^2 + 8x + 7 = 0$$
$$x^2 + 8x = -7$$

Step 3: The coefficient of x is 8, one-half of 8 is 4, and $4^2 = 16$. To complete the square, we add 16 to both sides.

$$x^2 + 8x + \mathbf{16} = \mathbf{16} - 7$$

1. $x^2 + 8x + 16 = 9$ $16 - 7 = 9.$

Step 4: Since the left-hand side of Equation 1 is a perfect square trinomial, we can factor it to get $(x + 4)^2$.

$$x^2 + 8x + 16 = 9$$

2. $(x + 4)^2 = 9$

Step 5: We then solve Equation 2 by using the square root property.

$$(x + 4)^2 = 9$$
$$x + 4 = \sqrt{9} \quad \text{or} \quad x + 4 = -\sqrt{9}$$
$$x + 4 = 3 \qquad\qquad x + 4 = -3$$
$$x = -1 \qquad\qquad x = -7$$

Verify that both solutions satisfy the equation.   ∎

EXAMPLE 5   Solve $6x^2 + 5x - 6 = 0$.

Solution   *Step 1:* To make the coefficient of x^2 equal to 1, we divide both sides of the equation by 6.

$$6x^2 + 5x - 6 = 0$$
$$\frac{6x^2}{6} + \frac{5}{6}x - \frac{6}{6} = \frac{0}{6} \qquad \text{Divide both sides by 6.}$$
$$x^2 + \frac{5}{6}x - 1 = 0 \qquad \text{Simplify.}$$

Step 2: We add 1 to both sides to place the constant on the right-hand side of the equal sign:

$$x^2 + \frac{5}{6}x = 1$$

Step 3: The coefficient of x is $\frac{5}{6}$, one-half of $\frac{5}{6}$ is $\frac{5}{12}$, and $\left(\frac{5}{12}\right)^2 = \frac{25}{144}$. To complete the square, we add $\frac{25}{144}$ to both sides.

$$x^2 + \frac{5}{6}x + \frac{\mathbf{25}}{\mathbf{144}} = 1 + \frac{\mathbf{25}}{\mathbf{144}}$$

3. $x^2 + \frac{5}{6}x + \frac{25}{144} = \frac{169}{144}$ $1 + \frac{25}{144} = \frac{144}{144} + \frac{25}{144} = \frac{169}{144}.$

Step 4: Since the left-hand side of Equation 3 is a perfect square trinomial, we can factor it to get $\left(x + \frac{5}{12}\right)^2$.

4. $\left(x + \dfrac{5}{12}\right)^2 = \dfrac{169}{144}$

Step 5: We can solve Equation 4 by using the square root property.

$$x + \frac{5}{12} = \sqrt{\frac{169}{144}} \quad \text{or} \quad x + \frac{5}{12} = -\sqrt{\frac{169}{144}}$$

$$x + \frac{5}{12} = \frac{13}{12} \qquad\qquad x + \frac{5}{12} = -\frac{13}{12}$$

$$x = -\frac{5}{12} + \frac{13}{12} \qquad\qquad x = -\frac{5}{12} - \frac{13}{12}$$

$$x = \frac{8}{12} \qquad\qquad x = -\frac{18}{12}$$

$$x = \frac{2}{3} \qquad\qquad x = -\frac{3}{2}$$

Verify that both solutions satisfy the original equation.   ∎

EXAMPLE 6   Solve $2x^2 + 4x + 1 = 0$.

Solution

$$2x^2 + 4x + 1 = 0$$

$$x^2 + 2x + \frac{1}{2} = \frac{0}{2}$$ Divide both sides by 2 to make the coefficient of x^2 equal to 1.

$$x^2 + 2x = -\frac{1}{2}$$ Subtract $\frac{1}{2}$ from both sides.

$$x^2 + 2x + 1 = 1 - \frac{1}{2}$$ Square half the coefficient of x and add it to both sides.

$$(x + 1)^2 = \frac{1}{2}$$ Factor and combine like terms.

$$x + 1 = \sqrt{\frac{1}{2}} \quad \text{or} \quad x + 1 = -\sqrt{\frac{1}{2}}$$

$$x + 1 = \frac{\sqrt{2}}{2} \qquad\qquad x + 1 = -\frac{\sqrt{2}}{2}$$ $\sqrt{\frac{1}{2}} = \frac{1}{\sqrt{2}} = \frac{1\cdot\sqrt{2}}{\sqrt{2}\sqrt{2}} = \frac{\sqrt{2}}{2}$.

$$x = -1 + \frac{\sqrt{2}}{2} \qquad\qquad x = -1 - \frac{\sqrt{2}}{2}$$

$$x = \frac{-2 + \sqrt{2}}{2} \qquad\qquad x = \frac{-2 - \sqrt{2}}{2}$$

Both values check.   ∎

Self Check

Answer

Solve $3x^2 + 6x + 1 = 0$.

$$\frac{-3 + \sqrt{6}}{3}, \frac{-3 - \sqrt{6}}{3}$$

■ THE QUADRATIC FORMULA

To develop a formula we can use to solve quadratic equations, we solve the general quadratic equation $ax^2 + bx + c = 0$ $(a \neq 0)$.

$$ax^2 + bx + c = 0$$

$$\frac{ax^2}{a} + \frac{bx}{a} + \frac{c}{a} = \frac{0}{a}$$ Since $a \neq 0$, we can divide both sides by a to make the coefficient of x^2 equal to 1.

$$x^2 + \frac{bx}{a} = -\frac{c}{a}$$ $\frac{0}{a} = 0$; subtract $\frac{c}{a}$ from both sides.

$$x^2 + \frac{b}{a}x + \left(\frac{b}{2a}\right)^2 = \left(\frac{b}{2a}\right)^2 - \frac{c}{a}$$ Complete the square on x by adding $\left(\frac{b}{2a}\right)^2$ to both sides.

$$x^2 + \frac{b}{a}x + \frac{b^2}{4a^2} = \frac{b^2}{4a^2} - \frac{4ac}{4aa}$$ Remove parentheses and get a common denominator on the right-hand side.

5. $$\left(x + \frac{b}{2a}\right)^2 = \frac{b^2 - 4ac}{4a^2}$$ Factor the left-hand side and add the fractions on the right-hand side.

We can solve Equation 5 by using the square root property.

$$x + \frac{b}{2a} = \sqrt{\frac{b^2 - 4ac}{4a^2}} \qquad \text{or} \qquad x + \frac{b}{2a} = -\sqrt{\frac{b^2 - 4ac}{4a^2}}$$

$$x + \frac{b}{2a} = \frac{\sqrt{b^2 - 4ac}}{2a} \qquad\qquad x + \frac{b}{2a} = -\frac{\sqrt{b^2 - 4ac}}{2a}$$

$$x = -\frac{b}{2a} + \frac{\sqrt{b^2 - 4ac}}{2a} \qquad\qquad x = -\frac{b}{2a} - \frac{\sqrt{b^2 - 4ac}}{2a}$$

$$= \frac{-b + \sqrt{b^2 - 4ac}}{2a} \qquad\qquad = \frac{-b - \sqrt{b^2 - 4ac}}{2a}$$

These two solutions give the **quadratic formula.**

The Quadratic Formula

The solutions of $ax^2 + bx + c = 0$ $(a \neq 0)$ are given by the formula

$$x = \frac{-b \pm \sqrt{b^2 - 4ac}}{2a}$$ Read the symbol $\pm$ as "plus or minus."

> **WARNING!** Be sure to draw the fraction bar under both parts of the numerator, and be sure to draw the radical sign exactly over $b^2 - 4ac$. Do not write the quadratic formula as
>
> $$x = -b \pm \frac{\sqrt{b^2 - 4ac}}{2a} \qquad \text{or as} \qquad x = -b \pm \sqrt{\frac{b^2 - 4ac}{2a}}$$

EXAMPLE 7   Solve $2x^2 - 3x - 5 = 0$.

Solution   In this equation $a = 2$, $b = -3$, and $c = -5$.

$$x = \frac{-b \pm \sqrt{b^2 - 4ac}}{2a}$$

$$= \frac{-(-3) \pm \sqrt{(-3)^2 - 4(2)(-5)}}{2(2)}$$ Substitute 2 for a, -3 for b, and -5 for c.

$$= \frac{3 \pm \sqrt{9 + 40}}{4}$$

$$= \frac{3 \pm \sqrt{49}}{4}$$

$$= \frac{3 \pm 7}{4}$$

$$x = \frac{3 + 7}{4} \quad \text{or} \quad x = \frac{3 - 7}{4}$$

$$x = \frac{10}{4} \qquad\qquad x = \frac{-4}{4}$$

$$x = \frac{5}{2} \qquad\qquad x = -1$$

Verify that both solutions satisfy the original equation.   ∎

Self Check   Solve $3x^2 - 5x - 2 = 0$.
Answer 2, $-\frac{1}{3}$

EXAMPLE 8   Solve $2x^2 + 1 = -4x$.

Solution   We write the equation in general form before identifying a, b, and c.

$$2x^2 + 4x + 1 = 0$$

In this equation, $a = 2$, $b = 4$, and $c = 1$.

$$x = \frac{-b \pm \sqrt{b^2 - 4ac}}{2a}$$

$$= \frac{-4 \pm \sqrt{4^2 - 4(2)(1)}}{2(2)} \qquad \text{Substitute 2 for } a, \text{ 4 for } b, \text{ and 1 for } c.$$

$$= \frac{-4 \pm \sqrt{16 - 8}}{4}$$

$$= \frac{-4 \pm \sqrt{8}}{4}$$

$$= \frac{-4 \pm 2\sqrt{2}}{4} \qquad \sqrt{8} = \sqrt{4 \cdot 2} = \sqrt{4}\sqrt{2} = 2\sqrt{2}.$$

$$= \frac{-2 \pm \sqrt{2}}{2} \qquad \frac{-4 \pm 2\sqrt{2}}{4} = \frac{2(-2 \pm \sqrt{2})}{4} = \frac{-2 \pm \sqrt{2}}{2}.$$

Thus, $x = \dfrac{-2 + \sqrt{2}}{2}$ or $x = \dfrac{-2 - \sqrt{2}}{2}$. ∎

Self Check Solve $3x^2 - 2x - 3 = 0$.

Answer $\dfrac{1 + \sqrt{10}}{3}, \dfrac{1 - \sqrt{10}}{3}$

■ PROBLEM SOLVING

EXAMPLE 9

Dimensions of a rectangle Find the dimensions of the rectangle shown in Figure 10-1, given that its area is 253 cm².

Solution If we let w represent the width of the rectangle, then $w + 12$ represents its length. Since the area of the rectangle is 253 square centimeters, we can form the equation

$$w(w + 12) = 253 \qquad \text{Area of a rectangle} = \text{width} \times \text{length}.$$

and solve it as follows:

$$w(w + 12) = 253$$
$$w^2 + 12w = 253 \qquad \text{Use the distributive property to remove parentheses.}$$
$$w^2 + 12w - 253 = 0 \qquad \text{Subtract 253 from both sides.}$$

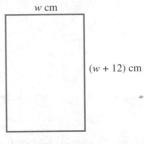

FIGURE 10-1

Solution by Factoring	*Solution by Formula*

$$(w - 11)(w + 23) = 0$$

$$w = \frac{-12 \pm \sqrt{12^2 - 4(1)(-253)}}{2(1)}$$

$$w - 11 = 0 \quad \text{or} \quad w + 23 = 0$$

$$= \frac{-12 \pm \sqrt{144 + 1{,}012}}{2}$$

$$w = 11 \qquad\qquad w = -23$$

$$= \frac{-12 \pm \sqrt{1{,}156}}{2}$$

$$= \frac{-12 \pm 34}{2}$$

$$w = 11 \quad \text{or} \quad w = -23$$

Since the rectangle cannot have a negative width, we discard the solution of -23. Thus, the only solution is $w = 11$. Since the rectangle is 11 centimeters wide and $(11 + 12)$ centimeters long, its dimensions are 11 centimeters by 23 centimeters.

Check: 23 is 12 more than 11, and the area of a rectangle with dimensions of 23 centimeters by 11 centimeters is 253 square centimeters.   ■

Orals   *Solve each equation.*

1. $x^2 = 49$ **2.** $x^2 = 10$

Find the number that must be added to the binomial to make it a perfect square trinomial.

3. $x^2 + 4x$ **4.** $x^2 - 6x$ **5.** $x^2 - 3x$ **6.** $x^2 + 5x$

Identify a, b, and c in each quadratic equation.

7. $3x^2 - 4x + 7 = 0$ **8.** $-2x^2 + x = 5$

EXERCISE 10.1

REVIEW   *Solve each equation or inequality.*

1. $\dfrac{t + 9}{2} + \dfrac{t + 2}{5} = \dfrac{8}{5} + 4t$

2. $\dfrac{1 - 5x}{2x} + 4 = \dfrac{x + 3}{x}$

3. $3(t - 3) + 3t - 5 \le 2(t + 1) + t - 4$

4. $-2(y + 4) - 3y + 3 \ge 3(2y - 3) - y - 5$

Solve for the indicated variable.

5. $Ax + By = C$ for B

6. $R = \dfrac{kL}{d^2}$ for L

VOCABULARY AND CONCEPTS *Fill in each blank to make a true statement.*

7. If $c > 0$, the solutions of $x^2 = c$ are _____ and _____.

8. To complete the square on x in $x^2 + 6x$, find one-half of __, square it to get __, and add __ to get _____.

9. The symbol $\pm$ is read as _____.

10. The solutions of $ax^2 + bx + c = 0$ $(a \neq 0)$ are given by the _____ formula, which is

$$x = \underline{\hspace{3cm}}$$

PRACTICE *In Exercises 11–22, use factoring to solve each equation.*

11. $6x^2 + 12x = 0$

12. $5x^2 + 11x = 0$

13. $2y^2 - 50 = 0$

14. $4y^2 - 64 = 0$

15. $r^2 + 6r + 8 = 0$

16. $x^2 + 9x + 20 = 0$

17. $7x - 6 = x^2$

18. $5t - 6 = t^2$

19. $2z^2 - 5z + 2 = 0$

20. $2x^2 - x - 1 = 0$

21. $6s^2 + 11s - 10 = 0$

22. $3x^2 + 10x - 8 = 0$

In Exercises 23–34, use the square root property to solve each equation.

23. $x^2 = 36$

24. $x^2 = 144$

25. $z^2 = 5$

26. $u^2 = 24$

27. $3x^2 - 16 = 0$

28. $5x^2 - 49 = 0$

29. $(x + 1)^2 = 1$

30. $(x - 1)^2 = 4$

31. $(s - 7)^2 - 9 = 0$

32. $(t + 4)^2 = 16$

33. $(x + 5)^2 - 3 = 0$

34. $(x + 3)^2 - 7 = 0$

In Exercises 35–48, use completing the square to solve each equation.

35. $x^2 + 2x - 8 = 0$

36. $x^2 + 6x + 5 = 0$

37. $x^2 - 6x + 8 = 0$

38. $x^2 + 8x + 15 = 0$

39. $x^2 + 5x + 4 = 0$

40. $x^2 - 11x + 30 = 0$

41. $x + 1 = 2x^2$

42. $-2 = 2x^2 - 5x$

43. $6x^2 + 11x + 3 = 0$

44. $6x^2 + x - 2 = 0$

45. $9 - 6r = 8r^2$

46. $11m - 10 = 3w^2$

47. $\dfrac{7x + 1}{5} = -x^2$

48. $\dfrac{3x^2}{8} = \dfrac{1}{8} - x$

In Exercises 49–60, use the quadratic formula to solve each equation.

49. $x^2 + 3x + 2 = 0$

50. $x^2 - 3x + 2 = 0$

51. $x^2 + 12x = -36$

52. $y^2 - 18y = -81$

53. $5x^2 + 5x + 1 = 0$

54. $4w^2 + 6w + 1 = 0$

55. $8u = -4u^2 - 3$

56. $4t + 3 = 4t^2$

57. $16y^2 + 8y - 3 = 0$

58. $16x^2 + 16x + 3 = 0$

59. $\dfrac{x^2}{2} + \dfrac{5}{2}x = -1$

60. $-3x = \dfrac{x^2}{2} + 2$

In Exercises 61–62, use the quadratic formula and a scientific calculator to solve each equation. Give all answers to the nearest hundredth.

61. $0.7x^2 - 3.5x - 25 = 0$

62. $-4.5x^2 + 0.2x + 3.75 = 0$

63. Integer problem   The product of two consecutive even positive integers is 288. Find the integers. (*Hint:* If one integer is x, the next consecutive even integer is $x + 2$.)

64. Integer problem   The product of two consecutive odd negative integers is 143. Find the integers. (*Hint:* If one integer is x, the next consecutive odd integer is $x + 2$.)

65. Integer problem   The sum of the squares of two consecutive positive integers is 85. Find the integers. (*Hint:* If one integer is x, the next consecutive positive integer is $x + 1$.)

66. Integer problem   The sum of the squares of three consecutive positive integers is 77. Find the integers. (*Hint:* If one integer is x, the next consecutive positive integer is $x + 1$, and the third is $x + 2$.)

In Exercises 67–70, note that a and b are solutions to the equation $(x - a)(x - b) = 0$.

67. Find a quadratic equation that has a solution set of $\{3, 5\}$.

68. Find a quadratic equation that has a solution set of $\{-4, 6\}$.

69. Find a third-degree equation that has a solution set of $\{2, 3, -4\}$.

70. Find a fourth-degree equation that has a solution set of $\{3, -3, 4, -4\}$.

APPLICATIONS

71. Dimensions of a rectangle   A rectangle is 4 feet longer than it is wide, and its area is 96 square feet. Find its dimensions.

72. Dimensions of a rectangle   One side of a rectangle is 3 times as long as another, and its area is 147 square meters. Find its dimensions.

73. Side of a square   The area of a square is numerically equal to its perimeter. Find the length of each side of the square.

74. Perimeter of a rectangle   A rectangle is 2 inches longer than it is wide. Numerically, its area exceeds its perimeter by 11. Find the perimeter.

75. Base of a triangle   The height of a triangle is 5 centimeters longer than three times its base. Find the base of the triangle if its area is 6 square centimeters.

76. Height of a triangle   The height of a triangle is 4 meters longer than twice its base. Find the height if the area of the triangle is 15 square meters.

77. Finding rates   A woman drives her snowmobile 150 miles at the rate of r mph. She could have gone the same distance in 2 hours less time if she had increased her speed by 20 mph. Find r.

78. Finding rates   Jeff bicycles 160 miles at the rate of r mph. The same trip would have taken 2 hours longer if he had decreased his speed by 4 mph. Find r.

79. Pricing concert tickets   Tickets to a rock concert cost $4, and the projected attendance is 300 persons. It is further projected that for every 10¢ increase in ticket price, the average attendance will decrease by 5. At what ticket price will the nightly receipts be $1,248?

80. Setting bus fares   A bus company has 3,000 passengers daily, paying a 25¢ fare. For each 5¢ increase in fare, the company estimates that it will lose 80 passengers. What increase in fare will produce a $994 daily revenue?

81. Computing profit   The *Gazette*'s profit is $20 per year for each of its 3,000 subscribers. Management estimates that the profit per subscriber will increase by 1¢ for each additional subscriber over the current 3,000. How many subscribers will bring a total profit of $120,000?

82. Finding interest rates   A woman invests $1,000 in a mutual fund for which interest is compounded annually at a rate r. After one year, she deposits an additional $2,000. After two years, the balance in the account is

$$\$1,000(1 + r)^2 + \$2,000(1 + r)$$

If this amount is $3,368.10, find r.

83. Framing a picture   The frame around the picture in Illustration 1 has a constant width. How wide is the frame if its area equals the area of the picture?

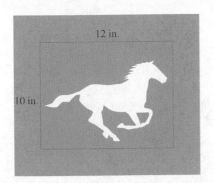

ILLUSTRATION 1

12 in.

10 in.

84. Metal fabrication   A box with no top is to be made by cutting a 2-inch square from each corner of the square sheet of metal shown in Illustration 2. After bending up the sides, the volume of the box is to be 200 cubic inches. How large should the piece of metal be?

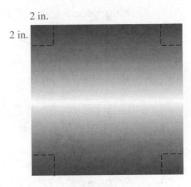

2 in.

2 in.

ILLUSTRATION 2

In Exercises 85–86, use a calculator.

85. Chemistry   A weak acid (0.1 M concentration) breaks down into free cations (the hydrogen ion, H^+) and anions (A^-). When this acid dissociates, the following equilibrium equation is established:

$$\frac{[H^+][A^-]}{[HA]} = 4 \times 10^{-4}$$

(where $[H^+]$, the hydrogen ion concentration, is equal to $[A^-]$, the anion concentration). $[HA]$ is the concentration of the undissociated acid itself. Find $[H^+]$ at equilibrium. (*Hint:* If $H^+ = x$, then $[HA] = 0.1 - x$.)

86. Chemistry   A saturated solution of hydrogen sulfide (0.1 M concentration) dissociates into cation H^+ and anion HS^-, where $H^+ = HS^-$. When this solution dissociates, the following equilibrium equation is established:

$$\frac{[H^+][HS^-]}{[HHS]} = 1.0 \times 10^{-7}$$

Find $[H^+]$. (*Hint:* If $H^+ = x$, then $[HHS] = 0.1 - x$.)

WRITING

87. Explain how to complete the square.

88. Tell why a cannot be 0 in the quadratic equation $ax^2 + bx + c = 0$.

SOMETHING TO THINK ABOUT

89. What number must be added to $x^2 + \sqrt{3}x$ to make it a perfect square trinomial?

90. Solve $x^2 + \sqrt{3}x - \frac{1}{4} = 0$ by completing the square.

10.2 Graphs of Quadratic Functions

■ QUADRATIC FUNCTIONS ■ GRAPHS OF $f(x) = ax^2$ ■ GRAPHS OF $f(x) = ax^2 + c$ ■ GRAPHS OF $f(x) = a(x - h)^2$ ■ GRAPHS OF $f(x) = a(x - h)^2 + k$ ■ GRAPHS OF $f(x) = ax^2 + bx + c$ ■ PROBLEM SOLVING ■ THE VARIANCE

Getting Ready   *If $y = f(x) = 3x^2 + x - 2$, find each value.*

1. $f(0)$ **2.** $f(1)$ **3.** $f(-1)$ **4.** $f(-2)$

If $x = -\dfrac{b}{2a}$, find x when a and b have the following values.

5. $a = 3$ and $b = -6$ **6.** $a = 5$ and $b = -40$

■ QUADRATIC FUNCTIONS

The graph shown in Figure 10-2 shows the height (in relation to time) of a toy rocket launched straight up into the air.

WARNING!   Note that the graph describes the height of the rocket, not the path of the rocket. The rocket goes straight up and comes straight down.

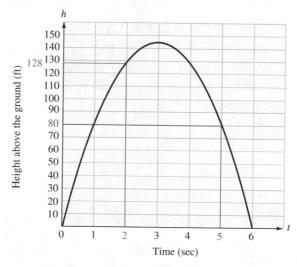

FIGURE 10-2

From the graph, we can see that the height of the rocket 2 seconds after it was launched is about 128 feet and that the height of the rocket 5 seconds after it was launched is 80 feet.

The parabola shown in Figure 10-2 is the graph of a *quadratic function,* a topic of this section.

> **Quadratic Functions**
> A **quadratic function** is a second-degree polynomial function of the form
> $$y = f(x) = ax^2 + bx + c \quad (a \neq 0)$$
> where a, b, and c are real numbers.

We begin the discussion of graphing quadratic functions by considering the graph of $f(x) = ax^2 + bx + c$, where $b = 0$ and $c = 0$.

■ GRAPHS OF $f(x) = ax^2$

EXAMPLE 1   Graph **a.** $f(x) = x^2$, **b.** $g(x) = 3x^2$, and **c.** $h(x) = \dfrac{1}{3}x^2$.

Solution   We can make a table of ordered pairs that satisfy each equation, plot each point, and join them with a smooth curve, as in Figure 10-3. We note that the graph of $h(x) = \frac{1}{3}x^2$ is wider than the graph of $f(x) = x^2$, and that the graph of $g(x) = 3x^2$ is narrower than the graph of $f(x) = x^2$. In the function $f(x) = ax^2$, the smaller the value of $|a|$, the wider the graph.

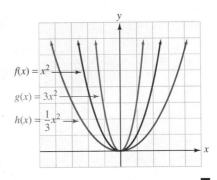

$f(x) = x^2$			$g(x) = 3x^2$			$h(x) = \frac{1}{3}x^2$		
x	$f(x)$	$(x, f(x))$	x	$g(x)$	$(x, g(x))$	x	$h(x)$	$(x, h(x))$
-2	4	$(-2, 4)$	-2	12	$(-2, 12)$	-2	$\frac{4}{3}$	$\left(-2, \frac{4}{3}\right)$
-1	1	$(-1, 1)$	-1	3	$(-1, 3)$	-1	$\frac{1}{3}$	$\left(-1, \frac{1}{3}\right)$
0	0	$(0, 0)$	0	0	$(0, 0)$	0	0	$(0, 0)$
1	1	$(1, 1)$	1	3	$(1, 3)$	1	$\frac{1}{3}$	$\left(1, \frac{1}{3}\right)$
2	4	$(2, 4)$	2	12	$(2, 12)$	2	$\frac{4}{3}$	$\left(2, \frac{4}{3}\right)$

FIGURE 10-3   ■

If we consider the graph of $f(x) = -3x^2$, we will see that it opens downward and has the same shape as the graph of $g(x) = 3x^2$.

    EXAMPLE 2   Graph $f(x) = -3x^2$.

Solution   We make a table of ordered pairs that satisfy the equation, plot each point, and join them with a smooth curve, as in Figure 10-4.

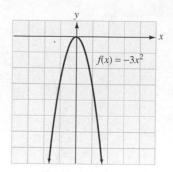

$$f(x) = -3x^2$$

x	$f(x)$	$(x, f(x))$
-2	-12	$(-2, -12)$
-1	-3	$(-1, -3)$
0	0	$(0, 0)$
1	-3	$(1, -3)$
2	-12	$(2, -12)$

FIGURE 10-4   ■

Self Check
Answer

Graph $f(x) = -\frac{1}{3}x^2$.

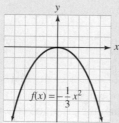

The graphs of quadratic functions are called **parabolas.** They open upward when $a > 0$ and downward when $a < 0$. The lowest point of a parabola that opens upward, or the highest point of a parabola that opens downward, is called the **vertex** of the parabola. The vertex of the parabola shown in Figure 10-4 is the point $(0, 0)$.

The vertical line, called an **axis of symmetry,** that passes through the vertex divides the parabola into two congruent halves. The axis of symmetry of the parabola shown in Figure 10-4 is the y-axis.

■ GRAPHS OF $f(x) = ax^2 + c$

EXAMPLE 3   Graph   **a.** $f(x) = 2x^2$,   **b.** $g(x) = 2x^2 + 3$,   and   **c.** $h(x) = 2x^2 - 3$.

Solution   We make a table of ordered pairs that satisfy each equation, plot each point, and join them with a smooth curve, as in Figure 10-5. We note that the graph of $g(x) = 2x^2 + 3$ is identical to the graph of $f(x) = 2x^2$, except that it has been translated 3 units upward. The graph of $h(x) = 2x^2 - 3$ is identical to the graph of $f(x) = 2x^2$, except that it has been translated 3 units downward.

$f(x) = 2x^2$		
x	$f(x)$	$(x, f(x))$
-2	8	$(-2, 8)$
-1	2	$(-1, 2)$
0	0	$(0, 0)$
1	2	$(1, 2)$
2	8	$(2, 8)$

$g(x) = 2x^2 + 3$		
x	$g(x)$	$(x, g(x))$
-2	11	$(-2, 11)$
-1	5	$(-1, 5)$
0	3	$(0, 3)$
1	5	$(1, 5)$
2	11	$(2, 11)$

$h(x) = 2x^2 - 3$		
x	$h(x)$	$(x, h(x))$
-2	5	$(-2, 5)$
-1	-1	$(-1, -1)$
0	-3	$(0, -3)$
1	-1	$(1, -1)$
2	5	$(2, 5)$

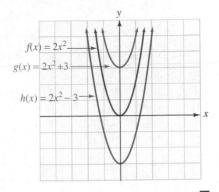

FIGURE 10-5

The results of Example 3 confirm the following facts, which we have previously discussed.

Vertical Translations of Graphs

If f is a function and k is a positive number, then

- The graph of $y = f(x) + k$ is identical to the graph of $y = f(x)$, except that it is translated k units upward.
- The graph of $y = f(x) - k$ is identical to the graph of $y = f(x)$, except that it is translated k units downward.

■ GRAPHS OF $f(x) = a(x - h)^2$

EXAMPLE 4

Graph   **a.** $f(x) = 2x^2$,   **b.** $g(x) = 2(x - 3)^2$,   and   **c.** $h(x) = 2(x + 3)^2$.

Solution

We make a table of ordered pairs that satisfy each equation, plot each point, and join them with a smooth curve, as in Figure 10-6. We note that the graph of $g(x) = 2(x - 3)^2$ is identical to the graph of $f(x) = 2x^2$, except that it has been translated 3 units to the right. The graph of $h(x) = 2(x + 3)^2$ is identical to the graph of $f(x) = 2x^2$, except that it has been translated 3 units to the left.

The results of Example 4 confirm the following facts, which we have previously discussed.

Horizontal Translations of Graphs

If f is a function and h is a positive number, then

- The graph of $y = f(x - h)$ is identical to the graph of $y = f(x)$, except that it is translated h units to the right.
- The graph of $y = f(x + h)$ is identical to the graph of $y = f(x)$, except that it is translated h units to the left.

$f(x) = 2x^2$		
x	$f(x)$	$(x, f(x))$
-2	8	$(-2, 8)$
-1	2	$(-1, 2)$
0	0	$(0, 0)$
1	2	$(1, 2)$
2	8	$(2, 8)$

$g(x) = 2(x - 3)^2$		
x	$g(x)$	$(x, g(x))$
1	8	$(1, 8)$
2	2	$(2, 2)$
3	0	$(3, 0)$
4	2	$(4, 2)$
5	8	$(5, 8)$

$h(x) = 2(x + 3)^2$		
x	$h(x)$	$(x, h(x))$
-5	8	$(-5, 8)$
-4	2	$(-4, 2)$
-3	0	$(-3, 0)$
-2	2	$(-2, 2)$
-1	8	$(-1, 8)$

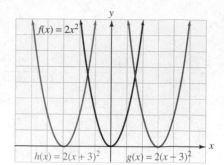

FIGURE 10-6

■ GRAPHS OF $f(x) = a(x - h)^2 + k$

EXAMPLE 5

Graph $f(x) = 2(x - 3)^2 - 4$.

Solution   The graph of $f(x) = 2(x - 3)^2 - 4$ is identical to the graph of $g(x) = 2(x - 3)^2$, except that it has been translated 4 units downward. The graph of $g(x) = 2(x - 3)^2$ is identical to the graph of $h(x) = 2x^2$, except that it has been translated 3 units to the right. Thus, to graph $f(x) = 2(x - 3)^2 - 4$, we can graph $h(x) = 2x^2$ and shift it 3 units to the right and then 4 units downward, as shown in Figure 10-7.

The vertex of the graph is the point $(3, -4)$, and the axis of symmetry is the line $x = 3$.

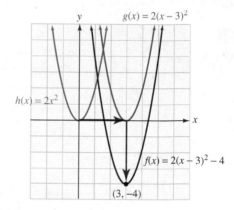

FIGURE 10-7

Self Check
Answer

Graph $f(x) = 2(x + 3)^2 + 1$.

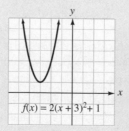

The results of Example 5 confirm the following facts, which we have previously discussed.

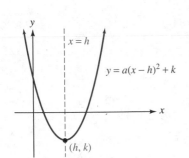

FIGURE 10-8

Vertex and Axis of Symmetry of a Parabola
The graph of the function

$$y = a(x - h)^2 + k \quad (a \neq 0)$$

is a parabola with vertex at (h, k). (See Figure 10-8.)
    The parabola opens upward when $a > 0$ and downward when $a < 0$. The axis of symmetry is the line $x = h$.

■ GRAPHS OF $f(x) = ax^2 + bx + c$

To graph functions of the form $f(x) = ax^2 + bx + c$, we can complete the square to write the function in the form $f(x) = a(x - h)^2 + k$.

EXAMPLE 6   Graph $f(x) = 2x^2 - 4x - 1$.

Solution   We complete the square on x to write the function in the form $f(x) = a(x - h)^2 + k$.

$$f(x) = 2x^2 - 4x - 1$$
$$f(x) = 2(x^2 - 2x) - 1 \qquad \text{Factor 2 from } 2x^2 - 4x.$$
$$f(x) = 2(x^2 - 2x + 1) - 1 - 2 \qquad \begin{array}{l}\text{Complete the square on } x. \text{ Since this adds} \\ \text{2 to the right-hand side, we also subtract 2} \\ \text{from the right-hand side.}\end{array}$$

1. $f(x) = 2(x - 1)^2 - 3 \qquad \text{Factor } x^2 - 2x + 1 \text{ and combine like terms.}$

From Equation 1, we can see that the vertex will be at the point $(1, -3)$. We can plot the vertex and a few points on either side of the vertex and draw the graph, which appears in Figure 10-9.

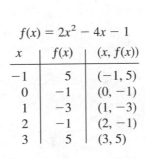

$$f(x) = 2x^2 - 4x - 1$$

x	$f(x)$	$(x, f(x))$
-1	5	$(-1, 5)$
0	-1	$(0, -1)$
1	-3	$(1, -3)$
2	-1	$(2, -1)$
3	5	$(3, 5)$

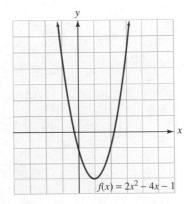

FIGURE 10-9   ■

Self Check

Answer

Graph $f(x) = 2x^2 - 4x + 1$.

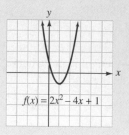

$f(x) = 2x^2 - 4x + 1$

■ ■ ■ ■ ■ ■ ■ ■ ■ ■ Graphing Quadratic Functions

GRAPHING
CALCULATORS

To graph $f(x) = 2x^2 + 6x - 3$ and find the coordinates of the vertex and the axis of symmetry of the parabola, we use a graphing calculator with window settings of $[-10, 10]$ for x and $[-10, 10]$ for y. If we enter the function, we will obtain the graph shown in Figure 10-10(a).

We then trace to move the cursor to the lowest point on the graph, as shown in Figure 10-10(b). By zooming in, we can determine that the vertex is the point $(-1.5, -7.5)$ and that the line $x = -1.5$ is the axis of symmetry.

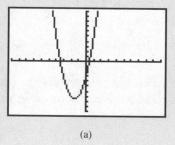

(a)

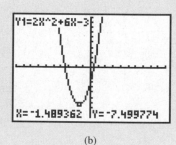

(b)

FIGURE 10-10

Because it is easy to graph quadratic functions with a graphing calculator, we can use graphing to find approximate solutions of quadratic equations. For example, the solutions of $0.7x^2 + 2x - 3.5 = 0$ are the numbers x that will make $y = 0$ in the quadratic function $y = f(x) = 0.7x^2 + 2x - 3.5$. To approximate these numbers, we graph the quadratic function and read the x-intercepts from the graph.

We can use the standard window settings of $[-10, 10]$ for x and $[-10, 10]$ for y and graph the function, as in Figure 10-11(a). We then trace to move the cursor to each *x*-intercept, as in Figures 10-11(b) and 10-11(c). From the graph, we can read the approximate value of the *x*-coordinate of each *x*-intercept. For better results, we can zoom in.

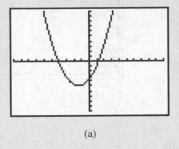

(a)

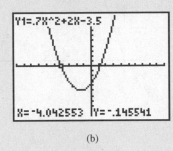

(b)

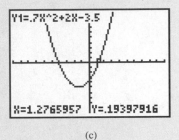

(c)

FIGURE 10-11

■ PROBLEM SOLVING

EXAMPLE 7

Ballistics   The ball shown in Figure 10-12(a) is thrown straight up with a velocity of 128 feet per second. The function $s = h(t) = -16t^2 + 128t$ gives the relation between *s* (the number of feet the ball is above the ground) and *t* (the time measured in seconds). How high does the ball travel, and when will it hit the ground?

Solution   The graph of $s = -16t^2 + 128t$ is a parabola. Since the coefficient of t^2 is negative, it opens downward, and the maximum height of the ball is given by the *s*-coordinate of the vertex of the parabola. We can find the coordinates of the vertex by completing the square:

$$s = -16t^2 + 128t$$
$$= -16(t^2 - 8t) \qquad\qquad \text{Factor out } -16.$$
$$= -16(t^2 - 8t \mathbf{+ 16 - 16}) \qquad \text{Add and subtract 16.}$$
$$= -16(t^2 - 8t + 16) + 256 \qquad (-16)(-16) = 256.$$
$$= -16(t - \mathbf{4})^2 + 256 \qquad \text{Factor } t^2 - 8t + 16.$$

From the result, we can see that the coordinates of the vertex are $(\mathbf{4}, \mathbf{256})$. Since $t = 4$ and $s = 256$ are the coordinates of the vertex, the ball reaches a maximum height of 256 feet in 4 seconds.

From the graph, we can see that the ball will hit the ground in 8 seconds, because the height is 0 when $t = 8$.

To solve this problem with a graphing calculator with window settings of [0, 10] for *x* and [0, 300] for *y*, we graph the function $h(t) = -16t^2 + 128t$ to get the graph in Figure 10-12(b). By using trace and zoom, we can determine that the ball reaches a height of 256 feet in 4 seconds and that the ball will hit the ground in 8 seconds.

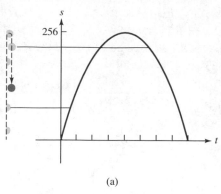

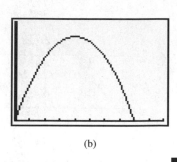

(a)

(b)

FIGURE 10-12

EXAMPLE 8   **Maximizing area**   A man wants to build the rectangular pen shown in Figure 10-13(a) to house his dog. If he uses one side of his barn, find the maximum area that he can enclose with 80 feet of fencing.

Solution   If we let the width of the area be w, the length is represented by $80 - 2w$.

We can find the maximum value of A as follows:

$$A = (80 - 2w)w \qquad\qquad A = lw.$$
$$= 80w - 2w^2 \qquad\qquad \text{Remove parentheses.}$$
$$= -2(w^2 - 40w) \qquad\qquad \text{Factor out } -2.$$
$$= -2(w^2 - 40w \mathbf{+ 400 - 400}) \qquad \text{Subtract and add 400.}$$
$$= -2(w^2 - 40w + 400) + 800 \qquad -2(-400) = 800.$$
$$= -2(w - 20)^2 + 800 \qquad\qquad \text{Factor } w^2 - 40w + 400.$$

Thus, the coordinates of the vertex of the graph of the quadratic function are (20, 800), and the maximum area is 800 square feet.

To solve this problem using a graphing calculator with window settings of [0, 50] for x and [0, 1,000] for y, we graph the function $s(t) = -2w^2 + 80w$ to get the graph in Figure 10-13(b). By using trace and zoom, we can determine that the maximum area is 800 square feet when the width is 20 feet.

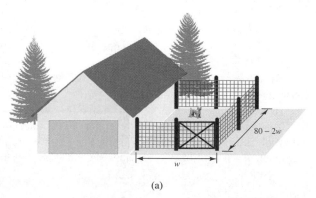

(a)

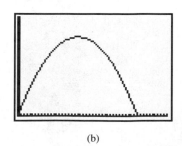

(b)

FIGURE 10-13

■ THE VARIANCE

In statistics, the square of the standard deviation is called the **variance.**

EXAMPLE 9

If p is the probability that a person selected at random has AIDS, then $1 - p$ is the probability that the person does not have AIDS. If 100 people in Minneapolis are randomly sampled, we know from statistics that the variance of this type of sample distribution will be $100p(1 - p)$. What value of p will maximize the variance?

Solution   The variance is given by the function

$$v(p) = 100p(1 - p) \quad \text{or} \quad v(p) = -100p^2 + 100p$$

Since all probabilities have values between 0 to 1, including 0 and 1, we use window settings of $[0, 1]$ for x when graphing the function $v(p) = -100p^2 + 100p$ on a graphing calculator. If we also use window settings of $[0, 30]$ for y, we will obtain the graph shown in Figure 10-14(a). After using trace and zoom to obtain Figure 10-14(b), we can see that a probability of 0.5 will give the maximum variance.

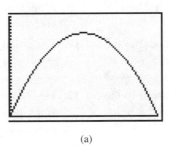
(a)

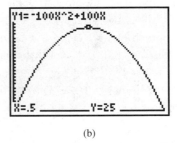
(b)

FIGURE 10-14   ■

Orals   *Tell whether the graph of each equation opens up or down.*

1. $y = -3x^2 + x - 5$ **2.** $y = 4x^2 + 2x - 3$

3. $y = 2(x - 3)^2 - 1$ **4.** $y = -3(x + 2)^2 + 2$

Find the vertex of the parabola determined by each equation.

5. $y = 2(x - 3)^2 - 1$ **6.** $y = -3(x + 2)^2 + 2$

EXERCISE 10.2

REVIEW   *In Exercises 1–2, find the value of x.*

1.

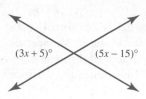

$(3x + 5)°$ $(5x − 15)°$

2. Lines r and s are parallel.

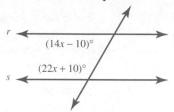

$(14x − 10)°$

$(22x + 10)°$

3. Madison and St. Louis are 385 miles apart. One train leaves Madison and heads toward St. Louis at the rate of 30 mph. Three hours later, a second train leaves Madison, bound for St. Louis. If the second train travels at the rate of 55 mph, in how many hours will the faster train overtake the slower train?

4. A woman invests $25,000, some at 7% annual interest and the rest at 8%. If the annual income from both investments is $1,900, how much is invested at the higher rate?

VOCABULARY AND CONCEPTS   *Fill in each blank to make a true statement.*

5. A quadratic function is a second-degree polynomial function that can be written in the form _____, where _____.

6. The graphs of quadratic functions are called _____.

7. The highest (or the lowest) point on a parabola is called the _____.

8. A vertical line that divides a parabola into two halves is called an _____ of symmetry.

9. The graph of $y = f(x) + k$ $(k > 0)$ is identical to the graph of $y = f(x)$, except that it is translated k units _____.

10. The graph of $y = f(x) − k$ $(k > 0)$ is identical to the graph of $y = f(x)$, except that it is translated k units _____.

11. The graph of $y = f(x − h)$ $(h > 0)$ is identical to the graph of $y = f(x)$, except that it is translated h units _____.

12. The graph of $y = f(x + h)$ $(h > 0)$ is identical to the graph of $y = f(x)$, except that it is translated h units _____.

13. The graph of $y = f(x) = ax^2 + bx + c$ $(a \neq 0)$ opens _____ when $a > 0$.

14. In statistics, the square of the standard deviation is called the _____.

PRACTICE   *In Exercises 15–26, graph each function.*

15. $f(x) = x^2$

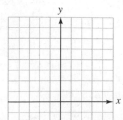

16. $f(x) = −x^2$

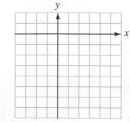

17. $f(x) = x^2 + 2$

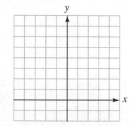

18. $f(x) = x^2 − 3$

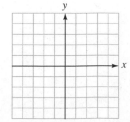

19. $f(x) = -(x - 2)^2$

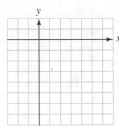

20. $f(x) = (x + 2)^2$

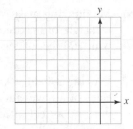

21. $f(x) = (x - 3)^2 + 2$

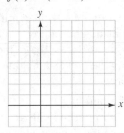

22. $f(x) = (x + 1)^2 - 2$

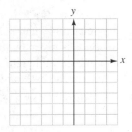

23. $f(x) = x^2 + x - 6$

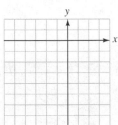

24. $f(x) = x^2 - x - 6$

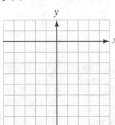

25. $f(x) = 12x^2 + 6x - 6$

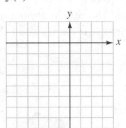

26. $f(x) = -2x^2 + 4x + 3$

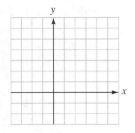

In Exercises 27–38, find the coordinates of the vertex and the axis of symmetry of the graph of each equation. If necessary, complete the square on x to write the equation in the form $y = a(x - h)^2 + k$. **Do not graph the equation.**

27. $y = (x - 1)^2 + 2$

28. $y = 2(x - 2)^2 - 1$

29. $y = 2(x + 3)^2 - 4$

30. $y = -3(x + 1)^2 + 3$

31. $y = -3x^2$

32. $y = 3x^2 - 3$

33. $y = 2x^2 - 4x$

34. $y = 3x^2 + 6x$

35. $y = -4x^2 + 16x + 5$

36. $y = 5x^2 + 20x + 25$

37. $y - 7 = 6x^2 - 5x$

38. $y - 2 = 3x^2 + 4x$

39. The equation $y - 2 = (x - 5)^2$ represents a quadratic function whose graph is a parabola. Find its vertex.

40. Show that $y = ax^2$, where $a \neq 0$, represents a quadratic function whose vertex is at the origin.

In Exercises 41–44, use a graphing calculator to find the coordinates of the vertex of the graph of each quadratic function. Give results to the nearest hundredth.

41. $y = 2x^2 - x + 1$

42. $y = x^2 + 5x - 6$

43. $y = 7 + x - x^2$

44. $y = 2x^2 - 3x + 2$

In Exercises 45–48, use a graphing calculator to solve each equation. If a result is not exact, give the result to the nearest hundredth.

45. $x^2 + x - 6 = 0$

46. $2x^2 - 5x - 3 = 0$

47. $0.5x^2 - 0.7x - 3 = 0$

48. $2x^2 - 0.5x - 2 = 0$

APPLICATIONS

49. Ballistics If a ball is thrown straight up with an initial velocity of 48 feet per second, its height s after t seconds is given by the equation $s = 48t - 16t^2$. Find the maximum height attained by the ball and the time it takes for the ball to return to earth.

50. Ballistics From the top of the building in Illustration 1, a ball is thrown straight up with an initial velocity of 32 feet per second. The equation $s = -16t^2 + 32t + 48$ gives the height s of the ball t seconds after it is thrown. Find the maximum height reached by the ball and the time it takes for the ball to hit the ground.

51. Maximizing area Find the dimensions of the rectangle of maximum area that can be constructed with 200 feet of fencing. Find the maximum area.

52. Fencing a field A farmer wants to fence in three sides of a rectangular field (shown in Illustration 2) with 1,000 feet of fencing. The other side of the rectangle will be a river. If the enclosed area is to be maximum, find the dimensions of the field.

ILLUSTRATION 1

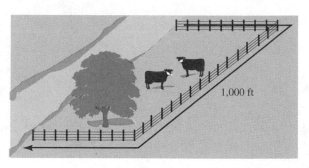

ILLUSTRATION 2

  In Exercises 53–60, use a graphing calculator to help solve each problem.

53. Maximizing revenue The revenue R received for selling x stereos is given by the equation

$$R = -\frac{x^2}{1,000} + 10x$$

Find the number of stereos that must be sold to obtain the maximum revenue.

54. Maximizing revenue In Exercise 53, find the maximum revenue.

55. Maximizing revenue The revenue received for selling x radios is given by the formula

$$R = -\frac{x^2}{728} + 9x$$

How many radios must be sold to obtain the maximum revenue? Find the maximum revenue.

56. Maximizing revenue The revenue received for selling x stereos is given by the formula

$$R = -\frac{x^2}{5} + 80x - 1,000$$

How many stereos must be sold to obtain the maximum revenue? Find the maximum revenue.

57. Maximizing revenue When priced at $30 each, a toy has annual sales of 4,000 units. The manufacturer estimates that each $1 increase in cost will decrease sales by 100 units. Find the unit price that will maximize total revenue. (*Hint:* Total revenue = price · the number of units sold.)

58. Maximizing revenue When priced at $57, one type of camera has annual sales of 525 units. For each $1 the camera is reduced in price, management expects to sell an additional 75 cameras. Find the unit price that will maximize total revenue. (*Hint:* Total revenue = price · the number of units sold.)

59. Finding the variance If p is the probability that a person sampled at random has high blood pressure, then $1 - p$ is the probability that the person doesn't. If 50 people are sampled at random, the variance of the sample will be $50p(1 - p)$. What two probabilities p will give a variance of 9.375?

60. Finding the variance If p is the probability that a person sampled at random smokes, then $1 - p$ is the probability that the person doesn't. If 75 people are sampled at random, the variance of the sample will be $75p(1 - p)$. What two probabilities p will give a variance of 12?

WRITING

61. The graph of $y = ax^2 + bx + c$ $(a \neq 0)$ passes the vertical line test. Explain why this shows that the equation defines a function.

62. The graph of $x = y^2 - 2y$ is a parabola. Explain why its graph does not represent a function.

SOMETHING TO THINK ABOUT

63. Can you use a graphing calculator to find solutions of the equation $x^2 + x + 1 = 0$? What is the problem? How do you interpret the result?

64. Complete the square on x in the equation $y = ax^2 + bx + c$ and show that the vertex of the parabolic graph is the point with coordinates of

$$\left(-\frac{b}{2a}, c - \frac{b^2}{4a} \right)$$

10.3 Complex Numbers

■ IMAGINARY NUMBERS ■ POWERS OF i ■ SIMPLIFYING IMAGINARY NUMBERS ■ COMPLEX NUMBERS ■ ARITHMETIC OF COMPLEX NUMBERS ■ RATIONALIZING THE DENOMINATOR ■ ABSOLUTE VALUE OF A COMPLEX NUMBER

Getting Ready *Do the following operations.*

1. $(3x + 5) + (4x - 5)$

2. $(3x + 5) - (4x - 5)$

3. $(3x + 5)(4x - 5)$

4. $(3x + 5)(3x - 5)$

■ IMAGINARY NUMBERS

So far, all of our work with quadratic equations has involved real numbers only. However, the solutions of many quadratic equations are not real numbers.

| EXAMPLE 1 | Solve $x^2 + x + 1 = 0$. |

Solution   Because the quadratic trinomial is prime and cannot be factored, we will use the quadratic formula, with $a = 1$, $b = 1$, and $c = 1$:

$$x = \frac{-b \pm \sqrt{b^2 - 4ac}}{2a}$$

$$= \frac{-1 \pm \sqrt{1^2 - 4(1)(1)}}{2(1)} \qquad \text{Substitute 1 for } a, \text{ 1 for } b, \text{ and 1 for } c.$$

$$= \frac{-1 \pm \sqrt{1 - 4}}{2}$$

$$= \frac{-1 \pm \sqrt{-3}}{2}$$

$$x = \frac{-1 \pm \sqrt{-3}}{2} \quad \text{or} \quad x = \frac{-1 - \sqrt{-3}}{2}$$ ∎

Self Check   Solve $a^2 + 3a + 3 = 0$.

Answer $\dfrac{-3 + \sqrt{-3}}{2}, \dfrac{-3 - \sqrt{-3}}{2}$

Each solution in Example 1 contains the number $\sqrt{-3}$. Since no real number squared is -3, $\sqrt{-3}$ is not a real number. For years, people believed that numbers such as

$$\sqrt{-1}, \sqrt{-3}, \sqrt{-4}, \text{ and } \sqrt{-9}$$

were nonsense. In the 17th century, René Descartes (1596–1650) called them **imaginary numbers.** Today, imaginary numbers have many important uses, such as describing the behavior of alternating current in electronics.

The imaginary number $\sqrt{-1}$ is often denoted by the letter i:

$$i = \sqrt{-1}$$

Because i represents the square root of -1, it follows that

$$i^2 = -1$$

■ POWERS OF i

The powers of i produce an interesting pattern:

$$i = \sqrt{-1} = i \qquad\qquad i^5 = i^4 i = 1i = i$$
$$i^2 = \left(\sqrt{-1}\right)^2 = -1 \qquad i^6 = i^4 i^2 = 1(-1) = -1$$
$$i^3 = i^2 i = -1i = -i \qquad i^7 = i^4 i^3 = 1(-i) = -i$$
$$i^4 = i^2 i^2 = (-1)(-1) = 1 \qquad i^8 = i^4 i^4 = (1)(1) = 1$$

The pattern continues: $i, -1, -i, 1, \ldots$.

EXAMPLE 2 Simplify i^{29}.

Solution We note that 29 divided by 4 gives a quotient of 7 and a remainder of 1. Thus, $29 = 4 \cdot 7 + 1$, and

$$
\begin{aligned}
i^{29} &= i^{4 \cdot 7 + 1} &&\quad 4 \cdot 7 + 1 = 29. \\
&= (i^4)^7 \cdot i &&\quad i^{4 \cdot 7 + 1} = i^{4 \cdot 7} \cdot i^1 = (i^4)^7 \cdot i. \\
&= 1^7 \cdot i &&\quad i^4 = 1. \\
&= i
\end{aligned}
$$

Self Check Simplify i^{31}.

Answer $-i$

■ ■ ■ ■ ■ ■ ■ ■ ■ PERSPECTIVE

The Pythagoreans (ca. 500 B.C.) understood the universe as a harmony of whole numbers. They did not classify fractions as numbers, and were upset that $\sqrt{2}$ was not the ratio of whole numbers. For 2,000 years, little progress was made in the understanding of the various kinds of numbers.

The father of algebra, François Vieta (1540–1603), understood the whole numbers, fractions, and certain irrational numbers. But he was unable to accept negative numbers, and certainly not imaginary numbers.

René Descartes (1596–1650) thought these numbers to be nothing more than figments of his imagination, so he called them *imaginary numbers*. Leonhard Euler (1707–1783) used the letter i for $\sqrt{-1}$; Augustin Cauchy (1789–1857) used the term *conjugate*; and Karl Gauss (1777–1855) first used the word *complex*.

Today, we accept complex numbers without question, and we use them in science, economics, medicine, and industry. But it took many centuries and the work of many mathematicians to make them respectable.

The results of Example 2 illustrate the following fact.

> **Powers of i**
>
> If n is a natural number that has a remainder of r when divided by 4, then
> $$i^n = i^r$$
> When n is divisible by 4, the remainder r is 0 and $i^0 = 1$.

EXAMPLE 3 Simplify i^{55}.

Solution We divide 55 by 4 and get a remainder of 3. Therefore,

$$i^{55} = i^3 = -i$$

Self Check	Simplify i^{62}.
Answer	-1

■ SIMPLIFYING IMAGINARY NUMBERS

If we assume that multiplication of imaginary numbers is commutative and associative, then

$$(2i)^2 = 2^2 i^2$$
$$= 4(-1) \qquad i^2 = -1.$$
$$= -4$$

Since $(2i)^2 = -4$, $2i$ is a square root of -4, and we can write

$$\sqrt{-4} = 2i$$

This result can also be obtained by using the multiplication property of radicals:

$$\sqrt{-4} = \sqrt{4(-1)} = \sqrt{4}\sqrt{-1} = 2i$$

We can use the multiplication property of radicals to simplify any imaginary number. For example,

$$\sqrt{-25} = \sqrt{25(-1)} = \sqrt{25}\sqrt{-1} = 5i$$

$$\sqrt{\frac{-100}{49}} = \sqrt{\frac{100}{49}(-1)} = \frac{\sqrt{100}}{\sqrt{49}}\sqrt{-1} = \frac{10}{7}i$$

These examples illustrate the following rule.

Properties of Radicals

If at least one of a and b is a nonnegative real number, then

$$\sqrt{ab} = \sqrt{a}\sqrt{b} \qquad \text{and} \qquad \sqrt{\frac{a}{b}} = \frac{\sqrt{a}}{\sqrt{b}} \quad (b \neq 0)$$

WARNING! If a and b are negative, then $\sqrt{ab} \neq \sqrt{a}\sqrt{b}$. For example, if $a = -16$ and $b = -4$,

$$\sqrt{(-16)(-4)} = \sqrt{64} = 8 \qquad \text{but} \qquad \sqrt{-16}\sqrt{-4} = (4i)(2i) = 8i^2 = 8(-1) = -8$$

The correct solution is -8.

■ COMPLEX NUMBERS

The imaginary numbers are a subset of a set of numbers called the **complex numbers.**

Complex Numbers

A **complex number** is any number that can be written in the form $a + bi$, where a and b are real numbers and $i = \sqrt{-1}$.

    In the complex number $a + bi$, a is called the **real part,** and b is called the **imaginary part.**

If $b = 0$, the complex number $a + bi$ is a real number. If $b \neq 0$ and $a = 0$, the complex number $0 + bi$ (or just bi) is an imaginary number.

Any imaginary number can be expressed in bi form. For example,

$$\sqrt{-1} = i$$
$$\sqrt{-9} = \sqrt{9(-1)} = \sqrt{9}\sqrt{-1} = 3i$$
$$\sqrt{-3} = \sqrt{3(-1)} = \sqrt{3}\sqrt{-1} = \sqrt{3}i$$

WARNING! The expression $\sqrt{3}i$ is often written as $i\sqrt{3}$ to make it clear that i is not part of the radicand. Do not confuse $\sqrt{3}i$ with $\sqrt{3i}$.

The relationship between the real numbers, the imaginary numbers, and the complex numbers is shown in Figure 10-15.

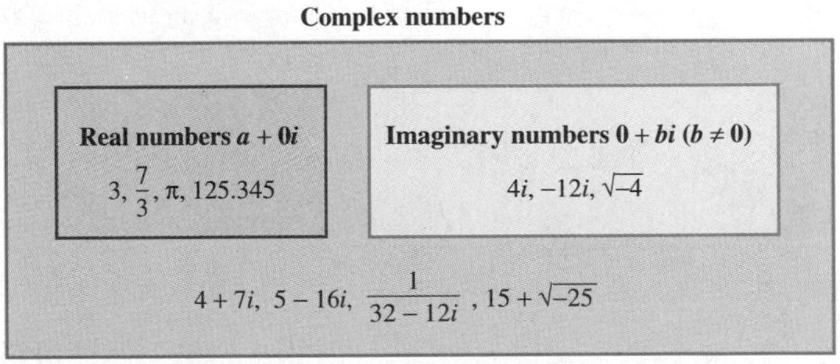

Complex numbers

Real numbers $a + 0i$	Imaginary numbers $0 + bi$ $(b \neq 0)$
$3, \dfrac{7}{3}, \pi, 125.345$	$4i, -12i, \sqrt{-4}$

$$4 + 7i,\ 5 - 16i,\ \frac{1}{32 - 12i},\ 15 + \sqrt{-25}$$

FIGURE 10-15

Equality of Complex Numbers

The complex numbers $a + bi$ and $c + di$ are equal if and only if

$$a = c \qquad \text{and} \qquad b = d$$

Because of the previous definition, complex numbers are equal when their real parts are equal and their imaginary parts are equal.

EXAMPLE 4

a. $2 + 3i = \sqrt{4} + \dfrac{6}{2}i$ because $2 = \sqrt{4}$ and $3 = \dfrac{6}{2}$.

b. $4 - 5i = \dfrac{12}{3} - \sqrt{25}i$ because $4 = \dfrac{12}{3}$ and $-5 = -\sqrt{25}$.

c. $x + yi = 4 + 7i$ if and only if $x = 4$ and $y = 7$.   ∎

■ ARITHMETIC OF COMPLEX NUMBERS

Addition and Subtraction of Complex Numbers
Complex numbers are added and subtracted as if they were binomials:

$$(a + bi) + (c + di) = (a + c) + (b + d)i$$

$$(a + bi) - (c + di) = (a + bi) + (-c - di) = (a - c) + (b - d)i$$

EXAMPLE 5   Do the operations.

a. $(8 + 4i) + (12 + 8i) = 8 + 4i + 12 + 8i$
$$= 20 + 12i$$

b. $(7 - 4i) + (9 + 2i) = 7 - 4i + 9 + 2i$
$$= 16 - 2i$$

c. $(-6 + i) - (3 - 4i) = -6 + i - 3 + 4i$
$$= -9 + 5i$$

d. $(2 - 4i) - (-4 + 3i) = 2 - 4i + 4 - 3i$
$$= 6 - 7i$$ ∎

Self Check   Do the operations: **a.** $(3 - 5i) + (-2 + 7i)$ and
b. $(3 - 5i) - (-2 + 7i)$.

Answers   **a.** $1 + 2i$,   **b.** $5 - 12i$

To multiply a complex number by an imaginary number, we use the distributive property to remove parentheses and then simplify. For example,

$$-5i(4 - 8i) = -5i(4) - (-5i)8i \qquad \text{Use the distributive property.}$$
$$= -20i + 40i^2 \qquad\qquad\quad \text{Simplify.}$$
$$= -20i + 40(-1) \qquad\qquad \text{Remember that } i^2 = -1.$$
$$= -40 - 20i$$

To multiply two complex numbers, we use the following definition.

> **Multiplying Complex Numbers**
> Complex numbers are multiplied as if they were binomials, with $i^2 = -1$:
> $$(a + bi)(c + di) = ac + adi + bci + bdi^2$$
> $$= ac + adi + bci + bd(-1)$$
> $$= (ac - bd) + (ad + bc)i$$

EXAMPLE 6 Multiply the complex numbers.

a. $(2 + 3i)(3 - 2i) = 6 - 4i + 9i - 6i^2$ Use the FOIL method.
$$= 6 + 5i + 6$$ $i^2 = -1$, and combine $-4i$ and $9i$.
$$= 12 + 5i$$

b. $(3 + i)(1 + 2i) = 3 + 6i + i + 2i^2$ Use the FOIL method.
$$= 3 + 7i - 2$$ $i^2 = -1$, and combine $6i$ and i.
$$= 1 + 7i$$

c. $(-4 + 2i)(2 + i) = -8 - 4i + 4i + 2i^2$ Use the FOIL method.
$$= -8 - 2$$ $i^2 = -1$, and combine $-4i$ and $4i$.
$$= -10$$ ∎

Self Check   Multiply $(-2 + 3i)(3 - 2i)$.
Answer $13i$

The next two examples show how to write complex numbers in $a + bi$ form. It is common to use $a - bi$ as a substitute for $a + (-b)i$.

EXAMPLE 7 Write each number in $a + bi$ form.

a. $7 = 7 + 0i$ **b.** $3i = 0 + 3i$

c. $4 - \sqrt{-16} = 4 - \sqrt{-1(16)}$ **d.** $5 + \sqrt{-11} = 5 + \sqrt{-1(11)}$
$$= 4 - \sqrt{16}\sqrt{-1} \qquad\qquad\qquad\qquad = 5 + \sqrt{11}\sqrt{-1}$$
$$= 4 - 4i \qquad\qquad\qquad\qquad\qquad\quad = 5 + \sqrt{11}i$$ ∎

Self Check   Write $3 - \sqrt{-25}$ in $a + bi$ form.
Answer $3 - 5i$

EXAMPLE 8 Simplify each expression.

a. $2i^2 + 4i^3 = 2(-1) + 4(-i)$

$\qquad\qquad = -2 - 4i$

b. $\dfrac{3}{2i} = \dfrac{3}{2i} \cdot \dfrac{i}{i}$ $\qquad \frac{i}{i} = 1.$

$\qquad = \dfrac{3i}{2i^2}$

$\qquad = \dfrac{3i}{2(-1)}$

$\qquad = \dfrac{3i}{-2}$

$\qquad = 0 - \dfrac{3}{2}i$

c. $-\dfrac{5}{i} = -\dfrac{5}{i} \cdot \dfrac{i^3}{i^3}$ $\qquad \frac{i^3}{i^3} = 1.$

$\qquad = -\dfrac{5(-i)}{1}$

$\qquad = 5i$

$\qquad = 0 + 5i$

d. $\dfrac{6}{i^3} = \dfrac{6i}{i^3 i}$ $\qquad \frac{i}{i} = 1.$

$\qquad = \dfrac{6i}{i^4}$

$\qquad = \dfrac{6i}{1}$

$\qquad = 6i$

$\qquad = 0 + 6i$ ∎

Self Check   Simplify   **a.** $3i^3 - 2i^2$ and   **b.** $\frac{2}{3i}$.

Answers   **a.** $2 - 3i$,   **b.** $0 - \frac{2}{3}i$

Complex Conjugates

The complex numbers $a + bi$ and $a - bi$ are called **complex conjugates.**

For example,

$\qquad 3 + 4i$ and $3 - 4i$ are complex conjugates.

$\qquad 5 - 7i$ and $5 + 7i$ are complex conjugates.

EXAMPLE 9 Find the product of $3 + i$ and its complex conjugate.

Solution   The complex conjugate of $3 + i$ is $3 - i$. We can find the product as follows:

$\qquad (3 + i)(3 - i) = 9 - 3i + 3i - i^2$ Use the FOIL method.

$\qquad\qquad\qquad = 9 - i^2$ Combine like terms.

$\qquad\qquad\qquad = 9 - (-1)$ $i^2 = -1.$

$\qquad\qquad\qquad = 10$ ∎

| Self Check | Multiply $(2 + 3i)(2 - 3i)$. |
| Answer | 13 |

The product of the complex number $a + bi$ and its complex conjugate $a - bi$ is the real number $a^2 + b^2$, as the following work shows:

$$(a + bi)(a - bi) = a^2 - abi + abi - b^2i^2 \qquad \text{Use the FOIL method.}$$
$$= a^2 - b^2(-1) \qquad\qquad i^2 = -1.$$
$$= a^2 + b^2$$

■ RATIONALIZING THE DENOMINATOR

To divide complex numbers, we often have to rationalize a denominator.

EXAMPLE 10 Divide and write the result in $a + bi$ form: $\dfrac{1}{3 + i}$.

Solution We can rationalize the denominator by multiplying the numerator and the denominator by the complex conjugate of the denominator.

$$\frac{1}{3 + i} = \frac{1}{3 + i} \cdot \frac{3 - i}{3 - i} \qquad \tfrac{3-i}{3-i} = 1.$$

$$= \frac{3 - i}{9 - 3i + 3i - i^2} \qquad \text{Multiply the numerators and multiply the denominators.}$$

$$= \frac{3 - i}{9 - (-1)} \qquad i^2 = -1.$$

$$= \frac{3 - i}{10}$$

$$= \frac{3}{10} - \frac{1}{10}i$$

| Self Check | Rationalize the denominator: $\frac{1}{5 - i}$. |
| Answer | $\frac{5}{26} + \frac{1}{26}i$ |

EXAMPLE 11 Write $\dfrac{3 - i}{2 + i}$ in $a + bi$ form.

Solution We multiply both the numerator and the denominator of the fraction by the complex conjugate of the denominator.

$$\frac{3-i}{2+i} = \frac{3-i}{2+i} \cdot \frac{2-i}{2-i}$$ $\frac{2-i}{2-i} = 1.$

$$= \frac{6 - 3i - 2i + i^2}{4 - 2i + 2i - i^2}$$ Multiply the numerators and multiply the denominators.

$$= \frac{5 - 5i}{4 - (-1)}$$ $i^2 = -1.$

$$= \frac{5(1 - i)}{5}$$ Factor out 5 in the numerator.

$$= 1 - i$$ Simplify.   ■

Self Check   Rationalize the denominator: $\frac{2+i}{5-i}$.

Answer $\frac{9}{26} + \frac{7}{26}i$

EXAMPLE 12   Write $\dfrac{4 + \sqrt{-16}}{2 + \sqrt{-4}}$ in $a + bi$ form.

Solution $$\frac{4 + \sqrt{-16}}{2 + \sqrt{-4}} = \frac{4 + 4i}{2 + 2i}$$ Write each number in $a + bi$ form.

$$= \frac{\overset{1}{\cancel{2}}(2 + 2i)}{\underset{1}{\cancel{2}} + 2i}$$ Factor out 2 in the numerator and simplify.

$$= 2 + 0i$$ ■

Self Check   Divide: $\dfrac{3 + \sqrt{-25}}{2 + \sqrt{-9}}$.

Answer $\frac{21}{13} + \frac{1}{13}i$

 WARNING!   To avoid mistakes, always put complex numbers in $a + bi$ form before doing any complex number arithmetic.

■ ABSOLUTE VALUE OF A COMPLEX NUMBER

Absolute Value of a Complex Number
The **absolute value** of the complex number $a + bi$ is $\sqrt{a^2 + b^2}$. In symbols,
$$|a + bi| = \sqrt{a^2 + b^2}$$

EXAMPLE 13   Find each absolute value.

a. $|3 + 4i| = \sqrt{3^2 + 4^2}$
$= \sqrt{9 + 16}$
$= \sqrt{25}$
$= 5$

b. $|3 - 4i| = \sqrt{3^2 + (-4)^2}$
$= \sqrt{9 + 16}$
$= \sqrt{25}$
$= 5$

c. $|-5 - 12i| = \sqrt{(-5)^2 + (-12)^2}$
$= \sqrt{25 + 144}$
$= \sqrt{169}$
$= 13$

d. $|a + 0i| = \sqrt{a^2 + 0^2}$
$= \sqrt{a^2}$
$= |a|$

∎

Self Check   Evaluate $|5 + 12i|$.
Answer   13

WARNING! Note that $|a + bi| = \sqrt{a^2 + b^2}$, not $|a + bi| = \sqrt{a^2 + (bi)^2}$.

Orals   *Simplify each power of i.*

1. i^3 **2.** i^2 **3.** i^4 **4.** i^5

Write each imaginary number in bi form.

5. $\sqrt{-49}$ **6.** $\sqrt{-64}$ **7.** $\sqrt{-100}$ **8.** $\sqrt{-81}$

Find each absolute value.

9. $|3 + 4i|$ **10.** $|5 - 12i|$

EXERCISE 10.3

REVIEW   *In Exercises 1–2, do each operation.*

1. $\dfrac{x^2 - x - 6}{9 - x^2} \cdot \dfrac{x^2 + x - 6}{x^2 - 4}$

2. $\dfrac{3x + 4}{x - 2} + \dfrac{x - 4}{x + 2}$

3. Wind speed   A plane that can fly 200 mph in still air makes a 330-mile flight with a tail wind and returns, flying into the same wind. Find the speed of the wind if the total flying time is $3\frac{1}{3}$ hours.

4. Finding rates   A student drove a distance of 135 miles at an average speed of 50 mph. How much faster would he have to drive on the return trip to save 30 minutes of driving time?

VOCABULARY AND CONCEPTS   *Fill in each blank to make a true statement.*

5. $\sqrt{-1}, \sqrt{-3},$ and $\sqrt{-4}$ are examples of _____ numbers.

6. $\sqrt{-1} = \_$

7. $i^2 = \_\_\_\_$

8. $i^3 = \_\_\_\_$

9. $i^4 = \_\_$

10. $\sqrt{ab} = _____,$ provided a and b are not both negative.

11. $\sqrt{\dfrac{a}{b}} = \_\_,$ provided a and b are not both negative.

12. $3 + 5i, 2 - 7i,$ and $5 - \dfrac{1}{2}i$ are examples of _____ numbers.

13. The real part of $5 + 7i$ is __. The imaginary part is __.

14. $a + bi = c + di$ if and only if $a = \_$ and $b = \_$.

15. $a + bi$ and $a - bi$ are called complex _____.

16. $|a + bi| = _____$

PRACTICE   *In Exercises 17–28, solve each equation. Write all roots in bi or a + bi form.*

17. $x^2 + 9 = 0$

18. $x^2 + 16 = 0$

19. $3x^2 = -16$

20. $2x^2 = -25$

21. $x^2 + 2x + 2 = 0$

22. $x^2 + 3x + 3 = 0$

23. $2x^2 + x + 1 = 0$

24. $3x^2 + 2x + 1 = 0$

25. $3x^2 - 4x = -2$

26. $2x^2 + 3x = -3$

27. $3x^2 - 2x = -3$

28. $5x^2 = 2x - 1$

In Exercises 29–36, simplify each expression.

29. i^{21}

30. i^{19}

31. i^{27}

32. i^{22}

33. i^{100}

34. i^{42}

35. i^{97}

36. i^{200}

In Exercises 37–42, tell whether the complex numbers are equal.

37. $3 + 7i, \sqrt{9} + (5 + 2)i$

38. $\sqrt{4} + \sqrt{25}i, 2 - (-5)i$

39. $8 + 5i, 2^3 + \sqrt{25}i^3$

40. $4 - 7i, -4i^2 + 7i^3$

41. $\sqrt{4} + \sqrt{-4}, 2 - 2i$

42. $\sqrt{-9} - i, 4i$

In Exercises 43–76, do the operations. Write all answers in a + bi form.

43. $(3 + 4i) + (5 - 6i)$

44. $(5 + 3i) - (6 - 9i)$

45. $(7 - 3i) - (4 + 2i)$

46. $(8 + 3i) + (-7 - 2i)$

47. $(8 + 5i) + (7 + 2i)$

48. $(-7 + 9i) - (-2 - 8i)$

49. $(1 + i) - 2i + (5 - 7i)$

50. $(-9 + i) - 5i + (2 + 7i)$

51. $(5 + 3i) - (3 - 5i) + \sqrt{-1}$

52. $(8 + 7i) - \left(-7 - \sqrt{-64}\right) + (3 - i)$

53. $\left(-8 - \sqrt{3}i\right) - \left(7 - 3\sqrt{3}i\right)$

54. $\left(2 + 2\sqrt{2}i\right) + \left(-3 - \sqrt{2}i\right)$

55. $3i(2 - i)$

56. $-4i(3 + 4i)$

57. $-5i(5 - 5i)$

58. $2i(7 + 2i)$

59. $(2 + i)(3 - i)$

60. $(4 - i)(2 + i)$

61. $(2 - 4i)(3 + 2i)$

62. $(3 - 2i)(4 - 3i)$

63. $\left(2 + \sqrt{2}i\right)\left(3 - \sqrt{2}i\right)$

64. $\left(5 + \sqrt{3}i\right)\left(2 - \sqrt{3}i\right)$

65. $\left(8 - \sqrt{-1}\right)\left(-2 - \sqrt{-16}\right)$

66. $\left(-1 + \sqrt{-4}\right)\left(2 + \sqrt{-9}\right)$

67. $(2 + i)^2$

68. $(3 - 2i)^2$

69. $(2 + 3i)^2$

70. $(1 - 3i)^2$

71. $i(5 + i)(3 - 2i)$

72. $i(-3 - 2i)(1 - 2i)$

73. $(2 + i)(2 - i)(1 + i)$

74. $(3 + 2i)(3 - 2i)(i + 1)$

75. $(3 + i)[(3 - 2i) + (2 + i)]$

76. $(2 - 3i)[(5 - 2i) - (2i + 1)]$

In Exercises 77–104, write each expression in a + bi form.

77. $\dfrac{1}{i}$

78. $\dfrac{1}{i^3}$

79. $\dfrac{4}{5i^3}$

80. $\dfrac{3}{2i}$

81. $\dfrac{3i}{8\sqrt{-9}}$

82. $\dfrac{5i^3}{2\sqrt{-4}}$

83. $\dfrac{-3}{5i^5}$

84. $\dfrac{-4}{6i^7}$

85. $\dfrac{5}{2 - i}$

86. $\dfrac{26}{3 - 2i}$

87. $\dfrac{13i}{5 + i}$

88. $\dfrac{2i}{5 + 3i}$

89. $\dfrac{-12}{7 - \sqrt{-1}}$

90. $\dfrac{4}{3 + \sqrt{-1}}$

91. $\dfrac{5i}{6 + 2i}$

92. $\dfrac{-4i}{2 - 6i}$

93. $\dfrac{3 - 2i}{3 + 2i}$

94. $\dfrac{2 + 3i}{2 - 3i}$

95. $\dfrac{3 + 2i}{3 + i}$

96. $\dfrac{2 - 5i}{2 + 5i}$

97. $\dfrac{\sqrt{5} - \sqrt{3}i}{\sqrt{5} + \sqrt{3}i}$

98. $\dfrac{\sqrt{3} + \sqrt{2}i}{\sqrt{3} - \sqrt{2}i}$

99. $\left(\dfrac{i}{3 + 2i}\right)^2$

100. $\left(\dfrac{5 + i}{2 + i}\right)^2$

101. $\dfrac{i(3 - i)}{3 + i}$

102. $\dfrac{5 + 3i}{i(3 - 5i)}$

103. $\dfrac{(2 - 5i) - (5 - 2i)}{5 - i}$

104. $\dfrac{5i}{(5 + 2i) + (2 + i)}$

In Exercises 105–112, find each value.

105. $|6 + 8i|$

106. $|12 + 5i|$

107. $|12 - 5i|$

108. $|3 - 4i|$

109. $|5 + 7i|$

110. $|6 - 5i|$

111. $\left|\dfrac{3}{5} - \dfrac{4}{5}i\right|$

112. $\left|\dfrac{5}{13} + \dfrac{12}{13}i\right|$

113. Show that $1 - 5i$ is a solution of $x^2 - 2x + 26 = 0$.

114. Show that $3 - 2i$ is a solution of $x^2 - 6x + 13 = 0$.

115. Show that i is a solution of $x^4 - 3x^2 - 4 = 0$.

116. Show that $2 + i$ is *not* a solution of $x^2 + x + 1 = 0$.

WRITING

117. Tell how to decide whether two complex numbers are equal.

118. Define the complex conjugate of a complex number.

SOMETHING TO THINK ABOUT

119. Rationalize the numerator of $\dfrac{3 - i}{2}$.

120. Rationalize the numerator of $\dfrac{2 + 3i}{2 - 3i}$.

10.4 The Discriminant and Equations That Can Be Written in Quadratic Form

■ THE DISCRIMINANT ■ EQUATIONS THAT CAN BE WRITTEN IN QUADRATIC FORM ■ SOLUTIONS OF A QUADRATIC EQUATION

Getting Ready *Evaluate $b^2 - 4ac$ for the following values.*

1. $a = 2$, $b = 3$, and $c = -1$

2. $a = -2$, $b = 4$, and $c = -3$

■ THE DISCRIMINANT

We can predict the type of solutions a quadratic equation will have without solving it. To see how, we suppose that the coefficients a, b, and c in the equation $ax^2 + bx + c = 0$ $(a \neq 0)$ are real numbers. Then the solutions of the equation are given by the quadratic formula

$$x = \frac{-b \pm \sqrt{b^2 - 4ac}}{2a} \quad (a \neq 0)$$

If $b^2 - 4ac \geq 0$, the solutions are real numbers. If $b^2 - 4ac < 0$, the solutions are nonreal complex numbers. Thus, the value of $b^2 - 4ac$, called the **discriminant,** determines the type of solutions for a particular quadratic equation.

The Discriminant

If a, b, and c are real numbers and

If $b^2 - 4ac$ is . . .	*then the solutions are . . .*
positive,	real numbers and unequal.
0,	real numbers and equal.
negative,	nonreal complex numbers and complex conjugates.

If a, b, and c are rational numbers and

If $b^2 - 4ac$ is . . .	*then the solutions are . . .*
a perfect square greater than 0,	rational numbers and unequal.
positive and not a perfect square,	irrational numbers and unequal.

EXAMPLE 1   Determine the type of solutions for the equation $x^2 + x + 1 = 0$.

Solution   We calculate the discriminant:

$$b^2 - 4ac = 1^2 - 4(1)(1) \qquad a = 1, b = 1, \text{ and } c = 1.$$
$$= -3$$

Since $b^2 - 4ac < 0$, the solutions are nonreal complex conjugates.   ■

Self Check   Determine the type of solutions for $x^2 + x - 1 = 0$.

Answer   real numbers that are irrational and unequal

EXAMPLE 2   Determine the type of solutions for the equation $3x^2 + 5x + 2 = 0$.

Solution   We calculate the discriminant:

$$b^2 - 4ac = 5^2 - 4(3)(2) \qquad a = 3, b = 5, \text{ and } c = 2.$$
$$= 25 - 24$$
$$= 1$$

Since $b^2 - 4ac > 0$ and $b^2 - 4ac$ is a perfect square, the solutions are rational and unequal.   ■

Self Check   Determine the type of solutions for $4x^2 - 10x + 25 = 0$.

Answer   nonreal numbers that are complex conjugates

EXAMPLE 3   What value of k will make the solutions of the equation $kx^2 - 12x + 9 = 0$ equal?

Solution   We calculate the discriminant:

$$b^2 - 4ac = (-12)^2 - 4(k)(9) \qquad a = k, b = -12, \text{ and } c = 9.$$
$$= 144 - 36k$$
$$= -36k + 144$$

Since the solutions are to be equal, we let $-36k + 144 = 0$ and solve for k.

$$-36k + 144 = 0$$
$$-36k = -144 \qquad \text{Subtract 144 from both sides.}$$
$$k = 4 \qquad \text{Divide both sides by } -36.$$

If $k = 4$, the solutions will be equal. Verify this by solving $4x^2 - 12x + 9 = 0$ and showing that the solutions are equal.   ■

Self Check

What value of k will make the solutions of $kx^2 - 20x + 25 = 0$ equal?

Answer   4

■ EQUATIONS THAT CAN BE WRITTEN IN QUADRATIC FORM

Many equations can be written in quadratic form and then solved with the techniques used for solving quadratic equations. For example, we can solve $x^4 - 5x^2 + 4 = 0$ as follows:

$$x^4 - 5x^2 + 4 = 0$$
$$(x^2)^2 - 5(x^2) + 4 = 0$$
$$y^2 - 5y + 4 = 0 \qquad \text{Let } y = x^2.$$
$$(y - 4)(y - 1) = 0 \qquad \text{Factor } y^2 - 5y + 4.$$
$$y - 4 = 0 \quad \text{or} \quad y - 1 = 0 \qquad \text{Set each factor equal to 0.}$$
$$y = 4 \qquad\qquad y = 1$$

Since $x^2 = y$, it follows that $x^2 = 4$ or $x^2 = 1$. Thus,

$$x^2 = 4 \qquad\qquad \text{or} \qquad\qquad x^2 = 1$$
$$x = 2 \quad \text{or} \quad x = -2 \qquad x = 1 \quad \text{or} \quad x = -1$$

This equation has four solutions: $1, -1, 2,$ and -2. Verify that each one satisfies the original equation. Note that this equation can be solved by factoring.

EXAMPLE 4

Solve $x - 7x^{1/2} + 12 = 0$.

Solution

If y^2 is substituted for x and y is substituted for $x^{1/2}$, the equation

$$x - 7x^{1/2} + 12 = 0$$

becomes a quadratic equation that can be solved by factoring:

$$y^2 - 7y + 12 = 0 \qquad \text{Substitute } y^2 \text{ for } x \text{ and } y \text{ for } x^{1/2}.$$
$$(y - 3)(y - 4) = 0 \qquad \text{Factor.}$$
$$y - 3 = 0 \quad \text{or} \quad y - 4 = 0 \qquad \text{Set each factor equal to 0.}$$
$$y = 3 \qquad\qquad y = 4$$

Because $x = y^2$, it follows that

$$x = 3^2 \quad \text{or} \quad x = 4^2$$
$$= 9 \qquad\qquad = 16$$

Verify that both solutions satisfy the original equation.   ■

Self Check

Solve $x + x^{1/2} - 6 = 0$. Be sure to check your solutions.

Answer   4

EXAMPLE 5   Solve $\dfrac{24}{x} + \dfrac{12}{x+1} = 11$.

Solution   Since the denominator cannot be 0, x cannot be 0 or -1. If either 0 or -1 appears as a suspected solution, it is extraneous and must be discarded.

$$\frac{24}{x} + \frac{12}{x+1} = 11$$

$$x(x+1)\left(\frac{24}{x} + \frac{12}{x+1}\right) = x(x+1)11 \qquad \text{Multiply both sides by } x(x+1).$$

$$24(x+1) + 12x = (x^2 + x)11 \qquad \text{Simplify.}$$

$$24x + 24 + 12x = 11x^2 + 11x \qquad \text{Use the distributive property to remove parentheses.}$$

$$36x + 24 = 11x^2 + 11x \qquad \text{Combine like terms.}$$

$$0 = 11x^2 - 25x - 24 \qquad \text{Subtract } 36x \text{ and } 24 \text{ from both sides.}$$

$$0 = (11x + 8)(x - 3) \qquad \text{Factor } 11x^2 - 25x - 24.$$

$$11x + 8 = 0 \quad \text{ or } \quad x - 3 = 0 \qquad \text{Set each factor equal to 0.}$$

$$x = -\frac{8}{11} \qquad\qquad x = 3$$

Verify that $-\frac{8}{11}$ and 3 satisfy the original equation. ∎

Self Check   Solve $\frac{12}{x} + \frac{6}{x+3} = 5$.
Answer $3, -\frac{12}{5}$

EXAMPLE 6   Solve the formula $s = 16t^2 - 32$ for t.

Solution   We proceed as follows:

$$s = 16t^2 - 32$$

$$s + 32 = 16t^2 \qquad \text{Add 32 to both sides.}$$

$$\frac{s+32}{16} = t^2 \qquad \text{Divide both sides by 16.}$$

$$t^2 = \frac{s+32}{16} \qquad \text{Write } t^2 \text{ on the left-hand side.}$$

$$t = \pm\sqrt{\frac{s+32}{16}} \qquad \text{Apply the square root property.}$$

$$t = \pm\frac{\sqrt{s+32}}{\sqrt{16}} \qquad \sqrt{\frac{a}{b}} = \frac{\sqrt{a}}{\sqrt{b}}.$$

$$t = \pm\frac{\sqrt{s+32}}{4}$$

∎

Self Check

Solve $a^2 + b^2 = c^2$ for a.

Answer

$a = \pm\sqrt{c^2 - b^2}$

■ SOLUTIONS OF A QUADRATIC EQUATION

Solutions of a Quadratic Equation

If r_1 and r_2 are the solutions of the quadratic equation $ax^2 + bx + c = 0$, with $a \neq 0$, then

$$r_1 + r_2 = -\frac{b}{a} \quad \text{and} \quad r_1 r_2 = \frac{c}{a}$$

Proof We note that the solutions to the equation are given by the quadratic formula

$$r_1 = \frac{-b + \sqrt{b^2 - 4ac}}{2a} \quad \text{and} \quad r_2 = \frac{-b - \sqrt{b^2 - 4ac}}{2a}$$

Thus,

$$r_1 + r_2 = \frac{-b + \sqrt{b^2 - 4ac}}{2a} + \frac{-b - \sqrt{b^2 - 4ac}}{2a}$$

$$= \frac{-b + \sqrt{b^2 - 4ac} - b - \sqrt{b^2 - 4ac}}{2a} \qquad \text{Keep the denominator and add the numerators.}$$

$$= -\frac{2b}{2a}$$

$$= -\frac{b}{a}$$

and

$$r_1 r_2 = \frac{-b + \sqrt{b^2 - 4ac}}{2a} \cdot \frac{-b - \sqrt{b^2 - 4ac}}{2a}$$

$$= \frac{b^2 - (b^2 - 4ac)}{4a^2} \qquad \text{Multiply the numerators and multiply the denominators.}$$

$$= \frac{b^2 - b^2 + 4ac}{4a^2}$$

$$= \frac{4ac}{4a^2} \qquad \qquad b^2 - b^2 = 0.$$

$$= \frac{c}{a} \qquad \qquad \qquad \square$$

It can also be shown that if

$$r_1 + r_2 = -\frac{b}{a} \quad \text{and} \quad r_1 r_2 = \frac{c}{a}$$

then r_1 and r_2 are solutions of $ax^2 + bx + c = 0$. We can use this fact to check the solutions of quadratic equations.

EXAMPLE 7   Show that $\frac{3}{2}$ and $-\frac{1}{3}$ are solutions of $6x^2 - 7x - 3 = 0$.

Solution   Since $a = 6$, $b = -7$, and $c = -3$, we have

$$-\frac{b}{a} = -\frac{-7}{6} = \frac{7}{6} \quad \text{and} \quad \frac{c}{a} = \frac{-3}{6} = -\frac{1}{2}$$

Since $\frac{3}{2} + \left(-\frac{1}{3}\right) = \frac{7}{6}$ and $\left(\frac{3}{2}\right)\left(-\frac{1}{3}\right) = -\frac{1}{2}$, these numbers are solutions. Solve the equation to see that the roots are $\frac{3}{2}$ and $-\frac{1}{3}$.   ■

Self Check   Are $-\frac{3}{2}$ and $\frac{1}{3}$ solutions of $6x^2 + 7x - 3 = 0$?

Answer   yes

Orals   *Find $b^2 - 4ac$ when*

1. $a = 1$, $b = 1$, $c = 1$ **2.** $a = 2$, $b = 1$, $c = 1$

Determine the type of solutions for

3. $x^2 - 4x + 1 = 0$ **4.** $8x^2 - x + 2 = 0$

Are the following numbers solutions of $x^2 - 7x + 6 = 0$?

5. 1, 5   **6.** 1, 6

EXERCISE 10.4

REVIEW   *Solve each equation.*

1. $\dfrac{1}{4} + \dfrac{1}{t} = \dfrac{1}{2t}$

2. $\dfrac{p - 3}{3p} + \dfrac{1}{2p} = \dfrac{1}{4}$

3. Find the slope of the line passing through $P(-2, -4)$ and $Q(3, 5)$.

4. Write the equation of the line passing through $P(-2, -4)$ and $Q(3, 5)$ in general form.

VOCABULARY AND CONCEPTS   *Consider the equation $ax^2 + bx + c = 0$ ($a \neq 0$), and fill in each blank to make a true statement.*

5. The discriminant is _____.

6. If $b^2 - 4ac < 0$, the solutions of the equation are nonreal complex _____.

7. If $b^2 - 4ac$ is a perfect square, the solutions are _____ numbers and _____.

8. If r_1 and r_2 are the solutions of the equation, then

$$r_1 + r_2 = \underline{\quad} \text{ and } r_1 r_2 = \underline{\quad}$$

PRACTICE   *In Exercises 9–16, use the discriminant to determine what type of solutions exist for each quadratic equation.* **Do not solve the equation.**

9. $4x^2 - 4x + 1 = 0$

10. $6x^2 - 5x - 6 = 0$

11. $5x^2 + x + 2 = 0$

12. $3x^2 + 10x - 2 = 0$

13. $2x^2 = 4x - 1$

14. $9x^2 = 12x - 4$

15. $x(2x - 3) = 20$

16. $x(x - 3) = -10$

In Exercises 17–24, find the value(s) of k that will make the solutions of each given quadratic equation equal.

17. $x^2 + kx + 9 = 0$

18. $kx^2 - 12x + 4 = 0$

19. $9x^2 + 4 = -kx$

20. $9x^2 - kx + 25 = 0$

21. $(k - 1)x^2 + (k - 1)x + 1 = 0$

22. $(k + 3)x^2 + 2kx + 4 = 0$

23. $(k + 4)x^2 + 2kx + 9 = 0$

24. $(k + 15)x^2 + (k - 30)x + 4 = 0$

25. Use the discriminant to determine whether the solutions of $1{,}492x^2 + 1{,}776x - 1{,}984 = 0$ are real numbers.

26. Use the discriminant to determine whether the solutions of $1{,}776x^2 - 1{,}492x + 1{,}984 = 0$ are real numbers.

27. Determine k such that the solutions of $3x^2 + 4x = k$ are nonreal complex numbers.

28. Determine k such that the solutions of $kx^2 - 4x = 7$ are nonreal complex numbers.

In Exercises 29–56, solve each equation.

29. $x^4 - 17x^2 + 16 = 0$

30. $x^4 - 10x^2 + 9 = 0$

31. $x^4 - 3x^2 = -2$

32. $x^4 - 29x^2 = -100$

33. $x^4 = 6x^2 - 5$

34. $x^4 = 8x^2 - 7$

35. $2x^4 - 10x^2 = -8$

36. $2x^4 + 24 = 26x^2$

37. $2x + x^{1/2} - 3 = 0$

38. $2x - x^{1/2} - 1 = 0$

39. $3x + 5x^{1/2} + 2 = 0$

40. $3x - 4x^{1/2} + 1 = 0$

41. $x^{2/3} + 5x^{1/3} + 6 = 0$

42. $x^{2/3} - 7x^{1/3} + 12 = 0$

43. $x^{2/3} - 2x^{1/3} - 3 = 0$

44. $x^{2/3} + 4x^{1/3} - 5 = 0$

45. $x + 5 + \dfrac{4}{x} = 0$

46. $x - 4 + \dfrac{3}{x} = 0$

47. $x + 1 = \dfrac{20}{x}$

48. $x + \dfrac{15}{x} = 8$

49. $\dfrac{1}{x - 1} + \dfrac{3}{x + 1} = 2$

50. $\dfrac{6}{x - 2} - \dfrac{12}{x - 1} = -1$

51. $\dfrac{1}{x + 2} + \dfrac{24}{x + 3} = 13$

52. $\dfrac{3}{x} + \dfrac{4}{x + 1} = 2$

53. $x^{-4} - 2x^{-2} + 1 = 0$

54. $4x^{-4} + 1 = 5x^{-2}$

55. $x + \dfrac{2}{x-2} = 0$

56. $x + \dfrac{x+5}{x-3} = 0$

In Exercises 57–64, solve each equation for the indicated variable.

57. $x^2 + y^2 = r^2$ for x

58. $x^2 + y^2 = r^2$ for y

59. $I = \dfrac{k}{d^2}$ for d

60. $V = \dfrac{1}{3}\pi r^2 h$ for r

61. $xy^2 + 3xy + 7 = 0$ for y

62. $kx = ay - x^2$ for x

63. $\sigma = \sqrt{\dfrac{\Sigma x^2}{N} - \mu^2}$ for μ^2

64. $\sigma = \sqrt{\dfrac{\Sigma x^2}{N} - \mu^2}$ for N

In Exercises 65–72, solve each equation and verify that the sum of the solutions is $-\frac{b}{a}$ and that the product of the solutions is $\frac{c}{a}$.

65. $12x^2 - 5x - 2 = 0$

66. $8x^2 - 2x - 3 = 0$

67. $2x^2 + 5x + 1 = 0$

68. $3x^2 + 9x + 1 = 0$

69. $3x^2 - 2x + 4 = 0$

70. $2x^2 - x + 4 = 0$

71. $x^2 + 2x + 5 = 0$

72. $x^2 - 4x + 13 = 0$

WRITING

73. Describe how to predict what type of solutions the equation $3x^2 - 4x + 5 = 0$ will have.

74. How is the discriminant related to the quadratic formula?

SOMETHING TO THINK ABOUT

75. Can a quadratic equation with integer coefficients have one real and one complex solution? Why?

76. Can a quadratic equation with complex coefficients have one real and one complex solution? Why?

10.5 Quadratic and Other Nonlinear Inequalities

■ SOLVING QUADRATIC INEQUALITIES ■ SOLVING OTHER INEQUALITIES ■ GRAPHS OF NONLINEAR INEQUALITIES IN TWO VARIABLES

Getting Ready *Factor each trinomial.*

1. $x^2 + 2x - 15$

2. $x^2 - 3x + 2$

■ SOLVING QUADRATIC INEQUALITIES

To solve the inequality $x^2 + x - 6 < 0$, we must find the values of x that make the inequality true. To find these values, we can factor the trinomial to obtain

$$(x + 3)(x - 2) < 0$$

Since the product of $x + 3$ and $x - 2$ is to be less than 0, their values must be opposite in sign. To find the intervals where this is true, we keep track of their signs by constructing the chart in Figure 10-16. The chart shows that

- $x - 2$ is 0 when $x = 2$, is positive when $x > 2$, and is negative when $x < 2$.
- $x + 3$ is 0 when $x = -3$, is positive when $x > -3$, and is negative when $x < -3$.

The only place where the values of the binomial are opposite in sign is in the interval $(-3, 2)$. Therefore, the product $(x + 3)(x - 2)$ will be less than 0 when

$$-3 < x < 2$$

The graph of the solution set is shown on the number line in Figure 10-16.

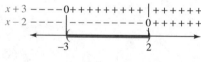

FIGURE 10-16

EXAMPLE 1   Solve $x^2 + 2x - 3 \geq 0$.

Solution   We factor the trinomial to get $(x - 1)(x + 3)$ and construct a sign chart, as in Figure 10-17.

- $x - 1$ is 0 when $x = 1$, is positive when $x > 1$, and is negative when $x < 1$.
- $x + 3$ is 0 when $x = -3$, is positive when $x > -3$, and is negative when $x < -3$.

The product of $x - 1$ and $x + 3$ will be greater than 0 when the signs of the binomial factors are the same. This occurs in the intervals $(-\infty, -3)$ and $(1, \infty)$. The numbers -3 and 1 are also included, because they make the product equal to 0. Thus, the solution set is

$$(-\infty, -3] \cup [1, \infty) \qquad \text{or} \qquad x \leq -3 \text{ or } x \geq 1$$

The graph of the solution set is shown on the number line in Figure 10-17.   ■

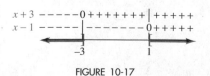

FIGURE 10-17

Self Check   Solve $x^2 + 2x - 15 > 0$ and graph the solution set.

Answer $(-\infty, -5) \cup (3, \infty)$

■ SOLVING OTHER INEQUALITIES

Making a sign chart is useful for solving many inequalities that are neither linear nor quadratic.

EXAMPLE 2   Solve $\dfrac{1}{x} < 6$.

Solution   We subtract 6 from both sides to make the right-hand side equal to 0. We then find a common denominator and add the fractions:

$$\frac{1}{x} < 6$$

$$\frac{1}{x} - 6 < 0 \qquad \text{Subtract 6 from both sides.}$$

$$\frac{1}{x} - \frac{6x}{x} < 0 \qquad \text{Get a common denominator.}$$

$$\frac{1 - 6x}{x} < 0 \qquad \text{Subtract the numerators and keep the common denominator.}$$

We now make a sign chart, as in Figure 10-18.

- The denominator x is 0 when $x = 0$, is positive when $x > 0$, and is negative when $x < 0$.
- The numerator $1 - 6x$ is 0 when $x = \frac{1}{6}$, is positive when $x < \frac{1}{6}$, and is negative when $x > \frac{1}{6}$.

The fraction $\frac{1-6x}{x}$ will be less than 0 when the numerator and denominator are opposite in sign. This occurs in the interval

$$\left(-\infty, 0\right) \cup \left(\frac{1}{6}, \infty\right) \qquad \text{or} \qquad x < 0 \text{ or } x > \frac{1}{6}$$

The graph of this interval is shown in Figure 10-18.   ■

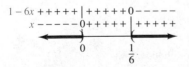

FIGURE 10-18

Self Check   Solve $\frac{3}{x} > 5$.

Answer $\left(0, \frac{3}{5}\right)$

> **WARNING!** Since we don't know whether x is positive, 0, or negative, multiplying both sides of the inequality $\frac{1}{x} < 6$ by x is a three-case situation:
>
> - If $x > 0$, then $1 < 6x$.
> - If $x = 0$, then the fraction $\frac{1}{x}$ is undefined.
> - If $x < 0$, then $1 > 6x$.
>
> If you multiply both sides by x and solve $1 < 6x$, you are only considering one case and will get only part of the answer.

EXAMPLE 3   Solve $\dfrac{x^2 - 3x + 2}{x - 3} \geq 0$.

Solution   We write the fraction with the numerator in factored form.

$$\frac{(x - 2)(x - 1)}{x - 3} \geq 0$$

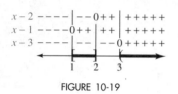

FIGURE 10-19

To keep track of the signs of the binomials, we construct the sign chart shown in Figure 10-19. The fraction will be positive in the intervals where all factors are positive, or where two factors are negative. The numbers 1 and 2 are included, because they make the numerator (and thus the fraction) equal to 0. The number 3 is not included, because it gives a 0 in the denominator.

The solution is the interval $[1, 2] \cup (3, \infty)$. The graph appears in Figure 10-19. ∎

Self Check   Solve $\dfrac{x + 2}{x^2 - 2x - 3} > 0$ and graph the solution set.

Answer $(-2, -1) \cup (3, \infty)$

EXAMPLE 4   Solve $\dfrac{3}{x - 1} < \dfrac{2}{x}$.

Solution   We subtract $\frac{2}{x}$ from both sides to get 0 on the right-hand side and proceed as follows:

$$\frac{3}{x - 1} < \frac{2}{x}$$

$$\frac{3}{x - 1} - \frac{2}{x} < 0 \qquad \text{Subtract } \tfrac{2}{x} \text{ from both sides.}$$

$$\frac{3x}{(x - 1)x} - \frac{2(x - 1)}{x(x - 1)} < 0 \qquad \text{Get a common denominator.}$$

$$\frac{3x - 2x + 2}{x(x - 1)} < 0 \qquad \text{Keep the denominator and subtract the numerators.}$$

$$\frac{x + 2}{x(x - 1)} < 0 \qquad \text{Combine like terms.}$$

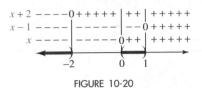

FIGURE 10-20

We can keep track of the signs of the three factors with the sign chart shown in Figure 10-20. The fraction will be negative in the intervals with either one or three negative factors. The numbers 0 and 1 are not included, because they give a 0 in the denominator, and the number -2 is not included, because it does not satisfy the inequality.

The solution is the interval $(-\infty, -2) \cup (0, 1)$, as shown in Figure 10-20. ∎

Self Check

Answer

Solve $\frac{2}{x+1} > \frac{1}{x}$ and graph the solution set.

$(-1, 0) \cup (1, \infty)$

Solving Inequalities

GRAPHING CALCULATORS

FIGURE 10-21

To approximate the solutions of the inequality $x^2 + 2x - 3 \geq 0$ (Example 1) by graphing, we can use window settings of $[-10, 10]$ for x and $[-10, 10]$ for y and graph the quadratic function $y = x^2 + 2x - 3$, as in Figure 10-21. The solution of the inequality will be those numbers x for which the graph of $y = x^2 + 2x - 3$ lies above or on the x-axis. We can trace to find that this interval is $(-\infty, -3] \cup [1, \infty)$.

To approximate the solutions of $\frac{3}{x-1} < \frac{2}{x}$ (Example 4), we first write the inequality in the form

$$\frac{3}{x-1} - \frac{2}{x} < 0$$

Then we use window settings of $[-5, 5]$ for x and $[-3, 3]$ for y and graph the function $y = \frac{3}{x-1} - \frac{2}{x}$, as in Figure 10-22(a). The solution of the inequality will be those numbers x for which the graph lies below the x-axis.

We can trace to see that the graph is below the x-axis when x is less than -2. Since we cannot see the graph in the interval $0 < x < 1$, we redraw the graph using window settings of $[-1, 2]$ for x and $[-25, 10]$ for y. See Figure 10-22(b).

We can now see that the graph is below the x-axis in the interval $(0, 1)$. Thus, the solution of the inequality is the union of two intervals:

$$(-\infty, -2) \cup (0, 1)$$

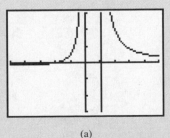

(a)

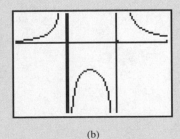

(b)

FIGURE 10-22

■ GRAPHS OF NONLINEAR INEQUALITIES IN TWO VARIABLES

We now consider the graphs of nonlinear inequalities in two variables.

EXAMPLE 5   Graph $y < -x^2 + 4$.

Solution   The graph of $y = -x^2 + 4$ is the parabolic boundary separating the region representing $y < -x^2 + 4$ and the region representing $y > -x^2 + 4$.

We graph $y = -x^2 + 4$ as a broken parabola, because equality is not permitted. Since the coordinates of the origin satisfy the inequality $y < -x^2 + 4$, the point $(0, 0)$ is in the graph. The complete graph is shown in Figure 10-23. ■

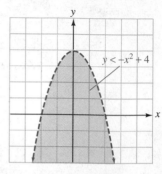

FIGURE 10-23

Self Check   Graph $y \geq -x^2 + 4$.
Answer

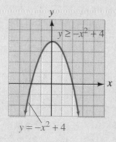

EXAMPLE 6   Graph $x \leq |y|$.

Solution   We first graph $x = |y|$ as in Figure 10-24(a), using a solid line because equality is permitted. Since the origin is on the graph, we cannot use the origin as a test point. However, another point, such as $(1, 0)$, will do. We substitute 1 for x and 0 for y into the inequality to get

$$x \leq |y|$$
$$1 \leq |0|$$
$$1 \leq 0$$

Since $1 \leq 0$ is a false statement, the point $(1, 0)$ does not satisfy the inequality and is not part of the graph. Thus, the graph of $x \leq |y|$ is to the left of the boundary. The complete graph is shown in Figure 10-24(b).

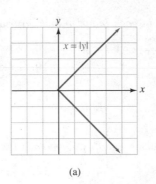

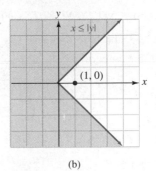

(a)   (b)

FIGURE 10-24

Self Check   Graph $x \geq -|y|$.
Answer

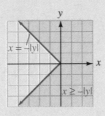

Orals   *Tell where $x - 2$ is*

1. 0   **2.** positive   **3.** negative

Tell where $x + 3$ is

4. 0   **5.** positive   **6.** negative

Multiply both sides of the equation $\frac{1}{x} < 2$ by x when x is

7. positive   **8.** negative

EXERCISE 10.5

REVIEW   *Write each expression as an equation.*

1. y varies directly with x.

2. y varies inversely with t.

3. t varies jointly with x and y.

4. d varies directly with t but inversely with u^2.

Find the slope of the graph of each equation.

5. $y = 3x - 4$

6. $\dfrac{2x - y}{5} = 8$

VOCABULARY AND CONCEPTS *Fill in each blank to make a true statement.*

7. When $x > 3$, the binomial $x - 3$ is _____ than zero.

8. When $x < 3$, the binomial $x - 3$ is _____ than zero.

9. If $x = 0$, the fraction $\frac{1}{x}$ is _____.

10. To keep track of the signs of factors in a product or quotient, we can use a _____ chart.

PRACTICE *In Exercises 11–50, solve each inequality. Give each result in interval notation and graph the solution set.*

11. $x^2 - 5x + 4 < 0$

12. $x^2 - 3x - 4 > 0$

13. $x^2 - 8x + 15 > 0$

14. $x^2 + 2x - 8 < 0$

15. $x^2 + x - 12 \leq 0$

16. $x^2 + 7x + 12 \geq 0$

17. $x^2 + 2x \geq 15$

18. $x^2 - 8x \leq -15$

19. $x^2 + 8x < -16$

20. $x^2 + 6x \geq -9$

21. $x^2 \geq 9$

22. $x^2 \geq 16$

23. $2x^2 - 50 < 0$

24. $3x^2 - 243 < 0$

25. $\dfrac{1}{x} < 2$

26. $\dfrac{1}{x} > 3$

27. $\dfrac{4}{x} \geq 2$

28. $-\dfrac{6}{x} < 12$

29. $-\dfrac{5}{x} < 3$

30. $\dfrac{4}{x} \geq 8$

31. $\dfrac{x^2 - x - 12}{x - 1} < 0$

32. $\dfrac{x^2 + x - 6}{x - 4} \geq 0$

33. $\dfrac{x^2 + x - 20}{x + 2} \geq 0$

34. $\dfrac{x^2 - 10x + 25}{x + 5} < 0$

35. $\dfrac{x^2 - 4x + 4}{x + 4} < 0$

36. $\dfrac{2x^2 - 5x + 2}{x + 2} > 0$

37. $\dfrac{6x^2 - 5x + 1}{2x + 1} > 0$

38. $\dfrac{6x^2 + 11x + 3}{3x - 1} < 0$

39. $\dfrac{3}{x - 2} < \dfrac{4}{x}$

40. $\dfrac{-6}{x + 1} \geq \dfrac{1}{x}$

41. $\dfrac{-5}{x + 2} \geq \dfrac{4}{2 - x}$

42. $\dfrac{-6}{x - 3} < \dfrac{5}{3 - x}$

43. $\dfrac{7}{x - 3} \geq \dfrac{2}{x + 4}$

44. $\dfrac{-5}{x - 4} < \dfrac{3}{x + 1}$

45. $\dfrac{x}{x + 4} \leq \dfrac{1}{x + 1}$

46. $\dfrac{x}{x + 9} \geq \dfrac{1}{x + 1}$

47. $\dfrac{x}{x + 16} > \dfrac{1}{x + 1}$

48. $\dfrac{x}{x + 25} < \dfrac{1}{x + 1}$

49. $(x + 2)^2 > 0$

50. $(x - 3)^2 < 0$

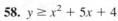 *In Exercises 51–54, use a graphing calculator to solve each inequality. Give the answer in interval notation.*

51. $x^2 - 2x - 3 < 0$ **52.** $x^2 + x - 6 > 0$ **53.** $\dfrac{x + 3}{x - 2} > 0$ **54.** $\dfrac{3}{x} < 2$

In Exercises 55–66, graph each inequality.

55. $y < x^2 + 1$ **56.** $y > x^2 - 3$ **57.** $y \leq x^2 + 5x + 6$ **58.** $y \geq x^2 + 5x + 4$

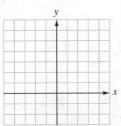 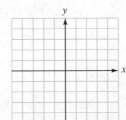 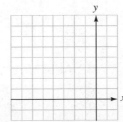 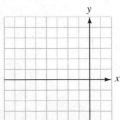

59. $y \geq (x-1)^2$

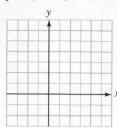

60. $y \leq (x+2)^2$

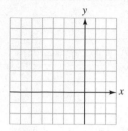

61. $-x^2 - y + 6 > -x$

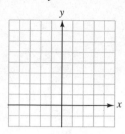

62. $y > (x+3)(x-2)$

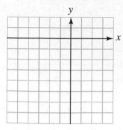

63. $y < |x+4|$

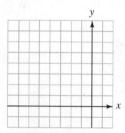

64. $y \geq |x-3|$

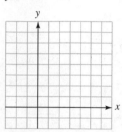

65. $y \leq -|x| + 2$

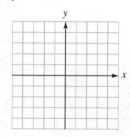

66. $y > |x| - 2$

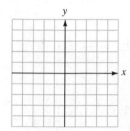

WRITING

67. Explain why $(x-4)(x+5)$ will be positive only when the signs of $x-4$ and $x+5$ are the same.

68. Tell how to find the graph of $y \geq x^2$.

SOMETHING TO THINK ABOUT

69. Under what conditions will the fraction $\frac{(x-1)(x+4)}{(x+2)(x+1)}$ be positive?

70. Under what conditions will the fraction $\frac{(x-1)(x+4)}{(x+2)(x+1)}$ be negative?

10.6 Algebra and Composition of Functions

■ ALGEBRA OF FUNCTIONS ■ COMPOSITION OF FUNCTIONS ■ THE IDENTITY FUNCTION
■ THE DIFFERENCE QUOTIENT ■ PROBLEM SOLVING

Getting Ready   *Assume that $P(x) = 2x + 1$ and $Q(x) = x - 2$. Find each expression.*

1. $P(x) + Q(x)$ **2.** $P(x) - Q(x)$

3. $P(x) \cdot Q(x)$ **4.** $\dfrac{P(x)}{Q(x)}$

■ ALGEBRA OF FUNCTIONS

We now consider how functions can be added, subtracted, multiplied, and divided.

Operations on Functions

If the domains and ranges of functions f and g are subsets of the real numbers, then

The **sum** of f and g, denoted as $f + g$, is defined by

$$(f + g)(x) = f(x) + g(x)$$

The **difference** of f and g, denoted as $f - g$, is defined by

$$(f - g)(x) = f(x) - g(x)$$

The **product** of f and g, denoted as $f \cdot g$, is defined by

$$(f \cdot g)(x) = f(x)g(x)$$

The **quotient** of f and g, denoted as f/g, is defined by

$$(f/g)(x) = \frac{f(x)}{g(x)} \quad (g(x) \neq 0)$$

The domain of each of these functions is the set of real numbers x that are in the domain of both f and g. In the case of the quotient, there is the further restriction that $g(x) \neq 0$.

EXAMPLE 1   Let $f(x) = 2x^2 + 1$ and $g(x) = 5x - 3$. Find each function and its domain: **a.** $f + g$ and   **b.** $f - g$.

Solution   **a.** $(f + g)(x) = f(x) + g(x)$

$$= (2x^2 + 1) + (5x - 3)$$

$$= 2x^2 + 5x - 2$$

The domain of $f + g$ is the set of real numbers that are in the domain of both f and g. Since the domain of both f and g is interval $(-\infty, \infty)$, the domain of $f + g$ is also the interval $(-\infty, \infty)$.

b. $(f - g)(x) = f(x) - g(x)$

$$= (2x^2 + 1) - (5x - 3)$$

$$= 2x^2 + 1 - 5x + 3 \qquad \text{Remove parentheses.}$$

$$= 2x^2 - 5x + 4 \qquad \text{Combine like terms.}$$

Since the domain of both f and g is $(-\infty, \infty)$, the domain of $f - g$ is also the interval $(-\infty, \infty)$. ∎

Self Check   Let $f(x) = 3x - 2$ and $g(x) = 2x^2 + 3x$. Find   **a.** $f + g$ and   **b.** $f - g$.
Answers   **a.** $2x^2 + 6x - 2$,   **b.** $-2x^2 - 2$

EXAMPLE 2   Let $f(x) = 2x^2 + 1$ and $g(x) = 5x - 3$. Find each function and its domain: **a.** $f \cdot g$ and   **b.** f/g.

Solution   **a.** $(f \cdot g)(x) = f(x)g(x)$

$$= (2x^2 + 1)(5x - 3)$$

$$= 10x^3 - 6x^2 + 5x - 3 \qquad \text{Multiply.}$$

The domain of $f \cdot g$ is the set of real numbers that are in the domain of both f and g. Since the domain of both f and g is the interval $(-\infty, \infty)$, the domain of $f \cdot g$ is also the interval $(-\infty, \infty)$.

b. $(f/g)(x) = \dfrac{f(x)}{g(x)}$

$$= \dfrac{2x^2 + 1}{5x - 3}$$

Since the denominator of the fraction cannot be 0, $x \neq \frac{3}{5}$. The domain of f/g is the interval $\left(-\infty, \frac{3}{5}\right) \cup \left(\frac{3}{5}, \infty\right)$.   ∎

Self Check   Let $f(x) = 2x^2 - 3$ and $g(x) = x^2 + 1$. Find   **a.** $f \cdot g$ and   **b.** f/g.

Answers   **a.** $2x^4 - x^2 - 3$,   **b.** $\dfrac{2x^2 - 3}{x^2 + 1}$

■ COMPOSITION OF FUNCTIONS

We have seen that a function can be represented by a machine: We put in a number from the domain, and a number from the range comes out. For example, if we put the number 2 into the machine shown in Figure 10-25(a), the number $f(2) = 5(2) - 2 = 8$ comes out. In general, if we put x into the machine shown in Figure 10-25(b), the value $f(x)$ comes out.

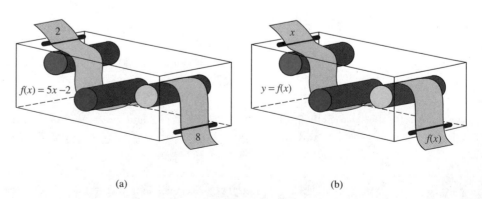

(a)   (b)

FIGURE 10-25

Often one quantity is a function of a second quantity that depends, in turn, on a third quantity. For example, the cost of a car trip is a function of the gasoline consumed. The amount of gasoline consumed, in turn, is a function of the number of miles driven. Such chains of dependence can be analyzed mathematically as **compositions of functions.**

Suppose that $y = f(x)$ and $y = g(x)$ define two functions. Any number x in the domain of g will produce the corresponding value $g(x)$ in the range of g. If $g(x)$ is in the domain of the function f, then $g(x)$ can be substituted into f, and a corresponding value $f(g(x))$ will be determined. This two-step process defines a new function, called a **composite function,** denoted by $f \circ g$.

The function machines shown in Figure 10-26 illustrate the composition $f \circ g$. When we put a number x into the function g, $g(x)$ comes out. The value $g(x)$ goes into function f, which transforms $g(x)$ into $f(g(x))$. If the function machines for g and f were connected to make a single machine, that machine would be named $f \circ g$.

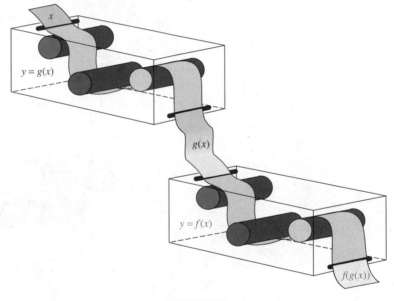

FIGURE 10-26

To be in the domain of the composite function $f \circ g$, a number x has to be in the domain of g. Also, the output of g must be in the domain of f. Thus, the domain of $f \circ g$ consists of those numbers x that are in the domain of g, and for which $g(x)$ is in the domain of f.

Composite Functions
The **composite function $f \circ g$** is defined by

$$(f \circ g)(x) = f(g(x))$$

For example, if $f(x) = 4x - 5$ and $g(x) = 3x + 2$, then

$$(f \circ g)(x) = f(g(x)) \qquad \text{or} \qquad (g \circ f)(x) = g(f(x))$$
$$= f(3x + 2) \qquad\qquad\qquad = g(4x - 5)$$
$$= 4(3x + 2) - 5 \qquad\qquad = 3(4x - 5) + 2$$
$$= 12x + 8 - 5 \qquad\qquad\quad = 12x - 15 + 2$$
$$= 12x + 3 \qquad\qquad\qquad = 12x - 13$$

WARNING!   Note that in the previous example, $(f \circ g)(x) \neq (g \circ f)(x)$. This shows that the composition of functions is not commutative.

 c

EXAMPLE 3   Let $f(x) = 2x + 1$ and $g(x) = x - 4$. Find   **a.** $(f \circ g)(9)$,   **b.** $(f \circ g)(x)$,   and **c.** $(g \circ f)(-2)$.

Solution   **a.** $(f \circ g)(9)$ means $f(g(9))$. In Figure 10-27(a), function g receives the number 9, subtracts 4, and releases the number $g(9) = 5$. The 5 then goes into the f function, which doubles 5 and adds 1. The final result, 11, is the output of the composite function $f \circ g$:

$$(f \circ g)(9) = f(g(9)) = f(5) = 2(5) + 1 = 11$$

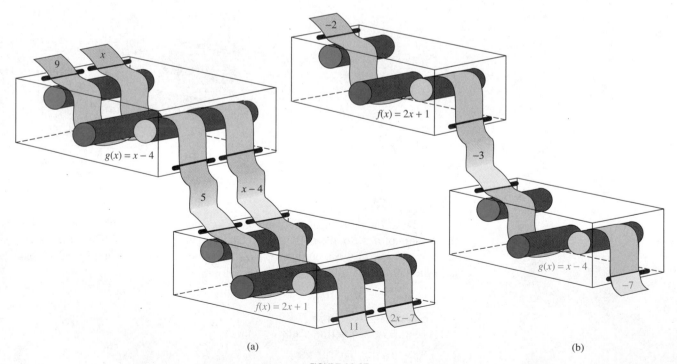

(a)   (b)

FIGURE 10-27

b. $(f \circ g)(x)$ means $f(g(x))$. In Figure 10-27(a), function g receives the number x, subtracts 4, and releases the number $x - 4$. The $x - 4$ then goes into the f function, which doubles $x - 4$ and adds 1. The final result, $2x - 7$, is the output of the composite function $f \circ g$.

$$(f \circ g)(x) = f(g(x)) = f(x - 4) = 2(x - 4) + 1 = 2x - 7$$

c. $(g \circ f)(-2)$ means $g(f(-2))$. In Figure 10-27(b), function f receives the number -2, doubles it and adds 1, and releases -3 into the g function. Function g subtracts 4 from -3 and releases a final result of -7. Thus,

$$(g \circ f)(-2) = g(f(-2)) = g(-3) = -3 - 4 = -7 \qquad \blacksquare$$

■ THE IDENTITY FUNCTION

The **identity function** is defined by the equation $I(x) = x$. Under this function, the value that corresponds to any real number x is x itself. If f is any function, the composition of f with the identity function is the function f:

$$(f \circ I)(x) = (I \circ f)(x) = f(x)$$

EXAMPLE 4   Let f be any function and I be the identity function, $I(x) = x$. Show that
a. $(f \circ I)(x) = f(x)$ and   **b.** $(I \circ f)(x) = f(x)$.

Solution   **a.** $(f \circ I)(x)$ means $f(I(x))$. Because $I(x) = x$, we have

$$(f \circ I)(x) = f(I(x)) = f(x)$$

b. $(I \circ f)(x)$ means $I(f(x))$. Because I passes any number through unchanged, we have $I(f(x)) = f(x)$ and

$$(I \circ f)(x) = I(f(x)) = f(x) \qquad \blacksquare$$

■ THE DIFFERENCE QUOTIENT

An important function in calculus, called the **difference quotient,** represents the slope of a line that passes through two given points on the graph of a function. The difference quotient is defined as follows:

$$\frac{f(x + h) - f(x)}{h}$$

EXAMPLE 5   If $f(x) = x^2 - 4$, evaluate the difference quotient.

Solution   First, we evaluate $f(x + h)$.

$$f(x) = x^2 - 4$$
$$f(x + h) = (x + h)^2 - 4 \qquad\qquad \text{Substitute } x + h \text{ for } x.$$
$$\qquad\quad = x^2 + 2xh + h^2 - 4 \qquad (x + h)^2 = x^2 + 2hx + h^2.$$

Then we note that $f(x) = x^2 - 4$. We can now substitute the values of $f(x + h)$ and $f(x)$ into the difference quotient and simplify.

$$\frac{f(x + h) - f(x)}{h} = \frac{(x^2 + 2xh + h^2 - 4) - (x^2 - 4)}{h}$$

$$= \frac{x^2 + 2xh + h^2 - 4 - x^2 + 4}{h} \qquad \text{Remove parentheses.}$$

$$= \frac{2xh + h^2}{h} \qquad \text{Combine like terms.}$$

$$= \frac{h(2x + h)}{h} \qquad \text{Factor out } h \text{ in the numerator.}$$

$$= 2x + h \qquad \text{Divide out } h; \frac{h}{h} = 1.$$

The difference quotient simplifies as $2x + h$. ■

■ PROBLEM SOLVING

EXAMPLE 6

Temperature change   A laboratory sample is removed from a cooler at a temperature of 15° Fahrenheit. Technicians are warming the sample at a controlled rate of 3° F per hour. Express the sample's Celsius temperature as a function of the time, t, since it was removed from refrigeration.

Solution   The temperature of the sample is 15° F when $t = 0$. Because it warms at 3° F per hour after that, it warms $3t°$, and the Fahrenheit temperature is given by the function

$$F(t) = 3t + 15$$

The Celsius temperature, C, is a function of this Fahrenheit temperature, F, given by the formula

$$C(F) = \frac{5}{9}(F - 32)$$

To express the sample's Celsius temperature as a function of time, we find the composition function

$$(C \circ F)(t) = C(F(t))$$

$$= \frac{5}{9}(F(t) - 32) \qquad \text{Substitute } F(t) \text{ for } F \text{ in } \frac{5}{9}(F - 32).$$

$$= \frac{5}{9}[(3t + 15) - 32] \qquad \text{Substitute } 3t + 15 \text{ for } F(t).$$

$$= \frac{5}{9}(3t - 17) \qquad \text{Simplify.}$$

■

Orals   *If $f(x) = 2x$, $g(x) = 3x$, and $h(x) = 4x$, find*

1. $f + g$ **2.** $h - g$ **3.** $f \cdot h$

4. g/f **5.** h/f **6.** $g \cdot h$

7. $(f \circ h)(x)$ **8.** $(f \circ g)(x)$ **9.** $(g \circ h)(x)$

EXERCISE 10.6

REVIEW   *Simplify each expression.*

1. $\dfrac{3x^2 + x - 14}{4 - x^2}$

2. $\dfrac{2x^3 + 14x^2}{3 + 2x - x^2} \cdot \dfrac{x^2 - 3x}{x}$

3. $\dfrac{8 + 2x - x^2}{12 + x - 3x^2} \div \dfrac{3x^2 + 5x - 2}{3x - 1}$

4. $\dfrac{x - 1}{1 + \dfrac{x}{x - 2}}$

VOCABULARY AND CONCEPTS   *Fill in each blank to make a true statement.*

5. $(f + g)(x) = $ _____

6. $(f - g)(x) = $ _____

7. $(f \cdot g)(x) = $ _____

8. $(f/g)(x) = $ ____ $(g(x) \neq 0)$

9. In Exercises 5–7, the domain of each function is the set of real numbers x that are in the _____ of both f and g.

10. $(f \circ g)(x) = $ _____

11. If I is the identity function, then $(f \circ I)(x) = $ ____.

12. If I is the identity function, then $(I \circ f)(x) = $ ____.

PRACTICE   *In Exercises 13–20, $f(x) = 3x$ and $g(x) = 4x$. Find each function and its domain.*

13. $f + g$ **14.** $f - g$ **15.** $f \cdot g$ **16.** f/g

17. $g - f$ **18.** $g + f$ **19.** g/f **20.** $g \cdot f$

In Exercises 21–28, $f(x) = 2x + 1$ and $g(x) = x - 3$. Find each function and its domain.

21. $f + g$ **22.** $f - g$ **23.** $f \cdot g$ **24.** f/g

25. $g - f$ **26.** $g + f$ **27.** g/f **28.** $g \cdot f$

In Exercises 29–32, $f(x) = 3x - 2$ and $g(x) = 2x^2 + 1$. Find each function and its domain.

29. $f - g$ **30.** $f + g$ **31.** f/g **32.** $f \cdot g$

In Exercises 33–36, $f(x) = x^2 - 1$ and $g(x) = x^2 - 4$. Find each function and its domain.

33. $f - g$ **34.** $f + g$ **35.** g/f **36.** $g \cdot f$

In Exercises 37–48, $f(x) = 2x + 1$ and $g(x) = x^2 - 1$. Find each value.

37. $(f \circ g)(2)$ **38.** $(g \circ f)(2)$ **39.** $(g \circ f)(-3)$ **40.** $(f \circ g)(-3)$

41. $(f \circ g)(0)$ **42.** $(g \circ f)(0)$ **43.** $(f \circ g)\left(\dfrac{1}{2}\right)$ **44.** $(g \circ f)\left(\dfrac{1}{3}\right)$

45. $(f \circ g)(x)$ **46.** $(g \circ f)(x)$ **47.** $(g \circ f)(2x)$ **48.** $(f \circ g)(2x)$

In Exercises 49–56, $f(x) = 3x - 2$ and $g(x) = x^2 + x$. Find each value.

49. $(f \circ g)(4)$ **50.** $(g \circ f)(4)$ **51.** $(g \circ f)(-3)$ **52.** $(f \circ g)(-3)$

53. $(g \circ f)(0)$ **54.** $(f \circ g)(0)$ **55.** $(g \circ f)(x)$ **56.** $(f \circ g)(x)$

In Exercises 57–68, find $\dfrac{f(x + h) - f(x)}{h}$.

57. $f(x) = 2x + 3$ **58.** $f(x) = 3x - 5$ **59.** $f(x) = x^2$

60. $f(x) = x^2 - 1$ **61.** $f(x) = 2x^2 - 1$ **62.** $f(x) = 3x^2$

63. $f(x) = x^2 + x$ **64.** $f(x) = x^2 - x$ **65.** $f(x) = x^2 + 3x - 4$

66. $f(x) = x^2 - 4x + 3$ **67.** $f(x) = 2x^2 + 3x - 7$ **68.** $f(x) = 3x^2 - 2x + 4$

In Exercises 69–80, find $\dfrac{f(x) - f(a)}{x - a}$.

69. $f(x) = 2x + 3$ **70.** $f(x) = 3x - 5$ **71.** $f(x) = x^2$

72. $f(x) = x^2 - 1$ **73.** $f(x) = 2x^2 - 1$ **74.** $f(x) = 3x^2$

75. $f(x) = x^2 + x$ **76.** $f(x) = x^2 - x$ **77.** $f(x) = x^2 + 3x - 4$

78. $f(x) = x^2 - 4x + 3$ **79.** $f(x) = 2x^2 + 3x - 7$ **80.** $f(x) = 3x^2 - 2x + 4$

81. If $f(x) = x + 1$ and $g(x) = 2x - 5$, show that $(f \circ g)(x) \neq (g \circ f)(x)$.

82. If $f(x) = x^2 + 1$ and $g(x) = 3x^2 - 2$, show that $(f \circ g)(x) \neq (g \circ f)(x)$.

83. If $f(x) = x^2 + 2x - 3$, find $f(a)$, $f(h)$, and $f(a + h)$. Then show that $f(a + h) \neq f(a) + f(h)$.

84. If $g(x) = 2x^2 + 10$, find $g(a)$, $g(h)$, and $g(a + h)$. Then show that $g(a + h) \neq g(a) + g(h)$.

85. If $f(x) = x^3 - 1$, find $\dfrac{f(x + h) - f(x)}{h}$.

86. If $f(x) = x^3 + 2$, find $\dfrac{f(x + h) - f(x)}{h}$.

APPLICATIONS

87. Alloys A molten alloy must be cooled slowly to control crystallization. When removed from the furnace, its temperature is 2,700° F, and it will be cooled at 200° per hour. Express the Celsius temperature as a function of the number of hours, t, since cooling began.

88. Weather forecasting A high pressure area promises increasingly warmer weather for the next 48 hours. The temperature is now 34° Celsius and is expected to rise 1° every 6 hours. Express the Fahrenheit temperature as a function of the number of hours from now. $\left(Hint: F = \frac{9}{5}C + 32.\right)$

WRITING

89. Explain how to find the domain of f/g.

90. Explain why the difference quotient represents the slope of a line passing through $(x, f(x))$ and $(x + h, f(x + h))$.

SOMETHING TO THINK ABOUT

91. Is composition of functions associative? Choose functions f, g, and h and determine whether $[f \circ (g \circ h)](x) = [(f \circ g) \circ h](x)$.

92. Choose functions f, g, and h and determine whether $f \circ (g + h) = f \circ g + f \circ h$.

10.7 Inverses of Functions

■ INTRODUCTION TO INVERSES OF FUNCTIONS ■ ONE-TO-ONE FUNCTIONS ■ THE HORIZONTAL LINE TEST ■ FINDING INVERSES OF FUNCTIONS

Getting Ready *Solve each equation for x.*

1. $y = 3x + 2$

2. $2x - 3y = 10$

■ INTRODUCTION TO INVERSES OF FUNCTIONS

The function defined by $C = \frac{5}{9}(F - 32)$ is the formula that we use to convert degrees Fahrenheit to degrees Celsius. If we substitute a Fahrenheit reading into the formula, a Celsius reading comes out. For example, if we substitute 41° for F, we obtain a Celsius reading of 5°:

$$C = \frac{5}{9}(F - 32)$$

$$= \frac{5}{9}(\mathbf{41} - 32) \qquad \text{Substitute 41 for } F.$$

$$= \frac{5}{9}(9)$$

$$= 5$$

If we want to find a Fahrenheit reading from a Celsius reading, we need a formula into which we can substitute a Celsius reading and have a Fahrenheit reading come out. Such a formula is $F = \frac{9}{5}C + 32$, which takes the Celsius reading of 5° and turns it back into a Fahrenheit reading of 41°.

$$F = \frac{9}{5}C + 32$$

$$= \frac{9}{5}(5) + 32 \qquad \text{Substitute 5 for } C.$$

$$= 41$$

The functions defined by these two formulas do opposite things. The first turns 41° F into 5° Celsius, and the second turns 5° Celsius back into 41° F. For this reason, we say that the functions are *inverses* of each other.

■ ONE-TO-ONE FUNCTIONS

Recall that for each input into a function, there is a single output. For some functions, different inputs have the same output, as shown in Figure 10-28(a). For other functions, different inputs have different outputs, as shown in Figure 10-28(b).

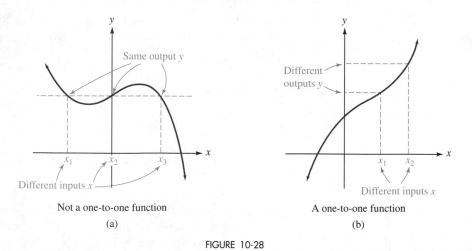

Not a one-to-one function
(a)

A one-to-one function
(b)

FIGURE 10-28

When every output of a function corresponds to exactly one input, we say that the function is *one-to-one*.

One-to-One Functions
A function is called **one-to-one** if each input value of x in the domain determines a different output value of y in the range.

EXAMPLE 1   Determine whether the functions **a.** $f(x) = x^2$ and **b.** $f(x) = x^3$ are one-to-one.

Solution   **a.** The function $f(x) = x^2$ is not one-to-one, because different input values x can determine the same output value y. For example, inputs of 3 and -3 produce the same output value of 9.

$$f(3) = 3^2 = 9 \quad \text{and} \quad f(-3) = (-3)^2 = 9$$

b. The function $f(x) = x^3$ is one-to-one, because different input values x determine different output values of y for all x. This is because different numbers have different cubes. ∎

Self Check   Determine whether $f(x) = 2x + 3$ is one-to-one.

Answer   yes

■ THE HORIZONTAL LINE TEST

The **horizontal line test** can be used to decide whether the graph of a function represents a one-to-one function. If any horizontal line that intersects the graph of a function does so only once, the function is one-to-one. Otherwise, the function is not one-to-one. See Figure 10-29.

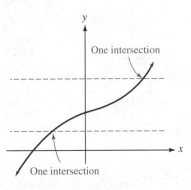
A one-to-one function

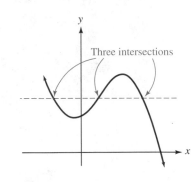
Not a one-to-one function

FIGURE 10-29

EXAMPLE 2   Use the horizontal line test to decide whether the graphs in Figure 10-30 represent one-to-one functions.

Solution   **a.** Because many horizontal lines intersect the graph shown in Figure 10-30(a) twice, the graph does not represent a one-to-one function.

b. Because each horizontal line that intersects the graph in Figure 10-30(b) does so exactly once, the graph does represent a one-to-one function.

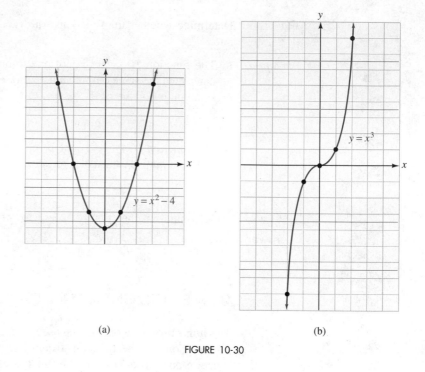

(a)   (b)

FIGURE 10-30

Self Check   Determine whether the following graph represents a one-to-one function.

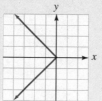

Answer   no

 WARNING!   Make sure to use the vertical line test to determine whether the graph represents a function. If it is, use the horizontal line test to determine whether the function is one-to-one.

■ FINDING INVERSES OF FUNCTIONS

If f is the function determined by the table shown in Figure 10-31(a), it turns the number 1 into 10, 2 into 20, and 3 into 30. Since the inverse of f must turn 10 back into 1, 20 back into 2, and 30 back into 3, it consists of the ordered pairs shown in Figure 10-31(b).

	Function f			**Inverse of** f	
x		y	x		y
1		10	10		1
2		20	20		2
3		30	30		3

Domain   Range   Domain   Range

Note that the inverse of f is also a function.

(a)   (b)

FIGURE 10-31

We note that the domain of f and the range of its inverse is $\{1, 2, 3\}$. The range of f and the domain of its inverse is $\{10, 20, 30\}$.

This example suggests that to form the inverse of a function f, we simply interchange the coordinates of each ordered pair that determines f. When the inverse of a function is also a function, we call it f **inverse** and denote it with the symbol f^{-1}.

WARNING!   The symbol $f^{-1}(x)$ is read as "the inverse of $f(x)$" or just "f inverse." The -1 in the notation $f^{-1}(x)$ is not an exponent. Remember that $f^{-1}(x) \neq \frac{1}{f(x)}$.

Finding the Inverse of a One-to-One Function
If a function is one-to-one, we find its inverse as follows:

1. Replace $f(x)$ with y, if necessary.
2. Interchange the variables x and y.
3. Solve the resulting equation for y.
4. This equation is $y = f^{-1}(x)$.

EXAMPLE 3   If $f(x) = 4x + 2$, find the inverse of f and tell whether it is a function.

Solution   We proceed as follows:

$$f(x) = 4x + 2$$
$$y = 4x + 2 \qquad \text{Replace } f(x) \text{ with } y.$$
$$x = 4y + 2 \qquad \text{Interchange the variables } x \text{ and } y.$$

To decide whether the inverse $x = 4y + 2$ is a function, we solve the equation for y.

$$x = 4y + 2$$
$$x - 2 = 4y \qquad \text{Subtract 2 from both sides.}$$

1. $$y = \frac{x - 2}{4} \qquad \text{Divide both sides by 4 and write } y \text{ on the left-hand side.}$$

Because each input x that is substituted into Equation 1 gives one output y, the inverse of f is a function, so we can express it in the form

$$f^{-1}(x) = \frac{x-2}{4}$$

∎

Self Check

Answers

If $y = f(x) = -5x - 3$, find the inverse of f and tell whether it is a function.

$y = -\frac{1}{5}x - \frac{3}{5}$, yes

To emphasize an important relationship between a function and its inverse, we substitute some number x, such as $x = 3$, into the function $f(x) = 4x + 2$ of Example 3. The corresponding value of y produced is

$$y = f(3) = 4(3) + 2 = 14$$

If we substitute 14 into the inverse function, f^{-1}, the corresponding value of y that is produced is

$$y = f^{-1}(14) = \frac{14-2}{4} = 3$$

Thus, the function f turns 3 into 14, and the inverse function f^{-1} turns 14 back into 3. In general, the composition of a function and its inverse is the identity function.

To prove that $f(x) = 4x + 2$ and $f^{-1}(x) = \frac{x-2}{4}$ are inverse functions, we must show that their composition (in both directions) is the identity function:

$$(f \circ f^{-1})(x) = f(f^{-1}(x)) \qquad\qquad (f^{-1} \circ f)(x) = f^{-1}(f(x))$$

$$= f\left(\frac{x-2}{4}\right) \qquad\qquad\qquad = f^{-1}(4x+2)$$

$$= 4\left(\frac{x-2}{4}\right) + 2 \qquad\qquad = \frac{4x+2-2}{4}$$

$$= x - 2 + 2 \qquad\qquad\qquad = \frac{4x}{4}$$

$$= x \qquad\qquad\qquad\qquad = x$$

Thus, $(f \circ f^{-1})(x) = (f^{-1} \circ f)(x) = x$, which is the identity function $I(x)$.

EXAMPLE 4

The set of all pairs (x, y) determined by $3x + 2y = 6$ is a function. Find its inverse function, and graph the function and its inverse on one coordinate system.

Solution

To find the inverse function of $3x + 2y = 6$, we interchange x and y to obtain

$$3y + 2x = 6$$

and then solve the equation for y.

$$3y + 2x = 6$$

$$3y = -2x + 6 \qquad \text{Subtract } 2x \text{ from both sides.}$$

$$y = -\frac{2}{3}x + 2 \qquad \text{Divide both sides by 3.}$$

Thus, $y = f^{-1}(x) = -\frac{2}{3}x + 2$. The graphs of $3x + 2y = 6$ and $y = f^{-1}(x) = -\frac{2}{3}x + 2$ appear in Figure 10-32.

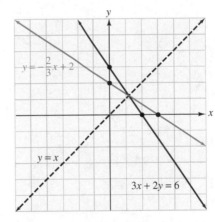

FIGURE 10-32   ■

Self Check   Find the inverse of the function defined by $2x - 3y = 6$. Graph the function and its inverse on one coordinate system.

Answer $f^{-1}(x) = \frac{3}{2}x + 3$

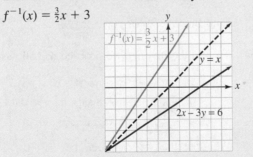

In Example 4, the graph of $3x + 2y = 6$ and $y = f^{-1}(x) = -\frac{2}{3}x + 2$ are symmetric about the line $y = x$. This is always the case, because when the coordinates (a, b) satisfy an equation, the coordinates (b, a) will satisfy its inverse.

In each example so far, the inverse of a function has been another function. This is not always true, as the following example will show.

EXAMPLE 5   Find the inverse of the function determined by $f(x) = x^2$.

Solution
$$y = x^2 \qquad \text{Replace } f(x) \text{ with } y.$$
$$x = y^2 \qquad \text{Interchange } x \text{ and } y.$$
$$y = \pm\sqrt{x} \qquad \text{Use the square root property and write } y \text{ on the left-hand side.}$$

When the inverse $y = \pm\sqrt{x}$ is graphed as in Figure 10-33, we see that the graph does not pass the vertical line test. Thus, it is not a function.

   The graph of $y = x^2$ is also shown in the figure. As expected, the graphs of $y = x^2$ and $y = \pm\sqrt{x}$ are symmetric about the line $y = x$.

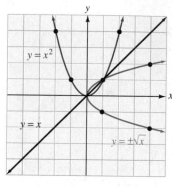

FIGURE 10-33

Self Check   Find the inverse of the function determined by $f(x) = 4x^2$.

Answer $y = \pm\dfrac{\sqrt{x}}{2}$

EXAMPLE 6   Find the inverse of $f(x) = x^3$.

Solution   To find the inverse, we proceed as follows:

$$y = x^3 \qquad \text{Replace } f(x) \text{ with } y.$$
$$x = y^3 \qquad \text{Interchange the variables } x \text{ and } y.$$
$$\sqrt[3]{x} = y \qquad \text{Take the cube root of both sides.}$$

We note that to each number x there corresponds one real cube root. Thus, $y = \sqrt[3]{x}$ represents a function. In $f^{-1}(x)$ notation, we have

$$f^{-1}(x) = \sqrt[3]{x}$$

Self Check   Find the inverse of $f(x) = x^5$.

Answer $f^{-1}(x) = \sqrt[5]{x}$

If a function is not one-to-one, we can often make it one-to-one by restricting its domain.

EXAMPLE 7   Find the inverse of the function defined by $y = x^2$ with $x \geq 0$. Then tell whether the inverse is a function. Graph the function and its inverse on one set of coordinate axes.

Solution   The inverse of the function $y = x^2$ with $x \geq 0$ is

$$x = y^2 \quad \text{with } y \geq 0 \qquad \text{Interchange the variables } x \text{ and } y.$$

Before considering the restriction, this equation can be written in the form

$$y = \pm\sqrt{x} \quad \text{with} \quad y \geq 0$$

Since $y \geq 0$, each number x gives only one value of y: $y = \sqrt{x}$. Thus, the inverse is a function.

The graphs of the two functions appear in Figure 10-34. The line $y = x$ is included so that we can see that the graphs are symmetric about the line $y = x$.

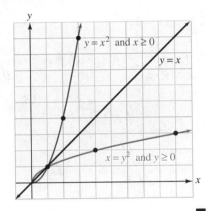

$y = x^2$ and $x \geq 0$			$x = y^2$ and $y \geq 0$		
x	y	(x, y)	x	y	(x, y)
0	0	$(0, 0)$	0	0	$(0, 0)$
1	1	$(1, 1)$	1	1	$(1, 1)$
2	4	$(2, 4)$	4	2	$(4, 2)$
3	9	$(3, 9)$	9	3	$(9, 3)$

FIGURE 10-34

Orals   *Find the inverse of each set of ordered pairs.*

1. $\{(1, 2), (2, 3), (5, 10)\}$ **2.** $\{(1, 1), (2, 8), (4, 64)\}$

Find the inverse function of each linear function.

3. $y = \dfrac{1}{2}x$ **4.** $y = 2x$

Tell whether each function is one-to-one.

5. $y = x^2 - 2$ **6.** $y = x^3$

EXERCISE 10.7

REVIEW *Write each complex number in a + bi form or find each value.*

1. $3 - \sqrt{-64}$

2. $(2 - 3i) + (4 + 5i)$

3. $(3 + 4i)(2 - 3i)$

4. $\dfrac{6 + 7i}{3 - 4i}$

5. $|6 - 8i|$

6. $\left| \dfrac{2 + i}{3 - i} \right|$

VOCABULARY AND CONCEPTS *Fill in each blank to make a true statement.*

7. A function is called _____ if each input determines a different output.

8. If every _____ line that intersects a graph of a function does so only once, the function is one-to-one.

9. If a one-to-one function turns an input of 2 into an output of 5, the inverse function will turn 5 into ___.

10. The symbol $f^{-1}(x)$ is read as _____ or _____.

11. $(f \circ f^{-1})(x) = $ ___

12. The graphs of a function and its inverse are symmetrical about the line _____.

PRACTICE *In Exercises 13–16, determine whether each function is one-to-one.*

13. $f(x) = 2x$

14. $f(x) = |x|$

15. $f(x) = x^4$

16. $f(x) = x^3 + 1$

In Exercises 17–24, each graph represents a function. Use the horizontal line test to decide whether the function is one-to-one.

17.
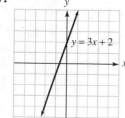
$y = 3x + 2$

18.
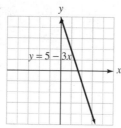
$y = 5 - 3x$

19.
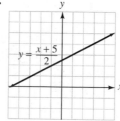
$y = \dfrac{x + 5}{2}$

20.
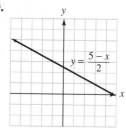
$y = \dfrac{5 - x}{2}$

21.
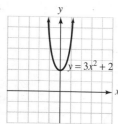
$y = 3x^2 + 2$

22.
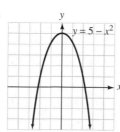
$y = 5 - x^2$

23.
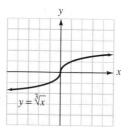
$y = \sqrt[3]{x}$

24.
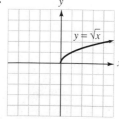
$y = \sqrt{x}$

In Exercises 25–30, find the inverse of each set of ordered pairs (x, y) and tell whether the inverse is a function.

25. $\{(3, 2), (2, 1), (1, 0)\}$

26. $\{(4, 1), (5, 1), (6, 1), (7, 1)\}$

27. $\{(1, 2), (2, 3), (1, 3), (1, 5)\}$

28. $\{(-1, -1), (0, 0), (1, 1), (2, 2)\}$ **29.** $\{(1, 1), (2, 4), (3, 9), (4, 16)\}$ **30.** $\{(1, 1), (2, 1), (3, 1), (4, 1)\}$

In Exercises 31–38, find the inverse of each function and express it in the form $y = f^{-1}(x)$. Verify each result by showing that $(f \circ f^{-1})(x) = (f^{-1} \circ f)(x) = I(x)$.

31. $f(x) = 3x + 1$

32. $y + 1 = 5x$

33. $x + 4 = 5y$

34. $x = 3y + 1$

35. $f(x) = \dfrac{x - 4}{5}$

36. $f(x) = \dfrac{2x + 6}{3}$

37. $4x - 5y = 20$

38. $3x + 5y = 15$

In Exercises 39–46, find the inverse of each function. Then graph the function and its inverse on one coordinate system. Find the equation of the line of symmetry.

39. $y = 4x + 3$ **40.** $x = 3y - 1$ **41.** $x = \dfrac{y - 2}{3}$ **42.** $y = \dfrac{x + 3}{4}$

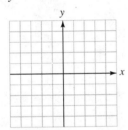   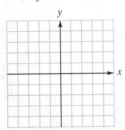   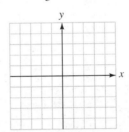   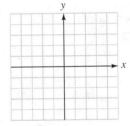

43. $3x - y = 5$ **44.** $2x + 3y = 9$ **45.** $3(x + y) = 2x + 4$ **46.** $-4(y - 1) + x = 2$

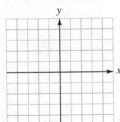   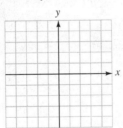   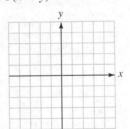   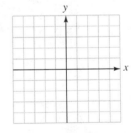

In Exercises 47–52, find the inverse of each function and tell whether it is a function.

47. $y = x^2 + 4$

48. $y = x^2 + 5$

49. $y = x^3$ **50.** $xy = 4$ **51.** $y = |x|$ **52.** $y = \sqrt[3]{x}$

In Exercises 53–54, show that the inverse of the function determined by each equation is also a function. Express it using $f^{-1}(x)$ notation.

53. $f(x) = 2x^3 - 3$

54. $f(x) = \dfrac{3}{x^3} - 1$

In Exercises 55–58, graph each equation and its inverse on one set of coordinate axes. Find the axis of symmetry.

55. $y = x^2 + 1$

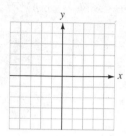

56. $y = \dfrac{1}{4}x^2 - 3$

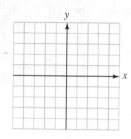

57. $y = \sqrt{x}$

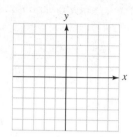

58. $y = |x|$

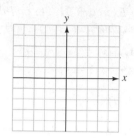

WRITING

59. Explain the purpose of the vertical line test.

60. Explain the purpose of the horizontal line test.

SOMETHING TO THINK ABOUT

61. Find the inverse of $y = \dfrac{x + 1}{x - 1}$.

62. Using the functions of Exercise 61, show that $(f \circ f^{-1})(x) = x$.

■ ■ ■ ■ ■ ■ ■ ■ ■ **PROJECTS**

PROJECT 1

Ballistics is the study of how projectiles fly. The general formula for the height above the ground of an object thrown straight up or straight down is given by the function

$$h(t) = -16t^2 + v_0 t + h_0$$

where h is the object's height (in feet) above the ground t seconds after it is thrown. The initial velocity v_0 is the velocity with which the object is thrown, measured in feet per second. The initial height h_0 is the object's height (in feet) above the ground when it is thrown. (If $v_0 > 0$, the object is thrown upward; if $v_0 < 0$, the object is thrown downward.)

This formula takes into account the force of gravity, but it disregards the force of air resistance. It is much more accurate for a smooth, dense ball than for a crumpled piece of paper.

One of the most popular acts of the Bungling Brothers Circus is the Amazing Glendo and his cannonball catching act. A cannon fires a ball vertically into the air; Glendo, standing on a platform above the cannon, uses his catlike reflexes to catch the ball as it passes by on its way toward the roof of the big top. As the balls fly past, they are within Glendo's reach only during a two-foot interval of their upward path.

(continued)

■ ■ ■ ■ ■ ■ ■ ■ ■ ■ **PROJECTS** *(continued)*

As an investigator for the company that insures the circus, you have been asked to find answers to the following questions. The answers will determine whether or not Bungling Brothers' insurance policy will be renewed.

a. In the first part of the act, cannonballs are fired from the end of a six-foot cannon with an initial velocity of 80 feet per second. Glendo catches one ball between 40 and 42 feet above the ground. Then he lowers his platform and catches another ball between 25 and 27 feet above the ground.

  i. Show that if Glendo missed a cannonball, it would hit the roof of the 56-foot-tall big top. How long would it take for a ball to hit the big top? To prevent this from happening, a special net near the roof catches and holds any missed cannonballs.

  ii. Find (to the nearest thousandth of a second) how long the cannonballs are within Glendo's reach for each of his catches. Which catch is easier? Why does your answer make sense? Your company is willing to insure against injuries to Glendo if he has at least 0.025 second to make each catch. Should the insurance be offered?

b. For Glendo's grand finale, the special net at the roof of the big top is removed, making Glendo's catch more significant to the people in the audience, who worry that if Glendo misses, the tent will collapse around them. To make it even more dramatic, Glendo's arms are tied to restrict his reach to a one-foot interval of the ball's flight, and he stands on a platform just under the peak of the big top, so that his catch is made at the very last instant (between 54 and 55 feet above the ground). For this part of the act, however, Glendo has the cannon charged with less gunpowder, so that the muzzle velocity of the cannon is 56 feet per second. Show work to prove that Glendo's big finale is in fact his *easiest* catch, and that even if he misses, the big top is never in any danger of collapsing, so insurance should be offered against injury to the audience.

PROJECT 2   The center of Sterlington is the intersection of Main Street (running east–west) and DueNorth Road (running north–south). The recreation area for the townspeople is Robin Park, a few blocks from there. The park is bounded on the south by Main Street and on every other side by Parabolic Boulevard, named for its distinctive shape. In fact, if Main Street and DueNorth Road were used as the axes of a rectangular coordinate system, Parabolic Boulevard would have the equation $y = -(x - 4)^2 + 5$, where each unit on the axes is 100 yards.

The city council has recently begun to consider whether or not to put two walkways through the park. (See Illustration 1.) The walkways would run from two points on Main Street and converge at the northernmost point of the park, dividing the area of the park exactly into thirds.

The city council is pleased with the esthetics of this arrangement but needs to know two important facts.

a. For planning purposes, they need to know exactly where on Main Street the walkways would begin.

b. In order to budget for the construction, they need to know how long the walkways will be.

Provide answers for the city council, along with explanations and work to show that your answers are correct. You will need to use the formula shown in Illustration 2, due to Archimedes (287–212 B.C.), for the area under a parabola but above a line perpendicular to the axis of symmetry of the parabola.

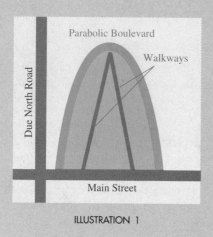

ILLUSTRATION 1

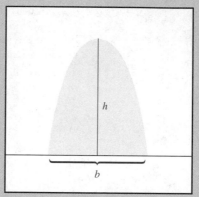

Shaded area $= \frac{2}{3} \cdot b \cdot h$

ILLUSTRATION 2

CHAPTER SUMMARY

CONCEPT

REVIEW EXERCISES

SECTION 10.1 — Completing the Square and the Quadratic Formula

Square root property:
If $c > 0$, the equation $x^2 = c$ has two real solutions:

$$x = \sqrt{c} \quad \text{and} \quad x = -\sqrt{c}$$

1. Solve each equation by factoring or by using the square root property.
   a. $12x^2 + x - 6 = 0$
   b. $6x^2 + 17x + 5 = 0$
   c. $15x^2 + 2x - 8 = 0$
   d. $(x + 2)^2 = 36$

To complete the square, add the square of one-half of the coefficient of x.

2. Solve each equation by completing the square.
   a. $x^2 + 6x + 8 = 0$
   b. $2x^2 - 9x + 7 = 0$

Quadratic formula:

$$x = \frac{-b \pm \sqrt{b^2 - 4ac}}{2a}$$

$(a \neq 0)$

3. Solve each equation by using the quadratic formula.
  a. $x^2 - 8x - 9 = 0$ **b.** $x^2 - 10x = 0$

  c. $2x^2 + 13x - 7 = 0$ **d.** $3x^2 - 20x - 7 = 0$

4. Dimensions of a rectangle   A rectangle is 2 centimeters longer than it is wide. If both the length and width are doubled, its area is increased by 72 square centimeters. Find the dimensions of the original rectangle.

5. Dimensions of a rectangle   A rectangle is 1 foot longer than it is wide. If the length is tripled and the width is doubled, its area is increased by 30 square feet. Find the dimensions of the original rectangle.

6. Ballistics   If a rocket is launched straight up into the air with an initial velocity of 112 feet per second, its height after t seconds is given by the formula

$$h = 112t - 16t^2$$

where h represents the height of the rocket in feet. After launch, how long will it be before it hits the ground?

7. Ballistics   What is the maximum height of the rocket discussed in Exercise 6?

SECTION 10.2	*Graphs of Quadratic Functions*

If f is a function and k is a positive number, then

The graph of $y = f(x) + k$ is identical to the graph of $y = f(x)$, except that it is translated k units upward.

The graph of $y = f(x) - k$ is identical to the graph of $y = f(x)$, except that it is translated k units downward.

8. Graph each function and give the coordinates of the vertex of the resulting parabola.
  a. $y = 2x^2 - 3$ **b.** $y = -2x^2 - 1$

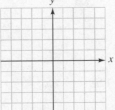

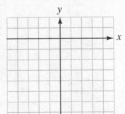

The graph of $y = f(x - h)$ is identical to the graph of $y = f(x)$, except that it is translated h units to the right.

The graph of $y = f(x + h)$ is identical to the graph of $y = f(x)$, except that it is translated h units to the left.

If $a \neq 0$, the graph of $y = a(x - h)^2 + k$ is a parabola with vertex at (h, k). It opens upward when $a > 0$ and downward when $a < 0$.

c. $y = -4(x - 2)^2 + 1$

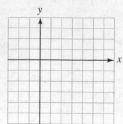

d. $y = 5x^2 + 10x - 1$

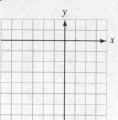

SECTION 10.3	*Complex Numbers*

Complex numbers: If a, b, c, and d are real numbers and $i^2 = -1$,

$a + bi = c + di$ if and only if $a = c$ and $b = d$

$(a + bi) + (c + di)$
$= (a + c) + (b + d)i$

$(a + bi)(c + di)$
$= (ac - bd) + (ad + bc)i$

$|a + bi| = \sqrt{a^2 + b^2}$

9. Do the operations and give all answers in $a + bi$ form.

a. $(5 + 4i) + (7 - 12i)$

b. $(-6 - 40i) - (-8 + 28i)$

c. $\left(-32 + \sqrt{-144}\right) - \left(64 + \sqrt{-81}\right)$

d. $\left(-8 + \sqrt{-8}\right) + \left(6 - \sqrt{-32}\right)$

e. $(2 - 7i)(-3 + 4i)$

f. $(-5 + 6i)(2 + i)$

g. $\left(5 - \sqrt{-27}\right)\left(-6 + \sqrt{-12}\right)$

h. $\left(2 + \sqrt{-128}\right)\left(3 - \sqrt{-98}\right)$

i. $\dfrac{3}{4i}$

j. $\dfrac{-2}{5i^3}$

k. $\dfrac{6}{2 + i}$

l. $\dfrac{7}{3 - i}$

m. $\dfrac{4 + i}{4 - i}$

n. $\dfrac{3 - i}{3 + i}$

o. $\dfrac{3}{5 + \sqrt{-4}}$

p. $\dfrac{2}{3 - \sqrt{-9}}$

q. $|9 + 12i|$

r. $|24 - 10i|$

The Discriminant and Equations That Can Be Written in Quadratic Form

The discriminant:
If $b^2 - 4ac > 0$, the solutions of $ax^2 + bx + c = 0$ are unequal real numbers.

If $b^2 - 4ac = 0$, the solutions of $ax^2 + bx + c = 0$ are equal real numbers.

If $b^2 - 4ac < 0$, the solutions of $ax^2 + bx + c = 0$ are complex conjugates.

If r_1 and r_2 are solutions of $ax^2 + bx + c = 0$, then

$$r_1 + r_2 = -\frac{b}{a}$$

$$r_1 r_2 = \frac{c}{a}$$

10. Use the discriminant to determine what type of solutions exist for each equation.
  a. $3x^2 + 4x - 3 = 0$

  b. $4x^2 - 5x + 7 = 0$

  c. Find the values of k that will make the solutions of $(k - 8)x^2 + (k + 16)x = -49$ equal.

  d. Find the values of k such that the solutions of $3x^2 + 4x = k + 1$ will be real numbers.

11. Solve each equation.
  a. $x - 13x^{1/2} + 12 = 0$

  b. $a^{2/3} + a^{1/3} - 6 = 0$

  c. $\dfrac{1}{x + 1} - \dfrac{1}{x} = -\dfrac{1}{x + 1}$

  d. $\dfrac{6}{x + 2} + \dfrac{6}{x + 1} = 5$

12. Find the sum of the solutions of the equation $3x^2 - 14x + 3 = 0$.

13. Find the product of the solutions of the equation $3x^2 - 14x + 3 = 0$.

Quadratic and Other Nonlinear Inequalities

To solve a quadratic inequality in one variable, make a sign chart.

14. Solve each inequality. Give each result in interval notation and graph the solution set.
  a. $x^2 + 2x - 35 > 0$

  b. $x^2 + 7x - 18 < 0$

To solve inequalities with rational expressions, get 0 on the right-hand side, add the fractions, and then factor the numerator and denominator. Then use a sign chart.

c. $\dfrac{3}{x} \le 5$

d. $\dfrac{2x^2 - x - 28}{x - 1} > 0$

15. Use a graphing calculator to solve each inequality. Compare the results with Review Exercise 14.

a. $x^2 + 2x - 35 > 0$ **b.** $x^2 + 7x - 18 < 0$

c. $\dfrac{3}{x} \le 5$ **d.** $\dfrac{2x^2 - x - 28}{x - 1} > 0$

To graph an inequality such as $y > 4x - 3$, first graph the equation $y = 4x - 3$. Then determine which half-plane represents the graph of $y > 4x - 3$.

16. Graph each inequality.

a. $y < \dfrac{1}{2}x^2 - 1$

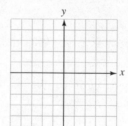

b. $y \ge -|x|$

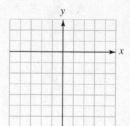

SECTION 10.6   *Algebra and Composition of Functions*

Operations with functions:

$(f + g)(x) = f(x) + g(x)$

$(f - g)(x) = f(x) - g(x)$

$(f \cdot g)(x) = f(x)g(x)$

$(f/g)(x) = \dfrac{f(x)}{g(x)} \quad (g(x) \ne 0)$

$(f \circ g)(x) = f(g(x))$

17. Let $f(x) = 2x$ and $g(x) = x + 1$. *Find each function or value.*

a. $f + g$ **b.** $f - g$

c. $f \cdot g$ **d.** f/g

e. $(f \circ g)(2)$ **f.** $(g \circ f)(-1)$

g. $(f \circ g)(x)$ **h.** $(g \circ f)(x)$

| **SECTION 10.7** | *Inverses of Functions* |

Horizontal line test:
If every horizontal line that intersects the graph of a function does so only once, the function is one-to-one.

18. Graph each function and use the horizontal line test to decide whether the function is one-to-one.

a. $f(x) = 2(x - 3)$

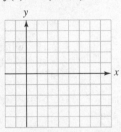

b. $f(x) = x(2x - 3)$

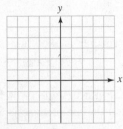

c. $f(x) = -3(x - 2)^2 + 5$

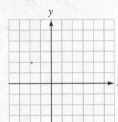

d. $f(x) = |x|$

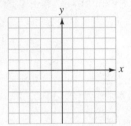

To find the inverse of a function, interchange the positions of variables x and y and solve for y.

19. Find the inverse of each function.

a. $f(x) = 6x - 3$

b. $f(x) = 4x + 5$

c. $y = 2x^2 - 1 \ (x \geq 0)$

d. $y = |x|$

■ Chapter Test

In Problems 1–2, solve each equation.

1. $x^2 + 3x - 18 = 0$

2. $x(6x + 19) = -15$

In Problems 3–4, determine what number must be added to each binomial to make it a perfect square.

3. $x^2 + 24x$

4. $x^2 - 50x$

In Problems 5–6, solve each equation.

5. $x^2 + 4x + 1 = 0$

6. $x^2 - 5x - 3 = 0$

In Problems 7–12, do the operations. Give all answers in a + bi form.

7. $(2 + 4i) + (-3 + 7i)$

8. $\left(3 - \sqrt{-9}\right) - \left(-1 + \sqrt{-16}\right)$

9. $2i(3 - 4i)$

10. $(3 + 2i)(-4 - i)$

11. $\dfrac{1}{i\sqrt{2}}$

12. $\dfrac{2 + i}{3 - i}$

13. Determine whether the solutions of $3x^2 + 5x + 17 = 0$ are real or nonreal numbers.

14. For what value(s) of k are the solutions of $4x^2 - 2kx + k - 1 = 0$ equal?

15. One leg of a right triangle is 14 inches longer than the other, and the hypotenuse is 26 inches. Find the length of the shorter leg.

16. Solve the equation $2y - 3y^{1/2} + 1 = 0$.

17. Graph the function $f(x) = \frac{1}{2}x^2 - 4$ and give the coordinates of its vertex.

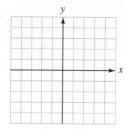

18. Graph the inequality $y \leq -x^2 + 3$

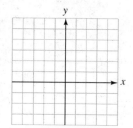

In Problems 19–20, solve each inequality and graph the solution set.

19. $x^2 - 2x - 8 > 0$

20. $\dfrac{x - 2}{x + 3} \leq 0$

In Problems 21–24, f(x) = 4x and g(x) = x − 1. Find each function.

21. $g + f$

22. $f - g$

23. $g \cdot f$

24. g/f

In Problems 25–28, f(x) = 4x and g(x) = x − 1. Find each value.

25. $(g \circ f)(1)$

26. $(f \circ g)(0)$

27. $(f \circ g)(-1)$

28. $(g \circ f)(-2)$

In Problems 29–30, f(x) = 4x and g(x) = x − 1. Find each function.

29. $(f \circ g)(x)$

30. $(g \circ f)(x)$

In Problems 31–32, find the inverse of each function.

31. $3x + 2y = 12$

32. $y = 3x^2 + 4 \ (x \leq 0)$

■ Cumulative Review Exercises

In Exercises 1–2, find the domain and range of each function.

1. $y = f(x) = 2x^2 - 3$

2. $y = f(x) = -|x - 4|$

In Exercises 3–4, write the equation of the line with the given properties.

3. $m = 3$, passing through $(-2, -4)$

4. Parallel to the graph of $2x + 3y = 6$ and passing through $(0, -2)$

In Exercises 5–6, do each operation.

5. $(2a^2 + 4a - 7) - 2(3a^2 - 4a)$

6. $(3x + 2)(2x - 3)$

In Exercises 7–8, factor each expression.

7. $x^4 - 16y^4$

8. $15x^2 - 2x - 8$

In Exercises 9–10, solve each equation.

9. $x^2 - 5x - 6 = 0$

10. $6a^3 - 2a = a^2$

In Exercises 11–18, simplify each expression. Assume that all variables represent positive numbers.

11. $\sqrt{25x^4}$

12. $\sqrt{48t^3}$

13. $\sqrt[3]{-27x^3}$

14. $\sqrt[3]{\dfrac{128x^4}{2x}}$

15. $8^{-1/3}$

16. $64^{2/3}$

17. $\dfrac{y^{2/3}y^{5/3}}{y^{1/3}}$

18. $\dfrac{x^{5/3}x^{1/2}}{x^{3/4}}$

In Exercises 19–20, graph each function and give the domain and the range.

19. $f(x) = \sqrt{x - 2}$

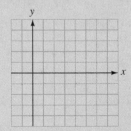

20. $f(x) = -\sqrt{x + 2}$

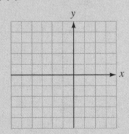

In Exercises 21–22, do the operations.

21. $(x^{2/3} - x^{1/3})(x^{2/3} + x^{1/3})$

22. $(x^{-1/2} + x^{1/2})^2$

In Exercises 23–28, simplify each statement.

23. $\sqrt{50} - \sqrt{8} + \sqrt{32}$

24. $-3\sqrt[4]{32} - 2\sqrt[4]{162} + 5\sqrt[4]{48}$

25. $3\sqrt{2}(2\sqrt{3} - 4\sqrt{12})$

26. $\dfrac{5}{\sqrt[3]{x}}$

27. $\dfrac{\sqrt{x} + 2}{\sqrt{x} - 1}$

28. $\sqrt[6]{x^3 y^3}$

In Exercises 29–30, solve each equation.

29. $5\sqrt{x + 2} = x + 8$

30. $\sqrt{x} + \sqrt{x + 2} = 2$

31. Find the length of the hypotenuse of the right triangle shown in Illustration 1.

32. Find the length of the hypotenuse of the right triangle shown in Illustration 2.

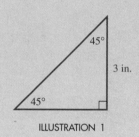

ILLUSTRATION 1

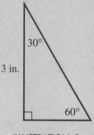

ILLUSTRATION 2

33. Find the distance between $P(-2, 6)$ and $Q(4, 14)$.

34. What number must be added to $x^2 + 6x$ to make a trinomial square?

35. Use the method of completing the square to solve $2x^2 + x - 3 = 0$.

36. Use the quadratic formula to solve $3x^2 + 4x - 1 = 0$.

37. Graph $y = f(x) = \frac{1}{2}x^2 + 5$ and find the coordinates of its vertex.

38. Graph $y \le -x^2 + 3$ and find the coordinates of its vertex.

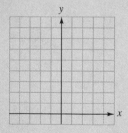

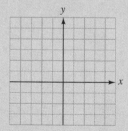

In Exercises 39–46, write each expression as a real number or as a complex number in $a + bi$ form.

39. $(3 + 5i) + (4 - 3i)$

40. $(7 - 4i) - (12 + 3i)$

41. $(2 - 3i)(2 + 3i)$

42. $(3 + i)(3 - 3i)$

43. $(3 - 2i) - (4 + i)^2$

44. $\dfrac{5}{3 - i}$

45. $|3 + 2i|$

46. $|5 - 6i|$

47. For what values of k will the solutions of $2x^2 + 4x = k$ be equal?

48. Solve $a - 7a^{1/2} + 12 = 0$

In Exercises 49–50, solve each inequality and graph the solution set on the number line.

49. $x^2 - x - 6 > 0$

50. $x^2 - x - 6 \leq 0$

In Exercises 51–54, $f(x) = 3x^2 + 2$ and $g(x) = 2x - 1$. Find each value or composite function.

51. $f(-1)$

52. $(g \circ f)(2)$

53. $(f \circ g)(x)$

54. $(g \circ f)(x)$

In Exercises 55–56, find the inverse of each function.

55. $f(x) = 3x + 2$

56. $f(x) = x^3 + 4$

11 Exponential and Logarithmic Functions

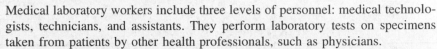

MATHEMATICS IN THE WORKPLACE

Medical Laboratory Worker

Medical laboratory workers include three levels of personnel: medical technologists, technicians, and assistants. They perform laboratory tests on specimens taken from patients by other health professionals, such as physicians.

Employment of medical laboratory workers is expected to expand faster than the average rate for all occupations through the year 2005, as physicians continue to use laboratory tests in routine physical checkups and in the diagnosis and treatment of disease.

SAMPLE APPLICATION ■ If a medium is inoculated with a bacterial culture that contains 1,000 cells per milliliter, how many generations will pass by the time the culture has grown to a population of 1 million cells per milliliter? (See Example 9 in Section 11.6.)

In this chapter, we will discuss two functions that are important in many applications of mathematics. **Exponential functions** are used to compute compound interest, find radioactive decay, and model population growth. **Logarithmic functions** are used to measure acidity of solutions, drug dosage, gain of an amplifier, intensity of earthquakes, and safe noise levels in factories.

11.1 Exponential Functions

■ IRRATIONAL EXPONENTS ■ EXPONENTIAL FUNCTIONS ■ GRAPHING EXPONENTIAL FUNCTIONS
■ VERTICAL AND HORIZONTAL TRANSLATIONS ■ COMPOUND INTEREST

Getting Ready   *Find each value.*

1. 2^3 **2.** $25^{1/2}$ **3.** 5^{-2} **4.** $\left(\dfrac{3}{2}\right)^{-3}$

The graph in Figure 11-1 shows the balance in a bank account in which $10,000 was invested in 1998 at 9% annual interest, compounded monthly. The graph shows that in the year 2008, the value of the account will be approximately $25,000, and in the year 2028, the value will be approximately $147,000. The curve shown in Figure 11-1 is the graph of a function called an *exponential function.*

Before we can discuss exponential functions, we must define irrational exponents.

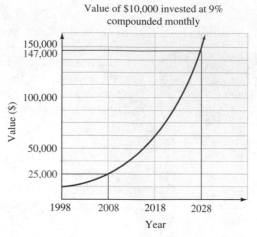

Value of $10,000 invested at 9% compounded monthly

FIGURE 11-1

■ IRRATIONAL EXPONENTS

We have discussed expressions of the form b^x, where x is a rational number.

$8^{1/2}$ means "the square root of 8."

$5^{1/3}$ means "the cube root of 5."

$3^{-2/5} = \dfrac{1}{3^{2/5}}$ means "the reciprocal of the fifth root of 3^2."

To give meaning to b^x when x is an irrational number, we consider the expression

$5^{\sqrt{2}}$ where $\sqrt{2}$ is the irrational number $1.414213562\ldots$

Each number in the following list is defined, because each exponent is a rational number.

$5^{1.4},\ 5^{1.41},\ 5^{1.414},\ 5^{1.4142},\ 5^{1.41421},\ \ldots$

Since the exponents are getting closer to $\sqrt{2}$, the numbers in this list are successively better approximations of $5^{\sqrt{2}}$. We can use a calculator to obtain a very good approximation.

■ ■ ■ ■ ■ ■ ■ ■ ■   **Evaluating Exponential Expressions**

CALCULATORS   To find the value of $5^{\sqrt{2}}$ with a scientific calculator, we enter these numbers and press these keys:

5   y^x 2 $\sqrt{}$ =

The display will read 9.738517742 .

■ ■ ■ ■ ■ ■ ■ ■ ■ ■ **Evaluating Exponential Expressions** *(continued)*

With a graphing calculator, we enter these numbers and press these keys:

5 ^ √ 2 Enter

The display will read 5^√(2)
$$9.738517742$$

In general, if b is positive and x is a real number, b^x represents a positive number. It can be shown that all of the familiar rules of exponents hold true for irrational exponents.

EXAMPLE 1 Use the rules of exponents to simplify **a.** $\left(5^{\sqrt{2}}\right)^{\sqrt{2}}$ and **b.** $b^{\sqrt{3}} \cdot b^{\sqrt{12}}$.

Solution **a.** $\left(5^{\sqrt{2}}\right)^{\sqrt{2}} = 5^{\sqrt{2}\sqrt{2}}$ Keep the base and multiply the exponents.
$$= 5^2 \qquad \sqrt{2}\sqrt{2} = \sqrt{4} = 2.$$
$$= 25$$

b. $b^{\sqrt{3}} \cdot b^{\sqrt{12}} = b^{\sqrt{3}+\sqrt{12}}$ Keep the base and add the exponents.
$$= b^{\sqrt{3}+2\sqrt{3}} \qquad \sqrt{12} = \sqrt{4}\sqrt{3} = 2\sqrt{3}.$$
$$= b^{3\sqrt{3}} \qquad \sqrt{3} + 2\sqrt{3} = 3\sqrt{3}. \qquad ■$$

Self Check Simplify **a.** $\left(3^{\sqrt{2}}\right)^{\sqrt{8}}$ and **b.** $b^{\sqrt{2}} \cdot b^{\sqrt{18}}$.

Answers **a.** 81, **b.** $b^{4\sqrt{2}}$

■ EXPONENTIAL FUNCTIONS

If $b > 0$ and $b \neq 1$, the function $y = f(x) = b^x$ is an **exponential function.** Since x can be any real number, its domain is the set of real numbers. This is the interval $(-\infty, \infty)$. Since b is positive, the value of $f(x)$ is positive, and the range is the set of positive numbers. This is the interval $(0, \infty)$.

Since $b \neq 1$, an exponential function cannot be the constant function $y = f(x) = 1^x$, in which $f(x) = 1$ for every real number x.

> **Exponential Functions**
> An **exponential function with base b** is defined by the equation
> $$y = f(x) = b^x \quad (b > 0, b \neq 1, \text{ and } x \text{ is a real number})$$
> The **domain of any exponential function** is the interval $(-\infty, \infty)$. The **range** is the interval $(0, \infty)$.

■ GRAPHING EXPONENTIAL FUNCTIONS

Since the domain and range of $y = f(x) = b^x$ are subsets of real numbers, we can graph exponential functions on a rectangular coordinate system.

EXAMPLE 2   Graph $y = f(x) = 2^x$.

Solution   To graph $y = f(x) = 2^x$, we find several points (x, y) whose coordinates satisfy the equation, plot the points, and join them with a smooth curve, as shown in Figure 11-2.

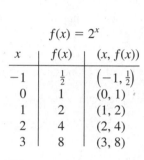

$f(x) = 2^x$

x	$f(x)$	$(x, f(x))$
-1	$\frac{1}{2}$	$\left(-1, \frac{1}{2}\right)$
0	1	$(0, 1)$
1	2	$(1, 2)$
2	4	$(2, 4)$
3	8	$(3, 8)$

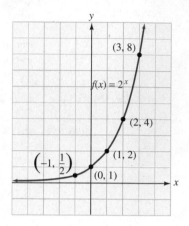

FIGURE 11-2

By looking at the graph, we can verify that the domain is the interval $(-\infty, \infty)$ and that the range is the interval $(0, \infty)$.

Note that as x decreases, the values of $f(x)$ decrease and approach 0. Thus, the x-axis is a horizontal asymptote of the graph.

Also note that the graph of $f(x) = 2^x$ passes through the points $(0, 1)$ and $(1, 2)$. ■

Self Check   Graph $y = f(x) = 4^x$.

Answer

EXAMPLE 3   Graph $y = f(x) = \left(\dfrac{1}{2}\right)^x$.

Solution   We find and plot pairs (x, y) that satisfy each equation. The graph of $y = f(x) = \left(\frac{1}{2}\right)^x$ appears in Figure 11-3.

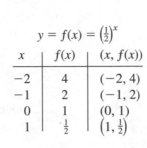

$$y = f(x) = \left(\tfrac{1}{2}\right)^x$$

x	$f(x)$	$(x, f(x))$
-2	4	$(-2, 4)$
-1	2	$(-1, 2)$
0	1	$(0, 1)$
1	$\frac{1}{2}$	$\left(1, \frac{1}{2}\right)$

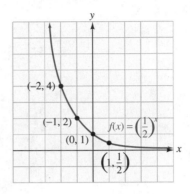

FIGURE 11-3

By looking at the graph, we can verify that the domain is the interval $(-\infty, \infty)$ and that the range is the interval $(0, \infty)$.

In this case, as x increases, the values of $f(x)$ decrease and approach 0. The x-axis is a horizontal asymptote. Note that the graph of $f(x) = \left(\frac{1}{2}\right)^x$ passes through the points $(0, 1)$ and $\left(1, \frac{1}{2}\right)$. ∎

Self Check   Graph $y = f(x) = \left(\frac{1}{4}\right)^x$.

Answer

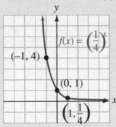

Examples 2 and 3 illustrate the following general properties of exponential functions.

Properties of Exponential Functions
The **domain** of the exponential function $y = f(x) = b^x$ is the interval $(-\infty, \infty)$.
The **range** is the interval $(0, \infty)$.
The graph has a y-intercept of $(0, 1)$.
The x-axis is an asymptote of the graph.
The graph of $y = f(x) = b^x$ passes through the point $(1, b)$.

EXAMPLE 4   From the graph of $y = f(x) = b^x$ shown in Figure 11-4, find the value of b.

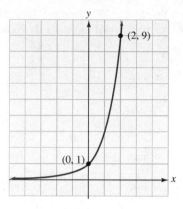

FIGURE 11-4

Solution   We first note that the graph passes through $(0, 1)$. Since the point $(2, 9)$ is on the graph, we substitute 9 for y and 2 for x in the equation $y = b^x$ to get

$$y = b^x$$
$$9 = b^2$$
$$3 = b \qquad \text{Take the positive square root of both sides.}$$

The base b is 3.   ■

Self Check   Is this the graph of an exponential function?

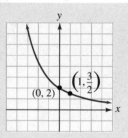

Answer   No; it doesn't pass through the point $(0, 1)$.

In Example 2 (where $b = 2$), the values of y increase as the values of x increase. Since the graph rises as we move to the right, we call the function an *increasing function*. When $b > 1$, the larger the value of b, the steeper the curve.

In Example 3 $\left(\text{where } b = \frac{1}{2}\right)$, the values of y decrease as the values of x increase. Since the graph drops as we move to the right, we call the function a *decreasing function*. When $0 < b < 1$, the smaller the value of b, the steeper the curve.

In general, the following is true.

Increasing and Decreasing Functions

If $b > 1$, then $f(x) = b^x$ is an **increasing function.**

If $0 < b < 1$, then $f(x) = b^x$ is a **decreasing function.**

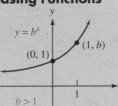

Increasing function

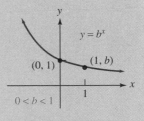

Decreasing function

An exponential function with base b is either increasing (for $b > 1$) or decreasing ($0 < b < 1$). Since different real numbers x determine different values of b^x, exponential functions are one-to-one.

The exponential function defined by

$$y = f(x) = b^x \quad \text{where } b > 0 \text{ and } b \neq 1$$

is one-to-one. Thus,

1. If $b^r = b^s$, then $r = s$.
2. If $r \neq s$, then $b^r \neq b^s$.

■ ■ ■ ■ ■ ■ ■ ■ ■ ■   **Graphing Exponential Functions**

GRAPHING
CALCULATORS

To use a graphing calculator to graph $y = f(x) = \left(\frac{2}{3}\right)^x$ and $y = f(x) = \left(\frac{3}{2}\right)^x$, we enter the right-hand sides of the equations. The screen will show the following equations.

$\backslash Y_1 = (2/3) \wedge X$

$\backslash Y_2 = (3/2) \wedge X$

If we use window settings of $[-10, 10]$ for x and $[-2, 10]$ for y and press the GRAPH key, we will obtain the graph shown in Figure 11-5.

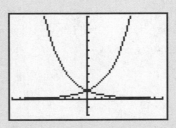

FIGURE 11-5

The graph of $y = f(x) = \left(\frac{2}{3}\right)^x$ passes through the points $(0, 1)$ and $\left(1, \frac{2}{3}\right)$. Since $\frac{2}{3} < 1$, the function is decreasing.

The graph of $y = f(x) = \left(\frac{3}{2}\right)^x$ passes through the points $(0, 1)$ and $\left(1, \frac{3}{2}\right)$. Since $\frac{3}{2} > 1$, the function is increasing.

Since both graphs pass the horizontal line test, each function is one-to-one.

■ VERTICAL AND HORIZONTAL TRANSLATIONS

We have seen that when $k > 0$ the graph of

$y = f(x) + k$ is the graph of $y = f(x)$ translated k units upward.

$y = f(x) - k$ is the graph of $y = f(x)$ translated k units downward.

$y = f(x - k)$ is the graph of $y = f(x)$ translated k units to the right.

$y = f(x + k)$ is the graph of $y = f(x)$ translated k units to the left.

EXAMPLE 5   On one set of axes, graph $y = f(x) = 2^x$ and $y = f(x) = 2^x + 3$.

Solution   The graph of $y = f(x) = 2^x + 3$ is identical to the graph of $y = f(x) = 2^x$, except that it is translated 3 units upward. (See Figure 11-6.)

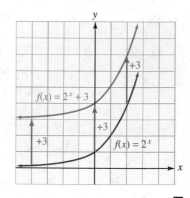

$y = f(x) = 2^x$		
x	y	(x, y)
-4	$\frac{1}{16}$	$\left(-4, \frac{1}{16}\right)$
0	1	$(0, 1)$
2	4	$(2, 4)$

$y = f(x) = 2^x + 3$		
x	y	(x, y)
-4	$3\frac{1}{16}$	$\left(-4, 3\frac{1}{16}\right)$
0	4	$(0, 4)$
2	7	$(2, 7)$

FIGURE 11-6

Self Check   Graph $y = f(x) = 4^x$ and $y = f(x) = 4^x - 3$.

Answer

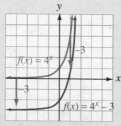

EXAMPLE 6   On one set of axes, graph $y = f(x) = 2^x$ and $y = f(x) = 2^{x+3}$.

Solution   The graph of $y = 2^{x+3}$ is identical to the graph of $y = f(x) = 2^x$, except that it is translated 3 units to the left. (See Figure 11-7.)

$y = f(x) = 2^x$				$y = f(x) = 2^{x+3}$		
x	y	(x, y)		x	y	(x, y)
0	1	(0, 1)		0	8	(0, 8)
2	4	(2, 4)		−1	4	(−1, 4)
4	16	(4, 16)		−2	2	(−2, 2)

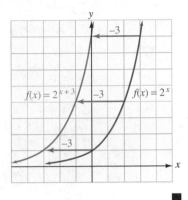

FIGURE 11-7

Self Check   On one set of axes, graph $y = f(x) = 4^x$ and $y = f(x) = 4^{x-3}$.

Answer

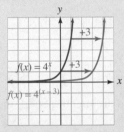

The graphs of $y = f(x) = kb^x$ and $y = f(x) = b^{kx}$ are vertical and horizontal stretchings of the graph of $y = f(x) = b^x$. To graph these functions, we can plot several points and join them with a smooth curve or use a graphing calculator.

Graphing Exponential Functions

GRAPHING CALCULATORS

To use a graphing calculator to graph the exponential function $y = f(x) = 3(2^{\frac{x}{3}})$, we enter the right-hand side of the equation. The display will show the equation

$$\backslash Y_1 = 3(2 \wedge (X/3))$$

If we use window settings of $[-10, 10]$ for x and $[-2, 18]$ for y and press the GRAPH key, we will obtain the graph shown in Figure 11-8.

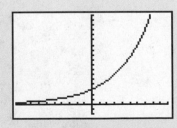

FIGURE 11-8

■ COMPOUND INTEREST

If we deposit $\$P$ in an account paying an annual simple interest rate r, we can find the amount A in the account at the end of t years by using the formula $A = P + Prt$, or $A = P(1 + rt)$.

Suppose that we deposit $\$500$ in an account that pays interest every six months. Then $P = 500$, and after six months $\left(\frac{1}{2}\text{ year}\right)$, the amount in the account will be

$$A = 500(1 + rt)$$
$$= 500\left(1 + r \cdot \frac{1}{2}\right) \qquad \text{Substitute } \tfrac{1}{2} \text{ for } t.$$
$$= 500\left(1 + \frac{r}{2}\right)$$

The account will begin the second six-month period with a value of $\$500\left(1 + \frac{r}{2}\right)$. After the second six-month period, the amount will be

$$A = P(1 + rt)$$
$$A = \left[500\left(1 + \frac{r}{2}\right)\right]\left(1 + r \cdot \frac{1}{2}\right) \qquad \text{Substitute } 500\left(1 + \tfrac{r}{2}\right) \text{ for } P \text{ and } \tfrac{1}{2} \text{ for } t.$$
$$= 500\left(1 + \frac{r}{2}\right)\left(1 + \frac{r}{2}\right)$$
$$= 500\left(1 + \frac{r}{2}\right)^2$$

At the end of a third six-month period, the amount in the account will be

$$A = 500\left(1 + \frac{r}{2}\right)^3$$

In this discussion, the earned interest is deposited back in the account and also earns interest. When this is the case, we say that the account is earning **compound interest.**

> **Formula for Compound Interest**
> If $\$P$ is deposited in an account and interest is paid k times a year at an annual rate r, the amount A in the account after t years is given by
> $$A = P\left(1 + \frac{r}{k}\right)^{kt}$$

EXAMPLE 7   **Saving for college**   To save for college, parents invest $\$12,000$ for their newborn child in a mutual fund that should average a 10% annual return. If the quarterly interest is reinvested, how much will be available in 18 years?

Solution   We substitute 12,000 for P, 0.10 for r, and 18 for t into the formula for compound interest and find A. Since interest is paid quarterly, $k = 4$.

$$A = P\left(1 + \frac{r}{k}\right)^{kt}$$

$$A = 12{,}000\left(1 + \frac{0.10}{4}\right)^{4(18)}$$

$$= 12{,}000(1 + 0.025)^{72}$$

$$= 12{,}000(1.025)^{72}$$

$$= 71{,}006.74 \qquad \text{Use a calculator, and press these keys:}$$

1.025 y^x 72 $=$ $\times$ 12,000 $=$.

In 18 years, the account will be worth $71,006.74. ■

Self Check

Answer

In Example 7, how much would be available if the parents invest $20,000?

$118,344.56

In business applications, the initial amount of money deposited is often called the **present value** (PV). The amount to which the money will grow is called the **future value** (FV). The interest rate used for each compounding period is the **periodic interest rate** (i), and the number of times interest is compounded is the number of **compounding periods** (n). Using these definitions, we have an alternate formula for compound interest.

Formula for Compound Interest

$$FV = PV(1 + i)^n$$

This alternate formula appears on business calculators. To use this formula to solve Example 7, we proceed as follows:

$$FV = PV(1 + i)^n$$

$$FV = 12{,}000(1 + 0.025)^{72} \qquad i = \tfrac{0.10}{4} = 0.025 \text{ and } n = 4(18) = 72.$$

$$= 71{,}006.74$$

■ ■ ■ ■ ■ ■ ■ ■ ■   **Solving Investment Problems**

GRAPHING
CALCULATORS

Suppose $1 is deposited in an account earning 6% annual interest, compounded monthly. To use a graphing calculator to estimate how much money will be in the account in 100 years, we can substitute 1 for P, 0.06 for r, and 12 for k into the formula

$$A = P\left(1 + \frac{r}{k}\right)^{kt}$$

$$A = 1\left(1 + \frac{0.06}{12}\right)^{12t}$$

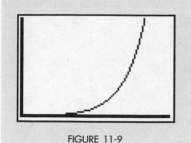

FIGURE 11-9

and simplify to get

$$A = (1.005)^{12t}$$

We now graph $A = (1.005)^{12t}$ using window settings of $[0, 120]$ for t and $[0, 400]$ for A to obtain the graph shown in Figure 11-9. We can then trace and zoom to estimate that $1 grows to be approximately $397 in 100 years. From the graph, we can see that the money grows slowly in the early years and rapidly in the later years.

Orals *If $x = 2$, evaluate each expression.*

1. 2^x **2.** 5^x **3.** $2(3^x)$ **4.** 3^{x-1}

If $x = -2$, evaluate each expression.

5. 2^x **6.** 5^x **7.** $2(3^x)$ **8.** 3^{x-1}

E X E R C I S E   11.1

REVIEW *In Illustration 1, lines r and s are parallel.*

1. Find x.

2. Find the measure of $\angle 1$.

3. Find the measure of $\angle 2$.

4. Find the measure of $\angle 3$.

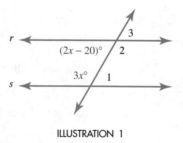

ILLUSTRATION 1

VOCABULARY AND CONCEPTS *Fill in each blank to make a true statement.*

5. If $b > 0$ and $b \neq 1$, $y = f(x) = b^x$ is called an _____ function.

6. The _____ of an exponential function is $(-\infty, \infty)$.

7. The range of an exponential function is the interval _____

8. The graph of $y = f(x) = 3^x$ passes through the points $(0, \_\_)$ and $(1, \_\_)$.

9. If $b > 1$, then $y = f(x) = b^x$ is an _____ function.

10. If $0 < b < 1$, then $y = f(x) = b^x$ is a _____ function.

11. The formula for compound interest is $A = $ _____.

12. An alternate formula for compound interest is $FV = $ _____.

PRACTICE   *In Exercises 13–16, find each value to four decimal places.*

13. $2^{\sqrt{2}}$

14. $7^{\sqrt{2}}$

15. $5^{\sqrt{5}}$

16. $6^{\sqrt{3}}$

In Exercises 17–20, simplify each expression.

17. $\left(2^{\sqrt{3}}\right)^{\sqrt{3}}$

18. $3^{\sqrt{2}}3^{\sqrt{18}}$

19. $7^{\sqrt{3}}7^{\sqrt{12}}$

20. $\left(3^{\sqrt{5}}\right)^{\sqrt{5}}$

In Exercises 21–28, graph each exponential function. Check your work with a graphing calculator.

21. $y = f(x) = 3^x$

22. $y = f(x) = 5^x$

23. $y = f(x) = \left(\dfrac{1}{3}\right)^x$

24. $y = f(x) = \left(\dfrac{1}{5}\right)^x$

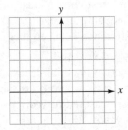 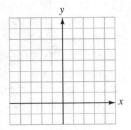 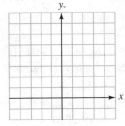 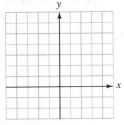

25. $y = f(x) = 3^x - 2$

26. $y = f(x) = 2^x + 1$

27. $y = f(x) = 3^{x-1}$

28. $y = f(x) = 2^{x+1}$

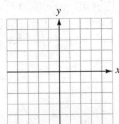 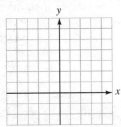 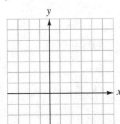 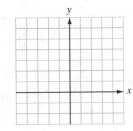

In Exercises 29–36, find the value of b, if any, that would cause the graph of $y = b^x$ to look like the graph indicated.

29.

30.

31.

32.

33.

34.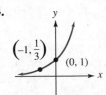

35.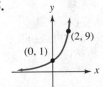

36.

In Exercises 37–40, use a graphing calculator to graph each function. Tell whether the function is an increasing or a decreasing function.

37. $y = f(x) = \dfrac{1}{2}\left(3^{x/2}\right)$ **38.** $y = f(x) = -3\left(2^{x/3}\right)$ **39.** $y = f(x) = 2\left(3^{-x/2}\right)$ **40.** $y = f(x) = -\dfrac{1}{4}(2^{-x/2})$

APPLICATIONS   *In Exercises 41–46, assume that there are no deposits or withdrawals.*

41. Compound interest   An initial deposit of $10,000 earns 8% interest, compounded quarterly. How much will be in the account after 10 years?

42. Compound interest   An initial deposit of $10,000 earns 8% interest, compounded monthly. How much will be in the account after 10 years?

43. Comparing interest rates   How much more interest could $1,000 earn in 5 years, compounded quarterly, if the annual interest rate were $5\frac{1}{2}$% instead of 5%?

44. Comparing savings plans   Which institution in Illustration 2 provides the better investment?

Fidelity Savings & Loan
Earn 5.25%
compounded monthly

Union Trust
Money Market Account
paying 5.35%,
compounded annually

ILLUSTRATION 2

45. Compound interest   If $1 had been invested on July 4, 1776, at 5% interest, compounded annually, what would it be worth on July 4, 2076?

46. Frequency of compounding $10,000 is invested in each of two accounts, both paying 6% annual interest. In the first account, interest compounds quarterly, and in the second account, interest compounds daily. Find the difference between the accounts after 20 years.

47. Radioactive decay   A radioactive material decays according to the formula $A = A_0\left(\frac{2}{3}\right)^t$, where A_0 is the initial amount present and t is measured in years. Find the amount present in 5 years.

48. Bacteria cultures   A colony of 6 million bacteria is growing in a culture medium. (See Illustration 3.) The population P after t hours is given by the formula $P = (6 \times 10^6)(2.3)^t$. Find the population after 4 hours.

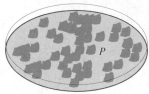

12:00 P.M.   4:00 P.M.

ILLUSTRATION 3

49. Discharging a battery   The charge remaining in a battery decreases as the battery discharges. The charge C (in coulombs) after t days is given by the formula $C = (3 \times 10^{-4})(0.7)^t$. Find the charge after 5 days.

50. Town population   The population of North Rivers is decreasing exponentially according to the formula $P = 3,745(0.93)^t$, where t is measured in years from the present date. Find the population in 6 years, 9 months.

51. Salvage value A small business purchases a computer for $4,700. It is expected that its value each year will be 75% of its value in the preceding year. If the business disposes of the computer after 5 years, find its salvage value (the value after 5 years).

52. Louisiana Purchase In 1803, the United States acquired territory from France in the Louisiana Purchase. The country doubled its territory by adding 827,000 square miles of land for $15 million. If the land has appreciated at the rate of 6% each year, what would one square mile of land be worth in 1996?

WRITING

53. If world population is increasing exponentially, why is there cause for concern?

54. How do the graphs of $y = b^x$ differ when $b > 1$ and $0 < b < 1$?

SOMETHING TO THINK ABOUT

55. In the definition of the exponential function, b could not equal 0. Why not?

56. In the definition of the exponential function, b could not be negative. Why not?

11.2 Base-e Exponential Functions

■ CONTINUOUS COMPOUND INTEREST ■ GRAPHING THE EXPONENTIAL FUNCTION ■ VERTICAL AND HORIZONTAL TRANSLATIONS ■ MALTHUSIAN POPULATION GROWTH ■ THE MALTHUSIAN THEORY

Getting Ready *Evaluate $\left(1 + \frac{1}{n}\right)^n$ for the following values. Round each answer to the nearest hundredth.*

1. $n = 1$ **2.** $n = 2$ **3.** $n = 4$ **4.** $n = 10$

■ CONTINUOUS COMPOUND INTEREST

If a bank pays interest twice a year, we say that interest is compounded semiannually. If it pays interest four times a year, we say that interest is compounded quarterly. If it pays interest continuously (infinitely many times in a year), we say that interest is compounded continuously.

To develop the formula for continuous compound interest, we start with the formula

$$A = P\left(1 + \frac{r}{k}\right)^{kt}$$ The formula for compound interest.

and substitute rn for k. Since r and k are positive numbers, so is n.

$$A = P\left(1 + \frac{r}{rn}\right)^{rnt}$$

We can then simplify the fraction $\frac{r}{rn}$ and use the commutative property of multiplication to change the order of the exponents.

$$A = P\left(1 + \frac{1}{n}\right)^{nrt}$$

Finally, we can use a property of exponents to write this formula as

1. $A = P\left[\left(1 + \frac{1}{n}\right)^{n}\right]^{rt}$ Use the property $a^{mn} = (a^{m})^{n}$.

To find the value of $\left(1 + \frac{1}{n}\right)^{n}$, we evaluate it for several values of n, as shown in Table 11-1.

n	$\left(1 + \dfrac{1}{n}\right)^{n}$
1	2
2	2.25
4	2.44140625. . .
12	2.61303529. . .
365	2.71456748. . .
1,000	2.71692393. . .
100,000	2.71826823. . .
1,000,000	2.71828046. . .

TABLE 11-1

The results suggest that as n gets larger, the value of $\left(1 + \frac{1}{n}\right)^{n}$ approaches the number 2.71828. . . . . This number is called e, which has the following value.

$e = 2.718281828459. . .$

In continuous compound interest, k (the number of compoundings) is infinitely large. Since k, r and n are all positive and $k = rn$, as k gets very large (approaches infinity), then so does n. Therefore, we can replace $\left(1 + \frac{1}{n}\right)^{n}$ in Equation 1 with e to get

$A = Pe^{rt}$

Formula for Exponential Growth

If a quantity P increases or decreases at an annual rate r, compounded continuously, then the amount A after t years is given by

$A = Pe^{rt}$

If time is measured in years, then r is called the *annual growth rate*. If r is negative, the "growth" represents a decrease.

To compute the amount to which $12,000 will grow if invested for 18 years at 10% annual interest, compounded continuously, we substitute 12,000 for P, 0.10 for r, and 18 for t in the formula for exponential growth:

$$A = Pe^{rt}$$

$$A = 12{,}000e^{0.10(18)}$$

$$= 12{,}000e^{1.8}$$

$$\approx 72{,}595.76957 \qquad \text{Use a scientific calculator and press these keys:}$$

$$1.8 \;\boxed{e^x}\; \boxed{\times}\; 12{,}000 \;\boxed{=}\;.$$

After 18 years, the account will contain $72,595.77. This is $1,589.03 more than the result in Example 7 in the previous section, where interest was compounded quarterly.

EXAMPLE 1   If $25,000 accumulates interest at an annual rate of 8%, compound continuously, find the balance in the account in 50 years.

Solution   We substitute 25,000 for P, 0.08 for r, and 50 for t.

$$A = Pe^{rt}$$

$$A = 25{,}000e^{(0.08)(50)}$$

$$= 25{,}000e^4$$

$$\approx 1{,}364{,}953.751 \qquad \text{Use a calculator.}$$

In 50 years, the balance will be $1,364,953.75—over one million dollars.   ■

Self Check   In Example 1, find the balance in 60 years.

Answer $3,037,760.44

The exponential function $y = f(x) = e^x$ is so important that it is often called **the exponential function.**

■ GRAPHING THE EXPONENTIAL FUNCTION

To graph the exponential function, we plot several points and join them with a smooth curve, as shown in Figure 11-10.

$$y = f(x) = e^x$$

x	$f(x)$	$(x, f(x))$
-2	0.1	$(-2, 0.1)$
-1	0.4	$(-1, 0.4)$
0	1	$(0, 1)$
1	2.7	$(1, 2.7)$
2	7.4	$(2, 7.4)$

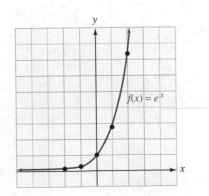

FIGURE 11-10

■ VERTICAL AND HORIZONTAL TRANSLATIONS

■ ■ ■ ■ ■ ■ ■ ■ ■   **Translations of the Exponential Function**

GRAPHING
CALCULATORS

Figure 11-11(a) shows the graphs of $y = f(x) = e^x$, $y = f(x) = e^x + 5$, and $y = f(x) = e^x - 3$. To graph these functions with window settings of $[-3, 6]$ for x and $[-5, 15]$ for y, we enter the right-hand sides of the equations after the symbols $Y_1 =$, $Y_2 =$, and $Y_3 =$. The display will show

$$Y_1 = e \char`\^ (x)$$
$$Y_2 = e \char`\^ (x) + 5$$
$$Y_3 = e \char`\^ (x) - 3$$

After graphing these functions, we can see that the graph of $y = f(x) = e^x + 5$ is 5 units above the graph of $y = f(x) = e^x$, and that the graph of $y = f(x) = e^x - 3$ is 3 units below the graph of $y = f(x) = e^x$.

Figure 11-11(b) shows the calculator graphs of $y = f(x) = e^x$, $y = f(x) = e^{x+5}$, and $y = f(x) = e^{x-3}$. To graph these functions with window settings of $[-7, 10]$ for x and $[-5, 15]$ for y, we enter the right-hand sides of the equations after the symbols $Y_1 =$, $Y_2 =$, $Y_3 =$. The display will show

$$Y_1 = e \char`\^ (x)$$
$$Y_2 = e \char`\^ (x + 5)$$
$$Y_3 = e \char`\^ (x - 3)$$

After graphing these functions, we can see that the graph of $y = f(x) = e^{x+5}$ is 5 units to the left of the graph of $y = f(x) = e^x$, and that the graph of $y = f(x) = e^{x-3}$ is 3 units to the right of the graph of $y = f(x) = e^x$.

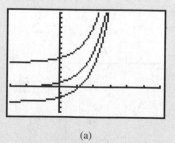   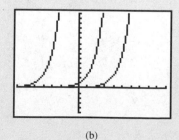

(a)   (b)

FIGURE 11-11

■ ■ ■ ■ ■ ■ ■ ■ ■ ■   **Graphing Exponential Functions**

GRAPHING
CALCULATORS

Figure 11-12 shows the calculator graph $y = f(x) = 3e^{-x/2}$. To graph this function with window settings of $[-7, 10]$ for x and $[-5, 15]$ for y, we enter the right-hand side of the equation after the symbol $Y_1 = $. The display will show the equation

$$Y_1 = 3(e \ \hat{} \ (-x/2))$$

The calculator graph appears in Figure 11-12. Explain why the graph has a y-intercept of $(0, 3)$.

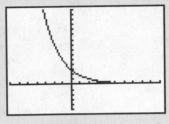

FIGURE 11-12

■ MALTHUSIAN POPULATION GROWTH

An equation based on the exponential function provides a model for **population growth.** In the **Malthusian model for population growth,** the future population of a colony is related to the present population by the formula $A = Pe^{rt}$.

EXAMPLE 2

City planning   The population of a city is currently 15,000, but changing economic conditions are causing the population to decrease by 2% each year. If this trend continues, find the population in 30 years.

Solution

Since the population is decreasing by 2% each year, the annual growth rate is -2%, or -0.02. We can substitute -0.02 for r, 30 for t, and 15,000 for P in the formula for exponential growth and find A.

$$A = Pe^{rt}$$
$$A = 15{,}000e^{-0.02(30)}$$
$$= 15{,}000e^{-0.6}$$
$$= 8{,}232.174541$$

In 30 years, city planners expect a population of approximately 8,232 persons.   ■

Self Check

In Example 2, find the population in 50 years.

Answer   5,518

■ THE MALTHUSIAN THEORY

The English economist Thomas Robert Malthus (1766–1834) pioneered in population study. He believed that poverty and starvation were unavoidable, because the human population tends to grow exponentially, but the food supply tends to grow linearly.

EXAMPLE 3   Suppose that a country with a population of 1,000 people is growing exponentially according to the formula

$$P = 1,000e^{0.02t}$$

where t is in years. Furthermore, assume that the food supply measured in adequate food per day per person is growing linearly according to the formula

$$y = 30.625x + 2,000$$

where x is in years. In how many years will the population outstrip the food supply?

Solution   We can use a graphing calculator, with window settings of [0, 100] for x and [0, 10,000] for y. After graphing the functions, we obtain Figure 11-13(a). If we trace, as in Figure 11-13(b), we can find the point where the two graphs intersect. From the graph, we can see that the food supply will be adequate for about 71 years. At that time, the population of approximately 4,200 people will begin to have problems.

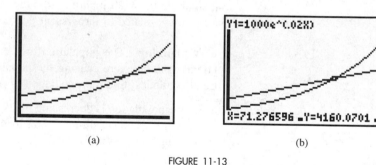

(a)   (b)

FIGURE 11-13   ■

Self Check   In Example 3, suppose that the population grows at a 3% rate. For how many years will the food supply be adequate?

Answer   about 38 years

Orals   *Use a calculator to find each value to the nearest hundredth.*

1. e^0 2. e^1 3. e^2 4. e^3

Fill in each blank to make a true statement.

5. The graph of $y = f(x) = e^x + 2$ is ___ units above the graph of $y = f(x) = e^x$.

6. The graph of $y = f(x) = e^{(x-2)}$ is ___ units to the right of the graph of $y = f(x) = e^x$.

EXERCISE 11.2

REVIEW   *Simplify each expression. Assume that all variables represent positive numbers.*

1. $\sqrt{240x^5}$

2. $\sqrt[3]{-125x^5y^4}$

3. $4\sqrt{48y^3} - 3y\sqrt{12y}$

4. $\sqrt[4]{48z^5} + \sqrt[4]{768z^5}$

VOCABULARY AND CONCEPTS   *Fill in each blank to make a true statement.*

5. To two decimal places, the value of *e* is _____.

6. The formula for continuous compound interest is $A = $ ____.

7. Since $e > 1$, the base-*e* exponential function is a(n) _____ function.

8. The graph of the exponential function $y = e^x$ passes through the points $(0, 1)$ and _____.

9. The Malthusian population growth formula is _____.

10. The Malthusian prediction is pessimistic, because _____ grows exponentially, but food supplies grow _____.

PRACTICE   *In Exercises 11–18, graph each function. Check your work with a graphing calculator. Compare each graph to the graph of $y = e^x$.*

11. $y = e^x + 1$

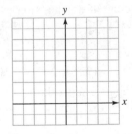

12. $y = e^x - 2$

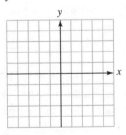

13. $y = e^{(x+3)}$

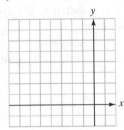

14. $y = e^{(x-5)}$

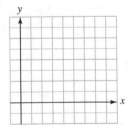

15. $y = -e^x$

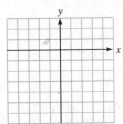

16. $y = -e^x + 1$

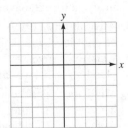

17. $y = 2e^x$

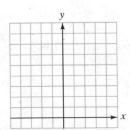

18. $y = \frac{1}{2}e^x$

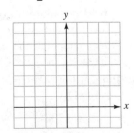

In Exercises 19–22, tell whether the graph of $y = e^x$ could look like the graph shown here.

19.

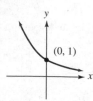
(0, 1)

20.

(1, 0)

21.

(0, 2)

22.

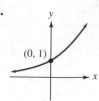
(0, 1)

APPLICATIONS *In Exercises 23–28, assume that there are no deposits or withdrawals.*

23. Continuous compound interest An investment of $5,000 earns 6% interest, compounded continuously. What will the investment be worth in 12 years?

24. Continuous compound interest An investment of $6,000 earns 7% interest, compounded continuously. What will the investment be worth in 35 years?

25. Determining the initial deposit An account now contains $12,000 and has been accumulating 7% annual interest, compounded continuously, for 9 years. Find the initial deposit.

26. Determining a previous balance An account now contains $8,000 and has been accumulating 8% annual interest, compounded continuously. How much was in the account 6 years ago?

27. Comparison of compounding methods An initial deposit of $5,000 grows at an annual rate of 8.5% for 5 years. Compare the final balances resulting from continuous compounding and annual compounding.

28. Comparison of compounding methods An initial deposit of $30,000 grows at an annual rate of 8% for 20 years. Compare the final balances resulting from continuous compounding and annual compounding.

 Solve each problem.

29. World population growth The population of the earth is approximately 6 billion people and is growing at an annual rate of 1.9%. Assuming a Malthusian growth model, find the world population in 30 years.

30. World population growth The population of the earth is approximately 6 billion people and is growing at an annual rate of 1.9%. Assuming a Malthusian growth model, find the world population in 40 years.

31. World population growth Assuming a Malthusian growth model and an annual growth rate of 1.9%, by what factor will the current population of the earth increase in 50 years? (See Exercise 29.)

32. Population growth The growth of a population is modeled by

$$P = 173e^{0.03t}$$

How large will the population be when $t = 30$?

33. Population decline The decline of a population is modeled by

$$P = 8,000e^{-0.008t}$$

How large will the population be when $t = 20$?

34. Epidemics The spread of ungulate fever through a herd of cattle can be modeled by the formula

$$P = P_0 e^{0.27t} \quad (t \text{ is in days})$$

If a rancher does not act quickly to treat two cases, how many cattle will have the disease in 10 days?

35. Alcohol absorption In one individual, the blood alcohol level t minutes after drinking two shots of whiskey is given by $P = 0.3(1 - e^{-0.05t})$. Find the blood alcohol level after 15 minutes.

36. Medicine The concentration, x, of a certain drug in an organ after t minutes is given by $x = 0.08(1 - e^{-0.1t})$. Find the concentration of the drug after 30 minutes.

37. Medicine Refer to Exercise 36. Find the initial concentration of the drug.

38. Skydiving Before her parachute opens, a skydiver's velocity v in meters per second is given by $v = 50(1 - e^{-0.2t})$. Find the initial velocity.

39. Skydiving Refer to Exercise 38 and find the velocity after 20 seconds.

40. Free-falling objects After t seconds, a certain falling object has a velocity v given by $v = 50(1 - e^{-0.3t})$. Which is falling faster after 2 seconds, this object or the skydiver in Exercise 38?

41. Depreciation A camping trailer originally purchased for $4,570 is continuously losing value at the rate of 6% per year. Find its value when it is $6\frac{1}{2}$ years old.

42. Depreciation A boat purchased for $7,500 has been continuously decreasing in value at the rate of 2% each year. It is now 8 years, 3 months old. Find its value.

In Exercises 43–44, use a graphing calculator to solve each problem.

43. In Example 3, suppose that better farming methods change the formula for food growth to $y = 31x + 2,000$. How long will the food supply be adequate?

44. In Example 3, suppose that a birth-control program changed the formula for population growth to $P = 1,000e^{0.01t}$. How long will the food supply be adequate?

WRITING

45. Explain why the graph of $y = f(x) = e^x - 5$ is 5 units below the graph of $y = f(x) = e^x$.

46. Explain why the graph of $y = f(x) = e^{(x+5)}$ is 5 units to the left of the graph of $y = f(x) = e^x$.

SOMETHING TO THINK ABOUT

47. The value of e can be calculated to any degree of accuracy by adding the first several terms of the following list:

$$1, 1, \frac{1}{2}, \frac{1}{2\cdot3}, \frac{1}{2\cdot3\cdot4}, \frac{1}{2\cdot3\cdot4\cdot5}, \dots$$

The more terms that are added, the closer the sum will be to e. Add the first six numbers in the preceding list. To how many decimal places is the sum accurate?

48. Graph the function defined by the equation

$$y = f(x) = \frac{e^x + e^{-x}}{2}$$

from $x = -2$ to $x = 2$. The graph will look like a parabola, but it is not. The graph, called a **catenary,** is important in the design of power distribution networks, because it represents the shape of a uniform flexible cable whose ends are suspended from the same height. The function is called the **hyperbolic cosine function.**

49. If $e^{t+5} = ke^t$, find k.

50. If $e^{5t} = k^t$, find k.

11.3 Logarithmic Functions

■ LOGARITHMS ■ GRAPHS OF LOGARITHMIC FUNCTIONS ■ VERTICAL AND HORIZONTAL TRANSLATIONS ■ BASE-10 LOGARITHMS ■ ELECTRONICS ■ SEISMOLOGY

Getting Ready *Find each value.*

1. 7^0 **2.** 5^2 **3.** 5^{-2} **4.** $16^{1/2}$

■ LOGARITHMS

Because an exponential function defined by $y = b^x$ is one-to-one, it has an inverse function that is defined by the equation $x = b^y$. To express this inverse function in the form $y = f^{-1}(x)$, we must solve the equation $x = b^y$ for y. For this, we need the following definition.

> **Logarithmic Functions**
> If $b > 0$ and $b \neq 1$, the **logarithmic function with base b** is defined by
>
> $\qquad y = \log_b x \quad$ if and only if $\quad x = b^y$
>
> The **domain of the logarithmic function** is the interval $(0, \infty)$. The **range** is the interval $(-\infty, \infty)$.

Since the function $y = \log_b x$ is the inverse of the one-to-one exponential function $y = b^x$, the logarithmic function is also one-to-one.

> **WARNING!**   Since the domain of the logarithmic function is the set of positive numbers, it is impossible to find the logarithm of 0 or the logarithm of a negative number.

The previous definition guarantees that any pair (x, y) that satisfies the equation $y = \log_b x$ also satisfies the equation $x = b^y$.

$\log_4 1 = 0$	because	$1 = 4^0$
$\log_5 25 = 2$	because	$25 = 5^2$
$\log_5 \dfrac{1}{25} = -2$	because	$\dfrac{1}{25} = 5^{-2}$
$\log_{16} 4 = \dfrac{1}{2}$	because	$4 = 16^{1/2}$
$\log_2 \dfrac{1}{8} = -3$	because	$\dfrac{1}{8} = 2^{-3}$
$\log_b x = y$	because	$x = b^y$

In each of these examples, the logarithm of a number is an exponent. In fact,

$\qquad$ *$\log_b x$ is the exponent to which b is raised to get x.*

In equation form, we write

$\qquad b^{\log_b x} = x$

EXAMPLE 1   Find y in each equation: **a.** $\log_6 1 = y$,   **b.** $\log_3 27 = y$,   and   **c.** $\log_5 \frac{1}{5} = y$.

Solution   **a.** We can change the equation $\log_6 1 = y$ into the equivalent exponential equation $6^y = 1$. Since $6^0 = 1$, it follows that $y = 0$. Thus,

$\qquad \log_6 1 = 0$

b. $\log_3 27 = y$ is equivalent to $3^y = 27$. Since $3^3 = 27$, it follows that $3^y = 3^3$, and $y = 3$. Thus,

$$\log_3 27 = 3$$

c. $\log_5 \frac{1}{5} = y$ is equivalent to $5^y = \frac{1}{5}$. Since $5^{-1} = \frac{1}{5}$, it follows that $5^y = 5^{-1}$, and $y = -1$. Thus,

$$\log_5 \frac{1}{5} = -1$$ ∎

Self Check Find y in each equation: **a.** $\log_3 9 = y$, **b.** $\log_2 64 = y$, and
c. $\log_5 \frac{1}{125} = y$.

Answers **a.** 2, **b.** 6, **c.** −3

 c

EXAMPLE 2 Find the value of x in each equation: **a.** $\log_3 81 = x$, **b.** $\log_x 125 = 3$, and
c. $\log_4 x = 3$.

Solution **a.** $\log_3 81 = x$ is equivalent to $3^x = 81$. Because $3^4 = 81$, it follows that $3^x = 3^4$. Thus, $x = 4$.

b. $\log_x 125 = 3$ is equivalent to $x^3 = 125$. Because $5^3 = 125$, it follows that $x^3 = 5^3$. Thus, $x = 5$.

c. $\log_4 x = 3$ is equivalent to $4^3 = x$. Because $4^3 = $ **64**, it follows that $x = 64$. ∎

Self Check Find x in each equation: **a.** $\log_2 32 = x$, **b.** $\log_x 8 = 3$, and
c. $\log_5 x = 2$.

Answers **a.** 5, **b.** 2, **c.** 25

EXAMPLE 3 Find the value of x in each equation: **a.** $\log_{1/3} x = 2$, **b.** $\log_{1/3} x = -2$, and
c. $\log_{1/3} \frac{1}{27} = x$.

Solution **a.** $\log_{1/3} x = 2$ is equivalent to $\left(\frac{1}{3}\right)^2 = x$. Thus, $x = \frac{1}{9}$.

b. $\log_{1/3} x = -2$ is equivalent to $\left(\frac{1}{3}\right)^{-2} = x$. Thus,

$$x = \left(\frac{1}{3}\right)^{-2} = 3^2 = 9$$

c. $\log_{1/3} \frac{1}{27} = x$ is equivalent to $\left(\frac{1}{3}\right)^x = \frac{1}{27}$. Because $\left(\frac{1}{3}\right)^3 = \frac{1}{27}$, it follows that $x = 3$. ∎

Self Check Find x in each equation: **a.** $\log_{1/4} x = 3$ and **b.** $\log_{1/4} x = -2$.

Answers **a.** $\frac{1}{64}$, **b.** 16

■ GRAPHS OF LOGARITHMIC FUNCTIONS

To graph the logarithmic function $y = \log_2 x$, we calculate and plot several points with coordinates (x, y) that satisfy the equation $x = 2^y$. After joining these points with a smooth curve, we have the graph shown in Figure 11-14(a).

To graph $y = \log_{1/2} x$, we calculate and plot several points with coordinates (x, y) that satisfy the equation $x = \left(\frac{1}{2}\right)^y$. After joining these points with a smooth curve, we have the graph shown in Figure 11-14(b).

$y = \log_2 x$

x	y	(x, y)
$\frac{1}{4}$	-2	$\left(\frac{1}{4}, -2\right)$
$\frac{1}{2}$	-1	$\left(\frac{1}{2}, -1\right)$
1	0	$(1, 0)$
2	1	$(2, 1)$
4	2	$(4, 2)$
8	3	$(8, 3)$

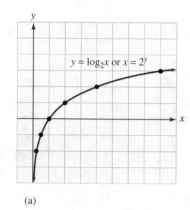

(a)

$y = \log_{1/2} x$

x	y	(x, y)
$\frac{1}{4}$	2	$\left(\frac{1}{4}, 2\right)$
$\frac{1}{2}$	1	$\left(\frac{1}{2}, 1\right)$
1	0	$(1, 0)$
2	-1	$(2, -1)$
4	-2	$(4, -2)$
8	-3	$(8, 3)$

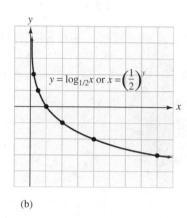

(b)

FIGURE 11-14

The graphs of all logarithmic functions are similar to those in Figure 11-15. If $b > 1$, the logarithmic function is increasing, as in Figure 11-15(a). If $0 < b < 1$, the logarithmic function is decreasing, as in Figure 11-15(b).

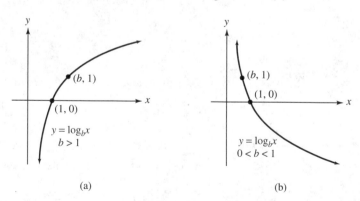

(a)   (b)

FIGURE 11-15

The graph of $y = f(x) = \log_b x$ has the following properties.

1. It passes through the point $(1, 0)$.
2. It passes through the point $(b, 1)$.
3. The y-axis is an asymptote.
4. The domain is $(0, \infty)$ and the range is $(-\infty, \infty)$.

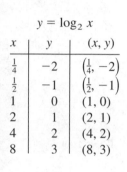

The exponential and logarithmic functions are inverses of each other and, therefore, have symmetry about the line $y = x$. The graphs of $y = \log_b x$ and $y = b^x$ are shown in Figure 11-16(a) when $b > 1$, and in Figure 11-16(b) when $0 < b < 1$.

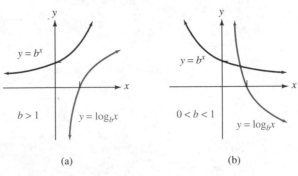

(a)   (b)

FIGURE 11-16

■ VERTICAL AND HORIZONTAL TRANSLATIONS

The graphs of many functions involving logarithms are translations of the basic logarithmic graphs.

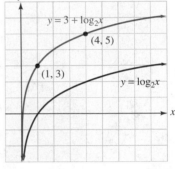

EXAMPLE 4   Graph the function defined by $y = 3 + \log_2 x$.

Solution   The graph of $y = 3 + \log_2 x$ is identical to the graph of $y = \log_2 x$, except that it is translated 3 units upward. (See Figure 11-17.)

FIGURE 11-17 ■

Self Check   Graph $y = \log_3 x - 2$.

Answer

EXAMPLE 5   Graph $y = \log_{1/2} (x - 1)$.

Solution   The graph of $y = \log_{1/2} (x - 1)$ is identical to the graph of $y = \log_{1/2} x$, except that it is translated 1 unit to the right. (See Figure 11-18.)

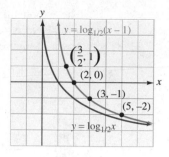

FIGURE 11-18

Self Check

Answer

Graph $y = \log_{1/3}(x + 2)$.

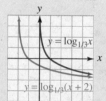

■ ■ ■ ■ ■ ■ ■ ■ ■ ■ **Graphing Logarithmic Functions**

GRAPHING
CALCULATORS

Graphing calculators can draw graphs of logarithmic functions directly if the base of the logarithmic function is 10 or e. To use a calculator to graph the logarithmic function $y = f(x) = -2 + \log_{10}\left(\frac{1}{2}x\right)$, we enter the right-hand side of the equation after the symbol $Y_1 =$. The display will show the equation

$$Y_1 = -2 + \log(1/2\text{*}x)$$

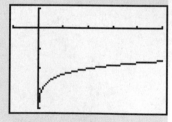

FIGURE 11-19

If we use window settings of $[-1, 5]$ for x and $[-4, 1]$ for y and press the GRAPH key, we will obtain the graph shown in Figure 11-19.

■ **BASE-10 LOGARITHMS**

For computational purposes and in many applications, we will use base-10 logarithms (also called **common logarithms**). When the base b is not indicated in the notation $\log x$, we assume that $b = 10$:

$\log x$ means $\log_{10} x$

Because base-10 logarithms appear so often, it is a good idea to become familiar with the following base-10 logarithms:

$$\log_{10} \frac{1}{100} = -2 \qquad \text{because} \qquad 10^{-2} = \frac{1}{100}$$

$$\log_{10} \frac{1}{10} = -1 \qquad \text{because} \qquad 10^{-1} = \frac{1}{10}$$

$$\log_{10} 1 = 0 \qquad \text{because} \qquad 10^{0} = 1$$

$$\log_{10} 10 = 1 \qquad \text{because} \qquad 10^{1} = 10$$

$$\log_{10} 100 = 2 \qquad \text{because} \qquad 10^{2} = 100$$

$$\log_{10} 1,000 = 3 \qquad \text{because} \qquad 10^{3} = 1,000$$

In general, we have

$$\log_{10} 10^x = x$$

■ ■ ■ ■ ■ ■ ■ ■ ■ ■ **Finding Logarithms**

CALCULATORS

Before calculators, extensive tables provided logarithms of numbers. Today, logarithms are easy to find with a calculator. For example, to find log 32.58 with a scientific calculator, we enter these numbers and press these keys:

32.58 LOG

The display will read `1.51295108`. To four decimal places, log 32.58 = 1.5130.

To use a graphing calculator, we enter these numbers and press these keys:

LOG 32.58 ENTER

The display will read `log (32.58`
`                              1.51295108`

EXAMPLE 6   Find x in the equation log $x = 0.3568$. Round to four decimal places.

Solution   The equation log $x = 0.3568$ is equivalent to $10^{0.3568} = x$. To find x with a scientific calculator, we enter these numbers and press these keys:

10 y^x .3568 =

The display will read `2.274049951`. To four decimal places,

$$x = 2.2740$$

If your calculator has a 10^x key, enter .3568 and press it to get the same result. ■

Self Check
Answer

Solve log $x = 2.7$. Round to four decimal places.

501.1872

■ ELECTRONICS

Common logarithms are used in electrical engineering to express the voltage gain (or loss) of an electronic device such as an amplifier. The unit of gain (or loss), called the **decibel,** is defined by a logarithmic relation.

Decibel Voltage Gain

If E_O is the output voltage of a device and E_I is the input voltage, the decibel voltage gain is given by

$$\text{db gain} = 20 \log \frac{E_O}{E_I}$$

EXAMPLE 7

Finding db gain   If the input to an amplifier is 0.4 volt and the output is 50 volts, find the decibel voltage gain of the amplifier.

Solution   We can find the decibel voltage gain by substituting 0.4 for E_I and 50 for E_O into the formula for db gain:

$$\text{db gain} = 20 \log \frac{E_O}{E_I}$$

$$\text{db gain} = 20 \log \frac{50}{0.4}$$

$$= 20 \log 125$$

$$\approx 42 \qquad \text{Use a calculator.}$$

The amplifier provides a 42-decibel voltage gain.   ■

■ SEISMOLOGY

In seismology, common logarithms are used to measure the intensity of earthquakes on the **Richter scale.** The intensity of an earthquake is given by the following logarithmic function.

Richter Scale

If R is the intensity of an earthquake, A is the amplitude (measured in micrometers), and P is the period (the time of one oscillation of the earth's surface, measured in seconds), then

$$R = \log \frac{A}{P}$$

EXAMPLE 8   **Measuring earthquakes**   Find the measure on the Richter scale of an earthquake with an amplitude of 10,000 micrometers (1 centimeter) and a period of 0.1 second.

Solution   We substitute 10,000 for A and 0.1 for P in the Richter scale formula and simplify:

$$R = \log \frac{A}{P}$$

$$R = \log \frac{10,000}{0.1}$$

$$= \log 100,000$$

$$= 5$$

The earthquake measures 5 on the Richter scale.   ∎

Orals   *Find the value of x in each equation.*

1. $\log_2 8 = x$ **2.** $\log_3 9 = x$ **3.** $\log_x 125 = 3$

4. $\log_x 8 = 3$ **5.** $\log_4 16 = x$ **6.** $\log_x 32 = 5$

7. $\log_{1/2} x = 2$ **8.** $\log_9 3 = x$ **9.** $\log_x \frac{1}{4} = -2$

EXERCISE 11.3

REVIEW   *Solve each equation.*

1. $\sqrt[3]{6x + 4} = 4$ **2.** $\sqrt{3x - 4} = \sqrt{-7x + 2}$

3. $\sqrt{a + 1} - 1 = 3a$ **4.** $3 - \sqrt{t - 3} = \sqrt{t}$

VOCABULARY AND CONCEPTS   *Fill in each blank to make a true statement.*

5. The equation $y = \log_b x$ is equivalent to _____.

6. The domain of the logarithmic function is the interval _____.

7. The _____ of the logarithmic function is the interval $(-\infty, \infty)$.

8. $b^{\log_b x} = $ __

9. Because an exponential function is one-to-one, it has an _____ function.

10. The inverse of an exponential function is called a _____ function.

11. $\log_b x$ is the _____ to which b is raised to get x.

12. The y-axis is an _____ to the graph of $y = f(x) = \log_b x$.

13. The graph of $y = f(x) = \log_b x$ passes through the points _____ and _____.

14. $\log_{10} 10^x = $ __.

15. db gain = _____

16. The intensity of an earthquake is measured by the formula $R = $ _____.

PRACTICE   *In Exercises 17–24, write each equation in exponential form.*

17. $\log_3 27 = 3$ **18.** $\log_8 8 = 1$ **19.** $\log_{1/2} \frac{1}{4} = 2$ **20.** $\log_{1/5} 1 = 0$

21. $\log_4 \dfrac{1}{64} = -3$ **22.** $\log_6 \dfrac{1}{36} = -2$

23. $\log_{1/2} \dfrac{1}{8} = 3$ **24.** $\log_{1/5} 1 = 0$

In Exercises 25–32, write each equation in logarithmic form.

25. $6^2 = 36$ **26.** $10^3 = 1{,}000$

27. $5^{-2} = \dfrac{1}{25}$ **28.** $3^{-3} = \dfrac{1}{27}$

29. $\left(\dfrac{1}{2}\right)^{-5} = 32$ **30.** $\left(\dfrac{1}{3}\right)^{-3} = 27$

31. $x^y = z$ **32.** $m^n = p$

In Exercises 33–72, find each value of x.

33. $\log_2 16 = x$ **34.** $\log_3 9 = x$ **35.** $\log_4 16 = x$ **36.** $\log_6 216 = x$

37. $\log_{1/2} \dfrac{1}{8} = x$ **38.** $\log_{1/3} \dfrac{1}{81} = x$ **39.** $\log_9 3 = x$ **40.** $\log_{125} 5 = x$

41. $\log_{1/2} 8 = x$ **42.** $\log_{1/2} 16 = x$ **43.** $\log_7 x = 2$ **44.** $\log_5 x = 0$

45. $\log_6 x = 1$ **46.** $\log_2 x = 4$ **47.** $\log_{25} x = \dfrac{1}{2}$ **48.** $\log_4 x = \dfrac{1}{2}$

49. $\log_5 x = -2$ **50.** $\log_3 x = -2$ **51.** $\log_{36} x = -\dfrac{1}{2}$ **52.** $\log_{27} x = -\dfrac{1}{3}$

53. $\log_{100} \dfrac{1}{1{,}000} = x$ **54.** $\log_{5/2} \dfrac{4}{25} = x$ **55.** $\log_{27} 9 = x$ **56.** $\log_{12} x = 0$

57. $\log_x 5^3 = 3$ **58.** $\log_x 5 = 1$ **59.** $\log_x \dfrac{9}{4} = 2$ **60.** $\log_x \dfrac{\sqrt{3}}{3} = \dfrac{1}{2}$

61. $\log_x \dfrac{1}{64} = -3$ **62.** $\log_x \dfrac{1}{100} = -2$ **63.** $\log_{2\sqrt{2}} x = 2$ **64.** $\log_4 8 = x$

65. $2^{\log_2 4} = x$ **66.** $3^{\log_3 5} = x$ **67.** $x^{\log_4 6} = 6$ **68.** $x^{\log_3 8} = 8$

69. $\log 10^3 = x$ **70.** $\log 10^{-2} = x$ **71.** $10^{\log x} = 100$ **72.** $10^{\log x} = \dfrac{1}{10}$

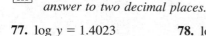 *In Exercises 73–76, use a calculator to find each value, if possible. Give answers to four decimal places.*

73. $\log 8.25$ **74.** $\log 0.77$ **75.** $\log 0.00867$ **76.** $\log 375.876$

In Exercises 77–84, use a calculator to find each value of y, if possible. If an answer is not exact, give the answer to two decimal places.

77. $\log y = 1.4023$ **78.** $\log y = 2.6490$ **79.** $\log y = 4.24$ **80.** $\log y = 0.926$

81. $\log y = -3.71$ **82.** $\log y = -0.28$ **83.** $\log y = \log 8$ **84.** $\log y = \log 7$

In Exercises 85–88, graph each function. Tell whether each function is an increasing or decreasing function.

85. $y = f(x) = \log_3 x$

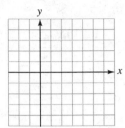

86. $y = f(x) = \log_{1/3} x$

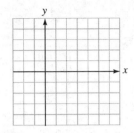

87. $y = f(x) = \log_{1/2} x$

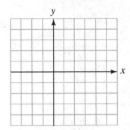

88. $y = f(x) = \log_4 x$

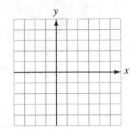

In Exercises 89–92, graph each function.

89. $y = f(x) = 3 + \log_3 x$

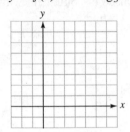

90. $y = f(x) = \log_{1/3} x - 1$

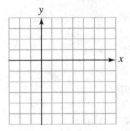

91. $y = f(x) = \log_{1/2} (x - 2)$

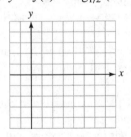

92. $y = f(x) = \log_4 (x + 2)$

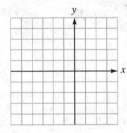

In Exercises 93–96, graph each pair of inverse functions on a single coordinate system.

93. $y = f(x) = 2^x$
$y = g(x) = \log_2 x$

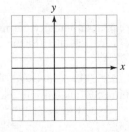

94. $y = f(x) = \left(\dfrac{1}{2}\right)^x$
$y = g(x) = \log_{1/2} x$

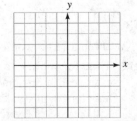

95. $y = f(x) = \left(\dfrac{1}{4}\right)^x$
$y = g(x) = \log_{1/4} x$

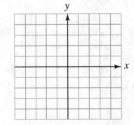

96. $y = f(x) = 4^x$
$y = g(x) = \log_4 x$

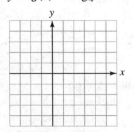

In Exercises 97–100, find the value of b, if any, that would cause the graph of f(x) = log_b x to look like the graph indicated.

97.

98.

99.

100.

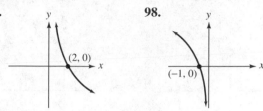

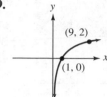

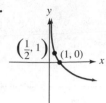

APPLICATIONS *In Exercises 101—112, if an answer is not exact, round to the nearest tenth.*

101. Finding the gain of an amplifier Find the db gain of an amplifier if input voltage is 0.71 volt when the output voltage is 20 volts.

102. Finding the gain of an amplifier Find the db gain of an amplifier if the output voltage is 2.8 volts when the input voltage is 0.05 volt.

103. db gain of an amplifier Find the db gain of the amplifier shown in Illustration 1.

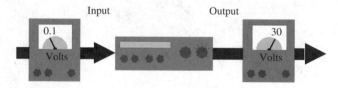

ILLUSTRATION 1

104. db gain of an amplifier An amplifier produces an output of 80 volts when driven by an input of 0.12 volts. Find the amplifier's db gain.

105. Earthquakes An earthquake has an amplitude of 5,000 micrometers and a period of 0.2 second. Find its measure on the Richter scale.

106. Earthquakes The period of an earthquake with amplitude of 80,000 micrometers is 0.08 second. Find its measure on the Richter scale.

107. Earthquakes An earthquake has a period of $\frac{1}{4}$ second and an amplitude of 2,500 micrometers. Find its measure on the Richter scale.

108. Earthquakes By what factor must the amplitude of an earthquake change to increase its severity by 1 point on the Richter scale? Assume that the period remains constant.

109. Depreciation Business equipment is often depreciated using the double declining-balance method. In this method, a piece of equipment with a life expectancy of N years, costing $\$C$, will depreciate to a value of $\$V$ in n years, where n is given by the formula

$$n = \frac{\log V - \log C}{\log \left(1 - \dfrac{2}{N} \right)}$$

A computer that cost \$17,000 has a life expectancy of 5 years. If it has depreciated to a value of \$2,000, how old is it?

110. Depreciation A printer worth \$470 when new had a life expectancy of 12 years. If it is now worth \$189, how old is it?

111. Time for money to grow If $\$P$ is invested at the end of each year in an annuity earning annual interest at a rate r, the amount in the account will be $\$A$ after n years, where

$$n = \frac{\log \left[\dfrac{Ar}{P} + 1 \right]}{\log (1 + r)}$$

If \$1,000 is invested each year in an annuity earning 12% annual interest, how long will it take for the account to be worth \$20,000?

112. Time for money to grow If \$5,000 is invested each year in an annuity earning 8% annual interest, how long will it take for the account to be worth \$50,000? (See Exercise 111.)

WRITING

113. Describe the appearance of the graph of $y = f(x) = \log_b x$ when $0 < b < 1$ and when $b > 1$.

114. Explain why it is impossible to find the logarithm of a negative number.

SOMETHING TO THINK ABOUT

115. Graph $y = f(x) = -\log_3 x$. How does the graph compare to the graph of $y = f(x) = \log_3 x$?

116. Find a logarithmic function that passes through the points $(1, 0)$ and $(5, 1)$.

11.4 Base-e Logarithms

■ BASE-e LOGARITHMS ■ GRAPHING BASE-e LOGARITHMS ■ DOUBLING TIME

Getting Ready   *Evaluate each expression.*

1. $\log_4 16$

2. $\log_2 \dfrac{1}{8}$

3. $\log_5 5$

4. $\log_7 1$

■ BASE-e LOGARITHMS

We have seen the importance of base-*e* exponential functions in mathematical models of events in nature. Base-*e* logarithms are just as important. They are called **natural logarithms** or **Napierian logarithms,** after John Napier (1550–1617), and are usually written as ln *x*, rather than $\log_e x$:

ln *x*   means $\log_e x$

As with all logarithmic functions, the domain of $y = f(x) = \ln x$ is the interval $(0, \infty)$, and the range is the interval $(-\infty, \infty)$.

To find the base-*e* logarithms of numbers, we can use a calculator.

■ ■ ■ ■ ■ ■ ■ ■ ■   **Evaluating Logarithms**

CALCULATORS   To use a scientific calculator to find the value of ln 9.87, we enter these numbers and press these keys:

9.87 LN

The display will read   `2.289499853` . To four decimal places, ln 9.87 = 2.2895.

To use a graphing calculator, we enter these numbers and press these keys:

LN 9.87 ENTER

The display will read ln 9.87
2.289499853

EXAMPLE 1 Use a calculator to find each value: **a.** ln 17.32 and **b.** ln (log 0.05).

Solution **a.** Enter these numbers and press these keys:

Scientific Calculator   *Graphing Calculator*
17.32 LN   LN 17.32 ENTER

Either way, the result is 2.851861903.

b. Enter these numbers and press these keys:

Scientific Calculator   *Graphing Calculator*
0.05 LOG LN   LN (LOG 0.05) ENTER

Either way, we obtain an error, because log 0.05 is a negative number, and we cannot take the logarithm of a negative number. ∎

Self Check Find each value to four decimal places: **a.** ln π and **b.** ln $\left(\log \dfrac{1}{2} \right)$.

Answers **a.** 1.1447, **b.** no value

EXAMPLE 2 Solve each equation: **a.** ln $x = 1.335$ and **b.** ln $x = \log 5.5$. Give each result to four decimal places.

Solution **a.** The equation ln $x = 1.335$ is equivalent to $e^{1.335} = x$. To use a scientific calculator to find x, enter these numbers and press these keys:

1.335 e^x

The display will read 3.799995946. To four decimal places,

$x = 3.8000$

b. The equation ln $x = \log 5.5$ is equivalent to $e^{\log 5.5} = x$. To use a scientific calculator to find x, press these keys:

5.5 LOG e^x

The display will read 2.096695826. To four decimal places,

$x = 2.0967$ ∎

Self Check	Solve **a.** $\ln x = 2.5437$ and **b.** $\log x = \ln 5$.
Answers	**a.** 12.7267, **b.** 40.6853

■ GRAPHING BASE-e LOGARITHMS

The equation $y = \ln x$ is equivalent to the equation $x = e^y$. To get the graph of $\ln x$, we can plot points that satisfy the equation $x = e^y$ and join them with a smooth curve, as shown in Figure 11-20(a). Figure 11-20(b) shows the calculator graph.

$y = \ln x$

x	y	(x, y)
$\dfrac{1}{e} \approx 0.4$	-1	$(0.4, -1)$
1	0	$(1, 0)$
$e \approx 2.7$	1	$(2.7, 1)$
$e^2 \approx 7.4$	2	$(7.4, 2)$

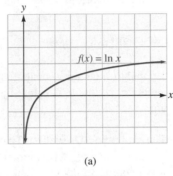

(a)

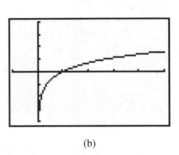

(b)

FIGURE 11-20

■ ■ ■ ■ ■ ■ ■ ■ ■ ■ **Graphing Logarithmic Functions**

GRAPHING
CALCULATORS

Many graphs of logarithmic functions involve translations of the graph of $y = f(x) = \ln x$. For example, Figure 11-21 shows calculator graphs of the functions $y = \ln x$, $y = \ln x + 2$, and $y = \ln x - 3$.

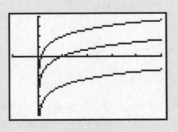

The graph of $y = \ln x + 2$ is 2 units above the graph of $y = \ln x$.

The graph of $y = \ln x - 3$ is 3 units below the graph of $y = \ln x$.

FIGURE 11-21

Figure 11-22 shows the calculator graphs of the functions $y = \ln x$, $y = \ln (x - 2)$, and $y = \ln (x + 3)$.

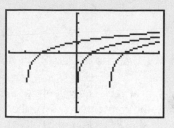

The graph of $y = \ln (x - 2)$ is 2 units to the right of the graph of $y = \ln x$.

The graph of $y = \ln (x + 3)$ is 3 units to the left of the graph of $y = \ln x$.

FIGURE 11-22

Base-e logarithms have many applications.

■ DOUBLING TIME

If a population grows exponentially at a certain annual rate, the time required for the population to double is called the **doubling time.** It is given by the following formula.

> **Formula for Doubling Time**
> If r is the annual rate (compounded continuously) and t is the time required for a population to double, then
> $$t = \frac{\ln 2}{r}$$

EXAMPLE 3   The population of the earth is growing at the approximate rate of 2% per year. If this rate continues, in how many years will the population double?

Solution   Because the population is growing at the rate of 2% per year, we substitute 0.02 for r into the formula for doubling time and simplify.

$$t = \frac{\ln 2}{r}$$

$$t = \frac{\ln 2}{0.02}$$

$$\approx 34.65735903$$

The population will double in about 35 years.   ■

Self Check   In Example 3, if the world population's annual growth rate could be reduced to 1.5% per year, what would be the doubling time?

Answer   46 years

EXAMPLE 4   **Doubling time**   How long will it take $1,000 to double at an annual rate of 8%, compounded continuously?

Solution   We substitute 0.08 for r and simplify:

$$t = \frac{\ln 2}{r}$$

$$t = \frac{\ln 2}{\mathbf{0.08}}$$

$$\approx 8.664339757$$

It will take about $8\frac{2}{3}$ years for the money to double.   ∎

Self Check   In Example 4, how long will it take at 9%, compounded continuously?

Answer   about 7.7 years

Orals   **1.** Write $y = \ln x$ as an exponential equation.
2. Write $e^a = b$ as a logarithmic equation.
3. Write the formula for doubling time.

EXERCISE 11.4

REVIEW   *Write the equation of the required line.*

1. Parallel to $y = 5x + 8$ and passing through the origin

2. Having a slope of 9 and a y-intercept of $(0, 5)$

3. Passing through the point $(3, 2)$ and perpendicular to the line $y = \frac{2}{3}x - 12$

4. Parallel to the line $3x + 2y = 9$ and passing through the point $(-3, 5)$

5. Vertical line through the point $(5, 3)$

6. Horizontal line through the point $(2, 5)$

Simplify each expression.

7. $\dfrac{2x + 3}{4x^2 - 9}$

8. $\dfrac{x + 1}{x} + \dfrac{x - 1}{x + 1}$

9. $\dfrac{x^2 + 3x + 2}{3x + 12} \cdot \dfrac{x + 4}{x^2 - 4}$

10. $\dfrac{1 + \dfrac{y}{x}}{\dfrac{y}{x} - 1}$

VOCABULARY AND CONCEPTS   *Fill in each blank to make a true statement.*

11. $\ln x$ means _____.

12. The domain of the function $y = f(x) = \ln x$ is the interval _____.

13. The range of the function $y = f(x) = \ln x$ is the interval _____.

14. The graph of $y = f(x) = \ln x$ has the _____ as an asymptote.

15. In the expression $\log x$, the base is understood to be ___.

16. In the expression $\ln x$, the base is understood to be ___.

17. If a population grows exponentially at a rate r, the time it will take the population to double is given by the formula $t =$ ___.

18. The logarithm of a negative number is _____.

PRACTICE   *In Exercises 19–26, use a calculator to find each value, if possible. Express all answers to four decimal places.*

19. $\ln 25.25$ **20.** $\ln 0.523$ **21.** $\ln 9.89$ **22.** $\ln 0.00725$

23. $\log (\ln 2)$ **24.** $\ln (\log 28.8)$ **25.** $\ln (\log 0.5)$ **26.** $\log (\ln 0.2)$

In Exercises 27–34, use a calculator to find y, if possible. Express all answers to four decimal places.

27. $\ln y = 2.3015$ **28.** $\ln y = 1.548$ **29.** $\ln y = 3.17$ **30.** $\ln y = 0.837$

31. $\ln y = -4.72$ **32.** $\ln y = -0.48$ **33.** $\log y = \ln 6$ **34.** $\ln y = \log 5$

In Exercises 35–38, tell whether the graph could represent the graph of $y = \ln x$.

35.
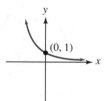

36.
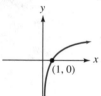

37.
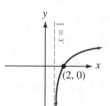

38.
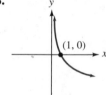

In Exercises 39–42, use a graphing calculator to graph each function.

39. $y = -\ln x$ **40.** $y = \ln x^2$ **41.** $y = \ln (-x)$ **42.** $y = \ln \left(\frac{1}{2}x\right)$

APPLICATIONS   *Use a calculator to solve each problem. Round each answer to the nearest tenth.*

43. Population growth   See Illustration 1. How long will it take the population of River City to double?

44. Doubling money   How long will it take $1,000 to double if it is invested at an annual rate of 5%, compounded continuously?

> **River City**
> *A growing community*
>
> • 6 parks   • 12% annual growth
> • 10 churches   • Low crime rate

ILLUSTRATION 1

45. Population growth A population growing at an annual rate r will triple in a time t given by the formula

$$t = \frac{\ln 3}{r}$$

How long will it take the population of a town growing at the rate of 12% per year to triple?

46. Tripling money Find the length of time for $25,000 to triple if invested at 6% annual interest, compounded continuously. (See Exercise 45.)

WRITING

47. Explain why an earthquake measuring 7 on the Richter scale is much worse than an earthquake measuring 6.

48. The time it takes money to double at an annual rate r, compounded continuously, is given by the formula $t = (\ln 2)/r$. Explain why money doubles more quickly as the rate increases.

SOMETHING TO THINK ABOUT

49. Use the formula $P = P_0 e^{rt}$ to verify that P will be twice P_0 when $t = \frac{\ln 2}{r}$.

50. Use the formula $P = P_0 e^{rt}$ to verify that P will be three times as large as P_0 when $t = \frac{\ln 3}{r}$.

51. Find a formula to find how long it will take money to quadruple.

52. Use a graphing calculator to graph

$$y = f(x) = \frac{1}{1 + e^{-2x}}$$

and discuss the graph.

11.5 Properties of Logarithms

■ PROPERTIES OF LOGARITHMS ■ THE CHANGE-OF-BASE FORMULA ■ CHEMISTRY ■ PHYSIOLOGY

Getting Ready *Simplify each expression.*

1. $x^m x^n$

2. x^0

3. $(x^m)^n$

4. $\dfrac{x^m}{x^n}$

■ PROPERTIES OF LOGARITHMS

Since logarithms are exponents, the properties of exponents have counterparts in the theory of logarithms. We begin with four basic properties.

Properties of Logarithms

If b is a positive number and $b \neq 1$, then

1. $\log_b 1 = 0$

2. $\log_b b = 1$

3. $\log_b b^x = x$

4. $b^{\log_b x} = x \quad (x > 0)$

Properties 1 through 4 follow directly from the definition of a logarithm.

1. $\log_b 1 = 0$, because $b^0 = 1$.
2. $\log_b b = 1$, because $b^1 = b$.
3. $\log_b b^x = x$, because $b^x = b^x$.
4. $b^{\log_b x} = x$, because $\log_b x$ is the exponent to which b is raised to get x.

Properties 3 and 4 also indicate that the composition of the exponential and logarithmic functions with the same base (in both directions) is the identity function. This is expected, because the exponential and logarithmic functions are inverse functions.

EXAMPLE 1   Simplify each expression: **a.** $\log_5 1$,   **b.** $\log_3 3$,   **c.** $\log_7 7^3$,   and   **d.** $b^{\log_b 7}$.

Solution   **a.** By Property 1, $\log_5 1 = 0$, because $5^0 = 1$.

b. By Property 2, $\log_3 3 = 1$, because $3^1 = 3$.

c. By Property 3, $\log_7 7^3 = 3$, because $7^3 = 7^3$.

d. By Property 4, $b^{\log_b 7} = 7$, because $\log_b 7$ is the power to which b is raised to get 7.   ∎

Self Check   Simplify   **a.** $\log_4 1$,   **b.** $\log_5 5$,   **c.** $\log_2 2^4$,   and   **d.** $5^{\log_5 2}$.
Answers   **a.** 0,   **b.** 1,   **c.** 4,   **d.** 2

The next two properties state that

The logarithm of a product is the sum of the logarithms.

The logarithm of a quotient is the difference of the logarithms.

Properties of Logarithms
If M, N, and b are positive numbers and $b \neq 1$, then

5. $\log_b MN = \log_b M + \log_b N$ 6. $\log_b \dfrac{M}{N} = \log_b M - \log_b N$

Proof   To prove Property 5, we let $x = \log_b M$ and $y = \log_b N$. We use the definition of logarithm to write each equation in exponential form.

$$M = b^x \qquad \text{and} \qquad N = b^y$$

Then $MN = b^x b^y$, and a property of exponents gives

$$MN = b^{x+y} \qquad b^x b^y = b^{x+y}\text{: Keep the base and multiply the exponents.}$$

We write this exponential equation in logarithmic form as

$$\log_b MN = x + y$$

Substituting the values of x and y completes the proof.

$$\log_b MN = \log_b M + \log_b N$$ $\square$

The proof of Property 6 is similar.

WARNING!   By Property 5 of logarithms, the logarithm of a *product* is equal to the *sum* of the logarithms. The logarithm of a sum or a difference usually does not simplify. In general,

$$\log_b (M + N) \neq \log_b M + \log_b N$$
$$\log_b (M - N) \neq \log_b M - \log_b N$$

By Property 6, the logarithm of a *quotient* is equal to the *difference* of the logarithms. The logarithm of a quotient is not the quotient of the logarithms:

$$\log_b \frac{M}{N} \neq \frac{\log_b M}{\log_b N}$$

■ ■ ■ ■ ■ ■ ■ ■ ■   **Verifying Properties**

CALCULATORS   We can use a calculator to illustrate Property 5 of logarithms by showing that

$$\ln [(3.7)(15.9)] = \ln 3.7 + \ln 15.9$$

We calculate the left- and right-hand sides of the equation separately and compare the results. To use a scientific calculator to find $\ln [(3.7)(15.9)]$, we enter these numbers and press these keys:

3.7 × 15.9 =   LN

The display will read 4.074651929 .
To find $\ln 3.7 + \ln 15.9$, we enter these numbers and press these keys:

3.7 LN   + 15.9 LN   =

The display will read 4.074651929 . Since the left- and right-hand sides are equal, the equation is true.

Two more properties state that

The logarithm of an expression to a power is the power times the logarithm of the expression.

If the logarithms of two numbers are equal, the numbers are equal.

Properties of Logarithms
If M, p, and b are positive numbers and $b \neq 1$, then
7. $\log_b M^p = p \log_b M$ **8.** If $\log_b x = \log_b y$, then $x = y$.

Proof To prove Property 7, we let $x = \log_b M$, write the expression in exponential form, and raise both sides to the pth power:

$$M = b^x$$

$$(M)^p = (b^x)^p \qquad \text{Raise both sides to the } p\text{th power.}$$

$$M^p = b^{px} \qquad \text{Keep the base and multiply the exponents.}$$

Using the definition of logarithms gives

$$\log_b M^p = px$$

Substituting the value for x completes the proof.

$$\log_b M^p = p \log_b M \qquad\qquad\qquad \square$$

Property 8 follows from the fact that the logarithmic function is a one-to-one function. Property 8 will be important in the next section when we solve logarithmic equations.

We can use the properties of logarithms to write a logarithm as the sum or difference of several logarithms.

EXAMPLE 2 Assume that b ($b \ne 1$), x, y, and z are positive numbers. Write each expression in terms of the logarithms of x, y, and z: **a.** $\log_b xyz$ and **b.** $\log_b \dfrac{xy}{z}$.

Solution **a.** $\log_b xyz$

$$= \log_b (xy)z$$

$$= \log_b (xy) + \log_b z \qquad \text{The log of a product is the sum of the logs.}$$

$$= \log_b x + \log_b y + \log_b z \qquad \text{The log of a product is the sum of the logs.}$$

b. $\log_b \dfrac{xy}{z}$

$$= \log_b (xy) - \log_b z \qquad \text{The log of a quotient is the difference of the logs.}$$

$$= (\log_b x + \log_b y) - \log_b z \qquad \text{The log of a product is the sum of the logs.}$$

$$= \log_b x + \log_b y - \log_b z \qquad \text{Remove parentheses.} \qquad \blacksquare$$

Self Check Find $\log_b \frac{x}{yz}$.
Answer $\log_b x - \log_b y - \log_b z$

 b

EXAMPLE 3 Assume that b ($b \ne 1$), x, y, and z are positive numbers. Write each expression in terms of the logarithms of x, y, and z.

a. $\log_b (x^2 y^3 z)$ and **b.** $\log_b \dfrac{\sqrt{x}}{y^3 z}$

Solution   **a.** $\log_b (x^2y^3z) = \log_b x^2 + \log_b y^3 + \log_b z$ The log of a product is the sum of the logs.

$$= 2 \log_b x + 3 \log_b y + \log_b z$$ The log of an expression to a power is the power times the log of the expression.

b. $\log_b \dfrac{\sqrt{x}}{y^3z} = \log_b \sqrt{x} - \log (y^3z)$ The log of a quotient is the difference of the logs.

$$= \log x^{1/2} - (\log y^3 + \log z)$$ $\sqrt{x} = x^{1/2}$; The log of a product is the sum of the logs.

$$= \frac{1}{2} \log x - (3 \log y + \log z)$$ The log of a power is the power times the log.

$$= \frac{1}{2} \log x - 3 \log y - \log z$$ Use the distributive property to remove parentheses.   ∎

Self Check   Find $\log_b \sqrt[4]{\dfrac{x^3y}{z}}$.

Answer $\frac{1}{4}(3 \log_b x + \log_b y - \log_b z)$

We can use the properties of logarithms to combine several logarithms into one logarithm.

EXAMPLE 4   Assume that b ($b \neq 1$), x, y, and z are positive numbers. Write each expression as one logarithm: **a.** $3 \log_b x + \frac{1}{2} \log_b y$ and   **b.** $\frac{1}{2} \log_b (x - 2) - \log_b y + 3 \log_b z$.

Solution   **a.** $3 \log_b x + \dfrac{1}{2} \log_b y = \log_b x^3 + \log_b y^{1/2}$ A power times a log is the log of the power.

$$= \log_b (x^3y^{1/2})$$ The sum of two logs is the log of a product.

b. $\dfrac{1}{2} \log_b (x - 2) - \log_b y + 3 \log_b z$

$$= \log_b (x - 2)^{1/2} - \log_b y + \log_b z^3$$ A power times a log is the log of the power.

$$= \log_b \frac{(x - 2)^{1/2}}{y} + \log_b z^3$$ The difference of two logs is the log of the quotient.

$$= \log_b \frac{z^3\sqrt{x - 2}}{y}$$ The sum of two logs is the log of a product.   ∎

Self Check   Write the expression as one logarithm: $2\log_b x + \frac{1}{2}\log_b y - 2\log_b (x - y)$.

Answer $\log_b \dfrac{x^2\sqrt{y}}{(x - y)^2}$

We summarize the properties of logarithms as follows.

Properties of Logarithms

If b, M, and N are positive numbers and $b \neq 1$, then

1. $\log_b 1 = 0$ 2. $\log_b b = 1$

3. $\log_b b^x = x$ 4. $b^{\log_b x} = x$

5. $\log_b MN = \log_b M + \log_b N$ 6. $\log_b \dfrac{M}{N} = \log_b M - \log_b N$

7. $\log_b M^p = p \log_b M$ 8. If $\log_b x = \log_b y$, then $x = y$.

EXAMPLE 5

Given that $\log 2 \approx 0.3010$ and $\log 3 \approx 0.4771$, find approximations for   **a.** $\log 6$, **b.** $\log 9$,   **c.** $\log 18$,   and   **d.** $\log 2.5$.

Solution   **a.** $\log 6 = \log (2 \cdot 3)$

$= \log 2 + \log 3$ The log of a product is the sum of the logs.

$\approx 0.3010 + 0.4771$ Substitute the value of each logarithm.

≈ 0.7781

b. $\log 9 = \log (3^2)$

$= 2 \log 3$ The log of a power is the power times the log.

$\approx 2(0.4771)$ Substitute the value of log 3.

≈ 0.9542

c. $\log 18 = \log (2 \cdot 3^2)$

$= \log 2 + \log 3^2$ The log of a product is the sum of the logs.

$= \log 2 + 2 \log 3$ The log of a power is the power times the log.

$\approx 0.3010 + 2(0.4771)$

≈ 1.2552

d. $\log 2.5 = \log \left(\dfrac{5}{2} \right)$

$= \log 5 - \log 2$ The log of a quotient is the difference of the logs.

$= \log \dfrac{10}{2} - \log 2$ Write 5 as $\frac{10}{2}$.

$= \log 10 - \log 2 - \log 2$ The log of a quotient is the difference of the logs.

$= 1 - 2 \log 2$ $\log_{10} 10 = 1$.

$\approx 1 - 2(0.3010)$

≈ 0.3980

Self Check — Use the values given in Example 5 and find   **a.** log 1.5   and   **b.** log 0.2.

Answers   **a.** 0.1761,   **b.** −0.6990

■ THE CHANGE-OF-BASE FORMULA

If we know the base-*a* logarithm of a number, we can find its logarithm to some other base *b* with a formula called the **change-of-base formula.**

Change-of-Base Formula
If *a*, *b*, and *x* are real numbers, then

$$\log_b x = \frac{\log_a x}{\log_a b}$$

Proof   To prove this formula, we begin with the equation $\log_b x = y$.

$$y = \log_b x$$
$$x = b^y \qquad \text{Change the equation from logarithmic to exponential form.}$$
$$\log_a x = \log_a b^y \qquad \text{Take the base-}a\text{ logarithm of both sides.}$$
$$\log_a x = y \log_a b \qquad \text{The log of a power is the power times the log.}$$
$$y = \frac{\log_a x}{\log_a b} \qquad \text{Divide both sides by } \log_a b.$$
$$\log_b x = \frac{\log_a x}{\log_a b} \qquad \text{Refer to the first equation and substitute } \log_b x \text{ for } y. \qquad \square$$

If we know logarithms to base *a* (for example, *a* = 10), we can find the logarithm of *x* to a new base *b*. We simply divide the base-*a* logarithm of *x* by the base-*a* logarithm of *b*.

WARNING! $\dfrac{\log_a x}{\log_a b}$ means that one logarithm is to be divided by the other. They are not to be subtracted.

EXAMPLE 6   Find $\log_4 9$ using base-10 logarithms.

Solution   We can substitute 4 for b, 10 for a, and 9 for x into the change-of-base formula:

$$\log_b x = \frac{\log_a x}{\log_a b}$$

$$\log_4 9 = \frac{\log_{10} 9}{\log_{10} 4}$$

$$\approx 1.584962501$$

To four decimal places, $\log_4 9 = 1.5850$.   ∎

Self Check   Find $\log_5 3$ using base-10 logarithms.
Answer   0.6826

■ CHEMISTRY

Common logarithms are used to express the acidity of solutions. The more acidic a solution, the greater the concentration of hydrogen ions. This concentration is indicated indirectly by the *pH scale,* or *hydrogen ion index.* The pH of a solution is defined by the following equation.

pH of a Solution
If $[H^+]$ is the hydrogen ion concentration in gram-ions per liter, then
$$pH = -\log [H^+]$$

EXAMPLE 7   **Finding pH of a solution**   Find the pH of pure water, which has a hydrogen ion concentration of 10^{-7} gram-ions per liter.

Solution   Since pure water has approximately 10^{-7} gram-ions per liter, its pH is

$$pH = -\log [H^+]$$
$$pH = -\log 10^{-7}$$
$$= -(-7) \log 10 \qquad \text{The log of a power is the power times the log.}$$
$$= -(-7) \cdot 1 \qquad \log 10 = 1.$$
$$= 7$$

∎

EXAMPLE 8   **Finding hydrogen ion concentration**   Find the hydrogen ion concentration of seawater if its pH is 8.5.

Solution   To find its hydrogen ion concentration, we substitute 8.5 for the pH and find $[H^+]$.

$$8.5 = -\log [H^+]$$
$$-8.5 = \log [H^+] \qquad \text{Multiply both sides by } -1.$$
$$[H^+] = 10^{-8.5} \qquad \text{Change the equation to exponential form.}$$

We can use a calculator to find that

$$[H^+] \approx 3.2 \times 10^{-9} \text{ gram-ions per liter.} \qquad ■$$

■ **PHYSIOLOGY**

In physiology, experiments suggest that the relationship between the loudness and the intensity of sound is a logarithmic one known as the **Weber–Fechner law.**

> **Weber-Fechner Law**
> If L is the apparent loudness of a sound, I is the actual intensity, and k is a constant, then
> $$L = k \ln I$$

EXAMPLE 9   **Weber-Fechner law**   Find the increase in intensity that will cause the apparent loudness of a sound to double.

Solution   If the original loudness L_O is caused by an actual intensity I_O, then

> **1.** $L_0 = k \ln I_0$

To double the apparent loudness, we multiply both sides of Equation 1 by 2 and use the power rule of logarithms:

$$2L_0 = 2k \ln I_0$$
$$= k \ln (I_0)^2$$

To double the loudness of a sound, the intensity must be squared.   ■

Self Check   In Example 9, what decrease in intensity will cause a sound to be half as loud?
Answer   Find the square root of the intensity.

Orals   *Find the value of x in each equation.*

1. $\log_3 9 = x$ **2.** $\log_x 5 = 1$ **3.** $\log_7 x = 3$

4. $\log_2 x = -2$ **5.** $\log_4 x = \dfrac{1}{2}$ **6.** $\log_x 4 = 2$

7. $\log_{1/2} x = 2$ **8.** $\log_9 3 = x$ **9.** $\log_x \dfrac{1}{4} = -2$

EXERCISE 11.5

REVIEW   *Consider the line that passes through P(−2, 3) and Q(4, −4).*

1. Find the slope of line *PQ*.   **2.** Find the distance *PQ*.

3. Find the midpoint of segment *PQ*.   **4.** Write the equation of line *PQ*.

VOCABULARY AND CONCEPTS   *Fill in each blank to make a true statement.*

5. $\log_b 1 = \underline{\quad}$ **6.** $\log_b b = \underline{\quad}$

7. $\log_b MN = \log_b \underline{\quad} + \log_b \underline{\quad}$ **8.** $b^{\log_b x} = \underline{\quad}$

9. If $\log_b x = \log_b y$, then $\underline{\quad} = \underline{\quad}$.   **10.** $\log_b \dfrac{M}{N} = \log_b M \underline{\quad} \log_b N$

11. $\log_b x^p = p \cdot \log_b \underline{\quad}$ **12.** $\log_b b^x = \underline{\quad}$

13. $\log_b (A + B) \underline{\quad} \log_b A + \log_b B$ **14.** $\log_b A + \log_b B \underline{\quad} \log_b AB$

15. $\log_4 1 = \underline{\quad}$ **16.** $\log_4 4 = \underline{\quad}$ **17.** $\log_4 4^7 = \underline{\quad}$ **18.** $4^{\log_4 8} = \underline{\quad}$

19. $5^{\log_5 10} = \underline{\quad}$ **20.** $\log_5 5^2 = \underline{\quad}$ **21.** $\log_5 5 = \underline{\quad}$ **22.** $\log_5 1 = \underline{\quad}$

23. $\log_7 1 = \underline{\quad}$ **24.** $\log_9 9 = \underline{\quad}$ **25.** $\log_3 3^7 = \underline{\quad}$ **26.** $5^{\log_5 8} = \underline{\quad}$

27. $8^{\log_8 10} = \underline{\quad}$ **28.** $\log_4 4^2 = \underline{\quad}$ **29.** $\log_9 9 = \underline{\quad}$ **30.** $\log_3 1 = \underline{\quad}$

PRACTICE   *In Exercises 31–36, use a calculator to verify each equation.*

31. $\log [(2.5)(3.7)] = \log 2.5 + \log 3.7$ **32.** $\ln \dfrac{11.3}{6.1} = \ln 11.3 - \ln 6.1$

33. $\ln (2.25)^4 = 4 \ln 2.25$ **34.** $\log 45.37 = \dfrac{\ln 45.37}{\ln 10}$

35. $\log \sqrt{24.3} = \dfrac{1}{2} \log 24.3$ **36.** $\ln 8.75 = \dfrac{\log 8.75}{\log e}$

In Exercises 37–48, assume that x, y, z, and b (b ≠ 1) are positive numbers. Use the properties of logarithms to write each expression in terms of the logarithms of x, y, and z.

37. $\log_b xyz$ **38.** $\log_b 4xz$

39. $\log_b \dfrac{2x}{y}$ **40.** $\log_b \dfrac{x}{yz}$

41. $\log_b x^3 y^2$ **42.** $\log_b xy^2 z^3$

43. $\log_b (xy)^{1/2}$

44. $\log_b x^3 y^{1/2}$

45. $\log_b x\sqrt{z}$

46. $\log_b \sqrt{xy}$

47. $\log_b \dfrac{\sqrt[3]{x}}{\sqrt[4]{yz}}$

48. $\log_b \sqrt[4]{\dfrac{x^3 y^2}{z^4}}$

In Exercises 49–56, assume that x, y, z, and b (b ≠ 1) are positive numbers. Use the properties of logarithms to write each expression as the logarithm of a single quantity.

49. $\log_b (x + 1) - \log_b x$

50. $\log_b x + \log_b (x + 2) - \log_b 8$

51. $2 \log_b x + \dfrac{1}{2}\log_b y$

52. $-2 \log_b x - 3 \log_b y + \log_b z$

53. $-3 \log_b x - 2 \log_b y + \dfrac{1}{2} \log_b z$

54. $3 \log_b (x + 1) - 2 \log_b (x + 2) + \log_b x$

55. $\log_b \left(\dfrac{x}{z} + x \right) - \log_b \left(\dfrac{y}{z} + y \right)$

56. $\log_b (xy + y^2) - \log_b (xz + yz) + \log_b z$

In Exercises 57–68, tell whether the given statement is true. If a statement is false, explain why.

57. $\log_b 0 = 1$

58. $\log_b (x + y) \neq \log_b x + \log_b y$

59. $\log_b xy = (\log_b x)(\log_b y)$

60. $\log_b ab = \log_b a + 1$

61. $\log_7 7^7 = 7$

62. $7^{\log_7 7} = 7$

63. $\dfrac{\log_b A}{\log_b B} = \log_b A - \log_b B$

64. $\log_b (A - B) = \dfrac{\log_b A}{\log_b B}$

65. $3 \log_b \sqrt[3]{a} = \log_b a$

66. $\dfrac{1}{3} \log_b a^3 = \log_b a$

67. $\log_b \dfrac{1}{a} = -\log_b a$

68. $\log_b 2 = \log_2 b$

In Exercises 69–80, assume that log 4 = 0.6021, log 7 = 0.8451, and log 9 = 0.9542. Use these values and the properties of logarithms to find each value. **Do not use a calculator.**

69. $\log 28$

70. $\log \dfrac{7}{4}$

71. $\log 2.25$

72. $\log 36$

73. $\log \dfrac{63}{4}$

74. $\log \dfrac{4}{63}$

75. $\log 252$

76. $\log 49$

77. $\log 112$

78. $\log 324$

79. $\log \dfrac{144}{49}$

80. $\log \dfrac{324}{63}$

In Exercises 81–88, use a calculator and the change-of-base formula to find each logarithm.

81. $\log_3 7$

82. $\log_7 3$

83. $\log_{1/3} 3$

84. $\log_{1/2} 6$

85. $\log_3 8$

86. $\log_5 10$

87. $\log_{\sqrt{2}} \sqrt{5}$

88. $\log_\pi e$

APPLICATIONS

89. pH of a solution Find the pH of a solution with a hydrogen ion concentration of 1.7×10^{-5} gram-ions per liter.

90. Hydrogen ion concentration Find the hydrogen ion concentration of a saturated solution of calcium hydroxide whose pH is 13.2.

91. Aquariums To test for safe pH levels in a freshwater aquarium, a test strip is compared with the scale shown in Illustration 1. Find the corresponding range in the hydrogen ion concentration.

92. pH of pickles The hydrogen ion concentration of sour pickles is 6.31×10^{-4}. Find the pH.

93. Change in loudness If the intensity of a sound is doubled, find the apparent change in loudness.

94. Change in loudness If the intensity of a sound is tripled, find the apparent change in loudness.

AquaTest pH Kit

Safe range

6.4   6.8   7.2   7.6   8.0

ILLUSTRATION 1

95. Change in intensity What change in intensity of sound will cause an apparent tripling of the loudness?

96. Change in intensity What increase in the intensity of a sound will cause the apparent loudness to be multiplied by 4?

WRITING

97. Explain why ln (log 0.9) is undefined.

98. Explain why $\log_b$ (ln 1) is undefined.

SOMETHING TO THINK ABOUT

99. Show that ln $(e^x) = x$.

100. If $\log_b 3x = 1 + \log_b x$, find b.

101. Show that $\log_{b^2} x = \dfrac{1}{2} \log_b x$.

102. Show that $e^{x \ln a} = a^x$.

11.6 Exponential and Logarithmic Equations

■ SOLVING EXPONENTIAL EQUATIONS ■ SOLVING LOGARITHMIC EQUATIONS ■ RADIOACTIVE DECAY ■ POPULATION GROWTH

Getting Ready   *Write each expression without using exponents.*

1. $\log x^2$

2. $\log x^{1/2}$

3. $\log x^0$

4. $\log a^b + b \log a$

An **exponential equation** is an equation that contains a variable in one of its exponents. Some examples of exponential equations are

$$3^x = 5, \qquad 6^{x-3} = 2^x, \qquad \text{and} \qquad 3^{2x+1} - 10(3^x) + 3 = 0$$

A **logarithmic equation** is an equation with logarithmic expressions that contain a variable. Some examples of logarithmic equations are

$$\log 2x = 25, \qquad \ln x - \ln (x - 12) = 24, \qquad \text{and} \qquad \log x = \log \frac{1}{x} + 4$$

In this section, we will learn how to solve many of these equations.

■ SOLVING EXPONENTIAL EQUATIONS

EXAMPLE 1

Solve $4^x = 7$.

Solution

Since logarithms of equal numbers are equal, we can take the common logarithm of each side of the equation. The power rule of logarithms then provides a way of moving the variable x from its position as an exponent to a position as a coefficient.

$$4^x = 7$$

$$\log 4^x = \log 7 \qquad \text{Take the common logarithm of each side.}$$

$$x \log 4 = \log 7 \qquad \text{The log of a power is the power times the log.}$$

1. $\qquad x = \dfrac{\log 7}{\log 4} \qquad \text{Divide both sides by } \log 4.$

$$\approx 1.403677461 \qquad \text{Use a calculator.}$$

To four decimal places, $x = 1.4037$. ■

Self Check   Solve $5^x = 4$.

Answer   0.8614

WARNING!   A careless reading of Equation 1 leads to a common error. The right-hand side of Equation 1 calls for a division, not a subtraction.

$$\frac{\log 7}{\log 4} \qquad \text{means} \qquad (\log 7) \div (\log 4)$$

It is the expression $\log \frac{7}{4}$ that means $\log 7 - \log 4$.

EXAMPLE 2   Solve $6^{x-3} = 2^x$.

Solution

$$6^{x-3} = 2^x$$

$$\log 6^{x-3} = \log 2^x$$ Take the common logarithm of each side.

$$(x - 3) \log 6 = x \log 2$$ The log of a power is the power times the log.

$$x \log 6 - 3 \log 6 = x \log 2$$ Use the distributive property.

$$x \log 6 - x \log 2 = 3 \log 6$$ Add 3 log 6 and subtract x log 2 from both sides.

$$x(\log 6 - \log 2) = 3 \log 6$$ Factor out x on the left-hand side.

$$x = \frac{3 \log 6}{\log 6 - \log 2}$$ Divide both sides by log 6 − log 2.

$$x \approx 4.892789261$$ Use a calculator. ∎

Self Check   Solve $5^{x-2} = 3^x$.

Answer $\dfrac{2 \log 5}{\log 5 - \log 3} \approx 6.301320206$

EXAMPLE 3   Solve $2^{x^2 + 2x} = \dfrac{1}{2}$.

Solution   Since $\frac{1}{2} = 2^{-1}$, we can write the equation in the form

$$2^{x^2+2x} = 2^{-1}$$

Since equal quantities with equal bases have equal exponents, we have

$$x^2 + 2x = -1$$

$$x^2 + 2x + 1 = 0$$ Add 1 to both sides.

$$(x + 1)(x + 1) = 0$$ Factor the trinomial.

$$x + 1 = 0 \quad \text{or} \quad x + 1 = 0$$ Set each factor equal to 0.

$$x = -1 \qquad\qquad x = -1$$

Verify that −1 satisfies the equation. ∎

Self Check   Solve $3^{x^2-2x} = \frac{1}{3}$.

Answer   1, 1

■ ■ ■ ■ ■ ■ ■ ■ ■ ■   **Solving Exponential Equations**

GRAPHING
CALCULATORS

To use a graphing calculator to approximate the solutions of $2^{x^2+2x} = \frac{1}{2}$ (see Example 3), we can subtract $\frac{1}{2}$ from both sides of the equation to get

$$2^{x^2+2x} - \frac{1}{2} = 0$$

and graph the corresponding function

$$y = f(x) = 2^{x^2+2x} - \frac{1}{2}$$

If we use window settings of $[-4, 4]$ for x and $[-2, 6]$ for y, we obtain the graph shown in Figure 11-23(a).

Since the solutions of the equation are its x-intercepts, we can approximate the solutions by zooming in on the values of the x-intercepts, as in Figure 11-23(b). Since $x = -1$ is the only x-intercept, -1 is the only solution. In this case, we have found an exact solution.

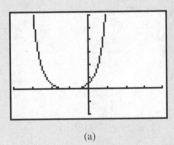

(a)

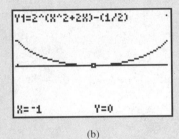

(b)

FIGURE 11-23

■ SOLVING LOGARITHMIC EQUATIONS

In each of the following examples, we use the properties of logarithms to change a logarithmic equation into an algebraic equation.

EXAMPLE 4

Solve $\log_b (3x + 2) - \log_b (2x - 3) = 0$.

Solution

$$\log_b (3x + 2) - \log_b (2x - 3) = 0$$

$$\log_b (3x + 2) = \log_b (2x - 3) \qquad \text{Add } \log_b (2x - 3) \text{ to both sides.}$$

$$3x + 2 = 2x - 3 \qquad \text{If } \log_b r = \log_b s, \text{ then } r = s.$$

$$x = -5 \qquad \text{Subtract } 2x \text{ and } 2 \text{ from both sides.}$$

Check:
$$\log_b (3x + 2) - \log_b (2x - 3) = 0$$
$$\log_b [3(-5) + 2] - \log_b [2(-5) - 3] \overset{?}{=} 0$$
$$\log_b (-13) - \log_b (-13) \overset{?}{=} 0$$

Since the logarithm of a negative number does not exist, the apparent solution of −5 must be discarded. This equation has no solutions. ∎

Self Check   Solve $\log_b (5x + 2) - \log_b (7x - 2) = 0$.

Answer   2

WARNING!   Example 4 illustrates that you must check the solutions of a logarithmic equation.

EXAMPLE 5   Solve $\log x + \log (x - 3) = 1$.

Solution

$\log x + \log (x - 3) = 1$	
$\log x(x - 3) = 1$	The sum of two logs is the log of a product.
$x(x - 3) = 10^1$	Use the definition of logarithms to change the equation to exponential form.
$x^2 - 3x - 10 = 0$	Remove parentheses and subtract 10 from both sides.
$(x + 2)(x - 5) = 0$	Factor the trinomial.
$x + 2 = 0$ or $x - 5 = 0$	Set each factor equal to 0.
$x = -2$ $x = 5$	

Check: The number −2 is not a solution, because it does not satisfy the equation (a negative number does not have a logarithm). We will check the remaining number, 5.

$\log x + \log (x - 3) = 1$	
$\log 5 + \log (5 - 3) \overset{?}{=} 1$	Substitute 5 for x.
$\log 5 + \log 2 \overset{?}{=} 1$	
$\log 10 \overset{?}{=} 1$	The sum of two logs is the log of a product.
$1 = 1$	$\log 10 = 1$.

Since 5 satisfies the equation, it is a solution. ∎

Self Check   Solve $\log x + \log (x + 3) = 1$.

Answer   2

EXAMPLE 6   Solve $\dfrac{\log (5x - 6)}{\log x} = 2$.

Solution   We can multiply both sides of the equation by $\log x$ to get

$$\log (5x - 6) = 2 \log x$$

and apply the power rule of logarithms to get

$$\log (5x - 6) = \log x^2$$

By Property 8 of logarithms, $5x - 6 = x^2$, because they have equal logarithms. Thus,

$$
\begin{aligned}
5x - 6 &= x^2 \\
0 &= x^2 - 5x + 6 \\
0 &= (x - 3)(x - 2) \\
x - 3 = 0 \quad &\text{or} \quad x - 2 = 0 \\
x = 3 \qquad\qquad & \qquad x = 2
\end{aligned}
$$

Verify that both 2 and 3 satisfy the equation.   ∎

Self Check   Solve $\dfrac{\log (5x + 6)}{\log x} = 2$.

Answer   6

■ ■ ■ ■ ■ ■ ■ ■ ■ ■   **Solving Logarithmic Equations**

GRAPHING CALCULATORS

To use a graphing calculator to approximate the solutions of $\log x + \log (x - 3) = 1$ (see Example 5), we can subtract 1 from both sides of the equation to get

$$\log x + \log (x - 3) - 1 = 0$$

and graph the corresponding function

$$y = f(x) = \log x + \log (x - 3) - 1$$

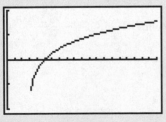

FIGURE 11-24

If we use window settings of $[0, 20]$ for x and $[-2, 2]$ for y, we obtain the graph shown in Figure 11-24. Since the solution of the equation is the x-intercept, we can find the solution by zooming in on the value of the x-intercept. The solution is $x = 5$.

■ RADIOACTIVE DECAY

Experiments have determined the time that it takes for half of a sample of a given radioactive material to decompose. This time is a constant, called the material's **half-life.**

When living organisms die, the oxygen/carbon dioxide cycle common to all living things stops, and carbon-14, a radioactive isotope with a half-life of 5,700 years, is no longer absorbed. By measuring the amount of carbon-14 present in an ancient object, archaeologists can estimate the object's age by using the radioactive decay formula.

> **Radioactive Decay Formula**
> If A is the amount of radioactive material present at time t, A_0 was the amount present at $t = 0$, and h is the material's half-life, then
> $$A = A_0 2^{-t/h}$$

EXAMPLE 7   **Carbon-14 dating**   How old is a wooden statue that retains only one-third of its original carbon-14 content?

Solution   To find the time t when $A = \frac{1}{3}A_0$, we substitute $\frac{A_0}{3}$ for A and 5,700 for h into the radioactive decay formula and solve for t:

$$A = A_0 2^{-t/h}$$

$$\frac{A_0}{3} = A_0 2^{-t/5,700}$$

$$1 = 3(2^{-t/5,700}) \qquad \text{Divide both sides by } A_0 \text{ and multiply both sides by 3.}$$

$$\log 1 = \log 3(2^{-t/5,700}) \qquad \text{Take the common logarithm of each side.}$$

$$0 = \log 3 + \log 2^{-t/5,700} \qquad \text{Log } 1 = 0, \text{ and the log of a product is the sum of the logs.}$$

$$-\log 3 = -\frac{t}{5,700} \log 2 \qquad \text{Subtract log 3 from both sides and use the power rule of logarithms.}$$

$$5,700 \left(\frac{\log 3}{\log 2}\right) = t \qquad \text{Multiply both sides by } -\frac{5,700}{\log 2}.$$

$$t \approx 9,034.286254 \qquad \text{Use a calculator.}$$

The statue is approximately 9,000 years old.   ■

Self Check   How old is a statue that retains 25% of its original carbon-14 content?

Answer   about 11,400 years

■ POPULATION GROWTH

Recall that when there is sufficient food and space, populations of living organisms tend to increase exponentially according to the Malthusian growth model.

> **Malthusian Growth Model**
> If P is the population at some time t, P_0 is the initial population at $t = 0$, and k depends on the rate of growth, then
> $$P = P_0 e^{kt}$$

EXAMPLE 8

Population growth   The bacteria in a laboratory culture increased from an initial population of 500 to 1,500 in 3 hours. How long will it take for the population to reach 10,000?

Solution   We substitute 500 for P_0, 1,500 for P, and 3 for t and simplify to find k:

$$P = P_0 e^{kt}$$

$\mathbf{1,500} = 500(e^{k3})$ Substitute 1,500 for P, 500 for P_0, and 3 for t.

$3 = e^{3k}$ Divide both sides by 500.

$3k = \ln 3$ Change the equation from exponential to logarithmic form.

$k = \dfrac{\ln 3}{3}$ Divide both sides by 3.

To find when the population will reach 10,000, we substitute 10,000 for P, 500 for P_0, and $\frac{\ln 3}{3}$ for k in the equation $P = P_0 e^{kt}$ and solve for t:

$$P = P_0 e^{kt}$$

$\mathbf{10,000} = 500 e^{[(\ln 3)/3]t}$

$20 = e^{[(\ln 3)/3]t}$ Divide both sides by 500.

$\left(\dfrac{\ln 3}{3} \right) t = \ln 20$ Change the equation to logarithmic form.

$t = \dfrac{3 \ln 20}{\ln 3}$ Multiply both sides by $\dfrac{3}{\ln 3}$.

≈ 8.180499084 Use a calculator.

The culture will reach 10,000 bacteria in about 8 hours.   ■

Self Check   In Example 8, how long will it take to reach 20,000?
Answer   about 10 hours

EXAMPLE 9

Generation time   If a medium is inoculated with a bacterial culture that contains 1,000 cells per milliliter, how many generations will pass by the time the culture has grown to a population of 1 million cells per milliliter?

Solution

During bacterial reproduction, the time required for a population to double is called the *generation time*. If b bacteria are introduced into a medium, then after the generation time of the organism has elapsed, there are $2b$ cells. After another generation, there are $2(2b)$, or $4b$ cells, and so on. After n generations, the number of cells present will be

1. $B = b \cdot 2^n$

To find the number of generations that have passed while the population grows from b bacteria to B bacteria, we solve Equation 1 for n.

$$\log B = \log (b \cdot 2^n)$$ Take the common logarithm of both sides.

$$\log B = \log b + n \log 2$$ Apply the product and power rules of logarithms.

$$\log B - \log b = n \log 2$$ Subtract $\log b$ from both sides.

$$n = \frac{1}{\log 2} (\log B - \log b)$$ Multiply both sides by $\dfrac{1}{\log 2}$.

2. $$n = \frac{1}{\log 2} \left(\log \frac{B}{b} \right)$$ Use the quotient rule of logarithms.

Equation 2 is a formula that gives the number of generations that will pass as the population grows from b bacteria to B bacteria.

To find the number of generations that have passed while a population of 1,000 cells per milliliter has grown to a population of 1 million cells per milliliter, we substitute 1,000 for b and 1,000,000 for B in Equation 2 and solve for n.

$$n = \frac{1}{\log 2} \log \frac{1,000,000}{1,000}$$

$$= \frac{1}{\log 2} \log 1,000$$ Simplify.

$$= 3.321928095(3)$$ $\dfrac{1}{\log 2} \approx 3.321928095$ and $\log 1,000 = 3$.

$$= 9.965784285$$

Approximately 10 generations will have passed.   ∎

Orals *Solve each equation for x.* **Do not simplify answers.**

1. $3^x = 5$ **2.** $5^x = 3$

3. $2^{-x} = 7$ **4.** $6^{-x} = 1$

5. $\log 2x = \log (x + 2)$ **6.** $\log 2x = 0$

7. $\log x^4 = 4$ **8.** $\log \sqrt{x} = \dfrac{1}{2}$

EXERCISE 11.6

REVIEW *Solve each equation.*

1. $5x^2 - 25x = 0$ **2.** $4y^2 - 25 = 0$

3. $3p^2 + 10p = 8$ **4.** $4t^2 + 1 = -6t$

VOCABULARY AND CONCEPTS *Fill in each blank to make a true statement.*

5. An equation with a variable in its exponent is called a(n) _____ equation.

6. An equation with a logarithmic expression that contains a variable is a(n) _____ equation.

7. The formula for carbon dating is $A =$ _____.

8. The formula for population growth is $P =$ _____.

PRACTICE ▦ *In Exercises 9–28, solve each exponential equation. If an answer is not exact, give the answer to four decimal places.*

9. $4^x = 5$ **10.** $7^x = 12$ **11.** $13^{x-1} = 2$ **12.** $5^{x+1} = 3$
13. $2^{x+1} = 3^x$ **14.** $5^{x-3} = 3^{2x}$ **15.** $2^x = 3^x$ **16.** $3^{2x} = 4^x$
17. $7^{x^2} = 10$ **18.** $8^{x^2} = 11$ **19.** $8^{x^2} = 9^x$ **20.** $5^{x^2} = 2^{5x}$

21. $2^{x^2-2x} = 8$ **22.** $3^{x^2-3x} = 81$ **23.** $3^{x^2+4x} = \dfrac{1}{81}$ **24.** $7^{x^2+3x} = \dfrac{1}{49}$

25. $4^{x+2} - 4^x = 15$ (*Hint:* $4^{x+2} = 4^x 4^2$.)

26. $3^{x+3} + 3^x = 84$ (*Hint:* $3^{x+3} = 3^x 3^3$.)

27. $2(3^x) = 6^{2x}$

28. $2(3^{x+1}) = 3(2^{x-1})$

▦ *In Exercises 29–32, use a calculator to solve each equation, if possible. Give all answers to the nearest tenth.*

29. $2^{x+1} = 7$ **30.** $3^{x-1} = 2^x$ **31.** $2^{x^2-2x} - 8 = 0$ **32.** $3^x - 10 = 3^{-x}$

In Exercises 33–62, solve each logarithmic equation. Check all solutions.

33. $\log 2x = \log 4$

34. $\log 3x = \log 9$

35. $\log (3x + 1) = \log (x + 7)$

36. $\log (x^2 + 4x) = \log (x^2 + 16)$

37. $\log (3 - 2x) - \log (x + 24) = 0$

38. $\log (3x + 5) - \log (2x + 6) = 0$

39. $\log \dfrac{4x + 1}{2x + 9} = 0$

40. $\log \dfrac{2 - 5x}{2(x + 8)} = 0$

41. $\log x^2 = 2$

42. $\log x^3 = 3$

43. $\log x + \log (x - 48) = 2$

44. $\log x + \log (x + 9) = 1$

45. $\log x + \log (x - 15) = 2$

46. $\log x + \log (x + 21) = 2$

47. $\log (x + 90) = 3 - \log x$

48. $\log (x - 90) = 3 - \log x$

49. $\log (x - 6) - \log (x - 2) = \log \dfrac{5}{x}$

50. $\log (3 - 2x) - \log (x + 9) = 0$

51. $\log x^2 = (\log x)^2$

52. $\log (\log x) = 1$

53. $\dfrac{\log (3x - 4)}{\log x} = 2$

54. $\dfrac{\log (8x - 7)}{\log x} = 2$

55. $\dfrac{\log (5x + 6)}{2} = \log x$

56. $\dfrac{1}{2} \log (4x + 5) = \log x$

57. $\log_3 x = \log_3 \left(\dfrac{1}{x}\right) + 4$

58. $\log_5 (7 + x) + \log_5 (8 - x) - \log_5 2 = 2$

59. $2 \log_2 x = 3 + \log_2 (x - 2)$

60. $2 \log_3 x - \log_3 (x - 4) = 2 + \log_3 2$

61. $\log (7y + 1) = 2 \log (y + 3) - \log 2$

62. $2 \log (y + 2) = \log (y + 2) - \log 12$

In Exercises 63–66, use a graphing calculator to solve each equation. If an answer is not exact, give all answers to the nearest tenth.

63. $\log x + \log (x - 15) = 2$

64. $\log x + \log (x + 3) = 1$

65. $\ln (2x + 5) - \ln 3 = \ln (x - 1)$

66. $2 \log(x^2 + 4x) = 1$

APPLICATIONS

67. Tritium decay The half-life of tritium is 12.4 years. How long will it take for 25% of a sample of tritium to decompose?

68. Radioactive decay In two years, 20% of a radioactive element decays. Find its half-life.

69. Thorium decay An isotope of thorium, ^{227}Th, has a half-life of 18.4 days. How long will it take for 80% of the sample to decompose?

70. Lead decay An isotope of lead, ^{201}Pb, has a half-life of 8.4 hours. How many hours ago was there 30% more of the substance?

71. Carbon-14 dating The bone fragment shown in Illustration 1 contains 60% of the carbon-14 that it is assumed to have had initially. How old is it?

ILLUSTRATION 1

72. Carbon-14 dating Only 10% of the carbon-14 in a small wooden bowl remains. How old is the bowl?

73. **Compound interest**   If $500 is deposited in an account paying 8.5% annual interest, compounded semiannually, how long will it take for the account to increase to $800?

74. **Continuous compound interest**   In Exercise 73, how long will it take if the interest is compounded continuously?

75. **Compound interest**   If $1,300 is deposited in a savings account paying 9% interest, compounded quarterly, how long will it take the account to increase to $2,100?

76. **Compound interest**   A sum of $5,000 deposited in an account grows to $7,000 in 5 years. Assuming annual compounding, what interest rate is being paid?

77. **Rule of seventy**   A rule of thumb for finding how long it takes an investment to double is called the **rule of seventy.** To apply the rule, divide 70 by the interest rate written as a percent. At 5%, it takes $\frac{70}{5} = 14$ years to double an investment. At 7%, it takes $\frac{70}{7} = 10$ years. Explain why this formula works.

78. **Bacterial growth**   A bacterial culture grows according to the formula

$$P = P_0 a^t$$

If it takes 5 days for the culture to triple in size, how long will it take to double in size?

79. **Rodent control**   The rodent population in a city is currently estimated at 30,000. If it is expected to double every 5 years, when will the population reach 1 million?

80. **Population growth**   The population of a city is expected to triple every 15 years. When can the city

planners expect the present population of 140 persons to double?

81. **Bacterial culture**   A bacterial culture doubles in size every 24 hours. By how much will it have increased in 36 hours?

82. **Oceanography**   The intensity I of a light a distance x meters beneath the surface of a lake decreases exponentially. From Illustration 2, find the depth at which the intensity will be 20%.

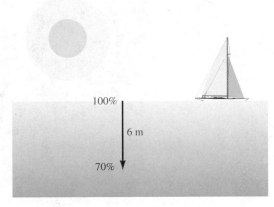

ILLUSTRATION 2

83. **Medicine**   If a medium is inoculated with a bacterial culture containing 500 cells per milliliter, how many generations will have passed by the time the culture contains 5×10^6 cells per milliliter?

84. **Medicine**   If a medium is inoculated with a bacterial culture containing 800 cells per milliliter, how many generations will have passed by the time the culture contains 6×10^7 cells per milliliter?

WRITING

85. Explain how to solve the equation $2^x = 7$.

86. Explain how to solve the equation $x^2 = 7$.

SOMETHING TO THINK ABOUT

87. Without solving the following equation, find the values of x that cannot be a solution:

$$\log (x - 3) - \log (x^2 + 2) = 0$$

88. Solve the equation $x^{\log x} = 10,000$.

■ ■ ■ ■ ■ ■ ■ ■ ■ ■ **PROJECTS**

PROJECT 1

When an object moves through air, it encounters air resistance. So far, all ballistics problems in this text have ignored air resistance. We now consider the case where an object's fall is affected by air resistance.

At relatively low velocities ($v < 200$ feet per second), the force resisting an object's motion is a constant multiple of the object's velocity:

Resisting force $= f_r = bv$

where b is a constant that depends on the size, shape, and texture of the object, and has units of kilograms per second. This is known as *Stokes' law of resistance*.

In a vacuum, the downward velocity of an object dropped with an initial velocity of 0 feet per second is

$v(t) = 32t$ (no air resistance)

t seconds after it is released. However, with air resistance, the velocity is given by the formula

$$v(t) = \frac{32m}{b}(1 - e^{-(b/m)t})$$

where m is the object's mass (in kilograms). There is also a formula for the distance an object falls (in feet) during the first t seconds after release, taking into account air resistance:

$$d(t) = \frac{32m}{b}t - \frac{32m^2}{b^2}(1 - e^{-(b/m)t})$$

Without air resistance, the formula would be

$$d(t) = 16t^2$$

a. Fearless Freda, a renowned skydiving daredevil, performs a practice dive from a hot-air balloon with an altitude of 5,000 feet. With her parachute on, Freda has a mass of 75 kg, so that $b = 15$ kg/sec. How far (to the nearest foot) will Freda fall in 5 seconds? Compare this with the answer you get by disregarding air resistance.

b. What downward velocity (to the nearest ft/sec) does Freda have after she has fallen for 2 seconds? For 5 seconds? Compare these answers with the answers you get by disregarding air resistance.

■ ■ ■ ■ ■ ■ ■ ■ ■ ■   **PROJECTS**   *(continued)*

c. Find Freda's downward velocity after falling for 20, 22, and 25 seconds (without air resistance, Freda would hit the ground in less than 18 seconds). Note that Freda's velocity increases only slightly. This is because for a large enough velocity, the force of air resistance almost counteracts the force of gravity; after Freda has been falling for a few seconds, her velocity becomes nearly constant. The constant velocity that a falling object approaches is called the *terminal velocity*.

$$\text{Terminal velocity} = \frac{32m}{b}$$

Find Freda's terminal velocity for her practice dive.

d. In Freda's show, she dives from a hot-air balloon with an altitude of only 550 feet, and pulls her ripcord when her velocity is 100 feet per second. (She can't tell her speed, but she knows how long it takes to reach that speed.) It takes a fall of 80 more feet for the chute to open fully, but then the chute increases the force of air resistance, making $b = 80$. After that, Freda's velocity approaches the terminal velocity of an object with this new b value.

To the nearest hundredth of a second, how long should Freda fall before she pulls the ripcord? To the nearest foot, how close is she to the ground when she pulls the ripcord? How close to the ground is she when the chute takes full effect? At what velocity will Freda hit the ground?

PROJECT 2   If an object at temperature T_0 is surrounded by a constant temperature T_s (for instance, an oven or a large amount of fluid that has a constant temperature), the temperature of the object will change with time t according to the formula

$$T(t) = T_s + (T_0 - T_s)e^{-kt}$$

This is *Newton's law of cooling and warming.* The number k is a constant that depends on how well the object absorbs and dispels heat.

In the course of brewing "yo ho! grog," the dread pirates of Hancock Isle have learned that it is important that their rather disgusting, soupy mash be heated slowly to allow all of the ingredients a chance to add their particular offensiveness to the mixture. However, after the mixture has simmered for several hours, it is equally important that the grog be cooled very quickly, so that it retains its potency. The kegs of grog are then stored in a cool spring.

By trial and error, the pirates have learned that by placing the mash pot into a tub of boiling water (100° C), they can heat the mash in the correct amount of time. They have also learned that they can cool the grog to the temperature of the spring by placing it in ice caves for 1 hour.

With a thermometer, you find that the pirates heat the mash from 20° C to 95° C and then cool the grog from 95° C to 7° C. Calculate how long the pirates cook the mash, and how cold the ice caves are. Assume that $k = 0.5$, and t is measured in hours.

C H A P T E R   S U M M A R Y

CONCEPT

REVIEW EXERCISES

SECTION 11.1   *Exponential Functions*

An exponential function with base b is defined by the equation

$$y = f(x) = b^x$$
$$(b > 0, b \neq 1)$$

1. Use properties of exponents to simplify.

    a. $5^{\sqrt{2}} \cdot 5^{\sqrt{2}}$ **b.** $\left(2^{\sqrt{5}}\right)^{\sqrt{2}}$

2. Graph the function defined by each equation.

    a. $y = 3^x$ **b.** $y = \left(\dfrac{1}{3}\right)^x$

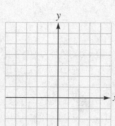 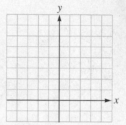

3. The graph of $f(x) = 6^x$ will pass through the points $(0, x)$ and $(1, y)$. Find x and y.

4. Give the domain and range of the function $f(x) = b^x$.

5. Graph each function by using a translation.

    a. $y = f(x) = \left(\dfrac{1}{2}\right)^x - 2$ **b.** $y = f(x) = \left(\dfrac{1}{2}\right)^{x+2}$

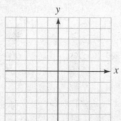 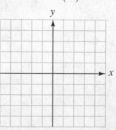

Compound interest

$$A = P\left(1 + \frac{r}{k}\right)^{kt}$$

6.   How much will \$10,500 become if it earns 9% per year for 60 years, compounded quarterly?

SECTION 11.2	*Base-e Exponential Functions*

$e = 2.71828182845904. . .$

Continuous compound interest

$A = Pe^{rt}$

7. If \$10,500 accumulates interest at an annual rate of 9%, compounded continuously, how much will be in the account in 60 years?

8. Graph each function.

    a. $y = f(x) = e^x + 1$

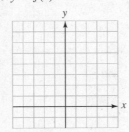

    b. $y = f(x) = e^{x-3}$

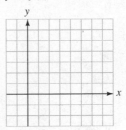

Malthusian population growth

$P = P_0 e^{kt}$

9. The population of the United States is approximately 275,000,000 people. Find the population in 50 years if $k = 0.015$.

SECTION 11.3	*Logarithmic Functions*

If $b > 0$ and $b \neq 1$, then

$y = \log_b x$ means $x = b^y$

10. Give the domain and range of the logarithmic function.

11. Find each value.

    a. $\log_3 9$ **b.** $\log_9 \dfrac{1}{3}$

    c. $\log_\pi 1$ **d.** $\log_5 0.04$

    e. $\log_a \sqrt{a}$ **f.** $\log_a \sqrt[3]{a}$

12. Find x.

    a. $\log_2 x = 5$ **b.** $\log_{\sqrt{3}} x = 4$

    c. $\log_{\sqrt{3}} x = 6$ **d.** $\log_{0.1} 10 = x$

    e. $\log_x 2 = -\dfrac{1}{3}$ **f.** $\log_x 32 = 5$

    g. $\log_{0.25} x = -1$ **h.** $\log_{0.125} x = -\dfrac{1}{3}$

    i. $\log_{\sqrt{2}} 32 = x$ **j.** $\log_{\sqrt{5}} x = -4$

    k. $\log_{\sqrt{3}} 9\sqrt{3} = x$ **l.** $\log_{\sqrt{5}} 5\sqrt{5} = x$

13. Graph each function.

a. $y = f(x) = \log(x - 2)$

b. $y = f(x) = 3 + \log x$

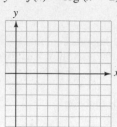 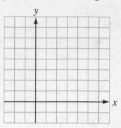

14. Graph each pair of equations on one set of coordinate axes.

a. $y = 4^x$ and $y = \log_4 x$

b. $y = \left(\dfrac{1}{3}\right)^x$ and $y = \log_{1/3} x$

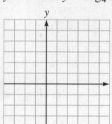 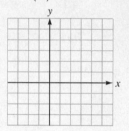

Decibel voltage gain

$$\text{db gain} = 20 \log \frac{E_O}{E_I}$$

15. An amplifier has an output of 18 volts when the input is 0.04 volt. Find the db gain.

Richter scale

$$R = \log \frac{A}{P}$$

16. An earthquake had a period of 0.3 second and an amplitude of 7,500 micrometers. Find its measure on the Richter scale.

SECTION 11.4	*Base-e* Logarithms

$\ln x$ means $\log_e x$.

17. 🔢 Use a calculator to find each value to four decimal places.

a. $\ln 452$

b. $\ln(\log 7.85)$

18. Solve each equation.

a. $\ln x = 2.336$

b. $\ln x = \log 8.8$

19. Graph each function.

a. $y = f(x) = 1 + \ln x$

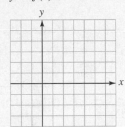

b. $y = f(x) = \ln(x + 1)$

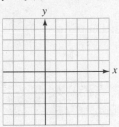

Population doubling time

$$t = \frac{\ln 2}{r}$$

20. How long will it take the population of the United States to double if the growth rate is 3% per year?

SECTION 11.5	*Properties of Logarithms*

Properties of logarithms

If b is a positive number and $b \ne 1$,

1. $\log_b 1 = 0$

2. $\log_b b = 1$

3. $\log_b b^x = x$

4. $b^{\log_b x} = x$

5. $\log_b MN = \log_b M + \log_b N$

6. $\log_b \dfrac{M}{N} = \log_b M - \log_b N$

7. $\log_b M^p = p \log_b M$

8. If $\log_b x = \log_b y$, then $x = y$.

21. Simplify each expression.

a. $\log_7 1$
b. $\log_7 7$
c. $\log_7 7^3$
d. $7^{\log_7 4}$

22. Simplify each expression.

a. $\ln e^4$
b. $\ln 1$
c. $10^{\log_{10} 7}$
d. $e^{\ln 3}$
e. $\log_b b^4$
f. $\ln e^9$

23. Write each expression in terms of the logarithms of x, y, and z.

a. $\log_b \dfrac{x^2 y^3}{z^4}$

b. $\log_b \sqrt{\dfrac{x}{yz^2}}$

24. Write each expression as the logarithm of one quantity.

a. $3 \log_b x - 5 \log_b y + 7 \log_b z$

b. $\dfrac{1}{2} \log_b x + 3 \log_b y - 7 \log_b z$

25. Assume that $\log a = 0.6$, $\log b = 0.36$, and $\log c = 2.4$. Find the value of each expression.

a. $\log abc$
b. $\log a^2 b$
c. $\log \dfrac{ac}{b}$
d. $\log \dfrac{a^2}{c^3 b^2}$

Change-of-base formula

$$\log_b y = \frac{\log_a y}{\log_a b}$$

26. 🖩 To four decimal places, find $\log_5 17$.

pH scale

$$pH = -\log[H^+]$$

27. pH of grapefruit The pH of grapefruit juice is about 3.1. Find its hydrogen ion concentration.

Weber-Fechner law

$$L = k \ln I$$

28. Find the decrease in loudness if the intensity is cut in half.

| SECTION 11.6 | *Exponential and Logarithmic Equations* |

29. Solve each equation for x, if possible.
  a. $3^x = 7$
  b. $5^{x+2} = 625$
  c. $2^x = 3^{x-1}$
  d. $2^{x^2+4x} = \dfrac{1}{8}$

30. Solve each equation for x.
  a. $\log x + \log (29 - x) = 2$
  b. $\log_2 x + \log_2 (x - 2) = 3$
  c. $\log_2 (x + 2) + \log_2 (x - 1) = 2$
  d. $\dfrac{\log (7x - 12)}{\log x} = 2$
  e. $\log x + \log (x - 5) = \log 6$
  f. $\log 3 - \log (x - 1) = -1$
  g. $e^{x \ln 2} = 9$
  h. $\ln x = \ln (x - 1)$
  i. $\ln x = \ln (x - 1) + 1$
  j. $\ln x = \log_{10} x$ (*Hint:* Use the change-of-base formula.)

Carbon dating

$$A = A_0 2^{-t/h}$$

31. Carbon-14 dating A wooden statue found in Egypt has a carbon-14 content that is two-thirds of that found in living wood. If the half-life of carbon-14 is 5,700 years, how old is the statue?

■ Chapter Test

In Problems 1–2, graph each function.

1. $f(x) = 2^x + 1$

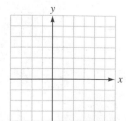

2. $f(x) = 2^{-x}$

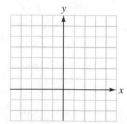

In Problems 3–4, solve each equation.

3. A radioactive material decays according to the formula $A = A_0(2)^{-t}$. How much of a 3-gram sample will be left in 6 years?

4. An initial deposit of \$1,000 earns 6% interest, compounded twice a year. How much will be in the account in one year?

5. Graph the function $f(x) = e^x$.

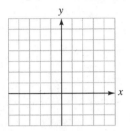

6. An account contains \$2,000 and has been earning 8% interest, compounded continuously. How much will be in the account in 10 years?

In Problems 7–12, find x.

7. $\log_4 16 = x$

8. $\log_x 81 = 4$

9. $\log_3 x = -3$

10. $\log_x 100 = 2$

11. $\log_{3/2} \dfrac{9}{4} = x$

12. $\log_{2/3} x = -3$

In Problems 13–14, graph each function.

13. $f(x) = -\log_3 x$

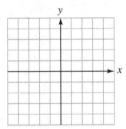

14. $f(x) = \ln x$

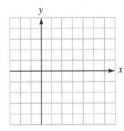

In Problems 15–16, write each expression in terms of the logarithms of a, b, and c.

15. $\log a^2bc^3$

16. $\ln \sqrt{\dfrac{a}{b^2c}}$

In Problems 17–18, write each expression as a logarithm of a single quantity.

17. $\dfrac{1}{2} \log (a + 2) + \log b - 3 \log c$

18. $\dfrac{1}{3} (\log a - 2 \log b) - \log c$

In Problems 19–20, assume that log 2 = 0.3010 and log 3 = 0.4771. Find each value. ***Do not use a calculator.***

19. $\log 24$

20. $\log \dfrac{8}{3}$

In Problems 21–22, use the change-of-base formula to find each logarithm. **Do not attempt to simplify the answer.**

21. $\log_7 3$

22. $\log_\pi e$

In Problems 23–26, tell whether each statement is true. If a statement is not true, explain why.

23. $\log_a ab = 1 + \log_a b$

24. $\dfrac{\log a}{\log b} = \log a - \log b$

25. $\log a^{-3} = \dfrac{1}{3 \log a}$

26. $\ln(-x) = -\ln x$

27. Find the pH of a solution with a hydrogen ion concentration of 3.7×10^{-7}. (*Hint:* pH $= -\log [\text{H}^+]$.)

28. Find the db gain of an amplifier when $E_O = 60$ volts and $E_I = 0.3$ volt. (*Hint:* db gain $= 20 \log (E_O/E_I)$.)

In Problems 29–30, solve each equation. **Do not simplify the logarithms.**

29. $5^x = 3$

30. $3^{x-1} = 100^x$

In Problems 31–32, solve each equation.

31. $\log(5x + 2) = \log(2x + 5)$

32. $\log x + \log(x - 9) = 1$

12 More Graphing and Conic Sections

■ ■ ■ ■ ■ ■ ■ ■ ■ ■   **TEACHER**

MATHEMATICS IN THE
WORKPLACE

Secondary school teachers help students delve more deeply into subjects introduced in elementary school and learn more about the world and about themselves. Teachers plan and evaluate lessons, prepare tests, grade papers, prepare
report cards, oversee study halls and homerooms, supervise extracurricular ac-
tivities, and meet with parents and school staff. In recent years, teachers have become more involved in curriculum design, such as choosing textbooks and
evaluating teaching methods.

SAMPLE APPLICATION ■ An art teacher wants to paint a mural on an elliptical background that is 10 feet wide and 6 feet high. To see how to do this construction, see page 890.

We have seen that the graphs of linear functions are straight lines, and that the
graphs of quadratic functions are parabolas. In this chapter, we will discuss some
special curves, called **conic sections**. We will then discuss **piecewise-defined functions** and **step functions**.

12.1   THE CIRCLE AND THE PARABOLA

■ INTRODUCTION TO THE CONIC SECTIONS ■ THE CIRCLE ■ PROBLEM SOLVING ■ THE PARABOLA
■ PROBLEM SOLVING

Getting Ready   *Square each binomial.*

1. $(x - 2)^2$ **2.** $(x + 4)^2$

What number must be added to each binomial to make it a perfect square trinomial?

3. $x^2 + 9x$ **4.** $x^2 - 12x$

■   INTRODUCTION TO THE CONIC SECTIONS

The graphs of second-degree equations in x and y represent figures that were fully
investigated in the 17th century by René Descartes (1596–1650) and Blaise Pascal
(1623–1662). Descartes discovered that graphs of second-degree equations fall into
one of several categories: a pair of lines, a point, a circle, a parabola, an ellipse, a
hyperbola, or no graph at all. Because all of these graphs can be formed by the intersection of a plane and a right-circular cone, they are called **conic sections**. See
Figure 12-1.

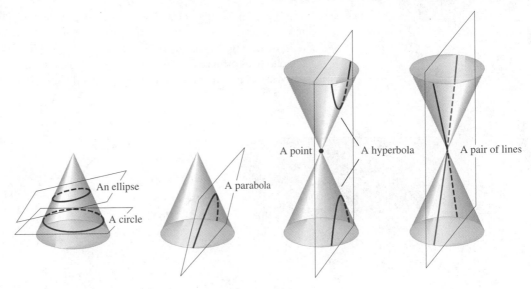

FIGURE 12-1

Conic sections have many applications. For example, everyone knows the importance of circular wheels and gears, pizza cutters, and ferris wheels.

Parabolas can be rotated to generate dish-shaped surfaces called *paraboloids*. Any light or sound placed at the *focus* of a paraboloid is reflected outward in parallel paths, as shown in Figure 12-2(a). This property makes parabolic surfaces ideal for flashlight and headlight reflectors. It also makes parabolic surfaces good antennas, because signals captured by such antennas are concentrated at the focus. Parabolic mirrors are capable of concentrating the rays of the sun at a single point and thereby generating tremendous heat. This property is used in the design of certain solar furnaces.

Any object thrown upward and outward travels in a parabolic path, as shown in Figure 12-2(b). In architecture, many arches are parabolic in shape, because this gives strength. Cables that support suspension bridges hang in the form of a parabola. (See Figure 12-2(c).)

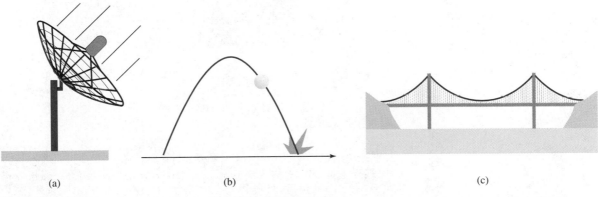

(a)   (b)   (c)

FIGURE 12-2

Ellipses have optical and acoustical properties that are useful in architecture and engineering. For example, many arches are portions of an ellipse, because the shape is pleasing to the eye. (See Figure 12-3(a).) The planets and many comets have elliptical orbits. (See Figure 12-3(b).) Gears are often cut into elliptical shapes to provide nonuniform motion. (See Figure 12-3(c).)

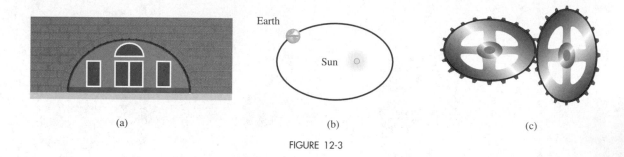

(a)   (b)   (c)

FIGURE 12-3

Hyperbolas serve as the basis of a navigational system known as LORAN (LOng RAnge Navigation). (See Figure 12-4.) They are also used to find the source of a distress signal, are the basis for the design of hypoid gears, and describe the orbits of some comets.

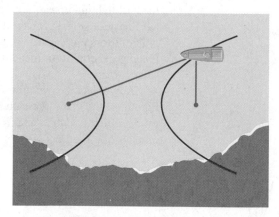

FIGURE 12-4

■ THE CIRCLE

The Circle

A **circle** is the set of all points in a plane that are a fixed distance from a point called its **center**.

The fixed distance is the **radius** of the circle.

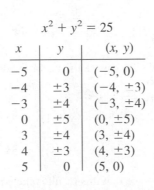

FIGURE 12-5

To develop the general equation of a circle, we must write the equation of a circle with a radius of r and with center at some point $C(h, k)$, as in Figure 12-5. This task is equivalent to finding all points $P(x, y)$ such that the length of line segment CP is r. We can use the distance formula to find r.

$$r = \sqrt{(x - h)^2 + (y - k)^2}$$

We then square both sides to obtain

1. $\quad r^2 = (x - h)^2 + (y - k)^2$

Equation 1 is called the **standard form of the equation of a circle** with a radius of r and center at the point with coordinates (h, k).

Standard Equation of a Circle with Center at (*h, k*)

Any equation that can be written in the form

$$(x - h)^2 + (y - k)^2 = r^2$$

has a graph that is a circle with radius r and center at point (h, k).

If $r = 0$, the graph reduces to a single point called a **point circle**. If $r^2 < 0$, a circle does not exist. If both coordinates of the center are 0, the center of the circle is the origin.

Standard Equation of a Circle with Center at (0, 0)

Any equation that can be written in the form

$$x^2 + y^2 = r^2$$

has a graph that is a circle with radius r and center at the origin.

EXAMPLE 1   Graph $x^2 + y^2 = 25$.

Solution   Because this equation can be written in the form $x^2 + y^2 = r^2$, its graph is a circle with center at the origin. Since $r^2 = 25 = 5^2$, the circle has a radius of 5. The graph appears in Figure 12-6.

$$x^2 + y^2 = 25$$

x	y	(x, y)
-5	0	$(-5, 0)$
-4	± 3	$(-4, \pm 3)$
-3	± 4	$(-3, \pm 4)$
0	± 5	$(0, \pm 5)$
3	± 4	$(3, \pm 4)$
4	± 3	$(4, \pm 3)$
5	0	$(5, 0)$

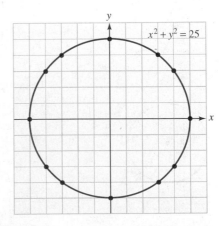

FIGURE 12-6

Self Check
Answer

Graph $x^2 + y^2 = 4$.

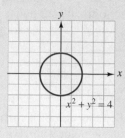

$x^2 + y^2 = 4$

EXAMPLE 2

Find the equation of the circle with radius 5 and center at $C(3, 2)$.

Solution

We substitute 5 for r, 3 for h, and 2 for k in the standard form of a circle and simplify.

$$(x - h)^2 + (y - k)^2 = r^2$$
$$(x - 3)^2 + (y - 2)^2 = 5^2$$
$$x^2 - 6x + 9 + y^2 - 4y + 4 = 25$$
$$x^2 + y^2 - 6x - 4y - 12 = 0$$

The equation is $x^2 + y^2 - 6x - 4y - 12 = 0$. ∎

Self Check
Answer

Find the equation of the circle with radius 6 and center at $(2, 3)$.
$x^2 + y^2 - 4x - 6y - 23 = 0$

EXAMPLE 3

Graph the equation $x^2 + y^2 - 4x + 2y = 20$.

Solution

To identify the curve, we complete the square on x and y and write the equation in standard form.

$$x^2 + y^2 - 4x + 2y = 20$$
$$x^2 - 4x + y^2 + 2y = 20$$

To complete the square on x and y, add 4 and 1 to both sides.

$$x^2 - 4x + \mathbf{4} + y^2 + 2y + \mathbf{1} = 20 + \mathbf{4} + \mathbf{1}$$
$$(x - 2)^2 + (y + 1)^2 = 25 \qquad \text{Factor } x^2 - 4x + 4 \text{ and } y^2 + 2y + 1.$$
$$(x - 2)^2 + [y - (-1)]^2 = 5^2$$

We can now see that this result is the equation of a circle with a radius of 5 and center at $h = 2$ and $k = -1$. If we plot the center and draw a circle with a radius of 5 units, we will obtain the circle shown in Figure 12-7. ∎

y

radius
5

$(2, -1)$

x

$x^2 + y^2 - 4x + 2y = 20$

FIGURE 12-7

Self Check

Write the equation $x^2 + y^2 + 2x - 4y - 11 = 0$ in standard form and graph it.

Answer

$(x + 1)^2 + (y - 2)^2 = 16$

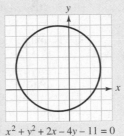

$x^2 + y^2 + 2x - 4y - 11 = 0$

■ ■ ■ ■ ■ ■ ■ ■ ■ ■ Graphing Circles

GRAPHING
CALCULATORS

Since the graphs of circles fail the vertical line test, their equations do not represent functions. It is somewhat more difficult to use a graphing calculator to graph equations that are not functions. For example, to graph the circle described by $(x - 1)^2 + (y - 2)^2 = 4$, we must split the equation into two functions and graph each one separately. We begin by solving the equation for y.

$$(x - 1)^2 + (y - 2)^2 = 4$$

$$(y - 2)^2 = 4 - (x - 1)^2 \qquad \text{Subtract } (x - 1)^2 \text{ from both sides.}$$

$$y - 2 = \pm \sqrt{4 - (x - 1)^2} \qquad \text{Take the square root of both sides.}$$

$$y = 2 \pm \sqrt{4 - (x - 1)^2} \qquad \text{Add 2 to both sides.}$$

This equation defines two functions. If we use window settings of $[-3, 5]$ for x and $[-3, 5]$ for y and graph the functions

$$y = 2 + \sqrt{4 - (x - 1)^2} \qquad \text{and} \qquad y = 2 - \sqrt{4 - (x - 1)^2}$$

we get the distorted circle shown in Figure 12-8(a). To get a better circle, graphing calculators have a squaring feature that gives an equal unit distance on both the x- and y-axes. After using this feature, we get the circle shown in Figure 12-8(b).

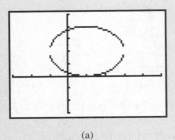

(a)

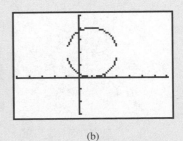

(b)

FIGURE 12-8

■ PROBLEM SOLVING

EXAMPLE 4

Radio translators   The broadcast area of a television station is bounded by the circle $x^2 + y^2 = 3{,}600$, where x and y are measured in miles. A translator station picks up the signal and retransmits it from the center of a circular area bounded by $(x + 30)^2 + (y - 40)^2 = 1{,}600$. Find the location of the translator and the greatest distance from the main transmitter that the signal can be received.

Solution   The coverage of the television station is bounded by $x^2 + y^2 = 60^2$, a circle centered at the origin with a radius of 60 miles, as shown in Figure 12-9. Because the translator is at the center of the circle $(x + 30)^2 + (y - 40)^2 = 1{,}600$, it is located at $(-30, 40)$, a point 30 miles west and 40 miles north of the television station. The radius of the translator's coverage is $\sqrt{1{,}600}$, or 40 miles.

As shown in Figure 12-9, the greatest distance of reception is the sum of A, the distance from the translator to the television station, and 40 miles, the radius of the translator's coverage.

To find A, we use the distance formula to find the distance between $(x_1, y_1) = (-30, 40)$ and the origin, $(x_2, y_2) = (0, 0)$.

$$A = \sqrt{(x_1 - x_2)^2 + (y_1 - y_2)^2}$$
$$A = \sqrt{(-30 - 0)^2 + (40 - 0)^2}$$
$$= \sqrt{(-30)^2 + 40^2}$$
$$= \sqrt{2{,}500}$$
$$= 50$$

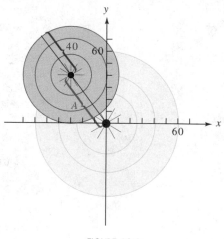

FIGURE 12-9

The translator is located 50 miles from the television station, and it broadcasts the signal an additional 40 miles. The greatest reception distance is $50 + 40$, or 90 miles.   ■

■ THE PARABOLA

We have seen that equations of the form $y = a(x - h)^2 + k$, with $a \neq 0$, represent parabolas with the vertex at the point (h, k). They open upward when $a > 0$ and downward when $a < 0$.

Equations of the form $x = a(y - k)^2 + h$ also represent parabolas with vertex at point (h, k). However, they open to the right when $a > 0$ and to the left when

$a < 0$. Parabolas that open to the right or left do not represent functions, because their graphs fail the vertical line test.

Several types of parabolas are summarized in the following chart.

Equations of Parabolas ($a > 0$)

Parabola opening	Vertex at origin	Vertex at (h, k)
Up	$y = ax^2$	$y = a(x - h)^2 + k$
Down	$y = -ax^2$	$y = -a(x - h)^2 + k$
Right	$x = ay^2$	$x = a(y - k)^2 + h$
Left	$x = -ay^2$	$x = -a(y - k)^2 + h$

EXAMPLE 5   Graph   **a.** $x = \frac{1}{2}y^2$ and   **b.** $x = -2(y - 2)^2 + 3$.

Solution   **a.** We make a table of ordered pairs that satisfy the equation, plot each pair, and draw the parabola, as in Figure 12-10(a). Because the equation is of the form $x = ay^2$ with $a > 0$, the parabola opens to the right and has its vertex at the origin.

b. We make a table of ordered pairs that satisfy the equation, plot each pair, and draw the parabola, as in Figure 12-10(b). Because the equation is of the form $x = -a(y - k)^2 + h$, the parabola opens to the left and has its vertex at the point with coordinates $(3, 2)$.

$x = \frac{1}{2}y^2$

x	y	(x, y)
0	0	$(0, 0)$
2	2	$(2, 2)$
2	-2	$(2, -2)$
8	4	$(8, 4)$
8	-4	$(8, -4)$

$x = -2(y - 2)^2 + 3$

x	y	(x, y)
-5	0	$(-5, 0)$
1	1	$(1, 1)$
3	2	$(3, 2)$
1	3	$(1, 3)$
-5	4	$(-5, 4)$

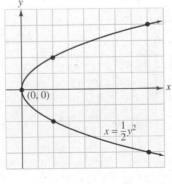

(a)

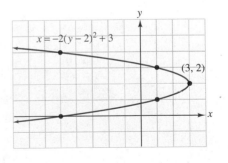

(b)

FIGURE 12-10

Self Check
Answer

Graph $x = \frac{1}{2}(y - 1)^2 - 2$.

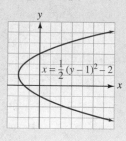

$x = \frac{1}{2}(y-1)^2 - 2$

EXAMPLE 6

Graph $y = -2x^2 + 12x - 15$.

Solution

Because the equation is not in standard form, the coordinates of the vertex of its graph are not obvious. To write the equation in standard form, we complete the square on x.

$$y = -2x^2 + 12x - 15$$
$$y = -2(x^2 - 6x) - 15 \qquad \text{Factor out } -2 \text{ from } -2x^2 + 12x.$$
$$y = -2(x^2 - 6x + 9) - 15 + 18 \qquad \text{Subtract and add 18; } -2(9) = -18.$$
$$y = -2(x - 3)^2 + 3$$

Because the equation is written in the form $y = -a(x - h)^2 + k$, we can see that the parabola opens downward and has its vertex at $(3, 3)$. The graph of the function is shown in Figure 12-11.

$y = -2x^2 + 12x - 15$

x	y	(x, y)
1	-5	$(1, -5)$
2	1	$(2, 1)$
3	3	$(3, 3)$
4	1	$(4, 1)$
5	-5	$(5, -5)$

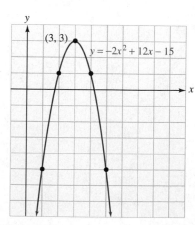

FIGURE 12-11

Self Check
Answer

Graph $y = 0.5x^2 - x - 1$.

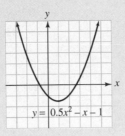

$y = 0.5x^2 - x - 1$

■ PROBLEM SOLVING

EXAMPLE 7

Gateway Arch   The shape of the Gateway Arch in St. Louis is approximately a parabola, as shown in Figure 12-12(a). How high is the arch 100 feet from its foundation?

Solution   We place the parabola in a coordinate system as in Figure 12-12(b), with ground level on the x-axis and the vertex of the parabola at the point $(h, k) = (0, 630)$. The equation of this downward-opening parabola has the form

$$y = -a(x - \boldsymbol{h})^2 + k$$
$$y = -a(x - \boldsymbol{0})^2 + 630 \qquad \text{Substitute } h = 0 \text{ and } y = 630.$$
$$y = -ax^2 + 630 \qquad \text{Simplify.}$$

Because the Gateway Arch is 630 feet wide at its base, the parabola passes through the point $\left(\frac{630}{2}, 0\right)$, or $(315, 0)$. To find a in the equation of the parabola, we proceed as follows:

$$y = -ax^2 + 630$$
$$0 = -a(315)^2 + 630 \qquad \text{Substitute 315 for } x \text{ and 0 for } y.$$
$$\frac{-630}{315^2} = -a \qquad \text{Subtract 630 from both sides and divide both sides by } 315^2.$$
$$\frac{2}{315} = a \qquad \text{Multiply both sides by } -1; \frac{630}{315^2} = \frac{2}{315}.$$

The equation of the parabola that approximates the shape of the Gateway Arch is

$$y = -\frac{2}{315} x^2 + 630$$

To find the height of the arch at a point 100 feet from its foundation, we substitute $315 - 100$, or 215, for x in the equation of the parabola and solve for y.

$$y = -\frac{2}{315}x^2 + 630$$

$$y = -\frac{2}{315}(215)^2 + 630$$

$$= 336.5079365$$

At a point 100 feet from the foundation, the height of the arch is about 337 feet.

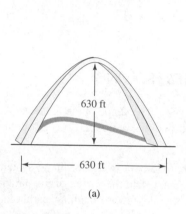

(a)

(b)

FIGURE 12-12

Orals *Find the center and the radius of each circle.*

1. $x^2 + y^2 = 144$ **2.** $x^2 + y^2 = 121$

3. $(x - 2)^2 + y^2 = 16$ **4.** $x^2 + (y + 1)^2 = 9$

Tell whether each parabola opens up or down or left or right.

5. $y = -3x^2 - 2$ **6.** $y = 7x^2 - 5$

7. $x = -3y^2$ **8.** $x = (y - 3)^2$

EXERCISE 12.1

REVIEW *Solve each equation.*

1. $|3x - 4| = 11$

2. $\left|\frac{4 - 3x}{5}\right| = 12$

3. $|3x + 4| = |5x - 2|$

4. $|6 - 4x| = |x + 2|$

VOCABULARY AND CONCEPTS *Fill in each blank to make a true statement.*

5. A _____ is the set of all points in a _____ that are a fixed distance from a given point.

6. The fixed distance in Exercise 5 is called the _____ of the circle, and the point is called its _____.

7. If _____ in the equation $x^2 + y^2 = r^2$, no circle exists.

8. The graph of $y = ax^2$ ($a > 0$) is a _____ with vertex at the _____ that opens _____.

9. The graph of $x = a(y - 2)^2 + 3$ ($a > 0$) is a _____ with vertex at _____ that opens to the _____.

10. The graph of $x = -a(y - 1)^2 - 3$ ($a > 0$) is a _____ with vertex at _____ that opens to the _____.

PRACTICE *In Exercises 11–20, graph each equation.*

11. $x^2 + y^2 = 9$

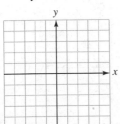

12. $x^2 + y^2 = 16$

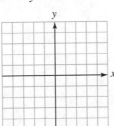

13. $(x - 2)^2 + y^2 = 9$

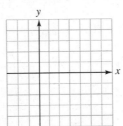

14. $x^2 + (y - 3)^2 = 4$

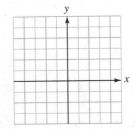

15. $(x - 2)^2 + (y - 4)^2 = 4$

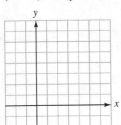

16. $(x - 3)^2 + (y - 2)^2 = 4$

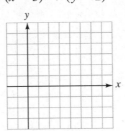

17. $(x + 3)^2 + (y - 1)^2 = 16$

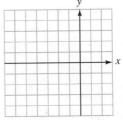

18. $(x - 1)^2 + (y + 4)^2 = 9$

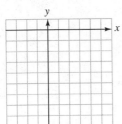

19. $x^2 + (y + 3)^2 = 1$

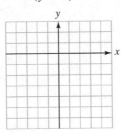

20. $(x + 4)^2 + y^2 = 1$

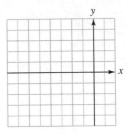

In Exercises 21–24, use a graphing calculator to graph each equation.

21. $3x^2 + 3y^2 = 16$

22. $2x^2 + 2y^2 = 9$

23. $(x + 1)^2 + y^2 = 16$

24. $x^2 + (y - 2)^2 = 4$

In Exercises 25–32, write the equation of the circle with the following properties.

25. Center at origin; radius 1

26. Center at origin; radius 4

27. Center at (6, 8); radius 5

28. Center at (5, 3); radius 2

29. Center at $(-2, 6)$; radius 12

30. Center at $(5, -4)$; radius 6

31. Center at the origin; diameter of $2\sqrt{2}$

32. Center at the origin; diameter of $8\sqrt{3}$

In Exercises 33–40, graph each circle. Give the coordinates of the center.

33. $x^2 + y^2 + 2x - 8 = 0$

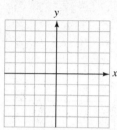

34. $x^2 + y^2 - 4y = 12$

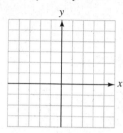

35. $9x^2 + 9y^2 - 12y = 5$

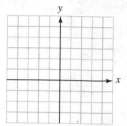

36. $4x^2 + 4y^2 + 4y = 15$

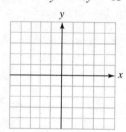

37. $x^2 + y^2 - 2x + 4y = -1$

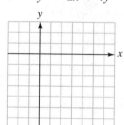

38. $x^2 + y^2 + 4x + 2y = 4$

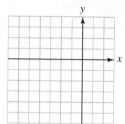

39. $x^2 + y^2 + 6x - 4y = -12$

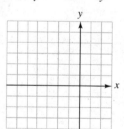

40. $x^2 + y^2 + 8x + 2y = -13$

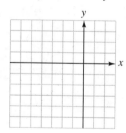

In Exercises 41–52, find the vertex of each parabola and graph it.

41. $x = y^2$

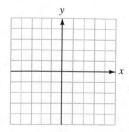

42. $x = -y^2 + 1$

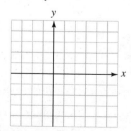

43. $x = -\dfrac{1}{4}y^2$

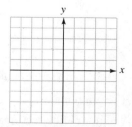

44. $x = 4y^2$

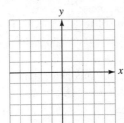

45. $y = x^2 + 4x + 5$

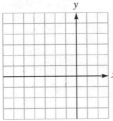

46. $y = -x^2 - 2x + 3$

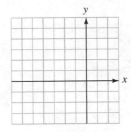

47. $y = -x^2 - x + 1$

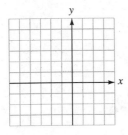

48. $x = \dfrac{1}{2}y^2 + 2y$

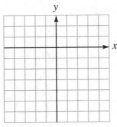

49. $y^2 + 4x - 6y = -1$

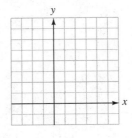

50. $x^2 - 2y - 2x = -7$

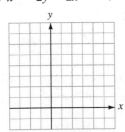

51. $y = 2(x - 1)^2 + 3$

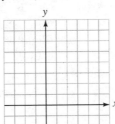

52. $y = -2(x + 1)^2 + 2$

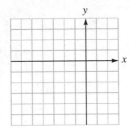

In Exercises 53–56, use a graphing calculator to graph each equation.

53. $x = 2y^2$ **54.** $x = y^2 - 4$ **55.** $x^2 - 2x + y = 6$ **56.** $x = -2(y - 1)^2 + 2$

APPLICATIONS

57. Meshing gears   For design purposes, the large gear in Illustration 1 is the circle $x^2 + y^2 = 16$. The smaller gear is a circle centered at $(7, 0)$ and tangent to the larger circle. Find the equation of the smaller gear.

58. Width of a walkway   The walkway in Illustration 2 is bounded by the two circles $x^2 + y^2 = 2{,}500$ and $(x - 10)^2 + y^2 = 900$, measured in feet. Find the

largest and the smallest width of the walkway.

59. Broadcast ranges   Radio stations applying for licensing may not use the same frequency if their broadcast areas overlap. One station's coverage is bounded by $x^2 + y^2 - 8x - 20y + 16 = 0$, and the other's by $x^2 + y^2 + 2x + 4y - 11 = 0$. May they be licensed for the same frequency?

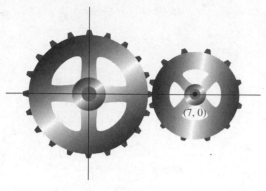

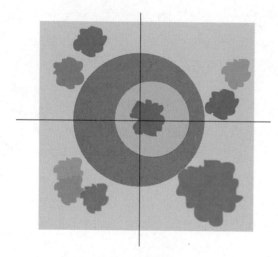

ILLUSTRATION 1

ILLUSTRATION 2

circle is $x^2 + y^2 - 16x - 20y + 155 = 0$, where distances are measured in kilometers. Find the locations (relative to the center of town) of the intersections of the highway with State and with Main.

61. Flight of a projectile   The cannonball in Illustration 4 follows the parabolic trajectory $y = 30x - x^2$. Where does it land?

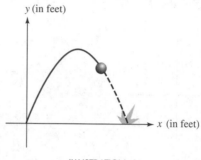

ILLUSTRATION 4

62. Flight of a projectile   In Exercise 61, how high does the cannonball get?

63. Orbit of a comet   If the orbit of the comet shown in Illustration 5 is given by the equation $2y^2 - 9x = 18$, how far is it from the sun at the vertex of the orbit? Distances are measured in astronomical units (AU).

60. Highway design   Engineers want to join two sections of highway with a curve that is one-quarter of a circle, as in Illustration 3. The equation of the

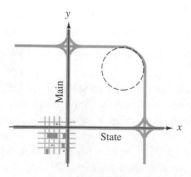

ILLUSTRATION 3

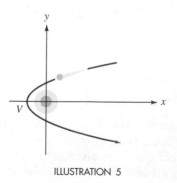

ILLUSTRATION 5

64. Satellite antenna The cross section of the satellite antenna in Illustration 6 is a parabola given by the equation $y = \frac{1}{16}x^2$, with distances measured in feet. If the dish is 8 feet wide, how deep is it?

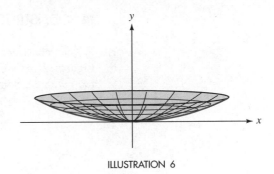

ILLUSTRATION 6

WRITING

65. Explain how to decide from its equation whether the graph of a parabola opens up, down, right, or left.

66. From the equation of a circle, explain how to determine the radius and the coordinates of the center.

SOMETHING TO THINK ABOUT

67. From the values of *a, h,* and *k,* explain how to determine the number of *x*-intercepts of the graph of $y = a(x - h)^2 + k$.

68. Under what conditions will the graph of $x = a(y - k)^2 + h$ have no *y*-intercepts?

12.2 The Ellipse

■ THE ELLIPSE ■ CONSTRUCTING AN ELLIPSE ■ GRAPHING ELLIPSES ■ PROBLEM SOLVING

Getting Ready   *Solve each equation for the indicated variable.*

1. $\dfrac{y^2}{b^2} = 1$ for *y*

2. $\dfrac{x^2}{a^2} = 1$ for *x*

■ THE ELLIPSE

The Ellipse

An **ellipse** is the set of all points *P* in the plane the sum of whose distances from two fixed points is a constant. See Figure 12-13, in which $d_1 + d_2$ is a constant.

Each of the two points is called a **focus**. Midway between the foci is the *center* of the ellipse.

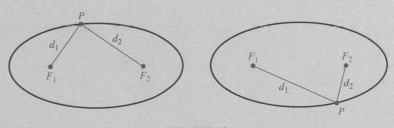

FIGURE 12-13

■ CONSTRUCTING AN ELLIPSE

We can construct an ellipse by placing two thumbtacks fairly close together, as in Figure 12-14. We then tie each end of a piece of string to a thumbtack, catch the loop with the point of a pencil, and, while keeping the string taut, draw the ellipse.

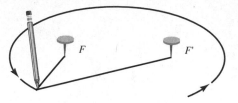

FIGURE 12-14

Using this method, we can construct an ellipse of any specific size. For example, to construct an ellipse that is 10 feet wide and 6 feet high, we must find the length of string to use and the distance between thumbtacks.

To do this, we will let a represent the distance between the center and vertex V, as shown in Figure 12-15(a). We will also let c represent the distance between the center of the ellipse and either focus. When the pencil is at vertex V, the length of the string is $c + a + (a - c)$, or just $2a$. Because $2a$ is the 10-foot width of the ellipse, the string needs to be 10 feet long. The distance $2a$ is constant for any point on the ellipse, including point B shown in Figure 12-15(b).

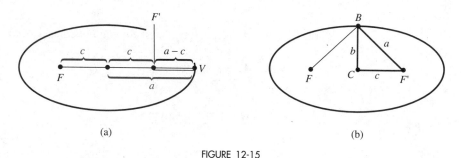

(a)   (b)

FIGURE 12-15

From right triangle CBF' and the Pythagorean theorem, we can find c as follows:

$$a^2 = b^2 + c^2 \qquad \text{or} \qquad c = \sqrt{a^2 - b^2}$$

Since distance b is one-half of the height of the ellipse, $b = 3$. Since $2a = 10$, $a = 5$. We can now substitute $a = 5$ and $b = 3$ into the formula to find c:

$$c = \sqrt{5^2 - 3^2}$$
$$= \sqrt{25 - 9}$$
$$= \sqrt{16}$$
$$= 4$$

Since $c = 4$, the distance between the thumbtacks must be 8 feet. We can construct the ellipse by tying a 10-foot string to thumbtacks that are 8 feet apart.

■ GRAPHING ELLIPSES

To graph ellipses, we can make a table of ordered pairs that satisfy the equation, plot them, and join the points with a smooth curve.

EXAMPLE 1   Graph $\dfrac{x^2}{36} + \dfrac{y^2}{9} = 1$.

Solution   First we note that the equation can be written in the form

$$\frac{x^2}{6^2} + \frac{y^2}{3^2} = 1 \qquad 36 = 6^2 \text{ and } 9 = 3^2.$$

After making a table of ordered pairs that satisfy the equation, plotting each of them, and joining the points with a curve, we obtain the ellipse shown in Figure 12-16.

We note that the center of the ellipse is the origin, the ellipse intersects the x-axis at points $(6, 0)$ and $(-6, 0)$, and the ellipse intersects the y-axis at points $(0, 3)$ and $(0, -3)$.

$$\frac{x^2}{36} + \frac{y^2}{9} = 1$$

x	y	(x, y)
-6	0	$(-6, 0)$
-4	± 2.2	$(-4, \pm 2.2)$
-2	± 2.8	$(-2, \pm 2.8)$
0	± 3	$(0, \pm 3)$
2	± 2.8	$(2, \pm 2.8)$
4	± 2.2	$(4, \pm 2.2)$
6	0	$(6, 0)$

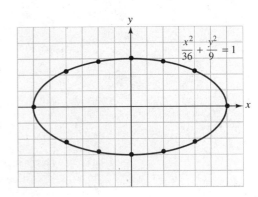

FIGURE 12-16   ■

Self Check   Graph $\dfrac{x^2}{4} + \dfrac{y^2}{16} = 1$.

Answer

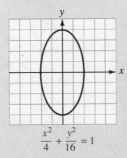

Example 1 illustrates that the graph of

$$\frac{x^2}{a^2} + \frac{y^2}{b^2} = 1$$

is an ellipse centered at the origin. To find the x-intercepts of the graph, we can let $y = 0$ and solve for x.

$$\frac{x^2}{a^2} + \frac{0^2}{b^2} = 1$$

$$\frac{x^2}{a^2} + 0 = 1$$

$$x^2 = a^2$$

$$x = a \quad \text{or} \quad x = -a$$

The x-intercepts are $(a, 0)$ and $(-a, 0)$.

To find the y-intercepts, we let $x = 0$ and solve for y.

$$\frac{0^2}{a^2} + \frac{y^2}{b^2} = 1$$

$$0 + \frac{y^2}{b^2} = 1$$

$$y^2 = b^2$$

$$y = b \quad \text{or} \quad y = -b$$

The y-intercepts are $(0, b)$ and $(0, -b)$.

In general, we have the following results.

Equations of an Ellipse Centered at the Origin

The equation of an ellipse centered at the origin, with x-intercepts at $V_1(a, 0)$ and $V_2(-a, 0)$ and with y-intercepts of $(0, b)$ and $(0, -b)$, is

$$\frac{x^2}{a^2} + \frac{y^2}{b^2} = 1 \quad (a > b > 0) \qquad \text{See Figure 12-17(a)}.$$

The equation of an ellipse centered at the origin, with y-intercepts at $V_1(0, a)$ and $V_2(0, -a)$ and x-intercepts at $(b, 0)$ and $(-b, 0)$, is

$$\frac{x^2}{b^2} + \frac{y^2}{a^2} = 1 \quad (a > b > 0) \qquad \text{See Figure 12-17(b)}.$$

In Figure 12-17, the points V_1 and V_2 are the **vertices** of the ellipse, the midpoint of V_1V_2 is the **center** of the ellipse, and the distance between the center of the ellipse and either vertex is a. The segment V_1V_2 is called the **major axis**, and the segment joining either $(0, b)$ and $(0, -b)$ or $(b, 0)$ and $(-b, 0)$ is called the **minor axis**.

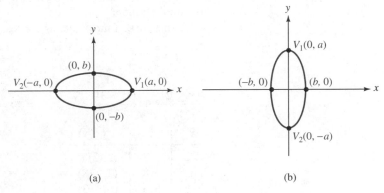

(a)   (b)

FIGURE 12-17

The equations for ellipses centered at (h, k) are as follows.

Equations of an Ellipse Centered at (h, k)

The equation of an ellipse centered at (h, k), with major axis parallel to the x-axis, is

1. $$\frac{(x - h)^2}{a^2} + \frac{(y - k)^2}{b^2} = 1 \quad (a > b > 0)$$

The equation of an ellipse centered at (h, k), with major axis parallel to the y-axis, is

2. $$\frac{(x - h)^2}{b^2} + \frac{(y - k)^2}{a^2} = 1 \quad (a > b > 0)$$

EXAMPLE 2   Graph the ellipse $\dfrac{(x - 2)^2}{16} + \dfrac{(y + 3)^2}{25} = 1$.

Solution   We first write the equation in the form

$$\frac{(x - 2)^2}{4^2} + \frac{[y - (-3)]^2}{5^2} = 1 \quad (5 > 4)$$

This is the equation of an ellipse centered at $(h, k) = (2, -3)$ with major axis parallel to the y-axis and with $b = 4$ and $a = 5$. We first plot the center, as shown in Figure 12-18. Since a is the distance from the center to the vertex, we can locate the vertices by counting 5 units above and 5 units below the center. The vertices are at points $(2, 2)$ and $(2, -8)$.

Since $b = 4$, we can locate two more points on the ellipse by counting 4 units to the left and 4 units to the right of the center. The points $(-2, -3)$ and $(6, -3)$ are also on the graph.

Using these four points as guides, we can draw the ellipse.

$$\frac{(x - 2)^2}{16} + \frac{(y + 3)^2}{25} = 1$$

x	y	(x, y)
2	2	$(2, 2)$
2	-8	$(2, -8)$
6	-3	$(6, -3)$
-2	-3	$(-2, -3)$

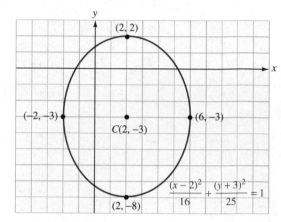

FIGURE 12-18

Self Check Graph $\dfrac{(x - 1)^2}{9} + \dfrac{(y + 2)^2}{16} = 1$.

Answer

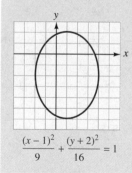

$$\frac{(x - 1)^2}{9} + \frac{(y + 2)^2}{16} = 1$$

■ ■ ■ ■ ■ ■ ■ ■ ■ **Graphing Ellipses**

GRAPHING CALCULATORS

To use a graphing calculator to graph

$$\frac{(x + 2)^2}{4} + \frac{(y - 1)^2}{25} = 1$$

■ ■ ■ ■ ■ ■ ■ ■ ■ **Graphing Ellipses** *(continued)*

we first clear the fractions by multiplying both sides by 100 and solving for y.

$$25(x + 2)^2 + 4(y - 1)^2 = 100$$ Multiply both sides by 100.

$$4(y - 1)^2 = 100 - 25(x + 2)^2$$ Subtract $25(x + 2)^2$ from both sides.

$$(y - 1)^2 = \frac{100 - 25(x + 2)^2}{4}$$ Divide both sides by 4.

$$y - 1 = \pm\frac{\sqrt{100 - 25(x + 2)^2}}{2}$$ Take the square root of both sides.

$$y = 1 \pm \frac{\sqrt{100 - 25(x + 2)^2}}{2}$$ Add 1 to both sides.

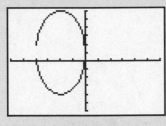

FIGURE 12-19

If we use window settings $[-6, 6]$ for x and $[-6, 6]$ for y and graph the functions

$$y = 1 + \frac{\sqrt{100 - 25(x + 2)^2}}{2} \quad \text{and} \quad y = 1 - \frac{\sqrt{100 - 25(x + 2)^2}}{2}$$

we will obtain the ellipse shown in Figure 12-19.

We can use completing the square to write the equations of many ellipses in standard form.

EXAMPLE 3   Write $4x^2 + 9y^2 - 16x - 18y = 11$ in standard form to show that the equation represents an ellipse.

Solution   We write the equation in standard form by completing the square on x and y:

$$4x^2 + 9y^2 - 16x - 18y = 11$$

$$4x^2 - 16x + 9y^2 - 18y = 11$$ Use the commutative property to rearrange terms.

$$4(x^2 - 4x) + 9(y^2 - 2y) = 11$$ Factor 4 from $4x^2 - 16x$ and factor 9 from $9y^2 - 18y$ to get coefficients of 1 for the squared terms.

$$4(x^2 - 4x + \mathbf{4}) + 9(y^2 - 2y + \mathbf{1}) = 11 + 16 + 9$$ Complete the square to make $x^2 - 4x$ and $y^2 - 2y$ perfect trinomial squares. Since $16 + 9$ is added to the left-hand side, add $16 + 9$ to the right-hand side.

$$4(x-2)^2 + 9(y-1)^2 = 36 \qquad \text{Factor } x^2 - 4x + 4 \text{ and } y^2 - 2y + 1.$$

$$\frac{(x-2)^2}{9} + \frac{(y-1)^2}{4} = 1 \qquad \text{Divide both sides by 36.}$$

Since this equation matches Equation 1 on page 893, it represents an ellipse. Its graph is shown in Figure 12-20. ■

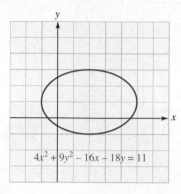

FIGURE 12-20

Self Check   Graph $4x^2 - 8x + 9y^2 - 36y = -4$.

Answer

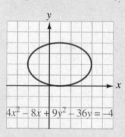

■ PROBLEM SOLVING

EXAMPLE 4   **Landscape design**   A landscape architect is designing an elliptical pool that will fit in the center of a 20-by-30-foot rectangular garden, leaving at least 5 feet of space on all sides. Find the equation of the ellipse.

Solution   We place the rectangular garden in a coordinate system, as in Figure 12-21. To maintain 5 feet of clearance at the ends of the ellipse, the vertices must be the points $V_1(10, 0)$ and $V_2(-10, 0)$. Similarly, the y-intercepts are the points $(0, 5)$ and $(0, -5)$.

The equation of the ellipse has the form

$$\frac{x^2}{a^2} + \frac{y^2}{b^2} = 1$$

with $a = 10$ and $b = 5$. Thus, the equation of the boundary of the pool is

$$\frac{x^2}{100} + \frac{y^2}{25} = 1$$

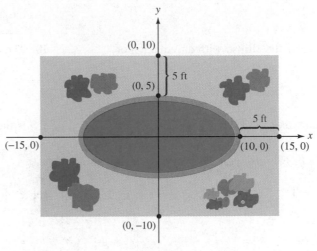

FIGURE 12-21

Orals *Find the x- and y-intercepts of each ellipse.*

1. $\dfrac{x^2}{9} + \dfrac{y^2}{16} = 1$

2. $\dfrac{x^2}{25} + \dfrac{y^2}{36} = 1$

Find the center of each ellipse.

3. $\dfrac{(x-2)^2}{9} + \dfrac{y^2}{16} = 1$

4. $\dfrac{x^2}{25} + \dfrac{(y+1)^2}{36} = 1$

EXERCISE 12.2

REVIEW *Find each product.*

1. $3x^{-2}y^2(4x^2 + 3y^{-2})$

2. $(2a^{-2} - b^{-2})(2a^{-2} + b^{-2})$

Write each expression without using negative exponents.

3. $\dfrac{x^{-2} + y^{-2}}{x^{-2} - y^{-2}}$

4. $\dfrac{2x^{-3} - 2y^{-3}}{4x^{-3} + 4y^{-3}}$

VOCABULARY AND CONCEPTS *Fill in each blank to make a true statement.*

5. An _____ is the set of all points in a plane the _____ of whose distances from two fixed points is a constant.

6. The fixed points in Exercise 5 are the _____ of the ellipse.

7. The midpoint of the line segment joining the foci of an ellipse is called the _____ of the ellipse.

8. The elliptical graph of

$$\frac{x^2}{a^2} + \frac{y^2}{b^2} = 1$$

has x-intercepts of _____ and y-intercepts of _____.

9. The center of the ellipse with an equation of

$$\frac{x^2}{a^2} + \frac{y^2}{b^2} = 1$$

is the point _____.

10. The center of the ellipse with an equation of

$$\frac{(x - h)^2}{a^2} + \frac{(y - k)^2}{b^2} = 1$$

is the point _____.

PRACTICE   *In Exercises 11–20, graph each equation.*

11. $\dfrac{x^2}{4} + \dfrac{y^2}{9} = 1$

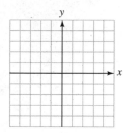

12. $x^2 + \dfrac{y^2}{9} = 1$

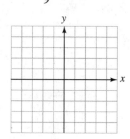

13. $x^2 + 9y^2 = 9$

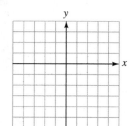

14. $25x^2 + 9y^2 = 225$

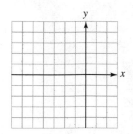

15. $16x^2 + 4y^2 = 64$

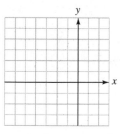

16. $4x^2 + 9y^2 = 36$

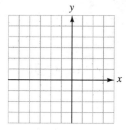

17. $\dfrac{(x - 2)^2}{9} + \dfrac{(y - 1)^2}{4} = 1$

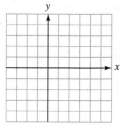

18. $\dfrac{(x - 1)^2}{9} + \dfrac{(y - 3)^2}{4} = 1$

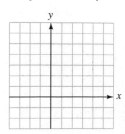

19. $(x + 1)^2 + 4(y + 2)^2 = 4$

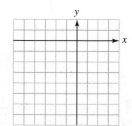

20. $25(x + 1)^2 + 9y^2 = 225$

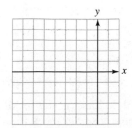

In Exercises 21–24, use a graphing calculator to graph each equation.

21. $\dfrac{x^2}{9} + \dfrac{y^2}{4} = 1$

22. $x^2 + 16y^2 = 16$

23. $\dfrac{x^2}{4} + \dfrac{(y-1)^2}{9} = 1$

24. $\dfrac{(x+1)^2}{9} + \dfrac{(y-2)^2}{4} = 1$

Write each equation in standard form and graph it.

25. $x^2 + 4y^2 - 4x + 8y + 4 = 0$

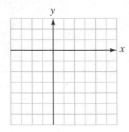

26. $x^2 + 4y^2 - 2x - 16y = -13$

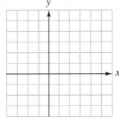

27. $9x^2 + 4y^2 - 18x + 16y = 11$

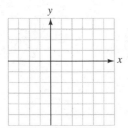

28. $16x^2 + 25y^2 - 160x - 200y + 400 = 0$

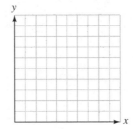

APPLICATIONS

29. Designing an underpass   The arch of the underpass in Illustration 1 is a part of an ellipse. Find the equation of the arch.

30. Calculating clearance   Find the height of the elliptical arch in Exercise 29 at a point 10 feet from the center of the roadway.

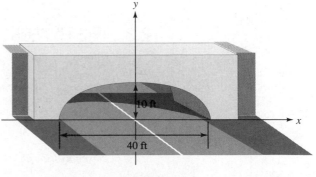

ILLUSTRATION 1

31. Area of an ellipse The area A of the ellipse

$$\frac{x^2}{a^2} + \frac{y^2}{b^2} = 1$$

is given by $A = \pi ab$. Find the area of the ellipse $9x^2 + 16y^2 = 144$.

32. Area of a track The elliptical track in Illustration 2 is bounded by the ellipses $4x^2 + 9y^2 = 576$ and $9x^2 + 25y^2 = 900$. Find the area of the track. (See Exercise 31.)

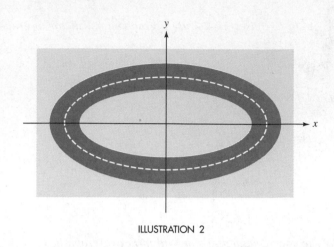

ILLUSTRATION 2

WRITING

33. Explain how to find the x- and the y-intercepts of the graph of the ellipse

$$\frac{x^2}{a^2} + \frac{y^2}{b^2} = 1$$

34. Explain the relationship between the center, focus, and vertex of an ellipse.

SOMETHING TO THINK ABOUT

35. What happens to the graph of

$$\frac{x^2}{a^2} + \frac{y^2}{b^2} = 1$$

when $a = b$?

36. Explain why the graph of $x^2 - 2x + y^2 + 4y + 20 = 0$ does not exist.

12.3 THE HYPERBOLA

■ THE HYPERBOLA ■ PROBLEM SOLVING

Getting Ready   *Find the value of y when* $\dfrac{x^2}{25} - \dfrac{y^2}{9} = 1$ *and x is the given value. Give each result to the nearest tenth.*

1. $x = 6$ **2.** $x = -7$

■ THE HYPERBOLA

The Hyperbola
A **hyperbola** is the set of all points P in the plane for which the difference of the distances of each point from two fixed points is a constant. See Figure 12-22, in which $d_1 - d_2$ is a constant.

Each of the two points is called a **focus**. Midway between the foci is the **center** of the hyperbola.

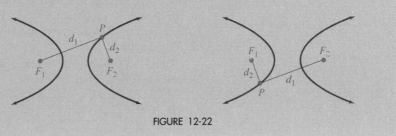

FIGURE 12-22

The graph of the equation

$$\frac{x^2}{25} - \frac{y^2}{9} = 1$$

is a hyperbola. To graph the equation, we make a table of ordered pairs that satisfy the equation, plot each pair, and join the points with a smooth curve as in Figure 12-23.

$$\frac{x^2}{25} - \frac{y^2}{9} = 1$$

x	y	(x, y)
-7	± 2.9	$(-7, \pm 2.9)$
-6	± 2.0	$(-6, \pm 2.0)$
-5	0	$(-5, 0)$
5	0	$(5, 0)$
6	± 2.0	$(6, \pm 2.0)$
7	± 2.9	$(7, \pm 2.9)$

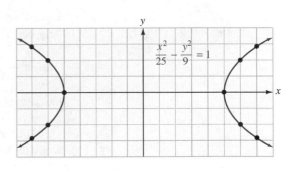

FIGURE 12-23

This graph is centered at the origin and intersects the x-axis at $\left(\sqrt{25}, 0\right)$ and $\left(-\sqrt{25}, 0\right)$. After simplifying, these points are $(5, 0)$ and $(-5, 0)$. We also note that the graph does not intersect the y-axis.

It is possible to draw a hyperbola without plotting points. For example, if we want to graph the hyperbola with an equation of

$$\frac{x^2}{a^2} - \frac{y^2}{b^2} = 1$$

we first look at the x- and y-intercepts. To find the x-intercepts, we let $y = 0$ and solve for x:

$$\frac{x^2}{a^2} - \frac{0^2}{b^2} = 1$$

$$x^2 = a^2$$

$$x = \pm a$$

Thus, the hyperbola crosses the x-axis at the points $V_1(a, 0)$ and $V_2(-a, 0)$, called the **vertices** of the hyperbola. See Figure 12-24.

To attempt to find the y-intercepts, we let $x = 0$ and solve for y:

$$\frac{0^2}{a^2} - \frac{y^2}{b^2} = 1$$

$$y^2 = -b^2$$

$$y = \pm\sqrt{-b^2}$$

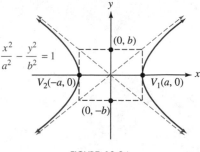

FIGURE 12-24

Since b^2 is always positive, $\sqrt{-b^2}$ is an imaginary number. This means that the hyperbola does not cross the y-axis.

If we construct a rectangle, called the **fundamental rectangle**, whose sides pass horizontally through $\pm b$ on the y-axis and vertically through $\pm a$ on the x-axis, the extended diagonals of the rectangle will be asymptotes of the hyperbola.

Equation of a Hyperbola Centered at the Origin

Any equation that can be written in the form

$$\frac{x^2}{a^2} - \frac{y^2}{b^2} = 1$$

has a graph that is a hyperbola centered at the origin, as in Figure 12-25. The x-intercepts are the vertices $V_1(a, 0)$ and $V_2(-a, 0)$. There are no y-intercepts.

The asymptotes of the hyperbola are the extended diagonals of the rectangle in the figure.

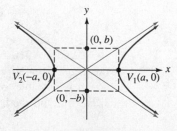

FIGURE 12-25

The branches of the hyperbola in previous discussions open to the left and to the right. It is possible for hyperbolas to have different orientations with respect to the x- and y-axes. For example, the branches of a hyperbola can open upward and downward. In that case, the following equation applies.

Equation of a Hyperbola Centered at the Origin

Any equation that can be written in the form

$$\frac{y^2}{a^2} - \frac{x^2}{b^2} = 1$$

has a graph that is a hyperbola centered at the origin, as in Figure 12-26. The y-intercepts are the vertices $V_1(0, a)$ and $V_2(0, -a)$. There are no x-intercepts.

The asymptotes of the hyperbola are the extended diagonals of the rectangle shown in the figure.

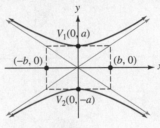

FIGURE 12-26

EXAMPLE 1   Graph $9y^2 - 4x^2 = 36$.

Solution   To write the equation in standard form, we divide both sides by 36 to obtain

$$\frac{9y^2}{36} - \frac{4x^2}{36} = 1$$

$$\frac{y^2}{4} - \frac{x^2}{9} = 1 \qquad \text{Simplify each fraction.}$$

We then find the y-intercepts by letting $x = 0$ and solving for y:

$$\frac{y^2}{4} - \frac{0^2}{9} = 1$$

$$y^2 = 4$$

Thus, $y = \pm 2$, and the vertices of the hyperbola are $V_1(0, 2)$ and $V_2(0, -2)$. (See Figure 12-27.)

Since $\pm \sqrt{9} = \pm 3$, we use the points $(3, 0)$ and $(-3, 0)$ on the x-axis to help draw the fundamental rectangle. We then draw its extended diagonals and sketch the hyperbola.

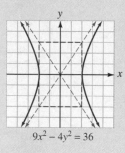

FIGURE 12-27   ∎

Self Check

Answer

Graph $9x^2 - 4y^2 = 36$.

If a hyperbola is centered at a point with coordinates (h, k), the following equations apply.

Equations of Hyperbolas Centered at (h, k)

Any equation that can be written in the form

$$\frac{(x - h)^2}{a^2} - \frac{(y - k)^2}{b^2} = 1$$

is a hyperbola with center at (h, k) that opens left and right.

Any equation of the form

$$\frac{(y - k)^2}{a^2} - \frac{(x - h)^2}{b^2} = 1$$

is a hyperbola with center at (h, k) that opens up and down.

EXAMPLE 2   Graph $\dfrac{(x - 3)^2}{16} - \dfrac{(y + 1)^2}{4} = 1$.

Solution   We write the equation in the form

$$\frac{(x - 3)^2}{4^2} - \frac{[y - (-1)]^2}{2^2} = 1$$

to see that its graph will be a hyperbola centered at the point $(h, k) = (3, -1)$. Its vertices are located at $a = 4$ units to the right and left of the center, at $(7, -1)$ and $(-1, -1)$. Since $b = 2$, we can count 2 units above and below the center to locate points $(3, 1)$ and $(3, -3)$. With these points, we can draw the fundamental rectangle along with its extended diagonals. We can then sketch the hyperbola, as shown in Figure 12-28.

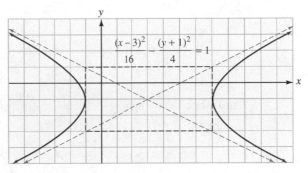

$$\frac{(x-3)^2}{16} - \frac{(y+1)^2}{4} = 1$$

FIGURE 12-28

■

Self Check   Graph $\dfrac{(x + 2)^2}{9} - \dfrac{(y - 1)^2}{4} = 1.$

Answer

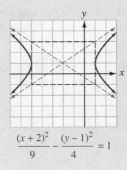

$$\frac{(x+2)^2}{9} - \frac{(y-1)^2}{4} = 1$$

EXAMPLE 3   Write the equation $x^2 - y^2 - 2x + 4y = 12$ in standard form to show that the equation represents a hyperbola. Then graph it.

Solution   We proceed as follows.

$$x^2 - y^2 - 2x + 4y = 12$$
$$x^2 - 2x - y^2 + 4y = 12 \qquad \text{Use the commutative property to rearrange terms.}$$
$$x^2 - 2x - (y^2 - 4y) = 12 \qquad \text{Factor } -1 \text{ from } -y^2 + 4y.$$

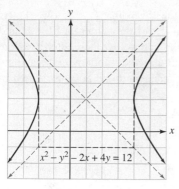

FIGURE 12-29

We then complete the square on x and y to make $x^2 - 2x$ and $y^2 - 4y$ perfect trinomial squares.

$$x^2 - 2x + 1 - (y^2 - 4y + 4) = 12 + 1 - 4$$

We then factor $x^2 - 2x + 1$ and $y^2 - 4y + 4$ to get

$$(x - 1)^2 - (y - 2)^2 = 9$$

$$\frac{(x - 1)^2}{9} - \frac{(y - 2)^2}{9} = 1 \qquad \text{Divide both sides by 9.}$$

This is the equation of a hyperbola with center at $(1, 2)$. Its graph is shown in Figure 12-29. ∎

Self Check
Answer

Graph $x^2 - 4y^2 + 2x - 8y = 7$.

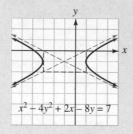

There is a special type of hyperbola (also centered at the origin) that does not intersect either the x- or the y-axis. These hyperbolas have equations of the form $xy = k$, where $k \neq 0$.

EXAMPLE 4

Graph $xy = -8$.

Solution

We make a table of ordered pairs, plot each pair, and join the points with a smooth curve to obtain the hyperbola in Figure 12-30.

$$xy = -8$$

x	y	(x, y)
1	-8	$(1, -8)$
2	-4	$(2, -4)$
4	-2	$(4, -2)$
8	-1	$(8, -1)$
-1	8	$(-1, 8)$
-2	4	$(-2, 4)$
-4	2	$(-4, 2)$
-8	1	$(-8, 1)$

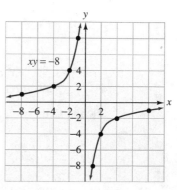

FIGURE 12-30

Self Check
Answer

Graph $xy = 6$.

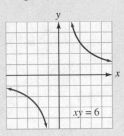

The result in Example 4 illustrates the following general equation.

Equations of Hyperbolas of the Form $xy = k$

Any equation of the form $xy = k$, where $k \neq 0$, has a graph that is a **hyperbola**, which does not intersect either the x- or the y-axis.

■ PROBLEM SOLVING

EXAMPLE 5

Atomic structure   In an experiment that led to the discovery of the atomic structure of matter, Lord Rutherford (1871–1937) shot high-energy alpha particles toward a thin sheet of gold. Because many were reflected, Rutherford showed the existence of the nucleus of a gold atom. The alpha particle in Figure 12-31 is repelled by the nucleus at the origin; it travels along the hyperbolic path given by $4x^2 - y^2 = 16$. How close does the particle come to the nucleus?

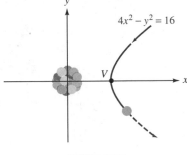

FIGURE 12-31

Solution   To find the distance from the nucleus at the origin, we must find the coordinates of the vertex V. To do so, we write the equation of the particle's path in standard form:

$$4x^2 - y^2 = 16$$

$$\frac{4x^2}{16} - \frac{y^2}{16} = \frac{16}{16} \qquad \text{Divide both sides by 16.}$$

$$\frac{x^2}{4} - \frac{y^2}{16} = 1 \qquad \text{Simplify.}$$

$$\frac{x^2}{2^2} - \frac{y^2}{4^2} = 1 \qquad \text{Write 4 as } 2^2 \text{ and 16 as } 4^2.$$

This equation is in the form

$$\frac{x^2}{a^2} - \frac{y^2}{b^2} = 1$$

with $a = 2$. Thus, the vertex of the path is $(2, 0)$. The particle is never closer than 2 units from the nucleus.   ∎

Orals   *Find the x- or y-intercepts of each hyperbola.*

1. $\dfrac{x^2}{9} - \dfrac{y^2}{16} = 1$

2. $\dfrac{y^2}{25} - \dfrac{x^2}{36} = 1$

EXERCISE 12.3

REVIEW   *Factor each expression.*

1. $-6x^4 + 9x^3 - 6x^2$

2. $4a^2 - b^2$

3. $15a^2 - 4ab - 4b^2$

4. $8p^3 - 27q^3$

VOCABULARY AND CONCEPTS   *Fill in each blank to make a true statement.*

5. A _____ is the set of all points in a plane for which the _____ of the distances from two fixed points is a constant.

6. The fixed points in Exercise 5 are the ____ of the hyperbola.

7. The midpoint of the line segment joining the foci of a hyperbola is called the _____ of the hyperbola.

8. The hyperbolic graph of

$$\frac{x^2}{a^2} - \frac{y^2}{b^2} = 1$$

has x-intercepts of _____. There are no _____.

9. The center of the hyperbola with an equation of

$$\frac{x^2}{a^2} - \frac{y^2}{b^2} = 1$$

is the point _____.

10. The center of the hyperbola with an equation of

$$\frac{(x - h)^2}{a^2} - \frac{(y - k)^2}{b^2} = 1$$

is the point _____.

PRACTICE   *In Exercises 11–22, graph each hyperbola.*

11. $\dfrac{x^2}{9} - \dfrac{y^2}{4} = 1$

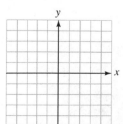

12. $\dfrac{x^2}{4} - \dfrac{y^2}{4} = 1$

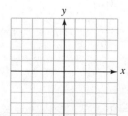

13. $\dfrac{y^2}{4} - \dfrac{x^2}{9} = 1$

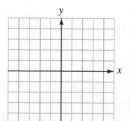

14. $\dfrac{y^2}{4} - \dfrac{x^2}{64} = 1$

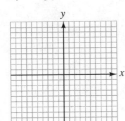

15. $25x^2 - y^2 = 25$

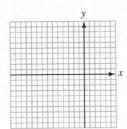

16. $9x^2 - 4y^2 = 36$

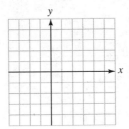

17. $\dfrac{(x-2)^2}{9} - \dfrac{y^2}{16} = 1$

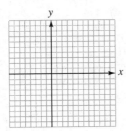

18. $\dfrac{(x+2)^2}{16} - \dfrac{(y-3)^2}{25} = 1$

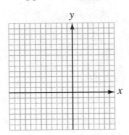

19. $4(x+3)^2 - (y-1)^2 = 4$

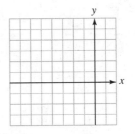

20. $(x+5)^2 - 16y^2 = 16$

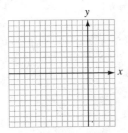

21. $xy = 8$

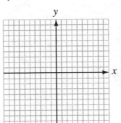

22. $xy = -10$

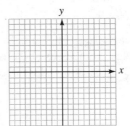

In Exercises 23–26, use a graphing calculator to graph each equation.

23. $\dfrac{x^2}{9} - \dfrac{y^2}{4} = 1$

24. $y^2 - 16x^2 = 16$

25. $\dfrac{x^2}{4} - \dfrac{(y-1)^2}{9} = 1$

26. $\dfrac{(y+1)^2}{9} - \dfrac{(x-2)^2}{4} = 1$

In Exercises 27–30, write each equation in standard form and graph it.

27. $4x^2 - y^2 + 8x - 4y = 4$

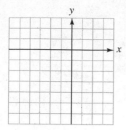

28. $x^2 - 9y^2 - 4x - 54y = 86$

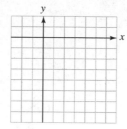

29. $4y^2 - x^2 + 8y + 4x = 4$

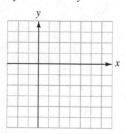

30. $y^2 - 4x^2 - 4y - 8x = 4$

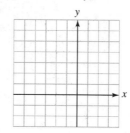

APPLICATIONS

31. Alpha particles The particle in Illustration 1 approaches the nucleus at the origin along the path $9y^2 - x^2 = 81$. How close does the particle come to the nucleus?

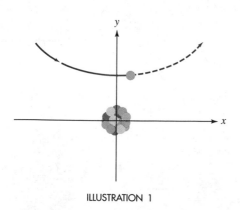

ILLUSTRATION 1

32. LORAN By determining the difference of the distances between the ship in Illustration 2 and two radio transmitters, the LORAN system places the ship on the hyperbola $x^2 - 4y^2 = 576$. If the ship is also 5 miles out to sea, find its coordinates.

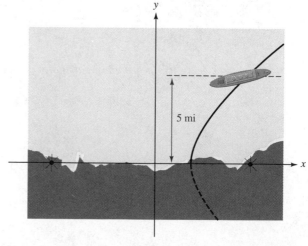

ILLUSTRATION 2

■ ■ ■ ■ ■ ■ ■ ■ ■ P E R S P E C T I V E

Focus on Conics

Satellite dishes and flashlight reflectors are familiar examples of a conic's ability to reflect a beam of light or to concentrate incoming satellite signals at one point. That property is shown in Illustration 1.

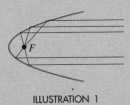

ILLUSTRATION 1

An ellipse has two foci, the points labeled *F* in Illustration 2. Any light or signal that starts at one focus will be reflected to the other. This property is the basis of whispering galleries, where a person standing at one focus can clearly hear another person speaking at the other focus.

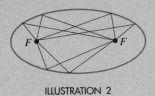

ILLUSTRATION 2

The focal property of the ellipse is also used in **lithotripsy**, a medical procedure for treating kidney stones. The patient is placed in an elliptical tank of water with the kidney stone at one focus. Shock waves from a small controlled explosion at the other focus are concentrated on the stone, pulverizing it.

The hyperbola also has two foci, the two points labeled *F* in Illustration 3. As in the ellipse, light aimed at one focus is reflected toward the other. Hyperbolic mirrors are used in some reflecting telescopes.

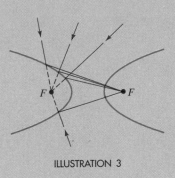

ILLUSTRATION 3

33. Sonic boom   The position of the sonic boom caused by the faster-than-sound aircraft in Illustration 3 is the hyperbola $y^2 - x^2 = 25$ in the coordinate system shown. How wide is the hyperbola 5 units from its vertex?

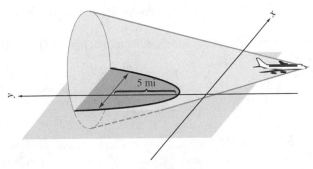

5 mi

ILLUSTRATION 3

34. Electrostatic repulsion   Two similarly charged particles are shot together for an almost head-on collision, as in Illustration 4. They repel each other and travel the two branches of the hyperbola given by $x^2 - 4y^2 = 4$. How close do they get?

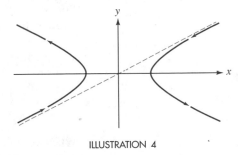

ILLUSTRATION 4

WRITING

35. Explain how to find the x- and the y-intercepts of the graph of the hyperbola

$$\frac{x^2}{a^2} - \frac{y^2}{b^2} = 1$$

36. Explain why the graph of the hyperbola

$$\frac{x^2}{a^2} - \frac{y^2}{b^2} = 1$$

has no y-intercept.

SOMETHING TO THINK ABOUT

37. Describe the fundamental rectangle of

$$\frac{x^2}{a^2} - \frac{y^2}{b^2} = 1$$

when $a = b$.

38. The hyperbolas $x^2 - y^2 = 1$ and $y^2 - x^2 = 1$ are called **conjugate** hyperbolas. Graph both on the same axes. What do they have in common?

12.5 Piecewise-Defined Functions and the Greatest Integer Function

■ PIECEWISE-DEFINED FUNCTIONS ■ INCREASING AND DECREASING FUNCTIONS ■ THE GREATEST INTEGER FUNCTION

Getting Ready

1. Is $f(x) = x^2$ positive or negative when $x > 0$?

2. Is $f(x) = -x^2$ positive or negative when $x > 0$?

3. What is the largest integer that is less than 98.6?

4. What is the largest integer that is less than -2.7?

■ PIECEWISE-DEFINED FUNCTIONS

In this section, we will discuss functions that are defined by using different equations for different parts of their domains. Such functions are called **piecewise-defined functions**.

A simple piecewise-defined function is the absolute value function, $f(x) = |x|$, which can be written in the form

$$f(x) = \begin{cases} x \text{ when } x \geq 0 \\ -x \text{ when } x < 0 \end{cases}$$

When x is in the interval $[0, \infty)$, we use the function $f(x) = x$ to evaluate $|x|$. However, when x is in the interval $(-\infty, 0)$, we use the function $f(x) = -x$ to evaluate $|x|$. The graph of the absolute value function is shown in Figure 12-32.

For $x \geq 0$			For $x \leq 0$		
x	y	(x, y)	x	y	(x, y)
0	0	(0, 0)	-4	4	$(-4, 4)$
1	1	(1, 1)	-3	3	$(-3, 3)$
2	2	(2, 2)	-2	2	$(-2, 2)$
3	3	(3, 3)	-1	1	$(-1, 1)$

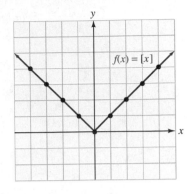

FIGURE 12-32

■ INCREASING AND DECREASING FUNCTIONS

We have seen that if the values of $f(x)$ increase as x increases on an interval, we say that the function is *increasing on the interval* (see Figure 12-33(a)). If the values of $f(x)$ decrease as x increases on an interval, we say that the function is *decreasing on the interval* (see Figure 12-33(b)). If the values of $f(x)$ remain constant as x increases on an interval, we say that the function is *constant on the interval* (see Figure 12-33(c)).

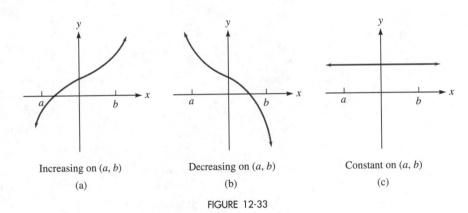

Increasing on (a, b)

(a)

Decreasing on (a, b)

(b)

Constant on (a, b)

(c)

FIGURE 12-33

The absolute value function, pictured in Figure 12-32, is decreasing on the interval $(-\infty, 0)$ and is increasing on the interval $(0, \infty)$.

EXAMPLE 1

Graph the piecewise-defined function given by

$$f(x) = \begin{cases} x^2 \text{ when } x \leq 0 \\ x \text{ when } 0 < x < 2 \\ -1 \text{ when } x \geq 2 \end{cases}$$

and tell where the function is increasing, decreasing, or constant.

Solution   For each number x, we decide which of the three equations will be used to find the corresponding value of y:

• For numbers $x \leq 0$, $f(x)$ is determined by $f(x) = x^2$, and the graph is the left half of a parabola. See Figure 12-34. Since the values of $f(x)$ decrease on this graph as x increases, the function is decreasing on the interval $(-\infty, 0)$.

• For numbers $0 < x < 2$, $f(x)$ is determined by $f(x) = x$, and the graph is part of a line. Since the values of $f(x)$ increase on this graph as x increases, the function is increasing on the interval $(0, 2)$.

• For numbers $x \geq 2$, $f(x)$ is the constant -1, and the graph is part of a horizontal line. Since the values of $f(x)$ remain constant on this line, the function is constant on the interval $(2, \infty)$.

The use of solid and open circles on the graph indicates that $f(x) = -1$ when $x = 2$.

Since every number x determines one value y, the domain of this function is the interval $(-\infty, \infty)$. The range is the interval $[0, \infty) \cup \{-1\}$.

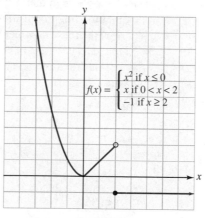

$$f(x) = \begin{cases} x^2 & \text{if } x \leq 0 \\ x & \text{if } 0 < x < 2 \\ -1 & \text{if } x \geq 2 \end{cases}$$

FIGURE 12-34 ■

Self Check   Graph the function
$$f(x) = \begin{cases} 2x & \text{when } x \leq 0 \\ \frac{1}{2}x & \text{when } x > 0 \end{cases}.$$

Answer

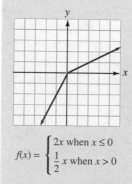

$$f(x) = \begin{cases} 2x & \text{when } x \leq 0 \\ \dfrac{1}{2}x & \text{when } x > 0 \end{cases}$$

■ THE GREATEST INTEGER FUNCTION

The **greatest integer function** is important in computer applications. It is a function determined by the equation

$$y = f(x) = [\![x]\!] \qquad \text{Read as ``}y\text{ equals the greatest integer in } x.\text{''}$$

where the value of y that corresponds to x is the greatest integer that is less than or equal to x. For example,

$$[\![4.7]\!] = 4, \quad \left[\!\left[2\tfrac{1}{2}\right]\!\right] = 2, \quad [\![\pi]\!] = 3, \quad [\![-3.7]\!] = -4, \quad [\![-5.7]\!] = -6$$

EXAMPLE 2 Graph $y = [\![x]\!]$.

Solution We list several intervals and the corresponding values of the greatest integer function:

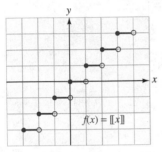

FIGURE 12-35

$[0, 1)$	$y = [\![x]\!] = 0$	For numbers from 0 to 1, not including 1, the greatest integer in the interval is 0.
$[1, 2)$	$y = [\![x]\!] = 1$	For numbers from 1 to 2, not including 2, the greatest integer in the interval is 1.
$[2, 3)$	$y = [\![x]\!] = 2$	For numbers from 2 to 3, not including 3, the greatest integer in the interval is 2.

In each interval, the values of y are constant, but they jump by 1 at integer values of x. The graph is shown in Figure 12-35. From the graph, we see that the domain is $(-\infty, \infty)$, and the range is the set of integers $\{. . ., -3, -2, -1, 0, 1, 2, 3,. . .\}$.

Since the greatest integer function is made up of a series of horizontal line segments, it is an example of a group of functions called **step functions**. ■

EXAMPLE 3 To print stationery, a printer charges \$10 for setup charges, plus \$20 for each box. The printer counts any portion of a box as a full box. Graph this step function.

Solution If we order stationery and cancel the order before it is printed, the cost will be \$10. Thus, the ordered pair (0, 10) will be on the graph.

If we purchase 1 box, the cost will be \$10 for setup plus \$20 for printing, for a total cost of \$30. Thus, the ordered pair (1, 30) will be on the graph.

The cost of $1\tfrac{1}{2}$ boxes will be the same as the cost of 2 boxes, or \$50. Thus, the ordered pairs (1.5, 50) and (2, 50) will be on the graph.

The complete graph is shown in Figure 12-36.

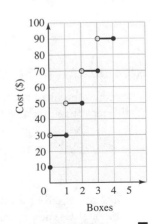

FIGURE 12-36 ■

Self Check	In Example 3, how much will $3\frac{1}{2}$ boxes cost?
Answer	$90

Orals   *Tell whether each function is increasing, decreasing, or constant on the interval* $(-2, 3)$.

1.

2.

3.

4.

EXERCISE 12.4

REVIEW   *Find the value of x. Assume that lines r and s are parallel.*

1.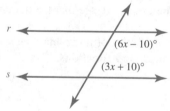

2.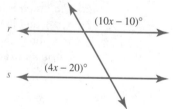

VOCABULARY AND CONCEPTS   *Fill in each blank to make a true statement.*

3. Piecewise-defined functions are defined by using different functions for different parts of their _____.

4. When the values of $f(x)$ increase as the values of x increase over an interval, we say that the function is an _____ function over that interval.

5. In a _____ function, the values of _____ are the same.

6. When the values of $f(x)$ decrease as the values of x _____ over an interval, we say that the function is a decreasing function over that interval.

7. When the graph of a function contains a series of horizontal line segments, the function is called a _____ function.

8. The function that gives the largest integer that is less than or equal to a number x is called the _____ function.

PRACTICE   *In Exercises 9–12, give the intervals on which each function is increasing, decreasing, or constant.*

9.

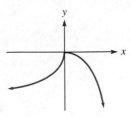

10.

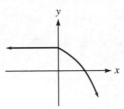

11.

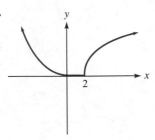

12.

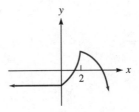

In Exercises 13–16, graph each function and give the intervals on which f is increasing, decreasing, or constant.

13. $f(x) = \begin{cases} -1 \text{ if } x \le 0 \\ x \text{ if } x > 0 \end{cases}$

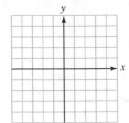

14. $f(x) = \begin{cases} -2 \text{ if } x \le 0 \\ x^2 \text{ if } x > 0 \end{cases}$

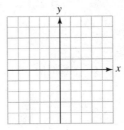

15. $f(x) = \begin{cases} -x \text{ if } x \le 0 \\ x \text{ if } 0 < x < 2 \\ -x \text{ if } x \ge 2 \end{cases}$

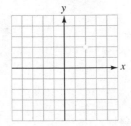

16. $f(x) = \begin{cases} -x \text{ if } x < 0 \\ x^2 \text{ if } 0 \le x \le 1 \\ 1 \text{ if } x > 1 \end{cases}$

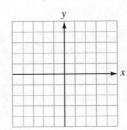

In Exercises 17–20, graph each function.

17. $f(x) = -[\![x]\!]$

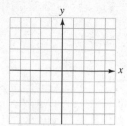

18. $f(x) = [\![x]\!] + 2$

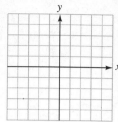

19. $f(x) = 2[\![x]\!]$

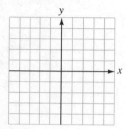

20. $f(x) = [\![\frac{1}{2}x]\!]$

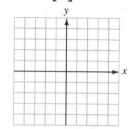

21. Signum function Computer programmers use a function denoted by $f(x) = \operatorname{sgn} x$, that is defined in the following way:

$$f(x) = \begin{cases} -1 \text{ if } x < 0 \\ 0 \text{ if } x = 0 \\ 1 \text{ if } x > 0 \end{cases}$$

Graph this function.

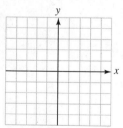

22. Heaviside unit step function This function, used in calculus, is defined by

$$f(x) = \begin{cases} 1 \text{ if } x > 0 \\ 0 \text{ if } x < 0 \end{cases}$$

Graph this function.

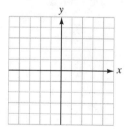

APPLICATIONS

23. Renting a jet ski A marina charges $20 to rent a jet ski for 1 hour, plus $5 for every extra hour (or portion of an hour). In Illustration 1, graph the ordered pairs (h, c), where h represents the number of hours and c represents the cost. Find the cost if the ski is used for 2.5 hours.

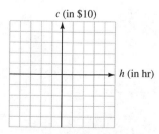

ILLUSTRATION 1

24. Riding in a taxi A cab company charges $3 for a trip up to 1 mile, and $2 for every extra mile (or portion of a mile). In Illustration 2, graph the ordered pairs (m, c), where m represents the number of miles traveled and c represents the cost. Find the cost to ride $10\frac{1}{4}$ miles.

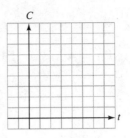

ILLUSTRATION 3

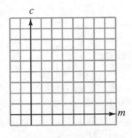

ILLUSTRATION 2

copies and 10% on sales thereafter. If the book sells for $10, express the royalty income, I, as a function of s, the number of copies sold, and use Illustration 4 to graph the function. (*Hint*: When sales are into the second 50,000 copies, how much was earned on the first 50,000?)

25. Information access Computer access to international data network A costs $10 per day plus $8 per hour or fraction of an hour. Network B charges $15 per day, but only $6 per hour or fraction of an hour. For each network, use Illustration 3 to graph the ordered pairs (t, C), where t represents the connect time and C represents the total cost. Find the minimal daily usage at which it would be more economical to use network B.

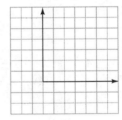

26. Royalties A publisher has agreed to pay the author of a novel 7% royalties on sales of the first 50,000

ILLUSTRATION 4

WRITING

27. Tell how to decide whether a function is increasing on the interval (a, b).

28. Describe the greatest integer function.

SOMETHING TO THINK ABOUT

29. Find a piecewise-defined function that is increasing on the interval $(-\infty, -2)$ and decreasing on the interval $(-2, \infty)$.

30. Find a piecewise-defined function that is constant on the interval $(-\infty, 0)$, increasing on the interval $(0, 5)$, and decreasing on the interval $(5, \infty)$.

■ ■ ■ ■ ■ ■ ■ ■ ■ ■ **PROJECT**

The zillionaire G.I. Luvmoney is known for his love of flowers. On his estate, he recently set aside a circular plot of land with a radius of 100 yards to be made into a flower garden. He has hired your landscape design firm to do the job. If Luvmoney is satisfied, he will hire your firm to do more lucrative jobs. Here is Luvmoney's plan.

The center of the circular plot of land is to be the origin of a rectangular coordinate system. You are to make 100 circles, all centered at the origin, with radii of 1 yard, 2 yards, 3 yards, and so on, up to the outermost circle, which will have a radius of 100 yards. Inside the innermost circle, he wants a fountain with a circular walkway around it. In the ring between the first and second circle, he wants to plant his favorite kind of flower; in the next ring his second favorite, and so on, until you reach the edge of the circular plot. Luvmoney provides you with a list ranking his 99 favorite flowers.

The first thing he wants to know is the area of each ring, so that he will know how many of each plant to order. Then he wants a simple formula that will give the area of any ring just by substituting in the number of the ring.

He also wants a walkway to go through the garden in the form of a hyperbolic path, following the equation

$$x^2 - \frac{y^2}{9} = 1$$

Luvmoney wants to know the x- and y-coordinates of the points where the path will intersect the circles, so that those points can be marked with stakes to keep gardeners from planting flowers where the walkway will later be built. He wants a formula (or two) that will enable him to put in the number of a circle and get out the intersection points.

Finally, although cost has no importance for Luvmoney, his accountants will want an estimate of the total cost of all of the flowers.

You go back to your office with Luvmoney's list. You find that because the areas of the rings grow from the inside of the garden to the outside, and because of Luvmoney's ranking of flowers, a strange thing happens. The first ring of flowers will cost $360, and the flowers in every ring after that will cost 110% as much as the flowers in the previous ring. That is, the second ring of flowers will cost $360(1.1) = $396, the third will cost $435.60, and so on.

Find answers to all of Luvmoney's questions, and show work that will convince him that you are right.

C H A P T E R   S U M M A R Y

CONCEPTS   **REVIEW EXERCISES**

| SECTION 12.1 | *The Circle and the Parabola* |

Equations of a circle:

$(x - h)^2 + (y - k)^2 = r^2$
  center (h, k), radius r

$x^2 + y^2 = r^2$
  center $(0, 0)$, radius r

1. Graph each equation.

a. $(x - 1)^2 + (y + 2)^2 = 9$ **b.** $x^2 + y^2 = 16$

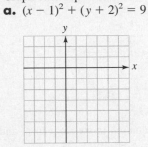   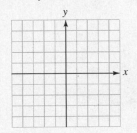

2. Write the equation in standard form and graph it.
$x^2 + y^2 + 4x - 2y = 4$

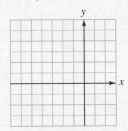

Equations of parabolas
($a > 0$):

Parabola opening	Vertex at origin
Up	$y = ax^2$
Down	$y = -ax^2$
Right	$x = ay^2$
Left	$x = -ay^2$

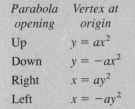

Parabola opening	Vertex at (h, k)
Up	$y = a(x - h)^2 + k$
Down	$y = -a(x - h)^2 + k$
Right	$x = a(y - k)^2 + h$
Left	$x = -a(y - k)^2 + h$

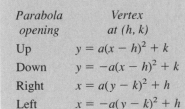

3. Graph each equation.

a. $x = -3(y - 2)^2 + 5$ **b.** $x = 2(y + 1)^2 - 2$

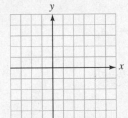   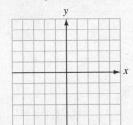

| SECTION 12.2 | *The Ellipse* |

Equations of an ellipse:

Center at $(0, 0)$

$$\frac{x^2}{a^2} + \frac{y^2}{b^2} = 1 \; a > b > 0$$

$$\frac{x^2}{b^2} + \frac{y^2}{a^2} = 1 \; a > b > 0$$

Center at (h, k)

$$\frac{(x-h)^2}{a^2} + \frac{(y-k)^2}{b^2} = 1$$

$$\frac{(x-h)^2}{b^2} + \frac{(y-k)^2}{a^2} = 1$$

4. Graph each ellipse.

a. $9x^2 + 16y^2 = 144$

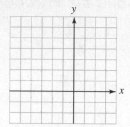

b. $\dfrac{(x-2)^2}{4} + \dfrac{(y-1)^2}{9} = 1$

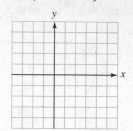

5. Write the equation in standard form and graph it.
$4x^2 + 9y^2 + 8x - 18y = 23$

| SECTION 12.3 | *The Hyperbola* |

Equations of a hyperbola:

Center at $(0, 0)$

$$\frac{x^2}{a^2} - \frac{y^2}{b^2} = 1$$

$$\frac{y^2}{a^2} - \frac{x^2}{b^2} = 1$$

Center at (h, k)

$$\frac{(x-h)^2}{a^2} - \frac{(y-k)^2}{b^2} = 1$$

$$\frac{(y-k)^2}{a^2} - \frac{(x-h)^2}{b^2} = 1$$

6. Graph each hyperbola.

a. $9x^2 - y^2 = -9$

b. $xy = 9$

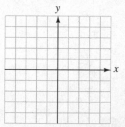

7. Write the equation $4x^2 - 2y^2 + 8x - 8y = 8$ in standard form and tell whether its graph will be an ellipse or a hyperbola.

8. Write the equation in standard form and graph it.
$$9x^2 - 4y^2 - 18x - 8y = 31$$

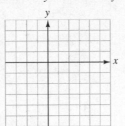

SECTION 12.4

Piecewise-Defined Functions and the Greatest Integer Function

A function is increasing on the interval (a, b) if the values of $f(x)$ increase as x increases from a to b.

A function is decreasing on the interval (a, b) if the values of $f(x)$ decrease as x increases from a to b.

A function is constant on the interval (a, b) if the value of $f(x)$ is constant as x increases from a to b.

9. Tell when the function is increasing, decreasing, or constant.

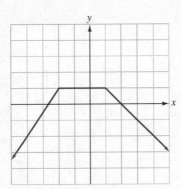

10. Graph each function.

a. $f(x) = \begin{cases} x \text{ if } x \le 1 \\ -x^2 \text{ if } x > 1 \end{cases}$

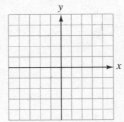

b. $f(x) = 3[\![x]\!]$

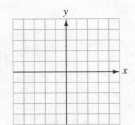

■ Chapter Test

1. Find the center and the radius of the circle $(x - 2)^2 + (y + 3)^2 = 4$.

2. Find the center and the radius of the circle $x^2 + y^2 + 4x - 6y = 3$.

In Problems 3–6, graph each equation.

3. $(x + 1)^2 + (y - 2)^2 = 9$

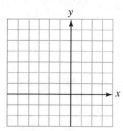

4. $x = (y - 2)^2 - 1$

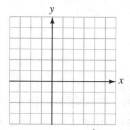

5. $9x^2 + 4y^2 = 36$

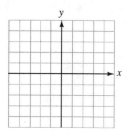

6. $\dfrac{(x - 2)^2}{9} - y^2 = 1$

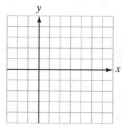

In Problems 7–8, write each equation in standard form and graph the equation.

7. $4x^2 + y^2 - 24x + 2y = -33$

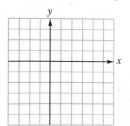

8. $x^2 - 9y^2 + 2x + 36y = 44$

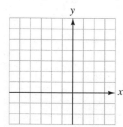

9. Tell where the function is increasing, decreasing, or constant.

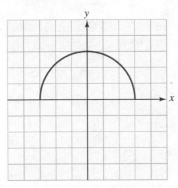

10. Graph $f(x) = \begin{cases} -x^2, & \text{when } x < 0 \\ -x, & \text{when } x \geq 0 \end{cases}$.

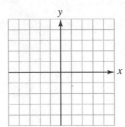

■ Cumulative Review Exercises

In Exercises 1–2, do the operations.

1. $(4x - 3y)(3x + y)$

2. $(a^n + 1)(a^n - 3)$

In Exercises 3–4, simplify each fraction.

3. $\dfrac{5a - 10}{a^2 - 4a + 4}$

4. $\dfrac{a^4 - 5a^2 + 4}{a^2 + 3a + 2}$

In Exercises 5–6, do the operations and simplify the result, if possible.

5. $\dfrac{a^2 - a - 6}{a^2 - 4} \div \dfrac{a^2 - 9}{a^2 + a - 6}$

6. $\dfrac{2}{a - 2} + \dfrac{3}{a + 2} - \dfrac{a - 1}{a^2 - 4}$

In Exercises 7–8, tell whether the graphs of the linear equations are parallel, perpendicular, or neither.

7. $3x - 4y = 12$, $y = \dfrac{3}{4}x - 5$

8. $y = 3x + 4$, $x = -3y + 4$

In Exercises 9–10, write the equation of each line with the following properties.

9. $m = -2$, passing through $(0, 5)$

10. Passing through $P(8, -5)$ and $Q(-5, 4)$

In Exercises 11–12, graph each inequality.

11. $2x - 3y < 6$

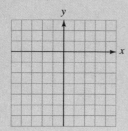

12. $y \geq x^2 - 4$

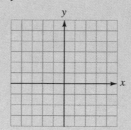

In Exercises 13–14, simplify each expression.

13. $\sqrt{98} + \sqrt{8} - \sqrt{32}$

14. $12\sqrt[3]{648x^4} + 3\sqrt[3]{81x^4}$

In Exercises 15–18, solve each equation.

15. $\sqrt{3a + 1} = a - 1$ /

16. $\sqrt{x + 3} - \sqrt{3} = \sqrt{x}$

17. $6a^2 + 5a - 6 = 0$

18. $3x^2 + 8x - 1 = 0$

19. If $f(x) = x^2 - 2$ and $g(x) = 2x + 1$, find $(f \circ g)(x)$.

20. Find the inverse function of $y = 2x^3 - 1$.

21. Graph $y = \left(\dfrac{1}{2}\right)^x$.

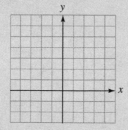

22. Write $y = \log_2 x$ as an exponential equation.

In Exercises 23–24, solve each equation.

23. $2^{x + 2} = 3^x$

24. $2 \log 5 + \log x - \log 4 = 2$

In Exercises 25–26, graph each equation.

25. $x^2 + (y + 1)^2 = 9$

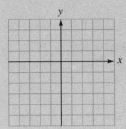

26. $x^2 - 9(y + 1)^2 = 9$

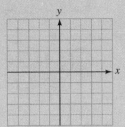

13 More on Systems of Equations and Inequalities

MATHEMATICS IN
THE WORKPLACE

Traffic Engineer

Traffic engineers supervise the design and construction of roads and gather and analyze data on traffic patterns. They determine efficient routes, designate the location of turn lanes and traffic control signals, and specify the timing of those signals to allow for smooth traffic flow.

SAMPLE APPLICATION ■ Before recommending traffic controls for the intersection of the two one-way streets shown in Illustration 1, a traffic engineer places counters across the roads to record traffic flow. The illustration shows the number of vehicles passing each of the four counters during one hour. From those data, find the number of vehicles passing straight through the intersection, and also the number that turn from one road to the other.
(See Project 2.)

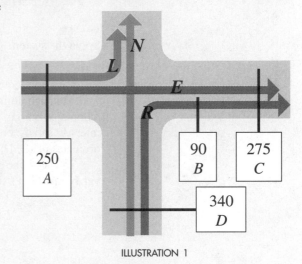

ILLUSTRATION 1

In Chapter 3, we solved systems of two linear equations in two variables by graphing, by substitution, and by addition. We also solved systems of two linear inequalities in two variables by graphing.

In this chapter, we will review the basic methods of solving systems of equations and inequalities and then discuss how to solve systems of three equations in three variables. We will also discuss how to solve linear systems by using matrices and determinants and how to solve systems of equations and inequalities that contain second-degree polynomials.

13.1   Solution of Two Equations in Two Variables

■ THE GRAPHING METHOD ■ THE SUBSTITUTION METHOD ■ THE ADDITION METHOD ■ PROBLEM SOLVING ■ SYSTEMS OF INEQUALITIES

Getting Ready   *Evaluate the function $f(x) = 3x + 2$ for the following values of x.*

1. $x = 0$ **2.** $x = 1$ **3.** $x = -2$ **4.** $x = -\dfrac{5}{3}$

■ THE GRAPHING METHOD

Recall that we follow these steps to solve a system of two equations in two variables by graphing.

Solving Systems of Equations by Graphing

1. On a single set of coordinate axes, graph each equation.

2. Find the coordinates of the point (or points) where the graphs intersect. These coordinates give the solution of the system.

3. Check the solution in both of the original equations.

4. If the graphs have no point in common, the system has no solution.

EXAMPLE 1   Solve the system $\begin{cases} x + 2y = 4 \\ 2x - y = 3 \end{cases}$.

Solution   We graph both equations on a single set of coordinate axes, as shown in Figure 13-1.

Although infinitely many pairs (x, y) satisfy $x + 2y = 4$ and infinitely many pairs (x, y) satisfy $2x - y = 3$, only the coordinates of the point where the graphs intersect satisfy both equations simultaneously. Since the intersection point has coordinates of $(2, 1)$, the solution is the pair $(2, 1)$, or $x = 2$ and $y = 1$.

To check the solution, we substitute 2 for x and 1 for y in each equation and verify that $(2, 1)$ satisfies each equation.

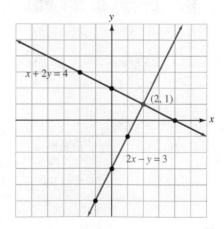

$x + 2y = 4$		
x	y	(x, y)
4	0	$(4, 0)$
0	2	$(0, 2)$
-2	3	$(-2, 3)$

$2x - y = 3$		
x	y	(x, y)
1	-1	$(1, -1)$
0	-3	$(0, -3)$
-1	-5	$(-1, -5)$

FIGURE 13-1

Self Check   Solve $\begin{cases} 2x + y = 0 \\ x - 2y = 5 \end{cases}$.

Answer

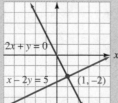

When a system of equations (as in Example 1) has a solution, the system is a **consistent system of equations.** A system with no solution is an **inconsistent system.**

EXAMPLE 2   Solve the system $\begin{cases} 2x + 3y = 6 \\ 4x + 6y = 24 \end{cases}$.

Solution   We graph both equations on the same set of coordinate axes, as shown in Figure 13-2. Since the lines do not intersect, the system does not have a solution. It is an inconsistent system.

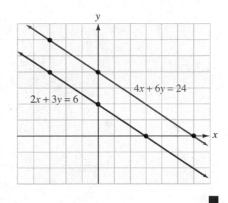

$2x + 3y = 6$				$4x + 6y = 24$		
x	y	(x, y)		x	y	(x, y)
3	0	$(3, 0)$		6	0	$(6, 0)$
0	2	$(0, 2)$		0	4	$(0, 4)$
-3	4	$(-3, 4)$		-3	6	$(-3, 6)$

FIGURE 13-2   ■

Self Check   Solve $\begin{cases} 2x - 5y = 15 \\ y = \dfrac{2}{5}x - 2 \end{cases}$.

Answer   no solution

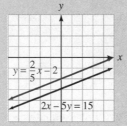

When the equations of a system have different graphs (as in Examples 1 and 2), the equations are **independent equations.** Two equations with the same graph are **dependent equations.**

EXAMPLE 3   Solve the system $\begin{cases} 2y - x = 4 \\ 2x + 8 = 4y \end{cases}$.

Solution   We graph each equation on the same set of coordinate axes, as shown in Figure 13-3. Since the graphs coincide, the system has infinitely many solutions. Any pair (x, y) that satisfies one equation satisfies the other also.

From the tables shown in the figure, we see that $(-4, 0)$ and $(0, 2)$ are solutions. We can find infinitely many more solutions by finding additional pairs (x, y) that satisfy either equation.

Because the two equations have the same graph, they are dependent equations.

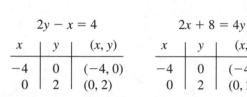

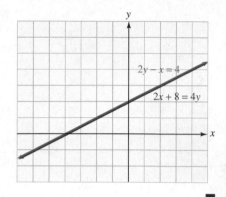

FIGURE 13-3

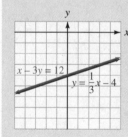

Self Check Solve $\begin{cases} x - 3y = 12 \\ y = \dfrac{1}{3}x - 4 \end{cases}.$

Answer   infinitely many solutions

When we graph two linear equations in two variables, the following possibilities can occur.

If the lines are different and intersect, the equations are independent and the system is consistent. *One solution exists.*

If the lines are different and parallel, the equations are independent and the system is inconsistent. *No solution exists.*

If the lines coincide, the equations are dependent and the system is consistent. *Infinitely many solutions exist.*

To solve a more difficult system such as

$$\begin{cases} \dfrac{3}{2}x - y = \dfrac{5}{2} \\ x + \dfrac{1}{2}y = 4 \end{cases}$$

we multiply both sides of the first equation by 2 and both sides of the second equation by 2 to eliminate the fractions:

$$\begin{cases} 3x - 2y = 5 \\ 2x + y = 8 \end{cases}$$

This new system is equivalent to the first. If we graph each equation in the new system, as in Figure 13-4, we see that the coordinates of the point where the two lines intersect are (3, 2).

For more examples of solving equations by graphing, refer to Section 3.3.

$3x - 2y = 5$		
x	y	(x, y)
0	$-\frac{5}{2}$	$\left(0, -\frac{5}{2}\right)$
$\frac{5}{3}$	0	$\left(\frac{5}{3}, 0\right)$

$2x + y = 8$		
x	y	(x, y)
4	0	$(4, 0)$
1	6	$(1, 6)$

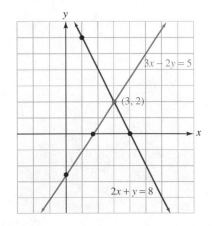

FIGURE 13-4

■ ■ ■ ■ ■ ■ ■ ■ ■ **Solving Systems of Equations**

GRAPHING
CALCULATORS

With a graphing calculator, we can obtain very good approximations of the solutions of a system of two linear equations in two variables. For example, to solve the system

$$\begin{cases} 3x + 2y = 12 \\ 2x - 3y = 12 \end{cases}$$

with a graphing calculator, we must first solve each equation for y to get the following equivalent system:

$$\begin{cases} y = -\dfrac{3}{2}x + 6 \\ y = \dfrac{2}{3}x - 4 \end{cases}$$

If we use window settings of $[-10, 10]$ for x and $[-10, 10]$ for y, the graphs of the equations will look like those in Figure 13-5(a). If we zoom in on the intersection point of the two lines as in Figure 13-5(b) and trace, we will find the approximate solution shown in the figure. To get better results, we can do additional zooms.

Verify that the exact solution is $x = \frac{60}{13}$ and $y = -\frac{12}{13}$.

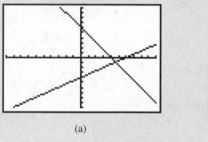

(a)

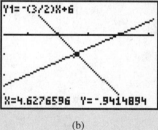

(b)

FIGURE 13-5

■ THE SUBSTITUTION METHOD

Recall that we use the following steps to solve a system of two equations in two variables by substitution.

Solving Systems of Equations by Substitution

1. If necessary, solve one equation for one of its variables.
2. Substitute the resulting expression for the variable obtained in Step 1 in the other equation and solve that equation.
3. Find the value of the other variable by substituting the value of the variable found in Step 2 in any equation containing both variables.
4. Check the solution in both of the original equations.

EXAMPLE 4   Solve the system $\begin{cases} \dfrac{4}{3}x + \dfrac{1}{2}y = -\dfrac{2}{3} \\ \dfrac{1}{2}x + \dfrac{2}{3}y = \dfrac{5}{3} \end{cases}$.

Solution   We first find an equivalent system without fractions by multiplying each side of each equation by 6 to obtain the system

1. $\begin{cases} 8x + 3y = -4 \\ 3x + 4y = 10 \end{cases}$
2.

Since no variable has a coefficient of 1, it is impossible to avoid fractions when solving for a variable. If we choose Equation 2 and solve it for x, we get

$$3x + 4y = 10$$

$$3x = -4y + 10 \qquad \text{Subtract } 4y \text{ from both sides.}$$

3. $\qquad\qquad x = -\dfrac{4}{3}y + \dfrac{10}{3} \qquad \text{Divide both sides by 3.}$

We can then substitute $-\frac{4}{3}y + \frac{10}{3}$ for x in Equation 1 and solve for y.

$$8x + 3y = -4$$

$$8\left(-\dfrac{4}{3}y + \dfrac{10}{3}\right) + 3y = -4 \qquad \text{Substitute } -\tfrac{4}{3}y + \tfrac{10}{3} \text{ for } x.$$

$$8(10 - 4y) + 9y = -12 \qquad \text{Multiply both sides by 3.}$$

$$80 - 32y + 9y = -12 \qquad \text{Use the distributive property to remove parentheses.}$$

$$-23y = -92 \qquad \text{Combine like terms and subtract 80 from both sides.}$$

$$y = 4 \qquad \text{Divide both sides by } -23.$$

We can find x by substituting 4 for y in Equation 3 and simplifying:

$$x = -\dfrac{4}{3}y + \dfrac{10}{3}$$

$$= -\dfrac{4}{3}(4) + \dfrac{10}{3} \qquad \text{Substitute 4 for } y.$$

$$= -2 \qquad\qquad -\tfrac{16}{3} + \tfrac{10}{3} = -\tfrac{6}{3} = -2.$$

The solution is the pair $(-2, 4)$. Verify that this solution satisfies each equation in the original system.   ∎

Self Check   Solve $\begin{cases} \dfrac{3}{2}x + \dfrac{1}{3}y = -5 \\ \dfrac{1}{2}x - \dfrac{2}{3}y = -4 \end{cases}$.

Answer $(-4, 3)$

For more examples of solving equations by substitution, refer to Section 3.4.

■ THE ADDITION METHOD

Recall that in the addition method, we combine the equations of the system in a way that will eliminate terms involving one of the variables.

> **Solving Systems of Equations by Addition**
> 1. Write both equations of the system in general form.
> 2. Multiply the terms of one or both of the equations by constants chosen to make the coefficients of x (or y) differ only in sign.
> 3. Add the equations and solve the equation that results, if possible.
> 4. Substitute the value obtained in Step 3 into either of the original equations and solve for the remaining variable.
> 5. The results obtained in Steps 3 and 4 are the solution of the system.
> 6. Check the solution in both of the original equations.

EXAMPLE 5   Solve the system $\begin{cases} \dfrac{4}{3}x + \dfrac{1}{2}y = -\dfrac{2}{3} \\ \dfrac{1}{2}x + \dfrac{2}{3}y = \dfrac{5}{3} \end{cases}$.

Solution   This system is the system discussed in Example 4. To solve it by addition, we find an equivalent system with no fractions by multiplying both sides of each equation by 6 to obtain

4. $\begin{cases} 8x + 3y = -4 \\ 3x + 4y = 10 \end{cases}$
5.

To make the y-terms drop out when we add the equations, we multiply both sides of Equation 4 by 4 and both sides of Equation 5 by -3 to get

$$\begin{cases} 32x + 12y = -16 \\ -9x - 12y = -30 \end{cases}$$

When these equations are added, the y-terms drop out, and we get

$23x = -46$

$x = -2$ Divide both sides by 23.

To find y, we substitute -2 for x in either Equation 4 or Equation 5. If we substitute -2 for x in Equation 5, we get

$3x + 4y = 10$

$3(-2) + 4y = 10$ Substitute -2 for x.

$-6 + 4y = 10$ Simplify.

$4y = 16$ Add 6 to both sides.

$y = 4$ Divide both sides by 4.

The solution is $(-2, 4)$.   ■

Self Check   Solve $\begin{cases} \frac{2}{3}x - \frac{2}{5}y = 10 \\ \frac{1}{2}x + \frac{2}{3}y = -7 \end{cases}$.

Answer $(6, -15)$

For more examples of solving equations by addition, refer to Section 3.5.

EXAMPLE 6   Solve the system $\begin{cases} y = 2x + 4 \\ 8x - 4y = 7 \end{cases}$.

Solution   Because the first equation is already solved for y, we use the substitution method.

$$8x - 4y = 7$$
$$8x - 4(2x + 4) = 7 \qquad \text{Substitute } 2x + 4 \text{ for } y.$$

We then solve this equation for x:

$$8x - 8x - 16 = 7 \qquad \text{Use the distributive property to remove parentheses.}$$
$$-16 \neq 7 \qquad \text{Combine like terms.}$$

This impossible result shows that the equations in the system are independent and that the system is inconsistent. Since the system has no solution, the graphs of the equations in the system will be parallel. ∎

Self Check   Solve $\begin{cases} 4x - 8y = 9 \\ y = \frac{1}{2}x - \frac{8}{9} \end{cases}$.

Answer   no solution

EXAMPLE 7   Solve the system $\begin{cases} 4x + 6y = 12 \\ -2x - 3y = -6 \end{cases}$.

Solution   Since the equations are written in general form, we use the addition method and multiply both sides of the second equation by 2 to get

$$\begin{cases} 4x + 6y = 12 \\ -4x - 6y = -12 \end{cases}$$

After adding the left-hand sides and the right-hand sides, we get

$$0x + 0y = 0$$
$$0 = 0$$

Here, both the x- and y-terms drop out. The true statement $0 = 0$ shows that the equations in this system are dependent and that the system is consistent.

Note that the equations of the system are equivalent, because when the second equation is multiplied by -2, it becomes the first equation.

The line graphs of these equations would coincide. Any ordered pair that satisfies one of the equations also satisfies the other. ∎

Self Check   Solve $\begin{cases} 2(x+y) - y = 12 \\ y = -2x + 12 \end{cases}$.

Answer   There are infinitely many solutions; any ordered pair that satisfies one of the equations also satisfies the other.

■ PROBLEM SOLVING

EXAMPLE 8   **Retail sales**   Hi-Fi Electronics advertises two types of car radios, one selling for $67 and the other for $100. If the receipts from the sale of 36 radios totaled $2,940, how many of each type were sold?

Analyze the problem   We can let x represent the number of radios sold for $67 and let y represent the number of radios sold for $100. Then the receipts for the sale of the less expensive radios are $67x$, and the receipts for the sale of the more expensive radios are $100y$.

Form two equations   The information in the problem gives the following two equations:

The number of less expensive radios	+	the number of more expensive radios	=	the total number of radios.
x	+	y	=	36

The value of the less expensive radios	+	the value of the more expensive radios	=	the total receipts.
$67x$	+	$100y$	=	2,940

Solve the resulting system of equations   We can solve the following system for x and y to find out how many of each type were sold:

6. $\begin{cases} x + y = 36 \\ 67x + 100y = 2{,}940 \end{cases}$
7.

We multiply both sides of Equation 6 by -100, add the resulting equation to Equation 7, and solve for x:

$$\begin{array}{r} -100x - 100y = -3{,}600 \\ 67x + 100y = 2{,}940 \\ \hline -33x = -660 \end{array}$$

$$x = 20 \qquad \text{Divide both sides by } -33.$$

To find y, we substitute 20 for x in Equation 6 and solve for y:

$$x + y = 36$$
$$20 + y = 36 \qquad \text{Substitute 20 for } x.$$
$$y = 16 \qquad \text{Subtract 20 from both sides.}$$

State the conclusion The store sold 20 of the less expensive radios and 16 of the more expensive radios.

Check the result If 20 of one type were sold and 16 of the other type were sold, a total of 36 radios were sold. Since the value of the less expensive radios is 20($67) = $1,340 and the value of the more expensive radios is 16($100) = $1,600, the total receipts are $2,940. ■

■ SYSTEMS OF INEQUALITIES

We now review the graphing method of solving systems of two linear inequalities in two variables. Recall that the solutions are usually the intersection of half-planes.

EXAMPLE 9 Graph the solution set of the system $\begin{cases} x + y \le 1 \\ 2x - y > 2 \end{cases}$.

Solution On the same set of coordinate axes, we graph each inequality, as in Figure 13-6.

The graph of the inequality $x + y \le 1$ includes the line graph of the equation $x + y = 1$ and all points below it. Since the boundary line is included, it is drawn as a solid line.

The graph of the inequality $2x - y > 2$ contains only those points below the graph of the equation $2x - y = 2$. Since the boundary line is not included, it is drawn as a broken line.

The area where the half-planes intersect represents the simultaneous solution of the given system of inequalities, because any point in that region has coordinates that will satisfy both inequalities.

$x + y = 1$		
x	y	(x, y)
0	1	$(0, 1)$
1	0	$(1, 0)$

$2x - y = 2$		
x	y	(x, y)
0	-2	$(0, -2)$
1	0	$(1, 0)$

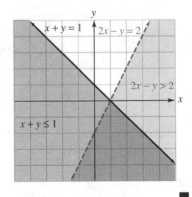

FIGURE 13-6 ■

Orals *Tell whether the following systems have one solution, no solutions, or infinitely many solutions.*

1. $\begin{cases} y = 2x \\ y = 2x + 5 \end{cases}$

2. $\begin{cases} y = 2x \\ y - x = x \end{cases}$

3. $\begin{cases} y = 2x \\ y = -2x \end{cases}$

4. $\begin{cases} y = 2x + 1 \\ 2x = y \end{cases}$

Solve each system for x.

5. $\begin{cases} y = 2x \\ x + y = 6 \end{cases}$

6. $\begin{cases} y = -x \\ 2x + y = 4 \end{cases}$

7. $\begin{cases} x - y = 6 \\ x + y = 2 \end{cases}$

8. $\begin{cases} x + y = 4 \\ 2x - y = 5 \end{cases}$

EXERCISE 13.1

REVIEW   *Simplify each expression. Write all answers without using negative exponents.*

1. $(a^2a^3)^2(a^4a^2)^2$

2. $\left(\dfrac{a^2b^3c^4d}{ab^2c^3d^4} \right)^{-3}$

3. $\left(\dfrac{-3x^3y^4}{x^{-5}y^3} \right)^{-4}$

4. $\dfrac{3t^0 - 4t^0 + 5}{5t^0 + 2t^0}$

Solve each formula for the given variable.

5. $A = p + prt$ for r

6. $A = p + prt$ for p

7. $\dfrac{1}{r} = \dfrac{1}{r_1} + \dfrac{1}{r_2}$ for r

8. $\dfrac{1}{r} = \dfrac{1}{r_1} + \dfrac{1}{r_2}$ for r_1

VOCABULARY AND CONCEPTS   *Fill in each blank to make a true statement.*

9. When a system of equations has a solution, it is called a _____ system of equations.

10. When a system of equations has no solution, it is called an _____ system of equations.

11. When the equations of a system have different graphs, the equations are called _____ equations.

12. Two equations with the same graph are called _____ equations.

PRACTICE   *In Exercises 13–24, solve each system by graphing, if possible. Check your graphs with a graphing calculator.*

13. $\begin{cases} x - y = 4 \\ 2x + y = 5 \end{cases}$

14. $\begin{cases} 2x + y = 1 \\ x - 2y = -7 \end{cases}$

15. $\begin{cases} x = 13 - 4y \\ 3x = 4 + 2y \end{cases}$

16. $\begin{cases} 3x = 7 - 2y \\ 2x = 2 + 4y \end{cases}$

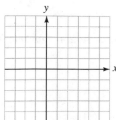 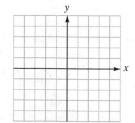 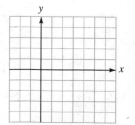 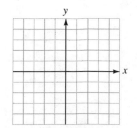

17. $\begin{cases} x = 3 - 2y \\ 2x + 4y = 6 \end{cases}$

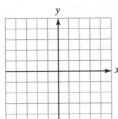

18. $\begin{cases} 3x = 5 - 2y \\ 3x + 2y = 7 \end{cases}$

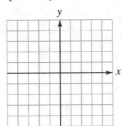

19. $\begin{cases} y = 3 \\ x = 2 \end{cases}$

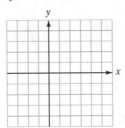

20. $\begin{cases} 2x + 3y = -15 \\ 2x + y = -9 \end{cases}$

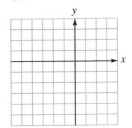

21. $\begin{cases} x = \dfrac{11 - 2y}{3} \\ y = \dfrac{11 - 6x}{4} \end{cases}$

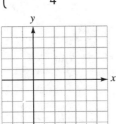

22. $\begin{cases} x = \dfrac{1 - 3y}{4} \\ y = \dfrac{12 + 3x}{2} \end{cases}$

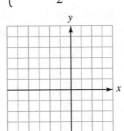

23. $\begin{cases} \dfrac{5}{2}x + y = \dfrac{1}{2} \\ 2x - \dfrac{3}{2}y = 5 \end{cases}$

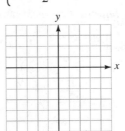

24. $\begin{cases} \dfrac{5}{2}x + 3y = 6 \\ y = \dfrac{24 - 10x}{12} \end{cases}$

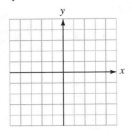

In Exercises 25–28, use a graphing calculator to solve each system. Give each answer to the nearest hundredth.

25. $\begin{cases} y = 3.2x - 1.5 \\ y = -2.7x - 3.7 \end{cases}$

26. $\begin{cases} y = -0.45x + 5 \\ y = 5.55x - 13.7 \end{cases}$

27. $\begin{cases} 1.7x + 2.3y = 3.2 \\ y = 0.25x + 8.95 \end{cases}$

28. $\begin{cases} 2.75x = 12.9y - 3.79 \\ 7.1x - y = 35.76 \end{cases}$

In Exercises 29–40, solve each system by substitution, if possible.

29. $\begin{cases} y = x \\ x + y = 4 \end{cases}$

30. $\begin{cases} y = x + 2 \\ x + 2y = 16 \end{cases}$

31. $\begin{cases} x - y = 2 \\ 2x + y = 13 \end{cases}$

32. $\begin{cases} x - y = -4 \\ 3x - 2y = -5 \end{cases}$

33. $\begin{cases} x + 2y = 6 \\ 3x - y = -10 \end{cases}$

34. $\begin{cases} 2x - y = -21 \\ 4x + 5y = 7 \end{cases}$

35. $\begin{cases} 3x = 2y - 4 \\ 6x - 4y = -4 \end{cases}$

36. $\begin{cases} 8x = 4y + 10 \\ 4x - 2y = 5 \end{cases}$

37. $\begin{cases} 3x - 4y = 9 \\ x + 2y = 8 \end{cases}$

38. $\begin{cases} 3x - 2y = -10 \\ 6x + 5y = 25 \end{cases}$

39. $\begin{cases} 2x + 2y = -1 \\ 3x + 4y = 0 \end{cases}$

40. $\begin{cases} 5x + 3y = -7 \\ 3x - 3y = 7 \end{cases}$

In Exercises 41–56, solve each system by addition, if possible.

41. $\begin{cases} x - y = 3 \\ x + y = 7 \end{cases}$

42. $\begin{cases} x + y = 1 \\ x - y = 7 \end{cases}$

43. $\begin{cases} 2x + y = -10 \\ 2x - y = -6 \end{cases}$

44. $\begin{cases} x + 2y = -9 \\ x - 2y = -1 \end{cases}$

45. $\begin{cases} 2x + 3y = 8 \\ 3x - 2y = -1 \end{cases}$

46. $\begin{cases} 5x - 2y = 19 \\ 3x + 4y = 1 \end{cases}$

47. $\begin{cases} 4x + 9y = 8 \\ 2x - 6y = -3 \end{cases}$

48. $\begin{cases} 4x + 6y = 5 \\ 8x - 9y = 3 \end{cases}$

49. $\begin{cases} 8x - 4y = 16 \\ 2x - 4 = y \end{cases}$

50. $\begin{cases} 2y - 3x = -13 \\ 3x - 17 = 4y \end{cases}$

51. $\begin{cases} x = \dfrac{3}{2}y + 5 \\ 2x - 3y = 8 \end{cases}$

52. $\begin{cases} x = \dfrac{2}{3}y \\ y = 4x + 5 \end{cases}$

53. $\begin{cases} \dfrac{x}{2} + \dfrac{y}{2} = 6 \\ \dfrac{x}{2} - \dfrac{y}{2} = -2 \end{cases}$

54. $\begin{cases} \dfrac{x}{2} - \dfrac{y}{3} = -4 \\ \dfrac{x}{2} + \dfrac{y}{9} = 0 \end{cases}$

55. $\begin{cases} \dfrac{3}{4}x + \dfrac{2}{3}y = 7 \\ \dfrac{3}{5}x - \dfrac{1}{2}y = 18 \end{cases}$

56. $\begin{cases} \dfrac{2}{3}x - \dfrac{1}{4}y = -8 \\ \dfrac{1}{2}x - \dfrac{3}{8}y = -9 \end{cases}$

In Exercises 57–60, graph the solution set of each system of inequalities.

57. $\begin{cases} y < 3x + 2 \\ y < -2x + 3 \end{cases}$

58. $\begin{cases} y \le x - 2 \\ y \ge 2x + 1 \end{cases}$

59. $\begin{cases} 3x + 2y > 6 \\ x + 3y \le 2 \end{cases}$

60. $\begin{cases} 3x + y \le 1 \\ -x + 2y \ge 6 \end{cases}$

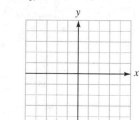
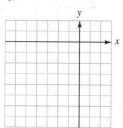
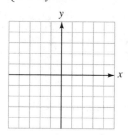
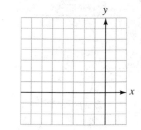

APPLICATIONS *Use two variables to solve each problem.*

61. Merchandising A pair of shoes and a sweater cost $98. If the sweater cost $16 more than the shoes, how much did the sweater cost?

62. Merchandising A sporting goods salesperson sells 2 fishing reels and 5 rods for $270. The next day, the salesperson sells 4 reels and 2 rods for $220. How much does each cost?

63. Electronics Two resistors in the voltage divider circuit in Illustration 1 have a total resistance of 1,375 ohms. To provide the required voltage, R_1 must be 125 ohms greater than R_2. Find both resistances.

64. Stowing baggage A small aircraft can carry 950 pounds of baggage, distributed between two storage

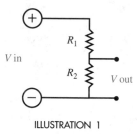

ILLUSTRATION 1

compartments. On one flight, the plane is fully loaded, with 150 pounds more baggage in one compartment than the other. How much is stowed in each compartment?

65. Geometry problem The rectangular field in Illustration 2 is surrounded by 72 meters of fencing. If the field is partitioned as shown, a total of 88 meters of fencing is required. Find the dimensions of the field.

66. Geometry problem In a right triangle, one acute angle is 15° greater than two times the other acute angle. Find the difference between the angles.

ILLUSTRATION 2

WRITING

67. Tell which method you would use to solve the following system. Why?

$$\begin{cases} y = 3x + 1 \\ 3x + 2y = 12 \end{cases}$$

68. Tell which method you would use to solve the following system. Why?

$$\begin{cases} 2x + 4y = 9 \\ 3x - 5y = 20 \end{cases}$$

69. When graphing a system of linear inequalities, explain how to decide which region to shade.

70. Explain how a system of two linear inequalities might have no solution.

SOMETHING TO THINK ABOUT

71. Under what conditions will a system of two equations in two variables be inconsistent?

72. Under what conditions will the equations of a system of two equations in two variables be dependent?

73. The solution of a system of inequalities in two variables is **bounded** if it is possible to draw a circle around it. Can the solution of a system of two linear inequalities be bounded?

74. The solution of $\begin{cases} y \ge |x| \\ y \le k \end{cases}$ has an area of 25. Find k.

13.2 Solutions of Three Equations in Three Variables

■ SOLVING THREE EQUATIONS IN THREE VARIABLES ■ A CONSISTENT SYSTEM ■ AN INCONSISTENT SYSTEM ■ SYSTEMS WITH DEPENDENT EQUATIONS ■ PROBLEM SOLVING ■ CURVE FITTING

Getting Ready *Tell whether the equation $x + 2y + 3z = 6$ is satisfied by the following values.*

1. $(1, 1, 1)$

2. $(-2, 1, 2)$

3. $(2, -2, -1)$

4. $(2, 2, 0)$

■ SOLVING THREE EQUATIONS IN THREE VARIABLES

We now extend the definition of a linear equation to include equations of the form $ax + by + cz = d$. The solution of a system of three linear equations with three variables is an ordered triple of numbers. For example, the solution of the system

$$\begin{cases} 2x + 3y + 4z = 20 \\ 3x + 4y + 2z = 17 \\ 3x + 2y + 3z = 16 \end{cases}$$

is the triple $(1, 2, 3)$, since each equation is satisfied if $x = 1$, $y = 2$, and $z = 3$.

$2x + 3y + 4z = 20$	$3x + 4y + 2z = 17$	$3x + 2y + 3z = 16$
$2(1) + 3(2) + 4(3) = 20$	$3(1) + 4(2) + 2(3) = 17$	$3(1) + 2(2) + 3(3) = 16$
$2 + 6 + 12 = 20$	$3 + 8 + 6 = 17$	$3 + 4 + 9 = 16$
$20 = 20$	$17 = 17$	$16 = 16$

The graph of an equation of the form $ax + by + cz = d$ is a flat surface called a **plane**. A system of three linear equations in three variables is consistent or inconsistent, depending on how the three planes corresponding to the three equations intersect. Figure 13-7 illustrates some of the possibilities.

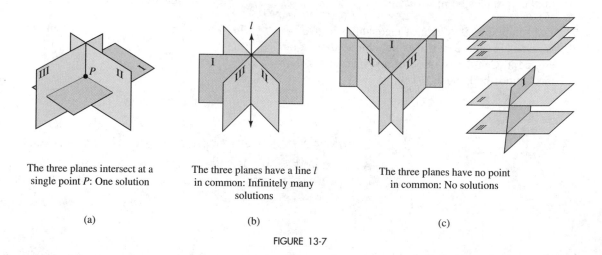

The three planes intersect at a single point P: One solution

(a)

The three planes have a line l in common: Infinitely many solutions

(b)

The three planes have no point in common: No solutions

(c)

FIGURE 13-7

To solve a system of three linear equations in three variables, we follow these steps.

Solving Three Equations in Three Variables

1. Pick any two equations and eliminate a variable.

2. Pick a different pair of equations and eliminate the same variable.

3. Solve the resulting pair of two equations in two variables.

4. To find the value of the third variable, substitute the values of the two variables found in Step 3 into any equation containing the unknown variable and solve the equation.

5. Check the solution in all three of the original equations.

■ A CONSISTENT SYSTEM

EXAMPLE 1 Solve the system $\begin{cases} 2x + y + 4z = 12 \\ x + 2y + 2z = 9 \\ 3x - 3y - 2z = 1 \end{cases}$.

Solution We are given the system

1. $\quad\begin{cases} 2x + y + 4z = 12 \\ x + 2y + 2z = 9 \\ 3x - 3y - 2z = 1 \end{cases}$
2.
3.

If we pick Equations 2 and 3 and add them, the variable z is eliminated:

$$
\begin{array}{ll}
\textbf{2.} & x + 2y + 2z = 9 \\
\textbf{3.} & 3x - 3y - 2z = 1 \\
\hline
\textbf{4.} & 4x - y = 10
\end{array}
$$

We now pick a different pair of equations (Equations 1 and 3) and eliminate z again. If each side of Equation 3 is multiplied by 2 and the resulting equation is added to Equation 1, z is again eliminated:

$$
\begin{array}{ll}
\textbf{1.} & 2x + y + 4z = 12 \\
& 6x - 6y - 4z = 2 \\
\hline
\textbf{5.} & 8x - 5y = 14
\end{array}
$$

Equations 4 and 5 form a system of two equations in two variables:

$$
\begin{array}{ll}
\textbf{4.} & \begin{cases} 4x - y = 10 \\ 8x - 5y = 14 \end{cases} \\
\textbf{5.} &
\end{array}
$$

To solve this system, we multiply Equation 4 by -5 and add the resulting equation to Equation 5 to eliminate y:

$$-20x + 5y = -50$$

5. $\quad\dfrac{8x - 5y = 14}{-12x = -36}$

$$x = 3 \qquad \text{Divide both sides by } -12.$$

To find y, we substitute 3 for x in any equation containing only x and y (such as Equation 5) and solve for y:

5. $\quad 8x - 5y = 14$

$8(3) - 5y = 14 \qquad$ Substitute 3 for x.

$24 - 5y = 14 \qquad$ Simplify.

${-5y = -10} \qquad$ Subtract 24 from both sides.

$y = 2 \qquad$ Divide both sides by -5.

To find z, we substitute 3 for x and 2 for y in an equation containing x, y, and z (such as Equation 1) and solve for z:

1. $\quad 2x + y + 4z = 12$

$2(3) + 2 + 4z = 12 \qquad$ Substitute 3 for x and 2 for y.

$8 + 4z = 12 \qquad$ Simplify.

$4z = 4 \qquad$ Subtract 8 from both sides.

$z = 1 \qquad$ Divide both sides by 4.

The solution of the system is $(x, y, z) = (3, 2, 1)$. Verify that these values satisfy each equation in the original system. ∎

Self Check   Solve $\begin{cases} 2x + y + 4z = 16 \\ x + 2y + 2z = 11 \\ 3x - 3y - 2z = -9 \end{cases}$.

Answer $(1, 2, 3)$

■ AN INCONSISTENT SYSTEM

EXAMPLE 2   Solve the system $\begin{cases} 2x + y - 3z = -3 \\ 3x - 2y + 4z = 2 \\ 4x + 2y - 6z = -7 \end{cases}$.

Solution   We are given the system of equations

1. $\begin{cases} 2x + y - 3z = -3 \\ 3x - 2y + 4z = 2 \\ 4x + 2y - 6z = -7 \end{cases}$
2.
3.

We can multiply Equation 1 by 2 and add the resulting equation to Equation 2 to eliminate y:

$$4x + 2y - 6z = -6$$
2. $\underline{3x - 2y + 4z = 2}$
4. $7x - 2z = -4$

We now add Equations 2 and 3 to eliminate y again:

2. $3x - 2y + 4z = 2$
3. $\underline{4x + 2y - 6z = -7}$
5. $7x - 2z = -5$

Equations 4 and 5 form the system

4. $\begin{cases} 7x - 2z = -4 \\ 7x - 2z = -5 \end{cases}$
5.

Since $7x - 2z$ cannot equal both -4 and -5, this system is inconsistent. Thus, it has no solution. ∎

Self Check   Solve $\begin{cases} 2x + y - 3z = 8 \\ 3x - 2y + 4z = 10 \\ 4x + 2y - 6z = -5 \end{cases}$.

Answer   no solution

■ SYSTEMS WITH DEPENDENT EQUATIONS

When the equations in a system of two equations in two variables were dependent, the system had infinitely many solutions. This is not always true for systems of three equations in three variables. In fact, a system can have dependent equations and still be inconsistent. Figure 13-8 illustrates the different possibilities.

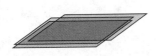

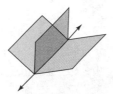

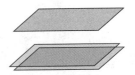

When three planes coincide, the equations are dependent, and there are infinitely many solutions.

When three planes intersect in a common line, the equations are dependent, and there are infinitely many solutions.

When two planes coincide and are parallel to a third plane, the system is inconsistent, and there are no solutions.

(a)

(b)

(c)

FIGURE 13-8

EXAMPLE 3   Solve the system $\begin{cases} 3x - 2y + z = -1 \\ 2x + y - z = 5 \\ 5x - y = 4 \end{cases}$.

Solution   We can add the first two equations to get

$$3x - 2y + z = -1$$
$$\underline{2x + \ y - z = \ \ \ 5}$$
1. $5x - \ y \ \ \ \ \ = \ \ \ 4$

Since Equation 1 is the same as the third equation of the system, the equations of the system are dependent, and there will be infinitely many solutions. From a graphical perspective, the equations represent three planes that intersect in a common line, as shown in Figure 13-8(b).

To write the general solution to this system, we can solve Equation 1 for y to get

$$5x - y = 4$$
$$-y = -5x + 4 \qquad \text{Subtract } 5x \text{ from both sides.}$$
$$y = 5x - 4 \qquad \text{Multiply both sides by } -1.$$

We can then substitute $5x - 4$ for y in the first equation of the system and solve for z to get

$$3x - 2y + z = -1$$
$$3x - 2(5x - 4) + z = -1 \qquad \text{Substitute } 5x - 4 \text{ for } y.$$
$$3x - 10x + 8 + z = -1 \qquad \text{Use the distributive property to remove parentheses.}$$
$$-7x + 8 + z = -1 \qquad \text{Combine like terms.}$$
$$z = 7x - 9 \qquad \text{Add } 7x \text{ and } -8 \text{ to both sides.}$$

Since we have found the values of y and z in terms of x, every solution to the system has the form $(x, 5x - 4, 7x - 9)$, where x can be any real number. For example,

If $x = 1$, a solution is $(1, 1, -2)$. $5(1) - 4 = 1$, and $7(1) - 9 = -2$.
If $x = 2$, a solution is $(2, 6, 5)$. $5(2) - 4 = 6$, and $7(2) - 9 = 5$.
If $x = 3$, a solution is $(3, 11, 12)$. $5(3) - 4 = 11$, and $7(3) - 9 = 12$.   ∎

Self Check   Solve $\begin{cases} 3x + 2y + z = -1 \\ 2x - y - z = 5 \\ 5x + y = 4 \end{cases}$.

Answer   infinitely many solutions; a general solution is $(x, 4 - 5x, -9 + 7x)$; three solutions are $(1, -1, -2)$, $(2, -6, 5)$, and $(3, -11, 12)$

■ PROBLEM SOLVING

EXAMPLE 4

Manufacturing hammers A company manufactures three types of hammers—good, better, and best. The cost of manufacturing each type of hammer is $4, $6, and $7, respectively, and the hammers sell for $6, $9, and $12. Each day, the cost of manufacturing 100 hammers is $520, and the daily revenue from their sale is $810. How many of each type are manufactured?

Analyze the problem   If we let x represent the number of good hammers, y represent the number of better hammers, and z represent the number of best hammers, we know that

The total number of hammers is $x + y + z$.

The cost of manufacturing the good hammers will be $4x$ ($4 times x hammers).

The cost of manufacturing the better hammers will be $6y$ ($6 times y hammers).

The cost of manufacturing the best hammers will be $7z$ ($7 times z hammers).

The revenue received by selling the good hammers is $6x$ ($6 times x hammers).

The revenue received by selling the better hammers is $9y$ ($9 times y hammers).

The revenue received by selling the best hammers is $12z$ ($12 times z hammers).

Form three equations   Since x represents the number of good hammers manufactured, y represents the number of better hammers manufactured, and z represents the number of best hammers manufactured, we have

The number of good hammers	+	the number of better hammers	+	the number of best hammers	=	the total number of hammers.
x	+	y	+	z	=	100

The cost of good hammers	+	the cost of better hammers	+	the cost of best hammers	=	the total cost.
$4x$	+	$6y$	+	$7z$	=	520

The revenue from good hammers	+	the revenue from better hammers	+	the revenue from best hammers	=	the total revenue.
$6x$	+	$9y$	+	$12z$	=	810

Solve the system We must now solve the system

$$
\begin{array}{rl}
\textbf{1.} & x + y + z = 100 \\
\textbf{2.} & 4x + 6y + 7z = 520 \\
\textbf{3.} & 6x + 9y + 12z = 810
\end{array}
$$

If we multiply Equation 1 by -7 and add the result to Equation 2, we get

$$
\begin{array}{rl}
 & -7x - 7y - 7z = -700 \\
 & \underline{4x + 6y + 7z = 520} \\
\textbf{4.} & -3x - y = -180
\end{array}
$$

If we multiply Equation 1 by -12 and add the result to Equation 3, we get

$$
\begin{array}{rl}
 & -12x - 12y - 12z = -1{,}200 \\
 & \underline{6x + 9y + 12z = 810} \\
\textbf{5.} & -6x - 3y = -390
\end{array}
$$

If we multiply Equation 4 by -3 and add it to Equation 5, we get

$$
\begin{array}{rl}
 & 9x + 3y = 540 \\
 & \underline{-6x - 3y = -390} \\
 & 3x = 150 \\
 & x = 50 \qquad \text{Divide both sides by 3.}
\end{array}
$$

To find y, we substitute 50 for x in Equation 4:

$$
\begin{array}{ll}
-3x - y = -180 & \\
-3(\textbf{50}) - y = -180 & \text{Substitute 50 for } x. \\
-y = -30 & \text{Add 150 to both sides.} \\
y = 30 & \text{Divide both sides by } -1.
\end{array}
$$

To find z, we substitute 50 for x and 30 for y in Equation 1:

$$
\begin{array}{ll}
x + y + z = 100 & \\
\textbf{50} + \textbf{30} + z = 100 & \\
z = 20 & \text{Subtract 80 from both sides.}
\end{array}
$$

State the conclusion The company manufactures 50 good hammers, 30 better hammers, and 20 best hammers each day.

Check the result Check the solution in each equation in the original system. ∎

■ CURVE FITTING

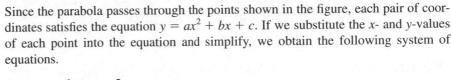

EXAMPLE 5

Curve fitting The equation of the parabola shown in Figure 13-9 is of the form $y = ax^2 + bx + c$. Find the equation of the parabola.

Solution   Since the parabola passes through the points shown in the figure, each pair of coordinates satisfies the equation $y = ax^2 + bx + c$. If we substitute the x- and y-values of each point into the equation and simplify, we obtain the following system of equations.

1. $a - b + c = 5$
2. $a + b + c = 1$
3. $4a + 2b + c = 2$

If we add Equations 1 and 2, we obtain $2a + 2c = 6$. If we multiply Equation 1 by 2 and add the result to Equation 3, we get $6a + 3c = 12$. We can then divide both sides of $2a + 2c = 6$ by 2 and divide both sides of $6a + 3c = 12$ by 3 to get the system

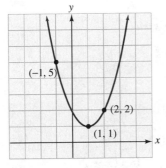

FIGURE 13-9

4. $\begin{cases} a + c = 3 \\ 2a + c = 4 \end{cases}$
5.

If we multiply Equation 4 by -1 and add the result to Equation 5, we get $a = 1$. To find c, we can substitute 1 for a in Equation 4 and find that $c = 2$. To find b, we can substitute 1 for a and 2 for c in Equation 2 and find that $b = -2$.

After we substitute these values of a, b, and c into the equation $y = ax^2 + bx + c$, we have the equation of the parabola.

$$y = \boldsymbol{a}x^2 + bx + c$$
$$y = \boldsymbol{1}x^2 + (\boldsymbol{-2})x + \boldsymbol{2}$$
$$y = x^2 - 2x + 2$$ ■

Orals   *Is the triple a solution of the system?*

1. $(1, 1, 1)$, $\begin{cases} 2x + y - 3z = 0 \\ 3x - 2y + 4z = 5 \\ 4x + 2y - 6z = 0 \end{cases}$ **2.** $(2, 0, 1)$, $\begin{cases} 3x + 2y - z = 5 \\ 2x - 3y + 2z = 4 \\ 4x - 2y + 3z = 10 \end{cases}$

EXERCISE 13.2

REVIEW   *Consider the line passing through $P(-2, -4)$ and $Q(3, 5)$.*

1. Find the slope of line PQ.   **2.** Write the equation of line PQ in general form

Let $f(x) = 2x^2 + 1$. Find each value.

3. $f(0)$ **4.** $f(-2)$ **5.** $f(s)$ **6.** $f(2t)$

VOCABULARY AND CONCEPTS *Fill in each blank to make a true statement.*

7. The graph of the equation $2x + 3y + 4z = 5$ is a flat surface called a _____.

8. When three planes coincide, the equations of the system are _____, and there are _____ many solutions.

9. When three planes intersect in a line, the system will have _____ many solutions.

10. When three planes are parallel, the system will have ____ solutions.

PRACTICE *In Exercises 11–12, tell whether the given triple is a solution of the given system.*

11. $(2, 1, 1)$, $\begin{cases} x - y + z = 2 \\ 2x + y - z = 4 \\ 2x - 3y + z = 2 \end{cases}$

12. $(-3, 2, -1)$, $\begin{cases} 2x + 2y + 3z = -1 \\ 3x + y - z = -6 \\ x + y + 2z = 1 \end{cases}$

In Exercises 13–24, solve each system.

13. $\begin{cases} x + y + z = 4 \\ 2x + y - z = 1 \\ 2x - 3y + z = 1 \end{cases}$

14. $\begin{cases} x + y + z = 4 \\ x - y + z = 2 \\ x - y - z = 0 \end{cases}$

15. $\begin{cases} 2x + 2y + 3z = 10 \\ 3x + y - z = 0 \\ x + y + 2z = 6 \end{cases}$

16. $\begin{cases} x - y + z = 4 \\ x + 2y - z = -1 \\ x + y - 3z = -2 \end{cases}$

17. $\begin{cases} a + b + 2c = 7 \\ a + 2b + c = 8 \\ 2a + b + c = 9 \end{cases}$

18. $\begin{cases} 2a + 3b + c = 2 \\ 4a + 6b + 2c = 5 \\ a - 2b + c = 3 \end{cases}$

19. $\begin{cases} 2x + y - z = 1 \\ x + 2y + 2z = 2 \\ 4x + 5y + 3z = 3 \end{cases}$

20. $\begin{cases} 4x + 3z = 4 \\ 2y - 6z = -1 \\ 8x + 4y + 3z = 9 \end{cases}$

21. $\begin{cases} 2x + 3y + 4z = 6 \\ 2x - 3y - 4z = -4 \\ 4x + 6y + 8z = 12 \end{cases}$

22. $\begin{cases} x - 3y + 4z = 2 \\ 2x + y + 2z = 3 \\ 4x - 5y + 10z = 7 \end{cases}$

23. $\begin{cases} x + \dfrac{1}{3}y + z = 13 \\ \dfrac{1}{2}x - y + \dfrac{1}{3}z = -2 \\ x + \dfrac{1}{2}y - \dfrac{1}{3}z = 2 \end{cases}$

24. $\begin{cases} x - \dfrac{1}{5}y - z = 9 \\ \dfrac{1}{4}x + \dfrac{1}{5}y - \dfrac{1}{2}z = 5 \\ 2x + y + \dfrac{1}{6}z = 12 \end{cases}$

APPLICATIONS

25. Integer problem The sum of three integers is 18. The third integer is four times the second, and the second integer is 6 more than the first. Find the integers.

26. Integer problem The sum of three integers is 48. If the first integer is doubled, the sum is 60. If the second integer is doubled, the sum is 63. Find the integers.

27. Geometry problem The sum of the angles in any triangle is 180°. In triangle *ABC*, angle *A* is 100° less than the sum of angles *B* and *C*, and angle *C* is 40° less than twice angle *B*. Find each angle.

28. Geometry problem The sum of the angles of any four-sided figure is 360°. In the quadrilateral shown in Illustration 1, angle *A* = angle *B*, angle *C* is 20° greater than angle *A*, and angle *D* = 40°. Find the angles.

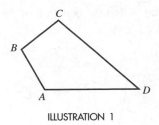

ILLUSTRATION 1

29. Nutritional planning One unit of each of three foods contains the nutrients shown in Illustration 2. How many units of each must be used to provide exactly 11 grams of fat, 6 grams of carbohydrates, and 10 grams of protein?

Food	Fat	Carbohydrates	Protein
A	1	1	2
B	2	1	1
C	2	1	2

ILLUSTRATION 2

30. Nutritional planning One unit of each of three foods contains the nutrients shown in Illustration 3. How many units of each must be used to provide exactly 14 grams of fat, 9 grams of carbohydrates, and 9 grams of protein?

Food	Fat	Carbohydrates	Protein
A	2	1	2
B	3	2	1
C	1	1	2

ILLUSTRATION 3

31. Making statues An artist makes three types of ceramic statues at a monthly cost of $650 for 180 statues. The manufacturing costs for the three types are $5, $4, and $3. If the statues sell for $20, $12, and $9, respectively, how many of each type should be made to produce $2,100 in monthly revenue?

32. Manufacturing footballs A factory manufactures three types of footballs at a monthly cost of $2,425 for 1,125 footballs. The manufacturing costs for the three

types of footballs are $4, $3, and $2. These footballs sell for $16, $12, and $10, respectively. How many of each type are manufactured if the monthly profit is $9,275? (*Hint:* Profit = income − cost.)

33. Concert tickets Tickets for a concert cost $5, $3, and $2. Twice as many $5 tickets were sold as $2 tickets. The receipts for 750 tickets were $2,625. How many of each price ticket were sold?

34. Mixing nuts The owner of a candy store mixed some peanuts worth $3 per pound, some cashews worth $9 per pound, and some Brazil nuts worth $9 per pound to get 50 pounds of a mixture that would sell for $6 per pound. She used 15 fewer pounds of cashews than peanuts. How many pounds of each did she use?

35. Chainsaw sculpting A north woods sculptor carves three types of statues with a chainsaw. The time required for carving, sanding, and painting a totem pole, a bear, and a deer are shown in Illustration 4. How many of each should be produced to use all available labor hours?

	Totem pole	Bear	Deer	Time available
Carving	2 hours	2 hours	1 hour	14 hours
Sanding	1 hour	2 hours	2 hours	15 hours
Painting	3 hours	2 hours	2 hours	21 hours

ILLUSTRATION 4

36. Making clothing A clothing manufacturer makes coats, shirts, and slacks. The time required for cutting, sewing, and packaging each item are shown in Illustration 5. How many of each should be made to use all available labor hours?

	Coats	Shirts	Slacks	Time available
Cutting	20 min	15 min	10 min	115 hr
Sewing	60 min	30 min	24 min	280 hr
Packaging	5 min	12 min	6 min	65 hr

ILLUSTRATION 5

37. Curve fitting Find the equation of the parabola shown in Illustration 6.

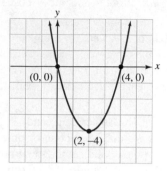

ILLUSTRATION 6

38. Curve fitting Find the equation of the parabola shown in Illustration 7.

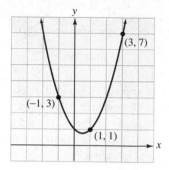

ILLUSTRATION 7

In Exercises 39–40, the equation of a circle is of the form $x^2 + y^2 + cx + dy + e = 0$.

39. Curve fitting Find the equation of the circle shown in Illustration 8.

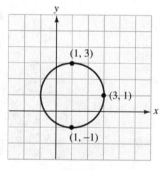

ILLUSTRATION 8

40. Curve fitting Find the equation of the circle shown in Illustration 9.

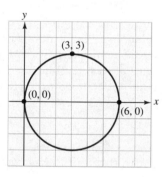

ILLUSTRATION 9

WRITING

41. What makes a system of three equations in three variables inconsistent?

42. What makes the equations of a system of three equations in three variables dependent?

SOMETHING TO THINK ABOUT

43. Solve the system

$$\begin{cases} x + y + z + w = 3 \\ x - y - z - w = -1 \\ x + y - z - w = 1 \\ x + y - z + w = 3 \end{cases}$$

44. Solve the system

$$\begin{cases} 2x + y + z + w = 3 \\ x - 2y - z + w = -3 \\ x - y - 2z - w = -3 \\ x + y - z + 2w = 4 \end{cases}$$

13.3 Solution by Matrices

■ MATRICES ■ GAUSSIAN ELIMINATION ■ SYSTEMS WITH MORE EQUATIONS THAN VARIABLES
■ SYSTEMS WITH MORE VARIABLES THAN EQUATIONS

Getting Ready *Multiply the first row by 2 and add the result to the second row.*

1. 2   3   5
    1   2   3

2. −1   0   4
    2   3   −1

Multiply the first row by −1 and add the result to the second row.

3. 2   3   5
    1   2   3

4. −1   0   4
    2   3   −1

■ MATRICES

Another method of solving systems of equations involves rectangular arrays of numbers called *matrices*.

> **Matrix**
> A **matrix** is any rectangular array of numbers.

Some examples of matrices are

$$A = \begin{bmatrix} 1 & 2 & 3 \\ 4 & 5 & 6 \end{bmatrix} \qquad B = \begin{bmatrix} 1 & 2 \\ 3 & 4 \\ 5 & 6 \end{bmatrix} \qquad C = \begin{bmatrix} 2 & 4 & 6 \\ 8 & 10 & 12 \\ 14 & 16 & 18 \end{bmatrix}$$

The numbers in each matrix are called **elements.** Because matrix A has two rows and three columns, it is called a 2×3 matrix (read "2 by 3" matrix). Matrix B is a 3×2 matrix, because the matrix has three rows and two columns. Matrix C is a 3×3 matrix (three rows and three columns).

Any matrix with the same number of rows and columns, like matrix C, is called a **square matrix.**

To show how to use matrices to solve systems of linear equations, we consider the system

$$\begin{cases} x - 2y - z = 6 \\ 2x + 2y - z = 1 \\ -x - y + 2z = 1 \end{cases}$$

which can be represented by the following matrix, called an **augmented matrix:**

$$\begin{bmatrix} 1 & -2 & -1 & 6 \\ 2 & 2 & -1 & 1 \\ -1 & -1 & 2 & 1 \end{bmatrix}$$

The first three columns of the augmented matrix form a 3×3 matrix called a **coefficient matrix**. It is determined by the coefficients of x, y, and z in the equations of the system. The 3×1 matrix in the last column is determined by the constants in the equations.

Coefficient matrix   ***Column of constants***

$$\begin{bmatrix} 1 & -2 & -1 \\ 2 & 2 & -1 \\ -1 & -1 & 2 \end{bmatrix} \qquad \begin{bmatrix} 6 \\ 1 \\ 1 \end{bmatrix}$$

Each row of the augmented matrix represents one equation of the system:

$$\begin{bmatrix} 1 & -2 & -1 & 6 \\ 2 & 2 & -1 & 1 \\ -1 & -1 & 2 & 1 \end{bmatrix} \begin{array}{l} \longleftrightarrow \\ \longleftrightarrow \\ \longleftrightarrow \end{array} \begin{cases} x - 2y - z = 6 \\ 2x + 2y - z = 1 \\ -x - y + 2z = 1 \end{cases}$$

■ GAUSSIAN ELIMINATION

To solve a 3×3 system of equations by **Gaussian elimination**, we transform an augmented matrix into the following matrix that has all 0's below its main diagonal, which is formed by the elements a, e, and h.

$$\begin{bmatrix} a & b & c & d \\ 0 & e & f & g \\ 0 & 0 & h & i \end{bmatrix} \quad (a, b, c, \dots, i \text{ are real numbers})$$

We can often write a matrix in this form, called **triangular form**, by using three operations called **elementary row operations**.

> **Elementary Row Operations**
>
> **1.** Any two rows of a matrix can be interchanged.
> **2.** Any row of a matrix can be multiplied by a nonzero constant.
> **3.** Any row of a matrix can be changed by adding a nonzero constant multiple of another row to it.

- A type 1 row operation corresponds to interchanging two equations of a system.
- A type 2 row operation corresponds to multiplying both sides of an equation by a nonzero constant.
- A type 3 row operation corresponds to adding a nonzero multiple of one equation to another.

None of these operations will change the solution of the given system of equations.

After we have written the matrix in triangular form, we can solve the corresponding system of equations by a back substitution process, as shown in Example 1.

EXAMPLE 1   Solve the system $\begin{cases} x - 2y - z = 6 \\ 2x + 2y - z = 1 \\ -x - y + 2z = 1 \end{cases}$.

Solution   We can represent the system with the following augmented matrix:

$$\begin{bmatrix} 1 & -2 & -1 & 6 \\ 2 & 2 & -1 & 1 \\ -1 & -1 & 2 & 1 \end{bmatrix}$$

To get 0's under the 1 in the first column, we use a type 3 row operation twice:

$$\begin{bmatrix} 1 & -2 & -1 & 6 \\ 2 & 2 & -1 & 1 \\ -1 & -1 & 2 & 1 \end{bmatrix} \approx \overset{\substack{\text{Multiply row 1 by } -2 \\ \text{and add to row 2.}}}{\begin{bmatrix} 1 & -2 & -1 & 6 \\ 0 & 6 & 1 & -11 \\ -1 & -1 & 2 & 1 \end{bmatrix}} \approx \overset{\substack{\text{Multiply row 1 by } 1 \\ \text{and add to row 3.}}}{\begin{bmatrix} 1 & -2 & -1 & 6 \\ 0 & 6 & 1 & -11 \\ 0 & -3 & 1 & 7 \end{bmatrix}}$$

The symbol "$\approx$" is read as "is row equivalent to." Each of the above matrices represents an equivalent system of equations.

To get a 0 under the 6 in the second column of the last matrix, we use another type 3 row operation:

$$\begin{bmatrix} 1 & -2 & -1 & 6 \\ 0 & 6 & 1 & -11 \\ 0 & -3 & 1 & 7 \end{bmatrix} \approx \overset{\substack{\text{Multiply row 2 by } \frac{1}{2} \\ \text{and add to row 3.}}}{\begin{bmatrix} 1 & -2 & -1 & 6 \\ 0 & 6 & 1 & -11 \\ 0 & 0 & \frac{3}{2} & \frac{3}{2} \end{bmatrix}}$$

We can use a type 2 row operation to simplify further.

$$\begin{bmatrix} 1 & -2 & -1 & 6 \\ 0 & 6 & 1 & -11 \\ 0 & 0 & \frac{3}{2} & \frac{3}{2} \end{bmatrix} \approx \overset{\substack{\text{Multiply row 3 by } \frac{2}{3}, \text{ which} \\ \text{is the reciprocal of } \frac{3}{2}.}}{\begin{bmatrix} 1 & -2 & -1 & 6 \\ 0 & 6 & 1 & -11 \\ 0 & 0 & 1 & 1 \end{bmatrix}}$$

The final matrix represents the system of equations

1. $\begin{cases} x - 2y - z = 6 \\ 0x + 6y + z = -11 \\ 0x + 0y + z = 1 \end{cases}$
2.
3.

From Equation 3, we can read that $z = 1$. To find y, we substitute 1 for z in Equation 2 and solve for y:

2. $6y + z = -11$

$\quad\quad 6y + \mathbf{1} = -11$ Substitute 1 for z.

$\quad\quad\quad\quad 6y = -12$ Subtract 1 from both sides.

$\quad\quad\quad\quad\quad y = -2$ Divide both sides by 6.

Thus, $y = -2$. To find x, we substitute 1 for z and -2 for y in Equation 1 and solve for x:

1. $x - 2y - z = 6$
$x - 2(-2) - 1 = 6$ Substitute 1 for z and -2 for y.
$x + 3 = 6$ Simplify.
$x = 3$ Subtract 3 from both sides.

Thus, $x = 3$. The solution of the given system is $(3, -2, 1)$. Verify that this triple satisfies each equation of the original system.   ∎

Self Check   Solve $\begin{cases} x - 2y - z = 2 \\ 2x + 2y - z = -5. \\ -x - y + 2z = 7 \end{cases}$

Answer $(1, -2, 3)$

■ SYSTEMS WITH MORE EQUATIONS THAN VARIABLES

We can use matrices to solve systems that have more equations than variables.

EXAMPLE 2   Solve the system $\begin{cases} x + y = -1 \\ 2x - y = 7 \\ -x + 2y = -8 \end{cases}$.

Solution   This system can be represented by a 3×3 augmented matrix:

$$\begin{bmatrix} 1 & 1 & -1 \\ 2 & -1 & 7 \\ -1 & 2 & -8 \end{bmatrix}$$

To get 0's under the 1 in the first column, we do a type 3 row operation twice:

Multiply row 1 by -2 and add to row 2.   Multiply row 1 by 1 and add to row 3.

$$\begin{bmatrix} 1 & 1 & -1 \\ 2 & -1 & 7 \\ -1 & 2 & -8 \end{bmatrix} \approx \begin{bmatrix} 1 & 1 & -1 \\ 0 & -3 & 9 \\ -1 & 2 & -8 \end{bmatrix} \approx \begin{bmatrix} 1 & 1 & -1 \\ 0 & -3 & 9 \\ 0 & 3 & -9 \end{bmatrix}$$

We can do other row operations to get a 0 under the -3 in the second column and then 1 in the second row, second column.

Multiply row 2 by 1 and add to row 3.   Multiply row 2 by $-\frac{1}{3}$.

$$\begin{bmatrix} 1 & 1 & -1 \\ 0 & -3 & 9 \\ 0 & 3 & -9 \end{bmatrix} \approx \begin{bmatrix} 1 & 1 & -1 \\ 0 & -3 & 9 \\ 0 & 0 & 0 \end{bmatrix} \approx \begin{bmatrix} 1 & 1 & -1 \\ 0 & 1 & -3 \\ 0 & 0 & 0 \end{bmatrix}$$

The final matrix represents the system

$$\begin{cases} x + y = -1 \\ 0x + y = -3 \\ 0x + 0y = 0 \end{cases}$$

The third equation can be discarded, because $0x + 0y = 0$ for all x and y. From the second equation, we can read that $y = -3$. To find x, we substitute -3 for y in the first equation and solve for x:

$$x + y = -1$$
$$x + (\mathbf{-3}) = -1 \qquad \text{Substitute } -3 \text{ for } y.$$
$$x = 2 \qquad \text{Add 3 to both sides.}$$

The solution is $(2, -3)$. Verify that this solution satisfies all three equations of the original system. ∎

Self Check   Solve $\begin{cases} x + y = 1 \\ 2x - y = 8 \\ -x + 2y = -7 \end{cases}$.

Answer $(3, -2)$

If the last row of the final matrix in Example 2 had been of the form $0x + 0y = k$, where $k \neq 0$, the system would not have a solution. No values of x and y could make the expression $0x + 0y$ equal to a nonzero constant k.

■ SYSTEMS WITH MORE VARIABLES THAN EQUATIONS

We can also solve many systems that have more variables than equations.

EXAMPLE 3   Solve the system $\begin{cases} x + y - 2z = -1 \\ 2x - y + z = -3 \end{cases}.$

Solution   We start by doing a type 3 row operation to get a 0 under the 1 in the first column.

$$\underset{\begin{array}{c}\text{Multiply row 1 by } -2 \\ \text{and add to row 2.}\end{array}}{} $$
$$\begin{bmatrix} 1 & 1 & -2 & -1 \\ 2 & -1 & 1 & -3 \end{bmatrix} \approx \begin{bmatrix} 1 & 1 & -2 & -1 \\ 0 & -3 & 5 & -1 \end{bmatrix}$$

We then do a type 2 row operation:

Multiply row 2 by $-\frac{1}{3}$.

$$\begin{bmatrix} 1 & 1 & -2 & -1 \\ 0 & -3 & 5 & -1 \end{bmatrix} \approx \begin{bmatrix} 1 & 1 & -2 & -1 \\ 0 & 1 & -\frac{5}{3} & \frac{1}{3} \end{bmatrix}$$

The final matrix represents the system

$$\begin{cases} x + y - 2z = -1 \\ y - \frac{5}{3}z = \frac{1}{3} \end{cases}$$

We add $\frac{5}{3}z$ to both sides of the second equation to obtain

$$y = \frac{1}{3} + \frac{5}{3}z$$

We have not found a specific value for y. However, we have found y in terms of z. To find a value of x in terms of z, we substitute $\frac{1}{3} + \frac{5}{3}z$ for y in the first equation and simplify to get

$$x + y - 2z = -1$$

$$x + \frac{1}{3} + \frac{5}{3}z - 2z = -1 \qquad \text{Substitute } \frac{1}{3} + \frac{5}{3}z \text{ for } y.$$

$$x + \frac{1}{3} - \frac{1}{3}z = -1 \qquad \text{Combine like terms.}$$

$$x - \frac{1}{3}z = -\frac{4}{3} \qquad \text{Subtract } \frac{1}{3} \text{ from both sides.}$$

$$x = -\frac{4}{3} + \frac{1}{3}z \qquad \text{Add } \frac{1}{3}z \text{ to both sides.}$$

A solution of this system must have the form

$$\left(-\frac{4}{3} + \frac{1}{3}z, \; \frac{1}{3} + \frac{5}{3}z, \; z\right) \qquad \begin{array}{l} \text{This solution is called "a general solution"} \\ \text{of the system.} \end{array}$$

for all values of z. This system has infinitely many solutions, a different one for each value of z. For example,

- If $z = 0$, the corresponding solution is $\left(-\frac{4}{3}, \frac{1}{3}, 0\right)$.
- If $z = 1$, the corresponding solution is $(-1, 2, 1)$.

Verify that both of these solutions satisfy each equation of the given system. ■

Self Check   Solve $\begin{cases} x + y - 2z = 11 \\ 2x - y + z = -2 \end{cases}$.

Answer   infinitely many solutions; a general solution is $\left(3 + \frac{1}{3}z, \; 8 + \frac{5}{3}z, \; z\right)$; two solutions are $(2, 3, -3)$ and $(3, 8, 0)$

■ ■ ■ ■ ■ ■ ■ ■ ■ PERSPECTIVE

Matrices with the same number of rows and columns can be added. We simply add their corresponding elements. For example,

$$\begin{bmatrix} 2 & 3 & -4 \\ -1 & 2 & 5 \end{bmatrix} + \begin{bmatrix} 3 & -1 & 0 \\ 4 & 3 & 2 \end{bmatrix}$$

$$= \begin{bmatrix} 2+3 & 3+(-1) & -4+0 \\ -1+4 & 2+3 & 5+2 \end{bmatrix}$$

$$= \begin{bmatrix} 5 & 2 & -4 \\ 3 & 5 & 7 \end{bmatrix}$$

To multiply a matrix by a constant, we multiply each element of the matrix by the constant. For example,

$$5 \cdot \begin{bmatrix} 2 & 3 & -4 \\ -1 & 2 & 5 \end{bmatrix}$$

$$= \begin{bmatrix} 5 \cdot 2 & 5 \cdot 3 & 5 \cdot (-4) \\ 5 \cdot (-1) & 5 \cdot 2 & 5 \cdot 5 \end{bmatrix}$$

$$= \begin{bmatrix} 10 & 15 & -20 \\ -5 & 10 & 25 \end{bmatrix}$$

Since matrices provide a good way to store information in computers, they are often used in applied problems. For example, suppose there are 66 security officers employed at either the downtown office or the suburban office:

Downtown Office

	Male	Female
Day shift	12	18
Night shift	3	0

Suburban Office

	Male	Female
Day shift	14	12
Night shift	5	2

The information about the employees is contained in the following matrices.

$$D = \begin{bmatrix} 12 & 18 \\ 3 & 0 \end{bmatrix} \quad \text{and} \quad S = \begin{bmatrix} 14 & 12 \\ 5 & 2 \end{bmatrix}$$

The entry in the first row-first column in matrix D gives the information that 12 males work the day shift at the downtown office. Company management can add the matrices D and S to find corporate-wide totals:

$$D + S = \begin{bmatrix} 12 & 18 \\ 3 & 0 \end{bmatrix} + \begin{bmatrix} 14 & 12 \\ 5 & 2 \end{bmatrix}$$

$$= \begin{bmatrix} 26 & 30 \\ 8 & 2 \end{bmatrix}$$

We interpret the total to mean:

	Male	Female
Day shift	26	30
Night shift	8	2

If one-third of the force in each category at the downtown location retires, the downtown staff would be reduced to $\frac{2}{3}D$ people. We can compute $\frac{2}{3}D$ by multiplying each entry by $\frac{2}{3}$.

$$\frac{2}{3}D = \frac{2}{3}\begin{bmatrix} 12 & 18 \\ 3 & 0 \end{bmatrix}$$

$$= \begin{bmatrix} 8 & 12 \\ 2 & 0 \end{bmatrix}$$

After retirements, downtown staff would be

	Male	Female
Day shift	8	12
Night shift	2	0

Orals   *Consider the system* $\begin{cases} 3x + 2y = 8 \\ 4x - 3y = 6 \end{cases}$

1. Find the coefficient matrix.   **2.** Find the augmented matrix.

Tell whether each matrix is in triangular form.

3. $\begin{bmatrix} 4 & 1 & 5 \\ 0 & 2 & 7 \\ 0 & 0 & 4 \end{bmatrix}$
     4. $\begin{bmatrix} 8 & 5 & 2 \\ 0 & 4 & 5 \\ 0 & 7 & 0 \end{bmatrix}$

EXERCISE 13.3

REVIEW *Write each number in scientific notation.*

1. 93,000,000   **2.** 0.00045   **3.** 63×10^3 **4.** 0.33×10^3

VOCABULARY AND CONCEPTS *Fill in each blank to make a true statement.*

5. A _____ is a rectangular array of numbers.

6. The numbers in a matrix are called its _____.

7. A 3×4 matrix has __ rows and 4 _____.

8. A _____ matrix has the same number of rows as columns.

9. An _____ matrix of a system of equations includes the coefficient matrix and the column of constants.

10. If a matrix has all 0's below its main diagonal, it is written in _____ form.

11. A _____ row operation corresponds to interchanging two equations in a system of equations.

12. A type 2 row operation corresponds to _____ both sides of an equation by a nonzero constant.

13. A type 3 row operation corresponds to adding a _____ multiple of one equation to another.

14. In the Gaussian method of solving systems of equations, we transform the _____ matrix into triangular form and finish the solution by using _____ substitution.

PRACTICE *In Exercises 15–18, use a row operation to find the missing number in the second matrix.*

15. $\begin{bmatrix} 2 & 1 & 1 \\ 5 & 4 & 1 \end{bmatrix}$
$\begin{bmatrix} 2 & 1 & 1 \\ 3 & 3 & \end{bmatrix}$

16. $\begin{bmatrix} -1 & 3 & 2 \\ 1 & -2 & 3 \end{bmatrix}$
$\begin{bmatrix} -1 & 3 & 2 \\ & 1 & 5 \end{bmatrix}$

17. $\begin{bmatrix} 3 & -2 & 1 \\ -1 & 2 & 4 \end{bmatrix}$
$\begin{bmatrix} 3 & -2 & 1 \\ -2 & 4 & \end{bmatrix}$

18. $\begin{bmatrix} 2 & 1 & -3 \\ 2 & 6 & 1 \end{bmatrix}$
$\begin{bmatrix} 6 & 3 & \\ 2 & 6 & 1 \end{bmatrix}$

In Exercises 19–30, use matrices to solve each system of equations. Each system has one solution.

19. $\begin{cases} x + y = 2 \\ x - y = 0 \end{cases}$

20. $\begin{cases} x + y = 3 \\ x - y = -1 \end{cases}$

21. $\begin{cases} x + 2y = -4 \\ 2x + y = 1 \end{cases}$

22. $\begin{cases} 2x - 3y = 16 \\ -4x + y = -22 \end{cases}$

23. $\begin{cases} 3x + 4y = -12 \\ 9x - 2y = 6 \end{cases}$

24. $\begin{cases} 5x - 4y = 10 \\ x - 7y = 2 \end{cases}$

25. $\begin{cases} x + y + z = 6 \\ x + 2y + z = 8 \\ x + y + 2z = 9 \end{cases}$

26. $\begin{cases} x - y + z = 2 \\ x + 2y - z = 6 \\ 2x - y - z = 3 \end{cases}$

27. $\begin{cases} 2x + y + 3z = 3 \\ -2x - y + z = 5 \\ 4x - 2y + 2z = 2 \end{cases}$　　　**28.** $\begin{cases} 3x + 2y + z = 8 \\ 6x - y + 2z = 16 \\ -9x + y - z = -20 \end{cases}$

29. $\begin{cases} 3x - 2y + 4z = 4 \\ x + y + z = 3 \\ 6x - 2y - 3z = 10 \end{cases}$　　　**30.** $\begin{cases} 2x + 3y - z = -8 \\ x - y - z = -2 \\ -4x + 3y + z = 6 \end{cases}$

In Exercises 31–38, use matrices to solve each system of equations. If a system has no solution, so indicate.

31. $\begin{cases} x + y = 3 \\ 3x - y = 1 \\ 2x + y = 4 \end{cases}$　　**32.** $\begin{cases} x - y = -5 \\ 2x + 3y = 5 \\ x + y = 1 \end{cases}$　　**33.** $\begin{cases} 2x - y = 4 \\ x + 3y = 2 \\ -x - 4y = -2 \end{cases}$　　**34.** $\begin{cases} 3x - 2y = 5 \\ x + 2y = 7 \\ -3x - y = -11 \end{cases}$

35. $\begin{cases} 2x + y = 7 \\ x - y = 2 \\ -x + 3y = -2 \end{cases}$　　**36.** $\begin{cases} 3x - y = 2 \\ -6x + 3y = 0 \\ -x + 2y = -4 \end{cases}$　　**37.** $\begin{cases} x + 3y = 7 \\ x + y = 3 \\ 3x + y = 5 \end{cases}$　　**38.** $\begin{cases} x + y = 3 \\ x - 2y = -3 \\ x - y = 1 \end{cases}$

In Exercises 39–42, use matrices to help find a general solution of each system of equations.

39. $\begin{cases} x + 2y + 3z = -2 \\ -x - y - 2z = 4 \end{cases}$　　　**40.** $\begin{cases} 2x - 4y + 3z = 6 \\ -4x + 6y + 4z = -6 \end{cases}$

41. $\begin{cases} x - y = 1 \\ y + z = 1 \\ x + z = 2 \end{cases}$　　　**42.** $\begin{cases} x + z = 1 \\ x + y = 2 \\ 2x + y + z = 3 \end{cases}$

In Exercises 43–46, remember these facts from geometry.

Two angles whose measures add up to 90° are complementary.

Two angles whose measures add up to 180° are supplementary.

The sum of the measures of the interior angles in a triangle is 180°.

43. Geometry　One angle is 46° larger than its complement. Find the angles.

44. Geometry　One angle is 28° larger than its supplement. Find the angles.

45. Geometry　In Illustration 1, angle B is 25° more than angle A, and angle C is 5° less than twice angle A. Find each angle in the triangle.

46. Geometry　In Illustration 2, angle A is 10° less than angle B, and angle B is 10° less than angle C. Find each angle in the triangle.

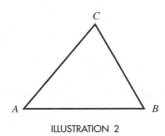

ILLUSTRATION 2

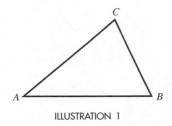

ILLUSTRATION 1

In Exercises 47–48, remember that the equation of a parabola is of the form $y = ax^2 + bx + c$.

47. Curve fitting Find the equation of the parabola passing through the points $(0, 1)$, $(1, 2)$, and $(-1, 4)$.

48. Curve fitting Find the equation of the parabola passing through the points $(0, 1)$, $(1, 1)$, and $(-1, -1)$.

WRITING

49. Explain how to check the solution of a system of equations.

50. Explain how to perform a type 3 row operation.

SOMETHING TO THINK ABOUT

51. If the system represented by

$$\begin{bmatrix} 1 & 1 & 0 & 1 \\ 0 & 0 & 1 & 2 \\ 0 & 0 & 0 & k \end{bmatrix}$$

has no solution, what do you know about k?

52. Is it possible for a system with fewer equations than variables to have no solution? Illustrate.

13.4 Solution by Determinants

■ DETERMINANTS ■ CRAMER'S RULE

Getting Ready   *Find each value.*

1. $3(-4) - 2(5)$

2. $5(2) - 3(-4)$

3. $2(2 - 5) - 3(5 - 2) + 2(4 - 3)$

4. $-3(5 - 2) + 2(3 + 1) - 2(5 + 1)$

■ DETERMINANTS

An idea closely related to the concept of matrix is the **determinant.** A determinant is a number that is associated with a square matrix. For any square matrix A, the symbol $|A|$ represents the determinant of A.

Value of a 2 × 2 Determinant

If a, b, c, and d are real numbers, the **determinant** of the matrix $\begin{bmatrix} a & b \\ c & d \end{bmatrix}$ is

$$\begin{vmatrix} a & b \\ c & d \end{vmatrix} = ad - bc$$

The determinant of a 2×2 matrix is the number that is equal to the product of the numbers on the major diagonal

$$\begin{vmatrix} a & b \\ c & d \end{vmatrix}$$

minus the product of the numbers on the other diagonal

$$\begin{vmatrix} a & b \\ c & d \end{vmatrix}$$

 b

EXAMPLE 1   Evaluate the determinants   **a.** $\begin{vmatrix} 3 & 2 \\ 6 & 9 \end{vmatrix}$ and   **b.** $\begin{vmatrix} -5 & \frac{1}{2} \\ -1 & 0 \end{vmatrix}$.

Solution **a.** $\begin{vmatrix} 3 & 2 \\ 6 & 9 \end{vmatrix} = 3(9) - 2(6)$ **b.** $\begin{vmatrix} -5 & \frac{1}{2} \\ -1 & 0 \end{vmatrix} = -5(0) - \frac{1}{2}(-1)$

$$= 27 - 12 \qquad\qquad\qquad\qquad = 0 + \frac{1}{2}$$

$$= 15 \qquad\qquad\qquad\qquad\qquad = \frac{1}{2} \qquad\blacksquare$$

Self Check   Evaluate $\begin{vmatrix} 4 & -3 \\ 2 & 1 \end{vmatrix}$.

Answer   10

A 3×3 determinant is evaluated by expanding by **minors**.

Value of a 3 × 3 Determinant

$$\begin{vmatrix} a_1 & b_1 & c_1 \\ a_2 & b_2 & c_2 \\ a_3 & b_3 & c_3 \end{vmatrix} = a_1 \overbrace{\begin{vmatrix} b_2 & c_2 \\ b_3 & c_3 \end{vmatrix}}^{\substack{\text{Minor} \\ \text{of } a_1}} - b_1 \overbrace{\begin{vmatrix} a_2 & c_2 \\ a_3 & c_3 \end{vmatrix}}^{\substack{\text{Minor} \\ \text{of } b_1}} + c_1 \overbrace{\begin{vmatrix} a_2 & b_2 \\ a_3 & b_3 \end{vmatrix}}^{\substack{\text{Minor} \\ \text{of } c_1}}$$

To find the minor of a_1, we find the determinant formed by crossing out the elements of the matrix that are in the same row and column as a_1:

$$\begin{vmatrix} a_1 & b_1 & c_1 \\ a_2 & b_2 & c_2 \\ a_3 & b_3 & c_3 \end{vmatrix} \qquad \text{The minor of } a_1 \text{ is } \begin{vmatrix} b_2 & c_2 \\ b_3 & c_3 \end{vmatrix}.$$

To find the minor of b_1, we cross out the elements of the matrix that are in the same row and column as b_1:

$$\begin{vmatrix} \cancel{a_1} & \cancel{b_1} & \cancel{c_1} \\ a_2 & \cancel{b_2} & c_2 \\ a_3 & \cancel{b_3} & c_3 \end{vmatrix}$$ The minor of b_1 is $\begin{vmatrix} a_2 & c_2 \\ a_3 & c_3 \end{vmatrix}$.

To find the minor of c_1, we cross out the elements of the matrix that are in the same row and column as c_1:

$$\begin{vmatrix} \cancel{a_1} & \cancel{b_1} & \cancel{c_1} \\ a_2 & b_2 & \cancel{c_2} \\ a_3 & b_3 & \cancel{c_3} \end{vmatrix}$$ The minor of c_1 is $\begin{vmatrix} a_2 & b_2 \\ a_3 & b_3 \end{vmatrix}$.

EXAMPLE 2   Evaluate the determinant $\begin{vmatrix} 1 & 3 & -2 \\ 2 & 1 & 3 \\ 1 & 2 & 3 \end{vmatrix}$.

Solution

$$\begin{vmatrix} \mathbf{1} & 3 & -2 \\ 2 & 1 & 3 \\ 1 & 2 & 3 \end{vmatrix} = \overset{\overset{\text{Minor}}{\text{of 1}}}{\mathbf{1}} \begin{vmatrix} 1 & 3 \\ 2 & 3 \end{vmatrix} - \overset{\overset{\text{Minor}}{\text{of 3}}}{3} \begin{vmatrix} 2 & 3 \\ 1 & 3 \end{vmatrix} + \overset{\overset{\text{Minor}}{\text{of } -2}}{(-2)} \begin{vmatrix} 2 & 1 \\ 1 & 2 \end{vmatrix}$$

$$= 1(3 - 6) - 3(6 - 3) - 2(4 - 1)$$
$$= -3 - 9 - 6$$
$$= -18$$

Self Check   Evaluate $\begin{vmatrix} 2 & -1 & 3 \\ 1 & 2 & -2 \\ 3 & 1 & 1 \end{vmatrix}$.

Answer   0

We can evaluate a 3×3 determinant by expanding it along any row or column. To determine the signs between the terms of the expansion of a 3×3 determinant, we use the following array of signs.

Array of Signs for a 3 × 3 Determinant

$$\begin{array}{ccc} + & - & + \\ - & + & - \\ + & - & + \end{array}$$

EXAMPLE 3 Evaluate $\begin{vmatrix} 1 & 3 & -2 \\ 2 & 1 & 3 \\ 1 & 2 & 3 \end{vmatrix}$ by expanding on the middle column.

Solution This is the determinant of Example 2. To expand it along the middle column, we use the signs of the middle column of the array of signs:

$$\begin{vmatrix} 1 & 3 & -2 \\ 2 & 1 & 3 \\ 1 & 2 & 3 \end{vmatrix} = \overset{\substack{\text{Minor} \\ \text{of 3} \\ \downarrow}}{-3} \begin{vmatrix} 2 & 3 \\ 1 & 3 \end{vmatrix} + \overset{\substack{\text{Minor} \\ \text{of 1} \\ \downarrow}}{1} \begin{vmatrix} 1 & -2 \\ 1 & 3 \end{vmatrix} - \overset{\substack{\text{Minor} \\ \text{of 2} \\ \downarrow}}{2} \begin{vmatrix} 1 & -2 \\ 2 & 3 \end{vmatrix}$$

$$= -3(6 - 3) + 1[3 - (-2)] - 2[3 - (-4)]$$
$$= -3(3) + 1(5) - 2(7)$$
$$= -9 + 5 - 14$$
$$= -18$$

As expected, we get the same value as in Example 2. ∎

Self Check Evaluate $\begin{vmatrix} 2 & -1 & 3 \\ 1 & 2 & 2 \\ 3 & 1 & 1 \end{vmatrix}$.

Answer -20

■ ■ ■ ■ ■ ■ ■ ■ ■ **Evaluating Determinants**

GRAPHING CALCULATORS It is possible to use a graphing calculator to evaluate determinants. For example, to evaluate the determinant in Example 3, we first enter the matrix by pressing the MATRIX key, selecting EDIT, and pressing the ENTER key. We then enter the dimensions and the elements of the matrix to get Figure 13-10(a). We then press 2nd QUIT to clear the screen. We then press MATRIX , select MATH, and press 1 to get Figure 13-10(b). We then press MATRIX , select NAMES, and press 1 to get Figure 13-10(c). To get the value of the determinant, we now press ENTER to get Figure 13-10(d), which shows that the value of the determinant is −18.

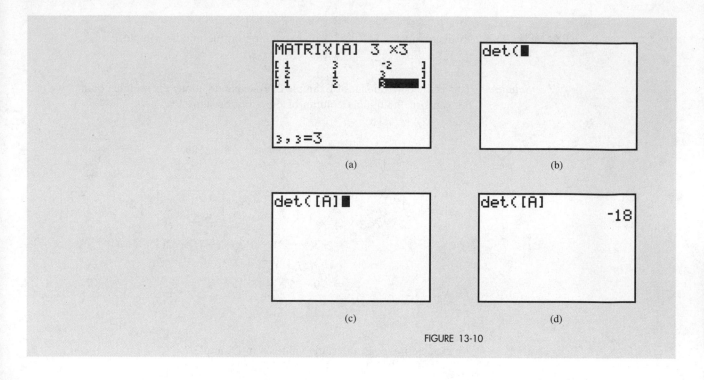

FIGURE 13-10

■ CRAMER'S RULE

The method of using determinants to solve systems of equations is called **Cramer's rule,** named after the 18th-century mathematician Gabriel Cramer. To develop Cramer's rule, we consider the system

$$\begin{cases} ax + by = e \\ cx + dy = f \end{cases}$$

where x and y are variables and a, b, c, d, e, and f are constants.

If we multiply both sides of the first equation by d and multiply both sides of the second equation by $-b$, we can add the equations and eliminate y:

$$\begin{array}{r} adx + bdy = \quad ed \\ \underline{-bcx - bdy = -bf} \\ adx - bcx \quad = ed - bf \end{array}$$

To solve for x, we use the distributive property to write $adx - bcx$ as $(ad - bc)x$ on the left-hand side and divide each side by $ad - bc$:

$$(ad - bc)x = ed - bf$$

$$x = \frac{ed - bf}{ad - bc} \quad (ad - bc \neq 0)$$

We can find y in a similar manner. After eliminating the variable x, we get

$$y = \frac{af - ec}{ad - bc} \quad (ad - bc \neq 0)$$

Determinants provide an easy way of remembering these formulas. Note that the denominator for both x and y is

$$\begin{vmatrix} a & b \\ c & d \end{vmatrix} = ad - bc$$

The numerators can be expressed as determinants also:

$$x = \frac{ed - bf}{ad - bc} = \frac{\begin{vmatrix} e & b \\ f & d \end{vmatrix}}{\begin{vmatrix} a & b \\ c & d \end{vmatrix}} \quad \text{and} \quad y = \frac{af - ec}{ad - bc} = \frac{\begin{vmatrix} a & e \\ c & f \end{vmatrix}}{\begin{vmatrix} a & b \\ c & d \end{vmatrix}}$$

If we compare these formulas with the original system

$$\begin{cases} ax + by = e \\ cx + dy = f \end{cases}$$

we note that in the expressions for x and y above, the denominator determinant is formed by using the coefficients a, b, c, and d of the variables in the equations. The numerator determinants are the same as the denominator determinant, except that the column of coefficients of the variable for which we are solving is replaced with the column of constants e and f.

Cramer's Rule for Two Equations in Two Variables

The solution of the system $\begin{cases} ax + by = e \\ cx + dy = f \end{cases}$ is given by

$$x = \frac{D_x}{D} = \frac{\begin{vmatrix} e & b \\ f & d \end{vmatrix}}{\begin{vmatrix} a & b \\ c & d \end{vmatrix}} \quad \text{and} \quad y = \frac{D_y}{D} = \frac{\begin{vmatrix} a & e \\ c & f \end{vmatrix}}{\begin{vmatrix} a & b \\ c & d \end{vmatrix}}$$

If every determinant is 0, the system is consistent but the equations are dependent.

If $D = 0$ and D_x or D_y is nonzero, the system is inconsistent.

If $D \neq 0$, the system is consistent and the equations are independent.

EXAMPLE 4   Use Cramer's rule to solve $\begin{cases} 4x - 3y = 6 \\ -2x + 5y = 4 \end{cases}$.

Solution   The value of x is the quotient of two determinants. The denominator determinant is made up of the coefficients of x and y:

$$D = \begin{vmatrix} 4 & -3 \\ -2 & 5 \end{vmatrix}$$

To solve for x, we form the numerator determinant from the denominator determinant by replacing its first column (the coefficients of x) with the column of constants (6 and 4).

To solve for y, we form the numerator determinant from the denominator determinant by replacing the second column (the coefficients of y) with the column of constants (6 and 4).

To find the values of x and y, we evaluate each determinant:

$$x = \frac{\begin{vmatrix} 6 & -3 \\ 4 & 5 \end{vmatrix}}{\begin{vmatrix} 4 & -3 \\ -2 & 5 \end{vmatrix}} = \frac{6(5) - (-3)(4)}{4(5) - (-3)(-2)} = \frac{30 + 12}{20 - 6} = \frac{42}{14} = 3$$

$$y = \frac{\begin{vmatrix} 4 & 6 \\ -2 & 4 \end{vmatrix}}{\begin{vmatrix} 4 & -3 \\ -2 & 5 \end{vmatrix}} = \frac{4(4) - 6(-2)}{4(5) - (-3)(-2)} = \frac{16 + 12}{20 - 6} = \frac{28}{14} = 2$$

The solution of this system is $(3, 2)$. Verify that $x = 3$ and $y = 2$ satisfy each equation in the given system.   ∎

Self Check   Solve $\begin{cases} 2x - 3y = -16 \\ 3x + 5y = 14 \end{cases}$.

Answer $(-2, 4)$

EXAMPLE 5   Use Cramer's rule to solve $\begin{cases} 7x = 8 - 4y \\ 2y = 3 - \frac{7}{2}x \end{cases}$.

Solution   We multiply both sides of the second equation by 2 to eliminate the fraction and write the system in the form

$$\begin{cases} 7x + 4y = 8 \\ 7x + 4y = 6 \end{cases}$$

When we attempt to use Cramer's rule to solve this system for x, we obtain

$$x = \frac{\begin{vmatrix} 8 & 4 \\ 6 & 4 \end{vmatrix}}{\begin{vmatrix} 7 & 4 \\ 7 & 4 \end{vmatrix}} = \frac{8}{0} \quad \text{which is undefined}$$

Since the denominator determinant is 0 and the numerator determinant is not 0, the system is inconsistent. It has no solutions.

We can see directly from the system that it is inconsistent. For any values of x and y, it is impossible that 7 times x plus 4 times y could be both 8 and 6. ■

Self Check Solve $\begin{cases} 3x = 8 - 4y \\ y = \dfrac{5}{2} - \dfrac{3}{4}x \end{cases}$.

Answer no solutions

Cramer's Rule for Three Equations in Three Variables

The solution of the system $\begin{cases} ax + by + cz = j \\ dx + ey + fz = k \\ gx + hy + iz = l \end{cases}$ is given by

$$x = \frac{D_x}{D}, \quad y = \frac{D_y}{D}, \quad \text{and} \quad z = \frac{D_z}{D}$$

where

$$D = \begin{vmatrix} a & b & c \\ d & e & f \\ g & h & i \end{vmatrix} \qquad D_x = \begin{vmatrix} j & b & c \\ k & e & f \\ l & h & i \end{vmatrix}$$

$$D_y = \begin{vmatrix} a & j & c \\ d & k & f \\ g & l & i \end{vmatrix} \qquad D_z = \begin{vmatrix} a & b & j \\ d & e & k \\ g & h & l \end{vmatrix}$$

If every determinant is 0, the system is consistent but the equations are dependent.

If $D = 0$ and D_x or D_y or D_z is nonzero, the system is inconsistent.

If $D \neq 0$, the system is consistent and the equations are independent.

EXAMPLE 6 Use Cramer's rule to solve $\begin{cases} 2x + y + 4z = 12 \\ x + 2y + 2z = 9 \\ 3x - 3y - 2z = 1 \end{cases}$.

Solution   The denominator determinant is the determinant formed by the coefficients of the variables. The numerator determinants are formed by replacing the coefficients of the variable being solved for by the column of constants. We form the quotients for x, y, and z and evaluate the determinants:

$$x = \frac{\begin{vmatrix} 12 & 1 & 4 \\ 9 & 2 & 2 \\ 1 & -3 & -2 \end{vmatrix}}{\begin{vmatrix} 2 & 1 & 4 \\ 1 & 2 & 2 \\ 3 & -3 & -2 \end{vmatrix}} = \frac{12\begin{vmatrix} 2 & 2 \\ -3 & -2 \end{vmatrix} - 1\begin{vmatrix} 9 & 2 \\ 1 & -2 \end{vmatrix} + 4\begin{vmatrix} 9 & 2 \\ 1 & -3 \end{vmatrix}}{2\begin{vmatrix} 2 & 2 \\ -3 & -2 \end{vmatrix} - 1\begin{vmatrix} 1 & 2 \\ 3 & -2 \end{vmatrix} + 4\begin{vmatrix} 1 & 2 \\ 3 & -3 \end{vmatrix}} = \frac{12(2) - (-20) + 4(-29)}{2(2) - (-8) + 4(-9)} = \frac{-72}{-24} = 3$$

$$y = \frac{\begin{vmatrix} 2 & 12 & 4 \\ 1 & 9 & 2 \\ 3 & 1 & -2 \end{vmatrix}}{\begin{vmatrix} 2 & 1 & 4 \\ 1 & 2 & 2 \\ 3 & -3 & -2 \end{vmatrix}} = \frac{2\begin{vmatrix} 9 & 2 \\ 1 & -2 \end{vmatrix} - 12\begin{vmatrix} 1 & 2 \\ 3 & -2 \end{vmatrix} + 4\begin{vmatrix} 1 & 9 \\ 3 & 1 \end{vmatrix}}{-24} = \frac{2(-20) - 12(-8) + 4(-26)}{-24} = \frac{-48}{-24} = 2$$

$$z = \frac{\begin{vmatrix} 2 & 1 & 12 \\ 1 & 2 & 9 \\ 3 & -3 & 1 \end{vmatrix}}{\begin{vmatrix} 2 & 1 & 4 \\ 1 & 2 & 2 \\ 3 & -3 & -2 \end{vmatrix}} = \frac{2\begin{vmatrix} 2 & 9 \\ -3 & 1 \end{vmatrix} - 1\begin{vmatrix} 1 & 9 \\ 3 & 1 \end{vmatrix} + 12\begin{vmatrix} 1 & 2 \\ 3 & -3 \end{vmatrix}}{-24} = \frac{2(29) - 1(-26) + 12(-9)}{-24} = \frac{-24}{-24} = 1$$

The solution of this system is $(3, 2, 1)$. ∎

Self Check   Solve $\begin{cases} x + y + 2z = 6 \\ 2x - y + z = 9 \\ x + y - 2z = -6 \end{cases}$.

Answer $(2, -2, 3)$

Orals   *Evaluate each determinant.*

1. $\begin{vmatrix} 2 & 1 \\ 1 & 1 \end{vmatrix}$ 2. $\begin{vmatrix} 0 & 2 \\ 1 & 1 \end{vmatrix}$ 3. $\begin{vmatrix} 0 & 1 \\ 0 & 1 \end{vmatrix}$

When using Cramer's rule to solve the system $\begin{cases} x + 2y = 5 \\ 2x - y = 4 \end{cases}$,

4. Find the denominator determinant for x.

5. Find the numerator determinant for x.

6. Find the numerator determinant for y.

EXERCISE 13.4

REVIEW *Solve each equation.*

1. $3(x + 2) - (2 - x) = x - 5$

2. $\dfrac{3}{7}x = 2(x + 11)$

3. $\dfrac{5}{3}(5x + 6) - 10 = 0$

4. $5 - 3(2x - 1) = 2(4 + 3x) - 24$

VOCABULARY AND CONCEPTS *Fill in each blank to make a true statement.*

5. A determinant is a _____ that is associated with a square matrix.

6. The value of $\begin{vmatrix} a & b \\ c & d \end{vmatrix}$ is _____.

7. The minor of b_1 in $\begin{vmatrix} a_1 & b_1 & c_1 \\ a_2 & b_2 & c_2 \\ a_3 & b_3 & c_3 \end{vmatrix}$ is _____.

8. We can evaluate a determinant by expanding it along any _____ or _____.

9. The denominator determinant for the value of x in the system $\begin{cases} 3x + 4y = 7 \\ 2x - 3y = 5 \end{cases}$ is _____.

10. If the denominator determinant for y in a system of equations is zero, the equations of the system are _____ or the system is _____.

PRACTICE *In Exercises 11–28, evaluate each determinant.*

11. $\begin{vmatrix} 2 & 3 \\ -2 & 1 \end{vmatrix}$

12. $\begin{vmatrix} 3 & -2 \\ -2 & 4 \end{vmatrix}$

13. $\begin{vmatrix} -1 & 2 \\ 3 & -4 \end{vmatrix}$

14. $\begin{vmatrix} -1 & -2 \\ -3 & -4 \end{vmatrix}$

15. $\begin{vmatrix} x & y \\ y & x \end{vmatrix}$

16. $\begin{vmatrix} x + y & y - x \\ x & y \end{vmatrix}$

17. $\begin{vmatrix} 1 & 0 & 1 \\ 0 & 1 & 0 \\ 1 & 1 & 1 \end{vmatrix}$

18. $\begin{vmatrix} 1 & 2 & 0 \\ 0 & 1 & 2 \\ 0 & 0 & 1 \end{vmatrix}$

19. $\begin{vmatrix} -1 & 2 & 1 \\ 2 & 1 & -3 \\ 1 & 1 & 1 \end{vmatrix}$

20. $\begin{vmatrix} 1 & 2 & 3 \\ 1 & 2 & 3 \\ 1 & 2 & 3 \end{vmatrix}$

21. $\begin{vmatrix} 1 & -2 & 3 \\ -2 & 1 & 1 \\ -3 & -2 & 1 \end{vmatrix}$

22. $\begin{vmatrix} 1 & 1 & 2 \\ 2 & 1 & -2 \\ 3 & 1 & 3 \end{vmatrix}$

23. $\begin{vmatrix} 1 & 2 & 3 \\ 4 & 5 & 6 \\ 7 & 8 & 9 \end{vmatrix}$

24. $\begin{vmatrix} 1 & 4 & 7 \\ 2 & 5 & 8 \\ 3 & 6 & 9 \end{vmatrix}$

25. $\begin{vmatrix} a & 2a & -a \\ 2 & -1 & 3 \\ 1 & 2 & -3 \end{vmatrix}$

26. $\begin{vmatrix} 1 & 2b & -3 \\ 2 & -b & 2 \\ 1 & 3b & 1 \end{vmatrix}$

27. $\begin{vmatrix} 1 & a & b \\ 1 & 2a & 2b \\ 1 & 3a & 3b \end{vmatrix}$

28. $\begin{vmatrix} a & b & c \\ 0 & b & c \\ 0 & 0 & c \end{vmatrix}$

In Exercises 29–54, use Cramer's rule to solve each system of equations, if possible.

29. $\begin{cases} x + y = 6 \\ x - y = 2 \end{cases}$

30. $\begin{cases} x - y = 4 \\ 2x + y = 5 \end{cases}$

31. $\begin{cases} 2x + y = 1 \\ x - 2y = -7 \end{cases}$

32. $\begin{cases} 3x - y = -3 \\ 2x + y = -7 \end{cases}$

33. $\begin{cases} 2x + 3y = 0 \\ 4x - 6y = -4 \end{cases}$

34. $\begin{cases} 4x - 3y = -1 \\ 8x + 3y = 4 \end{cases}$

35. $\begin{cases} y = \dfrac{-2x + 1}{3} \\ 3x - 2y = 8 \end{cases}$

36. $\begin{cases} 2x + 3y = -1 \\ x = \dfrac{y - 9}{4} \end{cases}$

37. $\begin{cases} y = \dfrac{11 - 3x}{2} \\ x = \dfrac{11 - 4y}{6} \end{cases}$

38. $\begin{cases} x = \dfrac{12 - 6y}{5} \\ y = \dfrac{24 - 10x}{12} \end{cases}$

39. $\begin{cases} x = \dfrac{5y - 4}{2} \\ y = \dfrac{3x - 1}{5} \end{cases}$

40. $\begin{cases} y = \dfrac{1 - 5x}{2} \\ x = \dfrac{3y + 10}{4} \end{cases}$

41. $\begin{cases} x + y + z = 4 \\ x + y - z = 0 \\ x - y + z = 2 \end{cases}$

42. $\begin{cases} x + y + z = 4 \\ x - y + z = 2 \\ x - y - z = 0 \end{cases}$

43. $\begin{cases} x + y + 2z = 7 \\ x + 2y + z = 8 \\ 2x + y + z = 9 \end{cases}$

44. $\begin{cases} x + 2y + 2z = 10 \\ 2x + y + 2z = 9 \\ 2x + 2y + z = 1 \end{cases}$

45. $\begin{cases} 2x + y - z = 1 \\ x + 2y + 2z = 2 \\ 4x + 5y + 3z = 3 \end{cases}$

46. $\begin{cases} 4x + 3z = 4 \\ 2y - 6z = -1 \\ 8x + 4y + 3z = 9 \end{cases}$

47. $\begin{cases} 2x + y + z = 5 \\ x - 2y + 3z = 10 \\ x + y - 4z = -3 \end{cases}$

48. $\begin{cases} 3x + 2y - z = -8 \\ 2x - y + 7z = 10 \\ 2x + 2y - 3z = -10 \end{cases}$

49. $\begin{cases} 2x + 3y + 4z = 6 \\ 2x - 3y - 4z = -4 \\ 4x + 6y + 8z = 12 \end{cases}$

50. $\begin{cases} x - 3y + 4z - 2 = 0 \\ 2x + y + 2z - 3 = 0 \\ 4x - 5y + 10z - 7 = 0 \end{cases}$

51. $\begin{cases} x + y = 1 \\ \dfrac{1}{2}y + z = \dfrac{5}{2} \\ x - z = -3 \end{cases}$

52. $\begin{cases} 3x + 4y + 14z = 7 \\ -\dfrac{1}{2}x - y + 2z = \dfrac{3}{2} \\ x + \dfrac{3}{2}y + \dfrac{5}{2}z = 1 \end{cases}$

53. $\begin{cases} 2x - y + 4z + 2 = 0 \\ 5x + 8y + 7z = -8 \\ x + 3y + z + 3 = 0 \end{cases}$

54. $\begin{cases} \dfrac{1}{2}x + y + z + \dfrac{3}{2} = 0 \\ x + \dfrac{1}{2}y + z - \dfrac{1}{2} = 0 \\ x + y + \dfrac{1}{2}z + \dfrac{1}{2} = 0 \end{cases}$

In Exercises 55–58, evaluate each determinant and solve the resulting equation.

55. $\begin{vmatrix} x & 1 \\ 3 & 2 \end{vmatrix} = 1$

56. $\begin{vmatrix} x & -x \\ 2 & -3 \end{vmatrix} = -5$

57. $\begin{vmatrix} x & -2 \\ 3 & 1 \end{vmatrix} = \begin{vmatrix} 4 & 2 \\ x & 3 \end{vmatrix}$

58. $\begin{vmatrix} x & 3 \\ x & 2 \end{vmatrix} = \begin{vmatrix} 3 & 2 \\ 1 & 1 \end{vmatrix}$

APPLICATIONS

59. Making investments   A student wants to average a 6.6% return by investing $20,000 in the three stocks listed in Illustration 1. Because HiTech is considered to be a high-risk investment, he wants to invest three times as much in SaveTel and HiGas combined as he invests in HiTech. How much should he invest in each stock?

Stock	Rate of return
HiTech	10%
SaveTel	5%
HiGas	6%

ILLUSTRATION 1

60. Making investments   See Illustration 2. A woman wants to average a $7\frac{1}{3}$% return by investing $30,000 in three certificates of deposit. She wants to invest five times as much in the 8% CD as in the 6% CD. How much should she invest in each CD?

Type of CD	Rate of return
12 month	6%
24 month	7%
36 month	8%

ILLUSTRATION 2

In Exercises 61–64, use a graphing calculator to evaluate each determinant.

61. $\begin{vmatrix} 2 & -3 & 4 \\ -1 & 2 & 4 \\ 3 & -3 & 1 \end{vmatrix}$

62. $\begin{vmatrix} -3 & 2 & -5 \\ 3 & -2 & 6 \\ 1 & -3 & 4 \end{vmatrix}$

63. $\begin{vmatrix} 2 & 1 & -3 \\ -2 & 2 & 4 \\ 1 & -2 & 2 \end{vmatrix}$

64. $\begin{vmatrix} 4 & 2 & -3 \\ 2 & -5 & 6 \\ 2 & 5 & -2 \end{vmatrix}$

WRITING

65. Tell how to find the minor of an element of a determinant.

66. Tell how to find x when solving a system of linear equations by Cramer's rule.

SOMETHING TO THINK ABOUT

67. Show that

$$\begin{vmatrix} x & y & 1 \\ -2 & 3 & 1 \\ 3 & 5 & 1 \end{vmatrix} = 0$$

is the equation of the line passing through $(-2, 3)$ and $(3, 5)$.

68. Show that

$$\frac{1}{2} \begin{vmatrix} 0 & 0 & 1 \\ 3 & 0 & 1 \\ 0 & 4 & 1 \end{vmatrix}$$

is the area of the triangle with vertices at $(0, 0)$, $(3, 0)$, and $(0, 4)$.

Determinants with more than 3 rows and 3 columns can be evaluated by expanding them by minors. The sign array for a 4 × 4 determinant is

$$\begin{array}{cccc} + & - & + & - \\ - & + & - & + \\ + & - & + & - \\ - & + & - & + \end{array}$$

Evaluate each determinant.

69. $\begin{vmatrix} 1 & 0 & 2 & 1 \\ 2 & 1 & 1 & 3 \\ 1 & 1 & 1 & 1 \\ 2 & 1 & 1 & 1 \end{vmatrix}$

70. $\begin{vmatrix} 1 & 2 & -1 & 1 \\ -2 & 1 & 3 & -1 \\ 0 & 1 & 1 & 2 \\ 2 & 0 & 3 & 1 \end{vmatrix}$

13.5 Solving Simultaneous Second-Degree Equations and Inequalities

■ SOLUTION BY GRAPHING ■ SOLUTION BY ELIMINATION ■ SOLUTIONS OF SYSTEMS OF INEQUALITIES

Getting Ready *Add the left-hand sides and the right-hand sides of the following equations.*

1. $3x^2 + 2y^2 = 12$
$\underline{4x^2 - 3y^2 = 32}$

2. $-7x^2 - 5y^2 = -17$
$\underline{12x^2 + 2y^2 = 25}$

■ SOLUTION BY GRAPHING

We now discuss ways to solve systems of two equations in two variables where at least one of the equations is of second degree.

EXAMPLE 1 Solve $\begin{cases} x^2 + y^2 = 25 \\ 2x + y = 10 \end{cases}$ by graphing.

Solution The graph of $x^2 + y^2 = 25$ is a circle with center at the origin and radius of 5. The graph of $2x + y = 10$ is a line. Depending on whether the line is a secant (intersecting the circle at two points) or a tangent (intersecting the circle at one point) or does not intersect the circle at all, there are two, one, or no solutions to the system, respectively.

After graphing the circle and the line, as shown in Figure 13-11, we see that there are two intersection points, $P(3, 4)$ and $P'(5, 0)$. These are the solutions of the system:

$$\begin{cases} x = 3 \\ y = 4 \end{cases} \quad \text{and} \quad \begin{cases} x = 5 \\ y = 0 \end{cases}$$

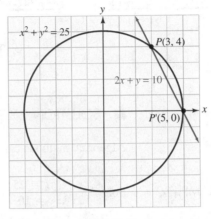

FIGURE 13-11 ■

Self Check   Solve $\begin{cases} x^2 + y^2 = 13 \\ y = -\dfrac{1}{5}x + \dfrac{13}{5}. \end{cases}$

Answer $(3, 2), (-2, 3)$

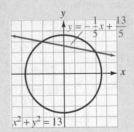

■ ■ ■ ■ ■ ■ ■ ■ ■ ■   **Solving Systems of Equations**

GRAPHING
CALCULATORS

To solve Example 1 with a graphing calculator, we graph the circle and the line on one set of coordinate axes (see Figure 13-12(a)). We then trace to find the co-ordinates of the intersection points of the graphs (see Figure 13-12(b) and Figure 13-12(c)).

We can zoom for better results.

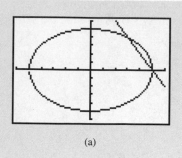
(a)

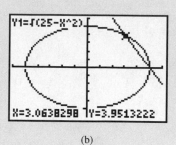
(b)

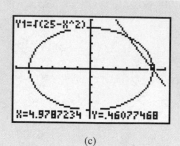
(c)

FIGURE 13-12

■ SOLUTION BY ELIMINATION

Algebraic methods can also be used to solve systems of equations.

EXAMPLE 2   Solve $\begin{cases} x^2 + y^2 = 25 \\ 2x + y = 10 \end{cases}$.

Solution   This system has one second-degree equation and one first-degree equation. We can solve this type of system by substitution. Solving the linear equation for y gives

$$2x + y = 10$$
1. $\qquad y = -2x + 10$

We can substitute $-2x + 10$ for y in the second-degree equation and solve the resulting quadratic equation for x:

$$x^2 + y^2 = 25$$
$$x^2 + (\mathbf{-2x + 10})^2 = 25$$
$$x^2 + 4x^2 - 40x + 100 = 25 \qquad\qquad \begin{array}{l}(-2x + 10)(-2x + 10) = \\ 4x^2 - 40x + 100.\end{array}$$

$$5x^2 - 40x + 75 = 0 \qquad\qquad \begin{array}{l}\text{Combine like terms and subtract 25} \\ \text{from both sides.}\end{array}$$

$$x^2 - 8x + 15 = 0 \qquad\qquad \text{Divide both sides by 5.}$$
$$(x - 5)(x - 3) = 0 \qquad\qquad \text{Factor } x^2 - 8x + 15.$$
$$x - 5 = 0 \quad \text{or} \quad x - 3 = 0 \qquad \text{Set each factor equal to 0.}$$
$$x = 5 \qquad\qquad x = 3$$

If we substitute 5 for x in Equation 1, we get $y = 0$. If we substitute 3 for x in Equation 1, we get $y = 4$. The two solutions are

$$\begin{cases} x = 5 \\ y = 0 \end{cases} \quad \text{or} \quad \begin{cases} x = 3 \\ y = 4 \end{cases}$$

■

Self Check   Solve by substitution: $\begin{cases} x^2 + y^2 = 13 \\ y = -\dfrac{1}{5}x + \dfrac{13}{5} \end{cases}$.

Answer $(3, 2), (-2, 3)$

EXAMPLE 3   Solve $\begin{cases} 4x^2 + 9y^2 = 5 \\ y = x^2 \end{cases}$.

Solution   We can solve this system by substitution.

$$4x^2 + 9y^2 = 5$$
$$4y + 9y^2 = 5 \qquad\qquad \text{Substitute } y \text{ for } x^2.$$
$$9y^2 + 4y - 5 = 0 \qquad\qquad \text{Subtract 5 from both sides.}$$
$$(9y - 5)(y + 1) = 0 \qquad\qquad \text{Factor } 9y^2 + 4y - 5.$$
$$9y - 5 = 0 \quad \text{or} \quad y + 1 = 0 \qquad \text{Set each factor equal to 0.}$$
$$y = \frac{5}{9} \qquad\qquad\quad y = -1$$

Since $y = x^2$, the values of x are found by solving the equations

$$x^2 = \frac{5}{9} \qquad \text{and} \qquad x^2 = -1$$

Because $x^2 = -1$ has no real solutions, this possibility is discarded. The solutions of $x^2 = \frac{5}{9}$ are

$$x = \frac{\sqrt{5}}{3} \qquad \text{or} \qquad x = -\frac{\sqrt{5}}{3}$$

The solutions of the system are

$$\left(\frac{\sqrt{5}}{3}, \frac{5}{9} \right) \qquad \text{and} \qquad \left(-\frac{\sqrt{5}}{3}, \frac{5}{9} \right) \qquad\qquad\blacksquare$$

Self Check   Solve $\begin{cases} x^2 + y^2 = 20 \\ y = x^2 \end{cases}$.

Answer $(2, 4), (-2, 4)$

EXAMPLE 4   Solve $\begin{cases} 3x^2 + 2y^2 = 36 \\ 4x^2 - y^2 = 4 \end{cases}$.

Solution   Since both equations are in the form $ax^2 + by^2 = c$, we can solve the system by addition.

We can copy the first equation and multiply the second equation by 2 to obtain the equivalent system

$$\begin{cases} 3x^2 + 2y^2 = 36 \\ 8x^2 - 2y^2 = 8 \end{cases}$$

We add the equations to eliminate y and solve the resulting equation for x:

$$11x^2 = 44$$
$$x^2 = 4$$
$$x = 2 \quad \text{or} \quad x = -2$$

To find y, we substitute 2 for x and then -2 for x in the first equation and proceed as follows:

For x = 2	*For x = -2*
$3x^2 + 2y^2 = 36$	$3x^2 + 2y^2 = 36$
$3(2)^2 + 2y^2 = 36$	$3(-2)^2 + 2y^2 = 36$
$12 + 2y^2 = 36$	$12 + 2y^2 = 36$
$2y^2 = 24$	$2y^2 = 24$
$y^2 = 12$	$y^2 = 12$
$y = +\sqrt{12}$ or $y = -\sqrt{12}$	$y = +\sqrt{12}$ or $y = -\sqrt{12}$
$y = 2\sqrt{3}$ $y = -2\sqrt{3}$	$y = 2\sqrt{3}$ $y = -2\sqrt{3}$

The four solutions of this system are

$$\left(2, 2\sqrt{3}\right), \quad \left(2, -2\sqrt{3}\right), \quad \left(-2, 2\sqrt{3}\right), \quad \text{and} \quad \left(-2, -2\sqrt{3}\right) \qquad \blacksquare$$

Self Check   Solve $\begin{cases} x^2 + 4y^2 = 16 \\ x^2 - y^2 = 1 \end{cases}$.

Answer $(2, \sqrt{3}), (2, -\sqrt{3}), (-2, \sqrt{3}), (-2, -\sqrt{3})$

■ SOLUTIONS OF SYSTEMS OF INEQUALITIES

EXAMPLE 5   Graph the solution set of the system $\begin{cases} y < x^2 \\ y > \dfrac{x^2}{4} - 2 \end{cases}$.

Solution   The graph of $y = x^2$ is the parabola shown in Figure 13-13, which opens upward and has its vertex at the origin. The points with coordinates that satisfy the inequality $y < x^2$ are those points below the parabola.

The graph of $y = \dfrac{x^2}{4} - 2$ is a parabola opening upward, with vertex at $(0, -2)$.

However, this time the points with coordinates that satisfy the inequality are those points above the parabola. The graph of the solution set of the system is the area between the parabolas.

$$y = x^2 \qquad\qquad y = \frac{x^2}{4} - 2$$

x	y	(x, y)
0	0	$(0, 0)$
1	1	$(1, 1)$
-1	1	$(-1, 1)$
2	4	$(2, 4)$
-2	4	$(-2, 4)$

x	y	(x, y)
0	-2	$(0, -2)$
2	-1	$(2, -1)$
-2	-1	$(-2, -1)$
4	2	$(4, 2)$
-4	2	$(-4, 2)$

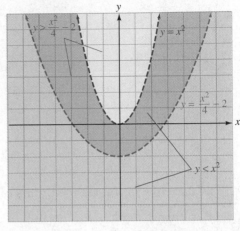

FIGURE 13-13

Orals *Give the possible number of solutions of a system when the graphs of the equations are*

1. A line and a parabola

2. A line and a hyperbola

3. A circle and a parabola

4. A circle and a hyperbola

EXERCISE 13.5

REVIEW *Simplify each radical expression. Assume that all variables represent positive numbers.*

1. $\sqrt{200x^2} - 3\sqrt{98x^2}$

2. $a\sqrt{112a} - 5\sqrt{175a^3}$

3. $\dfrac{3t\sqrt{2t} - 2\sqrt{2t^3}}{\sqrt{18t} - \sqrt{2t}}$

4. $\sqrt[3]{\dfrac{x}{4}} + \sqrt[3]{\dfrac{x}{32}} - \sqrt[3]{\dfrac{x}{500}}$

VOCABULARY AND CONCEPTS *Fill in each blank to make a true statement.*

5. We can solve systems of equations by _____, addition, or _____.

6. A line can intersect an ellipse in at most _____ points.

PRACTICE *In Exercises 7–14, solve each system of equations by graphing.*

7. $\begin{cases} 8x^2 + 32y^2 = 256 \\ x = 2y \end{cases}$

8. $\begin{cases} x^2 + y^2 = 2 \\ x + y = 2 \end{cases}$

9. $\begin{cases} x^2 + y^2 = 10 \\ y = 3x^2 \end{cases}$

10. $\begin{cases} x^2 + y^2 = 5 \\ x + y = 3 \end{cases}$

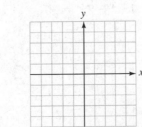

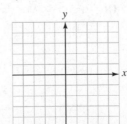

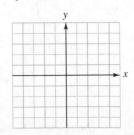

11. $\begin{cases} x^2 + y^2 = 25 \\ 12x^2 + 64y^2 = 768 \end{cases}$

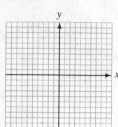

12. $\begin{cases} x^2 + y^2 = 13 \\ y = x^2 - 1 \end{cases}$

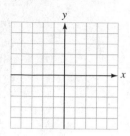

13. $\begin{cases} x^2 - 13 = -y^2 \\ y = 2x - 4 \end{cases}$

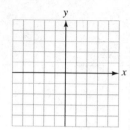

14. $\begin{cases} x^2 + y^2 = 20 \\ y = x^2 \end{cases}$

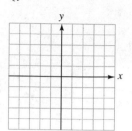

In Exercises 15–16, use a graphing calculator to solve each system.

15. $\begin{cases} x^2 - 6x - y = -5 \\ x^2 - 6x + y = -5 \end{cases}$

16. $\begin{cases} x^2 - y^2 = -5 \\ 3x^2 + 2y^2 = 30 \end{cases}$

In Exercises 17–42, solve each system of equations algebraically for real values of x and y.

17. $\begin{cases} 25x^2 + 9y^2 = 225 \\ 5x + 3y = 15 \end{cases}$

18. $\begin{cases} x^2 + y^2 = 20 \\ y = x^2 \end{cases}$

19. $\begin{cases} x^2 + y^2 = 2 \\ x + y = 2 \end{cases}$

20. $\begin{cases} x^2 + y^2 = 36 \\ 49x^2 + 36y^2 = 1{,}764 \end{cases}$

21. $\begin{cases} x^2 + y^2 = 5 \\ x + y = 3 \end{cases}$

22. $\begin{cases} x^2 - x - y = 2 \\ 4x - 3y = 0 \end{cases}$

23. $\begin{cases} x^2 + y^2 = 13 \\ y = x^2 - 1 \end{cases}$

24. $\begin{cases} x^2 + y^2 = 25 \\ 2x^2 - 3y^2 = 5 \end{cases}$

25. $\begin{cases} x^2 + y^2 = 30 \\ y = x^2 \end{cases}$

26. $\begin{cases} 9x^2 - 7y^2 = 81 \\ x^2 + y^2 = 9 \end{cases}$

27. $\begin{cases} x^2 + y^2 = 13 \\ x^2 - y^2 = 5 \end{cases}$

28. $\begin{cases} 2x^2 + y^2 = 6 \\ x^2 - y^2 = 3 \end{cases}$

29. $\begin{cases} x^2 + y^2 = 20 \\ x^2 - y^2 = -12 \end{cases}$

30. $\begin{cases} xy = -\dfrac{9}{2} \\ 3x + 2y = 6 \end{cases}$

31. $\begin{cases} y^2 = 40 - x^2 \\ y = x^2 - 10 \end{cases}$

32. $\begin{cases} x^2 - 6x - y = -5 \\ x^2 - 6x + y = -5 \end{cases}$

33. $\begin{cases} y = x^2 - 4 \\ x^2 - y^2 = -16 \end{cases}$

34. $\begin{cases} 6x^2 + 8y^2 = 182 \\ 8x^2 - 3y^2 = 24 \end{cases}$

35. $\begin{cases} x^2 - y^2 = -5 \\ 3x^2 + 2y^2 = 30 \end{cases}$

36. $\begin{cases} \dfrac{1}{x} + \dfrac{1}{y} = 5 \\ \dfrac{1}{x} - \dfrac{1}{y} = -3 \end{cases}$

37. $\begin{cases} \dfrac{1}{x} + \dfrac{2}{y} = 1 \\ \dfrac{2}{x} - \dfrac{1}{y} = \dfrac{1}{3} \end{cases}$

38. $\begin{cases} \dfrac{1}{x} + \dfrac{3}{y} = 4 \\ \dfrac{2}{x} - \dfrac{1}{y} = 7 \end{cases}$

39. $\begin{cases} 3y^2 = xy \\ 2x^2 + xy - 84 = 0 \end{cases}$

40. $\begin{cases} x^2 + y^2 = 10 \\ 2x^2 - 3y^2 = 5 \end{cases}$

41. $\begin{cases} xy = \dfrac{1}{6} \\ y + x = 5xy \end{cases}$

42. $\begin{cases} xy = \dfrac{1}{12} \\ y + x = 7xy \end{cases}$

43. Integer problem The product of two integers is 32, and their sum is 12. Find the integers.

44. Number problem The sum of the squares of two numbers is 221, and the sum of the numbers is 9. Find the numbers.

In Exercises 45–48, graph the solution set of each system of inequalities.

45. $\begin{cases} 2x - y > 4 \\ y < -x^2 + 2 \end{cases}$

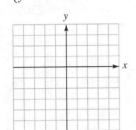

46. $\begin{cases} x \le y^2 \\ y \ge x \end{cases}$

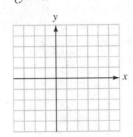

47. $\begin{cases} y > x^2 - 4 \\ y < -x^2 + 4 \end{cases}$

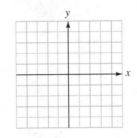

48. $\begin{cases} x \ge y^2 \\ y \ge x^2 \end{cases}$

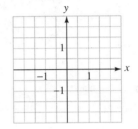

APPLICATIONS

49. Geometry problem The area of a rectangle is 63 square centimeters, and its perimeter is 32 centimeters. Find the dimensions of the rectangle.

50. Investing money Grant receives $225 annual income from one investment. Jeff invested $500 more than Grant, but at an annual rate of 1% less. Jeff's annual income is $240. What is the amount and rate of Grant's investment?

51. Investing money Carol receives $67.50 annual income from one investment. John invested $150 more than Carol at an annual rate of $1\frac{1}{2}$% more. John's annual income is $94.50. What is the amount and rate of Carol's investment? (*Hint:* There are two answers.)

52. Artillery The shell fired from the base of the hill in Illustration 1 follows the parabolic path $y = -\frac{1}{6}x^2 + 2x$, with distances measured in miles. The hill has a slope of $\frac{1}{3}$. How far from the gun is the point of impact? (*Hint:* Find the coordinates of the point and then the distance.)

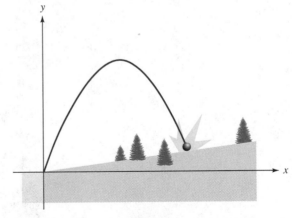

ILLUSTRATION 1

53. Driving rates Jim drove 306 miles. Jim's brother made the same trip at a speed 17 mph slower than Jim did and required an extra $1\frac{1}{2}$ hours. What was Jim's rate and time?

WRITING

54. Describe the benefits of the graphical method for solving a system of equations.

55. Describe the drawbacks of the graphical method.

SOMETHING TO THINK ABOUT

56. The graphs of the two independent equations of a system are parabolas. How many solutions might the system have?

57. The graphs of the two independent equations of a system are hyperbolas. How many solutions might the system have?

■ ■ ■ ■ ■ ■ ■ ■ ■   **PROJECTS**

PROJECT 1

In this project, you will explore two of the many uses of determinants. In the first, you will discover that the equation of a line can be written as a determinant equation. If you are given the coordinates of two fixed points, you can use a determinant to write the equation of the line passing through them.

• The equation of the line passing through the points $P(2, 3)$ and $Q(-1, 4)$ is

$$\begin{vmatrix} x & y & 1 \\ 2 & 3 & 1 \\ -1 & 4 & 1 \end{vmatrix} = 0$$

Verify this by expanding the determinant and graphing the resulting equation.

• In general, the **two-point form** of the equation of the line passing through the points $P(x_1, y_1)$ and $Q(x_2, y_2)$ is

$$\begin{vmatrix} x & y & 1 \\ x_1 & y_1 & 1 \\ x_2 & y_2 & 1 \end{vmatrix} = 0$$

Find the equation of the line passing through $P(-4, 5)$ and $Q(1, -3)$.

• Does this equation still work if the x-coordinates of the two points are equal? (For then the line would be vertical and therefore have no defined slope.)

As a second application, the formula for the area of a triangle can be written as a determinant.

• The vertices of the triangle in Illustration 1 are $A(-3, -2)$, $B(4, -2)$, and $C(4, 4)$. Clearly, ABC is a right triangle, and it is easy to find its area by the formula $A = \frac{1}{2}bh$. Show that the area is also given by

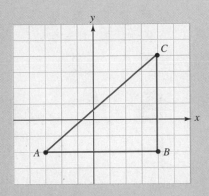

$$A = \frac{1}{2} \begin{vmatrix} -3 & -2 & 1 \\ 4 & -2 & 1 \\ 4 & 4 & 1 \end{vmatrix}$$

ILLUSTRATION 1

The formula works for any triangle, not just right triangles. The area of the triangle with vertices $A(x_1, y_1)$, $B(x_2, y_2)$, and $C(x_3, y_3)$ is

$$A = \frac{1}{2} \begin{vmatrix} x_1 & y_1 & 1 \\ x_2 & y_2 & 1 \\ x_3 & y_3 & 1 \end{vmatrix}$$

PROJECT 2   Before recommending traffic controls for the intersection of the two one-way streets shown in Illustration 2, a traffic engineer places counters across the roads to record traffic flow. The illustration shows the number of vehicles passing each of the four counters during one hour. To find the number of vehicles passing straight through the intersection and the number that turn from one road to the other, refer to the illustration and assign the following variables.

Let L represent the number of vehicles turning left.

Let R represent the number of vehicles turning right

Let N represent the number of vehicles headed north.

Let E represent the number of vehicles headed east.

Because the counter at A counts the total number of vehicles headed east and turning left, we have the equation

$$L + E = 250$$

From the data, explain why we can form the following system of equations:

$$\begin{aligned} L \quad\quad + E &= 250 \\ R \quad\quad &= 90 \\ R \quad + E &= 275 \\ R + N \quad &= 340 \end{aligned}$$

Solve the system and interpret the results.

(continued)

■ ■ ■ ■ ■ ■ ■ ■ ■ ■   **PROJECTS**   *(continued)*

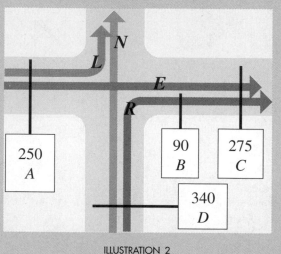

ILLUSTRATION 2

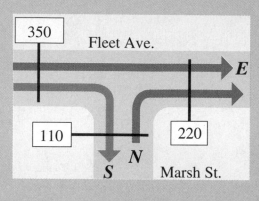

ILLUSTRATION 3

E X E R C I S E   The intersection of Marsh Street and one-way Fleet Avenue has three counters to record the traffic, as shown in Illustration 3.

1. Find *E,* the number of vehicles passing through the intersection headed east.
2. Find *S*, the number of vehicles turing south.
3. Find *N*, the number of northbound vehicles turning east.
4. The traffic engineer suspects that the counters are in error. Why?

C H A P T E R   S U M M A R Y

CONCEPTS

REVIEW EXERCISES

SECTION 13.1   *Solution of Two Equations in Two Variables*

If a system of equations has at least one solution, the system is a **consistent system**. Otherwise, the system is an **inconsistent system**.

If the graphs of the equations of a system are distinct, the equations are **independent equations**. Otherwise, the equations are **dependent equations**.

1. Solve each system by the graphing method.

a. $\begin{cases} 2x + y = 11 \\ -x + 2y = 7 \end{cases}$

b. $\begin{cases} 3x + 2y = 0 \\ 2x - 3y = -13 \end{cases}$

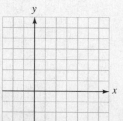

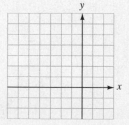

c. $\begin{cases} \dfrac{1}{2}x + \dfrac{1}{3}y = 2 \\ y = 6 - \dfrac{3}{2}x \end{cases}$

d. $\begin{cases} \dfrac{1}{3}x - \dfrac{1}{2}y = 1 \\ 6x - 9y = 2 \end{cases}$

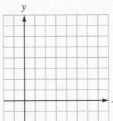

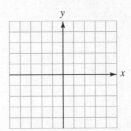

To solve a system by *substitution*, solve one equation for a variable. Then substitute the expression found for that variable into the other equation. Then solve for the other variable.

2. Solve each system by substitution.

a. $\begin{cases} y = x + 4 \\ 2x + 3y = 7 \end{cases}$

b. $\begin{cases} y = 2x + 5 \\ 3x - 5y = -4 \end{cases}$

c. $\begin{cases} x + 2y = 11 \\ 2x - y = 2 \end{cases}$

d. $\begin{cases} 2x + 3y = -2 \\ 3x + 5y = -2 \end{cases}$

To solve a system by *addition*, combine the equations of the system in a way that will eliminate one of the variables.

3. Solve each system by addition.

a. $\begin{cases} x + y = -2 \\ 2x + 3y = -3 \end{cases}$

b. $\begin{cases} 3x + 2y = 1 \\ 2x - 3y = 5 \end{cases}$

c. $\begin{cases} x + \dfrac{1}{2}y = 7 \\ -2x = 3y - 6 \end{cases}$

d. $\begin{cases} y = \dfrac{x - 3}{2} \\ x = \dfrac{2y + 7}{2} \end{cases}$

| **SECTION 13.2** | *Solutions of Three Equations in Three Variables* |

A system of three linear equations in three variables can be solved by using a combination of the addition method and the substitution method.

4. Solve each system.

a. $\begin{cases} x + y + z = 6 \\ x - y - z = -4 \\ -x + y - z = -2 \end{cases}$

b. $\begin{cases} 2x + 3y + z = -5 \\ -x + 2y - z = -6 \\ 3x + y + 2z = 4 \end{cases}$

| **SECTION 13.3** | *Solution by Matrices* |

A **matrix** is any rectangular array of numbers.

Systems of linear equations can be solved by using matrices and the method of **Gaussian elimination**.

5. Solve each system by using matrices.

a. $\begin{cases} x + 2y = 4 \\ 2x - y = 3 \end{cases}$

b. $\begin{cases} x + y + z = 6 \\ 2x - y + z = 1 \\ 4x + y - z = 5 \end{cases}$

c. $\begin{cases} x + y = 3 \\ x - 2y = -3 \\ 2x + y = 4 \end{cases}$

d. $\begin{cases} x + 2y + z = 2 \\ 2x + 5y + 4z = 5 \end{cases}$

| SECTION 13.4 | *Solution by Determinants* |

A **determinant of a square matrix** is a real number.

$$\begin{vmatrix} a & b \\ c & d \end{vmatrix} = ad - bc$$

$$\begin{vmatrix} a_1 & b_1 & c_1 \\ a_2 & b_2 & c_2 \\ a_3 & b_3 & c_3 \end{vmatrix} = a_1 \begin{vmatrix} b_2 & c_2 \\ b_3 & c_3 \end{vmatrix}$$

$$- b_1 \begin{vmatrix} a_2 & c_2 \\ a_3 & c_3 \end{vmatrix} + c_1 \begin{vmatrix} a_2 & b_2 \\ a_3 & b_3 \end{vmatrix}$$

6. Evaluate each determinant.

a. $\begin{vmatrix} 2 & 3 \\ -4 & 3 \end{vmatrix}$
b. $\begin{vmatrix} -3 & -4 \\ 5 & -6 \end{vmatrix}$

c. $\begin{vmatrix} -1 & 2 & -1 \\ 2 & -1 & 3 \\ 1 & -2 & 2 \end{vmatrix}$
d. $\begin{vmatrix} 3 & -2 & 2 \\ 1 & -2 & -2 \\ 2 & 1 & -1 \end{vmatrix}$

7. Use Cramer's rule to solve each system.

a. $\begin{cases} 3x + 4y = 10 \\ 2x - 3y = 1 \end{cases}$
b. $\begin{cases} 2x - 5y = -17 \\ 3x + 2y = 3 \end{cases}$

c. $\begin{cases} x + 2y + z = 0 \\ 2x + y + z = 3 \\ x + y + 2z = 5 \end{cases}$
d. $\begin{cases} 2x + 3y + z = 2 \\ x + 3y + 2z = 7 \\ x - y - z = -7 \end{cases}$

| SECTION 13.5 | *Solving Simultaneous Second-Degree Equations and Inequalities* |

8. Solve each system of equations.

a. $\begin{cases} 3x^2 + y^2 = 52 \\ x^2 - y^2 = 12 \end{cases}$
b. $\begin{cases} \dfrac{x^2}{16} + \dfrac{y^2}{12} = 1 \\ x^2 - \dfrac{y^2}{3} = 1 \end{cases}$

Systems of inequalities can be solved by graphing.

9. Graph the solution set in the system $\begin{cases} y \geq x^2 - 4 \\ y < x + 3 \end{cases}$.

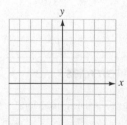

■ Chapter Test

1. Solve $\begin{cases} 2x + y = 5 \\ y = 2x - 3 \end{cases}$ by graphing.

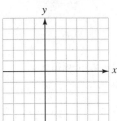

2. Use substitution to solve $\begin{cases} 2x - 4y = 14 \\ x = -2y + 7 \end{cases}$.

3. Use addition to solve $\begin{cases} 2x + 3y = -5 \\ 3x - 2y = 12 \end{cases}$.

4. Use any method to solve $\begin{cases} \dfrac{x}{2} - \dfrac{y}{4} = -4 \\ x + y = -2 \end{cases}$.

In Problems 5–6, consider the system $\begin{cases} 3(x + y) = x - 3 \\ -y = \dfrac{2x + 3}{3} \end{cases}$.

5. Are the equations of the system dependent or independent?

6. Is the system consistent or inconsistent?

In Problems 7–8, use an elementary row operation to find the missing number in the second matrix.

7. $\begin{bmatrix} 1 & 2 & -1 \\ 2 & -2 & 3 \end{bmatrix}, \begin{bmatrix} 1 & 2 & -1 \\ -1 & -8 & \blacksquare \end{bmatrix}$

8. $\begin{bmatrix} -1 & 3 & 6 \\ 3 & -2 & 4 \end{bmatrix}, \begin{bmatrix} -1 & 3 & 6 \\ 5 & \blacksquare & -8 \end{bmatrix}$

In Problems 9–10, consider the system $\begin{cases} x + y + z = 4 \\ x + y - z = 6 \\ 2x - 3y + z = -1 \end{cases}$.

9. Write the augmented matrix that represents the system.

10. Write the coefficient matrix that represents the system.

In Problems 11–12, use matrices to solve each system.

11. $\begin{cases} x + y = 4 \\ 2x - y = 2 \end{cases}$

12. $\begin{cases} x + y = 2 \\ x - y = -4 \\ 2x + y = 1 \end{cases}$

In Problems 13–16, evaluate each determinant.

13. $\begin{vmatrix} 2 & -3 \\ 4 & 5 \end{vmatrix}$
14. $\begin{vmatrix} -3 & -4 \\ -2 & 3 \end{vmatrix}$
15. $\begin{vmatrix} 1 & 2 & 0 \\ 2 & 0 & 3 \\ 1 & -2 & 2 \end{vmatrix}$
16. $\begin{vmatrix} 2 & -1 & 1 \\ 3 & 1 & 0 \\ 0 & 1 & 2 \end{vmatrix}$

In Problems 17–20, consider the system $\begin{cases} x - y = -6 \\ 3x + y = -6 \end{cases}$, *which is to be solved with Cramer's rule.*

17. When solving for x, what is the numerator determinant? **(Don't evaluate it.)**

18. When solving for y, what is the denominator determinant? **(Don't evaluate it.)**

19. Solve the system for x.

20. Solve the system for y.

In Problems 21–22, consider the system $\begin{cases} x + y + z = 4 \\ x + y - z = 6 \\ 2x - 3y + z = -1 \end{cases}$.

21. Solve for x.

22. Solve for z.

In Problems 23–24, solve each system.

23. $\begin{cases} x^2 + y^2 = 5 \\ x^2 - y^2 = 3 \end{cases}$

24. $\begin{cases} x^2 + y^2 = 25 \\ 4x^2 - 9y = 0 \end{cases}$

25. Solve the system $\begin{cases} y \geq x^2 \\ y < x + 3 \end{cases}$

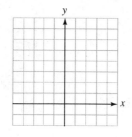

14 Miscellaneous Topics

MATHEMATICS IN THE WORKPLACE

Actuary

Why do young people pay more for automobile insurance than older people? How much should an insurance policy cost? How much should an organization contribute each year to its pension fund? Answers to these questions are provided by actuaries, who design insurance and pension plans and follow their performance to make sure that the plans are maintained on a sound financial basis.

SAMPLE APPLICATION ■ An *annuity* is a sequence of equal payments made periodically over a length of time. The sum of the payments and the interest earned during the *term* of the annuity is called the *amount* of the annuity.

After a sales clerk works six months, her employer will begin an annuity for her and will contribute $500 semiannually to a fund that pays 8% annual interest. After she has been employed for two years, what will be the amount of her annuity?

(See Example 7 in Section 14.4.)

In this chapter, we introduce several topics that have applications in advanced mathematics and in many occupational areas. The binomial theorem, permutations, and combinations are used in statistics. Arithmetic sequences and geometric sequences are used in the mathematics of finance.

14.1 The Binomial Theorem

■ RAISING BINOMIALS TO POWERS ■ PASCAL'S TRIANGLE ■ FACTORIAL NOTATION ■ THE BINOMIAL THEOREM

Getting Ready *Raise each binomial to the indicated power.*

1. $(x + 2)^2$ **2.** $(x - 3)^2$
3. $(x + 1)^3$ **4.** $(x - 2)^3$

■ RAISING BINOMIALS TO POWERS

We have discussed how to raise binomials to positive integral powers. For example, we know that

$$(a + b)^2 = a^2 + 2ab + b^2$$

and that

$$(a + b)^3 = (a + b)(a + b)^2$$
$$= (a + b)(a^2 + 2ab + b^2)$$
$$= a^3 + 2a^2b + ab^2 + a^2b + 2ab^2 + b^3$$
$$= a^3 + 3a^2b + 3ab^2 + b^3$$

To show how to raise binomials to positive-integer powers without doing the actual multiplications, we consider the following binomial expansions:

$$(a + b)^0 = 1$$
$$(a + b)^1 = a + b$$
$$(a + b)^2 = a^2 + 2ab + b^2$$
$$(a + b)^3 = a^3 + 3a^2b + 3ab^2 + b^3$$
$$(a + b)^4 = a^4 + 4a^3b + 6a^2b^2 + 4ab^3 + b^4$$
$$(a + b)^5 = a^5 + 5a^4b + 10a^3b^2 + 10a^2b^3 + 5ab^4 + b^5$$
$$(a + b)^6 = a^6 + 6a^5b + 15a^4b^2 + 20a^3b^3 + 15a^2b^4 + 6ab^5 + b^6$$

Several patterns appear in these expansions:

1. Each expansion has one more term than the power of the binomial.
2. The degree of each term in each expansion is equal to the exponent of the binomial that is being expanded.
3. The first term in each expansion is a, raised to the power of the binomial.
4. The exponents of a decrease by 1 in each successive term. The exponents of b, beginning with $b^0 = 1$ in the first term, increase by 1 in each successive term. Thus, the variables have the pattern

$$a^n, a^{n-1}b, a^{n-2}b^2, \ldots, ab^{n-1}, b^n$$

■ PASCAL'S TRIANGLE

To see another pattern, we write the coefficients of each expansion in the following triangular array:

$$
\begin{array}{ccccccccccccc}
& & & & & & 1 & & & & & & \\
& & & & & 1 & & 1 & & & & & \\
& & & & 1 & & 2 & & 1 & & & & \\
& & & 1 & & 3 & & 3 & & 1 & & & \\
& & 1 & & 4 & & 6 & & 4 & & 1 & & \\
& 1 & & 5 & & 10 & & 10 & & 5 & & 1 & \\
1 & & 6 & & 15 & & 20 & & 15 & & 6 & & 1
\end{array}
$$

In this array, called **Pascal's triangle**, each entry between the 1's is the sum of the closest pair of numbers in the line immediately above it. For example, the first 15 in the bottom row is the sum of the 5 and 10 immediately above it. Pascal's triangle continues with the same pattern forever. The next two lines are

$$1 \quad 7 \quad 21 \quad 35 \quad 35 \quad 21 \quad 7 \quad 1$$
$$1 \quad 8 \quad 28 \quad 56 \quad 70 \quad 56 \quad 28 \quad 8 \quad 1$$

EXAMPLE 1   Expand $(x + y)^5$.

Solution   The first term in the expansion is x^5, and the exponents of x decrease by 1 in each successive term. A y first appears in the second term, and the exponents on y increase by 1 in each successive term, concluding when the term y^5 is reached. Thus, the variables in the expansion are

$$x^5, \ x^4y, \ x^3y^2, \ x^2y^3, \ xy^4, \ y^5$$

The coefficients of these variables are given in Pascal's triangle in the row whose second entry is 5, the same as the exponent of the binomial being expanded:

$$1 \quad 5 \quad 10 \quad 10 \quad 5 \quad 1$$

Combining this information gives the following expansion:

$$(x + y)^5 = x^5 + 5x^4y + 10x^3y^2 + 10x^2y^3 + 5xy^4 + y^5 \qquad \blacksquare$$

Self Check   Expand $(x + y)^4$.
Answer $x^4 + 4x^3y + 6x^2y^2 + 4xy^3 + y^4$

EXAMPLE 2   Expand $(u - v)^4$.

Solution   We note that $(u - v)^4$ can be written in the form $[u + (-v)]^4$. The variables in this expansion are

$$u^4, \ u^3(-v), \ u^2(-v)^2, \ u(-v)^3, \ (-v)^4$$

and the coefficients are given in Pascal's triangle in the row whose second entry is 4:

$$1 \quad 4 \quad 6 \quad 4 \quad 1$$

Thus, the required expansion is

$$(u - v)^4 = u^4 + 4u^3(-v) + 6u^2(-v)^2 + 4u(-v)^3 + (-v)^4$$
$$= u^4 - 4u^3v + 6u^2v^2 - 4uv^3 + v^4 \qquad \blacksquare$$

Self Check   Expand $(x - y)^5$.
Answer $x^5 - 5x^4y + 10x^3y^2 - 10x^2y^3 + 5xy^4 - y^5$

■ FACTORIAL NOTATION

Although Pascal's triangle gives the coefficients of the terms in a binomial expansion, it is not the best way to expand a binomial. To develop a better way, we introduce **factorial notation.**

> **Factorial Notation**
> If n is a natural number, the symbol $n!$ (read as "n **factorial**" or as "**factorial** n") is defined as
> $$n! = n(n - 1)(n - 2)(n - 3) \cdots (3)(2)(1)$$

EXAMPLE 3　　Find　**a.** $2!$,　**b.** $5!$　**c.** $-9!$　and　**d.** $(n - 2)!$.

Solution　**a.** $2! = 2 \cdot 1 = 2$

b. $5! = 5 \cdot 4 \cdot 3 \cdot 2 \cdot 1 = 120$

c. $-9! = -9 \cdot 8 \cdot 7 \cdot 6 \cdot 5 \cdot 4 \cdot 3 \cdot 2 \cdot 1 = -362,880$

d. $(n - 2)! = (n - 2)(n - 3)(n - 4) \cdot \cdots \cdot 3 \cdot 2 \cdot 1$

> **WARNING!**　According to the previous definition, part **d** is meaningful only if $n - 2$ is a natural number.

■

Self Check　Find　**a.** $4!$　and　**b.** $x!$.
Answers　**a.** 24,　**b.** $x(x - 1)(x - 2) \cdot \cdots \cdot 3 \cdot 2 \cdot 1$

We define zero factorial as follows.

> **Zero Factorial**
> $$0! = 1$$

We note that

$$5 \cdot 4! = 5 \cdot 4 \cdot 3 \cdot 2 \cdot 1 = 5!$$
$$7 \cdot 6! = 7 \cdot 6 \cdot 5 \cdot 4 \cdot 3 \cdot 2 \cdot 1 = 7!$$
$$10 \cdot 9! = 10 \cdot 9 \cdot 8 \cdot 7 \cdot 6 \cdot 5 \cdot 4 \cdot 3 \cdot 2 \cdot 1 = 10!$$

These examples suggest the following property of factorials.

$n(n-1)!$
If n is a positive integer, then $n(n-1)! = n!$.

■ THE BINOMIAL THEOREM

We now state the binomial theorem.

The Binomial Theorem
If n is any positive integer, then

$$(a+b)^n = a^n + \frac{n!}{1!(n-1)!}a^{n-1}b + \frac{n!}{2!(n-2)!}a^{n-2}b^2 + \frac{n!}{3!(n-3)!}a^{n-3}b^3$$

$$+ \cdots + \frac{n!}{r!(n-r)!}a^{n-r}b^r + \cdots + b^n$$

In the binomial theorem, the exponents of the variables follow the familiar pattern:

- The sum of the exponents on a and b in each term is n,
- the exponents on a decrease, and
- the exponents on b increase.

Only the method of finding the coefficients is different. Except for the first and last terms, the numerator of each coefficient is $n!$. If the exponent of b in a particular term is r, the denominator of the coefficient of that term is $r!(n-r)!$.

EXAMPLE 4   Use the binomial theorem to expand $(a+b)^3$.

Solution   We can substitute directly into the binomial theorem and simplify:

$$(a+b)^3 = a^3 + \frac{3!}{1!(3-1)!}a^2b + \frac{3!}{2!(3-2)!}ab^2 + b^3$$

$$= a^3 + \frac{3!}{1! \cdot 2!}a^2b + \frac{3!}{2! \cdot 1!}ab^2 + b^3$$

$$= a^3 + \frac{3 \cdot 2 \cdot 1}{1 \cdot 2 \cdot 1}a^2b + \frac{3 \cdot 2 \cdot 1}{2 \cdot 1 \cdot 1}ab^2 + b^3$$

$$= a^3 + 3a^2b + 3ab^2 + b^3$$

■

Self Check   Use the binomial theorem to expand $(a+b)^4$.
Answer $a^4 + 4a^3b + 6a^2b^2 + 4ab^3 + b^4$

EXAMPLE 5   Use the binomial theorem to expand $(x - y)^4$.

Solution   We can write $(x - y)^4$ in the form $[x + (-y)]^4$, substitute directly into the binomial theorem, and simplify:

$$(x - y)^4 = [x + (-y)]^4$$

$$= x^4 + \frac{4!}{1!(4 - 1)!}x^3(-y) + \frac{4!}{2!(4 - 2)!}x^2(-y)^2 + \frac{4!}{3!(4 - 3)!}x(-y)^3 + (-y)^4$$

$$= x^4 - \frac{4 \cdot 3!}{1! \cdot 3!}x^3y + \frac{4 \cdot 3 \cdot 2!}{2! \cdot 2!}x^2y^2 - \frac{4 \cdot 3!}{3! \cdot 1!}xy^3 + y^4$$

$$= x^4 - 4x^3y + 6x^2y^2 - 4xy^3 + y^4 \qquad \blacksquare$$

Self Check   Use the binomial theorem to expand $(x - y)^3$.
Answer $x^3 - 3x^2y + 3xy^2 - y^3$

EXAMPLE 6   Use the binomial theorem to expand $(3u - 2v)^4$.

Solution   We write $(3u - 2v)^4$ in the form $[3u + (-2v)]^4$ and let $a = 3u$ and $b = -2v$. Then, we can use the binomial theorem to expand $(a + b)^4$.

$$(a + b)^4 = a^4 + \frac{4!}{1!(4 - 1)!}a^3b + \frac{4!}{2!(4 - 2)!}a^2b^2 + \frac{4!}{3!(4 - 3)!}ab^3 + b^4$$

$$= a^4 + 4a^3b + 6a^2b^2 + 4ab^3 + b^4$$

Now we can substitute $3u$ for a and $-2v$ for b and simplify:

$$(3u - 2v)^4 = (3u)^4 + 4(3u)^3(-2v) + 6(3u)^2(-2v)^2 + 4(3u)(-2v)^3 + (-2v)^4$$

$$= 81u^4 - 216u^3v + 216u^2v^2 - 96uv^3 + 16v^4 \qquad \blacksquare$$

Self Check   Use the binomial theorem to expand $(2a - 3b)^3$.
Answer $8a^3 - 36a^2b + 54ab^2 - 27b^3$

Orals   *Find each value.*

1. $1!$ **2.** $4!$ **3.** $0!$ **4.** $5!$

Expand each binomial.

5. $(m + n)^2$ **6.** $(m - n)^2$

7. $(p + 2q)^2$ **8.** $(2p - q)^2$

EXERCISE 14.1

REVIEW *Find each value of x.*

1. $\log_4 16 = x$

2. $\log_x 49 = 2$

3. $\log_{25} x = \dfrac{1}{2}$

4. $\log_{1/2} \dfrac{1}{8} = x$

VOCABULARY AND CONCEPTS *Fill in each blank to make a true statement.*

5. Every binomial expansion has _____ more term than the power of the binomial.

6. The first term in the expansion of $(a + b)^{20}$ is _____.

7. The triangular array that can be used to find the coefficients of a binomial expansion is called _____ triangle.

8. The symbol 5! is read as "_____."

9. $6 \cdot 5 \cdot 4 \cdot 3 \cdot 2 \cdot 1 =$ ___ (Write your answer in factorial notation.)

10. $8! = 8 \cdot$ ___

11. $0! =$ ___

12. According to the binomial theorem, the third term of the expansion of $(a + b)^n$ is _____ .

PRACTICE *In Exercises 13–32, evaluate each expression.*

13. $3!$

14. $7!$

15. $-5!$

16. $-6!$

17. $3! + 4!$

18. $2!(3!)$

19. $3!(4!)$

20. $4! + 4!$

21. $8(7!)$

22. $4!(5)$

23. $\dfrac{9!}{11!}$

24. $\dfrac{13!}{10!}$

25. $\dfrac{49!}{47!}$

26. $\dfrac{101!}{100!}$

27. $\dfrac{5!}{3!(5-3)!}$

28. $\dfrac{6!}{4!(6-4)!}$

29. $\dfrac{7!}{5!(7-5)!}$

30. $\dfrac{8!}{6!(8-6)!}$

31. $\dfrac{5!(8-5)!}{4! \cdot 7!}$

32. $\dfrac{6! \cdot 7!}{(8-3)!(7-4)!}$

In Exercises 33–48, use the binomial theorem to expand each expression.

33. $(x + y)^3$

34. $(x + y)^4$

35. $(x - y)^4$

36. $(x - y)^3$

37. $(2x + y)^3$

38. $(x + 2y)^3$

39. $(x - 2y)^3$

40. $(2x - y)^3$

41. $(2x + 3y)^3$

42. $(3x - 2y)^3$

43. $\left(\dfrac{x}{2} - \dfrac{y}{3}\right)^3$

44. $\left(\dfrac{x}{3} + \dfrac{y}{2}\right)^3$

45. $(3 + 2y)^4$

46. $(2x + 3)^4$

47. $\left(\dfrac{x}{3} - \dfrac{y}{2}\right)^4$

48. $\left(\dfrac{x}{2} + \dfrac{y}{3}\right)^4$

49. Without referring to the text, write the first ten rows of Pascal's triangle.

50. Find the sum of the numbers in each row of the first ten rows of Pascal's triangle. What is the pattern?

51. Find the sum of the numbers in the designated diagonal rows of Pascal's triangle shown in Illustration 1. What is the pattern?

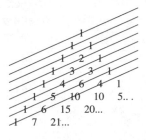

ILLUSTRATION 1

WRITING

52. Tell how to construct Pascal's triangle.

53. Tell how to find the variables of the terms in the expansion of $(r + s)^4$.

SOMETHING TO THINK ABOUT

54. If we apply the pattern of the coefficients to the coefficient of the first term in a binomial expansion, the coefficient would be $\frac{n!}{0!(n-0)!}$. Show that this expression is 1.

55. If we apply the pattern of the coefficients to the coefficient of the last term in a binomial expansion, the coefficient would be $\frac{n!}{n!(n-n)!}$. Show that this expression is 1.

14.2 The *n*th Term of a Binomial Expansion

■ FINDING A PARTICULAR TERM OF A BINOMIAL EXPANSION

Getting Ready   *Expand the binomial and find the coefficient of the third term in each expansion.*

1. $(r + s)^4$ **2.** $(r + s)^5$

Expand the binomial and find the coefficient of the fourth term in each expansion.

3. $(p - q)^6$ **4.** $(p - q)^4$

■ FINDING A PARTICULAR TERM OF A BINOMIAL EXPANSION

To find the fourth term of the expansion of $(a + b)^9$, we could raise the binomial $a + b$ to the 9th power and look at the fourth term. However, this task would be tedious. By using the binomial theorem, we can construct the fourth term without finding the complete expansion of $(a + b)^9$.

EXAMPLE 1   Find the fourth term in the expansion of $(a + b)^9$.

Solution   Since b^1 appears in the second term, b^2 appears in the third term, and so on, the exponent on b in the fourth term is 3. Since the exponent on b added to the exponent on a must equal 9, the exponent on a must be 6. Thus, the variables of the fourth term are

a^6b^3 The sum of the exponents must be 9.

The coefficient of these variables is

$$\frac{n!}{r!(n-r)!} = \frac{9!}{3!(9-3)!}$$

The complete fourth term is

$$\frac{9!}{3!(9-3)!}a^6b^3 = \frac{9 \cdot 8 \cdot 7 \cdot 6!}{3 \cdot 2 \cdot 1 \cdot 6!}a^6b^3$$
$$= 84a^6b^3$$

Self Check   Find the third term of the expansion of $(a + b)^9$.
Answer $36a^7b^2$

EXAMPLE 2   Find the sixth term in the expansion of $(x - y)^7$.

Solution   We first find the sixth term of $[x + (-y)]^7$. In the sixth term, the exponent on $(-y)$ is 5. Thus, the variables in the sixth term are

$x^2(-y)^5$ The sum of the exponents must be 7.

The coefficient of these variables is

$$\frac{n!}{r!(n-r)!} = \frac{7!}{5!(7-5)!}$$

The complete sixth term is

$$\frac{7!}{5!(7-5)!}x^2(-y)^5 = -\frac{7 \cdot 6 \cdot 5!}{5! \cdot 2 \cdot 1}x^2y^5$$
$$= -21x^2y^5$$

Self Check   Find the fifth term of the expansion of $(a - b)^7$.
Answer $35a^3b^4$

EXAMPLE 3   Find the fourth term of the expansion of $(2x - 3y)^6$.

Solution   We can let $a = 2x$ and $b = -3y$ and find the fourth term of the expansion of $(a + b)^6$:

$$\frac{6!}{3!(6-3)!} a^3 b^3 = \frac{6 \cdot 5 \cdot 4 \cdot 3!}{3! \cdot 3 \cdot 2 \cdot 1} a^3 b^3$$

$$= 20a^3 b^3$$

We can now substitute $2x$ for a and $-3y$ for b and simplify:

$$20a^3 b^3 = 20(2x)^3(-3y)^3$$

$$= -4,320x^3 y^3$$

The fourth term is $-4,320x^3 y^3$.   ∎

Self Check   Find the third term of the expansion of $(2a - 3b)^6$.
Answer $2,160a^4 b^2$

Orals   *In the expansion of $(x + y)^8$, find the exponent on y in the*

1. 3rd term   **2.** 4th term   **3.** 7th term

In the expansion of $(x + y)^8$, find the exponent on x in the

4. 3rd term   **5.** 4th term   **6.** 7th term

In the expansion of $(x + y)^8$, find the coefficient of the

7. 1st term   **8.** 2nd term

EXERCISE 14.2

REVIEW   *Solve each system of equations.*

1. $\begin{cases} 3x + 2y = 12 \\ 2x - y = 1 \end{cases}$

2. $\begin{cases} a + b + c = 6 \\ 2a + b + 3c = 11 \\ 3a - b - c = 6 \end{cases}$

Evaluate each determinant.

3. $\begin{vmatrix} 2 & -3 \\ 4 & -2 \end{vmatrix}$

4. $\begin{vmatrix} 1 & 2 & 3 \\ 4 & 5 & 0 \\ -1 & -2 & 1 \end{vmatrix}$

VOCABULARY AND CONCEPTS   *Fill in each blank to make a true statement.*

5. The exponent on b in the fourth term of the expansion of $(a + b)^6$ is ___.

6. The exponent on b in the fifth term of the expansion of $(a + b)^6$ is ___.

7. In the expansion of $(a + b)^7$, the sum of the exponents on a and b is ___.

8. The coefficient of the fourth term of the expansion of $(a + b)^9$ is 9! divided by _____.

PRACTICE   *In Exercises 9–36, use the binomial theorem to find the required term of each expansion.*

9. $(a + b)^3$; second term

10. $(a + b)^3$; third term

11. $(x - y)^4$; fourth term

12. $(x - y)^5$; second term

13. $(x + y)^6$; fifth term

14. $(x + y)^7$; fifth term

15. $(x - y)^8$; third term

16. $(x - y)^9$; seventh term

17. $(x + 3)^5$; third term

18. $(x - 2)^4$; second term

19. $(4x + y)^5$; third term

20. $(x + 4y)^5$; fourth term

21. $(x - 3y)^4$; second term

22. $(3x - y)^5$; third term

23. $(2x - 5)^7$; fourth term

24. $(2x + 3)^6$; sixth term

25. $(2x - 3y)^5$; fifth term

26. $(3x - 2y)^4$; second term

27. $\left(\sqrt{2}x + \sqrt{3}y\right)^6$; third term

28. $\left(\sqrt{3}x + \sqrt{2}y\right)^5$; second term

29. $\left(\dfrac{x}{2} - \dfrac{y}{3}\right)^4$; second term

30. $\left(\dfrac{x}{3} + \dfrac{y}{2}\right)^5$; fourth term

31. $(a + b)^n$; fourth term

32. $(a + b)^n$; third term

33. $(a - b)^n$; fifth term

34. $(a - b)^n$; sixth term

35. $(a + b)^n$; rth term

36. $(a + b)^n$; $(r + 1)$th term

WRITING

37. Tell how to find the coefficients in the expansion of $(x + y)^5$.

38. Explain why the signs alternate in the expansion of $(x - y)^9$.

SOMETHING TO THINK ABOUT

39. Find the constant term in the expansion of $\left(x + \frac{1}{x}\right)^{10}$.

40. Find the coefficient of a^5 in the expansion of $\left(a - \frac{1}{a}\right)^9$.

14.3 Arithmetic Sequences

■ SEQUENCES ■ ARITHMETIC SEQUENCES ■ ARITHMETIC MEANS ■ THE SUM OF THE FIRST n TERMS OF AN ARITHMETIC SEQUENCE ■ SUMMATION NOTATION

Getting Ready *Complete each table.*

1.

n	$2n + 1$
1	
2	
3	
4	

2.

n	$3n - 5$
3	
4	
5	
6	

■ SEQUENCES

A **sequence** is a function whose domain is the set of natural numbers. For example, the function $f(n) = 3n + 2$, where n is a natural number, is a sequence. Because a sequence is a function whose domain is the set of natural numbers, we can write its values as a list. If the natural numbers are substituted for n, the function $f(n) = 3n + 2$ generates the list

$$5, 8, 11, 14, 17, \ldots$$

It is common to call the list, as well as the function, a sequence. Each number in the list is called a **term** of the sequence. Other examples of sequences are

$1^3, 2^3, 3^3, 4^3, \ldots$	The ordered list of the cubes of the natural numbers.
$4, 8, 12, 16, \ldots$	The ordered list of the positive multiples of 4.
$2, 3, 5, 7, 11, \ldots$	The ordered list of prime numbers.
$1, 1, 2, 3, 5, 8, 13, 21, \ldots$	The Fibonacci sequence.

The **Fibonacci sequence** is named after the 12th-century mathematician Leonardo of Pisa—also known as Fibonacci. Beginning with the 2, each term of the sequence is the sum of the two preceding terms.

■ ARITHMETIC SEQUENCES

One common type of sequence is the **arithmetic sequence.**

> **Arithmetic Sequence**
> An **arithmetic sequence** is a sequence of the form
> $$a, a + d, a + 2d, a + 3d, \ldots, a + (n - 1)d, \ldots$$
> where a is the **first term**, $a + (n - 1)d$ is the **nth term,** and d is the **common difference.**

We note that the second term of an arithmetic sequence has an addend of $1d$, the third term has an addend of $2d$, the fourth term has an addend of $3d$, and the nth term has an addend of $(n-1)d$. We also note that the difference between any two consecutive terms in an arithmetic sequence is d.

EXAMPLE 1   An arithmetic sequence has a first term of 5 and a common difference of 4.

   a. Write the first six terms of the sequence.

   b. Write the 25th term of the sequence.

Solution   **a.** Since the first term is $a = 5$ and the common difference is $d = 4$, the first six terms are

$$5, \quad 5 + 4, \quad 5 + 2(4), \quad 5 + 3(4), \quad 5 + 4(4), \quad 5 + 5(4)$$

or

$$5, 9, 13, 17, 21, 25$$

   b. The nth term is $a + (n-1)d$. Since we want the 25th term, we let $n = 25$:

$$
\begin{aligned}
n\text{th term} &= a + (n-1)d \\
25\text{th term} &= 5 + (25-1)4 \qquad \text{Remember that } a = 5 \text{ and } d = 4. \\
&= 5 + 24(4) \\
&= 5 + 96 \\
&= 101
\end{aligned}
$$

Self Check   See Example 1.   **a.** Write the seventh term of the sequence.   **b.** Write the 30th term of the sequence.

Answers   **a.** 29,   **b.** 121

EXAMPLE 2   The first three terms of an arithmetic sequence are 3, 8, and 13. Find   **a.** the 67th term   and   **b.** the 100th term.

Solution   We first find d, the common difference. It is the difference between two successive terms:

$$d = 8 - 3 = 13 - 8 = 5$$

   a. We substitute 3 for a, 67 for n, and 5 for d in the formula for the nth term and simplify:

$$
\begin{aligned}
n\text{th term} &= a + (n-1)d \\
67\text{th term} &= 3 + (67-1)5 \\
&= 3 + 66(5) \\
&= 333
\end{aligned}
$$

b. We substitute 3 for a, 100 for n, and 5 for d in the formula for the nth term, and simplify:

$$\begin{aligned}
n\text{th term} &= a + (n - 1)d \\
100\text{th term} &= 3 + (100 - 1)5 \\
&= 3 + 99(5) \\
&= 498
\end{aligned}$$

■

Self Check   See Example 2. Find the 50th term.

Answer   248

EXAMPLE 3   The first term of an arithmetic sequence is 12, and the 50th term is 3,099. Write the first six terms of the sequence.

Solution   The key is to find the common difference. Because the 50th term of this sequence is 3,099, we can let $n = 50$ and solve the following equation for d:

$$\begin{aligned}
50\text{th term} &= a + (n - 1)d \\
3{,}099 &= 12 + (50 - 1)d \\
3{,}099 &= 12 + 49d && \text{Simplify.} \\
3{,}087 &= 49d && \text{Subtract 12 from both sides.} \\
63 &= d && \text{Divide both sides by 49.}
\end{aligned}$$

The first term of the sequence is 12, and the common difference is 63. Its first six terms are

12, 75, 138, 201, 264, 327

■

Self Check   The first term of an arithmetic sequence is 15, and the 12th term is 92. Write the first four terms of the sequence.

Answer   15, 22, 29, 36

■ **ARITHMETIC MEANS**

If numbers are inserted between two numbers a and b to form an arithmetic sequence, the inserted numbers are called **arithmetic means** between a and b.

If a single number is inserted between the numbers a and b to form an arithmetic sequence, that number is called **the arithmetic mean** between a and b.

EXAMPLE 4   Insert two arithmetic means between 6 and 27.

Solution   Here, the first term is $a = 6$, and the fourth term (or the last term) is $l = 27$. We must find the common difference so that the terms

$$6, \ \mathbf{6 + d}, \ \mathbf{6 + 2d}, \ 27$$

form an arithmetic sequence. To find d, we substitute 6 for a and 4 for n into the formula for the nth term:

$$\mathbf{n\text{th term}} = a + (n - 1)d$$
$$\mathbf{4\text{th term}} = 6 + (4 - 1)d$$

$27 = 6 + 3d$	Simplify.
$21 = 3d$	Subtract 6 from both sides.
$7 = d$	Divide both sides by 3.

The two arithmetic means between 6 and 27 are

$$\mathbf{6 + d} = 6 + 7 \quad \text{or} \quad \mathbf{6 + 2d} = 6 + 2(7)$$
$$= \mathbf{13} \qquad\qquad\qquad = 6 + 14$$
$$\qquad\qquad\qquad\qquad = \mathbf{20}$$

The numbers 6, 13, 20, and 27 are the first four terms of an arithmetic sequence.

■

Self Check   Insert two arithmetic means between 8 and 44.
Answer   20, 32

■ THE SUM OF THE FIRST n TERMS OF AN ARITHMETIC SEQUENCE

There is a formula that gives the sum of the first n terms of an arithmetic sequence. To develop this formula, we let S_n represent the sum of the first n terms of an arithmetic sequence:

$$S_n = a + [a + d] + [a + 2d] + \cdots + [a + (n - 1)d]$$

We write the same sum again, but in reverse order:

$$S_n = [a + (n - 1)d] + [a + (n - 2)d] + [a + (n - 3)d] + \cdots + a$$

We add these two equations together, term by term, to get

$$2S_n = [2a + (n - 1)d] + [2a + (n - 1)d] + [2a + (n - 1)d] + \cdots + [2a + (n - 1)d]$$

Because there are n equal terms on the right-hand side of the preceding equation, we can write

$$2S_n = n[2a + (n - 1)d]$$
$$2S_n = n[a + a + (n - 1)d]$$
$$2S_n = n[a + l] \qquad \text{Substitute } l \text{ for } a + (n - 1)d, \text{ because}$$
$$a + (n - 1)d \text{ is the last term of the sequence.}$$

$$S_n = \frac{n(a + l)}{2}$$

This reasoning establishes the following theorem.

> **Sum of the First *n* Terms of an Arithmetic Sequence**
> The sum of the first n terms of an arithmetic sequence is given by the formula
> $$S_n = \frac{n(a + l)}{2} \quad \text{with } l = a + (n - 1)d$$
> where a is the first term, l is the last (or nth) term, and n is the number of the terms in the sequence.

EXAMPLE 5 Find the sum of the first 40 terms of the arithmetic sequence 4, 10, 16, ...

Solution In this example, we let $a = 4$, $n = 40$, $d = 6$, and $l = 4 + (40 - 1)6 = 238$ and substitute these values into the formula for S_n:

$$S_n = \frac{n(a + l)}{2}$$
$$S_{40} = \frac{40(4 + 238)}{2}$$
$$= 20(242)$$
$$= 4,840$$

The sum of the first 40 terms is 4,840. ∎

Self Check Find the sum of the first 50 terms of the arithmetic sequence 3, 8, 13, ...
Answer 6,275

■ SUMMATION NOTATION

There is a shorthand notation for indicating the sum of a finite (ending) number of consecutive terms in a sequence. This notation, called **summation notation**, involves the Greek letter Σ (sigma). The expression

$$\sum_{k=2}^{5} 3k \qquad \text{Read as "the summation of } 3k \text{ as } k \text{ runs from 2 to 5."}$$

designates the sum of all terms obtained if we successively substitute the numbers 2, 3, 4, and 5 for k, called the **index of the summation**. Thus, we have

$$\sum_{k=2}^{5} 3k = 3(2) + 3(3) + 3(4) + 3(5)$$

$$= 6 + 9 + 12 + 15$$

$$= 42$$

EXAMPLE 6   Find each sum: **a.** $\displaystyle\sum_{k=3}^{5}(2k + 1)$,   **b.** $\displaystyle\sum_{k=2}^{5}k^2$,   and   **c.** $\displaystyle\sum_{k=1}^{3}(3k^2 + 3)$.

Solution   **a.** $\displaystyle\sum_{k=3}^{5}(2k + 1) = [2(3) + 1] + [2(4) + 1] + [2(5) + 1]$

$$= 7 + 9 + 11$$

$$= 27$$

b. $\displaystyle\sum_{k=2}^{5}k^2 = 2^2 + 3^2 + 4^2 + 5^2$

$$= 4 + 9 + 16 + 25$$

$$= 54$$

c. $\displaystyle\sum_{k=1}^{3}(3k^2 + 3) = [3(1^2) + 3] + [3(2^2) + 3] + [3(3^2) + 3]$

$$= 6 + 15 + 30$$

$$= 51$$ ∎

Self Check   Evaluate $\displaystyle\sum_{k=1}^{4}(2k^2 - 2)$.

Answer   52

Orals   *Find the next term in each arithmetic sequence.*

1. 2, 6, 10, . . .   **2.** 10, 7, 4, . . .

Find the common difference in each arithmetic sequence.

3. −2, 3, 8, . . .   **4.** 5, −1, −7, . . .

Find each sum.

5. $\displaystyle\sum_{k=1}^{2}k$ **6.** $\displaystyle\sum_{k=2}^{3}k$

EXERCISE 14.3

REVIEW *Do the operations and simplify, if possible.*

1. $3(2x^2 - 4x + 7) + 4(3x^2 + 5x - 6)$

2. $(2p + q)(3p^2 + 4pq - 3q^2)$

3. $\dfrac{3a + 4}{a - 2} + \dfrac{3a - 4}{a + 2}$

4. $2t - 3\overline{)8t^4 - 12t^3 + 8t^2 - 16t + 6}$

VOCABULARY AND CONCEPTS *Fill in each blank to make a true statement.*

5. A _____ is a function whose domain is the set of natural numbers.

6. The sequence 1, 1, 2, 3, 5, 8, 13, 21, . . . is called the _____ sequence.

7. The sequence 3, 9, 15, 21, . . . is an example of an _____ sequence with a common _____ of 6.

8. The last term of an arithmetic sequence is given by the formula _____.

9. If a number c is inserted between two numbers a and b to form an arithmetic sequence, then c is called the _____ between a and b.

10. The sum of the first n terms of an arithmetic sequence is given by the formula $S_n =$ _____.

11. $\displaystyle\sum_{k=1}^{5} k$ means _____.

12. In the symbol $\displaystyle\sum_{k=1}^{5}(2k - 5)$, k is called the _____ of the summation.

PRACTICE *In Exercises 13–26, write the first five terms of each arithmetic sequence with the given properties.*

13. $a = 3, d = 2$

14. $a = -2, d = 3$

15. $a = -5, d = -3$

16. $a = 8, d = -5$

17. $a = 5$, fifth term is 29

18. $a = 4$, sixth term is 39

19. $a = -4$, sixth term is -39

20. $a = -5$, fifth term is -37

21. $d = 7$, sixth term is -83

22. $d = 3$, seventh term is 12

23. $d = -3$, seventh term is 16

24. $d = -5$, seventh term is -12

25. The 19th term is 131 and the 20th term is 138.

26. The 16th term is 70 and the 18th term is 78.

27. Find the 30th term of the arithmetic sequence with $a = 7$ and $d = 12$.

28. Find the 55th term of the arithmetic sequence with $a = -5$ and $d = 4$.

29. Find the 37th term of the arithmetic sequence with a second term of -4 and a third term of -9.

30. Find the 40th term of the arithmetic sequence with a second term of 6 and a fourth term of 16.

31. Find the first term of the arithmetic sequence with a common difference of 11 and whose 27th term is 263.

32. Find the common difference of the arithmetic sequence with a first term of -164 if its 36th term is -24.

33. Find the common difference of the arithmetic sequence with a first term of 40 if its 44th term is 556.

34. Find the first term of the arithmetic sequence with a common difference of -5 and whose 23rd term is -625.

35. Insert three arithmetic means between 2 and 11.

36. Insert four arithmetic means between 5 and 25.

37. Insert four arithmetic means between 10 and 20.

38. Insert three arithmetic means between 20 and 30.

39. Find the arithmetic mean between 10 and 19.

40. Find the arithmetic mean between 5 and 23.

41. Find the arithmetic mean between -4.5 and 7.

42. Find the arithmetic mean between -6.3 and -5.2.

In Exercises 43–50, find the sum of the first n terms of each arithmetic sequence.

43. 1, 4, 7, . . . ; $n = 30$

44. 2, 6, 10, . . . ; $n = 28$

45. $-5, -1, 3, . . . ; n = 17$

46. $-7, -1, 5, . . . ; n = 15$

47. Second term is 7, third term is 12; $n = 12$

48. Second term is 5, fourth term is 9; $n = 16$

49. $f(n) = 2n + 1$, nth term is 31; n is a natural number

50. $f(n) = 4n + 3$, nth term is 23; n is a natural number

51. Find the sum of the first 50 natural numbers.

52. Find the sum of the first 100 natural numbers.

53. Find the sum of the first 50 odd natural numbers.

54. Find the sum of the first 50 even natural numbers.

In Exercises 55–60, find each sum.

55. $\displaystyle\sum_{k=1}^{4} 6k$

56. $\displaystyle\sum_{k=2}^{5} 3k$

57. $\displaystyle\sum_{k=3}^{4} (k^2 + 3)$

58. $\displaystyle\sum_{k=2}^{6} (k^2 + 1)$

59. $\displaystyle\sum_{k=4}^{4} (2k + 4)$

60. $\displaystyle\sum_{k=3}^{5} (3k^2 - 7)$

APPLICATIONS

61. Saving money Yasmeen puts $60 into a safety deposit box. Each month, she puts $50 more in the box. Write the first six terms of an arithmetic sequence that gives the monthly amounts in her savings, and find her savings after 10 years.

62. Installment loan Maria borrowed $10,000, interest-free, from her mother. She agreed to pay back the loan in monthly installments of $275. Write the first six terms of an arithmetic sequence that shows the balance due after each month, and find the balance due after 17 months.

63. Designing a patio Each row of bricks in the triangular patio in Illustration 1 (on the next page) is to have one more brick than the previous row, ending with the longest row of 150 bricks. How many bricks will be needed?

ILLUSTRATION 1

64. Falling object The equation $s = 16t^2$ represents the distance s in feet that an object will fall in t seconds. After 1 second, the object has fallen 16 feet. After 2 seconds, the object has fallen 64 feet, and so on. Find the distance that the object will fall during the second and third seconds.

65. Falling object Refer to Exercise 64. How far will the object fall during the 12th second?

66. Interior angles The sum of the angles of several polygons are given in the table shown in Illustration 2. Assuming that the pattern continues, complete the table.

Figure	Number of sides	Sum of angles
Triangle	3	180°
Quadrilateral	4	360°
Pentagon	5	540°
Hexagon	6	720°
Octagon	8	
Dodecagon	12	

ILLUSTRATION 2

WRITING

67. Define an arithmetic sequence.

68. Develop the formula for finding the sum of the first n terms of an arithmetic sequence.

SOMETHING TO THINK ABOUT

69. Write the first 6 terms of the arithmetic sequence given by

$$\sum_{n=1}^{k} \left(\frac{1}{2}n + 1 \right)$$

70. Find the sum of the first 6 terms of the sequence given in Exercise 69.

71. Show that the arithmetic mean between a and b is the average of a and b: $\dfrac{a + b}{2}$

72. Show that the sum of the two arithmetic means between a and b is $a + b$.

73. Show that $\displaystyle\sum_{k=1}^{5} 5k = 5 \sum_{k=1}^{5} k$.

74. Show that $\displaystyle\sum_{k=3}^{6} (k^2 + 3k) = \sum_{k=3}^{6} k^2 + \sum_{k=3}^{6} 3k$.

75. Show that $\displaystyle\sum_{k=1}^{n} 3 = 3n$. (*Hint:* Consider 3 to be $3k^0$.)

76. Show that $\displaystyle\sum_{k=1}^{3} \frac{k^2}{k} \neq \frac{\displaystyle\sum_{k=1}^{3} k^2}{\displaystyle\sum_{k=1}^{3} k}$.

14.4 Geometric Sequences

■ GEOMETRIC SEQUENCES ■ GEOMETRIC MEANS ■ THE SUM OF THE FIRST n TERMS OF A GEOMETRIC SEQUENCE ■ POPULATION GROWTH

Getting Ready   *Complete each table.*

1.

n	$5(2^n)$
1	
2	
3	

2.

n	$6(3^n)$
1	
2	
3	

■ GEOMETRIC SEQUENCES

Another common type of sequence is called a **geometric sequence**.

> **Geometric Sequence**
> A **geometric sequence** is a sequence of the form
> $$a, ar, ar^2, ar^3, \ldots, ar^{n-1}, \ldots$$
> where a is the **first term,** ar^{n-1} is the **nth term,** and r is the **common ratio.**

We note that the second term of a geometric sequence has a factor of r^1, the third term has a factor of r^2, the fourth term has a factor of r^3, and the nth term has a factor of r^{n-1}. We also note that the quotient obtained when any term is divided by the previous term is r.

EXAMPLE 1   A geometric sequence has a first term of 5 and a common ratio of 3.

a. Write the first five terms of the sequence.

b. Find the ninth term.

Solution   **a.** Because the first term is $a = 5$ and the common ratio is $r = 3$, the first five terms are

$$5, \quad 5(3), \quad 5(3^2), \quad 5(3^3), \quad 5(3^4)$$

or

$$5, 15, 45, 135, 405$$

b. The nth term is ar^{n-1} where $a = 5$ and $r = 3$. Because we want the ninth term, we let $n = 9$:

$$n\text{th term} = ar^{n-1}$$
$$9\text{th term} = 5(3)^{9-1}$$
$$= 5(3)^8$$
$$= 5(6{,}561)$$
$$= 32{,}805$$

■

Self Check   A geometric sequence has a first term of 3 and a common ratio of 4.   **a.** Write the first four terms.   **b.** Find the eighth term.

Answers   **a.** 3, 12, 48, 192,   **b.** 49,152

EXAMPLE 2   The first three terms of a geometric sequence are 16, 4, and 1. Find the seventh term.

Solution   We substitute 16 for a, $\frac{1}{4}$ for r, and 7 for n in the formula for the nth term and simplify:

$$n\text{th term} = ar^{n-1}$$
$$7\text{th term} = 16\left(\frac{1}{4}\right)^{7-1}$$
$$= 16\left(\frac{1}{4}\right)^{6}$$
$$= 16\left(\frac{1}{4{,}096}\right)$$
$$= \frac{1}{256}$$

■

Self Check   See Example 2. Find the tenth term.

Answer $\frac{1}{16{,}384}$

■ GEOMETRIC MEANS

If numbers are inserted between two numbers a and b to form a geometric sequence, the inserted numbers are called **geometric means** between a and b.

If a single number is inserted between the numbers a and b to form a geometric sequence, that number is called a **geometric mean** between a and b.

EXAMPLE 3   Insert two geometric means between 7 and 1,512.

Solution   Here the first term is $a = 7$, and the fourth term (or last term) is $l = 1,512$. To find the common ratio r so that the terms

$$7, \mathbf{7r}, \mathbf{7r^2}, 1,512$$

form a geometric sequence, we substitute 4 for n and 7 for a into the formula for the nth term of a geometric sequence and solve for r.

$$\textbf{\textit{n}th term} = ar^{n-1}$$
$$\textbf{4th term} = 7r^{4-1}$$
$$1,512 = 7r^3$$
$$216 = r^3 \qquad \text{Divide both sides by 7.}$$
$$6 = r \qquad \text{Take the cube root of both sides.}$$

The two geometric means between 7 and 1,512 are

$$\mathbf{7r} = 7(6) = \mathbf{42}$$

and

$$\mathbf{7r^2} = 7(6)^2 = 7(36) = \mathbf{252}$$

The numbers 7, 42, 252, and 1,512 are the first four terms of a geometric sequence.   ■

Self Check   Insert three positive geometric means between 1 and 16.

Answer   2, 4, 8

EXAMPLE 4   Find a geometric mean between 2 and 20.

Solution   We want to find the middle term of the three-term geometric sequence

$$2, \mathbf{2r}, 20$$

with $a = 2$, $l = 20$, and $n = 3$. To find r, we substitute these values into the formula for the nth term of a geometric sequence:

$$\textbf{\textit{n}th term} = ar^{n-1}$$
$$\textbf{3rd term} = 2r^{3-1}$$
$$20 = 2r^2$$
$$10 = r^2 \qquad \text{Divide both sides by 2.}$$
$$\pm\sqrt{10} = r \qquad \text{Take the square root of both sides.}$$

Because r can be either $\sqrt{10}$ or $-\sqrt{10}$, there are two values for the geometric mean. They are

$$2r = 2\sqrt{10}$$

and

$$2r = -2\sqrt{10}$$

The numbers 2, $2\sqrt{10}$, 20 and 2, $-2\sqrt{10}$, 20 both form geometric sequences. The common ratio of the first sequence is $\sqrt{10}$, and the common ratio of the second sequence is $-\sqrt{10}$.   ■

Self Check

Answer

Find the positive geometric mean between 2 and 200.

20

■ THE SUM OF THE FIRST n TERMS OF A GEOMETRIC SEQUENCE

There is a formula that gives the sum of the first n terms of a geometric sequence. To develop this formula, we let S_n represent the sum of the first n terms of a geometric sequence.

1. $S_n = a + ar + ar^2 + ar^3 + \ldots + ar^{n-1}$

We multiply both sides of Equation 1 by r to get

2. $S_n r = \quad ar + ar^2 + ar^3 + \ldots + ar^{n-1} + ar^n$

We now subtract Equation 2 from Equation 1 and solve for S_n:

$$S_n - S_n r = a - ar^n$$
$$S_n(1 - r) = a - ar^n \qquad \text{Factor out } S_n \text{ from the left-hand side.}$$
$$S_n = \frac{a - ar^n}{1 - r} \qquad \text{Divide both sides by } 1 - r.$$

This reasoning establishes the following formula.

Sum of the First n Terms of a Geometric Sequence
The sum of the first n terms of a geometric sequence is given by the formula

$$S_n = \frac{a - ar^n}{1 - r} \quad (r \neq 1)$$

where S_n is the sum, a is the first term, r is the common ratio, and n is the number of terms.

EXAMPLE 5   Find the sum of the first six terms of the geometric sequence 250, 50, 10, ...

Solution   Here $a = 250$, $r = \frac{1}{5}$, and $n = 6$. We substitute these values into the formula for the sum of the first n terms of a geometric sequence and simplify:

$$S_n = \frac{a - ar^n}{1 - r}$$

$$S_6 = \frac{250 - 250\left(\frac{1}{5}\right)^6}{1 - \frac{1}{5}}$$

$$= \frac{250 - 250\left(\frac{1}{15,625}\right)}{\frac{4}{5}}$$

$$= \frac{5}{4}\left(250 - \frac{250}{15,625}\right)$$

$$= \frac{5}{4}\left(\frac{3,906,000}{15,625}\right)$$

$$= 312.48$$

The sum of the first six terms is 312.48.   ∎

Self Check   Find the sum of the first five terms of the geometric sequence 100, 20, 4, ...
Answer   124.96

■ POPULATION GROWTH

EXAMPLE 6   **Growth of a town**   The mayor of Eagle River (population 1,500) predicts a growth rate of 4% each year for the next ten years. Find the expected population of Eagle River ten years from now.

Solution   Let P_0 be the initial population of Eagle River. After 1 year, there will be a different population, P_1. The initial population (P_0) plus the growth (the product of P_0 and the rate of growth, r) will equal this new population, P_1:

$$P_1 = P_0 + P_0r = P_0(1 + r)$$

The population after 2 years will be P_2, and

$$P_2 = P_1 + P_1r$$
$$= P_1(1 + r) \qquad \text{Factor out } P_1.$$
$$= P_0(1 + r)(1 + r) \qquad \text{Remember that } P_1 = P_0(1 + r).$$
$$= P_0(1 + r)^2$$

The population after 3 years will be P_3, and

$$P_3 = P_2 + P_2 r$$
$$\quad = P_2(1 + r) \qquad \text{Factor out } P_2.$$
$$\quad = P_0(1 + r)^2(1 + r) \qquad \text{Remember that } P_2 = P_0(1 + r)^2.$$
$$\quad = P_0(1 + r)^3$$

The yearly population figures

$$P_0, \quad P_1, \quad P_2, \quad P_3, \ldots$$

or

$$P_0, \quad P_0(1 + r), \quad P_0(1 + r)^2, \quad P_0(1 + r)^3, \ldots$$

form a geometric sequence with a first term of P_0 and a common ratio of $1 + r$. The population of Eagle River after 10 years is P_{10}, which is the 11th term of this sequence:

$$n\text{th term} = ar^{n-1}$$
$$P_{10} = 11\text{th term} = P_0(1 + r)^{10}$$
$$= 1,500(1 + 0.04)^{10}$$
$$= 1,500(1.04)^{10}$$
$$\approx 1,500(1.480244285)$$
$$\approx 2,220$$

The expected population ten years from now is 2,220 people. ∎

EXAMPLE 7

Amount of an annuity An **annuity** is a sequence of equal payments made periodically over a length of time. The sum of the payments and the interest earned during the *term* of the annuity is called the *amount* of the annuity.

After a sales clerk works six months, her employer will begin an annuity for her and will contribute $500 every six months to a fund that pays 8% annual interest. After she has been employed for two years, what will be the amount of the annuity?

Solution   Because the payments are to be made semiannually, there will be four payments of $500, each earning a rate of 4% per six-month period. These payments will occur at the end of 6 months, 12 months, 18 months, and 24 months. The first payment, to be made after 6 months, will earn interest for three interest periods. Thus, the amount of the first payment is $500(1.04)^3$. The amounts of each of the four payments after two years are shown in Figure 14-1.

The amount of the annuity is the sum of the amounts of the individual payments, a sum of $2,123.23.

Payment (at the end of period)	Amount of payment at the end of 2 years
1	$500(1.04)^3 = \$\ 562.43$
2	$500(1.04)^2 = \$\ 540.80$
3	$500(1.04)^1 = \$\ 520.00$
4	$\$500 = \$\ 500.00$
	$A_n = \$2{,}123.23$

FIGURE 14-1

Orals *Find the next term in each geometric sequence.*

1. 1, 3, 9, . . .   **2.** $1, \dfrac{1}{3}, \dfrac{1}{9}, \ldots$

Find the common ratio in each geometric sequence.

3. 0.2, 0.5, 1.25, . . .   **4.** $\sqrt{3}, 3, 3\sqrt{3}, \ldots$

Find x in each geometric sequence.

5. 2, x, 18, 54, . . .   **6.** $3, x, \dfrac{1}{3}, \dfrac{1}{9}, \ldots$

EXERCISE 14.4

REVIEW *Solve each inequality.*

1. $x^2 - 5x - 6 \leq 0$ **2.** $a^2 - 7a + 12 \geq 0$ **3.** $\dfrac{x-4}{x+3} > 0$ **4.** $\dfrac{t^2 + t - 20}{t + 2} < 0$

VOCABULARY AND CONCEPTS *Fill in each blank to make a true statement.*

5. A sequence of the form $a, ar, ar^2, \ldots$ is called a _____ sequence.

6. The formula for the nth term of a geometric sequence is _____.

7. In a geometric sequence, r is called the _____.

8. A number inserted between two numbers a and b to form a geometric sequence is called a geometric _____ between a and b.

9. The sum of the first n terms of a geometric sequence is given by the formula _____

10. In the formula for Exercise 9, a is the ____ term of the sequence.

PRACTICE *In Exercises 11–24, write the first five terms of each geometric sequence with the given properties.*

11. $a = 3, r = 2$

12. $a = -2, r = 2$

13. $a = -5, r = \dfrac{1}{5}$

14. $a = 8, r = \dfrac{1}{2}$

15. $a = 2, r > 0$, third term is 32

16. $a = 3$, fourth term is 24

17. $a = -3$, fourth term is -192

18. $a = 2, r < 0$, third term is 50

19. $a = -64, r < 0$, fifth term is -4

20. $a = -64, r > 0$, fifth term is -4

21. $a = -64$, sixth term is -2

22. $a = -81$, sixth term is $\dfrac{1}{3}$

23. The second term is 10, and the third term is 50.

24. The third term is -27, and the fourth term is 81.

25. Find the tenth term of the geometric sequence with $a = 7$ and $r = 2$.

26. Find the 12th term of the geometric sequence with $a = 64$ and $r = \frac{1}{2}$.

27. Find the first term of the geometric sequence with a common ratio of -3 and an eighth term of -81.

28. Find the first term of the geometric sequence with a common ratio of 2 and a tenth term of 384.

29. Find the common ratio of the geometric sequence with a first term of -8 and a sixth term of $-1,944$.

30. Find the common ratio of the geometric sequence with a first term of 12 and a sixth term of $\frac{3}{8}$.

31. Insert three positive geometric means between 2 and 162.

32. Insert four geometric means between 3 and 96.

33. Insert four geometric means between -4 and $-12,500$.

34. Insert three geometric means (two positive and one negative) between -64 and $-1,024$.

35. Find the negative geometric mean between 2 and 128.

36. Find the positive geometric mean between 3 and 243.

37. Find the positive geometric mean between 10 and 20.

38. Find the negative geometric mean between 5 and 15.

39. Find a geometric mean, if possible, between -50 and 10.

40. Find a negative geometric mean, if possible, between -25 and -5.

In Exercises 41–52, find the sum of the first n terms of each geometric sequence.

41. $2, 6, 18, \ldots ; n = 6$

42. $2, -6, 18, \ldots ; n = 6$

43. $2, -6, 18, \ldots ; n = 5$

44. $2, 6, 18, \ldots ; n = 5$

45. $3, -6, 12, \ldots ; n = 8$

46. $3, 6, 12, \ldots ; n = 8$

47. $3, 6, 12, \ldots ; n = 7$

48. $3, -6, 12, \ldots ; n = 7$

49. The second term is 1, and the third term is $\frac{1}{5}$; $n = 4$.

50. The second term is 1, and the third term is 4; $n = 5$.

51. The third term is -2, and the fourth term is 1; $n = 6$.

52. The third term is -3, and the fourth term is 1; $n = 5$.

APPLICATIONS   *In Exercises 53–62, use a calculator to help solve each problem.*

53. Population growth   The population of Union is predicted to increase by 6% each year. What will be the population of Union 5 years from now if its current population is 500?

54. Population decline   The population of Bensville is decreasing by 10% each year. If its current population is 98, what will be the population 8 years from now?

55. Declining savings   John has $10,000 in a safety deposit box. Each year he spends 12% of what is left in the box. How much will be in the box after 15 years?

56. Savings growth   Sally has $5,000 in a savings account earning 12% annual interest. How much will be in her account 10 years from now? (Assume that Sally makes no deposits or withdrawals.)

57. House appreciation   A house appreciates by 6% each year. If the house is worth $70,000 today, how much will it be worth 12 years from now?

58. Motorboat depreciation   A motorboat that cost $5,000 when new depreciates at a rate of 9% per year. How much will the boat be worth in 5 years?

59. Inscribed squares   Each inscribed square in Illustration 1 joins the midpoints of the next larger square. The area of the first square, the largest, is 1. Find the area of the 12th square.

60. Genealogy   The family tree in Illustration 2 spans 3 generations and lists 7 people. How many names would be listed in a family tree that spans 10 generations?

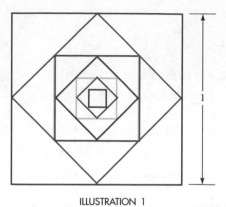

ILLUSTRATION 1

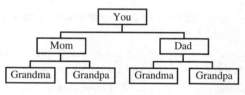

ILLUSTRATION 2

61. Annuities   Find the amount of an annuity if $1,000 is paid semiannually for two years at 6% annual interest. Assume that the first of the four payments is made immediately.

62. Annuities   Note that the amounts shown in Figure 14-1 form a geometric sequence. Verify the answer for Example 7 by using the formula for the sum of a geometric sequence.

WRITING

63. Define a geometric sequence.

64. Develop the formula for finding the sum of the first n terms of a geometric sequence.

SOMETHING TO THINK ABOUT

65. Show that the formula for the sum of the first n terms of a geometric sequence can be found by using the formula $S = \frac{a - lr}{1 - r}$.

66. Show that the geometric mean between a and b is $\sqrt{ab}$.

67. If $a > b > 0$, which is larger: the arithmetic mean between a and b or the geometric mean between a and b?

68. Is there a geometric mean between -5 and 5?

69. Show that the formula for the sum of the first n terms of a geometric sequence can be written in the form

$$S_n = \frac{lr - a}{r - 1} \quad \text{where } l = ar^{n-1}$$

70. Show that the formula for the sum of the first n terms of a geometric sequence can be written in the form

$$S_n = \frac{a(1 - r^n)}{1 - r}$$

14.5 Infinite Geometric Sequences

■ INFINITE GEOMETRIC SEQUENCES ■ THE SUM OF AN INFINITE GEOMETRIC SEQUENCE

Getting Ready *Evaluate each expression.*

1. $\dfrac{2}{1 - \dfrac{1}{2}}$ **2.** $\dfrac{3}{1 - \dfrac{1}{3}}$ **3.** $\dfrac{\dfrac{3}{2}}{1 - \dfrac{1}{2}}$ **4.** $\dfrac{\dfrac{5}{3}}{1 - \dfrac{1}{3}}$

■ INFINITE GEOMETRIC SEQUENCES

An **infinite geometric sequence** is a geometric sequence with infinitely many terms. Two examples of infinite geometric sequences are

$$2, 6, 18, 54, 162, \ldots \qquad (r = 3)$$

$$\frac{3}{2}, \frac{3}{4}, \frac{3}{8}, \frac{3}{16}, \frac{3}{32}, \ldots \qquad \left(r = \frac{1}{2}\right)$$

■ THE SUM OF AN INFINITE GEOMETRIC SEQUENCE

Under certain conditions, we can find the sum of all the terms of an infinite geometric sequence. To define this sum, we consider the geometric sequence

$$a, ar, ar^2, ar^3, \ldots, ar^{n-1}, \ldots$$

- The first **partial sum,** S_1, of the sequence is $S_1 = a$.
- The second partial sum, S_2, of the sequence is $S_2 = a + ar$.
- The third partial sum, S_3, of the sequence is $S_3 = a + ar + ar^2$.
- The nth partial sum, S_n, of the sequence is $S_n = a + ar + ar^2 + \cdots + ar^{n-1}$.

If the nth partial sum, S_n, approaches some number S as n approaches infinity, then S is called the **sum of the infinite geometric sequence**.

To develop a formula for finding the sum (if it exists) of all the terms in an infinite geometric sequence, we consider the formula that gives the sum of the first n terms of a geometric sequence.

$$S_n = \frac{a - ar^n}{1 - r} \quad (r \neq 1)$$

If $|r| < 1$ and a is constant, then the term ar^n in the above formula approaches 0 as n becomes very large. For example,

$$a\left(\frac{1}{4}\right)^1 = \frac{1}{4}a, \quad a\left(\frac{1}{4}\right)^2 = \frac{1}{16}a, \quad a\left(\frac{1}{4}\right)^3 = \frac{1}{64}a$$

and so on. When n is very large, the value of ar^n is negligible, and the term ar^n in the above formula can be ignored. This reasoning justifies the following theorem.

Sum of an Infinite Geometric Sequence

If a is the first term and r is the common ratio of an infinite geometric sequence, and if $|r| < 1$, the sum of the terms of the sequence is given by the formula

$$S = \frac{a}{1 - r}$$

EXAMPLE 1   Find the sum of the terms of the infinite geometric sequence 125, 25, 5, ...

Solution   Here $a = 125$ and $r = \frac{1}{5}$. Since $|r| = |\frac{1}{5}| = \frac{1}{5} < 1$, we can find the sum of all the terms of the sequence. We do this by substituting 125 for a and $\frac{1}{5}$ for r in the formula $S = \frac{a}{1-r}$ and simplifying:

$$S = \frac{a}{1-r} = \frac{\mathbf{125}}{1 - \dfrac{1}{5}} = \frac{125}{\dfrac{4}{5}} = \frac{5}{4}(125) = \frac{625}{4}$$

The sum of the terms of the sequence 125, 25, 5, . . . is 156.25. ∎

Self Check   Find the sum of the terms of the infinite geometric sequence 100, 20, 4, ...

Answer   125

EXAMPLE 2   Find the sum of the infinite geometric sequence 64, -4, $\frac{1}{4}$, ...

Solution   Here $a = 64$ and $r = -\frac{1}{16}$. Since $|r| = |-\frac{1}{16}| = \frac{1}{16} < 1$, we can find the sum of all the terms of the sequence. We substitute 64 for a and $-\frac{1}{16}$ for r in the formula $S = \frac{a}{1-r}$ and simplify:

$$S = \frac{a}{1-r} = \frac{\mathbf{64}}{1 - \left(-\dfrac{1}{16}\right)} = \frac{64}{\dfrac{17}{16}} = \frac{16}{17}(64) = \frac{1,024}{17}$$

The sum of the terms of the geometric sequence 64, -4, $\frac{1}{4}$, . . . is $\frac{1,024}{17}$. ∎

Self Check   Find the sum of the infinite geometric sequence 81, 27, 9, ...

Answer $\frac{243}{2}$

EXAMPLE 3   Change $0.\overline{8}$ to a common fraction.

Solution   The decimal $0.\overline{8}$ can be written as the sum of an infinite geometric sequence:

$$0.\overline{8} = 0.888 \ldots = \frac{8}{10} + \frac{8}{100} + \frac{8}{1,000} + \cdots$$

where $a = \frac{8}{10}$ and $r = \frac{1}{10}$. Because $|r| = \left|\frac{1}{10}\right| = \frac{1}{10} < 1$, we can find the sum as follows:

$$S = \frac{a}{1-r} = \frac{\dfrac{8}{10}}{1 - \dfrac{1}{10}} = \frac{\dfrac{8}{10}}{\dfrac{9}{10}} = \frac{8}{9}$$

Thus, $0.\overline{8} = \frac{8}{9}$. Long division will verify that $\frac{8}{9} = 0.888 \ldots$. ■

Self Check   Change $0.\overline{6}$ to a common fraction.
Answer $\frac{2}{3}$

EXAMPLE 4   Change $0.\overline{25}$ to a common fraction.

Solution   The decimal $0.\overline{25}$ can be written as the sum of an infinite geometric sequence:

$$0.\overline{25} = 0.252525 \ldots = \frac{25}{100} + \frac{25}{10,000} + \frac{25}{1,000,000} + \cdots$$

where $a = \frac{25}{100}$ and $r = \frac{1}{100}$. Since $|r| = \left|\frac{1}{100}\right| = \frac{1}{100} < 1$, we can find the sum as follows:

$$S = \frac{a}{1-r} = \frac{\dfrac{25}{100}}{1 - \dfrac{1}{100}} = \frac{\dfrac{25}{100}}{\dfrac{99}{100}} = \frac{25}{99}$$

Thus, $0.\overline{25} = \frac{25}{99}$. Long division will verify that this is true. ■

Self Check   Change $0.\overline{15}$ to a common fraction.
Answer $\frac{5}{33}$

Orals *Find the common ratio in each infinite geometric sequence.*

1. $\dfrac{1}{64}, \dfrac{1}{8}, 1, \ldots$ **2.** $1, \dfrac{1}{8}, \dfrac{1}{64}, \ldots$

3. $\dfrac{2}{3}, \dfrac{1}{3}, \dfrac{1}{6}, \ldots$ **4.** $64, 8, 1, \ldots$

Find the sum of the terms in each infinite geometric sequence.

5. $18, 6, 2, \ldots$ **6.** $12, 3, \dfrac{3}{4}, \ldots$

EXERCISE 14.5

REVIEW *Determine whether each equation determines y to be a function of x.*

1. $y = 3x^3 - 4$ **2.** $xy = 12$ **3.** $3x = y^2 + 4$ **4.** $x = |y|$

VOCABULARY AND CONCEPTS *Fill in each blank to make a true statement.*

5. If a geometric sequence has infinitely many terms, it is called an _____ geometric sequence.

6. The third partial sum of the sequence 2, 6, 18, 54, . . . is _____ = 26.

7. The formula for the sum of an infinite geometric sequence with $|r| < 1$ is _____.

8. Write $0.\overline{75}$ as the sum of an infinite geometric sequence. _____.

PRACTICE *In Exercises 9–20, find the sum of each infinite geometric sequence, if possible.*

9. $8, 4, 2, \ldots$ **10.** $12, 6, 3, \ldots$ **11.** $54, 18, 6, \ldots$

12. $45, 15, 5, \ldots$ **13.** $12, -6, 3, \ldots$ **14.** $8, -4, 2, \ldots$

15. $-45, 15, -5, \ldots$ **16.** $-54, 18, -6, \ldots$ **17.** $\dfrac{9}{2}, 6, 8, \ldots$

18. $-112, -28, -7, \ldots$ **19.** $-\dfrac{27}{2}, -9, -6, \ldots$ **20.** $\dfrac{18}{25}, \dfrac{6}{5}, 2, \ldots$

In Exercises 21–28, change each decimal to a common fraction. Then check the answer by doing a long division.

21. $0.\overline{1}$ **22.** $0.\overline{2}$ **23.** $-0.\overline{3}$ **24.** $-0.\overline{4}$

25. $0.\overline{12}$ **26.** $0.\overline{21}$ **27.** $0.\overline{75}$ **28.** $0.\overline{57}$

APPLICATIONS

29. Bouncing ball On each bounce, the rubber ball in Illustration 1 (on the next page) rebounds to a height one-half of that from which it fell. Find the total distance the ball travels.

30. Bouncing ball A golf ball is dropped from a height of 12 feet. On each bounce, it returns to a height two-thirds of that from which it fell. Find the total distance the ball travels.

10 m

ILLUSTRATION 1

31. Controlling moths To reduce the population of a destructive moth, biologists release 1,000 sterilized male moths each day into the environment. If 80% of these moths alive one day survive until the next, then after a long time, the population of sterile males is the sum of the infinite geometric sequence

$$1,000 + 1,000(0.8) + 1,000(0.8)^2 + 1,000(0.8)^3 + \cdots$$

Find the long-term population.

32. Controlling moths If mild weather increases the day-to-day survival rate of the sterile male moths in Exercise 31 to 90%, find the long-term population.

WRITING

33. Why must the absolute value of the common ratio be less than 1 before an infinite geometric sequence can have a sum?

34. Can an infinite arithmetic sequence have a sum?

SOMETHING TO THINK ABOUT

35. An infinite geometric sequence has a sum of 5 and a first term of 1. Find the common ratio.

36. An infinite geometric sequence has a common ratio of $-\frac{2}{3}$ and a sum of 9. Find the first term.

37. Show that $0.\overline{9} = 1$.

38. Show that $1.\overline{9} = 2$.

39. Does $0.999999 = 1$? Explain.

40. If $f(x) = 1 + x + x^2 + x^3 + x^4 + \cdots$, find $f\left(\frac{1}{2}\right)$ and $f\left(-\frac{1}{2}\right)$.

14.6 Permutations and Combinations

■ THE MULTIPLICATION PRINCIPLE FOR EVENTS ■ PERMUTATIONS ■ COMBINATIONS
■ ALTERNATIVE FORM OF THE BINOMIAL THEOREM

Getting Ready   *Evaluate each expression.*

1. $4 \cdot 3 \cdot 2 \cdot 1$

2. $5 \cdot 4 \cdot 3 \cdot 2 \cdot 1$

3. $\dfrac{6 \cdot 5 \cdot 4 \cdot 3 \cdot 2 \cdot 1}{4 \cdot 3 \cdot 2 \cdot 1}$

4. $\dfrac{8 \cdot 7 \cdot 6 \cdot 5 \cdot 4 \cdot 3 \cdot 2 \cdot 1}{2(5 \cdot 4 \cdot 3 \cdot 2 \cdot 1)}$

■ THE MULTIPLICATION PRINCIPLE FOR EVENTS

Steven goes to the cafeteria for lunch. He has a choice of three different sandwiches (hamburger, hot dog, or ham and cheese) and four different beverages (cola, root beer, orange, or milk). How many different lunches can he choose?

He has three choices of sandwich, and for any one of these choices, he has four choices of drink. The different options are shown in the *tree diagram* in Figure 14-2.

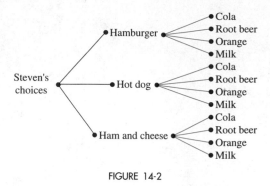

FIGURE 14-2

The tree diagram shows that there is a total of 12 different lunches to choose from. One of the possibilities is a hamburger with a cola, and another is a hot dog with milk.

A situation that can have several different outcomes—such as choosing a sandwich—is called an **event**. Choosing a sandwich and choosing a beverage can be thought of as two events. The preceding example illustrates the **multiplication principle for events.**

> **Multiplication Principle for Events**
> Let E_1 and E_2 be two events. If E_1 can be done in a_1 ways, and if—after E_1 has occurred—E_2 can be done in a_2 ways, the event "E_1 followed by E_2" can be done in $a_1 \cdot a_2$ ways.

EXAMPLE 1   Before studying for an exam, Heidi plans to watch the evening news and then a situation comedy on television. If she has a choice of four news broadcasts and two comedies, in how many ways can she choose to watch television?

Solution   Let E_1 be the event "watching the news" and E_2 be the event "watching a comedy." Because there are four ways to accomplish E_1 and two ways to accomplish E_2, the number of choices that she has is $4 \cdot 2 = 8$. ∎

Self Check   If Alex has 7 shirts and 5 pairs of pants, how many outfits could he wear?

Answer   35

The multiplication principle can be extended to any number of events. In Example 2, we use it to complete the number of ways that we can arrange objects in a row.

EXAMPLE 2   In how many ways can we arrange five books on a shelf?

Solution   We can fill the first space with any of the 5 books, the second space with any of the remaining 4 books, the third space with any of the remaining 3 books, the fourth space with any of the remaining 2 books, and the fifth space with the remaining 1 (or last) book. By the multiplication principle for events, the number of ways in which the books can be arranged is

$$5 \cdot 4 \cdot 3 \cdot 2 \cdot 1 = 120 \qquad\blacksquare$$

Self Check In how many ways can 4 boys line up in a row?

Answer 24

EXAMPLE 3 If a sailor has six flags, each of a different color, to hang on a flagpole, how many different signals can the sailor send by using four flags?

Solution The sailor must find the number of arrangements of 4 flags when there are 6 flags to choose from. The sailor can hang any one of the 6 flags in the top position, any one of the remaining 5 flags in the second position, any one of the remaining 4 flags in the third position, and any one of the remaining 3 flags in the lowest position. By the multiplication principle for events, the total number of signals that can be sent is

$$6 \cdot 5 \cdot 4 \cdot 3 = 360 \qquad\blacksquare$$

Self Check In Example 3, how many different signals can the sailor send if each signal uses three flags?

Answer 120

■ PERMUTATIONS

When computing the number of possible arrangements of objects such as books on a shelf or flags on a pole, we are finding the number of **permutations** of those objects. In Example 2, we found that the number of permutations of five books, using all five of them, is 120. In Example 3, we found that the number of permutations of six flags, using four of them, is 360.

The symbol $P(n, r)$, read as "the number of permutations of n things r at a time," is often used to express permutation problems. In Example 2, we found that $P(5, 5) = 120$. In Example 3, we found that $P(6, 4) = 360$.

EXAMPLE 4 If Sarah has seven flags, each of a different color, to hang on a flagpole, how many different signals can she send by using three flags?

Solution She must find $P(7, 3)$ (the number of permutations of 7 things 3 at a time). In the top position Sarah can hang any of the 7 flags, in the middle position any one of the remaining 6 flags, and in the bottom position any one of the remaining 5 flags. According to the multiplication principle for events,

$$P(7, 3) = 7 \cdot 6 \cdot 5 = 210$$

She can send 210 signals using only three of the seven flags. ■

Self Check

Answer
In Example 4, how many different signals can Sarah send using four flags?

840

Although it is correct to write $P(7, 3) = 7 \cdot 6 \cdot 5$, there is an advantage in changing the form of this answer to obtain a formula for computing $P(7, 3)$:

$$P(7, 3) = 7 \cdot 6 \cdot 5$$

$$= \frac{7 \cdot 6 \cdot 5 \cdot 4 \cdot 3 \cdot 2 \cdot 1}{4 \cdot 3 \cdot 2 \cdot 1} \qquad \text{Multiply both the numerator and denominator by } 4 \cdot 3 \cdot 2 \cdot 1.$$

$$= \frac{7!}{4!}$$

$$= \frac{7!}{(7 - 3)!}$$

The generalization of this idea gives the following formula.

Finding P(n, r)

The number of permutations of n things r at a time is given by the formula

$$P(n, r) = \frac{n!}{(n - r)!}$$

EXAMPLE 5 Compute **a.** $P(8, 2)$, **b.** $P(7, 5)$, **c.** $P(n, n)$, and **d.** $P(n, 0)$.

Solution

a. $P(8, 2) = \dfrac{8!}{(8 - 2)!}$

$$= \frac{8 \cdot 7 \cdot 6!}{6!}$$

$$= 8 \cdot 7$$

$$= 56$$

b. $P(7, 5) = \dfrac{7!}{(7 - 5)!}$

$$= \frac{7 \cdot 6 \cdot 5 \cdot 4 \cdot 3 \cdot 2!}{2!}$$

$$= 7 \cdot 6 \cdot 5 \cdot 4 \cdot 3$$

$$= 2{,}520$$

c. $P(n, n) = \dfrac{n!}{(n - n)!}$

$$= \frac{n!}{0!}$$

$$= \frac{n!}{1}$$

$$= n!$$

d. $P(n, 0) = \dfrac{n!}{(n - 0)!}$

$$= \frac{n!}{n!}$$

$$= 1$$

Self Check

Answers
Compute **a.** $P(10, 6)$ and **b.** $P(10, 0)$.

a. 151,200 **b.** 1

Parts **c** and **d** of Example 5 establish the following formulas.

> ### Finding $P(n, n)$ and $P(n, 0)$
> The number of permutations of n things n at a time and n things 0 at a time are given by the formulas
> $$P(n, n) = n! \quad \text{and} \quad P(n, 0) = 1$$

EXAMPLE 6

a. In how many ways can a television executive arrange the Saturday night lineup of 6 programs if there are 15 programs to choose from?

b. If there are only 6 programs to choose from?

Solution

a. To find the number of permutations of 15 programs 6 at a time, we use the formula $P(n, r) = \frac{n!}{(n-r)!}$ with $n = 15$ and $r = 6$.

$$\begin{aligned} P(15, 6) &= \frac{15!}{(15 - 6)!} \\ &= \frac{15 \cdot 14 \cdot 13 \cdot 12 \cdot 11 \cdot 10 \cdot 9!}{9!} \\ &= 15 \cdot 14 \cdot 13 \cdot 12 \cdot 11 \cdot 10 \\ &= 3,603,600 \end{aligned}$$

b. To find the number of permutations of 6 programs 6 at a time, we use the formula $P(n, n) = n!$ with $n = 6$.

$$P(6, 6) = 6! = 720$$ ∎

Self Check

In Example 6, how many ways are there if the executive has 20 programs to choose from?

Answer   27,907,200

■ COMBINATIONS

Suppose that Raul must read 4 books from a reading list of 10 books. The order in which he reads them is not important. For the moment, however, let's assume that order is important and find the number of permutations of 10 things 4 at a time:

$$\begin{aligned} P(10, 4) &= \frac{10!}{(10 - 4)!} \\ &= \frac{10 \cdot 9 \cdot 8 \cdot 7 \cdot 6!}{6!} \\ &= 10 \cdot 9 \cdot 8 \cdot 7 \\ &= 5,040 \end{aligned}$$

If order is important, there are 5,040 ways of choosing 4 books when there are 10 books to choose from. However, because the order in which Raul reads the books does not matter, the previous result of 5,040 is too big. Since there are 24 (or 4!) ways of ordering the 4 books that are chosen, the result of 5,040 is exactly 24 (or 4!) times too big. Therefore, the number of choices that Raul has is the number of permutations of 10 things 4 at a time, divided by 24:

$$\frac{P(10, 4)}{24} = \frac{5,040}{24} = 210$$

Raul has 210 ways of choosing four books to read from the list of ten books.

In situations where order is *not* important, we are interested in **combinations**, not permutations. The symbols $C(n, r)$ and $\binom{n}{r}$ both mean the number of combinations of n things r at a time.

If a selection of r books is chosen from a total of n books, the number of possible selections is $C(n, r)$ and there are $r!$ arrangements of the r books in each selection. If we consider the selected books as an ordered grouping, the number of orderings is $P(n, r)$. Therefore, we have

1. $r! \cdot C(n, r) = P(n, r)$

We can divide both sides of Equation 1 by $r!$ to get the formula for finding $C(n, r)$:

$$C(n, r) = \binom{n}{r} = \frac{P(n, r)}{r!} = \frac{n!}{r!(n - r)!}$$

Finding C(n, r)

The number of combinations of n things r at a time is given by

$$C(n, r) = \frac{n!}{r!(n - r)!}$$

EXAMPLE 7   Compute   **a.** $C(8, 5)$,   **b.** $\binom{7}{2}$,   **c.** $C(n, n)$,   and   **d.** $C(n, 0)$.

Solution   **a.** $C(8, 5) = \dfrac{8!}{5!(8 - 5)!}$ **b.** $\dbinom{7}{2} = \dfrac{7!}{2!(7 - 2)!}$

$\qquad\qquad\qquad = \dfrac{8 \cdot 7 \cdot 6 \cdot 5!}{5! \cdot 3!}$ $\qquad\quad = \dfrac{7 \cdot 6 \cdot 5!}{2 \cdot 1 \cdot 5!}$

$\qquad\qquad\qquad = 8 \cdot 7$ $\qquad\qquad = 21$

$\qquad\qquad\qquad = 56$

c. $C(n, n) = \dfrac{n!}{n!(n - n)!}$ **d.** $C(n, 0) = \dfrac{n!}{0!(n - 0)!}$

$\qquad\qquad = \dfrac{n!}{n!(0!)}$ $= \dfrac{n!}{0! \cdot n!}$

$\qquad\qquad = \dfrac{n!}{n!(1)}$ $= \dfrac{1}{0!}$

$\qquad\qquad = 1$ $= \dfrac{1}{1}$

$\qquad\qquad\qquad\qquad\qquad\qquad\qquad\qquad = 1$

The symbol $C(n, 0)$ indicates that we choose 0 things from the available n things. ∎

Self Check
Answers

Compute   **a.** $C(9, 6)$ and   **b.** $C(10, 10)$.
a. 84,   **b.** 1

Parts **c** and **d** of Example 7 establish the following formulas.

Finding $C(n, n)$ and $C(n, 0)$
The number of combinations of n things n at a time is 1. The number of combinations of n things 0 at a time is 1.

$\qquad C(n, n) = 1 \qquad$ and $\qquad C(n, 0) = 1$

EXAMPLE 8

If 15 students want to pick a committee of 4 students to plan a party, how many different committees are possible?

Solution

Since the ordering of people on each possible committee is unimportant, we find the number of combinations of 15 people 4 at a time:

$$C(15, 4) = \dfrac{15!}{4!(15 - 4)!}$$

$$= \dfrac{15 \cdot 14 \cdot 13 \cdot 12 \cdot 11!}{4 \cdot 3 \cdot 2 \cdot 1 \cdot 11!}$$

$$= \dfrac{15 \cdot 14 \cdot 13 \cdot 12}{4 \cdot 3 \cdot 2 \cdot 1}$$

$$= 1,365$$

There are 1,365 possible committees. ∎

Self Check

In how many ways can 20 students pick a committee of 5 students to plan a party?

Answer

15,504

■ ■ ■ ■ ■ ■ ■ ■ ■ PERSPECTIVE

Gambling is an occasional diversion for some and an obsession for others. Whether it is horse racing, slot machines, or state lotteries, the lure of instant riches is hard to resist.

Many states conduct lotteries, and the systems vary. One scheme is typical: For $1, you have two chances to match 6 numbers chosen from 55 numbers and win a grand prize of about $5 million. How likely are you to win? Is it worth $1 to play the game?

To match 6 numbers chosen from 55, you must choose the one winning combination out of $C(55, 6)$ possibilities:

$$C(n, r) = \frac{n!}{r!(n-r)!}$$

$$C(55, 6) = \frac{55!}{6!(55-6)!}$$

$$= \frac{55 \cdot 54 \cdot 53 \cdot 52 \cdot 51 \cdot 50}{6 \cdot 5 \cdot 4 \cdot 3 \cdot 2}$$

$$= 28,989,675$$

In this game, you have two chances in about 29 million of winning $5 million. Over the long haul, you will win $\frac{2}{29,000,000}$ of the time, so each ticket is worth $\frac{2}{29,000,000}$ of $5,000,000, or about 35¢, if you don't have to share the prize with another winner. For every dollar spent to play the game, you can expect to throw away 65¢. This state lottery is a poor bet. Casinos pay better than 50¢ on the dollar, with some slot machines returning 90¢. "You can't win if you're not in!" is the claim of the lottery promoters. A better claim would be "You won't regret if you don't bet!"

EXAMPLE 9   A committee in Congress consists of ten Democrats and eight Republicans. In how many ways can a subcommittee be chosen if it is to contain five Democrats and four Republicans?

Solution   There are $C(10, 5)$ ways of choosing the 5 Democrats and $C(8, 4)$ ways of choosing the 4 Republicans. By the multiplication principle for events, there are $C(10, 5) \cdot C(8, 4)$ ways of choosing the subcommittee:

$$C(10, 5) \cdot C(8, 4) = \frac{10!}{5!(10-5)!} \cdot \frac{8!}{4!(8-4)!}$$

$$= \frac{10 \cdot 9 \cdot 8 \cdot 7 \cdot 6 \cdot 5!}{120 \cdot 5!} \cdot \frac{8 \cdot 7 \cdot 6 \cdot 5 \cdot 4!}{24 \cdot 4!}$$

$$= \frac{10 \cdot 9 \cdot 8 \cdot 7 \cdot 6}{120} \cdot \frac{8 \cdot 7 \cdot 6 \cdot 5}{24}$$

$$= 17,640$$

There are 17,640 possible subcommittees.   ■

Self Check   See Example 9. In how many ways can a subcommittee be chosen if it is to contain four members from each party?

Answer   14,700

■ ALTERNATIVE FORM OF THE BINOMIAL THEOREM

We have seen that the expansion of $(x + y)^3$ is

$$(x + y)^3 = 1x^3 + 3x^2y + 3xy^2 + 1y^3$$

and that

$$\binom{3}{0} = 1, \binom{3}{1} = 3, \binom{3}{2} = 3, \text{ and } \binom{3}{3} = 1$$

Putting these facts together gives the following way of writing the expansion of $(x + y)^3$:

$$(x + y)^3 = \binom{3}{0} x^3 + \binom{3}{1} x^2y + \binom{3}{2} xy^2 + \binom{3}{3} y^3$$

Likewise, we have

$$(x + y)^4 = \binom{4}{0} x^4 + \binom{4}{1} x^3y + \binom{4}{2} x^2y^2 + \binom{4}{3} xy^3 + \binom{4}{4} y^4$$

The generalization of this idea allows us to state the binomial theorem in an alternative form using combinatorial notation.

The Binomial Theorem
If n is any positive integer, then

$$(a + b)^n = \binom{n}{0}a^n + \binom{n}{1}a^{n-1}b + \binom{n}{2}a^{n-2}b^2 + \cdots + \binom{n}{r}a^{n-r}b^r + \cdots + \binom{n}{n}b^n$$

EXAMPLE 10   Use the alternative form of the binomial theorem to expand $(x + y)^6$.

Solution
$$(x + y)^6 = \binom{6}{0}x^6 + \binom{6}{1}x^5y + \binom{6}{2}x^4y^2 + \binom{6}{3}x^3y^3 + \binom{6}{4}x^2y^4$$
$$+ \binom{6}{5}xy^5 + \binom{6}{6}y^6$$
$$= x^6 + 6x^5y + 15x^4y^2 + 20x^3y^3 + 15x^2y^4 + 6xy^5 + y^6 \qquad ■$$

Self Check   Use the alternative form of the binomial theorem to expand $(a + b)^2$.
Answer $a^2 + 2ab + b^2$

EXAMPLE 11   Use the alternative form of the binomial theorem to expand $(2x - y)^3$.

Solution
$$(2x - y)^3 = [2x + (-y)]^3$$
$$= \binom{3}{0}(2x)^3 + \binom{3}{1}(2x)^2(-y) + \binom{3}{2}(2x)(-y)^2 + \binom{3}{3}(-y)^3$$
$$= 1(2x)^3 + 3(4x^2)(-y) + 3(2x)(y^2) + (-y)^3$$
$$= 8x^3 - 12x^2y + 6xy^2 - y^3 \qquad ■$$

Self Check Answer	Use the alternative form of the binomial theorem to expand $(3a + b)^3$. $27a^3 + 27a^2b + 9ab^2 + b^3$

Orals
1. If there are 3 books and 5 records, in how many ways can you pick 1 book and 1 record?
2. In how many ways can 5 soldiers stand in line?
3. Find $P(3, 1)$
4. Find $P(3, 3)$
5. Find $C(3, 0)$
6. $C(3, 3)$

EXERCISE 14.6

REVIEW *Find each value of x.*

1. $|2x - 3| = 9$

2. $2x^2 - x = 15$

3. $\dfrac{3}{x - 5} = \dfrac{8}{x}$

4. $\dfrac{3}{x} = \dfrac{x - 2}{8}$

VOCABULARY AND CONCEPTS *Fill in each blank to make a true statement.*

5. If an event E_1 can be done in p ways and, after it occurs, a second event E_2 can be done in q ways, the event E_1 followed by E_2 can be done in _____ ways.

6. A _____ is an arrangement of objects.

7. The symbol _____ means the number of permutations of n things taken r at a time.

8. The formula for the number of permutations of n things taken r at a time is _____.

9. $P(n, n) =$ ___

10. $P(n, 0) =$ ___

11. The symbol $C(n, r)$ or _____ means the number of _____ of n things taken r at a time.

12. The formula for the number of combinations of n things taken r at a time is _____.

13. $C(n, n) =$ ___

14. $C(n, 0) =$ ___

PRACTICE *In Exercises 15–36, evaluate each permutation or combination.*

15. $P(3, 3)$
16. $P(4, 4)$
17. $P(5, 3)$
18. $P(3, 2)$

19. $P(2, 2) \cdot P(3, 3)$
20. $P(3, 2) \cdot P(3, 3)$
21. $\dfrac{P(5, 3)}{P(4, 2)}$
22. $\dfrac{P(6, 2)}{P(5, 4)}$

23. $\dfrac{P(6, 2) \cdot P(7, 3)}{P(5, 1)}$
24. $\dfrac{P(8, 3)}{P(5, 3) \cdot P(4, 3)}$

25. $C(5, 3)$
26. $C(5, 4)$
27. $\dbinom{6}{3}$
28. $\dbinom{6}{4}$

29. $\dbinom{5}{4}\dbinom{5}{3}$ **30.** $\dbinom{6}{5}\dbinom{6}{4}$ **31.** $\dfrac{C(38, 37)}{C(19, 18)}$ **32.** $\dfrac{C(25, 23)}{C(40, 39)}$

33. $C(12, 0)C(12, 12)$ **34.** $\dfrac{C(8, 0)}{C(8, 1)}$ **35.** $C(n, 2)$ **36.** $C(n, 3)$

In Exercises 37–42, use the alternative form of the binomial theorem to expand each expression.

37. $(x + y)^4$ **38.** $(x - y)^2$

39. $(2x + y)^3$ **40.** $(2x + 1)^4$

41. $(3x - 2)^4$ **42.** $(3 - x^2)^3$

In Exercises 43–46, find the indicated term of the binomial expansion.

43. $(x - 5y)^5$; fourth term   **44.** $(2x - y)^5$; third term

45. $(x^2 - y^3)^4$; second term   **46.** $(x^3 - y^2)^4$; fourth term

APPLICATIONS

47. Arranging an evening   Kristy plans to go to dinner and see a movie. In how many ways can she arrange her evening if she has a choice of five movies and seven restaurants?

48. Travel choices   Paula has five ways to travel from New York to Chicago, three ways to travel from Chicago to Denver, and four ways to travel from Denver to Los Angeles. How many choices are available to Paula if she travels from New York to Los Angeles?

49. Making license plates   How many six-digit license plates can be manufactured? Note that there are ten choices—0, 1, 2, 3, 4, 5, 6, 7, 8, 9—for each digit.

50. Making license plates   How many six-digit license plates can be manufactured if no digit can be repeated?

51. Making license plates   How many six-digit license plates can be manufactured if no license can begin with 0 and if no digit can be repeated?

52. Making license plates   How many license plates can be manufactured with two letters followed by four digits?

53. Phone numbers   How many seven-digit phone numbers are available in area code 815 if no phone number can begin with 0 or 1?

54. Phone numbers   How many ten-digit phone numbers are available if area codes of 000 and 911 cannot be used and if no local number can begin with 0 or 1?

55. Lining up   In how many ways can six people be placed in a line?

56. Arranging books   In how many ways can seven books be placed on a shelf?

57. Arranging books   In how many ways can four novels and five biographies be arranged on a shelf if the novels are placed on the left?

58. Making a ballot   In how many ways can six candidates for mayor and four candidates for the county board be arranged on a ballot if all of the candidates for mayor must be placed on top?

59. Combination locks   How many permutations does a combination lock have if each combination has three numbers, no two numbers of any combination are equal, and the lock has 25 numbers?

60. Combination locks   How many permutations does a combination lock have if each combination has three numbers, no two numbers of any combination are equal, and the lock has 50 numbers?

61. Arranging appointments   The receptionist at a dental office has only three appointment times available before Tuesday, and ten patients have toothaches. In how many ways can the receptionist fill those appointments?

62. Computers In many computers, a *word* consists of 32 *bits*—a string of thirty-two 1's and 0's. How many different words are possible?

63. Palindromes A palindrome is any word, such as *madam* or *radar*, that reads the same backward and forward. How many five-digit numerical palindromes (such as 13531) are there? (*Hint:* A leading 0 would be dropped.)

64. Call letters The call letters of U.S. commercial radio stations have 3 or 4 letters, and the first is always a W or a K. How many radio stations could this system support?

65. Planning a picnic A class of 14 students wants to pick a committee of 3 students to plan a picnic. How many committees are possible?

66. Choosing books Jeff must read 3 books from a reading list of 15 books. How many choices does he have?

67. Forming committees The number of three-person committees that can be formed from a group of persons is ten. How many persons are in the group?

68. Forming committees The number of three-person committees that can be formed from a group of persons is 20. How many persons are in the group?

69. Winning a lottery In one state lottery, anyone who picks the correct 6 numbers (in any order) wins. With the numbers 0 through 99 available, how many choices are possible?

70. Taking a test The instructions on a test read: "Answer any ten of the following fifteen questions. Then choose one of the remaining questions for homework, and turn in its solution tomorrow." In how many ways can the questions be chosen?

71. Forming a committee In how many ways can we select a committee of two men and two women from a group containing three men and four women?

72. Forming a committee In how many ways can we select a committee of three men and two women from a group containing five men and three women?

73. Choosing clothes In how many ways can we select 2 shirts and 3 neckties from a group of 12 shirts and 10 neckties?

74. Choosing clothes In how many ways can we select five dresses and two coats from a wardrobe containing nine dresses and three coats?

WRITING

75. State the multiplication principle for events.

76. Explain why *permutation lock* would be a better name for a combination lock.

SOMETHING TO THINK ABOUT

77. How many ways could five people stand in line if two people insist on standing together?

78. How many ways could five people stand in line if two people refuse to stand next to each other?

■ ■ ■ ■ ■ ■ ■ ■ ■ **PROJECTS**

PROJECT 1   Baytown is building an auditorium. The city council has already decided on the layout shown in Illustration 1. Each of the sections A, B, C, D, E is to be 60 feet in length from front to back. The aisle widths cannot be changed due to fire regulations. The one thing left to decide is how many rows of seats to put in each section. Based on the following information regarding each section of the auditorium, help the council decide on a final plan.

■ ■ ■ ■ ■ ■ ■ ■ ■ ■   **PROJECTS**   *(continued)*

Sections A and C each have four seats in the front row, five seats in the second row, six seats in the third row and so on, adding one seat per row as we count from front to back.

Section B has eight seats in the front row and adds one seat per row as we count from front to back.

Sections D and E each have 28 seats in the front row and add two seats per row as we count from front to back.

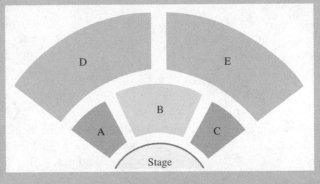

ILLUSTRATION 1

a. One plan calls for a distance of 36 inches, front to back, for each row of seats. Another plan allows for 40 inches (an extra four inches of legroom) for each row. How many seats will the auditorium have under each of these plans?

b. Another plan calls for the higher-priced seats (Sections A, B, and C) to have the extra room afforded by 40-inch rows, but for Sections D and E to have enough rows to make sure that the auditorium holds at least 2,700 seats. Determine how many rows Sections D and E would have to contain for this to work. (This answer should be an integer.) How much space (to the nearest tenth of an inch) would be allotted for each row in Sections D and E?

P R O J E C T   2   Pascal's triangle contains a wealth of interesting patterns. You have seen two in Exercises 50 and 51 of Section 14.1. Here are a few more.

a. Find the hockey-stick pattern in the numbers in Illustration 2. What would be the missing number in the rightmost hockey stick? Does this pattern work for larger hockey sticks? Experiment.

(continued)

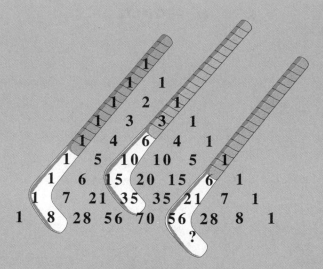

ILLUSTRATION 2

b. In Illustration 3, find the pattern in the sums of increasingly larger portions of Pascal's triangle. Find the sum of all of the numbers up to and including the row that begins 1   10   45 . . . .

$$
\begin{array}{ccccc}
 & & & & 1 \\
 & & & 1\ \ 1 \\
 & & 1 & 1\ 2\ 1 \\
 & 1\ 1 & 1\ 2\ 1 & 1\ 3\ 3\ 1 & 1\ 3\ 3\ 1 \\
1 & 1\ 1 & 1\ 2\ 1 & 1\ 3\ 3\ 1 & 1\ 4\ 6\ 4\ 1 \\
=1 & =3 & =7 & =? & =?
\end{array}
$$

ILLUSTRATION 3

c. In Illustration 4, find the pattern in the sums of the squares of the numbers in each row of the triangle. What is the sum of the squares of the numbers in the row that begins 1   10   45 . . . ? (*Hint:* Calculate $P(2, 1)$, $P(4, 2)$, $P(6, 3)$, ... Do these numbers appear elsewhere in the triangle?)

$$
\begin{aligned}
1^2 &= 1 \\
1^2 + 1^2 &= 2 \\
1^2 + 2^2 + 1^2 &= 6 \\
1^2 + 3^2 + 3^2 + 1^2 &= 20 \\
1^2 + 4^2 + 6^2 + 4^2 + 1^2 &= 70 \\
1^2 + 5^2 + 10^2 + 10^2 + 5^2 + 1^2 &= ? \\
1^2 + 6^2 + 15^2 + 20^2 + 15^2 + 6^2 + 1^2 &= ?
\end{aligned}
$$

ILLUSTRATION 4

■ ■ ■ ■ ■ ■ ■ ■ ■ ■ **PROJECTS**   *(continued)*

d. In 1653, Pascal described the triangle in *Treatise on the Arithmetic Triangle*, writing, "I have left out many more properties than I have included. It is remarkable how fertile in properties this triangle is. *Everyone can try his hand*." Accept Pascal's invitation. Find some of the triangle's patterns for yourself and share your discoveries with your class. Illustration 5 is an idea to get you started.

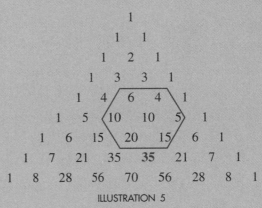

ILLUSTRATION 5

C H A P T E R   S U M M A R Y

CONCEPTS

REVIEW EXERCISES

SECTION 14.1

The Binomial Theorem

The symbol $n!$ (**n factorial**) is defined as

$$n! = n(n - 1)(n - 2) \cdots 2 \cdot 1$$

where n is a natural number.

$0! = 1$

$n(n - 1)! = n!$ (n is a natural number)

1. Evaluate each expression.

   a. $(4!)(3!)$ **b.** $\dfrac{5!}{3!}$

   c. $\dfrac{6!}{2!(6 - 2)!}$ **d.** $\dfrac{12!}{3!(12 - 3)!}$

   e. $(n - n)!$ **f.** $\dfrac{8!}{7!}$

The binomial theorem:

$$(a + b)^n =$$
$$a^n + \frac{n!}{1!(n - 1)!} a^{n-1}b$$
$$+ \frac{n!}{2!(n - 2)!} a^{n-2}b^2$$
$$+ \cdots + b^n$$

2. Use the binomial theorem to find each expansion.
   a. $(x + y)^5$
   b. $(x - y)^4$
   c. $(4x - y)^3$
   d. $(x + 4y)^3$

SECTION 14.2	The nth Term of a Binomial Expansion

The binomial theorem can be used to find the *n*th term of a binomial expansion.

3. Find the specified term in each expansion.
  a. $(x + y)^4$; third term
  b. $(x - y)^5$; fourth term
  c. $(3x - 4y)^3$; second term
  d. $(4x + 3y)^4$; third term

SECTION 14.3	Arithmetic Sequences

An **arithmetic sequence** is a sequence of the form

$$a, a + d, a + 2d, \ldots,$$
$$a + (n - 1)d$$

where *a* is the first term, $a + (n-1)d$ is the *n*th term, and *d* is the common difference.

If numbers are inserted between two given numbers *a* and *b* to form an arithmetic sequence, the inserted numbers are **arithmetic means** between *a* and *b*.

The sum of the first *n* terms of an arithmetic sequence is given by

$$S_n = \frac{n(a + l)}{2} \quad \text{with}$$
$$l = a + (n - 1)d$$

where *a* is the first term, *l* is the last (or *n*th) term, and *n* is the number of terms in the sequence.

$$\sum_{k=1}^{n} f(k) = f(1) + f(2)$$
$$+ \cdots + f(n)$$

4. Find the eighth term of an arithmetic sequence whose first term is 7 and whose common difference is 5.

5. Write the first five terms of the arithmetic sequence whose ninth term is 242 and whose seventh term is 212.

6. Find two arithmetic means between 8 and 25.

7. Find the sum of the first 20 terms of the sequence 11, 18, 25, ...

8. Find the sum of the first ten terms of the sequence 9, $6\frac{1}{2}$, 4, ...

9. Find each sum.

  a. $\sum_{k=4}^{6} \frac{1}{2}k$ **b.** $\sum_{k=2}^{5} 7k^2$

  c. $\sum_{k=1}^{4} (3k - 4)$ **d.** $\sum_{k=10}^{10} 36k$

| **SECTION 14.4** | *Geometric Sequences* |

A **geometric sequence** is a sequence of the form

$$a, ar, ar^2, ar^3, \ldots, ar^{n-1}$$

where a is the first term, ar^{n-1} is the nth term, and r is the common ratio.

10. Write the first five terms of the geometric sequence whose fourth term is 3 and whose fifth term is $\frac{3}{2}$.

11. Find the sixth term of a geometric sequence with a first term of $\frac{1}{8}$ and a common ratio of 2.

If numbers are inserted between a and b to form a geometric sequence, the inserted numbers are **geometric means** between a and b.

12. Find two geometric means between -6 and 384.

The sum of the first n terms of a geometric sequence is given by

$$S_n = \frac{a - ar^n}{1 - r} \quad (r \neq 1)$$

where S_n is the sum, a is the first term, r is the common ratio, and n is the number of terms in the sequence.

13. Find the sum of the first seven terms of the sequence $162, 54, 18, \ldots$.

14. Find the sum of the first eight terms of the sequence $\frac{1}{8}, -\frac{1}{4}, \frac{1}{2}, \ldots$.

| **SECTION 14.5** | *Infinite Geometric Sequences* |

If r is the common ratio of an infinite geometric sequence, and if $|r| < 1$, the sum of the terms of the infinite geometric sequence is given by

$$S = \frac{a}{1 - r}$$

where a is the first term and r is the common ratio.

15. Find the sum of the infinite geometric sequence $25, 20, 16, \ldots$.

16. Change the decimal $0.\overline{05}$ to a common fraction.

| SECTION 14.6 | *Permutations and Combinations* |

The multiplication principle for events:
If E_1 and E_2 are two events, and if E_1 can be done in a_1 ways and E_2 can be done in a_2 ways, then the event "E_1 followed by E_2" can be done in $a_1 \cdot a_2$ ways.

Formulas for permutations:

$$P(n, r) = \frac{n!}{(n - r)!}$$

$$P(n, n) = n! \quad \text{and}$$

$$P(n, 0) = 1$$

Formulas for combinations:

$$C(n, r) = \binom{n}{r} = \frac{n!}{r!(n - r)!}$$

$$C(n, n) = \binom{n}{n} = 1$$

$$C(n, 0) = \binom{n}{0} = 1$$

17. If there are 17 flights from New York to Chicago, and 8 flights from Chicago to San Francisco, in how many different ways could a passenger plan a trip from New York to San Francisco?

18. Evaluate each expression.
   a. $P(7, 7)$ **b.** $P(7, 0)$

   c. $P(8, 6)$ **d.** $\dfrac{P(9, 6)}{P(10, 7)}$

19. Evaluate each expression.
   a. $C(7, 7)$ **b.** $C(7, 0)$

   c. $\binom{8}{6}$ **d.** $\binom{9}{6}$

   e. $C(6, 3) \cdot C(7, 3)$ **f.** $\dfrac{C(7, 3)}{C(6, 3)}$

20. Car depreciation   A \$5,000 car depreciates at the rate of 20% of the previous year's value. How much is the car worth after 5 years?

21. Stock appreciation   The value of Mia's stock portfolio is expected to appreciate at the rate of 18% per year. How much will the portfolio be worth in 10 years if its current value is \$25,700?

22. Planting corn   A farmer planted 300 acres in corn this year. He intends to plant an additional 75 acres in corn in each successive year until he has 1,200 acres in corn. In how many years will that be?

23. Falling object   If an object is in free fall, the sequence 16, 48, 80, . . . represents the distance in feet that object falls during the first second, during the second second, during the third second, and so on. How far will the object fall during the first ten seconds?

24. Lining up   In how many ways can five people be arranged in a line?

25. Lining up In how many ways can three men and five women be arranged in a line if the women are placed ahead of the men?

26. Choosing people In how many ways can we pick three people from a group of ten?

27. Forming committees In how many ways can we pick a committee of two Democrats and two Republicans from a group containing five Democrats and six Republicans?

■ Chapter Test

1. Evaluate $\dfrac{7!}{4!}$

2. Evaluate $0!$

3. Find the second term in the expansion of $(x - y)^5$.

4. Find the third term in the expansion of $(x + 2y)^4$.

5. Find the tenth term of an arithmetic sequence with the first three terms of 3, 10, and 17.

6. Find the sum of the first 12 terms of the sequence $-2, 3, 8, \ldots$.

7. Find two arithmetic means between 2 and 98.

8. Evaluate $\displaystyle\sum_{k=1}^{3} (2k - 3)$.

9. Find the seventh term of the geometric sequence whose first three terms are $-\frac{1}{9}$, $-\frac{1}{3}$, and -1.

10. Find the sum of the first six terms of the sequence $\frac{1}{27}$, $\frac{1}{9}$, $\frac{1}{3}$, $\ldots$.

11. Find two geometric means between 3 and 648.

12. Find the sum of all of the terms of the infinite geometric sequence $9, 3, 1, \ldots$.

In Problems 13–20, find the value of each expression.

13. $P(5, 4)$

14. $P(8, 8)$

15. $C(6, 4)$

16. $C(8, 3)$

17. $C(6, 0) \cdot P(6, 5)$

18. $P(8, 7) \cdot C(8, 7)$

19. $\dfrac{P(6, 4)}{C(6, 4)}$

20. $\dfrac{C(9, 6)}{P(6, 4)}$

21. Choosing people In how many ways can we pick three people from a group of seven?

22. Choosing committees From a group of five men and four women, how many three-person committees can be chosen that will include two women?

■ Cumulative Review Exercises

1. Use graphing to solve $\begin{cases} 2x + y = 5 \\ x - 2y = 0 \end{cases}$.

2. Use substitution to solve $\begin{cases} 3x + y = 4 \\ 2x - 3y = -1 \end{cases}$.

3. Use addition to solve $\begin{cases} x + 2y = -2 \\ 2x - y = 6 \end{cases}$.

4. Use any method to solve $\begin{cases} \frac{x}{10} + \frac{y}{5} = \frac{1}{2} \\ \frac{x}{2} - \frac{y}{5} = \frac{13}{10} \end{cases}$.

5. Evaluate $\begin{vmatrix} 3 & -2 \\ 1 & -1 \end{vmatrix}$.

6. Use Cramer's rule and solve for y only:
$$\begin{cases} 4x - 3y = -1 \\ 3x + 4y = -7 \end{cases}$$

7. Solve $\begin{cases} x + y + z = 1 \\ 2x - y - z = -4 \\ x - 2y + z = 4 \end{cases}$.

8. Solve for z only: $\begin{cases} x + 2y + 3z = 6 \\ 3x + 2y + z = 6 \\ 2x + 3y + z = 6 \end{cases}$.

9. Solve $\begin{cases} 3x - 2y < 6 \\ y < -x + 2 \end{cases}$.

10. Solve $\begin{cases} y < x + 2 \\ 3x + y \le 6 \end{cases}$.

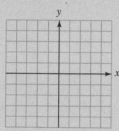

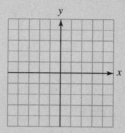

11. Graph $y = \left(\dfrac{1}{2}\right)^x$.

12. Write $y = \log_2 x$ as an exponential equation.

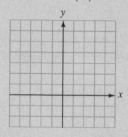

In Exercises 13–16, find x.

13. $\log_x 25 = 2$

14. $\log_5 125 = x$

15. $\log_3 x = -3$

16. $\log_5 x = 0$

17. Find the inverse of $y = \log_2 x$.

18. If $\log_{10} 10^x = y$, then y equals what quantity?

In Exercises 19–22, $\log 7 = 0.8451$ and $\log 14 = 1.1461$. Evaluate each expression without using a calculator or tables.

19. $\log 98$

20. $\log 2$

21. $\log 49$

22. $\log \dfrac{7}{5}$ (*Hint:* $\log 10 = 1$.)

23. Solve $2^{x+5} = 3^x$.

24. Solve $\log 5 + \log x - \log 4 = 1$.

In Exercises 25–26, use a calculator.

25. Boat depreciation How much will a $9,000 boat be worth after 9 years if it depreciates 12% per year?

26. Find $\log_6 8$.

27. Evaluate $\dfrac{6!7!}{5!}$.

28. Use the binomial theorem to expand $(3a - b)^4$.

29. Find the seventh term in the expansion of $(2x - y)^8$.

30. Find the 20th term of an arithmetic sequence with a first term of -11 and a common difference of 6.

31. Find the sum of the first 20 terms of an arithmetic sequence with a first term of 6 and a common difference of 3.

32. Insert two arithmetic means between -3 and 30.

33. Evaluate $\displaystyle\sum_{k=1}^{3} 3k^2$.

34. Evaluate $\displaystyle\sum_{k=3}^{5} (2k + 1)$.

35. Find the seventh term of a geometric sequence with a first term of $\frac{1}{27}$ and a common ratio of 3.

36. Find the sum of the first ten terms of the sequence $\frac{1}{64}, \frac{1}{32}, \frac{1}{16}, \cdots$

37. Insert two geometric means between -3 and 192.

38. Find the sum of all the terms of the sequence 9, 3, 1, ...

39. Evaluate $P(9, 3)$.

40. Evaluate $C(7, 4)$.

41. Evaluate $\dfrac{C(8, 4)C(8, 0)}{P(6, 2)}$.

42. If $n > 1$, which is smaller: $P(n, n)$ or $C(n, n)$?

43. Lining up In how many ways can seven people stand in a line?

44. Forming a committee In how many ways can a committee of three people be chosen from a group containing nine people?

SYMMETRIES OF GRAPHS

There are several ways that a graph can exhibit symmetry about the coordinate axes and the origin. It is often easier to draw graphs of equations if we first find the x- and y-intercepts and find any of the following symmetries of the graph:

1. **y-axis symmetry**: If the point $(-x, y)$ lies on a graph whenever the point (x, y) does, as in Figure I-1(a), we say that the graph is **symmetric about the y-axis**.

2. **Symmetry about the origin**: If the point $(-x, -y)$ lies on the graph whenever the point (x, y) does, as in Figure I-1(b), we say that the graph is **symmetric about the origin**.

3. **x-axis symmetry**: If the point $(x, -y)$ lies on the graph whenever the point (x, y) does, as in Figure I-1(c), we say that the graph is **symmetric about the x-axis**.

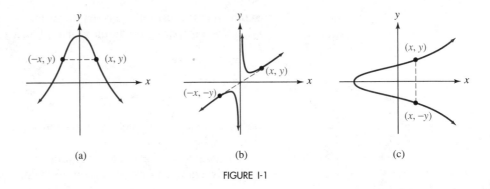

(a)   (b)   (c)

FIGURE I-1

Tests for Symmetry for Graphs in x and y

- To test a graph for y-axis symmetry, replace x with $-x$. If the new equation is equivalent to the original equation, the graph is symmetric about the y-axis. Symmetry about the y-axis will occur whenever x appears with only even exponents.

A-1

- To test a graph for symmetry about the origin, replace x with $-x$ and y with $-y$. If the resulting equation is equivalent to the original equation, the graph is symmetric about the origin.

- To test a graph for x-axis symmetry, replace y with $-y$. If the resulting equation is equivalent to the original equation, the graph is symmetric about the x-axis. The only function that is symmetric about the x-axis is $f(x) = 0$.

EXAMPLE 1   Find the intercepts and the symmetries of the graph of $y = f(x) = x^3 - 9x$. Then graph the function.

Solution   **x-intercepts**: To find the x-intercepts, we let $y = 0$ and solve for x:

$$y = x^3 - 9x$$
$$\mathbf{0} = x^3 - 9x \qquad \text{Substitute 0 for } y.$$
$$0 = x(x^2 - 9) \qquad \text{Factor out } x.$$
$$0 = x(x + 3)(x - 3) \qquad \text{Factor } x^2 - 9.$$
$$x = 0 \quad \text{or} \quad x + 3 = 0 \quad \text{or} \quad x - 3 = 0 \qquad \text{Set each factor equal to 0.}$$
$$x = -3 \qquad\qquad x = 3$$

Since the x-coordinates of the x-intercepts are 0, -3, and 3, the graph intersects the x-axis at $(0, 0)$, $(-3, 0)$, and $(3, 0)$.

y-intercepts: To find the y-intercepts, we let $x = 0$ and solve for y.

$$y = x^3 - 9x$$
$$y = \mathbf{0}^3 - 9(\mathbf{0}) \qquad \text{Substitute 0 for } x.$$
$$y = 0$$

Since the y-coordinate of the y-intercept is 0, the graph intersects the y-axis at $(0, 0)$.

Symmetry: We test for symmetry about the y-axis by replacing x with $-x$, simplifying, and comparing the result to the original equation.

1. $\quad y = x^3 - 9x$ $\qquad\qquad$ The original equation.
$\quad\quad y = (-x)^3 - 9(-x)$ $\qquad$ Replace x with $-x$.
2. $\quad y = -x^3 + 9x$ $\qquad\qquad$ Simplify.

Because Equation 2 is not equivalent to Equation 1, the graph is not symmetric about the y-axis.

We test for symmetry about the origin by replacing x and y with $-x$ and $-y$, respectively, and comparing the result to the original equation.

1. $\quad y = x^3 - 9x$ $\qquad\qquad$ The original equation.
$\quad\quad -y = (-x)^3 - 9(-x)$ $\qquad$ Replace x with $-x$, and y with $-y$.
$\quad\quad -y = -x^3 + 9x$ $\qquad\qquad$ Simplify.
3. $\quad y = x^3 - 9x$ $\qquad\qquad$ Multiply both sides by -1 to solve for y.

Because Equation 3 is equivalent to Equation 1, the graph is symmetric about the origin. Because the equation is the equation of a nonzero function, there is no symmetry about the x-axis.

To graph the equation, we plot the x-intercepts of $(-3, 0)$, $(0, 0)$, and $(3, 0)$ and the y-intercept of $(0, 0)$. We also plot other points for positive values of x and use the symmetry about the origin to draw the rest of the graph, as in Figure I-2(a). (Note that the scale on the x-axis is different from the scale on the y-axis.)

If we graph the equation with a graphing calculator, with window settings of $[-10, 10]$ for x and $[-10, 10]$ for y, we will obtain the graph shown in Figure I-2(b).

From the graph, we can see that the domain is the interval $(-\infty, \infty)$, and the range is the interval $(-\infty, \infty)$.

$y = x^3 - 9x$

x	y
0	0
1	-8
2	-10
3	0
4	28

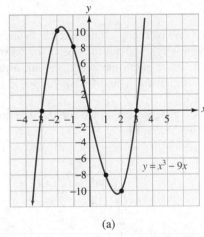

(a)

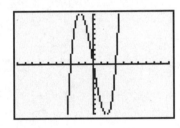

(b)

FIGURE I-2

EXAMPLE 2   Graph the function $y = f(x) = |x| - 2$.

Solution   **x-intercepts**: To find the x-intercepts, we let $y = 0$ and solve for x:

$$y = |x| - 2$$
$$0 = |x| - 2$$
$$2 = |x|$$
$$x = -2 \quad \text{or} \quad x = 2$$

Since -2 and 2 are solutions, the points $(-2, 0)$ and $(2, 0)$ are the x-intercepts, and the graph passes through $(-2, 0)$ and $(2, 0)$.

y-intercepts: To find the y-intercepts, we let $x = 0$ and solve for y:

$$y = |x| - 2$$
$$y = |0| - 2$$

Since $y = -2$, $(0, -2)$ is the y-intercept, and the graph passes through the point $(0, -2)$.

Symmetry: To test for y-axis symmetry, we replace x with $-x$.

4.	$y = \lvert x \rvert - 2$	The original equation.
	$y = \lvert -x \rvert - 2$	Replace x with $-x$.
5.	$y = \lvert x \rvert - 2$	$\lvert -x \rvert = \lvert x \rvert$.

Since Equation 5 is equivalent to Equation 4, the graph is symmetric about the y-axis. The graph has no other symmetries.

We plot the x- and y-intercepts and several other points (x, y), and use the y-axis symmetry to obtain the graph shown in Figure I-3(a).

If we graph the equation with a graphing calculator, with window settings of $[-10, 10]$ for x and $[-10, 10]$ for y, we will obtain the graph shown in Figure I-3(b).

From the graph, we see that the domain is the interval $(-\infty, \infty)$, and the range is the interval $[-2, \infty)$.

$y = \lvert x \rvert - 2$

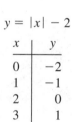

x	y
0	-2
1	-1
2	0
3	1
4	2

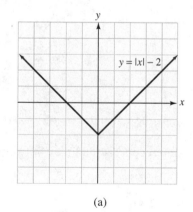

(a)

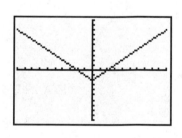

(b)

FIGURE I-3

EXERCISE I.1

In Exercises 1–12, find the symmetries of the graph of each relation. **Do not draw the graph.**

1. $y = x^2 - 1$ **2.** $y = x^3$ **3.** $y = x^5$ **4.** $y = x^4$

5. $y = -x^2 + 2$ **6.** $y = x^3 + 1$ **7.** $y = x^2 - x$ **8.** $y^2 = x + 7$

9. $y = -\lvert x + 2 \rvert$ **10.** $y = \lvert x \rvert - 3$ **11.** $\lvert y \rvert = x$ **12.** $y = 2\sqrt{x}$

In Exercises 13–24, graph each function and give its domain and range. Check each graph with a graphing calculator.

13. $y = x^4 - 4$

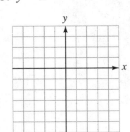

14. $y = \dfrac{1}{2}x^4 - 1$

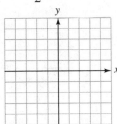

15. $y = -x^3$

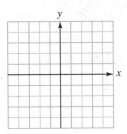

16. $y = x^3 + 2$

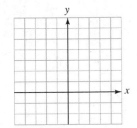

17. $y = x^4 + x^2$

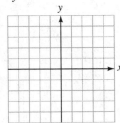

18. $y = 3 - x^4$

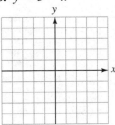

19. $y = x^3 - x$

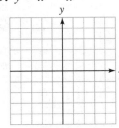

20. $y = x^3 + x$

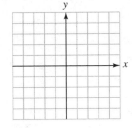

21. $y = \dfrac{1}{2}|x| - 1$

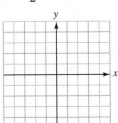

22. $y = -|x| + 1$

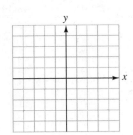

23. $y = -|x + 2|$

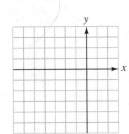

24. $y = |x - 2|$

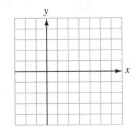

Table A   Powers and Roots

n	n^2	$\sqrt{n}$	n^3	$\sqrt[3]{n}$	n	n^2	$\sqrt{n}$	n^3	$\sqrt[3]{n}$
1	1	1.000	1	1.000	51	2,601	7.141	132,651	3.708
2	4	1.414	8	1.260	52	2,704	7.211	140,608	3.733
3	9	1.732	27	1.442	53	2,809	7.280	148,877	3.756
4	16	2.000	64	1.587	54	2,916	7.348	157,464	3.780
5	25	2.236	125	1.710	55	3,025	7.416	166,375	3.803
6	36	2.449	216	1.817	56	3,136	7.483	175,616	3.826
7	49	2.646	343	1.913	57	3,249	7.550	185,193	3.849
8	64	2.828	512	2.000	58	3,364	7.616	195,112	3.871
9	81	3.000	729	2.080	59	3,481	7.681	205,379	3.893
10	100	3.162	1,000	2.154	60	3,600	7.746	216,000	3.915
11	121	3.317	1,331	2.224	61	3,721	7.810	226,981	3.936
12	144	3.464	1,728	2.289	62	3,844	7.874	238,328	3.958
13	169	3.606	2,197	2.351	63	3,969	7.937	250,047	3.979
14	196	3.742	2,744	2.410	64	4,096	8.000	262,144	4.000
15	225	3.873	3,375	2.466	65	4,225	8.062	274,625	4.021
16	256	4.000	4,096	2.520	66	4,356	8.124	287,496	4.041
17	289	4.123	4,913	2.571	67	4,489	8.185	300,763	4.062
18	324	4.243	5,832	2.621	68	4,624	8.246	314,432	4.082
19	361	4.359	6,859	2.668	69	4,761	8.307	328,509	4.102
20	400	4.472	8,000	2.714	70	4,900	8.367	343,000	4.121
21	441	4.583	9,261	2.759	71	5,041	8.426	357,911	4.141
22	484	4.690	10,648	2.802	72	5,184	8.485	373,248	4.160
23	529	4.796	12,167	2.844	73	5,329	8.544	389,017	4.179
24	576	4.899	13,824	2.884	74	5,476	8.602	405,224	4.198
25	625	5.000	15,625	2.924	75	5,625	8.660	421,875	4.217
26	676	5.099	17,576	2.962	76	5,776	8.718	438,976	4.236
27	729	5.196	19,683	3.000	77	5,929	8.775	456,533	4.254
28	784	5.292	21,952	3.037	78	6,084	8.832	474,552	4.273
29	841	5.385	24,389	3.072	79	6,241	8.888	493,039	4.291
30	900	5.477	27,000	3.107	80	6,400	8.944	512,000	4.309
31	961	5.568	29,791	3.141	81	6,561	9.000	531,441	4.327
32	1,024	5.657	32,768	3.175	82	6,724	9.055	551,368	4.344
33	1,089	5.745	35,937	3.208	83	6,889	9.110	571,787	4.362
34	1,156	5.831	39,304	3.240	84	7,056	9.165	592,704	4.380
35	1,225	5.916	42,875	3.271	85	7,225	9.220	614,125	4.397
36	1,296	6.000	46,656	3.302	86	7,396	9.274	636,056	4.414
37	1,369	6.083	50,653	3.332	87	7,569	9.327	658,503	4.431
38	1,444	6.164	54,872	3.362	88	7,744	9.381	681,472	4.448
39	1,521	6.245	59,319	3.391	89	7,921	9.434	704,969	4.465
40	1,600	6.325	64,000	3.420	90	8,100	9.487	729,000	4.481
41	1,681	6.403	68,921	3.448	91	8,281	9.539	753,571	4.498
42	1,764	6.481	74,088	3.476	92	8,464	9.592	778,688	4.514
43	1,849	6.557	79,507	3.503	93	8,649	9.644	804,357	4.531
44	1,936	6.633	85,184	3.530	94	8,836	9.695	830,584	4.547
45	2,025	6.708	91,125	3.557	95	9,025	9.747	857,375	4.563
46	2,116	6.782	97,336	3.583	96	9,216	9.798	884,736	4.579
47	2,209	6.856	103,823	3.609	97	9,409	9.849	912,673	4.595
48	2,304	6.928	110,592	3.634	98	9,604	9.899	941,192	4.610
49	2,401	7.000	117,649	3.659	99	9,801	9.950	970,299	4.626
50	2,500	7.071	125,000	3.684	100	10,000	10.000	1,000,000	4.642

Table B   *(continued)*

N	0	1	2	3	4	5	6	7	8	9
5.5	7404	7412	7419	7427	7435	7443	7451	7459	7466	7474
5.6	7482	7490	7497	7505	7513	7520	7528	7536	7543	7551
5.7	7559	7566	7574	7582	7589	7597	7604	7612	7619	7627
5.8	7634	7642	7649	7657	7664	7672	7679	7686	7694	7701
5.9	7709	7716	7723	7731	7738	7745	7752	7760	7767	7774
6.0	7782	7789	7796	7803	7810	7818	7825	7832	7839	7846
6.1	7853	7860	7868	7875	7882	7889	7896	7903	7910	7917
6.2	7924	7931	7938	7945	7952	7959	7966	7973	7980	7987
6.3	7993	8000	8007	8014	8021	8028	8035	8041	8048	8055
6.4	8062	8069	8075	8082	8089	8096	8102	8109	8116	8122
6.5	8129	8136	8142	8149	8156	8162	8169	8176	8182	8189
6.6	8195	8202	8209	8215	8222	8228	8235	8241	8248	8254
6.7	8261	8267	8274	8280	8287	8293	8299	8306	8312	8319
6.8	8325	8331	8338	8344	8351	8357	8363	8370	8376	8382
6.9	8388	8395	8401	8407	8414	8420	8426	8432	8439	8445
7.0	8451	8457	8463	8470	8476	8482	8488	8494	8500	8506
7.1	8513	8519	8525	8531	8537	8543	8549	8555	8561	8567
7.2	8573	8579	8585	8591	8597	8603	8609	8615	8621	8627
7.3	8633	8639	8645	8651	8657	8663	8669	8675	8681	8686
7.4	8692	8698	8704	8710	8716	8722	8727	8733	8739	8745
7.5	8751	8756	8762	8768	8774	8779	8785	8791	8797	8802
7.6	8808	8814	8820	8825	8831	8837	8842	8848	8854	8859
7.7	8865	8871	8876	8882	8887	8893	8899	8904	8910	8915
7.8	8921	8927	8932	8938	8943	8949	8954	8960	8965	8971
7.9	8976	8982	8987	8993	8998	9004	9009	9015	9020	9025
8.0	9031	9036	9042	9047	9053	9058	9063	9069	9074	9079
8.1	9085	9090	9096	9101	9106	9112	9117	9122	9128	9133
8.2	9138	9143	9149	9154	9159	9165	9170	9175	9180	9186
8.3	9191	9196	9201	9206	9212	9217	9222	9227	9232	9238
8.4	9243	9248	9253	9258	9263	9269	9274	9279	9284	9289
8.5	9294	9299	9304	9309	9315	9320	9325	9330	9335	9340
8.6	9345	9350	9355	9360	9365	9370	9375	9380	9385	9390
8.7	9395	9400	9405	9410	9415	9420	9425	9430	9435	9440
8.8	9445	9450	9455	9460	9465	9469	9474	9479	9484	9489
8.9	9494	9499	9504	9509	9513	9518	9523	9528	9533	9538
9.0	9542	9547	9552	9557	9562	9566	9571	9576	9581	9586
9.1	9590	9595	9600	9605	9609	9614	9619	9624	9628	9633
9.2	9638	9643	9647	9652	9657	9661	9666	9671	9675	9680
9.3	9685	9689	9694	9699	9703	9708	9713	9717	9722	9727
9.4	9731	9736	9741	9745	9750	9754	9759	9763	9768	9773
9.5	9777	9782	9786	9791	9795	9800	9805	9809	9814	9818
9.6	9823	9827	9832	9836	9841	9845	9850	9854	9859	9863
9.7	9868	9872	9877	9881	9886	9890	9894	9899	9903	9908
9.8	9912	9917	9921	9926	9930	9934	9939	9943	9948	9952
9.9	9956	9961	9965	9969	9974	9978	9983	9987	9991	9996

Table B   Base-10 Logarithms

N	0	1	2	3	4	5	6	7	8	9
1.0	0000	0043	0086	0128	0170	0212	0253	0294	0334	0374
1.1	0414	0453	0492	0531	0569	0607	0645	0682	0719	0755
1.2	0792	0828	0864	0899	0934	0969	1004	1038	1072	1106
1.3	1139	1173	1206	1239	1271	1303	1335	1367	1399	1430
1.4	1461	1492	1523	1553	1584	1614	1644	1673	1703	1732
1.5	1761	1790	1818	1847	1875	1903	1931	1959	1987	2014
1.6	2041	2068	2095	2122	2148	2175	2201	2227	2253	2279
1.7	2304	2330	2355	2380	2405	2430	2455	2480	2504	2529
1.8	2553	2577	2601	2625	2648	2672	2695	2718	2742	2765
1.9	2788	2810	2833	2856	2878	2900	2923	2945	2967	2989
2.0	3010	3032	3054	3075	3096	3118	3139	3160	3181	3201
2.1	3222	3243	3263	3284	3304	3324	3345	3365	3385	3404
2.2	3424	3444	3464	3483	3502	3522	3541	3560	3579	3598
2.3	3617	3636	3655	3674	3692	3711	3729	3747	3766	3784
2.4	3802	3820	3838	3856	3874	3892	3909	3927	3945	3962
2.5	3979	3997	4014	4031	4048	4065	4082	4099	4116	4133
2.6	4150	4166	4183	4200	4216	4232	4249	4265	4281	4298
2.7	4314	4330	4346	4362	4378	4393	4409	4425	4440	4456
2.8	4472	4487	4502	4518	4533	4548	4564	4579	4594	4609
2.9	4624	4639	4654	4669	4683	4698	4713	4728	4742	4757
3.0	4771	4786	4800	4814	4829	4843	4857	4871	4886	4900
3.1	4914	4928	4942	4955	4969	4983	4997	5011	5024	5038
3.2	5051	5065	5079	5092	5105	5119	5132	5145	5159	5172
3.3	5185	5198	5211	5224	5237	5250	5263	5276	5289	5302
3.4	5315	5328	5340	5353	5366	5378	5391	5403	5416	5428
3.5	5441	5453	5465	5478	5490	5502	5514	5527	5539	5551
3.6	5563	5575	5587	5599	5611	5623	5635	5647	5658	5670
3.7	5682	5694	5705	5717	5729	5740	5752	5763	5775	5786
3.8	5798	5809	5821	5832	5843	5855	5866	5877	5888	5899
3.9	5911	5922	5933	5944	5955	5966	5977	5988	5999	6010
4.0	6021	6031	6042	6053	6064	6075	6085	6096	6107	6117
4.1	6128	6138	6149	6160	6170	6180	6191	6201	6212	6222
4.2	6232	6243	6253	6263	6274	6284	6294	6304	6314	6325
4.3	6335	6345	6355	6365	6375	6385	6395	6405	6415	6425
4.4	6435	6444	6454	6464	6474	6484	6493	6503	6513	6522
4.5	6532	6542	6551	6561	6571	6580	6590	6599	6609	6618
4.6	6628	6637	6646	6656	6665	6675	6684	6693	6702	6712
4.7	6721	6730	6739	6749	6758	6767	6776	6785	6794	6803
4.8	6812	6821	6830	6839	6848	6857	6866	6875	6884	6893
4.9	6902	6911	6920	6928	6937	6946	6955	6964	6972	6981
5.0	6990	6998	7007	7016	7024	7033	7042	7050	7059	7067
5.1	7076	7084	7093	7101	7110	7118	7126	7135	7143	7152
5.2	7160	7168	7177	7185	7193	7202	7210	7218	7226	7235
5.3	7243	7251	7259	7267	7275	7284	7292	7300	7308	7316
5.4	7324	7332	7340	7348	7356	7364	7372	7380	7388	7396

Table C   (continued)

N	0	1	2	3	4	5	6	7	8	9
5.5	1.7047	.7066	.7084	.7102	.7120	.7138	.7156	.7174	.7192	.7210
5.6	.7228	.7246	.7263	.7281	.7299	.7317	.7334	.7352	.7370	.7387
5.7	.7405	.7422	.7440	.7457	.7475	.7492	.7509	.7527	.7544	.7561
5.8	.7579	.7596	.7613	.7630	.7647	.7664	.7681	.7699	.7716	.7733
5.9	.7750	.7766	.7783	.7800	.7817	.7834	.7851	.7867	.7884	.7901
6.0	1.7918	.7934	.7951	.7967	.7984	.8001	.8017	.8034	.8050	.8066
6.1	.8083	.8099	.8116	.8132	.8148	.8165	.8181	.8197	.8213	.8229
6.2	.8245	.8262	.8278	.8294	.8310	.8326	.8342	.8358	.8374	.8390
6.3	.8405	.8421	.8437	.8453	.8469	.8485	.8500	.8516	.8532	.8547
6.4	.8563	.8579	.8594	.8610	.8625	.8641	.8656	.8672	.8687	.8703
6.5	1.8718	.8733	.8749	.8764	.8779	.8795	.8810	.8825	.8840	.8856
6.6	.8871	.8886	.8901	.8916	.8931	.8946	.8961	.8976	.8991	.9006
6.7	.9021	.9036	.9051	.9066	.9081	.9095	.9110	.9125	.9140	.9155
6.8	.9169	.9184	.9199	.9213	.9228	.9242	.9257	.9272	.9286	.9301
6.9	.9315	.9330	.9344	.9359	.9373	.9387	.9402	.9416	.9430	.9445
7.0	1.9459	.9473	.9488	.9502	.9516	.9530	.9544	.9559	.9573	.9587
7.1	.9601	.9615	.9629	.9643	.9657	.9671	.9685	.9699	.9713	.9727
7.2	.9741	.9755	.9769	.9782	.9796	.9810	.9824	.9838	.9851	.9865
7.3	.9879	.9892	.9906	.9920	.9933	.9947	.9961	.9974	.9988	2.0001
7.4	2.0015	.0028	.0042	.0055	.0069	.0082	.0096	.0109	.0122	.0136
7.5	2.0149	.0162	.0176	.0189	.0202	.0215	.0229	.0242	.0255	.0268
7.6	.0281	.0295	.0308	.0321	.0334	.0347	.0360	.0373	.0386	.0399
7.7	.0412	.0425	.0438	.0451	.0464	.0477	.0490	.0503	.0516	.0528
7.8	.0541	.0554	.0567	.0580	.0592	.0605	.0618	.0631	.0643	.0656
7.9	.0669	.0681	.0694	.0707	.0719	.0732	.0744	.0757	.0769	.0782
8.0	2.0794	.0807	.0819	.0832	.0844	.0857	.0869	.0882	.0894	.0906
8.1	.0919	.0931	.0943	.0956	.0968	.0980	.0992	.1005	.1017	.1029
8.2	.1041	.1054	.1066	.1078	.1090	.1102	.1114	.1126	.1138	.1150
8.3	.1163	.1175	.1187	.1199	.1211	.1223	.1235	.1247	.1258	.1270
8.4	.1282	.1294	.1306	.1318	.1330	.1342	.1353	.1365	.1377	.1389
8.5	2.1401	.1412	.1424	.1436	.1448	.1459	.1471	.1483	.1494	.1506
8.6	.1518	.1529	.1541	.1552	.1564	.1576	.1587	.1599	.1610	.1622
8.7	.1633	.1645	.1656	.1668	.1679	.1691	.1702	.1713	.1725	.1736
8.8	.1748	.1759	.1770	.1782	.1793	.1804	.1815	.1827	.1838	.1849
8.9	.1861	.1872	.1883	.1894	.1905	.1917	.1928	.1939	.1950	.1961
9.0	2.1972	.1983	.1994	.2006	.2017	.2028	.2039	.2050	.2061	.2072
9.1	.2083	.2094	.2105	.2116	.2127	.2138	.2148	.2159	.2170	.2181
9.2	.2192	.2203	.2214	.2225	.2235	.2246	.2257	.2268	.2279	.2289
9.3	.2300	.2311	.2322	.2332	.2343	.2354	.2364	.2375	.2386	.2396
9.4	.2407	.2418	.2428	.2439	.2450	.2460	.2471	.2481	.2492	.2502
9.5	2.2513	.2523	.2534	.2544	.2555	.2565	.2576	.2586	.2597	.2607
9.6	.2618	.2628	.2638	.2649	.2659	.2670	.2680	.2690	.2701	.2711
9.7	.2721	.2732	.2742	.2752	.2762	.2773	.2783	.2793	.2803	.2814
9.8	.2824	.2834	.2844	.2854	.2865	.2875	.2885	.2895	.2905	.2915
9.9	.2925	.2935	.2946	.2956	.2966	.2976	.2986	.2996	.3006	.3016

Table C   Base-e Logarithms

N	0	1	2	3	4	5	6	7	8	9
1.0	.0000	.0100	.0198	.0296	.0392	.0488	.0583	.0677	.0770	.0862
1.1	.0953	.1044	.1133	.1222	.1310	.1398	.1484	.1570	.1655	.1740
1.2	.1823	.1906	.1989	.2070	.2151	.2231	.2311	.2390	.2469	.2546
1.3	.2624	.2700	.2776	.2852	.2927	.3001	.3075	.3148	.3221	.3293
1.4	.3365	.3436	.3507	.3577	.3646	.3716	.3784	.3853	.3920	.3988
1.5	.4055	.4121	.4187	.4253	.4318	.4383	.4447	.4511	.4574	.4637
1.6	.4700	.4762	.4824	.4886	.4947	.5008	.5068	.5128	.5188	.5247
1.7	.5306	.5365	.5423	.5481	.5539	.5596	.5653	.5710	.5766	.5822
1.8	.5878	.5933	.5988	.6043	.6098	.6152	.6206	.6259	.6313	.6366
1.9	.6419	.6471	.6523	.6575	.6627	.6678	.6729	.6780	.6831	.6881
2.0	.6931	.6981	.7031	.7080	.7129	.7178	.7227	.7275	.7324	.7372
2.1	.7419	.7467	.7514	.7561	.7608	.7655	.7701	.7747	.7793	.7839
2.2	.7885	.7930	.7975	.8020	.8065	.8109	.8154	.8198	.8242	.8286
2.3	.8329	.8372	.8416	.8459	.8502	.8544	.8587	.8629	.8671	.8713
2.4	.8755	.8796	.8838	.8879	.8920	.8961	.9002	.9042	.9083	.9123
2.5	.9163	.9203	.9243	.9282	.9322	.9361	.9400	.9439	.9478	.9517
2.6	.9555	.9594	.9632	.9670	.9708	.9746	.9783	.9821	.9858	.9895
2.7	.9933	.9969	1.0006	.0043	.0080	.0116	.0152	.0188	.0225	.0260
2.8	1.0296	.0332	.0367	.0403	.0438	.0473	.0508	.0543	.0578	.0613
2.9	.0647	.0682	.0716	.0750	.0784	.0818	.0852	.0886	.0919	.0953
3.0	1.0986	.1019	.1053	.1086	.1119	.1151	.1184	.1217	.1249	.1282
3.1	.1314	.1346	.1378	.1410	.1442	.1474	.1506	.1537	.1569	.1600
3.2	.1632	.1663	.1694	.1725	.1756	.1787	.1817	.1848	.1878	.1909
3.3	.1939	.1969	.2000	.2030	.2060	.2090	.2119	.2149	.2179	.2208
3.4	.2238	.2267	.2296	.2326	.2355	.2384	.2413	.2442	.2470	.2499
3.5	1.2528	.2556	.2585	.2613	.2641	.2669	.2698	.2726	.2754	.2782
3.6	.2809	.2837	.2865	.2892	.2920	.2947	.2975	.3002	.3029	.3056
3.7	.3083	.3110	.3137	.3164	.3191	.3218	.3244	.3271	.3297	.3324
3.8	.3350	.3376	.3403	.3429	.3455	.3481	.3507	.3533	.3558	.3584
3.9	.3610	.3635	.3661	.3686	.3712	.3737	.3762	.3788	.3813	.3838
4.0	1.3863	.3888	.3913	.3938	.3962	.3987	.4012	.4036	.4061	.4085
4.1	.4110	.4134	.4159	.4183	.4207	.4231	.4255	.4279	.4303	.4327
4.2	.4351	.4375	.4398	.4422	.4446	.4469	.4493	.4516	.4540	.4563
4.3	.4586	.4609	.4633	.4656	.4679	.4702	.4725	.4748	.4770	.4793
4.4	.4816	.4839	.4861	.4884	.4907	.4929	.4951	.4974	.4996	.5019
4.5	1.5041	.5063	.5085	.5107	.5129	.5151	.5173	.5195	.5217	.5239
4.6	.5261	.5282	.5304	.5326	.5347	.5369	.5390	.5412	.5433	.5454
4.7	.5476	.5497	.5518	.5539	.5560	.5581	.5602	.5623	.5644	.5665
4.8	.5686	.5707	.5728	.5748	.5769	.5790	.5810	.5831	.5851	.5872
4.9	.5892	.5913	.5933	.5953	.5974	.5994	.6014	.6034	.6054	.6074
5.0	1.6094	.6114	.6134	.6154	.6174	.6194	.6214	.6233	.6253	.6273
5.1	.6292	.6312	.6332	.6351	.6371	.6390	.6409	.6429	.6448	.6467
5.2	.6487	.6506	.6525	.6544	.6563	.6582	.6601	.6620	.6639	.6658
5.3	.6677	.6696	.6715	.6734	.6752	.6771	.6790	.6808	.6827	.6845
5.4	.6864	.6882	.6901	.6919	.6938	.6956	.6974	.6993	.7011	.7029

Use the properties of logarithms and ln 10 = 2.3026 to find logarithms of numbers less than 1 or greater than 10.

Getting Ready (page 2)

1. 1, 2, 3, etc. **2.** $\frac{1}{2}, \frac{2}{3}$, etc. **3.** $-3, -21$, etc.

Orals (page 11)

11. -15 **12.** 25

Exercise 1.1 (page 11)

1. set **3.** whole **5.** subset **7.** prime **9.** is not equal to **11.** is greater than or equal to **13.** number **15.** 1, 2, 6, 9
17. 1, 2, 6, 9 **19.** $-3, -1, 0, 1, 2, 6, 9$ **21.** $-3, -\frac{1}{2}, -1, 0, 1, 2, \frac{5}{3}, \sqrt{7}, 3.25, 6, 9$ **23.** $-3, -1, 1, 9$ **25.** 6, 9
27. 9; natural, odd, composite, and whole number **29.** 0; even integer, whole number **31.** 24; natural, even, composite, and
whole number **33.** 3; natural, odd, prime, and whole number **35.** $=$ **37.** $<$ **39.** $>$ **41.** $=$ **43.** $=$ **45.** $<$ **47.** $=$
49. $7 > 3$ **51.** $8 \leq 8$ **53.** $3 + 4 = 7$ **55.** $7 \geq 3$ **57.** $0 < 6$ **59.** $8 < 3 + 8$ **61.** $10 - 4 > 6 - 2$ **63.** $3 \cdot 4 > 2 \cdot 3$
65. $\frac{24}{6} > \frac{12}{4}$ **67.** [number line: 3, 6] ; 6, 6 **69.** [number line: 6, 11] ; 11, 11 **71.** [number line: 0, 2] ; 2, 2
73. [number line: 0, 8] ; 8, 8 **75.** [number line: 1–8] **77.** [number line: 10–20]
79. [number line: 0] **81.** [number line: 15–25] **83.** [number line: 1, 5]
85. [number line: -3, 3] **87.** 36 **89.** 0 **91.** 230 **93.** 8

Getting Ready (page 14)

1. 250 **2.** 148 **3.** 16,606 **4.** 105

Orals (page 27)

1. $\frac{1}{2}$ **2.** $\frac{1}{2}$ **3.** $\frac{1}{2}$ **4.** $\frac{1}{3}$ **5.** $\frac{5}{12}$ **6.** $\frac{9}{20}$ **7.** $\frac{4}{9}$ **8.** $\frac{6}{25}$ **9.** $\frac{11}{9}$ **10.** $\frac{3}{7}$ **11.** $\frac{1}{6}$ **12.** $\frac{5}{4}$ **13.** 2.86 **14.** 1.24 **15.** 0.5 **16.** 3.9 **17.** 3.24
18. 3.25

Exercise 1.2 (page 27)

1. true **3.** false **5.** true **7.** true **9.** $=$ **11.** $=$ **13.** numerator **15.** simplify **17.** proper **19.** 1 **21.** multiply
23. numerators, denominator **25.** plus **27.** repeating **29.** $\frac{1}{2}$ **31.** $\frac{3}{4}$ **33.** $\frac{4}{3}$ **35.** $\frac{9}{8}$ **37.** $\frac{3}{10}$ **39.** $\frac{8}{5}$ **41.** $\frac{3}{2}$ **43.** $\frac{1}{4}$ **45.** 10

47. $\frac{20}{3}$ **49.** $\frac{9}{10}$ **51.** $\frac{5}{8}$ **53.** $\frac{1}{4}$ **55.** $\frac{14}{5}$ **57.** 28 **59.** $\frac{1}{5}$ **61.** $\frac{6}{5}$ **63.** $\frac{1}{13}$ **65.** $\frac{5}{24}$ **67.** $\frac{19}{15}$ **69.** $\frac{17}{12}$ **71.** $\frac{22}{35}$ **73.** $\frac{9}{4}$ **75.** $\frac{29}{3}$ **77.** $5\frac{1}{5}$
79. $1\frac{2}{3}$ **81.** $1\frac{1}{4}$ **83.** $\frac{5}{9}$ **85.** 158.65 **87.** 44.785 **89.** 44.88 **91.** 4.55 **93.** 350.49 **95.** 55.21 **97.** 3,337.52 **99.** 10.02
101. $121\frac{3}{5}$ m **103.** $53\frac{1}{6}$ ft **105.** 2,514,820 **107.** 270 lb **109.** 43.13 sec **111.** \$18,151.15 **113.** \$2,143.23
115. the high-capacity boards **117.** 205,200 lb **119.** the high-efficiency furnace

Getting Ready (page 31)

1. 4 **2.** 9 **3.** 27 **4.** 8 **5.** $\frac{1}{4}$ **6.** $\frac{1}{27}$ **7.** $\frac{8}{125}$ **8.** $\frac{27}{1,000}$

Orals (page 40)

1. 32 **2.** 81 **3.** 64 **4.** 125 **5.** 24 **6.** 36 **7.** 11 **8.** 1 **9.** 16 **10.** 24

Exercise 1.3 (page 40)

1. **3.** prime number **5.** exponent **7.** multiplication **9.** $P = 4s$
11. $P = 2l + 2w$ **13.** $P = a + b + c$ **15.** $P = a + b + c + d$ **17.** $C = \pi D$ **19.** $V = lwh$ **21.** $V = \frac{1}{3}Bh$ **23.** $V = \frac{4}{3}\pi r^3$
25. 16 **27.** 36 **29.** $\frac{1}{10,000}$ **31.** 493.039 **33.** 640.09 **35.** $x \cdot x$ **37.** $3 \cdot z \cdot z \cdot z \cdot z$ **39.** $5t \cdot 5t$ **41.** $5 \cdot 2x \cdot 2x \cdot 2x$ **43.** 36
45. 1,000 **47.** 18 **49.** 216 **51.** 11 **53.** 3 **55.** 28 **57.** 64 **59.** 13 **61.** 16 **63.** 2 **65.** 16 **67.** 21 **69.** 17 **71.** 9
73. 8 **75.** 8 **77.** $\frac{1}{144}$ **79.** 11 **81.** 1 **83.** $\frac{8}{9}$ **85.** 1 **87.** 4 **89.** 4 **91.** 12 **93.** 4 **95.** 11 **97.** 24 **99.** 12 **101.** 25
103. 1 **105.** 28 **107.** 35 **109.** 1 **111.** $(3 \cdot 8) + (5 \cdot 3)$ **113.** $(3 \cdot 8 + 5) \cdot 3$ **115.** 16 in. **117.** 15 m **119.** 25 m^2
121. 60 ft^2 **123.** 88 m **125.** 1,386 ft^2 **127.** 6 cm^3 **129.** 905 m^3 **131.** 1,056 cm^3 **133.** 40,764.51 ft^3 **135.** 480 ft^3 **137.** 8
141. bigger

Getting Ready (page 44)

1. 17.52 **2.** 2.94 **3.** 2 **4.** 1 **5.** 96 **6.** 382

Orals (page 52)

1. 5 **2.** -3 **3.** 3 **4.** -11 **5.** 4 **6.** -12 **7.** 2 **8.** 16 **9.** -6 **10.** -6

Exercise 1.4 (page 52)

1. 20 **3.** 24 **5.** arrows **7.** subtract, greater **9.** add, opposite **11.** 12 **13.** -10 **15.** 2 **17.** -2 **19.** 0.5 **21.** $\frac{12}{35}$ **23.** 1
25. 2.2 **27.** 7 **29.** -1 **31.** -7 **33.** -8 **35.** 3 **37.** 1.3 **39.** -1 **41.** 3 **43.** 10 **45.** -3 **47.** -1 **49.** 9 **51.** 1
53. 4 **55.** 12 **57.** 5 **59.** $\frac{1}{2}$ **61.** $-8\frac{3}{4}$ **63.** -4.2 **65.** 4 **67.** -7 **69.** 10 **71.** 0 **73.** 8 **75.** 3 **77.** 2.45 **79.** 9 **81.** -3
83. -15 **85.** 1 **87.** 3 **89.** -1 **91.** 9.9 **93.** -7.1 **95.** \$175 **97.** $+9$ **99.** $-4°$ **101.** 2,000 yr **103.** 1,325 m
105. 4,000 ft **107.** 5° **109.** 9,187 **111.** 700 **113.** \$422.66 **115.** \$83,425.57

Getting Ready (page 56)

1. 56 **2.** 54 **3.** 72 **4.** 63 **5.** 9 **6.** 6 **7.** 8 **8.** 8

Orals (page 62)

1. -3 **2.** 10 **3.** 18 **4.** -24 **5.** 24 **6.** -24 **7.** -2 **8.** 2 **9.** -9 **10.** -1

Exercise 1.5 (page 62)

1. 1,125 lb **3.** 53 **5.** positive **7.** positive **9.** positive **11.** a **13.** 0 **15.** 48 **17.** 56 **19.** -144 **21.** -16 **23.** 2 **25.** 1
27. 72 **29.** -24 **31.** -420 **33.** -96 **35.** 4 **37.** -9 **39.** -2 **41.** 5 **43.** -3 **45.** -8 **47.** -8 **49.** 6 **51.** 5 **53.** 7
55. -4 **57.** 2 **59.** -4 **61.** -20 **63.** 2 **65.** 1 **67.** -6 **69.** -30 **71.** 7 **73.** -10 **75.** -10 **77.** 14 **79.** -81 **81.** 88

83. $-\frac{1}{6}$ **85.** $-\frac{11}{12}$ **87.** $-\frac{7}{36}$ **89.** $-\frac{11}{48}$ **91.** $(+2)(+3) = +6$ **93.** $(-30)(15) = -450$ **95.** $(+23)(-120) = -2,760$
97. $\frac{-18}{-3} = +6$ **99.** 2-point loss per day   **101.** yes

Getting Ready (page 64)

1. sum   **2.** product   **3.** quotient   **4.** difference   **5.** quotient   **6.** difference   **7.** product   **8.** sum

Orals (page 70)

1. 1   **2.** -14 **3.** -11 **4.** 7   **5.** 16   **6.** 64   **7.** -12 **8.** 36

Exercise 1.6 (page 70)

1. 532   **3.** $\frac{1}{2}$ **5.** sum   **7.** multiplication   **9.** algebraic   **11.** variables   **13.** $x + y$ **15.** $x(2y)$ **17.** $y - x$ **19.** $\frac{y}{x}$ **21.** $z + \frac{x}{y}$
23. $z - xy$ **25.** $3xy$ **27.** $\frac{x+y}{y+z}$ **29.** $xy + \frac{y}{z}$ **31.** the sum of x and 3   **33.** the quotient obtained when x is divided by y
35. the product of 2, x, and y **37.** the quotient obtained when 5 is divided by the sum of x and y **39.** the quotient obtained when
the sum of 3 and x is divided by y **41.** the product of x, y, and the sum of x and y **43.** $x + z$; 10   **45.** $y - z$; 2   **47.** $yz - 3$; 5
49. $\frac{xy}{z}$; 16   **51.** 1; 6   **53.** 3; -1 **55.** 4; 3   **57.** 3; -4 **59.** 4; 3   **61.** 19 and x **63.** 29, x, y, and z **65.** 3, x, y, and z
67. 17, x, and z **69.** 5, 1, and 8   **71.** x and y **73.** 75   **75.** x and y **77.** $c + 4$ **79.** $\$9,987t$ **81.** $\frac{x}{5}$ **83.** $\$(3d + 5)$

Getting Ready (page 73)

1. 17   **2.** 17   **3.** 38.6   **4.** 38.6   **5.** 56   **6.** 56   **7.** 0   **8.** 1   **9.** 777   **10.** 777

Exercise 1.7 (page 79)

1. $x + y^2 \geq z$ **3.** 0   **5.** positive   **7.** real   **9.** a **11.** $(b + c)$ **13.** ac **15.** a **17.** element, multiplication   **19.** $\frac{1}{a}$ **21.** 10
23. -24 **25.** 144   **27.** 3   **29.** Both are 12.   **31.** Both are 29.   **33.** Both are 60.   **35.** Both are 0.   **37.** Both are -6.
39. Both are -12.   **41.** $3x + 3y$ **43.** $x^2 + 3x$ **45.** $-xa - xb$ **47.** $4x^2 + 4x$ **49.** $-5t - 10$ **51.** $-2ax - 2a^2$ **53.** $2, \frac{1}{2}$
55. $-\frac{1}{3}, 3$ **57.** 0, none   **59.** $\frac{5}{2}, -\frac{2}{5}$ **61.** 0.2, -5 **63.** $-\frac{4}{3}, \frac{3}{4}$ **65.** comm. prop. of add.   **67.** comm. prop. of mult.
69. distrib. prop.   **71.** comm. prop. of add.   **73.** identity for mult.   **75.** add. inverse   **77.** $3x + 3 \cdot 2$ **79.** xy^2 **81.** $(y + x)z$
83. $x(yz)$ **85.** x

Chapter 1 Summary (page 82)

1. a. 1, 2, 3, 4, 5   **b.** 2, 3, 5   **c.** 1, 3, 5   **d.** 4   **2. a.** $-6, 0, 5$ **b.** $-6, -\frac{2}{3}, 0, 2.6, 5$ **c.** 5   **d.** all of them   **e.** $-6, 0$ **f.** 5
g. $\sqrt{2}, \pi$ **3. a.** $<$ **b.** $<$ **c.** $=$ **d.** $>$ **4. a.** 8   **b.** -8 **5. a.**

14   15   16   17   18   19   20

b.

19   20   21   22   23   24   25

c.

-3 2

d.

-4 3

6. a. 11   **b.** 31   **7. a.** $\frac{5}{3}$ **b.** 11
8. a. $\frac{1}{3}$ **b.** $\frac{1}{3}$ **c.** 1   **d.** $\frac{5}{2}$ **e.** $\frac{4}{3}$ **f.** $\frac{1}{3}$ **g.** $\frac{10}{21}$ **h.** $\frac{73}{63}$ **i.** $\frac{11}{21}$ **j.** $\frac{2}{15}$ **k.** $8\frac{11}{12}$ **l.** $2\frac{11}{12}$ **9. a.** 48.61   **b.** 12.99   **c.** 18.55   **d.** 3.7
10. a. 4.70   **b.** 26.36   **c.** 3.57   **d.** 3.75   **11.** 6.85 hr   **12.** 57   **13.** 40.2 ft   **14. a.** 81   **b.** $\frac{4}{9}$ **c.** 0.25   **d.** 33   **15. a.** 81
b. 8   **16.** 15,133.6 ft^3 **17. a.** 32   **b.** 7   **c.** 6   **d.** 3   **e.** 98   **f.** 38   **g.** 3   **h.** 15   **18. a.** 58   **b.** 4   **c.** 7   **d.** 3   **19. a.** 22
b. 1   **20. a.** 15   **b.** -57 **c.** -6.5 **d.** $\frac{1}{2}$ **e.** -12 **f.** 16   **g.** 1.2   **h.** -3.54 **i.** 19   **j.** 1   **k.** -5 **l.** -7 **m.** $\frac{3}{2}$ **n.** 1   **o.** 1
p. $-\frac{1}{7}$ **21. a.** -4 **b.** -1 **c.** -2 **d.** 5   **e.** 4   **f.** 6   **22. a.** 12   **b.** 60   **c.** $\frac{1}{4}$ **d.** 1.3875   **e.** -35 **f.** -105 **g.** $-\frac{2}{3}$
h. -45.14 **i.** 5   **j.** 7   **k.** $\frac{7}{2}$ **l.** 6   **m.** -5 **n.** -2 **o.** 26   **p.** 7   **q.** 6   **r.** $\frac{3}{2}$ **23. a.** -6 **b.** 3   **c.** 2   **d.** 6   **e.** -7 **f.** 39
g. 6   **h.** -2 **24. a.** xz **b.** $x + 2y$ **c.** $2(x + y)$ **d.** $x - yz$ **25. a.** the product of 3, x, and y **b.** 5 decreased by the product
of y and z **c.** 5 less than the product of y and z **d.** the sum of x, y, and z, divided by twice their product   **26.** 3   **27.** 7   **28.** 1
29. 9   **30. a.** closure prop.   **b.** comm. prop. of mult.   **c.** assoc. prop. of add.   **d.** distrib. prop.   **e.** comm. prop. of add.
f. assoc. prop. of mult.   **g.** comm. prop. of add.   **h.** identity for mult.   **i.** add. inverse   **j.** identity for add.

Chapter 1 Test (page 88)

1. 31, 37, 41, 43, 47 **2.** 2 **3.**

4.

5. -23 **6.** 0
7. $=$ **8.** $<$ **9.** $>$ **10.** $=$ **11.** $\frac{13}{20}$ **12.** 1 **13.** $\frac{4}{5}$ **14.** $\frac{9}{2} = 4\frac{1}{2}$ **15.** -1 **16.** $-\frac{1}{13}$ **17.** 77.7 **18.** 301.57 ft^2 **19.** 64 cm^2
20. 1,539 in.^{3} **21.** -2 **22.** -14 **23.** -4 **24.** 12 **25.** 5 **26.** -23 **27.** $\frac{xy}{x+y}$ **28.** $5y - (x + y)$ **29.** $24x + 14y$
30. $\$(12a + 8b)$ **31.** 3 **32.** 4 **33.** 0 **34.** 5 **35.** comm. prop. of mult. **36.** distrib. prop. **37.** comm. prop. of add.
38. mult. inverse prop.

Getting Ready (page 91)

1. -3 **2.** 7 **3.** 4 **4.** 7 **5.** 17 **6.** x

Orals (page 98)

1. 20 **2.** 16 **3.** 2 **4.** 4 **5.** 0 **6.** 0 **7.** $\frac{3}{5}$ **8.** 1 **9.** 80° **10.** 100°

Exercise 2.1 (page 99)

1. 15; integer, composite **3.** -1; integer **5.** closure prop. of add. **7.** comm. prop. of add. **9.** 64 **11.** 27 **13.** equation
15. root **17.** equivalent **19.** x **21.** equal **23.** markup **25.** supplementary **27.** yes **29.** no **31.** yes **33.** yes **35.** yes
37. no **39.** yes **41.** yes **43.** yes **45.** yes **47.** 6 **49.** 19 **51.** 6 **53.** 519 **55.** 74 **57.** -28 **59.** 2 **61.** $\frac{5}{6}$ **63.** $-\frac{1}{5}$
65. $\frac{1}{2}$ **67.** \$9,345 **69.** \$90 **71.** \$260 **73.** \$53,000 **75.** \$195 **77.** \$145,149 **79.** 10° **81.** 159° **83.** 27° **85.** 53°
87. 130°

Getting Ready (page 102)

1. 1 **2.** $\frac{1}{5}$ **3.** 1 **4.** 4 **5.** 4 **6.** 3 **7.** 63 **8.** 72

Orals (page 108)

1. 1 **2.** 1 **3.** -2 **4.** 0 **5.** 10 **6.** -20 **7.** -12 **8.** -24 **9.** 0.30 **10.** 8%

Exercise 2.2 (page 108)

1. $\frac{22}{15}$ **3.** $\frac{25}{27}$ **5.** 14 **7.** -317 **9.** 25.2 ft^2 **11.** equal **13.** bc **15.** 100 **17.** 3 **19.** -9 **21.** 27 **23.** -11 **25.** 25
27. -64 **29.** 15 **31.** -33 **33.** -2 **35.** $-\frac{1}{2}$ **37.** 98 **39.** 5 **41.** $\frac{5}{2}$ **43.** 4,912 **45.** 1 **47.** 85 **49.** $-\frac{3}{2}$ **51.** 2.4 **53.** 80
55. 19 **57.** 320 **59.** 380 **61.** 150 **63.** 20% **65.** 8% **67.** 200% **69.** 117 **71.** 1,519 **73.** 55% **75.** \$270 **77.** 5,600
81. about 3.16

Getting Ready (page 110)

1. 22 **2.** 36 **3.** 5 **4.** $\frac{13}{2}$ **5.** -1 **6.** -1 **7.** $\frac{7}{9}$ **8.** $-\frac{19}{3}$

Orals (page 116)

1. add 7 **2.** subtract 3 **3.** add 3 **4.** multiply by 7 **5.** add 5 **6.** subtract 5 **7.** multiply by 3 **8.** subtract 2 **9.** 3 **10.** 13

Exercise 2.3 (page 117)

1. 50 cm **3.** 80.325 in.^{2} **5.** cost **7.** percent **9.** 1 **11.** -1 **13.** 3 **15.** -2 **17.** 2 **19.** -5 **21.** $\frac{3}{2}$ **23.** 2 **25.** 3
27. -54 **29.** -9 **31.** -33 **33.** 10 **35.** -4 **37.** 28 **39.** 5 **41.** 7 **43.** -8 **45.** 10 **47.** 4 **49.** 10 **51.** $\frac{17}{5}$ **53.** $-\frac{2}{3}$
55. 0 **57.** 6 **59.** $\frac{3}{5}$ **61.** 5 **63.** \$250 **65.** 7 days **67.** 29 min **69.** \$7,400 **71.** no chance; he needs 112 **73.** \$50
75. 15% to 6% **79.** $\frac{7x + 4}{22} = \frac{1}{2}$

Getting Ready (page 119)

1. $3x + 4x$ **2.** $7x + 2x$ **3.** $8w - 3w$ **4.** $10y - 4y$ **5.** $7x$ **6.** $9x$ **7.** $5w$ **8.** $6y$

Orals (page 125)

1. $8x$ **2.** y **3.** 0 **4.** $-2y$ **5.** 12 **6.** $6x$ **7.** 3 **8.** impossible **9.** 2 **10.** $\frac{1}{3}$

Exercise 2.4 (page 125)

1. 0 **3.** 2 **5.** $\frac{13}{56}$ **7.** $\frac{48}{35}$ **9.** variables, like **11.** identity **13.** $20x$ **15.** $3x^2$ **17.** $9x + 3y$ **19.** $7x + 6$ **21.** $7z - 15$
23. $12x + 121$ **25.** $6y + 62$ **27.** $-2x + 7y$ **29.** $2 + y$ **31.** $5x + 7$ **33.** $5x^2 + 24x$ **35.** -2 **37.** 3 **39.** 1 **41.** 1 **43.** $\frac{1}{3}$
45. 2 **47.** 6 **49.** 35 **51.** -9 **53.** 0 **55.** -20 **57.** -41 **59.** 9 **61.** -1 **63.** 8 **65.** 5 **67.** 4 **69.** -3 **71.** 1
73. identity **75.** impossible equation **77.** 16 **79.** impossible equation **81.** identity **83.** identity **89.** 0

Getting Ready (page 127)

1. $(2x + 2)$ ft **2.** $4x$ ft **3.** $P = 2l + 2w$ **5.** $\$840$ **6.** 385 mi **7.** 5.6 gal **8.** 9.5 lb

Orals (page 134)

1. $\$7d$ **2.** $\$18,000r$ **3.** $\frac{4}{6}$ ft **4.** $\left(\frac{P}{2} - 9\right)$ ft

Exercise 2.5 (page 135)

1. 200 cm^3 **3.** $7x - 6$ **5.** $-\frac{3}{2}$ **7.** $\$1,488$ **9.** $2l + 2w$ **11.** vertex **13.** $d = rt$ **15.** 4 ft and 8 ft **17.** 19 ft **19.** 29 m by 18 m
21. 17 in. by 39 in. **23.** $60°$ **25.** $\$4,500$ at 9% and $\$19,500$ at 14% **27.** $\$3,750$ **29.** $\$5,000$ **31.** 6% and 7% **33.** 3 hr
35. 6.5 hr **37.** 7.5 hr **39.** 500 mph **41.** 20 gal **43.** 50 gal **45.** 7.5 oz **47.** 40 lb lemon drops and 60 lb jelly beans
49. $\$1.20$ **51.** 80 lb

Getting Ready (page 139)

1. 3 **2.** -5 **3.** r **4.** $-a$ **5.** 7 **6.** 12 **7.** d **8.** s

Orals (page 143)

1. $a = \frac{d-c}{b}$ **2.** $b = \frac{d-c}{a}$ **3.** $c = d - ab$ **4.** $d = ab + c$ **5.** $a = \frac{c}{d} - b$ **6.** $b = \frac{c}{d} - a$ **7.** $c = d(a + b)$ **8.** $d = \frac{c}{a+b}$

Exercise 2.6 (page 144)

1. $5x - 5y$ **3.** $-x - 13$ **5.** literal **7.** isolate **9.** subtract **11.** $I = E/R$ **13.** $w = V/(lh)$ **15.** $b = P - a - c$
17. $w = (P - 2l)/2$ **19.** $t = (A - P)/(Pr)$ **21.** $r = C/(2\pi)$ **23.** $w = 2gK/v^2$ **25.** $R = P/I^2$ **27.** $g = wv^2/(2K)$
29. $M = Fd^2/(Gm)$ **31.** $d^2 = GMm/F$ **33.** $r = Gl/(2b) + 1$ or $r = (G + 2b)/(2b)$ **35.** $t = \frac{d}{r}$; $t = 3$ **37.** $t = \frac{i}{pr}$; $t = 2$
39. $c = P - a - b$; $c = 3$ **41.** $h = \frac{2K}{a+b}$; $h = 8$ **43.** $I = E/R$; $I = 4$ amp **45.** $r = C/(2\pi)$; $r = 2.28$ ft **47.** $R = P/I^2$;
$R = 13.78$ ohms **49.** $m = Fd^2/(GM)$ **51.** $D = (L - 3.25r - 3.25R)/2$; $D = 6$ ft **55.** $90{,}000{,}000{,}000$ joules

Getting Ready (page 146)

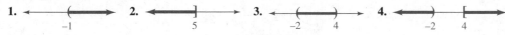

Orals (page 152)

1. $x < 2$ **2.** $x \geq 1$ **3.** $x \geq 2$ **4.** $x < -2$ **5.** $x < 6$ **6.** $x > -1$

Exercise 2.7   (page 152)

1. $5x^2 - 2y^2$ **3.** $-x + 14$ **5.** is less than   **7.** $\geq$ **9.** inequality   **11.** $x > 3$;   **13.** $x \geq -10$;   (−10)

15. $x < -1$; (−1)   **17.** $x \leq 4$; (4)   **19.** $x < -2$; (−2)

21. $x < -4$; (−4)   **23.** $x \leq -1$; (−1)   **25.** $x \geq -13$; (−13)

27. $x > -3$; (−3)   **29.** $x < -2$; (−2)   **31.** $x \geq 2$; (2)   **33.** $x > 3$; (3)

35. $x > -15$; (−15)   **37.** $x \leq 20$; (20)   **39.** $x \geq 3$; (3)

41. $x > -7$; (−7)   **43.** $x \geq 4$; (4)   **45.** $7 < x < 10$; (7 10)

47. $-9 < x \leq 3$; (−9 3)   **49.** $-10 \leq x \leq 0$; (−10 0)   **51.** $-5 < x < -2$; (−5 −2)

53. $-6 \leq x \leq 10$; (−6 10)   **55.** $2 \leq x < 3$; (2 3)   **57.** $-1 \leq x < 2$; (−1 2)

59. $-4 < x < 1$; (−4 1)   **61.** $-2 < x < 2$; (−2 2)   **63.** $98\% \leq s \leq 100\%$ **65.** $r \geq 27$ mpg

67. 0 ft $< s \leq 19$ ft   **69.** 0.1 mi $\leq x \leq 2.5$ mi   **71.** 3.3 mi $< x < 4.1$ mi   **73.** $66.2° < F < 71.6°$ **75.** 37.052 in. $< C < 38.308$ in.
77. 68.18 kg $< w < 86.36$ kg   **79.** 5 ft $< w < 9$ ft

Chapter 2 Summary   (page 157)

1. a. yes   **b.** no   **c.** no   **d.** yes   **e.** yes   **f.** no   **2. a.** 1   **b.** 9   **c.** 16   **d.** 0   **e.** 4   **f.** -2 **3.** $105.40   **4.** $97.70   **5.** $21°$
6. $111°$ **7. a.** 5   **b.** -2 **c.** $\frac{1}{2}$ **d.** $\frac{3}{2}$ **e.** 18   **f.** -35 **g.** $-\frac{1}{2}$ **h.** 6   **8. a.** 245   **b.** 1,300   **c.** 37%   **d.** 12.5%   **9.** about 81%
10. a. 3   **b.** 2   **c.** 1   **d.** 1   **e.** 1   **f.** -2 **g.** 2   **h.** 7   **i.** -2 **j.** -1 **k.** 5   **l.** 3   **m.** 13   **n.** -12 **o.** 5   **p.** 7   **q.** 8
r. 30   **s.** $\frac{15}{2}$ **t.** 44   **11.** $320   **12.** 6.5%   **13.** 96.4%   **14.** 53.8%   **15. a.** $14x$ **b.** $19a$ **c.** $5b$ **d.** $-2x$ **e.** $-2y$
f. not like terms   **g.** $9x$ **h.** $6 - 7x$ **i.** $4y^2 - 6$ **j.** 4   **16. a.** 7   **b.** 13   **c.** -3 **d.** -41 **e.** 9   **f.** -7 **g.** 7   **h.** 4   **i.** -8
j. -18 **17. a.** identity   **b.** contradiction   **c.** identity   **18.** 5 ft from one end   **19.** 13 in.   **20.** $16,000 at 7%, $11,000 at 9%
21. 20 min   **22.** 24 liters   **23.** 10 lb of each   **24.** 147 kwh   **25.** 85 ft   **26. a.** $R = \frac{E}{I}$ **b.** $t = \frac{i}{pr}$ **c.** $R = \frac{P}{I^2}$ **d.** $r = \frac{d}{t}$
e. $h = \frac{V}{lw}$ **f.** $m = \frac{y-b}{x}$ **g.** $h = \frac{V}{\pi r^2}$ **h.** $r = \frac{a}{2\pi h}$ **i.** $G = \frac{Fd^2}{Mm}$ **j.** $m = \frac{RT}{PV}$ **27. a.** (→ 1)   **b.** (→ −3)

c. (4)   **d.** (6)   **e.** (3)   **f.** (3)   **g.** (6 11)

h. (−1 1)

Chapter 2 Test   (page 161)

1. solution   **2.** solution   **3.** not a solution   **4.** solution   **5.** -36 **6.** 47   **7.** -12 **8.** -7 **9.** -2 **10.** 1   **11.** 7   **12.** -2
13. -3 **14.** 0   **15.** $6x - 15$ **16.** $8x - 10$ **17.** $-18x$ **18.** $-36x^2 + 13x$ **19.** $\frac{3}{5}$ hr   **20.** $7\frac{1}{2}$ liters   **21.** $t = \frac{d}{r}$ **22.** $l = \frac{P-2w}{2}$
23. $h = \frac{A}{2\pi r}$ **24.** $r = \frac{A-P}{Pt}$ **25.** (3)   **26.** (−28)   **27.** (−3 4)   **28.** (3/5 3)

Cumulative Review Exercises   (page 162)

1. integer, rational, real, positive   **2.** rational, real, negative   **3.** (1 2 3 4 5 6 7)   **4.** (2 7)   **5.** 0
6. $\frac{10}{3}$ **7.** $8\frac{1}{10}$ **8.** 35.65   **9.** 0   **10.** -2 **11.** 16   **12.** 0   **13.** 24.75   **14.** 5,275   **15.** 5   **16.** 37, y **17.** $-2x + 2y$ **18.** $x - 5$
19. x^2y^3 **20.** $4x^2$ **21.** 13   **22.** 41   **23.** $\frac{7}{4}$ **24.** -11 **25.** $22,814.56   **26.** $900   **27.** $12,650   **28.** 125 lb   **29.** no
30. 7.3 and 10.7 ft   **31.** $h = \frac{2A}{b+B}$ **32.** $x = \frac{y-b}{m}$ **33.** -9 **34.** 1   **35.** 280   **36.** -564 **37.** (→)
38. (−5 −1)   (−14)

Getting Ready   (page 165)

1.   2.   3.   4.

Orals   (page 174)

2. the origin   **3.** IV   **4.** y-axis

Exercise 3.1   (page 174)

1. 12   **3.** 8   **5.** 7   **7.** −49   **9.** ordered pair   **11.** origin   **13.** rectangular coordinate   **15.** no   **17.** origin, left, up   **19.** II
21. 3 or −3, 5 or −5, 4 or −4, 5 or −5, 3 or −3, 5 or −5, 4 or −4   **23.** 10 minutes before the workout, her heart rate was 60
beats per min.   **25.** 150 beats per min   **27.** approximately 5 min and 50 min after starting   **29.** 10 beats per min faster after
cooldown   **31.**

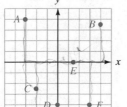

33. Carbondale (3, J), Champaign (4, D), Chicago (5, B), Peoria (3, C), Rockford (3, A),
Springfield (2, E), St. Louis (2, H)   **35. a.** 60°; 4 ft   **b.** 30°; 4 ft   **37. a.** $2   **b.** $4
c. $7   **d.** $9   **39.**

a. 35 mi
b. 4 gal
c. 32.5 mi

Distance (mi) vs Gasoline (gal)

41.

Value ($1,000s) vs Age of car (years)

a. A 3-yr-old car is worth $7,000.   **b.** $1,000   **c.** 6 yr

Getting Ready   (page 179)

1. 1   **2.** 5   **3.** −3   **4.** 2

Orals   (page 190)

1. 3

Exercise 3.2   (page 190)

1. −96   **3.** an expression   **5.** 1.25   **7.** 0.1   **9.** two   **11.** independent, dependent   **13.** linear   **15.** y-intercept   **17.** yes   **19.** no
21. −3, −2, −5   **23.** 0, −2, −6, 2, 4

25.

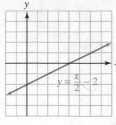

$y = 2x - 1$

27.

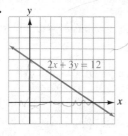

$y = \frac{x}{2} - 2$

29.

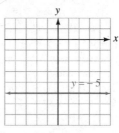

$x + y = 7$

31.

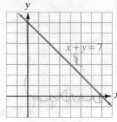

$x - y = 7$

33.

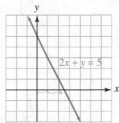

$2x + y = 5$

35.

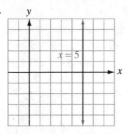

$2x + 3y = 12$

37.

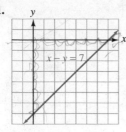

$y = -5$

39.

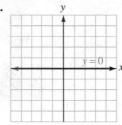

$x = 5$

41.

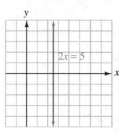

$y = 0$

43.

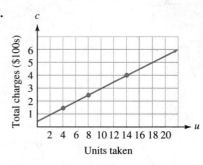

$2x = 5$

45.

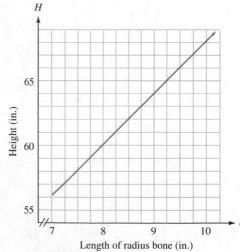

Total charges ($100s) vs Units taken

a. $c = 50 + 25u$
b. 150, 250, 400
c. The service fee is $50.
d. $850

47.

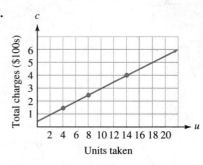

Height (in.) vs Length of radius bone (in.)

a. 56.2, 62.1, 64.0   **b.** taller the woman is   **c.** 58 in.
55. (6, 6)   **57.** $\left(-\frac{1}{2}, \frac{5}{2}\right)$ **59.** (7, 6)

Getting Ready (page 194)

1. -3 **2.** -2 **3.** 1   **4.** 6

Orals (page 203)

1. yes   **2.** yes   **3.** no   **4.** no

Exercise 3.3 (page 203)

1. 16   **3.** −18   **5.** system   **7.** independent   **9.** inconsistent   **11.** yes   **13.** yes   **15.** no   **17.** yes   **19.** no   **21.** no

23. 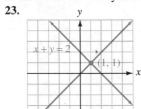   **25.** 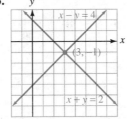   **27.** 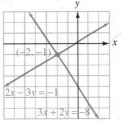   **29.**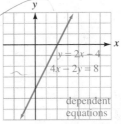

31. 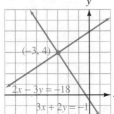   **33.** 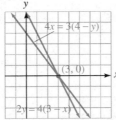   **35.** 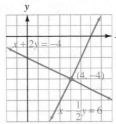   **37.**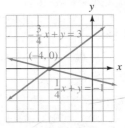

39. 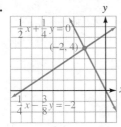   **41.**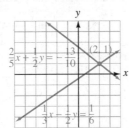

43. $(1, 3)$ **45.** $(0.67, -0.33)$ **47. a.** Donors outnumbered patients. **b.** 1994; 4,100   **c.** The patients outnumber the donors. **49. a.** Houston, New Orleans, St Augustine   **b.** St. Louis, Memphis, New Orleans   **c.** New Orleans

Getting Ready (page 207)

1. $6x + 4$ **2.** $-25 - 10x$ **3.** $2x - 4$ **4.** $3x - 12$

Orals (page 212)

1. $2z + 1$ **2.** $z + 2$ **3.** $3t + 3$ **4.** $\frac{t}{3} + 4$

Exercise 3.4 (page 212)

1. 5   **3.** 12   **5.** 10   **7.** y, terms   **9.** remove   **11.** infinitely many   **13.** $(2, 4)$ **15.** $(3, 0)$ **17.** $(-3, -1)$ **19.** inconsistent system   **21.** $(-2, 3)$ **23.** $(3, 2)$ **25.** $(3, -2)$ **27.** $(-1, 2)$ **29.** $(-1, -1)$ **31.** dependent equations   **33.** $(1, 1)$ **35.** $(4, -2)$ **37.** $(-3, -1)$ **39.** $(-1, -3)$ **41.** $\left(\frac{1}{2}, \frac{1}{3}\right)$ **43.** $(1, 4)$ **45.** $(4, 2)$ **47.** $(-5, -5)$ **49.** $(-6, 4)$ **51.** $\left(\frac{1}{5}, 4\right)$ **53.** $(5, 5)$

Getting Ready (page 213)

1. $5x = 10$ **2.** $y = 6$ **3.** $2x = 33$ **4.** $18y = 28$

Orals (page 219)

1. 1 **2.** 2 **3.** 3 **4.** 5

Exercise 3.5 (page 219)

1. 4 **3.** 4 **5.** $(-\infty, 2]$, 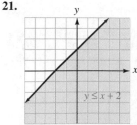 **7.** coefficient **9.** general **11.** 15 **13.** $(1, 4)$ **15.** $(-2, 3)$ **17.** $(-1, 1)$

19. $(2, 5)$ **21.** $(-3, 4)$ **23.** $(0, 8)$ **25.** $(2, 3)$ **27.** $(3, -2)$ **29.** $(2, 7)$ **31.** inconsistent system **33.** dependent equations
35. $(4, 0)$ **37.** $\left(\frac{10}{3}, \frac{10}{3}\right)$ **39.** $(5, -6)$ **41.** $(-5, 0)$ **43.** $(-1, 2)$ **45.** $\left(1, -\frac{5}{2}\right)$ **47.** $(-1, 2)$ **49.** $(0, 1)$ **51.** $(-2, 3)$ **53.** $(2, 2)$
57. $(1, 4)$

Getting Ready (page 221)

1. $x + y$ **2.** $x - y$ **3.** xy **4.** $\frac{x}{y}$ **5.** $A = lw$ **6.** $P = 2l + 2w$

Orals (page 230)

1. $2x$ **2.** $y + 1$ **3.** $2x + 3y$ **4.** $\$(3x + 2y)$ **5.** $\$(4x + 5y)$

Exercise 3.6 (page 230)

1. 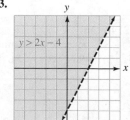 **3.** **5.** $8^3 c$ **7.** $a^2 b^2$ **9.** variable **11.** system **13.** 32, 64 **15.** 5, 8 **17.** 140
19. $\$15, \$5$ **21.** $\$5.40, \$6.20$ **23.** 10 ft, 15 ft **25.** $\$100,000$ **27.** causes: 24 min; outcome: 6 min **29.** 90,000 accidents;
540,000 cancer **31.** 25 ft by 30 ft **33.** 60 ft^2 **35.** 9.9 yr **37.** 80+ **39.** $\$2,000$ **41.** 250 **43.** 10 mph **45.** 50 mph
47. 5 L of 40% solution, 10 L of 55% solution **49.** 32 lb peanuts, 16 lb cashews **51.** 15 **53.** 9% **57.** 2

Getting Ready (page 234)

1. below **2.** above **3.** below **4.** on **5.** on **6.** above **7.** below **8.** above

Orals (page 243)

1. no **2.** no **3.** yes **4.** yes **5.** no **6.** yes **7.** no **8.** yes

Exercise 3.7 (page 243)

1. 3 **3.** $t = \frac{A - P}{Pr}$ **5.** $7a - 15$ **7.** $-2a + 7b$ **9.** inequality **11.** boundary **13.** inequalities **15.** doubly shaded **17. a.** yes
b. no **c.** yes **d.** no **19. a.** no **b.** yes
21. **23.** **25.** **27.**

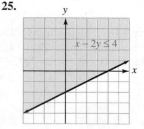
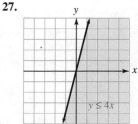

29.
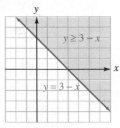

31.
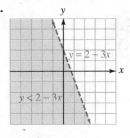

33.
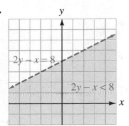

35.
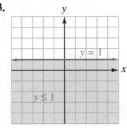

37.
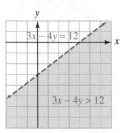

39.
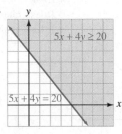

41.
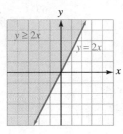

43.
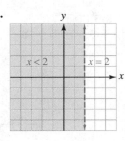

45. (10, 10), (20, 10), (10, 20)
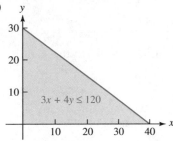

47. (50, 50), (30, 40), (40, 40)
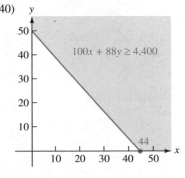

49. (80, 40), (80, 80), (120, 40)
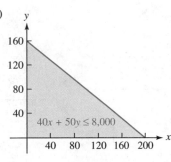

51.
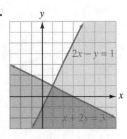

53.
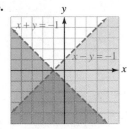

55.
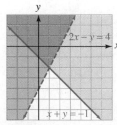

57.
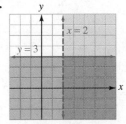

59.
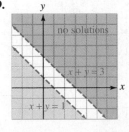

61.
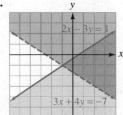

63.

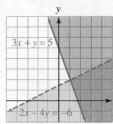

65.

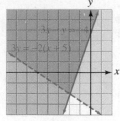

67.

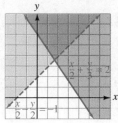

69.

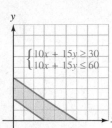

1 \$10 CD and 2 \$15 CDs;   **71.**
4 \$10 CDs and 1 \$15 CD

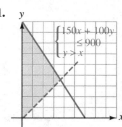

2 desk chairs and 4 side chairs;
1 desk chair and 5 side chairs

Chapter 3 Summary (page 250)

1.

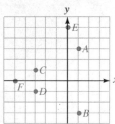

2. a. $(3, 1)$ **b.** $(-4, 5)$ **c.** $(-3, -4)$ **d.** $(2, -3)$ **e.** $(0, 0)$ **f.** $(0, 4)$ **g.** $(-5, 0)$ **h.** $(0, -3)$

3. a. no   **b.** yes

4. a.

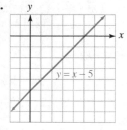

b.

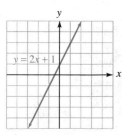

c.

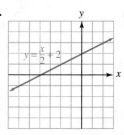

d.

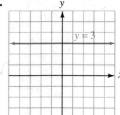

e.

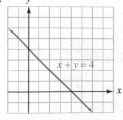

f.

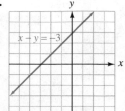

g.

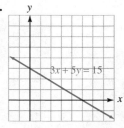

h.

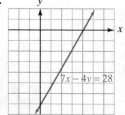

5. a. yes   **6. a.**
   b. no
   c. yes
   d. yes

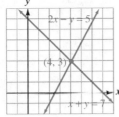   **b.** 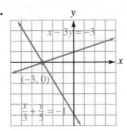   **c.** 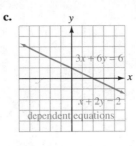   **d.**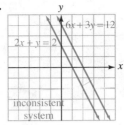

7. a. $(-1, -2)$ **b.** $(-2, 5)$ **c.** $(1, -1)$ **d.** $(-2, 1)$ **8. a.** $(3, -5)$ **b.** $\left(3, \frac{1}{2}\right)$ **c.** $(-1, 7)$ **d.** $\left(-\frac{1}{2}, \frac{7}{2}\right)$ **e.** $(0, 9)$
f. inconsistent system   **g.** dependent equations   **h.** $(0, 0)$ **9.** 3, 15   **10.** 3 ft by 9 ft   **11.** 50¢   **12.** $66   **13.** $1.69   **14.** $750

15. a. 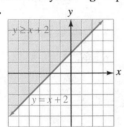   **b.**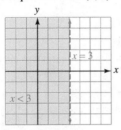

16. a. 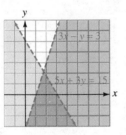   **b.** 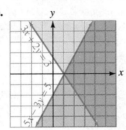   **c.** 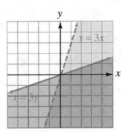   **d.**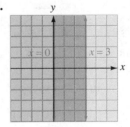

Chapter 3 Test   (page 255)

1. 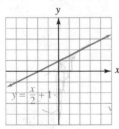   **2.** 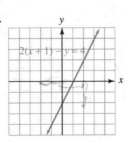   **3.** 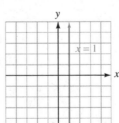   **4.** 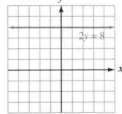   **5.** yes   **6.** no

7. 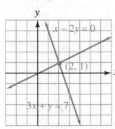   **8.** 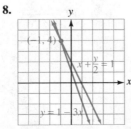   **9.** $(-3, -4)$ **10.** $(12, 10)$ **11.** $(2, 4)$ **12.** $(-3, 3)$ **13.** inconsistent
14. consistent   **15.** 65   **16.** $4,000

17. 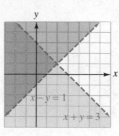   **18.**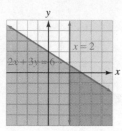

Getting Ready (page 258)
1. 8   **2.** 9   **3.** 6   **4.** 6   **5.** 12   **6.** 32   **7.** 18   **8.** 3

Orals (page 264)
1. base x, exponent 3   **2.** base 3, exponent x **3.** base b, exponent c **4.** base ab, exponent c **5.** 36   **6.** 36   **7.** 9   **8.** 27

Exercise 4.1 (page 265)
1. ⟵├──┼──•──┼──•──┼──•──┼──•──┼──•──┼──•──┼──⟶
      −4 −3 −2 −1 0 1 2 3   **3.** the product of 3 and the sum of x and y **5.** $|2x| + 3$ **7.** −5, 3
9. $(3x)(3x)(3x)(3x)$ **11.** $y \cdot y \cdot y \cdot y \cdot y$ **13.** $x^n y^n$ **15.** $a^{b \cdot c}$ **19.** base 4, exponent 3   **21.** base x, exponent 5   **23.** base $2y$, exponent 3   **25.** base x, exponent 4   **27.** base x, exponent 1   **29.** base x, exponent 3   **31.** $5 \cdot 5 \cdot 5$ **33.** $x \cdot x \cdot x \cdot x \cdot x \cdot x \cdot x$
35. $-4 \cdot x \cdot x \cdot x \cdot x \cdot x$ **37.** $(3t)(3t)(3t)(3t)(3t)$ **39.** 2^3 **41.** x^4 **43.** $(2x)^3$ **45.** $-4t^4$ **47.** 625   **49.** 13   **51.** 561   **53.** −725
55. x^7 **57.** x^{10} **59.** t^3 **61.** a^{12} **63.** y^9 **65.** $12x^7$ **67.** $-4y^5$ **69.** $6x^9$ **71.** 3^8 **73.** y^{15} **75.** a^{21} **77.** x^{25} **79.** $243z^{30}$
81. x^{31} **83.** r^{36} **85.** s^{33} **87.** $x^3 y^3$ **89.** $r^6 s^4$ **91.** $16a^2 b^4$ **93.** $-8r^6 s^9 t^3$ **95.** $\dfrac{a^3}{b^3}$ **97.** $\dfrac{x^{10}}{y^{15}}$ **99.** $\dfrac{-32a^5}{b^5}$ **101.** $\dfrac{b^6}{27a^3}$
103. x^2 **105.** y^4 **107.** $3a$ **109.** ab^4 **111.** $\dfrac{10r^{13}s^3}{3}$ **113.** $\dfrac{x^{12}y^{16}}{2}$ **115.** $\dfrac{y^3}{8}$ **117.** $-\dfrac{8r^3}{27}$ **119.** 2 ft   **121.** $16,000

Getting Ready (page 267)
1. $\frac{1}{3}$ **2.** $\frac{1}{y}$ **3.** 1   **4.** $\frac{1}{xy}$

Orals (page 271)
1. $\frac{1}{2}$ **2.** $\frac{1}{4}$ **3.** 2   **4.** 1   **5.** x **6.** $\frac{1}{y^7}$ **7.** 1   **8.** $\frac{y}{x}$

Exercise 4.2 (page 272)
1. 2   **3.** $\frac{6}{5}$ **5.** $s = \dfrac{f(P-L)}{i}$ or $s = \dfrac{fP - fL}{i}$ **7.** 1   **9.** 1   **11.** 8   **13.** 1   **15.** 1   **17.** 512   **19.** 2   **21.** 1   **23.** 1   **25.** − 2
27. $\dfrac{1}{x^2}$ **29.** $\dfrac{1}{b^5}$ **31.** $\dfrac{1}{16y^4}$ **33.** $\dfrac{1}{a^3 b^6}$ **35.** $\dfrac{1}{y}$ **37.** $\dfrac{1}{r^6}$ **39.** y^5 **41.** 1   **43.** $\dfrac{1}{a^2 b^4}$ **45.** $\dfrac{1}{x^6 y^3}$ **47.** $\dfrac{1}{x^3}$ **49.** $\dfrac{1}{y^2}$ **51.** $a^8 b^{12}$
53. $-\dfrac{y^{10}}{32x^{15}}$ **55.** a^{14} **57.** $\dfrac{1}{b^{14}}$ **59.** $\dfrac{256x^{28}}{81}$ **61.** $\dfrac{16y^{14}}{z^{10}}$ **63.** $\dfrac{x^{14}}{128y^{28}}$ **65.** $\dfrac{16u^4 v^8}{81}$ **67.** $\dfrac{1}{9a^2 b^2}$ **69.** $\dfrac{c^{15}}{216a^9 b^3}$ **71.** $\dfrac{1}{512}$
73. $\dfrac{17y^{27} z^5}{x^{35}}$ **75.** x^{3m} **77.** u^{5m} **79.** y^{2m+2} **81.** y^m **83.** $\dfrac{1}{x^{3n}}$ **85.** x^{2m+2} **87.** x^{8n-12} **89.** y^{4n-8} **91.** $6,678.04
93. $3,183.76

Getting Ready (page 274)
1. 100   **2.** 1,000   **3.** 10   **4.** $\frac{1}{100}$ **5.** 500   **6.** 8,000   **7.** 30   **8.** $\frac{7}{100}$

Orals (page 279)

1. 3.72×10^2 **2.** 37.2   **3.** 4.72×10^3 **4.** 3.72×10^3 **5.** 3.72×10^{-1} **6.** 2.72×10^{-2}

Exercise 4.3 (page 279)

1. 5   **3.** comm. prop. of add.   **5.** 6   **7.** scientific notation   **9.** 2.3×10^4 **11.** 1.7×10^6 **13.** 6.2×10^{-2} **15.** 5.1×10^{-6}
17. 4.25×10^3 **19.** 2.5×10^{-3} **21.** 230   **23.** 812,000   **25.** 0.00115   **27.** 0.000976   **29.** 25,000,000   **31.** 0.00051
33. 714,000   **35.** 30,000   **37.** 200,000   **39.** 2.57×10^{13} mi   **41.** 114,000,000 mi   **43.** 6.22×10^{-3} mi   **45.** 1.9008×10^{11} ft
47. 3.3×10^{-1} km/sec

Getting Ready (page 281)

1. $2x^2y^3$ **2.** $3xy^3$ **3.** $2x^2 + 3y^2$ **4.** $x^3 + y^3$ **5.** $6x^3y^3$ **6.** $5x^2y^2z^4$ **7.** $5x^2y^2$ **8.** $x^3y^3z^3$

Exercise 4.4 (page 291)

1. 8   **3.**
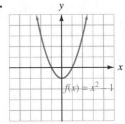
(number line with mark at -3)
   5. x^{18} **7.** y^9 **9.** algebraic   **11.** polynomial   **13.** trinomial   **15.** degree   **17.** function

19. domain   **21.** yes   **23.** yes   **25.** binomial   **27.** trinomial   **29.** monomial   **31.** binomial   **33.** trinomial   **35.** none of these
37. 4th   **39.** 3rd   **41.** 8th   **43.** 6th   **45.** 12th   **47.** 0th   **49.** 7   **51.** -8 **53.** -4 **55.** -5 **57.** 3   **59.** 11
61. $1, -2, -3, -2, 1$ **63.** $-6, 1, 2, 3, 10$ **65.**
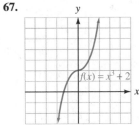
$f(x) = x^2 - 1$
   67. $f(x) = x^3 + 2$ **69.** 64 ft   **71.** 63 ft

Getting Ready (page 294)

1. $5x$ **2.** $2y$ **3.** $25x$ **4.** $5z$ **5.** $12r$ **6.** not possible   **7.** 0   **8.** not possible

Orals (page 298)

1. $4x^3$ **2.** $4xy$ **3.** $2y$ **4.** $2 - 2x$ **5.** $-2y^2$ **6.** $6x^2 + 4y$ **7.** $4x^2 + y$ **8.** $2y$

Exercise 4.5 (page 299)

1. -8 **3.** -9 **5.** (number line with mark at 3)   **7.** monomial   **9.** coefficients, variables   **11.** like terms   **13.** like terms, $7y$

15. unlike terms   **17.** like terms, $13x^3$ **19.** like terms, $8x^3y^2$ **21.** like terms, $65t^6$ **23.** unlike terms   **25.** $9y$ **27.** $-12t^2$
29. $16u^3$ **31.** $7x^5y^2$ **33.** $14rst$ **35.** $-6a^2bc$ **37.** $15x^2$ **39.** $4x^2y^2$ **41.** $95x^8y^4$ **43.** $7x + 4$ **45.** $2a + 7$ **47.** $7x - 7y$
49. $-19x - 4y$ **51.** $6x^2 + x - 5$ **53.** $7b + 4$ **55.** $3x + 1$ **57.** $5x + 15$ **59.** $3x - 3y$ **61.** $5x^2 - 25x - 20$
63. $5x^2 + x + 11$ **65.** $-7x^3 - 7x^2 - x - 1$ **67.** $2x^2y + xy + 13y^2$ **69.** $5x^2 + 6x - 8$ **71.** $-x^3 + 6x^2 + x + 14$
73. $-12x^2y^2 - 13xy + 36y^2$ **75.** $6x^2 - 2x - 1$ **77.** $t^3 + 3t^2 + 6t - 5$ **79.** $-3x^2 + 5x - 7$ **81.** $6x - 2$
83. $-5x^2 - 8x - 19$ **85.** $4y^3 - 12y^2 + 8y + 8$ **87.** $3a^2b^2 - 6ab + b^2 - 6ab^2$ **89.** $-6x^2y^2 + 4xy^2z - 20xy^3 + 2y$
91. \$114,000   **93.** \$132,000   **95. a.** \$263,000   **b.** \$263,000   **97.** $y = -1,100x + 6,600$ **99.** $y = -2,800x + 15,800$
103. $6x + 3h - 10$ **105.** 49

Getting Ready (page 302)

1. $6x$ **2.** $3x^4$ **3.** $5x^3$ **4.** $8x^5$ **5.** $3x + 15$ **6.** $x^2 + 5x$ **7.** $4y - 12$ **8.** $2y^2 - 6y$

Orals (page 310)

1. $6x^3 - 2x^2$ **2.** $10y^3 - 15y$ **3.** $7x^2y + 7xy^2$ **4.** $-4xy + 6y^2$ **5.** $x^2 + 5x + 6$ **6.** $x^2 - x - 6$ **7.** $2x^2 + 7x + 6$ **8.** $9x^2 - 1$
9. $x^2 + 6x + 9$ **10.** $x^2 - 10x + 25$

Exercise 4.6 (page 310)

1. distrib. prop. **3.** comm. prop. of mult. **5.** 0 **7.** monomial **9.** trinomial **11.** $6x^2$ **13.** $15x$ **15.** $12x^5$ **17.** $-24b^6$
19. $6x^5y^5$ **21.** $-3x^4y^7z^8$ **23.** $x^{10}y^{15}$ **25.** $a^5b^4c^7$ **27.** $3x + 12$ **29.** $-4t - 28$ **31.** $3x^2 - 6x$ **33.** $-6x^4 + 2x^3$
35. $3x^2y + 3xy^2$ **37.** $6x^4 + 8x^3 - 14x^2$ **39.** $2x^7 - x^2$ **41.** $-6r^3t^2 + 2r^2t^3$ **43.** $-6x^4y^4 - 6x^3y^5$ **45.** $a^2 + 9a + 20$
47. $3x^2 + 10x - 8$ **49.** $6a^2 + 2a - 20$ **51.** $6x^2 - 7x - 5$ **53.** $2x^2 + 3x - 9$ **55.** $6s^2 + 7st - 3t^2$ **57.** $x^2 + xz + xy + yz$
59. $u^2 + 2tu + uv + 2tv$ **61.** $-4r^2 - 20rs - 21s^2$ **63.** $4x^2 + 11x + 6$ **65.** $12x^2 + 14xy - 10y^2$ **67.** $x^3 - 1$
69. $2x^3 + 7x^2 + x - 1$ **71.** $x^2 + 8x + 16$ **73.** $t^2 - 6t + 9$ **75.** $r^2 - 16$ **77.** $x^2 + 10x + 25$ **79.** $4s^2 + 4s + 1$
81. $16x^2 - 25$ **83.** $x^2 - 4xy + 4y^2$ **85.** $4a^2 - 12ab + 9b^2$ **87.** $16x^2 - 25y^2$ **89.** $2x^2 - 6x - 8$ **91.** $3a^3 - 3ab^2$
93. $4t^3 + 11t^2 + 18t + 9$ **95.** $-3x^3 + 25x^2y - 56xy^2 + 16y^3$ **97.** $x^3 - 8y^3$ **99.** $5t^2 - 11t$ **101.** $x^2y + 3xy^2 + 2x^2$
103. $2x^2 + xy - y^2$ **105.** $8x$ **107.** $5s^2 - 7s - 9$ **109.** -3 **111.** -8 **113.** -1 **115.** 0 **117.** 1 **119.** 4 m **121.** 90 ft

Getting Ready (page 313)

1. $2xy^2$ **2.** y **3.** $\frac{3xy}{2}$ **4.** $\frac{x}{y}$ **5.** xy **6.** 3

Orals (page 318)

1. $2x^2$ **2.** $2y$ **3.** $5bc^2$ **4.** $-2pq$ **5.** 1 **6.** $3x$

Exercise 4.7 (page 318)

1. binomial **3.** none of these **5.** 2 **7.** polynomial **9.** two **11.** $\frac{a}{b}$ **13.** $\frac{1}{3}$ **15.** $-\frac{5}{3}$ **17.** $\frac{3}{4}$ **19.** 1 **21.** $-\frac{1}{4}$ **23.** $\frac{42}{19}$ **25.** $\frac{x}{z}$
27. $\frac{r^2}{s}$ **29.** $\frac{2x^2}{y}$ **31.** $-\frac{3u^3}{v^2}$ **33.** $\frac{4r}{y^2}$ **35.** $-\frac{13}{3rs}$ **37.** $\frac{x^4}{y^6}$ **39.** a^8b^8 **41.** $-\frac{3r}{s^9}$ **43.** $-\frac{x^3}{4y^3}$ **45.** $\frac{125}{8b^3}$ **47.** $\frac{xy^2}{3}$ **49.** a^8
51. z^3 **53.** $\frac{2}{y} + \frac{3}{x}$ **55.** $\frac{1}{5y} - \frac{2}{5x}$ **57.** $\frac{1}{y^2} + \frac{2y}{x^2}$ **59.** $3a - 2b$ **61.** $\frac{1}{y} - \frac{1}{2x} + \frac{2z}{xy}$ **63.** $3x^2y - 2x - \frac{1}{y}$ **65.** $5x - 6y + 1$
67. $\frac{10x^2}{y} - 5x$ **69.** $-\frac{4x}{3} + \frac{3x^2}{2}$ **71.** $xy - 1$ **73.** $\frac{x}{y} - \frac{11}{6} + \frac{y}{2x}$ **75.** 2 **77.** yes **79.** yes

Getting Ready (page 320)

1. 13 **2.** 21 **3.** 19 **4.** 13

Orals (page 325)

1. $2 + \frac{3}{x}$ **2.** $3 - \frac{5}{x}$ **3.** $2 + \frac{1}{x+1}$ **4.** $3 + \frac{2}{x+1}$ **5.** x **6.** x

Exercise 4.8 (page 325)

1. 21, 22, 24, 25, 26, 27, 28 **3.** 5 **5.** -5 **7.** $8x^2 - 6x + 1$ **9.** divisor, dividend **11.** remainder **13.** $4x^3 - 2x^2 + 7x + 6$
15. $6x^4 - x^3 + 2x^2 + 9x$ **17.** $0x^3$ and $0x$ **19.** $x + 2$ **21.** $y + 12$ **23.** $a + b$ **25.** $3a - 2$ **27.** $b + 3$ **29.** $x - 3y$
31. $2x + 1$ **33.** $x - 7$ **35.** $3x + 2y$ **37.** $2x - y$ **39.** $x + 5y$ **41.** $x - 5y$ **43.** $x^2 + 2x - 1$ **45.** $2x^2 + 2x + 1$

47. $x^2 + xy + y^2$ **49.** $x + 1 + \frac{-1}{2x+3}$ **51.** $2x + 2 + \frac{-3}{2x+1}$ **53.** $x^2 + 2x + 1$ **55.** $x^2 + 2x - 1 + \frac{6}{2x+3}$ **57.** $2x^2 + 8x + 14 + \frac{31}{x-2}$
59. $x + 1$ **61.** $2x - 3$ **63.** $x^2 - x + 1$ **65.** $a^2 - 3a + 10 + \frac{-30}{a+3}$ **67.** $5x^2 - x + 4 + \frac{16}{3x-4}$

Chapter Summary   (page 327)

1. a. $(-3x)(-3x)(-3x)(-3x)$ **b.** $\left(\frac{1}{2}pq\right)\left(\frac{1}{2}pq\right)\left(\frac{1}{2}pq\right)$ **2. a.** 125   **b.** 243   **c.** 64   **d.** -64 **e.** 13   **f.** 25   **3. a.** x^5 **b.** x^9 **c.** y^{21}
d. x^{42} **e.** $a^3 b^3$ **f.** $81x^4$ **g.** b^{12} **h.** $-y^2 z^5$ **i.** $256s^3$ **j.** $-3y^6$ **k.** x^{15} **l.** $4x^4 y^2$ **m.** x^4 **n.** $\frac{x^2}{y^2}$ **o.** $\frac{2y^2}{x^2}$ **p.** $5yz^4$

4. a. 1   **b.** 1   **c.** 9   **d.** $9x^4$ **e.** $\frac{1}{x^3}$ **f.** x **g.** y **h.** x^{10} **i.** $\frac{1}{x^2}$ **j.** $\frac{a^6}{b^3}$ **k.** $\frac{1}{x^5}$ **l.** $\frac{1}{9z^2}$ **5. a.** 7.28×10^2 **b.** 9.37×10^3
c. 1.36×10^{-2} **d.** 9.42×10^{-3} **e.** 7.73×10^0 **f.** 7.53×10^5 **g.** 1.8×10^{-4} **h.** 6×10^4 **6. a.** 726,000   **b.** 0.000391
c. 2.68   **d.** 57.6   **e.** 7.39   **f.** 0.000437   **g.** 0.03   **h.** 160   **7. a.** 7th, monomial   **b.** 2nd, binomial   **c.** 5th, trinomial
d. 5th, binomial   **8. a.** 11   **b.** 2   **c.** -4 **d.** 4   **9. a.** 402   **b.** 0   **c.** 82   **d.** 0.3405   **10. a.** -4 **b.** 21   **c.** 0   **d.** $-\frac{15}{4}$
11. a. 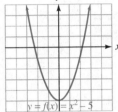   **b.** 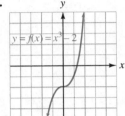   **12. a.** $7x$ **b.** in simplest terms   **c.** $4x^2 y^2$ **d.** $x^2 yz$ **e.** $8x^2 - 6x$

f. $4a^2 + 4a - 6$ **g.** $5x^2 + 19x + 3$ **h.** $6x^3 + 8x^2 + 3x - 72$ **13. a.** $10x^3 y^5$ **b.** $x^7 yz^5$ **14. a.** $5x + 15$ **b.** $6x + 12$
c. $3x^4 - 5x^2$ **d.** $2y^4 + 10y^3$ **e.** $-x^2 y^3 + x^3 y^2$ **f.** $-3x^2 y^2 + 3x^2 y$ **15. a.** $x^2 + 5x + 6$ **b.** $2x^2 - x - 1$ **c.** $6a^2 - 6$
d. $6a^2 - 6$ **e.** $2a^2 - ab - b^2$ **f.** $6x^2 + xy - y^2$ **16. a.** $x^2 + 6x + 9$ **b.** $x^2 - 25$ **c.** $y^2 - 4$ **d.** $x^2 + 8x + 16$
e. $x^2 - 6x + 9$ **f.** $y^2 - 2y + 1$ **g.** $4y^2 + 4y + 1$ **h.** $y^4 - 1$ **17. a.** $3x^3 + 7x^2 + 5x + 1$ **b.** $8a^3 - 27$ **18. a.** 1   **b.** -1
c. 7   **d.** 5   **e.** 1   **f.** 0   **19. a.** $\frac{3}{2y} + \frac{3}{x}$ **b.** $2 - \frac{3}{y}$ **c.** $-3a - 4b + 5c$ **d.** $-\frac{x}{y} - \frac{y}{x}$ **20. a.** $x + 1 + \frac{3}{x+2}$ **b.** $x - 5$ **c.** $2x + 1$
d. $x + 5 + \frac{3}{3x-1}$ **e.** $3x^2 + 2x + 1 + \frac{2}{2x-1}$ **f.** $3x^2 - x - 4$

Chapter 4 Test   (page 331)

1. $2x^3 y^4$ **2.** 134   **3.** y^6 **4.** $6b^7$ **5.** $32x^{21}$ **6.** $8r^{18}$ **7.** 3   **8.** $\frac{2}{y^3}$ **9.** y^3 **10.** $\frac{64a^3}{b^3}$ **11.** 2.8×10^4 **12.** 2.5×10^{-3}

13. 7,400   **14.** 0.000093   **15.** binomial   **16.** 10th degree   **17.** 0   **18.** 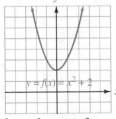   **19.** $-7x + 2y$ **20.** $-3x + 6$

21. $5x^3 + 2x^2 + 2x - 5$ **22.** $-x^2 - 5x + 4$ **23.** $-4x^5 y$ **24.** $3y^4 - 6y^3 + 9y^2$ **25.** $6x^2 - 7x - 20$ **26.** $2x^3 - 7x^2 + 14x - 12$
27. $\frac{1}{2}$ **28.** $\frac{y}{2x}$ **29.** $\frac{a}{4b} - \frac{b}{2a}$ **30.** $x - 2$

Cumulative Review Exercises   (page 332)

1. 11   **2.** 71   **3.** $-\frac{11}{10}$ **4.** 7   **5.** 15   **6.** 4   **7.** -10 **8.** -6 **9.** ←————(————→   **10.** ←—————)————→

11. ←——(————)——→
    -2 5
12. ←——[————]——→
    -2 4
13. $r = \frac{A - p}{pt}$ **14.** $h = \frac{2A}{b}$

15. 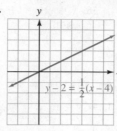   **16.**   **17.** 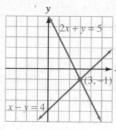   **18.**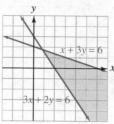

19. $(4, -3)$ **20.** $\left(\frac{1}{2}, \frac{2}{3}\right)$ **21.** y^{14} **22.** xy **23.** $\dfrac{a^7}{b^6}$ **24.** x^2y^2 **25.** $x^2 + 4x - 14$ **26.** $6x^2 + 10x - 56$ **27.** $x^3 - 8$ **28.** $2x + 1$
29. 4.8×10^{18} m   **30.** 4 in.   **31.** 879.6 in.^{2} **32.** \$512

Getting Ready (page 336)

1. $5x + 15$ **2.** $7y - 56$ **3.** $3x^2 - 2x$ **4.** $5y^2 + 9y$ **5.** $ab + 9a$ **6.** $3x + x^2 + xy$ **7.** $x^2y - 4xy$ **8.** $2x^2y^2 - 5xy^3$

Orals (page 342)

1. $2^2 \cdot 3^2$ **2.** 3^3 **3.** 3^4 **4.** $3^2 \cdot 5$ **5.** 3   **6.** $3ab$ **7.** $5(3xy + 2)$ **8.** $5xy(3 + 2y)$

Exercise 5.1 (page 342)

1. 7   **3.** 11   **5.** prime   **7.** largest   **9.** 0, 0   **11.** $2^2 \cdot 3$ **13.** $3 \cdot 5$ **15.** $2^3 \cdot 5$ **17.** $2 \cdot 7^2$ **19.** $3^2 \cdot 5^2$ **21.** $2^5 \cdot 3^2$ **23.** 4
25. r^2 **27.** $4, x$ **29.** $3(x + 2)$ **31.** $x(y - z)$ **33.** $t^2(t + 2)$ **35.** $r^2(r^2 - 1)$ **37.** $a^2b^3z^2(az - 1)$ **39.** $8xy^2z^3(3xyz + 1)$
41. $6uvw^2(2w - 3v)$ **43.** $3(x + y - 2z)$ **45.** $a(b + c - d)$ **47.** $2y(2y + 4 - x)$ **49.** $3r(4r - s + 3rs^2)$ **51.** $abx(1 - b + x)$
53. $2xyz^2(2xy - 3y + 6)$ **55.** $7a^2b^2c^2(10a + 7bc - 3)$ **57.** $-(a + b)$ **59.** $-(2x - 5y)$ **61.** $-(2a - 3b)$ **63.** $-(3m + 4n - 1)$
65. $-(3xy - 2z - 5w)$ **67.** $-(3ab + 5ac - 9bc)$ **69.** $-3xy(x + 2y)$ **71.** $-4a^2b^2(b - 3a)$ **73.** $-2ab^2c(2ac - 7a + 5c)$
75. $-7ab(2a^5b^5 - 7ab^2 + 3)$ **77.** $-5a^2b^3c(1 - 3abc + 5a^2)$ **79.** 2, -3 **81.** 4, -1 **83.** $\frac{5}{2}, -2$ **85.** 1, $-2, 3$ **87.** 0, 3
89. 0, $\frac{5}{2}$ **91.** 0, 7   **93.** 0, $-\frac{8}{3}$ **95.** 0, 2   **97.** 0, $-\frac{1}{5}$

Getting Ready (page 345)

1. $3x + 3y + ax + ay$ **2.** $xy + x + 5y + 5$ **3.** $5x + 5 - yx - y$ **4.** $x^2 + 2x - yx - 2y$ **5.** $3x^2 + 2xy - y^2$ **6.** $-y^2 + 12y - 35$

Orals (page 347)

1. $x + 3$ **2.** $a - 1$ **3.** $x - 2$ **4.** $y + 5$ **5.** $x - 7$ **6.** $2y + 9$

Exercise 5.2 (page 348)

1. u^9 **3.** $\frac{a}{b}$ **5.** $a + b$ **7.** $(a + b)$ **9.** $(p - q)$ **11.** $(x + y)(2 + b)$ **13.** $(x + y)(3 - a)$ **15.** $(r - 2s)(3 - x)$
17. $(x - 3)(x - 2)$ **19.** $2(a^2 + b)(x + y)$ **21.** $3(r + 3s)(x^2 - 2y^2)$ **23.** $(a + b + c)(3x - 2y)$ **25.** $7xy(r + 2s - t)(2x - 3)$
27. $(x + 1)(x + 3 - y)$ **29.** $(x^2 - 2)(3x - y + 1)$ **31.** $(x + y)(2 + a)$ **33.** $(r + s)(7 - k)$ **35.** $(r + s)(x + y)$
37. $(2x + 3)(a + b)$ **39.** $(b + c)(2a + 3)$ **41.** $(x + y)(2x - 3)$ **43.** $(v - 3w)(3t + u)$ **45.** $(3p + q)(3m - n)$
47. $(m - n)(p - 1)$ **49.** $(a - b)(x - y)$ **51.** $x^2(a + b)(x + 2y)$ **53.** $4a(b + 3)(a - 2)$ **55.** $(x^2 + 1)(x + 2)$
57. $y(x^2 - y)(x - 1)$ **59.** $(x + 2)(x + y + 1)$ **61.** $(m - n)(a + b + c)$ **63.** $(d + 3)(a - b - c)$ **65.** $(a + b + c)(x^2 - y)$
67. $(r - s)(2 + b)$ **69.** $(x + y)(a + b)$ **71.** $(a - b)(c - d)$ **73.** $r(r + s)(a - b)$ **75.** $(b + 1)(a + 3)$ **77.** $(r - s)(p - q)$

Getting Ready (page 350)

1. $a^2 - b^2$ **2.** $4r^2 - s^2$ **3.** $9x^2 - 4y^2$ **4.** $16x^4 - 9$

Orals (page 354)

1. $(x + 3)(x - 3)$ **2.** $(y + 6)(y - 6)$ **3.** $(z + 2)(z - 2)$ **4.** $(p + q)(p - q)$ **5.** $(5 + t)(5 - t)$ **6.** $(6 + r)(6 - r)$
7. $(10 + y)(10 - y)$ **8.** $(10 + y^2)(10 - y^2)$

Exercise 5.3 (page 354)

1. $p = w\left(k - h - \dfrac{v^2}{2g}\right)$ **3.** difference of two squares   **5.** $(p - q)$ **7.** $(x - 3)$ **9.** $(2m - 3n)$ **11.** $(x + 4)(x - 4)$
13. $(y + 7)(y - 7)$ **15.** $(2y + 7)(2y - 7)$ **17.** $(3x + y)(3x - y)$ **19.** $(5t + 6u)(5t - 6u)$ **21.** $(4a + 5b)(4a - 5b)$ **23.** prime
25. $(a^2 + 2b)(a^2 - 2b)$ **27.** $(7y + 15z^2)(7y - 15z^2)$ **29.** $(14x^2 + 13y)(14x^2 - 13y)$ **31.** $8(x + 2y)(x - 2y)$
33. $2(a + 2y)(a - 2y)$ **35.** $3(r + 2s)(r - 2s)$ **37.** $x(x + y)(x - y)$ **39.** $x(2a + 3b)(2a - 3b)$ **41.** $3m(m + n)(m - n)$
43. $x^2(2x + y)(2x - y)$ **45.** $2ab(a + 11b)(a - 11b)$ **47.** $(x^2 + 9)(x + 3)(x - 3)$ **49.** $(a^2 + 4)(a + 2)(a - 2)$
51. $(a^2 + b^2)(a + b)(a - b)$ **53.** $(9r^2 + 16s^2)(3r + 4s)(3r - 4s)$ **55.** $(a^2 + b^4)(a + b^2)(a - b^2)$
57. $(x^4 + y^4)(x^2 + y^2)(x + y)(x - y)$ **59.** $2(x^2 + y^2)(x + y)(x - y)$ **61.** $b(a^2 + b^2)(a + b)(a - b)$
63. $3n(4m^2 + 9n^2)(2m + 3n)(2m - 3n)$ **65.** $3ay(a^4 + 2y^4)$ **67.** $3a^2(a^4 + b^2)(a^2 + b)(a^2 - b)$
69. $2y^2(x^4 + 4y^2)(x^2 + 2y)(x^2 - 2y)$ **71.** $a^2b^2(a^2 + b^2c^2)(a + bc)(a - bc)$ **73.** $a^2b^3(b^2 + 25)(b + 5)(b - 5)$
75. $3rs(9r^2 + 4s^2)(3r + 2s)(3r - 2s)$ **77.** $(4x - 4y + 3)(4x - 4y - 3)$ **79.** $(a + 3)(a + 3)(a - 3)$ **81.** $(y + 4)(y - 4)(y - 3)$
83. $3(x + 2)(x - 2)(x + 1)$ **85.** $3(m + n)(m - n)(m + a)$ **87.** $2(m + 4)(m - 4)(mn^2 + 4)$ **89.** $5, -5$ **91.** $7, -7$ **93.** $\frac{1}{2}, -\frac{1}{2}$
95. $\frac{2}{3}, -\frac{2}{3}$ **97.** $7, -7$ **99.** $\frac{9}{2}, -\frac{9}{2}$

Getting Ready (page 356)

1. $x^2 + 12x + 36$ **2.** $y^2 - 14y + 49$ **3.** $a^2 - 6a + 9$ **4.** $x^2 + 9x + 20$ **5.** $r^2 - 7r + 10$ **6.** $m^2 - 4m - 21$
7. $a^2 + ab - 12b^2$ **8.** $u^2 - 8uv + 15v^2$ **9.** $x^2 - 2xy - 24y^2$

Orals (page 364)

1. 4   **2.** $-, -$ **3.** $-, 3$ **4.** $-, 2$ **5.** 6, 1   **6.** 6, 1

Exercise 5.4 (page 364)

1.   **3.**   **5.**   **7.**   **9.** $(x + y)^2$ **11.** 4, 2
13. $y, 2y$ **15.** $(x + 2)(x + 1)$ **17.** $(z + 11)(z + 1)$ **19.** $(a - 5)(a + 1)$ **21.** $(t - 7)(t - 2)$ **23.** prime   **25.** $(y - 6)(y + 5)$
27. $(a + 8)(a - 2)$ **29.** $(t - 10)(t + 5)$ **31.** prime   **33.** $(y + z)(y + z)$ **35.** $(x + 2y)(x + 2y)$ **37.** $(m + 5n)(m - 2n)$
39. $(a - 6b)(a + 2b)$ **41.** $(u + 5v)(u - 3v)$ **43.** $-(x + 5)(x + 2)$ **45.** $-(y + 5)(y - 3)$ **47.** $-(t + 17)(t - 2)$
49. $-(r - 10)(r - 4)$ **51.** $-(a + 3b)(a + b)$ **53.** $-(x - 7y)(x + y)$ **55.** $(x - 4)(x - 1)$ **57.** $(y + 9)(y + 1)$
59. $(c + 5)(c - 1)$ **61.** $-(r - 2s)(r + s)$ **63.** $(r + 3x)(r + x)$ **65.** $(a - 2b)(a - b)$ **67.** $2(x + 3)(x + 2)$ **69.** $3y(y + 1)(y + 1)$
71. $-5(a - 3)(a - 2)$ **73.** $3(z - 4t)(z - t)$ **75.** $4y(x + 6)(x - 3)$ **77.** $-4x(x + 3y)(x - 2y)$ **79.** $(x + 2)(ax + 2a + b)$
81. $(a + 5)(a + 3 + b)$ **83.** $(a + b + 2)(a + b - 2)$ **85.** $(b + y + 2)(b - y - 2)$ **87.** $(x + 3)(x + 3)$ **89.** $(y - 4)(y - 4)$
91. $(t + 10)(t + 10)$ **93.** $(u - 9)(u - 9)$ **95.** $(x + 2y)(x + 2y)$ **97.** $(r - 5s)(r - 5s)$ **99.** 12, 1   **101.** 5, -3 **103.** $-3, 7$
105. 8, 1   **107.** $-3, -5$ **109.** $-4, 2$ **111.** $0, -1, -2$ **113.** $0, 9, -3$ **115.** $1, -2, -3$

Getting Ready (page 367)

1. $6x^2 + 7x + 2$ **2.** $6y^2 - 19y + 10$ **3.** $8t^2 + 6t - 9$ **4.** $4r^2 + 4r - 15$ **5.** $6m^2 - 13m + 6$ **6.** $16a^2 + 16a + 3$

Orals (page 373)

1. 2, 3   **2.** 2   **3.** $+, -$ **4.** $+, -$ **5.** 3, 1   **6.** 3, 1

Exercise 5.5   (page 374)

1. $n = \frac{l - f + d}{d}$ **3.** descending   **5.** opposites   **7.** 2, 1   **9.** y, y **11.** $(2x - 1)(x - 1)$ **13.** $(3a + 1)(a + 4)$ **15.** $(z + 3)(4z + 1)$
17. $(3y + 2)(2y + 1)$ **19.** $(3x - 2)(2x - 1)$ **21.** $(3a + 2)(a - 2)$ **23.** $(2x + 1)(x - 2)$ **25.** $(2m - 3)(m + 4)$
27. $(5y + 1)(2y - 1)$ **29.** $(3y - 2)(4y + 1)$ **31.** $(5t + 3)(t + 2)$ **33.** $(8m - 3)(2m - 1)$ **35.** $(3x - y)(x - y)$
37. $(2u + 3v)(u - v)$ **39.** $(2a - b)(2a - b)$ **41.** $(3r + 2s)(2r - s)$ **43.** $(2x + 3y)(2x + y)$ **45.** $(4a - 3b)(a - 3b)$
47. $(3x + 2)(x - 5)$ **49.** $(2a - 5)(4a - 3)$ **51.** $(4y - 3)(3y - 4)$ **53.** prime   **55.** $(2a + 3b)(a + b)$ **57.** $(3p - q)(2p + q)$
59. prime   **61.** $(4x - 5y)(3x - 2y)$ **63.** $2(2x - 1)(x + 3)$ **65.** $y(y + 12)(y + 1)$ **67.** $3x(2x + 1)(x - 3)$
69. $3r^3(5r - 2)(2r + 5)$ **71.** $4(a - 2b)(a + b)$ **73.** $4(2x + y)(x - 2y)$ **75.** $-2mn(4m + 3n)(2m + n)$
77. $-2uv^3(7u - 3v)(2u - v)$ **79.** $(2x + 3)^2$ **81.** $(3x + 2)^2$ **83.** $(4x - y)^2$ **85.** $(2x + y + 4)(2x + y - 4)$
87. $(3 + a + 2b)(3 - a - 2b)$ **89.** $(2x + y + a + b)(2x + y - a - b)$ **91.** $\frac{1}{2}, 2$ **93.** $\frac{1}{5}, 1$ **95.** $-\frac{1}{3}, 3$ **97.** $\frac{2}{3}, -\frac{1}{5}$ **99.** $-\frac{3}{2}, \frac{2}{3}$
101. $\frac{1}{8}, 1$ **103.** $0, -3, -\frac{1}{3}$ **105.** $0, -3, -3$

Getting Ready   (page 376)

1. $3ax(x + a)$ **2.** $(x + 3y)(x - 3y)$ **3.** $(x - 5)(x + 2)$ **4.** $(2x - 3)(3x - 2)$ **5.** $2(3x - 1)(x - 2)$ **6.** $(a + b)(x + y)(x - y)$

Orals   (page 378)

1. common factor   **2.** difference of two squares   **3.** grouping   **4.** none, prime   **5.** common factor   **6.** difference of two squares

Exercise 5.6   (page 378)

1. $\frac{8}{3}$ **3.** 0, 7   **5.** factors   **7.** binomials   **9.** $3(2x + 1)$ **11.** $(x - 7)(x + 1)$ **13.** $(3t - 1)(2t + 3)$ **15.** $(2x + 5)(2x - 5)$
17. $(t - 1)(t - 1)$ **19.** $3(a + 2)(a - 2)$ **21.** $(y^2 - 2)(x + 1)(x - 1)$ **23.** $7p^4q^2(10q - 5 + 7p)$ **25.** $2a(b + 6)(b - 2)$
27. $-4p^2q^3(2pq^4 + 1)$ **29.** $(2a - b + 3)(2a - b - 3)$ **31.** prime   **33.** $-2x^2(x + 4)(x - 4)$ **35.** $2t^2(3t - 5)(t + 4)$
37. $(x - a)(a + b)(a - b)$ **39.** $2(x^2 + 4)(x + 2)(x - 2)$ **41.** $(x + 2)(x + 3)(x - 3)$ **43.** $-x^2y^2z(16x^2 - 24x^3yz^3 + 15yz^6)$
45. $(9p^2 + 4q^2)(3p + 2q)(3p - 2q)$ **47.** prime   **49.** $(x + y)(x - y)(x + 2)(x - 2)$ **51.** prime   **53.** $t(7t - 1)(3t - 1)$
55. $(x + y + 2)(x - y - 2)$ **57.** $2(a + b)(a - b)(c + 2d)$

Getting Ready   (page 380)

1. s^2 **2.** $2w + 4$ **3.** $x(x + 1)$ **4.** $w(w + 3)$

Orals   (page 383)

1. $A = lw$ **2.** $A = \frac{1}{2}bh$ **3.** $A = s^2$ **4.** $A = lwh$ **5.** $P = 2l + 2w$ **6.** $P = 4s$

Exercise 5.7   (page 384)

1. -10 **3.** 675 cm^2 **5.** analyze   **7.** 5, 7   **9.** 9   **11.** 9 sec   **13.** $\frac{15}{4}$ sec and 10 sec   **15.** 2 sec   **17.** 4 m by 9 m   **19.** 48 ft
21. $b = 4$ in., $h = 18$ in.   **23.** 18 sq units   **25.** 1 m   **27.** 3 cm   **29.** 4 cm by 7 cm

Chapter Summary   (page 388)

1. a. $5 \cdot 7$ **b.** $3^2 \cdot 5$ **c.** $2^5 \cdot 3$ **d.** $2 \cdot 3 \cdot 17$ **e.** $3 \cdot 29$ **f.** $3^2 \cdot 11$ **g.** $2 \cdot 5^2 \cdot 41$ **h.** 2^{12} **2. a.** $3(x + 3y)$ **b.** $5a(x^2 + 3)$
c. $7x(x + 2)$ **d.** $3x(x - 1)$ **e.** $2x(x^2 + 2x - 4)$ **f.** $a(x + y - z)$ **g.** $a(x + y - 1)$ **h.** $xyz(x + y)$ **3. a.** $0, -2$ **b.** $0, 3$
4. a. $(x + y)(a + b)$ **b.** $(x + y)(x + y + 1)$ **c.** $2x(x + 2)(x + 3)$ **d.** $3x(y + z)(1 - 3y - 3z)$ **e.** $(p + 3q)(3 + a)$
f. $(r - 2s)(a + 7)$ **g.** $(x + a)(x + b)$ **h.** $(y + 2)(x - 2)$ **i.** $(x + y)(a + b)$ **5. a.** $(x + 3)(x - 3)$ **b.** $(xy + 4)(xy - 4)$
c. $(x + 2 + y)(x + 2 - y)$ **d.** $(z + x + y)(z - x - y)$ **e.** $6y(x + 2y)(x - 2y)$ **f.** $(x + y + z)(x + y - z)$ **6. a.** $3, -3$ **b.** $5, -5$
7. a. $(x + 3)(x + 7)$ **b.** $(x - 3)(x + 7)$ **c.** $(x + 6)(x - 4)$ **d.** $(x - 6)(x + 2)$ **8. a.** $(2x + 1)(x - 3)$ **b.** $(3x + 1)(x - 5)$
c. $(2x + 3)(3x - 1)$ **d.** $3(2x - 1)(x + 1)$ **e.** $x(x + 3)(6x - 1)$ **f.** $x(4x + 3)(x - 2)$ **9. a.** 3, 4   **b.** 5, -3 **c.** $-4, 6$ **d.** 2, 8
e. 3, $-\frac{1}{2}$ **f.** 1, $-\frac{3}{2}$ **g.** $\frac{1}{2}, -\frac{1}{2}$ **h.** $\frac{2}{3}, -\frac{2}{3}$ **i.** 0, 3, 4   **j.** 0, $-2, -3$ **k.** 0, $\frac{1}{2}, -3$ **l.** 0, $-\frac{2}{3}, 1$ **10. a.** $y(3x - y)(x - 2)$
b. $5(x + 2)(x - 3y)$ **c.** $a(a + b)(2x + a)$ **d.** $(x + a + y)(x + a - y)$ **e.** $(x + 1)(ax + 3a - b)$
f. $a(x + y)(x^2 - xy + y^2)(x - y)(x^2 + xy + y^2)$
11. 5 and 7   **12.** $\frac{1}{3}$ **13.** 6 ft by 8 ft   **14.** 3 ft by 9 ft   **15.** 3 ft by 6 ft

Chapter 5 Test (page 391)

1. $2^2 \cdot 7^2$ **2.** $3 \cdot 37$ **3.** $5a(12b^2c^3 + 6a^2b^2c - 5)$ **4.** $3x(a + b)(x - 2y)$ **5.** $(x + y)(a + b)$ **6.** $(x + 5)(x - 5)$
7. $3(a + 3b)(a - 3b)$ **8.** $(4x^2 + 9y^2)(2x + 3y)(2x - 3y)$ **9.** $(x + 3)(x + 1)$ **10.** $(x - 11)(x + 2)$ **11.** $(x + 9y)(x + y)$
12. $6(x - 4y)(x - y)$ **13.** $(3x + 1)(x + 4)$ **14.** $(2a - 3)(a + 4)$ **15.** $(2x - y)(x + 2y)$ **16.** $(4x - 3)(3x - 4)$
17. $6(2a - 3b)(a + 2b)$ **18.** $(x^2 + 4y^2)(x + 2y)(x - 2y)$ **19.** $0, -3$ **20.** $-1, -\frac{3}{2}$ **21.** $3, -3$ **22.** $3, -6$ **23.** $\frac{9}{5}, -\frac{1}{2}$
24. $-\frac{9}{10}, 1$ **25.** $\frac{1}{5}, -\frac{9}{2}$ **26.** $-\frac{1}{10}, 9$ **27.** 12 sec **28.** 10 m

Getting Ready (page 393)

1. $\frac{1}{2}$ **2.** $\frac{2}{3}$ **3.** $-\frac{4}{5}$ **4.** $-\frac{5}{9}$

Orals (page 398)

1. $\frac{5}{7}$ **2.** $\frac{50}{1}$ **3.** $\frac{1}{3}$ **4.** $\frac{7}{10}$

Exercise 6.1 (page 398)

1. 17 **3.** 6 **5.** $2(x + 3)$ **7.** $(2x + 3)(x - 2)$ **9.** comparison, quotient **11.** equal **15.** $\frac{5}{7}$ **17.** $\frac{1}{2}$ **19.** $\frac{2}{3}$ **21.** $\frac{2}{7}$ **23.** $\frac{1}{3}$ **25.** $\frac{1}{5}$
27. $\frac{3}{7}$ **29.** $\frac{3}{4}$ **31.** \$1,825 **33.** $\frac{22}{365}$ **35.** \$8,725 **37.** $\frac{336}{1,745}$ **39.** $\frac{1}{16}$ **41.** $\frac{\$21.59}{17 \text{ gal}}$; \$1.27/gal **43.** 7¢/oz **45.** the 6-oz can **47.** the
first student **49.** $\frac{11,880 \text{ gal}}{27 \text{ min}}$; 440 gal/min **51.** 5% **53.** 65 mph **55.** the second car

Getting Ready (page 400)

1. 10 **2.** $\frac{7}{3}$ **3.** $\frac{20}{7}$ **4.** 5 **5.** $\frac{14}{3}$ **6.** 5 **7.** 21 **8.** $\frac{3}{2}$

Orals (page 409)

1. proportion **2.** not a proportion **3.** not a proportion **4.** proportion

Exercise 6.2 (page 410)

1. 90% **3.** $\frac{1}{3}$ **5.** 480 **7.** \$73.50 **9.** proportion, ratios **11.** means **13.** shape **15.** ad, bc **17.** triangle **19.** no **21.** yes
23. no **25.** yes **27.** 4 **29.** 6 **31.** -3 **33.** 9 **35.** 0 **37.** -17 **39.** $-\frac{3}{2}$ **41.** $\frac{83}{2}$ **43.** \$17 **45.** \$6.50 **47.** 24
49. about $4\frac{1}{4}$ **51.** 47 **53.** $7\frac{1}{2}$ gal **55.** \$309 **57.** 49 ft $3\frac{1}{2}$ in. **59.** 162 **61.** not exactly, but close **63.** 39 ft **65.** $46\frac{7}{8}$ ft
67. 6,750 ft **69.** 15,840 ft

Getting Ready (page 413)

1. $\frac{3}{4}$ **2.** 2 **3.** $\frac{5}{11}$ **4.** $\frac{1}{2}$

Orals (page 419)

1. $\frac{2}{3}$ **2.** 2 **3.** $\frac{z}{w}$ **4.** $2x$ **5.** $\frac{x}{y}$ **6.** $\frac{1}{y}$ **7.** 1 **8.** -1

Exercise 6.3 (page 419)

1. $(a + b) + c = a + (b + c)$ **3.** 0 **5.** $\frac{5}{3}$ **7.** numerator **9.** 0 **11.** negatives **13.** $\frac{a}{b}$ **15.** factor, common **17.** $\frac{4}{5}$ **19.** $\frac{4}{5}$
21. $\frac{2}{13}$ **23.** $\frac{2}{9}$ **25.** $-\frac{1}{3}$ **27.** $2x$ **29.** $-\frac{x}{3}$ **31.** $\frac{x}{a}$ **33.** $\frac{2}{z}$ **35.** $\frac{a}{3}$ **37.** $\frac{2}{3}$ **39.** $\frac{3}{2}$ **41.** in lowest terms **43.** $\frac{3x}{y}$ **45.** $\frac{7x}{8y}$ **47.** $\frac{1}{3}$ **49.** 5
51. $\frac{x}{2}$ **53.** $\frac{3x}{5y}$ **55.** $\frac{2}{3}$ **57.** -1 **59.** -1 **61.** -1 **63.** $\frac{x+1}{x-1}$ **65.** $\frac{x-5}{x+2}$ **67.** $\frac{2x}{x-2}$ **69.** $\frac{x}{y}$ **71.** $\frac{x+2}{x^2}$ **73.** $\frac{x-4}{x+4}$ **75.** $\frac{2(x+2)}{x-1}$
77. in lowest terms **79.** $\frac{3-x}{3+x}$ or $-\frac{x-3}{x+3}$ **81.** $\frac{4}{3}$ **83.** $x + 3$

Getting Ready (page 421)

1. $\frac{2}{3}$ **2.** $\frac{14}{3}$ **3.** 3 **4.** 6 **5.** $\frac{5}{2}$ **6.** 1 **7.** $\frac{3}{4}$ **8.** 2

Orals (page 428)

1. $\frac{3}{2}$ 2. $\frac{7}{5}$ 3. 5 4. 1 5. $\frac{1}{4}$ 6. x

Exercise 6.4 (page 428)

1. $-6x^5y^6z$ 3. $\dfrac{1}{81y^4}$ 5. $\dfrac{1}{x^m}$ 7. $4y^3 + 4y^2 - 8y + 32$ 9. numerator 11. numerators, denominators 13. 1

15. divisor, multiply 17. $\frac{45}{91}$ 19. $-\frac{3}{11}$ 21. $\frac{5}{7}$ 23. $\frac{3x}{2}$ 25. $\frac{yx}{z}$ 27. $\frac{14}{9}$ 29. x^2y^2 31. $2xy^2$ 33. $-3y^2$ 35. $\dfrac{b^3c}{a^4}$ 37. $\dfrac{r^3t^4}{s}$

39. $\dfrac{(z+7)(z+2)}{7z}$ 41. x 43. $\frac{x}{5}$ 45. $x+2$ 47. $\frac{3}{2x}$ 49. $x-2$ 51. x 53. $\dfrac{(x-2)^2}{x}$ 55. $\dfrac{(m-2)(m-3)}{2(m+2)}$ 57. 1

59. $\dfrac{c^2}{ab}$ 61. $\dfrac{x+1}{2(x-2)}$ 63. $\dfrac{2}{3}$ 65. $\dfrac{3}{5}$ 67. $\dfrac{3}{2y}$ 69. 3 71. $\dfrac{6}{y}$ 73. 6 75. $\dfrac{2x}{3}$ 77. $\dfrac{2}{y}$ 79. $\dfrac{2}{3x}$ 81. $\dfrac{2(z-2)}{z}$

83. $\dfrac{5z(z-7)}{z+2}$ 85. $\dfrac{x+2}{3}$ 87. 1 89. $\dfrac{x-2}{x-3}$ 91. $x+5$ 93. $\dfrac{9}{2x}$ 95. $\dfrac{x}{36}$ 97. $\dfrac{(x+1)(x-1)}{5(x-3)}$ 99. 2 101. $\dfrac{2x(1-x)}{5(x-2)}$

103. $\dfrac{y^2}{3}$ 105. $\dfrac{x+2}{x-2}$

Getting Ready (page 431)

1. $\frac{4}{5}$ 2. 1 3. $\frac{7}{8}$ 4. 2 5. $\frac{1}{9}$ 6. $\frac{1}{2}$ 7. $-\frac{2}{13}$ 8. $\frac{13}{10}$

Orals (page 441)

1. equal 2. equal 3. not equal 4. equal 5. equal 6. not equal 7. equal 8. equal

Exercise 6.5 (page 441)

1. 7^2 3. $2^3 \cdot 17$ 5. $2 \cdot 3 \cdot 17$ 7. $2^4 \cdot 3^2$ 9. LCD 11. numerators, common denominator 13. $\frac{2}{3}$ 15. $\frac{1}{3}$ 17. $\frac{4x}{y}$ 19. $\frac{2}{y}$

21. $\dfrac{2y+6}{5z}$ 23. 9 25. $\dfrac{1}{7}$ 27. $-\dfrac{1}{8}$ 29. $\dfrac{x}{y}$ 31. $\dfrac{y}{x}$ 33. $\dfrac{1}{y}$ 35. 1 37. $\dfrac{4x}{3}$ 39. $\dfrac{2x}{3y}$ 41. $\dfrac{2(2x-y)}{y+2}$ 43. $\dfrac{2(x+5)}{x-2}$

45. $\dfrac{125}{20}$ 47. $\dfrac{8xy}{x^2y}$ 49. $\dfrac{3x(x+1)}{(x+1)^2}$ 51. $\dfrac{2y(x+1)}{x^2+x}$ 53. $\dfrac{z(z+1)}{z^2-1}$ 55. $\dfrac{2(x+2)}{x^2+3x+2}$ 57. $6x$ 59. $18xy$ 61. x^2-1

63. $x^2 + 6x$ 65. $(x+1)(x+5)(x-5)$ 67. $\dfrac{7}{6}$ 69. $\dfrac{5y}{9}$ 71. $\dfrac{53x}{42}$ 73. $\dfrac{4xy+6x}{3y}$ 75. $\dfrac{2-3x^2}{x}$ 77. $\dfrac{4y+10}{15y}$

79. $\dfrac{x^2+4x+1}{x^2y}$ 81. $\dfrac{2x^2-1}{x(x+1)}$ 83. $\dfrac{2xy+x-y}{xy}$ 85. $\dfrac{x+2}{x-2}$ 87. $\dfrac{2x^2+2}{(x-1)(x+1)}$ 89. $\dfrac{2(2x+1)}{x-2}$ 91. $\dfrac{x}{x-2}$ 93. $\dfrac{5x+3}{x+1}$

95. $-\dfrac{1}{2(x-2)}$

Getting Ready (page 444)

1. 4 2. -18 3. 7 4. -8 5. $3 + 3x$ 6. $2 - y$ 7. $12x - 2$ 8. $3y + 2x$

Orals (4age 449)

1. $\frac{4}{3}$ 2. 4 3. $\frac{1}{4}$ 4. 3

Exercise 6.6 (page 449)

1. t^9 **3.** $-2r^7$ **5.** $\dfrac{81}{256r^8}$ **7.** $\dfrac{r^{10}}{9}$ **9.** complex fraction   **11.** single, divide   **13.** $\dfrac{8}{9}$ **15.** $\dfrac{3}{8}$ **17.** $\dfrac{5}{4}$ **19.** $\dfrac{5}{7}$ **21.** $\dfrac{x^2}{y}$ **23.** $\dfrac{5t^2}{27}$

25. $\dfrac{1-3x}{5+2x}$ **27.** $\dfrac{1+x}{2+x}$ **29.** $\dfrac{3-x}{x-1}$ **31.** $\dfrac{1}{x+2}$ **33.** $\dfrac{1}{x+3}$ **35.** $\dfrac{xy}{y+x}$ **37.** $\dfrac{y}{x-2y}$ **39.** $\dfrac{x^2}{(x-1)^2}$ **41.** $\dfrac{7x+3}{-x-3}$ **43.** $\dfrac{x-2}{x+3}$

45. -1 **47.** $\dfrac{y}{x^2}$ **49.** $\dfrac{x+1}{1-x}$ **51.** $\dfrac{a^2-a+1}{a^2}$ **53.** 2   **55.** $\dfrac{y-5}{y+5}$ **59.** $\frac{1}{2},\frac{2}{3},\frac{3}{5},\frac{5}{8}$

Getting Ready (page 451)

1. $3x+1$ **2.** $8x-1$ **3.** $3+2x$ **4.** $y-6$ **5.** 19   **6.** $7x+6$ **7.** y **8.** $3x+5$

Orals (page 456)

1. Multiply by 10.   **2.** Multiply by $x(x-1)$.   **3.** Multiply by 9.   **4.** Multiply by 15.

Exercise 6.7 (page 456)

1. $x(x+4)$ **3.** $(2x+3)(x-1)$ **5.** $(x^2+4)(x+2)(x-2)$ **7.** extraneous   **9.** LCD   **11.** xy **13.** 4   **15.** -20 **17.** 6   **19.** 60
21. -12 **23.** 0   **25.** -7 **27.** -1 **29.** 12   **31.** 0   **33.** -3 **35.** 3   **37.** no solution; 0 is extraneous   **39.** 1   **41.** 5   **43.** no
solution; -2 is extraneous   **45.** no solution; 5 is extraneous   **47.** -1 **49.** 6   **51.** 2   **53.** -3 **55.** 1   **57.** no solution; -2 is
extraneous   **59.** 1, 2   **61.** 3; -3 is extraneous   **63.** 3, -4 **65.** 1   **67.** 0   **69.** $-2, 1$ **71.** $a=\frac{b}{b-1}$ **73.** $f=\dfrac{d_1d_2}{d_1+d_2}$ **77.** 1

Getting Ready (page 459)

1. $\frac{1}{5}$ **2.** $\$(.05x)$ **3.** $\$\left(\frac{y}{.05}\right)$ **4.** $\frac{y}{52}$ hr

Orals (page 462)

1. $i=pr$ **2.** $d=rt$ **3.** $C=qd$

Exercise 6.8 (page 463)

1. $-1, 6$ **3.** $-2, -3, -4$ **5.** 0, 0, 1   **7.** 1, -1, 2, -2 **11.** 2   **13.** 5   **15.** $\frac{2}{3},\frac{3}{2}$ **17.** $2\frac{2}{9}$ hr   **19.** $2\frac{6}{11}$ days   **21.** $7\frac{1}{2}$ hr
23. 4 mph   **25.** 7% and 8%   **27.** 5   **29.** 30   **31.** 25 mph

Chapter Summary (page 466)

1. a. $\frac{1}{2}$ **b.** $\frac{4}{5}$ **c.** $\frac{2}{3}$ **d.** $\frac{5}{6}$ **2.** $\$2.93$ **3.** 568.75 kwh per week   **4. a.** no **b.** yes   **5. a.** $\frac{9}{2}$ **b.** 0 **c.** 7 **d.** 1   **6.** 20 ft   **7. a.** $\frac{2}{5}$
b. $-\frac{2}{3}$ **c.** $-\frac{1}{3}$ **d.** $\frac{7}{3}$ **e.** $\frac{1}{2x}$ **f.** $\frac{5}{2x}$ **g.** $\frac{x}{x+1}$ **h.** $\frac{1}{x}$ **i.** 2 **j.** 1 **k.** -1 **l.** $\frac{x+7}{x+3}$ **m.** $\frac{x}{x-1}$ **n.** in lowest terms   **8. a.** $\frac{3x}{y}$ **b.** $\frac{6}{x^2}$
c. 1 **d.** $\frac{2x}{x+1}$ **9. a.** $\frac{3y}{2}$ **b.** $\frac{1}{x}$ **c.** $x+2$ **d.** 1 **e.** $x+2$ **10. a.** 1 **b.** $\dfrac{2(x+1)}{x-7}$ **c.** $\dfrac{x^2+x-1}{x(x-1)}$ **d.** $\dfrac{x-7}{7x}$ **e.** $\dfrac{x-2}{x(x+1)}$
f. $\dfrac{x^2+4x-4}{2x^2}$ **g.** $\dfrac{x+1}{x}$ **h.** 0   **11. a.** $\frac{9}{4}$ **b.** $\frac{3}{2}$ **c.** $\dfrac{1+x}{1-x}$ **d.** $\dfrac{x(x+3)}{2x^2-1}$ **e.** x^2+3 **f.** $\dfrac{a(a+bc)}{b(b+ac)}$ **12. a.** 3 **b.** 1 **c.** 3
d. 4, $-\frac{3}{2}$ **e.** -2 **f.** 0   **13.** $r_1=\dfrac{rr_2}{r_2-r}$ **14.** $T_1=\dfrac{T_2}{1-E}$ or $T_1=\dfrac{-T_2}{E-1}$ **15.** $R=\dfrac{HB}{B-H}$ or $R=\dfrac{-HB}{H-B}$ **16.** $9\frac{9}{19}$ hr
17. $5\frac{5}{6}$ days   **18.** 5 mph   **19.** 40 mph

Chapter 6 Test (page 470)

1. $\frac{2}{3}$ **2.** yes **3.** $\frac{2}{3}$ **4.** 45 ft **5.** $\frac{8x}{9y}$ **6.** $\frac{x+1}{2x+3}$ **7.** 3 **8.** $\frac{5y^2}{4t}$ **9.** $\frac{x+1}{3(x-2)}$ **10.** $\frac{3t^2}{5y}$ **11.** $\frac{x^2}{3}$ **12.** $x+2$ **13.** $\frac{10x-1}{x-1}$

14. $\frac{13}{2y+3}$ **15.** $\frac{2x^2+x+1}{x(x+1)}$ **16.** $\frac{2x+6}{x-2}$ **17.** $\frac{2x^3}{y^3}$ **18.** $\frac{x+y}{y-x}$ **19.** -5 **20.** 6 **21.** 4 **22.** $B = \frac{RH}{R-H}$ **23.** $3\frac{15}{16}$ hr

24. 5 mph **25.** 8,050 ft

Cumulative Review Exercises (page 471)

1. x^7 **2.** x^{10} **3.** x^3 **4.** 1 **5.** $6x^3 - 2x - 1$ **6.** $2x^3 + 2x^2 + x - 1$ **7.** $13x^2 - 8x + 1$ **8.** $16x^2 - 24x + 2$ **9.** $-12x^5y^5$
10. $-35x^5 + 10x^4 + 10x^2$ **11.** $6x^2 + 14x + 4$ **12.** $15x^2 - 2xy - 8y^2$ **13.** $x+4$ **14.** $x^2 + x + 1$ **15.** $3xy(x - 2y)$
16. $(a+b)(3+x)$ **17.** $(a+b)(2+b)$ **18.** $(5p^2 + 4q)(5p^2 - 4q)$ **19.** $(x-12)(x+1)$ **20.** $(x-3y)(x+2y)$
21. $(3a+4)(2a-5)$ **22.** $(4m+n)(2m-3n)$ **23.** $(x^2 + 9)(x+3)(x-3)$ **24.** $(x+1+y)(x+1-y)$
25. 15 **26.** 4 **27.** $\frac{2}{3}$, $-\frac{1}{2}$ **28.** 0, 2 **29.** -1, -2 **30.** $\frac{3}{2}$, -4 **31.** ← —————— → **32.** ← —————— →

                                                       2   2

33. ← (——————) →   **34.** ← [——————] →   **35.**

         -2 5 -2 4

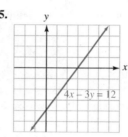

36.

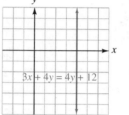

37. $(4, -3)$

38. $\left(\frac{1}{2}, \frac{2}{3}\right)$ **39.**

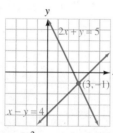

40.

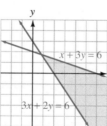

41. -3 **42.** 15 **43.** 5 **44.** $8x^2 - 3$ **45.** $\frac{x+1}{x-1}$

46. $\frac{x-3}{x-2}$ **47.** $\frac{(x-2)^2}{x-1}$ **48.** $\frac{(p+2)(p-3)}{3(p+3)}$ **49.** 1 **50.** $\frac{2(x^2+1)}{(x+1)(x-1)}$ **51.** $\frac{-1}{2(a-2)}$ **52.** $\frac{y+x}{y-x}$

Getting Ready (page 475)

1. 2 **2.** 8 **3.** 10 **4.** -7

Orals (page 487)

1. 1 **2.** 4 **3.** $x < 2$ **4.** $x \geq 3$ **5.** $x < -4$ **6.** $x \geq -8$

Exercise 7.1 (page 487)

1. $\frac{1}{t^{12}}$ **3.** 471 or more **5.** equation **7.** multiplied, divided **9.** impossible **11.** is greater than **13.** half-open **15.** positive

17. 6 **19.** 4 **21.** 2 **23.** 6 **25.** -4 **27.** -6 **29.** -11 **31.** 24 **33.** 6 **35.** 0 **37.** identity **39.** -6 **41.** impossible
43. identity **45.** $B = \frac{3V}{h}$ **47.** $w = \frac{p-2l}{2}$ **49.** $x = z\sigma + \mu$ **51.** $x = \frac{y-b}{m}$ **53.** $s = \frac{f(P-L)}{i}$ **55.** $(2, \infty)$ ← (——————→

                                                                                                  2

57. $[-2, \infty)$ **59.** $(-\infty, 1]$ **61.** $\left(-\infty, -\frac{8}{5}\right)$

63. $[-36, \infty)$ **65.** $(-2, 5)$ **67.** $(8, 11)$

69. $[-4, 6)$ **71.** no solution   **73.** $[-2, 4]$ **75.** $(-\infty, 2) \cup (7, \infty)$

77. $(-\infty, 1)$ **79.** no solution   **81.** 7 ft, 15 ft   **83.** 12 m by 24 m   **85.** 20 ft by 45 ft   **87.** $p < \$15$ **89.** 18

Getting Ready (page 491)

1. 6   **2.** 5   **3.** $2 - x$ **4.** $\pi - 2$

Orals (page 495)

1. 5   **2.** -5 **3.** -6 **4.** -4 **5.** 8 or -8 **6.** no solution   **7.** 5   **8.** 0, -2

Exercise 7.2 (page 496)

1. $\frac{3}{4}$ **3.** 6   **5.** x **7.** 0   **9.** $a = -b$ **11.** 8   **13.** 12   **15.** -2 **17.** -30 **19.** 50   **21.** $4 - \pi$ **23.** $|2|$ **25.** $|5|$ **27.** $|-2|$
29. $-|-4|$ **31.** $-|-7|$ **33.** $-x$ **35.** 8, -8 **37.** 9, -3 **39.** 4, -1 **41.** $\frac{14}{3}$, -6 **43.** no solution   **45.** 8, -4 **47.** 2, $-\frac{1}{2}$
49. -8 **51.** $-4, -28$ **53.** 0, -6 **55.** $\frac{20}{3}$ **57.** $-2, -\frac{4}{5}$ **59.** 3, -1 **61.** 0, -2 **63.** 0   **65.** $\frac{4}{3}$ **67.** no solution

Getting Ready (page 497)

1. $x > 1$ **2.** $x > -2$ **3.** $x \le 7$

Orals (page 501)

1. $-8 < x < 8$ **2.** $x < -8$ or $x > 8$ **3.** $x \le -4$ or $x \ge 4$ **4.** $-7 \le x \le 7$ **5.** $-3 < x < 1$ **6.** $x < -3$ or $x > 1$

Exercise 7.3 (page 501)

1. $t = \frac{A - p}{pr}$ **3.** $l = \frac{P - 2w}{2}$ **5.** $-k < x < k$ **7.** $x < -k$ or $x > k$ **9.** $(-4, 4)$,   **11.** $[-21, 3]$,

13. no solution   **15.** $\left[-\frac{3}{2}, 2\right]$,   **17.** $(-2, 5)$,   **19.** $(-\infty, -1) \cup (1, \infty)$,

21. $(-\infty, -12) \cup (36, \infty)$,   **23.** $\left(-\infty, -\frac{16}{3}\right) \cup (4, \infty)$,

25. $(-\infty, \infty)$,   **27.** $(-\infty, -2] \cup \left[\frac{10}{3}, \infty\right)$,   **29.** $(-\infty, -2) \cup (5, \infty)$,

31. $\left(-\infty, \frac{3}{8}\right) \cup \left(\frac{3}{8}, \infty\right)$,   **33.** $[-10, 14]$,   **35.** $\left(-\frac{5}{3}, 1\right)$,

37. $(-\infty, -4] \cup [-1, \infty)$,   **39.** no solution   **41.** $(-\infty, -24) \cup (-18, \infty)$,

43. $(-\infty, 25) \cup (25, \infty)$,   **45.** $\left(-\frac{65}{9}, -\frac{5}{9}\right)$,   **47.** $(-\infty, -4) \cup (-4, \infty)$,

49. $[-7, -7]$,   **51.** $[5, 5]$,   **55.** x and y must have different signs.

Getting Ready (page 503)

1. $6x^3y - 3x^2y^2$ **2.** $x^2 - 4$ **3.** $x^2 - x - 6$ **4.** $6x^2 + 7x - 3$

Orals (page 514)

1. $2x(x + 2)$ **2.** $3xy(y - 2x)$ **3.** $(x + 1)(x - 1)$ **4.** $(a^2 + 4)(a + 2)(a - 2)$ **5.** $(x + 6)(x - 1)$ **6.** $(2x + 1)(x - 1)$

Exercise 7.4 (page 514)

1. $x^3 + 1$ **3.** $r^3 - 8$ **5.** 12 **7.** $ab + ac$ **9.** perfect **11.** $2(x + 4)$ **13.** $2x(x - 3)$ **15.** $5x^2y(3 - 2y)$ **17.** $13ab^2c(c^2 - 2a^2)$
19. $3z(9z^2 + 4z + 1)$ **21.** $6s(4s^2 - 2st + t^2)$ **23.** $9x^7y^3(5x^3 - 7y^4 + 9x^3y^7)$ **25.** $-3(a + 2)$ **27.** $-3x(2x + y)$
29. $-7u^2v^3z^2(9uv^3z^7 - 4v^4 + 3uz^2)$ **31.** $x^2(x^n + x^{n+1})$ **33.** $y^n(2y^2 - 3y^3)$ **35.** $(x + y)(a + b)$ **37.** $(x + 2)(x + y)$
39. $(3 - c)(c + d)$ **41.** $r_1 = \dfrac{rr_2}{r_2 - r}$ **43.** $r = \dfrac{S - a}{S - l}$ **45.** $(x + 2)(x - 2)$ **47.** $(3y + 8)(3y - 8)$ **49.** $(9a^2 + 7b)(9a^2 - 7b)$
51. $(x + y + z)(x + y - z)$ **53.** $(x^2 + y^2)(x + y)(x - y)$ **55.** $2(x + 12)(x - 12)$ **57.** $2x(x + 4)(x - 4)$ **59.** $t^2(rs + x^2y)(rs - x^2y)$
61. $(x^m + y^{2n})(x^m - y^{2n})$ **63.** $(a + b)(a - b + 1)$ **65.** $(2x + y)(1 + 2x - y)$ **67.** $(x + 3)(x + 2)$ **69.** $(x - 2)(x - 5)$ **71.** prime
73. $3(x + 7)(x - 3)$ **75.** $b^2(a - 11)(a - 2)$ **77.** $-(a - 8)(a + 4)$ **79.** $-3(x - 3)(x - 2)$ **81.** $(3y + 2)(2y + 1)$
83. $(4a - 3)(2a + 3)$ **85.** prime **87.** $(4x - 3)(2x - 1)$ **89.** $(a + b)(a - 4b)$ **91.** $(2y - 3t)(y + 2t)$ **93.** $-(3a + 2b)(a - b)$
95. $x(3x - 1)(x - 3)$ **97.** $-x(2x - 3)^2$ **99.** $z(8x^2 + 6xy + 9y^2)$ **101.** $(x^2 + 5)(x^2 + 3)$ **103.** $(y^2 - 10)(y^2 - 3)$
105. $(a + 3)(a - 3)(a + 2)(a - 2)$ **107.** $(x^n + 1)^2$ **109.** $(2a^{3n} + 1)(a^{3n} - 2)$ **111.** $(x^{2n} + y^{2n})^2$ **113.** $(3x^n - 1)(2x^n + 3)$
115. $(x + 2 + y)(x + 2 - y)$ **117.** $(x + 1 + 3z)(x + 1 - 3z)$ **119.** $(c + 2a - b)(c - 2a + b)$ **121.** $(a - 16)(a - 1)$
123. $(2u + 3)(u + 1)$ **125.** $(5r + 2s)(4r - 3s)$ **127.** $(5u + v)(4u + 3v)$

Getting Ready (page 517)

1. $x^3 - 27$ **2.** $x^3 + 8$ **3.** $y^3 + 64$ **4.** $r^3 - 125$ **5.** $a^3 - b^3$ **6.** $a^3 + b^3$

Orals (page 521)

1. $(x - y)(x^2 + xy + y^2)$ **2.** $(x + y)(x^2 - xy + y^2)$ **3.** $(a + 2)(a^2 - 2a + 4)$ **4.** $(b - 3)(b^2 + 3b + 9)$
5. $(1 + 2x)(1 - 2x + 4x^2)$ **6.** $(2 - r)(4 + 2r + r^2)$ **7.** $(xy + 1)(x^2y^2 - xy + 1)$ **8.** $(5 - 2t)(25 + 10t + 4t^2)$

Exercise 7.5 (page 521)

1. 0.0000000000001 cm **3.** $(x^2 - xy + y^2)$ **5.** $(y + 1)(y^2 - y + 1)$ **7.** $(a - 3)(a^2 + 3a + 9)$ **9.** $(2 + x)(4 - 2x + x^2)$
11. $(s - t)(s^2 + st + t^2)$ **13.** $(3x + y)(9x^2 - 3xy + y^2)$ **15.** $(a + 2b)(a^2 - 2ab + 4b^2)$ **17.** $(4x - 3)(16x^2 + 12x + 9)$
19. $(3x - 5y)(9x^2 + 15xy + 25y^2)$ **21.** $(a^2 - b)(a^4 + a^2b + b^2)$ **23.** $(x^3 + y^2)(x^6 - x^3y^2 + y^4)$ **25.** $2(x + 3)(x^2 - 3x + 9)$
27. $-(x - 6)(x^2 + 6x + 36)$ **29.** $8x(2m - n)(4m^2 + 2mn + n^2)$ **31.** $xy(x + 6y)(x^2 - 6xy + 36y^2)$
33. $3rs^2(3r - 2s)(9r^2 + 6rs + 4s^2)$ **35.** $a^3b^2(5a + 4b)(25a^2 - 20ab + 16b^2)$ **37.** $yz(y^2 - z)(y^4 + y^2z + z^2)$
39. $2mp(p + 2q)(p^2 - 2pq + 4q^2)$ **41.** $(x + 1)(x^2 - x + 1)(x - 1)(x^2 + x + 1)$
43. $(x^2 + y)(x^4 - x^2y + y^2)(x^2 - y)(x^4 + x^2y + y^2)$ **45.** $(x + y)(x^2 - xy + y^2)(3 - z)$ **47.** $(m + 2n)(m^2 - 2mn + 4n^2)(1 + x)$
49. $(a + 3)(a^2 - 3a + 9)(a - b)$ **51.** $(y + 1)(y - 1)(y - 3)(y^2 + 3y + 9)$

Getting Ready (page 523)

1. $\frac{5}{3}$ **2.** $\frac{4}{15}$ **3.** $\frac{19}{6}$ **4.** $-\frac{11}{6}$

Orals (page 535)

1. $\frac{2}{3}$ **2.** $\frac{2}{3}$ **3.** $-\frac{5}{6}$ **4.** $-\frac{2}{5}$ **5.** $\frac{x}{y}$ **6.** 2 **7.** -1 **8.** $x - 1$

Exercise 7.6 (page 535)

1. **3.** $w = \dfrac{P - 2l}{2}$ **5.** $2, -2, 3, -3$ **7.** $\dfrac{a}{b}$ **9.** $\dfrac{ad}{bc}$ **11.** $4x^2$ **13.** $-\dfrac{4y}{3x}$ **15.** $\dfrac{3y}{7(y - z)}$ **17.** 1

19. $\dfrac{1}{x - y}$ **21.** $\dfrac{-3(x + 2)}{x + 1}$ **23.** $x + 2$ **25.** $\dfrac{x + 1}{x + 3}$ **27.** $\dfrac{m - 2n}{n - 2m}$ **29.** $\dfrac{x + 4}{2(2x - 3)}$ **31.** $\dfrac{3(x - y)}{x + 2}$ **33.** $\dfrac{1}{x^2 + xy + y^2 - 1}$

35. $\dfrac{xy^2d}{c^3}$ **37.** $-\dfrac{x^{10}}{y^2}$ **39.** $x+1$ **41.** 1 **43.** $\dfrac{x-4}{x+5}$ **45.** $\dfrac{(a+7)^2(a-5)}{12x^2}$ **47.** $\dfrac{t-1}{t+1}$ **49.** $\dfrac{n+2}{n+1}$ **51.** $x-5$ **53.** $\dfrac{x+y}{x-y}$

55. $-\dfrac{x+3}{x+2}$ **57.** $\dfrac{a+b}{(x-3)(c+d)}$ **59.** $-\dfrac{x+1}{x+3}$ **61.** $-\dfrac{2x+y}{3x+y}$ **63.** $-\dfrac{x^7}{18y^4}$ **65.** $x^2(x+3)$ **67.** $\dfrac{3x}{2}$ **69.** $\dfrac{x-7}{x+7}$ **71.** 1

73. $\dfrac{3-a}{a+b}$ **75.** 2 **77.** 3 **79.** $\dfrac{6x}{(x-3)(x-2)}$ **81.** $\dfrac{9a}{10}$ **83.** $\dfrac{17}{12x}$ **85.** $\dfrac{10a+4b}{21}$ **87.** $\dfrac{8x-2}{(x+2)(x-4)}$ **89.** $\dfrac{7x+29}{(x+5)(x+7)}$

91. $\dfrac{x^2+1}{x}$ **93.** 2 **95.** $\dfrac{2x^2+x}{(x+3)(x+2)(x-2)}$ **97.** $\dfrac{-4x^2+14x+54}{x(x+3)(x-3)}$ **99.** $\dfrac{x^2-5x-5}{x-5}$ **101.** $\dfrac{2}{x+1}$ **103.** $\dfrac{3x+1}{x(x+3)}$

105. $\dfrac{2y}{3z}$ **107.** $-\dfrac{1}{y}$ **109.** $\dfrac{b+a}{b}$ **111.** $y-x$ **113.** $\dfrac{-1}{a+b}$ **115.** $\dfrac{x+2}{x-3}$ **117.** $\dfrac{a-1}{a+1}$ **119.** $\dfrac{y+x}{y-x}$ **121.** xy **123.** $\dfrac{x^2(xy^2-1)}{y^2(x^2y-1)}$

125. $\dfrac{(b+a)(b-a)}{b(b-a-ab)}$ **127.** $\dfrac{3a^2+2a}{2a+1}$ **133.** yes **135.** a, d

Getting Ready (page 539)

1. $x+1$ with a remainder of 1, 1 **2.** $x+3$ with a remainder of 9, 9

Orals (page 544)

1. 9 **2.** -3 **3.** yes **4.** no

Exercise 7.7 (page 544)

1. 4 **3.** $12a^2+4a-1$ **5.** $8x^2+2x+4$ **7.** $P(r)$ **9.** $x+2$ **11.** $x-3$ **13.** $x+2$ **15.** $x-7+\frac{28}{x+2}$ **17.** $3x^2-x+2$
19. $2x^2+4x+3$ **21.** $6x^2-x+1+\frac{3}{x+1}$ **23.** $7.2x-0.66+\frac{0.368}{x-0.2}$ **25.** $2.7x-3.59+\frac{0.903}{x+1.7}$
27. $9x^2-513x+29{,}241+\frac{-1{,}666{,}762}{x+57}$ **29.** -1 **31.** -37 **33.** 23 **35.** -1 **37.** 2 **39.** -1 **41.** 18 **43.** 174 **45.** -8
47. 59 **49.** 44 **51.** $\frac{29}{32}$ **53.** yes **55.** no **57.** 64 **61.** 1

Chapter Summary (page 548)

1. a. 8 **b.** 7 **c.** 19 **d.** 8 **e.** identity **f.** impossible equation **2. a.** $h=\dfrac{3V}{\pi r^2}$ **b.** $x=\dfrac{6V}{ab}-y$ **3.** 5 ft from one end
4. 45 m² **5. a.** $(-\infty, -24]$, ←———┤——→ **b.** $\left(-\infty, -\frac{51}{11}\right)$, ←———)——→ **c.** $\left(-\frac{1}{3}, 2\right)$, ←(———)→
  -24 $-51/11$ $-1/3$ 2
d. $(2, \infty)$, ←(———→ **6. a.** 3, $-\frac{11}{3}$ **b.** $\frac{26}{3}$, $-\frac{10}{3}$ **c.** $\frac{1}{5}$, -5 **d.** -1, 1 **7. a.** $(-5, -2)$, ←(———)→
  2 -5 -2
b. $\left(-\infty, \frac{4}{3}\right] \cup [4, \infty)$, ←——┤—┤——→ **c.** $(-\infty, \infty)$, ←————→ **d.** no solutions **8. a.** $4(x+2)$ **b.** $5xy^2(xy-2)$
  $4/3$ 4 0
c. $-4x^2y^3z^2(2z^2+3x^2)$ **d.** $3a^2b^4c^2(4a^4+5c^4)$ **e.** $(x+2)(y+4)$ **f.** $(a+b)(c+3)$ **9.** $x^n(x^n+1)$ **10.** $y^{2n}(1-y^{2n})$
11. a. $(x^2+4)(x^2+y)$ **b.** $(a^3+c)(a^2+b^2)$ **c.** $(z+4)(z-4)$ **d.** $(y+11)(y-11)$ **e.** $2(x^2+7)(x^2-7)$
f. $3x^2(x^2+10)(x^2-10)$ **g.** $(y+20)(y+1)$ **h.** $(z-5)(z-6)$ **i.** $-(x+7)(x-4)$ **j.** $-(y-8)(y+3)$ **k.** $y(y+2)(y-1)$
l. $2a^2(a+3)(a-1)$ **m.** $3(5x+y)(x-4y)$ **n.** $5(6x+y)(x+2y)$ **o.** $(x+2+2p^2)(x+2-2p^2)$ **p.** $(y+2)(y+1+x)$
12. a. $(x+7)(x^2-7x+49)$ **b.** $(a-5)(a^2+5a+25)$ **c.** $8(y-4)(y^2+4y+16)$ **d.** $4y(x+3z)(x^2-3xz+9z^2)$
13. a. $\frac{31x}{72y}$ **b.** $\frac{x-7}{x+7}$ **14. a.** 1 **b.** 1 **c.** $\dfrac{5y-3}{x-y}$ **d.** $\dfrac{6x-7}{x^2+2}$ **e.** $\dfrac{5x+13}{(x+2)(x+3)}$ **f.** $\dfrac{4x^2+9x+12}{(x-4)(x+3)}$ **g.** $\dfrac{3x(x-1)}{(x-3)(x+1)}$ **h.** 1
i. $\dfrac{5x^2+11x}{(x+1)(x+2)}$ **j.** $\dfrac{2(3x+1)}{x-3}$ **k.** $\dfrac{5x^2+23x+4}{(x+1)(x-1)(x-1)}$ **l.** $\dfrac{-x^4-4x^3+3x^2+18x+16}{(x-2)(x+2)^2}$ **15. a.** $\dfrac{3y-2x}{x^2y}$ **b.** $\dfrac{y+2x}{2y-x}$
c. $\dfrac{2x+1}{x+1}$ **d.** $\dfrac{y-x}{y+x}$ **16. a.** 20 **b.** -1 **17. a.** yes **b.** no

Chapter 7 Test (page 551)

1. -12 **2.** 6   **3.** $i = \frac{f(P - L)}{s}$ **4.** $a = \frac{180n - 360}{n}$ **5.** $13\frac{1}{3}$ft   **6.** $36\,cm^2$ **7.** $(-\infty, -5]$,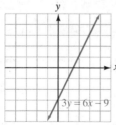

8. $(-2, 16)$, 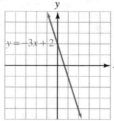   **9.** $4, -7$ **10.** $4, -4$ **11.** $[-7, 1]$,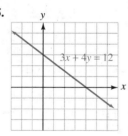

12. $(-\infty, -9) \cup (13, \infty)$, 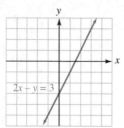   **13.** $3xy(y + 2x)$ **14.** $3abc(4a^2b - abc + 2c^2)$ **15.** $(a - y)(x + y)$

16. $(x + y)(a + b - c)$ **17.** $(x + 7)(x - 7)$ **18.** $2(x + 4)(x - 4)$ **19.** $4(y^2 + 4)(y + 2)(y - 2)$ **20.** $(b + 5)(b^2 - 5b + 25)$

21. $(b - 3)(b^2 + 3b + 9)$ **22.** $3(u - 2)(u^2 + 2u + 4)$ **23.** $(x + 5)(x + 3)$ **24.** $(3b + 2)(2b - 1)$ **25.** $3(u + 2)(2u - 1)$

26 $(x + 3 + y)(x + 3 - y)$ **27.** $\frac{-2}{3xy}$ **28.** $\frac{2x + 1}{4}$ **29.** $\frac{xz}{y^4}$ **30.** 1   **31.** $\frac{(x + y)^2}{2}$ **32.** $\frac{2x + 3}{(x + 1)(x + 2)}$ **33.** $\frac{u^2}{2vw}$ **34.** $\frac{2x + y}{xy - 2}$

35. -7 **36.** 47

Getting Ready (page 554)

1. 1   **2.** 9   **3.** 5   **4.** 2

Orals (page 562)

1. $(3, 0), (0, 3)$ **2.** $(2, 0), (0, 6)$ **3.** $(8, 0), (0, 2)$ **4.** $(4, 0), (0, -3)$ **5.** $(4, 6)$ **6.** $(0, -1)$

Exercise 8.1 (page 562)

1. 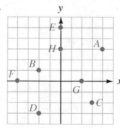   **3.** $x(x - 1)$ **5.** $(x - 1)(x^2 + x + 1)$ **7.** origin   **9.** y-coordinate   **11.** x-axis   **13.** horizontal

15–22. 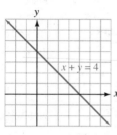   **23.** $(2, 4)$ **25.** $(-2, -1)$ **27.** $(4, 0)$ **29.** $(0, 0)$ **31.**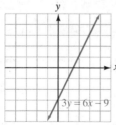

33. 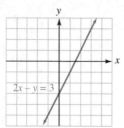   **35.**   **37.** 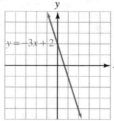   **39.**

41. 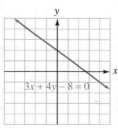   **43.** 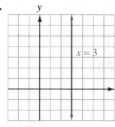   **45.** 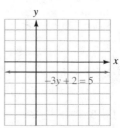   **47.** $(3, 4)$ **49.** $(9, 12)$ **51.** $\left(\frac{7}{2}, 6\right)$

53. $\left(\frac{1}{2}, -2\right)$ **55.** $(-4, 0)$ **57.** $\left(\frac{5a}{2}, 2b\right)$ **59.** $(a, 2b)$ **61.** $(4, 1)$ **63.** $162,500, $200,000 **65.** 200 **67.** 100 rpm
71. $a = 0, b > 0$

Getting Ready (page 565)

1. 1 **2.** -1 **3.** $\frac{13}{14}$ **4.** $\frac{3}{14}$

Orals (page 574)

1. 3 **2.** 2 **3.** yes **4.** 5 **5.** yes

Exercise 8.2 (page 574)

1. $x^9 y^6$ **3.** $\dfrac{x^{12}}{y^8}$ **5.** 1 **7.** y, x **9.** $\dfrac{y_2 - y_1}{x_2 - x_1}$ **11.** run **13.** vertical **15.** parallel **17.** 3 **19.** -1 **21.** $-\frac{1}{3}$ **23.** 0
25. undefined **27.** -1 **29.** $-\frac{3}{2}$ **31.** $\frac{3}{4}$ **33.** $\frac{1}{2}$ **35.** 0 **37.** negative **39.** positive **41.** undefined **43.** perpendicular
45. neither **47.** parallel **49.** parallel **51.** perpendicular **53.** neither **55.** not the same line **57.** not the same line
59. same line **61.** $y = 0, m = 0$ **69.** $\frac{1}{165}$ **71.** $\frac{18}{5}$ **73.** $20,000 per yr **79.** 4

Getting Ready (page 577)

1. 14 **2.** $-\frac{5}{3}$ **3.** $y = 3x - 4$ **4.** $x = \dfrac{-By - 3}{A}$

Orals (page 587)

1. $y - 3 = 2(x - 2)$ **2.** $y - 8 = 2(x + 3)$ **3.** $y = -3x + 5$ **4.** $y = -3x - 7$ **5.** parallel **6.** perpendicular

Exercise 8.3 (page 588)

1. 6 **3.** -1 **5.** 20 oz **7.** $y - y_1 = m(x - x_1)$. **9.** $Ax + By = C$ **11.** perpendicular **13.** $5x - y = -7$ **15.** $3x + y = 6$
17. $2x - 3y = -11$ **19.** $y = x$ **21.** $y = \frac{7}{3}x - 3$ **23.** $y = -\frac{9}{5}x + \frac{2}{5}$ **25.** $y = 3x + 17$ **27.** $y = -7x + 54$ **29.** $y = -4$
31. $y = -\frac{1}{2}x + 11$ **33.** 1, $(0, -1)$

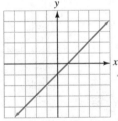

35. $\frac{2}{3}$, $(0, 2)$

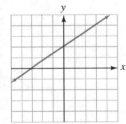

37. $-\frac{2}{3}$, $(0, 6)$

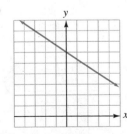

39. $\frac{3}{2}$, $(0, -4)$ **41.** $-\frac{1}{3}$, $\left(0, -\frac{5}{6}\right)$ **43.** $\frac{7}{2}$, $(0, 2)$ **45.** parallel **47.** perpendicular
49. parallel **51.** perpendicular **53.** perpendicular **55.** perpendicular **57.** $y = 4x$
59. $y = 4x - 3$ **61.** $y = \frac{4}{5}x - \frac{26}{5}$ **63.** $y = -\frac{1}{4}x$ **65.** $y = -\frac{1}{4}x + \frac{11}{2}$ **67.** $y = -\frac{5}{4}x + 3$
69. perpendicular **71.** parallel **73.** $x = -2$ **75.** $x = 5$ **77.** $y = -\frac{A}{B}x + \frac{C}{B}$
79. $y = -2,298x + 19,984$ **81.** $y = 50,000x + 250,000$ **83.** $y = -\frac{950}{3}x + 1,750$
85. $490 **87.** $154,000 **89.** $180 **99.** $a < 0, b > 0$

Getting Ready (page 592)

1. 1 **2.** 7 **3.** -20 **4.** $-\frac{11}{4}$

Orals (page 600)

1. yes   **2.** no   **3.** no   **4.** 1   **5.** 3   **6.** −3

Exercise 8.4 (page 600)

1. −2   **3.** 2   **5.** input   **7.** x   **9.** function, input, output   **11.** range   **13.** 0   **15.** mx + b   **17.** yes   **19.** yes   **21.** yes   **23.** no
25. 9, −3   **27.** 3, −5   **29.** 22, 2   **31.** 3, 11   **33.** 4, 9   **35.** 7, 26   **37.** 9, 16   **39.** 6, 15   **41.** 4, 4   **43.** 2, 2   **45.** $\frac{1}{5}$, 1
47. −2, $\frac{2}{5}$ **49.** 2w, 2w + 2   **51.** 3w − 5, 3w − 2   **53.** 12   **55.** 2b − 2a   **57.** 2b   **59.** 1   **61.** D = {−2, 4, 6}; R = {3, 5, 7}
63. D = (−∞, 4) ∪ (4, ∞); R = (−∞, 0) ∪ (0, ∞)   **65.** not a function   **67.** a function; D = (−∞, ∞); R = (−∞, ∞)
69.   D = (−∞, ∞); R = (−∞, ∞)   **71.**   D = (−∞, ∞); R = (−∞, ∞)   **73.** no

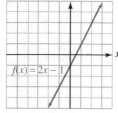

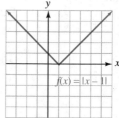

75. yes   **77.** 624 ft   **79.** 77° F   **81.** 192   **85.** yes

Getting Ready (page 603)

1. 2, (0, −3)   **2.** −3, (0, 4)   **3.** 6, −9   **4.** 4, $\frac{5}{2}$

Exercise 8.5 (page 614)

1. 41, 43, 47   **3.** a · b = b · a   **5.** 1   **7.** squaring   **9.** absolute value   **11.** horizontal   **13.** 2, down   **15.** 4, to the left
17. rational   **19.**   **21.**   **23.**   **25.**

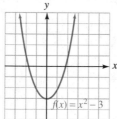   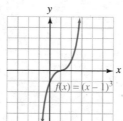   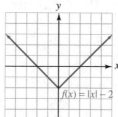   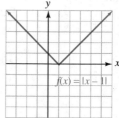

27.   **29.**   **31.**   **33.**

35.   **37.**   **39.**   **41.**

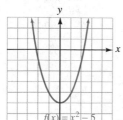   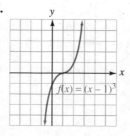   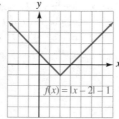   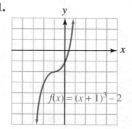

43.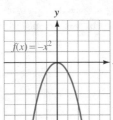

45.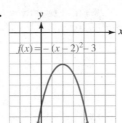

47. 20 hr   **49.** 12 hr   **51.** $5,555.56   **53.** $50,000

55. all reals but 2; all reals but 1   **57.** all reals but ±2; all reals   **59.** $c = f(x) = 1.25x + 700$ **61.** $1,325   **63.** $1.95
65. $c = f(n) = 0.09n + 7.50$ **67.** $77.25   **69.** 9.75¢   **71.** $c = f(x) = 350x + 5,000$ **73.** $47,000

Getting Ready   (page 618)

1. $\frac{3}{2}$ **2.** $\frac{10}{7}$ **3.** 4   **4.** 36

Orals   (page 624)

1. 1   **2.** 9   **3.** $\frac{14}{5}$ **4.** $a = kb$ **5.** $a = \frac{k}{b}$ **6.** $a = kbc$ **7.** $a = \frac{kb}{c}$

Exercise 8.6   (page 624)

1. x^{10} **3.** -1 **5.** 3.5×10^4 **7.** 0.0025   **9.** proportion   **11.** direct   **13.** rational   **15.** joint   **17.** direct   **19.** neither   **21.** 3
23. 5   **25.** -3 **27.** 5   **29.** 4, -1 **31.** 2, -2 **33.** 39   **35.** $-\frac{5}{2}$, -1 **37.** $A = kp^2$ **39.** $v = k/r^3$ **41.** $B = kmn$
43. $P = ka^2/j^3$ **45.** L varies jointly with m and n **47.** E varies jointly with a and the square of b **49.** X varies directly with x^2
and inversely with y^2 **51.** R varies directly with L and inversely with d^2 **53.** 36π in.^{2} **55.** 432 mi   **57.** 25 days   **59.** 12 in.^{3}
61. 85.3   **63.** 12   **65.** 26,437.5 gal   **67.** 3 ohms   **69.** 0.275 in.   **71.** 546 Kelvin

Chapter Summary   (page 630)

1. a.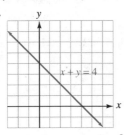

b.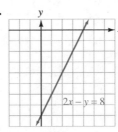

c.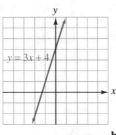

d.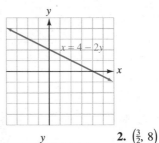

e.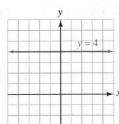

f.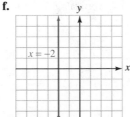

g.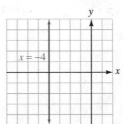

h.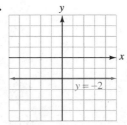

2. $\left(\frac{3}{2}, 8\right)$

3. a. 1   **b.** $\frac{14}{9}$ **c.** 5   **d.** $\frac{5}{11}$ **e.** 0   **f.** no defined slope   **4. a.** $\frac{2}{3}$ **b.** -2 **c.** undefined   **d.** 0   **5. a.** perpendicular

b. parallel   **c.** neither   **d.** perpendicular   **6.** $21,666.67   **7. a.** $3x - y = -29$ **b.** $13x + 8y = 6$ **c.** $3x - 2y = 1$
d. $2x + 3y = -21$ **8.**

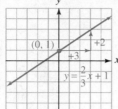

9. perpendicular   **10.** $y = -1,720x + 8,700$ **11. a.** yes   **b.** yes   **c.** no   **d.** no

12. a. -7 **b.** 60   **c.** 0   **d.** 17   **13. a.** $D = (-\infty, \infty)$; $R = (-\infty, \infty)$ **b.** $D = (-\infty, \infty)$; $R = (-\infty, \infty)$ **c.** $D = (-\infty, \infty)$;
$R = [1, \infty)$ **d.** $D = (-\infty, 2) \cup (2, \infty)$; $R = (-\infty, 0) \cup (0, \infty)$ **e.** $D = (-\infty, 3) \cup (3, \infty)$; $R = (-\infty, 0) \cup (0, \infty)$ **f.** $D = (-\infty, \infty)$;
$R = \{7\}$ **14. a.** a function   **b.** not a function   **c.** not a function   **d.** a function
15. a.

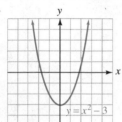

b.

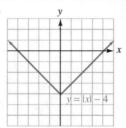

c.

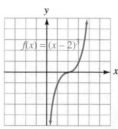

d.

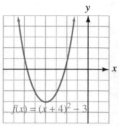

e.

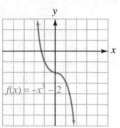

f.

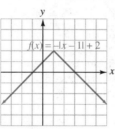

16. a.

b.

c.

d.

e.

f.

17. a.

D: $(-\infty, 2) \cup (2, \infty)$; R: $(-\infty, 0) \cup (0, \infty)$

b.

D: $(-\infty, -3) \cup (-3, \infty)$; R: $(-\infty, 1) \cup (1, \infty)$ **18. a.** $\frac{13}{3}$ **b.** $-4, -12$ **19.** 2   **20.** 6   **21.** $\frac{1}{32}$ **22.** 16

Chapter 8 Test   (page 635)

1.

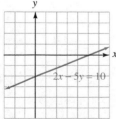

2. $\left(\frac{1}{2}, \frac{1}{2}\right)$ **3.** x-intercept $(3, 0)$, y-intercept $\left(0, -\frac{3}{5}\right)$ **4.** vertical   **5.** $\frac{1}{2}$ **6.** $\frac{2}{3}$ **7.** undefined   **8.** 0

9. $y = \frac{2}{3}x - \frac{23}{3}$ **10.** $8x - y = -22$ **11.** $m = -\frac{1}{3}, \left(0, -\frac{3}{2}\right)$ **12.** neither   **13.** perpendicular   **14.** $y = \frac{3}{2}x$ **15.** $y = \frac{3}{2}x + \frac{21}{2}$ **16.** no
17. $D = (-\infty, \infty)$; $R = [0, \infty)$ **18.** $D = (-\infty, \infty)$; $R = (-\infty, \infty)$ **19.** 10   **20.** -2 **21.** $3a + 1$ **22.** $x^2 - 2$ **23.** yes   **24.** no

25.

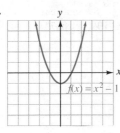

26.

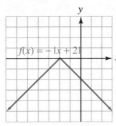

27. 6, −1   **28.** $\frac{6}{5}$ **29.** $\frac{44}{3}$ **30.** yes

Cumulative Review Exercises   (page 637)

1. 1, 2, 6, 7   **2.** 0, 1, 2, 6, 7   **3.** −2, 0, 1, 2, $\frac{13}{12}$, 6, 7   **4.** $\sqrt{5}$, π **5.** −2   **6.** −2, 0, 1, 2, $\frac{13}{12}$, 6, 7, $\sqrt{5}$, π **7.** 2, 7   **8.** 6
9. −2, 0, 2, 6   **10.** 1, 7   **11.** 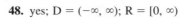   **12.** 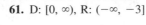   **13.** −2   **14.** −2   **15.** 22   **16.** −2
17. −4   **18.** −3   **19.** 4   **20.** −5   **21.** assoc. prop. of add.   **22.** distrib. prop.   **23.** comm. prop of add.

24. assoc. prop. of mult.   **25.** $x^8 y^{12}$ **26.** c^2 **27.** $-\dfrac{b^3}{a^2}$ **28.** 1   **29.** 4.97×10^{-6} **30.** 932,000,000   **31.** 8   **32.** −27   **33.** −1

34. 6   **35.** $a = \dfrac{2S}{n} - l$ **36.** $h = \dfrac{2A}{b_1 + b_2}$ **37.** 28, 30, 32   **38.** 14 cm by 42 cm   **39.** yes   **40.** $-\dfrac{7}{5}$ **41.** $y = -\dfrac{7}{5}x + \dfrac{11}{5}$
42. $y = -3x - 3$ **43.** 5   **44.** −1   **45.** $2t - 1$ **46.** $3r^2 + 2$
47. yes; D = (−∞, ∞); R = (−∞, 1]

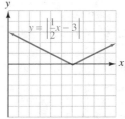

48. yes; D = (−∞, ∞); R = [0, ∞)

Getting Ready   (page 640)

1. 0   **2.** 16   **3.** 16   **4.** −16   **5.** $\frac{8}{125}$ **6.** $\frac{81}{256}$ **7.** $49x^2 y^2$ **8.** $343x^3 y^3$

Orals   (page 651)

1. 3   **2.** −4   **3.** −2   **4.** 2   **5.** $8|x|$ **6.** −3x   **7.** not a real number   **8.** $(x + 1)^2$

Exercise 9.1   (page 651)

1. $\dfrac{x + 3}{x - 4}$ **3.** 1   **5.** $\dfrac{3(m^2 + 2m - 1)}{(m + 1)(m - 1)}$ **7.** $(5x^2)^2$ **9.** positive   **11.** 3, up   **13.** $y^3 = x$ **15.** odd   **17.** 0   **19.** $3x^2$ **21.** $a^2 + b^3$
23. 11   **25.** −8   **27.** $\frac{1}{3}$ **29.** $-\frac{5}{7}$ **31.** not real   **33.** 0.4   **35.** 4   **37.** not real   **39.** 3.4641   **41.** 26.0624   **43.** $2|x|$
45. $|t + 5|$ **47.** $5|b|$ **49.** $|a + 3|$ **51.** 0   **53.** 4   **55.** 4.1231   **57.** 2.5539
59. D: [−4, ∞), R: [0, ∞)

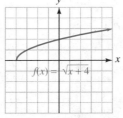

61. D: [0, ∞), R: (−∞, −3]

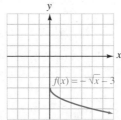

63. 1   **65.** −5

67. $-\frac{2}{3}$ **69.** 0.4 **71.** $2a$ **73.** $-10pq$ **75.** $-\frac{1}{2}m^2n$ **77.** $0.2z^3$ **79.** 3 **81.** -3 **83.** -2 **85.** $\frac{2}{5}$ **87.** $\frac{1}{2}$ **89.** not real **91.** $2|x|$ **93.** $2a$ **95.** $\frac{1}{2}|x|$ **97.** $|x^3|$ **99.** $-x$ **101.** $-3a^2$ **103.** $x+2$ **105.** $0.1x^2|y|$ **107.** 1.67 **109.** 11.8673 **111.** 3 units **113.** 4 sec **115.** about 7.4 amperes

Getting Ready (page 654)

1. 25 **2.** 169 **3.** 18 **4.** 11,236

Orals (page 659)

1. 5 **2.** 10 **3.** 13 **4.** 5 **5.** 10 **6.** 13 **7.** 4 **8.** 3 **9.** 5

Exercise 9.2 (page 659)

1. $12x^2 - 14x - 10$ **3.** $15t^2 + 2ts - 8s^2$ **5.** hypotenuse **7.** $a^2 + b^2 = c^2$ **9.** distance **11.** 10 ft **13.** 80 m **15.** 9.9 cm **17.** 5 **19.** 5 **21.** 13 **23.** 10 **25.** 10.2 **29.** $x = 7$ **31.** (7, 0) and (3, 0) **33.** 13 ft **35.** about 127 ft **37.** about 135 ft **39.** not quite **41.** yes **43.** 173 yd **45.** 0.05 ft **47.** 8 cm^3 **51.** 54.1 in.

Getting Ready (page 663)

1. a **2.** $5x$ **3.** $x+4$ **4.** $y-3$

Orals (page 669)

1. 7 **2.** 3 **3.** 0 **4.** 9 **5.** 17 **6.** 31

Exercise 9.3 (page 669)

1. 2 **3.** 6 **5.** $x^n = y^n$ **7.** square **9.** extraneous **11.** 2 **13.** 4 **15.** 0 **17.** 4 **19.** 8 **21.** $\frac{5}{2}, \frac{1}{2}$ **23.** 1 **25.** 16 **27.** 14, 6 **29.** 4, 3 **31.** 2, 7 **33.** 9, -25 **35.** 2, -1 **37.** -1, 1 **39.** 1, no solutions **41.** 0, 4 **43.** -3, no solutions **45.** 0 **47.** 1, 9 **49.** 4, 0 **51.** 2, 142 **53.** 2 **55.** 6, no solutions **57.** 0, $-\frac{12}{11}$ **59.** 1 **61.** 4, -9 **63.** 2, $\frac{21}{2}$ **65.** 2,010 ft **67.** about 29 mph **69.** \$5 **73.** 0, 4

Getting Ready (page 672)

1. x^7 **2.** a^{12} **3.** a^4 **4.** 1 **5.** $\dfrac{1}{x^4}$ **6.** a^3b^6 **7.** $\dfrac{b^6}{c^9}$ **8.** a^{10}

Orals (page 679)

1. 2 **2.** 3 **3.** 3 **4.** 1 **5.** 8 **6.** 4 **7.** $\frac{1}{2}$ **8.** 2 **9.** $2x$ **10.** $2x^2$

Exercise 9.4 (page 679)

1. $x < 3$ **3.** $r > 28$ **5.** $1\frac{2}{3}$ pints **7.** $a \cdot a \cdot a \cdot a$ **9.** a^{mn} **11.** $\dfrac{a^n}{b^n}$ **13.** $\dfrac{1}{a^n}$, 0 **15.** $\left(\dfrac{b}{a}\right)^n$ **17.** $|x|$ **19.** $\sqrt[3]{7}$ **21.** $\sqrt[4]{3x}$ **23.** $\sqrt[4]{\frac{1}{2}x^3y}$ **25.** $\sqrt{x^2 + y^2}$ **27.** $11^{1/2}$ **29.** $(3a)^{1/4}$ **31.** $\left(\frac{1}{7}abc\right)^{1/6}$ **33.** $(a^2 - b^2)^{1/3}$ **35.** 2 **37.** 2 **39.** 2 **41.** 2 **43.** $\frac{1}{2}$ **45.** $\frac{1}{2}$ **47.** -2 **49.** -3 **51.** not real **53.** 0 **55.** $5|y|$ **57.** $2|x|$ **59.** $3x$ **61.** not real **63.** 216 **65.** 27 **67.** 1,728 **69.** $\frac{1}{4}$ **71.** $125x^6$ **73.** $\dfrac{4x^2}{9}$ **75.** $\dfrac{1}{2}$ **77.** $\dfrac{1}{8}$ **79.** $\dfrac{1}{64x^3}$ **81.** $\dfrac{1}{9y^2}$ **83.** $\dfrac{1}{4p^2}$ **85.** 8 **87.** $\dfrac{16}{81}$ **89.** $-\dfrac{3}{2x}$ **91.** $5^{8/9}$ **93.** $4^{3/5}$ **95.** $9^{1/5}$ **97.** $7^{1/2}$ **99.** $\dfrac{1}{36}$ **101.** $2^{2/3}$ **103.** a **105.** $a^{2/9}$ **107.** $a^{3/4}b^{1/2}$ **109.** $\dfrac{n^{2/5}}{m^{3/5}}$ **111.** $\dfrac{2x}{3}$ **113.** $\dfrac{1}{3}x$ **115.** $y + y^2$

117. $x^2 - x + x^{3/5}$ **119.** $x - 4$ **121.** $x^{4/3} - x^2$ **123.** $x^{4/3} + 2x^{2/3}y^{2/3} + y^{4/3}$ **125.** $a^3 - 2a^{3/2}b^{3/2} + b^3$ **127.** $\sqrt{p}$ **129.** $\sqrt{5b}$
133. yes

Getting Ready (page 681)

1. 15   **2.** 24   **3.** 5   **4.** 7   **5.** $4x^2$ **6.** $\dfrac{8}{11}x^3$ **7.** $3ab^3$ **8.** $-2a^4$

Orals (page 689)

1. 7   **2.** 4   **3.** 3   **4.** $3\sqrt{2}$ **5.** $2\sqrt[3]{2}$ **6.** $\dfrac{\sqrt[3]{3x^2}}{4b^2}$ **7.** $7\sqrt{3}$ **8.** $3\sqrt{7}$ **9.** $5\sqrt[3]{9}$ **10.** $8\sqrt[5]{4}$

Exercise 9.5 (page 690)

1. $\dfrac{-15x^5}{y}$ **3.** $9t^2 + 12t + 4$ **5.** $3p + 4 + \dfrac{-5}{2p - 5}$ **7.** $\sqrt[n]{a}\sqrt[n]{b}$ **9.** 6   **11.** t **13.** $5x$ **15.** 10   **17.** $7x$ **19.** $6b$ **21.** 2   **23.** $3a$
25. $2\sqrt{5}$ **27.** $-10\sqrt{2}$ **29.** $2\sqrt[3]{10}$ **31.** $-3\sqrt[3]{3}$ **33.** $2\sqrt[4]{2}$ **35.** $2\sqrt[5]{3}$ **37.** $\dfrac{\sqrt{7}}{3}$ **39.** $\dfrac{\sqrt[3]{7}}{4}$ **41.** $\dfrac{\sqrt[4]{3}}{10}$ **43.** $\dfrac{\sqrt[5]{3}}{2}$ **45.** $5x\sqrt{2}$
47. $4\sqrt{2b}$ **49.** $-4a\sqrt{7a}$ **51.** $5ab\sqrt{7b}$ **53.** $-10\sqrt{3xy}$ **55.** $-3x^2\sqrt[3]{2}$ **57.** $2x^4y\sqrt[3]{2}$ **59.** $2x^3y\sqrt[4]{2}$ **61.** $\dfrac{z}{4x}$ **63.** $\dfrac{\sqrt[4]{5x}}{2z}$
65. $10\sqrt{2x}$ **67.** $\sqrt[5]{7a^2}$ **69.** $4\sqrt{3}$ **71.** $-\sqrt{2}$ **73.** $2\sqrt{2}$ **75.** $9\sqrt{6}$ **77.** $3\sqrt[3]{3}$ **79.** $-\sqrt[3]{4}$ **81.** -10 **83.** $-17\sqrt[4]{2}$
85. $16\sqrt[4]{2}$ **87.** $-4\sqrt{2}$ **89.** $3\sqrt{2} + \sqrt{3}$ **91.** $-11\sqrt[3]{2}$ **93.** $y\sqrt{z}$ **95.** $13y\sqrt{x}$ **97.** $12\sqrt[3]{a}$ **99.** $-7y^2\sqrt{y}$ **101.** $4x\sqrt[3]{xy^2}$
103. $2x + 2$ **105.** $h = 2.83, x = 2.00$ **107.** $x = 8.66, h = 10.00$ **109.** $x = 4.69, y = 8.11$ **111.** $x = 12.11, y = 12.11$
115. If $a = 0$, then b can be any real number. If $b = 0$, then a can be any real number.

Getting Ready (page 692)

1. a^7 **2.** b^3 **3.** $a^2 - 2a$ **4.** $6b^3 + 9b^2$ **5.** $a^2 - 3a - 10$ **6.** $4a^2 - 9b^2$

Orals (page 699)

1. 3   **2.** 2   **3.** $3\sqrt{3}$ **4.** $a^2|b|$ **5.** $6 + 3\sqrt{2}$ **6.** 1   **7.** $\dfrac{\sqrt{2}}{2}$ **8.** $\dfrac{\sqrt{3} + 1}{2}$

Exercise 9.6 (page 699)

1. 1   **3.** $\frac{1}{3}$ **5.** 2, $\sqrt{7}$, $\sqrt{5}$ **7.** FOIL   **9.** conjugate   **11.** 4   **13.** $5\sqrt{2}$ **15.** $6\sqrt{2}$ **17.** 5   **19.** 18   **21.** $2\sqrt[3]{3}$ **23.** ab^2
25. $5a\sqrt{b}$ **27.** $r\sqrt[3]{10s}$ **29.** $2a^2b^2\sqrt[3]{2}$ **31.** $x^2(x + 3)$ **33.** $3x(y + z)\sqrt[3]{4}$ **35.** $12\sqrt{5} - 15$ **37.** $12\sqrt{6} + 6\sqrt{14}$
39. $-8x\sqrt{10} + 6\sqrt{15x}$ **41.** $-1 - 2\sqrt{2}$ **43.** $8x - 14\sqrt{x} - 15$ **45.** $5z + 2\sqrt{15z} + 3$ **47.** $3x - 2y$ **49.** $6a + 5\sqrt{3ab} - 3b$
51. $18r - 12\sqrt{2r} + 4$ **53.** $-6x - 12\sqrt{x} - 6$ **55.** $\dfrac{\sqrt{7}}{7}$ **57.** $\dfrac{\sqrt{6}}{3}$ **59.** $\dfrac{\sqrt{10}}{4}$ **61.** 2   **63.** $\dfrac{\sqrt[3]{4}}{2}$ **65.** $\sqrt[3]{3}$ **67.** $\dfrac{\sqrt[3]{6}}{3}$
69. $2\sqrt{2x}$ **71.** $\dfrac{\sqrt{5y}}{y}$ **73.** $\dfrac{\sqrt[3]{2ab^2}}{b}$ **75.** $\dfrac{\sqrt[4]{4}}{2}$ **77.** $\dfrac{\sqrt[5]{2}}{2}$ **79.** $\sqrt{2} + 1$ **81.** $\dfrac{3\sqrt{2} - \sqrt{10}}{4}$ **83.** $2 + \sqrt{3}$ **85.** $\dfrac{9 - 2\sqrt{14}}{5}$
87. $\dfrac{2(\sqrt{x} - 1)}{x - 1}$ **89.** $\dfrac{x(\sqrt{x} + 4)}{x - 16}$ **91.** $\sqrt{2z} + 1$ **93.** $\dfrac{x - 2\sqrt{xy} + y}{x - y}$ **95.** $\dfrac{1}{\sqrt{3} - 1}$ **97.** $\dfrac{x - 9}{x(\sqrt{x} - 3)}$ **99.** $\dfrac{x - y}{\sqrt{x}(\sqrt{x} - \sqrt{y})}$
101. $f/4$ **105.** $x\sqrt[6]{x^5}$ **107.** $p\sqrt[10]{p}$ **109.** $\sqrt[6]{b^5}$

Chapter Summary (page 704)

1. a. 7 **b.** -11 **c.** -6 **d.** 15 **e.** -3 **f.** -6 **g.** 5 **h.** -2 **i.** $5|x|$ **j.** $|x+2|$ **k.** $3a^2b$ **l.** $4x^2|y|$

2. a.   **b.**   **c.**   **d.**

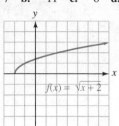 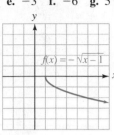 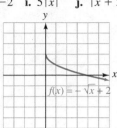 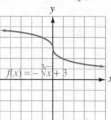

3. a. 12 **b.** about 5.7 **4.** 3 mi **5.** 8.2 ft **6.** 88 yd **7.** 16,000 yd, or about 9 mi **8.** 13 **9.** 2.83 units **10. a.** 22 **b.** 16, 9
c. 3, 9 **d.** $\frac{9}{16}$ **e.** 2 **f.** 0, -2 **11. a.** 5 **b.** -6 **c.** 27 **d.** 64 **e.** -2 **f.** -4 **g.** $\frac{1}{4}$ **h.** $\frac{1}{2}$ **i.** $-16,807$ **j.** $\frac{1}{3,125}$ **k.** 8 **l.** $\frac{27}{8}$
m. $3xy^{1/3}$ **n.** $3xy^{1/2}$ **o.** $125x^{9/2}y^6$ **p.** $\dfrac{1}{4u^{4/3}v^2}$ **12. a.** $5^{3/4}$ **b.** $a^{5/7}$ **c.** $u-1$ **d.** $v+v^2$ **e.** $x+2x^{1/2}y^{1/2}+y$
f. $a^{4/3}-b^{4/3}$ **13. a.** $\sqrt[3]{5}$ **b.** $\sqrt{x}$ **c.** $\sqrt[3]{3ab^2}$ **d.** $\sqrt{5ab}$ **14. a.** $4\sqrt{15}$ **b.** $3\sqrt[3]{2}$ **c.** $2\sqrt[4]{2}$ **d.** $2\sqrt[5]{3}$ **e.** $2|x|\sqrt{2x}$
f. $3x^2|y|\sqrt{2y}$ **g.** $2xy\sqrt[3]{2x^2y}$ **h.** $3x^2y\sqrt[3]{2x}$ **i.** $4|x|$ **j.** $2x$ **k.** $\dfrac{\sqrt[3]{2a^2b}}{3x}$ **l.** $\dfrac{\sqrt{17xy}}{8a^2}$ **15. a.** $3\sqrt{2}$ **b.** $\sqrt{5}$ **c.** 0 **d.** $8\sqrt[4]{2}$
e. $29x\sqrt{2}$ **f.** $32a\sqrt{3a}$ **g.** $13\sqrt[3]{2}$ **h.** $-4x\sqrt[4]{2x}$ **16.** $7\sqrt{2}$ m **17.** $6\sqrt{3}$ cm, 18 cm **18. a.** 7.07 in. **b.** 8.66 cm
19. a. $6\sqrt{10}$ **b.** 72 **c.** $3x$ **d.** 3 **e.** $-2x$ **f.** $-20x^3y^3\sqrt{xy}$ **g.** $4-3\sqrt{2}$ **h.** $2+3\sqrt{2}$ **i.** $\sqrt{10}-\sqrt{5}$ **j.** $3+\sqrt{6}$ **k.** 1
l. $5+2\sqrt{6}$ **m.** $x-y$ **n.** $6u-12+\sqrt{u}$ **20. a.** $\dfrac{\sqrt{3}}{3}$ **b.** $\dfrac{\sqrt{15}}{5}$ **c.** $\dfrac{\sqrt{xy}}{y}$ **d.** $\dfrac{\sqrt[3]{u^2}}{u^2v^2}$ **e.** $2(\sqrt{2}+1)$ **f.** $\dfrac{\sqrt{6}+\sqrt{2}}{2}$
g. $2(\sqrt{x}-4)$ **h.** $\dfrac{a+2\sqrt{a}+1}{a-1}$ **21. a.** $\dfrac{3}{5\sqrt{3}}$ **b.** $\dfrac{1}{\sqrt[3]{3}}$ **c.** $\dfrac{9-x}{2(3+\sqrt{x})}$ **d.** $\dfrac{a-b}{a+\sqrt{ab}}$

Chapter 9 Test (page 709)

1. 7 **2.** 4 **3.** $2|x|$ **4.** $2x$ **5.**   D $= [2, \infty)$, **6.**   D $= (-\infty, \infty)$,
                                              R $= [0, \infty)$ R $= (-\infty, \infty)$

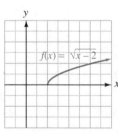 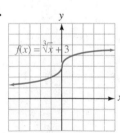

7. 10 **8.** about 2.5 **9.** 28 in. **10.** 1.25 m **11.** 10 **12.** 25 **13.** 10 **14.** 4, no solutions **15.** 2 **16.** 9 **17.** $\frac{1}{216}$ **18.** $\frac{9}{4}$
19. $2^{4/3}$ **20.** $8xy$ **21.** $4\sqrt{3}$ **22.** $5xy^2\sqrt{10xy}$ **23.** $2x^5y\sqrt[3]{3}$ **24.** $\frac{1}{4a}$ **25.** $2|x|\sqrt{3}$ **26.** $2|x^3|\sqrt{2}$ **27.** $3x\sqrt[3]{3}$
28. $3x^2y^4\sqrt{2y}$ **29.** $-\sqrt{3}$ **30.** $14\sqrt[3]{5}$ **31.** $2y^2\sqrt{3y}$ **32.** $6z\sqrt[4]{3z}$ **33.** 9.24 cm **34.** 8.67 cm **35.** $-6x\sqrt{y}-2xy^2$
36. $3-7\sqrt{6}$ **37.** $\dfrac{\sqrt{5}}{5}$ **38.** $2\sqrt[3]{3}$ **39.** $\sqrt{2}(\sqrt{5}-3)$ **40.** $\sqrt{3t}+1$ **41.** $\dfrac{3}{\sqrt{21}}$ **42.** $\dfrac{a-b}{a-2\sqrt{ab}+b}$

Getting Ready (page 713)

1. $(3x+2)(2x-1)$ **2.** $(2x-3)(2x+1)$ **3.** 7 **4.** 8

Orals (page 722)

1. ± 7 **2.** $\pm\sqrt{10}$ **3.** 4 **4.** 9 **5.** $\frac{9}{4}$ **6.** $\frac{25}{4}$ **7.** 3, -4, 7 **8.** -2, 1, -5

Exercise 10.1 (page 722)

1. 1 **3.** $t \le 4$ **5.** $B = \frac{-Ax + C}{y}$ **7.** $x = \sqrt{c}, x = -\sqrt{c}$ **9.** plus or minus **11.** 0, −2 **13.** 5, −5 **15.** −2, −4

17. 6, 1 **19.** 2, $\frac{1}{2}$ **21.** $\frac{2}{3}, -\frac{5}{2}$ **23.** ±6 **25.** ±$\sqrt{5}$ **27.** ±$\frac{4\sqrt{3}}{3}$ **29.** 0, −2 **31.** 4, 10 **33.** −5 ± $\sqrt{3}$ **35.** 2, −4 **37.** 2, 4

39. −1, −4 **41.** 1, −$\frac{1}{2}$ **43.** −$\frac{1}{3}$, −$\frac{3}{2}$ **45.** $\frac{3}{4}$, −$\frac{3}{2}$ **47.** $\frac{-7 \pm \sqrt{29}}{10}$ **49.** −1, −2 **51.** −6, −6 **53.** $\frac{-5 \pm \sqrt{5}}{10}$ **55.** −$\frac{3}{2}$, −$\frac{1}{2}$

57. $\frac{1}{4}$, −$\frac{3}{4}$ **59.** $\frac{-5 \pm \sqrt{17}}{2}$ **61.** 8.98, −3.98 **63.** 16, 18 **65.** 6, 7 **67.** $x^2 - 8x + 15 = 0$ **69.** $x^3 - x^2 - 14x + 24 = 0$

71. 8 ft by 12 ft **73.** 4 units **75.** $\frac{4}{3}$ cm **77.** 30 mph **79.** $4.80 or $5.20 **81.** 4,000 **83.** 2.26 in. **85.** about 6.13×10^{-3} M
89. $\frac{3}{4}$

Getting Ready (page 726)

1. −2 **2.** 2 **3.** 0 **4.** 8 **5.** 1 **6.** 4

Orals (page 735)

1. down **2.** up **3.** up **4.** down **5.** (3, −1) **6.** (−2, 2)

Exercise 10.2 (page 736)

1. 10 **3.** $3\frac{3}{5}$ hr **5.** $f(x) = ax^2 + bx + c, a \ne 0$ **7.** vertex **9.** upward **11.** to the right **13.** upward

15. 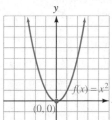 **17.** 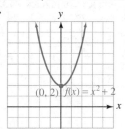 **19.** 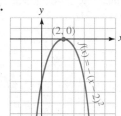 **21.**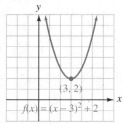

23. 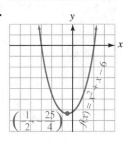 **25.**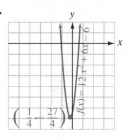

27. (1, 2), $x = 1$ **29.** (−3, −4), $x = -3$ **31.** (0, 0), $x = 0$
33. (1, −2), $x = 1$ **35.** (2, 21), $x = 2$ **37.** $\left(\frac{5}{12}, \frac{143}{24}\right)$, $x = \frac{5}{12}$ **39.** (5, 2)
41. (0.25, 0.88) **43.** (0.5, 7.25) **45.** 2, −3 **47.** −1.85, 3.25
49. 36 ft, 3 sec **51.** 50 ft by 50 ft, 2,500 ft^2 **53.** 5,000
55. 3,276, $14,742 **57.** $35 **59.** 0.25 and 0.75

Getting Ready (page 739)

1. $7x$ **2.** $-x + 10$ **3.** $12x^2 + 5x - 25$ **4.** $9x^2 - 25$

Orals (page 749)

1. $-i$ **2.** −1 **3.** 1 **4.** i **5.** $7i$ **6.** $8i$ **7.** $10i$ **8.** $9i$ **9.** 5 **10.** 13

Exercise 10.3 (page 749)

1. -1 **3.** 20 mph **5.** imaginary **7.** -1 **9.** 1 **11.** $\dfrac{\sqrt{a}}{\sqrt{b}}$ **13.** 5, 7 **15.** conjugates **17.** $\pm 3i$ **19.** $\pm\dfrac{4\sqrt{3}}{3}i$ **21.** $-1 \pm i$

23. $-\dfrac{1}{4} \pm \dfrac{\sqrt{7}}{4}i$ **25.** $\dfrac{2}{3} \pm \dfrac{\sqrt{2}}{3}i$ **27.** $\dfrac{1}{3} \pm \dfrac{2\sqrt{2}}{3}i$ **29.** i **31.** $-i$ **33.** 1 **35.** i **37.** yes **39.** no **41.** no **43.** $8 - 2i$

45. $3 - 5i$ **47.** $15 + 7i$ **49.** $6 - 8i$ **51.** $2 + 9i$ **53.** $-15 + 2\sqrt{3}i$ **55.** $3 + 6i$ **57.** $-25 - 25i$ **59.** $7 + i$ **61.** $14 - 8i$

63. $8 + \sqrt{2}i$ **65.** $-20 - 30i$ **67.** $3 + 4i$ **69.** $-5 + 12i$ **71.** $7 + 17i$ **73.** $5 + 5i$ **75.** $16 + 2i$ **77.** $0 - i$ **79.** $0 + \dfrac{4}{5}i$

81. $\dfrac{1}{8} - 0i$ **83.** $0 + \dfrac{3}{5}i$ **85.** $2 + i$ **87.** $\dfrac{1}{2} + \dfrac{5}{2}i$ **89.** $-\dfrac{42}{25} - \dfrac{6}{25}i$ **91.** $\dfrac{1}{4} + \dfrac{3}{4}i$ **93.** $\dfrac{5}{13} - \dfrac{12}{13}i$ **95.** $\dfrac{11}{10} + \dfrac{3}{10}i$

97. $\dfrac{1}{4} - \dfrac{\sqrt{15}}{4}i$ **99.** $-\dfrac{5}{169} + \dfrac{12}{169}i$ **101.** $\dfrac{3}{5} + \dfrac{4}{5}i$ **103.** $-\dfrac{6}{13} - \dfrac{9}{13}i$ **105.** 10 **107.** 13 **109.** $\sqrt{74}$ **111.** 1 **119.** $\dfrac{5}{3 + i}$

Getting Ready (page 752)

1. 17 **2.** -8

Orals (page 757)

1. -3 **2.** -7 **3.** irrational and unequal **4.** complex conjugates **5.** no **6.** yes

Exercise 10.4 (page 757)

1. -2 **3.** $\dfrac{9}{5}$ **5.** $b^2 - 4ac$ **7.** rational, unequal **9.** rational, equal **11.** complex conjugates **13.** irrational, unequal
15. rational, unequal **17.** 6, -6 **19.** 12, -12 **21.** 5 **23.** 12, -3 **25.** yes **27.** $k < -\dfrac{4}{3}$ **29.** 1, -1, 4, -4 **31.** 1, -1, $\sqrt{2}$,
$-\sqrt{2}$ **33.** 1, -1, $\sqrt{5}$, $-\sqrt{5}$ **35.** 1, -1, 2, -2 **37.** 1 **39.** no solution **41.** -8, -27 **43.** -1, 27 **45.** -1, -4

47. 4, -5 **49.** 0, 2 **51.** -1, $-\dfrac{27}{13}$ **53.** 1, 1, -1, -1 **55.** $1 \pm i$ **57.** $x = \pm\sqrt{r^2 - y^2}$ **59.** $d = \pm\sqrt{\dfrac{k}{I}} = \pm\dfrac{\sqrt{kI}}{I}$

61. $y = \dfrac{-3x \pm \sqrt{9x^2 - 28x}}{2x}$ **63.** $\mu^2 = \dfrac{\Sigma x^2}{N} - \sigma^2$ **65.** $\dfrac{2}{3}$, $-\dfrac{1}{4}$ **67.** $\dfrac{-5 \pm \sqrt{17}}{4}$ **69.** $\dfrac{1 \pm i\sqrt{11}}{3}$ **71.** $-1 \pm 2i$ **75.** no

Getting Ready (page 759)

1. $(x + 5)(x - 3)$ **2.** $(x - 2)(x - 1)$

Orals (page 765)

1. $x = 2$ **2.** $x > 2$ **3.** $x < 2$ **4.** $x = -3$ **5.** $x > -3$ **6.** $x < -3$ **7.** $1 < 2x$ **8.** $1 > 2x$

Exercise 10.5 (page 765)

1. $y = kx$ **3.** $t = kxy$ **5.** 3 **7.** greater **9.** undefined **11.**

(1, 4)

13.

$(-\infty, 3) \cup (5, \infty)$

15.

$[-4, 3]$

17.

$(-\infty, -5] \cup [3, \infty)$

19. no solutions **21.**

$(-\infty, -3] \cup [3, \infty)$

23.

$(-5, 5)$

25.

$(-\infty, 0) \cup (1/2, \infty)$

27.

$(0, 2]$

29.

$(-\infty, -5/3) \cup (0, \infty)$

31.

$(-\infty, -3) \cup (1, 4)$

33. $[-5, -2) \cup [4, \infty)$

35. $(-\infty, -4)$

37. $(-1/2, 1/3) \cup (1/2, \infty)$

39. $(0, 2) \cup (8, \infty)$

41. $(-\infty, -2) \cup (2, 18]$

43. $[-34/5, -4) \cup (3, \infty)$

45. $(-4, -2] \cup (-1, 2]$

47. $(-\infty, -16) \cup (-4, -1) \cup (4, \infty)$

49. $(-\infty, -2) \cup (-2, \infty)$ **51.** $(-1, 3)$ **53.** $(-\infty, -3) \cup (2, \infty)$ **55.**   **57.**

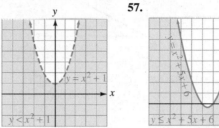

59.   **61.**   **63.**   **65.**

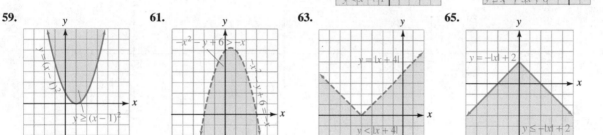

69. when 4 factors are negative, 2 factors are negative, or no factors are negative

Getting Ready (page 768)

1. $3x - 1$ **2.** $x + 3$ **3.** $2x^2 - 3x - 2$ **4.** $\dfrac{2x+1}{x-2}$

Orals (page 775)

1. $5x$ **2.** x **3.** $8x^2$ **4.** $\dfrac{3}{2}$ **5.** 2   **6.** $12x^2$ **7.** $8x$ **8.** $6x$ **9.** $12x$

Exercise 10.6 (page 775)

1. $-\dfrac{3x+7}{x+2}$ **3.** $\dfrac{x-4}{3x^2-x-12}$ **5.** $f(x) + g(x)$ **7.** $f(x)g(x)$ **9.** domain   **11.** $f(x)$ **13.** $7x, (-\infty, \infty)$ **15.** $12x^2, (-\infty, \infty)$
17. $x, (-\infty, \infty)$ **19.** $\frac{4}{3}, (-\infty, 0) \cup (0, \infty)$ **21.** $3x - 2, (-\infty, \infty)$ **23.** $2x^2 - 5x - 3, (-\infty, \infty)$ **25.** $-x - 4, (-\infty, \infty)$
27. $\frac{x-3}{2x+1}, \left(-\infty, -\frac{1}{2}\right) \cup \left(-\frac{1}{2}, \infty\right)$ **29.** $-2x^2 + 3x - 3, (-\infty, \infty)$ **31.** $(3x - 2)/(2x^2 + 1), (-\infty, \infty)$ **33.** $3, (-\infty, \infty)$
35. $(x^2 - 4)/(x^2 - 1), (-\infty, -1) \cup (-1, 1) \cup (1, \infty)$ **37.** 7   **39.** 24   **41.** -1 **43.** $-\frac{1}{2}$ **45.** $2x^2 - 1$ **47.** $16x^2 + 8x$ **49.** 58
51. 110   **53.** 2   **55.** $9x^2 - 9x + 2$ **57.** 2   **59.** $2x + h$ **61.** $4x + 2h$ **63.** $2x + h + 1$ **65.** $2x + h + 3$ **67.** $4x + 2h + 3$
69. 2   **71.** $x + a$ **73.** $2x + 2a$ **75.** $x + a + 1$ **77.** $x + a + 3$ **79.** $2x + 2a + 3$ **85.** $3x^2 + 3xh + h^2$
87. $C(t) = \frac{5}{9}(2{,}668 - 200t)$

Getting Ready (page 777)

1. $x = \dfrac{y-2}{3}$ **2.** $x = \dfrac{3y+10}{2}$

Orals (page 785)

1. $\{(2, 1), (3, 2), (10, 5)\}$ **2.** $\{(1, 1), (8, 2), (64, 4)\}$ **3.** $f^{-1}(x) = 2x$ **4.** $f^{-1}(x) = \frac{1}{2}x$ **5.** no   **6.** yes

Exercise 10.7   (page 786)

1. $3 - 8i$ **3.** $18 - i$ **5.** 10 **7.** one-to-one   **9.** 2 **11.** x **13.** yes   **15.** no

17.
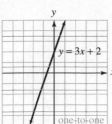
one-to-one

19.
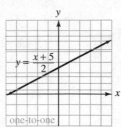
one-to-one

21.
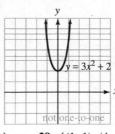
not one-to-one

23.
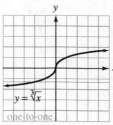
one-to-one

25. $\{(2, 3), (1, 2), (0, 1)\}$; yes   **27.** $\{(2, 1), (3, 2), (3, 1), (5, 1)\}$; no   **29.** $\{(1, 1), (4, 2), (9, 3), (16, 4)\}$; yes
31. $f^{-1}(x) = \frac{1}{3}x - \frac{1}{3}$ **33.** $f^{-1}(x) = 5x - 4$ **35.** $f^{-1}(x) = 5x + 4$ **37.** $f^{-1}(x) = \frac{5}{4}x + 5$

39.
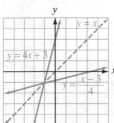

41.
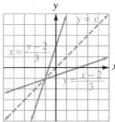

43.
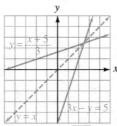

45.
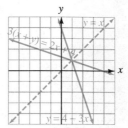

47. $y = \pm\sqrt{x - 4}$, no   **49.** $y = \sqrt[3]{x}$, yes   **51.** $x = |y|$, no   **53.** $f^{-1}(x) = \sqrt[3]{\frac{x+3}{2}}$

55.
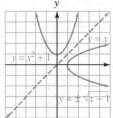

57.
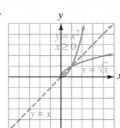

61. $f^{-1}(x) = \frac{x+1}{x-1}$

Chapter Summary   (page 790)

1. a. $\frac{2}{3}, -\frac{3}{4}$ **b.** $-\frac{1}{3}, -\frac{5}{2}$ **c.** $\frac{2}{3}, -\frac{4}{5}$ **d.** $4, -8$ **2. a.** $-4, -2$ **b.** $\frac{7}{2}, 1$ **3. a.** $9, -1$ **b.** $0, 10$ **c.** $\frac{1}{2}, -7$ **d.** $7, -\frac{1}{3}$
4. 4 cm by 6 cm   **5.** 2 ft by 3 ft   **6.** 7 sec   **7.** 196 ft
8. a.
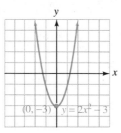

b.
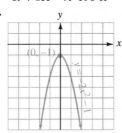

c.
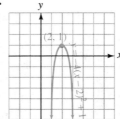

d.
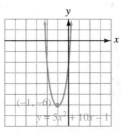

9. a. $12 - 8i$ **b.** $2 - 68i$ **c.** $-96 + 3i$ **d.** $-2 - 2\sqrt{2}i$ **e.** $22 + 29i$ **f.** $-16 + 7i$ **g.** $-12 + 28\sqrt{3}i$ **h.** $118 + 10\sqrt{2}i$
i. $0 - \frac{3}{4}i$ **j.** $0 - \frac{2}{5}i$ **k.** $\frac{12}{5} - \frac{6}{5}i$ **l.** $\frac{21}{10} + \frac{7}{10}i$ **m.** $\frac{15}{17} + \frac{8}{17}i$ **n.** $\frac{4}{5} - \frac{3}{5}i$ **o.** $\frac{15}{29} - \frac{6}{29}i$ **p.** $\frac{1}{3} + \frac{1}{3}i$ **q.** $15 + 0i$ **r.** $26 + 0i$
10. a. irrational, unequal   **b.** complex conjugates   **c.** 12, 152   **d.** $k \geq -\frac{7}{3}$ **11. a.** 1, 144   **b.** 8, -27   **c.** 1   **d.** 1, $-\frac{8}{5}$ **12.** $\frac{14}{3}$

13. 1   **14. a.** 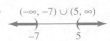 $(-\infty, -7) \cup (5, \infty)$ **b.** 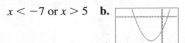 $(-9, 2)$ **c.** $(-\infty, 0) \cup [3/5, \infty)$ **d.** $(-7/2, 1) \cup (4, \infty)$

 -7 ⟶ 5 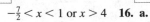 -9 2 0 $3/5$ $-7/2$ 1 4

15. a. 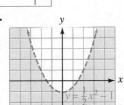 $x < -7$ or $x > 5$ **b.** 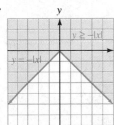 $-9 < x < 2$ **c.** $x < 0$ or $x \geq \frac{3}{5}$

d. 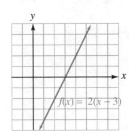 $-\frac{7}{2} < x < 1$ or $x > 4$ **16. a.** 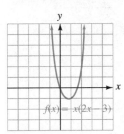   **b.**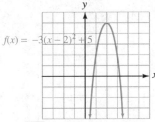

17. a. $(f + g)(x) = 3x + 1$
b. $(f - g)(x) = x - 1$
c. $(f \cdot g)(x) = 2x^2 + 2x$
d. $(f/g)(x) = \frac{2x}{x + 1}$ **e.** 6
f. -1 **g.** $2(x + 1)$
h. $2x + 1$

18. a. yes   **b.** no   **c.** no

$f(x) = 2(x - 3)$ $f(x) = x(2x - 3)$ $f(x) = -3(x - 2)^2 + 5$

d. no   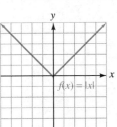   **19. a.** $f^{-1}(x) = \frac{x + 3}{6}$ **b.** $f^{-1}(x) = \frac{x - 5}{4}$ **c.** $y = \sqrt{\frac{x + 1}{2}}$ **d.** $x = |y|$

$f(x) = |x|$

Chapter 10 Test   (page 795)

1. 3, -6 **2.** $-\frac{3}{2}, -\frac{5}{3}$ **3.** 144   **4.** 625   **5.** $-2 \pm \sqrt{3}$ **6.** $\frac{5 \pm \sqrt{37}}{2}$ **7.** $-1 + 11i$ **8.** $4 - 7i$ **9.** $8 + 6i$ **10.** $-10 - 11i$

11. $0 - \frac{\sqrt{2}}{2}i$ **12.** $\frac{1}{2} + \frac{1}{2}i$ **13.** nonreal   **14.** 2   **15.** 10 in.   **16.** 1, $\frac{1}{4}$

17. 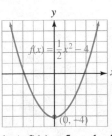   **18.** 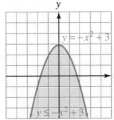   **19.** $(-\infty, -2) \cup (4, \infty)$ **20.** $(-3, 2]$

$f(x) = \frac{1}{2}x^2 - 4$ $y = -x^2 + 3$ $(0, -4)$ $y \leq -x^2 + 3$

 -2 4 -3 2

21. $(g + f)(x) = 5x - 1$ **22.** $(f - g)(x) = 3x + 1$ **23.** $(g \cdot f)(x) = 4x^2 - 4x$ **24.** $(g/f)(x) = \frac{x - 1}{4x}$ **25.** 3   **26.** -4 **27.** -8
28. -9 **29.** $4(x - 1)$ **30.** $4x - 1$ **31.** $y = \frac{12 - 2x}{3}$ **32.** $y = -\sqrt{\frac{x - 4}{3}}$

Cumulative Review Exercises (page 797)

1. $D = (-\infty, \infty)$; $R = [-3, \infty)$ **2.** $D = (-\infty, \infty)$; $R = (-\infty, 0]$ **3.** $y = 3x + 2$ **4.** $y = -\frac{2}{3}x - 2$ **5.** $-4a^2 + 12a - 7$
6. $6x^2 - 5x - 6$ **7.** $(x^2 + 4y^2)(x + 2y)(x - 2y)$ **8.** $(3x + 2)(5x - 4)$ **9.** $6, -1$ **10.** $0, \frac{2}{3}, -\frac{1}{2}$ **11.** $5x^2$ **12.** $4t\sqrt{3t}$ **13.** $-3x$
14. $4x$ **15.** $\frac{1}{2}$ **16.** 16 **17.** y^2 **18.** $x^{17/12}$ **19.** **20.** **21.** $x^{4/3} - x^{2/3}$

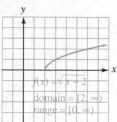

22. $\frac{1}{x} + 2 + x$ **23.** $7\sqrt{2}$ **24.** $-12\sqrt[4]{2} + 10\sqrt[4]{3}$ **25.** $-18\sqrt{6}$ **26.** $\dfrac{5\sqrt[3]{x^2}}{x}$ **27.** $\dfrac{x + 3\sqrt{x} + 2}{x - 1}$ **28.** $\sqrt{xy}$ **29.** $2, 7$ **30.** $\frac{1}{4}$

31. $3\sqrt{2}$ in. **32.** $2\sqrt{3}$ in. **33.** 10 **34.** 9 **35.** $1, -\frac{3}{2}$ **36.** $\dfrac{-2 \pm \sqrt{7}}{3}$

37. **38.** **39.** $7 + 2i$ **40.** $-5 - 7i$ **41.** 13 **42.** $12 - 6i$ **43.** $-12 - 10i$

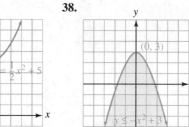

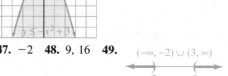

44. $\frac{3}{2} + \frac{1}{2}i$ **45.** $\sqrt{13}$ **46.** $\sqrt{61}$ **47.** -2 **48.** $9, 16$ **49. (−∞, −2) ∪ (3, ∞)** **50. [−2, 3]** **51.** 5 **52.** 27

53. $12x^2 - 12x + 5$ **54.** $6x^2 + 3$ **55.** $f^{-1}(x) = \frac{x-2}{3}$ **56.** $f^{-1}(x) = \sqrt[3]{x - 4}$

Getting Ready (page 801)

1. 8 **2.** 5 **3.** $\frac{1}{25}$ **4.** $\frac{8}{27}$

Orals (page 812)

1. 4 **2.** 25 **3.** 18 **4.** 3 **5.** $\frac{1}{4}$ **6.** $\frac{1}{25}$ **7.** $\frac{2}{9}$ **8.** $\frac{1}{27}$

Exercise 11.1 (page 812)

1. 40 **3.** $120°$ **5.** exponential **7.** $(0, \infty)$ **9.** increasing **11.** $P\left(1 + \frac{r}{k}\right)^{kt}$ **13.** 2.6651 **15.** 36.5548 **17.** 8 **19.** $7^{3\sqrt{3}}$
21. **23.** **25.** **27.** **29.** $b = \frac{1}{2}$

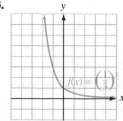

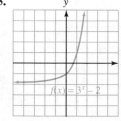

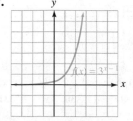

31. no value   **33.** $b = 2$ **35.** $b = 3$ **37.**   **39.** 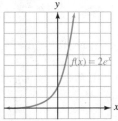   **41.** $22,080.40   **43.** $32.03   **45.** $2,273,996.13

47. $\frac{32}{243}A_0$ **49.** 5.0421×10^{-5} coulombs   **51.** $1,115.33

Getting Ready (page 815)

1. 2   **2.** 2.25   **3.** 2.44   **4.** 2.59

Orals (page 820)

1. 1   **2.** 2.72   **3.** 7.39   **4.** 20.09   **5.** 2   **6.** 2

Exercise 11.2 (page 821)

1. $4x^2\sqrt{15x}$ **3.** $10y\sqrt{3y}$ **5.** 2.72   **7.** increasing   **9.** $A = Pe^{rt}$

11. 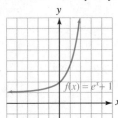   **13.** 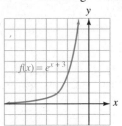   **15.**   **17.**

19. no   **21.** no   **23.** $10,272.17   **25.** $6,391.10   **27.** $7,518.28 from annual compounding; $7,647.95 from continuous compounding   **29.** 10.6 billion   **31.** 2.6   **33.** 6,817   **35.** 0.16   **37.** 0   **39.** 49 mps   **41.** $3,094.15   **43.** 72 yr   **49.** $k = e^5$

Getting Ready (page 823)

1. 1   **2.** 25   **3.** $\frac{1}{25}$ **4.** 4

Orals (page 831)

1. 3   **2.** 2   **3.** 5   **4.** 2   **5.** 2   **6.** 2   **7.** $\frac{1}{4}$ **8.** $\frac{1}{2}$ **9.** 2

Exercise 11.3 (page 831)

1. 10   **3.** 0; $-\frac{5}{9}$ does not check   **5.** $x = b^y$ **7.** range   **9.** inverse   **11.** exponent   **13.** $(b, 1), (1, 0)$

15. $20 \log \dfrac{E_O}{E_I}$ **17.** $3^3 = 27$ **19.** $\left(\frac{1}{2}\right)^2 = \frac{1}{4}$ **21.** $4^{-3} = \frac{1}{64}$ **23.** $\left(\frac{1}{2}\right)^3 = \frac{1}{8}$ **25.** $\log_6 36 = 2$ **27.** $\log_5 \frac{1}{25} = -2$ **29.** $\log_{1/2} 32 = -5$

31. $\log_x z = y$ **33.** 4   **35.** 2   **37.** 3   **39.** $\frac{1}{2}$ **41.** -3 **43.** 49   **45.** 6   **47.** 5   **49.** $\frac{1}{25}$ **51.** $\frac{1}{6}$ **53.** $-\frac{3}{2}$ **55.** $\frac{2}{3}$ **57.** 5   **59.** $\frac{3}{2}$

61. 4   **63.** 8   **65.** 4   **67.** 4   **69.** 3   **71.** 100   **73.** 0.9165   **75.** -2.0620 **77.** 25.25   **79.** 17,378.01   **81.** 0.00   **83.** 8

85. increasing   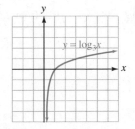   **87.** decreasing   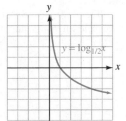   **89.**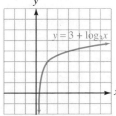

91.

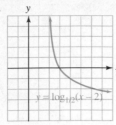

93.

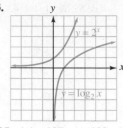

95.

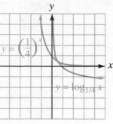

97. no value of b **99.** 3

101. 29.0 db **103.** 49.5 db **105.** 4.4 **107.** 4 **109.** 4.2 yr **111.** 10.8 yr

Getting Ready (page 835)

1. 2 **2.** −3 **3.** 1 **4.** 0

Orals (page 839)

1. $e^y = x$ **2.** $\ln b = a$ **3.** $t = \frac{\ln 2}{r}$

Exercise 11.4 (page 839)

1. $y = 5x$ **3.** $y = -\frac{3}{2}x + \frac{13}{2}$ **5.** $x = 5$ **7.** $\frac{1}{2x-3}$ **9.** $\frac{x+1}{3(x-2)}$ **11.** $\log_e x$ **13.** $(-\infty, \infty)$ **15.** 10 **17.** $\frac{\ln 2}{r}$ **19.** 3.2288
21. 2.2915 **23.** −0.1592 **25.** none **27.** 9.9892 **29.** 23.8075 **31.** 0.0089 **33.** 61.9098 **35.** no **37.** no
39.

41.

43. 5.8 yr **45.** 9.2 yr

Getting Ready (page 841)

1. x^{m+n} **2.** 1 **3.** x^{mn} **4.** x^{m-n}

Orals (page 850)

1. 2 **2.** 5 **3.** 343 **4.** $\frac{1}{4}$ **5.** 2 **6.** 2 **7.** $\frac{1}{4}$ **8.** $\frac{1}{2}$ **9.** 2

Exercise 11.5 (page 850)

1. $-\frac{7}{6}$ **3.** $\left(1, -\frac{1}{2}\right)$ **5.** 0 **7.** M, N **9.** x, y **11.** x **13.** $\neq$ **15.** 0 **17.** 7 **19.** 10 **21.** 1 **23.** 0 **25.** 7 **27.** 10 **29.** 1
37. $\log_b x + \log_b y + \log_b z$ **39.** $\log_b 2 + \log_b x - \log_b y$ **41.** $3 \log_b x + 2 \log_b y$ **43.** $\frac{1}{2}(\log_b x + \log_b y)$
45. $\log_b x + \frac{1}{2}\log_b z$ **47.** $\frac{1}{3} \log_b x - \frac{1}{4} \log_b y - \frac{1}{4} \log_b z$ **49.** $\log_b \dfrac{x+1}{x}$ **51.** $\log_b x^2 y^{1/2}$ **53.** $\log_b \dfrac{z^{1/2}}{x^3 y^2}$
55. $\log_b \dfrac{\frac{x}{z}+x}{\frac{y}{z}+y} = \log_b \dfrac{x}{y}$ **57.** false **59.** false **61.** true **63.** false **65.** true **67.** true **69.** 1.4472 **71.** 0.3521 **73.** 1.1972
75. 2.4014 **77.** 2.0493 **79.** 0.4682 **81.** 1.7712 **83.** −1.0000 **85.** 1.8928 **87.** 2.3219 **89.** 4.77 **91.** from 2.5119×10^{-8}
to 1.585×10^{-7} **93.** It will increase by $k \ln 2$. **95.** The intensity must be cubed.

Getting Ready (page 852)

1. $2 \log x$ **2.** $\frac{1}{2} \log x$ **3.** 0 **4.** $2b \log a$

Orals (page 861)

1. $x = \dfrac{\log 5}{\log 3}$ **2.** $x = \dfrac{\log 3}{\log 5}$ **3.** $x = -\dfrac{\log 7}{\log 2}$ **4.** $x = -\dfrac{\log 1}{\log 6} = 0$ **5.** $x = 2$ **6.** $x = \dfrac{1}{2}$ **7.** $x = 10$ **8.** $x = 10$

Exercise 11.6 (page 861)

1. $0, 5$ **3.** $\frac{2}{3}, -4$ **5.** exponential   **7.** $A_0 2^{-t/h}$ **9.** 1.1610 **11.** 1.2702 **13.** 1.7095 **15.** 0 **17.** ± 1.0878 **19.** $0, 1.0566$
21. $3, -1$ **23.** $-2, -2$ **25.** 0 **27.** 0.2789 **29.** 1.8 **31.** $3, -1$ **33.** 2 **35.** 3 **37.** -7 **39.** 4 **41.** $10, -10$ **43.** 50
45. 20 **47.** 10 **49.** 10 **51.** $1, 100$ **53.** no solution   **55.** 6 **57.** 9 **59.** 4 **61.** $1, 7$ **63.** 20 **65.** 8 **67.** 5.1 yr
69. 42.7 days   **71.** about $4,200$ yr   **73.** 5.6 yr   **75.** 5.4 yr   **77.** because $\ln 2 \approx 0.7$ **79.** 25.3 yr   **81.** 2.828 times larger
83. 13.3 **87.** $x \le 3$

Chapter Summary (page 866)

1. a. $5^{2\sqrt{2}}$ **b.** $2^{\sqrt{10}}$ **2. a.**

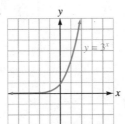

b.

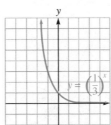

3. $x = 1, y = 6$ **4.** D: $(-\infty, \infty)$, R: $(0, \infty)$

5. a.

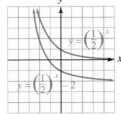

b.

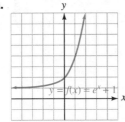

6. $\$2{,}189{,}703.45$ **7.** $\$2{,}324{,}767.37$

8. a.

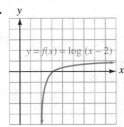

b.

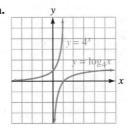

9. about $582{,}000{,}000$ **10.** $(0, \infty), (-\infty, \infty)$ **11. a.** 2 **b.** $-\frac{1}{2}$ **c.** 0
d. -2 **e.** $\frac{1}{2}$ **f.** $\frac{1}{3}$ **12. a.** 32 **b.** 9 **c.** 27 **d.** -1 **e.** $\frac{1}{8}$ **f.** 2
g. 4 **h.** 2 **i.** 10 **j.** $\frac{1}{25}$ **k.** 5 **l.** 3

13. a.

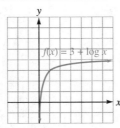

b.

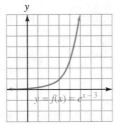

14. a.

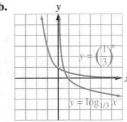

b.

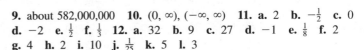

15. 53 db   **16.** 4.4   **17. a.** 6.1137   **b.** -0.1111 **18. a.** 10.3398   **b.** 2.5715
19. a.   **b.**

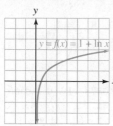

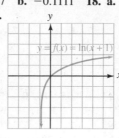

20. 23 yr   **21. a.** 0   **b.** 1   **c.** 3   **d.** 4   **22. a.** 4   **b.** 0   **c.** 7
d. 3   **e.** 4   **f.** 9   **23. a.** $2 \log_b x + 3 \log_b y - 4 \log_b z$
b. $\frac{1}{2}(\log_b x - \log_b y - 2 \log_b z)$ **24. a.** $\log_b \dfrac{x^3 z^7}{y^5}$ **b.** $\log_b \dfrac{y^3 \sqrt{x}}{z^7}$
25. a. 3.36   **b.** 1.56   **c.** 2.64   **d.** -6.72 **26.** 1.7604
27. about 7.94×10^{-4} gram-ions per liter   **28.** $k \ln 2$ less
29. a. $\frac{\log 7}{\log 3} \approx 1.7712$ **b.** 2   **c.** $\frac{\log 3}{\log 3 - \log 2} \approx 2.7095$ **d.** $-1, -3$
30. a. 25, 4   **b.** 4   **c.** 2   **d.** 4, 3   **e.** 6   **f.** 31   **g.** $\frac{\ln 9}{\ln 2} \approx 3.1699$
h. no solution   **i.** $\frac{e}{e-1} \approx 1.5820$ **j.** 1   **31.** about 3,300 yr

Chapter 11 Test   (page 870)

1.   **2.**   **3.** $\frac{3}{64}$ g   **4.** \$1,060.90   **5.**   **6.** \$4,451.08   **7.** 2

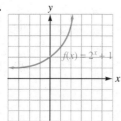

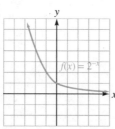

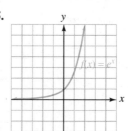

8. 3   **9.** $\frac{1}{27}$ **10.** 10   **11.** 2   **12.** $\frac{27}{8}$ **13.**   **14.**   **15.** $2 \log a + \log b + 3 \log c$

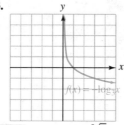

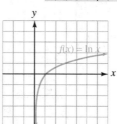

16. $\frac{1}{2}(\ln a - 2 \ln b - \ln c)$ **17.** $\log \dfrac{b\sqrt{a+2}}{c^3}$ **18.** $\log \dfrac{\sqrt[3]{a}}{c \sqrt[3]{b^2}}$ **19.** 1.3801   **20.** 0.4259   **21.** $\frac{\log 3}{\log 7}$ or $\frac{\ln 3}{\ln 7}$ **22.** $\frac{\log e}{\log \pi}$ or $\frac{\ln e}{\ln \pi}$
23. true   **24.** false   **25.** false   **26.** false   **27.** 6.4   **28.** 46   **29.** $\frac{\log 3}{\log 5}$ **30.** $\frac{\log 3}{(\log 3) - 2}$ **31.** 1   **32.** 10

Getting Ready   (page 874)

1. $x^2 - 4x + 4$ **2.** $x^2 + 8x + 16$ **3.** $\frac{81}{4}$ **4.** 36

Orals   (page 884)

1. (0, 0), 12   **2.** (0, 0), 11   **3.** (2, 0), 4   **4.** (0, -1), 3   **5.** down   **6.** up   **7.** left   **8.** right

Exercise 12.1   (page 884)

1. 5, $-\frac{7}{3}$ **3.** 3, $-\frac{1}{4}$ **5.** circle, plane   **7.** $r^2 < 0$ **9.** parabola, (3, 2), right   **11.**

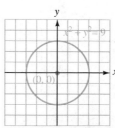

13.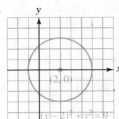

15.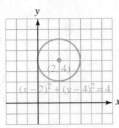

17.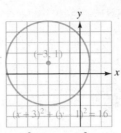

19.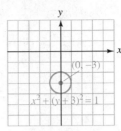

21.   **23.**   **25.** $x^2 + y^2 = 1$ **27.** $(x - 6)^2 + (y - 8)^2 = 25$ **29.** $(x + 2)^2 + (y - 6)^2 = 144$

31. $x^2 + y^2 = 2$ **33.**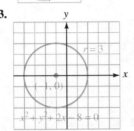

35.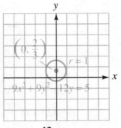

37.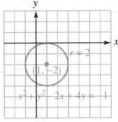

39.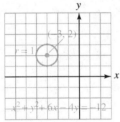

41.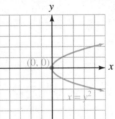

43.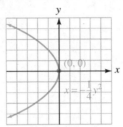

45.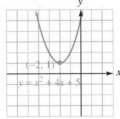

47.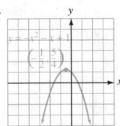

49.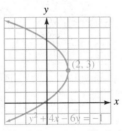

51.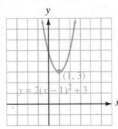

53.    **55.**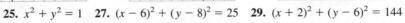

57. $(x - 7)^2 + y^2 = 9$ **59.** no   **61.** 30 ft away   **63.** 2 AU

Getting Ready   (page 889)

1. $y = \pm b$ **2.** $x = \pm a$

Orals   (page 897)

1. $(\pm 3, 0), (0, \pm 4)$ **2.** $(\pm 5, 0), (0, \pm 6)$ **3.** $(2, 0)$ **4.** $(0, -1)$

Exercise 12.2 (page 897)

1. $12y^2 + \dfrac{9}{x^2}$ **3.** $\dfrac{y^2 + x^2}{y^2 - x^2}$ **5.** ellipse, sum **7.** center **9.** $(0, 0)$ **11.**

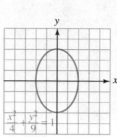

13.

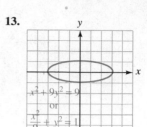

15.

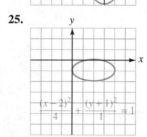

17.

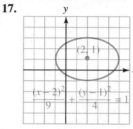

19.

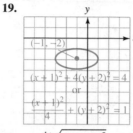

21.

23.

25.

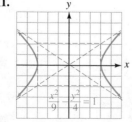

27.

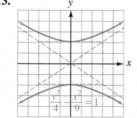

29. $y = \frac{1}{2}\sqrt{400 - x^2}$ **31.** $12\,\pi$ sq. units

Getting Ready (page 900)

1. $y = \pm 2.0$ **2.** $y = \pm 2.9$

Orals (page 908)

1. $(\pm 3, 0)$ **2.** $(0, \pm 5)$

Exercise 12.3 (page 908)

1. $-3x^2(2x^2 - 3x + 2)$ **3.** $(5a + 2b)(3a - 2b)$ **5.** hyperbola, difference **7.** center **9.** $(0, 0)$

11.

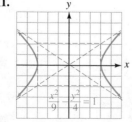

13.

15.

17.

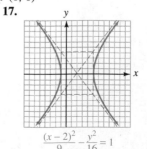

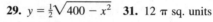

$$\dfrac{(x - 2)^2}{9} - \dfrac{y^2}{16} = 1$$

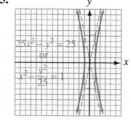

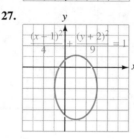

19.

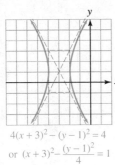

$$4(x+3)^2 - (y-1)^2 = 4$$
$$\text{or } (x+3)^2 - \frac{(y-1)^2}{4} = 1$$

21.

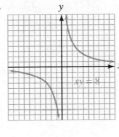

$xy = 8$

23.

25.

27.

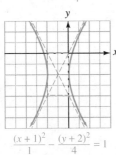

$$\frac{(x+1)^2}{1} - \frac{(y+2)^2}{4} = 1$$

29.

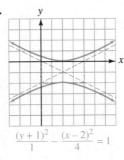

$$\frac{(y+1)^2}{1} - \frac{(x-2)^2}{4} = 1$$

31. 3 units   **33.** $10\sqrt{3}$ units

Getting Ready (page 912)

1. positive   **2.** negative   **3.** 98   **4.** -3

Orals (page 916)

1. increasing   **2.** decreasing   **3.** constant   **4.** increasing

Exercise 12.4 (page 916)

1. 20   **3.** domains   **5.** constant, $f(x)$ **7.** step   **9.** increasing on $(-\infty, 0)$, decreasing on $(0, \infty)$ **11.** decreasing on $(-\infty, 0)$, constant on $(0, 2)$, increasing on $(2, \infty)$ **13.** constant on $(-\infty, 0)$, increasing on $(0, \infty)$

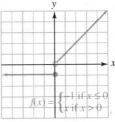

$$f(x) = \begin{cases} -1 \text{ if } x \le 0 \\ x \text{ if } x > 0 \end{cases}$$

15. decreasing on $(-\infty, 0)$, increasing on $(0, 2)$, decreasing on $(2, \infty)$

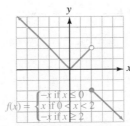

$$f(x) = \begin{cases} -x \text{ if } x \le 0 \\ x \text{ if } 0 < x < 2 \\ -x \text{ if } x \ge 2 \end{cases}$$

17.
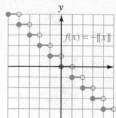

19.
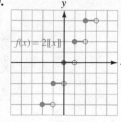

21.
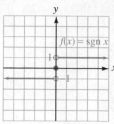

23. $30
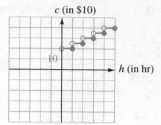

25. After 2 hours, network B is cheaper.
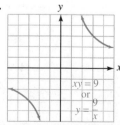

Chapter Summary   (page 921)

1. a.
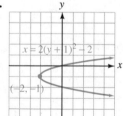

b.
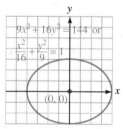

2.
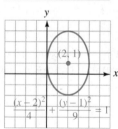

3. a.
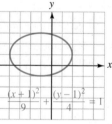

b.
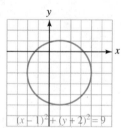

4. a.
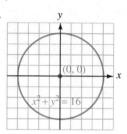

b.
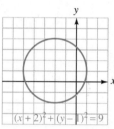

5.
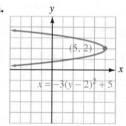

6. a.
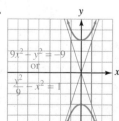

b.
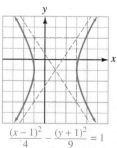

7. hyperbola   **8.**
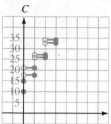

9. increasing on $(-\infty, -2)$, constant on $(-2, 1)$, decreasing on $(1, \infty)$ **10. a.**

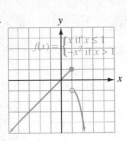

b.

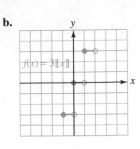

Chapter 12 Test   (page 924)

1. $(2, -3)$, 2   **2.** $(-2, 3)$, 4   **3.**

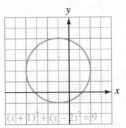

4.

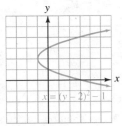

5.

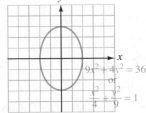

6.

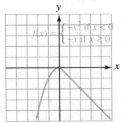

7.

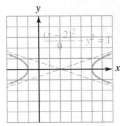

8.

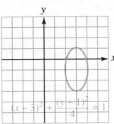

9. increasing on $(-3, 0)$, decreasing on $(0, 3)$

10.

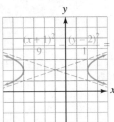

Cumulative Review Exercises   (page 925)

1. $12x^2 - 5xy - 3y^2$ **2.** $a^{2n} - 2a^n - 3$ **3.** $\frac{5}{a-2}$ **4.** $a^2 - 3a + 2$ **5.** 1   **6.** $\frac{4a-1}{(a+2)(a-2)}$ **7.** parallel   **8.** perpendicular
9. $y = -2x + 5$ **10.** $y = -\frac{9}{13}x + \frac{7}{13}$ **11.**

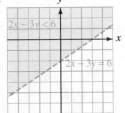

12.

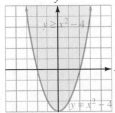

13. $5\sqrt{2}$ **14.** $81x\sqrt[3]{3x}$

15. $0, 5$ **16.** 0 **17.** $\frac{2}{3}, -\frac{3}{2}$ **18.** $\dfrac{-4 \pm \sqrt{19}}{3}$ **19.** $4x^2 + 4x - 1$ **20.** $f^{-1}(x) = \sqrt[3]{\dfrac{x+1}{2}}$

21.

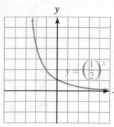

22. $2^y = x$ **23.** $\dfrac{2 \log 2}{\log 3 - \log 2}$ **24.** 16 **25.**

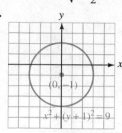

26.

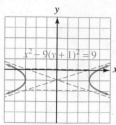

Getting Ready (page 929)

1. 2 **2.** 5 **3.** -4 **4.** -3

Orals (page 939)

1. no solution   **2.** infinitely many solutions   **3.** one solution   **4.** no solution   **5.** 2 **6.** 4 **7.** 4 **8.** 3

Exercise 13.1 (page 940)

1. a^{22} **3.** $\dfrac{1}{81x^{32}y^4}$ **5.** $r = \dfrac{A - p}{pt}$ **7.** $r = \dfrac{r_1 r_2}{r_2 + r_1}$ **9.** consistent   **11.** independent

13.

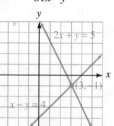

15.

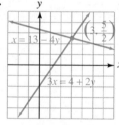

17.

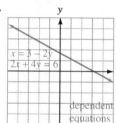

19.

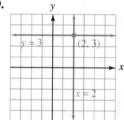

21.

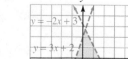
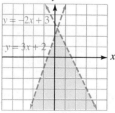

23.

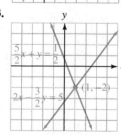
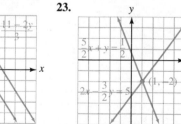

25. $(-0.37, -2.69)$ **27.** $(-7.64, 7.04)$ **29.** $(2, 2)$ **31.** $(5, 3)$
33. $(-2, 4)$ **35.** no solution   **37.** $\left(5, \frac{3}{2}\right)$ **39.** $\left(-2, \frac{3}{2}\right)$ **41.** $(5, 2)$
43. $(-4, -2)$ **45.** $(1, 2)$ **47.** $\left(\frac{1}{2}, \frac{2}{3}\right)$ **49.** dependent equations
51. no solution   **53.** $(4, 8)$ **55.** $(20, -12)$

57.

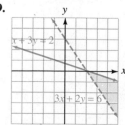

59.

61. $57   **63.** 625 ohms and 750 ohms   **65.** 16 m by 20 m   **73.** no

Getting Ready (page 943)

1. yes   **2.** yes   **3.** no   **4.** yes

Orals (page 951)

1. yes   **2.** no

Exercise 13.2 (page 951)

1. $\frac{9}{5}$ **3.** 1   **5.** $2s^2 + 1$ **7.** plane   **9.** infinitely   **11.** yes   **13.** (1, 1, 2)   **15.** (0, 2, 2)   **17.** (3, 2, 1)   **19.** inconsistent system
21. dependent equations   **23.** (2, 6, 9)   **25.** $-2, 4, 16$ **27.** $A = 40°, B = 60°, C = 80°$ **29.** 1, 2, 3   **31.** 30 expensive,
50 middle-priced, 100 inexpensive   **33.** 250 \$5 tickets, 375 \$3 tickets, 125 \$2 tickets   **35.** 3 poles, 2 bears, 4 deer
37. $y = x^2 - 4x$ **39.** $x^2 + y^2 - 2x - 2y - 2 = 0$ **43.** (1, 1, 0, 1)

Getting Ready (page 955)

1. 5 8 13   **2.** 0 3 7   **3.** -1 -1 -2 **4.** 3 3 -5

Orals (page 961)

1. $\begin{bmatrix} 3 & 2 \\ 4 & -3 \end{bmatrix}$ **2.** $\begin{bmatrix} 3 & 2 & 8 \\ 4 & -3 & 6 \end{bmatrix}$ **3.** yes   **4.** no

Exercise 13.3 (page 962)

1. 9.3×10^7 **3.** 6.3×10^4 **5.** matrix   **7.** 3, columns   **9.** augmented   **11.** type 1   **13.** nonzero   **15.** 0   **17.** 8   **19.** (1, 1)
21. (2, -3)   **23.** (0, -3)   **25.** (1, 2, 3)   **27.** ($-1, -1, 2$)   **29.** (2, 1, 0)   **31.** (1, 2)   **33.** (2, 0)   **35.** no solution   **37.** (1, 2)
39. ($-6 - z, 2 - z, z$)   **41.** ($2 - z, 1 - z, z$)   **43.** 22°, 68°   **45.** 40°, 65°, 75°   **47.** $y = 2x^2 - x + 1$ **51.** $k \neq 0$

Getting Ready (page 964)

1. -22 **2.** 22   **3.** -13 **4.** -13

Orals (page 972)

1. 1   **2.** -2 **3.** 0   **4.** $\begin{vmatrix} 1 & 2 \\ 2 & -1 \end{vmatrix}$ **5.** $\begin{vmatrix} 5 & 2 \\ 4 & -1 \end{vmatrix}$ **6.** $\begin{vmatrix} 1 & 5 \\ 2 & 4 \end{vmatrix}$

Exercise 13.4 (page 973)

1. -3 **3.** 0   **5.** number   **7.** $\begin{vmatrix} a_2 & c_2 \\ a_3 & c_3 \end{vmatrix}$ **9.** $\begin{vmatrix} 3 & 4 \\ 2 & -3 \end{vmatrix}$ **11.** 8   **13.** -2 **15.** $x^2 - y^2$ **17.** 0   **19.** -13 **21.** 26   **23.** 0
25. $10a$ **27.** 0   **29.** (4, 2)   **31.** ($-1, 3$)   **33.** $\left(-\frac{1}{2}, \frac{1}{3}\right)$ **35.** (2, -1)   **37.** no solution   **39.** $\left(5, \frac{14}{5}\right)$ **41.** (1, 1, 2)   **43.** (3, 2, 1)
45. no solution   **47.** (3, -2, 1)   **49.** dependent equations   **51.** ($-2, 3, 1$)   **53.** no solution   **55.** 2   **57.** 2   **59.** \$5,000 in
HiTech, \$8,000 in SaveTel, \$7,000 in HiGas   **61.** -23 **63.** 26   **69.** -4

Getting Ready (page 976)

1. $7x^2 - y^2 = 44$ **2.** $5x^2 - 3y^2 = 8$

Orals (page 981)

1. 0, 1, 2   **2.** 0, 1, 2   **3.** 0, 1, 2, 3, 4   **4.** 0, 1, 2, 3, 4

Exercise 13.5   (page 981)

1. $-11x\sqrt{2}$ **3.** $\frac{1}{2}$ **5.** graphing, substitution   **7.**

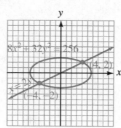   **9.** 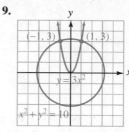   **11.**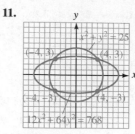

13.   **15.** (1, 0), (5, 0)   **17.** (3, 0), (0, 5)   **19.** (1, 1)   **21.** (1, 2), (2, 1)   **23.** (−2, 3), (2, 3)

25. $\left(\sqrt{5}, 5\right), \left(-\sqrt{5}, 5\right)$ **27.** (3, 2), (3, −2), (−3, 2), (−3, −2)

29. (2, 4), (2, −4), (−2, 4), (−2, −4)   **31.** $\left(-\sqrt{15}, 5\right), \left(\sqrt{15}, 5\right), (-2, -6),$ (2, −6)   **33.** (0, −4), (−3, 5), (3, 5)   **35.** (−2, 3), (2, 3), (−2, −3), (2, −3)

37. (3, 3)   **39.** (6, 2), (−6, −2), $\left(\sqrt{42}, 0\right), \left(-\sqrt{42}, 0\right)$ **41.** $\left(\frac{1}{2}, \frac{1}{3}\right), \left(\frac{1}{3}, \frac{1}{2}\right)$

43. 4 and 8

45. 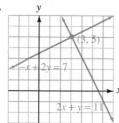   **47.** 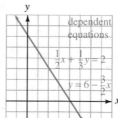   **49.** 7 cm by 9 cm   **51.** either $750 at 9% or $900 at 7.5%

53. 68 mph, 4.5 hr   **57.** 0, 1, 2, 3, 4

Chapter Summary   (page 986)

1. a. 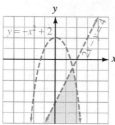   **b.** 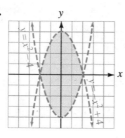   **c.** 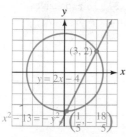   **d.**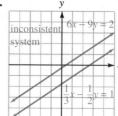

2. a. (−1, 3)   **b.** (−3, −1)   **c.** (3, 4)   **d.** (−4, 2)   **3. a.** (−3, 1)   **b.** (1, −1)   **c.** (9, −4)   **d.** $\left(4, \frac{1}{2}\right)$ **4. a.** (1, 2, 3)

b. inconsistent system   **5. a.** (2, 1)   **b.** (1, 3, 2)   **c.** (1, 2)   **d.** $(3z, 1 - 2z, z)$ **6. a.** 18   **b.** 38   **c.** −3   **d.** 28   **7. a.** (2, 1)

b. (−1, 3)   **c.** (1, −2, 3)   **d.** (−3, 2, 2)   **8. a.** (4, 2), (4, −2), (−4, 2), (−4, −2)   **b.** (2, 3), (2, −3), (−2, 3), (−2, −3)

9.

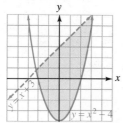

Chapter 13 Test   (page 989)

1.    **2.** (7, 0)   **3.** (2, −3)   **4.** (−6, 4)   **5.** dependent   **6.** consistent   **7.** 6   **8.** −8

9. $\begin{bmatrix} 1 & 1 & 1 & 4 \\ 1 & 1 & -1 & 6 \\ 2 & -3 & 1 & -1 \end{bmatrix}$ **10.** $\begin{bmatrix} 1 & 1 & 1 \\ 1 & 1 & -1 \\ 2 & -3 & 1 \end{bmatrix}$ **11.** (2, 2)   **12.** (−1, 3)   **13.** 22   **14.** −17   **15.** 4   **16.** 13

17. $\begin{vmatrix} -6 & -1 \\ -6 & 1 \end{vmatrix}$ **18.** $\begin{vmatrix} 1 & -1 \\ 3 & 1 \end{vmatrix}$ **19.** −3   **20.** 3   **21.** 3   **22.** −1   **23.** (2, 1), (2, −1), (−2, 1), (−2, −1)

24. (3, 4), (−3, 4)   **25.**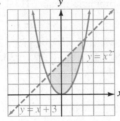

Getting Ready   (page 992)

1. $x^2 + 4x + 4$ **2.** $x^2 - 6x + 9$ **3.** $x^3 + 3x^2 + 3x + 1$ **4.** $x^3 - 6x^2 + 12x - 8$

Orals   (page 997)

1. 1   **2.** 24   **3.** 1   **4.** 120   **5.** $m^2 + 2mn + n^2$ **6.** $m^2 - 2mn + n^2$ **7.** $p^2 + 4pq + 4q^2$ **8.** $4p^2 - 4pq + q^2$

Exercise 14.1   (page 998)

1. 2   **3.** 5   **5.** one   **7.** Pascal's   **9.** 6!   **11.** 1   **13.** 6   **15.** −120   **17.** 30   **19.** 144   **21.** 40,320   **23.** $\frac{1}{110}$ **25.** 2,352   **27.** 10
29. 21   **31.** $\frac{1}{168}$ **33.** $x^3 + 3x^2y + 3xy^2 + y^3$ **35.** $x^4 - 4x^3y + 6x^2y^2 - 4xy^3 + y^4$ **37.** $8x^3 + 12x^2y + 6xy^2 + y^3$
39. $x^3 - 6x^2y + 12xy^2 - 8y^3$ **41.** $8x^3 + 36x^2y + 54xy^2 + 27y^3$ **43.** $\frac{x^3}{8} - \frac{x^2y}{4} + \frac{xy^2}{6} - \frac{y^3}{27}$

45. $81 + 216y + 216y^2 + 96y^3 + 16y^4$ **47.** $\frac{x^4}{81} - \frac{2x^3y}{27} + \frac{x^2y^2}{6} - \frac{xy^3}{6} + \frac{y^4}{16}$ **51.** 1, 1, 2, 3, 5, 8, 13, . . . ; beginning with 2,
each number is the sum of the previous two numbers.

Getting Ready   (page 999)

1. 6   **2.** 10   **3.** −20   **4.** −4

Orals   (page 1001)

1. 2   **2.** 3   **3.** 6   **4.** 6   **5.** 5   **6.** 2   **7.** 1   **8.** 8

(page 1001)

5. 3   **7.** 7   **9.** $3a^2b$ **11.** $-4xy^3$ **13.** $15x^2y^4$ **15.** $28x^6y^2$ **17.** $90x^3$ **19.** $640x^3y^2$ **21.** $-12x^3y$
$000x^4$ **25.** $810xy^4$ **27.** $180x^4y^2$ **29.** $-\frac{1}{6}x^3y$ **31.** $\frac{n!}{3!(n-3)!}a^{n-3}b^3$ **33.** $\frac{n!}{4!(n-4)!}a^{n-4}b^4$ **35.** $\frac{n!}{(r-1)!(n-r+1)!}a^{n-r+1}b^{r-1}$
3.

Getting Ready (page 1003)

1. 3, 5, 7, 9   **2.** 4, 7, 10, 13

Orals (page 1008)

1. 14   **2.** 1   **3.** 5   **4.** -6 **5.** 3   **6.** 5

Exercise 14.3 (page 1009)

1. $18x^2 + 8x - 3$ **3.** $\dfrac{6a^2 + 16}{(a+2)(a-2)}$ **5.** sequence   **7.** arithmetic, difference   **9.** arithmetic mean   **11.** $1 + 2 + 3 + 4 + 5$
13. 3, 5, 7, 9, 11   **15.** $-5, -8, -11, -14, -17$ **17.** 5, 11, 17, 23, 29   **19.** $-4, -11, -18, -25, -32$
21. $-118, -111, -104, -97, -90$ **23.** 34, 31, 28, 25, 22   **25.** 5, 12, 19, 26, 33   **27.** 355   **29.** -179 **31.** -23 **33.** 12
35. $\frac{17}{4}, \frac{13}{2}, \frac{35}{4}$ **37.** 12, 14, 16, 18   **39.** $\frac{29}{2}$ **41.** $\frac{5}{4}$ **43.** 1,335   **45.** 459   **47.** 354   **49.** 255   **51.** 1,275   **53.** 2,500   **55.** 60
57. 31   **59.** 12   **61.** 60, 110, 160, 210, 260, 310; $6,060   **63.** 11,325   **65.** 368 ft   **69.** $\frac{3}{2}, 2, \frac{5}{2}, 3, \frac{7}{2}, 4$

Getting Ready (page 1012)

1. 10, 20, 40   **2.** 18, 54, 162

Orals (page 1018)

1. 27   **2.** $\frac{1}{27}$ **3.** 2.5   **4.** $\sqrt{3}$ **5.** 6   **6.** 1

Exercise 14.4 (page 1018)

1. $[-1, 6]$ **3.** $(-\infty, -3) \cup (4, \infty)$ **5.** geometric   **7.** common ratio   **9.** $S_n = \dfrac{a - ar^n}{1 - r}$ **11.** 3, 6, 12, 24, 48
13. $-5, -1, -\frac{1}{5}, -\frac{1}{25}, -\frac{1}{125}$ **15.** 2, 8, 32, 128, 512   **17.** $-3, -12, -48, -192, -768$ **19.** $-64, 32, -16, 8, -4$
21. $-64, -32, -16, -8, -4$ **23.** 2, 10, 50, 250, 1,250   **25.** 3,584   **27.** $\frac{1}{27}$ **29.** 3   **31.** 6, 18, 54   **33.** $-20, -100, -500,$
$-2,500$ **35.** -16 **37.** $10\sqrt{2}$ **39.** No geometric mean exists.   **41.** 728   **43.** 122   **45.** -255 **47.** 381   **49.** $\frac{156}{25}$ **51.** $-\frac{21}{4}$
53. about 669 people   **55.** $1,469.74   **57.** $140,853.75   **59.** $\left(\frac{1}{2}\right)^{11} \approx 0.0005$ **61.** $4,309.14
67. arithmetic mean

Getting Ready (page 1021)

1. 4   **2.** 4.5   **3.** 3   **4.** 2.5

Orals (page 1024)

1. 8   **2.** $\frac{1}{8}$ **3.** $\frac{1}{2}$ **4.** $\frac{1}{8}$ **5.** 27   **6.** 16

Exercise 14.5 (page 1024)

1. yes **3.** no **5.** infinite **7.** $S = \frac{a}{1-r}$ **9.** 16 **11.** 81 **13.** 8 **15.** $-\frac{135}{4}$ **17.** no sum **19.** $-\frac{81}{2}$ **21.** $\frac{1}{9}$ **23.** $-\frac{1}{3}$ **25.** $\frac{4}{33}$ **27.** $\frac{25}{33}$ **29.** 30 m **31.** 5,000 **35.** $\frac{4}{5}$ **39.** no; $0.999999 = \frac{999,999}{1,000,000} < 1$

Getting Ready (page 1025)

1. 24 **2.** 120 **3.** 30 **4.** 168

Orals (page 1034)

1. 15 **2.** 120 **3.** 3 **4.** 6 **5.** 1 **6.** 1

Exercise 14.6 (page 1034)

1. 6, −3 **3.** 8 **5.** $p \cdot q$ **7.** $P(n, r)$ **9.** $n!$ **11.** $\binom{n}{r}$, combinations **13.** 1 **15.** 6 **17.** 60 **19.** 12 **21.** 5 **23.** 1,260
25. 10 **27.** 20 **29.** 50 **31.** 2 **33.** 1 **35.** $\frac{n!}{2!(n-2)!}$ **37.** $x^4 + 4x^3y + 6x^2y^2 + 4xy^3 + y^4$ **39.** $8x^3 + 12x^2y + 6xy^2 + y^3$
41. $81x^4 - 216x^3 + 216x^2 - 96x + 16$ **43.** $-1,250x^2y^3$ **45.** $-4x^6y^3$ **47.** 35 **49.** 1,000,000 **51.** 136,080 **53.** 8,000,000
55. 720 **57.** 2,880 **59.** 13,800 **61.** 720 **63.** 900 **65.** 364 **67.** 5 **69.** 1,192,052,400 **71.** 18 **73.** 7,920 **77.** 48

Chapter Summary (page 1039)

1. a. 144 **b.** 20 **c.** 15 **d.** 220 **e.** 1 **f.** 8 **2. a.** $x^5 + 5x^4y + 10x^3y^2 + 10x^2y^3 + 5xy^4 + y^5$
b. $x^4 - 4x^3y + 6x^2y^2 - 4xy^3 + y^4$ **c.** $64x^3 - 48x^2y + 12xy^2 - y^3$ **d.** $x^3 + 12x^2y + 48xy^2 + 64y^3$ **3. a.** $6x^2y^2$ **b.** $-10x^2y^3$
c. $-108x^2y$ **d.** $864x^2y^2$ **4.** 42 **5.** 122, 137, 152, 167, 182 **6.** $\frac{41}{3}, \frac{58}{3}$ **7.** 1,550 **8.** $-\frac{45}{2}$ **9. a.** $\frac{15}{2}$ **b.** 378 **c.** 14 **d.** 360
10. 24, 12, 6, 3, $\frac{3}{2}$ **11.** 4 **12.** 24, −96 **13.** $\frac{2,186}{9}$ **14.** $-\frac{85}{8}$ **15.** 125 **16.** $\frac{5}{99}$ **17.** 136 **18. a.** 5,040 **b.** 1 **c.** 20,160 **d.** $\frac{1}{10}$
19. a. 1 **b.** 1 **c.** 28 **d.** 84 **e.** 700 **f.** $\frac{7}{4}$ **20.** $1,638.40 **21.** $134,509.57 **22.** 12 yr **23.** 1,600 ft **24.** 120 **25.** 720
26. 120 **27.** 150

Chapter 14 Test (page 1043)

1. 210 **2.** 1 **3.** $-5x^4y$ **4.** $24x^2y^2$ **5.** 66 **6.** 306 **7.** 34, 66 **8.** 3 **9.** −81 **10.** $\frac{364}{27}$ **11.** 18, 108 **12.** $\frac{27}{2}$ **13.** 120
14. 40,320 **15.** 15 **16.** 56 **17.** 720 **18.** 322,560 **19.** 24 **20.** $\frac{7}{30}$ **21.** 35 **22.** 30

Cumulative Review Exercises (page 1043)

1. (2, 1) **2.** (1, 1) **3.** (2, −2) **4.** (3, 1) **5.** −1 **6.** −1 **7.** (−1, −1, 3) **8.** 1
9. 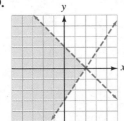 **10.** 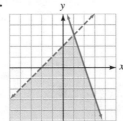 **11.**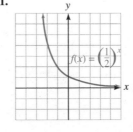

$f(x) = \left(\frac{1}{2}\right)^x$

12. $2^y = x$ **13.** 5 **14.** 3 **15.** $\frac{1}{27}$ **16.** 1
17. $y = 2^x$ **18.** x **19.** 1.9912
20. 0.3010 **21.** 1.6902 **22.** 0.1461
23. $\frac{5 \log 2}{\log 3 - \log 2}$ **24.** 8 **25.** $2,848.31
26. 1.16056 **27.** 30,240
28. $81a^4 - 108a^3b + 54a^2b^2 - 12ab^3 + b^4$
29. $112x^2y^6$ **30.** 103 **31.** 690
32. 8 and 19 **33.** 42 **34.** 27 **35.** 27
36. $\frac{1,023}{64}$ **37.** 12, −48 **38.** $\frac{27}{2}$ **39.** 504
40. 35 **41.** $\frac{7}{3}$ **42.** $C(n, n)$ **43.** 5,040
44. 84

Exercise I.1   (page A-4)

1. y-axis   **3.** origin   **5.** y-axis   **7.** none   **9.** none   **11.** x-axis   **13.**

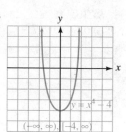

15.

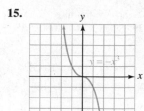

17.

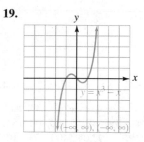

19.

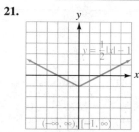

21.

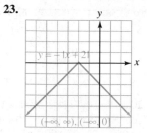

23.

■ ■ ■ ■ ■ ■ ■ ■ ■ ■ INDEX